SYNTHESIS OF FUSED HETEROCYCLES

PART 2

This is the second part of the Forty-seventh Volume in the Series

THE CHEMISTRY OF HETEROCYCLIC COMPOUNDS

THE CHEMISTRY OF HETEROCYCLIC COMPOUNDS

A SERIES OF MONOGRAPHS

EDWARD C. TAYLOR

Editor

SYNTHESIS OF FUSED HETEROCYCLES

PART 2

G. P. ELLIS

School of Chemistry and Applied Chemistry,
University of Wales,
Cardiff, Wales, UK

A Wiley–Interscience Publication

JOHN WILEY & SONS

Chichester · New York · Brisbane · Toronto · Singapore

Other Wiley Editorial Offices

John Wiley & Sons, Inc., 605 Third Avenue,
New York, NY 10158-0012, USA

Jacaranda Wiley Ltd, G.P.O. Box 859, Brisbane,
Queensland 4001, Australia

John Wiley & Sons (Canada) Ltd, 22 Worcester Road,
Rexdale, Ontario M9W 1L1, Canada

John Wiley & Sons (SEA) Pte Ltd, 37 Jalan Pemimpin #05-04,
Block B, Union Industrial Building, Singapore 2057

Library of Congress Cataloging-in-Publication Data
(Revised for volume 2)

Ellis G. P. (Gwynn Pennant)
Synthesis of fused heterocycles.
(The Chemistry of heterocyclic compounds;
v. 47)
'A Wiley–Interscience publication'.
Bibliography: p. 581–632; v. 2, p.
Includes indexes.
1. Heterocyclic chemistry. 2. Organic
compounds—Synthesis. 3. Heterocyclic
compounds. I. Title.
QD400.E45 1987 547′.59 86-28944
ISBN 0 471 91431 2 (v. 1)
ISBN 0 471 93070 9 (v. 2)

British Library Cataloguing in Publication Data

A catalogue record for this book is available from the British Library

ISBN 0 471 93070 9

Typeset in 10/12pt Times by Alden Multimedia, Northampton and
printed and bound in Great Britain by Biddles Ltd, Guildford, Surrey

Introduction to the Series

The series *The Chemistry of Heterocyclic Compounds*, published since 1950 under the initial editorship of Arnold Weissberger, and later, until Dr Weissberger's death in 1984, under our joint editorship, was organized according to compound classes. Each volume dealt with synthesis, reactions, properties, structure, physical chemistry and utility of compounds belonging to a specific ring system or class (e.g. pyridines, thiophenes, pyrimidines three-membered ring systems). This series, which has attempted to make the extraordinarily complex and diverse field of heterocyclic chemistry as readily accessible and organized as possible, has become the basic reference collection for information on heterocyclic compounds.

However, many broader aspects of heterocyclic chemistry are now recognized as disciplines of general significance which impinge on almost all aspects of modern organic and medicinal chemistry. For this reason we initiated several years ago a parallel series entitled *General Heterocyclic Chemistry* which treated such topics as nuclear mangetic resonance of heterocyclic compounds, mass spectra of heterocyclic compounds, photochemistry of heterocylic compounds, the utility of heterocyclic compounds in organic synthesis and the synthesis of heterocyclic compounds by means of 1,3-dipolar cycloaddition reactions. These volumes were intended to be of interest to all organic chemists, as well as to those whose particular concern is heterocyclic chemistry.

It became increasingly clear that this rather arbitrary distinction between the two series creates more problems than it solves. We have therefore elected to discontinue the more recently initiated series *General Heterocyclic Chemistry*, and to publish all forthcoming volumes in the general area of heterocyclic chemistry in *The Chemistry of Heterocyclic Compounds* series.

Edward C. Taylor

Department of Chemistry
Princeton University
Princeton, New Jersey 08544

Contents

Preface **xi**

1. Introduction 661
2. Acetal or Aldehyde and Amine or Carboxamide 670
3. Acetal and Ring-carbon or Ring-nitrogen 675
4. Acylamine and Aldehyde or Ketone 680
5. Acylamine or Carbamate and Amine or Hydrazine 683
6. Acylamine and Carboxamide 690
7. Acylamine and a Carboxylic Acid Derivative 693
8. Acylamine or Amine and Ether 697
9. Acylamine, Acylhydrazine, Amine or Isocyanate and Halogen 701
10. Acylamine or Amine and Hydroxy 712
11. Acylamine or Diazonium Salt and Lactam Carbonyl 726
12. Acylamine or Imine and Methylene or Alkene 733
13. Acylamine or Amine and Nitrile 739
14. Acylamine, Amine or Imine and Nitro 751
15. Acylamine or Amine and Nitroso or *N*-Oxide 756
16. Acylamine, Acyloxy, Amine or Hydroxy and Phosphorane 760
17. Acylamine or Acylhydrazine and Ring-carbon 767
18. Acylamine or Acylhydrazine and Ring-nitrogen 775
19. Acylamine or Amine and Sulphonamide or Thioureide 778
20. Acylamine or Amine and Thiocyanate 781
21. Acyl Halide and Ring-carbon or Ring-nitrogen 783
22. Aldehyde or Ketone and Alkene or Alkyne 786
23. Aldehyde or Ketone and Azide or Triazene 790
24. Aldehyde or Ketone and Carbamate, Isothiocyanate or Thiourea 792
25. Aldehyde or Ketone and Carboxamide or Sulphonamide 794
26. Aldehyde or Ketone and Carboxylic Acid or Ester 799
27. Aldehyde or Ketone and Ether 806
28. Aldehyde or Ketone and Halogen 809
29. Aldehyde, Ketone or Lactam Carbonyl and Hydroxy 817

30. Aldehyde and Ketone; Dialdehyde or Diketone 827
31. Aldehyde or Ketone and Methylene 839
32. Aldehyde or Ketone and Nitrile 845
33. Aldehyde or Ketone and Nitro, Nitroso or *N*-Oxide 848
34. Aldehyde or Other Carbonyl and Phosphorane 852
35. Aldehyde or Ketone and Ring-carbon 859
36. Aldehyde, Ketone or Lactam Carbonyl and Ring-nitrogen 872
37. Alkene or Alkyne and Amine, Azide or Nitro 884
38. Alkene or Alkyne and Carboxylic Acid or its Derivative 891
39. Alkene or Alkyne and Halogen 893
40. Alkene or Alkyne and Hydroxy or Ether 898
41. Alkene, Methylene or Ring-carbon and Lactam or Lactone Carbonyl 904
42. Alkene or Alkyne and Methylene, Ring-carbon or Ring-nitrogen 910
43. Amidine and Amine, Carboxylic Acid or its Derivative 921
44. Amidine and Ring-carbon or Ring-nitrogen 927
45. Amine and Azide, Azo or Diazo 932
46. Amine or Phosphorane and Carboxamide 935
47. Amine and Carboxylic Acid 940
48. Amine and Carboxylic Ester 946
49. Amine and Enamine or Oxime 957
50. Amine or Diazonium Salt and Hydrazide or Hydrazine 959
51. Amine and Hydrazone or Imine 964
52. Amine and Ketone 969
53. Amine and Ring-carbon 980
54. Amine and Ring-nitrogen 997
55. Azide and Azo or Nitro 1015
56. Azide or Isocyanate and a Carboxylic Acid Derivative 1018
57. Azide or Azo and Methyl or Methylene 1020
58. Azide or Azo and Ring-carbon 1023
59. Azide and Ring-nitrogen 1030
60. Azo and Carbamate 1033
61. Carbamate or Ureide and Ring-carbon or Ring-nitrogen 1035
62. Carbamate or Carboxamide and another Carboxylic Acid Derivative 1040
63. Carboxamide or Nitrile and Diazonium Salt 1045
64. Carboxamide, Nitrile or Ureide and Hydroxy or Ether 1047
65. Carboxamide or Nitrile and Ring-carbon or Ring-nitrogen 1054
66. Carboxylic Acid or its Derivative and Halogen 1063
67. Carboxylic Acid or Ester and Hydrazine or Hydroxylamine Derivative 1073
68. Carboxylic Acid, Acyl Chloride or Ester and Hydroxy or Ether 1076
69. Carboxylic Acid Derivative and Lactam Carbonyl, Cyanate or Isocyanate 1081
70. Carboxylic Acid Derivative, Lactam Carbonyl or Isocyanide and Methylene 1086

71. Carboxylic Acid Halide or Ester and Nitrile 1094
72. Carboxylic Acid, its derivative or Lactam Carbonyl and Nitro or Ureide 1096
73. Carboxylic Acid or Ester and Ring-carbon 1105
74. Carboxylic Acid or Ester and Ring-nitrogen 1112
75. Carboxylic Acid or its Derivative and a Sulphur-containing Group 1118
76. 1,2-Diamine 1123
77. 1,3-, 1,4- or 1,5-Diamine 1142
78. Diazo or Diazonium Salt and Halogen or Methylene 1147
79. Diazo or Diazonium Salt and Ring-carbon or Ring-nitrogen 1150
80. Dicarboxylic Acid or its Derivative or Diazide 1158
81. Dihalogen 1163
82. Dihydroxy 1170
83. Dinitrile or Dinitro 1177
84. Di-ring-carbon or Di-ring-nitrogen 1181
85. Enamine and Ester or Ketone 1200
86. Enamine and a Non-carbonyl Group 1204
87. Ether and Methylene, Ring-carbon or Ring-nitrogen 1208
88. Halogen and Ether or Hydroxy 1213
89. Halogen and Lactam Carbonyl or a Sulphur-containing Group 1219
90. Halogen and Methylene or Ring-carbon 1222
91. Halogen and Nitro 1233
92. Halogen and Ring-nitrogen 1236
93. Hydrazide, Hydrazine or Hydrazone and Nitro 1244
94. Hydrazine and Ring-carbon or Ring-nitrogen 1247
95. Hydrazone or Oxime and Ring-carbon or Ring-nitrogen 1260
96. Hydroxy or Ether and Hydrazone or Oxime 1267
97. Hydroxy and Methylene 1272
98. Hydroxy and Nitro or Ring-nitrogen 1276
99. Hydroxy and Ring-carbon 1280
100. Hydroxy, Ring-carbon or Ring-nitrogen and Imine or Iminophosphorane 1293
101. Isocyanate and Methylene, Ring-carbon or Ring-nitrogen 1303
102. Ketone and Lactam Carbonyl 1307
103. Methylene and Nitro or Nitroso 1312
104. Methylene and Ring-carbon or Ring-nitrogen 1316
105. Nitro or *N*-Oxide and Ring-carbon or Ring-nitrogen 1328
106. Ring-carbon or Ring-sulphur and Ring-Nitrogen 1335

List of Books and Monographs **1343**
References **1345**
General Index **1399**
Index of Ring Systems **1411**

Preface

The warm reception given to *Synthesis of Fused Heterocycles* by reviewers and readers since its publication in 1987 has encouraged me to prepare another classified selection of methods by which a new heterocyclic ring may be built on to a (carbocyclic or heterocyclic) ring. The 1987 book (which was reprinted with minor corrections in 1991) is referred to in this book as *Synthesis of Fused Heterocycles, Part 1*. The layout of the present volume is similar to that of Part 1 but some improvements have been made. These are described in the last two paragraphs of Chapter 1, Section II. Part 2 contains over two thousand references (including those numbered as 3037a, 3037b, 3059a, 3059b, 3067a, 3067b, etc.) not listed in Part 1.

I wish to thank reviewers and readers who made useful suggestions following the publication of Part 1; for example, Professor Brandsma suggested in a review of Part 1 that, in a product containing two or more heterocyclic rings, the index should show the newly formed ring in bold letters. The Index of Ring Systems at the end of this volume follows this suggestion.

I also thank Mr A. J. Cannard BSc for writing a computer program which overcomes the problem of alphabetizing the names of those fused heterocycles which begin with numbers and/or contain locants which need to be partially ignored during the production of an index which is useful to organic chemists. Thanks are also due to the staff of the libraries of the University of Wales colleges at Cardiff and Bangor.

I wish to thank my wife for enduring long periods of isolation while this book was being written, and for her advice and help when needed. Professor E. C. Taylor and the staff of John Wiley & Sons at Chichester have constantly provided encouragement and help for which I am grateful.

October 1991 G. P. Ellis

CHAPTER 1*

Introduction

I.	Aims	661
II.	Scope	662
III.	Use of the Book	664
	1. One- and Two-step Reactions	664
	2. Naming of Functional Groups	665
	3. Approved Names and Formulae	666
	4. Format	668
IV.	Abbreviations	669

I. AIMS

This book is intended to provide a convenient way of locating papers (and reviews) on methods for the synthesis of a heterocyclic ring which is fused to another ring. Figure 1.1 shows diagrammatically the types of cyclizations covered. The α-ring may be either homocyclic or heterocyclic, and may be of any size; the newly formed *β*-ring is heterocyclic and contains five to eight atoms, at least one of which (Z) is nitrogen, oxygen or sulphur. Atoms A and B may or may not be identical and may be carbon or nitrogen (or rarely sulphur, in which

Figure 1.1

*This is a modified and expanded version of Chapter 1, Part 1.

case it would not normally have a substituent X or Y attached to it). X and Y are usually functional groups; it is rare for both to be hydrogen. When A is a double-bonded or pyridine-type nitrogen, then X may be absent or is an *N*-oxide or an *N*-azinium (for example, a pyridinium) salt function.

The book is divided into chapters according to the identities of X and Y, the arrangement being alphabetical as far as possible. For example, reactants in which X and Y are amine and carboxamide, respectively, are considered in Chapter 46 while those where X and Y are hydroxy and methylene are contained in Chapter 99. Where only a few examples of a particular X and Y combination were found in the literature, these are included with cyclizations of closely related groups; for example, Chapter 20 contains ring closures of substrates having either acylamine or amine and thiocyanate.

Another aim of the book is to show the IUPAC-approved way of drawing each of the fused heterocycles whose synthesis is described. From this, it follows that the overall peripheral numbering of each ring system can be deduced by applying IUPAC nomenclature rules [B-1]. Although there is much to be said for drawing all ring systems according to IUPAC recommendations, a few of the simpler ones (for example, quinoline) and others (for example, β-lactam antibiotics) are often drawn differently in the literature. The formulae of reactants are drawn so as to require the minimum of alteration during their conversion into the product and are therefore not necessarily drawn in accordance with IUPAC rules. This feature makes it easier for the reader to see the structural relationship between the reactant and product, and is recommended to authors of papers and reviews.

Third, the approved name of the product ring system is given under the formula, but this omits dihydro-, tetrahydro-, etc., prefixes and functional groups except endocyclic carbonyl or thiocarbonyl. The basis of this distinction is that a heterocyclic ring containing one or more endocyclic carbonyl groups usually requires a radically different method of synthesis from its non-carbonyl parent, and compounds of this kind are often of considerable biological or industrial importance; the two types of ring systems are also separated in the index so that they may be more easily found. During the preparation of Parts 1 and 2 of this work, it was noticed that many ring systems were erroneously named in the literature. The correct names of such compounds are shown in this book; some notes on the application of IUPAC rules to polycyclic heterocycles are included in an Appendix to Part 1 (1987) (pp. 578–580).

II. SCOPE

In recent years, heterocyclic chemistry has greatly expanded its boundaries and complexity. The formation of a new fused heterocyclic ring is an important type of reaction which is often difficult to trace in abstracting journals. Parts 1 and

2 of this book contain references to a classified selection of papers, published since about 1970, which describe a wide variety of heterocyclizations. Each of these references is likely to contain citations of earlier relevant papers, and the reviews cited in this book should provide additional references. In this way, the reader may often locate suitable references more easily than by using the abstracting journals. Although Parts 1 and 2 of this book cannot cover all heterocyclic ring systems, the methods described can be applied to the synthesis of many more by varying the reactant and/or reagent.

Spiro and mesoionic products are not included, neither are reactions which have been shown to proceed by an ANRORC mechanism [551] or by other routes in which a heterocyclic ring is opened and closed again into a different one; these types of reactions are already covered by monographs [B-2, B-3, B-60]. However, an exception is made for reactions which involve opening of rings in reagents such as lactones, cyclic anhydrides, imides, epoxides, aziridines and diketene. Reactions in which two different rings are formed in one step are not included. Newly formed fused rings are usually unsaturated or contain one or more endocyclic carbonyl groups, but a few examples of the synthesis of saturated heterocyclic rings are included.

Some combinations of functional groups which were not in Part 1 are included in Part 2 and *vice versa*, and minor changes in the titles of a few chapters reflect this difference. Some rationalization has been made between the coverage of Chapters 29 and 96, and between Chapters 93 and 94; the types of cyclizations which were described in Chapter 107 of Part 1 have now been included in either Chapter 36 (lactam carbonyl and ring-nitrogen) or Chapter 98 (ring-nitrogen and thiol). In Part 2, functional groups which contain sulphur are included in chapters which discuss their oxygen analogues; their names do not usually appear in addition to those of the oxygen compounds in chapter headings. However, newly formed sulphur-containing rings are listed at the beginning of each relevant chapter. Other minor variations in the chapter titles are the result of the kinds of cyclizations found in the literature. Pagination, reference numbers and formulae numbers are continuous through Parts 1 and 2.

Reactions which are described in Part 1 are not repeated in Part 2 but an improvement or modification in the reagents, catalysts, mediators or reaction conditions are recorded, especially if this alteration gives a better yield or is more convenient to use. The general index has been improved in Part 2 to include all examples of reacting groups (X and Y in Figure 1.1); it also includes reviews. In the index of ring systems, the heteroring which is formed in a particular reaction is shown in bold, for example, oxazolo[5,4-*d*]**pyrimidine**. In names like **pyrano**[3,4-*b*]indol-3-**one**, it is important to note that the number in front of '-one' refers to the *peripheral* numbering of the ring system and not to that of the isolated pyranone ring, but in **pyrimido**[4,5-*d*]pyridazin-**2-one**, the newly formed ring is a pyrimidin-2-one. Ring systems which contain several carbonyl groups present additional problems; while the meaning of pyrimido[5,4-*e*]-**1,2,4-triazine-3**,5,7-tri**one** (the new ring is triazin-3-one) is clear, a newly formed pyrimidinedione in this ring system is represented by the less clear name **pyrimido**[5,4-*e*]-1,2,4-triazine-3,**5,7**-tri**one**. Trivial names such as purine or purinedione need a different method of indicating the newly formed ring: for example, purine (im) or purinedione (im2one) shows that imidazole or imidazol-2-one, respectively, is the newly formed ring. The abbreviations used are shown at the beginning of the index. Finally, pyrrolo[1,2-*a*]**quinazolin**-5-**one** means that the pyrimidinone ring is newly formed. If the synthesis of the compound from a quinazolinone is also described, the entry (**pyrrolo**[1,2-*a*]quinazolin-5-one) would be shown on a separate line.

III. USE OF THE BOOK

The aims of the book have been outlined in Section I; examples of how these may be achieved are given in this section.

1. One- and Two-step Reactions

Let us assume that the nitro ester (**1.1**) is to be converted into a novel tricyclic compound (**1.2**). No useful references are to be found in Chapter 72 (carboxylic acid or its derivative and nitro) because the nitro group has to be reduced before this kind of cyclization can occur. Chapter 48, which includes cyclization of amino esters, is more promising and Section II.1 (p. 228) offers at least one example of the direct conversion of a nitro ester into a pyridin-2-one ring; the conditions given [821] are also likely to be applicable to the synthesis of triazine (**1.2**).

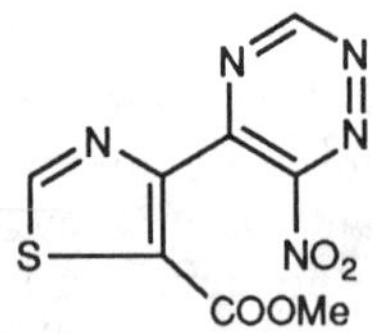

N N N N S NO_2 COOMe

(1.1)

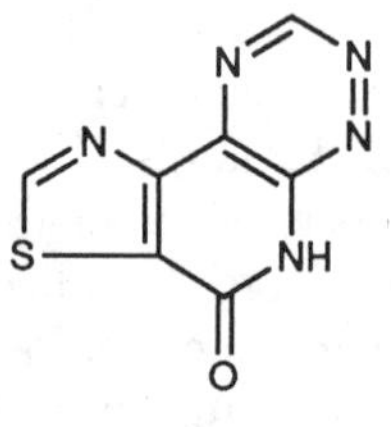

N N N N S NH O

(1.2)

There are two points to note regarding this example. First, two-step conversions such as this, where the intermediate is not isolated but its identity is known beyond reasonable doubt, are classified according to the functional groups of that intermediate. Cyclizations of this type are to be found amongst similar ones in which the intermediate had been isolated. Ring closures where the structure of an intermediate may be in some doubt are classified according to the functional groups of the compound actually placed in the reaction vessel.

Second, potentially relevant information may be gleaned by examining other cyclizations to be found in Chapter 48. For example, a glance at Section I.1 (p. 226) shows that variations in the reductive cyclization of nitro esters are possible; should ring closure of (**1.1**) give an unexpected product, knowledge of the alternative course taken by related reactions [348, 1221, 2778] may be useful. Minor changes in reaction conditions may affect yields, and Sections II.4 and III.3 of the same chapter show potentially useful variations.

2. Naming of Functional Groups

For most reactants, identification and the naming of the functions which participate in the ring closure is easy (as in the example in Section III.1), but several others presented difficulties. Brief comments on how the problems have been resolved may help the reader to appreciate the basis of classification. Several of the difficulties are collected in the (mythical?) compound (**1.3**) which has four functions, each of which must have an unambiguous name which determines in which chapter the cyclization is to appear.

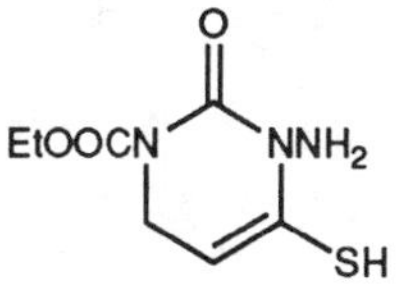

O
EtOOCN NNH$_2$
SH

(**1.3**)

The primary N-*amino group* may also be regarded as a disubstituted hydrazine but since its reactions resemble those of a primary amine and can only occur at one of the nitrogen atoms, it is classified as an amino group. For example, the cyclization of (**1.4**) appears under cyclizations of 1,2-diamines (Chapter 76).

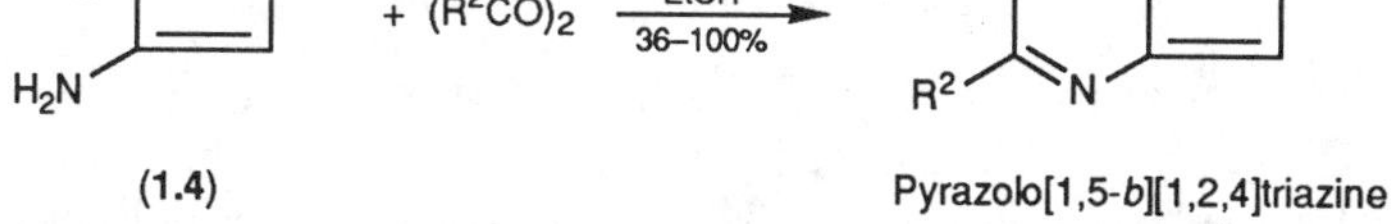

H$_2$NN N R^1 H$_2$N + $(R^2CO)_2$ $\xrightarrow[36–100\%]{EtOH}$ R^2 N N N R^1 R^2 N [1874]

(**1.4**) Pyrazolo[1,5-*b*][1,2,4]triazine

The ethoxycarbonyl group of (**1.3**), being attached to a nitrogen atom, differs in reactivity from that of a carboxylic ester and is classified as a carbamate.

The ring carbonyl group is part of a cyclic urea; it cannot therefore be regarded as a typical ketone or amide. A cyclic —NH—CO— group and its sulphur analogue are therefore classified as lactam carbonyl and lactam thiocarbonyl, respectively, whatever the size of the ring. Cyclization of lactam (**1.5**) is classified as an interaction between amine and lactam carbonyl; such reactions are to be found in Chapter 11. The difficulties of naming cyclization products which contain an endocyclic lactam carbonyl or thiocarbonyl group are mentioned in Section III.3 of this chapter.

xylene, $TiCl_4$, Δ
52–94%
[1959]

(1.5) Pyridazino[1,6-*a*]benzimidazole

Formulae (**1.6**) and (**1.7**) show other functions which present problems of classification. The former has a group which may be regarded as either a ureide or a carboxamide, and is classified as a carboxamide in this book. Since it is the ester group of (**1.7**) that reacts in the second example, the conversion is classified under carbamate and ring-nitrogen (Chapter 61).

$HC(OEt)_3$, Δ
62%
[58]

(1.6) [1,2,4]Triazolo[1,5-*a*][1,3,5]triazin-7-one

aq. NaOH
~94%
[1785]

(1.7)

Pyrazolo[1,5-*a*]-1,3,5-triazin-4-one

3. Approved Names and Formulae

Recommendations have been made by the International Union of Pure and Applied Chemistry (IUPAC) regarding the drawing and naming of fused heterocycles; these are followed by *Chemical Abstracts* in their indexes but some authors neglect some or all of the recommendations, especially those relating to the correct alignment of the formula. This is regrettable, because a formula which is properly drawn is more likely to have its correct peripheral numbering. There are a few common heterocycles in which the peripheral numbering is abnormal; these are shown in Figure 1.2. The IUPAC report [B-1] and a review [2014] give the basic rules and some examples; the Appendix in Part 1 (pp. 578–580) contains some additional guidance.

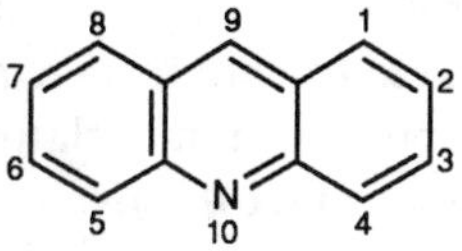

Acridine

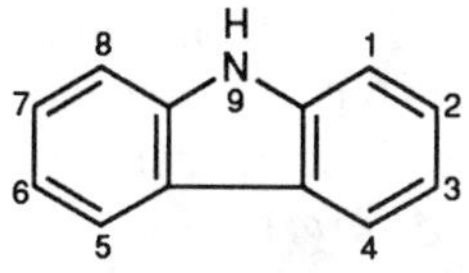

Carbazole

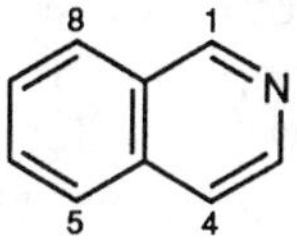

Isoquinoline

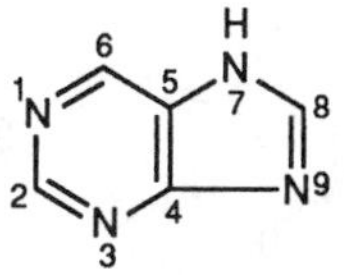

Purine

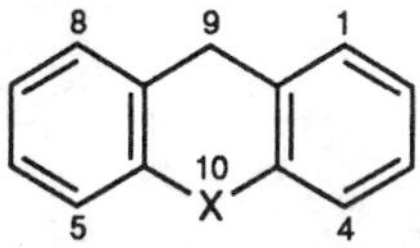

X = O Xanthene
X = S Thioxanthene

Figure 1.2

The confusion that can arise when the formula of a relatively simple fused heterocycle is wrongly drawn may be illustrated with a tricyclic compound which may be drawn in several ways, some of which are shown in (**1.8**) to (**1.12**). None of these is in accord with IUPAC recommendations; (**1.13**) is the correct formula and is the one that should be used. Peripheral numbering of this compound is shown and its name is pyrazino[2′,3′:4,5]thieno[3,2-*d*]pyrimidin-4(3H)-one.

Names of the heterocyclic ring systems (whose syntheses are shown) are given under the formulae; efforts have been made to ensure that, as far as possible, these comply with IUPAC recommendations, but it has not been possible to check every compound from recently published papers. As mentioned in Section I, the names are for the ring systems and do not include the names of substituents

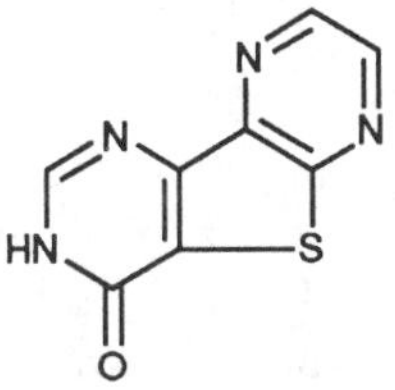

(**1.8**) 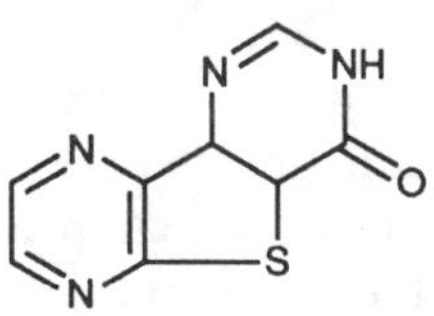 (**1.9**) 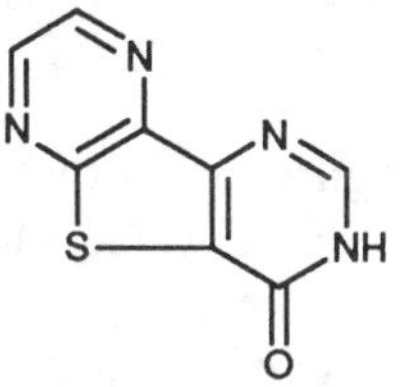 (**1.10**)

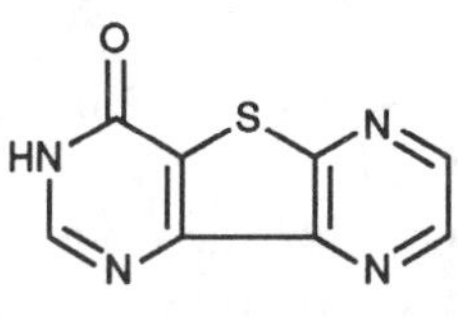

(**1.11**) 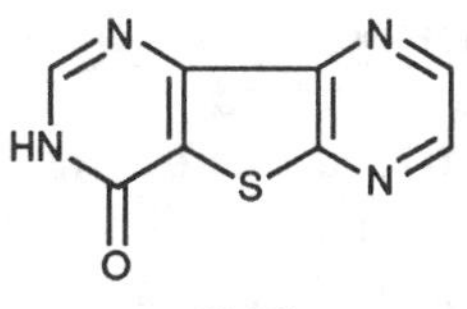 (**1.12**) 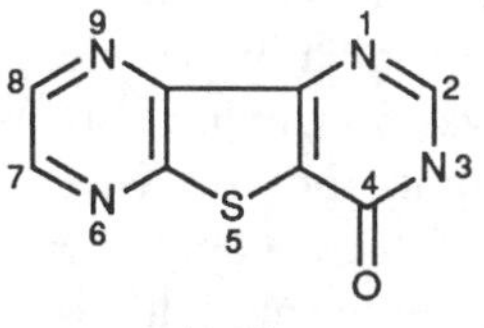 (**1.13**)

except for endocyclic carbonyl or thiocarbonyl groups. Some of these compounds are tautomeric but firm evidence for either keto or enol form is not always available. The oxo form is therefore often assumed for nomenclatural purposes (as in *Chemical Abstracts*). For example, compound (**1.14**) is named as its diketo tautomer (**1.15**), namely pteridine-2,4-dione (trivial name: lumazine), the designation of the added hydrogens being omitted.

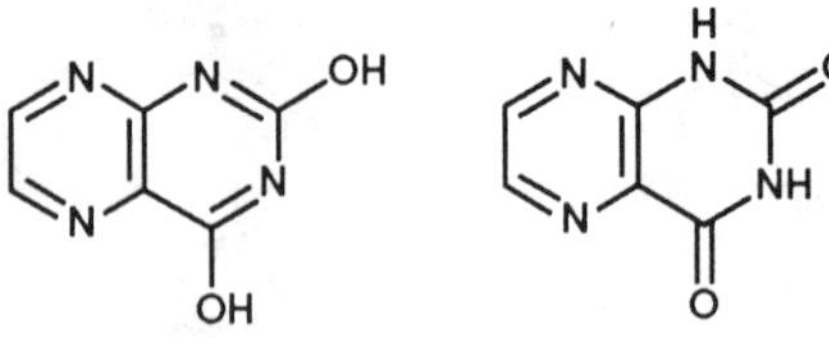

(1.14) (1.15)

4. Format

Each equation which shows a cyclization is intended to provide as much information as space allows. Figure 1.3 is a typical example; solvent(s), catalyst and promoter (acid, base or other compound) are shown above the arrow, and when the reaction temperature is higher than that of the surroundings, a delta (Δ) is added above the arrow. Its absence means that that reaction does not need externally applied heat. When the name of a solvent or reactant is abbreviated, reference should be made to the table in Section IV of this chapter or to the paragraph in front of the equation.

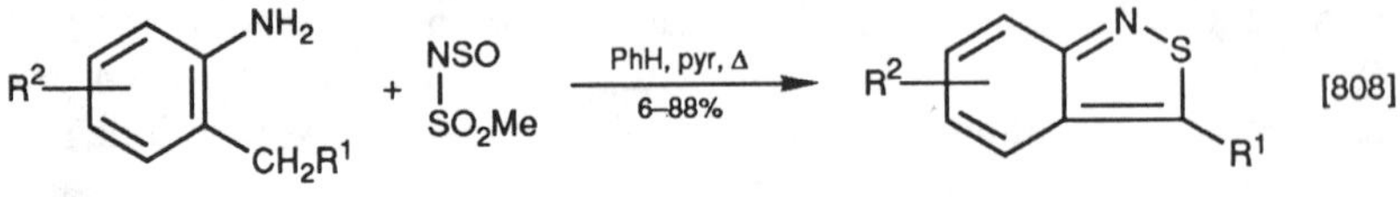

PhH, pyr, Δ; 6–88% [808]

R^1 = H, Me, Ph; R^2 = Me, Et, Cl, MeO, CN, COOMe 2,1-Benzisothiazole

Figure 1.3

The figures below the arrow indicate the yield quoted in the paper(s) cited; this consists of one value when the equation shows the synthesis of one particular compound, or an approximate value (for example, ~65%) when the yields for several related products fall within a range of ±5% of the value shown, or as in Figure 1.2, it shows the lowest and highest values for variations of the substituents R^1 and R^2 shown under the reactants. To the right of the equation or formula is the reference number; a list of references is to be found on the pages which have a black edge towards the end of the book. Books and monographs are listed separately just before the list of references.

With the compressed information supplied, it is hoped that the reader can judge the suitability of a particular reaction to his or her own work, and then need only consult the most promising papers.

IV. ABBREVIATIONS

The following abbreviations, are used in equations and occasionally in the text.

Ac	acetyl
AIBN	azobisisobutyronitrile
CDI	carbonyldi-imidazole
DBN	1,5-diazabicyclo[4.3.0]-nonan-5-one
DBO	1,4-diazabicyclo[2.2.2]-octane
DBU	1,5-diazabicyclo[5.4.0]-undec-5-ene
DCC	dicyclohexylcarbodi-imide
DDQ	2,3-dichloro-5,6-dicyano-1,4-benzoquinone
DEAD	diethyl acetylenedicarboxylate
DEG	diethylene glycol
diox	1,4-dioxane
DMA	*N,N*-dimethylacetamide
DMAD	dimethyl acetylenedicarboxylate
DMAP	*N,N*-dimethyl-4 pyridinamine
DME	dimethoxyethane
DMF	*N,N*-dimethylformamide
DMFDEA	*N,N*-dimethylformamide diethyl acetal
DMFDMA	*N,N*-dimethylformamide dimethyl acetal
DMSO	dimethyl sulphoxide
HMPT	hexamethylphosphoric triamide
LDA	lithium diisopropylamide
LTA	lead tetra-acetate
MCPBA	3-chloroperbenzoic acid
Mes	mesitylenesulphonyl
mor	morpholine
NBS	*N*-bromosuccinimide
NCS	*N*-chlorosuccinimide
NIS	*N*-iodosuccinimide
PEG	polyethylene glycol
pip	piperidine
Phth	phthaloyl
PPA	polyphosphoric acid
PPE	polyphosphoric acid ethyl ester
PTS	pyridinium 4-toluenesulphonate
pyr	pyridine
RaNi	Raney nickel
TCNE	tetracyanoethylene
TEA	triethylamine
tet	5-tetrazolyl
TFA	trifluoroacetic acid
THF	tetrahydrofuran
Ts	4-toluenesulphonyl

CHAPTER 2

Acetal or Aldehyde and Amine or Carboxamide

I.	Formation of a Five-membered Ring	670
	1. Pyrrole	670
II.	Formation of a Six-membered Ring	671
	1. Pyridine	671
	2. Pyridine 1-Oxide	673
	3. Pyridin-2-one	673
	4. Pyrazin-2-one	673
III.	Formation of a Seven-membered Ring	674
	1. 1,4-Diazepin-5-one	674
	2. 1,4-Thiazepin-5-one	674

I. FORMATION OF A FIVE-MEMBERED RING

1. Pyrrole

Catalytic hydrogenation of a nitro-(formylmethyl)pyrimidine is accompanied by cyclization and formation of a fused pyrrole ring. The report on page 8 of Part 1 of a paper which describes the cyclization of 2-(2-aminophenyl)acetaldehyde acetals is not cited in a 1986 paper [3189] which extends the method to the synthesis of 6-substituted indoles in which the substituent is electron-withdrawing; this type of indole is not easily accessible by the Fischer or Bischler synthesis.

Ph, OHCCH, N, N, O_2N, NH_2 — H_2, RaNi, EtOH, 54% → Ph, N, HN, N, NH_2 [2400]

Pyrrolo[3,2-*d*]pyrimidine

II. FORMATION OF A SIX-MEMBERED RING

1. Pyridine

Heating the aminoacetal (**2.1**) with ethanolic hydrogen chloride produces a quinoline [2612], but sulphuric acid causes the cyclization to follow another course (see Chapter 3, Section II.1). *o*-Aminoaldehydes are cyclized by heating with a reactive methylene-containing ketone, ester or nitrile but in the reaction with the pyrazole aldehyde (**2.2**), yields seem to depend greatly on the nature of the *N*-substituent; for example, when $R^2 = R^3 = CN$, the product is obtained in either 3.6% or 83% yields according to whether R^1 is methyl or phenyl [2798]. The chemistry of α-cyanothioacetamide (**2.4**), which is used in the conversion of 2-aminobenzaldehyde (**2.3**) to 2-aminoquinoline-3-thiocarboxamide [2926], has been reviewed [3331].

(**2.1**) —EtOH–HCl, Δ, 99%→ Quinoline [2612]

(**2.2**) + $R^2CH_2R^3$ —EtONa, Δ, 3–95%→ Pyrazolo[3,4-*b*]pyridine [2798]

R^1 = Me, Ph; $R^2 = R^3$ = Ac, CN or
R^2 = COOEt; R^3 = Ac or $R^2R^3 = CO(CH_2)_4$

(**2.3**) + (**2.4**) —1. EtOH, TEA 2. HCl, ~60%→ Quinoline [2926]

A Friedländer reaction (review: [3017]) between a 2-aminobenzaldehyde (**2.3**) and a 4-unsubstituted resorcinol in the presence of a base is a convenient method of synthesizing a fused pyridine, but the expected product (**2.6**) of a Friedländer reaction between 2-aminobenzaldehyde and *N*-acetyl-3-pyrrolidone in aqueous alkali was not obtained; instead, a low yield of the isomeric ring system (**2.5**) was isolated. Both types of products were obtained when the pyrrolidone was replaced by ethyl 3-oxopyrrolidine-1-carboxylate, the [3,2-*b*] isomer predominating. A mixture of isomers was also obtained when the reaction was attempted under acidic conditions.

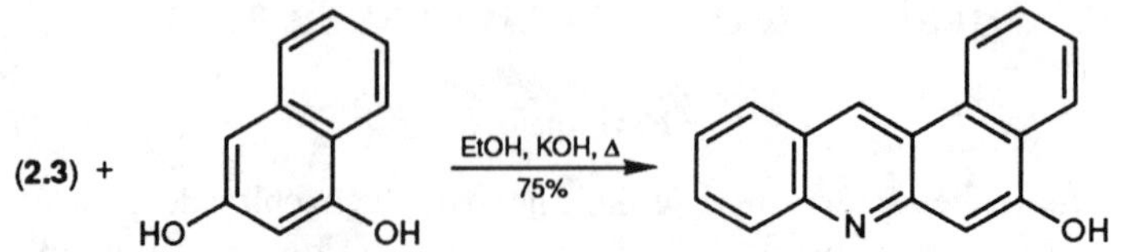

(2.3) + [naphthalene-1,3-diol] → EtOH, KOH, Δ, 75% → Benz[a]acridine [2481]

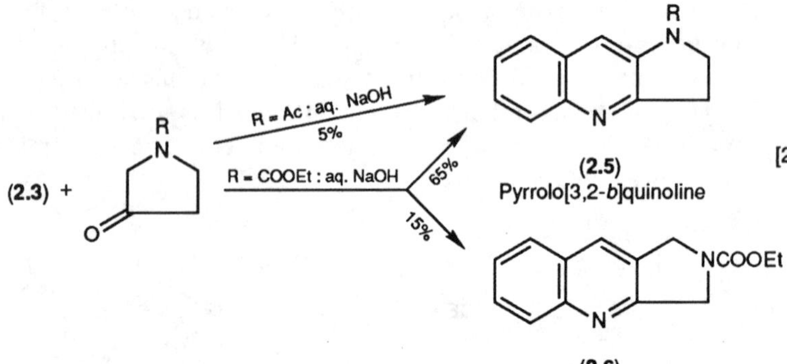

Benz[a]acridine

(2.3) + [1-R-pyrrolidin-3-one] → R = Ac : aq. NaOH, 5%; R = COOEt : aq. NaOH, 65% / 15% [2267]

(2.5)
Pyrrolo[3,2-b]quinoline

(2.6)
Pyrrolo[3,4-b]quinoline

A variation on the conversions of aminoaldehydes into fused pyridines (see p. 10) is their reaction with malono mono- or di-nitrile in the presence of DBN to yield a new fused 2-aminopyridine ring. Heating with a nitrile and ethanolic alkoxide is also effective [3497].

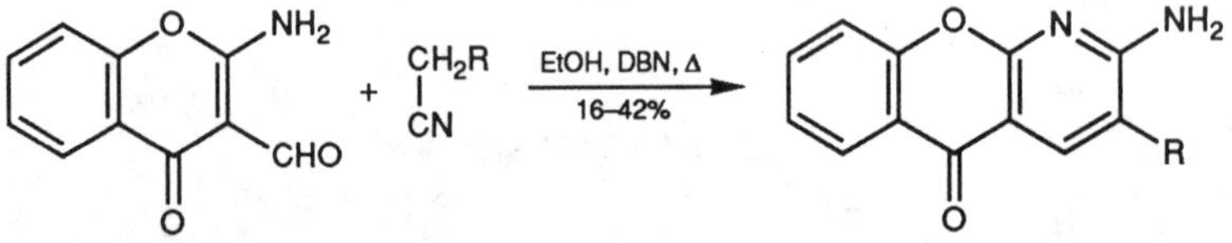

EtOH, DBN, Δ, 16–42% [2799]

R = CN, COOEt

[1]Benzopyrano[2,3-b]pyridin-5-one

Another ring system synthesized similarly:

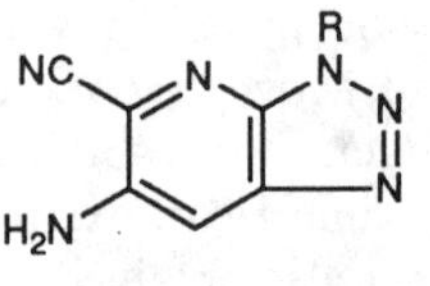

[3497]

1,2,3-Triazolo[4,5-b]pyridine

2. Pyridine 1-Oxide

An aldoxime and a suitably positioned 2,4-dinitrophenylamino group react in hot acid with elimination of 2,4-dinitroaniline and the formation of an isoquinoline 2-oxide in high yield.

CH=NOH / CH=CHNHAr → (EtOH, HCl, Δ; 81%) [2232]

Ar = 2,4-$(NO_2)_2C_6H_3$ — Isoquinoline 2-oxide

3. Pyridin-2-one

This ring may be constructed from an aminoaldehyde and either ethyl cyanoacetate or triethyl phosphonoacetate in the presence of a strong base.

H_2N, OHC + COOEt / CH_2CN → (EtOH, EtONa, Δ; ~43%) [3497]

R = $PhCH_2$, Ph — 1,2,3-Triazolo[4,5-*b*]pyridin-6-one

CF_3, NH_2, CHO, R → (NaH, $(EtO)_2POCH_2COOEt$, Δ; 52%) [3664]

R = 2,4-Me_2-imidazol-1-yl — 2-Quinolinone

4. Pyrazin-2-one

Side-chain acetal and nuclear carboxamide groups react in hot aqueous acid to form this fused ring.

RNHC(=O), $(MeO)_2CHCH_2N$, NH, S → (dil. HCl, Δ; ~83%) [3608]

R = R, alkyl, cyclopropyl — 1,2,4-Triazolo[4,3-*a*]pyrazin-8-one, 3-thioxo-

III. FORMATION OF A SEVEN-MEMBERED RING

1. 1,4-Diazepin-5-one

Cyclization of an aminoaldehyde [as its thioacetal (**2.7**)] was achieved in high yield and under mild conditions in the presence of mercury(II) chloride and calcium carbonate [3272, 3710]. Heating aminoacetals (review [3326]) such as (**2.8**) in TFA is an alternative procedure but may give lower yields [3124].

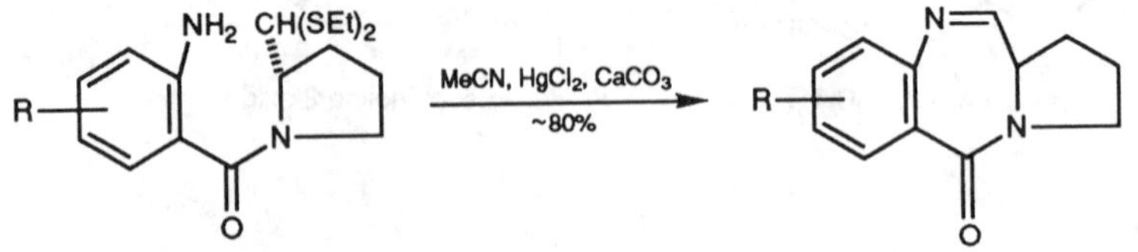

MeCN, $HgCl_2$, $CaCO_3$, ~80% [3272, 3710]

(**2.7**) R = H, MeO, Me, $PhCH_2O$

Pyrrolo[2,1-*c*][1,4]benzodiazepin-5-one

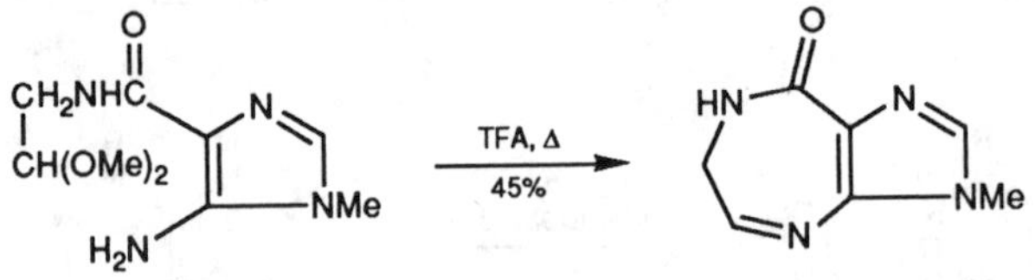

TFA, Δ, 45% [3124]

(**2.8**)

Imidazo[4,5-*e*][1,4]diazepin-8-one

3. 1,4-Thiazepin-5-one

An acetal attached through a sulphur-containing side-chain to a 2-carbamoylphenyl ring cyclizes on heating with 4-toluenesulphonic acid under anhydrous conditions.

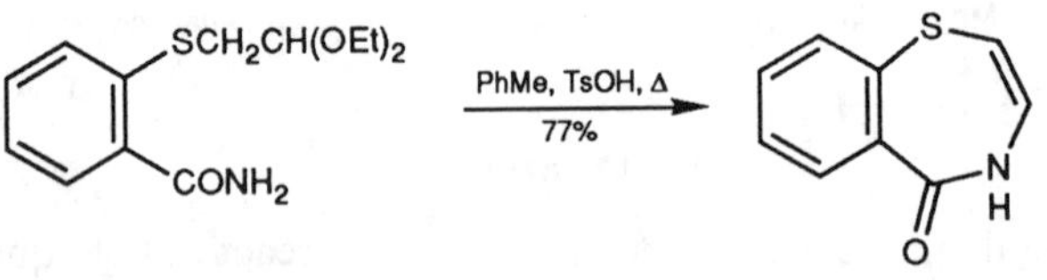

PhMe, TsOH, Δ, 77% [3391]

1,4-Benzothiazepin-5-one

CHAPTER 3

Acetal and Ring-carbon or Ring-nitrogen

I.	Formation of a Five-membered Ring	675
	1. Pyrrole	675
	2. Furan	676
	3. Thiophene	676
	4. Thiazole	677
II.	Formation of a Six-membered Ring	677
	1. Pyridine	677
	2. Pyridin-2-one	678
III.	Formation of a Seven-membered Ring	679
	1. Azepine	679
	2. Azepin-2-one	679

I. FORMATION OF A FIVE-MEMBERED RING

1. Pyrrole

One of the methylthio groups of a ketenedithioacetal is displaced by an anion derived from an active methylene group; simultaneously, the other MeS and R^3 (an activating group) are eliminated. A Pomeranz–Fritsch reaction, catalysed by boron trifluoride–trifluoroacetic anhydride (see also Section II.1), is an attractive synthesis of some 1-alkyl-2,3-unsubstituted indoles. Recent work on indoles has been reviewed [3893].

SMe; $N^+C{=}CSMe$; COOEt; R^1 + $R^2CH_2R^3$ —EtOH, TEA, Δ; ~53%→ R^1, R^2, SMe, COOEt, N [2880]

R^1 = H, Me; R^2 = H, CN; R^3 = NO_2, $PhSO_2$

Indolizine

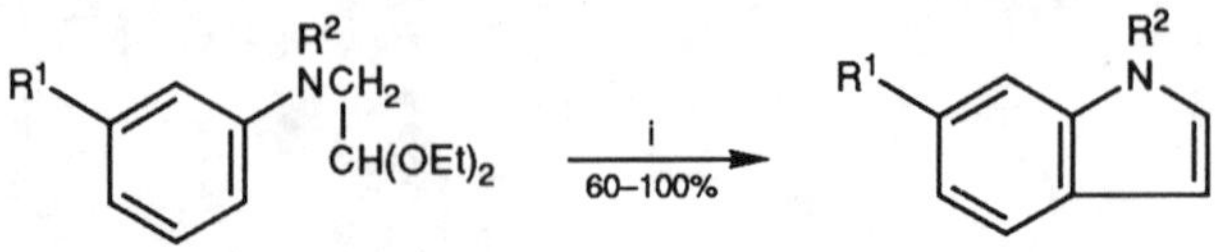

$R^1 = H, MeO$; $R^2 = Me, Et$
i, BF_3, $(CF_3CO)_2O$, AcOH

60–100%

Indole

[2378]

2. Furan

A 2,3-unsubstituted furo ring is formed when the acetal of a phenoxyacetaldehyde is heated with PPA. Recent developments in the chemistry of benzofurans, have been reviewed [3894, 3983].

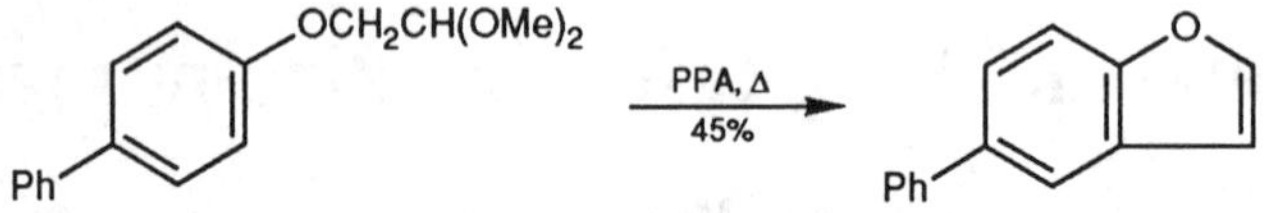

$OCH_2CH(OMe)_2$ PPA, Δ 45%

Benzofuran

[2468]

3. Thiophene

Synthesis of benzo[*b*]thiophene from a phenylthioethyl-acetal or -ketal often gives very good yields [2167], but a recent paper [3485] shows that in some instances, higher yields are obtained in a heterogeneous suspension in chlorobenzene containing a small amount of PPA.

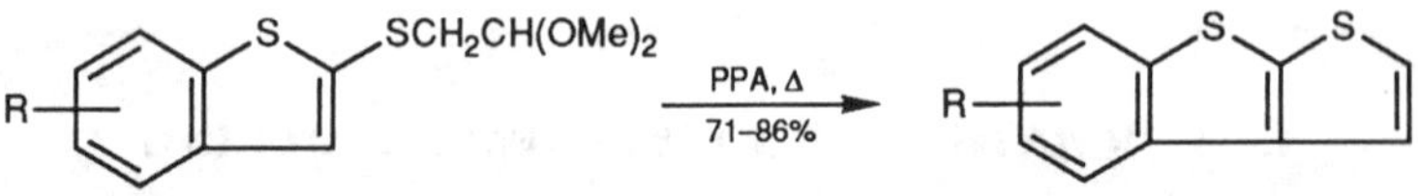

$SCH_2CH(OMe)_2$ PPA, Δ 71–86%

R = H, Cl, Br

Thieno[2,3-*b*][1]benzothiophene

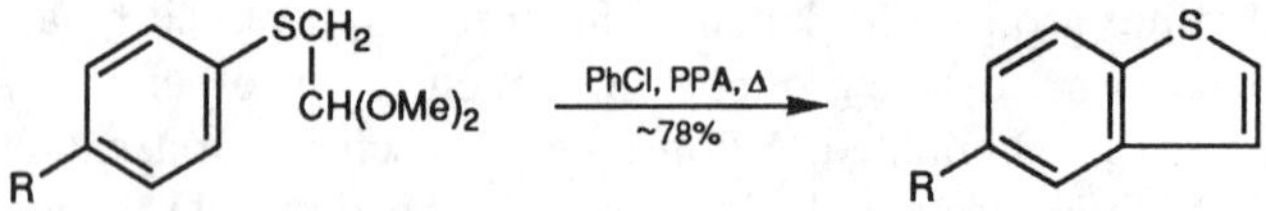

[2167]

$SCH_2CH(OMe)_2$ PhCl, PPA, Δ ~78%

R = Cl, Br, Me

Benzo[*b*]thiophene

[3485]

Another ring system synthesized similarly:

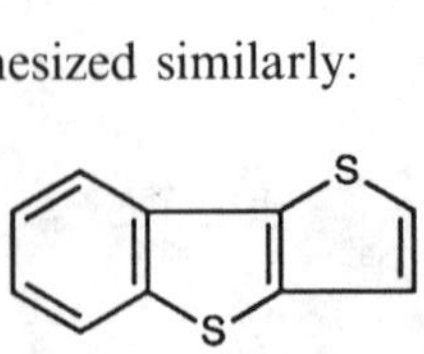

Thieno[3,2-*b*][1]benzothiophene

[2167]

4. Thiazole

An omega-acetal in a sulphur-containing side-chain can cyclize on to a ring-nitrogen in the presence of sulphuric acid.

CH_2S $(EtO)_2CH$ HN N $\xrightarrow[64\%]{H_2SO_4}$ S N N [2367]

Naphth[1',2':4,5]imidazo[2,1-*b*]-thiazole

II. FORMATION OF A SIX-MEMBERED RING

1. Pyridine

It has recently been shown that the Pomeranz–Fritsch reaction, on a methylthiobenzylamine gives, in additon to the product of *para* cyclization, a measurable amount of the *ortho* product [3063]. Consequently, a re-examination of the corresponding reaction on a 3-methoxy- and 3,4-dimethoxy-benzylamine has been made; this showed that whereas the methoxy substrate gave yields of 4 and 71% of quinolines from *ortho* and *para* attack, respectively, only the latter type of product was obtained from the dimethoxybenzylamine. This study also examined the *o*/*p* ratio of the products from related syntheses of isoquinolines. The effect of the MeS group was examined further in a number of cyclizations which lead to isoquinolines. This showed that (a) the MeS substituent has a strong activating effect and (b) the acidic catalyst chosen has a pronounced effect on the *ortho–para* ratio [3303]. Amongst the acid catalysts which are useful in the Pomeranz–Fritsch cyclization is boron trifluoride–trifluoroacetic anhydride, which is effective at ambient temperature, but substituents on the benzene ring have an important effect on yields. Heating the aminoacetal (**3.1**) (Chapter 2, Section I.1) with sulphuric acid converts it in high yield into an isoquinoline which is the result of annulation of the acetal group with a carbon in the unsubstituted phenyl ring.

MeO R^1 Ts CH_2NCH_2 $HC(OMe)_2$ $\xrightarrow[47–70\%]{\text{diox. HCl, }\Delta}$ MeO R^1 N [3302]

R^1 = H, MeO

Isoquinoline

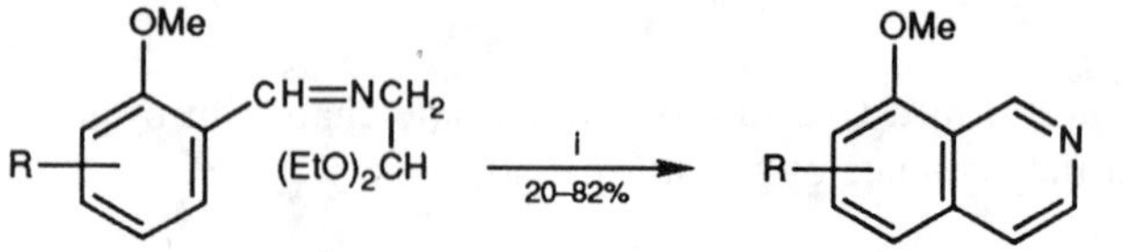

OMe CH=NCH$_2$ (EtO)$_2$CH → i, 20–82% → Isoquinoline [2378, 3418]

R = Me, MeO, HO, PhCH$_2$O
i, BF_3, $(CF_3CO)_2O$, AcOH

Isoquinoline

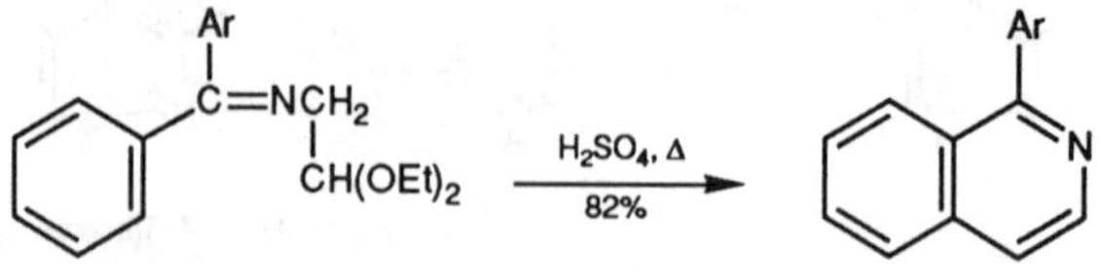

Ar C=NCH$_2$ CH(OEt)$_2$ → H_2SO_4, Δ, 82% → [2612]

(3.1) Ar = 2-NH_2-5-ClC_6H_3

Isoquinoline

When the *N*-tosylacetal (**3.2**) is heated with dilute hydrochloric acid and dioxan, a surprisingly facile detosylating–dehydrogenative cyclization occurs to give the isoquinolines in high yields—a valuable variation of the Pomeranz–Fritsch reaction. It has been used in a synthesis of the medicinally important ellipticine [2870].

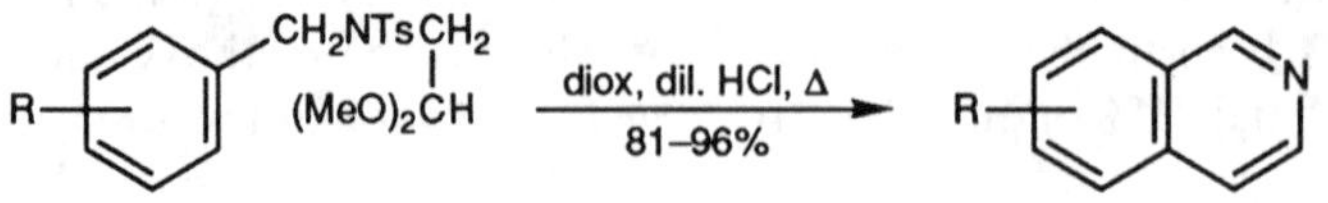

CH$_2$NTsCH$_2$ (MeO)$_2$CH → diox, dil. HCl, Δ, 81–96% → [2609]

(3.2)
R = HO, MeO, Cl

Isoquinoline

Another ring system synthesized similarly:

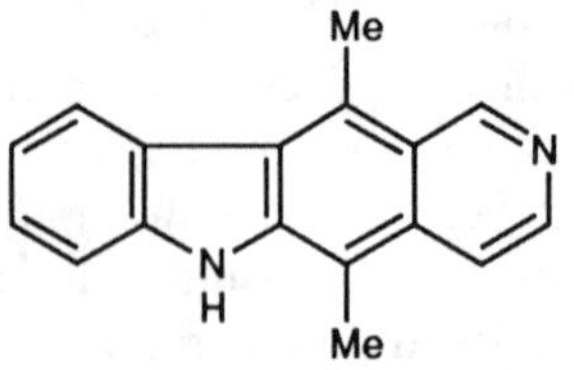

Me N H Me N [2870]

Pyrido[4,3-*b*]carbazole

2. Pyridin-2-one

4-Toluenesulphonic acid is a useful catalyst in the annulation of acetal and ring-carbon functions. This takes precedence over the expected reaction at the ring-nitrogen of the pyrrole (**3.3**). Di(methylthio)vinyl groups are susceptible to nucleophilic attack, for example, by anions derived from malonic acid derivatives.

PhMe, TsOH, Δ; 32% [3418]

(**3.3**) Pyrrolo[2,3-*c*]pyridin-7-one

Me_2SO_2, K_2CO_3; 40–85% [2515]

R = COOMe, COOEt, CN, 2-$MeC_6H_4SO_2$, 2-$NO_2C_6H_4$ 4-Quinolizinone

III. FORMATION OF A SEVEN-MEMBERED RING

1. Azepine

Methanesulphonic acid is an efficient catalyst for cyclizing omega-acetal side-chains on to ring-carbon atoms. An alternative method which gives comparable yields is to warm the acetal at about 50 °C for 4 h in a solvent which is saturated with hydrogen chloride.

AcOH, $MeSO_3H$; 79% [3574]

3-Benzazepine

2. Azepin-2-one

Electron-releasing substituents in the same benzene ring are needed for the success of the room temperature cyclization of the acetal group to a ring carbon of the carboxamide (**3.4**).

AcOH, HCl(g); 75% [3770]

(**3.4**) 3-Benzazepin-2-one

CHAPTER 4

Acylamine and Aldehyde or Ketone

I.	Formation of a Five-membered Ring	680
	1. Pyrrole	680
II.	Formation of a Six-membered Ring	681
	1. Pyridine	681
	2. Pyridin-2-one	681
	3. Pyridazine	681
	4. Pyrimidine	682
III.	Formation of a Seven-membered Ring	682
	1. Azepine	682

Cyclization of the monotosylhydrazone of a 1,2-diacylbenzene is exemplified.

I. FORMATION OF A FIVE-MEMBERED RING

1. Pyrrole

3-Acylamino-4-acylmethylpyridines are cyclized in high yield by heating with dilute hydrochloric acid; the *N*-acyl group is lost.

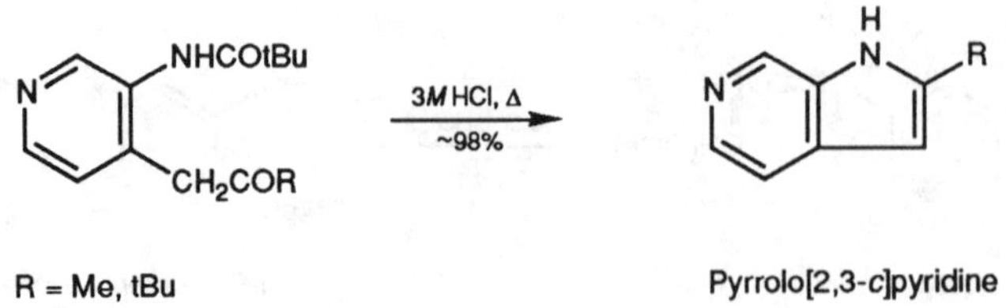

NHCOtBu, CH_2COR —(3*M* HCl, Δ; ~98%)→ [3414]

R = Me, tBu

Pyrrolo[2,3-*c*]pyridine

Other ring systems synthesized similarly:

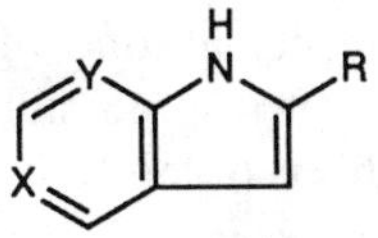

X = CH, Y = N Pyrrolo[2,3-*b*]pyridine
X = N, Y = CH Pyrrolo[3,2-*c*]pyridine

[3414]

II. FORMATION OF A SIX-MEMBERED RING

1. Pyridine

When the *o*-acylaminoacetophenone (**4.1**) is stirred with the chloroiminium chloride, cyclization proceeds in high yield to a quinoline which is an important intermediate in the synthesis of an antihypertensive agent.

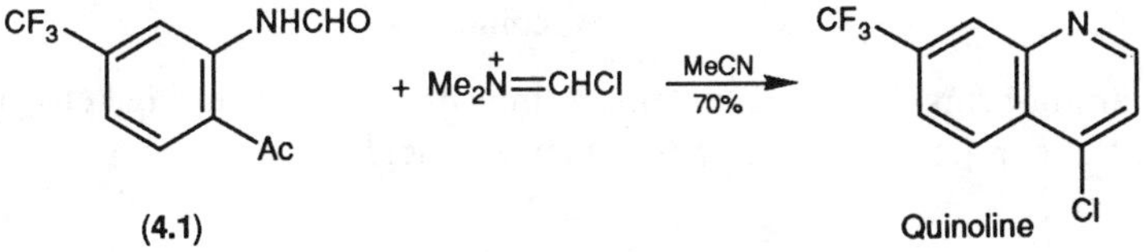

CF_3 NHCHO Ac (4.1) + $Me_2\overset{+}{N}{=}CHCl$ —MeCN, 70%→ Quinoline (CF_3, Cl)

[3859]

2. Pyridin-2-one

The acylamino ketone (**4.2**) is efficiently cyclized on heating for 2 h in DMF containing potassium carbonate.

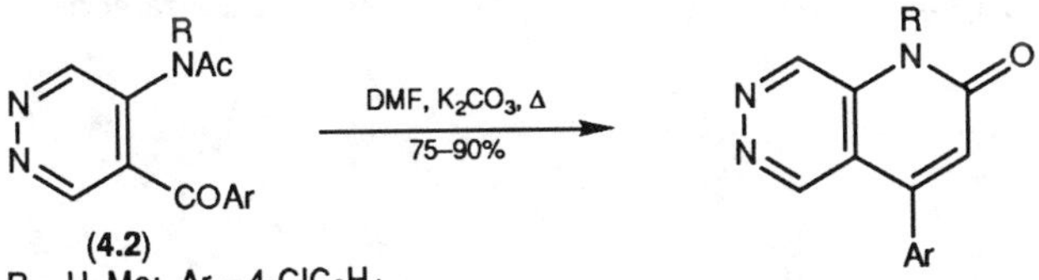

(4.2) R = H, Me; Ar = 4-ClC_6H_4 —DMF, K_2CO_3, Δ, 75–90%→ Pyrido[2,3-*d*]pyridazin-2-one

[3155]

3. Pyridazine

o-Acyl-tosylhydrazone (**4.3**) is cyclized on heating briefly in ethanol but when R = tBu, no cyclization occurs even on prolonged boiling.

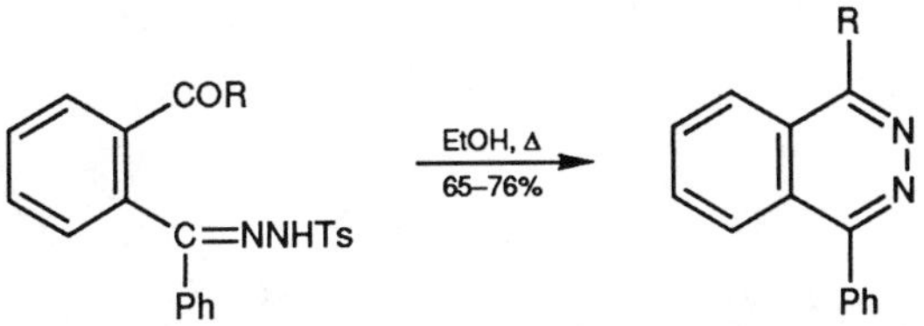

(4.3) R = Ph, 2-furyl —EtOH, Δ, 65–76%→ Phthalazine

[3186]

4. Pyrimidine

The bifunctional hydrazine, reacts at one of its amino groups with a 2-acetamidobenzophenone, and as it is impossible for dehydration to occur, the product is an amino alcohol. This is obtained in good yield only when an excess of hydrazine is available.

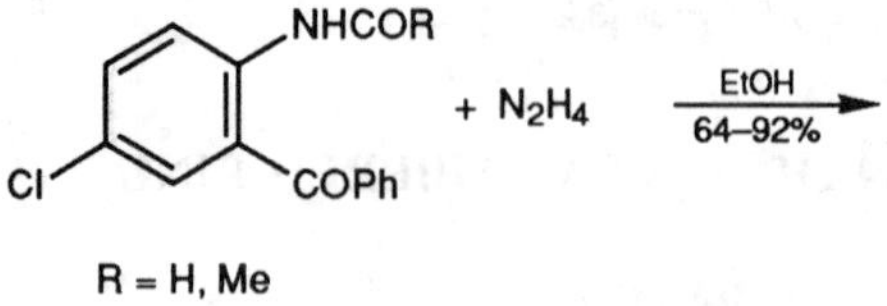

NHCOR, COPh, Cl + N_2H_4 → (EtOH, 64–92%) N, R, NNH_2, Cl, HO, Ph [2279a]

R = H, Me

Quinazoline

III. FORMATION OF A SEVEN-MEMBERED RING

1. Azepine

An acylaminopropyl side-chain forms a ring with a neighbouring keto group on heating the compound with 2*M* hydrochloric acid.

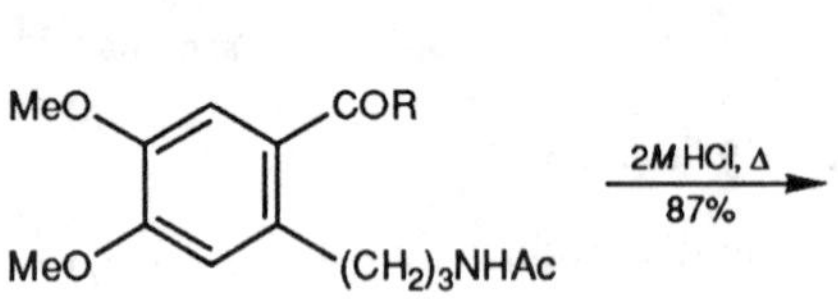

MeO, MeO, COR, $(CH_2)_3NHAc$ → (2*M* HCl, Δ, 87%) R, N, MeO, MeO [3348]

R = 3,4,5-$(MeO)_3C_6H_2(CH_2)_2$

2-Benzazepine

CHAPTER 5

Acylamine or Carbamate and Amine or Hydrazine

I. Formation of a Five-membered Ring 683
1. Imidazole 683
2. Imidazol-2-one 685
3. 1,2,4-Triazole 685
II. Formation of a Six-membered Ring 685
1. Pyrimidin-2-one 685
2. 1,2,4-Triazine 686
3. 1,2,4-Triazin-3-one 687
4. 1,3,5-Triazine-2,4-dione or 1,3,5-Triazin-2-one, 4-thioxo- 687
5. 1,2,4-Thiadiazine 1,1-Dioxide 687
6. 1,3,4-Thiadiazine 688
III. Formation of a Seven-membered Ring 688
1. 1,3-Diazepine 688
2. 1,2,4-Triazepine-3,7-dione 688

Examples of cyclizations of the following pairs of functions are also described: acylhydrazine–amine, acylamine–alkoxyamine and acylamine–imine.

I. FORMATION OF A FIVE-MEMBERED RING

1. Imidazole

The synthesis of purines from monoacyl pyrimidine-4,5-diamines is a commonly used method (for example, [2252]). Similarly, 8-hydroxymethylpurin-6-one is synthesized from 5-acylamino-6-aminopyrimidin-4-one and ethyl glycolate but not from the diamine and the keto ester [2506]. In contrast, the mono(chloroacetyl)aniline (**5.1**) undergoes simultaneous cyclization and hydrogenolysis when the nitro group is reduced catalytically [2046]. An *o*-acylamino-alkoxyamine is cyclized by heating with diethoxymethyl acetate, to give, in this example, a purine [3924].

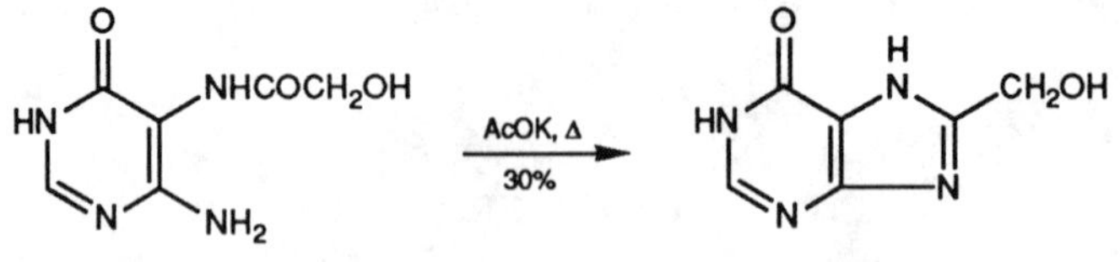

AcOK, Δ; 30% [2506]

6-Purinone

EtOH, H_2, RaNi; 78–86% [2046]

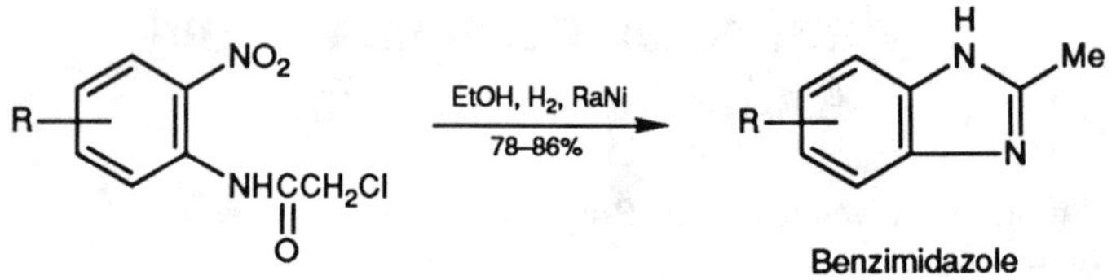

Benzimidazole

(**5.1**) R = H, MeO, Me

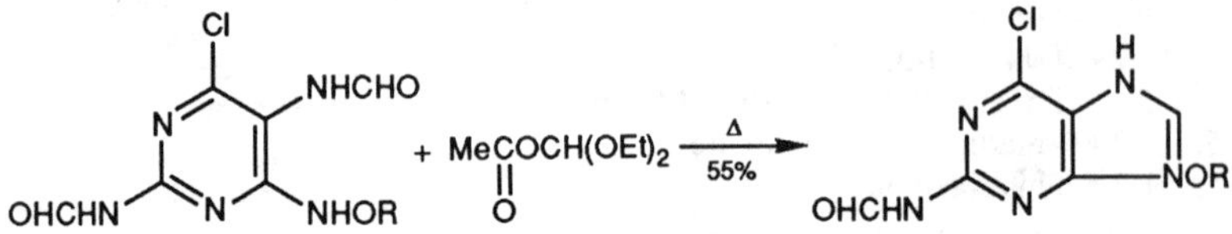

+ $MeCOCH(OEt)_2$ Δ; 55% [3924]

R = $CH_2{=}CH(CH_2)_2$

Purine

Annulation of an *o*-(acylamino)aniline has been induced by acidic or nearly neutral conditions (see p. 24), but an aqueous alkaline medium is also successful for this reaction. In another method, ethyl 2-ethoxyquinoline-1-carboxylate is used [2766].

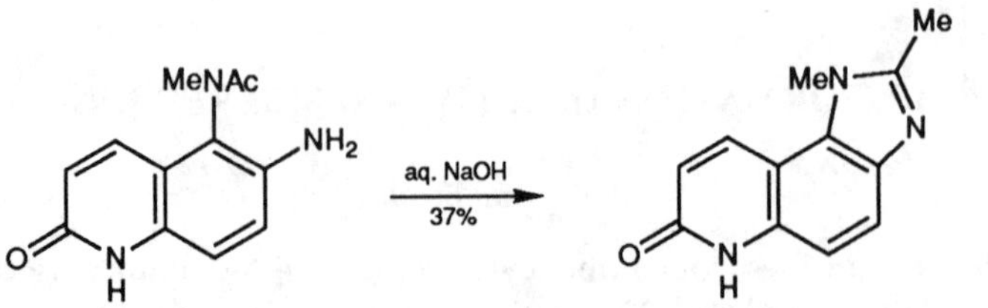

aq. NaOH; 37% [3354]

Imidazo[4,5-*f*]quinazolin-7-one

Another ring system synthesized similarly:

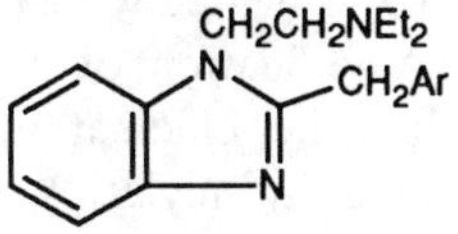

Benzimidazole [2766]

2. Imidazol-2-one

o-Aminocarbamates are cyclized under mild conditions with trichloromethyl chloroformate.

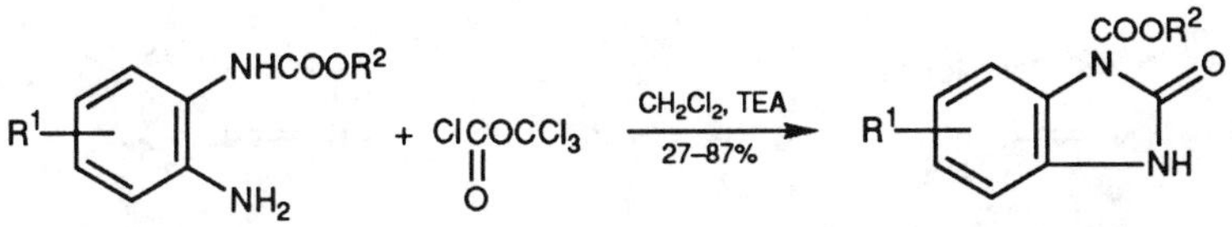

R^1 = Br, Cl, CF_3, Ac, MeO, Me;
R^2 = saturated 8-azabicyclo[3.2.1]octyl or similar group

CH_2Cl_2, TEA, 27–87%

2-Benzimidazolone [3773]

3. 1,2,4-Triazole

Treatment of an *o*-(*N*-acylamino)amine [such as (**5.2**)] with phosphorus oxychloride–PPA yields a fused triazole.

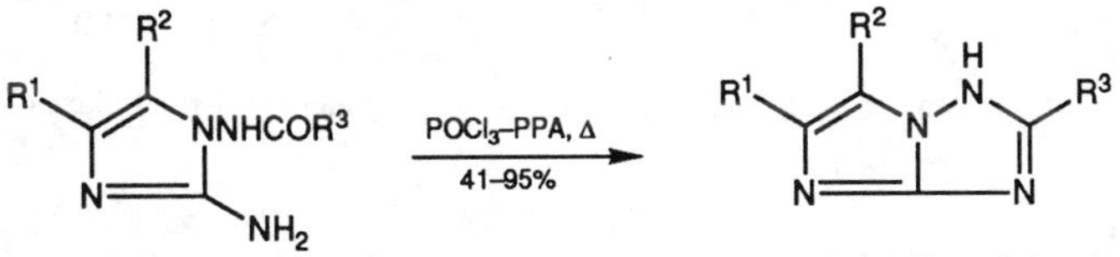

(**5.2**)
R^1 = Ph, 4-Cl-, 4-Br-, 4-$MeOC_6H_4$;
R^2 = H, Ph; R^3 = alkyl, $PhCH_2$

$POCl_3$–PPA, Δ, 41–95%

Imidazo[1,2-*b*][1,2,4]triazole [2097]

II. FORMATION OF A SIX-MEMBERED RING

1. Pyrimidin-2-one

This ring is formed by heating an amino-carbamate with a strong base.

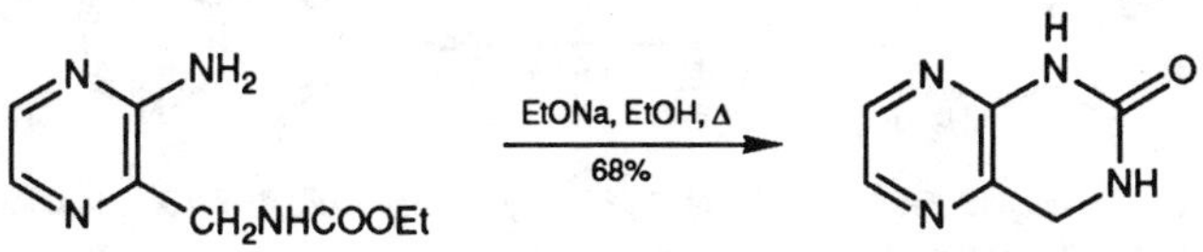

EtONa, EtOH, Δ, 68%

2-Pteridinone [2159]

2. 1,2,4-Triazine

When the (acylamino)amine (**5.3**) is heated with 1,2-diketones, a triazine ring is formed in high yield [2097]. Acidification of a 2-(acylhydrazino)amine with ethanolic hydrogen chloride is another efficient method [2319]. Moderate yields of a fused triazine are obtained by acylation of a 2-nitrophenylhydrazine followed by catalytic reduction and cyclization with mineral acid [3408]. The first cyclic product of the treatment of an *o*-acylhydrazino-amine with ethanolic hydrogen chloride is the unstable 1,2-dihydropyridotriazine which is dehydrogenated with either manganese dioxide or hexacyanoferrate(III) (ferricyanide) to the more stable triazine [3526].

Ph NNHCOMe NH$_2$ N + COR^1 COR^2 —EtOH, HCl, Δ / 76–91%→ Imidazo[1,2-*b*][1,2,4]triazine [2097]

(5.3)
R^1,R^2 = H, Me, Ph

R^1 NHAc NH$_2$ R^2 —EtOH–HCl / ~84%→ R^1 NH Me HCl [2319]

R^1 = H, Me; R^2 = H, Cl, COOMe   Pyrido[3,2-*e*]-1,2,4-triazine

Me NNHMe NO$_2$ R^1 —i – iii / ~50%→ Me NMe R^2 R^1 [3408]

R^1 = H, CF_3; R^2 = Me, Ph   1,2,4-Benzotriazine
i, R^2COCl, pyr; ii, EtOH, H_2, Pd–C; iii, HCl, Δ

NHNHCOR NH$_2$ HCl —i, ii / 9–76%→ R [3526]

R = H, alkyl, 2-F-, 4-CF_3, 4-CN-C_6H_4 Pyrido[3,4-*e*]-1,2,4-triazine
i, EtOH–HCl; Δ; ii, MnO_2, NaOH

3. 1,2,4-Triazin-3-one

Annulation of acylamino and hydrazino groups occurs in moderate yield on treatment with a strong base.

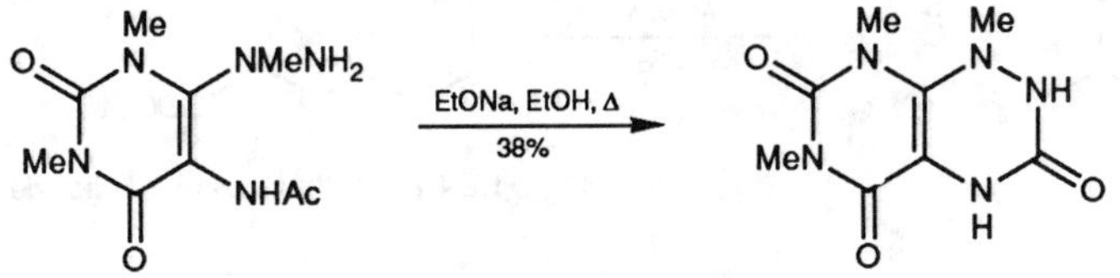

EtONa, EtOH, Δ 38% [2753]

Pyrimido[5,4-*e*]-1,2,4-triazine-3,5,7-trione

4. 1,3,5-Triazine-2,4-dione or 1,3,5-Triazin-2-one, 4-thioxo-

Amino and *N*-ethoxycarbonylthioureido (NHCSNHCOOEt) groups are cyclized by stirring the compound with a strong base at ambient temperature, or by heating in pyridine. (For classification purposes, the ethoxycarbonylthioureido group is regarded as a carbamate because it is the NHCOOEt group which reacts.)

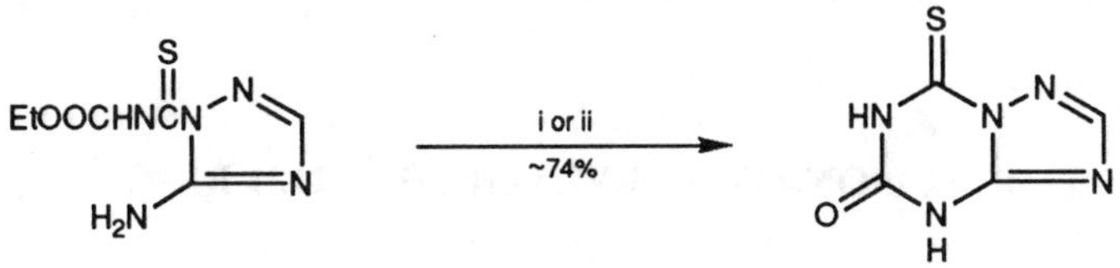

i or ii ~74% [2339, 2396]

i, EtOH–EtONa; ii, pyr, Δ

[1,2,4]Triazolo[1,5-*a*][1,3,5]triazin-5-one, 7-thioxo-

Other ring systems synthesized similarly:

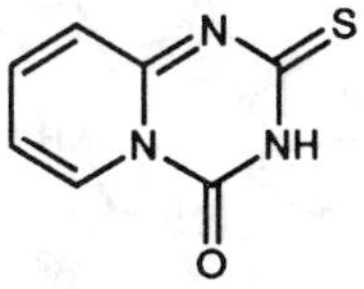

[2388]

Pyrido[1,2-*a*]-1,3,5-triazin-4-one, 2-thioxo-

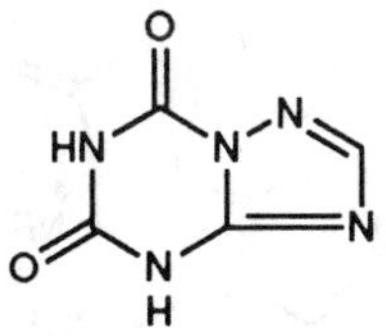

[2504]

[1,2,4]Triazolo[1,5-*a*][1,3,5]triazine-5,7-dione

5. 1,2,4-Thiadiazine 1,1-Dioxide

This ring is formed when a 2-aminobenzenesulphonamide, which carries an ethoxalyl group on the sulphonamide nitrogen, is treated with acetic anhydride.

Thiadiazines containing the 1,2-SO_2NH— moiety in the ring have been reviewed [3904].

Ac_2O, Δ, 32% [2805]

1,2,4-Benzothiadiazine 1,1-dioxide

6. 1,3,4-Thiadiazine

When the amino-thioacylhydrazine (**5.4**) is heated with dilute mineral acid, the amino group is eliminated and a thiadiazine ring is formed.

2*M* HCl, Δ, 10–73% [3382]

(**5.4**)
R^1 = Me, MeS, Ph, 4-Cl-, 4-Me-C_6H_4; R^2 = Me, Ph

[1,2,4]Triazino[5,6-*e*][1,3,4]thiadiazine

III. FORMATION OF A SEVEN-MEMBERED RING

1. 1,3-Diazepine

Acid-mediated annulation of acylamino and imino groups proceeds at ambient temperatures in moderate to high yields.

aq. HCl, 44-95% [2067]

R^1 = H, Me, MeO; R^2 = alkyl

Furo[3,4-*d*][1,3]benzodiazepin-3-one

2. 1,2,4-Triazepine-3,7-dione

In order to prevent an undesired cyclization, an *N*-methyl group was introduced to give the reactant (**5.5**). When heated with 4-nitrophenyl chloroformate, and

TEA, the triazepinedione was obtained. An alternative approach is to cyclize the amino-carbamate (**5.6**) with a strong base.

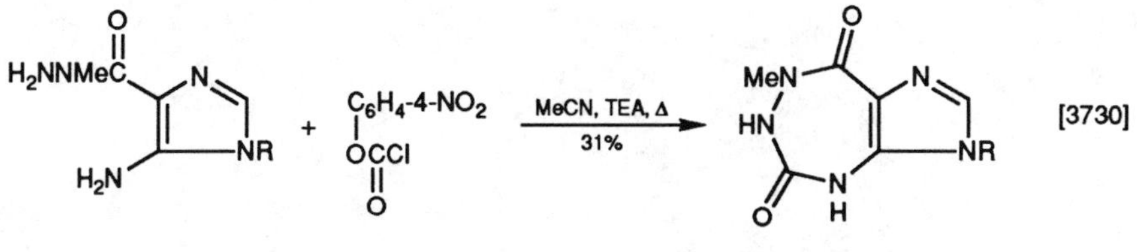

$$\text{(5.5) } (R = PhCH_2) + C_6H_4\text{-}4\text{-}NO_2OC(O)Cl \xrightarrow[31\%]{MeCN,\ TEA,\ \Delta} \text{Imidazo[4,5-}e\text{][1,2,4]-triazepine-5,8-dione} \qquad [3730]$$

(5.5)
R = $PhCH_2$

Imidazo[4,5-*e*][1,2,4]-triazepine-5,8-dione

$$\text{(5.6) } (R = PhCH_2) \xrightarrow[89\%]{MeOH,\ MeONa,\ \Delta} \qquad [3730]$$

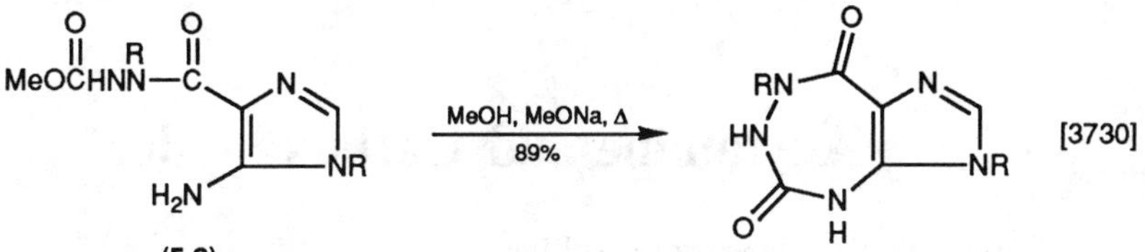

(5.6)
R = $PhCH_2$

CHAPTER 6

Acylamine and Carboxamide

I. Formation of a Five-membered Ring 690
1. Imidazole 690
II. Formation of a Six-membered Ring 691
1. Pyrimidine 691
2. Pyrimidine 1-oxide 691
3. Pyrimidin-2-one or -4-one 691

Examples of the cyclization of *O*- and *N*-acyl derivatives of an *o*-aminoamidroxamine are included in this chapter.

I. FORMATION OF A FIVE-MEMBERED RING

1. Imidazole

When the *O*-benzoyl derivative of 2-aminobenzamide oxime is kept in boiling water for about 5 h and its pH adjusted to about 9 with aqueous alkali, 2-aminobenzimidazole is obtained in almost quantitative yield.

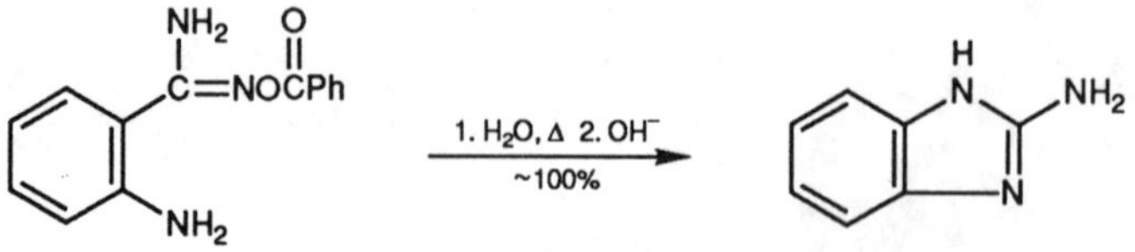

NH2 O C=NOCPh NH2 —(1. H_2O, Δ 2. OH^-; ~100%)→ H N NH2 N [3429]

Benzimidazole

II. FORMATION OF A SIX-MEMBERED RING

1. Pyrimidine

Most cyclizations of acylaminocarboxamides yield pyrimidinone rings, but when treated with primary amines in the presence of phosphorus pentoxide and TEA hydrochloride, a new pyrimidine ring is formed.

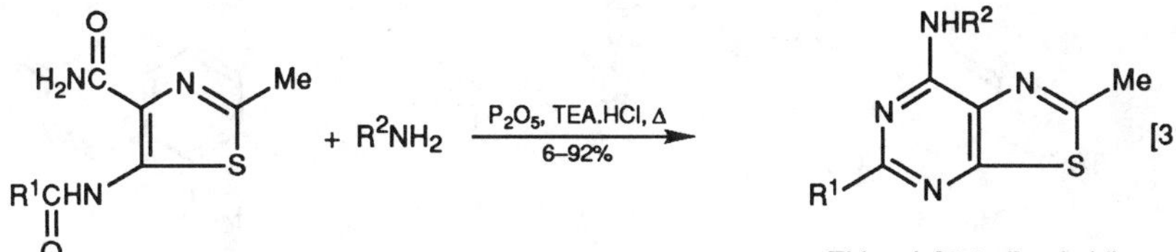

$H_2NC(=O)$, Me, N, S, $R^1C(=O)HN$ + R^2NH_2 —P_2O_5, TEA.HCl, Δ; 6–92%→ NHR^2, N, N, Me, S, R^1 [3107]

Thiazolo[5,4-*d*]pyrimidine

R^1 = H, Me; R^2 = Ph, 4-Cl-, 4-F-, 4-Me-, 3-CF_3-C_6H_4, $Me_2C_6H_3$

2. Pyrimidine 1-Oxide

The *o*-(acylamino)amidroxamine (**6.1**) is a versatile intermediate and is cyclized to the *N*-oxide by heating with ethanol and hydrochloric acid for 40 min (see also Section II.3). Annulation of the acylamino and oximino functions in the *peri* positions of the thioxanthene (**6.2**) is an efficient acid-induced process.

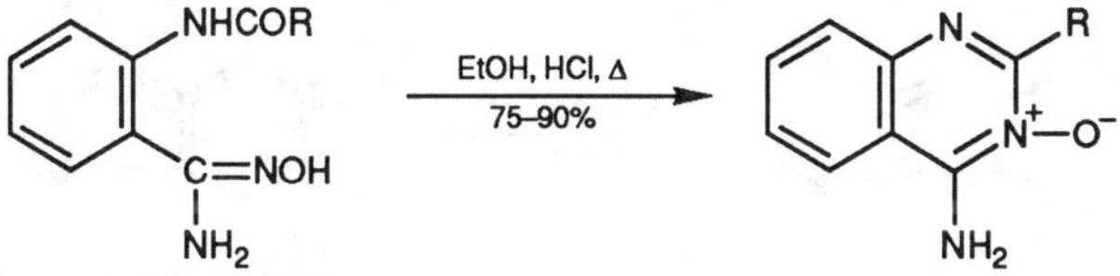

NHCOR, C=NOH, NH_2 —EtOH, HCl, Δ; 75–90%→ N, R, N^+–O^-, NH_2 [3429]

(6.1)
R = Me, Ph

Quinazoline 3-oxide

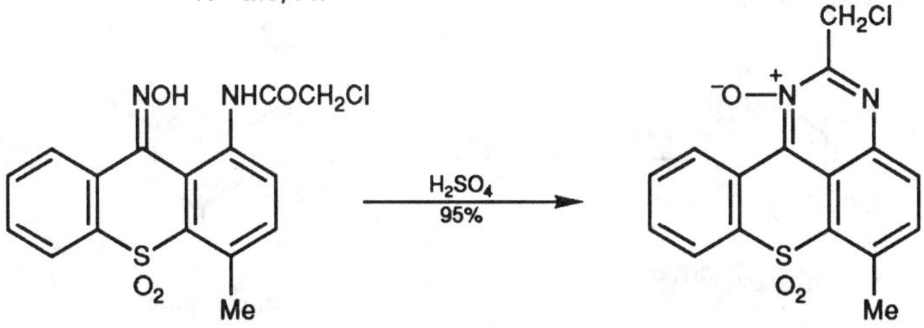

NOH, $NHCOCH_2Cl$, S, O_2, Me —H_2SO_4; 95%→ CH_2Cl, ^-O–N^+, N, S, O_2, Me [2443]

(6.2)

[1]Benzothiopyrano[4,3,2-*de*]quinazoline 1,7,7-trioxide

3. Pyrimidin-2-one or -4-one

Although the yield in this synthesis of 2-quinazolinone is low, it provides a route to a 4-substituted ring system; it deserves further study of its scope [3105].

Attempts to hydrolyse the acylamino group of an acylamino-carboxamide with aqueous alkali resulted in the formation of a pyrimidinone ring [2311] which is also formed in a variant of a well-known reaction (see p. 29) but using sodium methoxide–DMF as the cyclizing agent [3215, 3661].

A 2-acylaminobenzonitrile is converted into a quinazolinone under basic and oxidizing conditions [3771], while the amidroxamine **(6.1)** gives a high yield of a 4-pyrimidinone oxime by treatment with hot ethanolic alkali [3429].

NHCOOEt, CN, MeO, MeO + RMgBr —(Et_2O, diox, Δ; ~11%)→ 2-Quinazolinone [3105]

R = Me, Et

2-Quinazolinone

R^1R^2N, Ph, N, N, N, $R^3HNC(=O)$ —(aq. NaOH, Δ; 80–98%)→ [2311]

R^1 = Ac, EtCO, $PhCH_2CO$; R^2 = H, Ac;
R^3 = H, Ac, EtCO; R^4 = H, Me, Et, $PhCH_2$

1,2,3-Triazolo[4,5-*d*]pyrimidin-4-one

$H_2NC(=O)$, NCH_2Ar, RCHN(C=O) —(MeONa–DMF, Δ; 24–77%)→ [3215, 3661]

Ar = Ph, 4-MeC_6H_4;
R = H, Me, CF_3, $PhCH_2$

Imidazo[4,5-*d*]pyrimidin-7-one

NHC(=O)R, Cl, CN —(EtOH, NaOH, H_2O_2, Δ; 9–76%)→ [3771]

R = $PhCH_2$, $Ph(CH_2)_2$, PhC≡C

4-Quinazolinone

(6.1) —(EtOH, NaOH, Δ; 75%)→ [3429]

R = Ph

4-Quinazolinone oxime

CHAPTER 7

Acylamine and a Carboxylic Acid Derivative

I.	Formation of a Five-membered Ring	693
	1. Pyrrole	693
	2. Pyrazol-3-one	694
II.	Formation of a Six-membered Ring	694
	1. Pyridin-2-one	694
	2. Pyrimidin-4-one	695
	3. Pyrimidin-4-one, 2-thioxo-	695
	4. 1,3-Oxazin-6-one	695
	5. 1,2-Thiazin-5-one 1,1-Dioxide	696
	6. 1,3,4-Oxadiazin-6-one	696

Examples of substrates in which the acylamino group is replaced by either an acylhydrazino or a sulphonylamino are included in this chapter; the second group may be a carboxylic acid.

I. FORMATION OF A FIVE-MEMBERED RING

1. Pyrrole

A synthesis of a partially reduced indolizine which allows the introduction of a substituent at any of the C-3 to C-6 atoms utilizes an *N*-acylpiperidine-2-carboxylic acid which is cyclized on heating with ethyl propynate. A lactone intermediate is proposed for this interesting reaction.

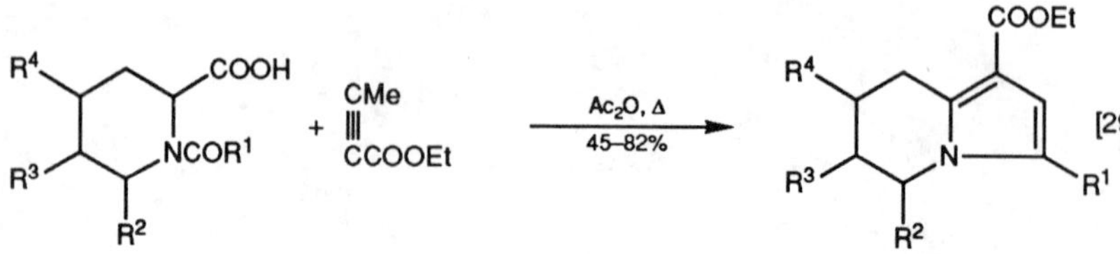

R^1 = H, Me; R^2 = H, COOEt; R^3 = R^4 = H or R^3R^4 = benzo

Ac_2O, Δ 45–82%

Indolizine
Benz[*f*]Indolizine

[2903]

2. Pyrazol-3-one

Cyclization followed by migration of the acyl group from N-2 to N-1 probably occurs when 2-acylhydrazinobenzoic acid is heated.

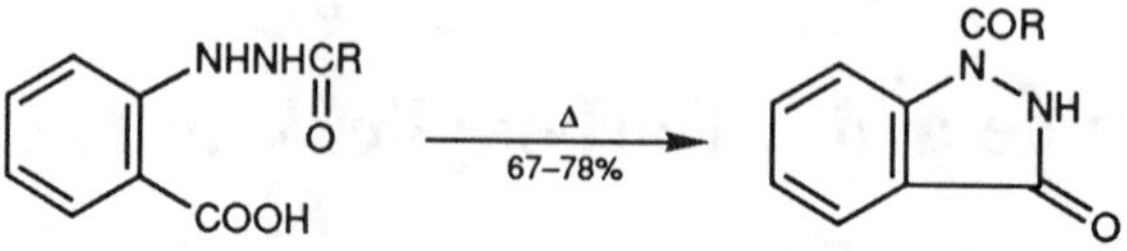

R = Me, Ph

Δ 67–78%

3-Indazolone

[3080]

II. FORMATION OF A SIX-MEMBERED RING

1. Pyridin-2-one

A new and efficient annulation of acylamino and carboxylate ester groups offers a convenient synthetic route to naphthyridinones and 2-quinolinone. The intermediate (**7.2**) is obtained in two stages from 2-pivaloylaminopyridine; very good yields are obtained at each of the three stages.

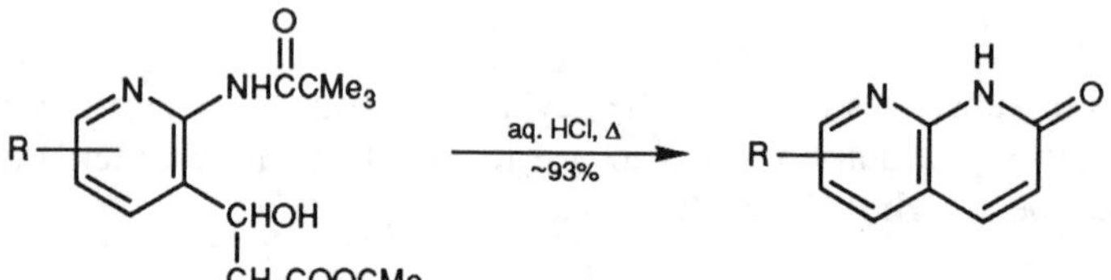

(**7.2**)
R = H, Me, Cl

aq. HCl, Δ ~93%

1,8-Naphthyridin-2-one

[3847]

Other ring systems synthesized similarly:

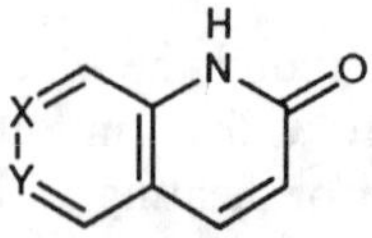

X = Y = CH 2-Quinolinone
X = CH; Y = N 1,6-Naphthyridin-2-one
X = N; Y = CH 1,7-Naphthyridin-2-one

[3847]

2. Pyrimidin-4-one

2-Arylquinazolin-4-ones may be synthesized from acylamino-carboxylic acids, primary arylamines and either phosphorus trichloride or PPA.

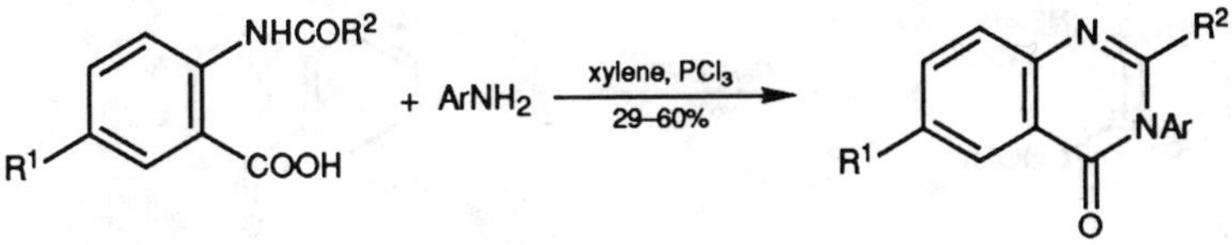

NHCOR^2 + ArNH$_2$ —(xylene, PCl_3; 29–60%)→ [2689]

R^1 = H, NO_2; R^2 = 2-pyridyl;
Ar = Cl-, NO_2-C_6H_4

4-Quinazolinone

3. Pyrimidin-4-one, 2-thioxo-

Thiocarbamate and carboxylic ester functions are annulated under mild conditions—stirring with methylamine for 30 min. When the reactants are heated, the product is a 3-alkyl-2-alkylaminopyrimidin-4-one.

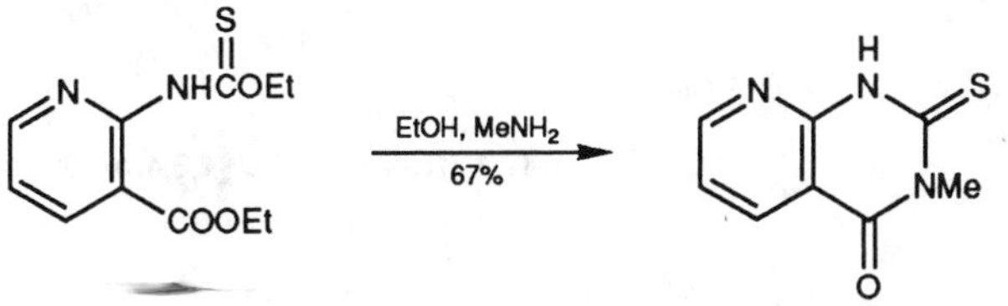

EtOH, MeNH$_2$; 67% [3681]

Pyrido[2,3-*d*]pyrimidin-4-one, 2-thioxo-

4. 1,3-Oxazin-6-one

Amides of 4-oxo-3,1-benzoxazine-2-carboxylic acid may be synthesized by heating oxalamides (**7.3**) with an excess of acetic anhydride.

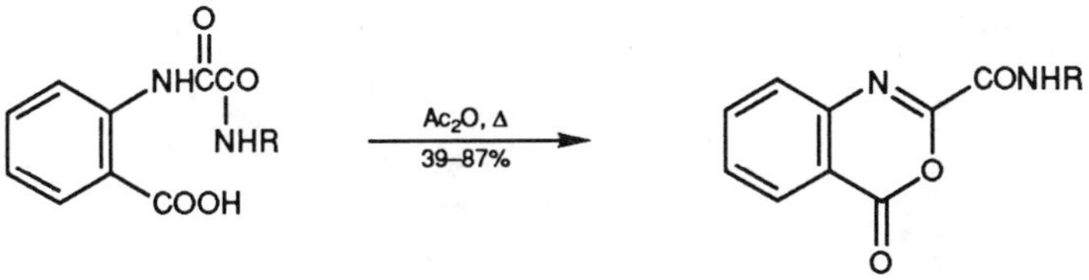

Ac_2O, Δ; 39–87% [2642]

(**7.3**) R = H, alkyl, $PhCH_2$

3,1-Benzoxazin-4-one

5. 1,2-Thiazin-5-one 1,1-Dioxide

A sulphonylamino-carboxylic ester is cyclized by treatment at ambient temperature with sodium hydride–DMF.

NMeSO$_2$Me, COOMe; NaH–DMF, 95%; Me, N, SO$_2$, O [2398]

2,1-Benzothiazin-4-one 2,2-dioxide

6. 1,3,4-Oxadiazin-6-one

This ring is formed when an *N*-acylaminoazole-2-carboxylic acid is heated with thionyl chloride–pyridine. Lower yields are recorded when DCC–THF is used.

Ar^{1}CONHN, N, N, HOOC, Ar^2; pyr-SOCl$_2$, Δ, 71–88%; Ar^1, N, N, N, N, O, O, Ar^2 [3221]

Ar^1 = Ph, 2- or 4-NO$_2$-, 4-MeO-, 2- or 4-Cl-C$_6$H$_4$;
Ar^2 = Ph, 4-MeOC$_6$H$_4$

[1,2,3]Triazolo[1,5-*d*][1,3,4]oxadiazin-4-one

CHAPTER 8

Acylamine or Amine and Ether

I.	Formation of a Five-membered Ring	697
	1. Pyrrole	697
	2. 1,2,4-Triazole	698
	3. Thiazole	699
	4. Thiazol-2-one	699
II.	Formation of a Six-membered Ring	699
	1. Pyridin-4-one	699
III.	Formation of a Seven-membered Ring	700
	1. 1,4-Diazepine	700

I. FORMATION OF A FIVE-MEMBERED RING

1. Pyrrole

Catalytic reduction of the nitro-epoxide (**8.1**, R = COOMe) leads directly to the indole carboxylic ester but when R = H, reduction of the nitro-epoxide does not give the indole. However, cleavage of the epoxide and reduction of the nitro group gives the amine (**8.2**) which is cyclized by a strong base.

An attempt to demethylate the psychotomimetic phenylethylamine (**8.3**) resulted in a cyclized product being isolated [2629]. Another pharmacologically important indole (butotenin) may be synthesized by *in situ* demethylation of the diether (**8.3**) followed by dehydrogenation under alkaline conditions [2696]. Palladium-catalysed coupling of bromoarenes with a stannylethene provides ready access to β-ethoxyvinylbenzenes (**8.4**) which cyclize at room temperature in acid [3929].

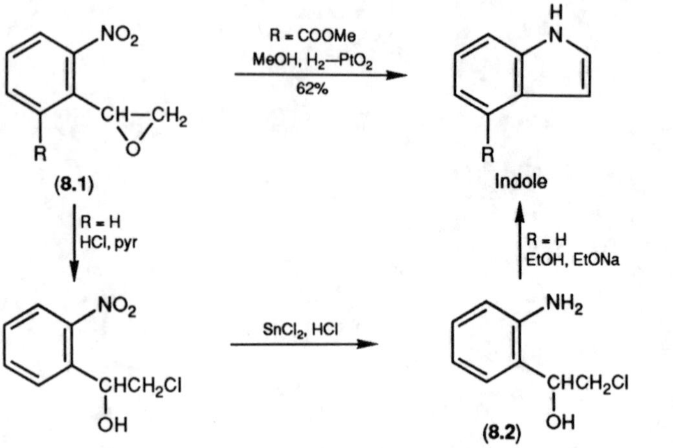

$R = COOMe$; MeOH, H_2—PtO_2; 62%

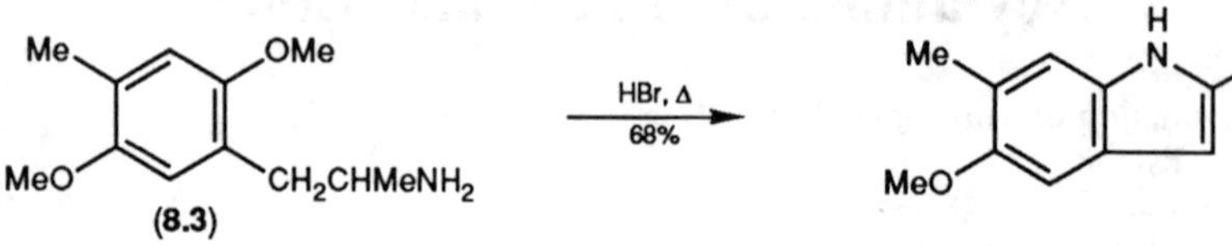

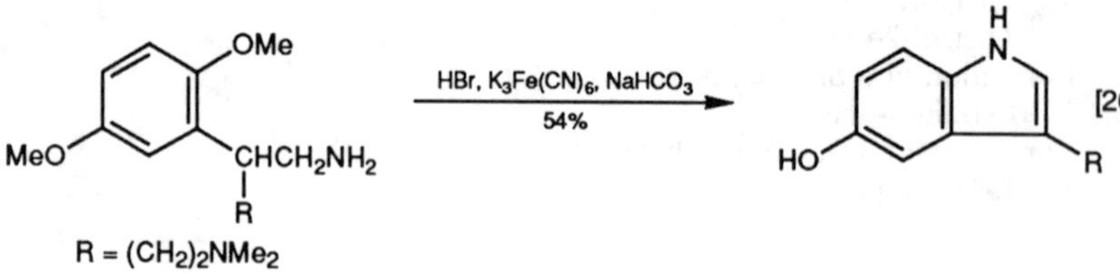

(8.1) → Indole

$R = H$; HCl, pyr

$SnCl_2$, HCl

(8.2)

$R = H$; EtOH, EtONa [2426]

(8.3) HBr, Δ; 68% [2629]

HBr, $K_3Fe(CN)_6$, $NaHCO_3$; 54% [2696]

$R = (CH_2)_2NMe_2$

(8.4) $R = H, Ac$ PhH, TsOH; 29–70% [3929]

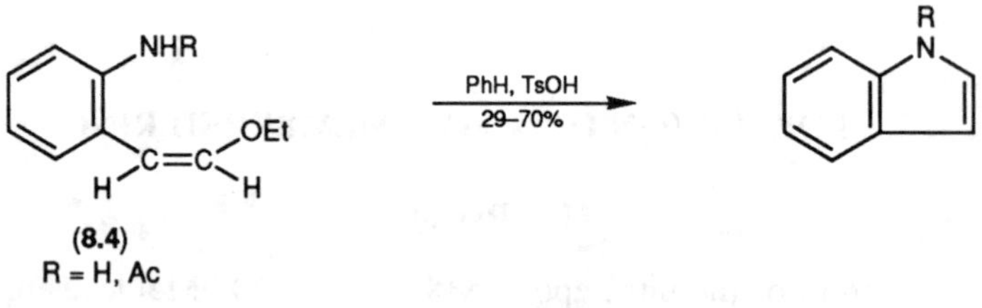

2. 1,2,4-Triazole

Carbodiimides are often used as condensing agents but they are sometimes useful as reactants, for example, in their reaction over many hours with *o*-methylthio-*N*-amines.

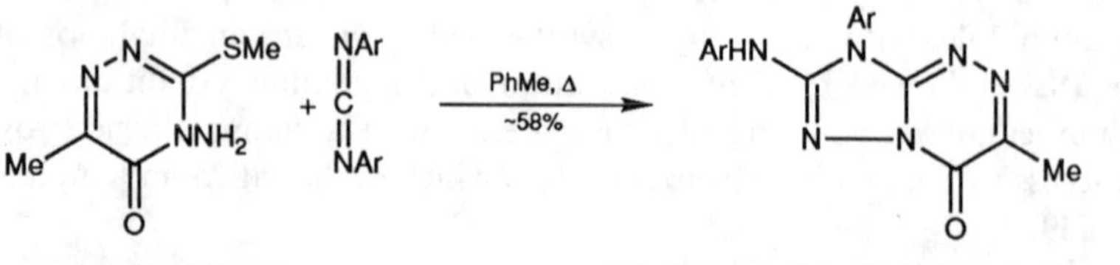

PhMe, Δ; ~58% [3136]

Ar = Ph, 4-BrC_6H_4

[1,2,4]Triazolo[5,1-*c*][1,2,4]-triazin-4-one

3. Thiazole

This cyclization of a 2-(methylthio)anilide was discovered when an attempt was being made to prepare imidoyl chlorides by reaction with phosgene.

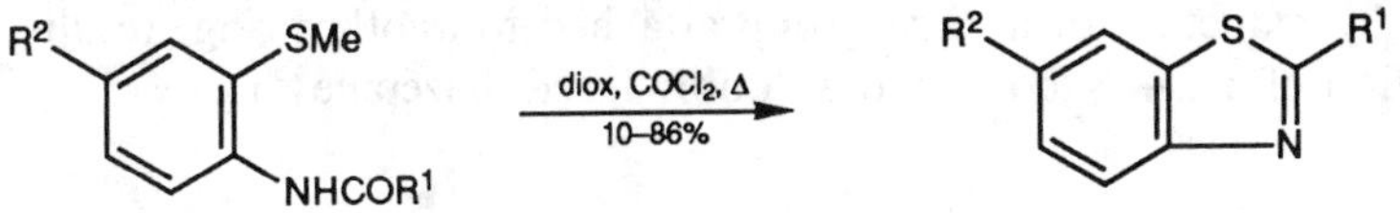

diox, $COCl_2$, Δ
10–86%

[2623]

R^1 = Me, CF_3, cyclopropyl; R^2 = H, Cl

Benzothiazole

4. Thiazol-2-one

Phenols may be converted in thiocarbamates ($ArSCONR_2$) by a Newman rearrangement; amino-thiocarbamates, are thus accessible from aminophenols. This provides a convenient route to 2-benzothiazolones.

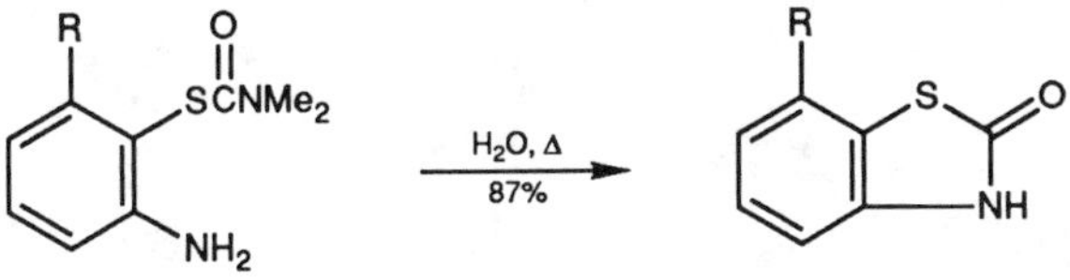

H_2O, Δ
87%

[3358]

R = $CF_3CONH(CH_2)_2$

2-Benzothiazolone

II. FORMATION OF A SIX-MEMBERED RING

1. Pyridin-4-one

Base-induced cyclization of a 2-amino- or 2-acetamido-2′-methoxybenzophenone gives a good yield of the acridinone but variations in the base used, substituent R^2, solvent and temperature have a considerable effect on the yield. Replacing the ether by a tosyloxy group improves the yield of 9-acridinone to 91%.

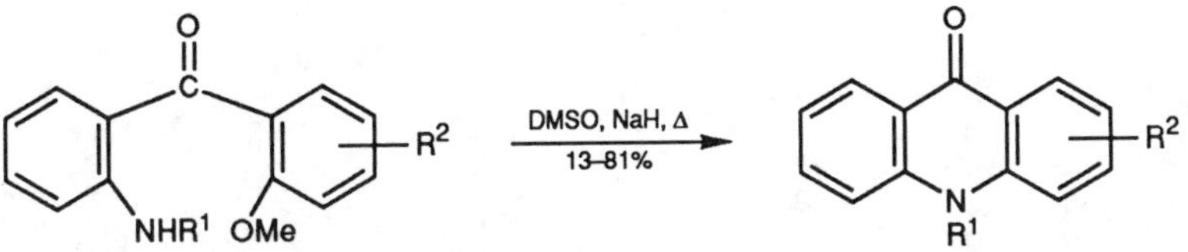

DMSO, NaH, Δ
13–81%

[2850]

R^1 = H, Ac; R^2 = H, MeO

9-Acridinone

III. FORMATION OF A SEVEN-MEMBERED RING

1. 1,4-Diazepine

Displacement of a methylthio attached to one ring by an amino group (formed *in situ* by reduction of a nitro group attached to another ring) results in the formation of a C—N bond and a doubly fused diazepine ring.

NO_2 MeS H N^+ N $(CH_2)_2$ Zn–AcOH, Δ 40% H N N N [2560]

Imidazo[2,1-*b*][1,3]benzodiazepine

CHAPTER 9

Acylamine, Acylhydrazine, Amine or Isocyanate and Halogen

I. Formation of a Five-membered Ring 702
1. Pyrrole 702
2. Pyrrol-2-one 702
3. Pyrazole 703
4. Imidazole 703
5. Furan 703
6. Oxazole 704
7. Oxazol-2-one 704
II. Formation of a Six-membered Ring 704
1. Pyridine 704
2. Pyridin-4-one 706
3. Pyridazine 707
4. Pyridazin-3-one 707
5. Pyridazin-4-one 708
6. Pyrazine 708
7. 1,2,4-Triazine 708
8. 1,2,4-Triazin-3-one 708
9. 1,3-Oxazine 709
10. 1,4-Thiazine 709
11. 1,3,4-Oxadiazine or 1,3,4-Thiadiazine 710
III. Formation of a Seven-membered Ring 710
1. 1,4-Diazepin-5-one 710
2. 1,4-Oxazepine 711
3. 1,4-Thiazepin-4-one 711
4. 1,2,5-Thiadiazepine 1,1-Dioxide 711

Examples of the cyclization of a hydrazine, a semicarbazide or of a ureido group with halogen are included in this chapter.

I. FORMATION OF A FIVE-MEMBERED RING

1. Pyrrole

A useful degree of regioselectivity is observed in the synthesis of 1-, 2- and/or 3-substituted indoles from 2-iodoanilines and a variety of alkynes in a reaction which is mediated by palladium(II) acetate, a small amount of triphenylphosphine and lithium chloride (which is preferred to tetrabutylammonium chloride for several syntheses). Palladium-catalysed reactions have been reviewed [B-41, 3501, 3505]. A nuclear fluorine is displaced by a side-chain amino group; this cyclization is accompanied by thermal dehydration. Indole is also prepared in good yield from a 2-(2-chloroethyl)aniline (see Chapter 8, Section I.1) in which halogen and amine functions have been interchanged. Recent work on indoles has been reviewed [3893, 3982].

NHR^1, I + $R^2C{\equiv}CR^3$ —i, 51–98%→ [indole with R^1, R^2, R^3] [3998]

R^1 = H, Me, Ac, Ts; R^2 = Pr, tBu, Me_3Si, Ph;
R^3 = alkyl, $H_2C{=}CMe$, CH_2OH, Ph
i, $Pd(OAc)_2$, PPh_3, LiCl

[tetrafluorophenyl with $CR^2(OH)CHR^3NHR^1$] —DMF, Δ, 32–90%→ [tetrafluoroindole with R^1, R^3, R^2] [2220, 2221]

R^1 = H, alkyl, $PhCH_2$; R^2,R^3 = H, Me

Indole

2. Pyrrol-2-one

A *peri*-annulation between acylamino and chloro functions occurs in the presence of sodamide, and proceeds through a benzyne.

[Cl, COR, N-tetrahydroquinoline] —liq. NH_3–Na, 40–96%→ [R, O, N tricyclic product] [2235]

R = Me, Et, $AcCH_2$, $PhCH_2$

Pyrrolo[3,2,1-*ij*]quinolin-2-one

3. Pyrazole

The chlorine-containing salt formed during a Vilsmeier reaction loses dimethylamine on treatment with hydrazine. Fused pyrazoles have been reviewed [3902].

$PO_2Cl_2^-$ — EtOH, N_2H_4, Δ; 85% → Pyrazolo[4,3-*b*][1,4]benzoxazine [2357]

4. Imidazole

An attempt to diacylate an *o*-chlorophenylhydroxylamine with acetic anhydride–pyridine gave instead a cyclized product in which pyridine had been incorporated. A cleaner sample of the same product was obtained by stirring the diacetate in pyridine at ambient temperature. 2-Amino-3-chloropyrazines may react with a 2-unsubstituted pyridine at ambient temperature to form a fused imidazole system in which the pyridine is also fused. The presence of substituents on the pyridine ring may necessitate the use of DMF as solvent and heat to be supplied.

pyr; 51–63% → Pyrido[1,2-*a*]benzimidazole [3097]

R = NO_2, CF_3

DMF, Δ; 14–73% → Pyrido[1',2' : 1,2]imidazo[4,5-*b*]pyrazine [3103a]

R = H, alkyl, $CONH_2$, COPh, COOMe, Ph

5. Furan

An isobenzofuran is obtained in good yield when the bromo-hydrazone (**9.2**) is treated successively with methyllithium, butyllithium, benzoyl chloride and an excess of lithium bromide. Recent developments in the chemistry of benzofurans have been reviewed [3894, 3983].

(9.2) → Isobenzofuran [3186]

i, THF, MeLi, BuLi, PhCOCl, LiBr, Δ

6. Oxazole

Aliphatic iodo and ureido groups cyclize when the compound is stirred with silver(I) perchlorate in the absence of light. A similar result is obtained by stirring with either water and then potassium hydroxide or MCPBA in chloroform; these procedures gave slightly lower yields. Fully aromatized 2-aryloxazolopyridines are prepared in high yield by heating an acylamino-chloropyridine with trimethylsilyl polyphosphoric ester (PPSE).

Et_2O, $AgClO_4$, 66% → Benzoxazole [3043]

PPSE, o-$Cl_2C_6H_4$, Δ, 75–99% → Oxazolo[5,4-*b*]pyridine [3709]

R = H, Cl; Ar = Ph, halo-, 4-MeO-, 2-NO_2-, 4-Me-, 4-CN-C_6H_4

7. Oxazol-2-one

A 2-chloro-isocyanate cyclizes on standing in a desiccator to a product which contains this ring.

32% → Oxazolo[5,4-*d*]pyrimidine-2,5,7-trione [2753]

II. FORMATION OF A SIX-MEMBERED RING

1. Pyridine

Addition of bromine to a 5-allylpyrimidin-6-amine followed by heating in DMF gives a new unsubstituted 1,4-dihydropyridine ring [3034]. Coupling a 2-

bromoaniline or its more reactive acetyl derivative with a formylbenzeneboronic acid, with catalysis by tetrakis(triphenylphosphine)palladium, gives a new pyridine ring [3195]. The bromoamine was later modified to the more reactive bromo-t-butyl carbamate [3432, 3577] and the boronic acid, replaced by a tributylstannyl [3433] or a trimethylstannyl group [3577].

1. $CHCl_3$, Br_2 2. DMF, Δ; 82% [3034]

Pyrido[2,3-*d*]pyrimidine

$(MeOCH_2)_2$, $Pd(PPh_3)_4$, Δ; 46% [3195]

Phenanthridine

Other ring systems synthesized similarly:

Thieno[3,4-*c*]quinoline [3577]

Dithieno[3,2-*b* : 3',2'-*d*]pyridine [3432]

Dithieno[34-*b* : 3',2'-*d*]pyridine [3432]

Dithieno[3,4-*b* : 3',4'-*d*]pyridine [3432]

Dithieno[3,2-*b* : 2',3'-*d*]pyridine [3433, 3577, 3596]

Dithieno[2,3-*b* : 2',3'-*d*]pyridine [3433, 3577, 3596]

Dithieno[3,4-*b* : 2',3'-*d*]pyridine [3433, 3596]

Thieno[2,3-*c*]quinoline [3195]

Benzo[*c*][1,8]naphthyridine [3195]

Thieno[3,2-*c*][1,8]naphthyridine [3195]

Another palladium-mediated cyclization enables 2-iodoaniline to be converted at 140 °C to quinoline by reaction with a substituted allyl alcohol in the presence of triphenylphosphine, HMPT and hydrogencarbonate. Allyl alcohol itself gives poor results in this reaction.

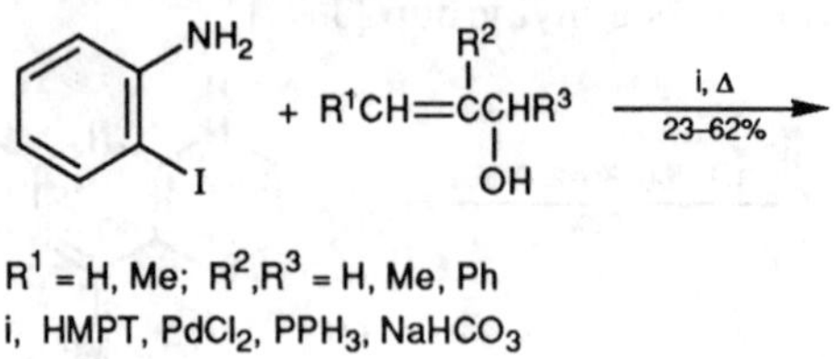

NH2
I
+ R1CH=CCHR3
R2
OH
i, Δ
23–62%
R3
R2
R1
N
Quinoline
R1 = H, Me; R2,R3 = H, Me, Ph
i, HMPT, PdCl2, PPH3, NaHCO3

[3969]

2. Pyridin-4-one

The course of the cyclization of amino (or acylamino) ketones to pyrido[2,3-*d*]pyridazin-2-one (see Chapter 4, Section II.2) changes when the phenyl group is replaced by 2-fluorophenyl. Under mildly alkaline conditions, the amino and fluoro functions react to form a sandwiched 4-pyridinone ring. This compound is methylated (MeI–NaH–DMF) at N-2. A side-chain amino group which may be regarded as a vinylogous carboxamide displaces a bz-fluorine atom on heating with sodium hydride.

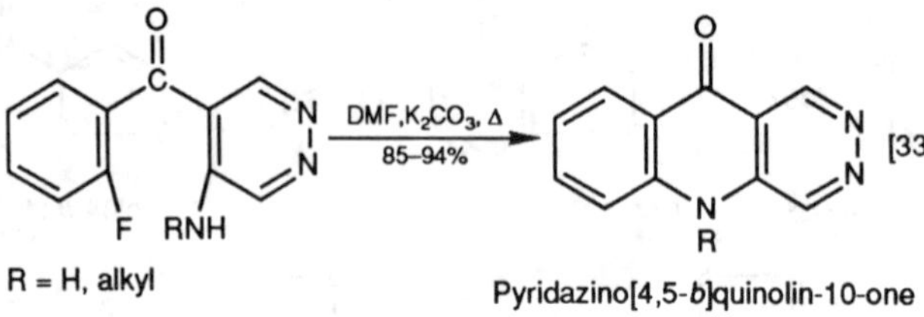

O
C
N
N
F
RNH
DMF,K2CO3, Δ
85–94%
O
N
N
N
R
R = H, alkyl
Pyridazino[4,5-b]quinolin-10-one

[3393, 3645, 3647]

Another ring system synthesized similarly:

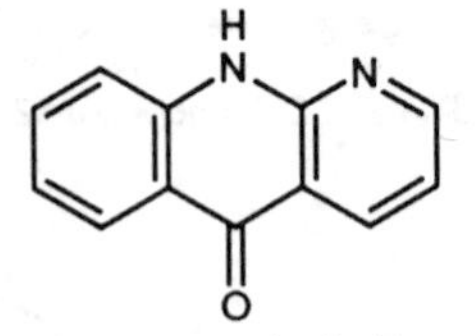

H
N
N
O
Pyrido[2,3-b]quinolin-5-one

[3751]

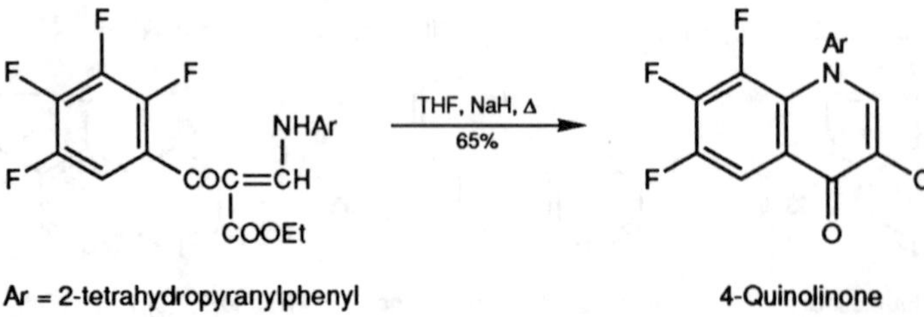

F
F
F
F
NHAr
COC=CH
COOEt
THF, NaH, Δ
65%
F
Ar
N
F
F
COOEt
O
Ar = 2-tetrahydropyranylphenyl
4-Quinolinone

[3207]

o-Halogenoanilines are cyclized with a terminal alkyne in an atmosphere of carbon monoxide with palladium [as $PdCl_2(PPh_3)$]; a pressure of $20\,kg\,cm^{-2}$ increases the yield. A mixture of (*E*)- and (*Z*)-2-acetamido-α-bromochalcones is cyclized on heating in acetic acid and orthophosphoric acid—a type of cyclization which is not found in the corresponding 2-acetyloxychalcone.

+ RC≡CH + $PdCl_2(PPh_3)_2$ —(Et_2NH, CO, Δ; 62–95%)→ [3968]

R = C_6H_{13}, Ph, $AcO(CH_2)_3$, $MeOOC(CH_2)_3$, 4-MeO-, 4-COOEt-C_6H_4; X = Br, I

4-Quinolinone

—(AcOH, H_3PO_4, Δ; 77%)→ [3838]

3. Pyridazine

The bromohydrazine (**9.2a**) adds on to DMAD at room temperature to give an unexpected product.

+ DMAD —(DMF; 28–39%)→ [3940]

(**9.2a**)
R = H, Et, $PhCH_2$, Ph

Pyridazino[4,5-*c*]pyridazin-8-one

4. Pyridazin-3-one

Dilithiation [B-47] of the bromohydrazone (**9.2**) to form the dianion, followed by addition of oxalyl chloride at −78 °C gives a good yield of the phthalazone.

(**9.2**) —(THF, MeLi, BuLi, $(COCl)_2$; 65%)→ [3186]

1-Phthalazinone

5. Pyridazin-4-one

Attempts to hydrolyse the carbamate (**9.3**) under alkaline conditions gave an unexpected product which is formed by displacement of the fluorine (or less readily, chlorine) atom and formation of a pyridazinone ring. Optimum yields of this product were developed.

aq. EtOH, NaOH, Δ 41–98% [3478]

(**9.3**)
R^1 = F, Cl; R^2 = H, Cl; R^3 = H, Me

Pyrrolo[1,2-*b*]cinnolin-10-one

6. Pyrazine

Prolonged heating with TEA is required to annulate side-chain chloroethylenamino and bz-amino groups.

TEA, Δ ~29% [2839]

R = H, Me

Quinoxaline

7. 1,2,4-Triazine

Intramolecular displacement of an iodine atom by nucleophilic nitrogen in a side-chain leads to the formation of a fused 1,2,4-triazine ring.

MeCN, Ag_2O ~47% [3531]

1,3,4-Triazabicyclo[4.2.0]oct-2-en-8-one

8. 1,2,4-Triazin-3-one

Thermolysis of an *o*-halo-semicarbazide (**9.4**, R = H) at 175–180 °C produced a new reduced triazinone ring, but when N^3 of the semicarbazide group carried a methyl (as in **9.5**, R = Me), cyclization was accompanied by migration of the methyl group.

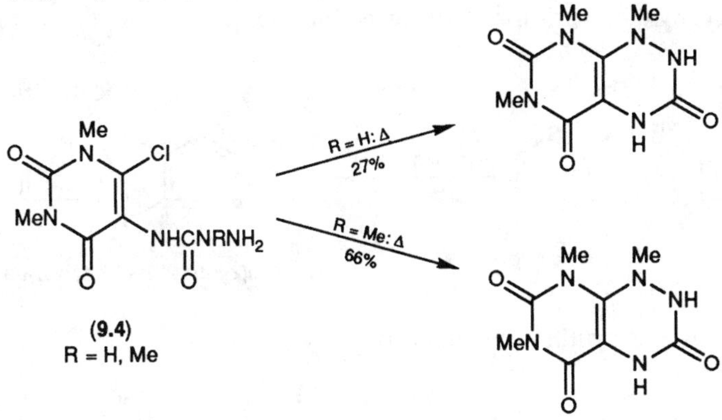

(9.4)
R = H, Me

R = H: Δ, 27%

R = Me: Δ, 66%

Pyrimido[5,4-*d*]-1,2,4-triazine-3,5,7-trione

[2753]

9. 1,3-Oxazine

Base-induced reaction of ureido carbonyl and reactive bromine leads to the formation of a 3,1-benzoxazine ring system in good yield.

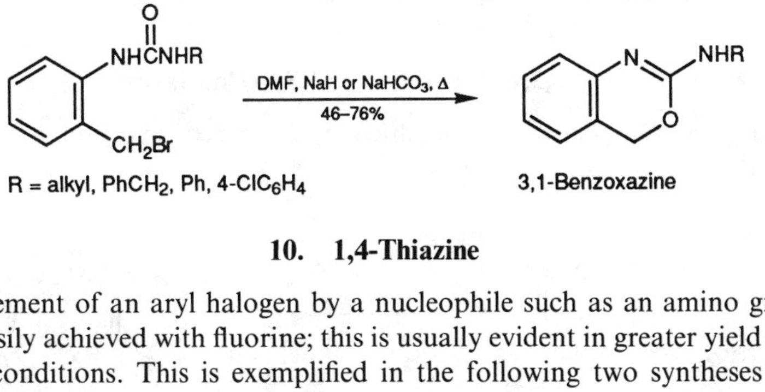

DMF, NaH or $NaHCO_3$, Δ, 46–76%

R = alkyl, $PhCH_2$, Ph, 4-ClC_6H_4

3,1-Benzoxazine

[3781]

10. 1,4-Thiazine

Displacement of an aryl halogen by a nucleophile such as an amino group is more easily achieved with fluorine; this is usually evident in greater yield and/or milder conditions. This is exemplified in the following two syntheses of the antipsychotic drug thioridazine (**9.5**) [3730]; the related antiemetic metopimazine may also be synthesized in this way. Phenothiazines have been reviewed [B-44].

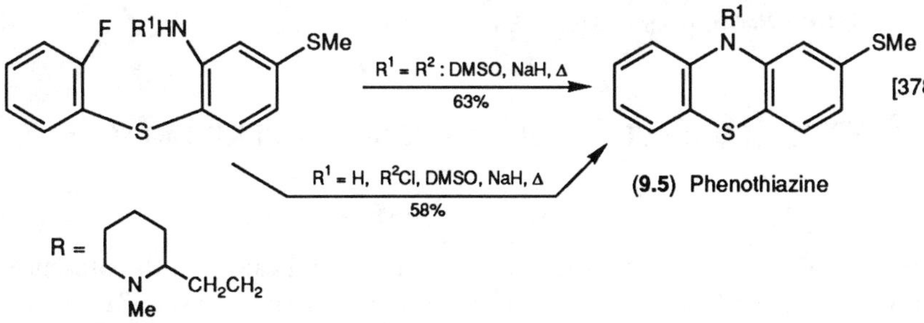

$R^1 = R^2$: DMSO, NaH, Δ, 63%

R^1 = H, R^2Cl, DMSO, NaH, Δ, 58%

(9.5) Phenothiazine

[3783]

Acylamino and bromo functions attached to different rings (joined through a sulphur atom) react on heating with potassium acetate and copper bronze; this is accompanied by elimination of hydrogen chloride.

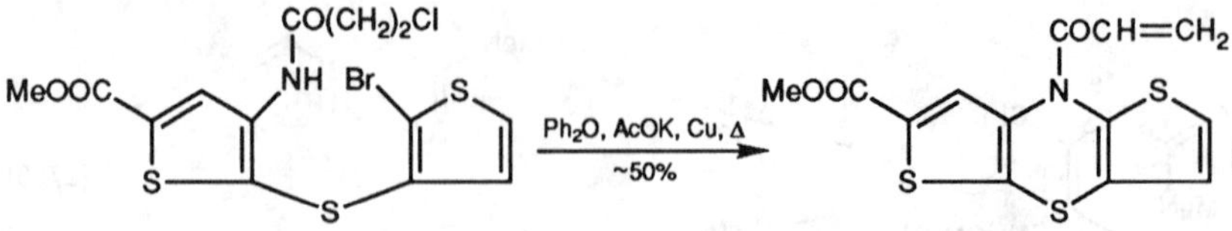

CO(CH2)2Cl
NH Br
MeOOC
S
S
S
Ph2O, AcOK, Cu, Δ
~50%
COCH=CH2
N
MeOOC
S
S
S

[2656]

Dithieno[2,3-*b* : 2',3'-*e*][1,4]thiazine

Other ring systems synthesized similarly:

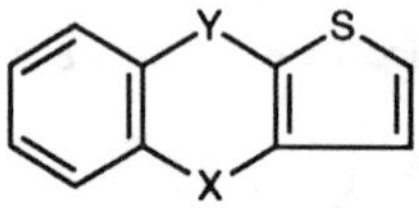

Y
S
X

[2769]

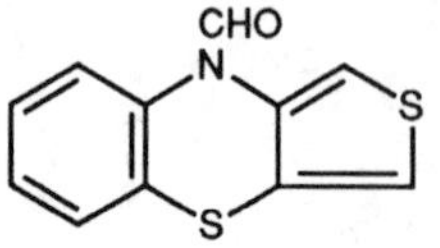

CHO
N
S
S

[2769]

X = NCHO, Y = S Thieno[2,3-*b*][1,4]benzothiazine
X = S, Y = NCHO Thieno[3,2-*b*][1,4]benzothiazine

Thieno[3,4-*b*][1,4]benzothiazine

11. 1,3,4-Oxadiazine or 1,3,4-Thiadiazine

An *N*-acyl or *N*-thioacyl 2-halohydrazine is cyclized on heating in a basic medium.

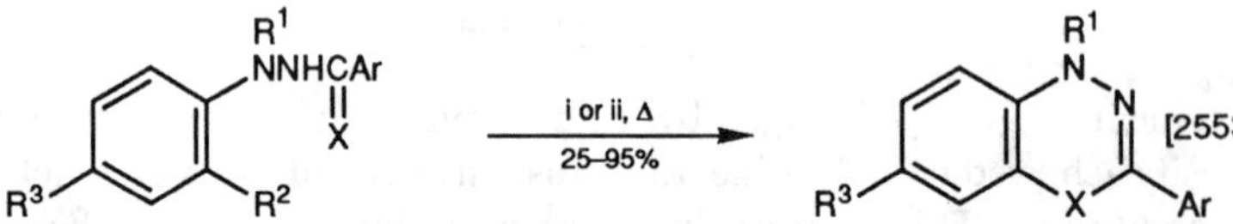

R1
NNHCAr
X
R3
R2
i or ii, Δ
25–95%
R1
N
N
R3
X
Ar

[2553, 2683]

R^1 = H, Ac, 4-$NO_2C_6H_4$; R^2 = Br, F, I
R^3 = Br, F, I, NO_2, CF_3, COOEt, Me_2NSO_2;
X = O, S; Ar = Ph, 4-Cl-, 4-MeO-C_6H_4
i, DMF, NaOH, TEA; ii, MeCN, TEA

4,1,2-Benzoxadiazine
4,1,2-Benzothiadiazine

III. FORMATION OF A SEVEN-MEMBERED RING

1. 1,4-Diazepin-5-one

A reactive haloamine reacts with an amino ester on heating in the presence of a strong base. Tricyclic 1,4-benzodiazepines have been reviewed [3135].

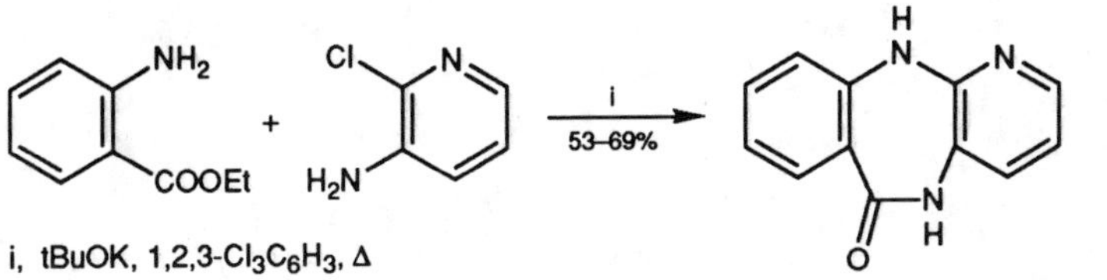

[3523]

i, tBuOK, 1,2,3-$Cl_3C_6H_3$, Δ

i; 53–69%

Pyrido[2,3-*b*][1,4]benzodiazepin-6-one

2. 1,4-Oxazepine

Thermally induced interaction between an aryl bromide and an arylamino group or its acetyl derivative (as in **9.6**, R = H or Ac) sometimes fails but in this example, the formamido homologue reacts in a basic medium.

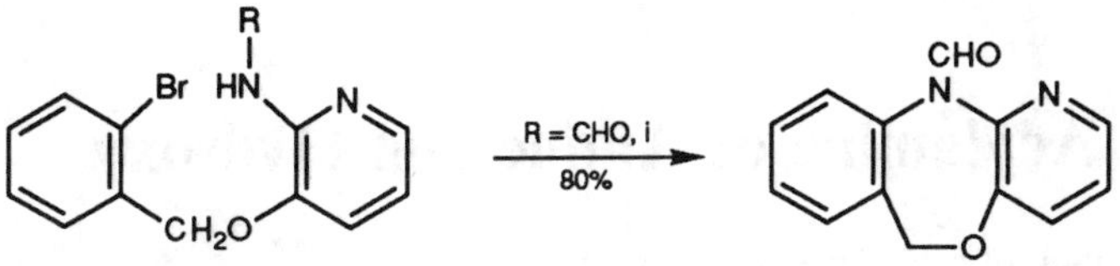

(**9.6**)

i, EtOH, aq. NaOH, Δ

R = CHO, i; 80%

[2228]

Pyrido[3,2-*b*][4,1]benzoxazepine

3. 1,4-Thiazepin-4-one

This ring is formed by treating a 2-chloro(2-acylamino)pyridine with LDA and a thiocarboxylate ester under argon.

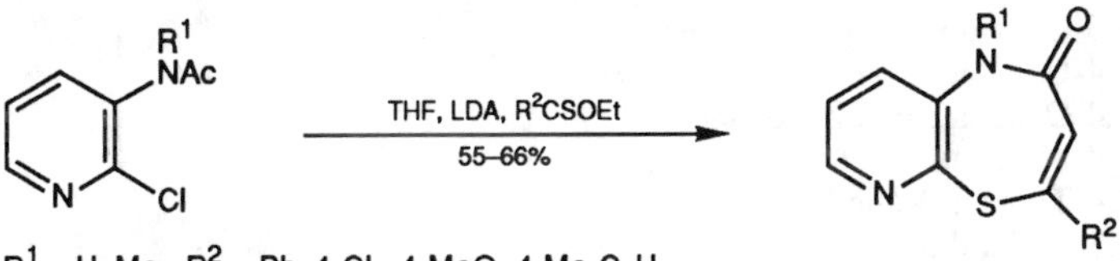

THF, LDA, R^2CSOEt; 55–66%

[3844]

R^1 = H, Me; R^2 = Ph, 4-Cl-, 4-MeO-,4-Me-C_6H_4, 3,4-$OCH_2OC_6H_3$, 2-furyl, 2-thienyl

Pyrido[2,3-*b*][1,4]thiazepin-2-one

4. 1,2,5-Thiadiazepine 1,1-Dioxide

2-Acylamino and 2′-halo functions attached to two different rings of *N*-arylbenzenesulphonamides react on heating the compounds with copper powder and copper(I) bromide in DMF.

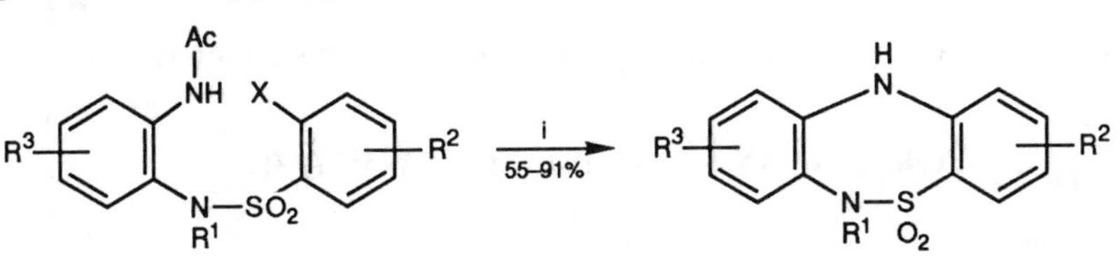

i; 55–91%

[3926]

R^1 = alkyl; R^2 = H, Cl, MeO, CF_3; R^3 = H, Cl, Me; X = Br, Cl

i, DMF, CuBr, Cu, Δ

Dibenzo[*c,f*][1,2,5]thiadiazepine 5,5-dioxide

CHAPTER 10

Acylamine or Amine and Hydroxy

I.	Formation of a Five-membered Ring	712
	1. Pyrrole	712
	2. Oxazole	713
	3. Oxazol-2-one	715
	4. Thiazole	716
	5. Thiazol-2-one	719
	6. 1,2,3-Thiadiazole	720
II.	Formation of a Six-membered ring	720
	1. 1,3-Oxazine	720
	2. 1,4-Oxazine	720
	3. 1,4-Oxazin-2-one or -3-one	721
	4. 1,4-Thiazine	722
	5. 1,4-Thiazin-3-one	723
	6. 1,3,4-Thiadiazine	723
III.	Formation of a Seven-membered ring	723
	1. 1,4-Oxazepine or 1,4-Oxazepin-5-one	723
	2. 1,4-Oxazepine-5,7-dione	724
	3. 1,4-Thiazepine	724
	4. 1,2,5-Oxathiazepin-4-one 2-Oxide	725

Cyclization of 2-aminophenyl disulphide is included in this chapter.

I. FORMATION OF A FIVE-MEMBERED RING

1. Pyrrole

Hydrogenation of a nitro-alcohol in the presence of copper-on-silica at 250 °C is an effective method of reductive cyclization. Nitro and amino alcohols have recently been cyclized in high yield by heating in toluene, xylene or diglyme with ruthenium(II) chloride–triphenylphosphine. Substituents present in the aniline included halogen, methoxy and methyl [3835].

H_2, Cu–SiO_2, Δ; 99%; Indole [2601]

2. Oxazole

Cyclization of aminophenols with carboxylic acids (or esters) often requires high temperatures even in the presence of PPA. Lower temperatures (10–30 °C) suffice when PPA is replaced by PPE [2111, 3084, 3625]. Alternatively, a dithioic ester may be stirred with the aminophenol at pH 8 [2860]. Annulation of amino and phenolic groups is accomplished by treatment with a nitrile at ambient temperature in a basic environment [2770]. The use of nitriles in the synthesis of heterocycles has been reviewed [3332a].

i or ii; 44–85%; Benzoxazole, Benzothiazole [2111, 2860]

i, R^1COOH, PPE; R^1 = Me, Ph, 4-Cl-, 4-NO_2-, 4-AcNH-C_6H_4, 2-furyl, 3-pyridyl; X = O
ii, MeOH, (aq. NaOH), pH 8, R^1CSSR^2; X = O, S; R^1 = Me, Ph; R^2 = Et, CH_2COOH

MeONa–MeOH; 43–60%; Benzoxazole [2770]

R = COOH, CHMeCOOH

An anhydride [2309, 2323] (catalysed by mineral acid) or ethyl hippurate (ethyl 2-benzoylglycinate) [2186] cyclizes an *o*-aminophenol on prolonged refluxing. Chloral hydrate and hydroxylamine convert an aminophenol into a benzoxazole-2-carboxaldehyde oxime [2104].

Ac_2O, H_2SO_4, Δ; 65% [2309, 2323]

Another ring system synthesized similarly:

[2323]

Oxazolo[5,4-*d*]pyrimidin-7-one

+ $Cl_3CH(OH)_2$ + NH_2OH —AcONa, Δ; 58–76%→ [2104]

R = H, Me, Cl, NO_2

In a basic environment, *o*-aminophenols add on to 1,1-dichloro-2-nitroethene to form a 2-nitromethyloxazole ring [2717]. Both hydroximoyl chlorides [3468] and 2-chloro-1,1,1-triethoxyethane [3652] are versatile reagents because of their high reactivity even at low temperatures, for example, an oxazole ring is formed at ambient temperature.

+ $Cl_2C{=}CH(NO_2)$ —EtOH–EtONa; ~58%→ [2717]

R = H, Me

EtOH, ArC(=NOH)Cl; ~87% → [3468]

$(EtO)_3CCH_2Cl$; 46–90% → [3652]

Thioaroylcarbamates are readily prepared compounds which react as monofunctional compounds with aminophenols without addition of an acid or a base [2899]. Cyanohydrins can provide the 2-carbon atom of this ring when aminophenols are substrates, and a second molecule of aminophenol is attached at C-2 methyl group [3705, 3707].

Many of the methods available for the synthesis of fused oxazoles lead to 2-substituted products. A method which leads to 2-unsubstituted 5-R-benzoxazoles uses triethyl orthoformate and aqueous hydrochloric acid at moderate temperatures [3889]; sulphuric acid (used in an earlier paper) tends to cause charring and to lower yields. Both 2-substituted and 2-unsubstituted

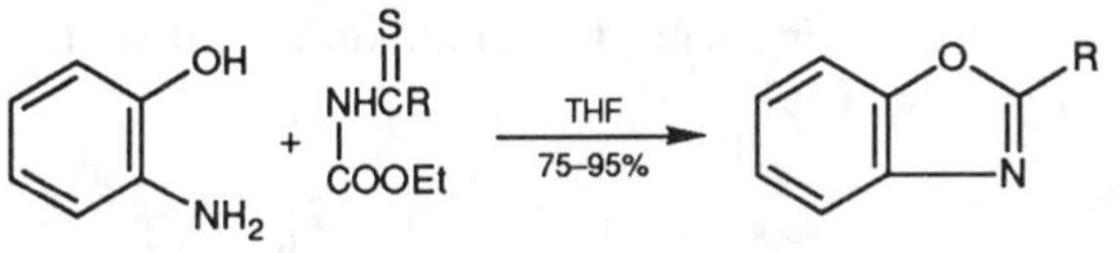

[2899]

R = 4-Me-, 4-MeO-C_6H_4, 2-thienyl

Benzoxazole

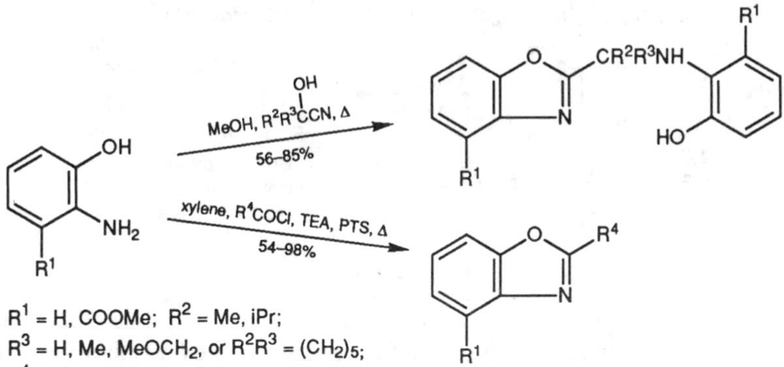

[3705]

[3707]

R^1 = H, COOMe; R^2 = Me, iPr;
R^3 = H, Me, $MeOCH_2$, or $R^2R^3 = (CH_2)_5$;
R^4 = H, Me, tBu, Ph, 4-MeO-, 4-NO_2-C_6H_4,
2-furyl, 3-pyridyl

oxazoloazines may be obtained in moderate yield by heating the aminoazinol at 140–150 °C with an orthoester and 4-toluenesulphonic acid [3860].

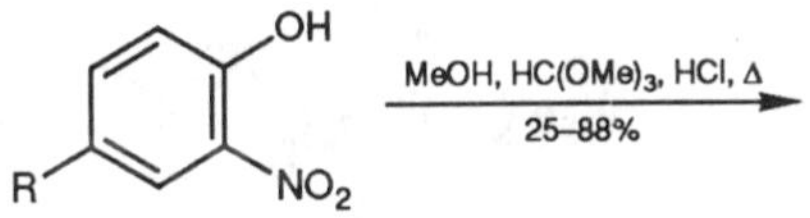

[3889]

R = halo, CN, MeO, COOEt, AcNH, PhCO, NO_2

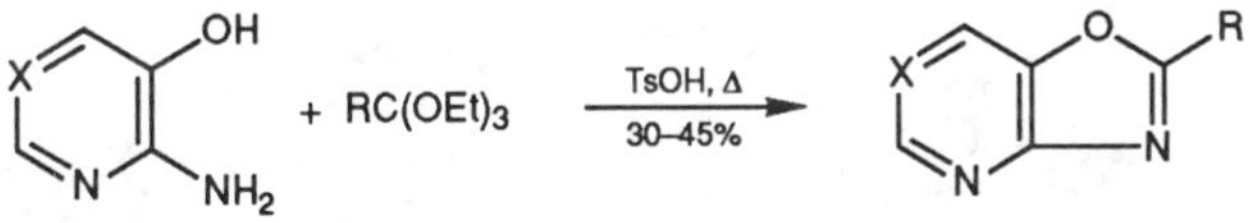

[3860]

R = H, alkyl, Ph; X = CH, N

Oxazolo[4,5-*b*]pyridine
Oxazolo[4,5-*d*]pyrimidine

3. Oxazol-2-one

Bis(trichloromethyl) carbonate ('triphosgene') is described as a more convenient reagent than phosgene and other reagents as a source of a CO group. This is well exemplified in the synthesis of a 2-benzoxazolone from a 2-aminophenol at ambient temperature [3617]. Carbonyl sulphide is another reagent which is worth considering in this synthesis [3096]. An aminophenol may be cyclized under mild conditions by stirring with disuccinimido carbonate (DSC), which supplies a carbonyl group [3019].

R = H, MeO; i, $(CCl_3O)_2CO$; ii, COS

i or ii, TEA, THF, 75–97%

2-Benzoxazolone [3096, 3617]

R = H, Cl; X = CH, N; Y = O, S

MeCN, DSC, 70–91%

2-Benzoxazolone
2-Benzothiazolone
Oxazolo[4,5-*b*]pyridin-2-one [3019]

4. Thiazole

2-Aminothiophenol is cyclized by reaction with prop-2-ynyltriphenylphosphonium bromide, but some 2-substituted anilines give the phosphonium salt (a possible intermediate in the thiophenol reaction) which may be cyclized separately (see Chapter 16, Section I.5) [2736]. Reaction of thiophenols with the lactonic 2-oxazolin-5-ones can give any one of three products: in acetic acid–sodium acetate, cyclization to the thiazole occurs while in some other solvents, acylamino and acylthio products are formed, and may be cyclized by acetic acid–sodium acetate [3088].

(10.1) + $CH_2C{\equiv}CH$ / $^+PPh_3Br^-$ —MeCN, 62%→ Benzothiazole [2736]

(10.1) + oxazolone: EtOH, Δ, 58–78%; AcOH–AcONa; AcOH–AcONa, Δ, 56–75%; PhH, pip, ~56%; AcOH–AcONa [3088]

Ar = Ph, 4-$NO_2C_6H_4$;
R = Ph, 4-Cl-, 4-MeO-C_6H_4,
2-thienyl

Several methods of cyclizing acylamino-ols or -thiols are synthetically useful; heating for a few minutes with phosphorus pentasulphide is usually sufficient although yields may be low [2196].

P_2S_5, Δ, 26%

Thieno[2,3-*d*]thiazole [2196]

Another ring system synthesized similarly:

Thieno[2',3' : 4,5]thieno[2,3-*d*]thiazole [2817]

When the reaction of an aminophenol with *N*-ethoxycarbonylthioamide or a dithioate ester is applied to an aminothiophenol, a high yield of benzothiazole is obtained [2860, 2899]. The same substrate is converted into a benzothiazole efficiently at ambient temperature with hydroxamoyl chloride (the chloride of a hydroxamic acid) [3468]. Similar compounds in which the 2-phenyl group is varied are of interest as dye intermediates and are synthesized from quinonoid aminothiols [3477]; the chemistry of quinonoid heterocycles has been reviewed [2947, 3650].

(10.1) + NHCSR / COOEt → (THF, Δ, 46–92%) Benzothiazole [2899]

R = Et, 4-Me-, 4-MeO-C_6H_4, 2-thienyl

(10.1) + ArC(Cl)=NOH → (EtOH, ~90%) [3468]

Ar = Ph, 4-Cl-, 3-NO_2-C_6H_4

+ ArCHO → (AcOH, Δ, 65–82%) [3477]

Ar = 4-$Me_2NC_6H_4$, 2- or 4-pyridyl

Naphtho[2,3-*d*]thiazole-4,9-dione

An acyl- (or ethoxycarbonyl-) aminothiophenol, in which the thiol is present as its dithiocarbamate, is cyclized by heating with formic acid but the corresponding amino-dithiocarbamate cyclizes with retention of the amino moiety of the carbamate when treated with trifluoroacetic anhydride.

R = tBuCO, HCOOH, Δ 81%
R = H; THF, $(CF_3CO)_2O$ 65%
R = COOEt, HCOOH, Δ 34%

[3677]

Thiazolo[5,4-*c*]pyridine

Acetic formic anhydride is a versatile reagent in organic synthesis (review [3823]), for example, when it is stirred with an *o*-aminothiophenol, a 2-unsubstituted benzothiazole is formed. Aminothiophenols, like their oxygen counterparts [2111], are cyclized by carboxylic acids or their esters, and also by dichloromethyleneammonium salts (review [2549]) to give good yields of benzothiazoles [2548]. The use of a dithioic ester is shown in Section I.2 [2860].

H_2O 49% [3010a]

(**10.1**) + RCOOEt MeOH, pH 8 85–98% [2860]

R = Me, Ph

Benzothiazole

Reductive cyclization occurs when a nitrodisulphide is heated with aqueous sodium sulphide (which reduces the nitro group) and carbon disulphide (review [B-52]). An acylated thiol forms a thiazole ring with an *N*-carbonyl function when compound (**10.2**) is reduced with triethyl phosphite (the formula of the product is incorrectly shown in the paper). Microbiologically useful *β*-lactams can also be synthesized from the disulphide (**10.3**) by treatment with a two-molar proportion of certain phosphoranes at −50 °C in an intramolecular

Wittig reaction which generates a fused thiazole ring carrying a protected carboxylic acid group.

CS_2, aq. Na_2S, Δ; 45% [2724]

Benzothieno[2,3-*d*]thiazole

$(EtO)_3P$, Δ; ~48% [3534]

(10.2)

R^1 = Me, Pr; R^2 = Me_3Si, H_4 pyranyl;
R^3 = $CH_2{=}CHCH_2$

4-Thia-1-azabicyclo[3.2.0]hept-2-en-7-one

(10.3) + $CHR^3{=}PPh_3$; 50–70% [3873]

R^1 = 4-$NO_2C_6H_4$; R^2 = benzothiazol-2-yl;
R^3 = Ph, PhS, PhCH=CH (*t*); R^4 = $tBuMe_2SiO$

5. Thiazol-2-one

This ring is formed in high yield from a mercapto-amine and either phosgene or disuccinimido carbonate (see Section I.3). Ethyl potassium xanthate is also effective [3038]. An aminophenol in which the hydroxy group is masked as its dimethyl aminothiocarbamate cyclizes under conditions which generate the amine from the nitro group. The product is the result of a Newman–Kwart rearrangement [B-40, Vol. 3, p. 11].

i, PhH, Δ; 94% [3038, 3358]

R = $(CH_2)_2NPr_2.HCl$
i, disuccinimido carbonate or $COCl_2$

2-Benzothiazolone

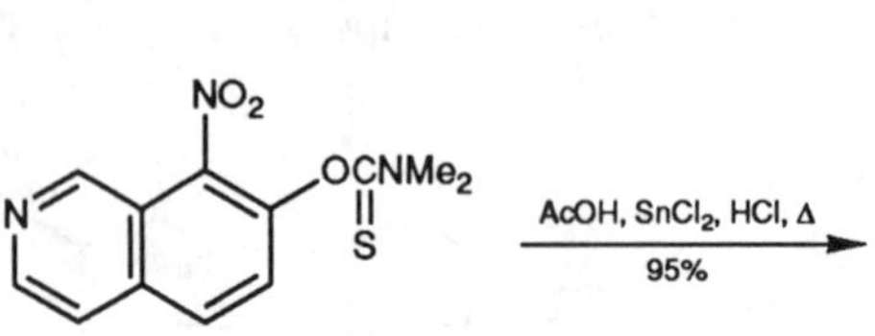

AcOH, $SnCl_2$, HCl, Δ
95%

[3522]

Thiazolo[5,4-*h*]isoquinolin-2-one

6. 1,2,3-Thiadiazole

Diazotization of an aminothiophenol is a convenient route to this ring; if the thiol is benzylated for protection at an earlier stage, it can be freed after diazotization.

(10.1) $NaNO_2$, AcOH, 77% [2109]

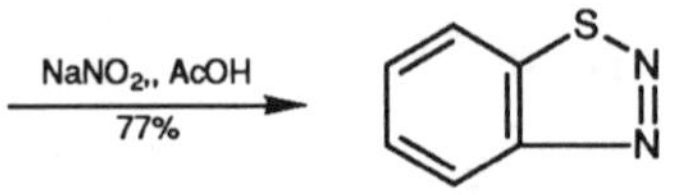

1,2,3-Benzothiadiazole

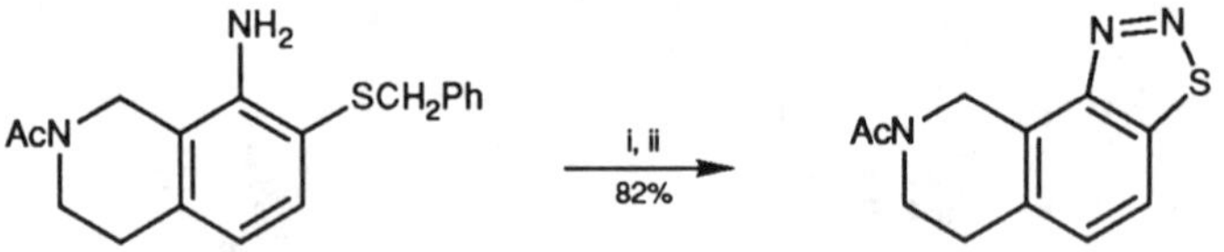

i, ii, 82% [3522]

i, AcOH, HCl, $NaNO_2$; ii, CuCl, HCl

[1,2,3]Thiadiazolo[5,4-*h*]isoquinoline

II. FORMATION OF A SIX-MEMBERED RING

1. 1,3-Oxazine

The imine of an *o*-acylphenol reacts with an orthoester to form an oxazine in good yield, especially when R^1 = Ar.

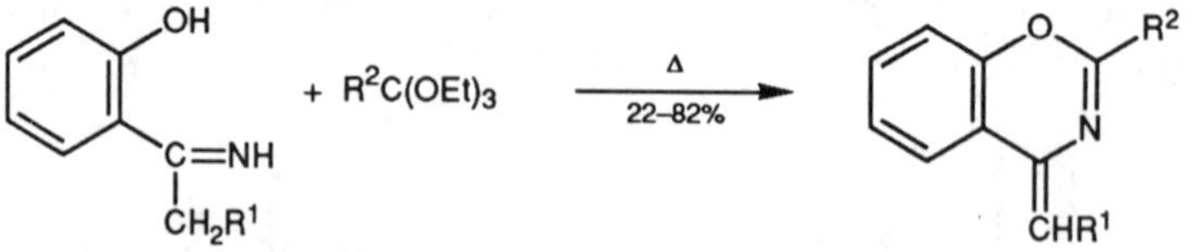

+ $R^2C(OEt)_3$, Δ, 22–82% [2090]

R^1 = H, Ph; R^2 = H, Me

1,3-Benzoxazine

2. 1,4-Oxazine

Oxidative cyclization of an aminophenol can give rise to a doubly fused oxazine in good yield from two molecules of aminophenol [2331]. A non-oxidative method using orthophosphoric acid is an alternative approach to a phenoxazine [2372]. 2-Chloroacetoacetate ester condenses with an aminohydroxypyrimidine to form a fused oxazine ring [2546]. Synthesis of a tetracyclic benzoxazine is described in Chapter 35, Section II.6.

EtO_2, HCl, $K_3Fe(CN)_6$; 62–71%

R = MeO, $NHCH_2COOMe$, NHCHMeCOOMe

3-Phenoxazinone [2331]

H_3PO_4, $ZnCl_2$, Δ; 70%

Phenoxazine [2372]

ClCHAc(COOEt); EtOH, TEA, Δ; ~82%

R = H, Me

Pyrimido[5,4-*b*][1,4]oxazine [2546]

3. 1,4-Oxazin-2-one or -3-one

A 4-aminopyrimidin-5-ol reacts exothermally with DMAD with the formation of a fused oxazin-3-one ring in high yield [2786], but prolonged heating of an aminophenol with DMAD yields the oxazin-2-one as the main but not only product [3125]. Another approach to a fused oxazin-3-one is demonstrated by the reaction of ethyl α-chloroalkanoates with 4-aminopyrimidin-5-ol under basic conditions [2459, 3810].

EtOH, DMAD; 98%

Pyrimido[5,4-*b*][1,4]oxazin-7-one [2786]

diox, DMAD, Δ; 13–68%

R = H, Cl, Me

1,4-Benzoxazin-2-one [3125]

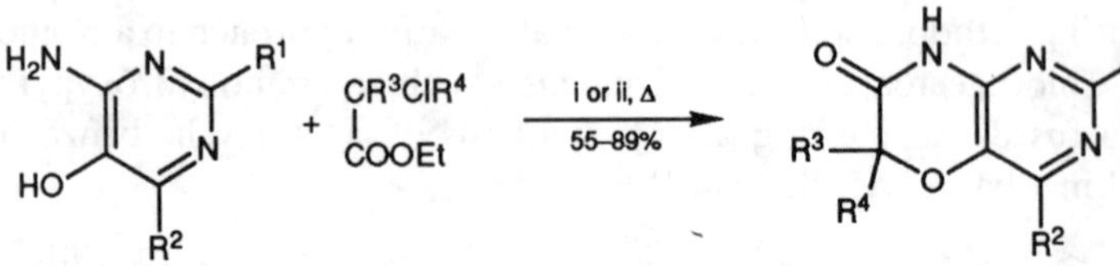

[2459, 3810]

R^1 = H, Me; R^2 = H, OH, Me, Cl;
R^3 = H, Me, Et; R^4 = H, Me
i, TEA; ii, EtOH–EtONa

Pyrimido[5,4-*b*][1,4]oxazin-7-one

4. 1,4-Thiazine

Some substituted *o*-quinones react with mercaptoamines to give 1-phenothiazinones but others (including *o*-benzoquinone) give more complex products [2635, 3179, 3332]. Cyclic 1,3-diketones (in their tautomeric form) react with a 2-mercaptoaniline to give phenothiazine-like compounds [3285]. The chemistry of phenothiazines has been reviewed [B-44, 3332]. *o*-Mercaptoaniline is annulated by heating with a hydroxycyclopentadienone [2635].

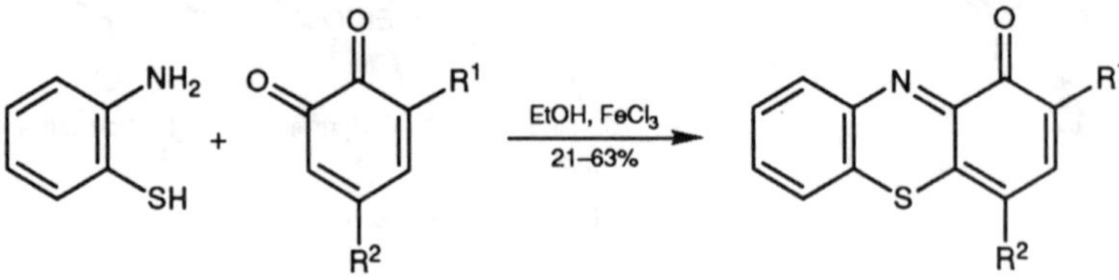

[3179]

R^1 = Me, tBu; R^2 = Me, tBu, Ph

1-Phenothiazinone

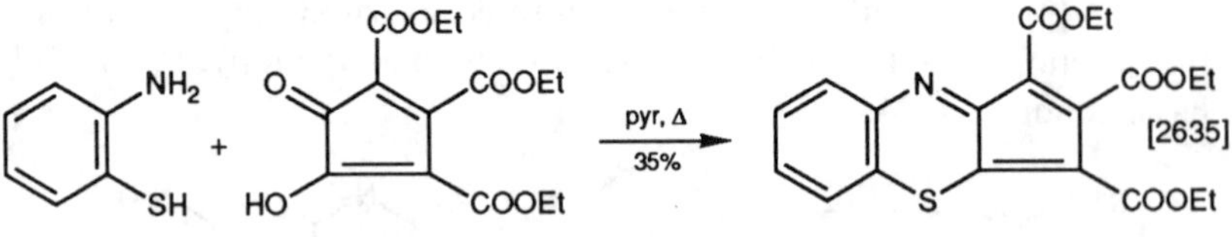

[2635]

Cyclopenta[*b*][1,4]benzothiazine

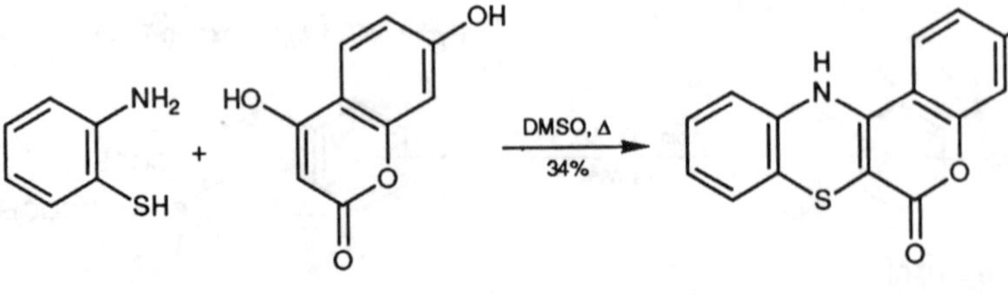

[3285]

[1]Benzopyrano[3,4-*b*][1,4]benzothiazin-6-one

Another ring system synthesized similarly:

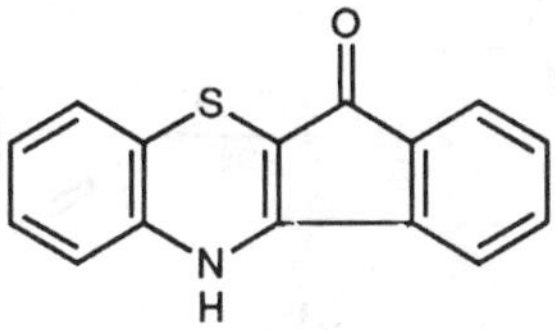

[3285]

Benz[*b*]indeno[1,2-*e*][1,4]thiazin-11-one

5. 1,4-Thiazin-3-one

A versatile synthesis of 1,4-benzothiazin-3-ones is that in which 2-aminophenyl disulphide is heated with methyl (or ethyl) arylacetate and a strong base.

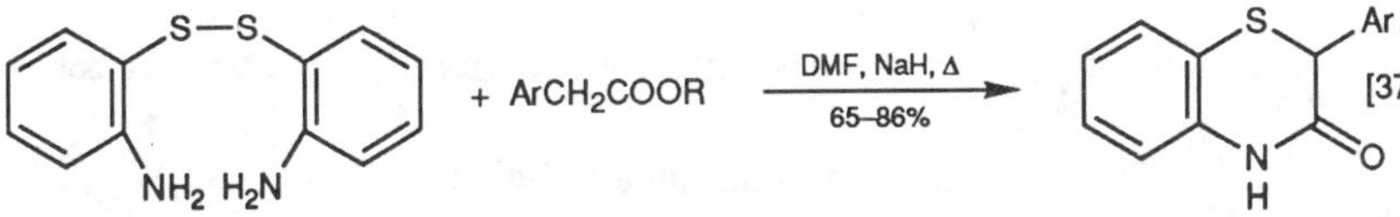

+ $ArCH_2COOR$ —DMF, NaH, Δ; 65–86%→ [3727]

Ar = Ph, halo-, CF_3-, NO_2-, MeO-, Me-, iPr-C_6H_4, 3-Br-4-MeO-, 3,4-$(MeO)_2$-, 3,4-OCH_2O-C_6H_3, 1-naphthyl-, 2 or 3-pyridyl; R = Me, Et

1,4-Benzothiazin-3-one

6. 1,3,4-Thiadiazine

The hydrazinium salt (**10.2**) of a 1-amino-1,2,3-triazole-5-thiol is cyclized by heating with an α-chloro ketone [B-45].

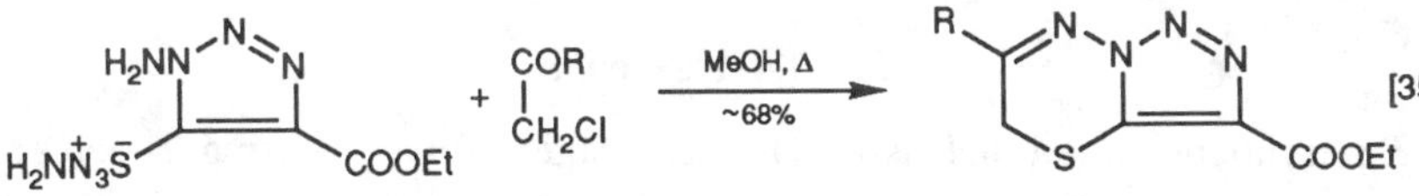

+ COR / CH_2Cl —MeOH, Δ; ~68%→ [3590]

(**10.2**) R = Me, Ph, 4-ClC_6H_4

[1,2,3]Triazolo[5,1-*b*][1,3,4]thiadiazine

III. FORMATION OF A SEVEN-MEMBERED RING

1. 1,4-Oxazepine or 1,4-Oxazepin-5-one

o-Aminophenol (**10.3**) condenses in an acidic medium with an *o*-aminonitrile or an amino ester to form a fused 1,4-oxazepine or a 1,4-oxazepin-5-one, respectively.

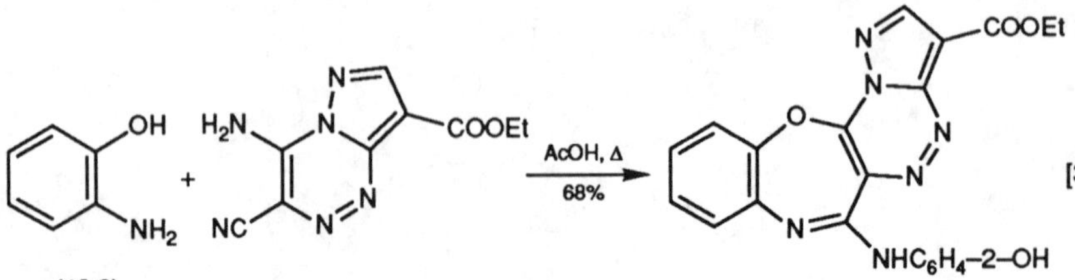

(10.3) + 7-amino-6-cyanopyrazolo[5,1-c][1,2,4]triazine-3-COOEt —AcOH, Δ, 68%→ [3484]

NHC_6H_4–2–OH

Pyrazolo[5',1' : 3,4][1,2,4]triazino[5,6-*b*][1,5]benzoxazepine

(10.3) + H_2N / EtOOC pyrazolotriazine COOEt —AcOH, Δ, 66%→ [3484]

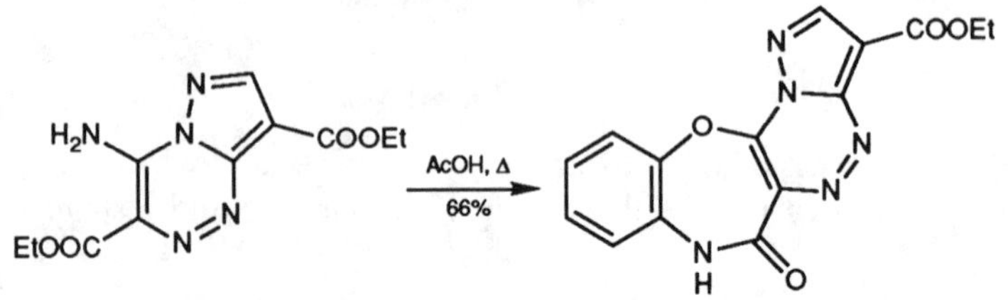

Pyrazolo[5',1' : 3,4][1,2,4]triazino[5,6-*b*][1,5]benzoxazepin-6-one

2. 1,4-Oxazepine-5,7-dione

2-Aminophenol is cyclized at 5 °C with diphenyl thiomalonate in the presence of DMAP.

(10.3) + $CH_2(CSPh)_2$ (C=O) —i, 57%→ [3641]

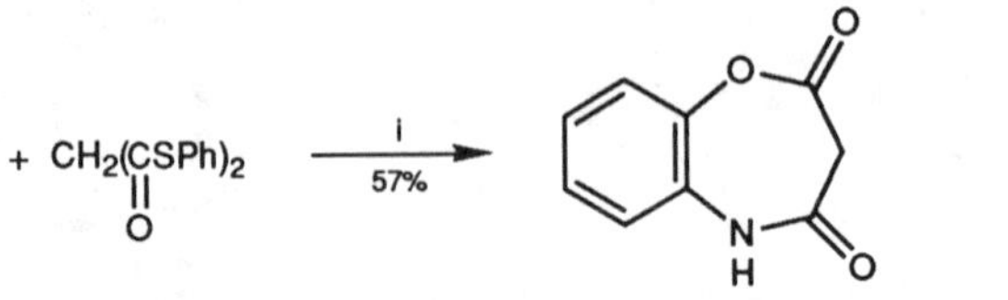

i, Me_2CHCH_2Ac, DMAP

1,5-Benzoxazepine-2,4-dione

3. 1,4-Thiazepine

The 3-arylideneimide (**10.4**) is an efficient reagent for converting 2-mercaptoaniline into a benzothiazepine. The use of lactams in heterocyclic synthesis has been reviewed [2387].

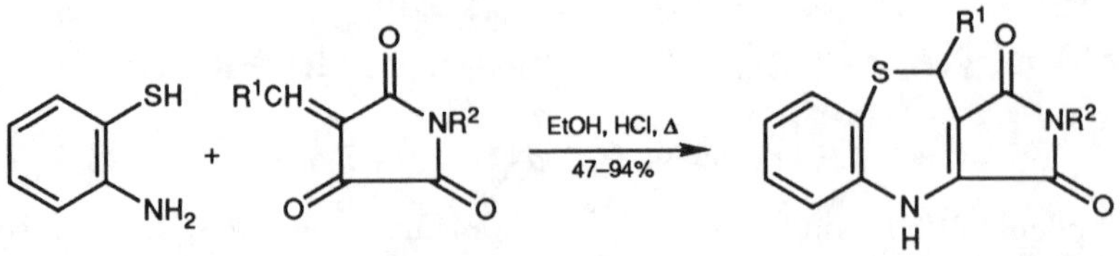

2-mercaptoaniline + R^1CH= (10.4) NR^2 —EtOH, HCl, Δ, 47–94%→ [3247]

(**10.4**)

R^1 = Ph, 4-Cl, 3-NO_2-C_6H_4; R^2 = 4-MeC_6H_4

Pyrrolo[3,4-*c*][1,5]benzothiazepine-1,3-dione

4. 1,2,5-Oxathiazepin-4-one 2-Oxide

Cyclization of 2-aminophenol with ketene in liquid sulphur dioxide at or below 0 °C gives a modest yield of this fused ring compound.

$$\textbf{(10.3)} + CH_2{=}CO + SO_2 \xrightarrow[25\%]{\text{liq. } SO_2} \quad [2068]$$

1,2,5-Benzoxathiazepin-4-one 2-oxide

CHAPTER 11

Acylamine or Diazonium Salt and Lactam Carbonyl

I. Formation of a Five-membered Ring 726
  1. Pyrazole 726
  2. Imidazole 727
  3. 1,2,4-Triazole 727
  4. Thiazole 728
  5. 1,3,4-Thiadiazole 729
  6. 1,3,4-Thiadiazol-4-ium 730
II. Formation of a Six-membered Ring 730
  1. Pyran-2-one 730
  2. 1,4-Thiazine 730
  3. 1,3,4-Oxadiazine 731
  4. 1,3,4-Thiadiazine 731
  5. 1,3,4-Thiadiazinium 732
III. Formation of a Seven-membered Ring 732
  1. 1,4-Diazepin-5-one 732

Cyclization of the following additional pairs of functions are also described: acylhydrazino-lactam carbonyl, amidino-lactam carbonyl.

I. FORMATION OF A FIVE-MEMBERED RING

1. Pyrazole

A 1-aminopyridin-2-one is cyclized by heating it with a reagent which has a reactive methylene group, and anhydrous zinc chloride.

DMF, $ZnCl_2$, Δ, ~67% [3238]

R^1 = Me, Ph, 3-coumarinyl;
R^2 = H, Ac, COOEt

Pyrazolo[1,5-*a*]pyridine

2. Imidazole

Lawesson's reagent was originally used to convert a carbonyl into a thiocarbonyl group but it can also promote cyclization, possibly through the corresponding thiocarbonyl function (cf. p. 35). In this example, it is used in the presence of (+)-10-camphorsulphonic acid (CSA) to annulate side-chain amino and lactam carbonyl groups [3212]. Dehydration of an amidino-lactam yields a fused imidazole ring [2402]. Synthesis of heterocycles from lactams has been reviewed [2387].

xylene, i, Δ, 38% [3212]

i, Lawesson's reagent, (+)–CSA

Benz[*g*]imidazo[2,1-*a*]isoindole

pyr, $POCl_3$, Δ, 22–50% [2402]

Ar = Ph, 4-Me-, 4-MeO-C_6H_4, 3,4,5-$(MeO)_3C_6H_2$

Imadazo[4,5-*d*]imidazole

3. 1,2,4-Triazole

Carboxamides cyclize *N*-aminopyridin-2-ones (**11.1**) on heating with zinc chloride under anhydrous conditions [3154]. A novel ring system was synthesized by warming the 3-amino-2-thioxothiazol-4-one (a rhodanine derivative) with an imidoyl chloride [3024]. The two carbonyl groups of the oxadiazole (**11.2**) condense with hydrazine to form a fused triazole ring [3945].

+ $RCONH_2$ — DMF, $ZnCl_2$, Δ; 64–79% — [3154]

(**11.1**)
R = H, Me, Ph, PhCH=CH

[1,2,4]Triazolo[1,5-*a*]pyridine

+ $ClC(R^3)=NR^2$ — Δ; 20–64% — [3024]

R^1,R^2 = Me, Ph; R^3 = Ph, 4-MeO-, 4-NO_2-C_6H_4

Thiazolo[3,4-*b*][1,2,4]triazole-5-thione

+ N_2H_4 — EtOH, Δ; 55–70% — [3945]

(**11.2**)
Ar = 2,4-$Cl_2C_6H_3$;
R = Me, Bu, 2,4-$Cl_2C_6H_3$, $PhCH_2$, 4-MeC_6H_4

1,2,4-Triazolo[3,4-*b*][1,3,4]oxadiazole

4. Thiazole

Acylamido and lactam carbonyl groups may be annulated by heating with phosphorus pentasulphide in pyridine.

pyr, P_2S_5, Δ; 60% — [2823]

Thiazolo[5,4-*b*]indole

5. 1,3,4-Thiadiazole

N-Amino and lactam thiocarbonyl functions may be joined together by reaction with cyanogen bromide or by heating with an aldehyde. An *N*-acylaminopyrimidine-2-thione is cyclized by heating with phosphorus oxychloride.

aq. EtOH, BrCN, Δ; 46–80% [3219]

R = H, Br, Cl, 4-Me, 4-MeO

1,2,4-Triazolo[3,4-*b*][1,3,4]thiadiazole

+ ArCHO; EtOH, Δ; 56% [2119]

Ar = 4-MeOC_6H_4

[1,3,4]Thiadiazolo[2,3-*b*]quinazolin-9-one

$POCl_3$, Δ; 66% [2245]

1,3,4-Thiadiazolo[3,2-*a*]pyrimidin-5-one

An unexpected cyclization occurred when the thiazolone (**11.3**) was heated with phosphorus pentasulphide. Several bicyclic products may be obtained from the *N*-amino-thiolactam (shown in its enol form) (**11.4**); with a carboxylic acid and phosphorus oxychloride, a new thiadiazole ring is formed.

PhMe, P_2S_5, Δ; 17–40% [2809]

(**11.3**) R = Me, Ph

Thiazolo[4,3-*b*]-1,3,4-thiadiazole-5-thione

+ RCOOH; $POCl_3$, Δ; 76–94% [3809]

(**11.4**)
R = Ph, NH_2-, HO-, MeO-C_6H_4, 2,4-Cl_2-, 2,4-$(HO)_2$-C_6H_3, $PhCH_2CH_2$; Ar = 2,4-$Cl_2C_6H_3$

1,2,4-Triazolo[3,4-*b*][1,3,4]thiadiazole

6. 1,3,4-Thiadiazol-4-ium

1,4,6-Triaminopyrimidine-2-thione reacts with the Vilsmeier reagent to give one or more of several products according to the temperature of the reaction. In this way, thiadiazolopyrimidines are formed at 0 °C.

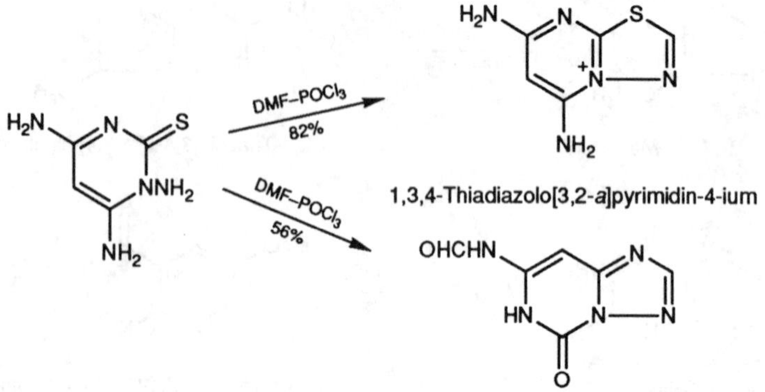

DMF–$POCl_3$ 82%

1,3,4-Thiadiazolo[3,2-*a*]pyrimidin-4-ium [3684]

DMF–$POCl_3$ 56%

[1,2,4]Triazolo[1,5-*c*]pyrimidin-5-one

II. FORMATION OF A SIX-MEMBERED RING

1. Pyran-2-one

In the presence of a strong base, malononitrile reacts with a dimethylamino-methylene group and forms a ring with a neighbouring lactam carbonyl function, the dimethylamino group being eliminated. The usefulness of nitriles in the synthesis of heterocycles as been reviewed [3332a].

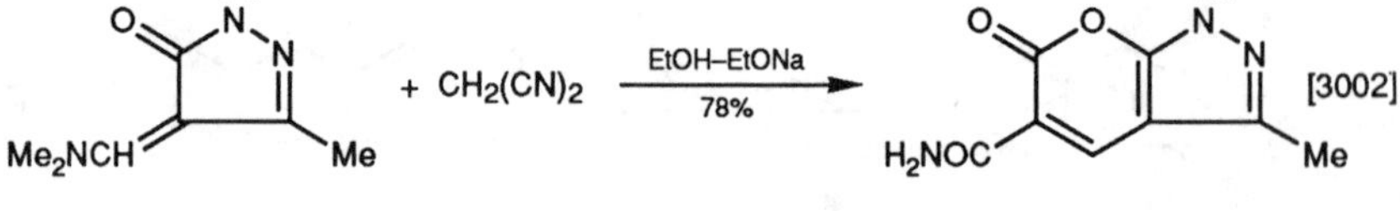

+ $CH_2(CN)_2$ EtOH–EtONa 78% [3002]

Pyrano[2,3-*c*]pyrazol-6-one

2. 1,4-Thiazine

A pyridothiazine may be synthesized efficiently by stirring a 2-mercapto-3-aminopyridine in an alkaline medium. When an organic sulphide such as (**11.5**) is heated with ethanolic hydrogen bromide under argon, cyclization occurs in good yield.

EtOH, KOH 78% [2195]

Pyrido[2,3-*b*][1,4]thiazine

EtOH–HCl, Δ 55-81% [2950]

(11.5)
R = CH_2=$CHCH_2$, $Me_2N(CH_2)_3$

Pyrimido[5,4-*b*][1,4]benzothiazine-2,4-dione

3. 1,3,4-Oxadiazine

Acylhydrazine and lactam carbonyl groups interact to form a ring on treatment with boiling thionyl chloride.

$SOCl_2$, Δ 83–98% [3229]

Ar^1 = Ph, 4-Me-, 4-Br-C_6H_4
Ar^2 = Ph, 4-Me-, 4-MeO-C_6H_4

[1,2,4]Triazino[5,6-*e*][1,3,4]oxadiazinium

4. 1,3,4-Thiadiazine

An α-bromo ketone reacts with a cyclic thiosemicarbazide (containing a lactam thiocarbonyl) on heating in ethanol; the base is freed from the resulting hydrochloride salt with carbonate. Annulation of *N*-benzylideneamino and lactam groups by heating with phenacyl bromide gives a new thiadiazine ring with elimination of the benzylidene group.

+ $RCOCH_2Br$ — i, ii 81% [2128]

R = 5-NO_2-furyl
i, EtOH, Δ; ii, aq. Na_2CO_3

Imidazo[2,1*b*][1,3,4]thiadiazine

MeOH, Δ 40% [3270]

1,2,4-Triazolo[3,4-*b*][1,3,4]thiadiazin-1-ium

Benzoin undergoes a cyclocondensation with the aminothiolactam (**11.4**) to form a new thiadiazine ring while the reactive chlorine atoms of 2,3-dichloroquinoxaline under mildly basic conditions react to form a doubly fused thiadiazine.

(**11.4**) — EtOH, RCOCHOHR, Δ; 30–50% → 1,2,4-Triazolo[3,4-*b*][1,3,4]thiadiazine [3809]

(**11.4**) — EtOH, AcONa, *i*; 90% → 1,2,4-Triazolo[3′,4′ : 2,3][1,3,4]thiadiazino[5,6-*b*]quinoxaline

R = Ph, 2-furyl;
Ar = 2,4-$Cl_2C_6H_3$
i, 2,3-dichloroquinoxaline

5. 1,3,4-Thiadiazinium

Phenacyl bromides cyclize amino-lactam thiocarbonyl groups (the latter is a cyclic thiosemicarbazide in this example) on heating in ethanol. The charge is shown at N-8 but may also be at N-1.

+ COAr–CH_2Br — EtOH, Δ; 74–96% → [1,2,4]Triazolo[5,1-*b*][1,3,4]thiadiazin-8-ium [3725]

R = 4-Br-, 4-NO_2-$C_6H_4CH_2$, 2,4-$Cl_2C_6H_3CH_2$
Ar = 4-Cl-, 4-Me_2N-C_6H_4, 2-naphthyl

III. FORMATION OF A SEVEN-MEMBERED RING

1. 1,4-Diazepin-5-one

Amino and lactam thiocarbonyl groups may be converted into a diazepinone ring by reaction with the cyclic carbamate-anhydride (**11.6**).

+ (**11.6**) — DMF, Δ; 43–64% → [3082]

R¹ = Me, Ph, C_6H_4, 4-Cl-, 4-NO_2-, 4-Me-C_6H_4;
R² = R³ = H or R²R³ = benzo

1,2,4-Triazino[5,6-*b*][1,4]diazepin-8-one
1,2,4-Triazino[5,6-*b*][1,4]benzodiazepin-10-one

CHAPTER 12

Acylamine or Imine and Methylene or Alkene

I. Formation of a Five-membered Ring 733
1. Pyrrole 733
2. Pyrrol-2-one 735
3. Pyrazole 736
4. Imidazole 736
5. Isothiazole 737
II. Formation of a Six-membered Ring 737
1. Pyridine 737
2. Pyridin-4-one 737
3. Pyrimidine 738
4. 1,3-Oxazine 738
III. Formation of a Seven-membered Ring 738
1. 1,3-Oxazepine 738

The following pairs of functional groups are also mentioned in this chapter: carbamate and methylene, acylamine or carbamate and trimethylsilylmethyl, alkene and tosylamino, alkyne and imine.

I. FORMATION OF A FIVE-MEMBERED RING

1. Pyrrole

Treatment of an imidate which is placed *ortho* to a reactive methylene group with sodium hydroxide–DMSO results in the formation of a new pyrrole ring. Good-to-excellent yields of 4-nitroindoles are obtained when 3-nitro-2-methylanilines (or acetanilides) are treated with diethyl oxalate in the presence of a strong base (cf. Madelung synthesis of indoles, p. 68).

R^2 | N=COR^3 | R^4 X $CH_2SO_2R^1$ → DMSO, NaOH, 52–84% → H N R^2 | R^4 X SO_2R^1 [3149]

R^1 = Ph, 4-MeC_6H_4, $N[(CH_2)_2]_2O$; R^2 = H, Me; R^3 = Me, Et; R^4 = H, Cl, MeO, Ph

X = CH Indole
X = N Pyrrolo[3,2-*b*]pyridine

R^1 | NCOR^2 | R^3 Me | NO_2 + $(COOEt)_2$ → EtOK, DMF, 40–99% → R^1 N R^2 | R^3 | NO_2 [3832]

R^1 = H, Me; R^2 = H, Me, Et, Ph; R^3 = H, alkyl, Br, MeO

Indole

When the iminophenylacetic acid (**12.1**) in acetonitrile is heated or allowed to stand for two days at ambient temperature, it is converted into an indole-3-carboxylic acid [2999]. Oxidative cyclization of the aminoalkene (**12.2**) occurs under mild conditions (temperature of 23 °C or lower) to give the hydrazone of an indole-3-carboxaldehyde [3843]. A number of 2,5-disubstituted indoles are accessible from the dilithium derivative of a 2-trimethylsilylmethylbenzanilide and an ester; the latter is synthesized from a 4-substituted nitrobenzene in three stages [3824].

N=CHR | CH_2COOH → MeCN, Δ, 52–86% → H N R | COOH [2999]

(**12.1**)

R = PhCO, PhCH=CH, Ph, 2-MeC_6H_4, 2-pyridyl

NHR | CH=CHCH=$NNMe_2$ → i, 52% → R N | CH=$NNMe_2$ [3843]

(**12.2**)

R = Ac, PhCO

i, CH_2Cl_2, 1,4-benzoquinone, $BF_3.Et_2O$

Li | NCOR^2 | R^1 $CHSiMe_3$ | Li + R^3COOR^4 → i, 52–74% → H N R^3 | R^1 [3824]

R^1 = MeO, Me, Ph; R^2 = Ph, 4-MeC_6H_4; R^3 = Ph, 4-MeO-, 4-Cl-C_6H_4; R^4 = alkyl

i, THF, 2,2,6,6-Me_4 piperidide Li

A mixed copper and zinc reagent (prepared in three steps from an *o*-toluidine via the benzyl bromide) reacts with various electrophilic reagents such as acyl halides at low temperature (~ −20 °C) to yield 2-*R*-indoles [3637]. 2-Vinylanilines [protected as their *N*-tosyl derivatives (**12.3**)] are cyclized to *N*-tosylindoles in a palladium-mediated reaction (review [B-41]) in the presence of benzoquinone and lithium chloride. This method gives good yields with variously substituted anilines, and is especially useful in the synthesis of 4-substituted indoles, but no variations in the vinyl group are reported [3410].

NHTs, CH=CH₂ —i, 47–88%→ Ts N [3410]

(**12.3**)

R = H, MeO, AcO, Cl, Me, COOMe, $CH(OAc)_2$, CH_2=CH, CF_3SO_2O

i, $PdCl_2.(MeCN)_2$, benzoquinone, LiCl, DMF

Application of the Vilsmeier reaction (review [1676]) to an *o*-methyl-amine may result in the formation of a fused pyrrole-3-carboxaldehyde ring. The same functional groups are annulated by reaction with trimethyl orthoformate to give a 2,3-unsubstituted pyrrole ring.

Me, N, Me, NH, H_2N, O, HCl —$POCl_3$, DMF, Δ, 67%→ CHO, N, HN, NH, O [2897]

Pyrrolo[3,2-*d*]pyrimidin-4-one

NH_2, N, Me + $HC(OEt)_3$ —EtOH–HCl, Δ, 63%→ H N, N [2851]

Pyrrolo[3,2-*b*]quinoline

2. Pyrrol-2-one

Some pyrrolo[2,3-*b*]pyridin-2-ones exhibit anti-inflammatory activity; this ring system may be formed by cyclizing the (phenylthiomethyl)amine (**12.4**) using butyllithium (review [B-47]) followed by carbon dioxide and trifluoroacetic acid.

(12.4)
R^1 = iPr, tBu, $PhCH_2$, Ph, 3-Cl-, 4-iPr-, 3-CF_3-C_6H_4, 3,4- Cl_2-,3-Cl-2-Me-C_6H_3; R^2 = Me, Ph
i, BuLi, THF; ii, CO_2; iii, CH_3COOH, CH_2Cl_2

i, ii, iii 60% [3775]

Pyrrolo[3,2-*b*]pyridin-2-one

3. Pyrazole

1-Aminobenzimidazoles are useful intermediates, and their alkylation may occur at either NH_2 or the =N— atom. On heating with an iodoalkane, only the 3-quaternary salt is obtained. This is cyclized on heating with an anhydride and sodium carbonate; a lower yield is obtained when the carbonate is replaced by pyridine. When a 2-hydroxymethyl-1-aminopyridinium chloride is heated with acetic anhydride, the *N,O*-diacetate is formed but in the presence of a base, this reaction yields the pyrazolopyridine.

+ $(R^2CO)_2O$ — Na_2CO_3, Δ 73–85% → [2982]

R^1 = Me, Et, Pr; R^2 = Me, Et, Ph

Pyrazolo[1,5-*a*]benzimidazole

+ $(R^2CO)_2O$ — pyr or AcONa 52–89% → [2519]

R^1 = H, MeO; R^2 = alkyl, CF_3

Pyrazolo[1,5-*a*]pyridine

4. Imidazole

2-Aminobenzimidazoles, quaternized with 1-bromoprop-2-yne and converted with ammonia into the 2-imines, cyclize on heating with a strong base; the quaternary salt behaves similarly. The enzyme catalase promotes the cyclization of the amino-alkene (**12.5**) to a benzimidazole under comparatively mild conditions (review [B-53]).

EtONa, Δ ~94% → [2713]

R = Me, Et, $PhCH_2$

Imidazo[1,2-*a*]benzimidazole

(12.5)
R = H, Me, PhCO

catalase, EtOH 75–88%

Benzimidazole

[3796]

5. Isothiazole

When *o*-toluidines are heated with thionyl chloride in xylene, this ring is formed in good yield.

R = H, Me, NO_2

xylene, $SOCl_2$, Δ 50–80%

2,1-Benzisothiazole

[2064, 2485]

II. FORMATION OF A SIX-MEMBERED RING

1. Pyridine

An *N*-(*o*-vinyl)imino ester is probably an intermediate in the two-step thermal cyclization of 2-vinyl-*N*-acylanilines. A photochemical method also gives high yields except when R^1 = COOMe; the thermal method gives none of this product.

R^1 = H, Me; R^2 = Et, C_5H_{11}, Ph, 2-furyl
i, $Et_3\overset{+}{O}\overset{-}{B}F_4$, CH_2Cl_2, Δ; ii, Ph_2O, Δ

i, ii 71–92%

Quinoline

[3440]

2. Pyridin-4-one

A methylene group activated by carbonyl and sulphonamido groups reacts with a neighbouring amino group on heating with orthoformate ester.

R^1 = H, Me, MeO, Cl; R^2 = H, Et;
X = CH, N

$HC(OEt)_3$, Δ 14–92%

4-Quinolinone
1,8-Naphthyridin-4-one

[2517]

3. Pyrimidine

Imino and methyl groups may be annulated by prolonged heating with crotonic anhydride.

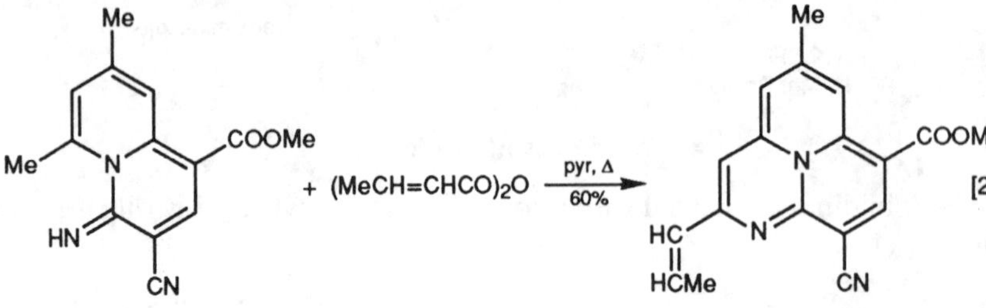

+ $(MeCH{=}CHCO)_2O$ —pyr, Δ; 60%→ [2734]

Pyrimido[2,1,6-*de*]quinolizine

4. 1,3-Oxazine

The methyl(bismethylthio)sulphonium salt **(12.6)**, a mild methylthiolating agent for alkenes and alkynes, converts *o*-vinyl-acylamines into fused oxazines.

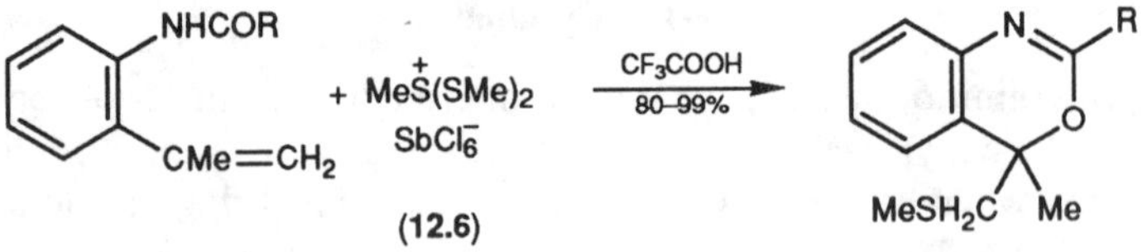

+ $Me\overset{+}{S}(SMe)_2$ $SbCl_6^-$ **(12.6)** —CF_3COOH; 80–99%→ [3138]

3,1-Benzoxazine

III. FORMATION OF A SEVEN-MEMBERED RING

1. 1,3-Oxazepine

o-Allylbenzanilide is annulated by stirring with a bis(methylthio)sulphonium salt **(12.6)** followed by basification to release the free base.

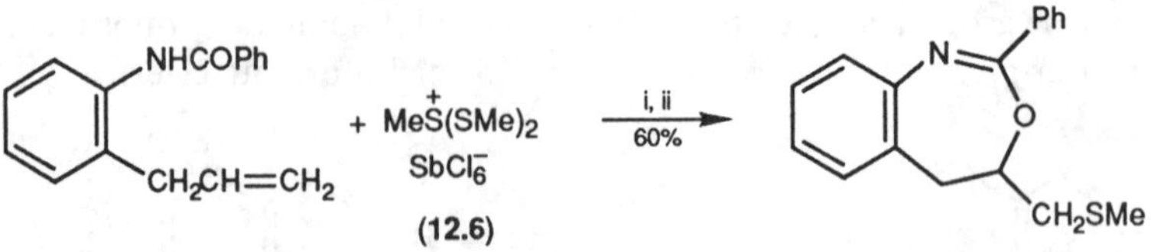

+ $Me\overset{+}{S}(SMe)_2$ $SbCl_6^-$ **(12.6)** —i, ii; 60%→ [3329]

i, CH_2Cl_2, C_5H_{12}; ii, aq. $NaHCO_3$

1,3-Benzoxazepine

CHAPTER 13

Acylamine or Amine and Nitrile

I. Formation of a Five-membered Ring 739
   1. Pyrrole 739
   2. Pyrazole 740
   3. Imidazole 740
   4. 1,2,4-Triazole 741
II. Formation of a Six-membered Ring 741
   1. Pyridine 741
   2. Pyridin-2-one 743
   3. Pyrimidine 743
   4. Pyrimidin-2-one or -4-one 746
   5. 1,2,3-Triazin-4-one 2-Oxide 747
   6. 1,2,4-Triazine or 1,2,4-Triazin-5-one 747
   7. 1,3,5-Triazine or 1,3,5-Triazin-2-one 748
   8. 1,4-Oxazine or 1,4-Thiazine 748
   9. 1,2,4-Thiadiazine 1,1-Dioxide 749
III. Formation of a Seven-membered Ring 749
   1. 1,4-Diazepine 749
   2. 1,4-Thiazepine 750
   3. 1,3,4-Thiadiazepine 750

An example of the cyclization of an amino-iminoether (often prepared from a nitrile) is included in this chapter.

I. FORMATION OF A FIVE-MEMBERED RING

1. Pyrrole

Reductive cyclization of a nitro-nitrile yields a fused pyrrole ring.

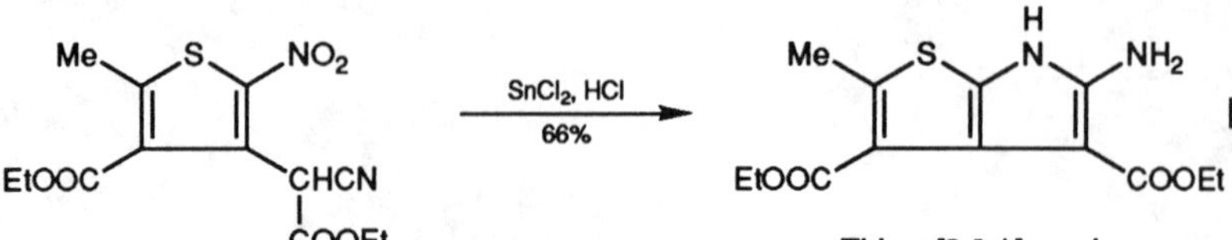

SnCl$_2$, HCl, 66% [2545]

Thieno[2,3-*b*]pyrrole

2. Pyrazole

A 3-aminopyrazole ring is formed on stirring an aminonitrile at ambient temperature with hydrazine hydrate.

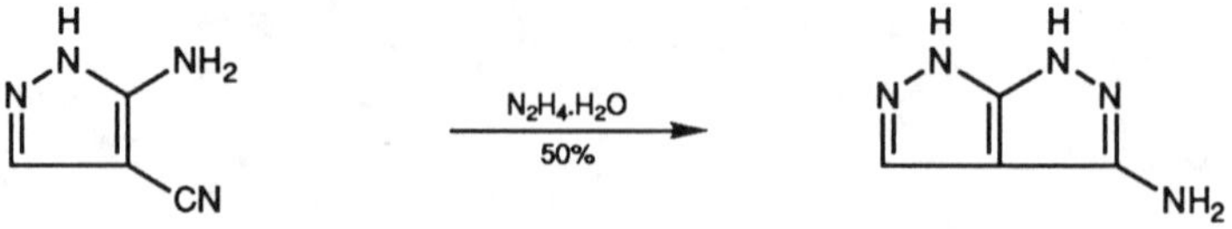

$N_2H_4.H_2O$, 50% [3780]

Pyrazolo[3,4-*c*]pyrazole

3. Imidazole

An iminoether (formed *in situ* from a nitrile group) and an acylamino function react in the presence of a mild base to form a fused imidazole ring [3137]. Nucleophilic attack by an amino group on an iminoether (formed *in situ*) gives high yields of fused imidazoles [2446]. Irradiation (review [B-50]) of a 2-cyanocyclohexylamine in THF gives a high yield of a reduced benzimidazole [2822].

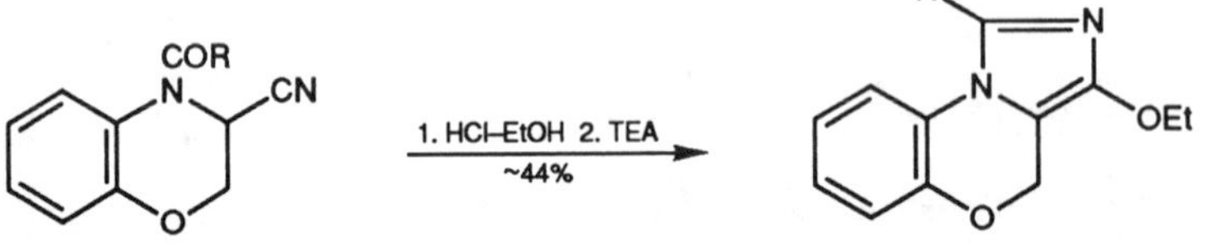

1. HCl–EtOH 2. TEA, ~44% [3137]

R = Me, Ph

Imidazo[5,1-*c*][1,4]benzoxazine

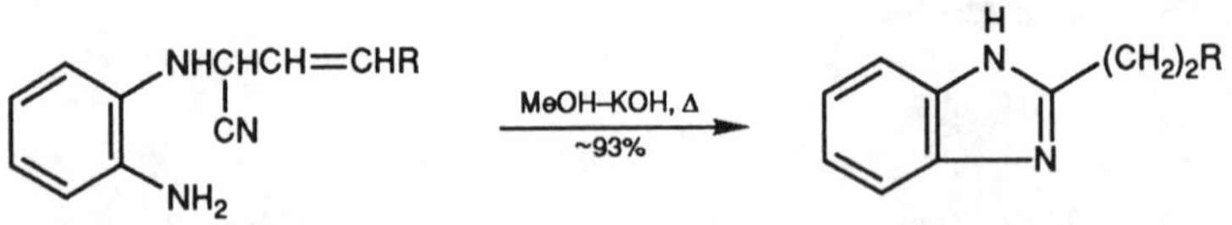

MeOH–KOH, Δ, ~93% [2446]

R = Ph, 2-furyl

Benzimidazole

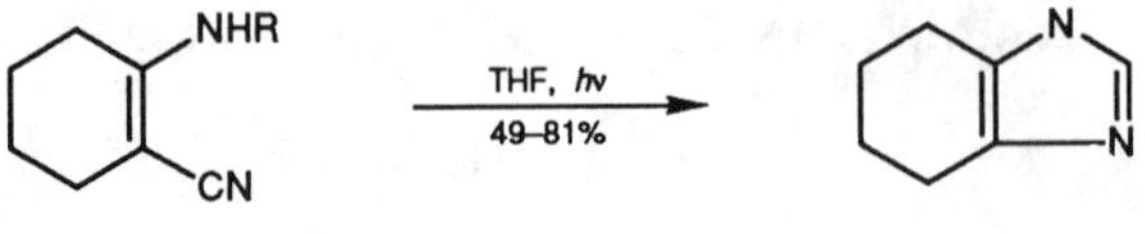

THF, *hν*, 49–81% [2822]

R = H, Me

Another ring system synthesized similarly:

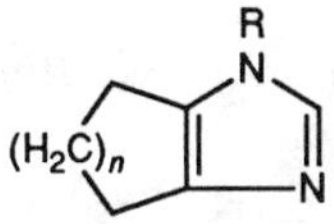

[2821]

R = H, Me; n = 1 or 3
Cyclopent[*d*]imidazole
Cyclohept[*d*]imidazole

4. 1,2,4-Triazole

The amino-cyanoiminothiazole (**13.3**) [formed *in situ* by amination of the thiazole (**13.2**)] cyclizes on stirring with *O*-(2,4-dinitrophenyl)hydroxylamine.

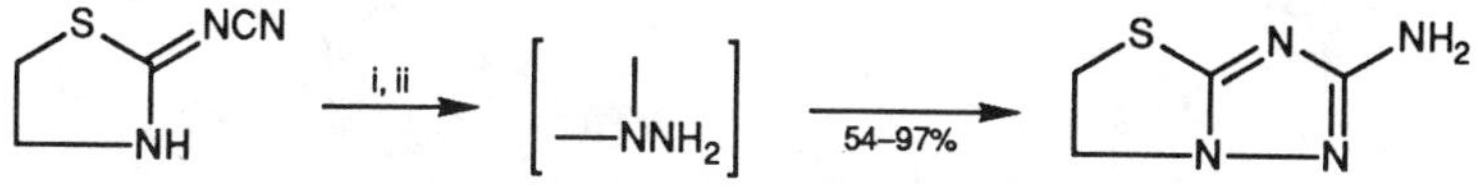

[3374]

(**13.2**) (**13.3**) Thiazolo[3,2-*b*][1,2,4]triazole
i, THF, NaH; ii, 2,4-$(NO_2)_2C_6H_3ONH_2$

II. FORMATION OF A SIX-MEMBERED RING

1. Pyridine

Amino and dicyano- or cyanoethoxycarbonyl-ethylidene functions are annulated by heating with t-butoxide, TEA or acetic anhydride, or by stirring with methanolic sodium hydroxide. The 4-nitrophenyl derivative gave the highest yield, the 4-methoxyphenyl the lowest. Condensed pyrazoles have been reviewed [3902].

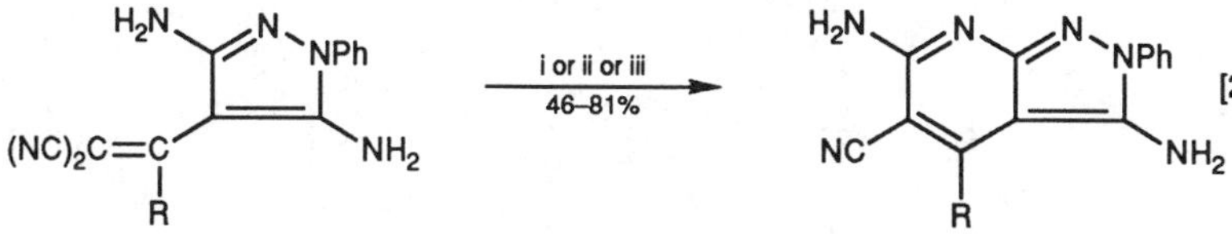

[2986, 3127]

R = Me, Et, Ph, 4-ClC_6H_4, 4-$NO_2C_6H_4$
i, tBuOH, TEA, Δ; ii, Ac_2O, Δ; iii, MeOH–NaOH

Pyrazolo[3,4-*b*]pyridine

Other ring systems synthesized similarly:

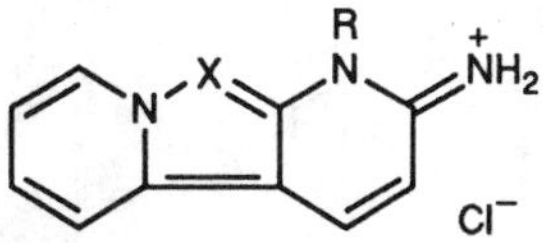

[2986]

X = N Dipyrido[1,2-*b*:3',2'-*d*]pyrazole
X = CH Pyrido[3,2-*a*]indolizine

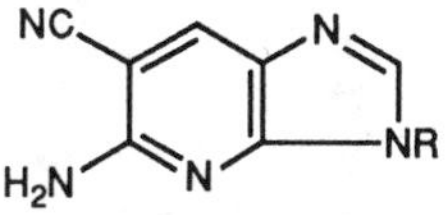

[3668]

Imadazo[4,5-*b*]pyridine

The powerful Lewis acidity of a titanium(IV) salt is demonstrated in its room-temperature catalysis of the reaction of an aminonitrile with a lactone, δ-valerolactone or its 6-methyl homologue. γ-Butyrolactone gave poor yields [3638]. Tin(IV) chloride also promotes cyclization of aminobenzonitriles with β-keto esters to give the medicinally useful 4-aminoquinolines [3871].

CH_2Cl_2, $TiCl_4$, TEA; 50–62% [3638]

R^1 = H, F; R^2 = H, Me

Pyrano[2,3-*b*]quinoline

PhMe, $SnCl_4$, Δ; 60–80% [3871]

(13.4)
R = H, NO_2

Quinoline

$\alpha\beta$-Unsaturated nitriles are widely used in the synthesis of heterocycles (review [1096, 3123]); benzylidenemalononitrile in the presence of sodium converts an aminonitrile into a fused 4-amino-3-cyanopyridine ring. When a milder base (TEA) is used, both nitrile groups are retained [3897, 3353]. When the 4*H*-pyran (**13.5**) was stirred with diethyl benzylidenemalonate and TEA, none of the expected product was isolated but another derivative of this ring system was obtained, namely, that which is formed by reaction of the pyran (**13.5**) with benzylidenemalononitrile. The pyran ring has opened, reacted with itself and cyclized to give the pyrano[2,3-*b*]pyridine [3819].

+ PhCH=C(CN)$_2$ → (diox, Na, Δ; ~48%) [3897]

R = CN, COOEt

Thieno[2,3–*b*]pyridine

+ PhCH=C(CN)$_2$ → (i; 60%); PhCH=C(COOEt)$_2$, ii; 19% [3353, 3819]

(13.5)
i, EtOH, TEA, Δ; ii, EtOH, Δ

Pyrano[2,3-*b*]pyridine

2-Aminocinnamonitriles may be obtained from benzaldimines by a palladium-mediated reaction with acrylonitrile; these are then cyclized (without purification) by heating in mesitylene. When acrylonitrile is replaced by styrene, and mercury(II) oxide added (to oxidize the imine group to the aldehyde), cyclization gives 3-phenylisoquinolin-1-ones in good yield.

CH=NtBu; i, ii; CH=NtBu, CH=CHCN; iii, 10–56%; [3398]

Isoquinoline

R = H, 6- or 7-MeO, 6,7-$(MeO)_2$

i, AcOH, AcONa, $PdCl_2$, Δ;

ii, PhMe, CH_2=CHCN, TEA, Δ; iii, $C_6H_3Me_3$, Δ

2. Pyridin-2-one

Cyclization of aminonitriles with malonic esters is mediated by tin(IV) chloride (cf. previous section) but the yields may be low [3871]. Amino and cyano groups attached to different rings interact to form a fused pyrimidin-2-one on heating the compound with potassium hydroxide [3899].

(13.4) + $CH_2(COOR^2)_2$ — PhMe, $SnCl_4$, Δ; ~25% → [3871]

R = H, NO_2; R^2 = Me, Et

2-Quinolinone

EtOCH_2CH_2OH, KOH, Δ; 57% [3899]

Benzofuro[2,3-*c*]quinolin-6-one

3. Pyrimidine

A number of reactions involve addition of hydrogen chloride to a nitrile group (review [2547]) and the formation of a chlorimine which later loses either a Cl or NH_2 as HCl or NH_3, respectively. Both of these possible routes are observed when anthranilonitrile or a heterocyclic analogue is heated with another nitrile. The (single) product formed depends on the nature of the aminonitrile and is difficult to predict. This reaction is also successful when hydrogen chloride is replaced by a strong base.

+ R^2CN → diox, HCl(g), 11–87% [3670]

R^1 = H, Me, Ph; R^2 = Me, $ClCH_2$, Cl_2CH, Cl_3C, $PhCH_2$, COOEt, CH_2COOEt, 4-$NO_2C_6H_4CH_2$, 4-$NO_2C_6H_4$, $PhOCH_2$, $PhSO_2CH_2$; X = CH=CH, O, S

Quinazoline
Furo[2,3-*d*]pyrimidine
Thieno[2,3-*d*]pyrimidine

+ R^2CN → iPrOH, MeONa, Δ, 20–98% [3205]

R^1 = H, Ph, PhMeCH; R^2 = Me, Ph, Ph_2CH, 2- or 4-ClC_6H_4

Pyrrolo[2,3-*d*]pyrimidine

A side-chain cyano group reacts with a ring-amino to form a pyrimidine ring; this cyclization occurs on heating the substrate either alone or with acetic acid [2670]. A fused 4-aminopyrimidine ring is a constituent of several compounds of medicinal interest and may be constructed from an aminonitrile and a number of reagents (see p. 76); formamide [2636] or diethoxymethyl acetate [2746] frequently give good results. Recent work on polyazines has been reviewed [3985].

→ Δ, ~80% [2670]

R = H, Me

Pyrazolo[1,5-*a*]pyrimidine

+ $HCONH_2$ → Δ, ~95% [2636]

Ar = Ph, 4-Cl-, 4-Me-C_6H_4

Thieno[2,3-*d*]pyrimidine

Other ring systems synthesized similarly:

[3898]

Pyrimido[4′,5′ : 6,5]pyrano[3,2-*h*]quinoline

(13.5) + $OCH(OMe)_2$–COMe → ($MeOH$, NH_3; 74%) [2746]

Pyrazolo[3,4-*d*]pyrimidine

Cyclization of an aminonitrile under Vilsmeier conditions but employing t-carboxamides other than DMF leads to moderate yields of fused 4-chloropyrimidines [3919]. When an aminonitrile (**13.5**) is heated with a secondary formamide, cyclization is accompanied by migration of the *N*-alkyl group from the ring-nitrogen to the amino-nitrogen (a Dimroth rearrangement) [2507]. A variant of the same ring system has been synthesized without rearrangement using *N*-methylformamide; an *N*-substituted acetamide gives a comparable yield but some unreacted aminonitrile remains [2746].

+ $POCl_3$ + $RCONMe_2$ → (Δ; 30–52%) [3919]

R = Me, Ph

Benzofuro[3,2-*d*]pyrimidine

(13.5) + R^1CONHR^2 → (Δ; 54–77%) → (Δ) [2507]

R^1 = H, Me;
R^2 = alkyl, $CH_2{=}CHCH_2$

Pyrazolo[3,4-*d*]pyrimidine

A fused 4-hydrazinopyrimidine ring is obtained when neighbouring imino and nitrile groups are annulated by means of an excess of a hydrazine [3372]. Two molecules of an aminonitrile sometimes react together under strongly basic conditions, and under more severe or prolonged conditions, a trimer may form. The example shows the product obtained from 2-aminoisophthalonitrile at ambient temperature over 7 days [3559].

MeOCH=N, NC, X, Y, NR + MeNHNH$_2$ → (PhMe, MeOH, TFA, Δ; 65–86%) NHNHMe [3372]

R = Me, PhCH$_2$

X = N; Y = CH Purine
X = CH; Y = N Pyrazolo[3,4-*d*]pyrimidine

CN, NH$_2$, CN → (DMSO, tBuOK; ~60%) [3559]

Quinazoline

4. Pyrimidin-2-one or -4-one

An efficient cyclization of an aminonitrile under mild conditions (without external heat) is obtained by reacting it with chlorosulphonyl isocyanate (review [3340]) to give a high yield of pyrimidin-2-one [3368]; heating with urea is also effective [2774]. The corresponding 4-one is formed in lower yield by heating an (acylamino)nitrile with a primary amine hydrochloride, a t-amine and phosphorus pentoxide [3789], or with acetic anhydride–pyridine [3898].

NHR, CN + SO$_2$Cl–NCO → (CH_2Cl_2; ~91%) [3368]

R = H, Me, Et, PhCH$_2$CH$_2$

2-Quinazolinone

→ (i, Δ; 18–51%) [3789]

R^1 = H, PhCH$_2$; R^2 = Me, Ph;
Ar = Ph, 4-Cl-, 3-F-, 4-alkyl-, 3-CF$_3$-C$_6$H$_4$
i, ArNH$_2$·HCl, C$_6$H$_{11}$NMe$_2$HCl, P$_2$O$_5$, H$_2$O

Pyrrolo[2,3-*d*]pyrimidine-4-one

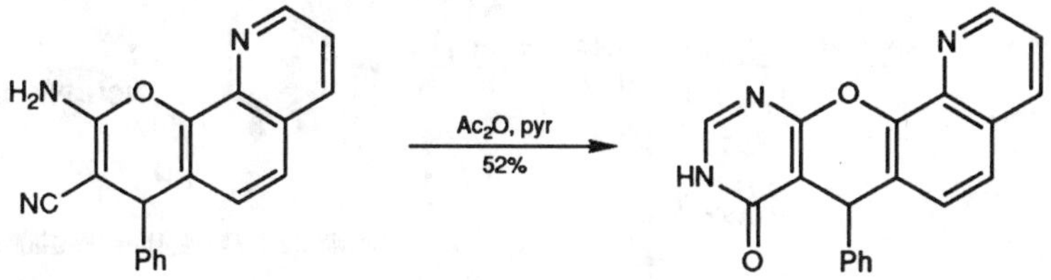

[3898]

Pyrimido[4',5' : 6,5]pyrano[3,2-*h*]quinolin-8-one

5. 1,2,3-Triazin-4-one 2-Oxide

Treatment of an aminonitrile with a nitric acid–sulphuric acid mixture at about 0 °C produces this ring, probably through the formation of the *N*-nitro compound.

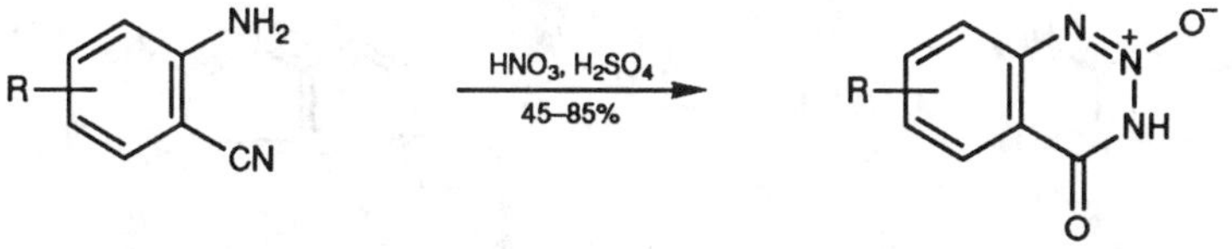

[3381]

1,2,3-Benzotriazin-4-one 2-oxide

Another ring system synthesized similarly:

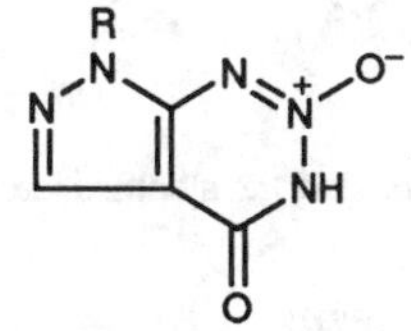

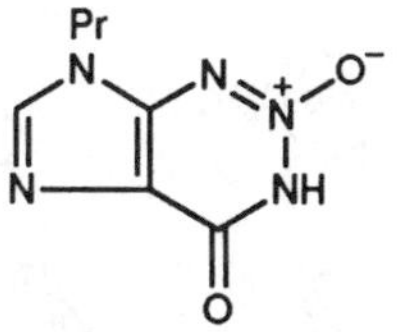

[3381]

Pyrazolo[3,4-*d*]-1,2,3-triazin-4-one 2-oxide

[3381]

Imadazo[4,5-*d*]-1,2,3-triazin-4-one 2-oxide

6. 1,2,4-Triazine or 1,2,4-Triazin-5-one

N-Aminopyridinium salts (review [1870]) are useful substrates for the synthesis of a number of fused heterocycles. With ethyl orthoformate, *N*-ethoxyimino derivatives are formed and react with benzylamine at ambient temperature to give the readily cyclized aminomethylene derivative which, with piperidine and then mineral acid, yields the triazinone.

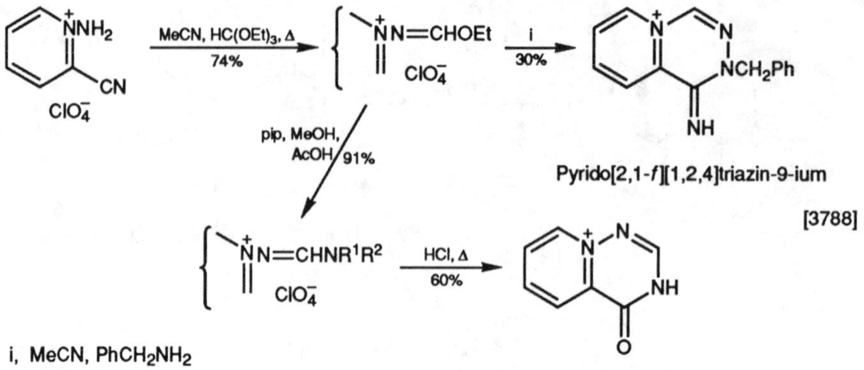

MeCN, $HC(OEt)_3$, Δ 74%; i 30%; pip, MeOH, AcOH 91%; HCl, Δ 60%

Pyrido[2,1-*f*][1,2,4]triazin-9-ium [3788]

i, MeCN, $PhCH_2NH_2$
$NR^1R^2 = N(CH_2)_5$

Pyrido[2,1-*f*][1,2,4]triazin-9-ium, 4-oxo-

7. 1,3,5-Triazine or 1,3,5-Triazin-2-one

Neighbouring amino and *N*-iminoether groups are cyclized by heating with a carboxylic acid to give a triazine while aryl isocyanates yield the triazinone.

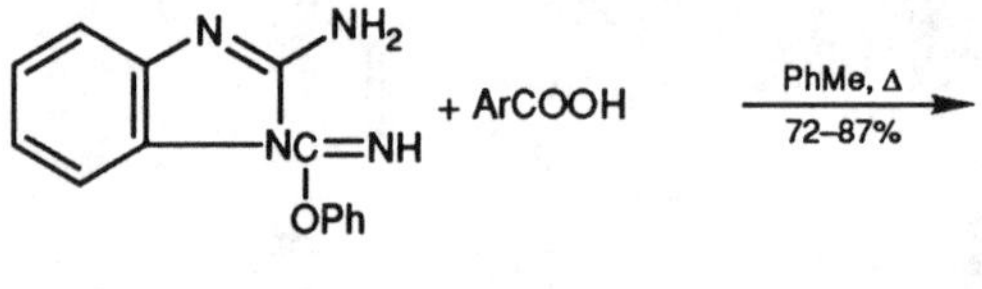

+ ArCOOH → PhMe, Δ 72–87% [2603]

Ar = Ph, 4-Cl-, 4-Br-, 4-NO_2-C_6H_4

1,3,5-Triazino[1,2-*a*]benzimidazole

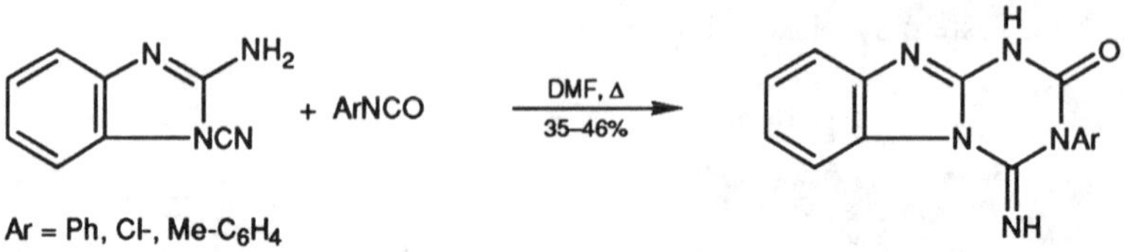

+ ArNCO → DMF, Δ 35–46% [2603]

Ar = Ph, Cl-, Me-C_6H_4

1,3,5-Triazino[1,2-*a*]benzimidazol-2-one

8. 1,4-Oxazine or 1,4-Thiazine

Cyanomethyl-oxy or -thio and nuclear amino groups are annulated by heating with potassium hydroxide [2456] or by stirring with sodium methoxide [2377].

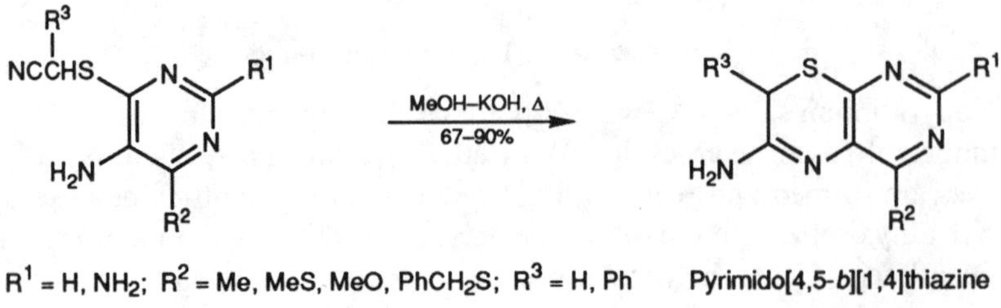

MeOH–KOH, Δ 67–90% [2456]

R^1 = H, NH_2; R^2 = Me, MeS, MeO, $PhCH_2S$; R^3 = H, Ph

Pyrimido[4,5-*b*][1,4]thiazine

Other ring systems synthesized similarly:

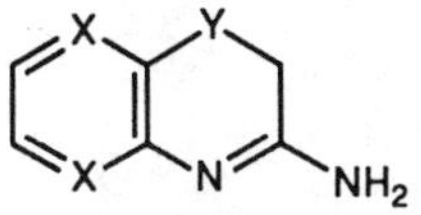

[2377, 2456]

X = CH; Y = O 1,4-Benzoxazine
X = N; Y = S Pyrazino[2,3-*b*]-1,4-thiazine

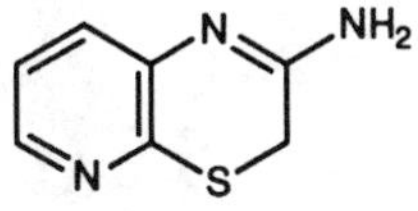

[2456]

Pyrido[2,3-*b*][1,4]thiazine

9. 1,2,4-Thiadiazine 1,1-Dioxide

This ring is formed efficiently when an *o*-amino-(*N*-cyanosulphonamide) (obtained in high yield as its sodium salt from the sulphonyl chloride and sodium cyanamide) is stirred with dilute mineral acid without external heating.

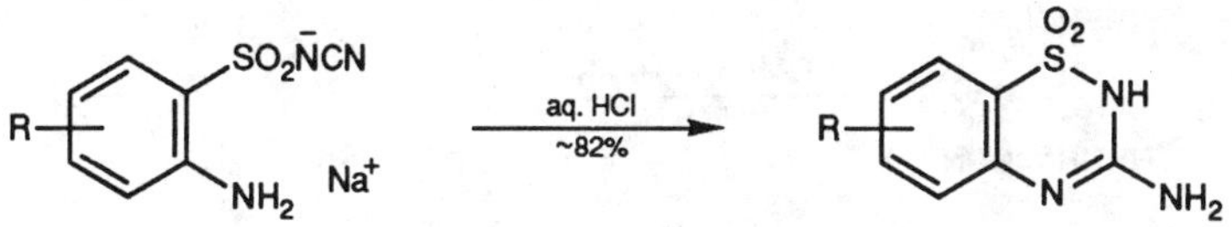

aq. HCl, ~82% [2554]

R = Cl, CF_3

1,2,4-Benzothiadiazine 1,1-dioxide

III. FORMATION OF A SEVEN-MEMBERED RING

1. 1,4-Diazepine

When a compound containing amino (possibly formed *in situ*) and cyano functions (attached to different rings) is heated in either an acidic or a basic medium, a 5-aminodiazepine ring is formed.

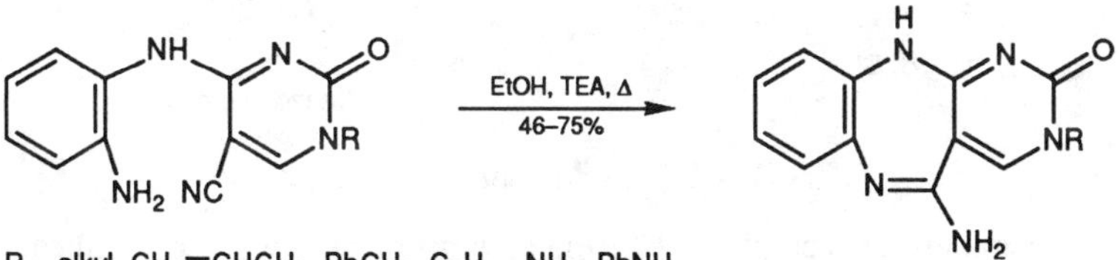

EtOH, TEA, Δ, 46–75% [3101, 3150]

R = alkyl, $CH_2{=}CHCH_2$, $PhCH_2$, C_6H_{11}, NH_2, PhNH

Pyrimido[4,5-*b*][1,5]benzodiazepin-2-one

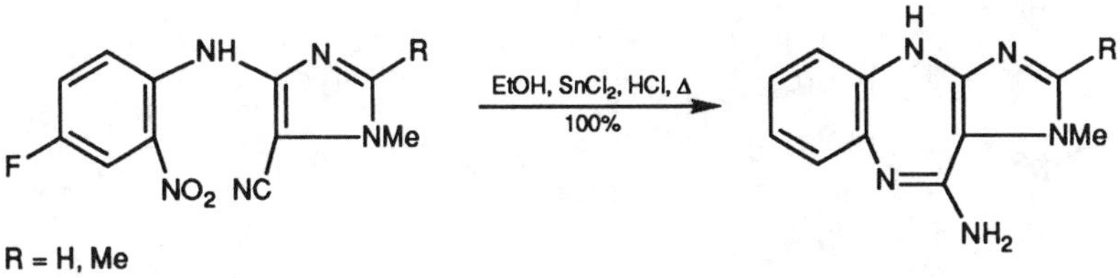

EtOH, $SnCl_2$, HCl, Δ, 100% [3525]

R = H, Me

Imidazo[4,5-*b*][1,5]benzodiazepine

Another ring system synthesized similarly:

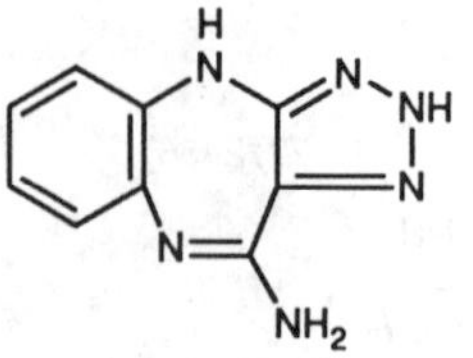

[3525]

1,2,3-Triazolo[4,5-*b*][1,5]benzodiazepine

A high yield of a 1,5-benzodiazepine is obtained under chemically mild conditions by refluxing the malononitrile derivative (**13.6**) in ethanol. The product, being an *o*-aminonitrile, would be expected to react with guanidine to form a new 2,4-diaminopyrimidine ring (cf. Part 1, p. 78), but this reaction failed.

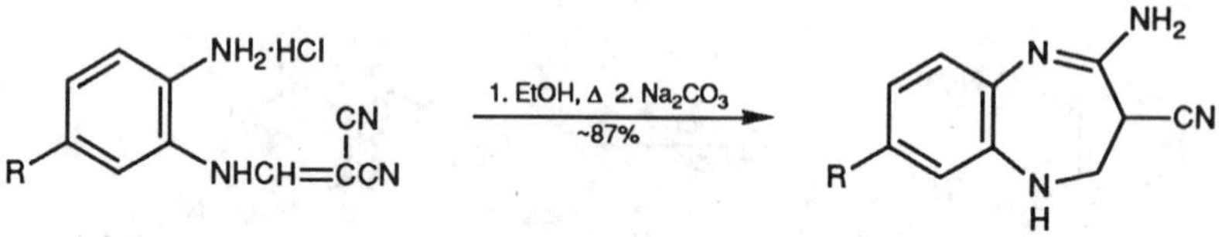

1. EtOH, Δ 2. Na_2CO_3
~87%

[2730]

(**13.6**)
R = H, Cl

1,5-Benzodiazepine

2. 1,4-Thiazepine

This ring may be formed in an acid-catalysed cyclization of a 2-β-cyanoethylthioaniline; the use of this type of reaction for the synthesis of heterocycles has been reviewed [2547].

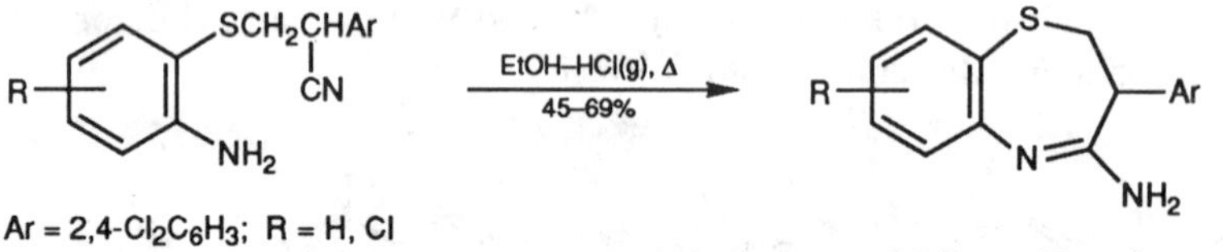

EtOH–HCl(g), Δ
45–69%

[2324]

Ar = 2,4-$Cl_2C_6H_3$; R = H, Cl

1,5-Benzothiazepine

3. 1,3,4-Thiadiazepine

This ring was formed unexpectedly by heating an aminonitrile with thiosemicarbazide hydrochloride in acetic acid.

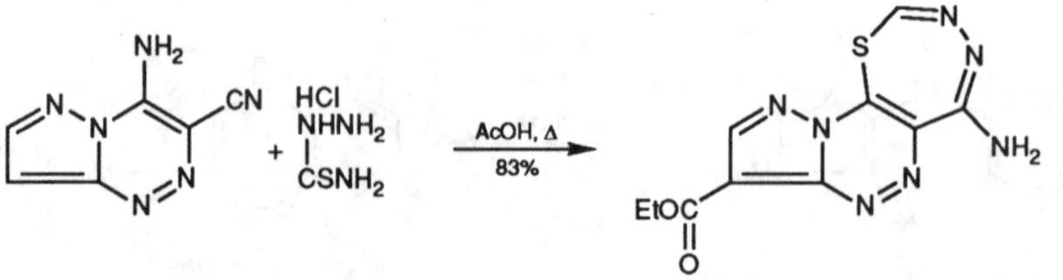

AcOH, Δ
83%

[3576]

Pyrazolo[5',1' : 3,4][1,2,4]triazino[6,5-*f*][1,3,4]thiadiazepine

CHAPTER 14

Acylamine, Amine or Imine and Nitro

I. Formation of a Five-membered Ring 751
 1. Pyrazole or Pyrazole 1-Oxide 751
 2. Pyrazol-3-one 752
 3. Imidazole or Imidazole 1-Oxide 752
 4. Furazan 1-Oxide 753
II. Formation of a Six-membered Ring 753
 1. Pyridin-4-one 753
 2. Pyridazine 1-Oxide 754
 3. Pyrazine 1-Oxide 754
 4. 1,2,4-Triazine 1-Oxide 754

Cyclizations of a nitro-imine and a nitro-guanidine are also described in this chapter.

I. FORMATION OF A FIVE-MEMBERED RING

1. Pyrazole or Pyrazole 1-Oxide

Base-catalysed cyclization of the nitrobenzylaniline (**14.1**) produces the strongly coloured yellow *N*-oxide. The unoxidized pyrazole is produced (often together with the cinnoline *N*-oxide, see Section II.2) when 2-nitrobenzylidenearylamines (**14.2**) are heated with potassium cyanide.

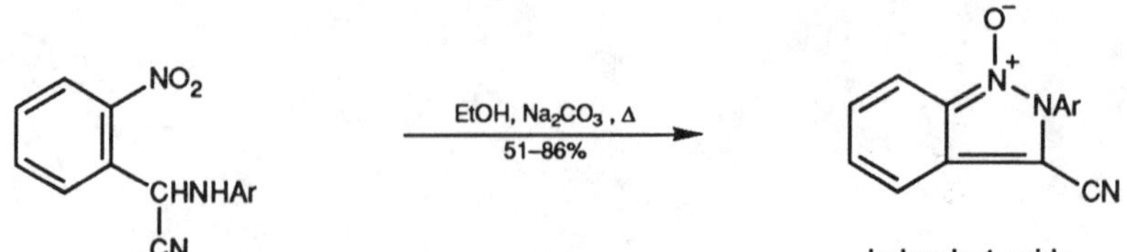
NO2
CHNHAr
CN
EtOH, Na2CO3, Δ
51–86%
O⁻
N+
NAr
CN

[3262]

Indazole 1-oxide

(14.1)
Ar = Ph, 4-Br-, 4-Cl-, 4-I-, 2-Me-, 4-Me-, 4-MeO-C_6H_4

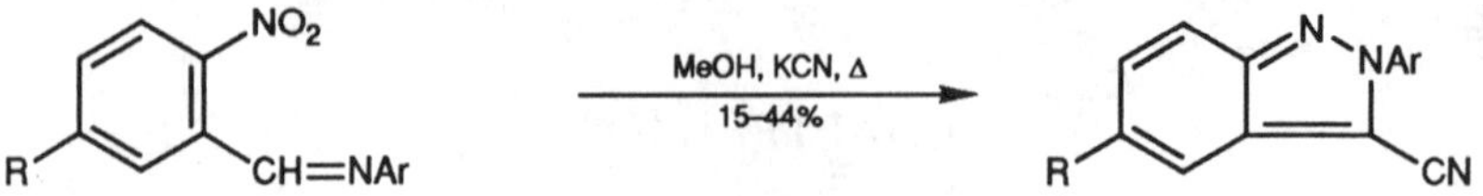
NO2
R
CH=NAr
MeOH, KCN, Δ
15–44%
N
NAr
R
CN

[3262]

(14.2)
Ar = Ph, 4-Br-, 4-Cl-, 4-I-, 2-Me-, 4-Me-, 4-MeO-C_6H_4; R = H

Indazole

2. Pyrazol-3-one

Annulation of methylene and nitro groups of 2-arylaminomethyl-3-nitropyridine by heating with a primary arylamine in acetic acid leads to the pyrazolone together with a small amount of the corresponding pyrazole in some experiments.

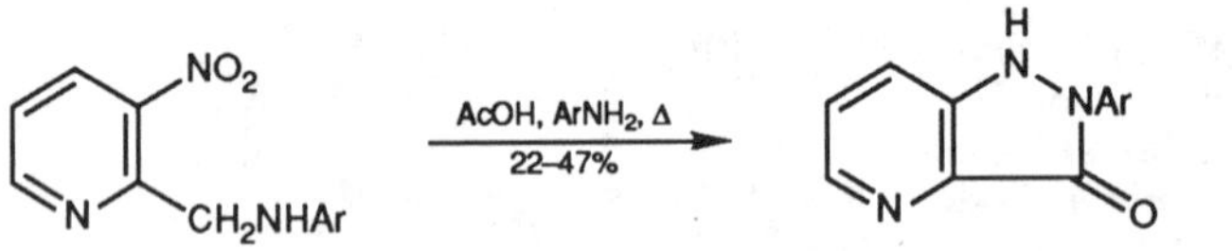
NO2
N
CH2NHAr
AcOH, ArNH2, Δ
22–47%
H
N
NAr
N
O

[2495]

Ar = Ph, 4-Cl-, 4-MeO-, 4-EtOOC-C_6H_4

Pyrazolo[4,3-*b*]pyridin-3-one

3. Imidazole or Imidazole 1-Oxide

N-Benzylidene-2-nitroanilines are useful substrates: heating with potassium cyanide–methanol gives the 1-hydroxy-2-phenylbenzimidazole while the *o*-nitrobenzylidene-2-nitroanilines lead to cinnoline 1-oxide in low yields (see Section II.2) [2266, 2848]. Thermolysis of a mixture of an *o*-nitroaniline and a benzaldehyde gives a 2-arylbenzimidazole [2848] while reduction or photolysis of *o*-nitro-t-amines converts them into benzimidazoles or their *N*-oxides, depending on steric and electronic characteristics [2197, 2224].

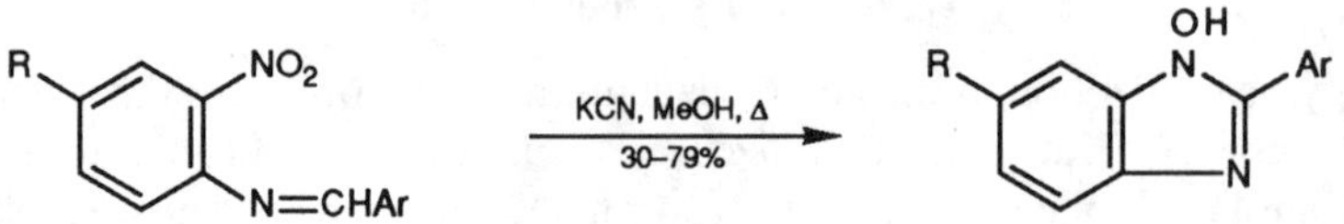
R
NO2
N=CHAr
KCN, MeOH, Δ
30–79%
OH
R
N
Ar
N

[2266, 2848]

R = H, Me; Ar = Ph, 2-Br-, 2-Cl-, 2-MeO-C_6H_4

Benzimidazole

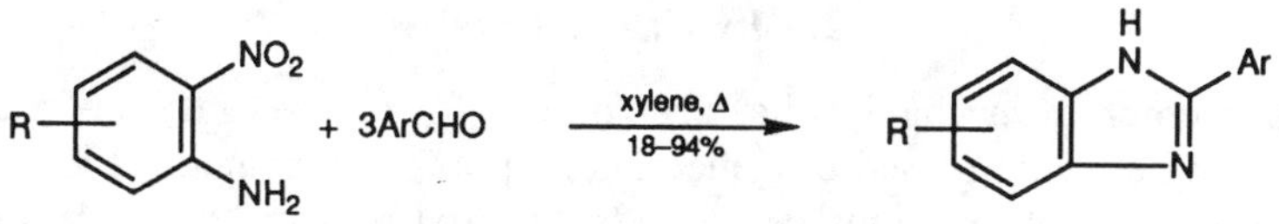

R = H, Me, MeO; Ar = Ph, 4-MeC_6H_4, 4-$MeOC_6H_4$ [2848]

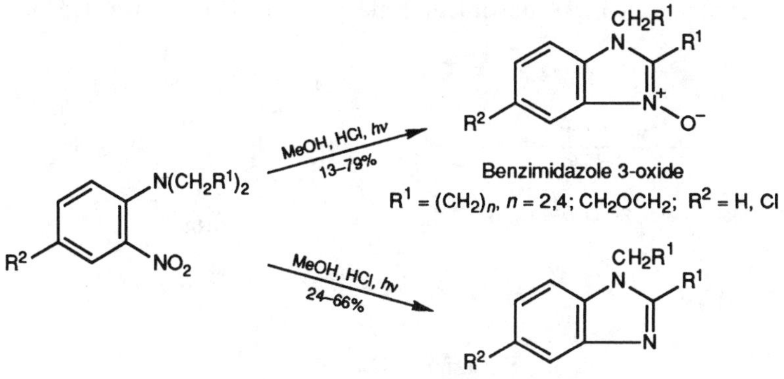

Benzimidazole 3-oxide

$R^1 = (CH_2)_n$, n = 2,4; CH_2OCH_2; R^2 = H, Cl [2197, 2224]

$R^1 = (CH_2)_n$, n = 3,4; R^2 = H

4. Furazan 1-Oxide

Oxidation of a nitroamine with (diacetoxyiodo)benzene (review [3714]) often gives better results in this cyclization than are obtained with hypochlorite.

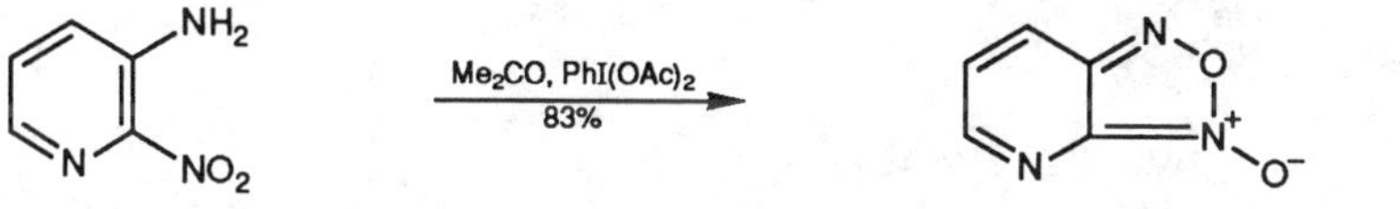

[2162]

[1,2,5]Oxadiazolo[4,5-*b*]pyridine 3-oxide

II. FORMATION OF A SIX-MEMBERED RING

1. Pyridin-4-one

Displacement of a nitro group by a side-chain amine-derived anion can lead to the formation of a pyridinone derivative which may have antibacterial properties.

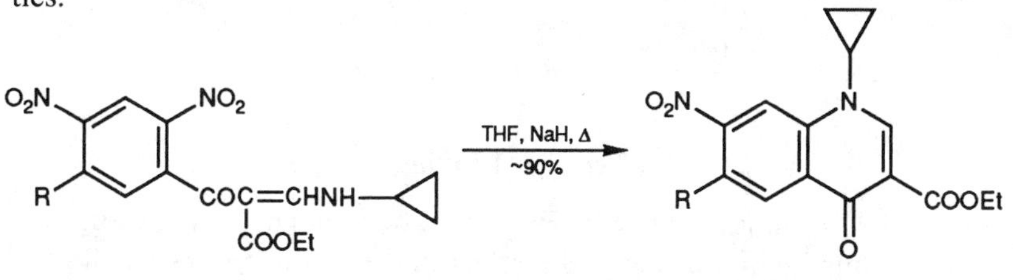

[3200]

R = F, Cl

4-Quinolinone

2. Pyridazine 1-Oxide

A rather complex mechanism has been offered for the cyclization of a 2-nitroaniline imine in the presence of methanolic potassium cyanide. Yields in this type of cyclization may not be high [2848]. 2-Nitrobenzylidenarylamines are cyclized by the same reagents. Variable amounts of indazoles are also produced but when Ar contains an *ortho*-substituent, little or no indazole is formed [3262].

(14.2) MeOH, KCN, Δ / 9–55% → [2848, 3262]

Ar = Ph, Me-, MeO-, halo-, 2-COOMe-C_6H_4;
R = H, MeO, Me

Cinnoline 1-oxide

3. Pyrazine 1-Oxide

A 5-unsubstituted 4-benzyliminopyrimidinone (**14.3**) undergoes successive nitration and cyclization to give a good yield of a pyrazinopyrimidine *N*-oxide which may also be synthesized from the nitrobenzyliminopyrimidine in high yield.

(14.3) AcOH, KNO_3, Δ / 67% → [2667]

AcOH, KNO_3, H_2SO_4 / 90% →

Pyrimido[4,5-*b*]quinoxaline-2,4-dione 5-oxide

4. 1,2,4-Triazine 1-Oxide

Amino and nitro groups in different rings may be annulated by heating with aqueous potassium hydroxide in pyridine. The triazine oxide ring is also formed when an *o*-nitro-guanidine is heated with aqueous carbonate.

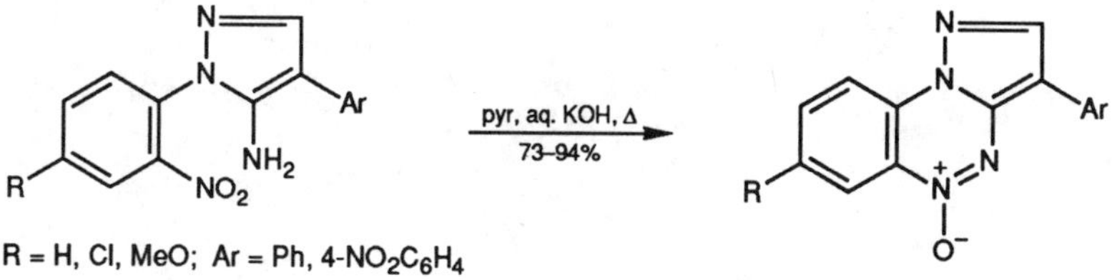

N
N
Ar
NH₂
R
NO₂

pyr, aq. KOH, Δ
73–94%

N
N
Ar
R
N
N
O⁻

[2292]

R = H, Cl, MeO; Ar = Ph, 4-$NO_2C_6H_4$

Pyrazolo[5,1-*c*][1,2,4]benzotriazine 5-oxide

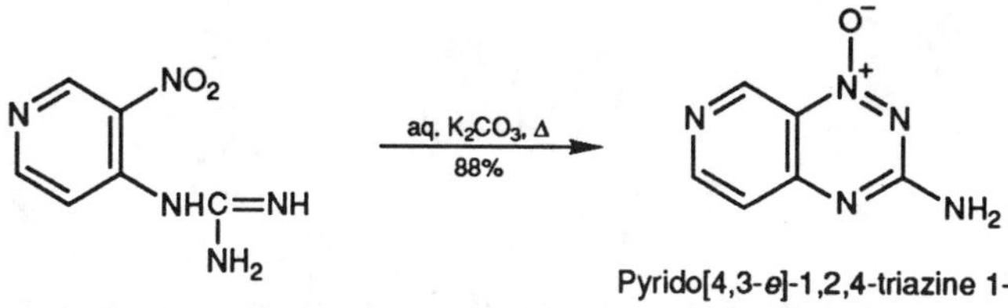

N
NO_2
NHC=NH
NH_2

aq. K_2CO_3, Δ
88%

O⁻
N
N
N
N
NH_2

[2320]

Pyrido[4,3-*e*]-1,2,4-triazine 1-oxide

CHAPTER 15

Acylamine or Amine and Nitroso or *N*-Oxide

I. Formation of a Five-membered Ring 756
1. Imidazole 756
2. 1,2,4-Oxadiazole 757
II. Formation of a Six-membered Ring 758
1. Pyrazine 758
2. 1,2,4-Triazine 759

I. FORMATION OF A FIVE-MEMBERED RING

1. Imidazole

Reductive cyclization of a nitrosoacylamine in acetic acid can give high yields of a fused 2-substituted imidazole [2765]. A nitroso-aminopyrimidinedione reacts with a variety of aldehyde hydrazones and with the Vilsmeier reagent to give good yields of purinediones [2225, 2447]. When *o*-nitroso-*N*-alkylpyrazolamines are heated for 10–90 min in pyridine, a fused imidazole ring is formed [3449, 3831].

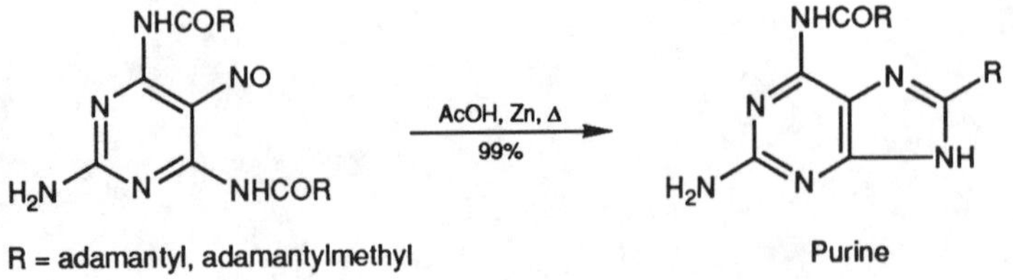

NHCOR, NO, N, N, H_2N, NHCOR — AcOH, Zn, Δ, 99% → NHCOR, N, N, R, NH, H_2N, N [2765]

R = adamantyl, adamantylmethyl

Purine

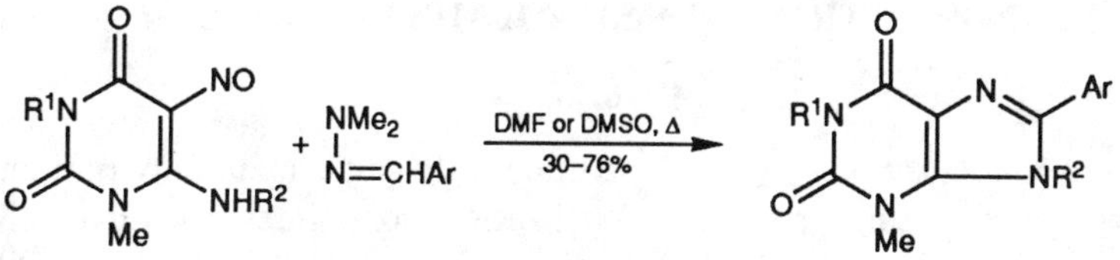

O
R^1N NO
O N NHR^2
Me
+ NMe_2
N=CHAr
DMF or DMSO, Δ
30–76%
O
R^1N N Ar
O N NR^2
Me

[2225]

2,6-Purinedione

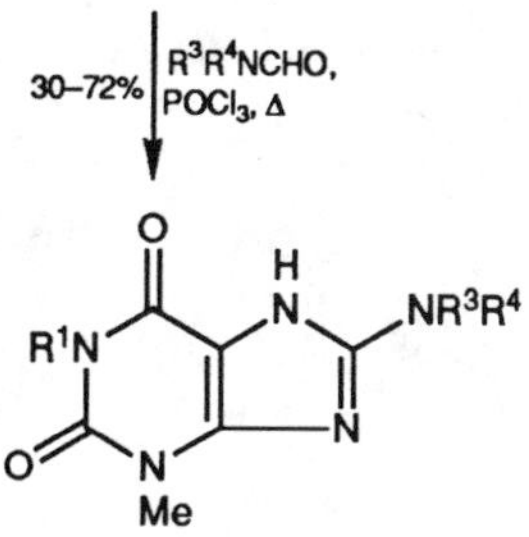

30–72% R^3R^4NCHO, $POCl_3$, Δ

O
H
R^1N N NR^3R^4
N
O N
Me

R^1,R^2 = H, Me; R^3 = H, Me, Et;
R^4 = Me, Et, Ph; Ar = Ph, 4-Cl-, 4-Br-C_6H_4, 2-thienyl

[2447, 2474]

Another ring system synthesized similarly:

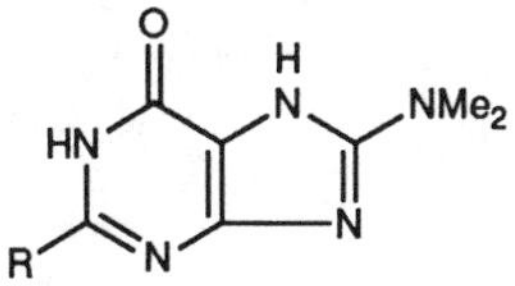

O
H
HN N NMe_2
N
R N

[2474]

6-Purinone

R^1
R^2CH_2HN N N
ON Me
pyr, Δ
58–88%
H R^1
R^2 N N N
N Me

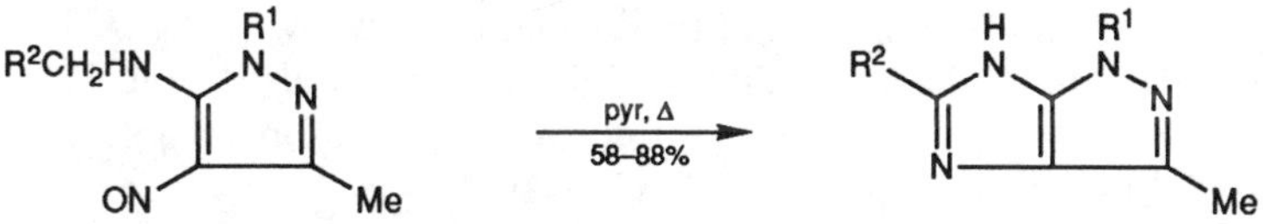

[3449, 3831]

R^1 = Me, Ph, Cl-, F-C_6H_4; R^2 = Me, Ph

Imidazo[4,5-*c*]pyrazole

2. 1,2,4-Oxadiazole

Cyanogen bromide reacts with 2-aminoazine *N*-oxides to give a fused oxadiazole in good yield.

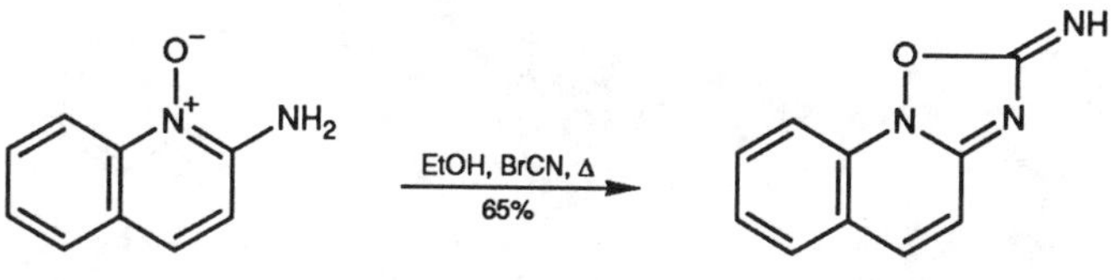

O^-
N^+ NH_2
EtOH, BrCN, Δ
65%
NH
O
N N

[2665]

1,2,4-Oxadiazolo[2,3-*a*]quinoline

II. FORMATION OF A SIX-MEMBERED RING

1. Pyrazine

Acid-catalysed condensation of 1,3-indanedione with a nitrosoaminopyrimidine gives a new pyrazine ring—a type of cyclization sometimes known as the Timmis synthesis. Another indanedione derivative, 2,2-dihydroxy-1,3-indanedione, condenses with a nitrosoamine with even better results and without the need for acid to be present.

AcOH, HCl, Δ ~45% [2644]

Indeno[2,1-*g*]pteridin-6-one

EtOH, Δ 91% [2397]

Indeno[1,2-*b*]pyrido[3,2-*e*]pyrazin-6-one

Base-induced condensation of a 2-nitroamine with an arylacetonitrile gives a member of the biologically important pteridine family (reviews [3454, 3594]) in good yield and without the possibility of 7-arylpteridines also being formed [2987]. Reduction of an *o*-nitrosoamine followed by addition of polyglyoxal produces a 2,3-unsubstituted pyrazine ring [2082]. A 2-acetoxypyrazine ring is formed by heating a nitrosoamine with acetic anhydride [2052] or with an enamine [2233].

+ RCH_2CN EtOH–EtONa, Δ 65% [2987]

R = 4-pyridyl

Pteridine

+ $(CHO)_2$ $Na_2S_2O_4$, HCl, Δ ~56% [2082]

R = H, Me

Pteridine

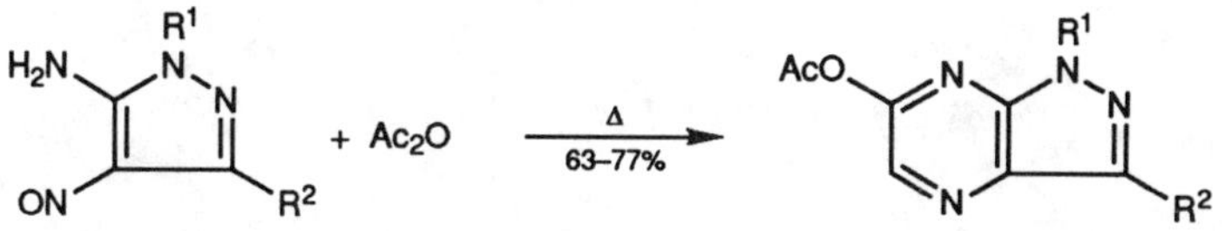

R^1 = Me, Ph; R^2 = Me, Ph

+ Ac_2O —Δ, 63–77%→ [2052]

Pyrazolo[3,4-*b*]pyrazine

Another ring system synthesized similarly:

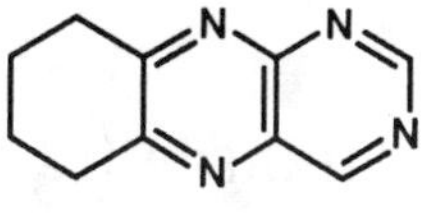

[2233]

Benzo[*g*]pteridine

2. 1,2,4-Triazine

Nitroso and amino groups may be annulated by heating the compound with an acylhydrazine (a hydrazide) in an aprotic solvent but acethydrazide did not give a characteristic triazine.

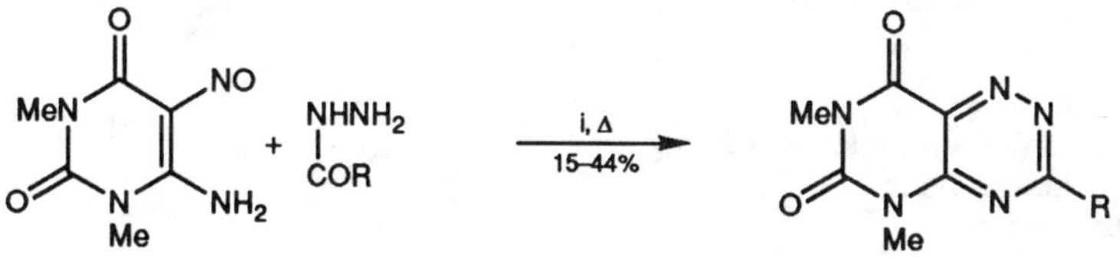

—i, Δ, 15–44%→ [2928]

R = H, Ph, 2-furyl, 2-thienyl, 3- or 4-pyridyl
i, DMF, DMSO or sulpholene

Pyrimido[4,5-*e*]-1,2,4-triazine-6,8-dione

CHAPTER 16

Acylamine, Acyloxy, Amine or Hydroxy and Phosphorane

I. Formation of a Five-membered Ring 760
1. Pyrazole 760
2. 1,2,4-Triazole 761
3. Furan 762
4. Thiophene 762
5. Oxazole 762
II. Formation of a Six-membered ring 763
1. Pyridazin-4-one 763
2. Pyrimidine 763
3. Pyrimidine-2-thione or Pyrimidin-2-one 763
4. 1,2,4,5-Tetrazine 764
5. Pyran or Thiopyran 765
6. Pyran-4-one 765

Examples of the cyclization of compounds carrying a phosphorane and one of the following functional groups are shown in this chapter: acylhydrazine, hydrazono, thioxy and imino.

I. FORMATION OF A FIVE-MEMBERED RING

1. Pyrazole

An iminophosphorane function (formed *in situ*) is annulated to a neighbouring imino group by reaction in dry dichloromethane at or below room temperature. The first product isolated is a 2-(triphenylphosphoranylidene)aminoindazole which can be converted into the phosphorus-free 2-aminoindazole by acid hydrolysis.

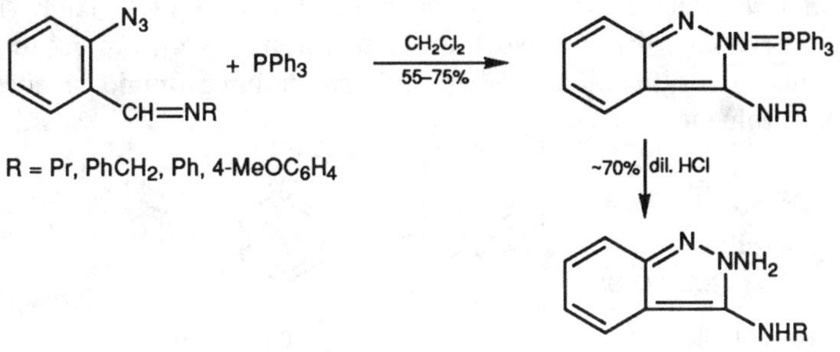

R = Pr, $PhCH_2$, Ph, 4-$MeOC_6H_4$

CH_2Cl_2 55–75%

~70% dil. HCl

Indazole

[3846]

2. 1,2,4-Triazole

Isocyanates react at ambient temperature with a hydrazono-iminophosphorane to yield a 7-substituted aminotriazolo compound. The initial reaction is probably between the isocyanate and the iminophosphorane to form a carbodi-imide. Heat is needed to induce a hydrazono-iminophosphorane to react with benzoyl chloride and to form a new triazole ring, but the yields are then usually high.

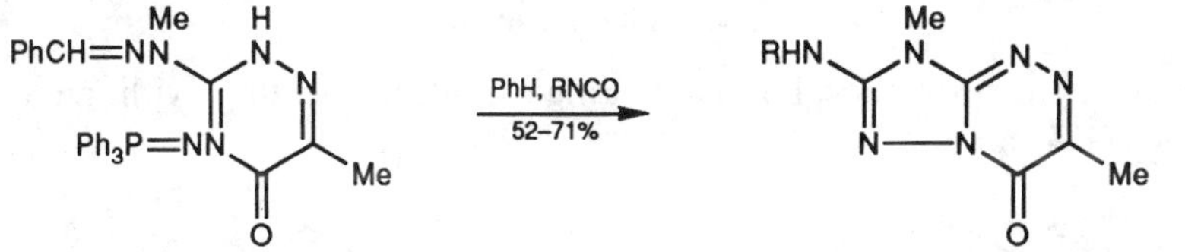

PhH, RNCO 52–71%

[3529]

R = alkyl, C_6H_{11}, $PhCH_2$, $CH_2{=}CHCH_2$

[1,2,4]Triazolo[5,1-*c*][1,2,4]triazin-4-one

Another ring system synthesized similarly:

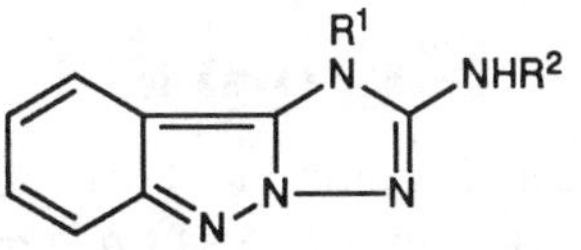

[3639, 3846]

[1,2,4]Triazolo[1,5-*b*]indazole

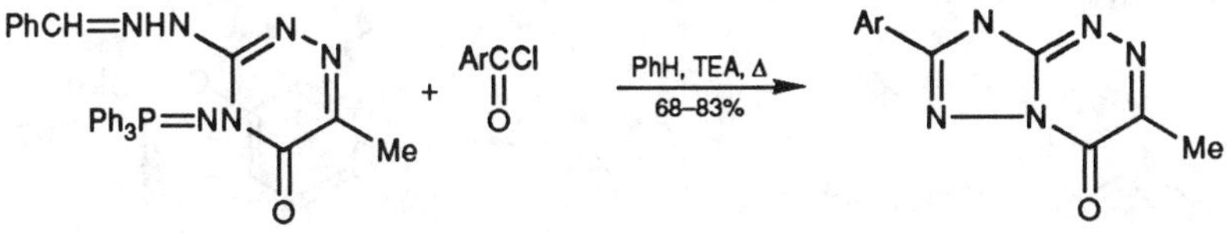

PhH, TEA, Δ 68–83%

[3176]

Ar = Ph, 4-Me-, 4-MeO-, 4-NO_2-C_6H_4, 2-naphthyl

[1,2,4]Triazolo[5,1-*c*][1,2,4]triazine-4-one

3. Furan

Annulation of a phosphonium salt and the carbonyl of an acyloxy group (formed *in situ*) proceeds in good yield on heating with TEA. High yields of benzofurans (review of recent work on benzofurans [3894]) are obtainable by heating a mixture of the α-hydroxybenzylphosphonium bromide, an acyl halide and TEA in toluene.

RCOCl, TEA, Δ; 50–93% → Benzofuran [2972]

R = alkyl, alkenyl, 2-furyl, Ph

i, ii; > 60% [3072]

R^1 = H, MeO; R^2 = H, ArCO; Ar = 3,5-$(AcO)_2C_6H_3$

i, MeCN, Δ; ii, PhMe, TEA, Δ

4. Thiophene

An *o*-mercaptobenzylphosphorane is ring-closed by heating with an acyl chloride–TEA mixture.

RCOCl, PhMe, TEA, Δ; 40–95% → Benzo[*b*]thiophene [3371]

R = Me, *n*-C_7H_{15}, Ph, PhCH=CH, 2-furyl

5. Oxazole

When the reaction between 2-aminothiophenol and phosphonium bromide (Chapter 10, Section I.4) is applied to a closely related phenol, the phosphonium salt (**16.1**) is isolated in 87% yield and is cyclized by a catalytic amount of sodium hydride.

(**16.1**) → MeCN, NaH, Δ; 80% → Benzoxazole [2736]

II. FORMATION OF A SIX-MEMBERED RING

1. Pyridazin-4-one

A primary amino group, after diazotization, reacts with an α-ketophosphorane function to form a pyridazinone ring but the triphenylphosphine moiety is retained and can be removed by heating with aqueous alkali.

NH_2 ; $CCH{=}PPh_3$; O ; i, ii ; 53–88% ; H ; N ; N ; O ; [3233]

R = H, Cl, Me

i, $C_5H_{11}NO_2$, MeOH; ii, aq. NaOH, Δ

4-Cinnolinone

2. Pyrimidine

The amino-iminophosphorane (**16.2**) is cyclized by heating with an isocyanate. A closely related iminophosphorane (**16.3**) reacts with an isocyanate in unheated dichloromethane to form a new fused 2-iminopyrimidine. When the imino is replaced with a hydrazono function (R^1 = ArNH), a similar 2-imino-pyrimidine ring is produced. The chemistry of pyrazolopyrimidines has been reviewed [3360].

$Ph_3P{=}N$; Ph ; R^1NHCH_2 ; Me ; + R^2NCO ; PhMe, Δ ; 50–77% ; R^2HN ; R^1N ; [3716]

(16.2)

R^1 = alkyl, $HC{\equiv}CCH_2$, Ph, $4\text{-}MeC_6H_4$;

R^2 = Et, Ph, $4\text{-}MeC_6H_4$

Pyrazolo[3,4-*d*]pyrimidine

$Ph_3P{=}N$; Ph ; $R^1N{=}CH$; Me ; CH_2Cl_2, PhMe, R^2NCO ; 43–72% ; R^2N ; R^1N ; [3419]

(16.3)

R^1 = Ph, $4\text{-}MeC_6H_4$, $4\text{-}NO_2C_6H_4NH$; R^2 = Et, $CH_2{=}CHCH_2$, Ph, 4-MeO-, 3- or 4-Cl-C_6H_4

3. Pyrimidine-2-thione or Pyrimidin-2-one

Carbon disulphide reacts at room temperature with the iminophosphorane (**16.3**) to give a high yield of the thione. It is necessary to heat the iminophosphorane with carbon dioxide in order to synthesize the corresponding pyrimidinone.

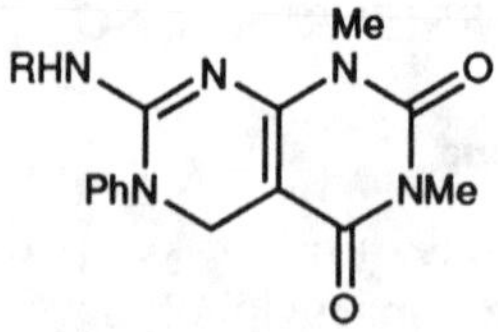

[3833]

Pyrimido[4,5-*d*]pyrimidine-2,4-dione

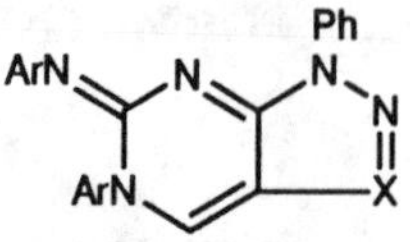

[3317]

X = CMe Pyrazolo[3,4-*d*]pyrimidine
X = N 1,2,3-Triazolo[4,5-*d*]pyrimidine

(16.3) $\xrightarrow[75\text{–}97\%]{CH_2Cl_2,\ CS_2}$ [3419]

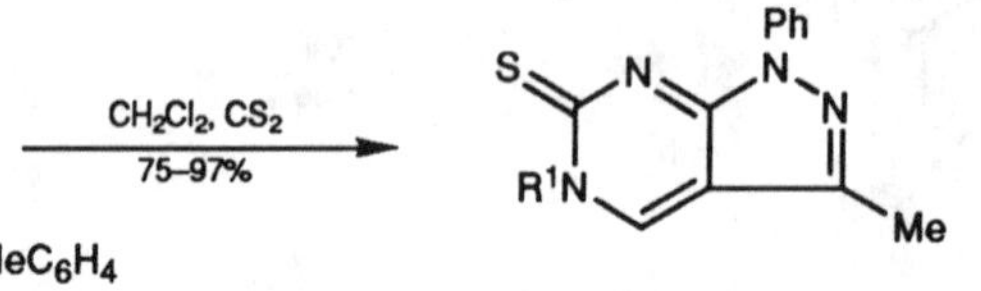

R^1 = Ph, 4-MeC_6H_4

Pyrazolo[3,4-*d*]pyrimidine-6-thione

Other ring systems synthesized similarly:

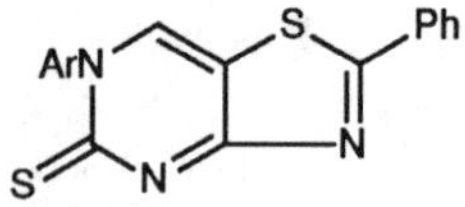

[3419]

Thiazolo[4,5-*d*]pyrimidine-5-thione

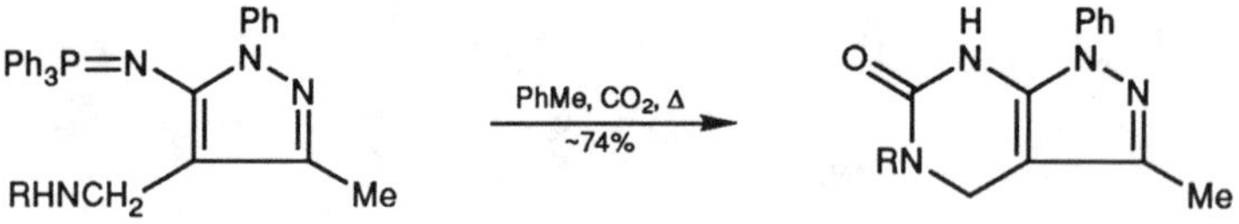

$\xrightarrow[\sim 74\%]{PhMe,\ CO_2,\ \Delta}$ [3716]

R = Ph, 4-MeC_6H_4

Pyrazolo[3,4-*d*]pyrimidin-6-one

4. 1,2,4,5-Tetrazine

Acylhydrazino-iminophosphoranes are cyclized by stirring in dichloromethane with an isocyanate for 24 h.

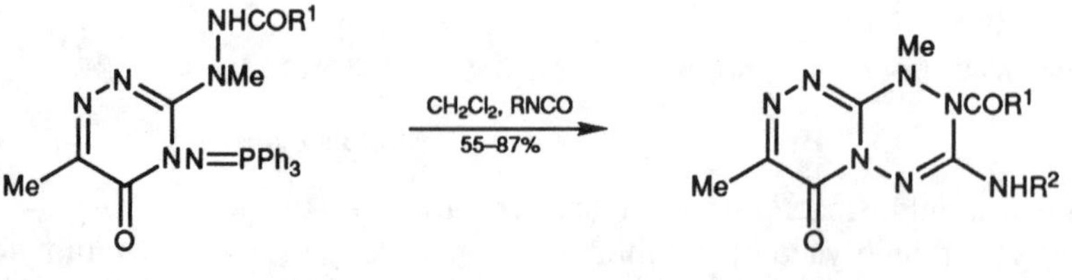

$\xrightarrow[55\text{–}87\%]{CH_2Cl_2,\ RNCO}$ [3529]

R^1 = Me, 4-MeO-, 4-Me-C_6H_4;
R^2 = Me, iPr, Ph, 4-MeO-, 4-Me-C_6H_4

[1,2,4]Triazino[4,3-*b*]-1,2,4,5-tetrazin-6-one

5. Pyran or Thiopyran

2-Acyloxy-phosphonium salts (**16.4**) are cyclized to a 2-benzopyran on heating in toluene with t-butyl methoxide. The acyl group may be provided by a reagent rather than have been attached to the oxygen or sulphur atom; in this example, the thiophenol is treated with the acid chloride–TEA in boiling toluene.

PhMe, $tBuCH_2ONa$, 54–73% [2973]

(**16.4**)
R = H, Me, Ph

2-Benzopyran

i, ii, 40–51% [3371]

R^1 = Me, Ph, 4-$MeOC_6H_4$, 3-$PhCH_2O$-5-$MeOC_6H_3$; R^2 = H, iPr
i, MeOH-MeONa, PhMe; ii, Δ

2*H*-1-Benzothiopyran

6. Pyran-4-one

Acyloxy and phosphorane groups interact when the compound is heated in an inert solvent, and a chromone (not a flavone as stated in the paper) is formed [3827]. *O*-Acylsalicylic acids are cyclized to flavones in a similar type of reaction involving carbon tetrachloride and triphenylphosphine [3827]. Higher yields may be obtained by treating a phenolic phosphorane first with an acyl chloride and then with methoxide [3193a].

o-$Cl_2C_6H_4$, Δ, 65% [3827]

1-Benzopyran-4-one

+ PPh_3 + CCl_4; CH_2Cl_2, TEA, 47–60% [3827]

R = Ph, 4-ClC_6H_4, 3-pyridyl

$$\text{2-hydroxy-}R^1R^2R^3\text{-}C_6H\text{-}C(=O)CH{=}PPh_3 \xrightarrow[\text{70–92\%}]{\text{i, ii}} \text{5-}R^1\text{-7-}R^2\text{-8-}R^3\text{-2-}R^4\text{-chromone} \quad [3193a]$$

R^1–R^3 = H, OH; R^4 = Me, Ph, 4-MeO-, 4-NO_2-C_6H_4, 3,4-$(MeO)_2C_6H_3$

i, R^4COCl, pyr, Δ; ii, MeOH–MeONa

CHAPTER 17

Acylamine or Acylhydrazine and Ring-carbon

I.	Formation of a Five-membered Ring	767
	1. Pyrrole	767
	2. Pyrrol-2-one	768
	3. Pyrazole	768
	4. 1,2,4-Triazole	768
	5. Oxazole	769
	6. Thiazole	769
II.	Formation of a Six-membered Ring	769
	1. Pyridine	769
	2. Pyridin-2-one	771
	3. Pyrimidine	771
	4. Pyrazine	772
	5. 1,3,4-Thiadiazine	772
III.	Formation of a Seven-membered Ring	772
	1. 1,4-Diazepine	772
	2. 1,2,5-Thiadiazepine 1,1-Dioxide	773
IV.	Formation of an Eight-membered Ring	774
	1. 1,5-Diazocine	774

In addition to the functional groups in the title, cyclization of a sulphonylimino group and ring-carbon is described.

I. FORMATION OF A FIVE-MEMBERED RING

1. Pyrrole

An *N*-arylsulphonylimino-1,4-quinone is cyclized by reaction with 3-aminobut-2-enoate esters (3-aminocrotonate esters) to produce the highly substituted indole-3-carboxylates. Formation of an azomethine ylide is believed to occur in the cyclization of the diazo ketone (**17.1**) with methyl acrylate in the presence of a catalytic amount of rhodium(II) acetate (review of such reactions [3928].

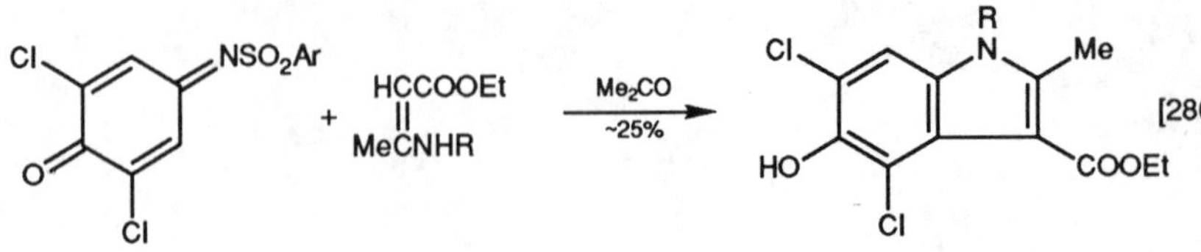

R = Me, Et

Indole

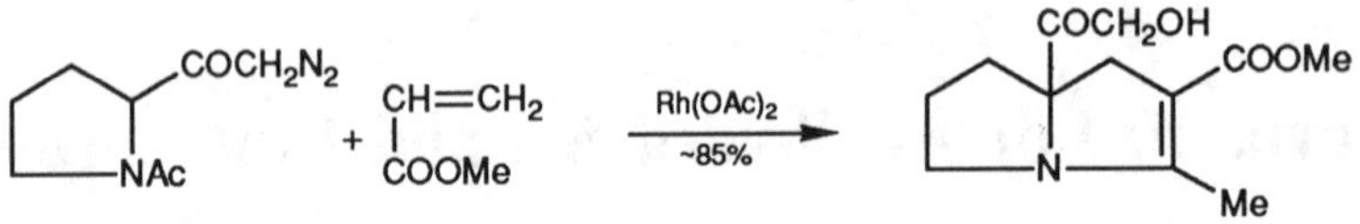

[2806]

(17.1)

Pyrrolo[1,2-*a*]pyrrole

[3666]

2. Pyrrol-2-one

Acid-catalysed cyclization of a 3-acylaminomethylquinoline through a Reissert compound gives a pyrrolone-fused derivative (review [2959a]).

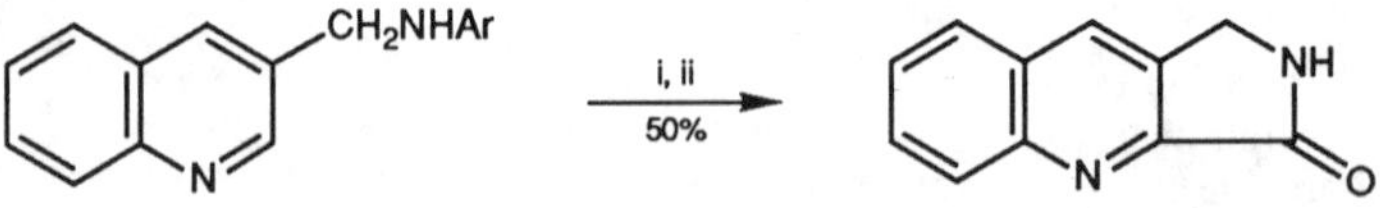

i, KCN, PhCOCl, H_2O; ii, $CHCl_3$–diox, HCl(g)

Pyrrolo[3,4-*b*]quinolin-3-one

[2243]

3. Pyrazole

Addition of an *N*-acetamidopyridazinium ylide to benzyne in the presence of butyl nitrite gives a pyridazinoindazole in good yield and under mild conditions, but pyridazine *N*-oxide reacts with benzyne differently, a 1-benzoxepine and a pyridazine being formed.

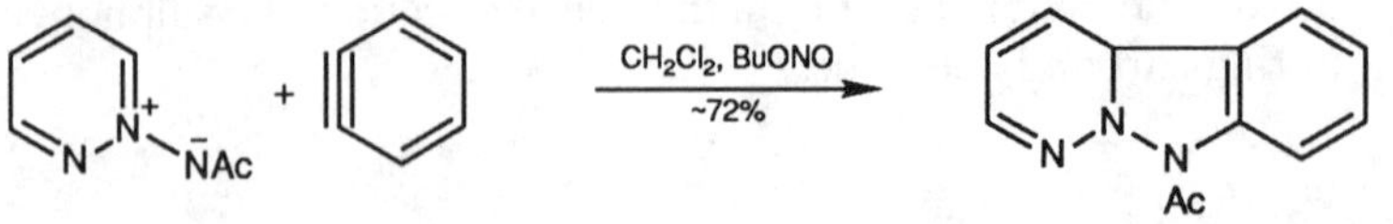

Pyridazino[1,6-*b*]indazole

[2735]

4. 1,2,4-Triazole

2-Thioformylaminopyridine is converted in low yield into its triazolo analogue by stirring with hydroxylamine.

NHCHS — MeOH, $NH_2OH.HCl$, 9% → [3325]

[1,2,4]Triazolo[1,5-*a*]pyridine

5. Oxazole

Photolysis of *o*-thallated acylarylamines produces a benzoxazole in good yield.

$Tl(COCF_3)_2$, $NHCOR^1$ — C_6H_{12}, $h\nu$, 33–60% → [3181]

R^1 = Me, Ph, 4-ClC_6H_4; R^2 = H, Me

Benzoxazole

6. Thiazole

Mild oxidation with potassium hexacyanoferrate(III) under aqueous alkaline conditions induces thioacylanilines to cyclize.

NHC(S)CH(CN)COPh — i, ~56% → CH(CN)COPh [2640]

Benzothiazole

R = H, Me, NH_2, NO_2, Cl i, EtOH, aq. NaOH, $K_3Fe(CN)_6$

II. FORMATION OF A SIX-MEMBERED RING

1. Pyridine

A side-chain methylsulphonylimino group attached to C-3 of a pyran-2-one ring acts as a 1-azadiene in an inverse electron Diels–Alder reaction with 1,1-dimethoxy-1-propene at ambient temperature. The first-formed product is more conveniently isolated after removal of the N-SO_2Me and MeO functions with t-butoxide at $-30\,°C$.

NSO_2Me, CPh + $(MeO)_2C{=}CHMe$ — i, ii, 52% → OMe, Me, N, Ph [3953]

i, PhH; ii, THF, tBuOK

[1]Benzopyrano[3,4-*c*]pyridin-5-one

A suitably positioned acylamino group is annulated to a ring-carbon in a Bischler–Napieralski reaction with phosphorus oxychloride; comparison of a number of reagents (phosphorus pentoxide–toluene, phosphorus pentoxide–tetralin, phosphorus oxychloride, PPA) for their efficacy in this reaction showed that PPA was the most useful.

$R^1 = Me$, $R^2 = OH$; PhMe, $POCl_3$, Δ; 96% → Isoquinoline [2178]

$R^1 = PhCH_2$, $R^2 = H$; PhMe, $POCl_3$, Δ; 73%

Ar = 3,4-$(PhCH_2))_2C_6H_3$

Other ring systems synthesized similarly:

[2153, 2308]

X = O [1]Benzofuro[3,2-*c*]pyridine
X = S [1]Benzothieno[3,2-*c*]pyridine

Benz[*f*]isoquinoline [2307]

Cyclization of an acylamino group on to a neighbouring ring-carbon by the Bischler–Napieralski reaction usually gives a reduced pyridine ring, but similar treatment of an acyl*imine* gives good yields of the fully aromatized pyridine, as also does heating *N*-acylarylamines at a high temperature with DMF and HMPT. Yields fell as R^1 changed from H (72%) to iPr (0%).

$POCl_3$, Δ; 32–61% [3857]

R = H, Ph, 4-Cl-, 4-NO_2-C_6H_4, 3-pyridyl
$m = 1,2$; $n = 1,2$

$m = n = 1$ Dipyrrolo[2,3-*b* : 2′,3′-*d*]pyridine
$m = 2$; $n = 1$ Pyrrolo[3,2-*c*][1,8]naphthyridine
$m = n = 2$ Pyrido[2,3-*h*]-1,6-naphthyridine

DMF, HMPT, Δ; 26–76% → Quinoline [2602]

R^1 = H, alkyl; R^2 = H, Me, MeO

Sometimes, a phenyl group undergoes a C—C migration during a Pictet–Gams synthesis; this occurs through an oxazoline which ring-opens to a compound such as MeCAr=CHNHCOPh. The rearranged oxazoline has recently been isolated in a cyclization promoted by phosphorus pentoxide.

PhMe, PPA, $POCl_3$, Δ; 53% [2277]

Isoquinoline

P_2O_5, decalin, Δ; 2–58% [3431]

R = H, Me, MeO

It has recently been shown that the use of a Vilsmeier reagent to annulate an acylamino group to an adjacent ring-carbon (p. 95) can be extended to acylanilines which contain various substituents in the ring, but regiospecificity is not always high. Electron-attracting groups usually lower the yield, for example, a nitro group reduces the yield to 20% while an activating substituent can result in yields of about 85% [3888].

2. Pyridin-2-one

N-Methylacetanilides cyclize on treatment with phosgene; the quaternary salt formed may be converted into the 2-quinolinone by heating with aqueous carbonate.

i, $COCl_2$; ii, aq. K_2CO_3, Δ; 57–66% [2698]

R = Me, Ph

2-Quinolinone

3. Pyrimidine

This ring is formed from an acylaminoazole by conversion (with phosphorus pentachloride) into the imidoyl halide which is cyclized by reaction with a nitrile (through the nitrilium salt) and Friedel–Crafts catalyst. Yields are high for several nitriles but the acylamino group was not varied.

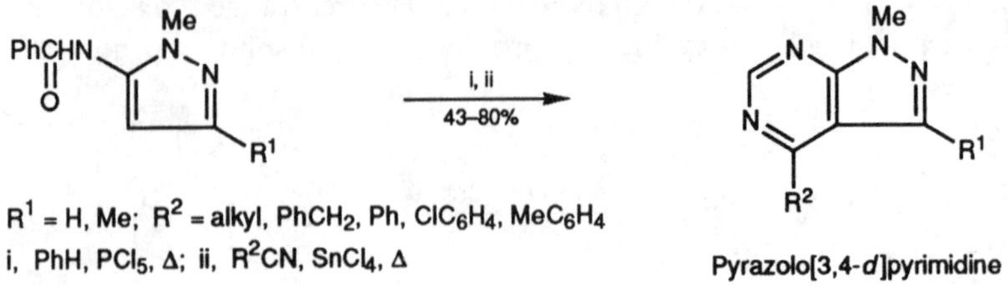

i, ii; 43–80%; [3242]

R^1 = H, Me; R^2 = alkyl, $PhCH_2$, Ph, ClC_6H_4, MeC_6H_4
i, PhH, PCl_5, Δ; ii, R^2CN, $SnCl_4$, Δ

Pyrazolo[3,4-*d*]pyrimidine

4. Pyrazine

Annulation of an acetamido group to a ring-carbon atom occurs under conditions which were expected to hydrolyse the amide group, that is, ethanolic potassium hydroxide.

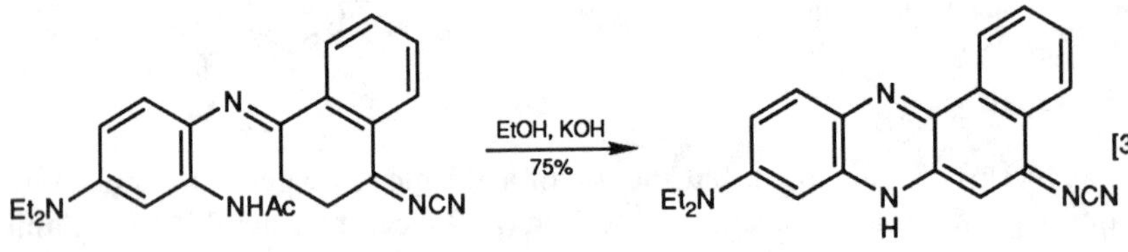

EtOH, KOH; 75%; [3747]

Benzo[*a*]phenazine

5. 1,3,4-Thiadiazine

Oxidative cyclization of a thioacylhydrazine converts it into a fused thiadiazine in low yield.

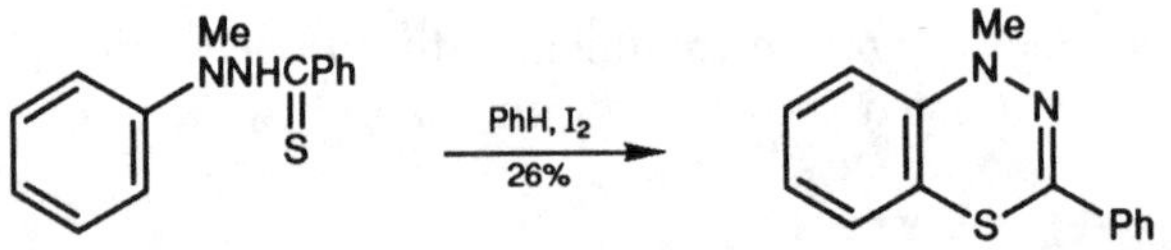

PhH, I_2; 26%; [2844]

4,1,2-Benzothiadiazine

III. FORMATION OF A SEVEN-MEMBERED RING

1. 1,4-Diazepine

Treatment of an acylamino group with phosphorus oxychloride is a widely used method of annulating the carbonyl group to a carbon of another ring [2258] (see also p. 96), but when this method fails, heating with phosphorus oxychloride–phosphorus pentoxide is sometimes successful [3005]. The amide (**17.2**) was resistant to other methods and was eventually cyclized by converting it first (with phosphorus pentachloride) into its imidoyl chloride which was then subjected to Friedel–Crafts conditions. Another synthesis of this ring is described in Section IV.1.

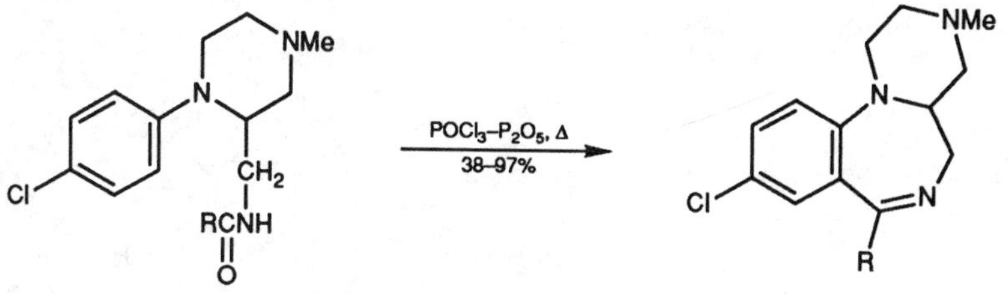

$POCl_3$–P_2O_5, Δ
38–97%

[3005]

R = H, Me, Ac, $PhCH_2$, Ph, 2- or 4-ClC_6H_4

Pyrazino[1,2-*a*][1,4]benzodiazepine

i, ii
~62%

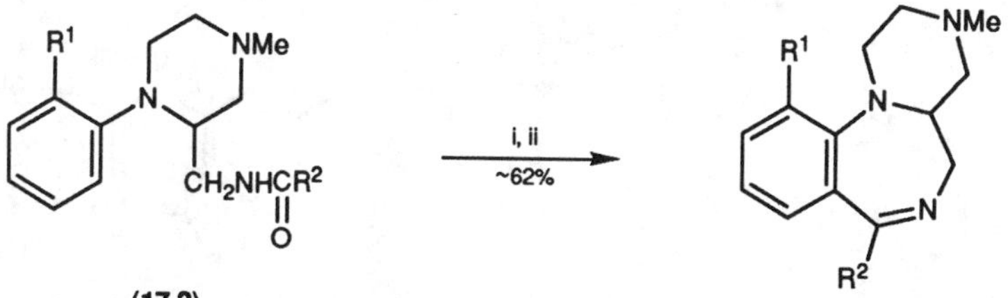

[3511]

(17.2)
R^1 = F, Me, MeO; R^2 = 2-thienyl
i, CH_2Cl_2, PCl_5, Δ; ii, $PhNO_2$, $AlCl_3$, Δ

Pyrazino[1,2-*a*][1,4]benzodiazepine

2. 1,2,5-Thiadiazepine 1,1-Dioxide

Heating the 2-(acylamino)sulphonamide **(17.3)** with phosphorus oxychloride gives a good yield of a tricyclic thiadiazepine; a new C—C bond is formed between the acyl carbonyl and the pyrrole C-2 atom.

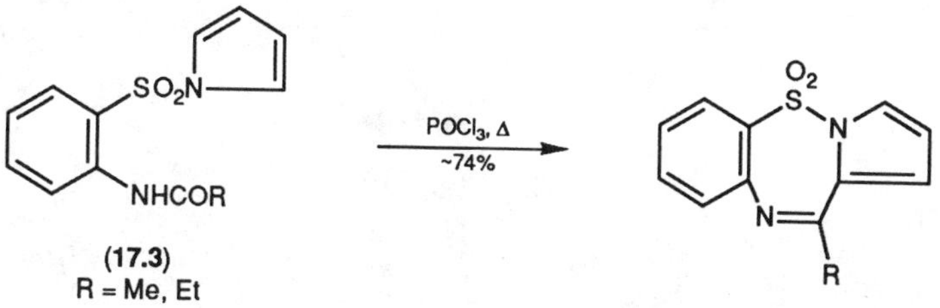

$POCl_3$, Δ
~74%

[2610]

(17.3)
R = Me, Et

Pyrrolo[1,2-*b*][1,2,5]benzothiadiazepine 5,5-dioxide

IV. FORMATION OF AN EIGHT-MEMBERED RING

1. 1,5-Diazocine

When the acylamine (**17.4**) is heated at 80 °C with phosphorus oxychloride, a diazocine ring is formed in good yield but at higher temperatures (preferably 130 °C), the product is a benzodiazepine.

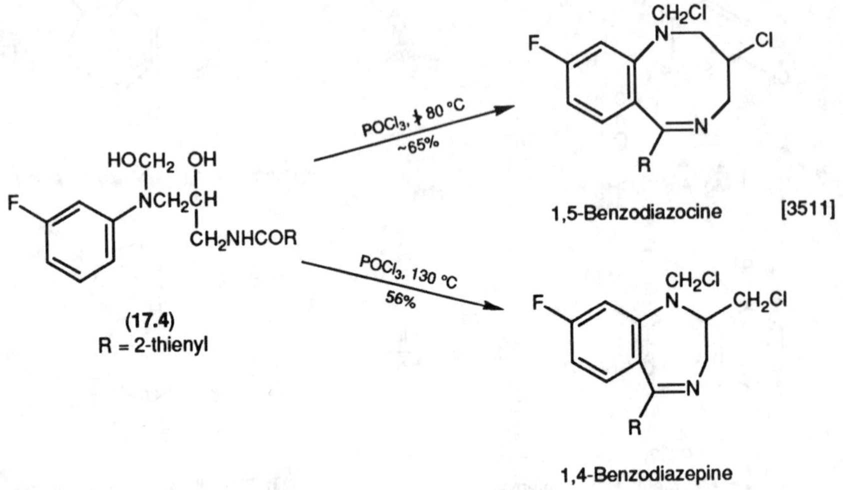

HOCH$_2$ OH
NCH$_2$CH
CH$_2$NHCOR

(17.4)
R = 2-thienyl

$POCl_3$, ≯ 80 °C
~65%

1,5-Benzodiazocine [3511]

$POCl_3$, 130 °C
56%

1,4-Benzodiazepine

CHAPTER 18

Acylamine or Acylhydrazine and Ring-nitrogen

I. Formation of a Five-membered Ring 775
1. Imidazole 775
2. 1,2,4-Triazole 776
3. 1,2,4-Thiadiazole 776
4. 1,2,3,5-Thiatriazole 1-Oxide 777

Ring closure of a sulphenamide group on to a ring-nitrogen is shown in this chapter.

I. FORMATION OF A FIVE-MEMBERED RING

1. Imidazole

Prolonged heating of an α-acylaminoazole with phosphorus oxychloride leads to the formation of a new imidazole ring. When the acylamino group is attached to one of the *peri* positions of a bicyclic molecule, the imidazole ring may be formed at the *peri* positions as in this example of a quinoxaline where catalytic reduction has also affected the pyrazine ring.

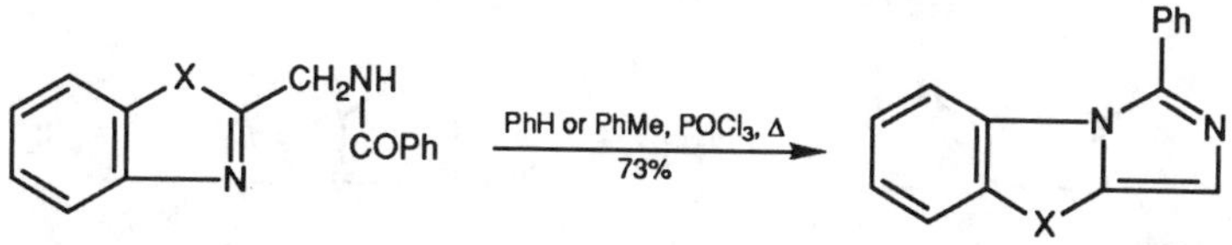

CH_2NH / COPh —[PhH or PhMe, $POCl_3$, Δ; 73%]→ [2186, 2189]

X = O, NMe

Imidazo[5,1-*b*]benzoxazole
Imidazo[5,1-*b*]benzimidazole

[2514]

R[1] = H, MeO; R[2] = Me, Ph

H_2,Pd–C, Δ
70–95%

Imidazo[1,5,4-*de*]quinoxaline

2. 1,2,4-Triazole

Several examples of the conversion of 2-(acylhydrazino)azines into a triazoloazine are mentioned on pp. 99–101; another variation on these experimental conditions is to heat the substrate in refluxing toluene to which 4-toluenesulphonic acid monohydrate is added. The synthesis of triazoles fused at their 3,4-bond has been reviewed [3903].

PhMe, TsOH, H_2O, Δ
~74%

[3774, 3947]

R = iBu, $MeS(CH_2)_2$

1,2,4-Triazolo[4,3-*a*]pyrazine

3. 1,2,4-Thiadiazole

A sulphenamido group placed α to a ring-nitrogen reacts with an arylamine (such as a nitroaniline) to form a fused thiadiazole ring.

+ $ArNH_2$

$CHCl_3$, TEA
12–62%

[2755]

Ar = $O_2NC_6H_4$, 2,5-$Cl_2C_6H_3$

Isoxazolo[3,2-*c*][1,2,4]thiadiazole

Other ring systems synthesized similarly:

[2755]

[2755]

X = O [1,2,4]Thiadiazolo[3,4-*b*][1,3,4]oxadiazole
X = S [1,3,4]Thiadiazolo[2,3-*c*][1,2,4]thiadiazole

Thiazolo[2,3-*c*][1,2,4]thiadiazole

4. 1,2,3,5-Thiatriazole 1-Oxide

Thionyl chloride at ambient temperature converts a 2-acylhydrazinoazine into a thiatriazole oxide ring in high yield.

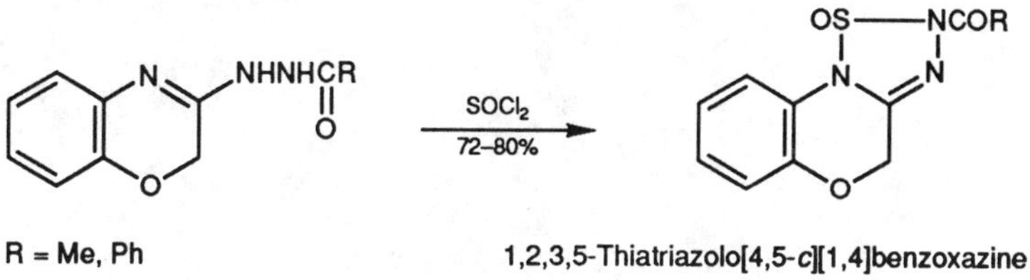

NHNHCR
O
$SOCl_2$
72–80%
OS
NCOR
N
N
[3567]

R = Me, Ph

1,2,3,5-Thiatriazolo[4,5-*c*][1,4]benzoxazine 1-oxide

CHAPTER 19

Acylamine or Amine and Sulphonamide or Thioureide

I.	Formation of a Five-membered Ring	778
	1. Imidazole	778
	2. Imidazole-2-thione	779
II.	Formation of a Six-membered Ring	779
	1. 1,2,4-Thiadiazine 1,1-Dioxide	779
	2. 1,2,4-Thiadiazine 1-Oxide	780
III.	Formation of a Seven-membered Ring	780
	1. 1,2,4-Thiadiazepin-3-one 1-Oxide	780

I. FORMATION OF A FIVE-MEMBERED RING

1. Imidazole

Simultaneous desulphurization and cyclization of an amino-thioureide (**19.2**) is brought about by heating gently with methyl iodide or mercury(II) oxide, and a trace of sulphur. The usefulness of mercury(II) oxide in organic chemistry has been reviewed [B-22]. A similar cyclization is effected by heating with mercury(II) chloride, in chloroform [2473a], or to a temperature above its melting point (without a solvent) [2559].

NH_2 / NHCNHR (C=S) + MeI —(i or ii, Δ; 30–76%)→ H, N, NHR, N [2030, 2591]

(19.2)

Benzimidazole

R = Et, Bu, $PhCH_2$, Ph, 2-MeC_6H_4

i, MeI, EtOH; ii, HgO, S, $CHCl_3$

2. Imidazole-2-thione

Thiophosgene cyclizes 2-thioureidoaniline at ambient temperature to the potentially interesting intermediate 2-benzimidazolethione (**19.3**).

(**19.2**) $\xrightarrow[40–74\%]{Me_2CO,\ CSCl_2}$ [3052]

R = alkyl, $PhCH_2$, Ph

(**19.3**)
2-Benzimidazolethione

II. FORMATION OF A SIX-MEMBERED RING

1. 1,2,4-Thiadiazine 1,1-Dioxide

The 2-*N*-hydroxysulphonamidoaniline (**19.4**) reacts with DMAD or with 4-nitrobenzaldehyde to give this fused ring; the chemistry of 1,2,*x*-thiadiazines, has recently been reviewed [3904].

(**19.4**) $\xrightarrow[45\%]{MeOH,\ DMAD,\ \Delta}$; $\xrightarrow[89\%]{MeOH,\ 4\text{-}NO_2C_6H_4CHO,\ \Delta}$ [2430]

1,2,4-Benzothiadiazine 1,1-dioxide (review [3364])

1,3-Cyclobutanediones react with *o*-aminosulphonamides to form a thiadiazine dioxide ring but with *o*-aminocarbohydrazides, two new rings are formed. Annulation of aminosulphonamides with orthoesters (p. 103) may be improved by heating the reactants in an excess of orthoester for 30 min, removing the excess of it and heating the residue in diphenyl ether. *N*-Substituted sulphonamides may be used in this synthesis.

$\xrightarrow[21–62\%]{xylene,\ TsOH,\ \Delta}$ [2916]

$R^1 = R^2 = Me$ or $R^1R^2 = (CH_2)_4$ or $(CH_2)_5$; $R^3 = H, Cl$

SO_2NHR^1, NH_2 + $R^2C(OR^3)_3$ —(i, ii, Δ; 65–91%)→ O_2, S, NR^1, N, R^2 [3182]

R^1 = Ph, 1-naphthyl, 2- or 4-Me-, 4-Meo-, 4-COOMe-C_6H_4;

R^2 = H, Me; R^3 = Me, Et

i, Δ; ii, Ph_2O, Δ

1,2,4-Benzothiadiazine 1,1-dioxide

2. 1,2,4-Thiadiazine 1-Oxide

The amino-sulphoximide (**19.5**) is cyclized by refluxing in acetonitrile with cyanogen bromide.

R^1, S, O, NH, R^2, NH_2 —(MeCN, BrCN, Δ; 65–79%)→ R^1, S, O, N, R^2, N, NH_2 [2837]

(19.5)

R^1 = Me, Ph; R^2 = H, Me, Cl

1,2,4-Benzothiadiazine 1-oxide

III. FORMATION OF A SEVEN-MEMBERED RING

1. 1,2,4-Thiadiazepin-3-one 1-Oxide

N,N′-Carbonyldi-imidazole annulates an amino and a sulphoximide on heating the compound in an inert solvent.

R, SO=NH, CH_2NH_2 + CO(imid)$_2$ —(THF; ~77%)→ R, O, S, N, O, N, H [2450]

R = Me, Et

1,2,4-Benzothiadiazepin-3-one 1-oxide

CHAPTER 20

Acylamine or Amine and Thiocyanate

I. Formation of a Five-membered Ring 781
 1. Pyrrole 781
 2. Pyrrol-2-one 782
 3. Thiazole 782
II. Formation of a Six-membered Ring 782
 1. Pyrimidine-2-thione 782

Examples of the cyclization of an *o*-(trimethylsilylmethyl)acylamine–carbamate and of a (ring-nitrogen)-thiocyanate are included in this chapter.

1. FORMATION OF A FIVE-MEMBERED RING

1. Pyrrole

2-Trimethylsilylmethyl *N*-methylanilides (**20.1**) are conveniently synthesized from a nitrobenzene by successive treatments with trimethylsilylmethyl magnesium chloride at 0 °C, reduction of the nitro group and *N*-acylation. The anilide cyclizes in high yield at −20 °C.

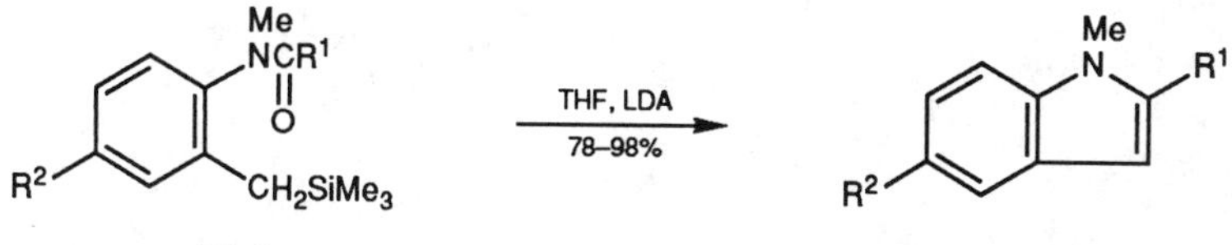

Me, NCR[1], O, CH_2SiMe_3, R^2 — THF, LDA, 78–98% → Me, N, R^1, R^2 [3493]

(20.1) Indole

R^1 = Ph, 4-ClC_6H_4, PhCH=CH; R^2 = MeO, Cl

2. Pyrrol-2-one

When the *N*-acyl group in (**20.1**) is a methoxycarbonyl, the reaction with LDA gives high yields of the pyrrolone.

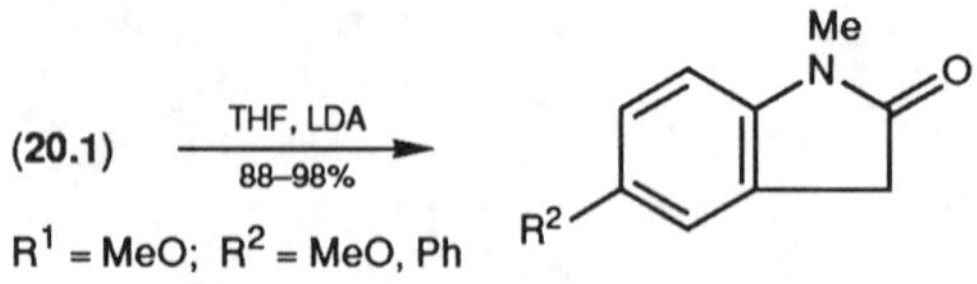

(**20.1**) → THF, LDA; 88–98% [3493]

R^1 = MeO; R^2 = MeO, Ph

Indol-2-one

3. Thiazole

This ring may be formed from aminothiocyanates by heating in a suitable solvent such as ethanol, but substitution in the parent ring can inhibit cyclization. In such cases, heating in acetic acid sometimes overcomes this problem.

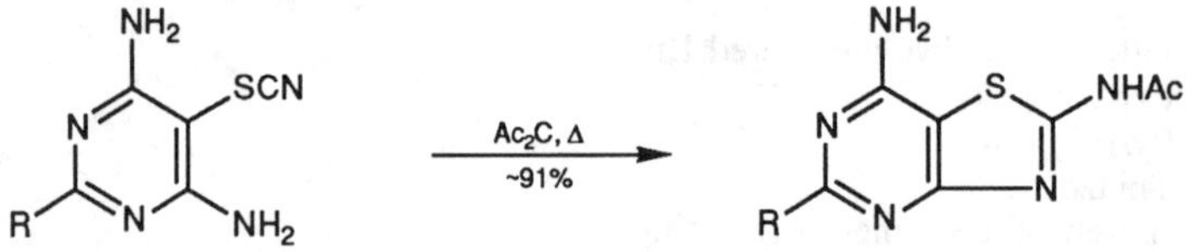

Ac_2O, Δ; ~91% [2169]

R = MeS, NH_2

Thiazolo[4,5-*d*]pyrimidine

1. FORMATION OF A SIX-MEMBERED RING

1. Pyrimidine-2-thione

A chloroenamino group reacts with potassium thiocyanate under mild conditions to form a pyrimidinethione ring with the *peri*-nitrogen atom.

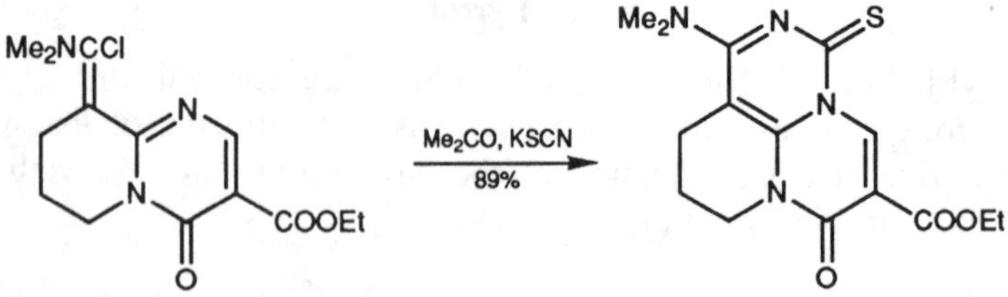

Me_2CO, KSCN; 89% [3070a]

2,3a,6a-Triazaphenalen-6-one

CHAPTER 21

Acyl Halide and Ring-carbon or Ring-nitrogen

I. Formation of a Five-membered Ring 783
1. Pyrrol-2-one 783
2. Furan-3-one 784
3. Thiophene 784
II. Formation of a Six-membered Ring 784
1. Pyrazine-2,5-dione 784
2. Thiopyran-2-thione 785

I. FORMATION OF A FIVE-MEMBERED RING

1. Pyrrol-2-one

One of the chlorocarbonyl groups of the malonyl chloride formed from **(21.2)** reacts with a ring-carbon atom under mild conditions to give a new fused 2-oxopyrrole-3-carbonyl chloride ring. A 4,5-fused pyrrol-2-one ring is formed when α-chloro-α,α-diphenylacetyl chloride is heated first with *N*,*N*-dimethylhydrazine and then with *N*-ethylmorpholine (NEM) (a more efficient base than TEA). When the hydrazine is replaced by *O*-methylhydroxylamine, 3-methoxy-3-phenyl-2-indolone is the main product [2381].

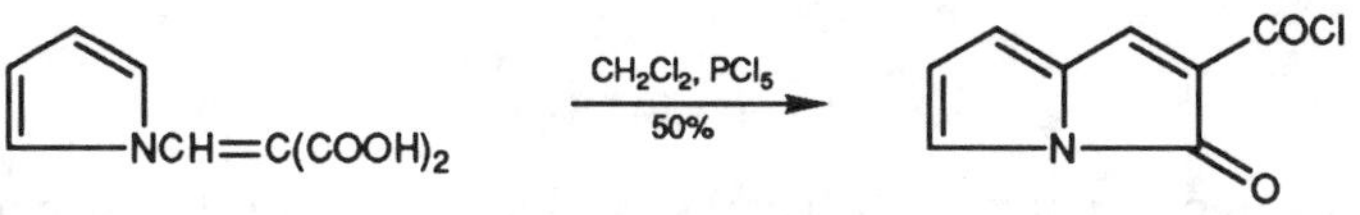

(21.2)

3-Pyrrolizinone

[2989]

$PhC(Cl)(Ph)COCl$ + RNMe–NH_2 $\xrightarrow[32–79\%]{i, ii}$ 2-Indolone (N-NMeR, 3-Ph) [2525]

R = Me, $PhCH_2$, $3,4\text{-}(MeO)_2C_6H_3CH_2$
i, THF, Δ; ii, NEM, Δ

2. Furan-3-one

Friedel–Crafts intramolecular acylation of an aryloxyacetic acid (through its acyl chloride) at low temperature usually gives high yields of this ring.

Me–C_6H_4–OCH_2COCl $\xrightarrow[71–91\%]{PhH, AlCl_3}$ Me–Benzofuran-3-one [2409]

Benzofuran-3-one

3. Thiophene

Treatment of alkanoic acids with thionyl chloride and a catalytic amount of pyridine causes oxidation at the α-methylene group and the formation of an α-chloro-α-(chlorosulphenyl)acyl chloride, $RCH_2C(SCl)ClCOCl$. This usually reacts further and when R contains a benzene ring, this may be attacked with consequent formation of a fused ring system. Unsaturated aralkanoic acids behave similarly. Extension of this work to related compounds suggests that a concerted elimination–cyclization operates.

$Ph(CH_2)_2COOH$ $\xrightarrow[31\%]{i}$ 3-Cl-benzo[b]thiophene-2-COCl $\xleftarrow{i}$ PhCH=CHOOH [2759]

i, pyr, $SOCl_2$

Benzo[*b*]thiophene

II. FORMATION OF A SIX-MEMBERED RING

1. Pyrazine-2,5-dione

Attempts to prepare imidazol-4(5)-carbonyl chloride give the tricyclic pyrazinedione—a type of reaction which also occurs with other azolecarbonyl chlorides.

$$\text{Imidazole-4-carboxylic acid} \xrightarrow[\sim 95\%]{SOCl_2,\ \Delta} \text{product} \qquad [2631]$$

Diimidazo[1,5-*a* : 1',5'-*d*]pyrazine-5,10-dione

2. Thiopyran-2-thione

2-Biphenylyl chlorodithioformate undergoes a Friedel–Crafts cyclization to give a quantitative yield of the thiopyranthione.

$$\xrightarrow[\sim 100\%]{CS_2,\ AlCl_3} \qquad [2832]$$

Dibenzo[*b*,*d*]thiopyran-6-thione

CHAPTER 22

Aldehyde or Ketone and Alkene or Alkyne

I.	Formation of a Five-membered Ring	786
	1. Pyrrole	786
	2. Thiophene	787
II.	Formation of a Six-membered Ring	787
	1. Pyridine or Pyridin-2-one	787
	2. Pyridine 1-Oxide	787
	3. Pyrimidin-2-one	788
	4. Pyran	788
III.	Formation of an Eight-membered Ring	789
	1. 1,5-Oxathiocin	789

I. FORMATION OF A FIVE-MEMBERED RING

1. Pyrrole

Benzyne (from benzenediazonium 2-carboxylate) reacts with a 2-ethylidene-*N*-formylpiperidine to form a doubly fused pyrrole ring in low yield.

CHMe; R^1O; R^2O; NCHO; + ; 26%; Me; N; [3307]

$R^1 = R^2 = Me$ or $R^1R^2 = CH_2$

Indolo[2,1-*a*]isoquinoline

2. Thiophene

Side-chain vinyl and lactam thiocarbonyl groups react with bromine at 0 °C to form a thiophene ring, probably by addition, *S*-alkylation and elimination of hydrogen bromide.

Ph N S CH=CHCOAr $\xrightarrow[38\text{–}48\%]{CHCl_3,\ Br_2}$ Ph N^+ S COAr Br^- [3147]

Ar = 4-Cl-, 4-MeO-, 4-Me-C_6H_4

Thieno[2,3*b*]pyridinium

II. FORMATION OF A SIX-MEMBERED RING

1. Pyridine or Pyridin-2-one

3-Benzylidenethiochromanones (**22.2**) react with malononitrile and solid sodium hydroxide in methanol without external heating to form the thiopyranopyridine in moderate-to-good yield. The [4 + 2] addition of a benzylidene ketone can give one of several possible products (p. 111) depending on the nature of the substrates and on the two-carbon reactant. The benzothiopyranone (**22.2**) also gives two different products depending on the reagent and the temperature (see also Section II.4).

R O CHAr S + $CH_2(CN)_2$ $\xrightarrow[27\text{–}69\%]{MeOH,\ NaOH}$ R N OMe CN Ar S [3425]

(**22.2**)
R = H, MeO, Cl; Ar = Ph, MeO-, Cl-C_6H_4

[1]Benzothiopyrano[4,3-*b*]pyridine

(**22.2**) + CH_2CN / COOMe $\xrightarrow[17\%]{AcONH_4,\ \Delta}$ R HN O CN Ph S [3935]

R = H, Me; Ar = Ph

[1]Benzothiopyrano[4,3-*b*]pyridin-2-one

2. Pyridine 1-Oxide

Cyclization of an *o*-alkynyl-aldoxime occurs readily on heating with a mild base.

R^1 = Bu, Ph, $SiMe_3$; R^2 = H, MeO

EtOH, K_2CO_3, Δ 35–78% [3118]

Isoquinoline 2-oxide

3. Pyrimidin-2-one

Condensaton of an enone with urea results in cyclodehydration and the formation of a partially reduced pyrimidinone ring.

+ $CO(NH_2)_2$ AcOH, Δ 83% [2033]

Ar = 3-$MeOC_6H_4$

Indeno[1,2-*d*]pyrimidine-2,5-dione

4. Pyran

The α,β-unsaturated ketone (**22.2**) reacts with malononitrile (and piperidine) at room temperature to form the thiopyranopyran in a [4 + 2] cycloaddition. 1-Benzopyran-5,8-diones were unknown until their recent synthesis from acetylenic quinone (**22.3**) by heating in toluene until the disappearence of the alkyne. A Claisen rearrangement followed by enolization and a 1,5-hydrogen shift probably preceded cyclization. The chemistry of heterocyclic quinones has been reviewed [2947, 3650].

(**22.2**) + $CH_2(CN)_2$ pip, EtOH 4–52% [3425, 3935]

R = H, MeO, Me; Ar = Ph, $MeOC_6H_4$

[1]Benzothiopyrano[4,3-*b*]pyran

PhMe, Δ 68–96% [3245]

(**22.3**)

R^1 = H, Me, C_5H_{11}, Ph; R^2 = H, Me

2*H*-1-Benzopyran-5,8-dione

III. FORMATION OF AN EIGHT-MEMBERED RING

1. 1,5-Oxathiocin

o-Allyloxythiobenzophenones cyclize on heating in xylene to give high yields of this eight-membered ring product. When R^1 = Cl the reaction time becomes much longer than the 0.05–1 h needed for the analogues.

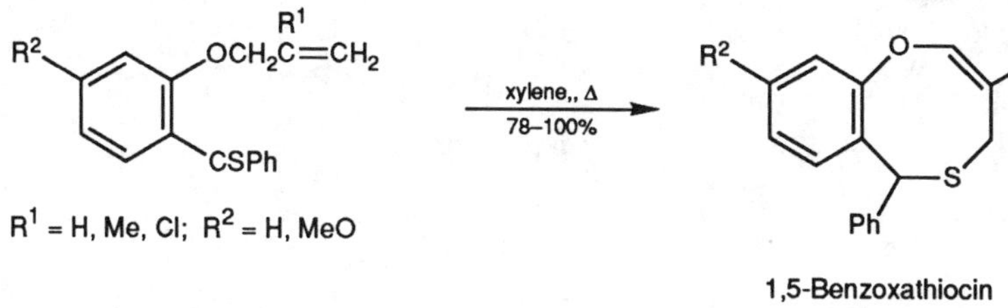

R^1 = H, Me, Cl; R^2 = H, MeO

xylene,, Δ

78–100%

[3622]

1,5-Benzoxathiocin

CHAPTER 23

Aldehyde or Ketone and Azide or Triazene

I.	Formation of a Five-membered Ring	790
	1. Pyrazole	790
II.	Formation of a Six-membered Ring	790
	1. Pyridin-2-one	790
III.	Formation of a Seven-membered Ring	791
	1. 1,4-Diazepin-2-one	791

I. FORMATION OF A FIVE-MEMBERED RING

1. Pyrazole

The triazenophosphorane (**23.1**) cyclizes when treated with a primary arylamine in acetic acid at room temperature.

(23.1) [2-(N=N–N=PPh$_3$)C$_6$H$_4$CHO] + 4-MeC$_6$H$_4$NH$_2$ —(EtOH, AcOH; 40%)→ Indazole [N–N=PPh$_3$, NHC$_6$H$_4$-4-Me] [3979]

II. FORMATION OF A SIX-MEMBERED RING

1. Pyridin-2-one

Thermolysis of an azido-aldehyde can produce a fused pyridine ring in high yield.

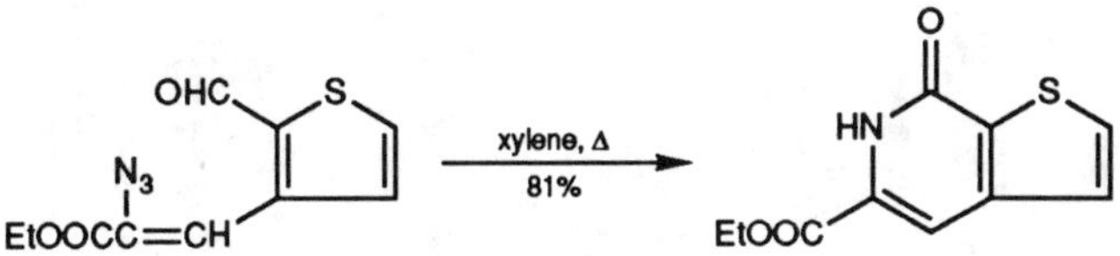

OHC, S, N_3, EtOOCC=CH → xylene, Δ, 81% → HN, O, S, EtOOC [2781]

Thieno[2,3-*c*]pyridin-7-one

Another ring system synthesized similarly:

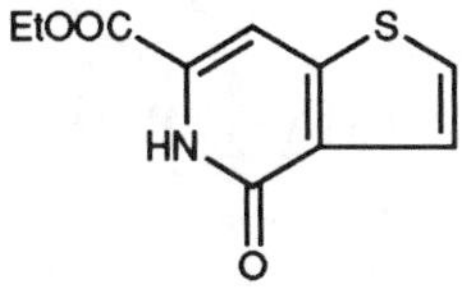

EtOOC, S, HN, O [2781]

Thieno[3,2-*c*]pyridin-4-one

III. FORMATION OF A SEVEN-MEMBERED RING

1. 1,4-Diazepin-2-one

In the synthesis of the medicinally important benzodiazepine, an example of the use of an azide as an unisolated intermediate is given in Part 1 (p. 260). The azide may be isolated and then converted by a reductive cyclization into the diazepinone.

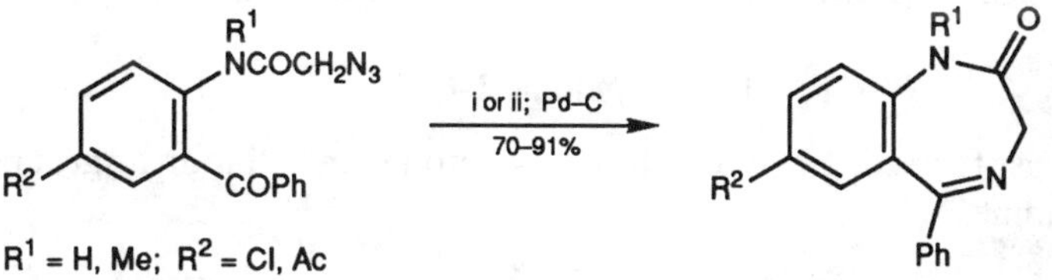

R^1, $NCOCH_2N_3$, R^2, COPh → i or ii; Pd–C, 70–91% → R^1, N, O, N, R^2, Ph [2075, 2647]

R^1 = H, Me; R^2 = Cl, Ac

i, H_2; ii, N_2H_4

1,4-Benzodiazepin-2-one

CHAPTER 24

Aldehyde or Ketone and Carbamate, Isothiocyanate or Thiourea

I. Formation of a Six-membered Ring 792
1. Pyrimidine 1-Oxide 792
2. Pyrimidine-2-thione 792
3. Pyrimidin-2-one 793
4. Pyran-2-one 793

I. FORMATION OF A SIX-MEMBERED RING

1. Pyrimidine 1-Oxide

This ring is formed by interaction of ketone and thiourea groups with hydroxylamine.

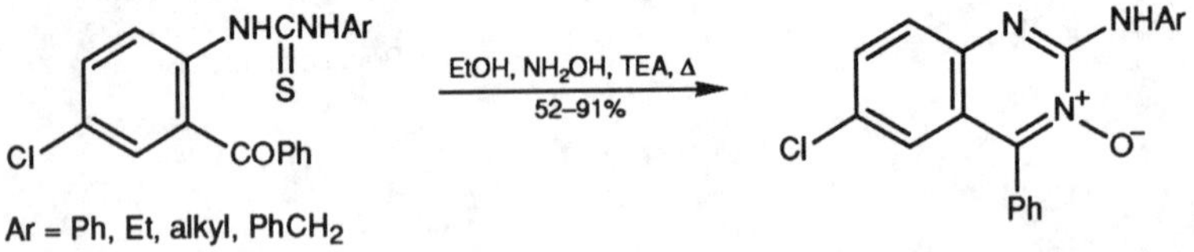

NHC(=S)NHAr / COPh / Cl — EtOH, NH_2OH, TEA, Δ; 52–91% → Quinazoline 3-oxide (NHAr, N^+–O^-, Ph, Cl) [2772]

Ar = Ph, Et, alkyl, $PhCH_2$

Quinazoline 3-oxide

2. Pyrimidine-2-thione

Cyclization of an *o*-benzoyl-isothiocyanate occurs at low temperatures on treatment with a primary amine or hydrazine, the latter behaving as a monofunctional reagent. The use of isothiocyanates in heterocyclic synthesis has been reviewed [3593, 3990].

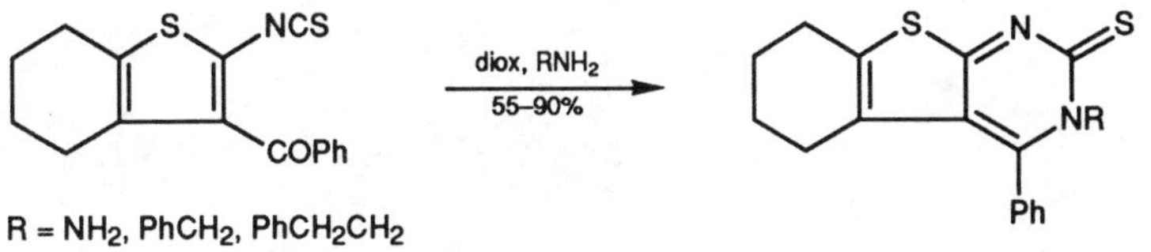

diox, RNH_2, 55–90% [3706]

$R = NH_2, PhCH_2, PhCH_2CH_2$

[1]Benzothieno[2,3-*d*]pyrimidine-2-thione

3. Pyrimidin-2-one

A ketonic group annulates a neighbouring isocyanate function (formed *in situ* by thermolysis of an azidocarbonyl) to form a fused pyrimidinone in high yield.

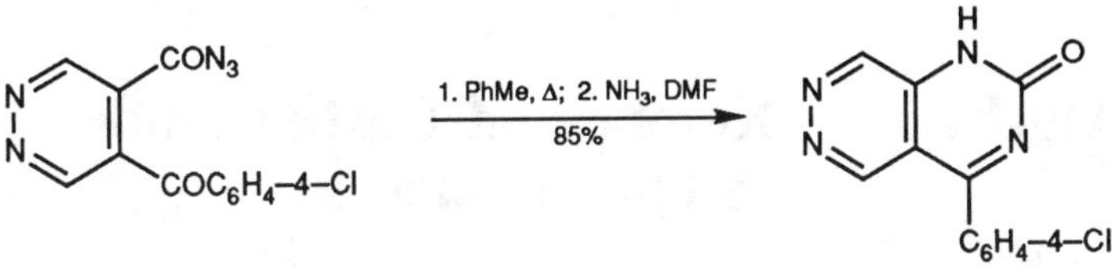

1. PhMe, Δ; 2. NH_3, DMF, 85% [3085]

Pyrimido[4,5-*d*]pyridazin-2-one

4. Pyran-2-one

Urea is eliminated from the ureidomethylene ketone (**24.1**) when the compound is heated with acetic anhydride in a Perkin-type reaction.

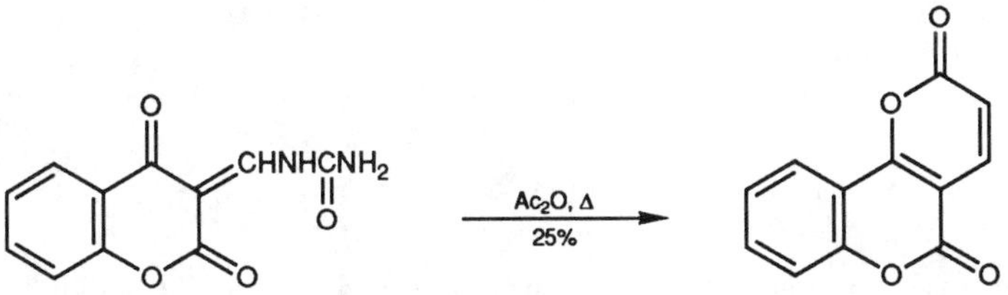

Ac_2O, Δ, 25% [3003]

(**24.1**)

Pyrano[3,2-*c*][1]benzopyran-2,5-dione

CHAPTER 25

Aldehyde or Ketone and Carboxamide or Sulphonamide

I.	Formation of a Five-membered Ring	794
	1. Pyrazole	794
	2. Thiophene	795
	3. Isoxazole	795
	4. Isothiazole 1,1-Dioxide	795
II.	Formation of a Six-membered Ring	795
	1. Pyridin-2-one	795
	2. Thiopyran-4-one	796
III.	Formation of a Seven-membered Ring	797
	1. 1,2-Diazepine	797
	2. 1,4-Oxazepin-5-one	798

Cyclizations of a hydrazono ketone and of 2-acyl *O*- or *S*-(2-acylphenyl)-carbamates are included in this chapter.

I. FORMATION OF A FIVE-MEMBERED RING

1. Pyrazole

Although a fused isoxazole (**25.1**) is formed in high yield from the ketonic carboxamide and hydroxylamine (see Section I.3), replacing the reagent with phenylhydrazine and heating the mixture for 12 h gives a very low yield of the fused pyrazole.

Ar = 4-MeOC$_6$H$_4$ + PhNHNH$_2$ → (AcOH, Δ; 4%) [2055]

Pyrazolo[3,4-*b*]quinoxaline

2. Thiophene

The formation of a fused thiophene ring from an *S*-(2-acetylphenyl)thiocarbamate is shown in Section II.2.

3. Isoxazole

A ketoximino group (formed *in situ*) adjacent to a carboxamide displaces the latter and forms this ring on heating in ethanol–pyridine.

Ar = 4-MeOC$_6$H$_4$ + NH$_2$OH excess → (EtOH, pyr, Δ; 80%) [2055]

(25.1)
Isoxazolo[4,5-*b*]quinoxaline

4. Isothiazole 1,1-Dioxide

Reductive cyclization occurs when an *o*-acylsulphonamide is treated with sodium dithionite in pyridine.

→ (pyr, Na$_2$S$_2$O$_4$; 10–48%) [2561, 2764]

R^1 = AlkO, NO$_2$, AlkS; R^2 = NH$_2$, AlkS;
Ar = Ph, Me-, Cl-C$_6$H$_4$

1,2,Benzisothiazole 1,1-dioxide

II. FORMATION OF A SIX-MEMBERED RING

1. Pyridin-2-one

Reaction of a keto ester with ammonia (to form the keto-amide *in situ*) leads, on further heating, to cyclization and the formation of a pyridin-2-one ring [2040]. The same ring is produced under Dean and Stark conditions from the cyclohexanone (**25.2**) [2218]. In contrast to many cyclizations which involve a carboxamide, ring closure of a 2-formyl-t-acetamide proceeds with loss of the amide nitrogen [2375]. When methylamine is replaced by hydrazine or hydroxylamine, these reagents behave as monofunctional compounds to give (**25.2a**, R = NH_2 or OH). The azine (**25.2b**) (formed as a byproduct in the preparation of the hydrazone) is efficiently cyclized to the isoquinolinone (**25.2c**, R^2 = NH_2) by hot mineral acid.

COR^1, $CHCOOH$, R^2 — 1. CH_2N_2; 2. MeOH, NH_3, Δ; 52–89% — [2040]

R^1 = H, Me, Ph; R^2 = H, Ph

3-Isoquinolinone

$(CH_2)_2CNHR^2$ — PhMe, TsOH, Δ; 80–90% — [2218]

(**25.2**)
R^1 = H, MeO; R^2 = H, Me

Benzo[*h*]quinolin-2-one

R^1O, CHO, CH_2CONH_2 — EtOH, R^2NH_2, Δ; 73–85% — [2375]

R^1 = Me; R^2 = Me, NH_2, OH or R^1R^1 = CH_2

(**25.2a**)
3-Isoquinolinone

R^3=N–N=R^3 (**25.2b**) — 6M HCl, Δ; 92% — (**25.2a**) R^1R^1 = CH_2, R^2 = NH_2 — [2375]
1,3-Dioxolo[*g*]isoquinolin-7-one

R^3 = =CH–, Me_2NCCH_2 (C=O)

2. Thiopyran-4-one

Treatment of an *S*-(2-acetylphenyl)thiocarbamate (**25.3**) with sodium hydride–DMF under nitrogen yields a 2-hydroxythiochromone (**25.4**), but in air, a significant amount of a benzothiophene (**25.5**) is also formed when R = H or

MeO. The *S*-arylthiocarbamates (**25.3**) are prepared by heating the *O*-analogues at 175 °C without a solvent; when 1,2-dichlorobenzene is present as solvent, an appreciable amount of a cyclic product, namely, the 5-hydroxythiochromone (**25.6**) is sometimes formed alongside the desired product.

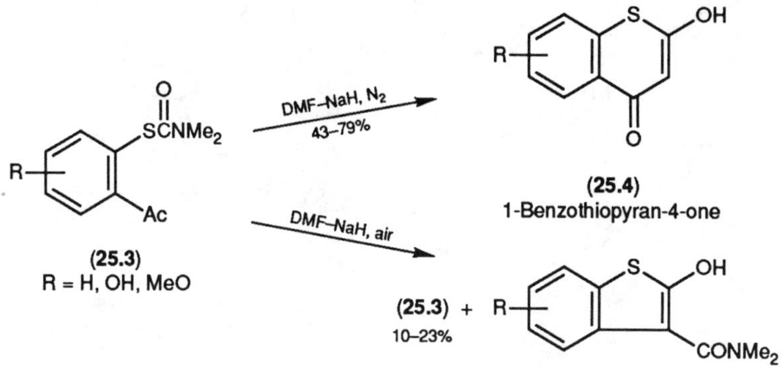

(25.3)
R = H, OH, MeO

DMF–NaH, N_2
43–79%

(25.4)
1-Benzothiopyran-4-one

[3274]

DMF–NaH, air

(25.3) +
10–23%

(25.5)
Benzo[*b*]thiophene
7–39%

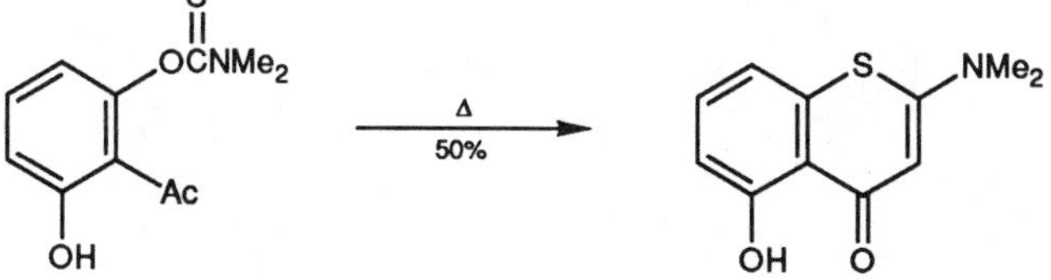

Δ
50%

(25.6)
1-Benzothiopyran-4-one

[3274]

III. FORMATION OF A SEVEN-MEMBERED RING

1. 1,2-Diazepine

A hydrazone group which contains a terminal NH_2 group and is separated from a carbonyl by four carbon atoms may cyclize to form a diazepine ring on stirring the compound in methanol containing hydrogen chloride.

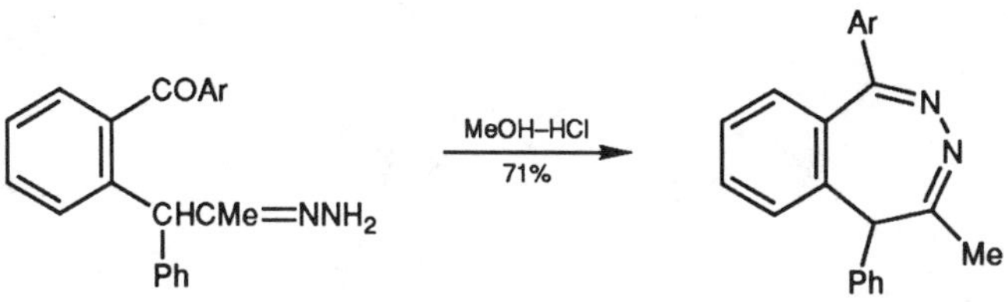

MeOH–HCl
71%

[2579]

Ar = 3,4-$(MeO)_2C_6H_3$

2,3-Benzodiazepine

2. 1,4-Oxazepin-5-one

Dehydrative cyclization of the keto-carboxamide (**25.7**) under Dean and Stark conditions gives a new fused 1,4-oxazepinone ring.

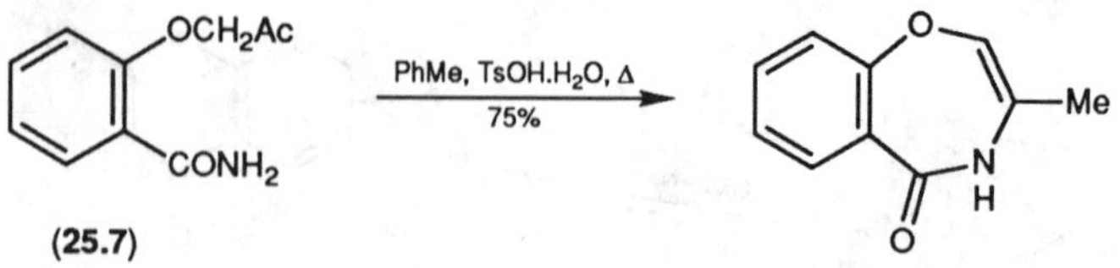

OCH$_2$Ac
CONH$_2$
(**25.7**)

PhMe, TsOH.H$_2$O, Δ
75%

O
Me
N
H
O

[3671]

1,4-Benzoxazepin-5-one

CHAPTER 26

Aldehyde or Ketone and Carboxylic Acid or Ester

I. Formation of a Five-membered Ring 799
1. Pyrrol-2-one 799
2. Pyrazole 800
3. Furan 800
4. Furan-2-one 801
5. Isoxazole 801
II. Formation of a Six-membered Ring 801
1. Pyridine 801
2. Pyridin-2-one 801
3. Pyran-2-one 802
III. Formation of a Seven-membered Ring 804
1. Azepine-2,7-dione 804
2. 1,2-Diazepin-3-one 804
3. 1,4-Diazepin-5-one 805

I. FORMATION OF A FIVE-MEMBERED RING

1. Pyrrol-2-one

Passing a current of dry ammonia for 10 h through a solution of the keto ester (**26.4**) converts it into a lactam while hydroxylamine in methanol gives a 6,7-dihydroxy derivative.

(**26.4**) —PhH, NH_3, 68%→ Pyrrolo[3,4-*d*]pyridazin-5-one [3081]

2. Pyrazole

Treatment of a β-ketolactone with hydrazine leads to a new pyrazole ring being formed, and replacement of the lactone by a pyrrolone ring. A carboxylic acid group is displaced by a neighbouring phenylhydrazono group on heating in ethanolic hydrogen chloride; **(26.5)** is thus formed (cf. the formation of an isoxazole ring, Chapter 25, Section I.3).

+ N_2H_4 —TEA, Δ; 38–88%→ [2752]

$CH_2\overset{O}{\overset{\|}{C}}Ar$

Ar = 4-MeO-, 4-F-, 4-CF_3-, 3-Me-C_6H_4, 2,4,6-$Me_3C_6H_2$

Pyrazolo[5,1-*a*]isoindol-8-one

COOH, COAr + $PhNHNH_2$ —EtOH, HCl, Δ; 83%→ [2055]

Ar = 4-$MeOC_6H_4$

(26.5)
Pyrazolo[3,4-*b*]quinoxaline

3. Furan

The formation of a ketene (some are too unstable to be isolated) from a CHCOCl group and TEA is believed to be a step in the cyclization of *o*-formyl(or *o*-acyl)phenoxyacetic acid. Conversion of this into a furan ring is accompanied by decarboxylation. Two other methods of producing the ketene were applied successfully to the synthesis of a benzofuran. A ketene may also be generated *in situ* by heating a 4-toluenesulphonyl ester with TEA. Recent work on benzofurans has been reviewed [3894].

R^1, OCHCOOH, COR^2 —i, ii, Δ; 53–82%→ [3283]

R^1 = H, Me, Et, Ph; R^2 = H, Me, Ph
i, $(COCl)_2$, PhH; ii, PhH, TEA

Benzofuran

Ph, OCHCOOH, COR^2, R^1 —PhH, TsCl, TEA, Δ; ~78%→ [3236]

R^1 = H, MeO; R^2 = H, Me

4. Furan-2-one

o-Aroyl carboxylic acids react with refluxing thionyl chloride over 4 h to give high yields of fused furan-2-ones.

$SOCl_2$, Δ
~93%

[3081]

Ar = Ph, 2-Cl-, 4-Cl-, 2-F-C_6H_4

Furo[3,4-*d*]pyridazin-5-one

5. Isoxazole

The synthesis of this ring from carboxamido and keto groups (Chapter 25, Section I.3) also occurs with carboxylic acid and keto groups; in this way, the isoxazolo[4,5-*b*]quinoxaline (**25.1**) is synthesized in 70% yield [2055].

II. FORMATION OF A SIX-MEMBERED RING

1. Pyridine

Keto and carboxylic ester groups in side-chains are annulated by reaction with ammonia. In this example, the reactants are heated in molten ammonium acetate.

NH_4OAc, NH_3, Δ
29–62%

[3488]

R = H, Et, COOMe

Thieno[2,3-*c*]pyridine

2. Pyridin-2-one

An attempt to prepare the oxime of the dibenzyl ketone (**26.6**) by heating it with hydroxylamine hydrochloride–sodium acetate gave the *N*-hydroxyisoquinoline as the only product [2120]. In a similar cyclization of a keto acid, ammonium acetate provides the nitrogen atom [2140]. Ammonium carbonate in acetic acid may also be used [2257].

(26.6) + NH_2OH → (EtOH, AcONa, Δ; 57%) 1-Isoquinolinone [2120]

Ar = 3,4-$(MeO)_2C_6H_3$ → (AcOH, $AcONH_4$, Δ; 85%) 3-Isoquinolinone [2140]

→ (AcOH, $(NH_4)_2CO_3$; 25%) Fluoreno[1,9,9a-*cd*]pyridin-3-one [2257]

3. Pyran-2-one

Heating the keto ester (**26.7**) with sodium hydride and a catalytic amount of t-butyl alcohol converts it into the isocoumarin in good yield. A 2-acyl(or formyl)indole-3-acetic acid is cyclized by heating in acetic anhydride or on stirring with the anhydride for 36 h [3741, 3937]. The latter process is sometimes shortened by addition of boron trifluoride dietherate [3742].

(26.7) → (PhMe, NaH, tBuOH, Δ; 55–92%) 2-Benzopyran-1-one [3156]

R^1 = H, MeO; R^2 = Me, Pr, MeCH=CH, Ph

→ (Ac_2O, Δ; 40%) Pyrano[3,4-*b*]indol-3-one [3399]

Other ring systems synthesized similarly:

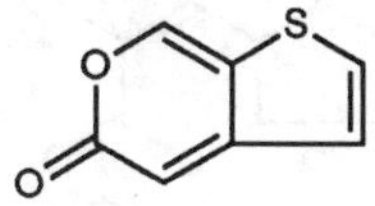

Thieno[2,3-*c*]pyran-5-one [3755]

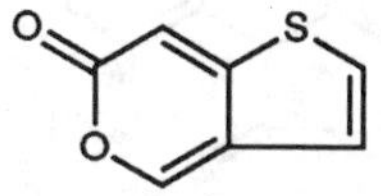

Thieno[3,2-*c*]pyran-6-one [3755]

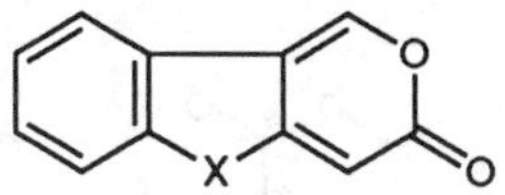

X = NH Pyrano[4,3-*b*]indol-3-one [3741, 3742]
X = S Benzothieno[3,2-*c*]pyran-3-one

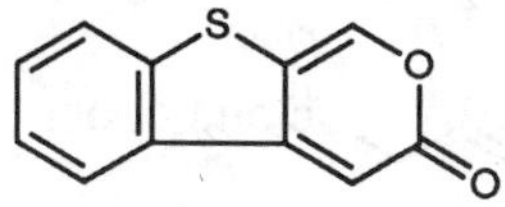

Benzothieno[2,3-*c*]pyran-3-one [3742]

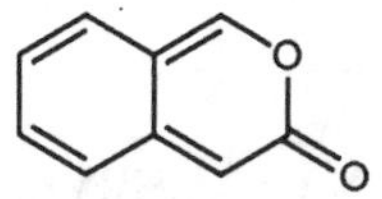

2-Benzopyran-3-one [3937]

A degradation product of the naturally occurring isophellopterin is synthesized by reductive cyclization of the acetylenic quinone (**26.8**) using a slightly poisoned catalyst [3416] (review of heterocyclic quinones [3650]). Acid-promoted dehydrative cyclization of a 2-oxophenylacetic acid gives the 4,5-fused pyranone in high yield [2102]. This process may be brought about thermally [2463] or in the presence of DBU [3915]. The chemistry of DBU has been reviewed [3364].

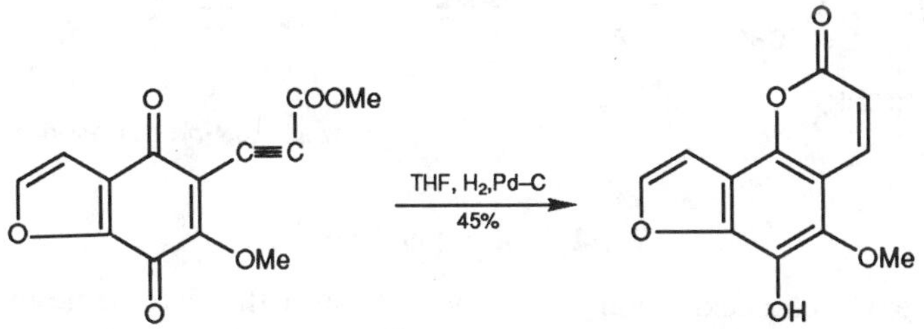

(**26.8**) → THF, H_2,Pd–C, 45% → Furo[2,3-*h*][1]benzopyran-2-one [3416]

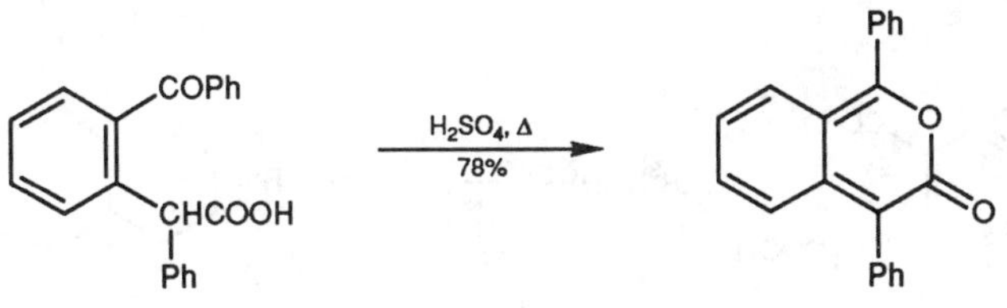

→ H_2SO_4, Δ, 78% → 2-Benzopyran-3-one [2102]

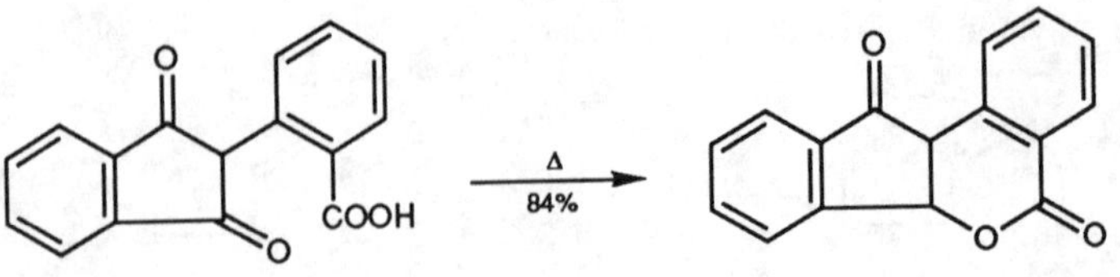

Δ, 84% [2463]

Indeno[2,1-*c*][2]benzopyran-5,11-dione

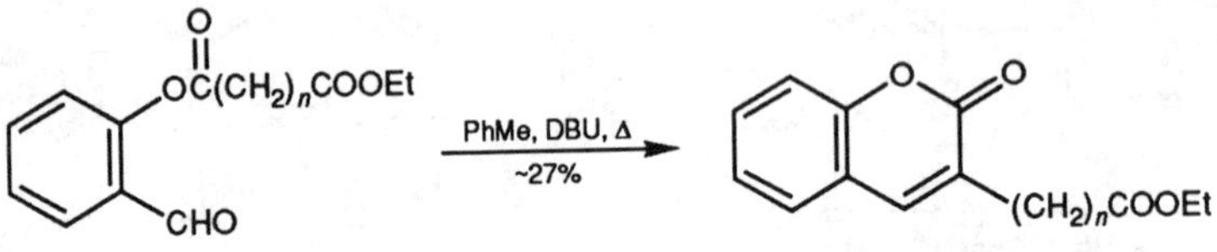

PhMe, DBU, Δ, ~27% [3915]

n = 3-6

1-Benzopyran-2-one

III. FORMATION OF A SEVEN-MEMBERED RING

1. Azepine-2,7-dione

An intramolecular Knoevenagel reaction can be used to annulate aldehyde and carboxylate groups using cyanoacetamide or malondiamide to complete the azepinedione ring.

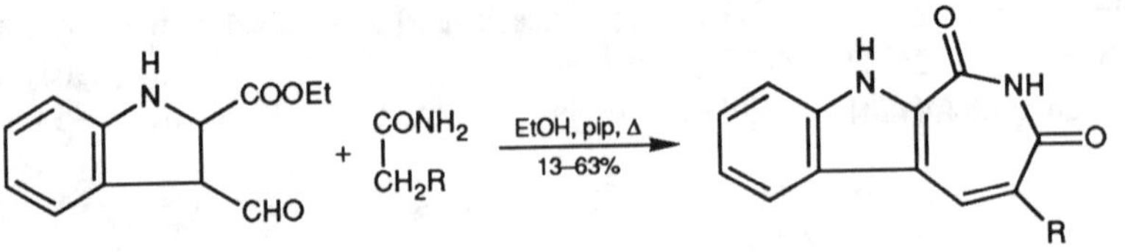

EtOH, pip, Δ, 13–63% [2953]

R = CN, $CONH_2$

Azepino[3,4-*b*]indole-1,3-dione

2. 1,2-Diazepin-3-one

2-Aroylphenylacetic acid is annulated by reaction with a hydrazine in boiling butanol or xylene—a more convenient reaction than that of the less readily obtained aroylphenylacetic ester hydrazone previously used.

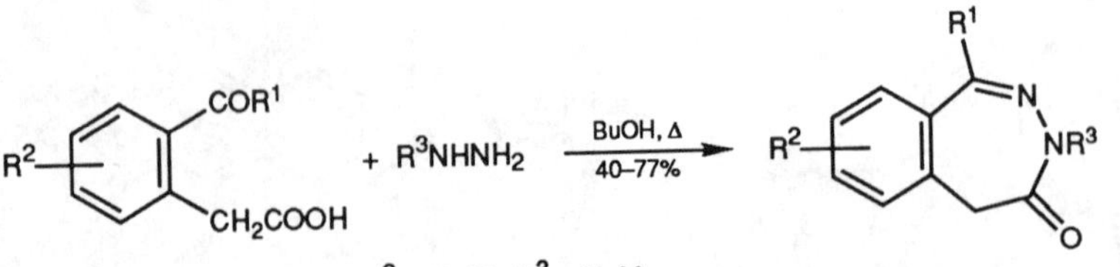

+ R^3NHNH_2 — BuOH, Δ, 40–77% [2264]

R^1 = Ph, 4-$MeOC_6H_4$; R^2 = H, Cl; R^3 = H, Me

2,3-Benzodiazepin-4-one

3. 1,4-Diazepin-5-one

A cyclic 3-oxo carboxylic ester and *o*-phenylenediamine undergo a thermal cyclization in moderate yield.

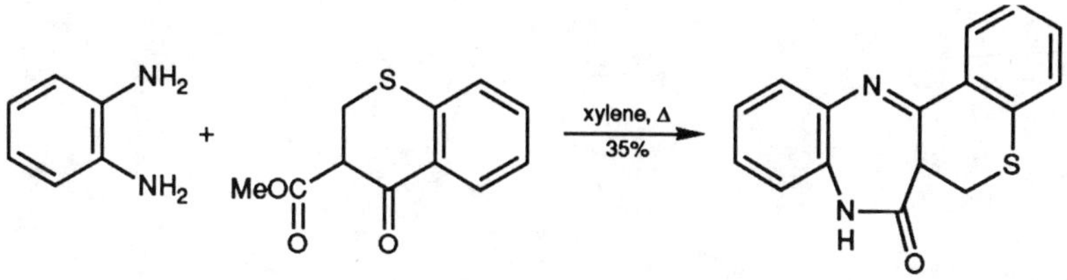

xylene, Δ
35%

[2954]

[1]Benzothiopyrano[4,3-*b*][1,4]benzodiazepin-7-one

CHAPTER 27

Aldehyde or Ketone and Ether

I. Formation of a Five-membered Ring 806
1. Pyrrole 806
2. Pyrazole or Imidazole 807
3. Oxazole 807
II. Formation of a Six-membered Ring 808
1. Pyrimidine 808
2. Pyran-4-one 808
3. 1,2-Thiazine 1-Oxide 808

Cyclization of an *o*-acyl-sulphoximine is described in this chapter.

I. FORMATION OF A FIVE-MEMBERED RING

1. Pyrrole

The labile alkoxy group of an alkoxyquinone is displaced by a primary alkylamine at ambient temperature, and a side-chain keto group reacts with the amino function to form this ring. Reviews of the chemistry of heterocyclic quinones are available [2947, 3650].

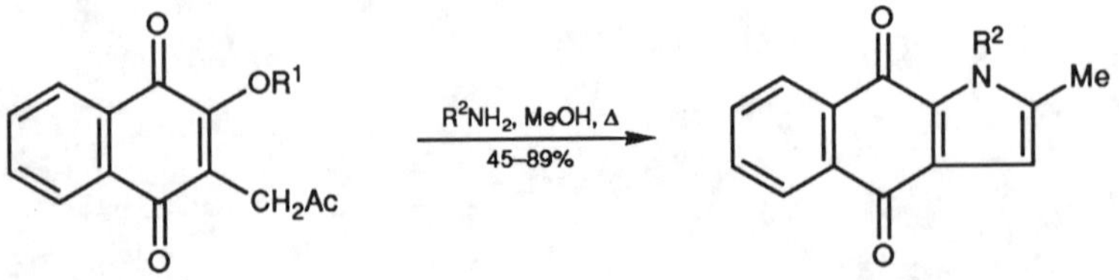

OR^1, CH_2Ac → (R^2NH_2, MeOH, Δ; 45–89%) → Me, R^2 [3606]

R^1 = alkyl; R^2 = alkyl, allyl

Benz[*f*]indole-4,9-dione

2. Pyrazole or Imidazole

The 2-methoxytropone (**27.2**) reacts with an amidine and a strong base to give a fused imidazole which was the only cyclized product; guanidine cyclizes the tropone efficiently over 24 h without a base. Although ureas failed to effect cyclization, thiourea under similar conditions gave an 81% yield of the imidazole (R^2 = HS) over 2 h. The isomeric 3-R^1-2-methoxytropone gave lower yields of imidazoles.

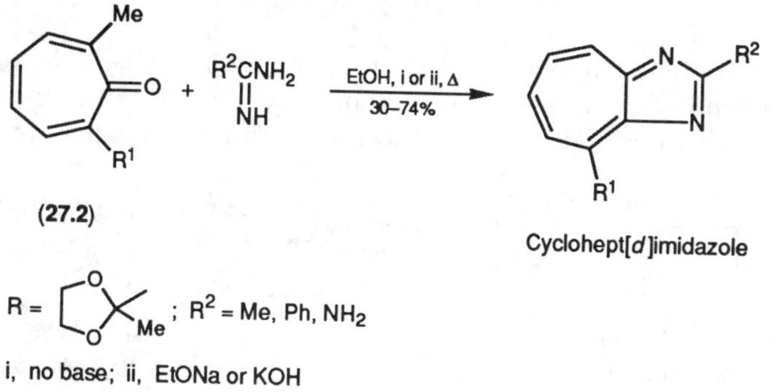

(27.2) + $R^2C(=NH)NH_2$ —EtOH, i or ii, Δ; 30–74%→ Cyclohept[*d*]imidazole [3296]

R = 2-methyl-1,3-dioxolan-2-yl; R^2 = Me, Ph, NH_2

i, no base; ii, EtONa or KOH

Another ring system synthesized similarly:

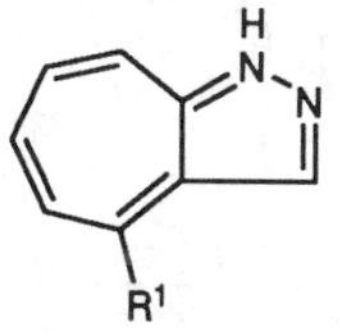

Cyclohepta[*c*]pyrazole [3296]

3. Oxazole

N-Phenacyl-2-pyridinium salts cyclize on heating in pyridine, the methylthio group being lost (cf. Chapter 36, Section I.3). Simultaneous Beckmann rearrangement, demethylation and cyclization converts an *o*-alkoxy-ketoxime into a fused oxazole ring.

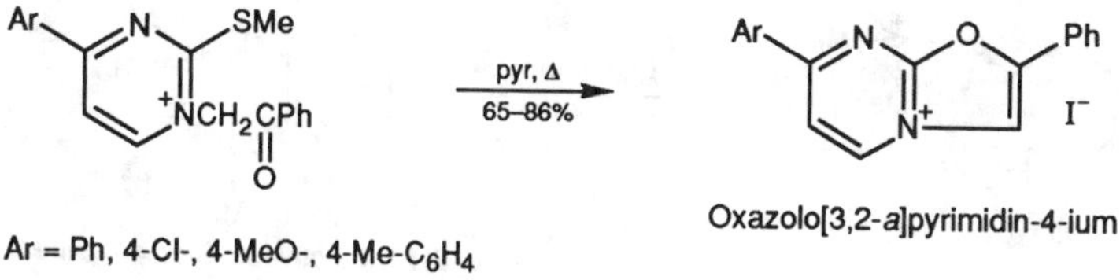

Ar, SMe, $^+NCH_2C(=O)Ph$ —pyr, Δ; 65–86%→ Oxazolo[3,2-*a*]pyrimidin-4-ium (Ar, Ph, I^-) [3383]

Ar = Ph, 4-Cl-, 4-MeO-, 4-Me-C_6H_4

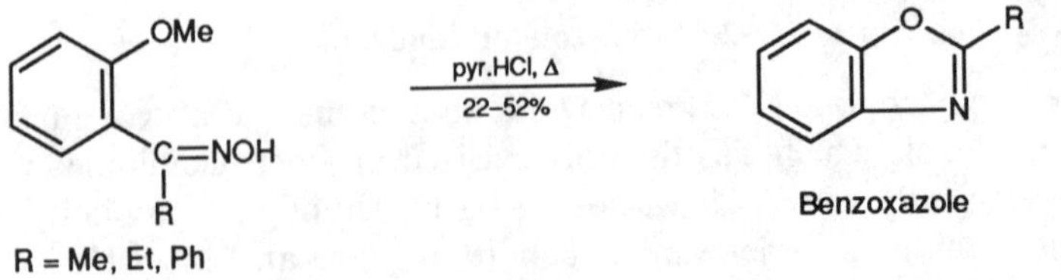

R = Me, Et, Ph

pyr.HCl, Δ 22–52% [2048]

Benzoxazole

2. FORMATION OF A SIX-MEMBERED RING

1. Pyrimidine

Displacement of an alkoxy by a basic nitrogen group results in the formation of a C—N bond; when the nitrogen is part of an amidine, the other nitrogen may attack a suitably positioned carbonyl to form a fused pyrimidine ring.

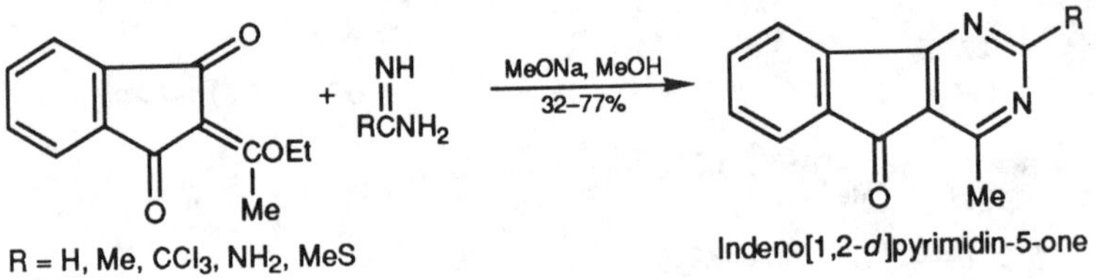

MeONa, MeOH 32–77% [2395]

R = H, Me, CCl_3, NH_2, MeS

Indeno[1,2-*d*]pyrimidin-5-one

2. Pyran-4-one

One of the less common methods of forming this ring is by heating an *o*-ethoxypropyl-1,3-dione with TEA for 12 h.

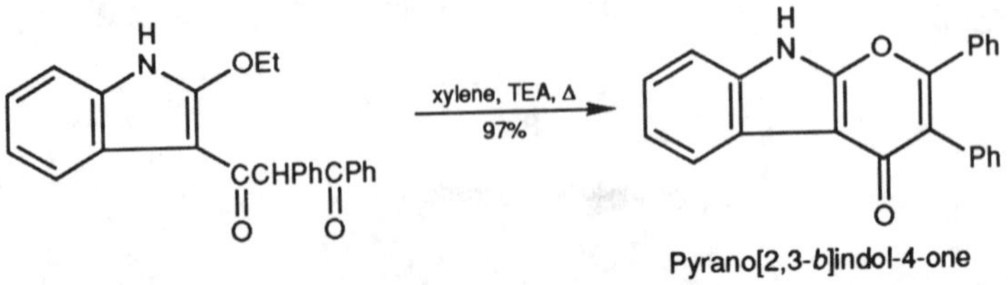

xylene, TEA, Δ 97% [2840]

Pyrano[2,3-*b*]indol-4-one

3. 1,2-Thiazine 1-Oxide

Treatment of a 2-(dimethylsulphoximino) ketone with sodium hydride can give one of two anions but the product formed suggests that a hydrogen ion is removed from the MeS group.

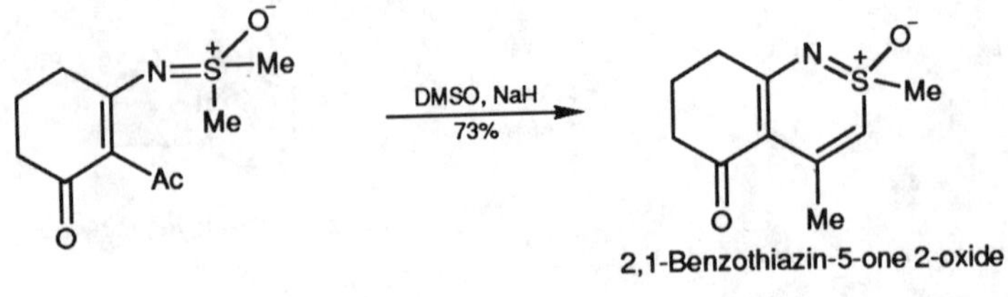

DMSO, NaH 73% [2902]

2,1-Benzothiazin-5-one 2-oxide

CHAPTER 28

Aldehyde or Ketone and Halogen

I.	Formation of a Five-membered Ring	809
	1. Pyrrole	809
	2. Pyrazole	810
	3. Imidazole	811
	4. Furan	811
	5. Thiophene	812
II.	Formation of a Six-membered Ring	812
	1. Pyridine	812
	2. Pyridin-4-one	813
	3. Pyridazine	813
	4. Pyrazin-2-one	814
	5. 1,2,4-Triazine	814
	6. Pyran-2-one	814
	7. Pyran-4-one or Thiopyran-4-one	815
III.	Formation of a Seven-membered Ring	815
	1. 1,4-Diazepine	815
	2. 1,4-Diazepin-2-one	816
	3. 1,4-Oxazepine	816
IV.	Formation of an Eight-membered Ring	816
	1. 1,5-Diazocin-2-one	816

I. FORMATION OF A FIVE-MEMBERED RING

1. Pyrrole

2-Halogenotropones (or 2-tosyloxytropone) (**28.1**) may be cyclized to cycloheptapyrroles by reaction with either a cyclic or an acyclic β-aminoenone, or an *N*-vinyliminophosphorane in the presence of a mild base such as TEA or TEA–potassium carbonate [3942, 3528]. Prolonged heating of the chloroketopyrimidine (**28.2**) with a primary amine in ethanol gives good yields of a fused pyrrole [2049].

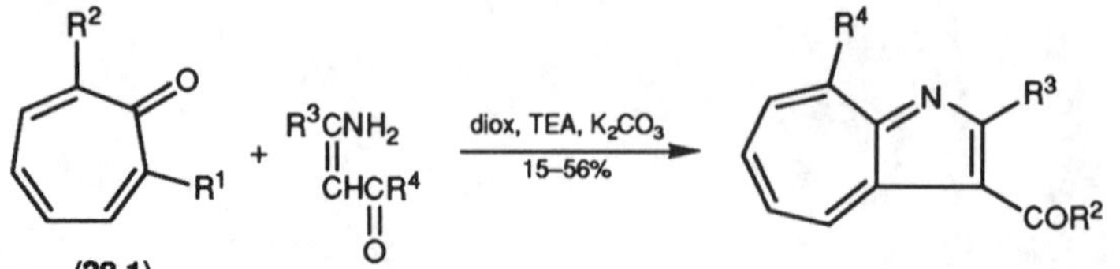

(**28.1**) + R^3CNH_2 / $CHCR^4$ / O —diox, TEA, K_2CO_3; 15–56%→ Cyclohepta[*b*]pyrrole [3942]

R^1 = Br, Cl, TsO; R^2 = H;
R^3 = Me or R; R = $(CH_2)_n$, $CH_2CMe_2CH_2$;
n = 3,4; R^4 = MeO

(**28.1**) + $CR^3{=}CH_2$ / $N{=}PPh_3$ —PhH, TEA, Δ; 32–77%→ [3528]

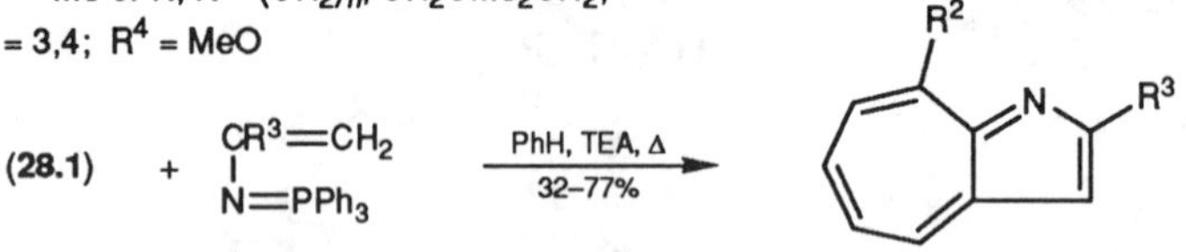

R^1 = Br, Cl; R^2 = H, Br; R^3 = Me, Bu, Ph

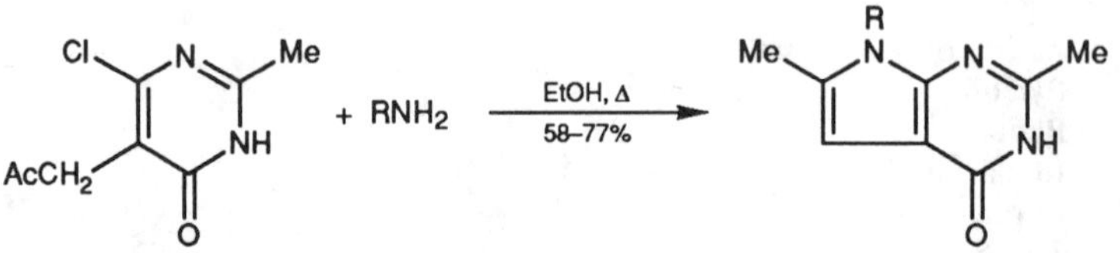

(**28.2**) + RNH_2 —EtOH, Δ; 58–77%→ Pyrrolo[2,3–*d*]pyrimidin-4-one [2049]

R = $HO(CH_2)_2$, 2-furyl

Another ring system synthesized similarly:

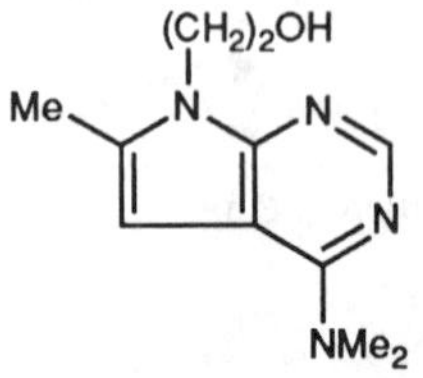

Pyrrolo[2,3–*d*]pyrimidine [2049]

2. Pyrazole

Under acidic or weakly basic conditions, an *o*-chloro-aldehyde or -ketone reacts with a phenylhydrazine to form a fused pyrazole ring in good yield [2055].

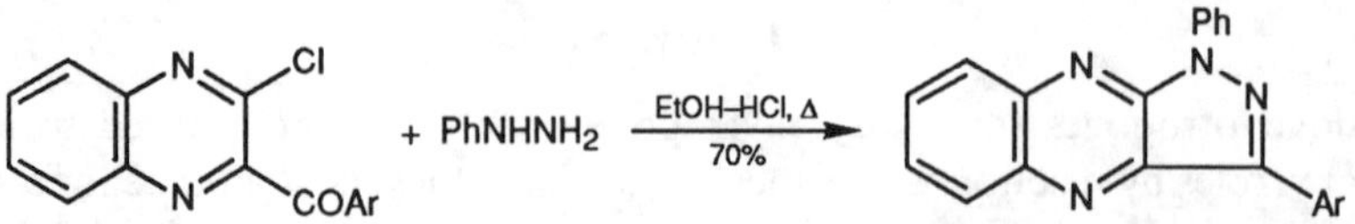

+ $PhNHNH_2$ —EtOH–HCl, Δ; 70%→ Pyrazolo[3,4-*b*]quinoxaline [2055]

Ar = 4-$MeOC_6H_4$

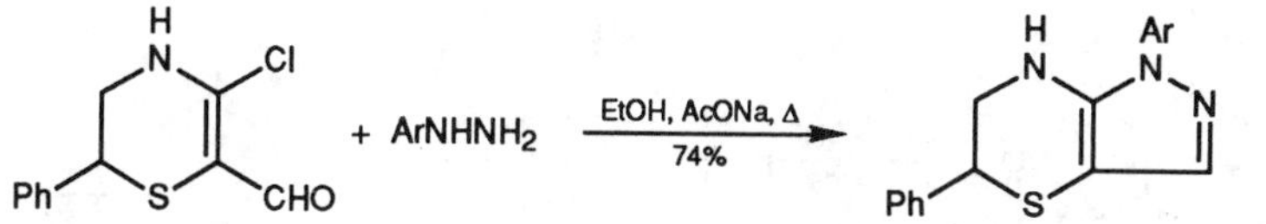

+ ArNHNH2 —EtOH, AcONa, Δ / 74%→ [2418]

Ar = 4-ClC6H4

Pyrazolo[3,2-*c*][1,4]thiazine

Other ring systems synthesized similarly:

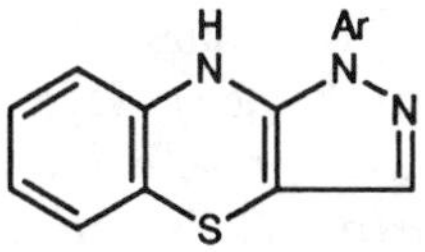

[2418] [2418]

Pyrazolo[4,3-*b*][1,4]benzothiazine

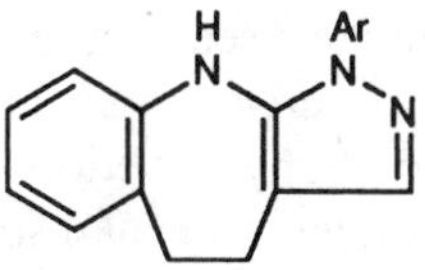

Pyrazolo[3,4-*b*][1]benzazepine

3. Imidazole

A reactive nuclear 2-bromoazole carrying an *N*-oxomethyl chain reacts with a primary amine to form this ring.

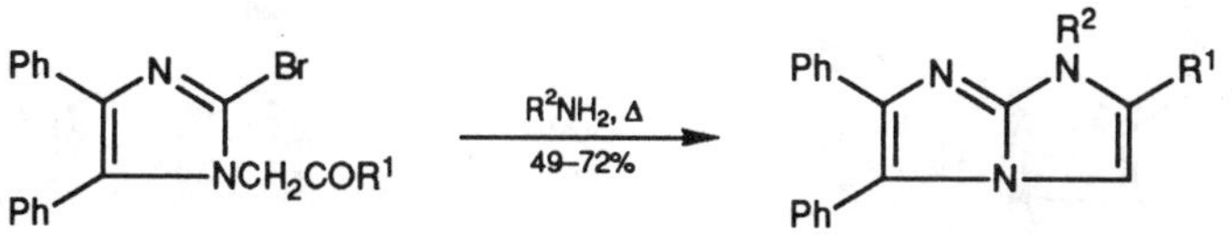

R^2NH_2, Δ / 49–72% [2368]

R^1 = Me, Ph; R^2 = Ph, MeC_6H_4

Imidazo[1,2-*a*]imidazole

4. Furan

Bromine reacts at room temperature and in the absence of light with the acylmethylquinone (**28.3**) to give good-to-excellent yields of the quinonoid heterocycle—a class of compounds which has been reviewed [2947, 3650]. An oxygen enolate ion intramolecularly displaces a fluorine atom in pentafluorophenyl compounds with the formation of fused furan ring.

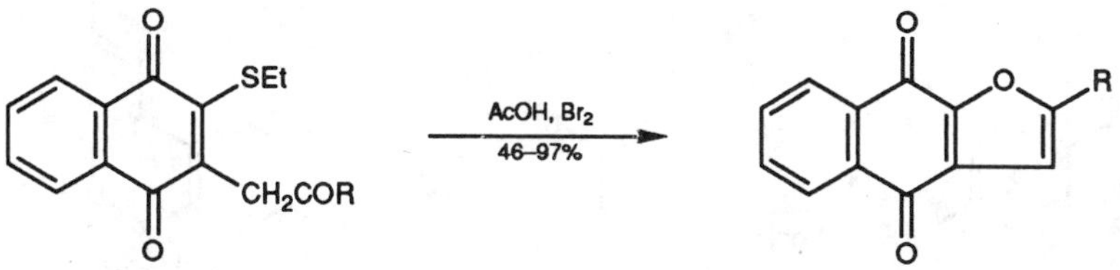

AcOH, Br_2 / 46–97% [3740]

(**28.3**)
R = tBu, Ph, Cl-, 4-F-, Br-, 4-MeO-, 4-NO_2-C_6H_4

Naphtho[2,3-*b*]furan-4,9-dione

R = H, Me

NaH–DMF, 13–76% → [2269]

Benzofuran

5. Thiophene

When the amine used in the cyclization of a pyrimidine related to (**28.2**) is replaced by thiourea, sulphur is incorporated in the newly formed ring [2050]. Bromo and carbonyl functions may be annulated by reacting with a thioamide [2933]. A stronger base than sodium carbonate (p. 131) is sometimes preferred in the cyclization of a haloaldehyde [3267].

$+ CS(NH_2)_2$ — EtOH, Δ, 25–66% → [2050]

R^1 = H, MeS, Me; R^2 = SH, Me, Me_2N

Thieno[2,3-*d*]pyrimidine

EtOH, Δ, 92% → [2933]

Thieno[3,4-*d*]-1,2,3-thiadiazole

II. FORMATION OF A SIX-MEMBERED RING

1. Pyridine

2-Bromobenzaldehydes are coupled to arylboronic acids (**28.4**) by heating with triphenylphosphine; coupling is probably followed by loss of the *N*-protecting group. The aldehyde group of the pyrimidinedione (**28.5**) forms a bond with the electron-rich carbon of the 3-hydroxyaniline while the chlorine is displaced by the amino group.

(**28.4**) — i, 45–89% → [3447]

Phenanthridine

R^1 = H, MeO; R^2 = H, MeO, NO_2; R^3 = H, MeO

i, PhMe, PPh_3, N_2, Δ

Another ring system synthesized similarly:

[3447]

[1,3]Dioxolo[4,5-*j*]phenanthridine

[3583]

(28.5)
R^1 = Me, Ph; R^2 = alkyl

Pyrimido[4,5-*b*]quinoline-2,4-dione

2. Pyridin-4-one

Ring closure by reaction between an amino and a fluoro function is a frequently used method (see Chapter 9); it is sometimes convenient to use a heterocyclic carbonyl group as a precursor of the amine, and thus reduce the number of compounds that have to be isolated and purified in a multistep synthesis. Pyrimidinediones are susceptible to this type of conversion as is demonstrated in the synthesis of a diaza-acridinone.

[3647]

(28.6)
i, pyr, $POCl_3$, Δ; ii, CH_2Cl_2, $BuNH_2$, Δ;
iii, DMF, K_2CO_3, Δ

Pyridazino[4,5-*b*]quinolin-10-one

3. Pyridazine

Complexes of chlorobenzenes with iron (cyclopentadienyl)hexafluorophosphate have useful reactivities, for example, in cyclization. The complex **(28.7)** is cyclized with hydrazine to a dihydrocinnoline complex; the metallophosphate is removed and the ring is aromatized by sodamide.

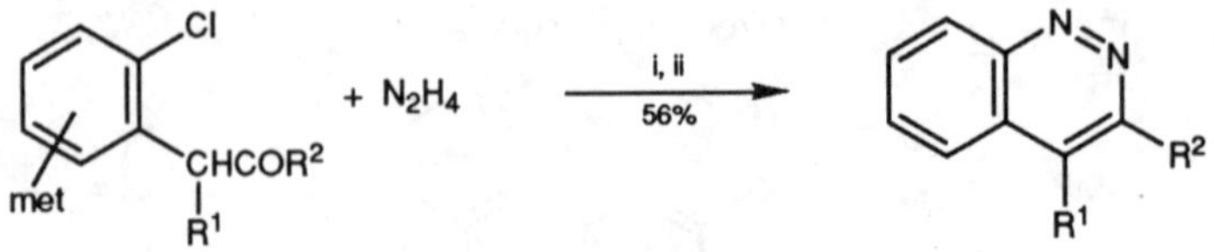

(28.7) + N_2H_4 —(i, ii; 56%)→ Cinnoline [3481]

R^1 = Me, Ph; R^2 = H, Me; met = Fe-cyclopentadienyl-PF_6
i, CH_2Cl_2, DMF, N_2H_4; ii, CH_2Cl_2, $NaNH_2$

4. Pyrazin-2-one

Trichloromethyl and carbonyl groups react at or below room temperature when the compound is stirred with methylamine, and the former substituent is lost.

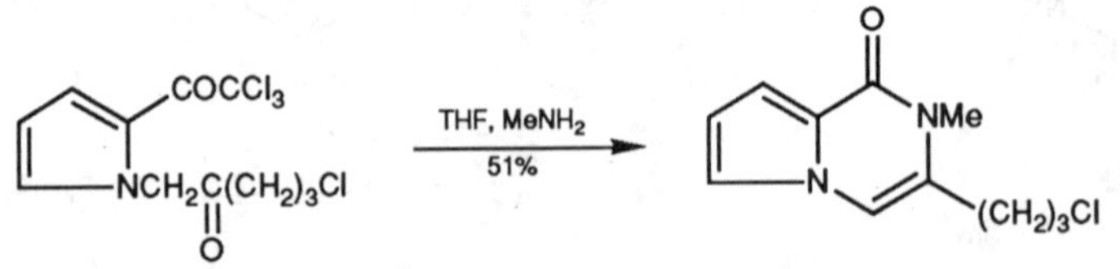

—(THF, $MeNH_2$; 51%)→ Pyrrolo[1,2-*a*]pyrazin-1-one [3418]

5. 1,2,4-Triazine

A side-chain carbonyl is annulated to an activated nuclear bromine or chlorine by reaction with hydrazine hydrate or methylhydrazine.

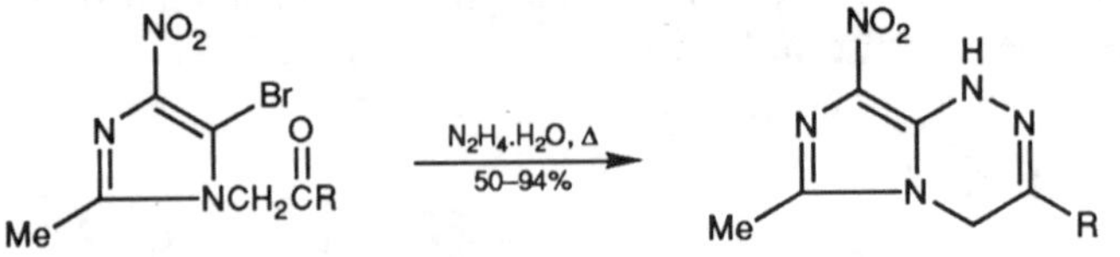

—($N_2H_4.H_2O$, Δ; 50–94%)→ Imidazo[5,1-*c*][1,2,4]triazine [2716]

R = Ph, 4-ClC_6H_4

Other ring systems synthesized similarly:

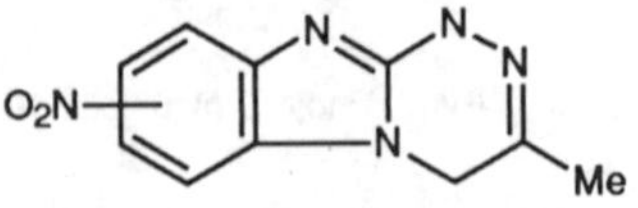

[1,2,4]Triazino[4,3-*a*]benzimidazole [3257]

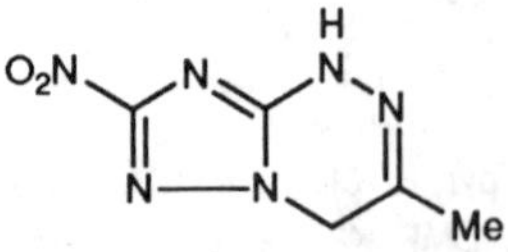

[1,2,4]Triazolo[5,1-*c*][1,2,4]triazine [3257]

6. Pyran-2-one

A fluorine atom adjacent to a carbonyl group in a benzenoid compound is labile, as is exemplified in the synthesis of a coumarin from 2-fluorobenzaldehyde and a reactive methylene-containing carboxylic acid.

$ArCH_2COOH$ + 2-fluorobenzaldehyde —(Ac_2O, TEA, Δ; 27–47%)→ [3357]

Ar = Ph, 4-Cl-, 4-Me-C_6H_4, 2- or 3-thienyl

1-Benzopyran-2-one

7. Pyran-4-one or Thiopyran-4-one

Annulation of a ring-ketone and a fluorine atom (attached to different rings) is achieved by heating the compound (**28.6**) in either DMF containing potassium carbonate or pyridine–phosphorus pentasulphide.

(**28.6**) —(i or ii, Δ; ~83%)→ [3393]

i, DMF–K_2CO_3; ii, pyr–P_2S_5

X = O [1]Benzopyrano[2,3-*d*]pyridazin-10-one
X = S [1]Benzothiopyrano[2,3-*d*]pyridazin-10-one

III. FORMATION OF A SEVEN-MEMBERED RING

1. 1,4-Diazepine

Neighbouring aldehyde and halogen functions react with an *o*-phenylenediamine to form a new doubly fused diazepine ring, but this may not be the major product. For example, the chloroaldehyde (**28.8**) reacts on heating with the diamine in ethanol to give a 63% yield of 2-(2-chlorocyclohepta[*b*]pyrrol-3-yl)benzimidazole and a 29% yield of the fused diazepine. Hexamine (review [2966a]) supplies the second nitrogen atom in the cyclization of a 2-(ω-bromoethylamino)benzophenone.

(**28.8**) + *o*-phenylenediamine —(EtOH, Δ; 29%)→ [3765]

Cyclohepta[*d*]pyrrolo[2,3-*b*][1,5]benzodiazepine

+ $(CH_2)_6N_4$ —(EtOH, Δ; 75–88%)→ [2328]

R = H, Me

1,4-Benzodiazepine

2. 1,4-Diazepin-2-one

Annulation of haloacylamino and keto groups to produce a 1,4-benzodiazepinone (p. 135) is improved when ammonia is replaced by ammonium carbonate [2992] and when hydrochloric acid is present [2993]; a different mechanism is proposed on the basis of work using [^{14}C]hexamethylenetetramine [2993]. The role of R^1 and R^2 (see p. 135) in controlling the amount of the imidazolidine byproduct is discussed [2993]. The relative usefulness of ammonia (under pressure), ammonium carbonate and hexamethylenetetramine (hexamine) in this cyclization is considered [2992].

3. 1,4-Oxazepine

2-Chloroquinoline-3-carboxaldehydes condense with 2-aminophenol on heating first in THF and then with carbonate–DMF.

[3352]

R = H, MeO, Me
i, THF, Δ; ii, DMF–K_2CO_3, Δ

i, ii
92–100%

Quino[2,3-*b*][1,5]benzoxazepine

IV. FORMATION OF AN EIGHT-MEMBERED RING

1. 1,5-Diazocin-2-one

A benzophenone carrying a δ-chloro side-chain cyclizes on stirring with methanolic ammonia. The chemistry of benzodiazocines has been reviewed [3355, 3366].

[2462]

R = Br, Cl

MeOH–NH_3
~48%

1,5-Benzodiazocin-2-one

CHAPTER 29

Aldehyde, Ketone or Lactam Carbonyl and Hydroxy

I. Formation of a Five-membered Ring 817
  1. Pyrrole 817
  2. Furan 818
  3. 1,3-Dioxol-2-one 819
  4. 1,3-Oxathiol-2-one 820
  5. Oxazole 820
II. Formation of a Six-membered Ring 820
  1. Pyran 820
  2. Pyran-2-one 822
  3. Pyran-4-one 824
  4. 1,3-Oxazin-3-one 825
  5. 1,4-Oxazin-3-one 826
  6. 1,2-Oxathiazine 2,2-Dioxide 826

Cyclization of a sulphonamide containing a modified bis(methylthio)ketal is included in this chapter.

I. FORMATION OF A FIVE-MEMBERED RING

1. Pyrrole

When the dithioacetal *S*-oxide (**29.1**) is stirred in dichloromethane containing hydrogen sulphide and hydrochloric acid, it is efficiently cyclized to a 1-arylsulphonyl-2-methylthioindole.

(29.1) $R = H, Br, Cl, Me$; $Ar = Ph, 4\text{-}MeC_6H_4$ —(CH_2Cl_2, H_2S, HCl; 62–80%)→ Indole [3996]

2. Furan

Salicylaldehydes such as (**29.2**) condense with ethyl bromomalonate in the presence of a mild base to give a fused furan ring [2468]. Heating a methoxyaldehyde (**29.3**) at about 175 °C with PPA causes demethylation and cyclization [2491]. A 2-(β-oxoalkyl)phenol is cyclized by heating with 4-toluenesulphonic acid to a 2-substituted benzofuran [2089], but when the carbonyl group is attached directly to the phenolic ring, heating with chloroacetonitrile or chloroacetamide can give high yields of a 2- or 2,3-substituted benzofuran [2276]. 2-Benzyloxychalcones are readily prepared, and when treated with thallium(III) nitrate, they are oxidatively rearranged to the ketoacetal (**29.4**). When the latter is stirred with hydrobromic (or hydrochloric) acid, or is catalytically debenzylated and then stirred with mineral acid, it is cyclized to the benzofuran [3911]. Recent progress in the chemistry of fused furans has been reviewed [3894, 3983].

(29.2) + $BrCH(COOEt)_2$ —(AcEt, K_2CO_3, Δ; 83%)→ Benzofuran [2468]

(29.3) —(PPA, Δ; 29%)→ Furo[3,2-*c*]quinolin-4-one [2491]

—(PhH, TsOH, Δ; 63%)→ [2089]

+ $ClCH_2R^3$ —(DMF, K_2CO_3, Δ; 18–98%)→ Benzofuran [2276]

$R^1 = H, Me, MeO, Cl, Br$; $R^2 = H, Me, Et, Ph$; $R^3 = CN, CONH_2$

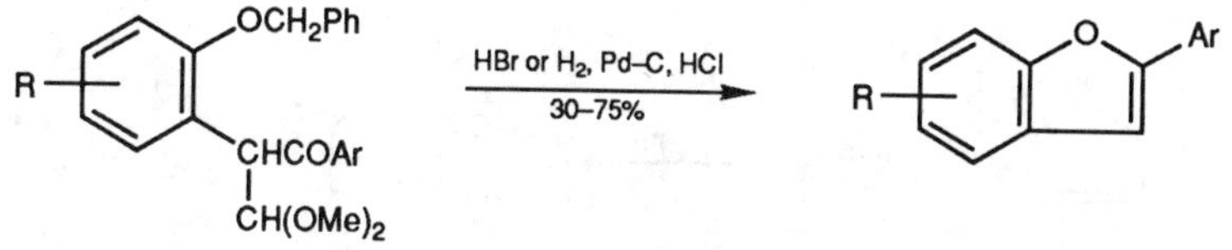

HBr or H_2, Pd–C, HCl
30–75%
[4006]

(29.4)
R = H, MeO; Ar = Ph, MeO-, HO-C_6H_4, $(MeO)_2C_6H_3$

Hydroxy and lactam carbonyl functions attached to different rings are readily converted into an ether linkage joining the two rings. The hydroxy is sometimes generated *in situ* from a methoxy by the action of pyridine hydrochloride or other dealkylating agent [3001, 3568]. Another variation is to heat the hydroxy-lactam with aniline for several hours, but the nucleophilic properties of the amine may affect other parts of the ring system, as in the example [3482]. The usefulness of pyridine hydrochloride as a demethylating agent has been reviewed [2891], as also has the chemistry of condensed furans [3779, 3894].

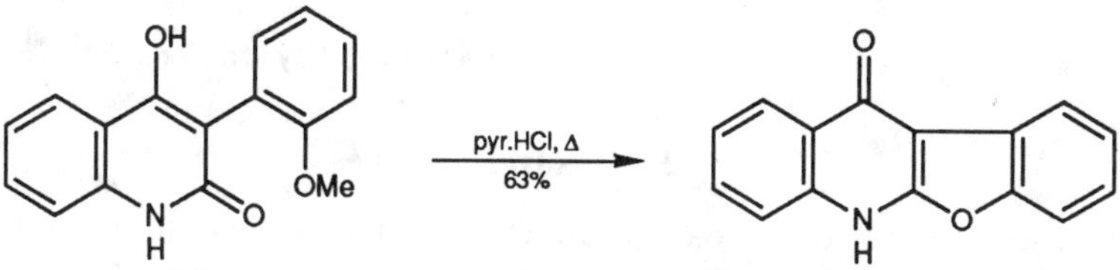

pyr.HCl, Δ
63%
[3001]

Benzofuro[2,3-*b*]quinolin-11-one

Another ring system synthesized similarly:

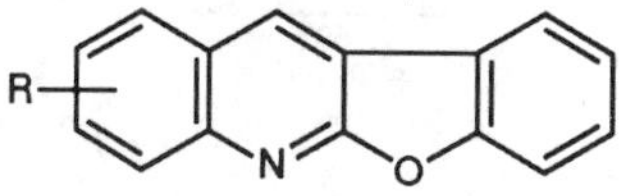

[3568]

Benzofuro[2,3-*b*]quinoline

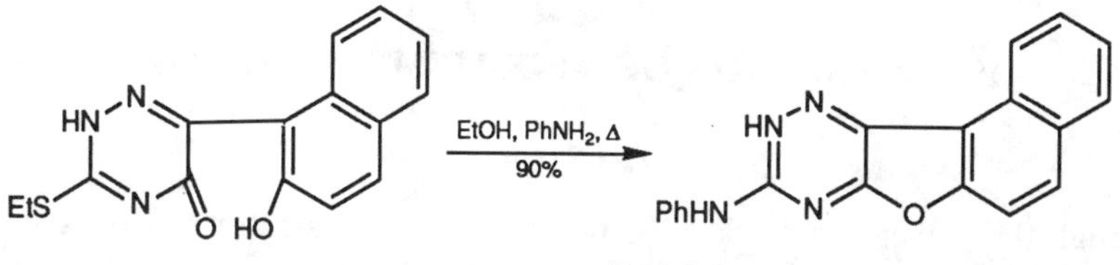

EtOH, $PhNH_2$, Δ
90%
[3482]

Naphtho[1',2' : 4,5]furo[2,3-*e*][1,2,4]triazine

3. 1,3-Dioxol-2-one

When a cyclic 2-hydroxy ketone reacts with phosgene at room temperature, a high yield of the fused dioxolone is obtained.

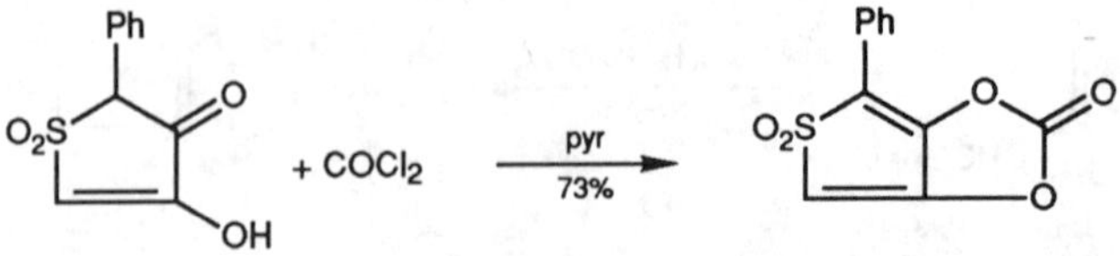

[2778]

Thieno[3,4-*d*][1,3]dioxol-2-one 5,5-dioxide

4. 1,3-Oxathiol-2-one

This ring is formed by reaction of neighbouring hydroxy and lactam thiocarbonyl groups with carbonyldi-imidazole and sodium imidazolide (NaIm) at or below room temperature. The product is a cyclic thiocarbonate which acylates amines under mild conditions.

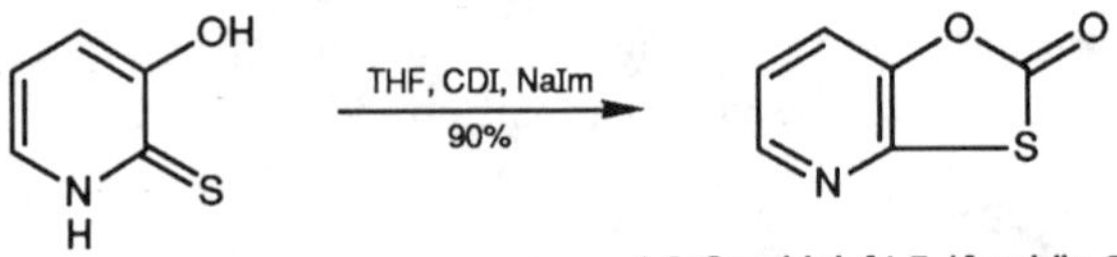

[3954]

1,3-Oxathiolo[4,5-*b*]pyridin-2-one

5. Oxazole

Neighbouring acetyl and hydroxy groups are annulated when the compound is treated with sodium azide in a Schmidt reaction which is followed by cyclization.

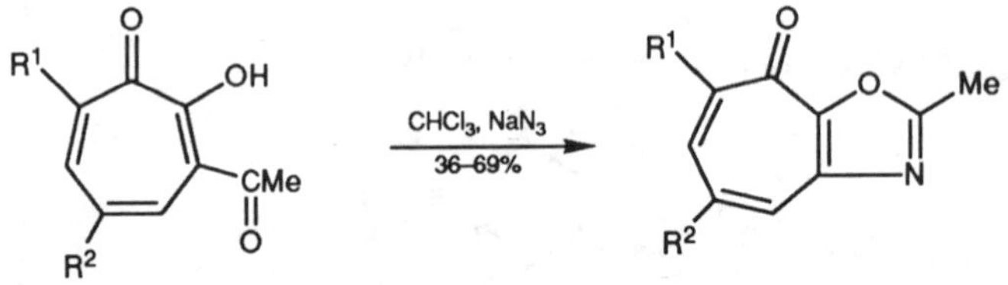

[3211]

R = Br, I; R^2 = H, Br

Cyclohept[*d*]oxazol-8-one

II. FORMATION OF A SIX-MEMBERED RING

1. Pyran

Although the pioneering work of Schweizer [B-12] demonstrated that a Wittig reaction may be applied to the synthesis of 2*H*-chromenes, synthesis of other chromones and pyranopyridines by this method has only fairly recently been accomplished; this procedure is possible largely as a result of another discovery of a method of synthesizing 1-aryl-3-alkoxyprop-1-enes from a benzaldehyde, 2-hydroxytriphenylphosphonium chloride, potassium carbonate and an alcohol.

3-Nitrobenzopyrans are potentially valuable intermediates in the synthesis of other 3-substituted derivatives, and are formed in good yield from the reaction of *o*-hydroxybenzaldehydes with β-nitrostyrenes [3087]. The synthetic uses of nitroalkenes have been reviewed [3123]. An analogous method using acrolein or a vinyl ketone leads to a benzopyran-3-aldehyde or -3-ketone [2618]; on the other hand, α-cyanothioacetamide (reviews [3114, 3331]) gives 2-imino-1-benzopyran-3-thiocarboxamide [2926].

$Ph_3\overset{+}{P}(CH_2)_2OH$ Cl^- — PhCl, K_2CO_3, Δ; 15–50% [3970]

R = H, MeO; X = N or CH; Y = CH or N

X = Y = CH 2*H*-1-Benzopyran
X = CH; Y = N Pyrano[3,2-*b*]pyridine
X = N; Y = CH Pyrano[2,3-*c*]pyridine

+ $ArCH{=}CHNO_2$ — TEA; 51–82% [3087]

R = H, MeO, Cl, NO_2; Ar = Ph, $EtOC_6H_4$

2*H*-1-Benzopyran, 2-aryl-

(29.4) + $CH_2{=}CHC(O)R$ — aq. NaOH; 72–85% [2618]

R = H, Ac, Ph

2*H*-1-Benzopyran

Enaminones react with 2-hydroxybenzaldehydes on prolonged heating under Dean and Stark conditions to give chromenes. *Peri*-cyclization of aldehyde and hydroxy groups by reaction with ethyl chloroacetate in an alkaline medium produces a pyrano derivative in high yield. *Peri*-annulated naphthalenes and similar compounds are discussed in a review [3905].

+ $CHCOR^2{=}CHNR^3_2$ — PhMe, mor, Δ; 13–59% [3369]

R^1 = H, halogen, MeO, NO_2; R^2 = Me, Ph; NR^3 = morpholine

+ CH_2Cl–COOEt — DMF, K_2CO_3, Δ; 98% [3702]

[1]Benzopyrano[6,5,4-*def*][1]benzopyran

2. Pyran-2-one

An *o*-hydroxyaraldehyde reacts with a cumulative phosphorane or acetic anhydride–trimethylamine (TMA) to form a new pyranone ring [2777, 2413]. Heating salicylaldehyde with either cyanoacetamide or cyanothioacetamide (reviews [3114, 3331]) often gives high yields of the corresponding pyran-2-one or -2-thione [3907]. *N,N*-Dimethyl(dichlorophosphoryloxymethylene)ammonium chloride (DDA) phosphorylates and activates carboxylic acids to give compounds of the type $RCOOPOCl_2$; these, in the presence of TEA, cyclize *o*-acylphenols in high yield [3159]. DDA is prepared from DMF and phosphorus oxychloride.

PhH, Δ; 70% [2777]

Naphtho[2,1-*b*]pyran-3-one

Ac_2O, TMA, Δ; 31% [2413]

1-Benzopyran-2-one

(29.4) + $NH_2C(=X)CH_2CN$; EtOH, $MeCOONH_4$, Δ; ~84% [3907]

X = O, S

1-Benzopyran-2-one
1-Benzopyran-2-thione

+ CH_2R^2COOH + DDA; CH_2Cl_2, TEA, Δ; 72–98% [3159]

R^1 = H, Me; R^2 = $PhCH_2$, COOEt, Ph, PhO, $PhCOCH_2$

1-Benzopyran-2-one

2-Hydroxy-3-carbonylated pyridines are available from the *o*-metallation of 2-halopyridines followed by reaction with a nucleophile. These hydroxy or methoxy pyridines, under Perkin reaction conditions, are converted into pyranopyridines. Replacing the CHO group by a 2-methoxyaroyl group can lead to the synthesis of tricyclic compounds such as benzopyranopyridinones [3751].

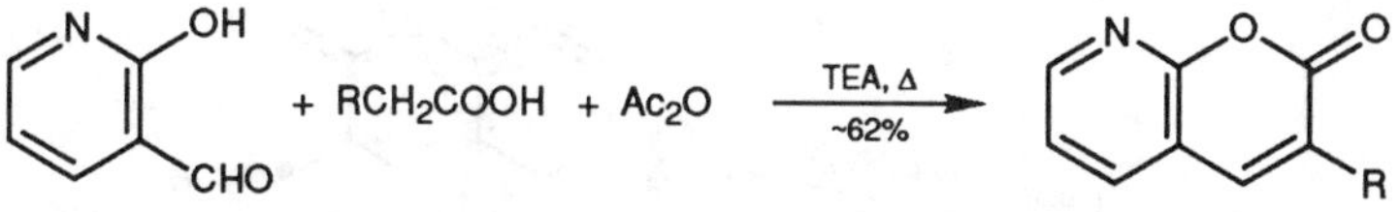

$+ RCH_2COOH + Ac_2O$ —TEA, Δ, ~62%→ [3751]

R = Ph, 4-FC_6H_4, 3,4,5-$(MeO)_3C_6H_2$, pyridin-4-yl

Pyrano[2,3-*b*]pyridin-2-one

Another ring system synthesized similarly:

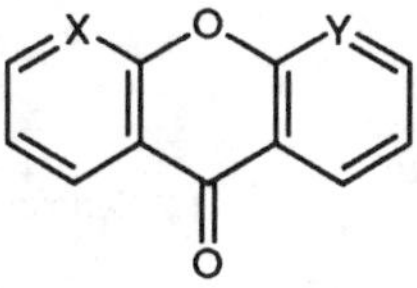

[3751]

X = CH; Y = N [1]Benzopyrano[2,3-*b*]pyridin-5-one
X = Y = N Pyrano[2,3-*b* : 6,5-*b*']dipyridin-5-one

Malonic acid derivatives have been widely used to synthesize fused pyran-2-ones from 2-hydroxyacetophenones [3337]. The uses of cyanoacetamide have been reviewed [3114]. An interesting alternative to these malonic acid-based reagents is diphenyl sulphoacetate (**29.5a**) which reacts in a mildly basic environment. Variations (**29.5b–d**) on its structure (see table) have been studied and some of these lead to sultones (**29.6**) instead of coumarins [2741].

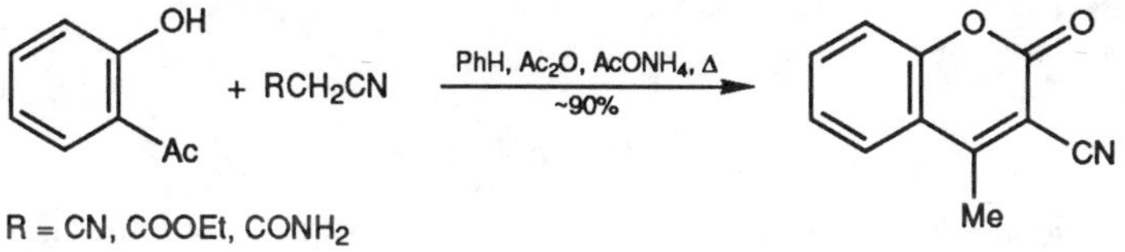

$+ RCH_2CN$ —PhH, Ac_2O, $AcONH_4$, Δ, ~90%→ [3337]

R = CN, COOEt, $CONH_2$

1-Benzopyran-2-one

XSO_2CH_2Y
(**29.5**)

(**29.5**)		X	Y
	a	COOPh	PhO
	b	COOPh	PhNH
	c	CONHPh	PhNH
	d	COPh	PhO

Another ring system synthesized similarly:

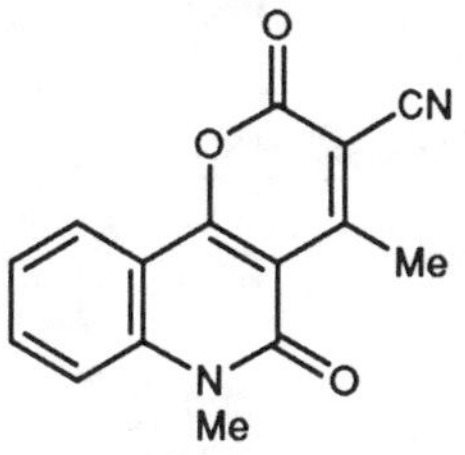

[3337]

Pyrano[3,2-*c*]quinoline-2,5-dione

(29.4) —(29.5a), K_2CO_3, Me_2CO, 77%→ 1-Benzopyran-2-one (SO_2OH)

(29.4) —(29.5d), pip, Δ, 65%→ (29.6) 1,2-Benoxathiazine 2,2-dioxide

[2741]

Meldrum's acid (review [3949]) and DMF supply three carbon atoms in the two-stage low-temperature conversion of a 2-methoxybenzaldehyde into a coumarin-3-carboxylic acid.

R = H, Br, MeO —1. DMF 2. H_2SO_4, 55–80%→ COOH

[3455]

3. Pyran-4-one

The phenolic aldehydes (**29.7**) or their acetals are cyclized by mineral acid; this leads to 2-unsubstituted chromones (R^2 = H) or isoflavones (R^2 = Ar) [2181, 2147]. The corresponding phenolic aryl ketones similarly give flavones in high yield [3120]; these substrates may also be cyclized under weakly basic conditions [2138, 2325]. DBU (review [3364]) may be used instead of potassium hydroxide [2340] in the Baker–Venkataraman synthesis of chromones (see p. 531) but yields may be low [3915] as compared with the cyclization of 2-aroyloxyacetophenones to give flavones [3660]. Suitably placed lactam carbonyl and hydroxy groups may react by heating with methanolic hydrogen chloride [2117].

(29.7) —aq. H_2SO_4 or HCl, Δ, 60–71%→ 1-Benzopyran-4-one

(29.7)
R^1 = H, MeO; R^2 = H, 4-$MeOC_6H_4$

[2147, 2181]

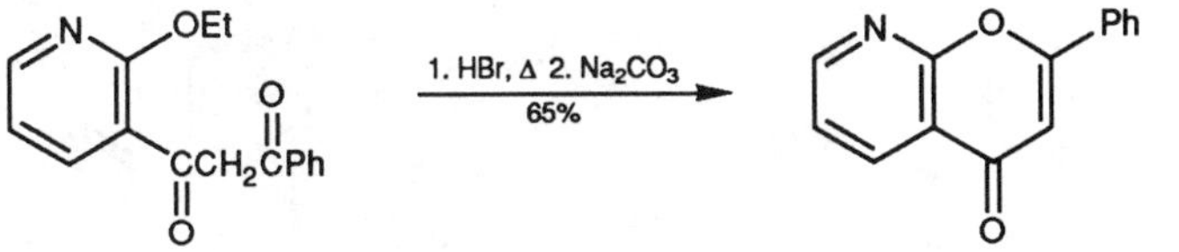

1. HBr, Δ 2. Na_2CO_3
65%

[2138]

Pyrano[2,3-*b*]pyridin-4-one

Another ring system synthesized similarly:

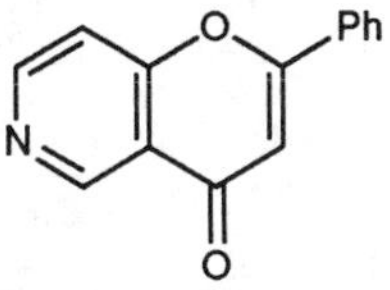

[2325]

Pyrano[3,2-*c*]pyridin-4-one

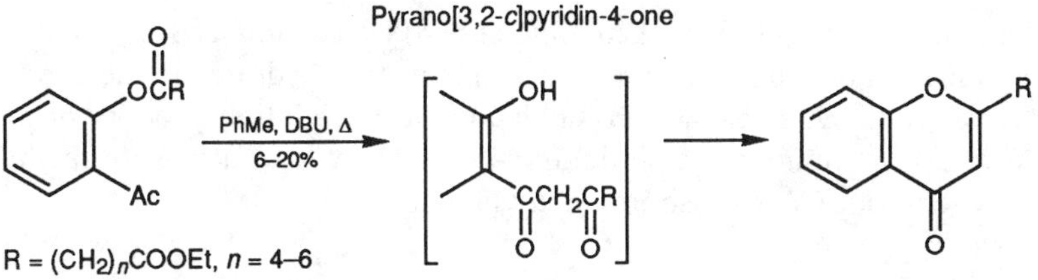

PhMe, DBU, Δ
6–20%

[3915]

R = $(CH_2)_nCOOEt$, n = 4–6

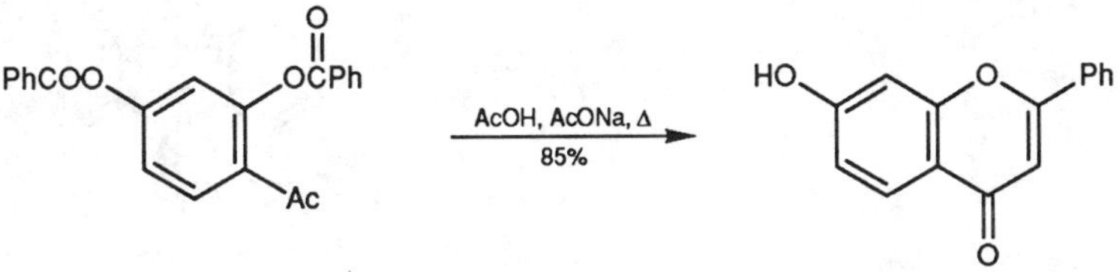

AcOH, AcONa, Δ
85%

[3660]

1-Benzopyran-4-one, 2-aryl-

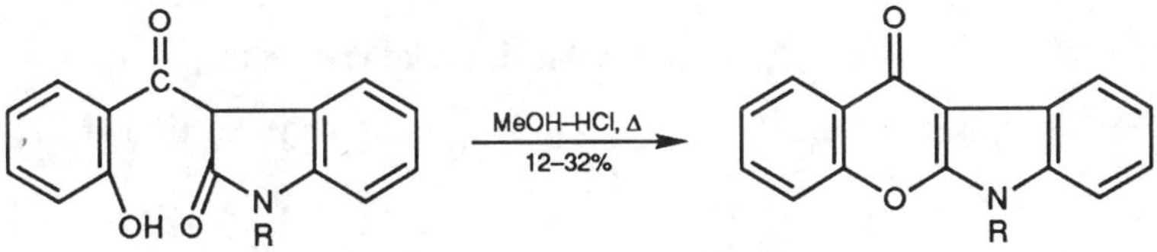

MeOH–HCl, Δ
12–32%

[2117]

R = H, Me

[1]Benzopyrano[2,3-*b*]indol-11-one

4. 1,3-Oxazin-2-one

Chlorosulphonyl isocyanate (reviews [324, 3147a, 3340] cyclizes 2-hydroxyaldehydes and -acetophenones in good yields; the corresponding benzophenones are not cyclized in this way. Methyl isocyanate in alkaline solution also reacts with salicylaldehyde to give a high yield of a fused oxazinone.

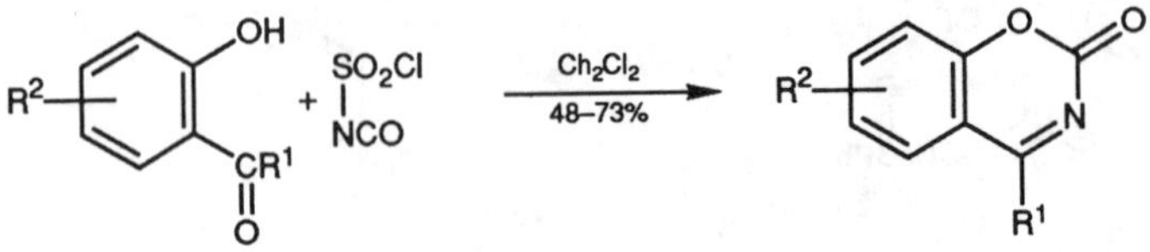

R^1 = H, Me; R^2 = H, Cl, Me

CH_2Cl_2, 48–73%

1,3-Benzoxazin-2-one [3018]

(**29.4**) + MeNCO → (EtOH, KOH, ~95%) [2095]

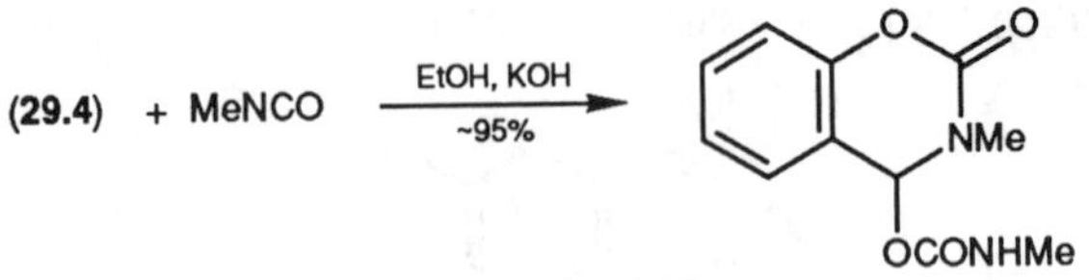

5. 1,4-Oxazin-3-one

Catalytic debenzylation of the ketoamide (**29.8**) is accompanied by cyclization and, if desired, the hydroxylactam produced may be dehydrated to the 2-methylene derivative (**29.9**, R^1R^2 = CH_2) (which is a degradation product of auromomycin, an antitumour antibiotic) in 94% yield by treatment with methanesulphonyl chloride–*N*,*N*-di-isopropylethylamine.

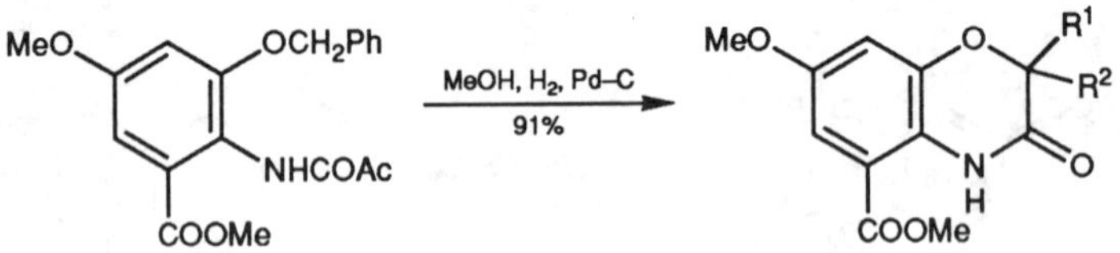

MeOH, H_2, Pd–C, 91% [3187]

(**29.8**)
R^1 = Me; R^2 = OH or R^1R^2 = CH_2

(**29.9**)
1,4-Benzoxazin-3-one

6. 1,2-Oxathiazine 2,2-Dioxide

A synthesis of this fused oxathiazine 2,2-dioxide is shown in Section II.2 [2741].

CHAPTER 30

Aldehyde and Ketone; Dialdehyde or Diketone

I. Formation of a Five-membered Ring 827
1. Pyrrole 827
2. Pyrrol-2-one 828
3. Pyrazole 829
4. Imidazole 829
5. Furan 830
6. Thiophene 830
7. Isoxazole 831
8. Furazan or Furazan 2-Oxide 831
9. 1,2,5-Thiadiazole or 1,2,5-Thiadiazole *N*- or *S*-Oxide 833
II. Formation of a Six-membered Ring 834
1. Pyridine 834
2. Pyridazine 834
3. Pyrimidine 835
4. Pyrazine 835
5. Pyrazin-2-one or Pyrazin-2-one 4-Oxide 836
6. 1,2,4-Triazine 836
7. Pyran-2-one 837
8. 1,3-Dioxin-2-one 837
9. 1,2,6-Thiadiazine 1,1-Dioxide 837
III. Formation of a Seven-membered Ring 838
1. 1,4-Diazepine 838

Examples of the cyclization of dioximes are included in this chapter.

I. FORMATION OF A FIVE-MEMBERED RING

1. Pyrrole

A 1,4-diketone [such as (**30.1**)] cyclizes to a pyrrole on being heated with ammonium acetate–acetic acid. Several possible mechanisms have been proposed for the rearrangement which occurs durng this reaction. When the triketone (**30.1**) is dissolved in liquid ammonia, a 3*H*-indoline is formed and may be thermally converted to the 5-indolizinone [3392]. Indoles in which the benzene ring is reduced may be synthesized from 2-(2-oxoalkyl)cyclohexanone and an amine [2130]. Diels–Alder reaction of a butadiene and 1,2-dibenzoylacetylene gives the diketone (**30.2**) which is a useful intermediate for the formation of fused five-membered rings. Ammonia or a primary amine yields isoindoles [2281].

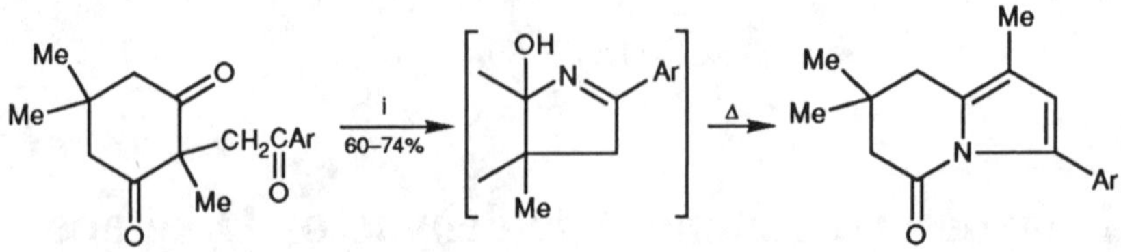

(**30.1**) → 5-Indolizinone [3392]

Ar = Ph, 4-Br-, 4-Me-C_6H_4; i, AcOH, $AcONH_4$ (i, 60–74%; Δ)

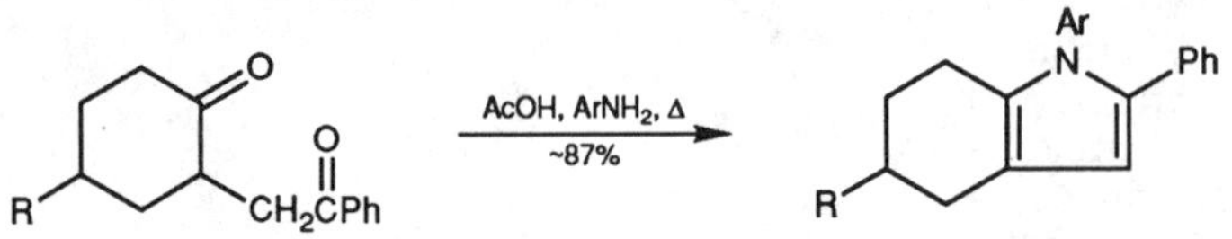

Indole [2130]

AcOH, $ArNH_2$, Δ, ~87%

R = H, MeO; Ar = 4-HOC_6H_4

(**30.2**) + R^2NH_2 → Isoindole [2281]

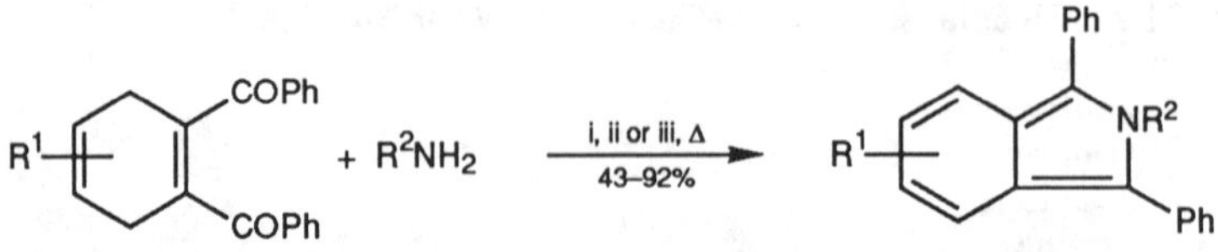

i, ii or iii, Δ, 43–92%

R^1 = H, Me, Ph; R^2 = H, Me, Ph

i, $AcONH_4$, EtOH; ii, aq. $MeNH_2$; iii, AcOH, $PhNH_2$

2. Pyrrol-2-one

Phthalaldehyde reacts with an isocyanate, an alkylamine or a carbodi-imide to give good-to-excellent yields of an isoindolone.

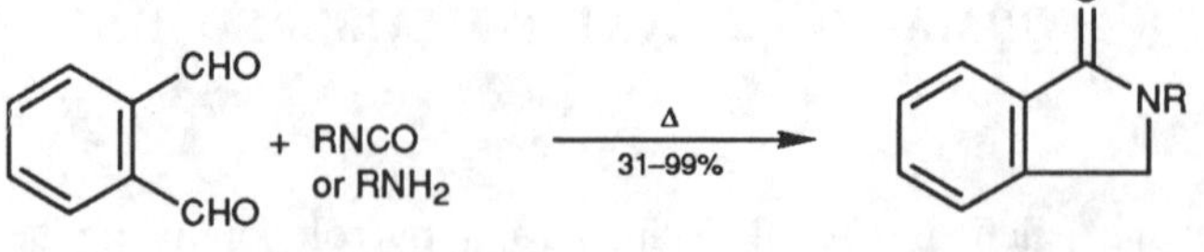

+ RNCO or RNH_2 → 1-Isoindolone [2360, 3121]

Δ, 31–99%

R = Ph, Me-, Cl-C_6H_4, $HOCH_2CMe_2$

3. Pyrazole

The product obtained by heating 2-acylindanediones for 24 h with hydrazine depends on the ratio of hydrazine to triketone. A molar ratio gives the oxopyrazole, a 2.5 molar excess leads to the hydrazone while a larger excess surprisingly gives the methylene-pyrazole—a reaction which is reminiscent of a Wolff–Kishner reaction—but this is not observed for all indandiones. In a correction of earlier work, it has been shown that it is the side-chain hydrazone (**30.3**) (as opposed to the 4-hydrazono isomer) which cyclizes under Dean and Stark conditions to give a furo- or pyrrolo-pyrazolone.

1 : 1 N_2H_4, Δ, 16–88% → Indeno[1,2-*c*]pyrazol-4-one

1 : 2.5 N_2H_4, Δ, 64–91% → Indeno[1,2-*c*]pyrazole

1 : >2.5 N_2H_4, Δ, 45–82%

R = cycloalkyl, 2- or 3-thienyl, 2-pyrrolyl

[2329]

xylene, TsOH, Δ, 29–50% → Furo[3,4-*c*]pyrazol-4-one, Pyrrolo[3,4-*c*]pyrazol-4-one

(**30.3**)
X = O, NH; R^1 = H, Me; R^2 = Me, Ph

[3020a]

4. Imidazole

The basis for a test for arginine and peptides containing it is a cyclization of a diketone such as 1,2-phenanthraquinone by the guanidine group. Similar fluorescent products are obtained from hypoglycaemic biguanides such as methoformin (metformin), butylbiguanide (buformin) and phenformin (2-phenylethylbiguanide).

+ $H_2NC(=NH)NHC(=NH)NR^1R^2$ —EtOH, NaOH, 27–62%→ [2928a]

R^1 = H, Me; R^2 = Me, Bu, $PhCH_2CH_2$

Phenanthro[9,10-*d*]imidazole

5. Furan

Borohydride reduction of a diketone followed by dehydrative thermolysis is a good method of producing an isobenzofuran free of the diketone [2391]. Acid-catalysed dehydrative cyclization of the diketone (**30.2**) is an alternative route [2281] to isobenzofurans (review [3508]. An *o*-acylbenzaldehyde may be cyclized by reaction with potassium cyanide–acetic acid [2100].

COPh, COPh —1. MeOH, NaOH, Δ 2. Ac_2O, Δ, 70%→ Ph, Ph [2391]

Isobenzofuran

(**30.2**) —PhH, TsOH, Δ, ~66%→ R^1, Ph, Ph [2281]

R^1 = H, Me

Isobenzofuran

CHO, COPh —AcOH, KCN, 75%→ CN, Ph [2100]

6. Thiophene

The diketone (**30.2**) may be converted into a benzothiophene by heating with either sulphur or phosphorus pentasulphide in tetralin. A similar method but one which uses milder conditions is a recently published synthesis using acetonitrile–sodium hydrogencarbonate. This reaction is run at 30 °C for 4 h, and with a wide variety of substituents at C-1, C-3, C-5 and C-6, yields of 50–95% are obtainable [3722].

(30.2) —S or P_2S_5–tetralin, Δ; 61–78%→ Benzo[*c*]thiophene [2281]

R = H, Me

7. Isoxazole

When 2-oxocyclohexanecarboxaldehyde reacts with hydroxylamine, a mixture of the 1,2- and the 2,1-isoxazole is formed; these can be difficult to separate, but when the ketal-carboxaldehyde is used, the product is the 1,2-isoxazole. A similar method was effective on 2-acetylcyclohexanone but some of the 2,1-isomer was formed. A high yield of a *peri*-fused oxazole is obtained by treating the acridinone (**30.4**) with PPA at 80 °C for 4 h. The product is a derivative of an antimalarial drug.

+ NH_2OH —aq. EtOH, K_2CO_3, Δ; ~72%→ 1,2-Benzisoxazole [2821]

(30.4) —PPA, Δ; ~84%→ Isoxazolo[3,4,5-*kl*]acridine [3923]

R = HO, MeO; Ar = 4-Cl-, 4-F-, 4-CF_3-C_6H_4

8. Furazan or Furazan 2-Oxide

The diketone dioxime (**30.5**) undergoes an oxidative ring closure to the 2-oxide, but warm aqueous alkali converts it into the oxadiazole [2202]. Heating a di-amidoxime with ethyl chloroformate for a short time converts it into a furazan ring in low yield. Comparison of the relative reactivities of the NH and two OH groups of the benzoxazine (**30.6**) towards electrophilic reagents shows that two products are produced under the experimental conditions used. The product (**30.7**, R^3 = H) is converted into the carbamate (**30.7**, R^3 = COOEt) by further heating with the electrophile [3922].

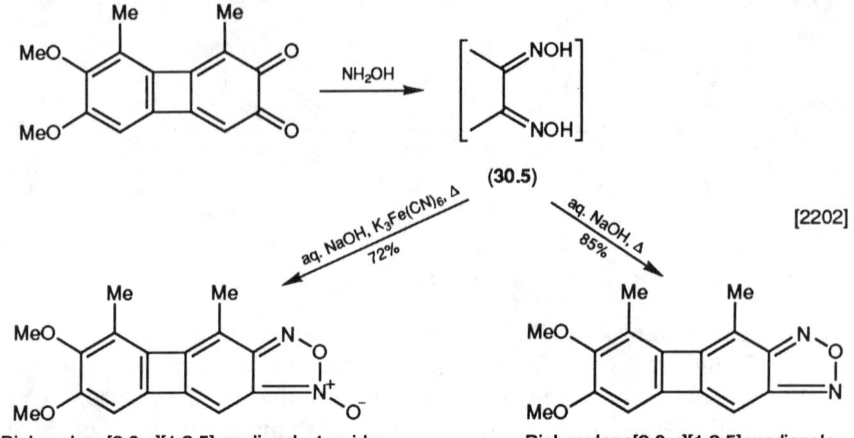

NH_2OH

(30.5)

aq. NaOH, $K_3Fe(CN)_6$, Δ 72%

aq. NaOH, Δ 85%

[2202]

Biphenyleno[2,3-*c*][1,2,5]oxadiazole 1-oxide

Biphenyleno[2,3-*c*][1,2,5]oxadiazole

PhH, ClCOOEt, Δ ~23%

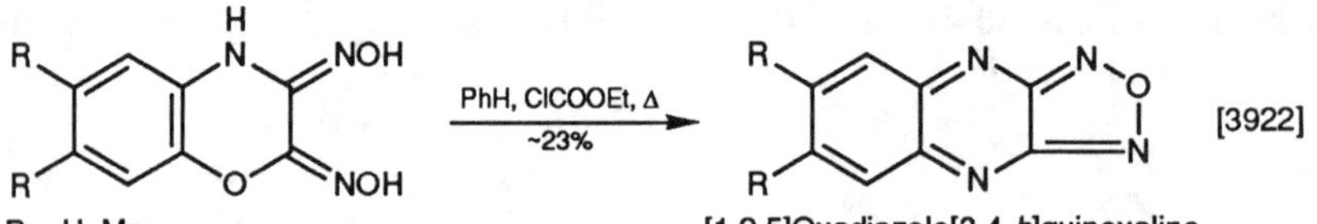

R = H, Me

[1,2,5]Oxadiazolo[3,4-*b*]quinoxaline

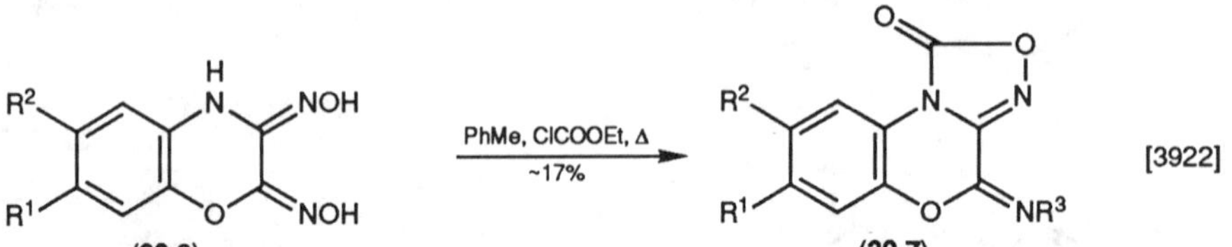

[3922]

PhMe, ClCOOEt, Δ ~17%

(30.6)

R^1 = H, Me; R^2 = H, Me, Cl;

R^3 = HO, COOEt

(30.7)

[1,2,4]Oxadiazolo[3,4-*c*][1,4]benzoxazin-1-one

[3922]

The dioxime of a cyclic 1,2-diketone (prepared in this example by successive nitrosation and oximation of the 1,3-cyclohexanedione) is oxidized to a furoxan (1,2,5-oxadiazole 2-oxide); the isomeric *N*-oxide is formed by mild treatment of the dioxime with thionyl chloride [2496].

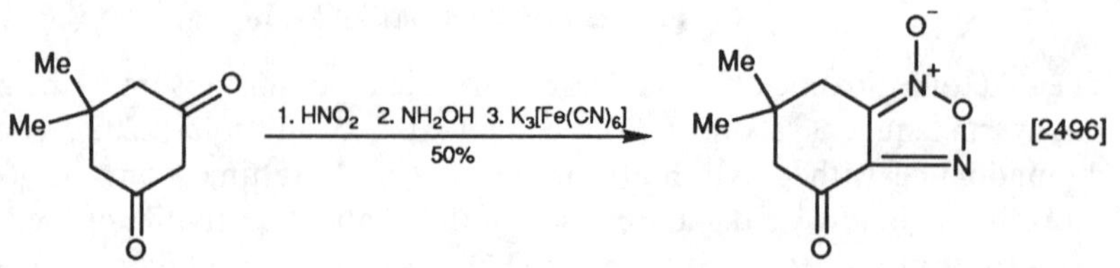

1. HNO_2 2. NH_2OH 3. $K_3[Fe(CN)_6]$ 50%

[2496]

4-Benzofurazanone 1-oxide

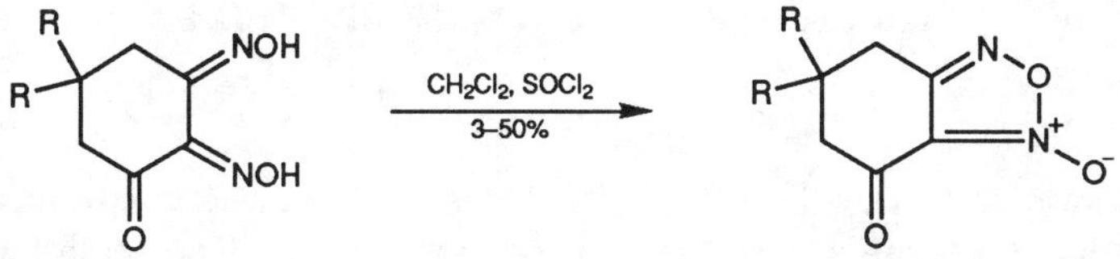

R = H, Me

CH_2Cl_2, $SOCl_2$ 3–50%

2,1,3-Benzodiazol-4-one 3-oxide

[2496]

9. 1,2,5-Thiadiazole or 1,2,5-Thiadiazole *N*- or *S*-Oxide

Treatment of *o*-benzoquinone dioxime with sulphur dichloride gives a mixture of 2,1,3-benzothiadiazole (38%) and its 1-*N*-oxide (34%). The presence of three chlorine atoms in the benzene ring reduces the yield [2177]. 9,10-Phenanthrenequinone dioxime gives a low yield of mainly the thiadiazole; the *S*-oxide of this may be prepared from the quinone di(trimethylsilylimine) and thionyl chloride.

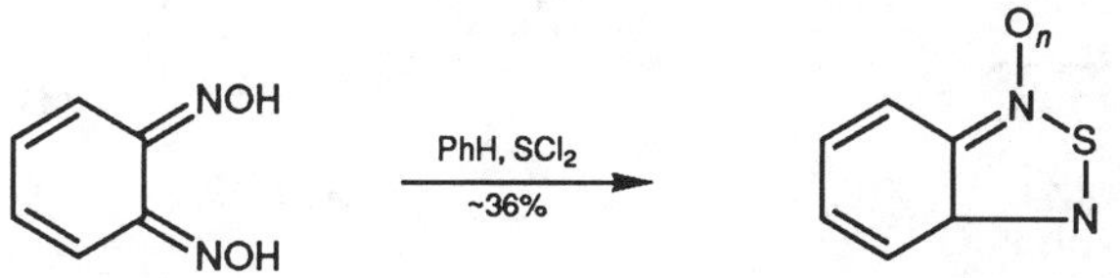

n = 0 or 1

PhH, SCl_2 ~36%

2,1,3-Benzothiadiazole (*N*-oxide)

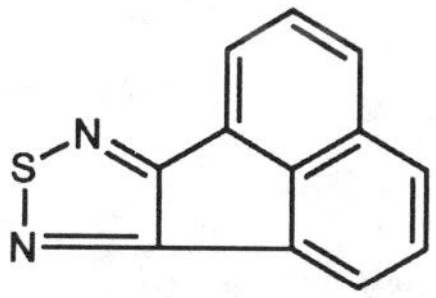

[2177]

Acenaphtho[1,2-*c*][1,2,5]thiadiazole

[2177]

Another ring system synthesized similarly:

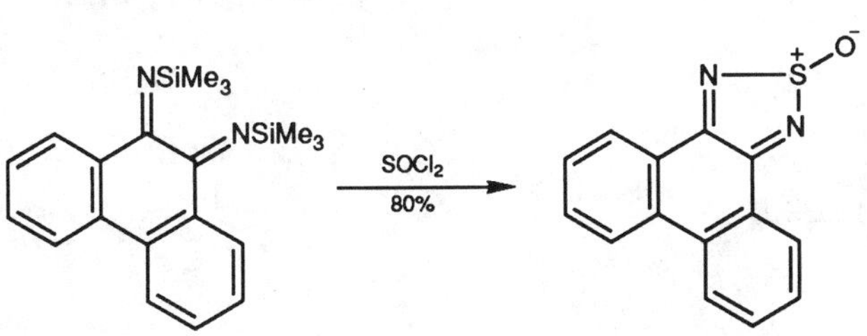

$SOCl_2$ 80%

Phenanthro[9,10-*c*][1,2,5]thiadiazole 2-oxide

[2035, 2177]

II. FORMATION OF A SIX-MEMBERED RING

1. Pyridine

Cyanoacetamide (review [3114]) reacts under weakly basic conditions with a 1,3-diketone to form a fused pyridine-4-carbonitrile. Ketene dithioacetals (review of their reactions [3712, 3830]) are versatile synthons; in this example, the dithioacetal is part of the reactant (**30.8**) which is stirred at −78 °C.

$(H_2C)_n$ cyclic 2-acetylketone + CH_2CN / $CONH_2$ → (EtOH, Et_2NH; 50–85%) → fused pyridine with Me, N, OH, CN [2066]

n = 5-12

(30.8) n = 2: $(H_2C)_n$ cyclic ketone with $=C(SMe)_2$ → (THF, $LiCH_2CN$; 76%) → SMe, N, SMe [3450]

Quinoline

2. Pyridazine

When a 4,5-diacylpyridazine is heated with a sixfold excess of semicarbazide hydrochloride in methanol, the yellow dipyridazine is obtained in high yield. 1,4-Diketones are cyclized by heating briefly with hydrazine hydrate.

4,5-di(COR)pyridazine + $NHNH_2$ / $CONH_2$ → (MeOH, Δ; ~87%) → R, R [2925]

R = Ph, 4-$MeOC_6H_4$, 3,4-$(MeO)_2C_6H_3$

Pyridazino[4,5-*d*]pyridazine

3,4-diacetyl-2-(CH_2OH)-5-methylfuran (Ac, Ac, CH_2OH, Me) + $N_2H_4.H_2O$ → (EtOH, Δ; 56%) → Me, CH_2OH, Me, Me [2158]

Furo[3,4-*d*]pyridazine

Another ring system synthesized similarly:

[2470]

[1]Benzothieno[2,3-*d*]pyridazine

3. Pyrimidine

1,3-Diketones react with 5-aminotetrazole to form a fused pyrimidine ring when the group R in (**30.9**) is not too bulky; when R is a t-butyl or phenyl, the product is the isomeric azide.

EtOH, AcOH, Δ
51–63%
[3726]

(**30.9**)
R = Me, Et, iBu

Indeno[1,2-*d*]tetrazolo[1,5-*a*]pyrimidin-9-one

4. Pyrazine

Although the *o*-quinone (**30.10**) condenses with a 1,2-diaminobenzene (**30.11**) as expected, some *o*-quinones do not always react at these functional groups; for example, in methanol, the two otherwise unsubstituted reactants (with an excess of quinone) at ambient temperature give, on treatment with acetic anhydride–pyridine, the phenazine diacetate (**30.12**), but in diethyl ether, the unsubstituted phenazine is formed [2305]. Heating a 1,5-diketone with ammonium acetate leads to the formation of a fused pyrazine in high yield [2819].

AcOH, Δ
63%
[2091]

(**30.10**) (**30.11**) Benz[*a*]imidazo[4,5-*c*]phenazin-2-one

(30.11) + [scheme] → 1. MeOH 2. Ac_2O, ~75 → (30.12) Phenazine [2305]

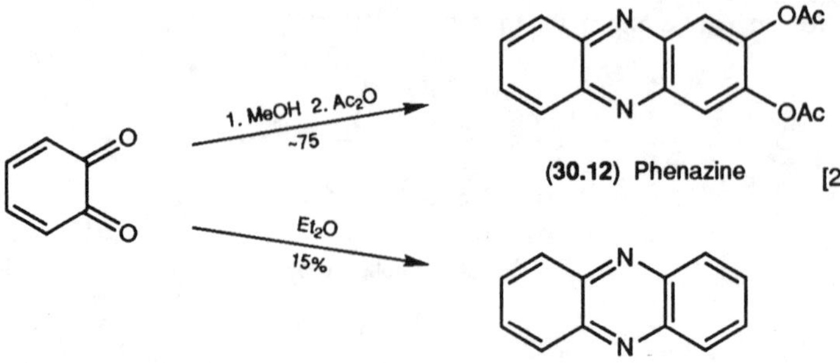

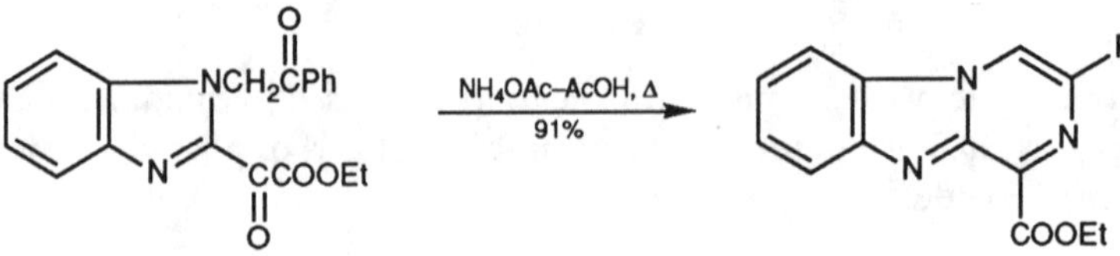

Et_2O, 15%

NH_4OAc–AcOH, Δ, 91% → Pyrazino[1,2-*a*]benzimidazole [2819]

5. Pyrazin-2-one or Pyrazin-2-one 4-Oxide

o-Benzoquinone dioxime condenses with 2-oxoaldehydes to give good yields of the 2-quinoxalinone 4-oxides. The corresponding quinone di(phenylsulphonyl-imine) is cyclized by heating with diphenylketene.

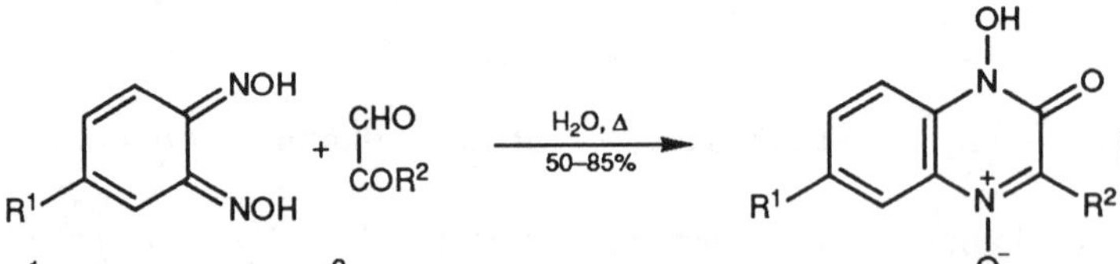

H_2O, Δ, 50–85% [2229, 2737]

R^1 = H, Me, COOMe; R^2 = H, Me

2-Quinoxalinone 4-oxide

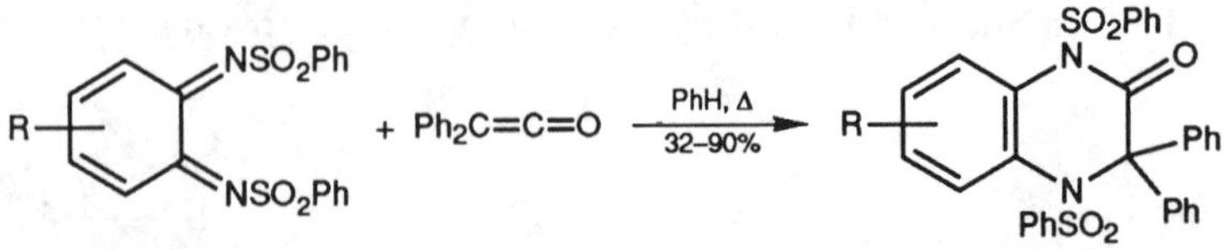

+ $Ph_2C{=}C{=}O$ → PhH, Δ, 32–90% [2691]

R = Me, Cl

2-Quinoxalinone

6. 1,2,4-Triazine

This ring is formed when 1,2-diones are heated with the hydrazones of carbox-amides.

ON– + H_2NN=$RCNH_2$ —MeOH, Δ, 28–75%→ HON= [3046]

R = H, Me, COOMe, C_6H_{11}, Ph, cC_3H_5

Cyclohepta[*e*]-1,2,4-triazine

7. Pyran-2-one

Stirring the dithioacetals (**30.8**) with t-butyl lithioacetate and LDA at −78 °C and then with t-butyl acetate gives a high yield of the pyran-2-one.

(**30.8**) + $LiCH_2COOtBu$ —i, ii, ~89%→ [3190a]

i, THF, LDA; ii, MeCOOtBu

SMe, $(H_2C)_n$

n = 1 Cyclopenta[*d*]pyran-3-one
n = 2 2-Benzopyran-3-one

8. 1,3-Dioxin-2-one

A ring of this type is formed when 1-acylindan-2-one is stirred with either phosgene or phenyl chloroformate; the corresponding 2-thione is obtained when *O*-phenyl chlorothioformate is used.

COR —i, ii or iii, 20–51%→ X [2692]

R = Me, $PhCH_2$, Ph
i, $COCL_2$, PhH, $EtNiPr_2$; ii, ClCOOPh, $EtNiPr_2$;
iii, ClCSOPh, $EtNiPr_2$

X = O Indeno[2,1-*d*]-1,3-dioxin-2-one
X = S Indeno[2,1-*d*]-1,3-dioxin-2-thione

9. 1,2,6-Thiadiazine 1,1-Dioxide

Sulphamide annulates adjacent aldehyde and ring ketone groups in the presence of a strong base.

CHO + NH_2–SO_2NH_2 —MeOH–MeONa, Δ, 85%→ [2868]

[1]Benzopyrano[4,3-*c*][1,2,6]thiadiazine 2,2-dioxide

III. FORMATION OF A SEVEN-MEMBERED RING

1. 1,4-Diazepine

Prolonged boiling of ethylenediamine with some 1,3-diketones (usually in the presence of a small amount of formic acid) leads to a diazepine, but other diketones fail to produce cyclized products. The substituent (R) in the diketone seems to have a role in the selectivity of this reaction.

COR + CH_2NH_2–CH_2NH_2 → (EtOH, HCOOH, Δ; 60–75%) [2175]

R = Et, Ph_2CH, Ph

Indeno[1,2-*e*]-1,4-diazepin-6-one

CHAPTER 31

Aldehyde or Ketone and Methylene

I. Formation of a Five-membered Ring 839
1. Pyrrole 839
2. Furan 840
3. Thiophene 841
II. Formation of a Six-membered Ring 842
1. Pyridine 842
2. Pyridin-2- or -3-one 842
3. Pyran 843
4. 1,3-Thiazine 1,1-Dioxide 843
III. Formation of a Seven-membered Ring 843
1. Azepine or Azepin-2-one 843

I. FORMATION OF A FIVE-MEMBERED RING

1. Pyrrole

Syntheses of indolizines carrying substituents in the pyridine ring are not numerous but substituted α-picolines quaternized (review [1870]) with an α-halo ketone (review [B-45]) are convenient substrates for cyclization to indolizines or related compounds; the reaction is mediated by ammonia on its own at 5 °C or with hydrogencarbonate, TEA or methanolic sodium hydroxide. 2-Alkylpyridines may be *N*-alkylated and cyclized in a one-pot reaction (see Chapter 104, Section I.1).

i, aq. NH_3, Δ; 31–93% [2814, 2946]

Indolizine

R^1 = Me, tBu, Ph, 4-MeO-, 4-NO_2-, 4-Me-C_6H_4;
R^2 = H, COOEt
i, $NaHCO_3$ or TEA–EtOH

Another ring system synthesized similarly:

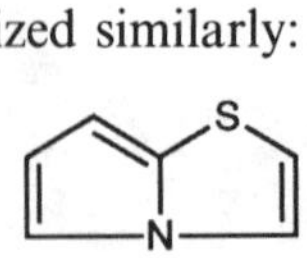

Pyrrolo[2,1,-*b*]thiazole [3106, 3172]

Indenetriones (**31.1**) react in an unexpected fashion when stirred at 50 °C with tosyl chloride in anhydrous pyridine. A possible mechanism through tosylation of the hydroxy group and then formation of a bond between pyridine-N and C-1 of indanone is suggested. Pyridines may be replaced by isoquinoline or phthalazine in this cyclization [3071, 3785]. A 2-phenacylidenepyridine may be cyclized to an indolizine by heating with an anhydride [2458].

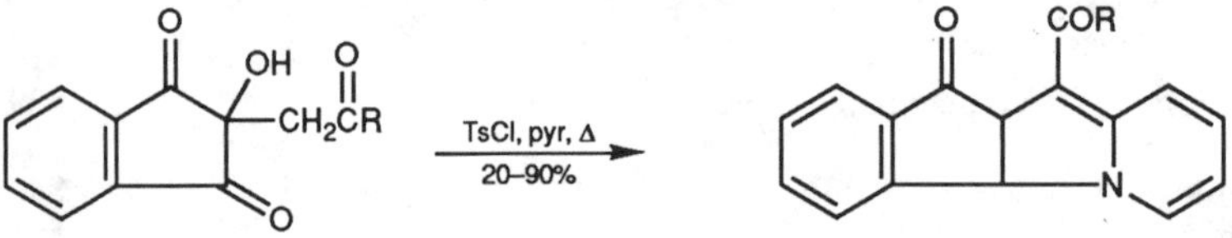

TsCl, pyr, Δ; 20–90% [3071, 3785]

(**31.1**)
R = Ph, 3-NO_2, 4-NO_2, 4-NH_2, 4-$PhCH_2OOCNH$, iBu

Indeno[2,1-*b*]indolizin-10-one

Another ring system synthesized similarly:

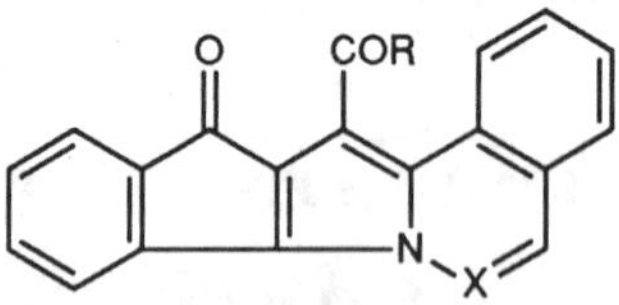

[3785]

X = CH Indeno[2',1' : 4,5]pyrrolo[2,1-*a*]isoquinolin-12-one
X = N Indeno[2',1' : 4,5]pyrrolo[2,1-*a*]phthalazin-12-one

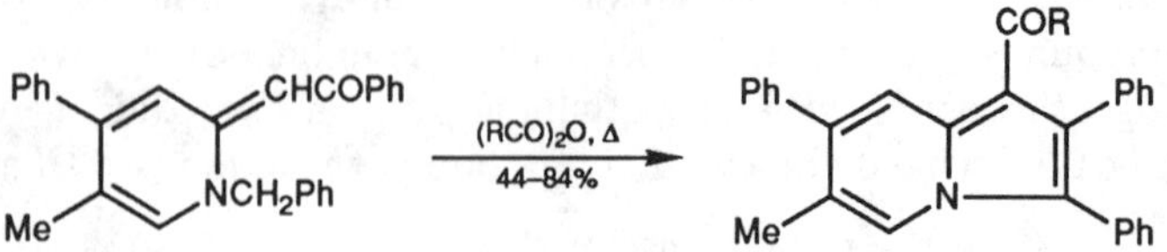

$(RCO)_2O$, Δ; 44–84% [2458]

R = Me, Et, Ph

Indolizine

2. Furan

Base-catalysed cyclization and simultaneous decarboxylation of an *o*-acylphenoxyacetic acid leads to the formation of a furan ring [2499] but the corresponding phenoxyacetonitrile gives a 2-benzofurancarbonitrile [3899]. A recent study of the former reaction shows that in the presence of TEA (without any acetic acid present) decarboxylation does not occur, but when acetic acid is present, a mixture of products (**31.2**, R = H and R = COOH) is obtained. The key intermediate is shown to be a 3-hydroxy-3-methyl-2,3-dihydrobenzofuran-2-carboxylic acid; its behaviour in the presence of various acids has been examined [3216]. A furan ring may be formed in the *peri* positions of a bicyclic ring system by this method [3921].

Ac_2O, AcONa, Δ; 75% [2499]

Furo[3,2-*g*][1]benzopyran-5-one

EtOH, KOH, Δ; 73% [3899]

Ar = 2-$NH_2C_6H_4$

Benzofuran

TEA, Ac_2O, Δ; 89%

(**31.2**, R = COOH) [3216]

AcOH, AcONa, Ac_2O

(**31.2**, R = COOH) + (**31.2**, R = H)
9% 85%

Ac_2O, AcONa, Δ; 38% [3921]

Furo[4,3,2-*de*][1]benzopyran

3. Thiophene

An anion derived from a side-chain methylene (SCH_2COOR) attacks a neighbouring aldehyde (p. 155) or an acyl group, the latter course leading to a 3-substituted thiophene ring.

EtOH, EtONa, Δ
~97%

[2349]

R = H, Me

Thieno[2,3-*b*]thiophene

II. FORMATION OF A SIX-MEMBERED RING

1. Pyridine

When 3-acyl-2-methylchromones are heated with ammonium acetate in acetic acid, a pyridine ring is formed, probably through an intermediate enamine, the 2-CH=CHNH$_2$ derivative. Some *o*-methyl aldehydes are cyclized in good-to-excellent yields under weakly basic conditions by stirring with a primary alkylamine.

AcOH, AcONH$_4$, Δ
~93%

[3400]

R = H, Me, Ph

[1]Benzopyrano[3,2-*c*]pyridin-10-one

+ R^{1}NH$_2$

THF
56–96%

[3466]

R^1 = Bu, PhCH$_2$, CH$_2$=CHCH$_2$, EtOOCCH$_2$, CNCH$_2$, 3- or 4-pyridylmethyl; R^2 = 2,6-dioxo-1,3,5-Me$_3$-pyrimidin-5-yl

Pyrido[3,4-*d*]pyrimidine-2,4-dione

2. Pyridin-2- or -3-one

A very efficient cyclization of *o*-acyl-*ω*-substituted acetanilides is effected by heating them with a base; the *ω*-substituent should be electron-withdrawing. Suitably positioned acetyl and benzoyl functions undergo an aldol condensation; alumina as a base gave better results than potassium carbonate.

i or ii, Δ
~95%

[2075, 3960]

R^1 = H, Me; R^2 = N$_3$, COOEt; R^3 = iPr, Ph
i, EtOH, aq. NaOH; ii, EtOH, EtONa

2-Quinolinone

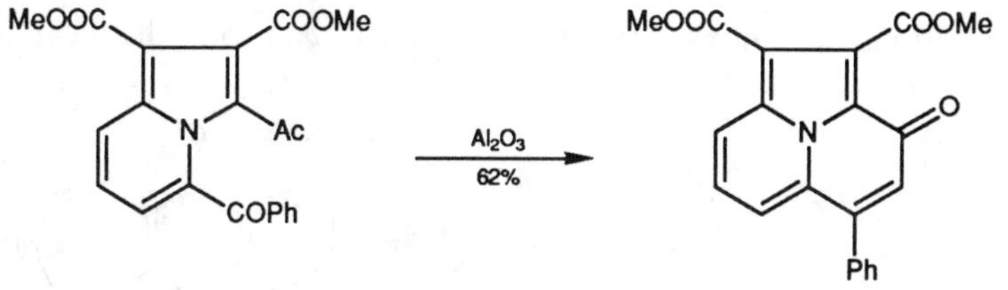

Al_2O_3, 62% [3330]

Pyrrolo[2,1,5-*de*]quinolizin-3-one

3. Pyran

Treatment of the ketone (**31.3**) successively with bromine in the dark and TEA without external heating leads to the formation of the naphthopyran by bromination and dehydrobromination.

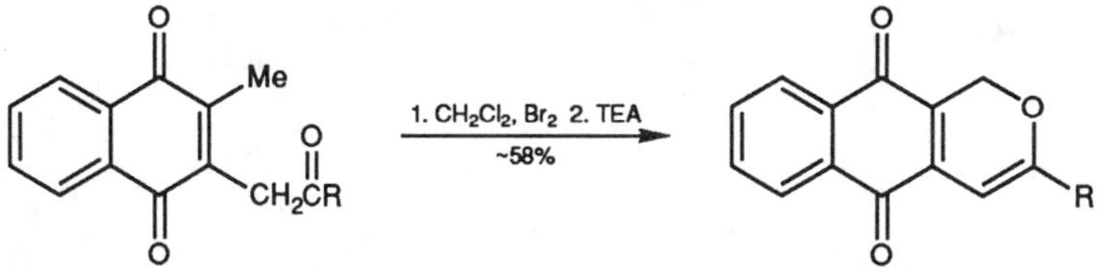

1. CH_2Cl_2, Br_2 2. TEA, ~58% [3750]

(**31.3**)
R = 2-furyl, 2-thienyl

Naphtho[2,3-*c*]pyran-5,10-dione

4. 1,3-Thiazine 1,1-Dioxide

Reduction of a 1,6-diketone with trimethyl phosphite followed by cyclization in the boiling solvent is a method which has been used by several workers to synthesize cephalosporin-type antibiotics. In this example, the diketone is prepared *in situ* by ozonolysis on an alkene side-chain.

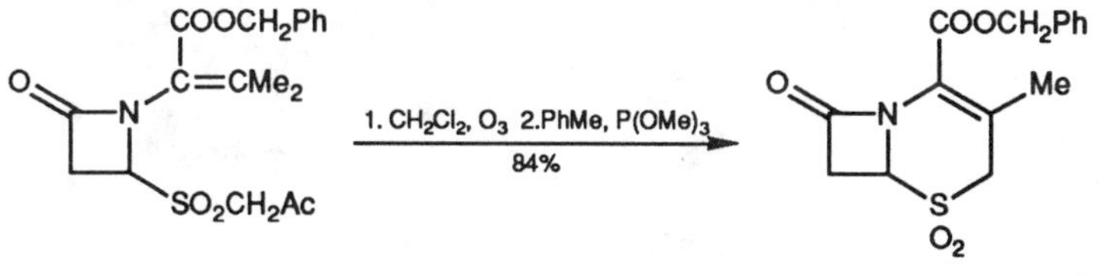

1. CH_2Cl_2, O_3 2. PhMe, $P(OMe)_3$, 84% [3407]

5-Thia-1-azabicyclo[4.2.0]oct-2-en-8-one 5,5-dioxide

III. FORMATION OF A SEVEN-MEMBERED RING

1. Azepine or Azepin-2-one

A carbanion from a side-chain methylene reacts with a neighbouring carbonyl (but not carboxaldehyde) group to form an azepine ring. Under the strongly basic conditions, an ethoxycarbonyl group is lost. Aqueous alkali can also be used in this type of reaction.

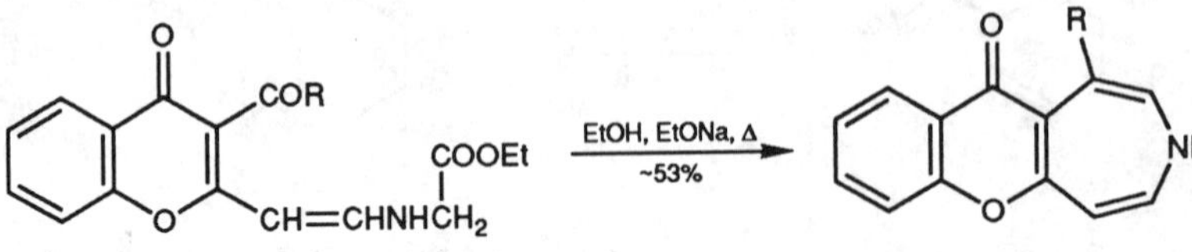

O

COR

COOEt

CH=CHNHCH$_2$

EtOH, EtONa, Δ

~53%

O R

NH [3400]

R = Me, Ph

[1]Benzopyrano[2,3-*d*]azepin-11-one

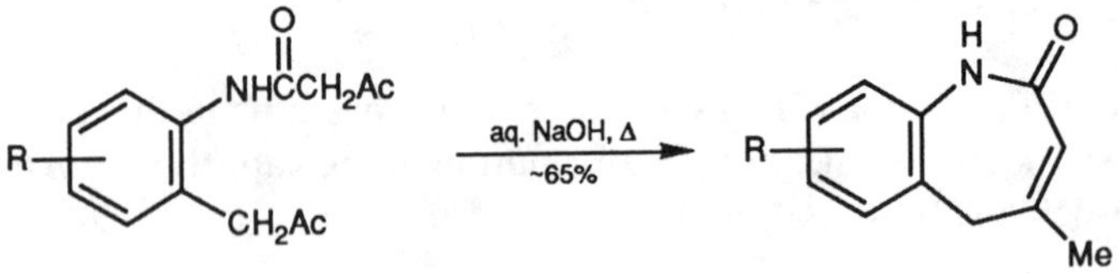

O

NHCCH$_2$Ac

R

CH$_2$Ac

aq. NaOH, Δ

~65%

H O

N

R

Me

[2788]

R = H, Me

1-Benzazepin-2-one

CHAPTER 32

Aldehyde or Ketone and Nitrile

I.	Formation of a Five-membered Ring	845
	1. Pyrrol-2-one	845
	2. Isoxazole	846
II.	Formation of a Six-membered Ring	846
	1. Pyridin-2-one	846
	2. Pyridazine	846
	3. Pyridazin-3-one	847
	4. Pyrimidine	847
	5. Pyrazine	847

1. FORMATION OF A FIVE-MEMBERED RING

1. Pyrrol-2-one

A 2-formylbenzonitrile reacts with hydrazine hydrate to give the monohydrazone of phthalimide (with simultaneous loss of a second CN present in the substrate), and with primary or secondary amines at room temperature to form an isoindol-1-one.

CN, CN, CHO — $N_2H_4 \cdot H_2O$, Δ, 40% → 1-Isoindolone (O, NH, NNH_2)

RNH$_2$, 37–100% → 1-Isoquinolinone (CN, O, NH, NHR)

R = Pr, iPr, tBu or NHR = morphdinyl, piperidinyl

[3763]

2. Isoxazole

A nitrile group is displaced by a neighbouring ketoxime group when the compound is heated with pyridine.

+ NH_2OH —EtOH, pyr, Δ; 90%→ [2055]

Ar = 4-$MeOC_6H_4$

Isoxazolo[4,5-*b*]quinoxaline

II. FORMATION OF A SIX-MEMBERED RING

1. Pyridin-2-one

This ring is formed from the reaction of cyanoethylidene and acetyl functions by heating the compound with phosphoric acid.

—H_3PO_4, Δ; 73%→ [3234]

1,6-Naphthyridin-7-one

Another ring system synthesized similarly:

[3234]

2,7-Naphthyridin-1-one

2. Pyridazine

Hydrazine hydrate annulates carbonyl and nitrile functions on heating the reactants for 4 h.

+ $N_2H_4 \cdot H_2O$ —Δ; ~63%→ [3380]

R^1 = Me, Ph, Ph; R^2 = OH, NH_2

Pyrido[2,3-*d*]pyridazine

3. Pyridazin-3-one

This ring is formed in good yield when a 2-cyanobenzaldehyde is heated with hydrazine hydrate in hydrochloric acid.

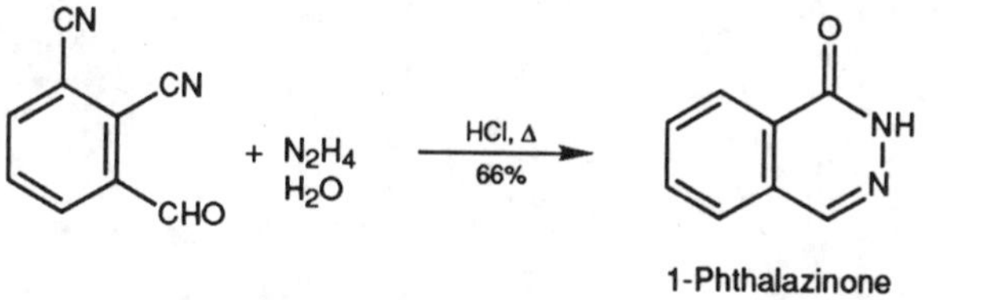

+ N_2H_4 H_2O → HCl, Δ 66% [3763]

1-Phthalazinone

4. Pyrimidine

A fivefold excess of formamide at about 145 °C in a stream of ammonia converts the 2-cyano ketone (**32.1**) efficiently into a fused 4-aminopyrimidine.

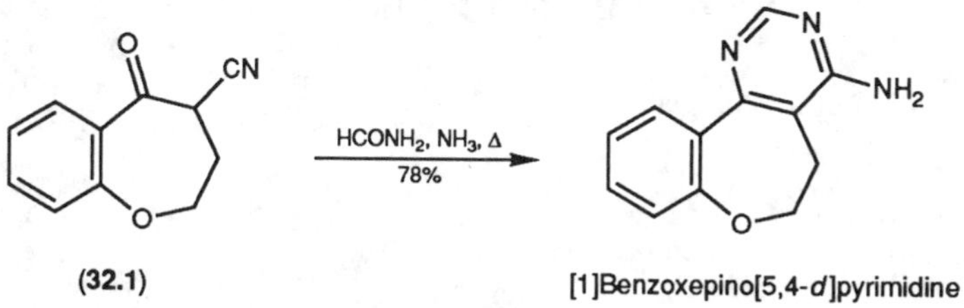

(**32.1**) → $HCONH_2$, NH_3, Δ 78% [3966]

[1]Benzoxepino[5,4-*d*]pyrimidine

5. Pyrazine

Adjacent ketone and nitrile groups are annulated by heating the compound with formamide and either phosphorus oxychloride or (often for a higher yield) ammonia.

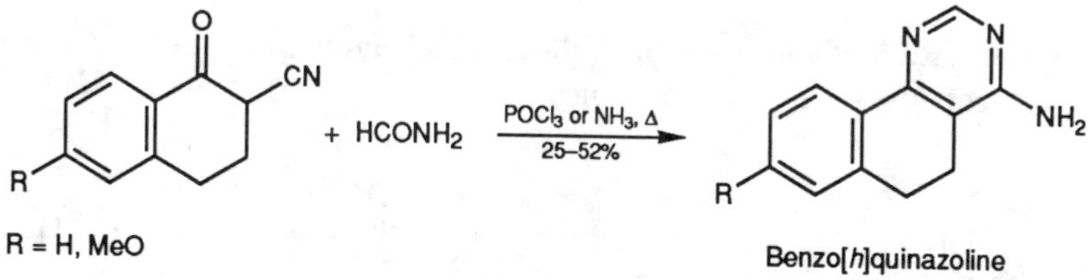

R = H, MeO + $HCONH_2$ → $POCl_3$ or NH_3, Δ 25–52% [2663]

Benzo[*h*]quinazoline

CHAPTER 33

Aldehyde or Ketone and Nitro, Nitroso or *N*-Oxide

I.	Formation of a Five-membered Ring	848
	1. Pyrrole	848
	2. Pyrazole	848
	3. Isoxazole	849
II.	Formation of a Six-membered Ring	849
	1. Pyridine 1-Oxide	849
III.	Formation of a Seven-membered Ring	850
	1. 1,4-Oxazepine	850

I. FORMATION OF A FIVE-MEMBERED RING

1. Pyrrole

Mild hydrogenation of a nitro-ketone followed by heating the resulting cyclized solid in toluene gives a 1-hydroxyindole.

NO_2 / $CH_2CCH_2SO_2Ph$ (C=O) → (i, ii; 77%) OH, N, CH_2SO_2Ph — Indole [3273]

i, AcOEt, H_2, PtO_2; ii, PhMe, Δ

2. Pyrazole

Reductive cyclization of the nitro-thienylideneaniline (**33.1**) (a masked nitro-aldehyde) with triethyl phosphite gives low yields of the tricyclic pyrazole but higher yields may be obtainable from the azides (Chapter 59, Section I.2).

(33.1) → [1]Benzothieno[3,2-*c*]pyrazole [2493]

$P(OEt)_2$, Δ; ~26%

Ar = Ph, 2-Cl-, 4-MeO-, 4-Me_2N-C_6H_4

3. Isoxazole

2-Nitrosopropiophenone reacts with hydrogen bromide in benzene at ambient temperature to form a 5-bromo-2,1-benzisoxazole as the main product while the corresponding 5- and 7-chloro derivatives may be formed when hydrogen chloride is used. A similar substrate may be deoxygenatively cyclized with triphenylphosphine in good-to-excellent yields.

HBr, PhH; 39% → 2,1-Benzisoxazole [2813]

PhH, Ph_3P; 61–90% [2720]

R^1 = Me, Et; R^2 = H, Br, Ac, tBu, NO_2

II. FORMATION OF A SIX-MEMBERED RING

1. Pyridine 1-Oxide

Catalytic hydrogenation of a nitro ketone under mild conditions leads to an azine *N*-oxide, but the nature of the ketone moiety has an effect on the relative amounts of *N*-oxide and the free base which is otherwise a minor byproduct.

H_2, Pd–C; 38–86% → Quinoline 1-oxide [2729]

R^1,R^2 = H, OCH_2O; R^3 = Me, Ph

Two thienyl rings joined at various positions and having a formyl group attached to one ring and a nitro to the other undergo cyclization when subjected to partial reduction with iron(II) sulphate and ammonia.

$FeSO_4$, aq. NH_3, Δ; 95% [3197]

Dithieno[3,2-*b* : 3′,2′-*d*]pyridine 4-oxide

Other ring systems synthesized similarly:

[3197]

Dithieno[3,4-*b* : 3′,2′-*d*]pyridine 4-oxide

[3197]

Dithieno[3,2-*b* : 3′,4′-*d*]pyridine 4-oxide

[3687, 3688]

X = S Thieno[3,2-*c*]quinoline 5-oxide
X = O Furo[3,2-*c*]quinoline 5-oxide

[3196]

Phenanthridine 5-oxide

[3196, 3651]

X = O Furo[2,3-*c*]quinoline 5-oxide
X = S Thieno[2,3-*c*]quinoline 5-oxide

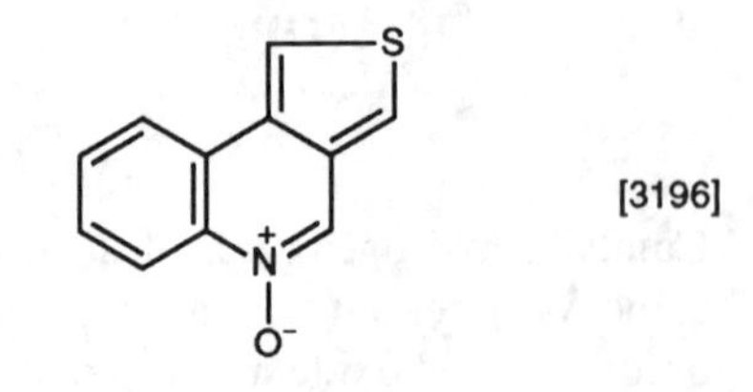

[3196]

Thieno[3,4-*c*]quinoline 5-oxide

III. FORMATION OF A SEVEN-MEMBERED RING

1. 1,4-Oxazepine

Partial reduction of the nitro-aldehyde (**33.2**) with iron(II) sulphate–ammonia results in the formation of the N=CH linkage of an oxazepine ring.

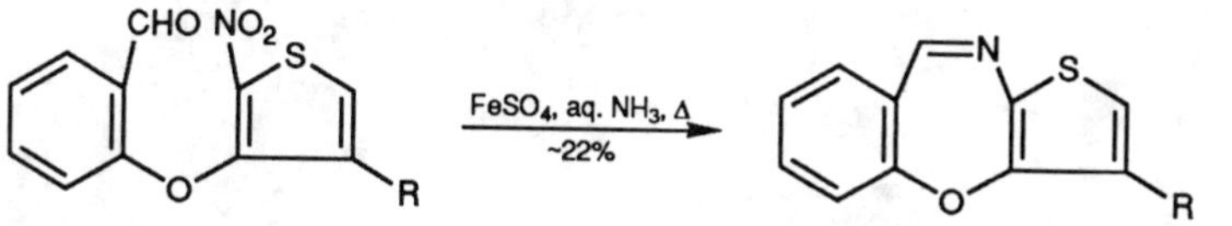

CHO NO_2 S O R

$FeSO_4$, aq. NH_3, Δ
~22%

N S O R

[3060]

(**33.2**) R = H, Br

Thieno[3,2-*b*][1,4]benzoxazepine

Other ring systems synthesized similarly:

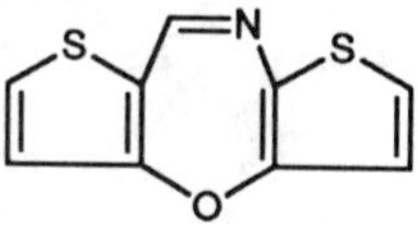

S N S O [3060]

N N S O [3060]

Dithieno[3,2-*b* : 2',3'-*f*][1,4]oxazepine

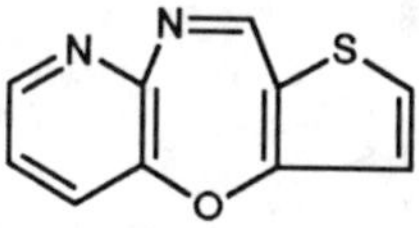

Pyrido[3,2-*b*]thieno[2,3-*f*][1,4]oxazepine

CHAPTER 34

Aldehyde or Other Carbonyl and Phosphorane

I. Formation of a Five-membered Ring 852
  1. Pyrrole 852
  2. Thiophene 854
  3. Thiazole 854
II. Formation of a Six-membered Ring 855
  1. Pyridine 855
  2. Pyridazine 855
  3. Pyrimidin-2- or -4-one 856
  4. Pyran 857
III. Formation of a Seven-membered Ring 857
  1. 1,4-Diazepin-2-one 857
  2. 1,2,4-Triazepine 858

Cyclizations of compounds containing neighbouring (a) thiocyanato and phosphorane, and (b) side-chain carbonyl and phosphorodiamidothioate groups are included in this chapter.

I. FORMATION OF A FIVE-MEMBERED RING

1. Pyrrole

Carbapenem esters are obtained by heating a triethoxyphosphorane thioester in toluene or xylene. The synthesis of heterocycles using phosphoranes has been reviewed [2673]. Phosphonate and aldehyde groups (the latter being formed *in situ* by ozonization of the styryl group) react in the presence of dimethyl sulphide and DBU (review [3364]), a C—C double bond being formed.

PhMe or xylene, Δ 40–88% [3067a]

R^1 = Me_3Si, $tBuMe_2Si$;
R^2 = iPr, $4\text{-}NO_2C_6H_4CH_2OC(O)NH(CH_2)_2$;
R^3 = $4\text{-}NO_2C_6H_4CH_2$

1-Azabicyclo[3.2.0]hept-2-en-7-one

Another ring systems synthesized similarly:

[3301]

n = 0,1; 1-Azabicyclo[3.2.0]hept-2-en-7-one

It is sometimes desirable to avoid the use of alkoxide during the formation of C—C double bonds in the synthesis of β-lactams [3324]. An iminophosphorane (formed *in situ* from an azide and triphenylphosphine) reacts with a carboxylic ester group in a neighbouring malonylidene side-chain to form this ring [3792]. The classic Wittig reaction has been extended to the synthesis of 2-vinylindoles [3603], and other indoles have been obtained from isothiocyanato phosphoranes [3297].

i, Δ 42% [3324]

R^1 = allyl; R^2 = $tBuSiMe_2$; R^3 = Cl, Br
i, PPh_3, 2,6-lutidine, THF

1-Azabicyclo[3.2.0]hept-2-en-7-one

1. THF, Ph_3P 2. H_2O 48% [3792]

R = $4\text{-}MeOC_6H_4SO_2$

Pyrrolo[3,4-*b*]pyrrole

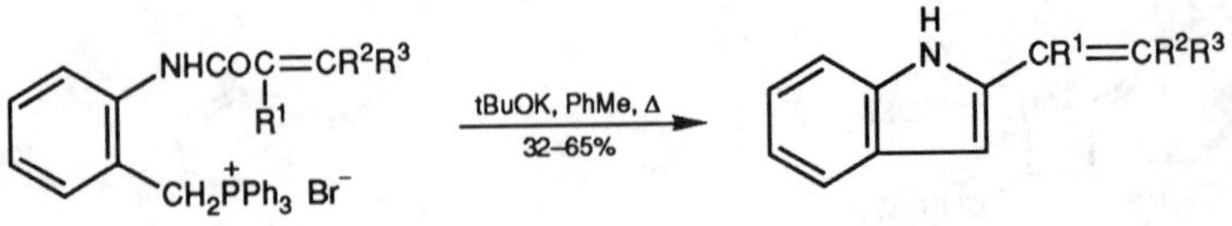

R^1,R^3 = H, Me; R^2 = H, Me, Ph

tBuOK, PhMe, Δ, 32–65%

Indole

[3603]

MeCN, TEA, 94%

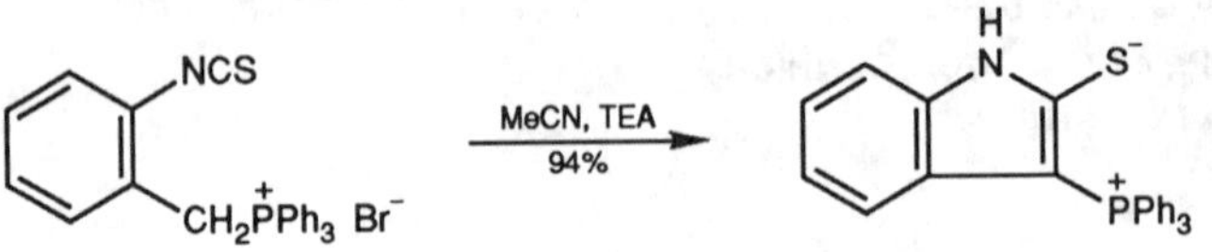

[3297]

2. Thiophene

The *N,N,N′,N′*-tetramethylphosphorodiamidothioate (**34.2**) is cyclized to a benzothiophene by heating in formic acid.

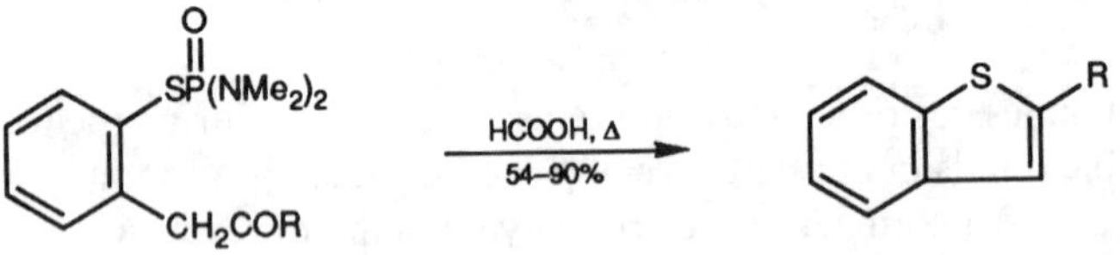

HCOOH, Δ, 54–90%

(34.2)
R = Me, Ph, 3,4-$(MeO)_2C_6H_4$, 4-pyridyl

Benzo[*b*]thiophene

[3916]

3. Thiazole

A thioxo group of a trithiocarbonate reacts with a triphenylphosphorane in an efficient thermal cyclization of a penem ring system; this gives 74% of the *trans* and 15% of the *cis* isomers. The same product is obtained by treating the oxalylimide (R^4 = O) with triethyl phosphite.

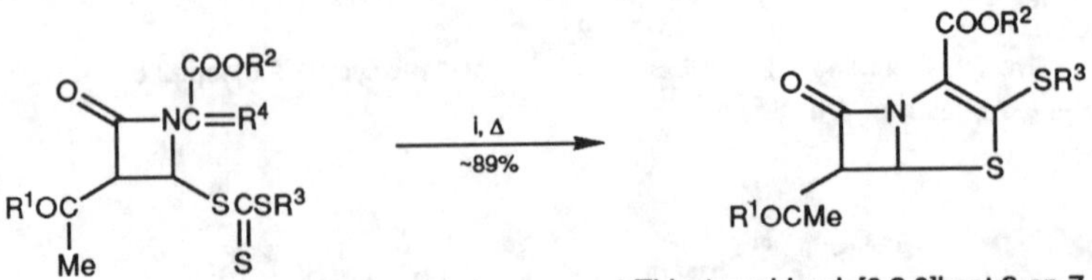

i, Δ, ~89%

4-Thia-1-azabicyclo[3.2.0]hept-2-en-7-one

[3321]

R^1 = $tBuSiMe_2$; R^2 = 4-$NO_2C_6H_4CH_2$;
R^3 = $F(CH_2)_2$; R^4 = PPh_3
i, hydroquinone

II. FORMATION OF A SIX-MEMBERED RING

1. Pyridine

An iminophosphorane (from an azide and triphenylphosphine) reacts with a neighbouring keto group to give a fused pyridine ring in high yield [3264]; when the keto group is part of a carboxylic ester, a similar thermal cyclization leads to a fused 2-ethoxypyridine ring [3561]. Chloromethyl and aldehyde groups attached to different rings may be annulated using a Wittig reaction. Once formed, the phosphorane is treated with ethoxide at room temperature [3220]. One of the limitations of such a reaction is that base-labile functions may be affected in which case a procedure similar to that mentioned in Section I.1 [3324] may be of interest.

COR^1; $CH{=}CN_3$; COOEt — C_6H_{12}, PPh_3, ~93% → R^1, N, $COOR^2$ — Isoquinoline [3264]

R^1 = H, Me, Ph; R^2 = Me, Et

$N{=}PPh_3$; $CH{=}CHCOOEt$ — tetralin, Δ, 70% → N, OEt — Cyclopenta[*b*]pyridine [3561]

PhN—N; N; CH_2Cl; CHO — 1. PPh_3, Δ 2. EtONa, 52% → PhN—N; N — Pyrazolo[3,4-*e*]indolizine [3220]

2. Pyridazine

Pyridazines in which both nitrogen atoms are at bridgeheads may be obtained in moderate yields by cyclizing an aldehydo phosphonate in the presence of DBU (review [3364]). A thermal method of forming a pyridazine ring gives similar yields. The keto-azo-phosphorane (**34.3**) undergoes a thermally induced cyclization to produce a 3-substituted cinnoline.

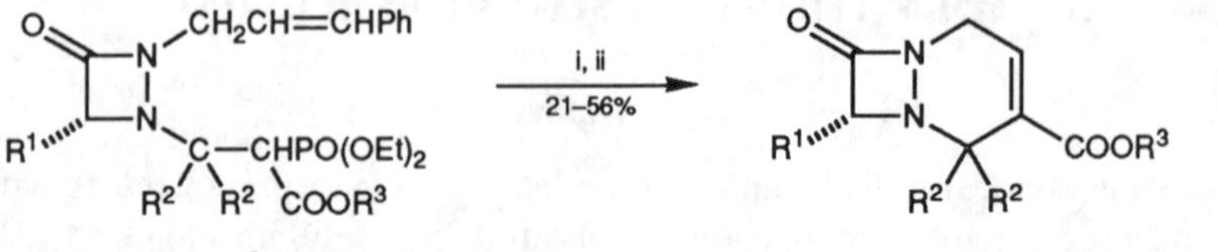

R^1,R^2 = H, Me; R^3 = $SiMe_3CH_2CH_2$
i, CH_2Cl_2, O_3; ii, Me_2S, DBu

i, ii
21–56%

1,6-Diazabicyclo[4.2.0]oct-3-en-8-one

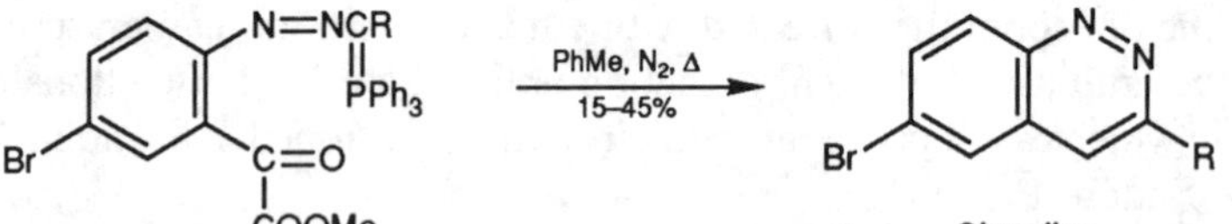

[3180]

PhMe, N_2, Δ
15–45%

(34.3)
R = Ac, COOMe

Cinnoline

[3686]

3. Pyrimidin-2- or -4-one

Formyl and iminophosphorane functions react with an isocyanate either in dichloromethane at about 10 °C or in refluxing toluene to give high yields of the pyrimidin-2-one. Replacing the formyl by an ester group produces the corresponding pyrimidin-4-one.

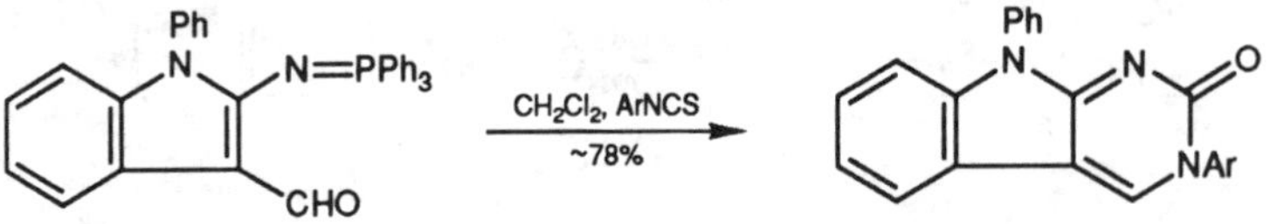

CH_2Cl_2, ArNCS
~78%

Ar = Ph, 4-Cl-, 4-MeO-C_6H_4

Pyrimido[4,5-*d*]indol-2-one

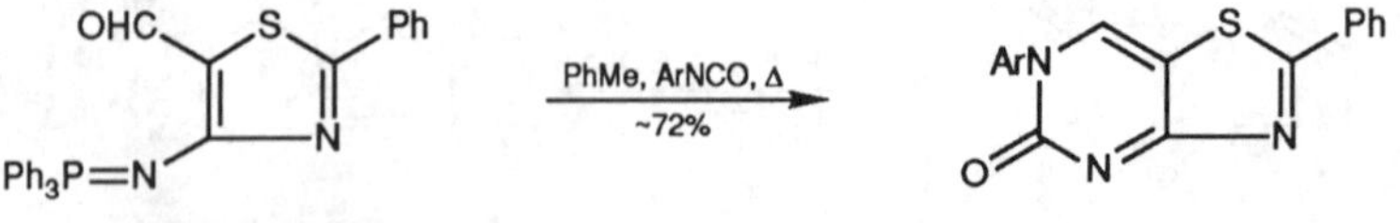

[3401]

PhMe, ArNCO, Δ
~72%

Ar = Ph, 4-MeO-, 4-Me-C_6H_4

Thiazolo[4,5-*d*]pyrimidin-5-one

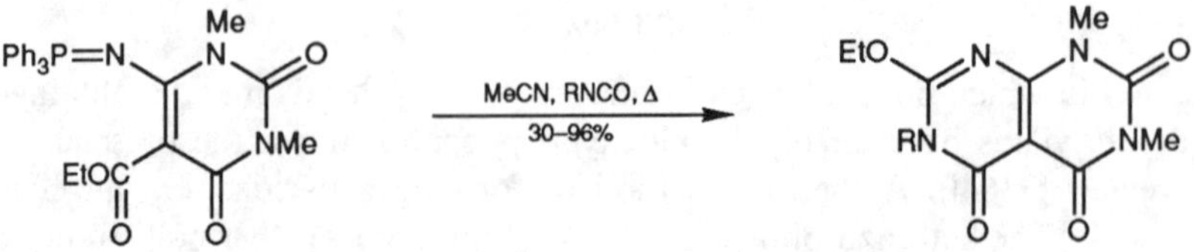

[3419]

MeCN, RNCO, Δ
30–96%

R = $PhCH_2$, Ph, 1-naphthyl, 3-Cl-, 3-CF_3-, 4-Me-C_6H_4

Pyrimido[4,5-*d*]pyrimidine-2,4,5-trione

[3388]

Other ring systems synthesized similarly:

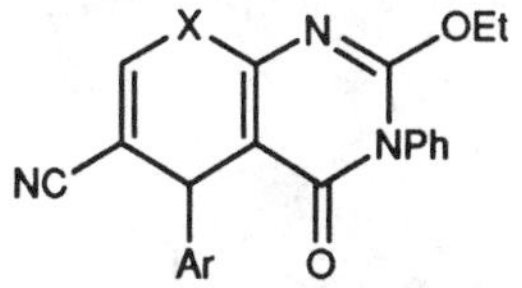

X N OEt
NPh
NC
Ar O

[3734]

X = O Pyrano[2,3-*d*]pyrimidin-4-one
X = N Pyrido[2,3-*d*]pyrimidin-4-one

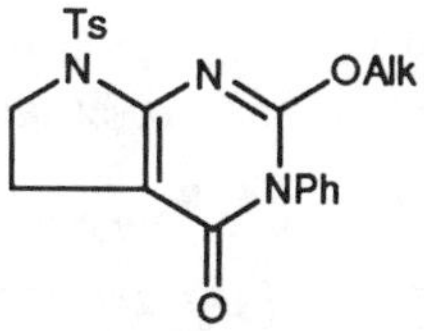

Ts
N N OAlk
NPh
O

[3733]

Pyrrolo[2,3-*d*]pyrimidin-4-one

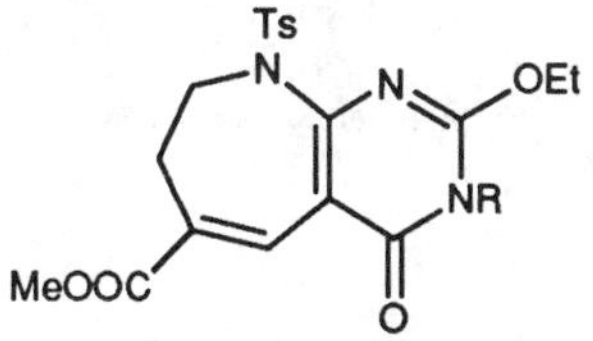

Ts
N N OEt
NR
MeOOC
O

[3733]

Pyrimido[4,5-*b*]azepin-4-one

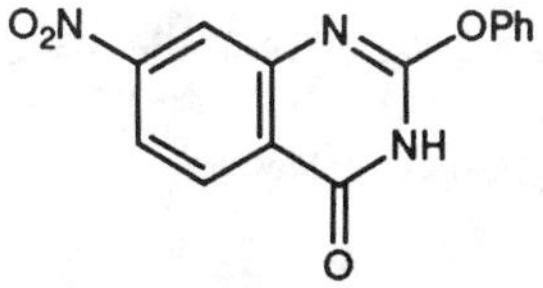

O_2N N OPh
NH
O

[3733]

4-Quinazolinone

4. Pyran

The keto-phosphorane (**34.4**) is cyclized in high yield on stirring it at ambient temperature with sodium ethoxide.

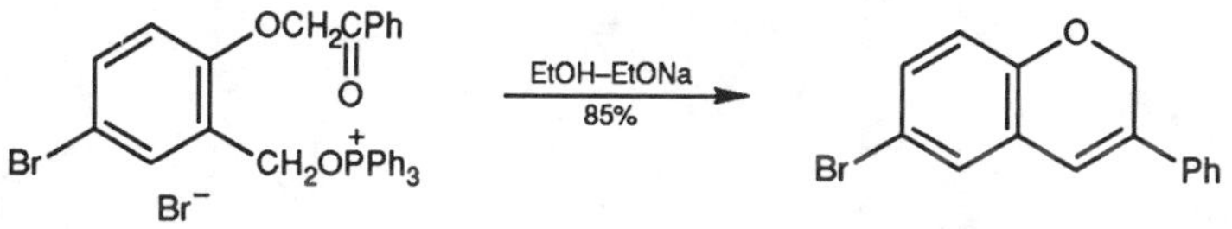

OCH_2CPh (C=O)
Br $CH_2O\overset{+}{P}Ph_3$
Br^-

(**34.4**)

EtOH–EtONa
85%

O
Br Ph

2*H*-1-Benzopyran, 3-aryl-

[3759]

III. FORMATION OF A SEVEN-MEMBERED RING

1. 1,4-Diazepin-2-one

2-Azidoacetylamidobenzophenone can be converted into a benzodiazepinone in good yield through the iminophosphorane which undergoes a thermal cyclization.

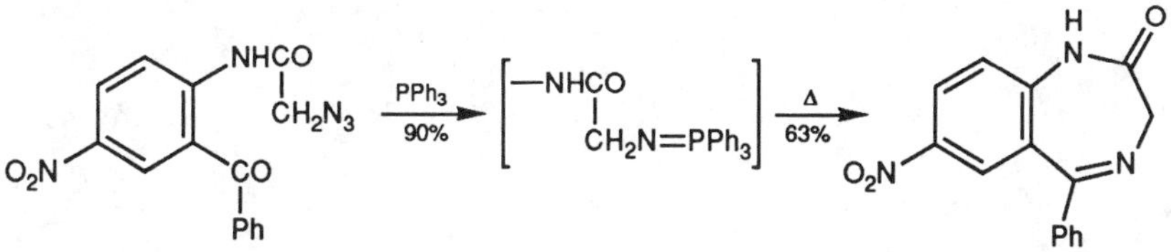

NHCO
CH_2N_3
O_2N CO
Ph

PPh_3
90%

[—NHCO
$CH_2N{=}PPh_3$]

Δ
63%

H O
N
N
O_2N
Ph

[3565]

1,4-Benzodiazepin-2-one

2. 1,2,4-Triazepine

This ring may also be constructed from a suitable acyl iminophosphorane (**34.5**) by heating in methoxyethanol or acetonitrile.

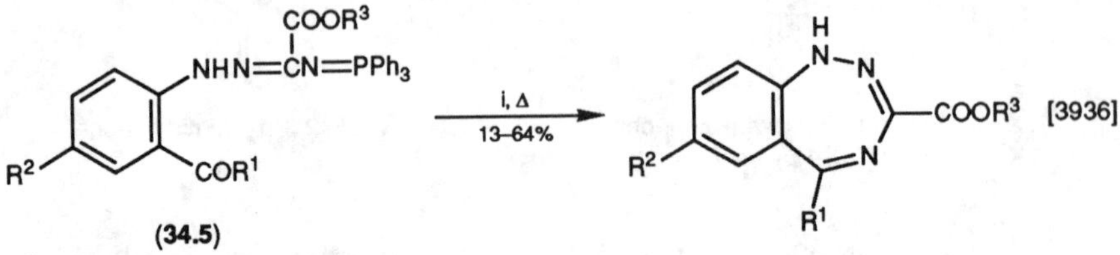

i, Δ; 13–64%; [3936]

(**34.5**)

R^1 = Me, COOMe, Ph, 4-Cl-, 2-F-C_6H_4, 3- or 4-pyridyl; R^2 = H, Br, Cl; R^3 = Me, Et

i, $MeO(CH_2)_2OH$ or MeCN

1,2,4-Benzotriazepine

CHAPTER 35

Aldehyde or Ketone and Ring-carbon

I. Formation of a Five-membered Ring 859
   1. Pyrrole 859
   2. Pyrrol-2-one 861
   3. Pyrazole 862
   4. Imidazole 862
   5. Furan 862
   6. Thiophene or Thiophen-2-one 864
   7. Thiazole 865
II. Formation of a Six-membered Ring 866
   1. Pyridine 866
   2. Pyridin-2-one 867
   3. Pyridazin-3-one 868
   4. Pyran 869
   5. Thiopyran-4-one 870
   6. 1,4-Oxazine 871

I. FORMATION OF A FIVE-MEMBERED RING

1. Pyrrole

In the Nenitzescu synthesis of 5-hydroxyindoles, an *N*-aryl substituted amino-propenoic ester may be used, albeit with a low yield of product [2343]. Better yields may be obtained by reacting the quinone at ambient temperature with an enaminonitrile [3246]. The methylthio group may be removed by heating with Raney nickel. Magnesium methyl carbonate is sometimes used to carboxylate a methyl ketone but when it is warmed with the methyl ketone (**35.3**), it causes cyclization of the keto side-chain [2747]. A dioxopyrimidine aldehyde (**35.4**) is cyclized in moderate yield by heating with carbonate in DMF [2668].

Me$_2$CO, Δ
20%

Indole [2343]

40–92% MeSC=CHCN, NHR, AcOH

[3246]

R = Me, PhCH$_2$, Ph, 4-Cl-, 4-MeO-C$_6$H$_4$;
Ar = Ph, 4-ClC$_6$H$_4$

i, ii
71%

(35.3)

Pyrrolizino[2,1-*b*]quinolin-5-one [2747]

i, MeOH, [O=C(OMe)$_2$]$_2$Mg, Δ; ii, HCl

DMF, K$_2$CO$_3$, Δ
56%

(35.4)

Pyrrolo[2,3-*d*]pyrimidine-2,4-dione [2668]

One modification of the Knorr synthesis of pyrroles uses an isonitroso ketone but the more readily available monophenylhydrazone of 1,2-diketones may be a better substrate; its condensation with cyclohexanone, cycloheptanone, or cyclohexane-1,3-dione gives reduced indoles [2453, 3309]. A C—C bond is formed from an aldehyde group to a carbon atom in another ring by treatment with a secondary amine hydrochloride in a Mannich reaction [2298]. The carbonyl group of tropone and an adjacent ring-carbon are annulated by reaction with an *N*-vinyliminophosphorane. The dihydroaza-azulenes may be converted into the fully aromatic analogues with either nickel or manganese peroxide [3528].

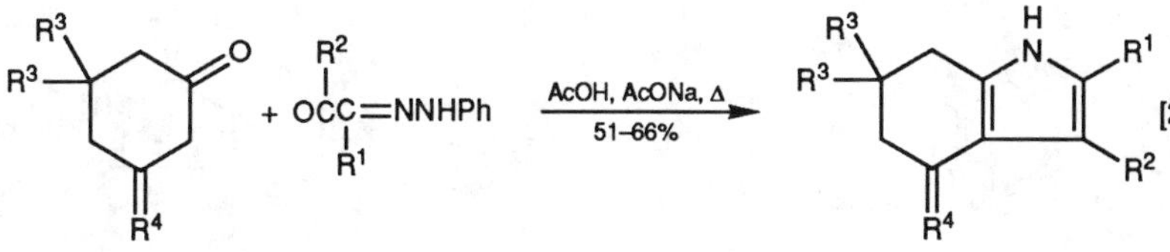

R^1 = Me, Ac, COOEt; R^2 = Me, EtO or R^1R^2 = $(CH_2)_4$; R^3 = H, Me; R^4 = H_2, O

AcOH, AcONa, Δ 51–66% [2453]

Indole

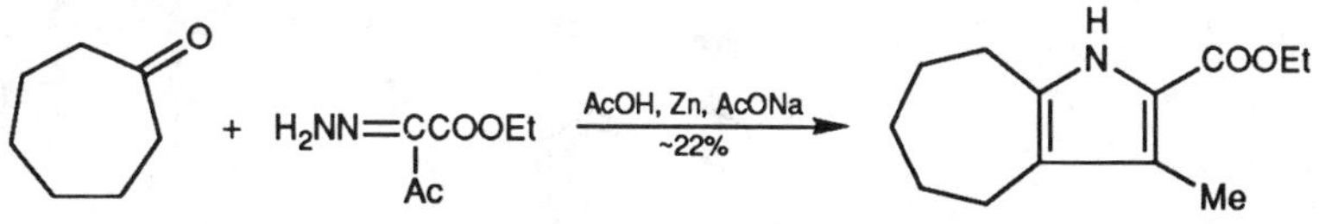

AcOH, Zn, AcONa ~22% [3309]

Cyclohepta[*b*]pyrrole

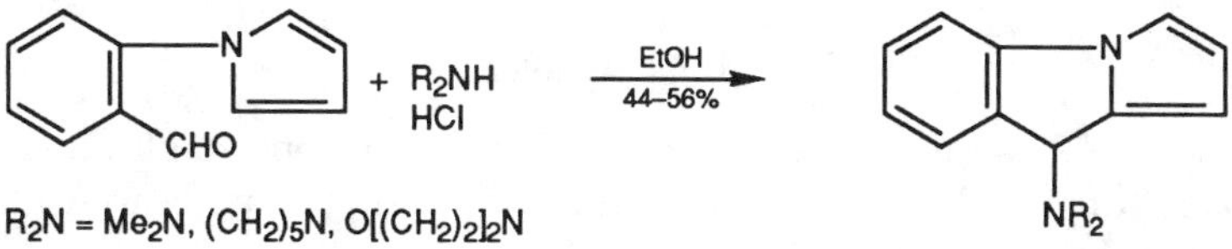

R_2N = Me_2N, $(CH_2)_5N$, $O[(CH_2)_2]_2N$

EtOH 44–56% [2298]

Pyrrolo[1,2-*a*]indole

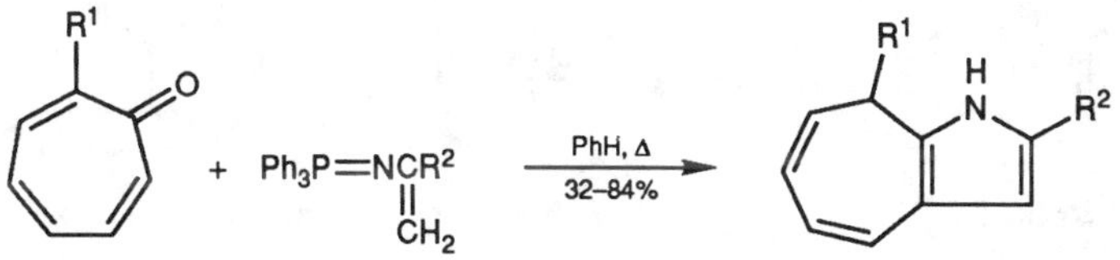

R^1 = H, Me; R^2 = H, Ph

PhH, Δ 32–84% [3528]

Cyclohepta[*b*]pyrrole

2. Pyrrol-2-one

N-Acetoacetylindoles are cyclized by Lewis acids at ambient temperature by attack at the 2-position, but when this position is occupied, the reaction takes another course (see Section II.2). In a nonacidic medium, this ring may be formed by heating *o*-manganated acetophenones with isocyanates.

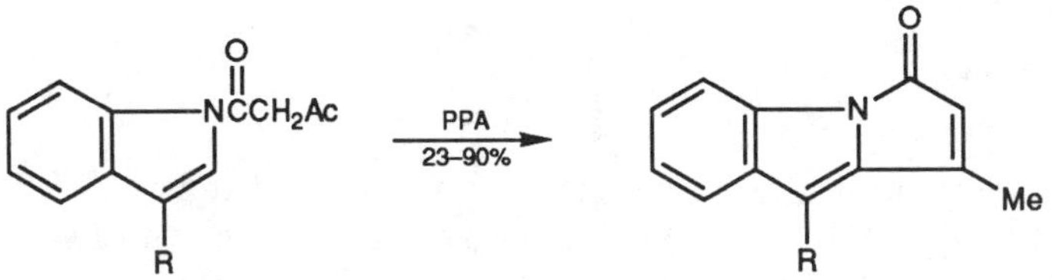

PPA 23–90% [2855]

R = Me, $PhCH_2$, H_2NCOCH_2, $Ac(CH_2)_2$, Ph

Pyrrolo[1,2-*a*]indol-1-one

R^1 = H, F, Cl, Br, MeO, CF_3; R^2 = Me, Et, 4-MeC_6H_4 + R^2NCO —(diox, Δ; 13–76%)→ 1-Isoindolone [3875]

Another ring system synthesized similarly:

Cyclohept[1,2,3-*cd*]isoindol-2-one [3875]

3. Pyrazole

Conversion of 2-aroylquinoxaline into its phenylhydrazone and heating this in ethanolic hydrogen chloride results in annulation of the phenyl-bearing nitrogen to a ring-carbon.

Ar = 4-$MeOC_6H_4$ + $PhNHNH_2$ —(EtOH–HCl, Δ; 70%)→ Pyrazolo[3,4-*b*]quinoxaline [2055]

4. Imidazole

Cyclization is accompanied by rearrangement when the hydrazone (**35.5**) is heated in ethanol for 5 min.

(**35.5**) R = 4-$NO_2C_6H_4$ —(EtOH, Δ; 70%)→ Pyrrolo[1,2-*a*]imidazole [3891]

5. Furan

Cyclodehydration of a side-chain ketone (p. 168) is also feasible in an acidic medium [2301]. *S,N*-Acetals which are also enamines react with 1,4-benzoquinones to form fused furans. Of several solvents tried for this reaction, toluene and THF seem to give the best yield [3268]. Annulation of a cyclic keto group to a neighbouring carbon atom occurs when the ketone is heated with ethyl 2-chloroacetoacetate [2812]. 1,3-Cyclohexanediones are cyclized to benzofurans by reaction with an allenic sulphonium bromide in the presence of a strong base. The intermediate 3-methylene is isomerized by acid to the 3-methylfuran [3798].

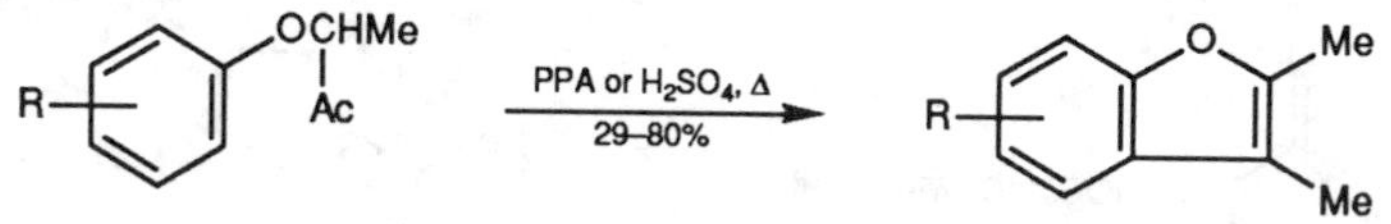

OCHMe Ac — PPA or H_2SO_4, Δ; 29–80% → Me, Me [2308]

R = AcNH, NO_2

Benzofuran

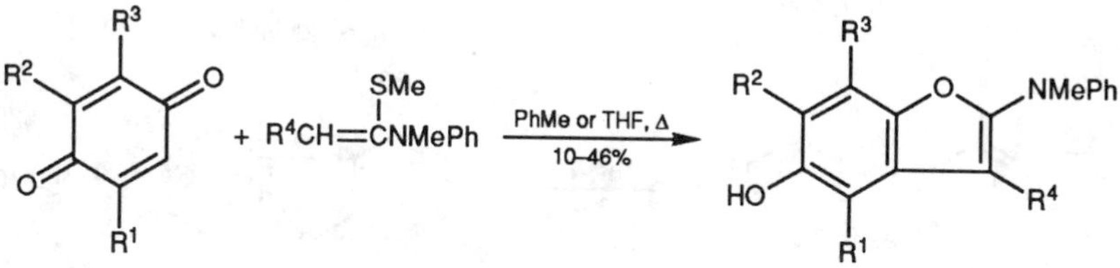

+ $R^4CH{=}C(SMe)NMePh$ — PhMe or THF, Δ; 10–46% → NMePh, R^4, HO [3268]

R^1,R^2 = H, Cl, CN; R^3 = H, CN; R^4 = H, Me

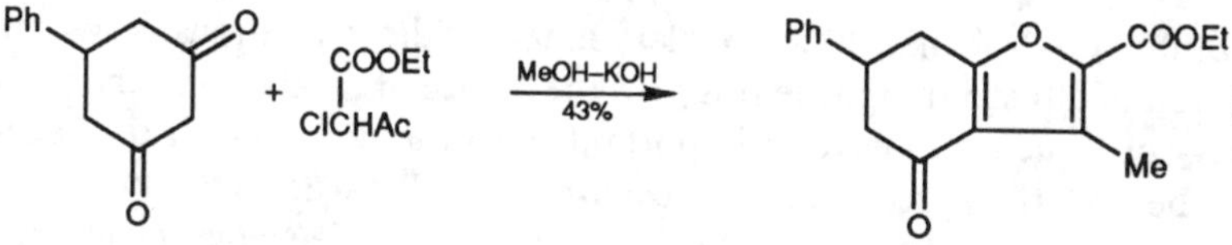

Ph + COOEt ClCHAc — MeOH–KOH; 43% → Ph, COOEt, Me [2812]

4-Benzofuranone

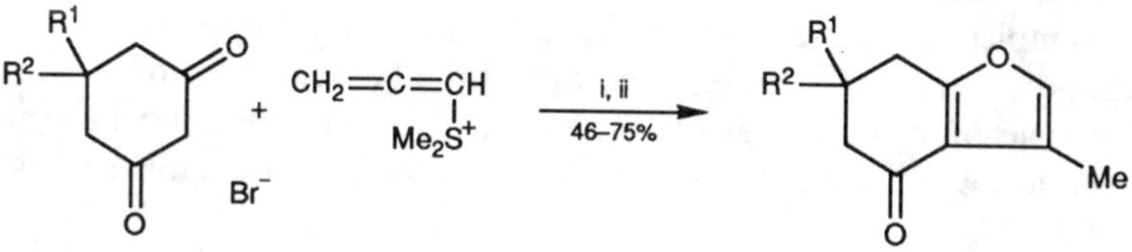

+ $CH_2{=}C{=}CH$ Me_2S^+ Br^- — i, ii; 46–75% → Me [3798]

R^1,R^2 = H, Me

i, EtOH–EtONa, Δ; ii, PhH, TsOH

Another ring system synthesized similarly:

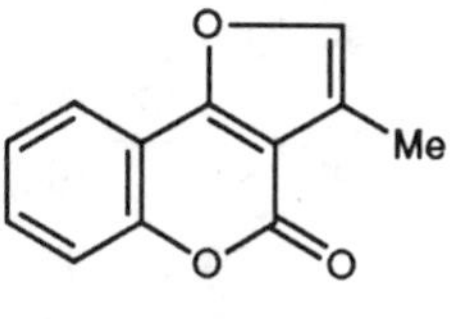

Me [3798]

Furo[3,2-*c*][1]benzopyran-4-one

Regiospecific cycloaddition of the hydrazone (**35.6**) to the 5,8-quinolinedione occurs at ambient temperature. The isomeric 5,8-isoquinolinedione, however, reacts differently (see Chapter 84, Section II.1). An inter-ring C—C bond is formed when a 2-oxocyclohexyloxybenzene is treated with sulphuric acid.

EtOC=CHMe / HC=NNMe$_2$ (**35.6**) + quinolinedione (R = H, Me) —$CHCl_3$, N_2, ~42%→ Furo[5,6-*f*]quinoline (CH=NNMe$_2$, Me, OH) [3973]

Ac-substituted 2-oxocyclohexyloxybenzene —H_2SO_4, 19%→ Dibenzofuran [2415]

6. Thiophene or Thiophen-2-one

Cyclization of 2-phenacylthiopyrroles under acidic conditions leads to the formation of a fused thiophene ring, but the choice of catalyst and the presence of an *N*-substituent can have an important effect on the course of this reaction. When the 1-methylpyrrole (**35.7**) is cyclized with PPA at 90 °C, two isomeric thienopyrroles are formed, one (**35.8**) of which was independently synthesized from the pyrrole (**35.9**) (by the method described on p. 299), followed by methylation [3068a, 3094]. The formation of the [3,2-*b*] isomer is consistent with other examples of rearrangements of 2-thiopyrroles. Copper-catalysed ring formation between a C=S group and one of the benzene rings of a thiobenzophenone is accomplished using ditosyldiazomethane and the copper complex of acetoacetate [3995]. Recent progress in benzothiophene chemistry has been reviewed [3981].

(**35.7**) (Me, SCH$_2$COPh) —PPA, Δ→ Thieno[3,2-*b*]pyrrole (13%) and (**35.8**) (27%) [3068a, 3094]

(**35.9**) (H, SCH$_2$COPh) —1. PPA 2. MeI–DMF–NaH, 74%→ (**35.8**) Thieno[3,2-*b*]pyrrole

Other ring systems synthesized similarly:

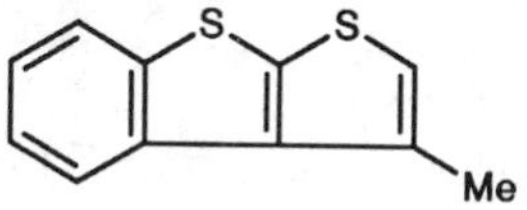

Thieno[2,3-*b*][1]benzothiophene [2167]

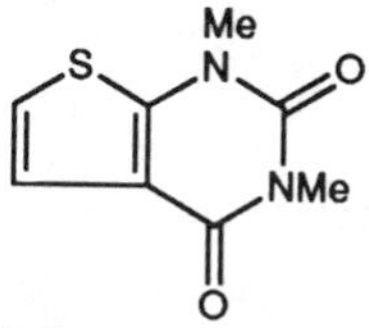

Thieno[2,3-*d*]pyrimidine-2,4-dione [2416]

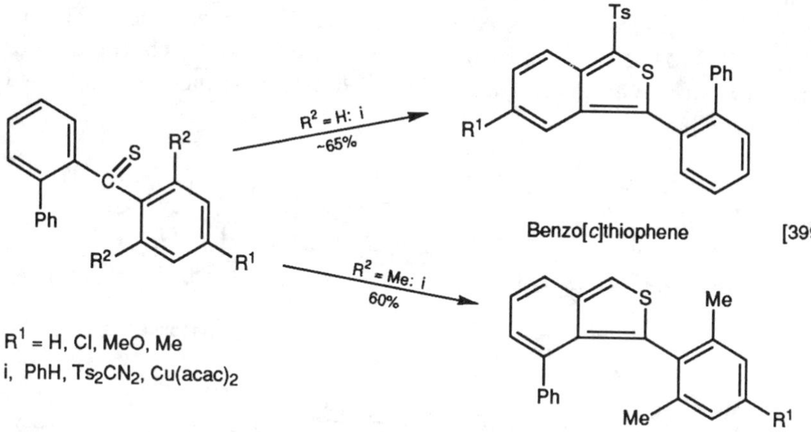

R^2 = H: i, ~65%

R^2 = Me: i, 60%

R^1 = H, Cl, MeO, Me

i, PhH, Ts_2CN_2, $Cu(acac)_2$

Benzo[*c*]thiophene [3995]

An [8 + 2] cycloaddition of diphenylketene to tropothione proceeds in high yield at a temperature below 10 °C.

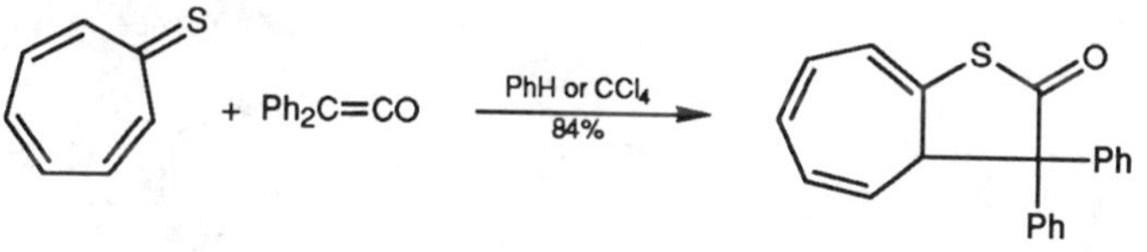

+ $Ph_2C{=}CO$ — PhH or CCl_4, 84% →

2-Cyclohepta[*b*]thiophenone [3872]

7. Thiazole

Thiourea reacts with 1,4-benzoquinone to give several products depending on the conditions. One of the most useful cyclizations is that in which an excess of the quinone reacts at ambient temperature in mineral acid to give 6-hydroxybenzothiazol-2-amines.

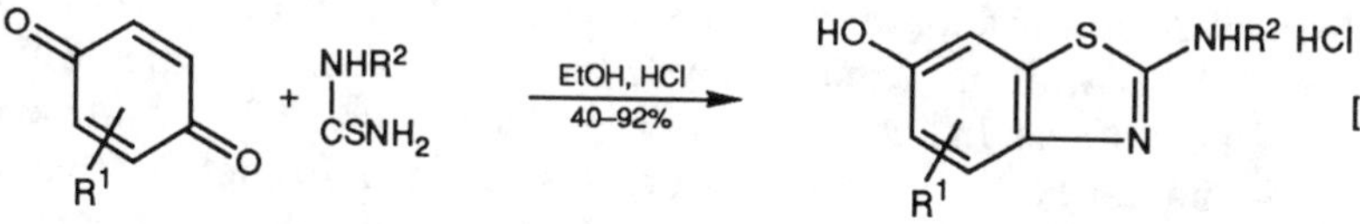

+ NHR^2–$CSNH_2$ — EtOH, HCl, 40–92% → NHR^2 HCl [2227]

R^1 = H, Me, iPr, Cl, PhS; R^2 = H, Me, Ph

Benzothiazole

II. FORMATION OF A SIX-MEMBERED RING

1. Pyridine

When the Gattermann aldehyde synthesis (hydrogen cyanide–zinc chloride) is applied to some benzyl alkyl or aryl ketones, isoquinolines are formed [2268]. Application of the Schmidt reaction to *trans*-dibenzoylstilbene at a temperature slightly above that of the surroundings converts it into a 4-benzoylquinoline [2216]. Care should be exercised with the use of hydrazoic acid. The acetal (**35.10**) failed to cyclize when treated with such Lewis acids as boron trichloride or tin(IV) chloride but hydrolysis to the aldehyde was immediately followed by cyclization [3488].

R^1 = H, Me; R^2 = alkyl, aryl — Et_2O, HCN, $ZnCl_2$, 20–40% → Isoquinoline [2268]

+ HN_3 — $CHCl_3$, H_2SO_4, Δ, 53% → Quinoline [2216]

(**35.10**) — aq. HBr–AcOH, Δ, 34% → Thieno[3,2-*c*]pyridine [3488]

In the Combes quinoline synthesis (see Chapter 53), the intermediate monoketone may be isolated, and on heating this with sulphuric acid, the quinoline is formed in moderate yield [2135]. A trifluoromethyl group often enhances pharmacological activity in some types of biologically active compounds, and so a method of incorporating this group into heterocyclic rings is useful. The synthesis of 2- or 4-trifluoromethylquinolines may be effected by treatment of enamine (**35.11**) with an acidic catalyst, but regiospecificity varies with the nature of the enamine. In one case, an inseparable mixture of 2- and 4-trifluoromethyl-4- and -2-hexyl-6-bromoquinolines was obtained. In contrast, treatment of the arylamine with a trifluoroacetylethyne is regiospecific (see Chapter 54, Section II.1).

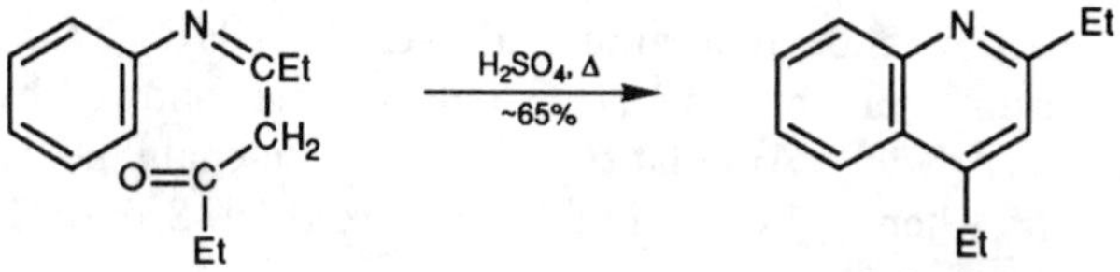

[2135]

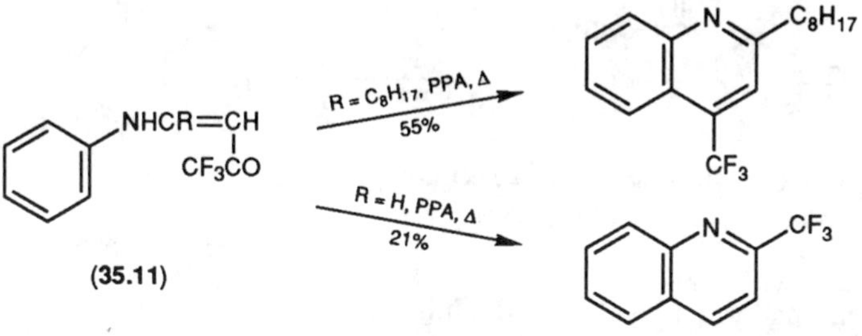

Quinoline

[3867]

(35.11)

2. Pyridin-2-one

Compounds containing an —$NHCOCH_2COR$ side-chain are cyclized to this ring by mineral acid (the Knorr reaction) but the course of this reaction varies according to the substituent R; loss of a methyl group may also occur. A double Knorr cyclization is a step in the synthesis of the antibiotic diazaquinomycin A (**35.12**).

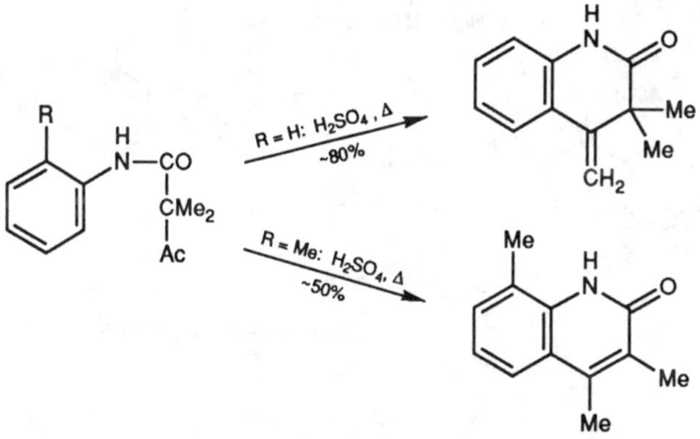

[2120]

2-Quinolinone

Another ring system synthesized similarly:

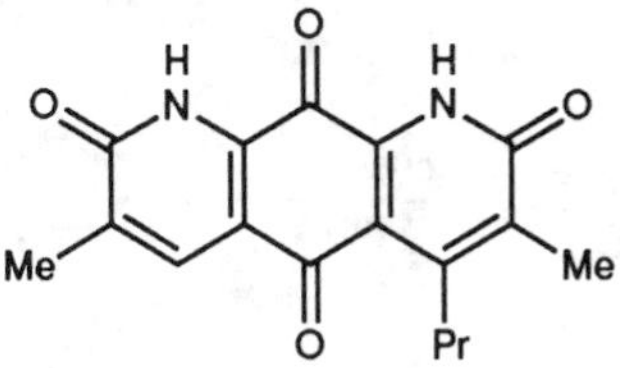

[3498]

(**35.12**)
Pyrido[3,2-*g*]quinoline-2,5,8,10-tetraone

A reversed carboxamide (containing a —CONHCHR[1]COR[2] chain) also cyclizes to a pyridin-2-one but this needs mildly basic followed by cyclodehydrating conditions [3790]. Annulation of an *N*-side-chain keto group to a *peri*-ring-C occurs when C-2 of an indole is blocked (cf. Section I.2).

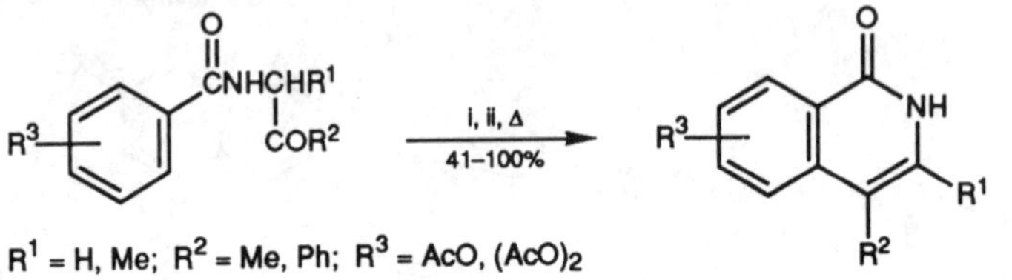

R^1 = H, Me; R^2 = Me, Ph; R^3 = AcO, $(AcO)_2$
i, MeOH, K_2CO_3; ii, $CHCl_3$, Ac_2O, pyr

i, ii, Δ 41–100%

1-Isoquinolinone [3790]

Another ring system synthesized similarly:

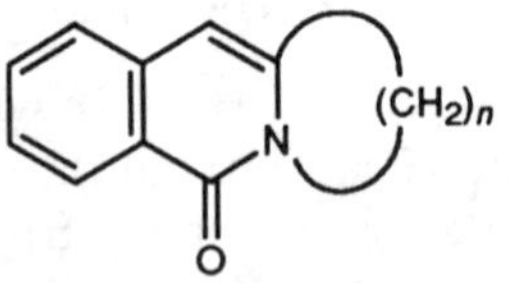

[3790]

n = 3 Pyrrolo[1,2-*b*]isoquinolin-5-one
n = 4 Benzo[*b*]quinolizin-6-one

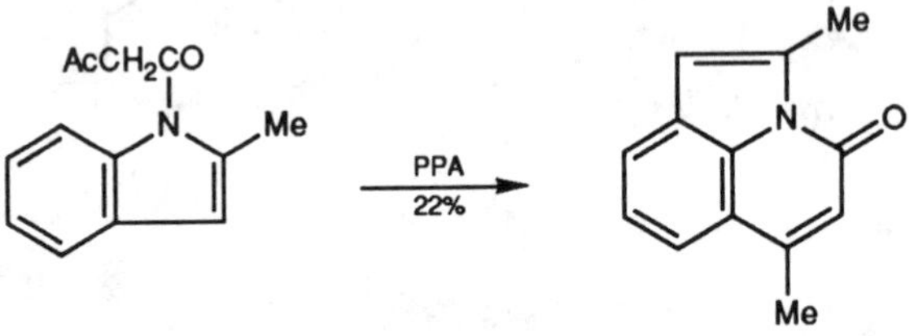

PPA 22%

Pyrrolo[3,2,1-*ij*]quinolin-4-one [2856]

3. Pyridazin-3-one

Cyclization of a *δ*-keto side-chain on to a ring carbon is catalysed by hydrogen chloride and usually proceeds without external heating.

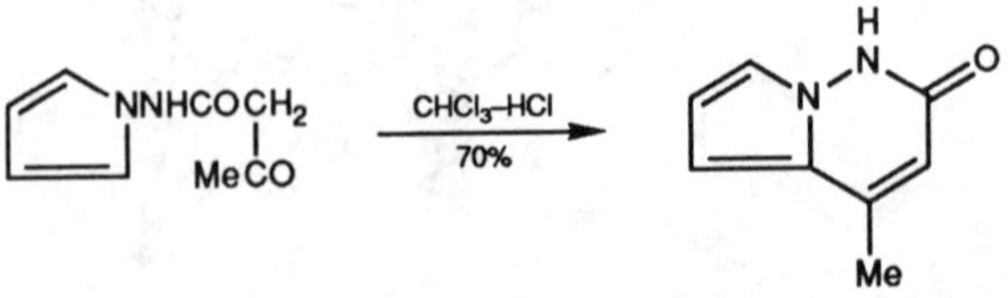

$CHCl_3$–HCl 70%

Pyrrolo[1,2-*b*]pyridazin-2-one [2108]

4. Pyran

3-Arylpyruvic acid esters, on successive treatment with an anhydride and perchloric acid, yield 2-benzopyrylium salts, which may also be prepared from benzyl ketones and acetals in the presence of triphenylmethyl perchlorates. A cation is probably formed from the acetal, and this attacks the reactive benzene ring.

R^1 = COOAlk, $(R^2CO)_2O$, $HClO_4$, 16–41% [2544]

R^1 = Me, $R^3CH(OAlk)_2$, Ph_3CClO_4, 13–63% [2541]

R^2 = Me, Et; R^3 = H, alkyl, Ph

2-Benzopyrylium

The —CH=C—CH=O portion of chromone-3-carboxaldehydes acts as a heterodiene and adds on alkoxyethenes or dihydropyran to give usually (but not always) a mixture of stereoisomers differing in the configuration of C-3. The ratio of isomers varies but the *endo*-product predominates. Prolonged reaction at ambient temperature for 15 days may be necessary but in a sealed system, reaction progresses at 115 °C over a couple of days [3164, 3054, 3055]. Heterodienes and their uses have been reviewed [B-42, 2455, 3067a, 3880].

+ $R^3OC(R^2)=CH_2$ —i, 67–96%→ [3164]

R^1 = H, OH, AcO, MeO, NO_2; R^2 = H, Me; R^3 = alkyl
i, CH_2Cl_2, $CHCl_3$, THF, DMF

Pyrano[4,3-*b*][1]benzopyran-10-one

Another ring system synthesized similarly:

[3164]

Pyrano[3',2' : 5,6]pyrano[4,3-*b*][1]benzopyran-7-one

2-Benzoyldiaryl ethers are cyclized in good-to-excellent yield by treatment with hot sulphuric acid–acetic acid [3027, 3029]. A 2-aminopyran ring is formed when a heterocyclic ketone is heated with a cinnamonitrile and piperidine [3095]. Reduced chromen-5-ones are also formed in a basic medium when acroleins are heated with 1,3-cyclohexanediones [2763a].

AcOH, H_2SO_4, Δ; 40–88% [3027, 3029]

R^1 = H, Br, Cl, F, Me; R^2 = H, Me; R^3 = H, Me, Cl, Br

[1]Benzopyran[2,3-*b*]pyridine

EtOH, pip, Δ; ~40% [3095]

R = CN, COOEt

Pyrano[2,3-*d*]pyridazine

pyr, $MgSO_4$, Δ; 67–82% [2763a]

R^1 = H, Me; R^2 = H, Me, $Me_2C{=}C(CH_2)_2$

2*H*-1-Benzopyran-5-one

5. Thiopyran-4-one

This ring may be built from a ring-carbonyl and its adjacent carbon atom with 3-mercaptopropanoic acid. The enolacetate of indoxyl (3-indolone) or thioindoxyl may also be cyclized likewise.

i, ii; 33–61% [3882]

X = NH, O, S

i, PPA, Δ; ii, Ph_3CClO_4

X = NH Thiopyrano[3,2-*b*]indol-4-one

X = O Thiopyrano[3,2-*b*]benzofuran-4-one

X = S Thiopyrano[3,2-*b*]benzothiophen-4-one

6. 1,4-Oxazine

o-Aminophenol condenses with the carbonyl and the adjacent ring-carbon of a 1,4-quinone on heating the reactants in pyridine for 24 h.

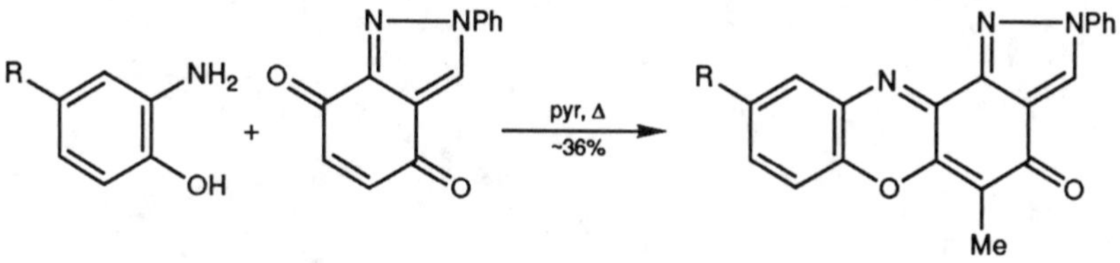

pyr, Δ
~36%

[3487]

Pyrazolo[3,4-*a*]phenoxazin-4-one

CHAPTER 36

Aldehyde, Ketone or Lactam Carbonyl and Ring-nitrogen

I. Formation of a Five-membered Ring 872
 1. Pyrrole 872
 2. Pyrrol-2-one 873
 3. Imidazole 874
 4. 1,2,4-Triazole 875
 5. 1,2,4-Triazol-3-one 875
 6. Thiazole 876
 7. Thiazol-4-one 878
II. Formation of a Six-membered Ring 879
 1. Pyridine 879
 2. Pyridin-2-one 879
 3. Pyrimidine 880
 4. Pyrimidin-4-one 880
 5. Pyrazine 881
 6. Pyrazin-2-one 881
 7. 1,2,4-Triazine 881
 8. 1,3-Oxazine 881
 9. 1,4-Oxazine 882
 10. 1,3-Thiazine 882
 11. 1,3-Thiazin-4-one 882
III. Formation of a Seven-membered Ring 883
 1. 1,4-Thiazepine 883

1. FORMATION OF A FIVE-MEMBERED RING

1. Pyrrole

In a synthesis of an analogue of the antibacterial carbapenems, the keto-γ-lactam (**36.2**) is converted in four steps (including carbene insertion, diphenylphosphorylation and aminoethylthiolation; the intermediates were not isolated) to a pyrrolizinone which is related to thienamycin [3404]. Thermal dehydration of a 2-imidazolylbenzophenone under acidic conditions can lead to the formation of a fused pyrrole [2200]. 2-Acylazoles may be cyclized by reaction with a cumulative phosphorane at ambient temperature or on heating in benzene for a shorter time [2777]. The chemistry of pyrrolizine has been reviewed [3225].

$R^2OOCCH_2CCH_2$ (C=O) HN NHR^1 O → i–iv, ~57% → R^2OOC N NHR^1 O [3404]

(36.2)
$R^1 = 4\text{-}NO_2C_6H_4CH_2OCO$ or COOtBu;
$R^2 = 4\text{-}NO_2C_6H_4CH_2$; R^3 = Ac, $4\text{-}NO_2C_6H_4CH_2$
i, TsN_3, TEA, MeCN; ii, THF, $Rh_2(OAc)_4$, Ar;
iii, MeCN, Ph_2PO_3Cl, iPrNEt; iv, iPr_2NEt, MeCN, $R^3NH(CH_2)_2SH$

3-Pyrrolizinone

N HN COPh → xylene, TsOH, Δ, 76% → N N Ph [2200]

Imidazo[2,1-*a*]isoindole

COPh NH + $Ph_3P{=}C{=}C{=}R$ → PhH, Δ, ~84% → Ph N R [2777]

R = O, PhN

Pyrrolizine
3-Pyrrolizinone

2. Pyrrol-2-one

Annulation of a 2-acylazole under basic conditions produces the fluorescent pyrroloimidazoles (**36.4**). Sometimes, the N^2-acyl derivative is also isolated.

O Et R^1C NH HN O + $(R^2CH_2C)_2O$ (C=O) → i, 6–80% → R^1 R^2 NH O N O [3952]

R^1 = Me, Ph; R^2 = H, Me, Ph
i, DMF, MeOH, MeONa, Δ

(36.4)
Pyrrolo[1,2-*c*]imidazole-3,5-dione

3. Imidazole

A side-chain carbonyl group reacts with a ring-nitrogen (usually at ambient temperature) on treatment with either acetic anhydride–perchloric acid or an excess of thionyl chloride. The first method yields a perchlorate salt [3148] whereas the latter procedure leads to a chlorinated product [3719]. An attempt to prepare the imine of 2-aroylpyridine with benzylamine and boron trifluoride etherate yielded instead the imidazopyridine [3190]. Another reaction which gave an unexpected product was the attempt to form a fused diazepine from the benzoylamidomethyltriazole and PPE; this gave good yields of the imidazotriazole [2599]. Activation of a lactam NH with diethyl chlorophosphate followed by treatment with ethyl isocyanoacetate under mild conditions gives a new imidazole-4-carboxylate ring [3567].

Ac_2O, $HClO_4$ ~63% [3148]

Ar = Ph, 4-Cl-, 4-NO_2-, 4-Me-, 4-MeO-C_6H_4 Imidazo[1,2-*d*][1,2,4]triazin-4-ium, 8-oxo-

CCl_4, $SOCl_2$, Δ 42–90% [3719]

R = H, Me,Cl; Ar = Ph, 4-Cl-, 4-MeO-, 4-Me-C_6H_4 X = CH Imidazo[1,2-*a*]pyridine
X = N Imidazo[1,2-*a*]pyrimidine

+ $PhCH_2NH_2$ $BF_3.Et_2O$, Δ ~43% [3190]

Ar = Ph, 2-pyridyl   Imidazo[1,5-*a*]pyridine

PPE, Δ 67–76% [2599]

Ar = Ph, 4-ClC_6H_4 Imidazo[1,5-*b*][1.2,4]triazole

i, ii 56% [3567]

i, DMF, tBuOH, $ClPO_3Et_2$; ii, $CNCH_2COOEt$ Imidazo[5,1-*c*][1,4]benzoxazine

4. 1,2,4-Triazole

Reaction of a lactam with dimorpholinophosphoric chloride gives an *O*-substituted derivative which is susceptible to reaction at ambient temperature with acethydrazide. It is not necessary to isolate the intermediate. A lactam thiocarbonyl and its ring-nitrogen are annulated on heating the compound first with hydrazine hydrate and then with trimethyl orthoacetate.

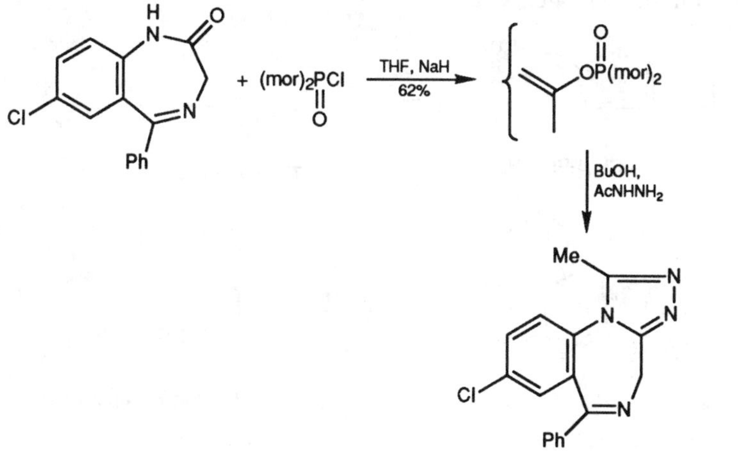

+ $(mor)_2P(O)Cl$ — THF, NaH, 62% → intermediate $OP(O)(mor)_2$ — BuOH, $AcNHNH_2$ →

[2833]

[1,2,4]Triazolo[4,3-*a*][1,4]benzodiazepine

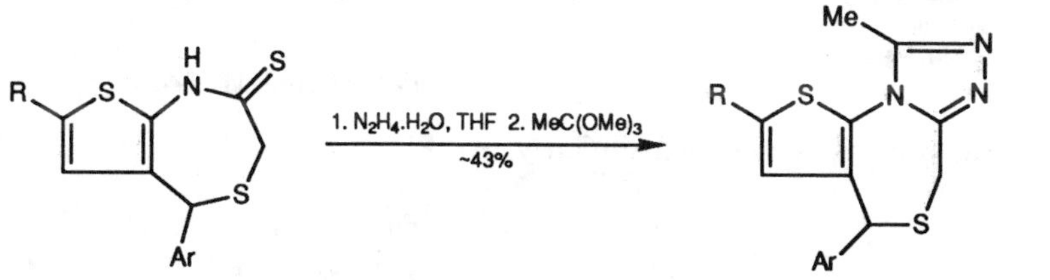

1. $N_2H_4.H_2O$, THF 2. $MeC(OMe)_3$, ~43%

[3801]

R = H, Br; Ar = 2-FC_6H_4

Thieno[2,3-*e*][1,2,4]triazolo[3,4-*c*][1,4]thiazepine

5. 1,2,4-Triazol-3-one

This ring is formed when a triazine-3-thione is heated with ethyl carbazate; simultaneously, the amino group is hydrolysed to a carbonyl.

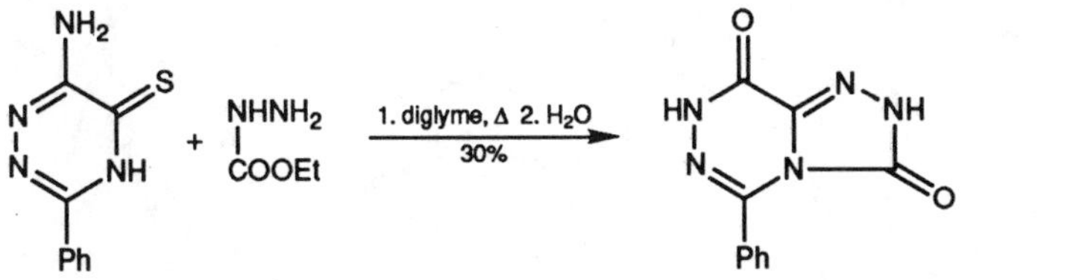

+ $NHNH_2$–COOEt — 1. diglyme, Δ 2. H_2O, 30% →

[3382]

1,2,4-Triazolo[4,3-*d*][1,2,4]triazine-3,8-dione

6. Thiazole

PPA, phosphorus pentoxide–orthophosphoric acid, acetic anhydride or ethanolic hydrogen chloride may be used to produce this ring in moderate-to-good yields from an acyl- or aroyl-methylthio side-chain of an azole or azine. Heating the ketone with 48% hydrobromic acid gives the same result. A 4-phenacylthioimidazole and similar compounds cyclize under mild conditions to a thiazolium or related salt.

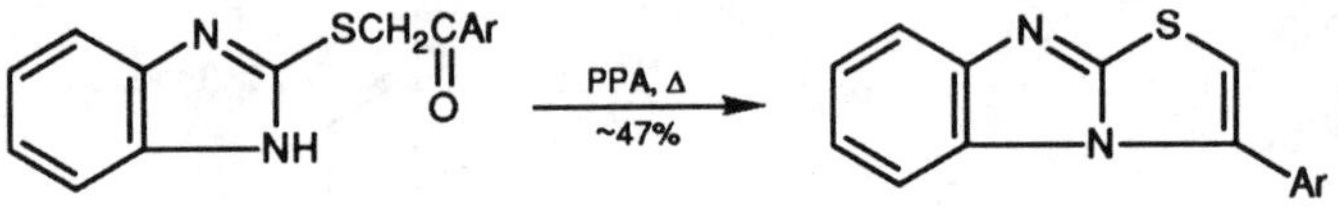

PPA, Δ; ~47% [2118, 3912]

Ar = Ph, 4-Cl-, 4-Br-C_6H_4

Thiazolo[3,2-*a*]benzimidazole

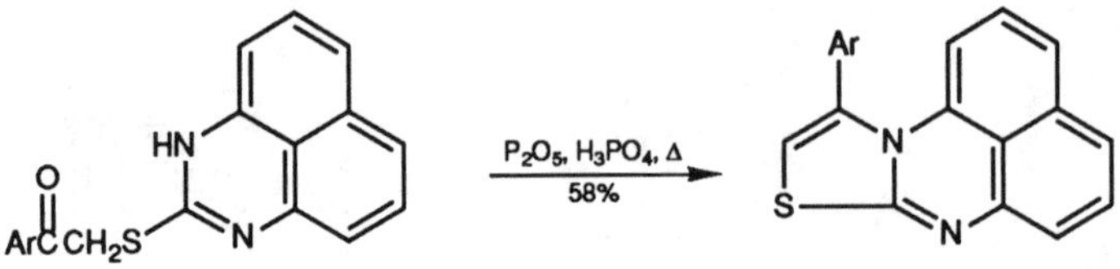

P_2O_5, H_3PO_4, Δ; 58% [2121]

Ar = 4-Br-C_6H_4

Thiazolo[3,2-*a*]perimidine

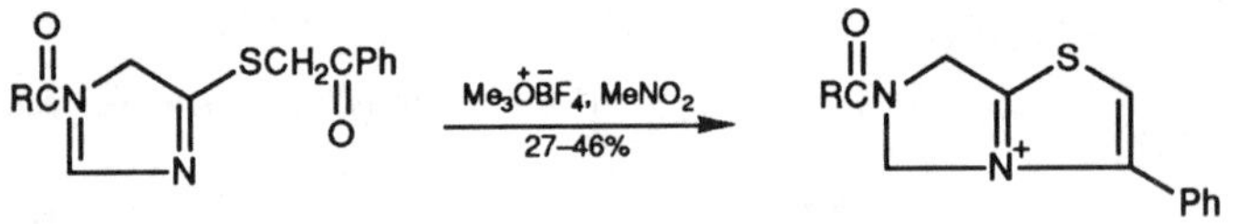

$Me_3\overset{+}{O}\overset{-}{B}F_4$, $MeNO_2$; 27–46% [2898]

R = Me, MeO

Imidazo[2,1-*b*]thiazol-4-ium

Other ring systems synthesized similarly:

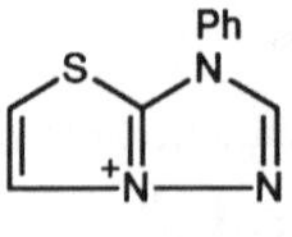

[2711]

Thiazolo[3,2-*b*][1,2,4]triazol-4-ium

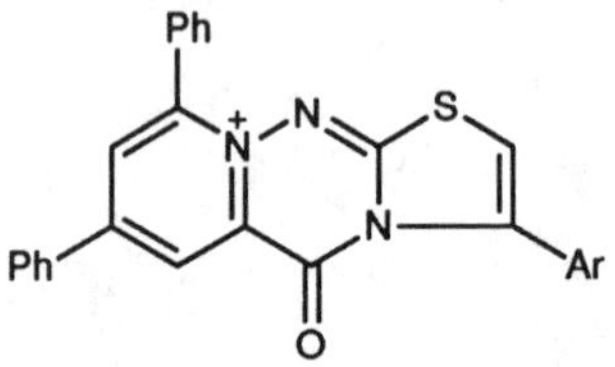

[3347]

Thiazolo[2,1-*f*]pyrido[2,3-*c*][1,2,4]triazin-10-ium, 5-oxo-

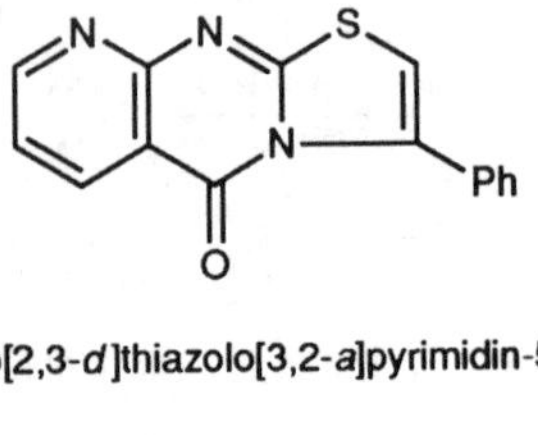

[3333]

Pyrido[2,3-*d*]thiazolo[3,2-*a*]pyrimidin-5-one

A thiazole ring may also be constructed from a lactam thiocarbonyl and one of the following reagents: DMAD [3177], a 2-bromoacrylic acid or its nitrile [2051, 3509], an aryl isothiocyanate and an alkyne [2706, 2915], a reactive α-methylene ketone such as dimedone [3944], an acyl halide [2433] or a halo ketone which is prepared *in situ* [2475].

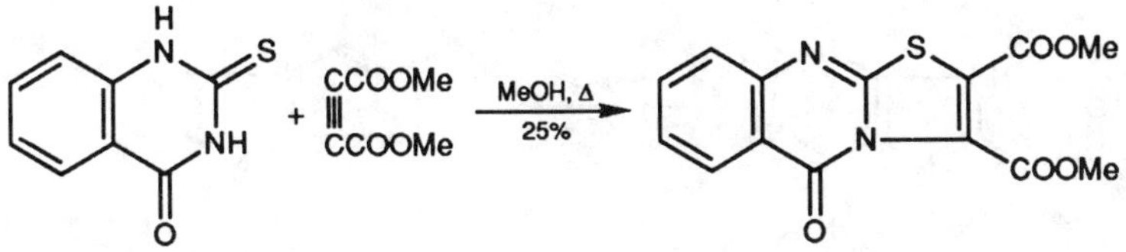

MeOH, Δ; 25% [3177]

Thiazolo[2,3-*b*]quinazolin-5-one

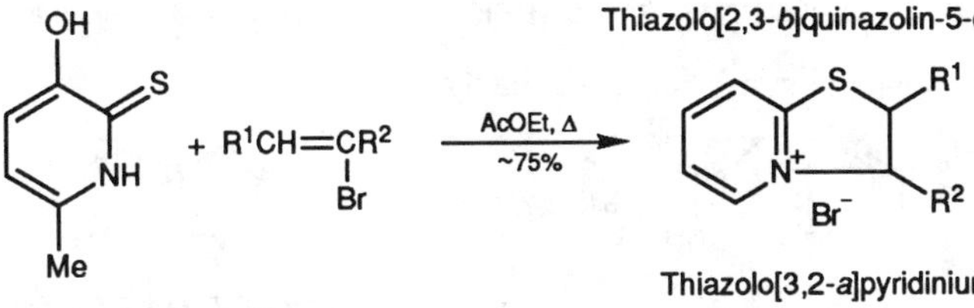

AcOEt, Δ; ~75% [2051, 3509]

Thiazolo[3,2-*a*]pyridinium

R^1 = H, Me, Ph; R^2 = COOH, CN

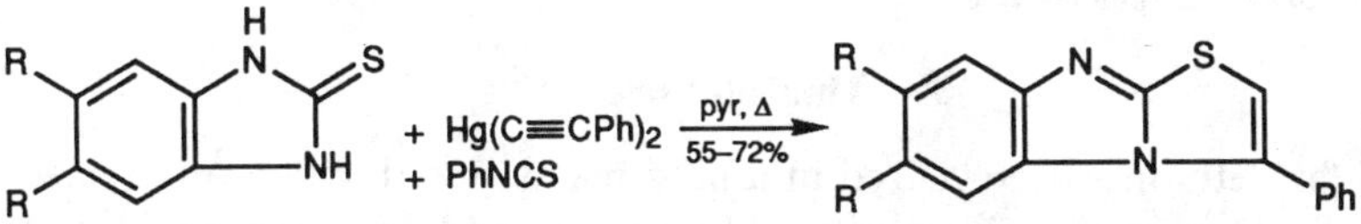

pyr, Δ; 55–72% [2706, 2915]

R = H, Me

Thiazolo[3,2-*a*]benzimidazole

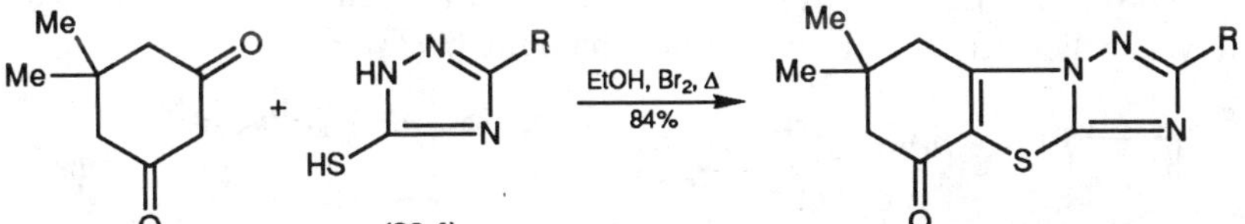

EtOH, Br_2, Δ; 84% [3944]

(36.4)

R = 2,4-$Cl_2C_6H_3$

[1,2,4]Triazolo[5,1-*b*]benzothiazol-5-one

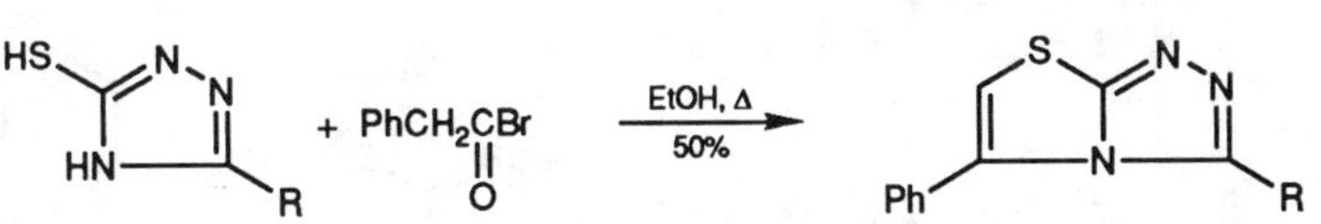

EtOH, Δ; 50% [2433, 2512]

R = 2-furyl

Thiazolo[3,2-*c*]-1,2,4-triazole

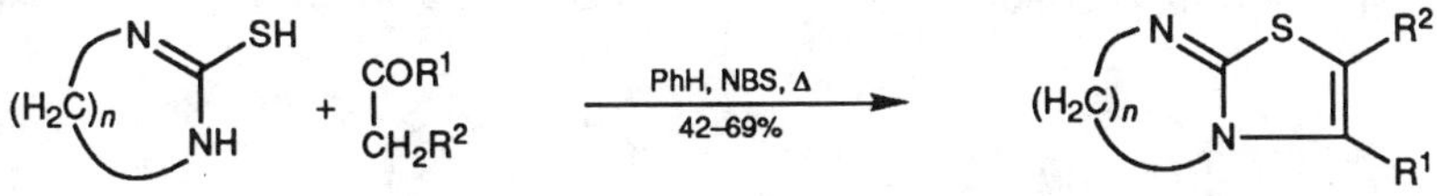

PhH, NBS, Δ; 42–69% [2475]

R^1 = Me, Ph, 4-Cl-, 4-Me-C_6H_4; R^2 = H, Me, Ph; n = 3 or 4

Thiazolo[3,2-*a*]pyrimidine
Thiazolo[3,2-*a*][1,3]diazepine

A keto-containing side-chain at position 8 of an adenine derivative may cyclize on to either of its ring-nitrogen neighbours, but under acid catalysis it closes on to *N*-9 exclusively [2510]. However, a 2,6-purinedione cyclizes under similar conditions (using chloroacetone) to *N*-7 [2512].

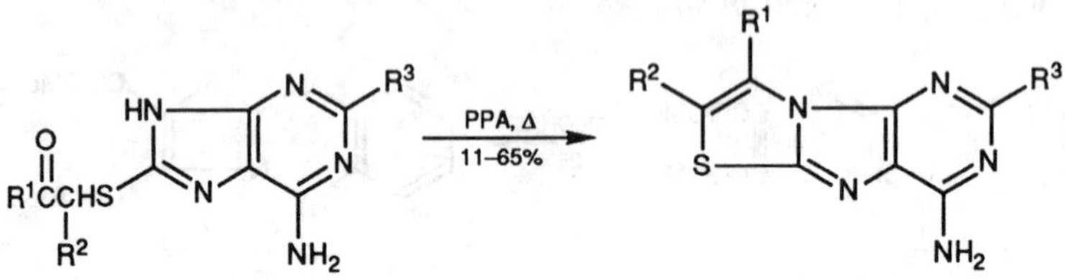

PPA, Δ 11–65% [2510]

R^1 = H, Cl, MeS; R^2 = H, Me; R^3 = alkyl, Ph

Thiazolo[3,2-*e*]purine

Other ring systems synthesized similarly:

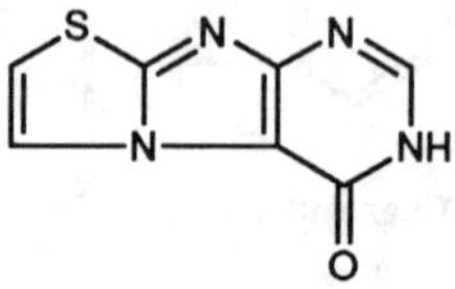

[2512, 2725]

Thiazolo[3,2-*f*]purin-4-one

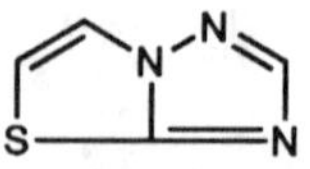

[3103]

Thiazolo[3,2-*b*][1,2,4]triazole

7. Thiazol-4-one

2-Benzimidazolethiones are cyclized in a mild reaction with either an arylazochloroacetyl chloride [2366] or with chloroacetic acid and a benzaldehyde [2454]. Imidazole-2-thione adds on to DMAD also without the need for external heating [2270]; the epoxide of arylidenemalononitrile cyclizes 2-benzimidazolethione very efficiently to the thiazolone analogue [3933].

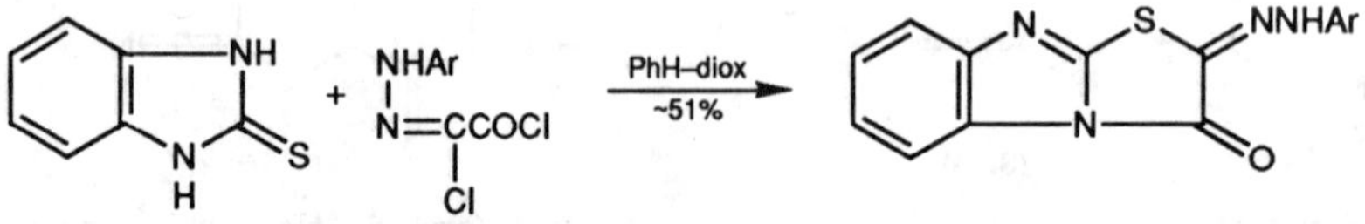

PhH–diox ~51% [2366]

Ar = 3-Me-, 2-NO_2-, 2-MeO-C_6H_4

Thiazolo[3,2-*a*]benzimidazol-3-one

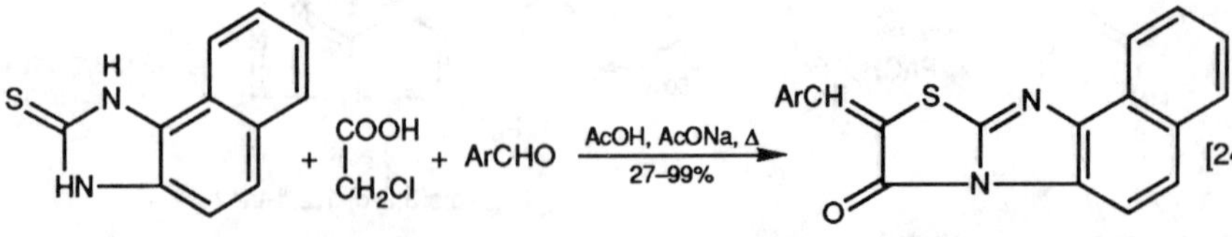

AcOH, AcONa, Δ 27–99% [2454]

Ar = Ph, Me-, NO_2-C_6H_4, PhCH=CH

Naphth[1,2-*d*]imidazo[2,1-*b*]thiazol-8-one

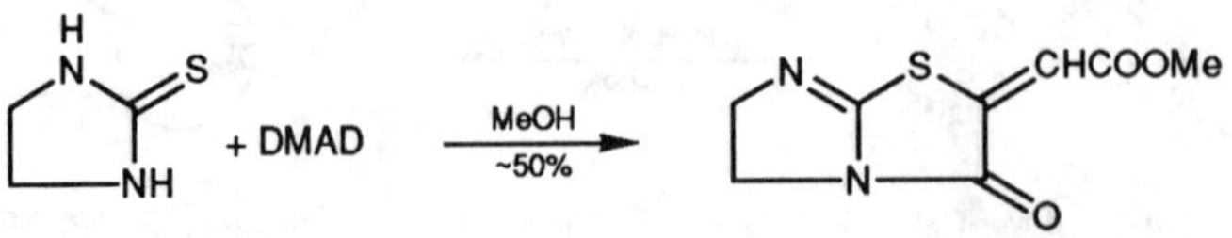

MeOH ~50% [2270]

Imidazo[2,1-*b*]thiazol-3-one

H N S NH + ArCH—C(CN)$_2$ O

Me_2CO, N_2 90%

N S Ar N O [3933]

Ar = Ph, 4-ClC_6H_4

Thiazolo[3,2-*a*]benzimidazol-3-one

II. FORMATION OF A SIX-MEMBERED RING

1. Pyridine

Ring closure of 2-phenacylimidazole by stirring with DMAD is an exothermic reaction; a lower yield is obtained when the alkyne is replaced by propynoic acid.

$PhCCH_2$ (C=O) H N N + C≡CR^1 COOMe

MeOH 25–70%

Ph N N HOOC R^2 [2739]

R^1 = H, COOMe; R^2 = H, COOH

Imidazo[1,2-*a*]pyridine

2. Pyridin-2-one

In a strongly basic medium, the β-keto ester (**36.5**) reacts with an acyl chloride to form 5-indolizinone in good yield [2353]. *Peri* cyclization of NH and a CHO-carbon proceeds under either mildly or strongly basic conditions depending on the type of reagent used [2044, 3602]; the pyrrolo[3,2,1-*ij*]quinolinone ring system is present in several alkaloids. A fused pyridinone is formed in high yield when a compound containing a CHO group and an NH in an adjacent ring is heated with dimethyl malonate and a mild base [2776].

COOMe ArC HN + $ArCH_2CCl$ O

PhH, NaH 65%

COOMe Me Ar N O [2353]

(**36.5**)

Ar = 4-$MeOC_6H_4$

5-Indolizinone

CHO H N N O + CH_2COOH RNCPh O

AcONa, Ac_2O, Δ 59–89%

R O N N O [2044]

R = Me, Ac, CN, Ph

Benzimidazo[1,2,3-*ij*]benzo[*c*][1,8]-naphthyridine-3,10-dione

+ MeCOOEt —(EtONa, Δ; 71–92%)→ [3602]

R = H, Me, Ph, 4-BrC_6H_4

Pyrrolo[3,2,1-*ij*]quinoilin-4-one

+ $CH_2(COOMe)_2$ —(TEA, Δ; 83%)→ [2776]

Indolo[3,2,1-*de*][1,5]naphthyridin-6-one

3. Pyrimidine

The isonitrile (**36.6**) cyclizes the pyrrole-2-carboxaldehyde on heating in THF containing DBU (review [3364]).

+ COOMe–CH_2NC (**36.6**) —(THF, DBU, Δ; 50–69%)→ [2826]

R^1 = H, Me, COOEt, 3-MeOC_6H_4; R^2,R^3 = H, Me, COOEt

Pyrrolo[1,2-*c*]pyrimidine

4. Pyrimidin-4-one

A new fused pyrimidinone ring is formed when the oxazinone (**36.7**) is treated successively with phosphorus oxychloride and anthranilic acid.

(**36.7**) + anthranilic acid —(i, ii; 60–90%)→ [3376]

R = H, Cl, NO_2, MeO, Me; X = O, S

i, $(CH_2Cl)_2$, $POCl_3$; ii, TEA, Δ

Quinazolino[2,3-*c*][1,4]benzoxazin-12-one

Quinazolino[2,3-*c*][1,4]benzothiazin-12-one

5. Pyrazine

Annulation of a ketone group to an adjacent ring-nitrogen to form this ring is effected by heating with the bifunctional 2-bromoacetaldehyde dibutyl acetal and ammonium acetate.

CH_2Br–$CH(OBu)_2$ + $AcONH_4$ —EtONa, DMF, Δ; 78%→ [2063]

Pyrazino[1,2-*a*]benzimidazole

6. Pyrazin-2-one

The oxocarboxamide (**36.8**) is cyclized by heating with thionyl chloride, a triazolopyrazinium chloride being formed in high yield.

(**36.8**) —$SOCl_2$; 71–96%→ [3362]

Ar^1 = Ph, 4-Me-, 4-MeO-, 4-Cl-C_6H_4
Ar^2 = Ph, 4-Br-, 4-Cl-C_6H_4

[1,2,4]Triazolo[1,5-*a*]pyrazin-1-ium, 8-oxo-

7. 1,2,4-Triazine

A side-chain *β*-carbonylazo group is incorporated into a triazine ring either thermally or in an acidic medium.

—HCl, Δ; ~91%→ [2616]

R^1 = Me, Ph; R^2 = Ac, COOEt

1,2,4-Triazino[4,3-*b*]indazole

8. 1,3-Oxazine

2-Phenacylimidazole reacts with an isocyanate or isothiocyanate (review of reactions [3990]) to give high yields of fused oxazine.

PhC(O)CH2-imidazoline (X = O, S) + PhNCX —THF, Δ; 67–90%→ [2739]

Imidazo[1,2-*c*][1,3]oxazine

9. 1,4-Oxazine

The N=C—C=O system of a 1-phenothiazinone undergoes an intramolecular Diels–Alder reaction [B-42, 2455, 3067a, 3880] with diethyl acetylenedicarboxylate, but with tetracyanoethylene (reviews [3146, 3244]), a mixture of products is obtained.

+ DEAD —PhCl, Δ; 40%→ [3179]

[1,4]Oxazino[2,3,4-*kl*]phenothiazine

10. 1,3-Thiazine

When a thiolactam adds on to a cyanoalkyne under basic conditions and at room temperature a fused thiazinimine ring is formed in high yield.

R = H, Me; + PhC≡CCN —diox, TEA; 95%→ [3030]

[1,2,4]Triazolo[5,1-*b*][1,3]thiazine

11. 1,3-Thiazin-4-one

2-Thiohydantoin and allene-1,3-dicarboxylic ester react under argon to form a fused thiazinone ring [3405]. 5-Methyl-1,2,4-triazole-3-thione (or its tautomer) adds on to DMAD in refluxing methanol to give a high yield of a triazolo[3,4-*b*]thiazin-5-one [3920], but the 5-phenyl homologue in methanol containing a trace of acetic acid is reported to lead to the [5,1-*b*]thiazin-7-one [3944]. 2-Mercaptobenzimidazoles are cyclized in low-to-moderate yield by heating with a 3-chlorocrotonic ester and a strong base [2869].

$$\text{2-thioxoimidazolidin-4-one} + \text{CHCOOMe=C=CHCOOMe} \xrightarrow[45\%]{\text{MeOH, Ar, }\Delta}$$ [3405]

Imidazo[2,1-*b*][1,3]thiazine-2,5-dione

(36.4) + DMAD $\xrightarrow[40\%]{\text{MeOH, AcOH, }\Delta}$ [3944]

R = 2,4-$Cl_2C_6H_3$

[1,2,4]Triazolo[5,1-*b*][1,3]thiazin-7-one

(36.4) + DMAD $\xrightarrow[75\%]{\text{MeOH, }\Delta}$ [3920]

R = Me

1,2,4-Triazolo[3,4-*b*][1,3]thiazin-5-one

2-mercaptobenzimidazole + COOEt–CH=CMe–Cl $\xrightarrow[15\text{–}44\%]{\text{MeOH–MeONa, }\Delta}$ [2869]

R = H, Me, Cl, NO_2, COOH

[1,3]Thiazino[3,2-*a*]benzimidazol-4-one

III. FORMATION OF A SEVEN-MEMBERED RING

1. 1,4-Thiazepine

Thermal or base-induced cyclization of an oxomethylthio side-chain to a ring-NH is accomplished by dehydration in DMF.

$\xrightarrow[60\text{–}90\%]{\text{DMF, }\Delta}$ [2709]

R^1 = H, Cl, Br, NO_2, PhCO, H_2NSO_2, Me_2NCO;
R^2 = Me, Ph, 4-Cl-, 4-F-, 4-HO-, 4-NO_2-C_6H_4

Imidazo[1,2-*d*][1,4]benzothiazepine

CHAPTER 37

Alkene or Alkyne and Amine, Azide or Nitro

I.	Formation of a Five-membered Ring	884
	1. Pyrrole	884
	2. Imidazole	886
II.	Formation of a Six-membered Ring	887
	1. Pyridine	887
	2. Pyridine 1-Oxide	889
	3. Pyridazine	889
	4. Pyrimidine	889
	5. 1,3-Oxazine	890
III.	Formation of a Seven-membered Ring	890
	1. Azepine	890
IV.	Formation of an Eight-membered Ring	890
	1. 1,4-Diazocine	890

Cyclizations of compounds containing the following pairs of functions are also included in this chapter: acylamine and alkene, alkene and imidate, alkene and imine, and alkyne and carbamate.

I. FORMATION OF A FIVE-MEMBERED RING

1. Pyrrole

Thermolysis of an alkenyl-azide can give high yields of a fused pyrrole; the method also has the advantages of requiring moderate temperature and of being conducted in a neutral medium. Functional groups such as an *N*-oxide [as in **(37.1)**] thus often survive the reaction [2436], but when this kind of reaction is induced photolytically, some substituents on the terminal carbon atom may migrate [3163].

(37.1) → PhMe, Δ, 95% → Pyrrolo[3,2-*c*]pyridine 5-oxide [2436]

MeCN, *hν*, 26–90% → Thieno[3,2-*b*]pyrrole [3163]

R^1 = H, Ac, COOEt, RS, RSO, RSO_2;
R^2 = Ac, COOEt, MeS

An early report of annulation of an iminophosphorane (formed *in situ*) to a malonylidene function describes the synthesis of a thienopyrrole in high yield [3084a]. A different procedure gives the isomeric [3,4-*c*] product [3492a]. A styryl phenyl ketone is converted into a 1-hydroxyindole by warming with potassium cyanide [2163]; the chemistry of 1-hydroxyindoles has been reviewed [3906].

THF, PPH_3, Δ, 78% → Thieno[2,3-*c*]pyrrole [3084a]

1. TsOH 2. Na_2CO_3, ~70% → Thieno[3,4-*c*]pyrrole [3499]

Metal-catalysed cyclization is an increasingly important method of synthesis (reviews [B-41, 2998, 3069, 3501, 3505]). Bis(acetonitrile)dichloropalladium promotes cyclization of an alkynyl carbamate with epoxypropane and a vinyl chloride [3438]; an alkynylaniline is converted into an indole by heating with a palladium chloride–triphenylphosphine catalyst [3631]. Mercury(II) acetate is an effective promotor of the cyclization of the alkenylaniline (**37.2**) to a reduced carbazole [3861].

i, 31–82% → Indole [3438]

R^1 = Bu, Ph; R^2 = H, Cl, Me; R^3 = H, Me, Et
i, THF, bis(acetonitrile)dichloropalladium, Ar

$\xrightarrow[\sim 90\%]{PdCl_2(PPh_3)_2,\ TEA,\ \Delta}$ [3631]

R = CH_2=CH, aryl, heteroaryl

$\xrightarrow[\sim 80\%]{ROH,\ AcOH,\ Hg(OAc)_2}$ [3861]

(37.2)
R = alkyl

Carbazole

2. Imidazole

Thermolysis of a vinyl-azide (**37.3**) results in the formation of an imidazole ring; the product consisted of a separable mixture of epimers. Baker's yeast (*Saccharomyces cerevisiae*) converts *N*-allylbenzoxazinone into its imidazo analogue under aerobic conditions at 37 °C in 5 days. Pretreatment of the yeast by ultrasonification increased the yield. Similar results are obtainable from quinazoline substrates. The use of ultrasonic irradiation in organic chemistry has been reviewed [B-58, 3615, 3646].

$\xrightarrow[\sim 81\%]{yeast,\ 37\ ^\circ C}$ [3754]

X = O, S; Y = O, NH, MeN, PhN

Imidazo[1,2-*c*][1,3]benzoxazin-5-one
Imidazo[1,2-*c*][1,3]benzoxazine-5-thione
Imidazo[1,2-*c*]quinazolin-5-one

$\xrightarrow[58\%]{PhMe,\ \Delta}$ [3167]

(37.3)

1,3-Diazabicyclo[3.2.0]hept-3-en-7-one

II. FORMATION OF A SIX-MEMBERED RING

1. Pyridine

A nitrovinyl-iminophosphorane may be cyclized to a fused pyridine ring by heating it in toluene for about 12 h with an isocyanate; yields are usually very good.

RNCO, PhMe, Δ; 79–97% [3618]

R = Et, Ph, 4-Cl-, 4-MeO-, 4-Me-C_6H_4

Pyrido[2,3-*b*]indole

Other ring systems synthesized similarly:

Pyrazolo[3,4-*b*]pyridine [3619]

Pyrido[2,3-*d*]pyrimidine-2,4-dione [3717]

This ring may also by obtained by heating either a 2-vinylaniline with phosgene [2209a] or a 2-vinylacetanilide with phosphorus oxychloride [3345]. *o*-Vinylimines undergo oxidative cyclization when heated (in air) at 180 °C in an inert solvent such as 1,2,4-trichlorobenzene (TCB). When a methoxycarbonyl group is attached at the β-vinyl carbon, a higher temperature (200 °C) is required [3417].

+ $COCl_2$; PhMe, Δ; 77–88% [2209a]

Ar = Ph, 4-Me-, 2- or 4-Cl-C_6H_4

Quinoline

PhMe, $POCl_3$, Δ; 34–90% [3345]

R^1 = H or COR^2; R^2 = Me, Ph; R^3 = H, Me; R^4 = H, MeO

Thermolytic annulation of azide and alkene groups is a widely used method of ring formation which often gives good yields. In compounds such as the α-azido-*o*-vinylcinnamic ester (**37.4**), it is possible for the nitrene to attack either a ring carbon (see Chapter 58) or the vinyl group. In practice, the latter reaction takes precedence but a substituent (R) on the β-carbon of the vinyl group has an influence of the product(s) formed, as is shown in the scheme [3169]. Epoxidation of the vinyl group prevents this cyclization (see Chapter 58, Section I.1). Another factor which controls the course of this cyclization is the stereochemistry of the ester (**37.4**, R = Ph). The *Z*-isomer yields the isoquinoline as the main product while the other isomer is mostly converted into the benzazepine [3756]. Replacing the vinyl group of (**37.4**) by an allyl chain and heating the compound in toluene gives the benzazepine as the main product but in other solvents, a mixture of products is obtained, none of which is an isoquinoline [3169].

CH=CHR
CH=CN$_3$
COOEt
(37.4)
PhMe, Δ
R = H
75%
CH$_2$R
N
COOEt
(37.5)
Isoquinoline
[3169]
R = Ph
(**37.5**, R = Ph)
(36%)
+
COOEt
N
(37%)
Ph
3-Benzazepine

H
R[1]
C=C
R[2]
N$_3$
CH=CCOOMe
Ar = 2-BrC$_6$H$_4$
R[1] = H; R[2] = Ar; PhMe, Δ
39%
MeO
MeO
CH$_2$R
N
COOMe
Isoquinoline
[3756]
R[1] = Ar; R[2] = H; PhMe, Δ
55%
MeO
MeO
COOMe
N
Ar
3-Benzazepine

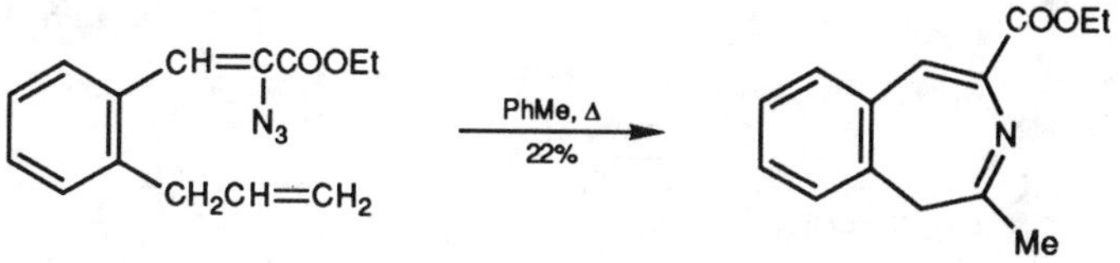

PhMe, Δ
22%
[3169]

2. Pyridine 1-Oxide

A *β*-subtituted vinyl group carrying two electron-withdrawing groups reacts under mild conditions with a neighbouring nitro function and potassium cyanide to produce a good yield of a pyridine oxide.

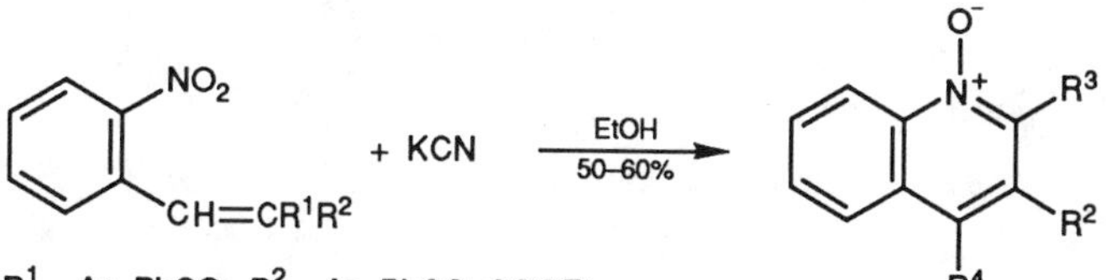

+ KCN
EtOH
50–60%
[2163]

R^1 = Ac, PhCO; R^2 = Ac, PhCO, COOEt;
R^3 = Me, Ph; R^4 = $CONH_2$, CN

Quinoline 1-oxide

3. Pyridazine

Cinnoline-3- and -4-carbaldehyde are not readily available; the 3,4-dialdehyde was first synthesized in 1987 from a 2-vinylbenzenediazonium salt followed by oxidation of the hydroxymethyl group.

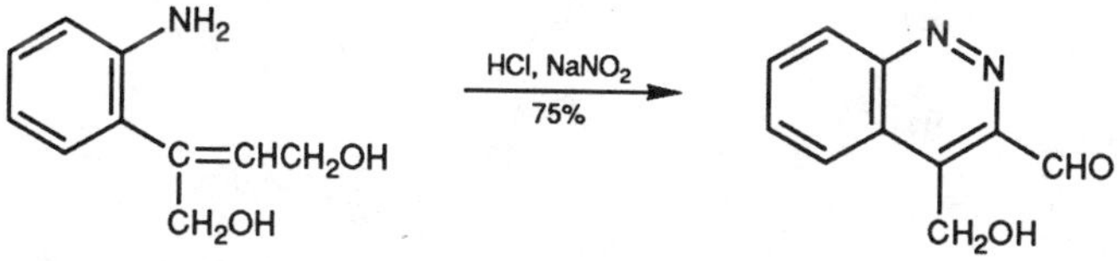

HCl, $NaNO_2$
75%
[3269]

Cinnoline

4. Pyrimidine

An alkynyl azide such as **(37.5)** is thermolytically converted into a pyrimidine; alkenyl azides similarly yield dihydro products.

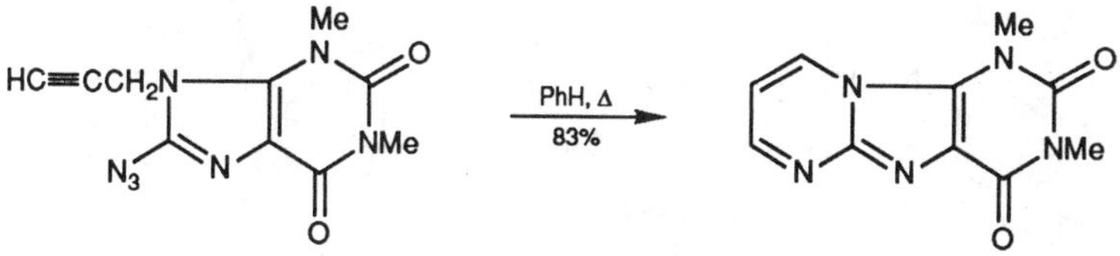

PhH, Δ
83%
[3610]

(37.5)

Pyrimido[1,2-*e*]purine-2,4-dione

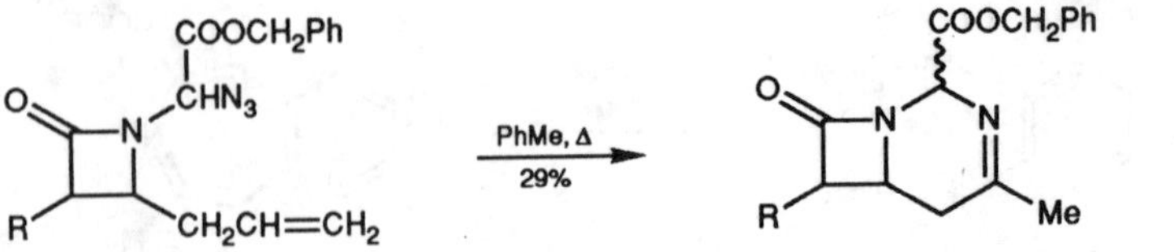

$R = PhOCH_2CONH$

$PhMe, \Delta$ / 29%

1,3-Diazabicyclo[4.2.0]oct-3-en-8-one [3167]

5. 1,3-Oxazine

When the enamine (**37.6**) is treated with t-butyl hypochlorite, the oxazine is formed.

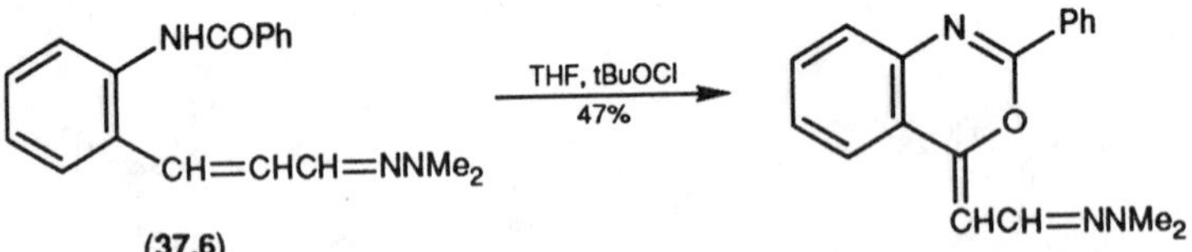

(**37.6**) THF, tBuOCl / 47%

3,1-Benzoxazine [3843]

III. FORMATION OF A SEVEN-MEMBERED RING

1. Azepine

Thermolysis of 2′-styryl-2-azidocinnamic ester (**37.4**, R = Ph) in toluene or xylene gives a moderate yield of the benzazepine (see also Section II.1).

(**37.4**) R = H PhMe or xylene, Δ / 37%

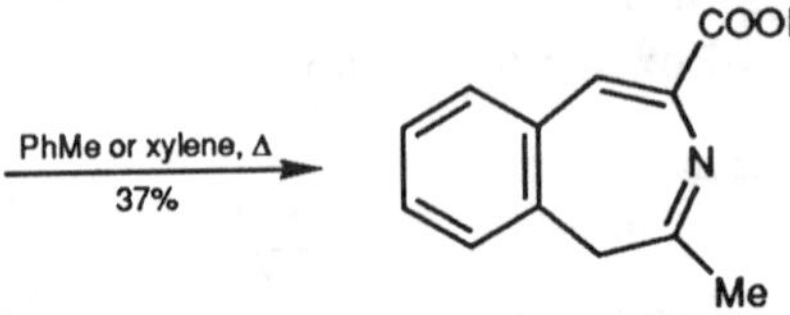

3-Benzazepine [3169, 3319]

IV. FORMATION OF AN EIGHT-MEMBERED RING

1. 1,4-Diazocine

A reductive cyclization of *o*-alkenylaminoanilines (**37.7**) in the presence of rhodium acetate in an atmosphere of carbon monoxide and hydrogen under pressure gives the diazocine and a smaller amount of the pyrrolo[1,2-*a*]benzimidazole

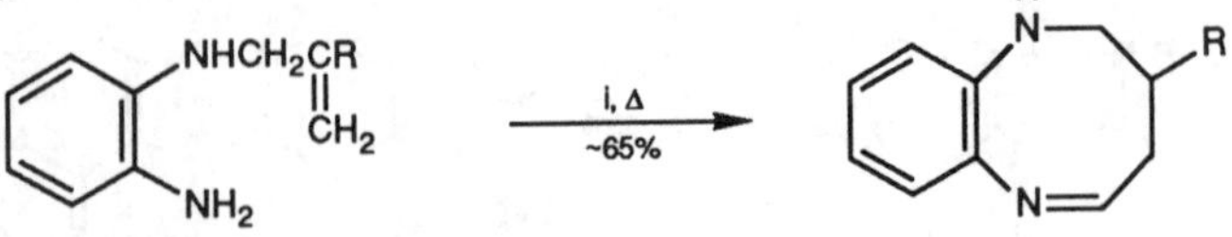

(**37.7**) i, Δ / ~65%

1,6-Benzodiazocine [3977]

R = H, Me

i, H_2, CO, $[Rh(OAc)_2]_2PPh_3$

CHAPTER 38

Alkene or Alkyne and Carboxylic Acid or its Derivative

I. Formation of a Six-membered Ring 891
   1. Pyridin-2-one 891
   2. Pyran-2-one 891
   3. Thiopyran-4-one 892

I. FORMATION OF A SIX-MEMBERED RING

1. Pyridin-2-one

2-Vinylbenzamides (prepared from 2-bromobenzamides and ethene) cyclize to the isoquinolinones in high yield when heated with dilute hydrochloric acid.

R = H, Cl, MeO, Me — dil. HCl, Δ, 73–90% → 1-Isoquinolinone [3691]

2. Pyran-2-one

The nitrogen atom is lost (as ammonia) when the 2-alkynyl-3-pyridinecarbonitrile (**38.1**) is heated with PPA.

(**38.1**) R = Bu, Ph — PPA, Δ, ~61% → Pyrano[4,3-*b*]pyridin-5-one [3496]

3. Thiopyran-4-one

Both stereoisomers of a 2-(vinylthio)benzoic acid (**38.2**) undergo cyclization on being heated with PPA. The diester formed is a useful intermediate; for example, it may be converted into the 2-methylthiochromone-3-carboxylic ester when an attempt is made to hydrolyse the diester with 50% acetic acid.

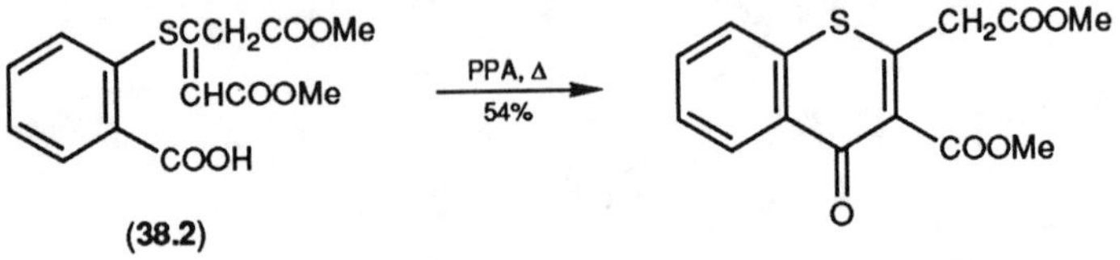

SCCH$_2$COOMe
||
CHCOOMe
COOH

(**38.2**)

PPA, Δ
54%

S CH$_2$COOMe
COOMe
O

[3405]

1-Benzothiopyran-4-one

CHAPTER 39

Alkene or Alkyne and Halogen

I.	Formation of a Five-membered Ring	893
	1. Pyrrole	893
	2. Pyrrol-2-one	894
	3. Furan	895
	4. Thiophene	895
II.	Formation of a Six-membered Ring	896
	1. Pyridin-2-one	896
III.	Formation of a Seven-membered Ring	896
	1. 1,2-Diazepine	896

I. FORMATION OF A FIVE-MEMBERED RING

1. Pyrrole

This ring may be built by the formation of a new C—C bond between alkyne and halogen functions, a reaction which is assisted by palladium dichloride–triphenylphosphine (reviews of palladium catalysis [B-41, 3069, 3505] in the presence of piperidine (as base) and formic acid (to remove hydride ion). The reaction is regio- and stereo-selective. Replacement of the piperidine and formic acid by R-substituted tributyltin enables the R group to be transferred to the indole stereospecifically [3448].

$NAcCH_2C{\equiv}CR^1$, R^2, I → (i, Δ; 40–60%) → Ac, N, R^2, CHR^1 [3442, 3448]

R^1 = H, Me, $(CH_2)_5NCH_2$; R^2 = H, COOEt
i, MeCN, $PdCl_2$, PPh_3, pip, HCOOH

Indole

Homolytic cyclization of a bromo-alkene yields dihydropyrroles but bromo-alkynes by the same mechanism form a mixture of pyrrole and 3-methylene-pyrrole in high yield (the ratio is not stated) [3434]. Aryl radicals produced by the action of samarium di-iodide on an aryl bromide react with an *o*-propynyl-amino (or propynyloxy—see Section I.3) side-chain to form a new fused pyrrole (or furan) under mild conditions. The corresponding allylamino bromides give 2,3-dihydroindoles in good yields [3974]. An activated C—C double bond reacts with ammonia or a primary amine and a bromomethyl function to form a pyrrole or an *N*-substituted pyrrole. Malonate ester is eliminated [3494, 3840].

PhH, Bu_3SnH, AIBN, 99% [3434]

THF, SmI_2, HMPA, ~52% [3974]

R = Ac, HC≡CCH_2

$(EtOOC)_2C$=CH … + RNH_2 → EtOH, Δ, 30–86% [3494, 3840]

R = H, alkyl, $MeOOCCH_2$, $PhCH_2$, Ph, C_6H_{11}

Thieno[2,3-*c*]pyrrole

Other ring systems synthesized similarly:

Pyrrolo[3,4-*c*]pyridine [3494]

Pyrrolo[3,4-*b*]pyrazine [3494]

Benzofuro[2,3-*c*]pyrrole [3494, 3840]

2. Pyrrol-2-one

In a homolytic cyclization, a side-chain alkene displaces a neighbouring nuclear bromine atom on heating the compound in toluene with tributyltin hydride and AIBN. An optical purity of 39% is observed when an asymmetric *N*-substituent is present. Palladium diacetate-catalysed cyclization (reviews [B-41, 3069, 3505] of the bromoester (**39.1**) gives the oxindole in moderate yield; the product may consist of a mixture of stereoisomers; for example, when R = Ph, one isomer predominates in the ratio 60:13.

PhMe, Bu_3SiH, AIBN ~61%

2-Indolone [3632]

i, MeCN, $Pd(OAc)_2$, PPh_3, Bu_3N

47–63% [2959, 3505]

3. Furan

Cyclization of an allyloxy-iodobenzene under conditions which are broadly similar to those described in Section I.1 gives a good-to-excellent yield of the corresponding oxygen heterocycle. A reduced benzofuran is formed in a homolytic process when the bromoacetal (**39.1**) is heated with tributyltin hydride and a catalytic amount of AIBN; the first-formed product is dehydrogenated at room temperature in the presence of 4-toluenesulphonic acid to give the naturally occurring evodone.

R = H, alkyl, cyclopentyl

i, DMF, Bu_4NCl, $Pd(OAc)_2$, HCOONa, 80 °C, 48 h

45–83% Benzofuran [3443]

(**39.1**) 4-Benzofuranone [3444]

i, Bu_3SnH, AIBN, $NC(CH_2)_3N{=}N(CH_2)_3CN$;

ii, TsOH

45%

4. Thiophene

Neighbouring chloro and ethynyl functions are converted into a thiophene ring by either sodium sulphide–DMF or morpholine and sulphur.

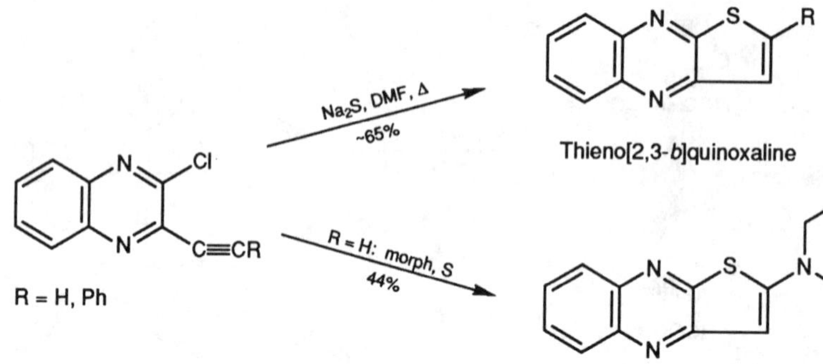

Na_2S, DMF, Δ ~65%

Thieno[2,3-*b*]quinoxaline [3074]

R = H, Ph

R = H: morph, S 44%

II. FORMATION OF A SIX-MEMBERED RING

1. Pyridin-2-one

Palladium-catalysed cyclization of suitable iodo-alkynes (**39.1**) is improved by the presence of thallium(I) carbonate to give the aminovinylisoquinoline (**39.2**); otherwise, a poor yield of the vinylisoquinoline (**39.3**) is obtained.

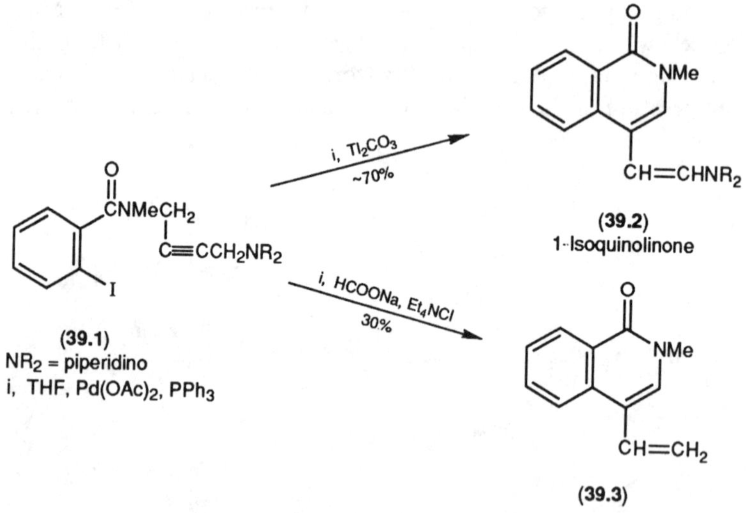

i, Tl_2CO_3 ~70%

(**39.2**) 1-Isoquinolinone [3879]

(**39.1**)
NR_2 = piperidino
i, THF, $Pd(OAc)_2$, PPh_3

i, HCOONa, Et_4NCl 30%

(**39.3**)

III. FORMATION OF A SEVEN-MEMBERED RING

1. 1,2-Diazepine

Treatment of *o*-vinylphenylhydrazones (**40.5**) with sodium acetate or sodium cyanide in benzene at 25 or 40 °C gives a mixture of the 1,2-benzodiazepine and its 4,5-dihydro derivative.

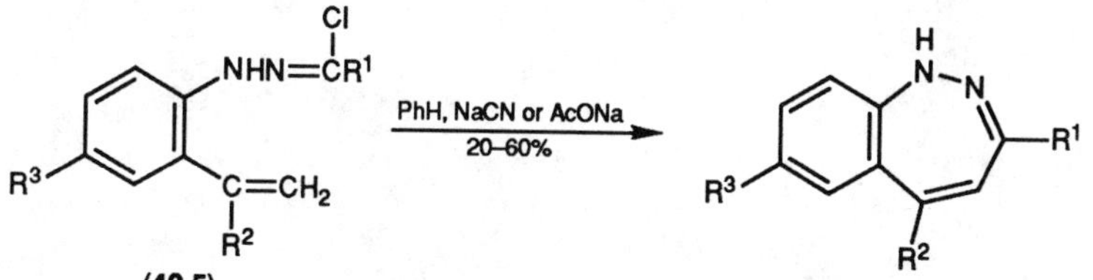

Cl
NHHN=CR[1]
R[3]
C=CH$_2$
R[2]
(40.5)

PhH, NaCN or AcONa
20–60%

H
N–N
R[1] [3009a]
R[3]
R[2]

R^1 = Ac, COOEt, Ph; R^2 = H, Ph; R^3 = H, Cl

1,2-Benzodiazepine

CHAPTER 40

Alkene or Alkyne and Hydroxy or Ether

I. Formation of a Five-membered Ring 898
1. Pyrazole 898
2. Furan 899
3. Furan-3-one 899
II. Formation of a Six-membered Ring 900
1. Pyridin-2-one 900
2. Pyran 900
3. Pyran-4-one 901
4. Thiopyran-4-one 902
5. 1,3-Oxazin-2-one 903
6. 1,4-Oxazine 903

I. FORMATION OF A FIVE-MEMBERED RING

1. Pyrazole

Disrotatory cyclization of the reduced pyridine (**40.1**) occurs when it is refluxed in toluene.

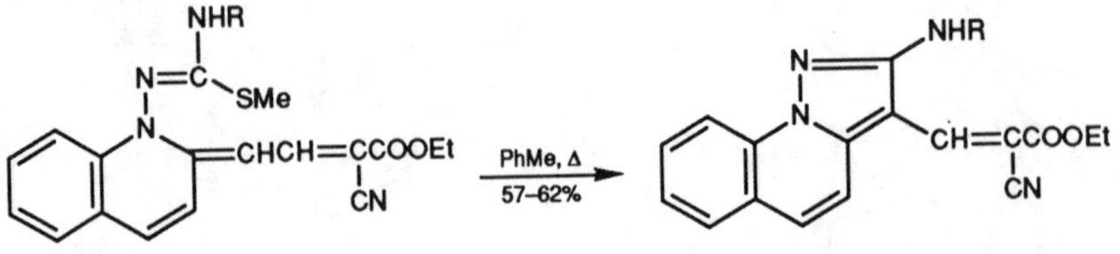

NHR, N=C, SMe, N, =CHCH=CCOOEt, CN → PhMe, Δ, 57–62% → N, N, NHR, CH=CCOOEt, CN [2986]

(**40.1**)
R = Me, Et

Pyrazolo[1,5-*a*]quinoline

2. Furan

Selective demethylation by pyridine hydrochloride *in situ* of an *o*-alkynylanisole in the presence of sodium hydrosulphite gives a high yield of a benzofuran [2444]. A 3-propynylpyridin-2-ol is cyclized in high yield on heating in TFA for 6 h [3836]. Several oxidative ring closures of alkenylphenols are mentioned in this chapter in Part 1; bromine in acetic acid–sodium acetate at ambient temperature is worth mentioning, but in this example, simultaneous bromination occurs [3678].

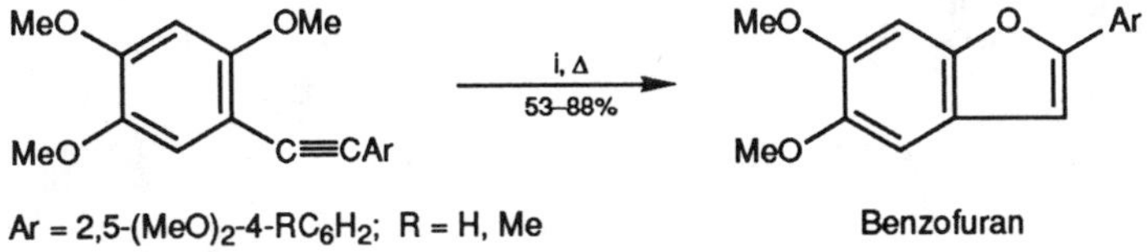
MeO
OMe
MeO
C≡CAr
i, Δ
53–88%
MeO
O
Ar
MeO
Ar = 2,5-(MeO)2-4-RC6H2; R = H, Me
Benzofuran

[2444]

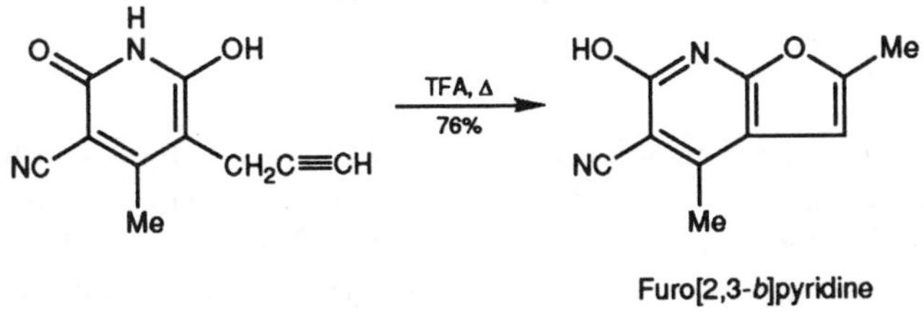
H
O
N
OH
NC
CH2C≡CH
Me
TFA, Δ
76%
HO
N
O
Me
NC
Me
Furo[2,3-b]pyridine

[3836]

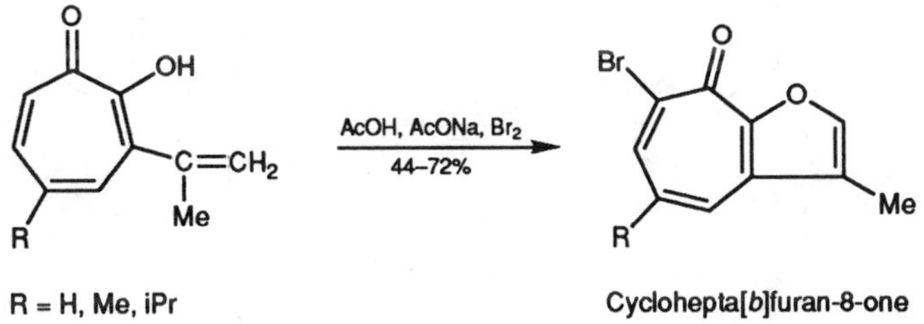
O
OH
C=CH2
Me
R
AcOH, AcONa, Br2
44–72%
Br
O
O
Me
R
R = H, Me, iPr
Cyclohepta[b]furan-8-one

[3678]

3. Furan-3-one

This ring is efficiently synthesized by stirring an arylpropynoylphenol with silver nitrate in methanol. In some cyclizations of this type, an appreciable amount of the isomeric flavone is produced. An azide reacts with triphenylphosphine to form the iminophosphorane which, by reaction with one of several functions (see Chapters 16 and 100 for examples), forms a new nitrogen-containing ring. However, the iminophosphorane group derived from the azide (**40.2**) is retained while the ring is formed by interaction of the alkene and phenol functions. This corrects an earlier suggestion.

MeOH, $AgNO_3$ ~90% [3737]

3-Benzofuranone

R = H, MeO, iPrO; Ar = Ph, 4-$MeOC_6H_4$

Et_2O, PPh_3 52–90% [3720]

(40.2)
Ar = Ph, 4-Cl-, 4-MeO-C_6H_4

II. FORMATION OF A SIX-MEMBERED RING

1. Pyridin-2-one

Photocyclization of an enamide (review [2595]) containing a methoxy group leads to the formation of a pyridinone ring of an intermediate in the synthesis of (−)-sincitine. Reviews of numerous syntheses of isoquinoline alkaloids [2487] and of the use of Reissert compounds in such syntheses [2488] are available.

MeOH, *hν* ~44% [2486]

R = H, Me

Benzo[*a*]-1,3-benzodioxolo[4,5-*g*]quinolizin-14-one

2. Pyran

Palladium-induced cyclization of *o*-allylphenols (at about 65 °C) usually leads to a mixture of chromenes, benzofurans and their dihydro derivatives, but by judicious choice of carboxylic acid sodium salt, moderately good yields of the predominating 2*H*-benzopyran may be obtained [2864]. Such reactions have been reviewed [B-41, 3069, 3501, 3505]. 3-Chloropropynyloxybenzene undergoes a Claisen rearrangement and subsequent cyclization when the ether is heated in *N,N*-diethylaniline, but when the chlorine atom is replaced by a bromine, a mixture of products is formed [3821]. Formation of a pyran ring from a 2-allylphenol normally needs a dehydrogenating agent such as DDQ, but a 2-(1-hydroxyallyl)phenol undergoes a combined dehydration–cyclization on heating in diglyme [2445].

OH, $CH_2CH{=}CMe_2$ + RCOONa —(i, Δ; 17–40%)→ Me, Me, O [2864]

R = Me, CH_2F, CF_3, Et, tBu
i, MeOH, $PdCl_2$, $CuCl_2$

2*H*-1-Benzopyran

R^1, $OCR^2R^3C{\equiv}CCl$ —($PhNEt_2$, Δ; 60–80%)→ R^1, O, R^2, R^3 [3821]

R^1 = Cl, MeO, Me; R^2,R^3 = H, Cl, MeO, Me

OH, $C(HO)(R)CH{=}CH_2$ —(diglyme, Δ; 22–75%)→ O, R [2445]

R = H, $CH_2{=}CH$, Ph

3. Pyran-4-one

2′-Hydroxychalcone is readily cyclized with hot aqueous alkali; this is the basis of a widely applicable synthesis of flavanones [B-4]. In addition to the examples of dehydrogenative cyclization of chalcones described in Part 1 (pp. 191, 192), that produced by DDQ should be mentioned. In boiling dioxan, the reaction takes 20–72 h and produces flavones together with smaller amounts of flavanones and aurones.

R, OH, $COCH{=}CHAr$ —(aq, NaOH, Δ; 64–92%)→ R, O, Ar, O [2201]

R = H, MeO; Ar = Ph, 4-$MeOC_6H_4$

1-Benzopyran-4-one, 2-aryl-

R, OH, $C(=O)CH{=}CHAr$ —(diox, DDQ, Δ; 28–42%)→ R, O, Ar, O [3231]

R = H, MeO; Ar = Ph, MeO-, Cl-C_6H_4

The Algar–Flynn–Oyamada reaction is a convenient and efficient method of converting a chalcone into a 3-hydroxyflavone [2392] but chalcones which contain a 6′-substituent (R ≠ H) give benzofuran-3-ones (coumarones) instead [2261]. A dehydrogenative high-yielding cyclization of a chalcone may be effected by heating with selenium dioxide in isopentyl alcohol [2521].

EtOH, NaOH, Δ; 56% [2392]

Ar = 4-$MeOC_6H_4$

aq. NaOH. H_2O_2; 52–80% [2261]

Ar = $MeOC_6H_4$, Ph

3-Benzofuranone

An acryloylphenol in which the hydroxy group is protected as its methoxymethyl ether cyclizes on being heated in refluxing acetic acid for 6 h. When thallium(III) nitrate (TN) causes ring closure of acyloxychalcones such as **(40.3)**, a rearrangement also occurs to form the isoflavone in good yield.

AcOH, Δ; ~56% [3855]

R^1 = $MeOCH_2$; R^2 = H, Me, Ph

Naphtho[1,2-*b*]pyran-4-one

MeOH, TN, Δ; 70% [3564]

(40.3)
Ar = 4′-$PhCH_2O$-2′,5′-$(MeO)_2C_6H_4$

Benzo[1,2-*b* : 5,4-*b*′]dipyran-4,6-dione

4. Thiopyran-4-one

A 3-alkenoylcoumarin such as **(40.4)** may be thiated at C-4 and cyclized when the compound is heated with hydrogen sulphide and sodium acetate.

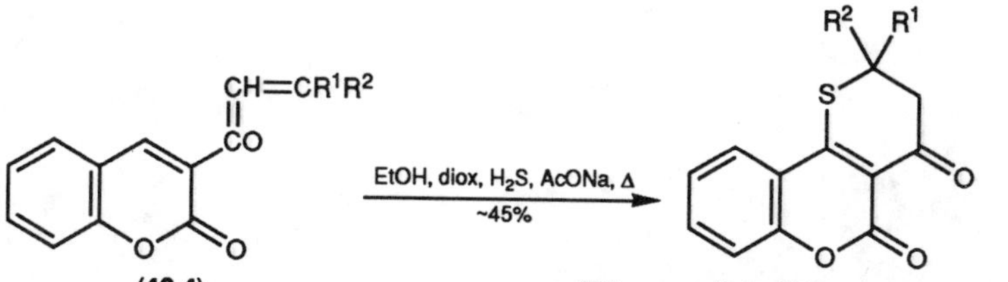

EtOH, diox, H_2S, AcONa, Δ ~45% [3235]

(40.4)
R^1 = Me, 4-MeC_6H_4; R^2 = H, Me

Thiopyrano[3,2-*c*][1]benzopyran-4,5-dione

5. 1,3-Oxazin-2-one

Chalcones may be ring-closed to oxazinones by reaction with alkyl or aryl isocyanates in boiling benzene containing a catalytic amount of potassium hydroxide.

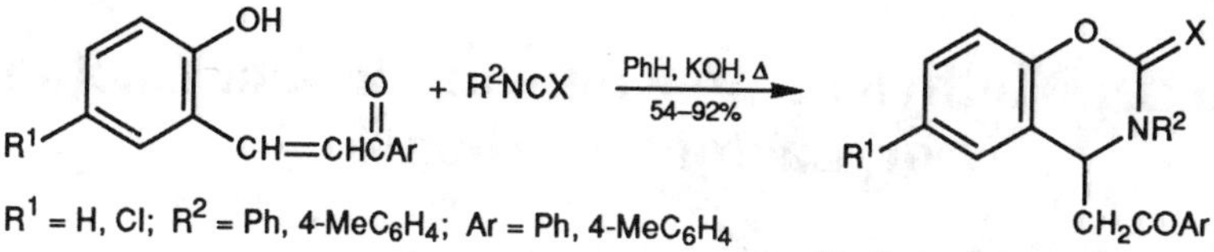

+ R^2NCX PhH, KOH, Δ 54–92% [3373]

R^1 = H, Cl; R^2 = Ph, 4-MeC_6H_4; Ar = Ph, 4-MeC_6H_4

X = O 1,3-Benzoxazin-2-one
X = S 1,3-Benzoxazine-2-thione

6. 1,4-Oxazine

Phase-transfer catalysis has been used successfully to annulate side-chain alkyne and hydroxy functions.

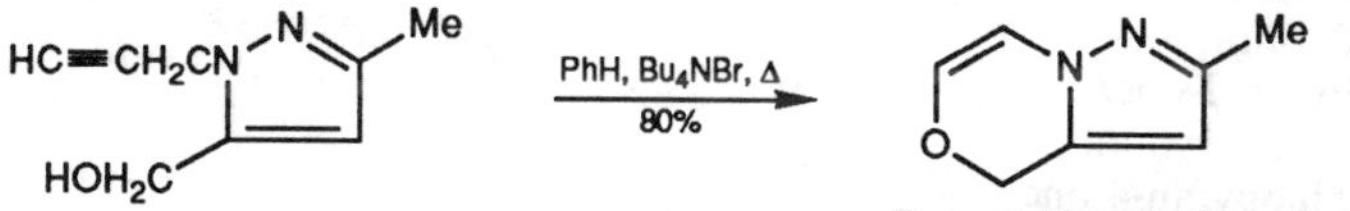

PhH, Bu_4NBr, Δ 80% [3020]

Pyrazolo[5,1*c*][1,4]oxazine

CHAPTER 41

Alkene, Methylene or Ring-carbon and Lactam or Lactone Carbonyl

I. Formation of a Five-membered Ring 904
  1. Pyrrole 904
  2. Pyrrole-2-thione 905
  3. Thiazole 905
II. Formation of a Six-membered Ring 906
  1. Pyridine 906
  2. Pyridazine 907
  3. Pyrimidin-4-one 907
  4. Pyran 908
  5. Pyran-2-one 908
  6. Thiopyrylium 909
  7. Thiopyran-4-one 909

A review on the use of lactams (and lactones) and their sulphur analogues in the synthesis of heterocyclic compounds [2387] is relevant to reactions described in this chapter.

1. FORMATION OF A FIVE-MEMBERED RING

1. Pyrrole

When the lactam (**41.1**) is heated with acetic anhydride and DMAD, both methyl and carboxyl groups are lost during the formation of a fused pyrrole ring [3542]. Heating a side-chain ester (**41.2**) with soda-lime brings about annulation of the α-methylene group to the lactam carbonyl [2794].

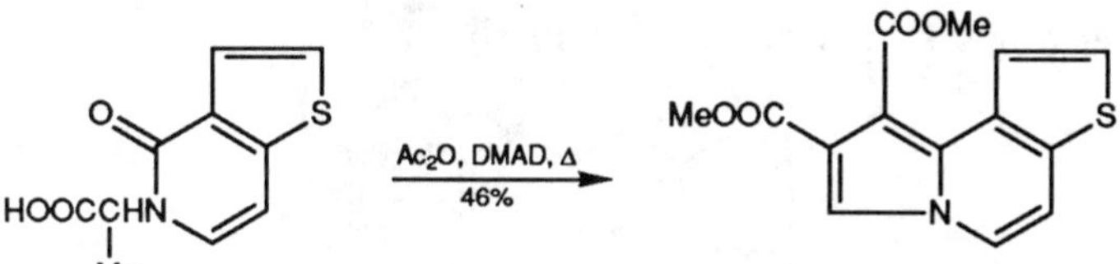

Ac_2O, DMAD, Δ 46%

(41.1)

Thieno[2,3-*g*]indolizine

[3542]

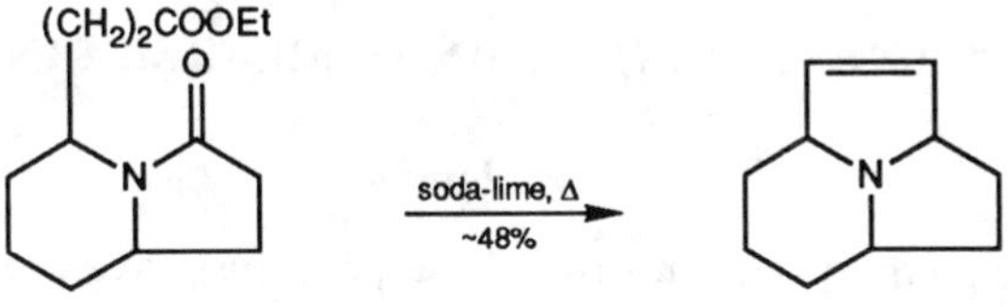

soda-lime, Δ ~48%

(41.2)

Pyrrolo[2,1,5-*cd*]indolizine

[2794]

2. Pyrrole-2-thione

Cyanothioacetamide (reviews [3114, 3331, 3458]) reacts with a maleimide to give a new pyrrolethione ring in good yield.

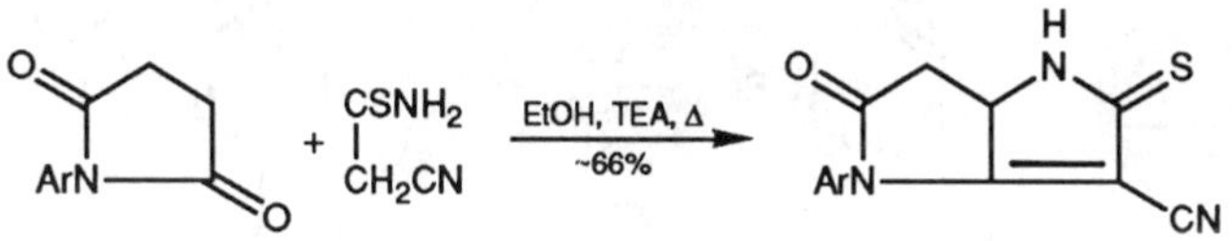

EtOH, TEA, Δ ~66%

Ar = Ph, 2- or 4-Cl-C_6H_4

Pyrrolo[3,2-*b*]pyrrol-5-one, 2-thioxo-

[3134]

3. Thiazole

NBS–sulphuric acid at 70–100 °C is an efficient reagent for the dehydrogenative cyclization of a substrate containing a thiolactam and a neighbouring benzene ring. *N*-Chlorosuccinimide–sulphuric acid may give better results on 6-thioxopyrazolo[3,4-*d*]pyrimidin-4-one [3349].

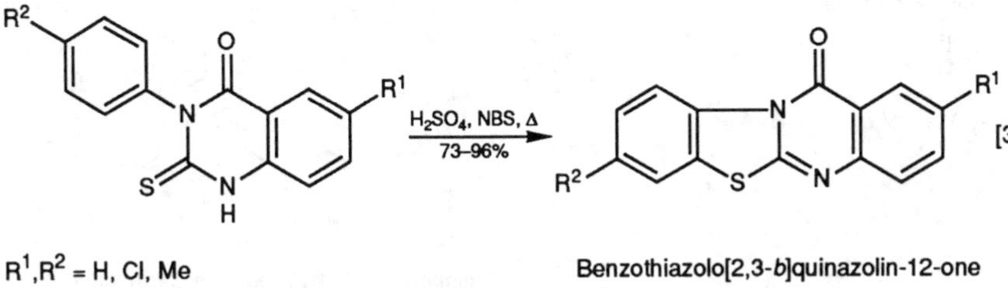

H_2SO_4, NBS, Δ 73–96%

R^1,R^2 = H, Cl, Me

Benzothiazolo[2,3-*b*]quinazolin-12-one

[3349]

Another ring system synthesized similarly:

[3349]

Pyrazolo[3',4' : 4,5]pyrimido[2,1-*b*]benzothiazol-4-one

II. FORMATION OF A SIX-MEMBERED RING

1. Pyridine

A more versatile method than most of those previously available for the synthesis of triaryloxazolopyridine is based on the cyclization of 4-benzylidene-oxazol-5-one with *N*-phenacylpyridinium bromide [3152]. Hot phosphorus oxychloride promotes C—C bond formation between a lactam carbonyl and a ring-carbon and the pyrimidinone ring is chlorinated [3093]. Malononitrile in a basic medium converts an alkenyl-lactam into a fused 3-cyanopyridine ring. Yields in this reaction are low and heating for 16 h is necessary [3427].

CH_2COPh ... Br^- —i, 45–70%→ [3152]

Ar = Ph, 2-NO_2-, 4-Cl-, 4-MeO-, 3-NH_2-C_6H_4, 2,5-$(MeO)_2C_6H_3$

Oxazolo[5,4-*b*]pyridine

$POCl_3$, Δ 60–75%

Thieno[2',3' : 3,4]pyrido[1,2-*c*]pyrimidin-7-one

[3093]

X = S; Y = CH or X = CH; Y = S

$POCl_3$, Δ ~68%

Thieno[3',2' : 3,4]pyrido[1,2-*c*]pyrimidin-4-one

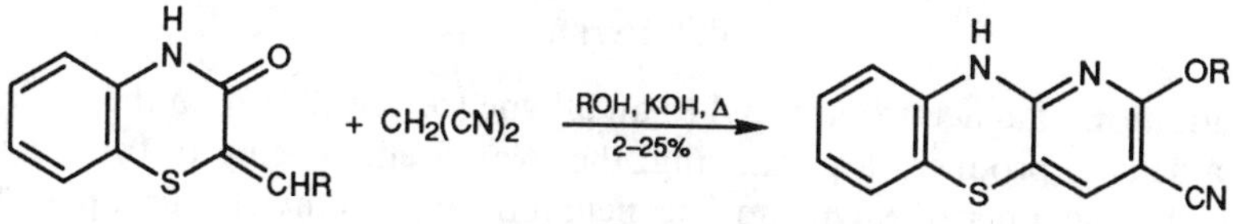

+ $CH_2(CN)_2$ → (ROH, KOH, Δ; 2–25%) [3427]

R = Me, Et

Pyrido[3,2-*b*][1,4]benzothiazine

2. Pyridazine

A 1,3-diene in which one double bond is endocyclic undergoes a [4 + 2] cycloaddition (resembling a Diels–Alder reaction) to the N=N bond of 1,2,4-triazole-3,5-dione; no external heating is required in this reaction.

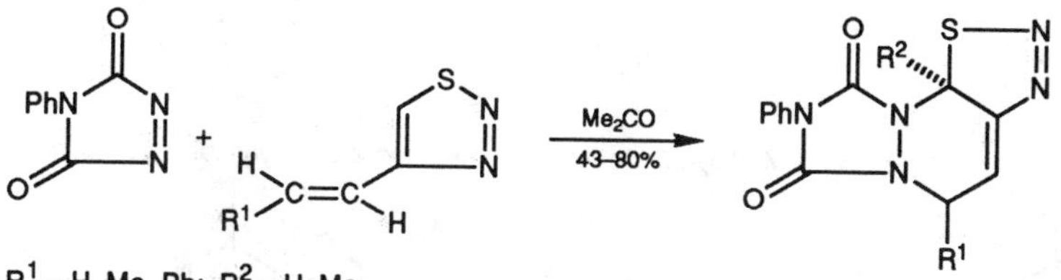

→ (Me_2CO; 43–80%) [3109]

R^1 = H, Me, Ph; R^2 = H, Me

1,2,3-Thiadiazolo[5,4-*c*][1,2,4]triazolo[1,2-*a*]pyridazine-7,9-dione

3. Pyrimidin-4-one

In a recent synthesis of tryptanthrin, one step consisted of annulating the carbonyl group of 2-indolone to its C-3 atom by reaction with isatoic anhydride (review [3008]) in the presence of tetramethylethylenediamine (TMEDA).

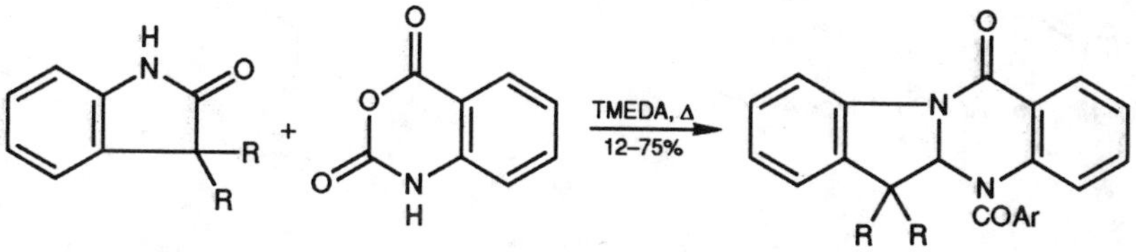

→ (TMEDA, Δ; 12–75%) [3452]

Ar = 2-$NH_2C_6H_4$; R = H, Me

Indolo[2,1-*b*]quinazolin-12-one

Another ring system synthesized similarly:

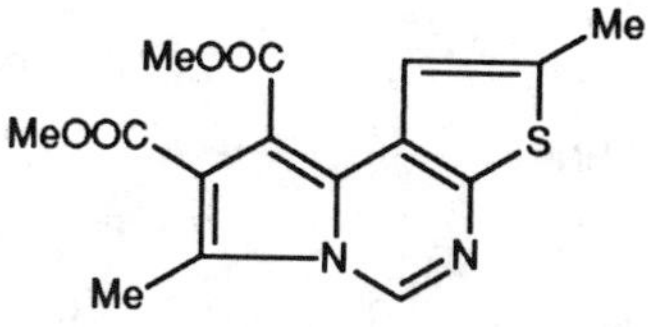

[3542]

Pyrrolo[1,2-*c*]thieno[3,2-*e*]pyrimidine

4. Pyran

Cycloaddition of a lactam carbonyl to an alkyne (review [2816]) in the presence of an aldehyde produces a pyran ring, the overall effect being to form a ring joining the lactam carbonyl and the neighbouring methylene group [2085]. Heating a lactam with an arylidenemalononitrile (p. 194) [3047] (or a near relative of it [3013]) and a base achieves the same overall effect. Malononitrile also annulates efficiently a lactam carbonyl to a neighbouring benzylidene group under basic conditions [2583].

+ PhC≡CH + R^3CHO —AcOH, Δ, 33–92%→ [2085]

R^1,R^2 = H, Me; R^3 = H, Et, Ph, 4-Cl-, 4-NO_2-, 3-MeO-C_6H_4

Pyrano[2,3-*d*]pyrimidine-2,4-dione

+ ArCH=C(R)CN —EtOH, pip, Δ, 50–75%→ [3047]

R = COOEt, PhCO, CN; Ar = 2-furyl, 2-thienyl

Pyrano[2,3-*c*]pyrazole

+ $CH_2(CN)_2$ —MeOH, MeONa, Δ, ~90%→ [2583]

R = Me, Ph; Ar = Ph, 4-Me-, 4-MeO-C_6H_4

5. Pyran-2-one

A lactam carbonyl is annulated to its neighbouring methylene by heating with either the enamine **(41.3)** or the acetonedicarboxylic ester **(41.4)**

+ MeOOC–C(=CHNMe₂)–NHCOPh **(41.3)** —AcOH, Δ, ~87%→ [3584]

R = H, Me

Pyrano[2,3-*d*]pyrimidine-2,4,7-trione

Other ring systems synthesized similarly:

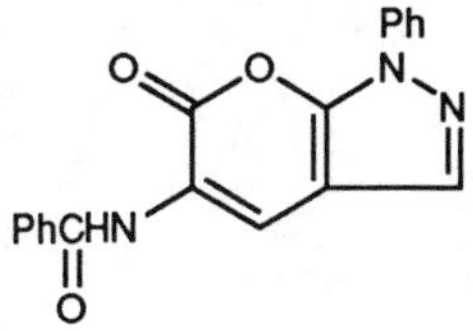

[3584] [3584]

Pyrano[2,3-*c*]pyrazol-6-one   1-Benzopyran-2-one

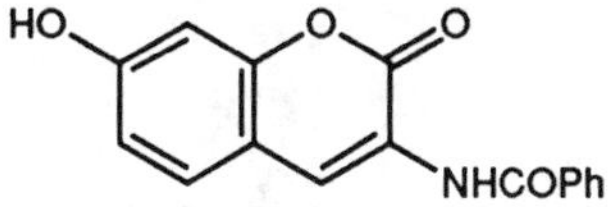

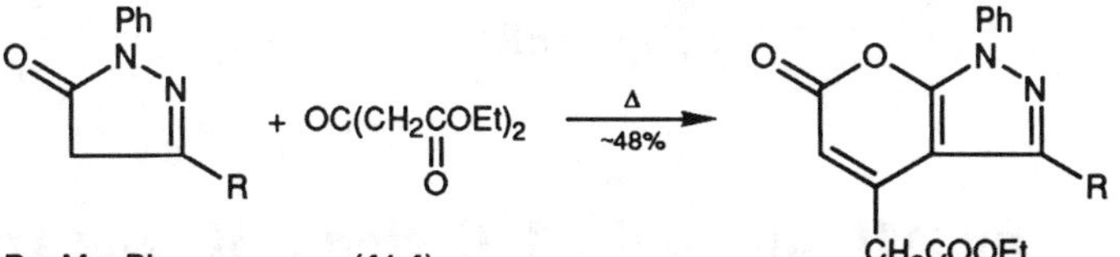

R = Me, Ph   (41.4) $\xrightarrow[\sim 48\%]{\Delta}$ [2110]

Pyrano[2,3-*c*]pyrazol-6-one

6. Thiopyrylium

A lactam thiocarbonyl becomes part of a thiopyrylium ring on reaction at ambient temperature with a 1,3-diketone.

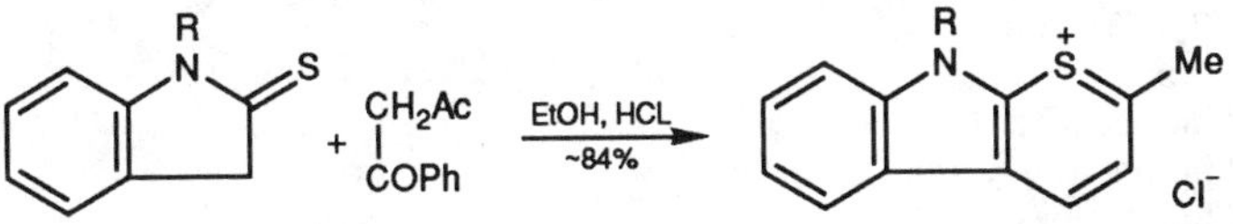

R = H, Me $\xrightarrow[\sim 84\%]{\text{EtOH, HCL}}$ [2142]

Thiopyrano[2,3-*b*]indol-1-ium

7. Thiopyran-4-one

Ethoxymethylidenemalonate reacts with a thiolactam with the formation of a new fused thiopyrimidinone ring in good yield.

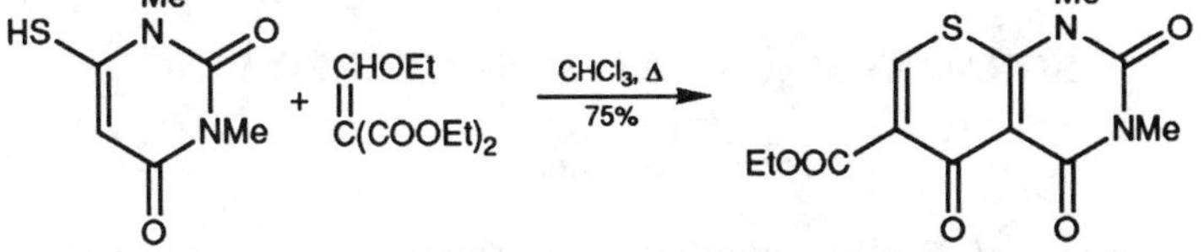

$\xrightarrow[75\%]{CHCl_3, \Delta}$ [2787]

Thiopyrano[2,3-*d*]pyrimidine-2,4,5-trione

CHAPTER 42

Alkene or Alkyne and Methylene, Ring-carbon or Ring-nitrogen

I. Formation of a Five-membered Ring 910
1. Pyrrole 910
2. Pyrrol-2-one 913
3. Pyrazole 913
4. Imidazole 913
5. 1,2,3-Triazole 913
6. Furan 914
7. Thiophene 914
8. Thiophene 1-Oxide 914
9. Thiazole 915
II. Formation of a Six-membered Ring 916
1. Pyridine 916
2. Pyridin-2-one 917
3. Pyridin-4-one 918
4. Pyrimidine 919
5. Pyrimidin-4-one 919
6. Pyran or Thiopyran 920
7. 1,2-Oxathiin 2,2-Dioxide 920

I. FORMATION OF A FIVE-MEMBERED RING

1. Pyrrole

The high reactivity of *N*-substituted pyridines (review [1870]) is exemplified in the mild conditions needed to cyclize the *N*-allylic side-chain of the pyridinium salt (**42.1**). Evidence for the participation of pyridinium ylides in this reaction is provided by adding ethyl propynoate and carbonate to the pyridinium salt [2469, 2471, 2518]. The pyridinium salt (**42.2**, R = Et) reacts with ketene dithioacetals (review [3712]) to give 3-substituted indolizines [2881]; the behaviour of 2-picolinium ylides towards the thioacetals is mentioned in Chapter 104, Section I.1. Another heterocycle containing a bridgehead nitrogen may be synthesized from the acetylenic benzimidazole (**42.3**) by heating with sodium hydrogensulphite–hydrogencarbonate [2815].

K_2CO_3, 42–95% [2469, 2471]

Indolizine

(**42.1**)
R^1 = H, Me; R^2 = Me, Ph; R^3 = MeO, Me

EtOH, ~22% [2518]

(**42.2**)
R = Me, Et

(**42.2**) + $C(SMe)_2$=$C(CN)_2$ — EtOH, TEA, Δ, 32% [2881]

R = Et

$NaHSO_3$, $NaHCO_3$, Δ, 30–76% [2815]

Pyrrolo[1,2-*a*]benzimidazole

(**42.3**)
R^1 = Me, Et; R^2 = H, Me, Ph

An indolizine may also be synthesized from a 2-allylpyridine by heating the compound at 120 °C [3753]. A good yield of a tricyclic indolizine is obtained by oxidative cyclization of the alkyne (**42.4**) at about 5 °C with MCPBA; the first-formed *N*-oxide undergoes a rearrangement of the alkynyl chain. Addition of a nucleophile towards the end of this one-pot synthesis leads to a *peri*-fused quinoline in good yield [3535]. *N*-Benzyl-*N*-propynylanilines undergo oxidative cyclization on treatment with a perbenzoic acid and a nucleophile (such as thiophenol or cyanide ion) under mild conditions [2784].

OAc
CHCCOOMe
CH_2
N
Δ
68%
COOMe
N
Indolizine

[3753]

$ArOCH_2C{\equiv}CCH_2$
N
1. CH_2Cl_2, MCPBA 2. Nu, Δ
55–75%
NuH_2C
CH_2OAr
N
Pyrrolo[3,2,1-*ij*]quinoline

[3535]

(42.4)
R^1,R^2 = H, Cl, Me; Nu = CN, 4-ClC_6HS;
Ar = 2-R^1-4-$R^2C_6H_3$

CH_2Ph
NCH_2C
CR^3
R^1
i, ii
7–71%
CH_2Ph
N
R^2
R^1
CH_2R^3
Indole

[2784]

R^1 = H, MeO, Cl; R^2 = H, Me; R^3 = PhS, CN, N_3
i, 3-ClC_6H_4COOH (C=O), $CHCl_3$, NaOH; ii, PhSH or KCN or NaN_3

The alkylidene group of (**42.5**) is joined to the ring-nitrogen by heating with bromoacetal [3598]; a less predictable reaction is that of the *N*-aryl-*S*-alkenyl-sulphinamides (**42.6**) (prepared from *N*-sulphinylanilines and an alkenyl Grignard reagent) on being heated in benzene or toluene. The C—S bond is probably cleaved and a C—N bond formed with loss of hydrogen sulphide and other sulphur-containing molecules. The sulphinamides may also be cyclized to indoles by low-temperature treatment with triethyloxonium tetrafluoroborate [3185].

RCOCH
S
HN
+
CH_2Br
$CH(OEt)_2$
DMF, Δ
61–73%
COR
S
N
Pyrrolo[2,1-*b*]thiazole

[3598]

(42,5)
R = Me, Ph, 4-ClC_6H_4

H
N
$SOCR^2$
CH_2
R^1
i or ii
20–75%
H
N
R^2
R^1
Indole

[3185]

(42.6)
R^1 = H, Cl, Br, MeO, Me; R^2 = H, Me
i, PhMe or PhH, Δ; ii, CH_2Cl_2, $Et_3\overset{+}{O}\overset{-}{B}F_4$

2. Pyrrol-2-one

α-Alkenylazoles react with maleic anhydride so as to annulate the two functions into a ring in high yield.

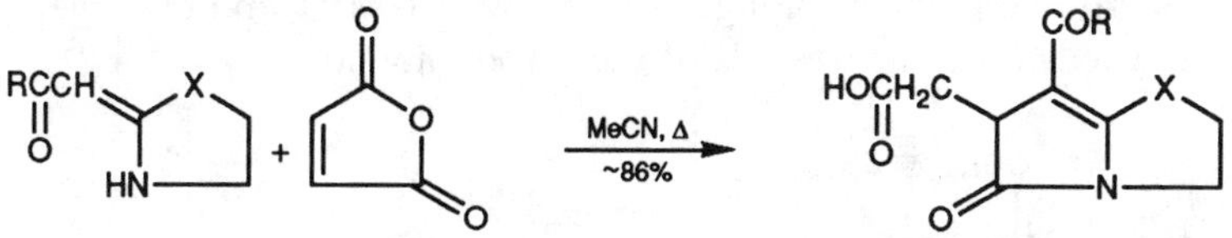

MeCN, Δ ~86% [3375]

R = Me, Ph, 4-Cl-, 4-MeO-, 4-Me- C_6H_4

X = S Pyrrolo[2,1-*b*]thiazol-5-one
X = NH Pyrrolo[1,2-*a*]imidazol-5-one

3. Pyrazole

A C—N bond is formed (in an acid-catalysed reaction) between an activated carbon—carbon double bond and a ring-nitrogen.

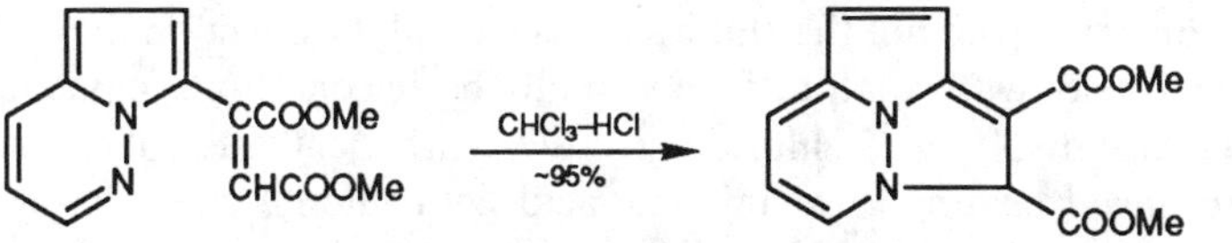

$CHCl_3$–HCl ~95% [2108]

4a,7b-Diazacyclopent[*cd*]indene

4. Imidazole

Annulation of a terminal alkyne group to a neighbouring ring-nitrogen atom occurs when the compound is refluxed with formic acid.

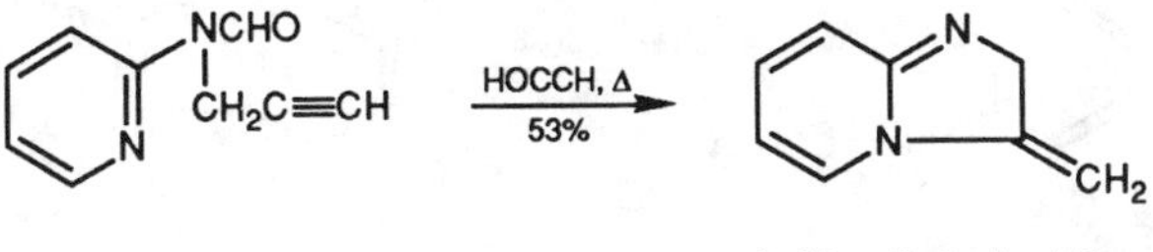

HOCCH, Δ 53% [3470]

Imidazo[1,2-*a*]pyridine

5. 1,2,3-Triazole

Tosyl azide reacts with the CH of an acyl ketene *NS*-cyclic ketal in refluxing dioxane to form a fused triazole ring.

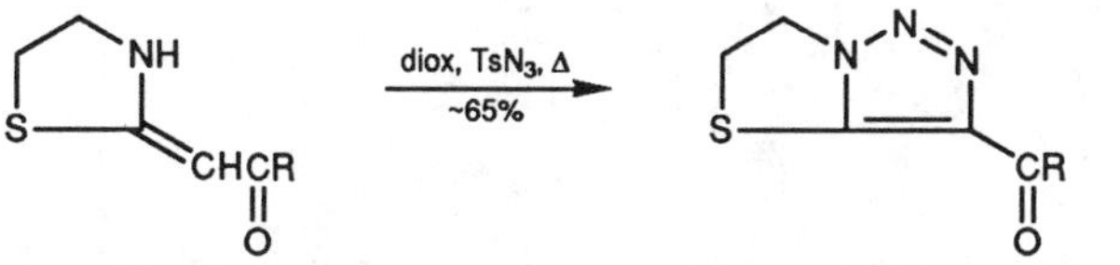

diox, TsN_3, Δ ~65% [3384]

R = Ph, 4-Cl-, 4-Me-C_6H_4

Thiazolo[3,2-*c*][1,2,3]triazole

6. Furan

Several methods of converting a propynyloxy side-chain into a fused furan ring have been described (see p. 199 for some of them). When this side-chain is attached to a 2-quinolinone, prolonged heating with HMPT and hydrogencarbonate gives good-to-excellent yields of the furoquinolinone.

$OCH_2C{\equiv}CH$ — HMPT, $NaHCO_3$, Δ; 51–94% → [3597]

R^1 = Me, Ph; R^2 = H, Br, NO_2

Furo[3,2-*c*]quinolin-4-one

7. Thiophene

Hydrogen bromide (but not the chloride) has a catalytic effect on the cyclization of alkenylbenzenes with sulphur. Under mildly basic conditions, cinnamonitriles on heating with disulphur dichloride give the synthetically useful chloronitriles (**42.7**). Prolonged heating of a cinnamic acid with thionyl chloride–pyridine is a useful synthesis of 3-chlorobenzo[*b*]thiophene-2-carbonyl chlorides.

CH=CHPh — S, HBr, Δ; 30% → [2187]

Benzo[*b*]thiophene

CH=C(R)CN — pyr, S_2Cl_2, Δ; 25–54% → [2210]

(**42.7**)

R = CN, COOEt

CH=CHCOOH — pyr, $SOCl_2$, PhCl, Δ; 26–90% → COCl, Cl [2326]

R = H, Me, Cl, F, NO_2

8. Thiophene 1-Oxide

Thionyl chloride–aluminium chloride cyclizes ethynylbenzene in good yield and without external heating.

$SOCl_2$, $AlCl_3$ 71% [2032]

Benzo[*b*]thiophene 1-oxide

9. Thiazole

Thiazolotriazoles are of considerable medicinal interest and are accessible from 2-propynyl-1,2,4-triazole by cyclization in either an alkaline or an acidic medium. In a basic medium, the main product is the [3,2-*b*] isomer whilst the [2,3-*c*] compound is the major component of acid-induced cyclization, although the yield is low [2965]. 4-(Prop-2-ynylthio)pyrimidines are cyclized in the presence of palladium(II) chloride (reviews of palladium-catalysis in organic synthesis [B-41, 3501, 3505]). In the absence of the palladium salt, no product is obtained when $R^1 = R^2 = R^3 = R^4 = H$, but the presence of a methyl group (especially at R^2) increases the yield so that moderate or even high yields are sometimes attained in the absence of palladium catalyst [3144].

$HClO_4$, H_2SO_4, Δ 19–28%

EtONa, EtOH, Δ ~67%

Ar = Ph, 2- or 4-MeO-C_6H_4

Thiazolo[2,3-*c*]-1,2,4-triazole [2965]

Thiazolo[3,2-*b*][1,2,4]triazole

i 35–93% [3144]

R^1 = H, F, Br, Me; R^2,R^3,R^4 = H, Me

i, MeCN or DME, $PdCl_2(PhCN)_2$, N_2

Thiazolo[3,2-*c*]pyrimidin-5-one

Benzyloxynitromethane is a useful reagent in the synthesis of the precursors of some *β*-lactam antibiotic molecules. The nitrovinyl thioether (**42.8**) and others are thus synthesized, and may be cyclized with tetrabutylammonium fluoride, ozone and DBU (review of the chemistry of DBU [3364]. 2-Vinylthiopyridine forms a fused thiazolium ring on stirring with bromine–carbon tetrachloride [2811].

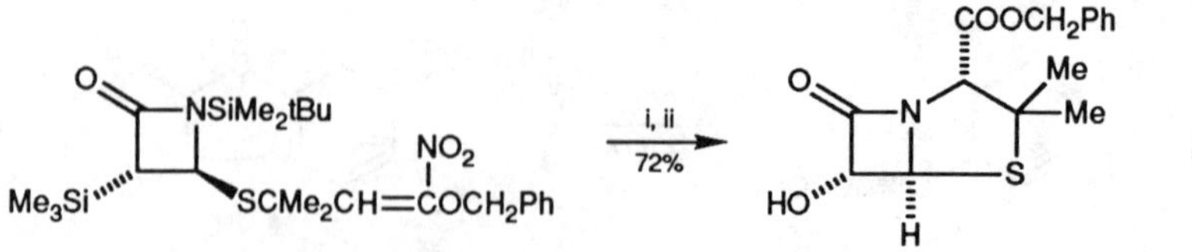

(42.8)

i, THF, Bu_4NF; O_3, CH_2Cl_2; ii, DBU

4-Thia-1-azabicyclo[3.2.0]heptan-7-one [3850]

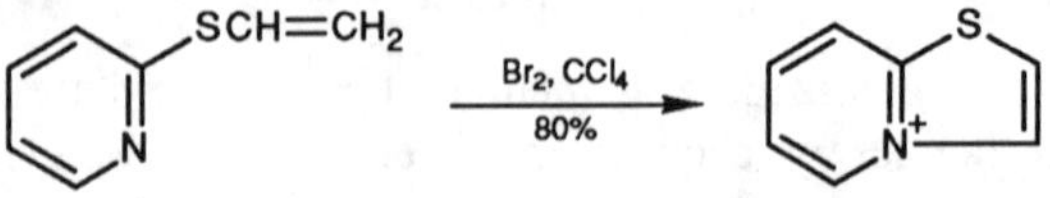

Thiazolo[3,2-*a*]pyridinium [2811]

II. FORMATION OF A SIX-MEMBERED RING

1. Pyridine

N^1-Aryl-N^2-phenylalkynylamidines are cyclized to quinolines by treatment with hot PPA; the main product is the 2-(*N*-methyl-*N*-phenylamino)quinoline but cyclization on to the other benzene ring may occur to some extent. 3-Phenylaminobut-2-enoate esters are cyclized (on to a ring-carbon) at or below room temperature by the application of Vilsmeier reagents.

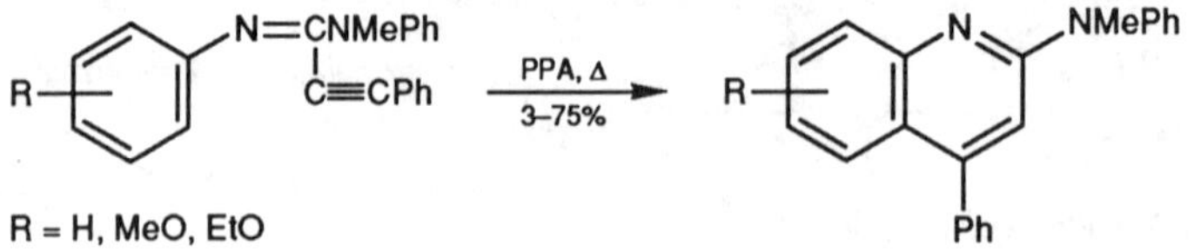

R = H, MeO, EtO

Quinoline [2836]

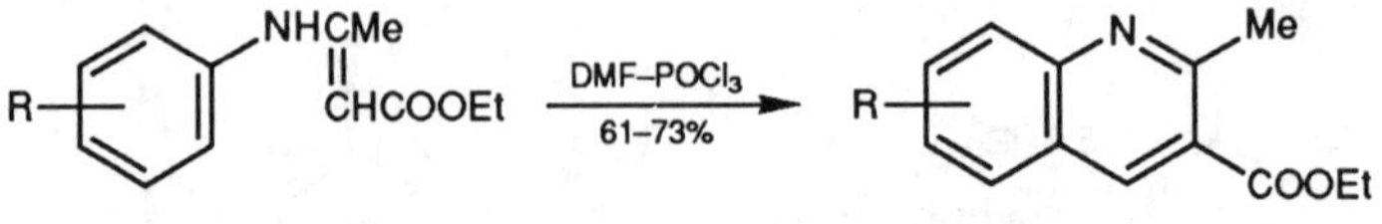

R = 3- or 4-MeO-, 2- or 4-NO_2-

[3044]

When the heterylidenemalononitrile (**42.9**) is heated in butanol, cyclization {as in the 't-amino effect' (reviews [1618, 3885])} occurs in good yield. An electron-rich alkyne adds across an allene-like C=N bond at low temperature.

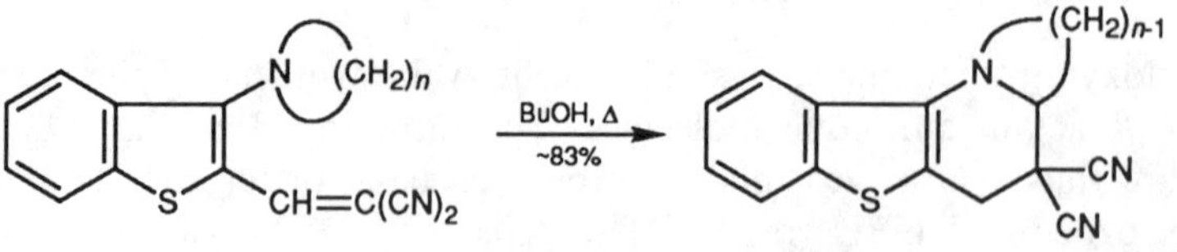

(42.9)

n = 4,5

n = 3 Benzothieno[3,2-*e*]indolizine

n = 4 Benzothieno[2,3-*c*]quinolizine [3886]

Other ring systems synthesized similarly:

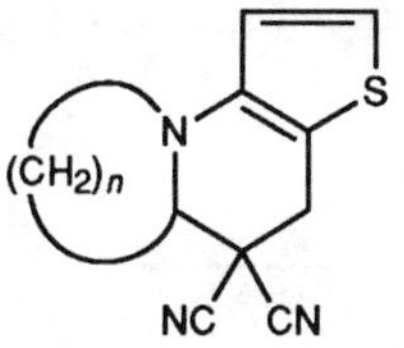

[3886]

$n = 3$ Thieno[3,2-*e*]indolizine
$n = 4$ Thieno[2,3-*c*]quinolizine

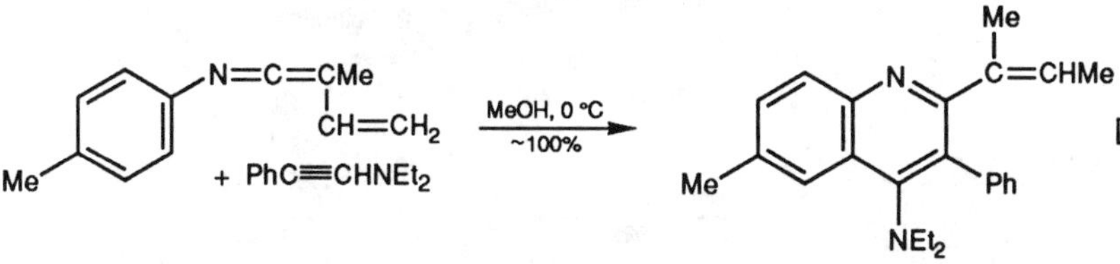

[2562]

Quinoline

2. Pyridin-2-one

The phosphate of an enol (**42.10**) converts a 2-methylidenepyrrolidine into a pyrrolopyridinone on stirring the compound with ethanolic TEA. Replacing the phosphate by an *N*-propynoylimidazole and heating the compounds in toluene gives an excellent yield.

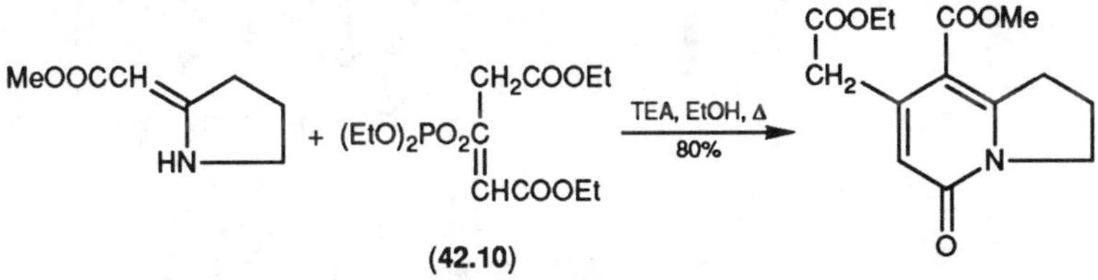

[2624]

(**42.10**)

5-Indolizinone

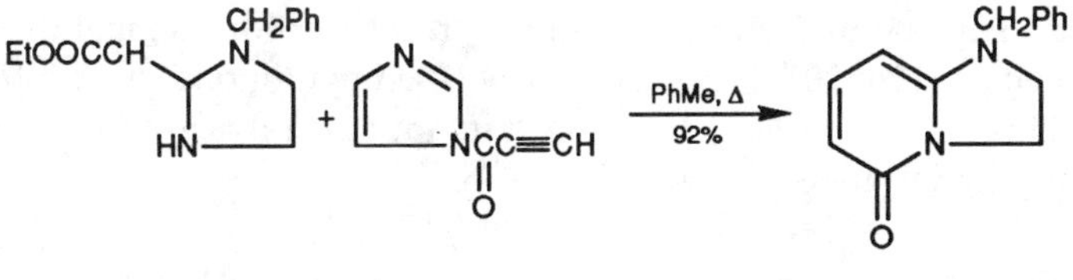

[3445]

Imidazo[1,2-*a*]pyridin-5-one

Thermolysis of the ω,ω-diphenylallene derivative (**42.11**) for 6–7 h yields the quinolinone, but when the two phenyl groups are replaced by hydrogen or alkyl groups, the reaction produces a bridged indol-2-one [3657]. The 2-arylidenoxazolidine (**42.12**) adds on to a reactive alkyne on refluxing with DMAD in methanol for 48 h [3658]. Prolonged heating (but at 220 °C) is also required for the enehydrazine (**42.13**) to react with DMAD to produce the pyridopyridazinone in a Diels–Reese reaction [3940].

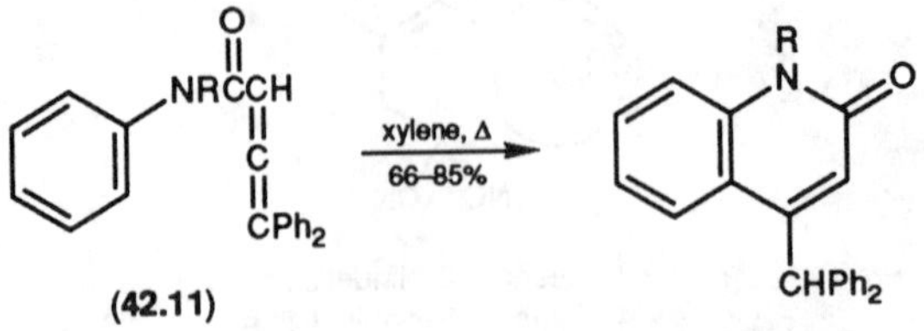

xylene, Δ
66–85%

[3657]

(**42.11**)
R = H, Ph

2-Quinolinone

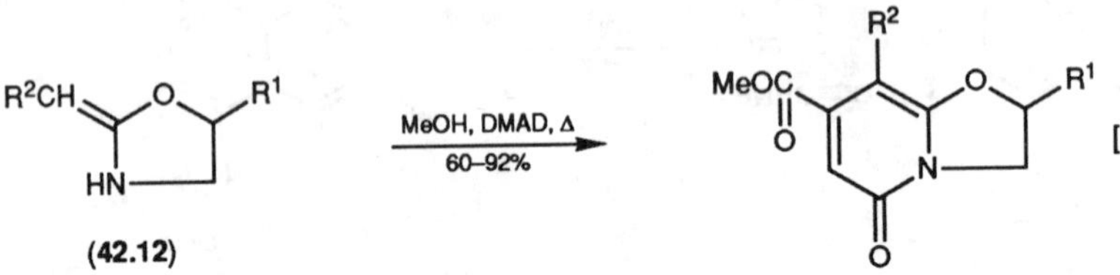

MeOH, DMAD, Δ
60–92%

[3658]

(**42.12**)
R^1 = H, Me; R^2 = PhCO, 4-MeO-, 4-Cl-, 4-Me-C_6H_4

Oxazolo[3,2-*a*]pyridin-5-one

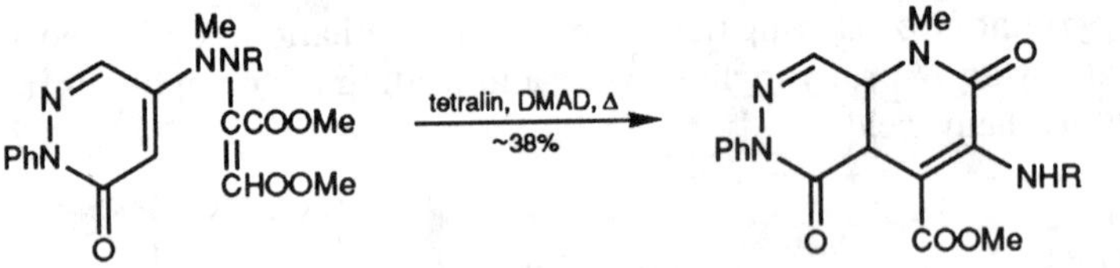

tetralin, DMAD, Δ
~38%

[3940]

(**42.13**)
R = Et, Pr

Pyrido[2,3-*d*]pyridazine-2,5-dione

3. Pyridin-4-one

2-Nitromethylideneimidazolidines (**42.14**) [synthesized from 1-nitro-2,2-bis(methylthio)ethene and ethylenediamines] react at room temperature with diketene (reviews [116, 2018]) to form a new fused pyridinone ring [3816]. The readily available chalcones may be cyclized in hot mineral acid to form a fused pyran-4-one ring in moderate yield [3517].

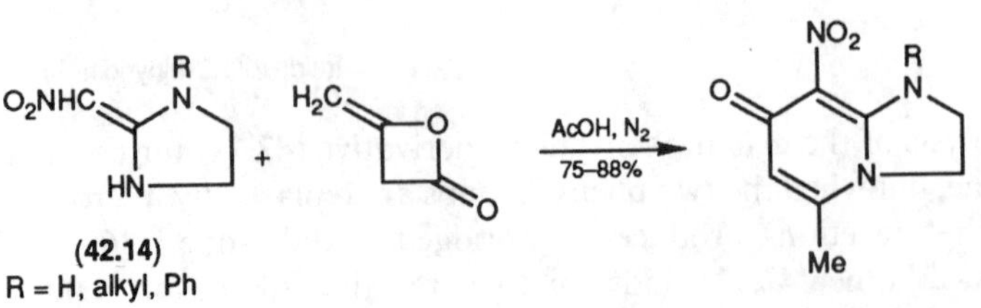

AcOH, N_2
75–88%

[3816]

(**42.14**)
R = H, alkyl, Ph

Imidazo[1,2-*a*]pyridin-7-one

Another ring system synthesized similarly:

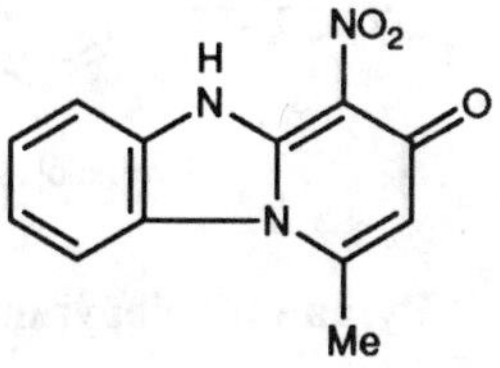

NO_2
H
N
O
N
Me

[3816]

Pyrido[1,2-*a*]benzimidazol-2-one

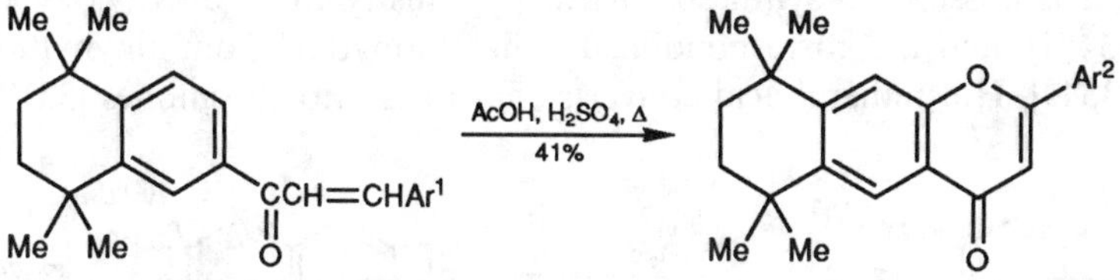

Me Me
Me Me
CCH=CHAr[1]
O

AcOH, H_2SO_4, Δ
41%

Me Me
O
Ar^2
Me Me
O

[3517]

Ar^1 = 4-$COOMeC_6H_4$; Ar^2 = 4-$COOHC_6H_4$

Naphtho[2,3-*b*]pyran-4-one

4. Pyrimidine

Heating an *N*-allylbenzamidine with PPA gives a dihydroquinazoline but the more unsaturated cinnamylideneamino-azole (**42.15**) gives a high yield of an aromatized pyrazolopyrimidine.

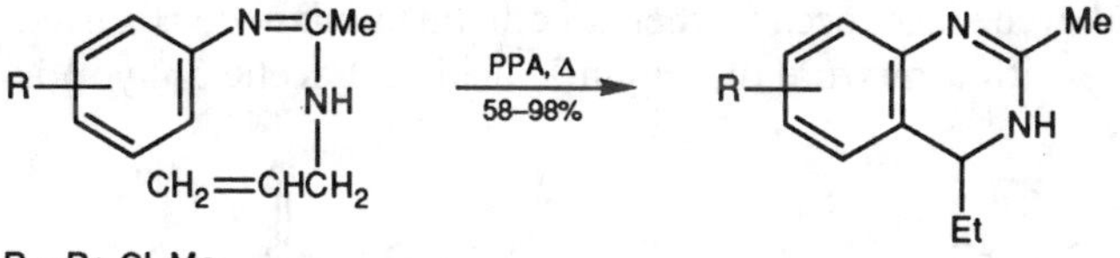

R
N=CMe
NH
CH_2=$CHCH_2$

PPA, Δ
58–98%

R
N
Me
NH
Et

[2498]

R = Br, Cl, Me

Quinazoline

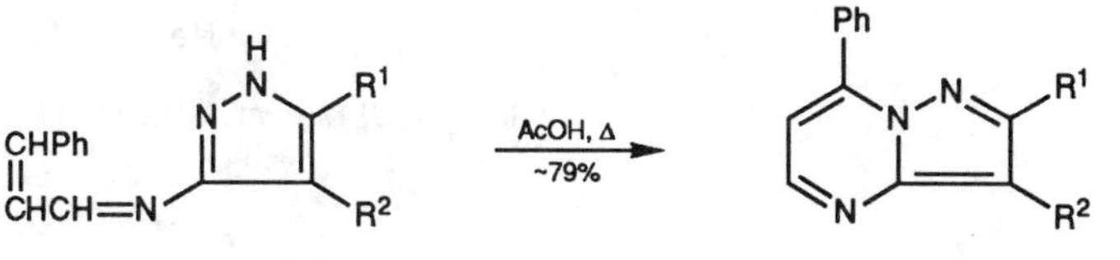

CHPh
CHCH=N
H
N
N
R^1
R^2

AcOH, Δ
~79%

Ph
N
N
R^1
N
R^2

[3767]

(**42.15**)
R^1 = Me, Ph; R^2 = H, Ph

Pyrazolo[1,5-*a*]pyrimidine

5. Pyrimidin-4-one

Heating a 2-acryloylaminoazine in xylene is a very efficient way of linking the C—C double bond to a ring-nitrogen.

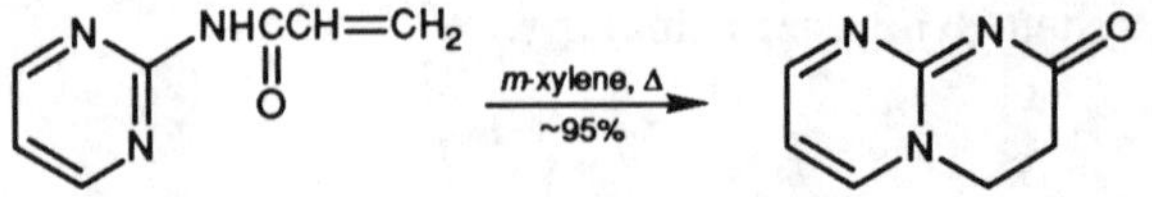

$\xrightarrow[\sim 95\%]{m\text{-xylene}, \Delta}$ [2279]

Pyrimido[1,2-*a*]pyrimidin-2-one

6. Pyran or Thiopyran

Alkynyloxybenzenes cyclize on heating (p. 201); when this reaction is applied to ω-chloropropynyloxybenzenes, 4-chlorochromenes are obtained in good yields [3439]. It is possible to synthesize 1-benzopyrans from an alkynyloxybenzene [2376, 3451], and the corresponding 1-benzothiopyrans from alkynylthiobenzenes [3544]. Hot mineral acid converts chalcones into chromones [3517].

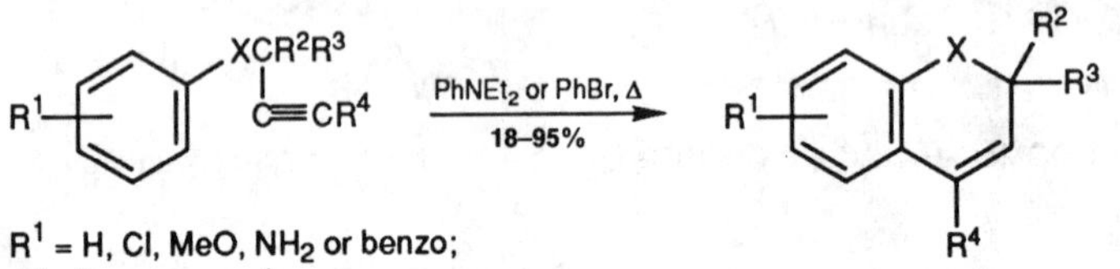

$\xrightarrow[18\text{–}95\%]{PhNEt_2 \text{ or } PhBr, \Delta}$ [2376, 3439, 3451, 3544]

R^1 = H, Cl, MeO, NH_2 or benzo;
R^2,R^3 = H, Me; R^4 = H, Cl; X = O, S

2*H*-1-Benzopyran
2*H*-1-Benzothiopyran

7. 1,2-Oxathiin 2,2-Dioxide

An unusual cyclization occurs when an ethylidenindanone is stirred at ambient temperature with a mixture of sulphuric acid and acetic anhydride.

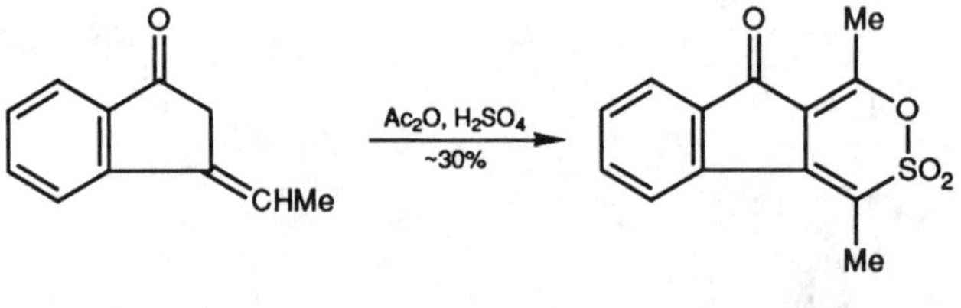

$\xrightarrow[\sim 30\%]{Ac_2O, H_2SO_4}$ [2657]

Indeno[1,2-*d*][1,2]oxathiin 3,3-dioxide

CHAPTER 43

Amidine and Amine, Carboxylic Acid or its Derivative

I. Formation of a Five-membered Ring 921
  1. Pyrrole 921
  2. Imidazole 922
  3. 1,2,4-Triazole 922
II. Formation of a Six-membered Ring 923
  1. Pyrimidine 923
  2. Pyrimidine 1-Oxide 924
  3. Pyrimidin-4-one 924
  4. 1,3,5-Triazine 925
  5. 1,2,4-Thiadiazine 1,1-Dioxide 925
III. Formation of a Seven-membered Ring 925
  1. 1,3,4-Triazepin-5-one 925
  2. 1,3,4-Thiadiazepine 926

Examples of annulation of amidine with methylene and with methylthio groups are included in this chapter.

I. FORMATION OF A FIVE-MEMBERED RING

1. Pyrrole

2-Methylbenzamidines react at room temperature with dimethyl oxalate and t-butoxide in DMF; the pyruvate intermediate may be isolated if desired or heated further (with or without ethanol as a solvent) to give the indole. Flash vacuum photolysis of 2-alkylbenzamidines at about 900 °C gives indole in moderate-to-good yields, but as both the 2-methyl- and the 2-ethyl-amidine give indole, it is not always easy to predict the structure of the product; some amidines give quinolines.

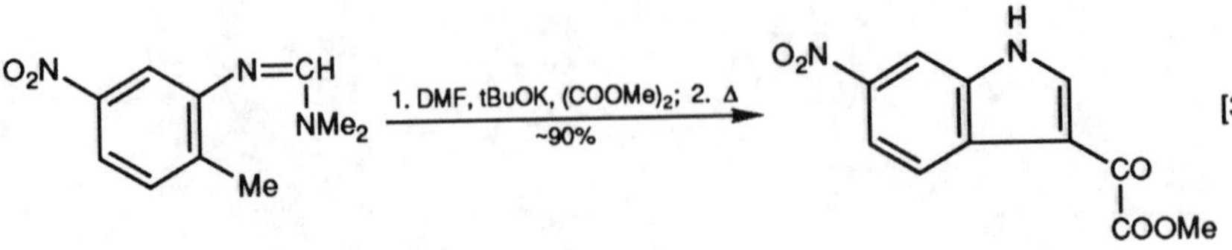

1. DMF, tBuOK, $(COOMe)_2$; 2. Δ ~90% [3832]

Indole

900 °C, 10^{-2} Torr 51–76% [3318]

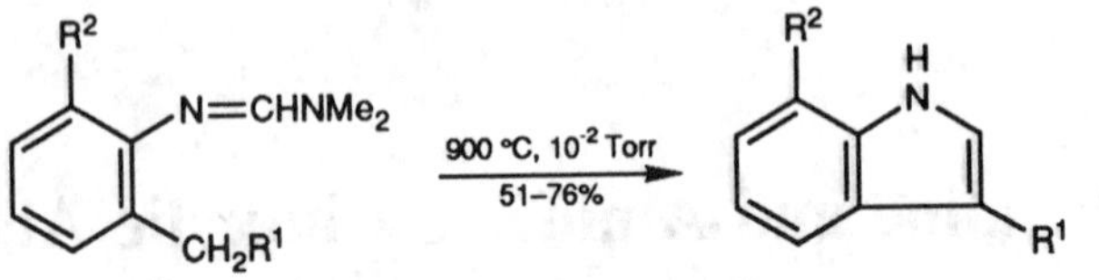

R^1 = H, Me; R^2 = Me, Et

2. Imidazole

High yields of benzimidazole are obtained by heating *o*-aminoamidines with hot mineral acid.

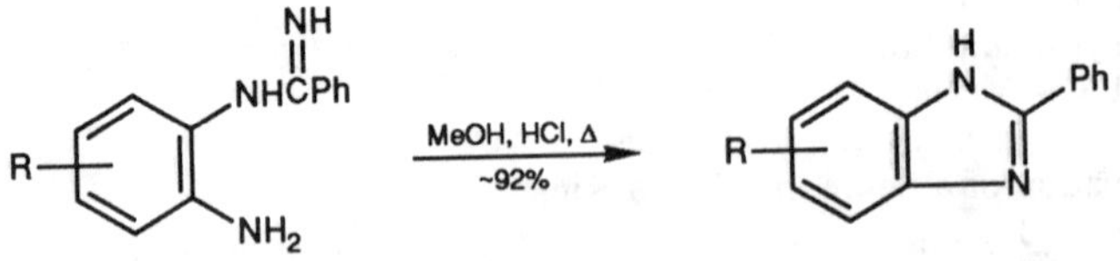

MeOH, HCl, Δ ~92% [3048]

R = H, Cl, Me, COOMe

Benzimidazole

3. 1,2,4-Triazole

N-Guanidino and methylthio groups interact in the presence of a strong base; after the loss of the MeS group, a fused triazole is obtained in moderate-to-high yields.

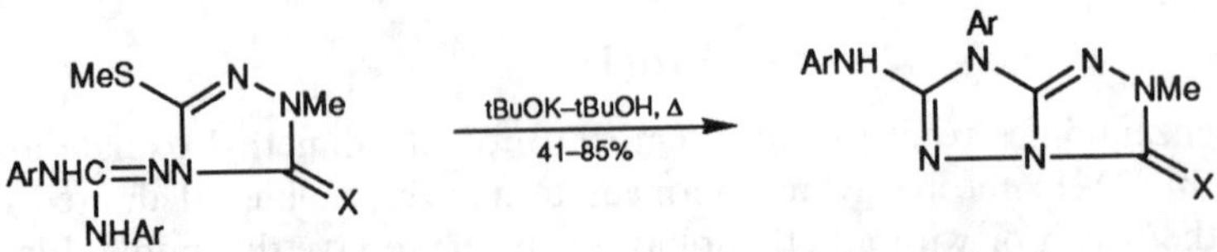

tBuOK–tBuOH, Δ 41–85% [3271]

Ar = Ph, 4-Br-, 4-Cl-, 4-MeO-C_6H_4; X = O, S

1,2,4-Triazolo[4,3-*b*][1,2,4]triazol-3-one
1,2,4-Triazolo[4,3-*b*][1,2,4]triazole-3-thione

II. FORMATION OF A SIX-MEMBERED RING

1. Pyrimidine

Amidino and amino groups are converted in high yield into a fused pyrimidine ring on stirring the compound with (in this example, labelled) acetic formic anhydride (review of this compound [3823]. Amidino-nitriles react with either ammonium acetate or methylamine to give fairly good yields of the amino-pteridines (reviews of pteridines: [3454, 3594, 3669]). Formic acid provides a carbon atom to complete the pyrimidine ring of a purine in this example of cyclization of an amino-amidine [2431]. No additional atoms are needed in the cyclization of the pyrazine (**43.1**) to a dihydropteridine [2159].

AcO^{14}CHO, 81% [2949a]

1,2,3-Triazolo[4,5-*d*]pyrimidine

i, Δ, 65–96% [2271]

R^1 = H, Cl; R^2 = H, Me; i, aq. $AcONH_4$ or $AcONH_3Me$

Pteridine

Another ring system synthesized similarly:

[2405]

1,2,3-Triazolo[4,5-*d*]pyrimidine

HCOOH, Δ, 57% [2431]

R = cyclopentyl

Purine

i, ii, 61% [2159]

(**43.1**)

i, EtOH–EtONa, Δ; ii, TsOH

Pteridine

2. Pyrimidine 1-Oxide

Ethyl orthoformate converts an *N*-hydroxyamidino-amine into a fused pyrimidine *N*-oxide.

$\xrightarrow[25\text{–}44\%]{HC(OEt)_3,\ \Delta}$ [2798]

R = Me, Ph

Pyrazolo[3,4-*d*]pyrimidine 5-oxide

3. Pyrimidin-4-one

A 2-amidinobenzoate (prepared from the amino ester and either a nitrilium salt or DMFDEA) is cyclized on heating with t-butoxide–t-butanol or a primary amine. Both methods yield a 3-substituted product. The internal amidinium salt (**43.2**) cyclizes when heated at 220 °C for 5 min.

N=CMeNHtBu, COOMe $\xrightarrow[50\%]{tBuOH,\ tBuONa,\ \Delta}$ Me, NtBu, O [2477]

4-Quinazolinone

N=CHNMe$_2$, COOEt $\xrightarrow[72\%]{PhCH_2NH_2,\ \Delta}$ NCH$_2$Ph [2983]

Pyrimido[4',5' : 4,5]pyrrolo[2,3-*c*]azepine-4,6-dione

COO$^-$, NHC(Me)=$\overset{+}{N}$HBu, OAr $\xrightarrow[63\%]{\Delta}$ Bu, Me, OAr [3653]

(**43.2**)
Ar = 4-MeC$_6$H$_4$

Pyrimido[4,5-*c*]isoquinolin-1-one

4. 1,3,5-Triazine

An *N*-amidino-aminopyrazole may be cyclized by reaction with an orthoester, acetic formic anhydride, formic acid, phenyl isocyanate dichloride or dimethyl oxalate according to the substituent that is required at position 2 of the product.

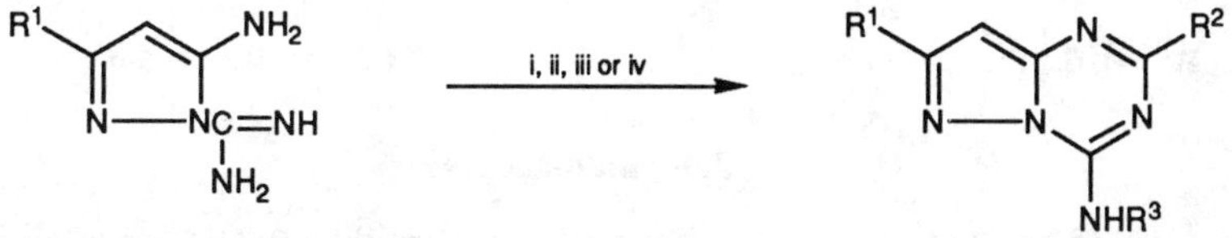

[2773]

i, $MeC(OEt)_3$, diox, Δ; $R^1 = R^2 = Me$; $R^3 = H$, 90%;
ii, Ac_2O–HCOOH, Δ; $R^1 = R^2 = H$; $R^3 = CHO$, 93%;
iii, diox, $PhN{=}CCl_2$; $R^1 = R^3 = H$; $R^2 = PhNH$, 12%;
iv, $(COOMe)_2$, Δ; $R^1 = R^3 = H$; $R^2 = COOMe$, 47%

Pyrazolo[1,5-*a*]-1,3,5-triazine

Another ring system synthesized similarly:

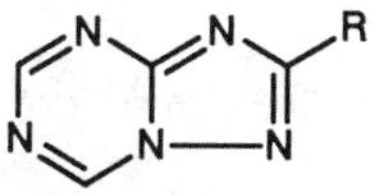

[2190]

[1,2,4]Triazolo[1,5-*a*][1,3,5]triazine

5. 1,2,4-Thiadiazine 1,1-Dioxide

A modest yield of the benzothiadiazine dioxide is obtained when an *o*-amino-*N*-amidinosulphonamide is heated at about 225 °C for several hours.

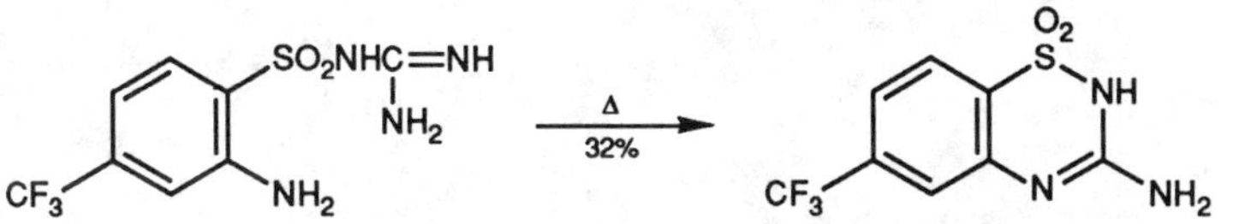

[2554]

1,2,4-Benzothiadiazine 1,1-dioxide

III. FORMATION OF A SEVEN-MEMBERED RING

1. 1,3,4-Triazepin-5-one

Amidine and amine groups may be joined together to form this triazepinone ring by heating the compound with methanolic sulphuric acid for 3 h. The reactant is obtained by stirring isatoic anhydride (review of reactions [3008]) with an amidrazone (formally, a hydrazone of a carboxamide).

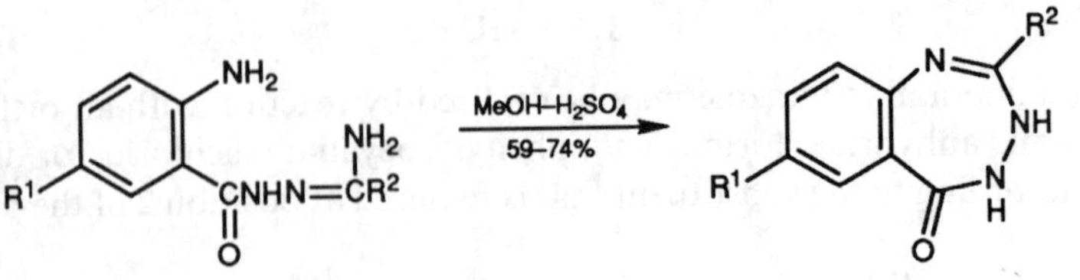

MeOH–H_2SO_4, 59–74% [2643]

R^1 = H, Br; R^2 = pyrid-2-yl

1,3,4-Benzotriazepin-5-one

2. 1,3,4-Thiadiazepine

An unexpected cyclization occurs when the aminoamidine (**43.3**) is heated with thiosemicarbazide hydrochloride.

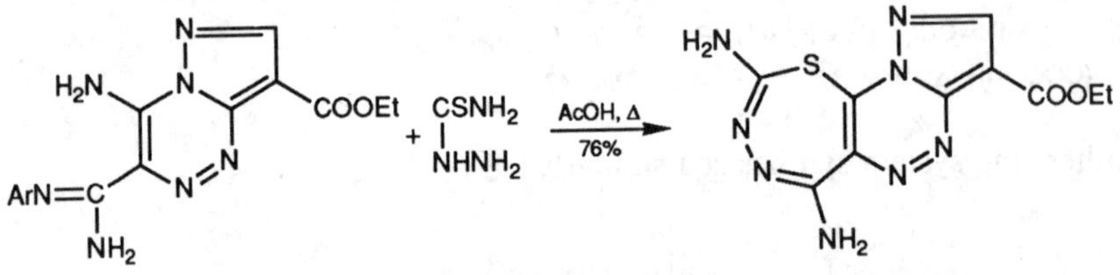

AcOH, Δ, 76% [3576]

(**43.3**)

Pyrazolo[5',1' : 3,4][1,2,4]triazino[6,5-*f*][1,3,4]thiadiazepine

CHAPTER 44

Amidine and Ring-carbon or Ring-nitrogen

I. Formation of a Five-membered Ring 927
  1. Imidazole 927
  2. 1,2,4-Triazole 928
II. Formation of a Six-membered Ring 929
  1. Pyridin-2-one 929
  2. Pyrimidine 929
  3. Pyrimidin-4-one or Pyrimidine-4-thione 930
  4. 1,3,5-Triazine 930
  5. 1,3,5-Triazin-2-one 931
  6. 1,3,5-Triazine-2,4-dione 931

Cyclization of compounds containing a sulphimidoamidine function is included in this chapter.

I. FORMATION OF A FIVE-MEMBERED RING

1. Imidazole

A synthesis of a benzimidazole in which the final step consists of the formation of a bond between a nitrogen atom and a ring-carbon atom was used to synthesize 6-hydroxy-1,2-diarylbenzimidazoles from the benzamidine (**44.1**). Naphthylguanidines undergo oxidative cyclization at temperatures below 0 °C to form a new 2-iminoimidazole ring. In this example, simultaneous oxidation and bromination occurs elsewhere in the molecule. LTA has also been used as the oxidant.

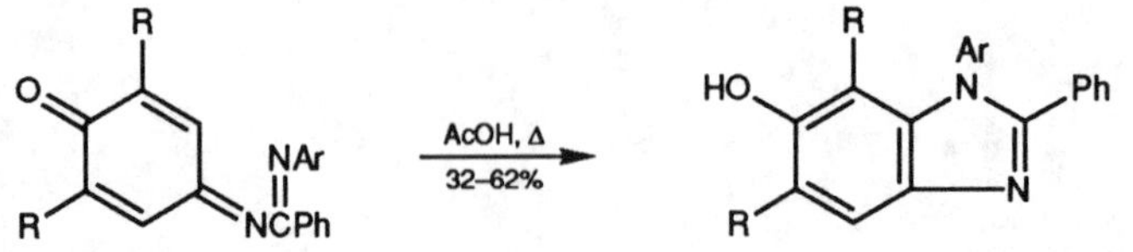

(44.1)

R = H, Cl, Me; Ar = Ph, 4-Cl-, 4-Me-, 4-MeO-C_6H_4, 2,6-Me_2-, 2,5-Me_2-C_6H_3

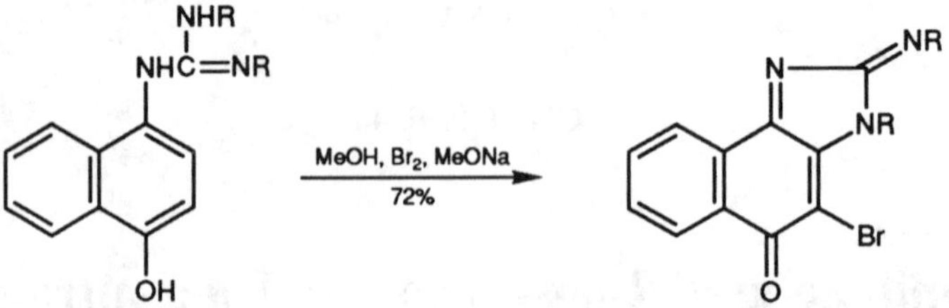

AcOH, Δ, 32–62% → Benzimidazole [3479]

MeOH, Br_2, MeONa, 72% → Naphth[1,2-*d*]imidazol-5-one [2802]

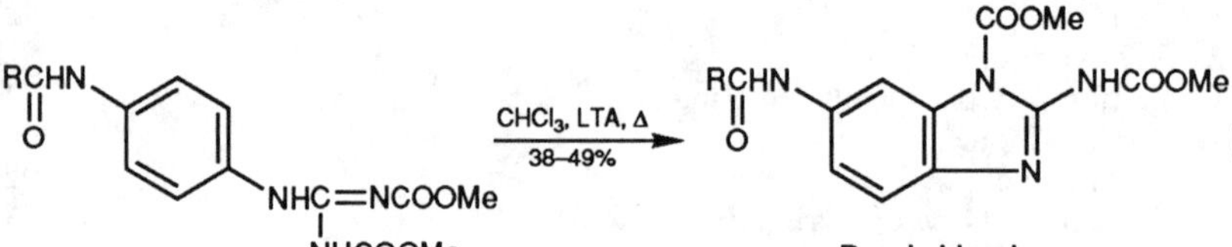

R = tBu, cyclohexyl

$CHCl_3$, LTA, Δ, 38–49% → Benzimidazole [3057]

R = Me, tBu, Ph

2. 1,2,4-Triazole

Oxidative cyclization of an amidinoazine with LTA can lead to a good yield of a triazole. A formazan (which contains an N=C—N linkage) is cyclized by acid which also cleaves the NAr group.

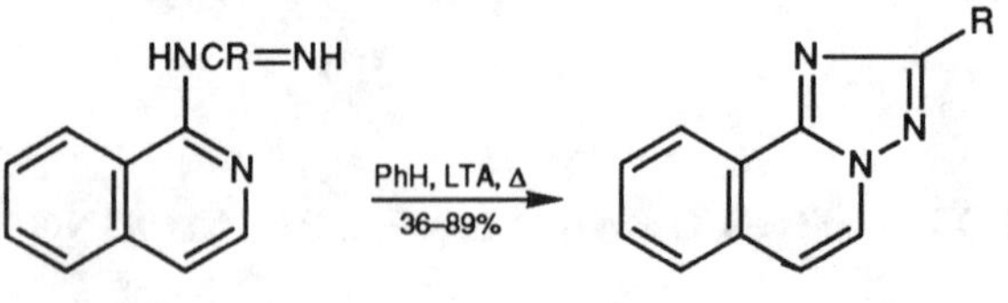

PhH, LTA, Δ, 36–89% → [1,2,4]Triazolo[5,1-*a*]isoquinoline [2098, 2341, 2835]

R = Me, CF_3, CF_2Cl, Ph

Another ring system synthesized similarly:

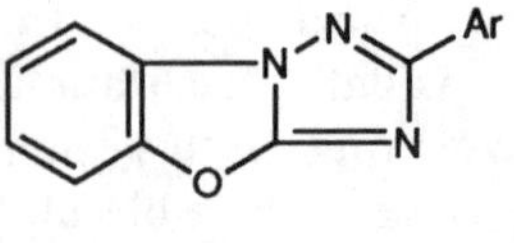

[1,2,4]Triazolo[5,1-*b*]benzoxazole [3713]

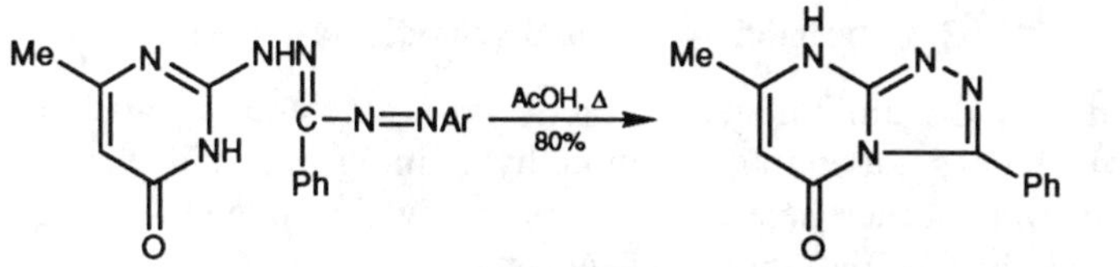

[2895]

Ar = Ph, 4-MeC6H4

1,2,4-Triazolo[4,3-*a*]pyrimidin-5-one

II. FORMATION OF A SIX-MEMBERED RING

1. Pyridin-2-one

When the amidine (**44.3**) is heated in an inert atmosphere, the dimethylamino group is eliminated and a bond is made to the ring-carbon atom.

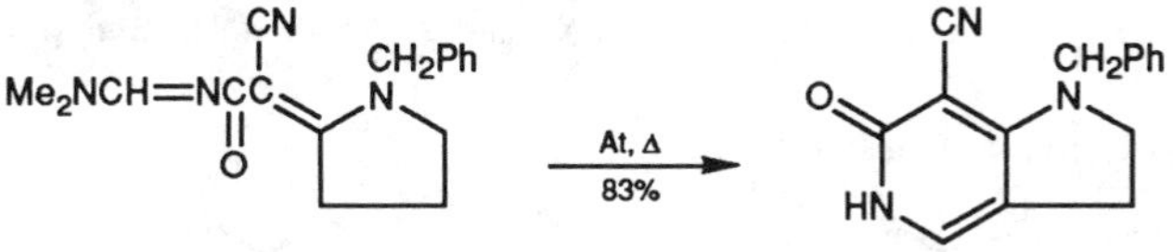

[2944]

(**44.3**)

Pyrrolo[3,2-*c*]pyridin-6-one

2. Pyrimidine

N-Aryl-*N*′-sulphimidoamindines react with enamines to form quinazolines in good yields. Further studies on this reaction [using the $S(CD_3)_2$ analogue] showed that the product acquires a hydrogen from the SMe_2 group and that a 1,3-diaza-1,3-diene is probably an intermediate. When the enamine carries an α-substituent (other than hydrogen), a 3,4-dihydroquinazoline is obtained. The corresponding 2-pyridyl sulphimide is photocyclized to a 1,2,4-triazolo[1,5-*a*]pyridine in high yield [2686].

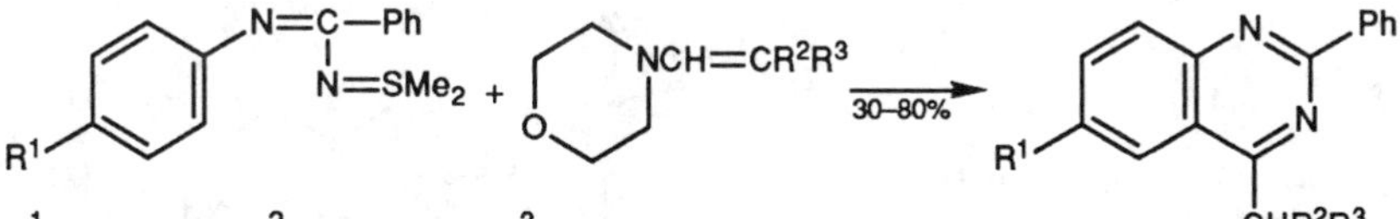

[3600]

R^1 = Me, MeO; R^2 = H, Me, Ph; R^3 = Me, Ph

Quinazoline

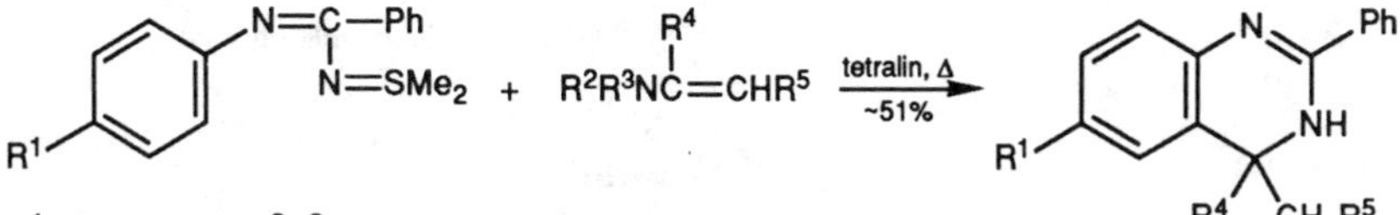

[3828]

R^1 = Me, MeO; R^2R^3N = morpholino;
R^4 = Et, Ph; R^5 = H, Me or R^4R^5 = $(CH_2)_3$ or $(CH_2)_4$

3. Pyrimidin-4-one or Pyrimidine-4-thione

An *N,N*-dimethyl-α-amidinoazole reacts with diketene to give a cyclized product which does not contain the dimethylamino group. This reaction can be applied to several other heterocycles containing this side-chain. Heating the amidine (**44.4**) briefly in quinoline effects its cyclization in high yield; replacing the ethoxycarbonyl group by a chlorine atom and heating the compound with potassium thiocyanate can give high yields of the corresponding thione.

PhH, Δ
27–72%
[2886]

X = O 1,3,4-Oxadiazolo[3,2-*a*]pyrimidin-5-one
X = S 1,3,4-Thiadiazolo[3,2-*a*]pyrimidin-5-one

Other ring systems synthesized similarly:

[2886] [2886]

X = O Pyrimido[2,1-*b*]benzoxazol-4-one
X = NH Pyrimido[1,2-*a*]benzimidazol-4-one

Pyrido[1,2-*a*]pyrimidin-4-one

R^3 = COOEt: quinoline, Δ
73%

R^3 = Cl: $(MeOCH_2)_2$, KSCN, Δ
30–90%

(**44.4**)

R^1 = H, Me, NO_2; R^2 = H, Me, NO_2, 2-thienyl-, Me_2N, morpholino

4-Quinazolinone [2620]

4-Quinazolinethione [2566]

4. 1,3,5-Triazine

Guanidinobenzimidazole reacts with benzoyl isothiocyanate in a basic medium to give a high yield of the fused triazine.

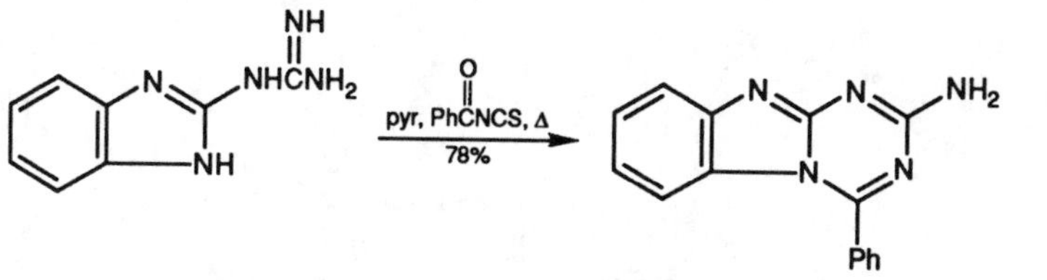

pyr, PhC(O)NCS, Δ; 78% [2577]

1,3,5-Triazino[1,2-*a*]benzimidazole

5. 1,3,5-Triazin-2-one

Phosgene at ambient temperature cyclizes 1-amidinoisoquinoline to its triazino derivative in good yields.

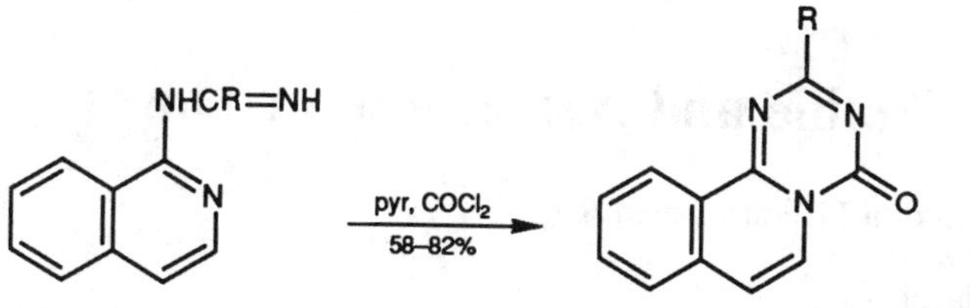

pyr, $COCl_2$; 58–82% [2338, 2835]

R = CF_3, Ph

1,3,5-Triazino[2,1-*a*]isoquinolin-4-one

6. 1,3,5-Triazine-2,4-dione

An aryl isocyanate annulates an amidino-nitrogen to a neighbouring ring-nitrogen.

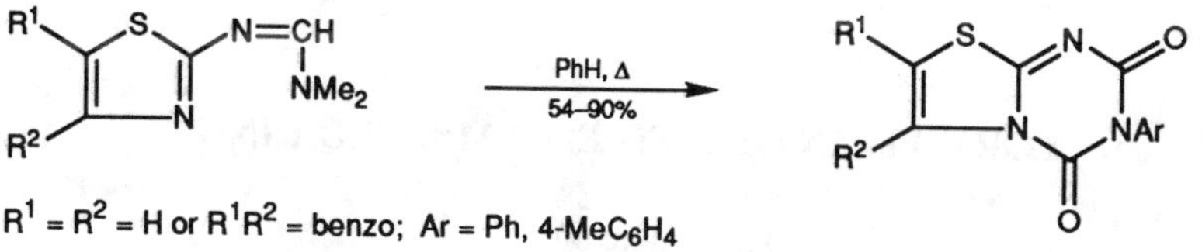

PhH, Δ; 54–90% [2096]

$R^1 = R^2 = H$ or R^1R^2 = benzo; Ar = Ph, 4-MeC_6H_4

Thiazolo[3,2-*a*]-1,3,5-triazine-2,4-dione
1,3,5-Triazino[1,2-*b*]benzothiazole-2,4-dione

CHAPTER 45

Amine and Azide, Azo or Diazo

I. Formation of a Five-membered Ring 932
1. Imidazole 932
2. Imidazol-2-one 933
3. 1,2,3-Triazole 933

A rare example of the cyclization of an oxime of an acylamino and an amino group is described.

I. FORMATION OF A FIVE-MEMBERED RING

1. Imidazole

Amino and azo functions are annulated to form an imidazole ring by fusing the compound with a benzaldehyde at 220 °C. An azido-imine is cyclized (via a nitrenium complex) by brief treatment at room temperature with boron trifluoride etherate.

ArCHO, Δ
70–95%

[2580]

Ar = Ph, 4-MeO-, 4-Cl-, 4-Me_2N-C_6H_4

2,6-Purinedione

Another ring system synthesized similarly:

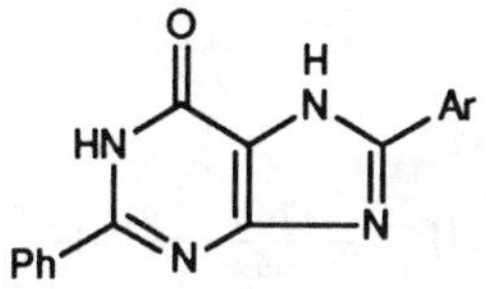

6-Purinone

[2580]

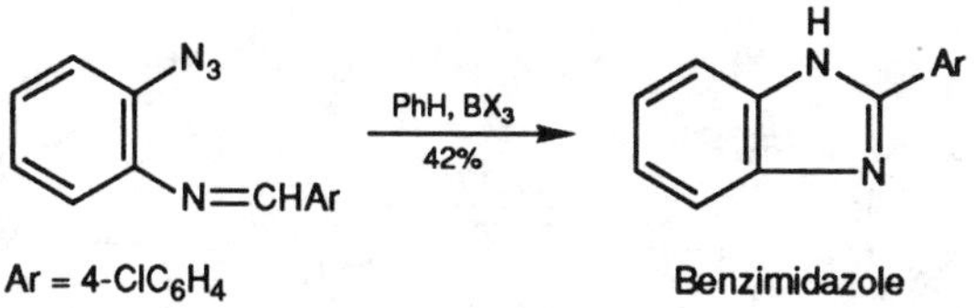

PhH, BX_3
42%

Ar = 4-ClC_6H_4

Benzimidazole

[3453]

2. Imidazol-2-one

An azidocarbonyl group undergoes a thermally induced Curtius reaction to an isocyanate, and this reacts readily with a neighbouring primary or secondary amino group to form a fused imidazolone ring. Toluene or diphenyl ether may be used as solvent according to the temperature needed for the reaction to proceed in a reasonable time but without causing decomposition.

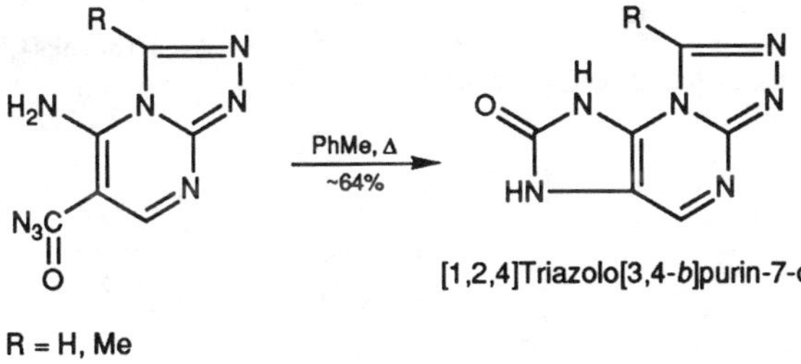

PhMe, Δ
~64%

R = H, Me

[1,2,4]Triazolo[3,4-*b*]purin-7-one

[3090]

Another ring system synthesized similarly:

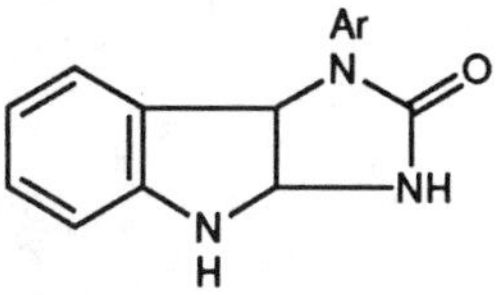

Imidazo[4,5-*b*]indol-2-one

[3363]

3. 1,2,3-Triazole

A high yield of a fused triazole is obtained when the oxime of a 2-aminobenz-anilide is treated with nitrous acid [3048]. Cyclization of amino-azo compounds with LTA is often carried out in dichloromethane but acetic acid also gives good results [2417]. Pyridine as solvent and copper(II) sulphate as oxidant can also give high yields [2476].

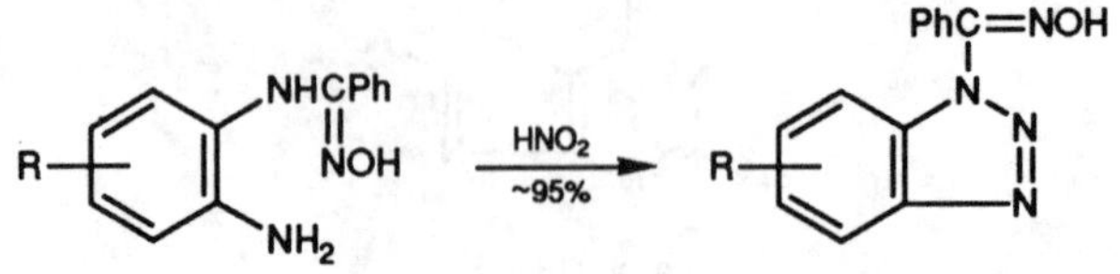

HNO_2, ~95% [3048]

R = H, Me, Cl, COOMe

Benzotriazole

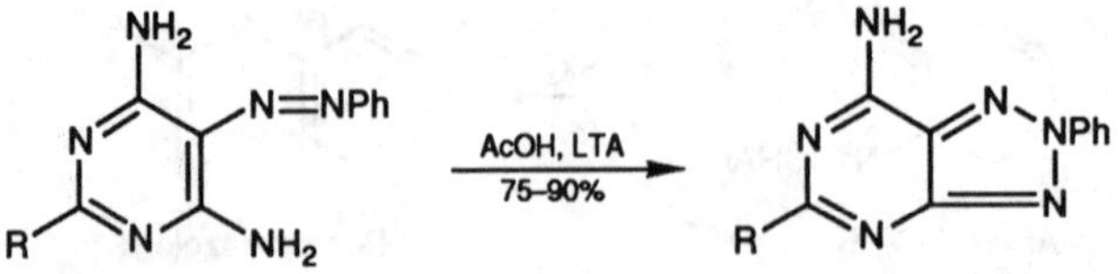

AcOH, LTA, 75–90% [2417]

R = H, Ph, NH_2

1,2,3-Triazolo[4,5-*d*]pyrimidine

Other ring systems synthesized similarly:

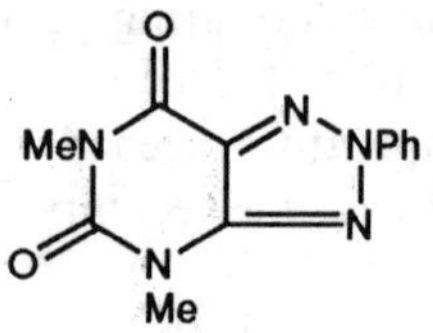

[2417]

1,2,3-Triazolo[4,5-*d*]pyrimidine-5,7-dione

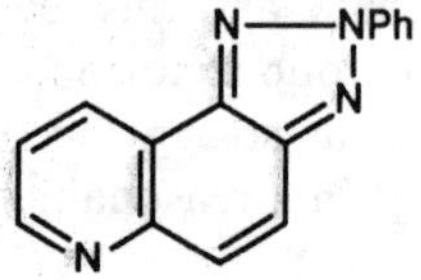

[2476]

1,2,3-Triazolo[4,5-*f*]quinoline

CHAPTER 46

Amine or Phosphorane and Carboxamide

I.	Formation of a Six-membered Ring	935
	1. Pyridin-4-one	935
	2. Pyrimidin-4-one or Pyrimidin-4-one, 2-thioxo-	935
	3. Pyrimidine-2,4-dione	938
II.	Formation of a Seven-membered Ring	939
	1. 1,3-Diazepin-4-one	939
	2. 1,4-Diazepine-2,3,5-trione	939

I. FORMATION OF A SIX-MEMBERED RING

1. Pyridin-4-one

o-Amino-t-benzamides are cyclized to 4-quinolinones by successive treatment with a ketone containing at least one α-methylene group with LDA in THF at temperatures varying from 0 °C to refluxing temperature of the solvent.

NH_2 / $CONR^1_2$ + $R^3COCH_2R^2$ —(i, ii; 45–93%)→ 4-Quinolinone [3193]

R^1,R^3 = Me, Et; R^2 = Me, Et, C_5H_{11}, COOEt, Ph or R^2R^3 = cC_6H_{11}

i, Me_2CO or PhH, 20–80 °C; ii, THF, LDA

4-Quinolinone

2. Pyrimidin-4-one or Pyrimidin-4-one, 2-Thioxo-

Annulation of a side-chain phosphonium bromide and a carboxamide group [as in compound (**46.1**)] occurs under base catalysis [2736]. A pyrimidinone ring is also formed in high yield and under mild experimental conditions when an iminophosphorane (formed *in situ* from an azide and triphenylphosphine) reacts with an *N*-acylcarboxamide (that is, a mixed imide) at room temperature [3636]. Other products can be prepared from these two pairs of reacting groups; for instance, an isocyanate converts **(46.2)** into the 2-aminoquinazolinone whereas carbon disulphide converts it into 2-thioxoquinazolin-4-one, both in high yield [3441].

NHCMe=CH$\overset{+}{P}$Ph$_3$ Br$^-$ / CONH$_2$ **(46.1)** —MeCN, NaH, Δ, 75%→ 4-Quinazolinone [2736]

Another ring system synthesized similarly:

Pyrrolo[2,1-*b*]quinazolin-9-one [3636]

N$_3$ / CONMeCOR —1. xylene, PPh$_3$ 2. stir, ~97%→ [3636]

R = Me, cC$_3$H$_5$, PhCH=CH

N=PPh$_3$ / CONHR^1 **(46.2)** —R^{2}NCO, ~95%→ (NHR^2, NR^1)

—R^1 = Me: CS$_2$, 93%→ 4-Quinazolinone, 2-thioxo- [3441]

R^1 = Me, 4-MeC$_6$H$_4$;
R^2 = Me, 4-Cl-, 4-MeO-C$_6$H$_4$

A cyclic hydroxamic acid is formed when an amino-(*N*-hydroxycarboxamide) is heated with an acylating agent such as an anhydride or acid chloride [2399]. 1,3-Bis(methoxycarbonyl)-*S*-methylisothiourea (**46.3**) converts aminocarboxamides into fused 4-pyrimidinones which carry a 2-carbamated function. A 2,2-dicarbamate is formed first (under mildly acid conditions), and briefly warming this with alkoxide yields the monocarbamate [3711].

NH_2, CONHOH + R^1COX or $(R^2CO)_2O$ —(i, Δ; 75–88%)→ 4-Quinazolinone [2399]

R^1 = $ClCH_2$, Ph, 4-$NO_2C_6H_4$, 2-$COOHC_6H_4$, CH=CHCOOH, $BrCH_2$;
$(R^2CO)_2O$ = succinic, maleic, phthalic; X = Cl, Br
i, EtOH or DMF

(**46.3**) —(i, ii, iii; 38–50%)→ [3711]

R = H, MeO
i, RaNi–N_2H_4, MeOH; ii, $MeO_2CN{=}C(SMe)NHCOOMe$, AcOH, Δ; iii, MeOH–MeONa

Many reagents may be used to cyclize an aminocarboxamide into a fused pyrimidinone (see pp. 211–216); the following may be added to the list: triformamidomethane, DMFDEA and thiourea. Another example is shown in Section III.2.

—($HC(NHCHO)_3$, Δ; 66%)→ Pyrimido[4,5-*d*]pyrimidin-4-one [3090]

—(DMFDMA, Δ; 39%)→

Another ring system synthesized similarly:

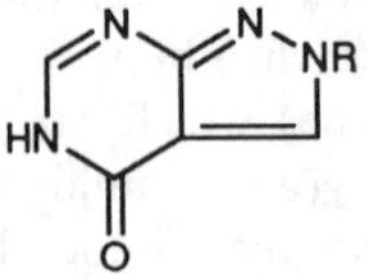

R = D-ribosyl
Pyrazolo[3,4-*d*]pyrimidin-4-one

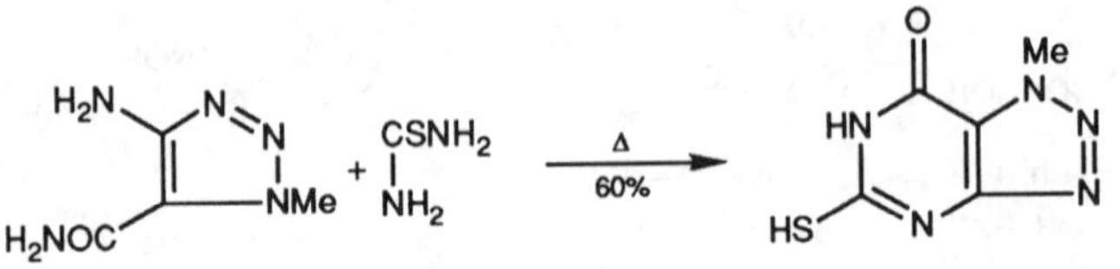

[2411]

[2404]

1,2,3-Triazolo[4,5-*d*]pyrimidin-7-one

3. Pyrimidine-2,4-dione

To the list of reagents mentioned in Part 1 (p. 216) used to synthesize this fused ring from an aminocarboxamide, phosgene [2564] and ethyl chloroformate [2913] may be added. The iminophosphorane (**46.2**) is also a substrate as it reacts with carbon dioxide to give the dione in excellent yield.

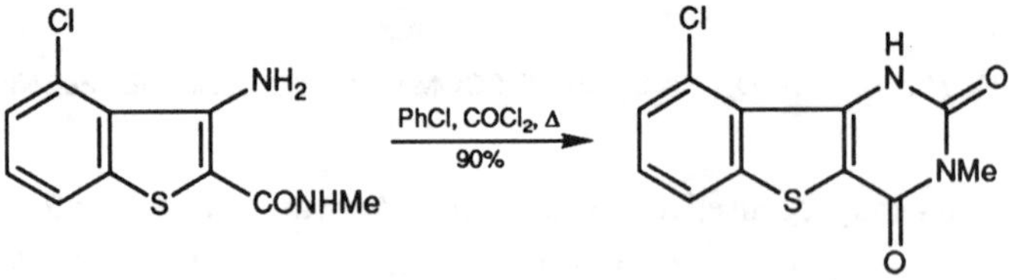

[2564]

[1]Benzothieno[3,2-*d*]pyrimidine-2,4-dione

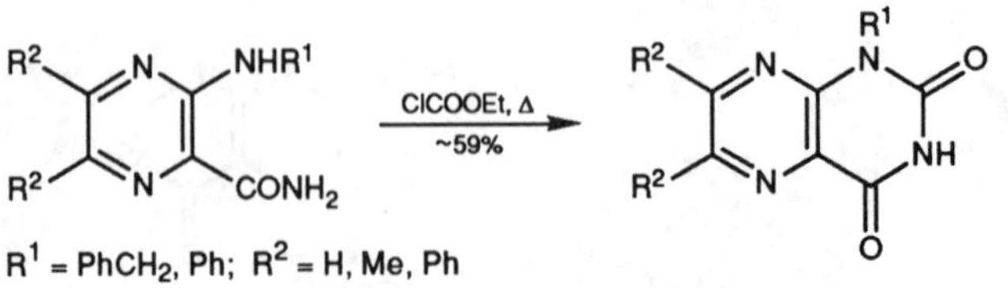

R^1 = PhCH$_2$, Ph; R^2 = H, Me, Ph

[2913]

2,4-Pteridinedione

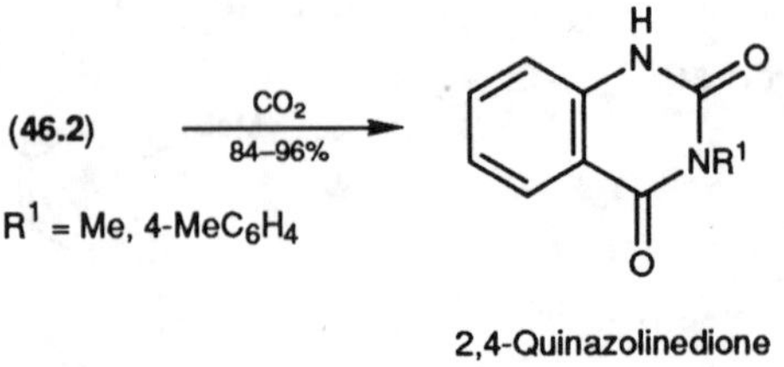

[3441]

R^1 = Me, 4-MeC$_6$H$_4$

2,4-Quinazolinedione

II. FORMATION OF A SEVEN-MEMBERED RING

1. 1,3-Diazepin-4-one

Formaldehyde supplies the methylene group to join the amino and carboxamido functions of (**46.4**) to form this ring.

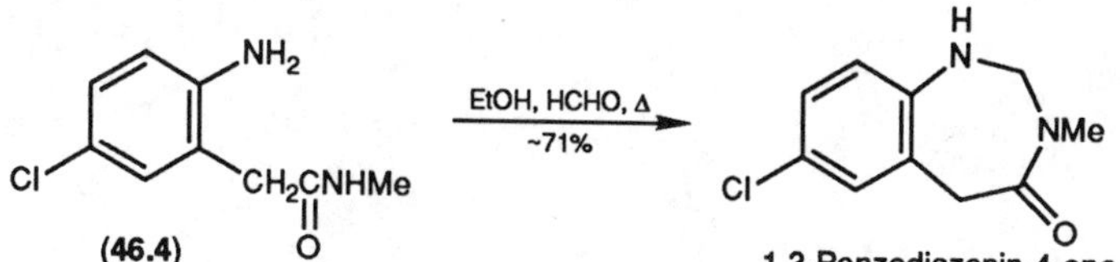

(46.4) —EtOH, HCHO, Δ; ~71%→ 1,3-Benzodiazepin-4-one [2058]

2. 1,4-Diazepine-2,3,5-trione

Anthranilamide and many other aminocarboxamides are converted into the fused pyrimidin-4-one analogues in a base-catalysed reaction with diethyl oxalate [182]. When this reaction was applied to the chromone analogues (**46.5**), the expected product was obtained when R = COOEt, but the diazepinetrione was the only product isolated from the chromone (**46.5**, R = H).

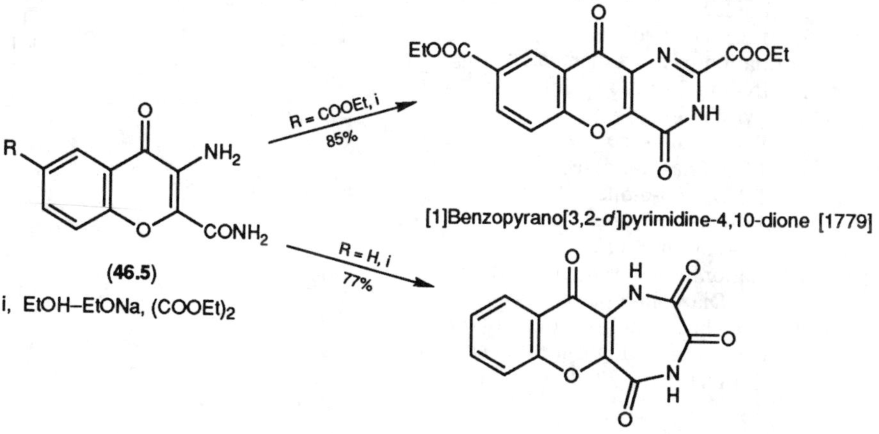

(46.5) —R = COOEt, i; 85%→ [1]Benzopyrano[3,2-*d*]pyrimidine-4,10-dione [1779]

(46.5) —R = H, i; 77%→ [1]Benzopyrano[3,2-*e*]-1,4-diazepine-2,3,5,11-tetraone

i, EtOH–EtONa, $(COOEt)_2$

CHAPTER 47

Amine and Carboxylic Acid

I.	Formation of a Five-membered Ring	940
	1. Imidazole	940
	2. 1,2,3-Triazole	941
II.	Formation of a Six-membered Ring	941
	1. Pyridin-2-one	941
	2. Pyridin-4-one	941
	3. Pyrimidin-4-one	942
	4. 1,2,3-Triazin-4-one	943
	5. 1,3-Oxazin-6-one	943
	6. 1,4-Oxazin-3-one	944
	7. 1,3-Thiazin-6-one	944
III.	Formation of a Seven-membered Ring	944
	1. 1,4-Diazepin-5-one	944
	2. 1,4-Thiazepin-3-one or -5-one	945
IV.	Formation of an Eight-membered Ring	945
	1. 1,5-Diazocine-2,6-dione	945

Cyclization of an imino-carboxylic acid is included in this chapter.

I. FORMATION OF A FIVE-MEMBERED RING

1. Imidazole

Cyclization combined with chlorination of the newly formed ring occurs when an imino-carboxylic acid is heated with phosphorus oxychloride; the yield is low when R = Me.

HN, S, HOOCH$_2$N, R → $POCl_3$, Δ; 10–87% → Cl, N, S, N, R [2069]

R = H, Me

Imidazo[2,1-*b*]thiazole

2. 1,2,3-Triazole

A benzyne (generated from anthranilic acid) reacts with an aryl azide and isopentyl nitrite to give a good yield of a fused triazole.

NH_2, COOH + ArN_3 → i, $C_5H_{11}ONO$, CH_2Cl_2, Δ; 57–75% → Ar, N, N, N [3260a]

Benzotriazole

II. FORMATION OF A SIX-MEMBERED RING

1. Pyridin-2-one

2-Aminobenzoic acid undergoes cycloacylation to give moderate yields of quinolin-2-ones.

R^1, NHR^2, COOH → AcOH, Ac_2O, Δ; 15–67% → R^1, R^2, N, O, OH [2767]

(47.1)

R^1 = H, Me, Et, Cl, Br; R^2 = H, Me, Et, Ph

2-Quinolinone

2. Pyridin-4-one

This ring is formed by reaction of a 3-amino-5-chloroindole-2-carboxylic acid with DEAD (review of addition to alkynes [2816]) in alkaline solution; when the chlorine atom was replaced by other substituents, a mixture of products was obtained.

Cl, NH_2, N, H, COOK + DEAD → aq. KOH, Δ; 46% → Cl, H, N, COOEt, N, H, O [2888]

Pyrido[3,2-*b*]indol-4-one

3. Pyrimidin-4-one

A 2-fluoromethylquinazolin-2-one may be synthesized by stirring an anthranilic acid and a base with the adduct from fluoroacetonitrile, ethanol and gaseous hydrogen chloride [3777]. The use of nitrile–hydrogen chloride adducts in the synthesis of heterocycles has been reviewed [2547], as also has the cyclization of *o*-aminocarboxylic acids with an amide (by a modified Niementowski quinazoline synthesis) [2484].

(47.2) R^1 = 5-Me + FCH_2CN —1. EtOH–HCl(g) 2. EtONa, 67%→ 4-Quinazolinone [3777]

Application of a modified Willgerodt reaction to the synthesis of 3-heteryl-quinazolinones gives low yields [2427]. Cyanogen [2996] or imino esters [2106] may be used to cyclize anthranilic acids; 1,3,5-triazine is an effective amino-methylating agent for the same kind of substrate [2442].

(47.2) R^1 = H + $ArNH_2$ + R^2Me —S, Δ, 18–45%→ [2427]

R^2 = 2- or 4-pyridyl; Ar = Ph, MeO-, Me-C_6H_4

+ $(CN)_2$ —ROH, ~60%→ Pyrazolo[4,3-*g*]quinazolin-5-one [2996]

R = Me, Et

(47.2) + 1,3,5-triazine —pip, Δ, 50–85%→ 4-Quinazolinone [2442]

R^1 = H, Hal, NO_2, SO_3H

4. 1,2,3-Triazin-4-one

1,2,3-Benzotriazin-4-ones are useful reagents in peptide synthesis and may be synthesized from the anthranilic acid by successive reaction with thionyl chloride (which forms an unstable sulphinamide anhydride), *O*-(trimethylsilyl)hydroxylamine and a nitrous acid–hydrochloric acid mixture at low temperature.

(47.2) —— i, ii, iii; 54–79% ——> [3728]

R^1 = H, Cl, I, Me, MeO, NO_2

i, PhH, $SOCl_2$, Δ; ii, PhH, Me_3SiONH_2; iii, HCl, $NaNO_2$

1,2,3-Benzotriazin-4-one

5. 1,3-Oxazin-6-one

Addition of ethyl chloroformate to an anthranilic acid stirred at ambient temperature in aqueous hydrogencarbonate or in pyridine results in the formation of a 2-ethoxybenzoxazinone in high yield [3770]. Cyanogen bromide in a similar low-temperature reaction yields a 2-aminobenzoxazinone while thiophosgene gives a high yield of the benzoxazinethione. Formaldehyde annulates an anthranilic acid to a 2-unsubstituted benzoxazinone [2213].

(47.1)

R^1 = Me, Et;
R^2 = H or Ar;
Ar = 3-$CF_3C_6H_4$;
R^3 = Me, Et

$ClCOOR^3$, pyr; 50–89% → 3,1-Benzoxazin-4-one

$CSCl_2$, Me_3SiCl; ~95% → 3,1-Benzoxazin-4-one, 2-thioxo- [3769]

aq. NaOH, CNBr; 54–90%

R^1 = H; R^2 = Ar; EtOH, HCHO, Δ; ~50% [2213]

6. 1,4-Oxazin-3-one

A cyclic amide is formed intramolecularly when a compound containing suitably positioned amino and carboxylic groups is heated in acetic acid–acetic anhydride.

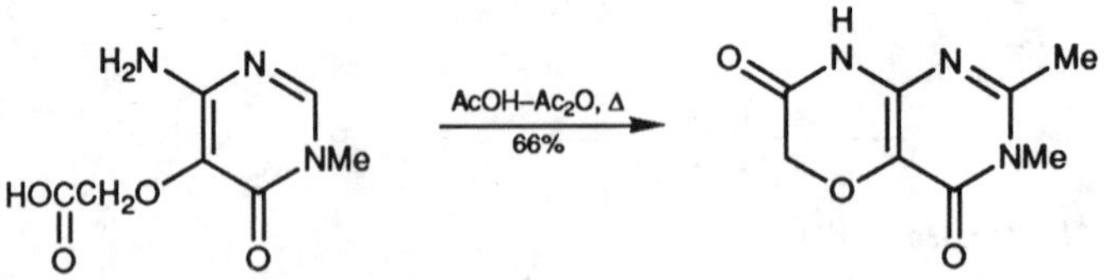

H_2N, N, NMe, $HOCCH_2O$, O, O → AcOH–Ac_2O, Δ; 66% → O, H, N, N, Me, NMe, O, O [3703]

Pyrimido[5,4-*b*][1,4]oxazine-4,7-dione

7. 1,3-Thiazin-6-one

An excess of thioacetic acid converts anthranilic acid into the benzothiazinone in theoretical yield.

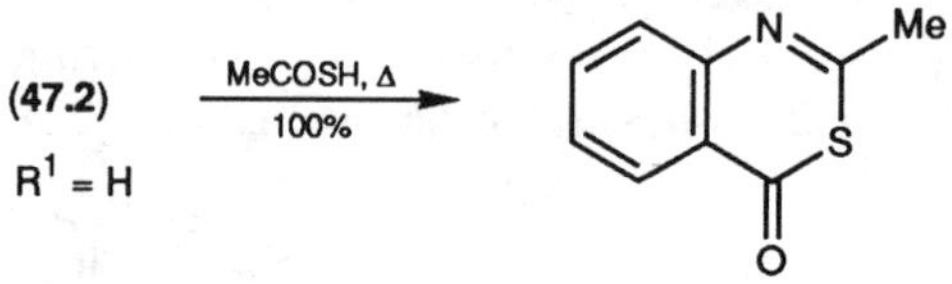

(47.2) R^1 = H → MeCOSH, Δ; 100% → N, Me, S, O [2037]

3,1-Benzothiazin-4-one

III. FORMATION OF A SEVEN-MEMBERED RING

1. 1,4-Diazepin-5-one

Suitably positioned amino and carboxylic acid groups attached to different rings usually react in the presence of methoxide or sodamide, or on refluxing in a high-boiling solvent or on heating with PPA, but DCC is sometimes the most useful reagent even though the yields may not be high (see also Chapter 48, Section III.3).

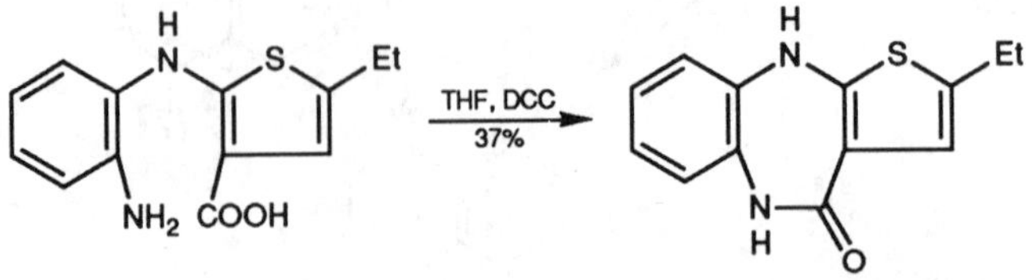

H, N, S, Et, NH_2, COOH → THF, DCC; 37% → H, N, S, Et, N, H, O [2942]

Thieno[2,3-*b*][1,5]benzodiazepin-4-one

2. 1,4-Thiazepin-3-one or -5-one

This ring may be constructed by annulating amino and carboxylic groups attached to different rings with oxalyl chloride-DMF or in a high-boiling solvent such as xylene.

Ar = 2-FC_6H_4
i, $(COCl)_2$, CF_3COOH, DMF, CH_2Cl_2

i, 67% — Thieno[2,3-*e*][1,4]thiazepin-2-one [3801]

Δ, 75% — Dibenzo[*b*,*f*][1,4]thiazepin-11-one [2589]

IV. FORMATION OF AN EIGHT-MEMBERED RING

1. 1,5-Diazocine-2,6-dione

Two molecules of *N*-methylanthranilic acid react together when heated with thionyl chloride to give a high yield of the diazocinedione.

(47.1) — PhH, $SOCl_2$, Δ, 84% → [2874]

R^1 = H; R^2 = Me

Dibenzo[*b*,*f*][1,5]diazocin-6,12-dione

CHAPTER 48

Amine and Carboxylic Ester

I. Formation of a Five-membered Ring 946
1. Pyrrol-2-one 946
2. Pyrazol-3-one 947
3. Imidazol-4-one 947
II. Formation of a Six-membered Ring 948
1. Pyridin-2-one 948
2. Pyridin-4-one 949
3. Pyrimidin-4-one 951
4. Pyrimidin-4-one, 2-Thioxo- or Pyrimidine-2,4-dione 951
5. 1,3-Oxazin-6-one 952
6. 1,4-Oxazin-3-one 953
III. Formation of a Seven-membered Ring 953
1. 1,3-Diazepin-4-one 953
2. 1,4-Diazepin-5-one 953
3. 1,4-Diazepine-2,5-dione 954
4. 1,4-Oxazepin-5-one 955
5. 1,4-Thiazepin-5-one 955
IV. Formation of an Eight-membered Ring 956
1. 1,3,6-Thiadiazocin-7-one 956

I. FORMATION OF A FIVE-MEMBERED RING

1. Pyrrol-2-one

A 2-aminophenylacetate spontaneously cyclizes in acetic acid or with PPA [2416] to give an oxindole. An unexpected cyclization observed when the amino-pyridinium salt (**48.1**) is heated with benzylamine is a Dimroth rearrangement.

PhCH$_2$O, NO$_2$, MeO, CH$_2$COOEt → (H_2–PdC, AcOH, 57%) HO, MeO, H N, O [2112]

2-Indolone

CH$_2$, N$^+$, COMe, O, NH$_2$, Br$^-$ (48.1) → ($PhCH_2NH_2$, Δ, 86%) O, N, N [2253]

1-Isoindolone

Another ring system synthesized similarly:

H, Me, O, N, N, O, NMe, O [2416]

Pyrrolo[2,3-*d*]pyrimidine-2,4,6-trione

2. Pyrazol-3-one

Heating an amino ester with hydrazine hydrate leads to the formation of a pyrazolone ring, but in this particular example, a simultaneous replacement of sulphur by hydrazine also occurs in the substrate ring with the loss of hydrogen sulphide, methanol and ammonia.

Me, S, NH$_2$, COOEt + $2N_2H_4.H_2O$ → (Δ, 50%) H, N, N, N, NH, Me, O [3389]

Pyrazolo[3,4-*c*]pyridazin-3-one

3. Imidazol-4-one

Amino and ester groups interact on heating the compound with methoxide. When base-sensitive functions are present, resort may have to be made either to thermolysis under more drastic conditions, for example, heating at 160 °C under vacuum for several hours, or to the possibility that a weaker base will be effective.

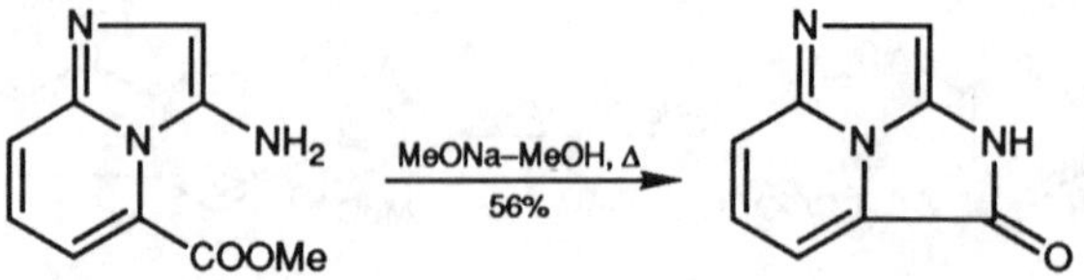

MeONa–MeOH, Δ 56% [3266]

1,3,7b-Triazacyclopent[*cd*]inden-4-one

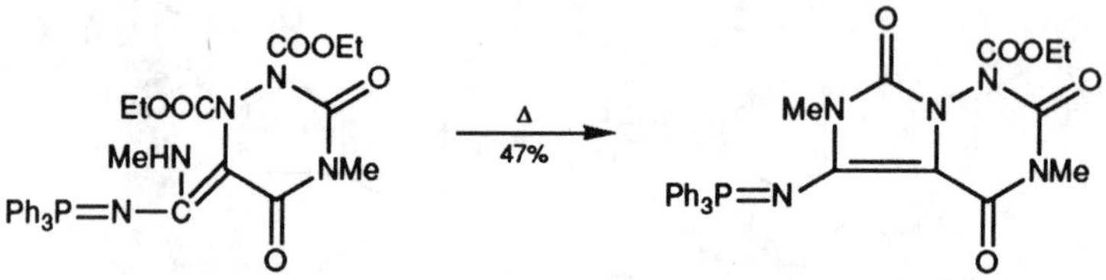

Δ 47% [3183]

Imidazo[5,1-*f*]1,2,4]triazine-2,4,7-trione

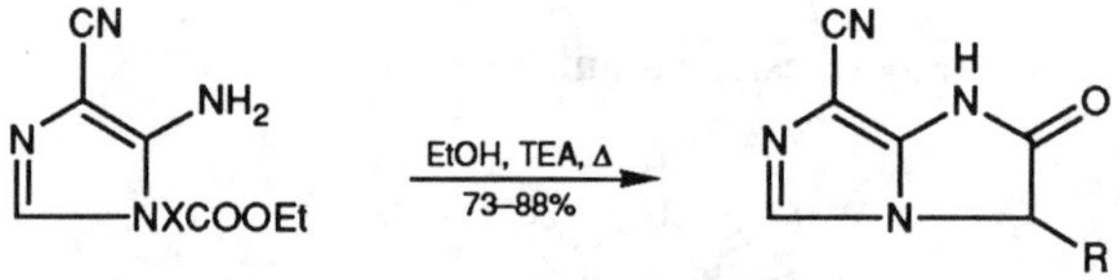

EtOH, TEA, Δ 73–88% [3961]

X = CH_2 or MeCH; R = H or Me

Imidazo[1,2-*a*]imidazol-2-one

II. FORMATION OF A SIX-MEMBERED RING

1. Pyridin-2-one

The first report of the synthesis of an imidazo[4,5-*b*]pyridin-5-one by closure of a pyridine ring is a base-catalysed cyclization of the amino-malonylidene diester. Complementary to this method is that in which a monoester mononitrile of malonic acid is heated in acetic acid to give a high yield of the fused pyridinone.

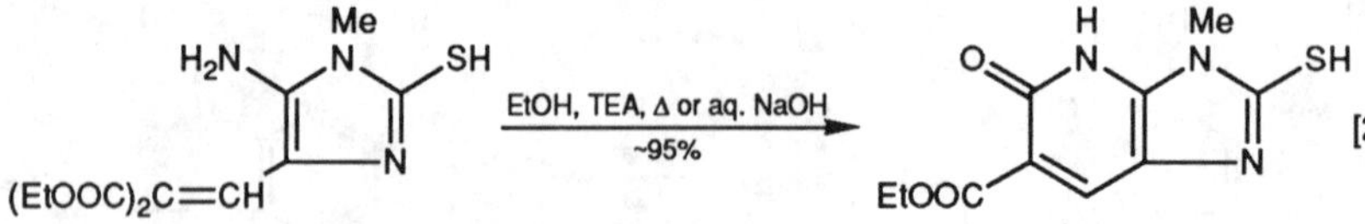

EtOH, TEA, Δ or aq. NaOH ~95% [2931]

Imidazo[4,5-*b*]pyridin-5-one

Another ring system synthesized similarly:

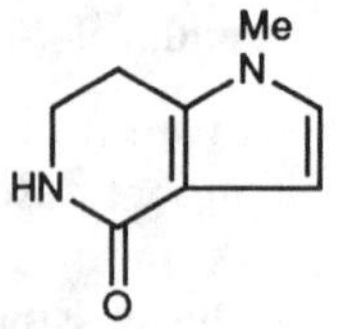

[2144]

Pyrrolo[3,2-*c*]pyridin-4-one

R^1 = H, Me; R^2 = Me, Et; X = N, CCOOEt

AcOH, Δ 87–97%

X = CH Pyrido[3,2-*a*]indolizin-2-one
X = N Dipyrido[1,2-*b* : 3',2'-*d*]pyrazol-2-one

[2986]

Another ring system synthesized similarly:

Pyrido[2',3' : 3,4]pyrazolo[1,5-*a*]quinolin-9-one

[2986]

Catalytic hydrogenation of a 2-nitro-α-oxocarboxylic ester with a large amount of platinum oxide present reduces both nitro and oxo groups; a 3-hydroxypyridin-2-one ring is formed. A smaller proportion of catalyst tends to favour the formation of a five-membered ring.

H_2, PtO_2, EtOH 75%

1,7-Naphthyridin-2-one

[2278]

Another ring system synthesized similarly:

1,5-Naphthyridin-2-one

[2278]

2. Pyridin-4-one

The use of dimethyl penta-2,3-dienedioate (dimethyl allenedicarboxylate) in cyclizing salicylic acid (p. 139) has been extended to that of anthranilic esters (**48.2**); the intermediate addition product of the amino group to the allene may be isolated [3142]. An *o*-amino carboxylic ester (as its hydrochloride salt) adds on to DMAD to form a fused pyridine-2,3-dicarboxylate (which may exist as its tautomer) [3556]. When an amino ester is heated with butoxide and dimethylacetamide diethyl acetal, the dimethylamine group is retained in the pyridinone ring [2949].

R (ring) NH_2, COOMe (48.2) + CHCOOMe ‖ CH=CHCOOMe —tBuOK–MeOH, Δ; 45–91%→ R (ring) H N, CH_2COOMe, COOMe, O [3142]

(48.2)
R = H or 4-Cl-5-F

4-Quinolinone

S (ring) NH_2HCl, COOMe —AcOH, DMAD, Δ; 54%→ S (ring) N, COOMe, COOMe, OH [3556]

Thieno[3,4-*b*]pyridine

Other ring systems synthesized similarly:

$MeOOCCH_2$, H N, S, MeOOC, O [3142, 3405]

Thieno[2,3-*b*]pyridin-4-one

S, H N, CH_2COOMe, COOMe, O [3405]

[1]Benzothieno[2,3-*b*]pyridin-4-one

H_2N, N, R, N, EtOOC + $MeC(OEt)_2$ | NMe_2 —i, ii; 42%→ Me_2N, H N, N, R, N, O [2949]

R = H, MeS
i, PhH, BuONa; ii, H^+

Pyrido[2,3-*d*]pyrimidin-5-one

A recent synthesis of acridinones uses lithiated ethyl anthranilate and benzyne (from a bromobenzene) at temperatures well below 0 °C. In this reaction, lithium *N*-isopropylcyclohexylamide (LiICA, prepared from butyl-lithium and *N*-isopropylcyclohexylamine) was more effective than LDA or lithium 2,2,6,6-tetramethylcyclohexylamide.

COOEt, NHR^1 + (benzyne) R^2 —THF, i; 31–68%→ O, N R^1, R^2 [3068, 3869]

R^1 = H, Me; R^2 = H, MeO, OCH_2O
i, LiICA or LDA

9-Acridinone

3. Pyrimidin-4-one

Benzoylcyanamide in DMF reacts with anthranilic acid, its allyl ester or amide to form 4-quinazolinone in moderate-to-good yield, but an electron-withdrawing substituent in the amino acid prevented cyclization [2625]. Cyclization of anthranilic esters with an imino ether proceeds without external heating and can give high yields [2574]. An almost theoretical yield of a 2-amino-4-quinazolinone is reported in a reaction between an anthranilic ester and chloroformamidine hydrochloride [3772].

NH_2 + NHCN / COPh —DMF, Δ; 24–62%→ [2625]

R^1 = HO, MeO, EtO, NH_2; R^2 = H, Me, MeO

4-Quinazolinone

Another ring system synthesized similarly:

[2625]

Pyrido[3,4-*d*]pyrimidin-4-one

(48.2) + CF_2CHCl_2–C(=NH)–OMe —AcOH; 29–88%→ [2574]

R = H, HO, Cl, Br

4-Quinazolinone

4. Pyrimidin-4-one, 2-thioxo- or Pyrimidine-2,4-dione

An —NH—CS group is incorporated in a new ring when ethoxycarbonyl isothiocyanate (review [2688]) reacts with an amino ester (cf. the reaction of its methyl homologue with an amino-imino ether, p. 77) [2388]; other isothiocyanates give 3-substituted products. When an isocyanate is used, the product is a dione. In alkaline solution, an amino ester reacts with a dithiocarbamate to give a fused thioxopyrimidinone in good yield [3342].

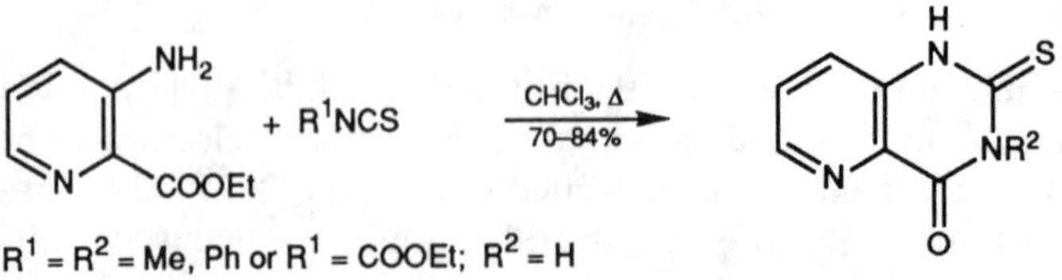

$R^1 = R^2 = Me, Ph$ or $R^1 = COOEt$; $R^2 = H$

+ R^1NCS → ($CHCl_3$, Δ; 70–84%) [2078, 2388]

Pyrido[3,2-*d*]pyrimidin-4-one, 2-thioxo-

Another ring system synthesized similarly:

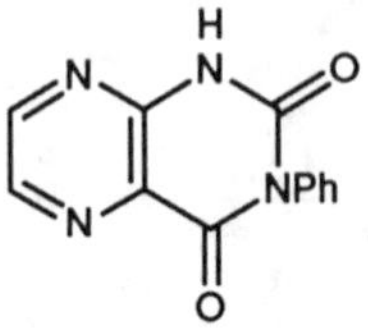

[2144]

2,4-Pteridinedione

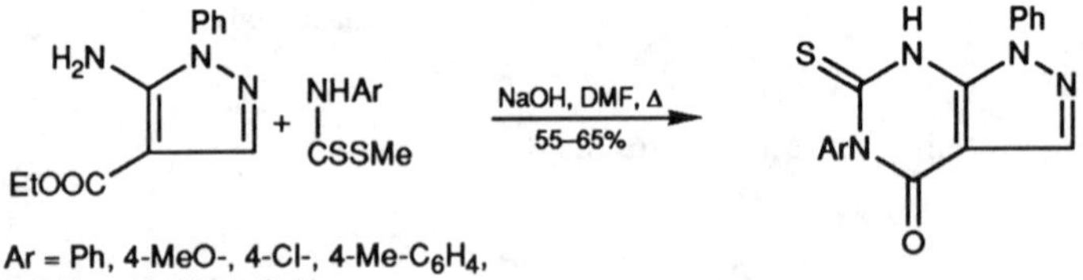

+ NHAr–CSSMe → (NaOH, DMF, Δ; 55–65%) [3342]

Ar = Ph, 4-MeO-, 4-Cl-, 4-Me-C_6H_4, 2,4-Me_2-, 3,5-Me_2-C_6H_3

Pyrazolo[3,4-*d*]pyrimidin-4-one, 6-thioxo-

5. 1,3-Oxazin-6-one

When an amino ester is heated with an anhydride, water and ethanol are lost during the formation of an oxazinone ring. Triphenyl phosphite or dibromotriphenylphosphorane may be used to cyclize an amino ester.

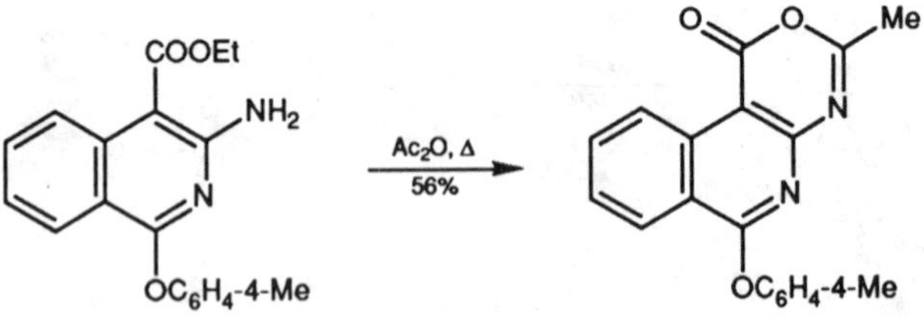

→ (Ac_2O, Δ; 56%) [3653]

OC_6H_4-4-Me

[1,3]Oxazino[4,5-*c*]isoquinolin-1-one

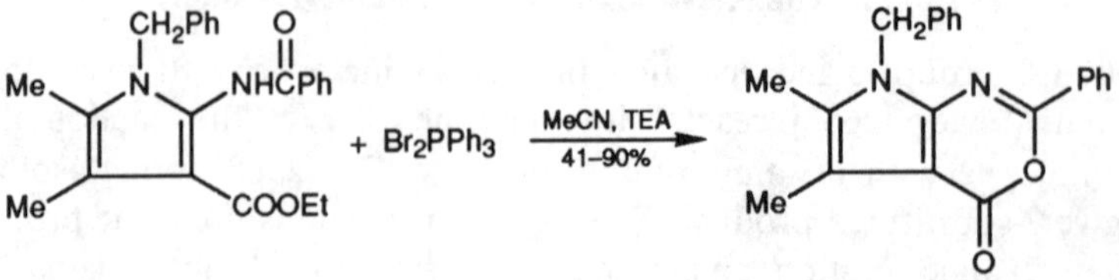

+ Br_2PPh_3 → (MeCN, TEA; 41–90%) [3300]

Pyrrolo[2,3-*d*][1,3]oxazin-4-one

Other ring systems synthesized similarly:

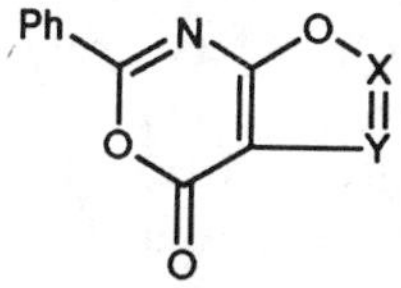

[3300] [3300]

X = N; Y = CPh Isoxazolo[5,4-*d*][1,3]oxazin-4-one
X = CPh; Y = N Oxazolo[5,4-*d*][1,3]oxazin-4-one

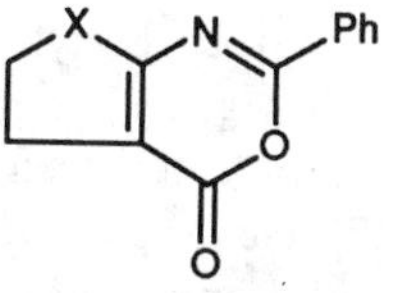

X = O Furo[2,3-*d*][1,3]oxazin-4-one
X = NR Pyrrolo[2,3-*d*][1,3]oxazin-4-one

6. 1,4-Oxazin-3-one

N-Hydroxy-1,4-oxazin-3-ones (which are cyclic hydroxamic acids and may have insecticidal action in plants) may be synthesized by a combined reduction and cyclization of 2-nitrophenoxyacetates.

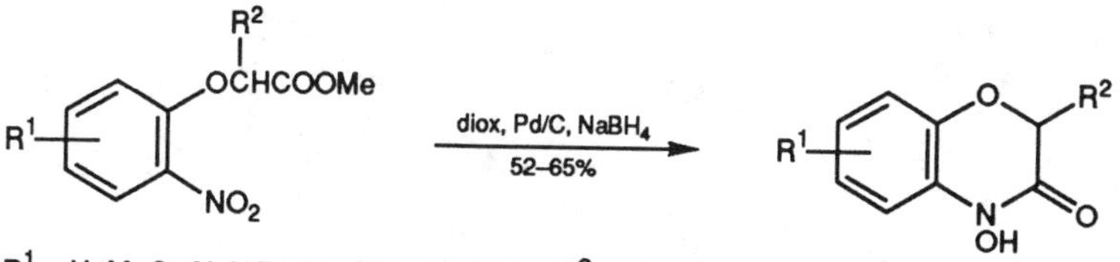

diox, Pd/C, $NaBH_4$ 52–65% [3958]

R^1 = H, MeO, Cl, NO_2, Me, CF_3, COOMe; R^2 = H, MeO

1,4-Benzoxazin-3-one

III. FORMATION OF A SEVEN-MEMBERED RING

1. 1,3-Diazepin-4-one

Amino and carboxylate ester groups attached to different rings may be annulated by heating with sodium methoxide.

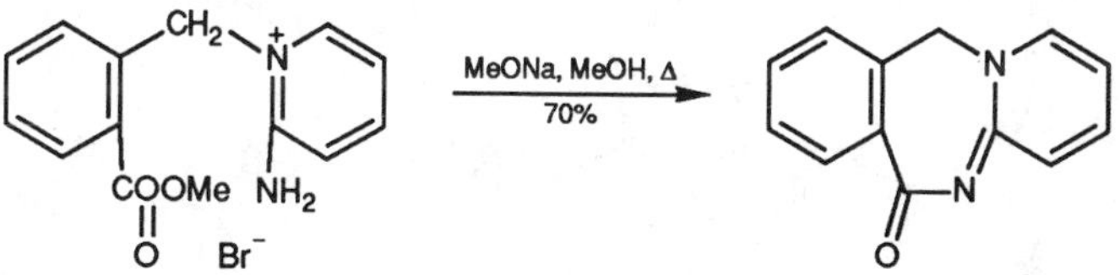

MeONa, MeOH, Δ 70% [2253]

Pyrido[1,2-*b*][2,4]benzodiazepin-6-one

2. 1,4-Diazepin-5-one

In contrast to the condensation of diamine with acetoacetate (Chapter 76), base-catalysed cyclization of the product of acetoacetate with one amine group gives the benzodiazepine-2-one [2327]. The anilinopyrrole (**48.3**) is cyclized to form a new imidazole ring when heated with PPA (Chapter 54), but when PPA is replaced by 4-toluenesulphonic acid, the imidazole (39%) is accompanied by an appreciable amount of a fused diazepinone [3766]. Nevertheless, annulation of amino and carboxylic ester groups sometimes presents problems. One method which gave better yields than several others used sodium methylsulphinylmethanide; a threefold amount of this was heated at 80 °C with the amino ester [2942]. Heating an amino ester in xylene containing a catalytic amount of 4-toluenesulphonic acid can also give excellent results [3721].

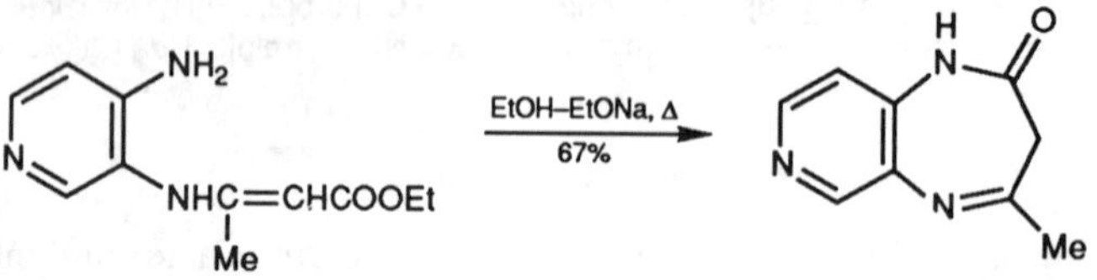

EtOH–EtONa, Δ 67% [2327]

Pyrido[3,4-*b*][1,4]diazepin-2-one

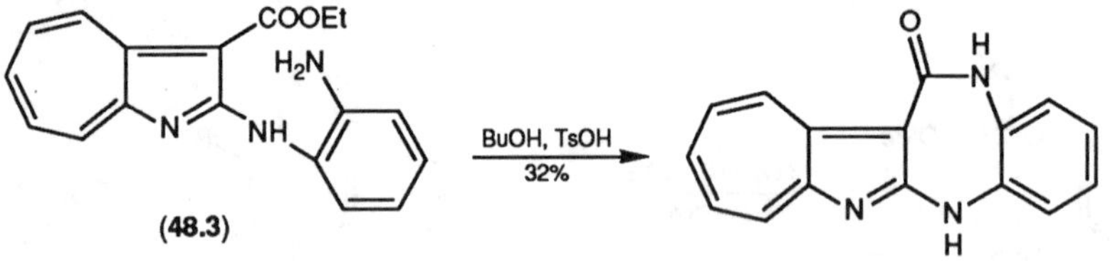

(48.3) BuOH, TsOH 32% [3766]

Cyclohepta[*d*]pyrrolo[2,3-*b*][1,5]benzodiazepin-12-one

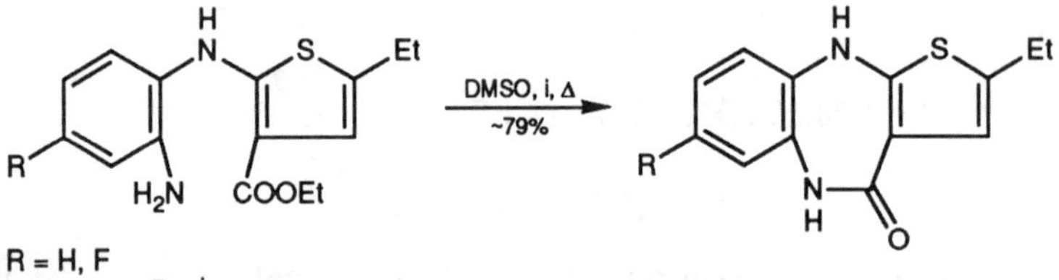

DMSO, i, Δ ~79% [2942]

R = H, F
i, $MeSOCH_2^-Na^+$

Thieno[2,3-*b*][1,5]benzodiazepin-4-one

Another ring system synthesized similarly:

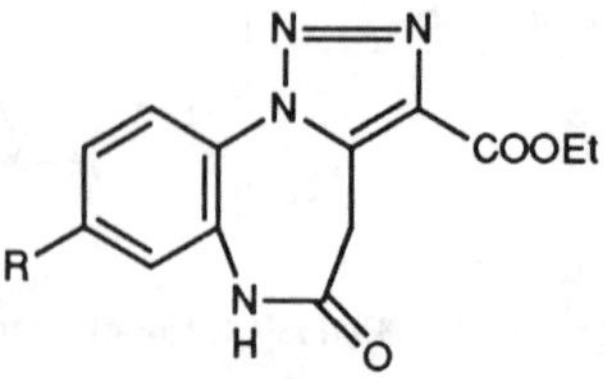

[3721]

[1,2,3]Triazolo[1,5-*a*]benzodiazepin-5-one

3. 1,4-Diazepine-2,5-dione

A substituent at C-3 of a benzodiazepine ring may be introduced by cyclizing an *N*-substituted carboxamide such as (**48.4**) in either acetic acid or methanolic piperidine. Heating an amino ester in pyridine is sometimes a useful procedure.

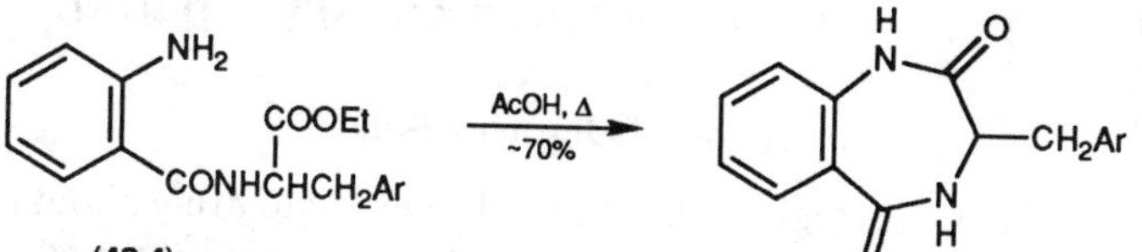

(48.4)
Ar = Ph, 4-HOC_6H_4

AcOH, Δ ~70%

1,4-Benzodiazepine-2,5-dione [2251, 2910]

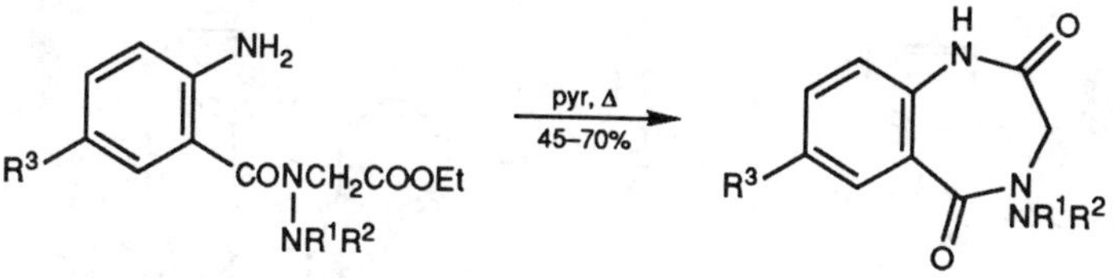

pyr, Δ 45–70%

R^1 = Me; R^2 = Ac or R^1 = R^2 = Me or R^1R^2 = $(CH_2)_5$ or NR^1R^2 = morpholine; R^3 = H, Me, Cl

1,4-Benzodiazepine-2,5-dione [3204]

4. 1,4-Oxazepin-5-one

4-Toluenesulphonic acid promotes the formation of this ring also (see Section III.3).

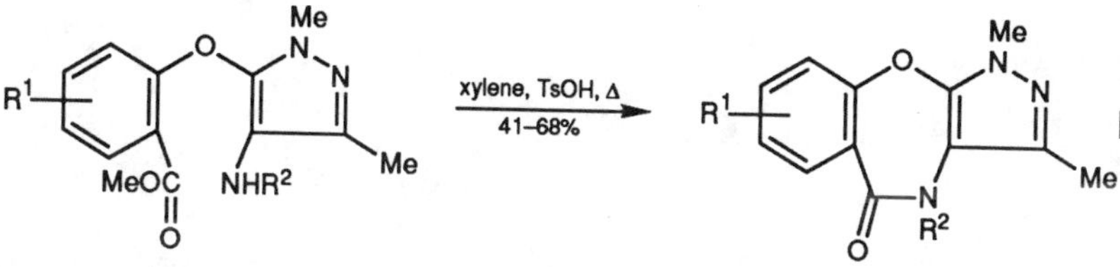

xylene, TsOH, Δ 41–68%

R^1 = H, Br, MeO, EtO; R^2 = H, Me

Pyrazolo[3,4-*b*][1,4]benzoxazepin-5-one [2429]

5. 1,4-Thiazepin-5-one

This ring is a significant feature of the potassium channel blocker diltiazem (review [3294]) in which the stereochemistry of the substituents at C-2 and C-3 is important. Analogues of diltiazem are synthesized from the amino-dicarboxylic ester (**48.5**) in a strongly basic medium.

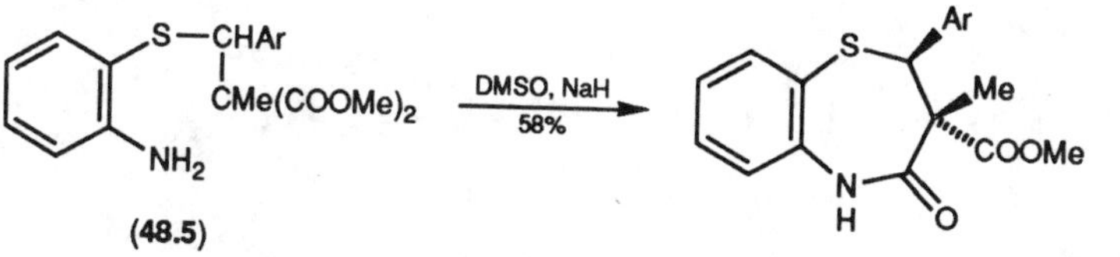

(48.5)
Ar = 4-$MeOC_6H_4$

DMSO, NaH 58%

1,5-Benzothiazepin-4-one [3852]

IV. FORMATION OF AN EIGHT-MEMBERED RING

1. 1,3,6-Thiadiazocin-7-one

A wide variety of reagents and conditions has been used to bring about reaction between amino and carboxylic ester groups. A less commonly used reagent, trimethylaluminium, was effective at ambient temperature in forming this ring in good yield.

SCH_2COOEt, NH_2 — C_6H_{14}, $AlMe_3$, 62% → [3218]

Pyrrolo[1,2-*a*][3,1,6]benzothiadiazocin-6-one

CHAPTER 49

Amine and Enamine or Oxime

I.	Formation of a Five-membered Ring	957
	1. Pyrrole	957
II.	Formation of a Six-membered Ring	958
	1. Pyrimidine 1-Oxide	958

I. FORMATION OF A FIVE-MEMBERED RING

1. Pyrrole

An enamine (formed *in situ*) reacts with a neighbouring amino group (also formed *in situ* by sodium borohydride–nickel boride–hydrazine hydrate reduction of a nitro group) to give a high yield of the precursor of a biologically important indole.

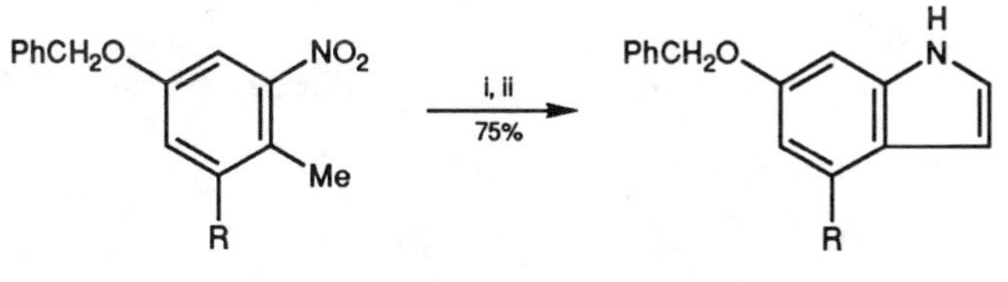

PhCH$_2$O, NO$_2$, Me, R → (i, ii; 75%) → PhCH$_2$O, H, N, R [3799]

R = Pr$_2$N(CH$_2$)$_2$ Indole

i, tripiperidinomethane, Ni(OAc)$_2$, NaBH$_4$;

ii, N$_2$H$_4$, Δ

A valuable modification of the Batcho–Leimgruber reaction (p. 239), in which the α-carbon of the enamine is substituted, enables tryptophan derivatives to be synthesized in very good yields.

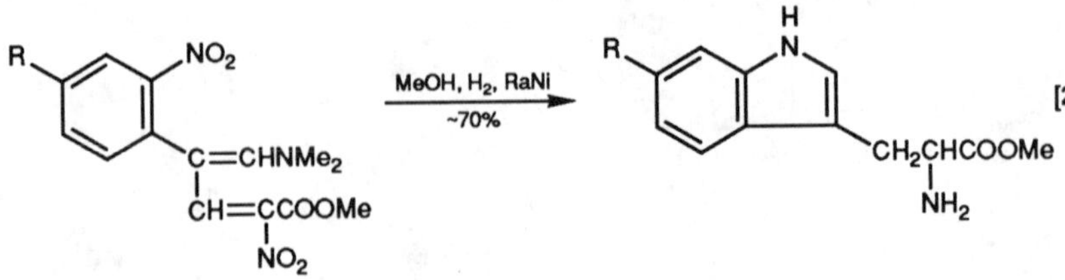

MeOH, H_2, RaNi ~70% [2962]

Of several reducing agents which have been used to cyclize the nitro-oxime (**49.1**), iron–acetic acid gave the best yield although the most convenient to use was Pd/C–$NaBH_4$ which gave slightly lower yields.

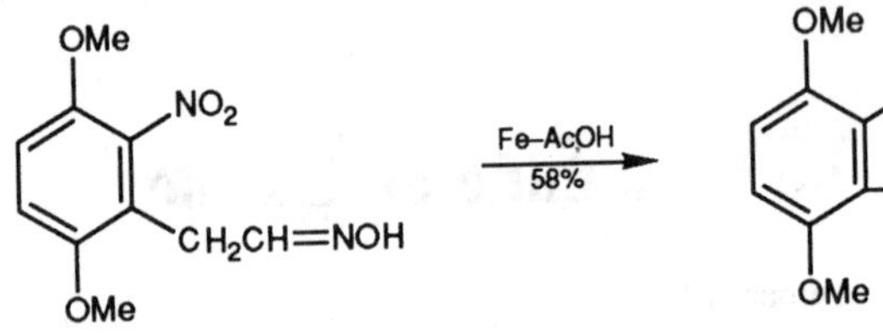

Fe–AcOH 58% [3812]

(49.1)

II. FORMATION OF A SIX-MEMBERED RING

2. Pyrimidine 1-Oxide

Heating the oxime of 2-aminobenzophenone with ethyl orthoformate gives quinazoline *N*-oxide. A pyrazolopyrimidine *N*-oxide has been similarly synthesized; the chemistry of pyrazolopyrimidines has been reviewed [3360].

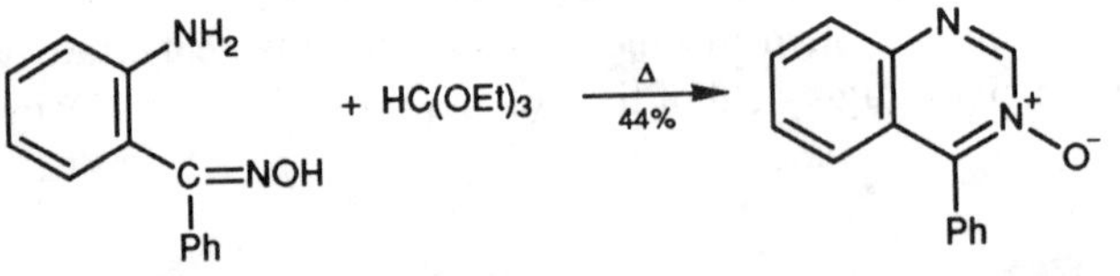

+ $HC(OEt)_3$ Δ 44% [2425]

Quinazoline 3-oxide

Another ring system synthesized similarly:

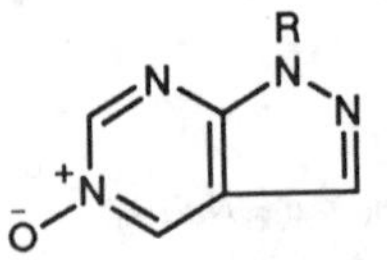

[2798]

Pyrazolo[3,4-*d*]pyrimidine 5-oxide

CHAPTER 50

Amine or Diazonium Salt and Hydrazide or Hydrazine

I. Formation of a Five-membered Ring 959
  1. Imidazole 959
  2. Imidazol-2-one 960
  3. 1,2,4-Triazole 960
II. Formation of a Six-membered Ring 961
  1. Pyrazin-2-one 961
  2. 1,2,4-Triazine 961
  3. 1,2,4-Triazin-3-one 961
  4. 1,2,4-Triazin-5-one 962
III. Formation of a Seven-membered Ring 962
  1. 1,2,4-Triazepin-7-one 962
  2. 1,2,4-Triazepine-3,7-dione 963

Cyclizations of *o*-acylhydrazino-amines are dicussed in Chapter 5.

I. FORMATION OF A FIVE-MEMBERED RING

1. Imidazole

2-(Acylhydrazinyl)anilines are cyclized in good-to-excellent yields by heating with an alkanoic acid.

NHNHC(O)R^1 / NH_2 + R^2COOH —Δ, 42–92%→ NHCOR^1 / R^2 [2500]

R^1,R^2 = H, Me, Et

Benzimidazole

2. Imidazol-2-one

Diazotization of an *o*-amino-hydrazide at low temperatures leads to the formation of a pyrazole ring (p. 241), but under more drastic conditions, a fused imidazolinone ring may be formed.

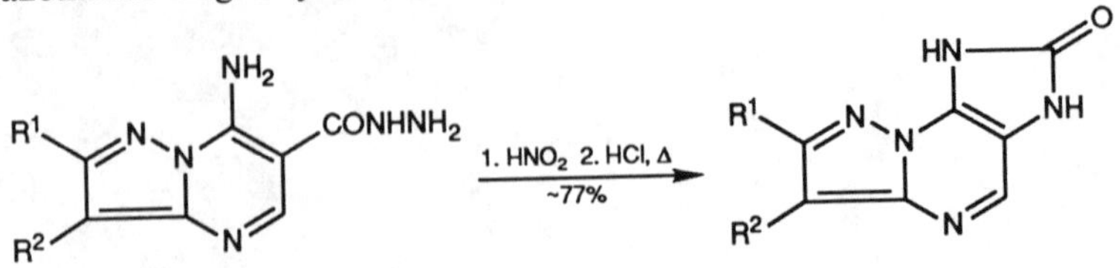

1. HNO_2 2. HCl, Δ ~77% [2028]

R^1,R^2 = H, Me

Pyrazolo[5,1-*f*]purin-2-one

Another ring system synthesized similarly:

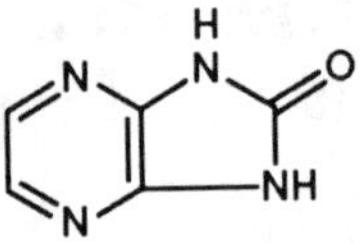

[2532]

Imidazo[4,5-*b*]pyrazin-2-one

3. 1,2,4-Triazole

When the *N*-amino-hydrazine (**50.1**) is treated with an orthoester, two products are possible; triethyl orthoacetate yields only the linear molecule whereas the orthoformate gives a mixture of linear and angular isomers.

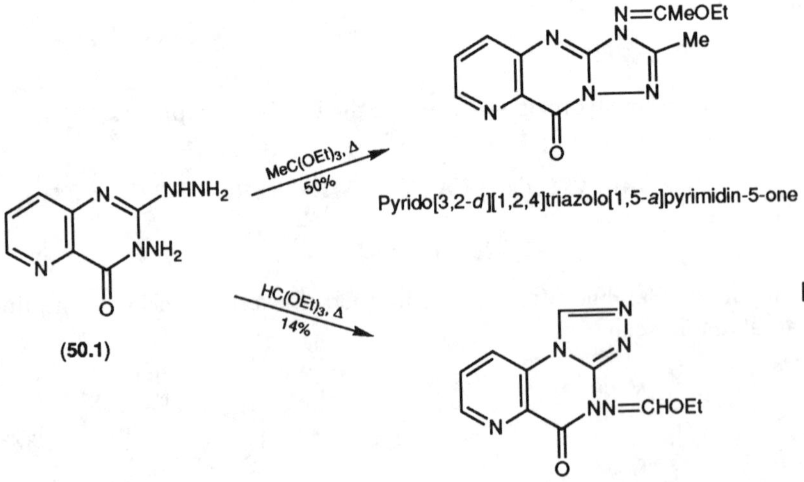

$MeC(OEt)_3$, Δ 50%

Pyrido[3,2-*d*][1,2,4]triazolo[1,5-*a*]pyrimidin-5-one

[3674]

$HC(OEt)_3$, Δ 14%

(50.1)

Pyrido[2,3-*e*][1,2,4]triazolo[4,3-*a*]pyrimidin-5-one

II. FORMATION OF A SIX-MEMBERED RING

1. Pyrazin-2-one

Of the several possible ways in which the amino-hydrazine (**50.2**) may react with an unsymmetrical ketone, that which gives the pyrazinone as the major product predominates.

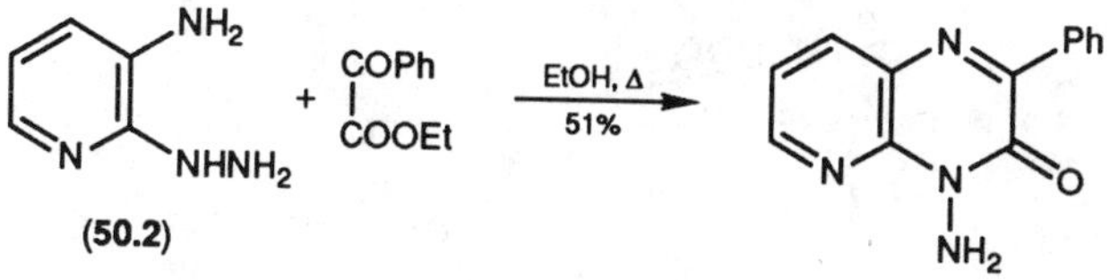

(50.2) + COPh–COOEt → (EtOH, Δ; 51%) Pyrido[2,3-*b*]pyrazin-3-one [3251]

2. 1,2,4-Triazine

Annulation of amino and hydrazino groups with an orthoester is a convenient low-temperature method of forming a dihydrothiazine ring, or, in some instances, the triazine ring [2751].

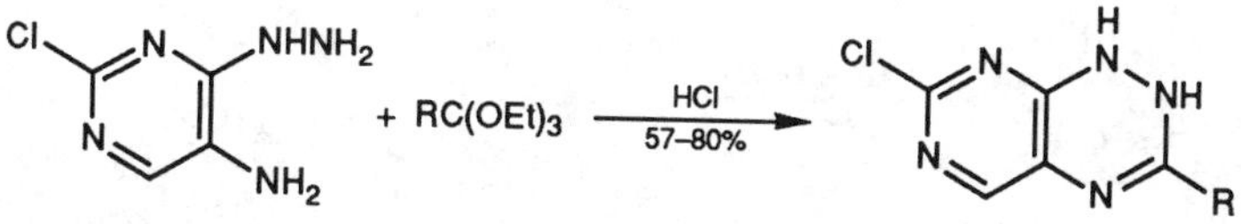

+ $RC(OEt)_3$ → (HCl; 57–80%) Pyrimido[5,4-*e*]-1,2,4-triazine [2255, 2293]

R = H, Me, $MeOCH_2$, $ClCH_2$, $EtOOCCH_2$

Other ring systems synthesized similarly:

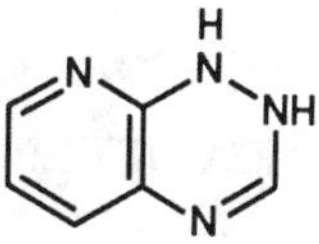

Pyrido[3,2-*e*]-1,2,4-triazine [2319]

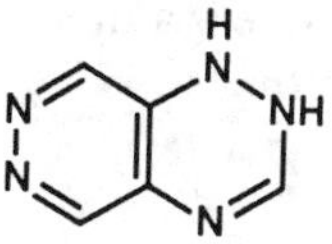

Pyridazino[4,5-*e*]-1,2,4-triazine [2649]

3. 1,2,4-Triazin-3-one

LTA oxidatively cyclizes an *o*-amino-*N*,*N*′-di(ethoxycarbonyl)hydrazine to this ring in good yield, as also does a strong base such as ethoxide.

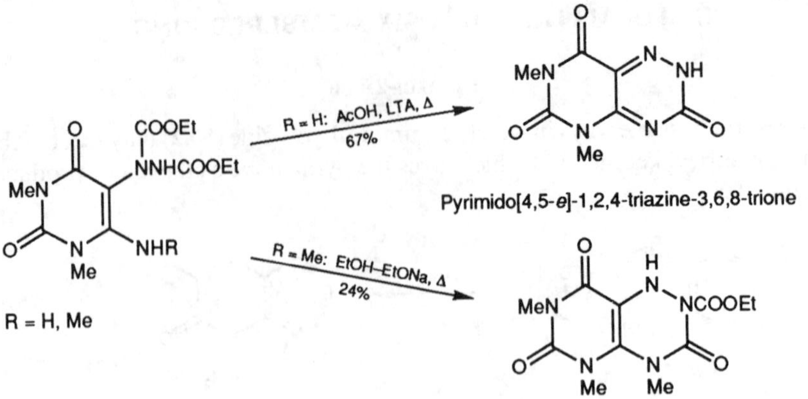

R = H: AcOH, LTA, Δ 67%

Pyrimido[4,5-*e*]-1,2,4-triazine-3,6,8-trione [2753]

R = Me: EtOH–EtONa, Δ 24%

R = H, Me

4. 1,2,4-Triazin-5-one

The pyrimidine (**50.3**) may formally be a dihydrazine but in the cyclization with DMAD, only three of the four nitrogens are in the newly formed ring. The addition of heterocycles to acetylenes has been reviewed [2816].

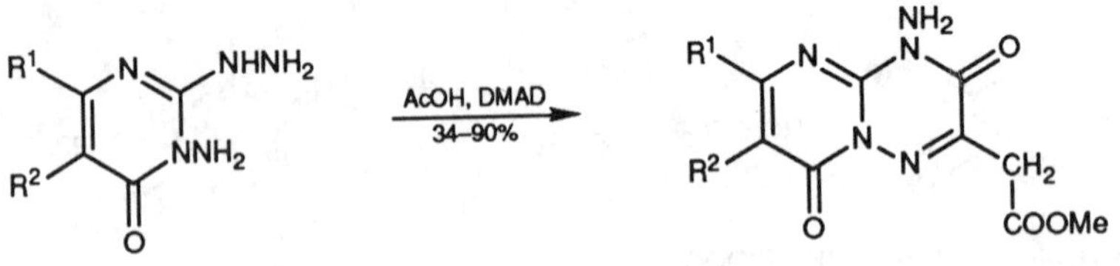

AcOH, DMAD 34–90% [3473]

(**50.3**)

R^1 = Me, Ph; R^2 = H, Et

Pyrimido[1,2-*b*][1,2,4]triazine-2,6-dione

III. FORMATION OF A SEVEN-MEMBERED RING

1. 1,2,4-Triazepin-7-one

The terminal nitrogen of a hydrazide is annulated to an amino group by heating the compound with a dialkyl ketone.

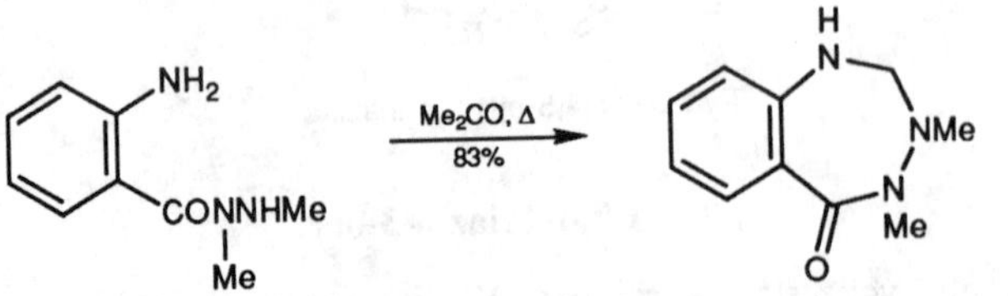

Me_2CO, Δ 83% [2059]

1,3,4-Benzotriazepin-5-one

2. 1,2,4-Triazepine-3,7-dione

Amino and hydrazide groups combine to form this ring when the compound is heated with ethyl chloroformate–TEA.

$$\text{2-}(H_2N)C_6H_4\text{-CON(Me)NH}_2 \xrightarrow[26\%]{ClCOOEt,\ TEA,\ \Delta} \text{1,3,4-Benzotriazepine-2,5-dione} \qquad [2834]$$

1,3,4-Benzotriazepine-2,5-dione

CHAPTER 51

Amine and Hydrazone or Imine

I.	Formation of a Five-membered Ring	964
	1. Pyrazole	964
	2. Imidazole	965
	3. 1,2,4-Triazole	965
II.	Formation of a Six-membered Ring	966
	1. Pyridine	966
	2. Pyrimidine	966
	3. 1,2,4-Triazine	966
	4. 1,2,4-Triazine-5,6-dione	967
III.	Formation of a Seven-membered Ring	968
	1. 1,3,5-Triazepine	968

I. FORMATION OF A FIVE-MEMBERED RING

1. Pyrazole

The quinonoid hydrazones (**51.1**) are readily cyclized to indazolediones at ambient temperature in dilute mineral acid. When the mixture is warmed, the Ts group is cleaved as it is also in the thermal cyclization of (**51.1**) at 180 °C [2879]. This unexpected reaction provides a convenient synthesis of these indazoles and a mechanism involving displacement of the t-amino group by the tosyl-carrying nitrogen is proposed. The chemistry of heterocyclic quinones has been reviewed [2947, 3650].

EtOH, HCl 13–88%

[2790, 2879]

(51.1)

R^1 = t-amino; R^2 = H, t-amino

4,7-Indazoledione

2. Imidazole

Imino ethers which have a neighbouring amino group are cyclized on heating under anhydrous conditions; this provides a method of synthesizing purinediones with a particular stereoconfiguration at the α-carbon of a chain at C-8.

PhH, Δ 80%

[3778]

2,6-Purinedione

3. 1,2,4-Triazole

Annulation of a nuclear amino group with a side-chain hydrazone function [as in **(51.2)**] proceeds in moderate yield on heating the compound with sodium acetate–acetic acid for 10–20 h under anhydrous conditions. Ethoxymethylene-malonates also convert imino-*N*-amines into triazoles with a similar loss of carboxyl-derived functions.

AcONa–AcOH, Δ 39–57%

[3298]

(51.2)

Ar^1 = 4-halogeno-, 4-MeO-, 4-Bu-, 4-tBu-, 4-MeS-C_6H_4;
Ar^2 = 4-F-, 4-NO_2-C_6H_4, 2,4-$Cl_2C_6H_3$

[1,2,4]Triazolo[3,4-*c*]-1,2,4-triazole

EtOH, Δ 84%

[3724]

(51.3)

R = CN, COOEt

Pyrazolo[4′,3′ : 5,6]pyrano[3,2-*e*][1,2,4]triazolo[1,5-*c*]pyrimidine

II. FORMATION OF A SIX-MEMBERED RING

1. Pyridine

Intramolecular cycloaddition occurs when the aminoimine (**51.5**) [formed *in situ* from the imine (**51.4**) and DMFDMA] is heated.

DMFDMA, Δ 48–93% [3007]

(**51.4**) (**51.5**) Pyrido[3,2-*d*]pyrimidine-2,4-dione

2. Pyrimidine

The imino-amine (**51.6**) reacts at room temperature with aldehydes or ketones with the formation of purines. The products obtained from aldehydes are slowly oxidized (by loss of two hydrogen atoms from the pyrimidine ring) at room temperature, but the ketone-derived purines are stable and are accompanied by smaller amounts of an imidazo[1,5-*c*]imidazole. During cyclization, the nitrile group is converted into a carboxamide. Pentane-2,4-dione, 2-furfuraldehyde and but-2-enal give the fully aromatized purine as the main or only product isolated (in 38–49% yield). Stirring an imino-nitrile for several hours or heating it for a few minutes with an anhydride converts it into a fused pyrimidine.

+ R^1R^2CO MeOH 35–87% [3744]

(**51.6**) Purine

R^1 = H; R^2 = Me, Et or R^1 = R^2 = Me, Bu or R^1R^2 = $(CH_2)_5$

$(RCO)_2O$, $CHCl_3$ 52–92% [3377]

R = Me, Et, CF_3, CH=CHCOOH, $(CH_2)_2COOH$

3. 1,2,4-Triazine

An imino-amine is cyclized to a 6-hydroxy-1,2,4-triazine (rather than its keto tautomer, according to spectral evidence) by heating with ethyl cyanoacetate in ethanol for 10 h but heating with malononitrile causes loss of the pyranopyrazole fragment and the formation of a triazine ring [3724]. Amino and hydrazono functions are annulated thermally by orthoesters or with an acylnitrile and a trace of sodium in dioxane [3385]. When R = NH_2, the reaction takes a different course [3643].

(51.3) → (CH_2(COOEt)CN, EtOH, Δ; 61%) [3724]

Pyrazolo[4",3" : 5',6']pyrano[2',3' : 4,5]pyrimido[1,6-*b*][1,2,4]triazine

(51.3) → (EtOH, $CH_2(CN)_2$, Δ; 69%)

Pyrimido[1,6-*b*][1,2,4]triazine

HC(OEt)$_3$, Δ; 63% [3385]

R = Me, MeS, Ph, 4-Cl-, 4-Me-C_6H_4

[1,2,4]Triazino[6,5-*e*]-1,2,4-triazine

4. 1,2,4-Triazine-5,6-dione

Oxalyl dichloride in benzene converts an imino-amine (**51.3**) into a fused triazinedione ring.

(**51.3**) → (PhH, $(COCl)_2$, Δ; 71%) [3724]

Pyrazolo[4",3" : 5',6']pyrano[2',3' : 4,5]pyrimido[1,6-*b*][1,2,4]triazine-2,3-dione

III. FORMATION OF A SEVEN-MEMBERED RING

1. 1,3,5-Triazepine

A thermally induced ring closure of an imino-amine with an alkanoic acid proceeds in refluxing toluene containing a trace of acetic acid.

+ R^3COOH → PhMe, AcOH, Δ; 58–80% [2862]

R^1 = H, MeO, Me; R^2 = H, Me; R^3 = H, Me, Et

Thiazolo[3,2-*b*][1,3,5]benzotriazepine

CHAPTER 52

Amine and Ketone

I. Formation of a Five-membered Ring 969
1. Pyrrole 969
2. Pyrazole 970
3. Imidazole 971
4. Imidazol-2-one 971
5. Isoxazole 971
6. Oxazol-2-one or Oxazole-2-thione 972
7. Isothiazole 1,1-Dioxide 972
II. Formation of a Six-membered Ring 973
1. Pyridine 973
2. Pyridin-2-one 973
3. Pyrimidine 974
4. Pyrimidin-2-one 974
5. Pyrimidine-2-thione 975
6. Pyrazine 975
III. Formation of a Seven-membered Ring 975
1. 1,3-Diazepin-4-one 975
2. 1,4-Diazepine 976
3. 1,4-Diazepin-2-one 977

I. FORMATION OF A FIVE-MEMBERED RING

1. Pyrrole

Reduction of ethyl 2-nitrophenyl-3-oxobutanoate with titanium(III) chloride at ambient temperature is accompanied by cyclization in high yield. Recent developments in indole chemistry have been reviewed [3893, 3982].

$$o\text{-}NO_2C_6H_4CH_2COCH_2COOEt \xrightarrow[75\%]{Me_2CO,\ AcONH_4,\ TiCl_3} \text{2-}(CH_2COOEt)\text{-indole}$$ [3741]

Indole

2. Pyrazole

In the reaction of an *o*-aminoketone with hydrazine hydrate (p. 252), the amino group is displaced; in this example, the reaction proceeds in high yield in acetic acid [2490] or in PPA [2332]. When a substituted hydrazine is used, two isomeric products are possible and the ratio of isomers formed depends on the conditions and the nature of R [3718]. Heterocyclic quinones have been reviewed [2947, 3650].

COOEt, NHBu, COMe (pyridine) $\xrightarrow[92\%]{N_2H_4,\ AcOH,\ \Delta}$ EtOOC, N-H, Me [2490]

Pyrazolo[4,3-*c*]pyridine

$C{=}NNH_2$ / R $\xrightarrow[\sim 92\%]{PPA,\ \Delta}$ [2332]

R = Me, Et, $PhCH_2$, Ph_2CH

Indeno[1,2-*c*]pyrazol-4-one

Other ring systems synthesized similarly:

{2332]

X = CH; Y = N Pyrazolo[3',4' : 3,4]cyclopenta[1,2-*b*]pyridin-4-one
X = N; Y = CH Pyrazolo[3',4' : 3,4]cyclopenta[1,2-*c*]pyridin-4-one

NMe_2, $COCF_3$, $COCF_3$ + $RNHNH_2$ $\xrightarrow[19\text{–}100\%]{R = H,\ Me:\ MeCN,\ \Delta}$ RN—N, CF_3, $COCF_3$

$\xrightarrow[40\text{–}100\%]{R = alkyl,\ Ar:\ MeCN,\ \Delta}$ N—NR, CF_3, $COCF_3$ [3718]

(52.1)

Ar = Ph, 4-$NO_2C_6H_4$

Benz[*g*]indazole

3. Imidazole

Annulation of an imine and a side-chain keto group in boiling mineral acid can give a high yield of the imidazo derivative [2192]. Alternatively, the reaction may proceed in a basic medium without external heat [2435]. Another variation on this cyclization is shown by the behaviour of the aminothiazolium salt (**52.2**) [3533].

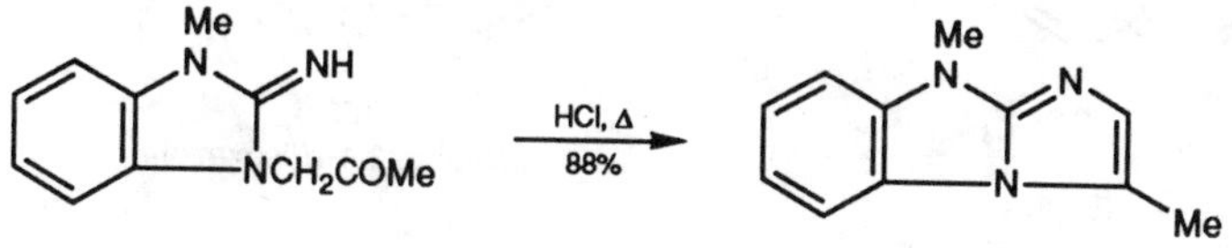

HCl, Δ
88%

Imidazo[1,2-*a*]benzimidazole [2192]

NaOH
~90%

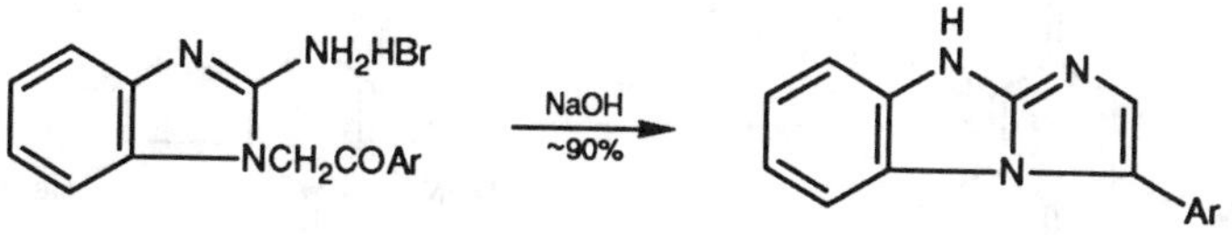

Ar = Ph, 4-Br-, 4-Cl-C_6H_4

Imidazo[1,2-*a*]benzimidazole [2435]

DMF, Δ
83%

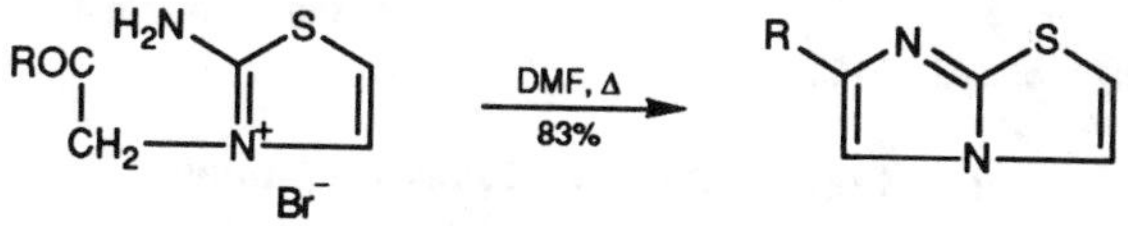

(**52.2**)
R = 4-$NO_2C_6H_4$

Imidazo[2,1-*b*]thiazole [3533]

4. Imidazol-2-one

Reaction of the 2-aminobenzoquinone (**52.3**) with phenyl isocyanate was expected to give the 1-phenylnaphth[1,2-*d*]imidazol-2-one but produced the 3-phenyl isomer; a four-membered spiran intermediate may have been involved.

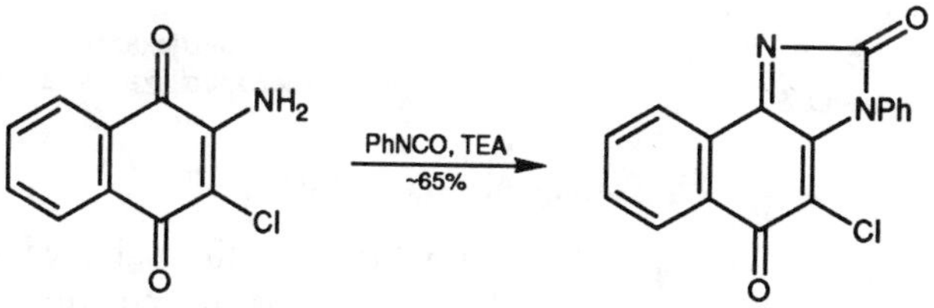

PhNCO, TEA
~65%

(**52.3**)

Naphth[1,2-*d*]imidazol-2-one [2091]

5. Isoxazole

When the naphthalene (**52.1**) is treated with a small excess of hydroxylamine, the isoxazole (**52.4**) is formed. Its [2,1-*d*] isomer (**52.5**) is produced in low yield by heating 2,4-di(trifluoroacetyl)-1-naphthylamine with hydroxylamine and TEA. Hydroxylamine reacts with quinonoid amino-ketones such as (**52.6**) under milder conditions than usual to give high yields of the isoxazole.

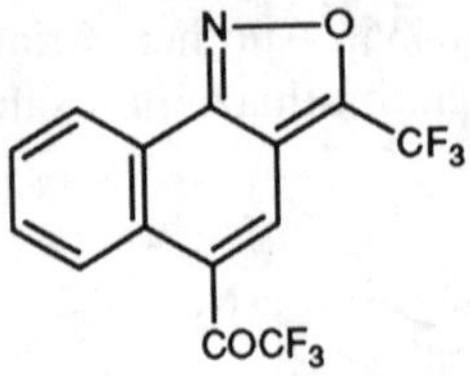

[3716] [3716]

(**52,4**)
Naphth[1,2-*c*]isoxazole

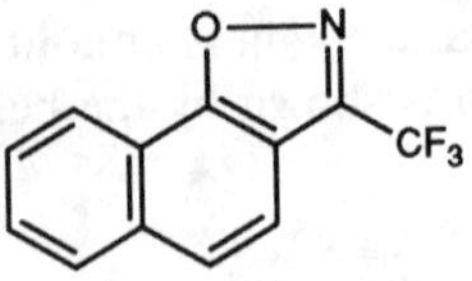

(**52.5**)
Naphth[2,1-*d*]isoxazole

NH_2OH, MeOH, Δ
75–96%

[2590]

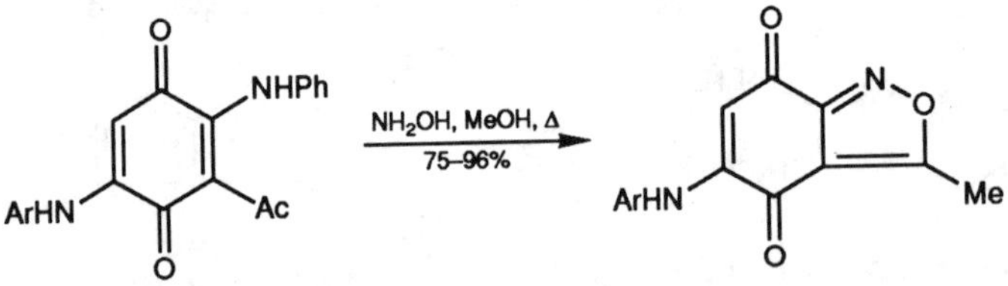

(**52.6**)
Ar = Ph, 2- or 3-MeC_6H_4, 4-ClC_6H_4

2,1-Benzisoxazole-4,7-dione

6. Oxazol-2-one or Oxazole-2-thione

When colchineinamide (**52.7**) is treated with phosgene or thiophosgene at about 0 °C, the amino and carbonyl groups react to give a moderate yield of the oxazolone.

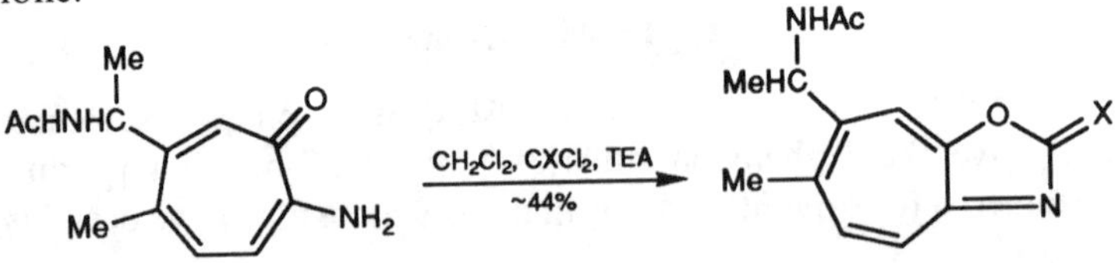

CH_2Cl_2, $CXCl_2$, TEA
~44%

[3808]

(**52.7**)
X = O, S

Cyclohept[*d*]oxazol-2-one
Cyclohept[*d*]oxazole-2-thione

7. Isothiazole 1,1-Dioxide

Conversion of an amino ketone into this ring proceeds through the intermediate formation of the (crude) sulphonyl chloride which is treated with ammonia.

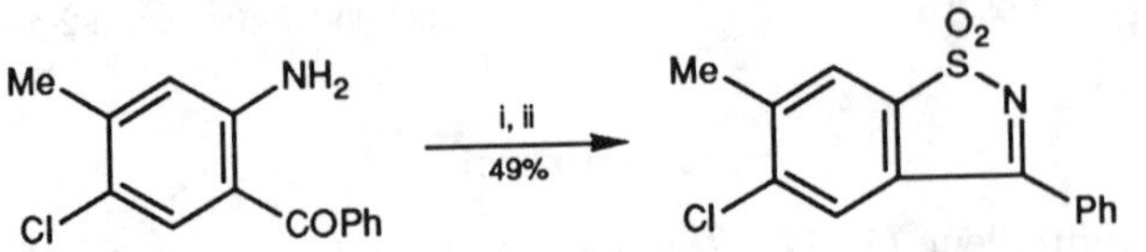

i, ii
49%

[2528]

i, HNO_2, SO_2, $CuCl_2$; ii, NH_3

1,2-Benzisothiazole 1,1-dioxide

II. FORMATION OF A SIX-MEMBERED RING

1. Pyridine

2-Benzoylarylamines may be cyclized using either a carboxylic acid or *N,N*-dimethylacetamide dimethylacetal; in the former method, two molecular proportions of amino ketone and of one the carboxylic acid are heated with PPA [2137]. In an unusual reaction which also involves a rearrangement, the enaminone (**52.9**) is heated with a cycloalkanone and ammonium acetate [2906].

+ R^3CH_2COOH PPA, Δ 71–92% [2137]

(52.8)

R^1 = H, Me, Cl; R^2 = H; R^3 = H, Me, Et, Br; Ar = Ph

Quinoline

+ $MeC(OMe)_2NMe_2$ Δ 98% [3155]

Pyrido[2,3-*d*]pyridazine

NH_4OAc, Δ 8–50% [2906]

(52.9)

m = 3,4; *n* = 3,4

Dicyclopenta[*b,e*]pyridine

Acridine

2. Pyridin-2-one

This ring may be formed by condensation of an amino ketone with an orthoester to give the 2-ethoxypyridine which is hydrolysed to the pyridinone with aqueous acid. A more highly substituted pyridinone ring is formed when the amino ketone (**52.10**) is heated with malonic ester [3155]. Heating 2-aminobenzophenone (review of its synthesis [3009]) with a γ-keto ester is a variation of Knorr's quinoline synthesis which gives high yields [3171].

i ~89% HCl, Δ 78–95% [3155]

Ar = 4-ClC_6H_4; R = H, Me

i, $RCH_2C(OEt)_3$

Pyrido[2,3-*d*]pyridazin-2-one

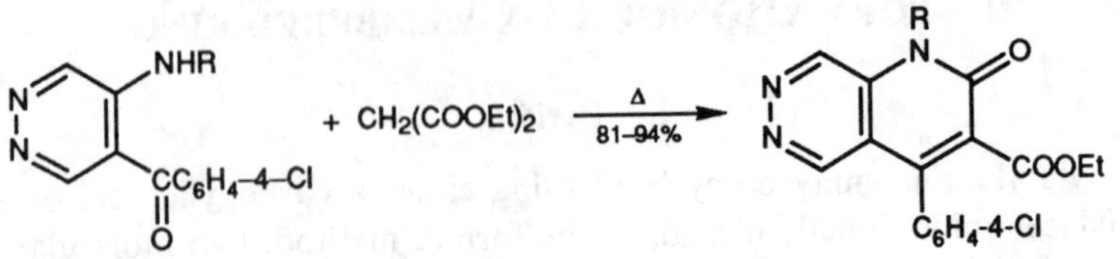

(52.10) R = H, Me + $CH_2(COOEt)_2$ —Δ, 81–94%→ Pyrido[2,3-*d*]pyridazin-2-one [3155]

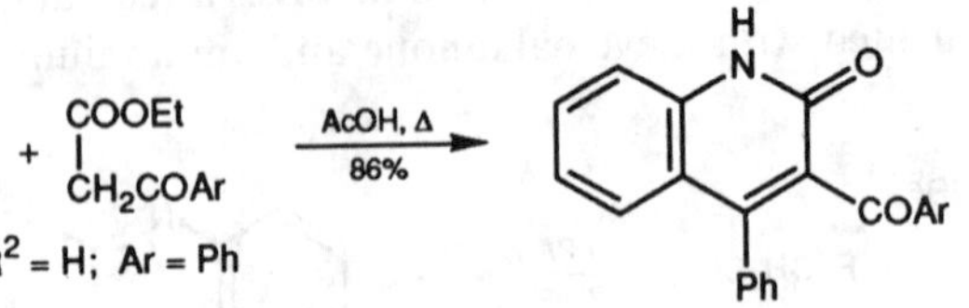

(52.8) + CH_2COAr / COOEt —AcOH, Δ, 86%→ 2-Quinolinone [3171]

$R^1 = R^2 = H$; Ar = Ph

3. Pyrimidine

Substituents in a 1,4-quinone may be displaced by nucleophiles in the presence of strong bases; an amidine supplies the necessary atoms for pyrimidine formation. When the amino and carbonyl functions are in *peri* positions, a doubly fused ring is formed by reaction with formamide.

(52.6) + RC=NH(NH_2) —EtOH–EtONa, Δ, 16–82%→ 5,8-Quinazolinedione [2800]

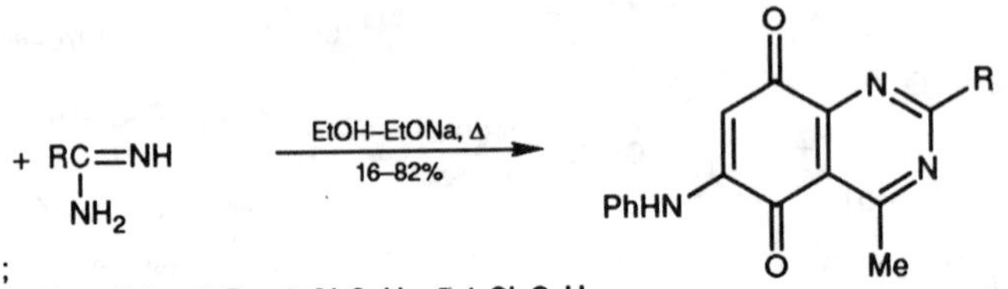

Ar = Ph;
R = alkyl, Ph, 4-NO_2-, 3-Br-, 3-Cl-C_6H_4, 3,4-$Cl_2C_6H_3$

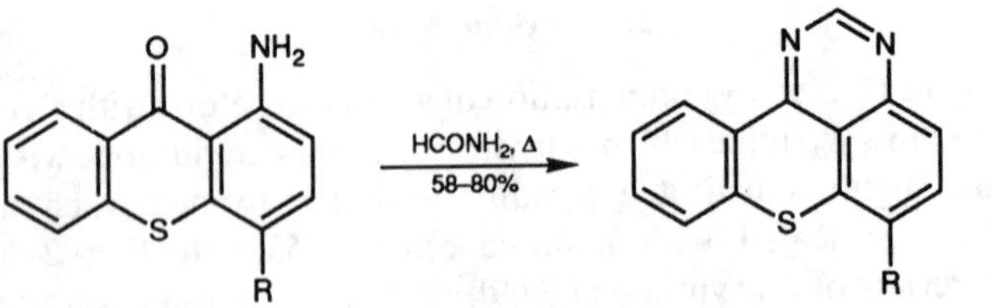

—$HCONH_2$, Δ, 58–80%→ [1]Benzothiopyrano[4,3,2-*de*]quinazoline [2584]

R = Cl, NO_2, Me

4. Pyrimidin-2-one

In the synthesis of this ring from 2-aminobenzophenone (review of synthesis [3009]) or a heteroanalogue, ethyl carbamate or chlorosulphonyl isocyanate (reviews [3147a, 3340] may be used.

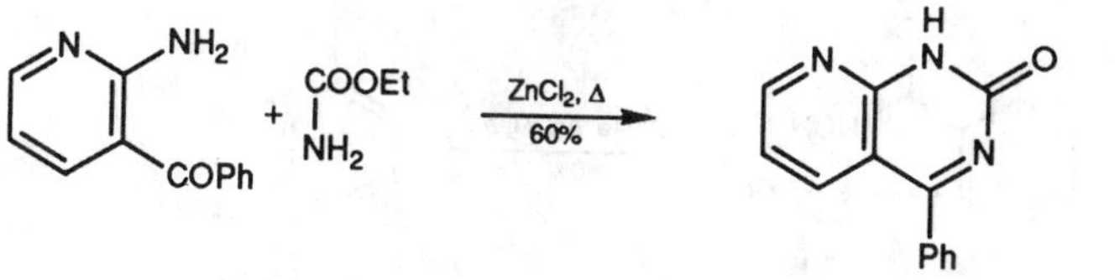

NH$_2$, COPh + COOEt / NH$_2$ —ZnCl$_2$, Δ; 60%→ Pyrido[2,3-*d*]pyrimidin-2-one [2628]

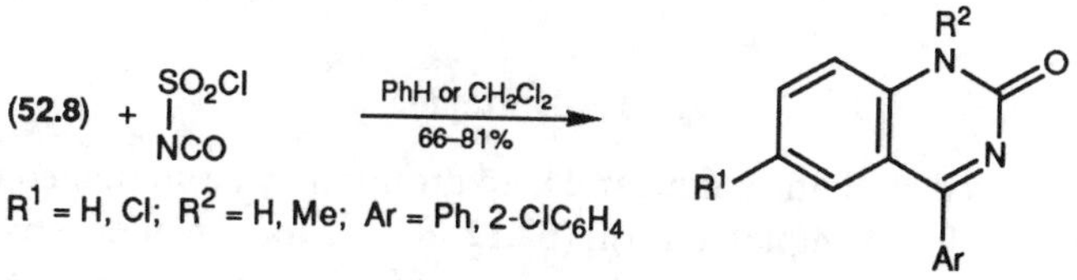

(52.8) + SO$_2$Cl / NCO —PhH or CH$_2$Cl$_2$; 66–81%→ 2-Quinazolinone [2976]

R^1 = H, Cl; R^2 = H, Me; Ar = Ph, 2-ClC$_6$H$_4$

5. Pyrimidine-2-thione

At low temperature (−40 to 20 °C) and under non-acidic conditions, 2-amino-acetophenone or 2-aminobenzophenones (review [3009] are cyclized with triphenylphosphine thiocyanate in high yields.

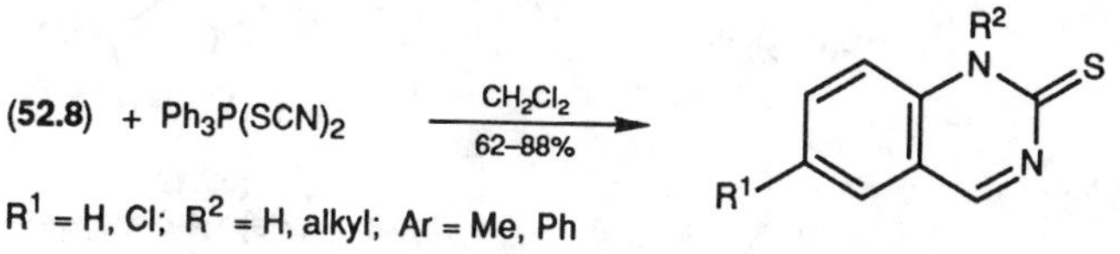

(52.8) + $Ph_3P(SCN)_2$ —CH$_2$Cl$_2$; 62–88%→ 2-Quinazolinethione [2966]

R^1 = H, Cl; R^2 = H, alkyl; Ar = Me, Ph

6. Pyrazine

Amino and aldehyde groups attached to different joined rings condense dehydratively in aqueous ammonia to form this ring. A 4-pteridinone has been synthesized in high yield from an amino-ketone by heating it with a deaerated pyridine–pyridine hydrochloride buffer [2745].

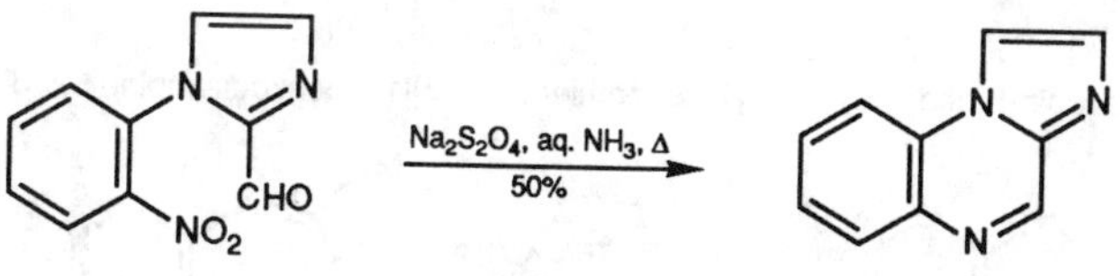

CHO, NO$_2$ —Na$_2$S$_2$O$_4$, aq. NH$_3$, Δ; 50%→ Imadazo[1,2-*a*]quinoxaline [2365]

III. FORMATION OF A SEVEN-MEMBERED RING

1. 1,3-Diazepin-4-one

The lactam ring of (**52.11**) is opened by traces of acetic acid in boiling toluene, and closes to a seven-membered ring in moderate yield.

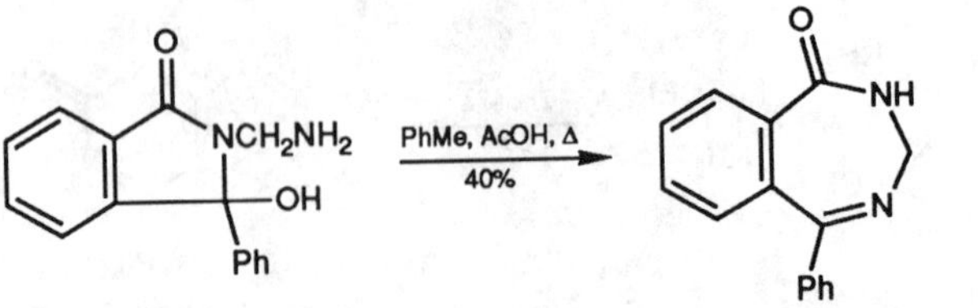

(52.11) $\xrightarrow[40\%]{\text{PhMe, AcOH, }\Delta}$ 2,4-Benzodiazepin-1-one [2958]

2. 1,4-Diazepine

Suitably separated aliphatic amino and keto groups may be annulated thermally without solvent, for example, to form (**52.12**, R = Me). An alternative method is to use a solvent containing a small amount of formic acid, as in the synthesis of (**52.12**, R = H), with dilute hydrochloric acid or 4-toluenesulphonic acid [2057, 2575].

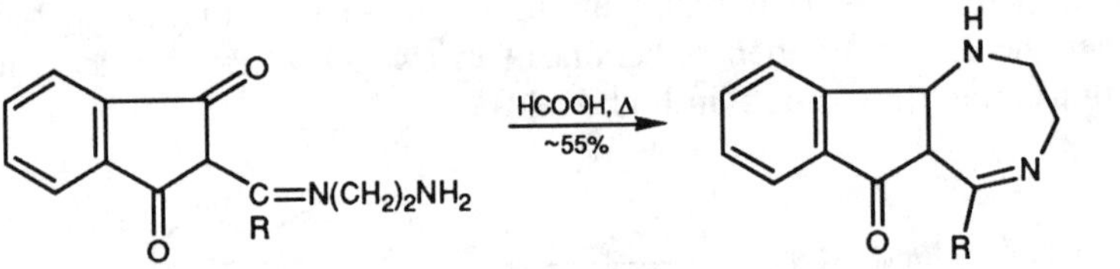

$\xrightarrow[\sim 55\%]{\text{HCOOH, }\Delta}$ [2175]

R = H, Me

(52.12)
Indeno[1,2-*e*]-1,4-diazepin-6-one

Other ring systems synthesized similarly:

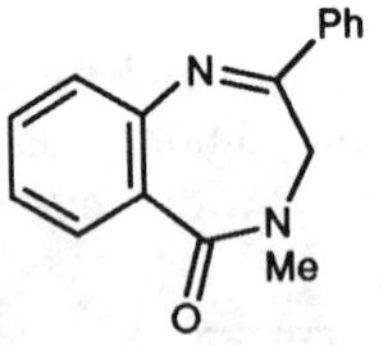

[2057]

1,4-Benzodiazepin-5-one

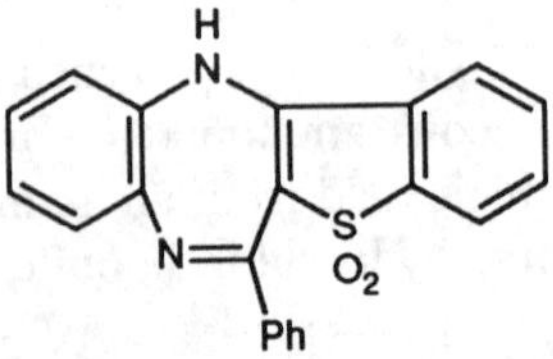

[2575]

[1]Benzothieno[3,2-*b*][1,5]benzodiazepine 5,5-dioxide

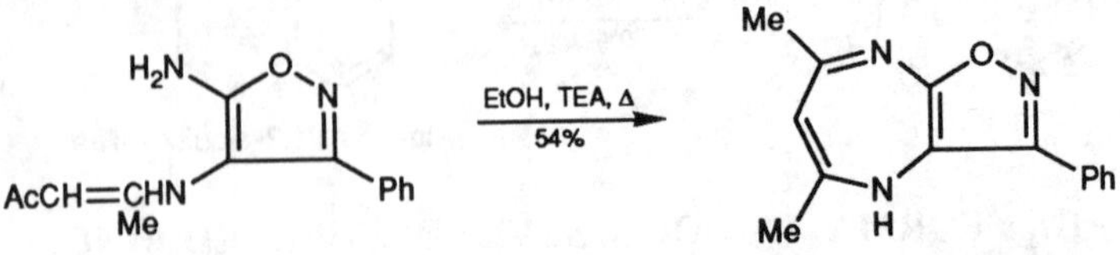

$\xrightarrow[54\%]{\text{EtOH, TEA, }\Delta}$ [2143]

Isoxazolo[4,5-*b*][1,4]diazepine

Aromatic amino and side-chain ketone groups are annulated by heating in a base. An *N*-protected amino acid reacts with a 2-aminobenzophenone (review of the synthesis of 2-aminobenzophenones [3009]) on successive treatments with DCC, mineral acid and alkali [3359]. The protected amino acid may be replaced with the corresponding acyl chloride hydrochloride but the yields are usually low [3359]. Reduction of the nitro group of (**52.13**) catalytically or with sodium hydrosulphite in water or ethanol gave poor results, but when the hydrosulphite was used in aqueous DMF containing sodium hydroxide under nitrogen at 70 °C, good yields were obtained but the unstable amine was not isolated. Acidification caused cyclization and the methyl ketone was obtained under these conditions in a good overall yield from the nitro compound. Reduction with iron–acid gave similar results [3748].

(**52.7**) + $BocNR^3CH(R^4)COOH$ —(i, ii, iii; 15–79%)→ 1,4-Benzodiazepin-2-one [3359]

R^1 = H, Cl; R^2 = H; R^3 = H, Me; R^4 = $PhCH_2$, iBu, 1- or 2-naphthylmethyl, 3-indolylmethyl; Ar = Ph, 2-Cl-, 2-F-C_6H_4

i, CH_2CH_2, THF, DCC; ii, HCl; iii, NaOH

(**52.13**) —(i or ii; ~53%)→ Pyrimido[4,5,-*b*][1,4]diazepin-4-one [3748]

i, DMF, $Na_2S_2O_4$, NaOH; ii, DMF, Fe, HCl

3. 1,4-Diazepin-2-one

1,4-Benzodiazepine with a particular stereochemistry at C-3 is obtained by deprotecting the side-chain amino group of the diphenyl ketone (**52.14**) and treating the product with di-isopropylethylamine. A 3-amino-1,4-benzodiazepin-2-one may be synthesized by addition of ammonium acetate either alone [3308] or with [3283] acetic acid to the 2-(α-aminoglycinoylamino)diphenyl ketone at 23 °C. The protecting benzyloxycarbonyl group can then be removed.

(**52.14**) —(i, ii; 70%)→ 1,4-Benzodiazepin-2-one [3680]

i, CH_2Cl_2, HBr(g); ii, MeOH, iPr_2NEt

(52.8) $\xrightarrow[>75\%]{\text{AcOH–AcONH}_4}$ [3282]

R^1 = H; R^2 = C(=O)CH(NH$_2$)NHCbz; Ar = Ph

Another method which usually gives good yields is to heat the amino ketone (**52.15**) in a mixture of pyridine, acetic acid and benzene so that the water formed may be removed azeotropically [2529]. Heating an ethanolic or pyridine solution of an amino ketone under reflux is sometimes successful [2552], the amino group being introduced as a first step but the amine is not isolated [2917]. A 2-aminobenzophenone may be converted into a 1,4-benzodiazepin-2-one by successive treatment with bromoacetyl bromide and ammonia (cf. Chapter 28) without isolating the bromide [2647].

$NR^2COCH_2NH_2$, COAr $\xrightarrow[64–88\%]{\text{AcOH, pyr, PhH, }\Delta}$ [2529]

(**52.15**)

R^1 = H, alkyl; R^2 = H, Me, $CH_2{=}CHCH_2$, $PhCH_2$;
Ar = Ph, 2- or 4-Me-, -Cl-, MeO-C_6H_4

Thieno[2,3-*e*]-1,4-diazepin-2-one

(**52.15**) $\xrightarrow[77\%]{\text{EtOH, }\Delta}$ [2552]

$R^1 = R^2$ = H; Ar = Ph

Me, $NHC(=O)CH_2Cl$, COR $\xrightarrow[54–74\%]{\text{1. DMF, NH}_3\text{ 2. pyr, }\Delta}$ [2917]

R = Me, Ph, 3-Me-, 3-Cl-, 4-Cl-C_6H_4

Thieno[3,4-*e*]-1,4-diazepin-2-one

Deprotection of the substituted glycylamino group of **(52.16)** under mild conditions enabled this ring to be formed efficiently. Protecting groups which needed more drastic treatment were not useful in this cyclization.

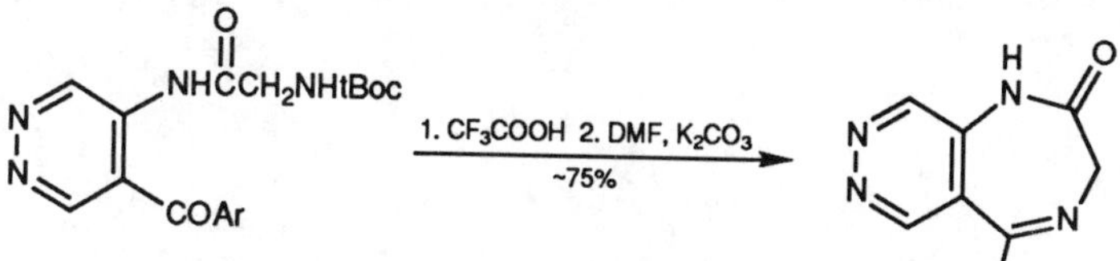

$NHCOCH_2NHtBoc$ ... COAr

1. CF_3COOH 2. DMF, K_2CO_3

~75%

[3495]

(52.16)
Ar = Ph, 4-Cl-, 2-F-C_6H_4

Ar

Pyridazino[4,5-*e*]-1,4-diazepin-2-one

CHAPTER 53

Amine and Ring-carbon

I. Formation of a Five-membered Ring 980
1. Pyrrole 980
2. Pyrazole 983
3. Imidazole or Imidazole 3-Oxide 983
4. 1,2,3-Triazole 984
5. Furan 984
6. 1,2-Dithiole-3-thione 985
7. Isothiazole 985
8. Thiazol-2-one 985
9. 1,2,3-Thiadiazole 986
10. 1,2,4-Thiadiazole 986
II. Formation of a Six-membered Ring 986
1. Pyridine 986
2. Pyridin-2-one or -4-one 992
3. Pyridazine or Pyridazin-3-one 993
4. Pyrazine or Pyrazin-2-one 993
5. 1,2,4-Triazine 994
6. Pyran 994
7. Pyran-2-one 995
8. 1,3-Thiazine-2-thione 995
9. 1,2,4-Oxadiazine 995
III. Formation of a Seven-membered Ring 996
1. 1,4-Diazepine 996
2. 1,4-Oxazepine or 1,4-Thiazepine 996

I. FORMATION OF A FIVE-MEMBERED RING

1. Pyrrole

A useful alternative to the Fischer synthesis of indoles is that in which a primary arylamine is treated at −70 °C successively with t-butyl hypochlorite, methylthioacetone and then TEA at 0 °C. Desulphurization, if desired, may then be achieved catalytically in high yield [2909]. 6-Aminopyrimidin-4-ones may be cyclized by heating with chloroacetaldehyde in aqueous sodium acetate [3685]. In another synthesis, the amine and an α-hydroxy ketone are heated in xylene with an acid catalyst [3091].

NH_2 → i, ii, iii 76% → HN, Me, SMe [2909]

i, THF, tBuOCl, N_2, –70 °C;
ii, $MeSCH_2Ac$, –70 °C; iii, TEA, 0 °C

Imidazo[4,5-*f*]quinoline

Another ring system synthesized similarly:

NH [2909]

Benz[*e*]indole

H_2N, N, OMe, NR, O + CH_2Cl / CHO → aq. AcONa, Δ ~65% → H N, N, OMe, NR, O [3685]

R = H, Me

Pyrrolo[2,3-*d*]pyrimidin-4-one

NH_2, O, O, O, R + PhCHOH / PhCO → xylene, TsOH, Δ ~49% → HN, O, O, O, R [3091]

R = H, Me

Benzofuro[3',2' : 6,7][1]benzopyrano[4,3-*b*]pyrrol-4-one

In the presence of ruthenium trichloride–triphenylphosphine, arylamines condense under pressure with 1,2-diols to give indoles. The solvent, the size of the *N*-alkyl group and the ratio of reactants affects yields [3275]. Phenylsulphonylalkynes react at ambient temperature with *C,N*-diphenylnitrone (*Chemical Abstracts* name: *N*-(phenylmethylene)benzenamine *N*-oxide) (**53.1**) [B-43, B-51] to give an indole as the main product. This reaction takes a complex course and among the minor products (28% yield) are two 4-phenylsulphonyl-isoxazoles, one of which changes into the major indolic product on standing. When R^1 = Me, no indole is formed. Reaction of the nitrone with 1-phenyl-sulphonylallene in chloroform gives a small yield of a sulphonyl-containing indole and a benzazepinone as the major product. In ethanol, the latter was the only product (43%) [3273]. A similar study using the allene methyl 2-methyl-buta-2,3-dienoate has given both indoles and benzazepin 4-ones.

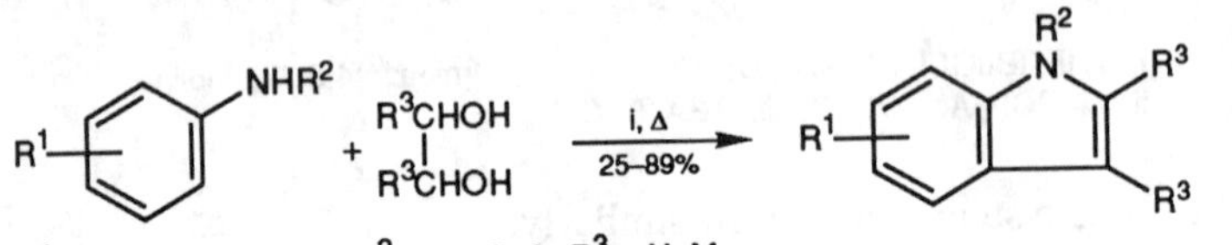

R^1 = H, Cl, MeO, Me; R^2 = H, alkyl; R^3 = H, Me

i, $(CH_2OH)_2$, $RuCl_3.nH_2O.(PPh_3)_3$, Ar, pressure

i, Δ; 25–89% — Indole [3275]

(**53.1**) R^1 = H; R^2 = Me, Ph

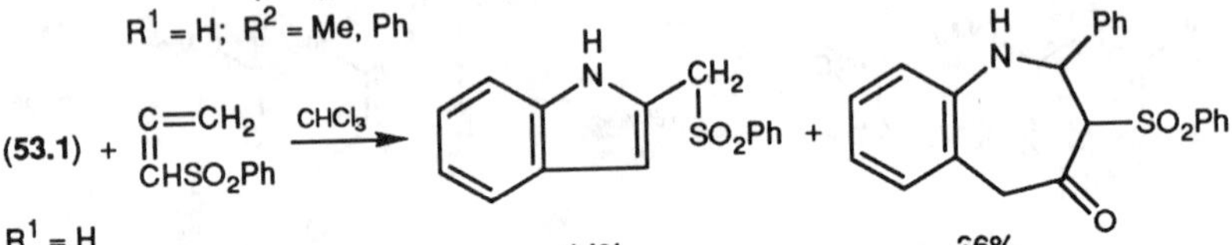

$CHCl_3$, 20 h; 30–48% [3273]

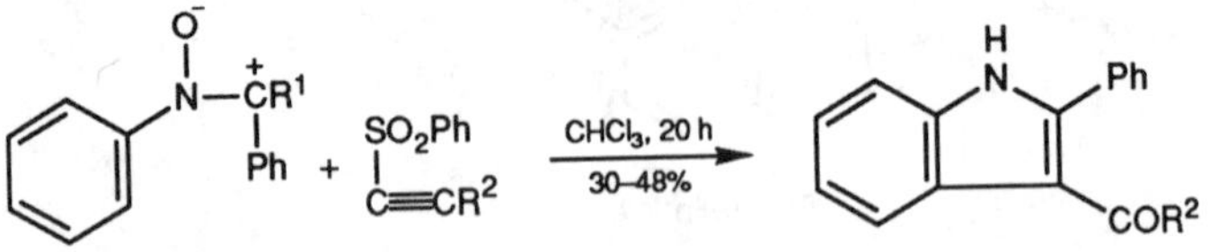

(**53.1**) + CH_2=C=CHSO$_2$Ph (R^1 = H), $CHCl_3$ → 14% + 36% 1-Benzazepin-4-one [3273]

t-Butyl groups are useful blockers when incoming substituents need to be guided to particular positions (review [2971]), and they can later be removed efficiently under Friedel–Crafts conditions. Reductive cyclization of the 2-nitro-diphenyl gives a carbazole[2964]. The readily oxidized catecholamine 1-(3′,4′-dihydroxyphenyl)-2-aminoethanol is cyclized by mushroom tyrosinase; in the absence of the secondary alcohol group, the yield is low [3315]. Enamines of cyclic ketones react with α-ketols to give reduced indol-4-one or benz[*f*]indolones [2344].

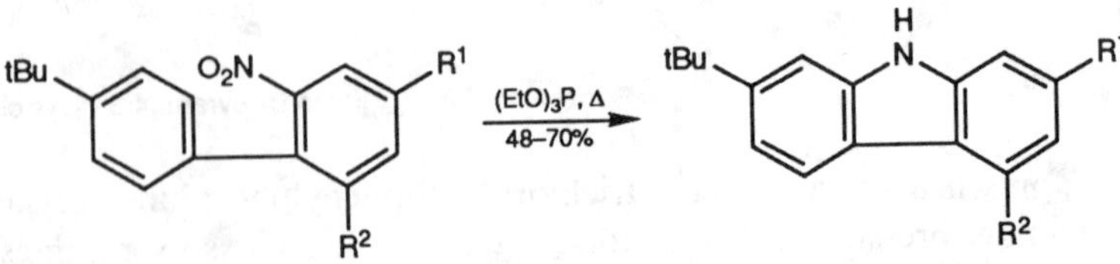

R^1 = H, tBu; R^2 = H, Me

$(EtO)_3P$, Δ; 48–70% — Carbazole [2964]

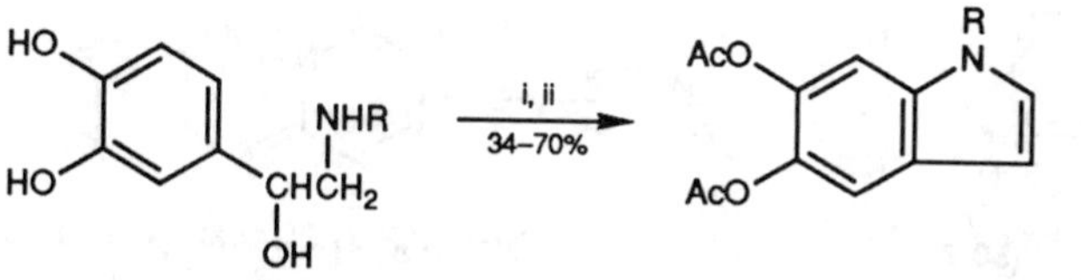

[3315]

R = H, Me, iPr
i, tyrosinase, pH 6·8; ii, Ac_2O, DMAP

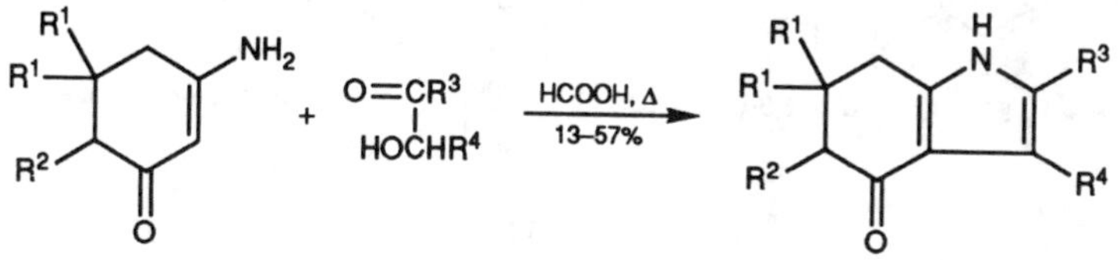

[2344]

R^1 = H, Me; R^2 = H, $PhCH_2$; R^3 = Me, Ph;
R^4 = Me, Ph or R^3R^4 = $(CH_2)_4$

4-Indolone

2. Pyrazole

N-Amino groups are converted into pyrazole rings on heating the compound with a methyl ketone under acidic conditions; an intermediate imine may be involved. 1-Aminopyridinium iodide reacts with diketene to give the aceto-acetylimino ylide which cyclizes under the influence of a base.

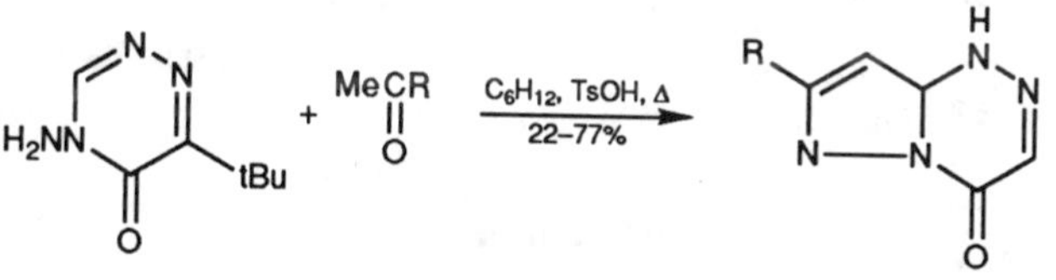

[2801]

R = Me, Ph

Pyrazolo[3,2-*c*][1,2,4]triazin-4-one

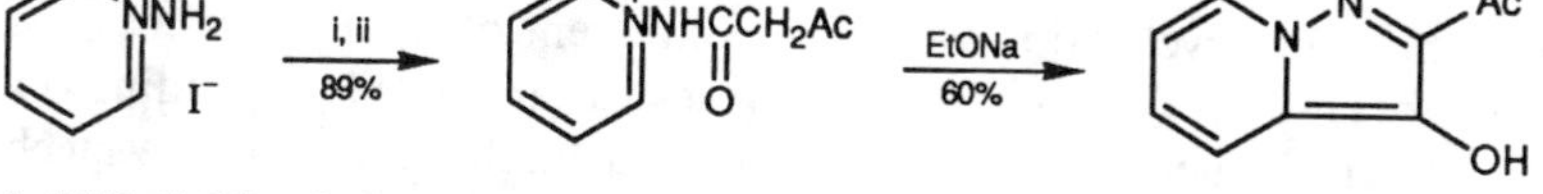

[2726]

i, DMF, K_2CO_3; ii, diketene

Pyrazolo[1,5-*a*]pyridine

3. Imidazole or Imidazole 3-Oxide

The reactivity of *N*-substituted pyridinium salts is shown in the cyclization of the salt (**53.2**). *N*-Phenylhydroxylamine reacts with benzonitrile oxide (review [B-43]) at ambient temperature to form a fused 1-hydroxyimidazole 3-oxide.

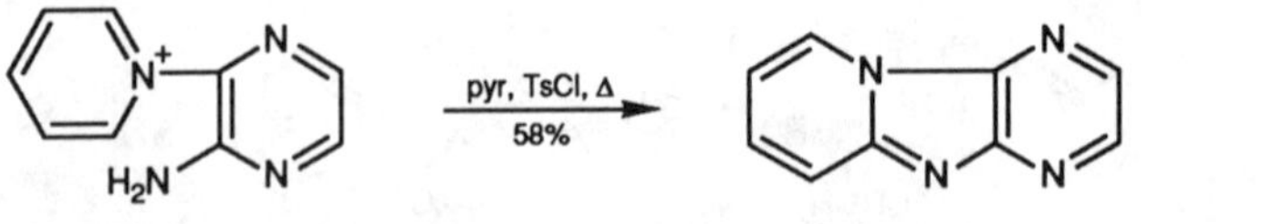

pyr, TsCl, Δ; 58% [2424]

(53.2)

Pyrido[1′,2′ : 1,2]imidazo[4,5-*b*]pyrazine

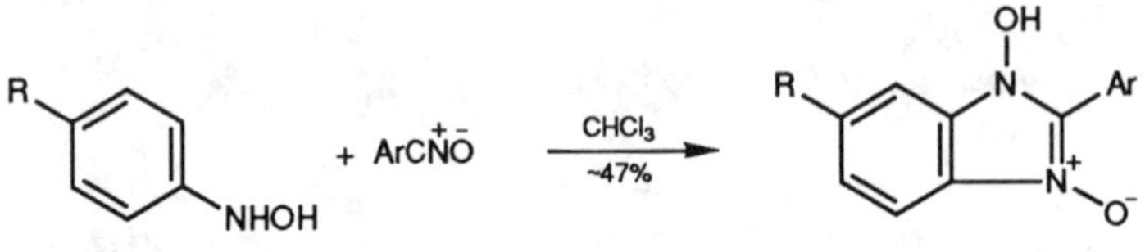

+ ArCNO; $CHCl_3$; ~47% [2699]

R = H, tBu; Ar = Ph, 2,4,6-$(MeO)_3C_6H_2$

Benzimidazole 3-oxide

4. 1,2,3-Triazole

Transfer of an azide group from an azidothiazolium salt followed by cyclization in an acid medium can be a useful method of building a triazole ring (p. 263); tosyl azide is possibly a more convenient source of azide, and cyclization at ambient temperature in a basic medium is complete in about 15 min.

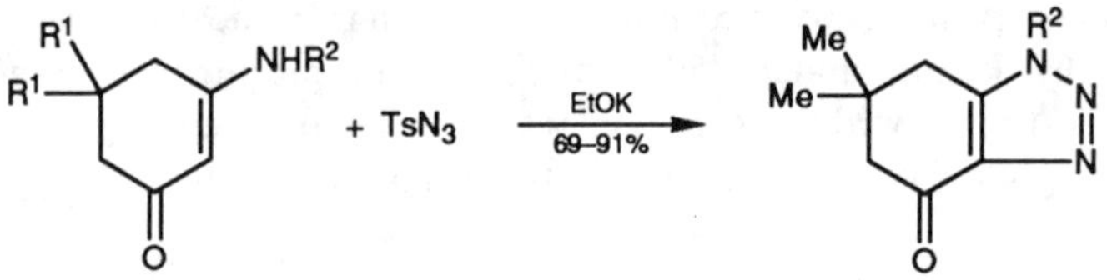

+ TsN_3; EtOK; 69–91% [2043]

(53.3)

R^1 = Me; R^2 = Ph, 4-Cl-, MeO-, NO_2-C_6H_4

4-Benzotriazolone

5. Furan

O-Aryloximes (formed *in situ*) rearrange on heating in ethanol containing hydrogen chloride to give benzofurans. *N*-Acetoacetyl-*O*-arylhydroxylamines (prepared by reacting the *O*-phenylhydroxylamine with diketene) undergoes a rearrangement and cyclization which is reminiscent of the Fischer indole synthesis. The reaction proceeds at 0 °C and gives variable yields depending on the substituents on the benzene ring. It may have a Fischer reaction-like mechanism; when the benzene ring is unsymmetrically substituted, two products are formed.

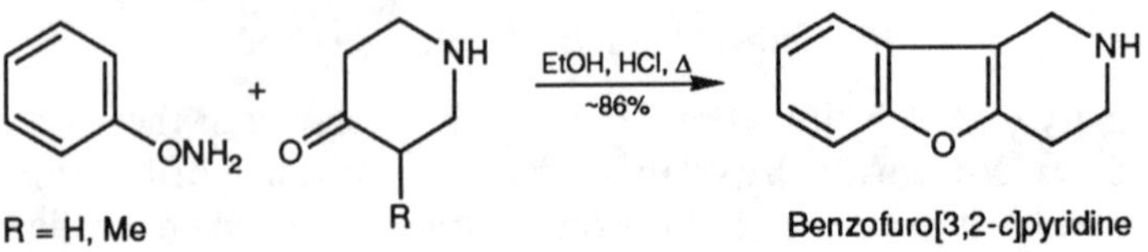

EtOH, HCl, Δ; ~86% [2542]

R = H, Me

Benzofuro[3,2-*c*]pyridine

CF_3SO_3OH, CF_3COOH, 34–79% [3192]

R = H, Me, tBu, MeO, Ph

Benzofuran

6. 1,2-Dithiole-3-thione

The enaminone (**53.4**) reacts with carbon disulphide in alkali to form the title ring, and the arylamino group is lost. A spiro intermediate may be involved.

DMSO, CS_2, NaOH, ~80% [3784]

(**53.4**)
Ar = Ph, 4-MeC_6H_4

Indeno[1,2-*c*]-1,2-dithiole-3-thione

Another ring system synthesized similarly:

[3784]

1,2-Benzodithiol-4-one, 3-thioxo-

7. Isothiazole

When 6-aminouracils are treated with DMF and thionyl chloride, one of the main products is the isothiazolopyrimidinedione.

+ DMF + $SOCl_2$ Δ, 9–43% [2785]

R = H, Et

Isothiazolo[3,4-*d*]pyrimidine-4,6-dione

8. Thiazol-2-one

2-Benzothiazolones are useful intermediates in the synthesis of plant growth regulators, and may be synthesized from substituted anilines by first converting the latter (with formaldehyde) into 1,3,5-(triaminoaryl)-1,3,5-triazines which then break down on stirring successively with chlorocarbonylsulphenyl chloride and aluminium chloride.

R^1 = H, Me; R^2 = H, Cl, Me
i, HCHO, PhMe; ii, ClC(O)SCl, CH_2Cl_2, $AlCl_3$

i, ii; 35–65%

2-Benzothiazolone [3379]

9. 1,2,3-Thiadiazole

When primary aromatic amines are treated with sulphur monochloride (S_2Cl_2, also known as disulphur dichloride)—the Herz reaction—one or more products may be formed. One of these, the benzothiazathiolium salt, on treatment with nitrous acid gives the benzothiadiazole. Depending on the type of substitution in the ring, it is possible to obtain a good yield. The benzene ring may be chlorinated simultaneously, and the thiadiazole may possibly rearrange to another five-membered ring.

i, ii; ~50%

1,2,3-Benzothiadiazole [2165]

R = H, F
i, S_2Cl_2, AcOH; ii, H_2SO_4, $NaNO_2$

10. 1,2,4-Thiadiazole

This ring is obtained in one step by reaction of a 2-aminopyridine at room temperature with trichloromethanesulphenyl chloride (perchloromethylmercaptan). A bromine atom in the amine may be lost in the reaction.

+ Cl_3CSCl ; $CHCl_3$, TEA; 35–90%

[1,2,4]Thiadiazolo[4,3-*a*]pyridine [2179]

R = Me, Cl, Br

II. FORMATION OF A SIX-MEMBERED RING

1. Pyridine

2-Amino-azoles or -azines react with 1,3-dicarbonyl compounds in the presence of a mild base [2361] or an acid [2088] to give a fused pyridine ring.

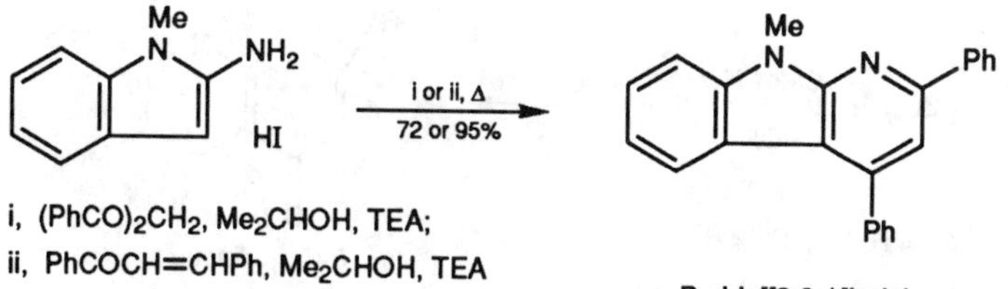

i, $(PhCO)_2CH_2$, Me_2CHOH, TEA;
ii, PhCOCH=CHPh, Me_2CHOH, TEA

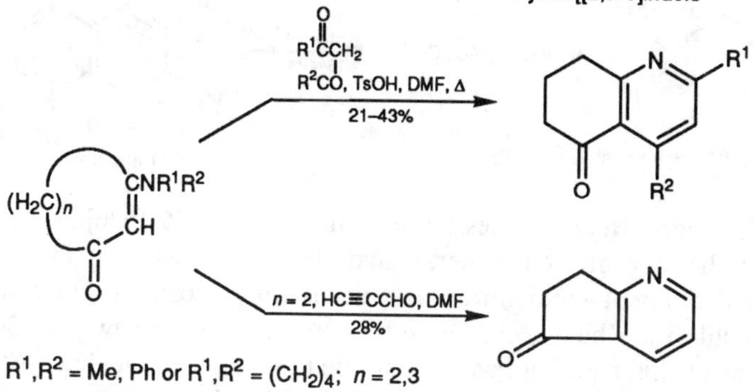

i or ii, Δ 72 or 95%

Pyrido[[2,3-*b*]indole [2361]

$R^1COCH_2COR^2$, TsOH, DMF, Δ 21–43%

n = 2, HC≡CCHO, DMF 28%

R^1,R^2 = Me, Ph or R^1,R^2 = $(CH_2)_4$; *n* = 2,3

Cyclopenta[*b*]pyridin-5-one [2088]

2,3-Unsaturated aldehydes or ketones also condense with the amines in an acidic [2569, 2364, 3012] or basic [2361] medium; some of these reactions are known by the names of their discoverers, for example, the Doebner–Miller synthesis [2364] or the Doebner quinoline synthesis [3214]. The latter needs neither acid nor base.

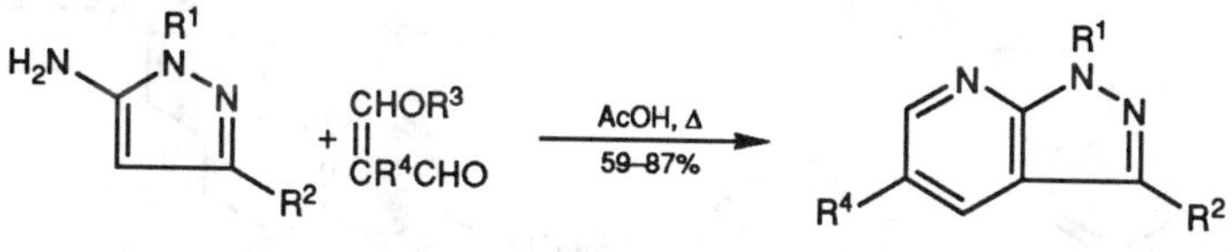

AcOH, Δ 59–87%

R^1 = Me, Ph; R^2 = H, Me;
R^3 = H, alkyl; R^4 = alkyl

Pyrazolo[3,4-*b*]pyridine [2569]

Other ring systems synthesized similarly:

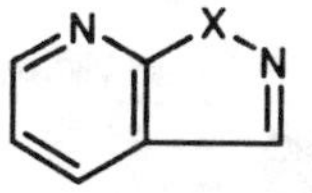

[2967, 2968]

X = O Isoxazolo[5,4-*b*]pyridine
X = S Isothiazolo[5,4-*b*]pyridine

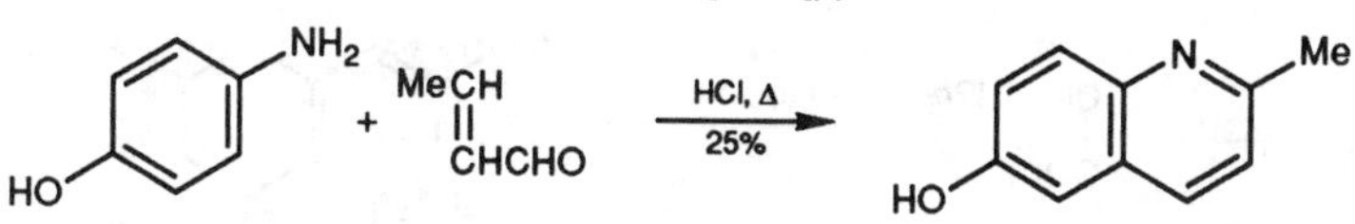

HCl, Δ 25%

Quinoline [2364]

(53.3) + CMeCHO=CHOEt —EtCOOH, Δ; 30–53%→ 5-Quinolinone [3012]

R^1 = H, Me; R^2 = H

5-Quinolinone

R-, MeO-substituted aniline + ArCHO + AcCOOH —EtOH, Δ; 16–42%→ Quinoline [3214]

R = H, Cl; Ar = 4-Me-C_6H_4, 2-thienyl

Quinoline

The Doebner–Miller synthesis (illustrated on p. 265 [526]) usually needs heating with concentrated mineral acid, but a particularly facile example proceeds at ambient temperature and needs only a catalytic amount of dry hydrogen chloride which cyclizes and then aromatizes the new ring [3014]. The mechanism of this reaction has been studied [3543]. The Doebner–Miller synthesis of 2-methylquinolines often gives low yields, but an improvement in the method of isolating the product may give better results: addition of an equimolecular proportion of zinc chloride to the reaction mixture enabled one group of products to be isolated more efficiently [2904]. An arylamine is converted into a 2-trifluoromethylquinoline by heating with 1,1,1-trifluorobut-3-yn-2-one in trifluoroacetic acid (cf. Chapter 35, Section II.1) [3867].

Methyl 6-amino-5-methoxyindole-2-carboxylate —CH_2Cl_2, i, ii; >90%→ Pyrrolo[2,3-*f*]quinoline [3014]

i, CHCOOH=CHCOCOOH; ii, HCl(g)

Pyrrolo[2,3-*f*]quinoline

(53.5) + CHMe=CHCHO —HCl, $ZnCl_2$, Δ; 40–50%→ Quinoline [2904]

(53.5)

R^1 = H, halogeno, Me, MeO

Quinoline

(53.5) + C≡CHPh / $COCF_3$ —CF_3COOH, Δ; 48%→ 5,7-(MeO)$_2$-4-Ph-2-CF_3-quinoline [3867]

R^1 = 3,5-(MeO)$_2$

1,8-Naphthyridines may be synthesized by means of a Skraup reaction on 2-aminopyridines. Best yields are obtained in the presence of boric acid and iron(II) sulphate [2241], but arsenic acid, 85% phosphoric acid [2527] and 65% sulphuric acid [2978] improve the yields in some Skraup reactions. When 2-aminocarbazole is subjected to this reaction, the angular isomer is produced. The linear isomer of this and of benzoquinoline may be synthesized by heating the amine with 2,4-pentanedione and sulphuric acid or PPA (the Beyer–Combes reaction) [2428, 2494]. Although the Skraup cyclization is a vigorous reaction, the presence of an imidazole ring in the substrate does not interfere [2585], and provided that the reaction temperature is kept below about 140 °C, an *N*-oxide function should survive the reaction [2791].

R = H, Me (53.6) 1,8-Naphthyridine [2241]

i, $3\text{-}NO_2C_6H_4SO_3Na$, H_2SO_4, H_3BO_3, $FeSO_4.6H_2O$

i, Δ 38–87%

Other ring systems synthesized similarly:

[2241] [2791]

X = CH; Y = N 1,5-Naphthyridine
X = N; Y = CH 1,6-Naphthyridine

1,5-Naphthyridine 1-oxide

(53.5) + $ArCOHCH_2OEt$ / CH_2OEt — H_2SO_4, Δ 15–36% → Quinoline [2527, 2978]

R^1 = 3-COOEt, Me-3-COOEt; Ar = Ph, 4-ClC_6H_4

Pyrido[2,3-*c*]carbazole [2494]

Pyrido[3,2-*b*]carbazole [2428, 2494]

+ (53.6) — H_2SO_4, Δ 47% → Imidazo[4,5-*f*]quinoline [2585]

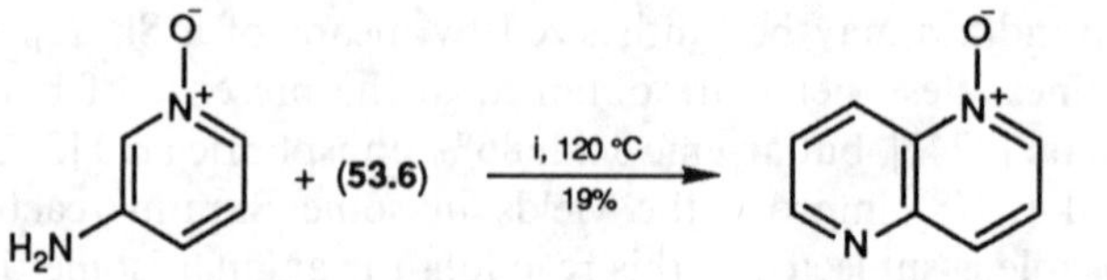
+ (53.6)
i, 120 °C
19%
H2N

[2791]

i, $H_2SO_4.SO_3$, $FeSO_4$, H_3BO_3 1,5-Naphthyridine 1-oxide

Although the Combes synthesis sometimes gives rather low yields, it is attractive because the reactants are usually readily available and only one step is involved [2587]. The enaminone from cyclopentane-1,3-dione is a useful synthon which reacts with 1,3-diketones or α-ynals with the formation of a pyridine ring. The homologous aminocyclohexanone has similar properties [2088]. Heating the enamine with a 6-aminopyrimidinedione generates the medicinally interesting 5-deazapteridinedione [2101, 2558] (see next paragraph).

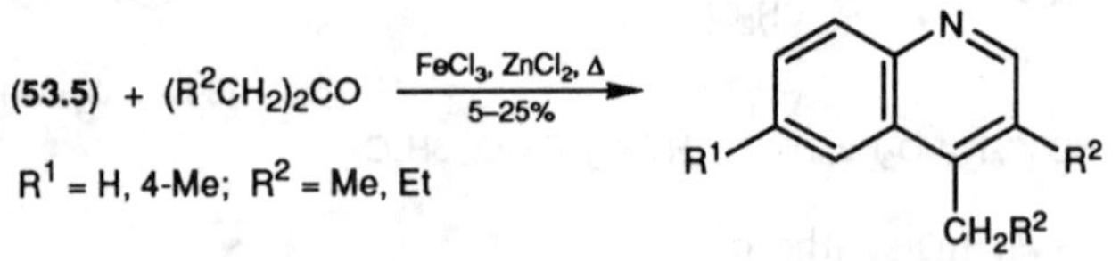
(53.5) + (R2CH2)2CO
FeCl3, ZnCl2, Δ
5–25%
R1 = H, 4-Me; R2 = Me, Et
R1
R2
CH2R2

[2587]

Quinoline

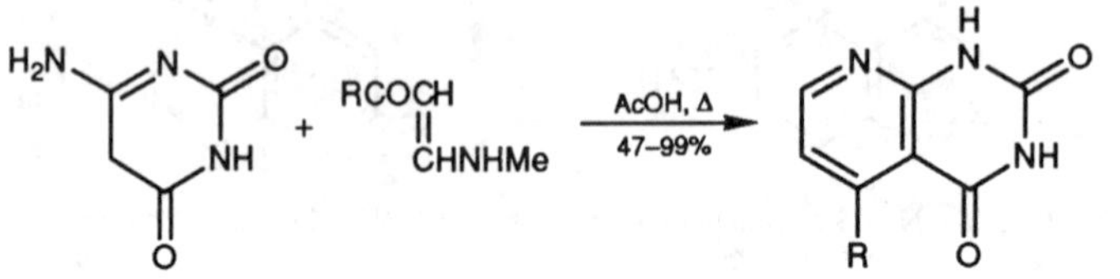
H2N
NH
RCOCH
CHNHMe
AcOH, Δ
47–99%
NH
R

[2101]

R = Me, Ph, 4-Me-, 4-MeO-C_6H_4 Pyrido[2,3-*d*]pyrimidine-2,4-dione

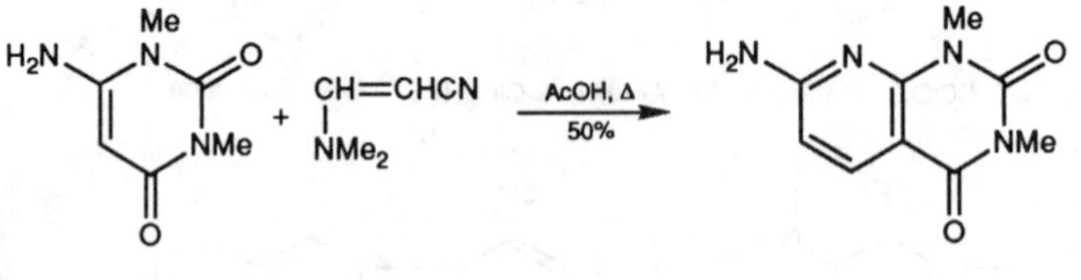
Me
H2N
NMe
CH=CHCN
NMe2
AcOH, Δ
50%
H2N
Me
NMe

[2558]

Pyrido[2,3-*d*]pyrimidine-2,4-dione

Several malonic acid derivatives are useful for annulating an amino group to the neighbouring ring-carbon atom. An acidic catalyst is often used [3527] but the reactants may be heated in propanol [3527, 3834]. Medicinally important 5-deazapteridines (reviews [3454, 3594, 3669]) are available from a pyrimidinone and a malononitrile derivative [3842]. The formation of a mixture of a reduced quinolinone and acridinedione from cyclohexanediones and 3-aminoacrolein (Chapter 99, Section II.1) is averted by using an aminocyclohexenone and a 3-ethoxyacrolein [3012].

$$\text{5-amino-1-pentylpyrazole} + AcCH(COOEt)_2 \xrightarrow[80\%]{PPA,\ \Delta} \text{4-hydroxy-1-pentylpyrazolo[3,4-}b\text{]pyridine} \quad [3527]$$

Pyrazolo[3,4-*b*]pyridine

$$\text{6-amino-3-methyl-1-}R^1\text{-uracil} + R^2CH{=}C(R^3)CN \xrightarrow[\sim 90\%]{PrOH,\ \Delta} \text{pyrido[2,3-}d\text{]pyrimidine-2,4-dione} \quad [3834]$$

R^1 = H, Me; R^2 = Ph, 2-furyl, 2-thienyl; R^3 = CN, COOEt, $CONH_2$

Pyrido[2,3-*d*]pyrimidine-2,4-dione

$$\text{2,6-diaminopyrimidin-4(3H)-one} + CHOEt{=}CR{-}HC{=}C(CN)_2 \xrightarrow[\sim 64\%]{AcOH,\ \Delta} \text{2-amino-6-R-pyrido[2,3-}d\text{]pyrimidin-4-one} \quad [3842]$$

R = $4\text{-COOMeC}_6H_4(CH_2)_n$; n = 2,3

Pyrido[2,3-*d*]pyrimidin-4-one

A synthesis of 7,8-dioxygenated isoquinolines (found in cularine alkaloids) requires *o*-metallation of the *N*-protected *β*-phenylethylamine. The primary amine is protected by reaction with two equivalents of trimethylsilyl chloride. Phenylhydroxylamine and the allenic nitrile (**53.7**) react in boiling ethanol over 48 h to give a high yield of 4-aminoquinoline—an important intermediate for the synthesis of some antimalarial drugs.

$$\text{3-MeO-4-OR-C}_6H_3(CH_2)_2NH_2 \xrightarrow[\sim 40\%]{i,\ ii} \text{7-MeO-8-OR-3,4-dihydroisoquinoline} \quad [3513]$$

R = Me, $MeOCH_2$

i, Me_3SiCl, THF; ii, BuLi, DMF

Isoquinoline

$$PhNHOH + \underset{(\mathbf{53.7})}{CHCN{=}C{=}CR^1R^2} \xrightarrow[\sim 84\%]{EtOH,\ \Delta} \text{4-amino-2-(}CR^1R^2OH\text{)quinoline} \quad [3530]$$

R^1, R^2 = Me, Et

Quinoline

Cyclizations catalysed by transition metals often need less drastic conditions than traditional methods. A catalytic quantity of ruthenium trichloride and a larger amount of triphenylphosphine promote the cyclization of an arylamine by 1,3-propanediol in refluxing diglyme to give good yields of quinolines. Palladium or rhodium complexes were much less effective in this reaction. Variations in substituents on the arylamine have an effect on yield: 4-methoxy-aniline gave a good yield but its 2-isomer failed to react, probably because it coordinates with the catalyst. Unfortunately, amines in which R is a strongly electron-attracting substituent do not seem to have been studied. Unsymmetrically substituted amines or diols usually give a mixture of positional isomers [3275].

(**53.5**) + CH_2OH–CH_2CH_2OH → (i, Δ; 0–59%) R^1-quinoline [3275]

R^1 = H, Cl, MeO, Me

i, diglyme, $RuCl_3.nH_2O$–PPh_3, Ar

2. Pyridin-2-one or -4-one

A pyridin-2-one ring is formed by reacting a 3-aminodiazine with a reactive malonate ester such the 2,4,6-trichlorophenyl ester; this cyclization needs a temperature of about 240 °C to be maintained for about 20 min [3676]. The ease of displacement of a methylthio group in compounds such as (**53.8**) demonstrates their usefulness; a second molecule of the arylamine reacts with the methoxycarbonyl group [3690]. An alternative method (to that described in Chapter 9, Section II.2) of synthesizing pyridazino[4,5-*b*]quinolin-10-ones which avoids the use of fluorophenyl ketones is oxidative cyclization of 2-aminophenyl ketones (**53.9**) [3642].

COOMe, NH_2, N, PhN, O + $R^1CH(COOR^2)_2$ → (Δ; ~80%) O=COMe, H, N, O, N, PhN, R^1, O, OH [3676]

R^1 = Bu, Ph; R^2 = 2,4,6-$Cl_3C_6H_2$

Pyrido[2,3-*d*]pyridazine-2,5-dione

Another ring system synthesized similarly:

O, Ph, Me, RN, Me, OH, O [2415]

Furo[2,3-*h*]quinolin-2-one

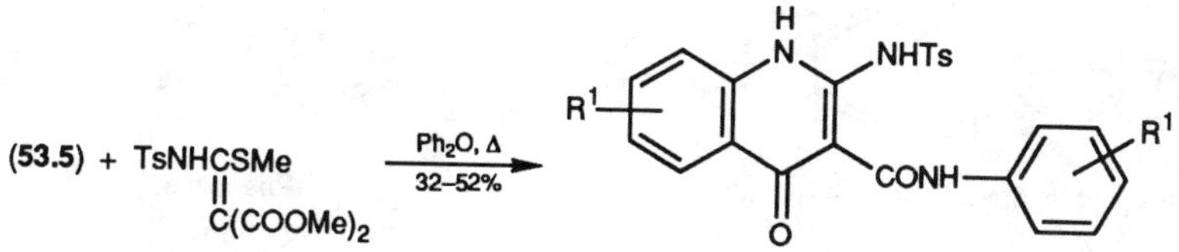

(53.5) + TsNHC(SMe)=C(COOMe)$_2$ —Ph$_2$O, Δ; 32–52%→ 4-Quinolinone [3690]

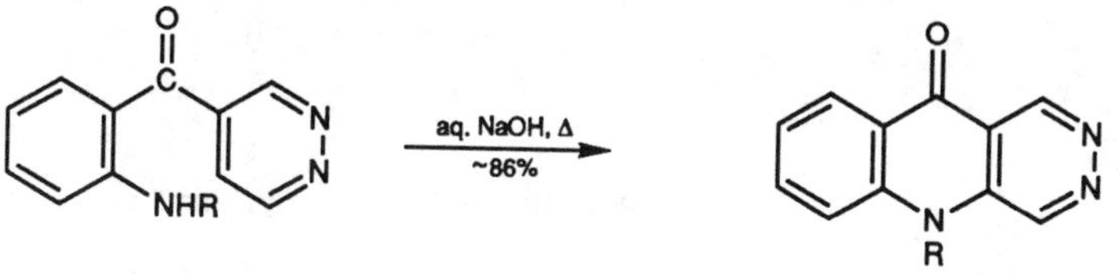

(53.8)

R^1 = H, MeO, Cl

(53.9) —aq. NaOH, Δ; ~86%→ Pyridazino[4,5-*b*]quinolin-10-one [3642]

R = Bu, CH_2=CHCH$_2$, PhNH$_2$

3. Pyridazine or Pyridazine-3-one

1-Aminopyrazoles react with 1,3-diketones or 3-ethoxyacrolein diethylacetal to give a pyrrolopyridazine in low yield.

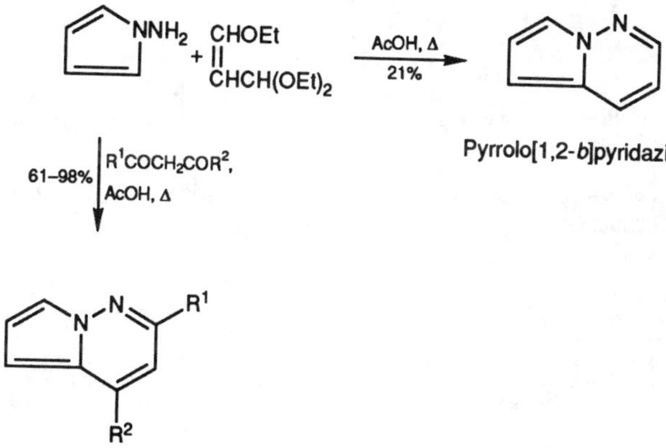

1-Aminopyrrole + CHOEt=CHCH(OEt)$_2$ —AcOH, Δ; 21%→ Pyrrolo[1,2-*b*]pyridazine [2108,2317]

—$R^1COCH_2COR^2$, AcOH, Δ; 61–98%→ [2108,2317]

R^1 = R^2 = Me, Ph
or R^1 = Me, R^2 = Ph

4. Pyrazine or Pyrazin-2-one

2-Nitrodiphenylamine is reductively cyclized with sodium borohydride in the presence of a base. A primary amino group and a carbon of another ring may be annulated by reaction with ethyl chloroformate to give the pyrazinone.

EtOH, EtONa, $NaBH_4$, Δ 32–80% [2206]

R = H, Me, MeO, COOH, NH_2, SO_2NH_2

Phenazine

ClCOOEt, Δ 70% [3133]

Pyrido[2,3-*e*]pyrrolo[1,2-*a*]pyrazin-6-one

5. 1,2,4-Triazine

This cyclization is another example of the synthetic value of 1-aminopyridinium salts (review [1870]); its base-catalysed reaction with acetophenone *N,N*-dimethylhydrazone methiodide without external heat leads to a new fused triazine ring. 2-Chloroacrylic esters react similarly but in the presence of a weaker base; two moles of the pyridinium salt react and an N—N bond is cleaved to supply the third nitrogen.

THF, tBuOK 10–53% [2905]

Pyrido[1,2-*b*][1,2,4]triazine

$R^2CH{=}C(Cl)COOR^3$, $CHCl_3$, K_2CO_3 4–87% [2615]

Pyrido[1,2-*b*][1,2,4]triazine
R^1 = H, Me or benzo;
R^2 = Me, Ph; R^3 = Me, Et;
Ar = Ph, 4-MeC_6H_4, 2-naphthyl

6. Pyran

A pyran ring is formed from a vinologous carbonyl group and a tertiary amino function on heating the compound with a 1,3-diketone.

$R^1 = NEt_2, \overset{+}{N}HMe_2\overset{-}{Cl}$; R^2 = Me, Ph

Ac_2O, AcOH, Δ 66–78%

Pyrano[3,2-*c*]quinoline [2854]

7. Pyran-2-one

O-Arylhydroxylamines react at low temperatures with DMAD and a rearrangement occurs to give high yields of 3-aminocoumarins. The yield is much lower when R is an electron-withdrawing group.

EtOH, DMAD ~75%

R = H, Me

1-Benzopyran-2-one [2472]

8. 1,3-Thiazine-2-thione

Thiocarbamoyl isothiocyanate reacts with cyclic enamines to form a fused thiazinethione under mild conditions; a mechanism has been suggested for this unusual cyclization. The reactions of isothiocyanates have been reviewed [3990].

Et_2O 47%

R = Et, iPr, cyclohexyl

1,3-Benzothiazine-4-thione [2094]

9. 1,2,4-Oxadiazine

S,*S*-Dimethylsulphimides (readily prepared from the primary amine, dimethyl sulphide and *N*-chlorosuccinimide) react with nitrile oxides (review [B-43]) under mild conditions to form a fused oxadiazine.

CH_2Cl_2 27–64%

R^1 = H, Cl, Ph, NO_2; R^2 = 4-MeC_6H_4

1,2,4-Benzoxadiazine [2852]

III. FORMATION OF A SEVEN-MEMBERED RING

1. 1,4-Diazepine

Lithiation of 2-aminodiphenylamine (**53.10**, X = NH) with butyl-lithium followed by stirring with DMF and decomposition with water results in the formation of a dibenzodiazepine in good yield.

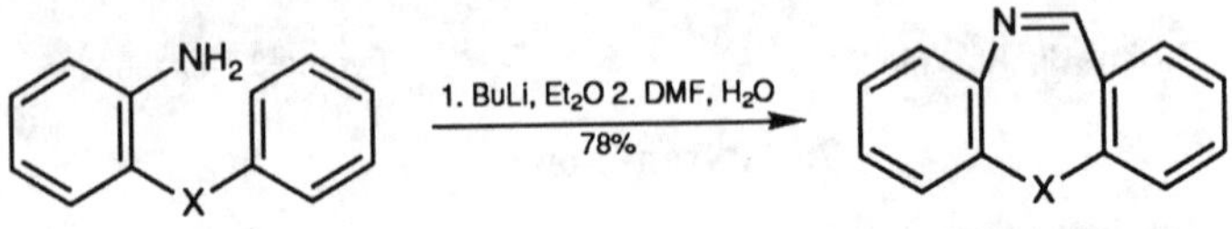

1. BuLi, Et_2O 2. DMF, H_2O
78%

[2969]

(53.10)
X = NH, O or S

X = NH Dibenzo[*b,e*][1,4]diazepine
X = O Dibenz[*b,e*][1,4]oxazepine
X = S Dibenzo[*b,e*][1,4]thiazepine

2. 1,4-Oxazepine or 1,4-Thiazepine

When the method described in Section III.1 is applied to either 2-aminodiphenyl ether (**53.10,** X = O) or thio ether (**53.10,** X = S), a new oxazepine or thiazepine ring is formed.

CHAPTER 54

Amine and Ring-nitrogen

I. Formation of a Five-membered Ring 997
1. Imidazole 997
2. Imidazol-2-one or -4-one 1000
3. Imidazole-2-thione 1001
4. 1,2,4-Triazole or 1,2,4-Triazole 1-Oxide 1001
5. 1,2,4-Thiadiazole or 1,2,4-Thiadiazol-5-one 1001
II. Formation of a Six-membered Ring 1002
1. Pyridine 1002
2. Pyrimidine 1003
3. Pyrimidin-2-one 1007
4. Pyrimidin-4-one 1007
5. Pyrazine 1012
6. Pyrazine-2,3-dione 1012
7. 1,3,5-Triazine 1013
8. 1,3,5-Triazin-2-one 1013
9. 1,3,5-Triazine-2,4-dione 1014
III. Formation of a Seven-membered Ring 1014
1. 1,4-Diazepin-5-one 1014

I. FORMATION OF A FIVE-MEMBERED RING

1. Imidazole

The use of α-halogenoketones for cyclizing 2-aminoazines or 2-aminoazoles is well-known (see p. 276 and [B-45, 3533]), but the same result may be obtained by heating the heterocycle with a ketone, iodine or bromine and sodium hydrogencarbonate in ethanol [2710]. In an extension of the reaction of 2-aminothiazoles with α-bromoketones (p. 277), 1-bromomethyl 2-t-butyl ketone has been shown to give high yields of 6-t-butylimidazo[2,1-*b*]thiazoles, the presence of the bulky alkyl group having no adverse effect on the yield [3939]. A variation of this reaction is described in Section II.2. Propargylamine converts an α-aminodiazepine into an imidazodiazepine; α-bromoketones perform a similar function but in this example they give lower yields [2885]. 2-Aminoacetal may also be useful for preparing an intermediate imine which is cyclized by hot formic acid. An α-bromoketone cyclizes 2-aminopyridine—the Tschitshibabin synthesis of imidazo[1,2-*a*]pyrimidines [3931]. Some cyclizations with α-haloketones need basic conditions; ketols (acyloins) fulfil a similar role under acidic conditions, for example, in acetic and toluenesulphonic acids [2449].

+ R^3COMe — EtOH, Br_2 or I_2, $NaHCO_3$, Δ; 34–94% — [2710]

R^1 = H, MeO; R^2 = H, Br; R^3 = Me, Ph, 4-Br-, 4-NO_2-C_6H_4

Imidazo[1,2-*c*]pyrimidine

i; 33–94% — [2885]

R = H,Cl,CF_3;
Ar = Ph, 2-FC_6H_4
i, HC≡CCH_2NH_2, BuOH, TsOH,Δ

Imidazo[1,2-*a*][1,5]benzodiazepin-5-one

+ MeCHOH–MeCO — AcOH, TsOH, Δ; 36% — [2449]

Imidazo[1,2-*b*][1,2,4]triazole

Annulation of aminomethyl and ring-nitrogen functions with ethyl acetimidate hydrochloride gives a good yield of a dihydroimidazo compound [3736]. Glycidylaldehyde reacts with guanine derivatives to give the linear tricyclic product [3066]. 6-Aminopyrimidine-2,4-diones are converted into purines by reaction with diethyl azodicarboxylate [2914]. The chemistry of α-haloaldehydes and -ketones has been reviewed [B-45, 2348].

HCl + MeC(=NH)OEt — EtOH, Δ; 66% — [3736]

Imidazo[1,5-*a*]quinoline

R = ribosyl

pH10
59%

[3066]

Imidazo[1,2-*a*]purin-9-one

Δ
30–83%

[2914]

R^1 = H, Me; R^2 = alkyl, $PhCH_2$, Ph

2,6-Purinedione

An unexpected cyclization occurred when the pyridylaminoaniline (**54.1**) was heated with dilute hydrochloric acid: the product was a pyridobenzimidazole [2156]. PPA brought about a similar annulation of a primary amino group and a ring-nitrogen in (**54.2**) even in the presence of an ethoxycarbonyl group, but replacement of PPA by 4-toluenesulphonic acid in boiling butanol produced two kinds of cyclizations simultaneously. The major product (**54.3**, 39%) was a fused imidazole, and the byproduct (**54.4**, 32%) had been formed through the involvement of the ethoxycarbonyl group. The ratio of these two compounds varied with the catalyst (see also Section III.1) [3766].

EtOH, aq. HCl, Δ
65%

[2156]

(**54.1**)

Pyrido[1,2-*a*]benzimidazole

R = H, COOEt: PPA, Δ
48–81%

[3766]

(**54.2**)

(**54.3**)
Cyclohepta[*d*]pyrrolo[1,2-*a*]benzimidazole

R = COOEt:
TsOH, BuOH, Δ
see text

(**54.3**) +

[3766]

(**54.4**)
Cyclohepta[*d*]pyrrolo[2,3-*b*[1,5]benzodiazepin-12-one

2. Imidazol-2-one or -4-one

When an amino group is in a *peri* position with respect to the heteroatom [as in (**54.5** after reduction)], an imidazole ring may be built to join the two by reaction with a carboxylic acid and PPE [2513]. Replacing the carboxylic acid by urea in the cyclization of (**54.5**) leads to the formation of a doubly fused imidazol-2-one ring [2513]. An imidazol-4-one may be constructed by the reaction of either phenylglyoxal [3039, 3188] or the phenylazo derivative of chloroacetyl chloride [2366] with a 2-aminoazine.

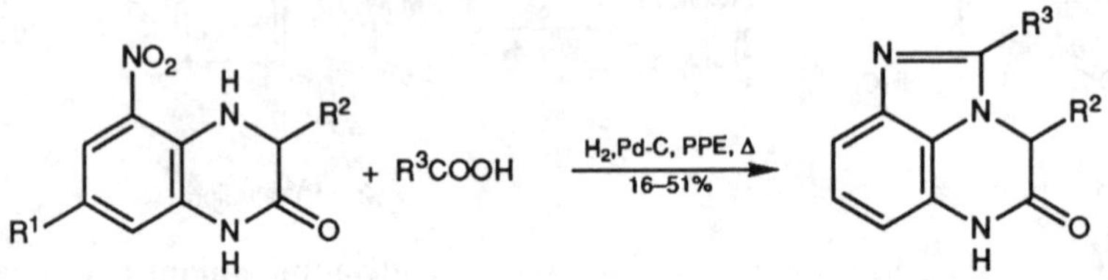

(**54.5**) + R^3COOH —(H_2, Pd-C, PPE, Δ; 16–51%)→ Imidazo[1,5,4-*de*]quinoxalin-5-one [2513]

R^1 = H, Cl; R^2 = H, Me; R^3 = H, Me, Ph

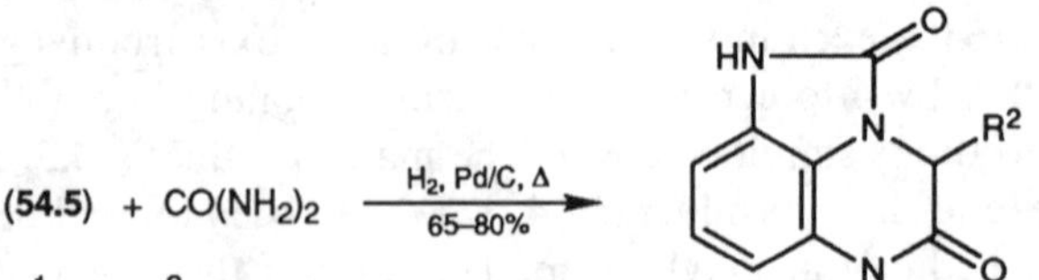

(**54.5**) + $CO(NH_2)_2$ —(H_2, Pd/C, Δ; 65–80%)→ Imidazo[1,5,4-*de*]quinoxaline-2,5-dione [2513]

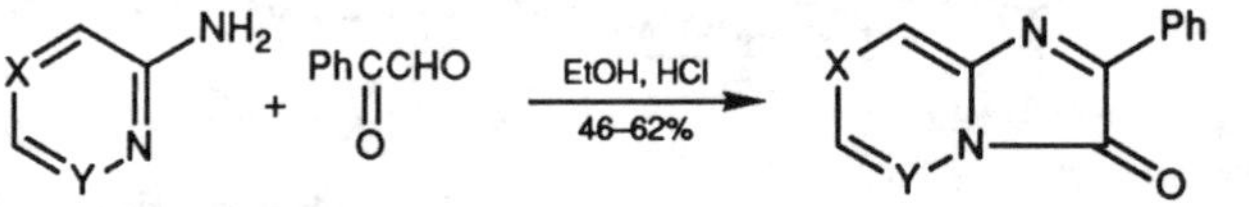

R^1 = H; R^2 = Me

2-Aminoazine + PhCOCHO —(EtOH, HCl; 46–62%)→ [3039]

X = CH, Y = N Imidazo[1,2-*b*]pyridazin-3-one
X = N, Y = CH Imidazo[1,2-*a*]pyrazin-3-one

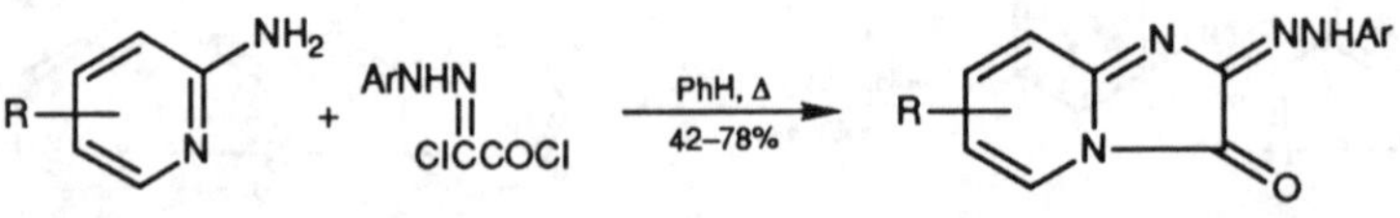

2-Aminopyridine + ArNHN=C(Cl)COCl —(PhH, Δ; 42–78%)→ Imidazo[1,2-*a*]pyridin-3-one [2366]

R = H, Me; Ar = MeO-, NO_2-C_6H_4

Other ring systems synthesized similarly:

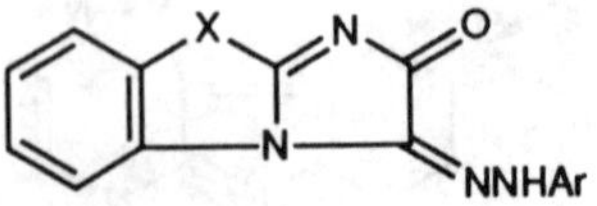

[2366]

X = O Imidazo[2,1-*b*]benzoxazol-2-one
X = S Imidazo[2,1-*b*]benzothiazol-2-one
X = NH Imidazo[1,2-*b*]benzimidazol-2-one

3. Imidazole-2-thione

The use of carbon disulphide in organic synthesis [reviews B-51, 3582] is demonstrated by its cyclization of a 2-aminomethyldiazine in excellent yield.

N N CH_2NH_2 NH_2 —CS_2, pyr, Δ; 89%→ S N NH N NH_2 [2159]

Imidazo[1,5-*a*]pyrazine-1-thione

4. 1,2,4-Triazole or 1,2,4-Triazole 1-Oxide

Pyridine *N*-oxide cyclizes 3-aminopyrazoles to give a low yield of a tricyclic product. The reactive *S,S*-dimethylsulphimides (see Chapter 53, Section II.9) are cyclized in a mild reaction with a nitrile oxide (reviews [B-43, 2115] to produce the triazole 1-oxide in good yield. The chemistry of [1,2,4]triazolo[1,5-*a*]pyrimidines has been reviewed [3890].

$N^+–O^-$ + H N N H_2N CN —diox, Δ; 15%→ N N N N H CN [3780]

Pyrazolo[2',3' : 2,3][1,2,4]triazolo[1,5-*a*]pyridine

X N R^1 $\bar{N}\overset{+}{S}Me_2$ + $R^2C\overset{+}{N}–O^-$ —CH_2Cl_2; 31–77%→ X N R^2 N N^+ O^- R^1 [2853]

R^1 = H, Me; R^2 = 4-MeC_6H_4, COOEt;
X = CH, N

X = CH [1,2,4]Triazolo[1,5-*a*]pyridine 3-oxide
X = N [1,2,4]Triazolo[1,5-*a*]pyrazine 3-oxide

Another ring system synthesized similarly:

O^- N^+ N N N [2853]

[1,2,4]Triazolo[1,5-*a*]pyrimidine 1-oxide

5. 1,2,4-Thiadiazole or 1,2,4-Thiadiazol-5-one

1-Chloro-1-phenyliminomethanesulphenyl chloride reacts without external heat with aminoazines to form a fused thiadiazole ring. Chlorothioformyl chloride reacts likewise with 2-aminoazoles or 2-aminoazines to form a new fused thiadiazol-5-one ring; this reaction is sensitive to the solvent used.

+ PhN=CSCl(Cl) —CHCl₃, TEA / 22–44%→ [3158]

X = Y = CH [1,2,4]Thiadiazolo[4,3-*a*]pyridine
X = N, Y = CH [1,2,4]Thiadiazolo[4,3-*a*]pyrazine
X = CH, Y = N [1,2,4]Thiadiazolo[4,3-*a*]pyrimidine

Another ring system synthesized similarly:

[3158] [3158]

X = CH Thiazolo[2,3-*c*][1,2,4]thiadiazole
X = N [1,3,4]Thiadiazolo[2,3-*c*][1,2,4]thiadiazole

[1,2,4]Thiadiazolo[3,4-*b*]benzothiazole

+ ClCSCl(=O) —CHCl₃,TEA / 41%→ [2677]

Thiazolo[2,3-*c*][1,2,4]thiadiazol-3-one

Other ring systems synthesized similarly:

[2677]

X = N, Y = CH [1,2,4]Thiadiazolo[4,3-*a*]pyrimidin-3-one
X = CH, Y = N [1,2,4]Thiadiazolo[4,3-*c*]pyrimidin-3-one

II. FORMATION OF A SIX-MEMBERED RING

1. Pyridine

2-Phenyl-3-dimethylaminoacrolein is a useful synthon as it reacts with 2-aminoazoles or 2-aminoazines with loss of the t-amino group. Hydrogenation converts it into indole-3-carboxaldehyde in low (15%) yield. Another synthetically useful acrolein derivative is the cyanoacetal (**54.6**), which cyclizes amino-*N*-heterocycles on heating in methanol–hydrochloric acid.

+ PhC(CHO)=CHNMe₂ —aq. AcOH, Δ / 55%→ [2645]

Pyrazolo[3,4-*b*]pyridin-3-one

Another ring system synthesized similarly:

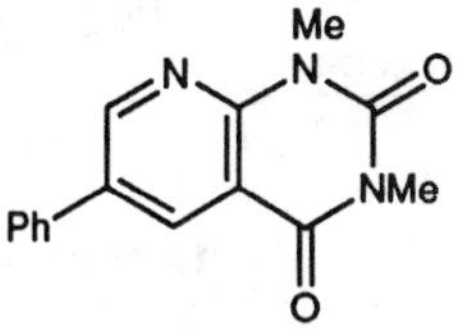

Pyrido[2,3-*d*]pyrimidine-2,3-dione [2645]

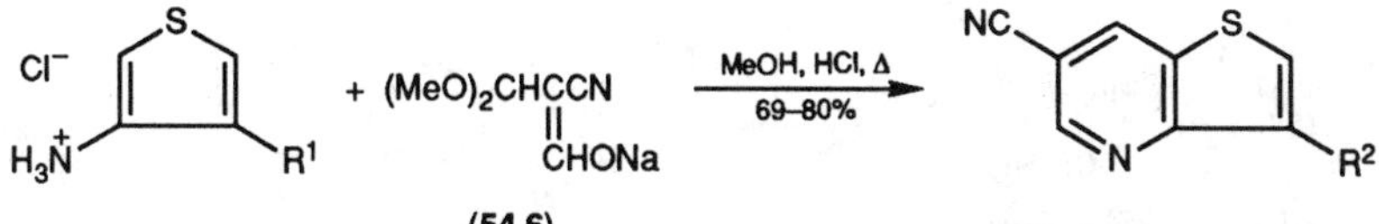

(54.6) + $(MeO)_2CHC(CN)=CHONa$ —MeOH, HCl, Δ; 69–80%→ Thieno[3,2-*b*]pyridine [3249]

$R^1 = R^2 = H$ or $R^1 = \overset{+}{N}H_3\bar{Cl}$, $R^2 = NH_2$

Other ring systems synthesized similarly:

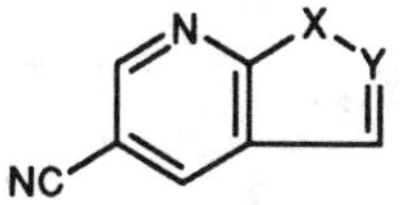

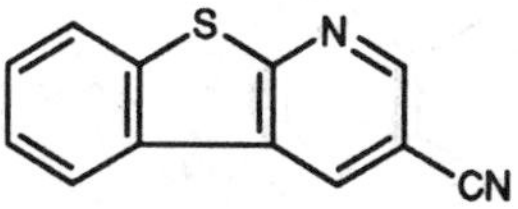

X = S, Y = CH Thieno[2,3-*b*]pyridine [3249]
X = $PhCH_2N$, Y = CH Pyrrolo[2,3-*b*]pyridine
X = PhN, Y = N Pyrazolo[3,4-*b*]pyridine

[1]Benzothieno[2,3-*b*]pyridine [3249]

2. Pyrimidine

Many methods of converting an α-aminoazine or α-aminoazole into a fused pyrimido ring are available. 3-Aminocrotononitrile joins an amino group to the ring nitrogen in high yields [2651]; the value of nitriles in heterocyclic synthesis has been reviewed [3332a]. Several malonic acid derivatives may be applied to the synthesis of a fused pyrimidine ring. Ethoxymalononitrile [2123] and benzylidene malononitrile [3226], for example, react under relatively mild conditions. Fluoromalonaldehyde (review of halomalondialdehydes [2771]) reacts with a 3-aminopyrazole to give a fluoropyrimido derivative [2707] while several other monosubstituted malonaldehydes react in an acidic medium with amino heterocycles to give 5-substituted fused pyrimidines [2824]. Tetracyanoethylene (reviews [3146, 3244] in a similar reaction gives a fused pyrimidine dinitrile [3767].

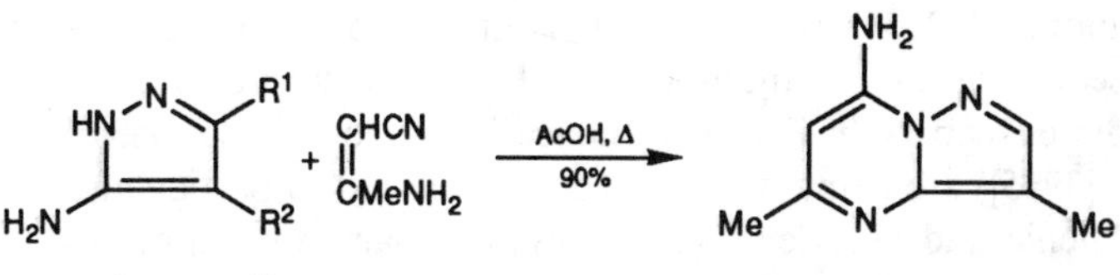

(54.7) R^1 = Me; R^2 = H + $CHCN{=}CMeNH_2$ —AcOH, Δ; 90%→ Pyrazolo[1,5-*a*]pyrimidine [2651]

(**54.7**) + CHOEt=C(CN)$_2$ —AcOH, Δ; 89%→ [2123]

$R^1 = R^2 = H$

[1,2,4]Triazolo[1,5-*a*]pyrimidine

(**54.7**) + PhCH=C(CN)$_2$ —R = CH; EtOH; 75–87%→ [3226]

$R^1 = NH_2$; $R^2 = ArCH_2$;
Ar = Ph, 4-Cl-, 4-Me-C_6H_4

Pyrazolo[1,5-*a*]pyrimidine

+ CHF(CHO)$_2$ —MeOH, AcOH, Δ; 42%→ [2707]

Pyrimido[1,2-*b*]indazole

HCl + RCH(CHO)$_2$ —HCl, Δ; ~46%→ [2824]

R = Me, MeO, Ph, 4-MeO-, 4-Me_2N-C_6H_4,
3,4,5-$(MeO)_3)C_6H_2$

Pyrimido[1,2-*a*]purin-10-one

(**54.7**) + C(CN)$_2$=C(CN)$_2$ —MeCN, Δ; ~65%→ [3767]

R^1 = Me, NH_2, Ph;
R^2 = H, Ph, PhN=N

Pyrazolo[1,5-*a*]pyrimidine

Enaminones [3213] and ω-dimethylaminopropiophenones [3767] provide further scope for the synthesis of specifically substituted fused pyrimidines. Intermediates such as the ketene dithioacetals (reviews [3712, 3830] and a recent synthesis [3965]) are being widely used and tailored to particular requirements [3822]. Allenic and acetylenic nitriles cyclize 2-aminopyridines on heating in dichloromethane [3397].

[3213]

R = Me, Ph; Ar = Ph, Cl-, 4-MeO-, 4-Me-, 3-CF_3-C_6H_4

AcOH, Δ 38–72%

Pyrimido[1,2-*a*]benzimidazole

(54.7) + R^4CCOR^3 = $C(SMe)_2$ → AcOH, pip, Δ 76–93%

[3822]

R^1 = H, MeS; R^2 = H, Ph; R^3 = Me, Et, Ph, 4-Cl-, 4-MeO-C_6H_4, 2-furyl-, 2-thienyl, PhCH=CH; R^4 = H, Me, Bu, PhCO

Pyrazolo[1,5-*a*]pyrimidine

Other ring systems synthesized similarly:

[3822]

n = 3 Cyclopenta[*e*]pyrazolo[1,5-*a*]pyrimidine
n = 4 Pyrazolo[1,5-*a*]quinazoline
n = 5 Cyclohepta[*e*]pyrazolo[1,5-*a*]pyrimidine

[3965]

[2]Benzothiopyrano[3,4-*e*]pyrazolo[1,5-*a*]pyrimidine

CH_2Cl_2, Δ 41–80%

[3397]

R^1 = H, Me, HN_2; R^2, R^3 = alkyl

Pyrido[1,2-*a*]pyrimidine

Cyclization with 1,3-dicarbonyl compounds can give high yields of fused pyrimidines [2249, 2722]; the halogen of 3-bromo-2,4-pentanedione (which would be expected to be labile) survives the reaction with a 3-aminopyrazole to give a fused bromodimethylpyrimidine in good yield [3426].

AcOH, Δ 40–92%

[2722]

R^1 = H, Me, $PhCH_2$, Ph; R^2–R^4 = H, Me

[1,3,4]Oxadiazolo[3,2-*a*]pyrimidin-4-ium

(54.8) + Ac_2CH_2 —EtOH, pip or NaOH, Δ; 91%→ [1,2,3]Triazolo[1,5-*a*]pyrimidine [2249]

+ AcCHAc(Br) —EtOH, Δ; 65%→ Pyrazolo[1,5-*a*:3,4-*d'*]dipyrimidine [3426]

A weak Lewis acid promotes the cyclization of heteroamines with acetophenones (to give a mixture of isomers) and related ketones [3963]. Finally, two methods which are useful for annulating a ring-nitrogen to an amino group attached to another ring use an anhydride [3132] or an aldehyde [3611].

ArC(O)Me, $ZnCl_2$, Δ; 71–92% → [1,2,4]Triazolo[1,5-*a*]pyrimidine [3963]

PhCMe=CHC(O)Ph, $ZnCl_2$, Δ; 60%

Ar = Ph, 4-MeO-, 4-Me-C_6H_4

Ac_2O, Δ; 59–80% → Pyrazolo[1',5' : 1,6]pyrimido[4,5-*d*]pyridazin-4-one [3132]

R^1 = H, Me; R^2 = Me, Ph;
Ar = Ph, 4-ClC_6H_4

+ R^2CHO —PhH, Δ; 38–56%→ Imidazo[1,2-*c*]-1,2,3-triazolo[4-5-*e*]pyrimidine [3611]

R^1 = $PhCH_2$, 2,6-$Cl_2C_6H_3CH_2$;
R^2 = H, 4-Cl-, 3-NO_2-C_6H_4,
2,6-$Cl_2C_6H_3$, 3,4,5-$(MeO)_3C_6H_2$

3. Pyrimidin-2-one

Annulation of amino and ring-nitrogen functions in different rings may be achieved under very mild conditions by stirring the compound with either phosgene [2652] or trichloromethyl chloroformate with TEA [3994].

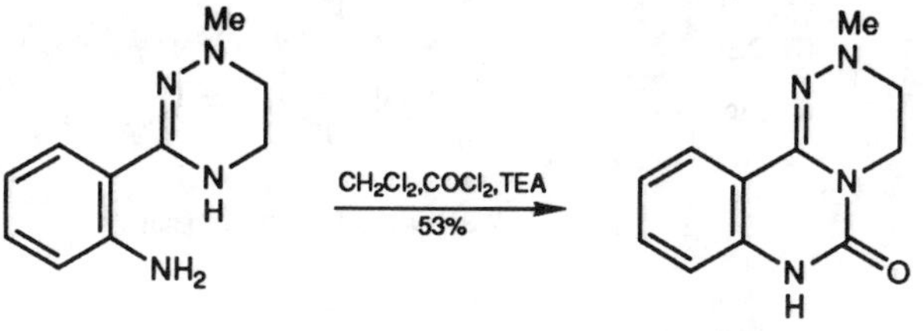

$CH_2Cl_2, COCl_2$, TEA 53% [2652]

[1,2,4]Triazino[4,3-*c*]quinazolin-6-one

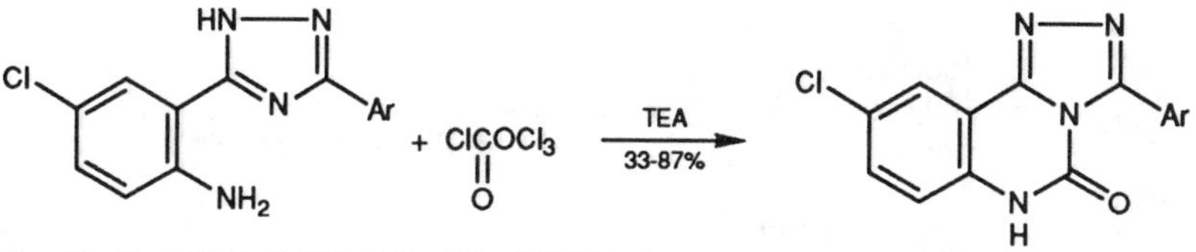

TEA 33-87% [3994]

Ar = Ph, F-, 4-MeO-, 3-CF_3-C_6H_4, 2-furyl, 2-thienyl

1,2,4-Triazolo[4,3-*c*]quinazolin-5-one

4. Pyrimidin-4-one

Cyclization of a 2-aminoheterocycle by reaction with a 1,3-dicarbonyl compound is a widely used method for the synthesis of a fused pyrimidin-4-one (see also pp. 284–286). The conversion of 2-aminopyridine into a pyridopyrimidinone by heating with an acetoacetate in the presence of a Friedel–Crafts reaction-type catalyst is typical; PPA and PPE vary in their effectiveness from one reaction to another [2250, 2289, 2296]. This reaction is also catalysed by bases such as piperidine [2501].

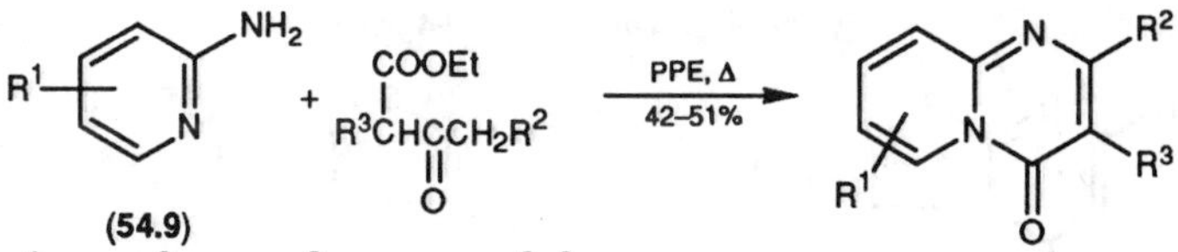

PPE, Δ 42–51% [2250]

(54.9)

R^1 = H, R^2 = Me, R^3 = H, Me or R^2R^3 = $(CH_2)_4$

Pyrido[1,2-*a*]pyrimidin-4-one
Pyrido[2,1-*b*]quinazolin-5-one

Another ring system synthesized similarly:

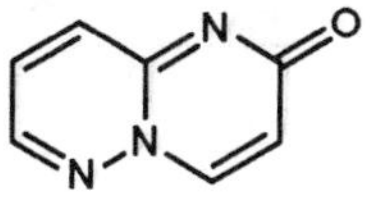

[2289, 2296]

Pyrimido[1,2-*b*]pyridazin-2-one

Variations from the simple ethyl acetoacetate [2211] are numerous and extend the usefulness of the reaction greatly, for example, 2- and 4-chloroacetoacetate [2857, 2866], 3-acyllactones [2533] and cyclic acetoacetates [3263, 3592].

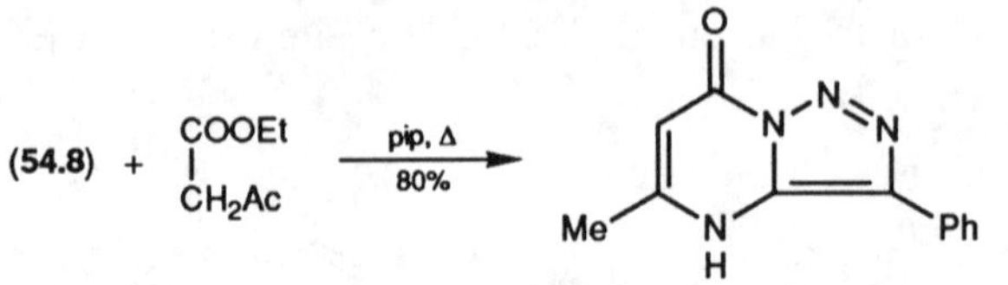

(54.8) + COOEt / CH_2Ac —pip, Δ, 80%→ [2501]

[1,2,3]Triazolo[3,4,*a*]pyrimidin-7-one

(54.9) + CHClAc / COOEt —PPA, Δ, 59%→ [2857]
R^1 = H

Pyrido[1,2-*a*]pyrimidin-4-one

58%, $COCH_2Cl$ / CH_2COOEt, PPA, Δ

[2866]

Pyrido[1,2-*a*]pyrimidin-4-one

Other ring systems synthesized similarly:

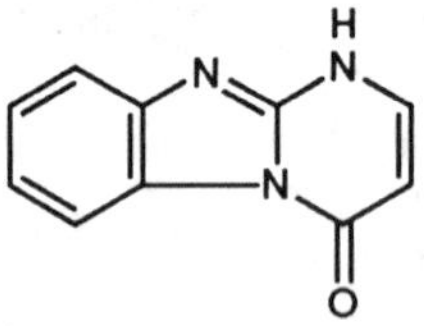

[2866]

Pyrimido[1,2-*a*]benzimidazol-4-one

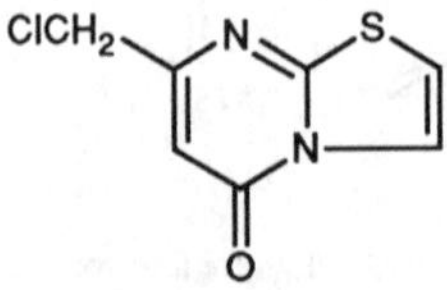

[2866]

Thiazolo[3,2-*a*]pyrimidin-5-one

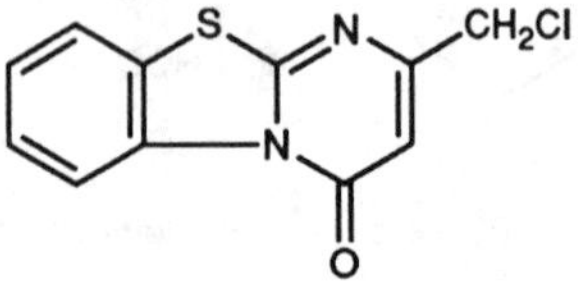

[2866]

Pyrimido[2,1-*b*]benzothiazol-4-one

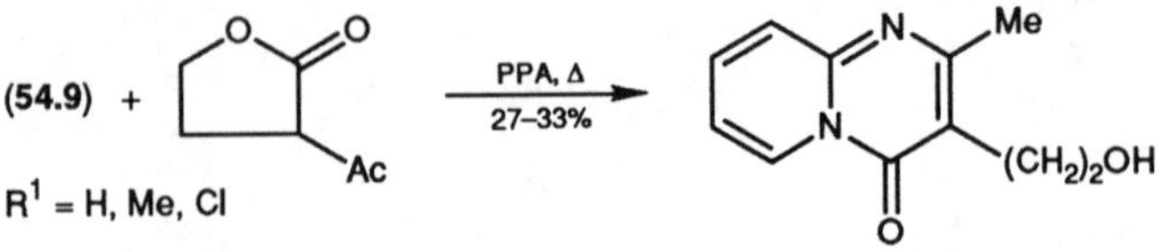

(54.9) + —PPA, Δ, 27–33%→ [2533]
R^1 = H, Me, Cl

Pyrido[1,2-*a*]pyrimidin-4-one

Other ring systems synthesized similarly:

[2533]

$n = 2,3$
Imidazo[1,2-*a*]pyrimidin-4-one
Pyrimido[1,2-*a*]pyrimidin-4-one

PPA, Δ 55–75% [3263]

R^1 = H, Me, Cl, Br; R^2 = H, iPr, $PhCH_2$

Dipyrido[1,2-*a* : 4,3-*d*]pyrimidin-11-one

$R^1 = PhCH_2$: Δ ~70%

[3592]

R^2 = MeS,O[CH_2CH_2]N
[1,2,4]Triazolo[1,5-*a*]quinazolin-5-one

R^1 = H: AcOH, Δ ~83%

[1,2,4]Triazolo[5,1-*b*]quinazolin-9-one

The acetoacetates have sometimes been replaced by ethoxymethylenemalonates [2289, 3786] under different experimental conditions from those given on p. 285, diketene [2041] (review [2130]) and several kinds of unsaturated esters [2092, 2093, 3405, 3644]; an atmosphere of argon is recommended for some of these.

PPA, Δ ~78% [2289]

R = H, Cl

Pyrimido[1,2-*b*]pyridazin-4-one

Another ring system synthesized similarly:

[3786]

Pyrimido[1,2-*a*]benzimidazol-4-one

(54.9) + ; $R^1 = H$ — H_2O, 97% → Pyrido[1,2-*a*]pyrimidin-2-one [2041]

(54.7) — $R^3C{\equiv}CCOOMe$, EtOH, Δ, 15–65% → Pyrazolo[1,5-*a*]pyrimidin-7-one + Pyrazolo[1,5-*a*]pyrimidin-5-one [2092]

$R^1 = R^2 = H$;
R^3 = H, Me, Ph, COOMe

+ $HC{\equiv}C{-}COOMe$ — EtOH,H_2O, 22% → 1,2,4-Triazolo[4,3-*a*]pyrimidin-7-one [2093]

(54.10) X = S, NH + CHCOOMe / C=CHCOOMe — MeOH,Ar, 63–76% → [3405]

X = S Pyrimido[1,2-*a*]benzothiazol-2-one
X = NH Pyrimido[1,2-*a*]benzimidazol-2-one

Other ring systems synthesized similarly:

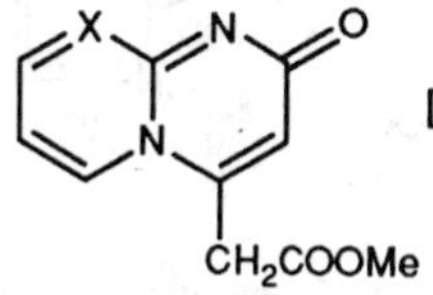

[3405] [3405]

X = CH Pyrido[1,2-*a*]pyrimidin-2-one
X = N Pyrimido[1,2-*a*]pyrimidin-2-one

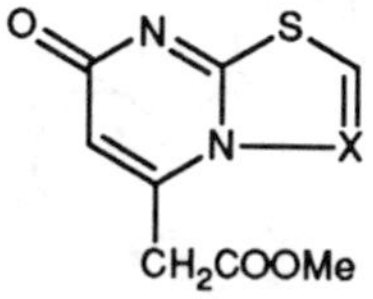

X = CH Thiazolo[3,2-*a*]pyrimidin-7-one
X = N 1,3,4-Thiadiazolo[3,2-*a*]pyrimidin-7-one

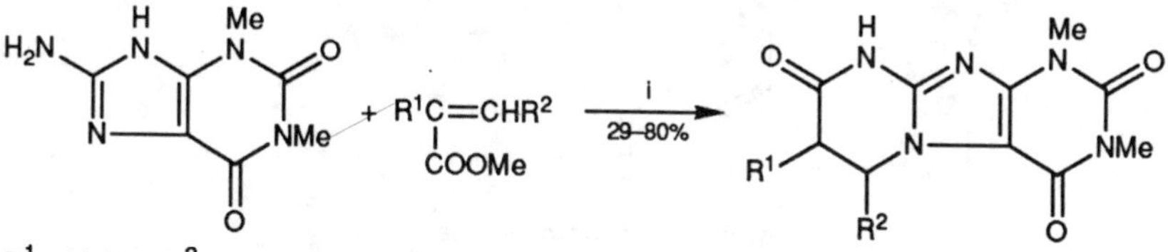

+ $R^1C{=}CHR^2$ (COOMe) —i, 29–80%→ [3644]

R^1 = H, Me; R^2 = H, Me, Ph
i, $MeO(CH_2)_2OH$, Triton B or KOH, Δ

Pyrimido[2,1-*f*]purine-2,4,8-trione

2-Chloro-3-oxobutanonitrile reacts in the presence of PPA and gives the fused pyrimidin-4-one [2839].

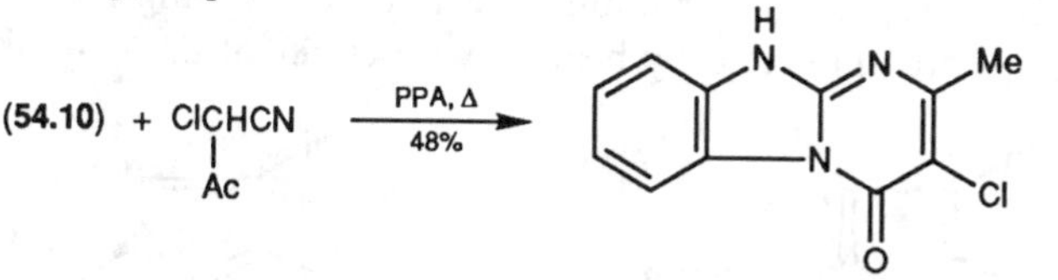

(**54.10**) + ClCHCN (Ac) —PPA, Δ, 48%→ [2839]

Pyrimido[1,2-*a*]benzimidazol-4-one

Other ring systems synthesized similarly:

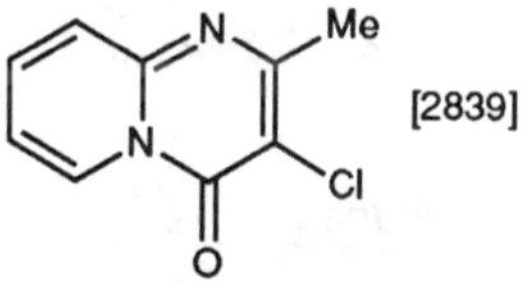

[2839] [2839]

Pyrido[1,2-*a*]pyrimidin-4-one

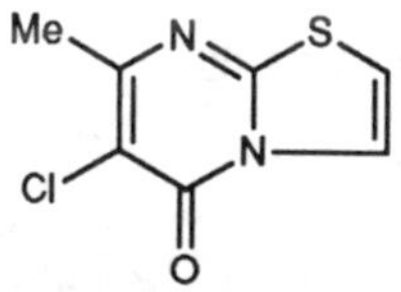

Thiazolo[3,2-*a*]pyrimidin-5-one

Carbon suboxide (review [100]) and 2-methylaminopyridine give an excellent yield of an inner salt [2288]. A bis(methylthio)methylene derivative of Meldrum's acid (review of its properties [3949]) is a versatile source of three ring-carbon atoms and an attached carboxyl group in this reaction [4000].

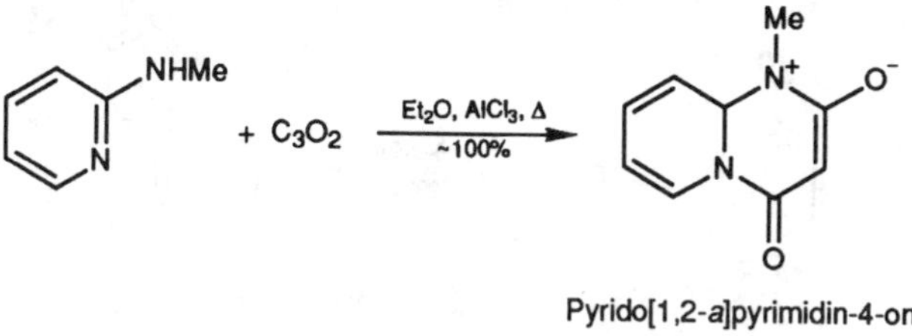

+ C_3O_2 —Et_2O, $AlCl_3$, Δ, ~100%→ [2288]

Pyrido[1,2-*a*]pyrimidin-4-one

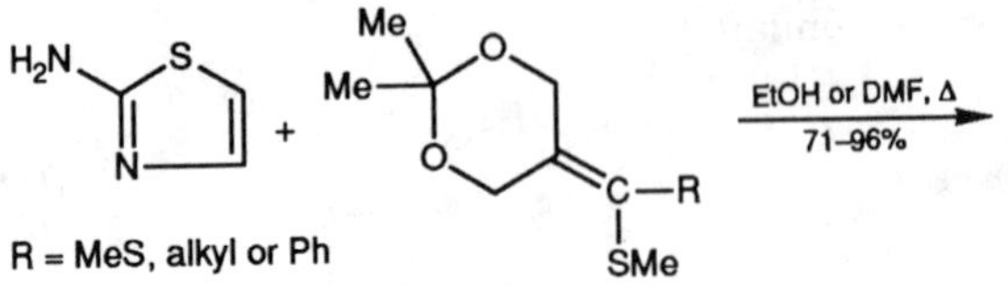

R = MeS, alkyl or Ph

EtOH or DMF, Δ
71–96%

[4000]

Thiazolo[3,2-*a*]pyrimidin-5-one

Another ring system synthesized similarly:

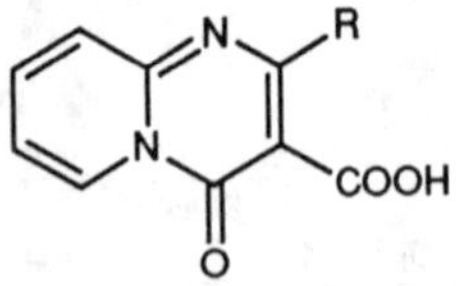

[4000]

Pyrido[1,2-*a*]pyrimidin-4-one

5. Pyrazine

Dibenzoylacetylene and DMAD (review [2816] are useful reagents for cyclizing aminoheterocycles; in this example, the amino and ring-nitrogen atoms in different rings are annulated by heating with either of the two alkynes [3768].

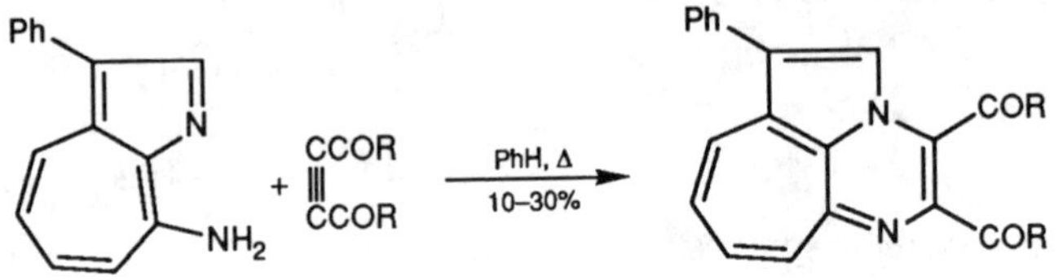

PhH, Δ
10–30%

[3768]

R = MeO, Ph

2a,5-Diazabenz[*cd*]azulene

6. Pyrazine-2,3-dione

1-Tosylaminophenothiazine (R = Ts) and oxalyl chloride gave a better yield of cyclized product than did the 1-amine compound (R = H) [2236]. A thermally induced cyclization of the methylaminomethylazepine (**54.11**) with diethyl oxalate gave a good yield of product [2125].

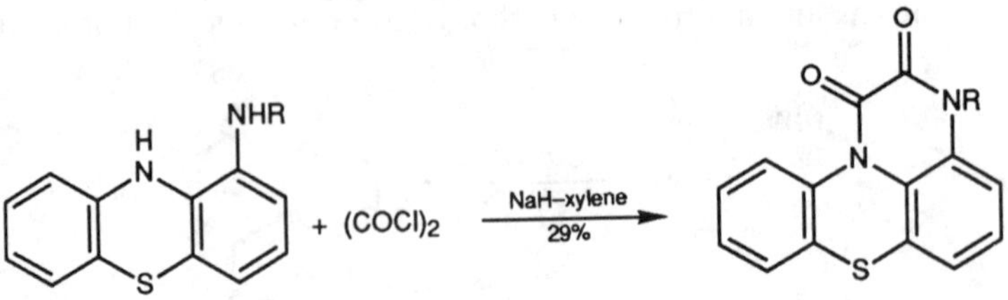

+ $(COCl)_2$

NaH–xylene
29%

[2236]

R = H, Ts

Pyrazino[3,2,1-*kl*]phenothiazine-1,2-dione

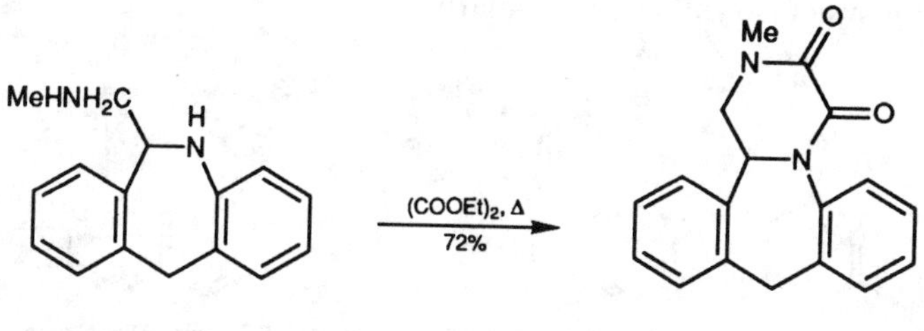

MeHNH$_2$C H N (COOEt)$_2$, Δ 72% Me N O O N

(54.11)

Dibenzo[*cf*]pyrazino[1,2-*a*]azepine-3,4-dione [2125]

7. 1,3,5-Triazine

Two moles of a 2-aminopyridine react with di-iodomethane to give a tricyclic quaternary salt. Electron-withdrawing substituents in the pyridine ring inhibit the reaction; of several aminopyrimidines, aminopyrazine, aminoimidazoles and aminopyrazoles, only 4-aminopyrimidine reacted [3089].

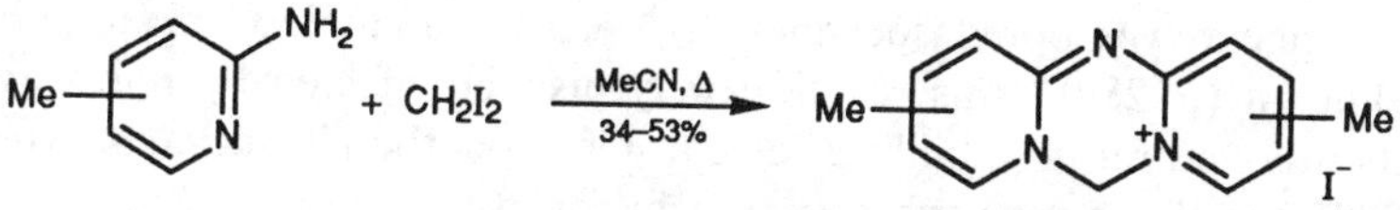

Me NH$_2$ N + CH_2I_2 MeCN, Δ 34–53% Me N N N$^+$ Me I^- [3089]

Dipyrido[1,2-*a* : 2',1'-*d*][1,3,5]-triazin-5-ium

Another ring system synthesized similarly:

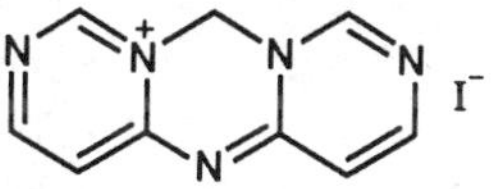

N N$^+$ N N N I^- [3089]

Dipyrimido[1,6-*a* : 6',1'-*d*][1,3,5]-triazin-10-ium

8. 1,3,5-Triazin-2-one

2-Aminoazines react with *N*-(dichloromethylene)carboxamides and a mild base to form this ring.

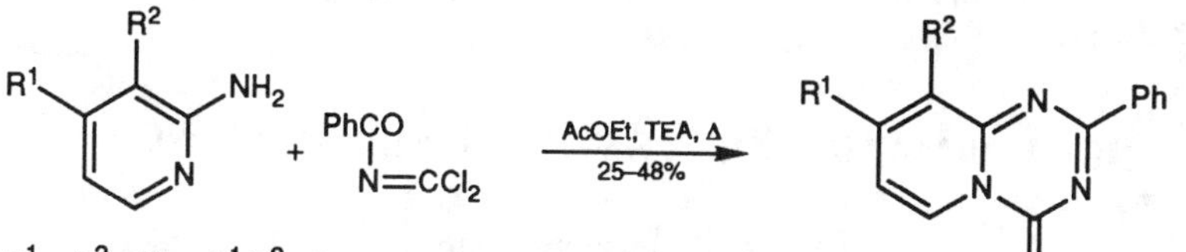

R^2 R^1 NH$_2$ N + PhCO N=CCl$_2$ AcOEt, TEA, Δ 25–48% R^2 R^1 N Ph N N O [2338]

$R^1 = R^2 = H$ or $R^1 R^2$ = benzo

Pyrido[1,2-*a*]-1,3,5-triazin-4-one
1,3,5-Triazino[2,1-*a*]isoquinolin-4-one

Other ring systems synthesized similarly:

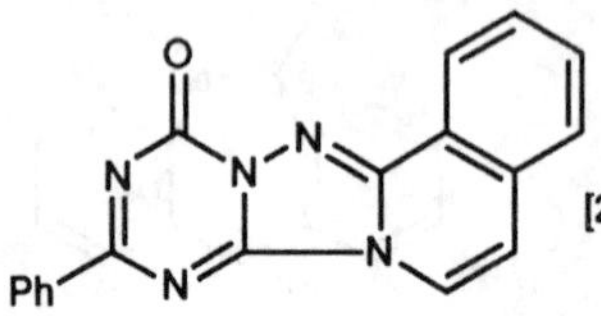

[2338]

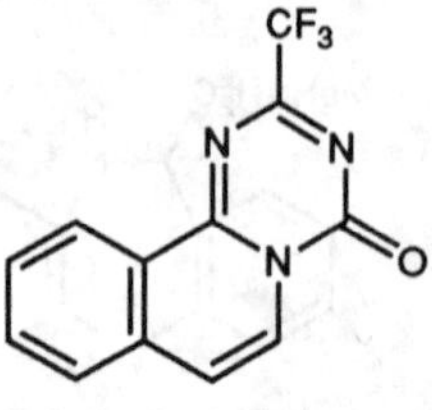

[1,3,5]Triazino[1',2' : 1,5][1,2,4]triazolo[3,4-*a*]isoquinolin-11-one

[2835]

1,3,5-Triazino[2,1-*a*]isoquinolin-4-one

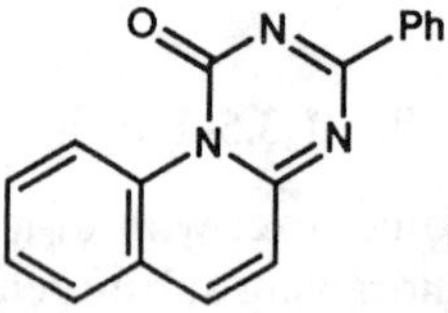

[2835]

1,3,5-Triazino[1,2-*a*]quinolin-1-one

9. 1,3,5-Triazine-2,4-dione

The use of phenoxycarbonyl isocyanate for cyclizing an aminoazole in pyridine is well known (p. 289). This reaction may also be effected by refluxing the reactants in a neutral high-boiling solvent. Chlorocarbonyl isocyanate (review [2929, 3802]) is also a very effective reagent for cyclizing such amines.

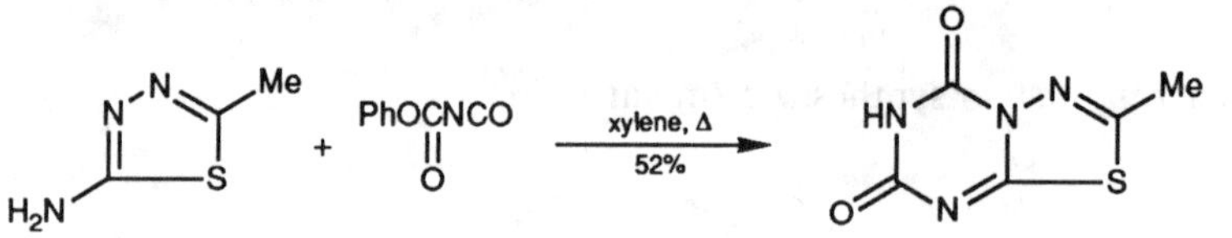

+ PhOCNCO, xylene, Δ, 52% [3208]

[1,3,4]Thiadiazolo[3,2-*a*][1,3,5]triazine-5,7-dione

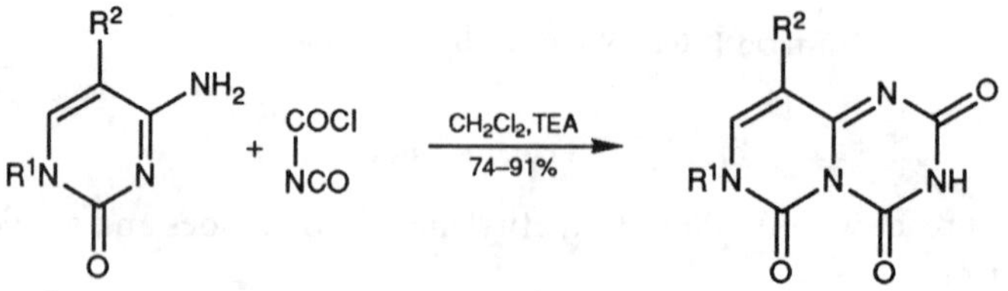

+ COCl / NCO, CH_2Cl_2,TEA, 74–91% [3415]

R^1 = D-ribosyl derivative
R^2 = H, Me

Pyrimido[1,6-*a*]-1,3,5-triazine-2,4,6-trione

III. FORMATION OF A SEVEN-MEMBERED RING

1. 1,4-Diazepin-5-one

With trifluoroacetic acid as catalyst, the cyclohepta[*b*]pyrrole (**54.2**) is cyclized into the diazepinone (**54.4**) in 96% yield.

CHAPTER 55

Azide and Azo or Nitro

I. Formation of a Five-membered Ring 1015
1. 1,2,3-Triazole 1015
2. Furazan *N*-Oxide 1016
II. Formation of a Six-membered Ring 1017
1. 1,2,3-Triazine 1017

Cyclizations which involve azides have been reviewed [3515]. Examples of ring formation from a nitro-carbodi-imide and an amino hydrazone are included in this chapter.

I. FORMATION OF A FIVE-MEMBERED RING

1. 1,2,3-Triazole

Boron trihalide promotes cyclization of *o*-azido-azo compounds under mild conditions with the formation of a benzotriazole in high yield. A nitrenium ion–BX_3 complex may be involved. The conversion of 2-azidoazobenzenes (p. 292) to benzotriazole is promoted by boron trichloride so that the conversion proceeds at room temperature and in high yield [3056].

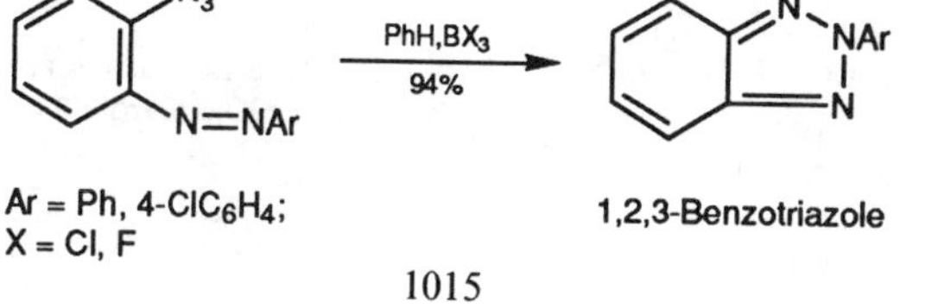

N_3 ; N=NAr $\xrightarrow[94\%]{PhH, BX_3}$ N, NAr, N [3453]

Ar = Ph, 4-ClC_6H_4;
X = Cl, F

1,2,3-Benzotriazole

The nitro-carbodi-imides (**55.1**) are cyclized into the triazole by heating in bromobenzene.

N=C=NAr / NO_2 (**55.1**) —PhBr, Δ, 75%→ benzotriazole N-NAr [2991]

2. Furazan *N*-Oxide

Thermolysis of nitro-azides is an efficient method of forming a furazan ring; its *N*-oxide may be prepared by heating the same precursor in acetic acid. Photochemical cyclization of the nitro-azide in acetic acid is a less convenient method [1491].

—AcOH, Δ, 80%→ [1491]

Imidazo[4,5-*f*]-2,1,3-benzoxadiazole 1-oxide

—$(CH_2OH)_2$, Δ, 81%→ [1491]

Pyrido[1',2':1,2]imidazo[4,5-*f*]-2,1,3-benzoxadiazole 1-oxide

It is not always necessary to isolate the nitroazides; for example, the 2-azido-3-nitropyridines may be prepared *in situ* from the chloronitropyridine and sodium azide [2432]. A recent improvement in this process by the use of phase-transfer catalysis means that lower temperatures are used and better yields are obtained for a wider range of substituted products [3241].

+ NaN_3 —i, ~95%→ [2432]

R^1 = NO_2, Ac, CHO, COOH, CF_3, PhCO, Cl
X = CR^2 or N; R^2 = H, NO_2
i, $PhCH_2\overset{+}{N}Bu_3Br^-$, $(CH_2Cl)_2$, 30-55°C

X = CH 2,1,3-Benzoxadiazole 1-oxide
X = N [1,2,5]Oxadiazolo[3,4-*b*]pyridine 1-oxide

II. FORMATION OF A SIX-MEMBERED RING

1. 1,2,3-Triazine

When the amino-hydrazone (**55.2**) is treated with LTA, the benzotriazine is obtained.

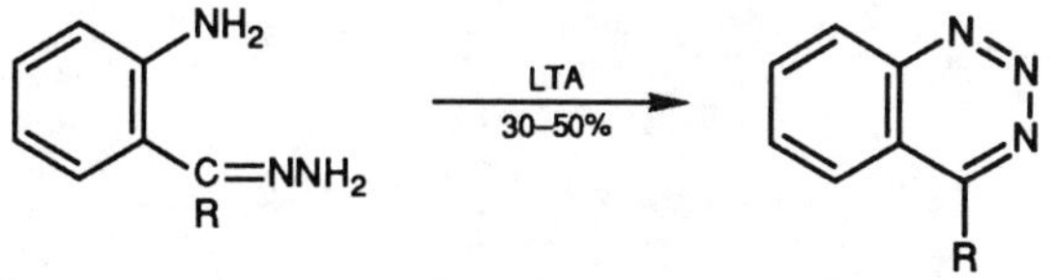

NH_2 C=NNH_2 R — LTA, 30–50% → N N N R [2379]

(55.2)
R = Me, Ph, 4-MeO

1,2,3-Benzotriazine

CHAPTER 56

Azide or Isocyanate and a Carboxylic Acid Derivative

I.	Formation of a Five-membered Ring	1018
	1. Pyrrole	1018
	2. Pyrazol-2-one	1019
	3. Oxazole	1019
	4. Tetrazole	1019
II.	Formation of a Six-membered Ring	1019
	1. Pyrimidine-2,4-dione	1019

A review of cyclizations which involve azides [3515] is relevant to many of the reactions discussed in this chapter. A rare type of cyclization resulting from interaction of a chlorine and a side-chain azide is included.

I. FORMATION OF A FIVE-MEMBERED RING

1. Pyrrole

Thermolysis of an *o*-halogeno-(β-azidovinyl)benzene sometimes produces a mixture of products; from a tetrahydrobenzenoid analogue, a newly formed pyrrole-containing product was isolated in low yield.

Cl; N_3; CH=CCOOEt; PhBr, Δ; 17%; H; N; [3075]

Indole

2. Pyrazol-3-one

An azido-carboxylate ester, on being heated for 1 h in ethanol with hydrazine hydrate, was unexpectedly converted into a fused pyrazolone.

N_3, COOMe + N_2H_4 H_2O → EtOH, Δ 44% [2608]

Benz[*f*]indazol-3-one

3. Oxazole

Cyclization of 2-azidophenyl esters with triethyl phosphite probably involves the formation of an intermediate phosphorimidate.

OCR, O, N_3 → C_6H_{12},P(OEt)$_3$, ~70% [2352]

R = Me, Ph

Benzoxazole

4. Tetrazole

An azido-nitrile (formed *in situ* from the reactive chloropyridazine and sodium azide) is efficiently converted into a fused tetrazole ring on heating in DMF.

CN, Cl → DMF,NaN_3 80% [3463]

Tetrazolo[1,5-*b*]pyridazine

II. FORMATION OF A SIX-MEMBERED RING

1. Pyrimidine-2,4-dione

Carboxamide and isocyanate (formed *in situ* from a hydrazide) interact on heating the compound in toluene to form this ring.

$CONHNH_2$, $CONH_2$, Me → 1. HNO_2 2. PhMe, Δ 84% [2937]

Pyrrolo[2,3-*d*]pyrimidine-2,4-dione

CHAPTER 57

Azide or Azo and Methyl or Methylene

I. Formation of a Five-membered Ring 1020
1. Pyrrole 1020
2. Pyrrol-3-one 1021
II. Formation of a Six-membered Ring 1021
1. 1,2-Thiazine 1021
III. Formation of a Seven-membered Ring 1022
1. 1,4-Thiazepine 1022
2. 1,2,4-Triazepine 1022

Included in this chapter are ring-forming reactions between an azide and a thio ether. Cyclizations involving azides have been reviewed [3515].

I. FORMATION OF A FIVE-MEMBERED RING

1. Pyrrole

Thermolysis of azides (pp. 294, 297) whose stereochemistry is not favourable to cyclization may be effected in dimethyl sulphoxide. For example, the azide **(57.1)** is unchanged on heating in toluene whereas its *E*-isomer readily forms a tricyclic product in high yields. The conversion of **(57.1)** into an indole is unlikely to involve the formation of a nitrene (which reacts with the sulphoxide).

N_3; R^1; C=C—H; R^2 CN

(57.1)

R^1 = H, Cl; R^2 = H, Me, Ph

DMSO, Δ → 31–60%

H N CN; R^1; R^2

Indole

[3006]

2. Pyrrol-3-one

A comparison of several base-induced cyclizations of 2-acylbenzazides by several bases showed that ethanolic potassium hydroxide at room temperature was an efficient process. Some evidence is available that nitrogen loss assisted by an enolate ion rather than nitrene formation occurs under these conditions.

EtOH-KOH, 67–96% [3168]

3-Indolone

R^1 = H, Me; R^2 = Me, Et, Ph
or $R^1R^2 = (CH_2)_5$

II. FORMATION OF A SIX-MEMBERED RING

1. 1,2-Thiazine

Thermolysis of an azide which contains a thio ether group in a neighbouring side-chain may result in the formation of an N^-—S^+ bond in a thiazine. Yields are dependent on the character of substituents in the side-chain and of the heteroatom X in (**57.2**). Reversing the positions of azide and methylthio groups and using toluene as solvent gives a high yield of (**57.3**). A similar cyclization of the benzenoid azide (**57.4**) needed a higher reaction temperature.

PhH or xylene, Δ, 4–90% [3161]

(**57.2**)
R^1 = H, COOEt; R^2 = Me, Et, Ph;
X = S, O

2λ⁴-Thieno[3,2-*c*][1,2]thiazine
2λ⁴-Furo[3,2-*c*][1,2]thiazine

[3161]

(**57.3**)
Thieno[2,3-*e*][1λ⁴,2]thiazine

xylene, Δ, 13–61% [3162]

(**57.4**)
R = Cl, PhS

1,2-Benzothiazine

III. FORMATION OF A SEVEN-MEMBERED RING

1. 1,4-Thiazepine

Ring closure of the azide (**57.5**) required a high-boiling solvent. This may account for the rearrangement which accompanied the cyclization.

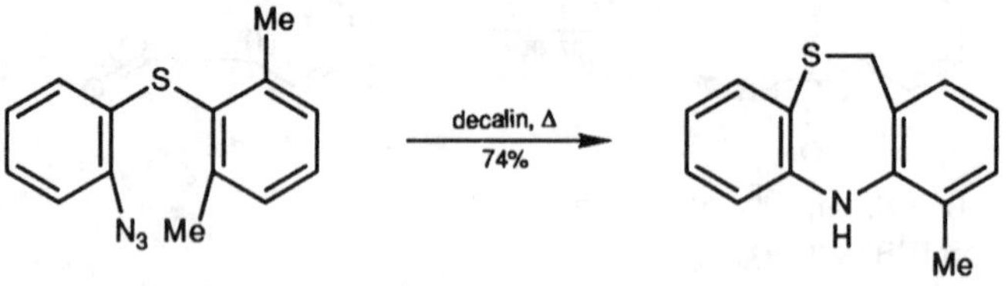

decalin, Δ
74%

(57.5)

Dibenzo[*b*,*e*][1,4]thiazepine

[2139]

2. 1,2,4-Triazepine

Thermolysis of compounds which contain side-chain azoimine groups which are adjacent to a bz-methyl or -methylene group leads to cyclization which is facilitated by a methoxycarbonyl or nitrile function attached to the carbon which separates the azo and imine groups.

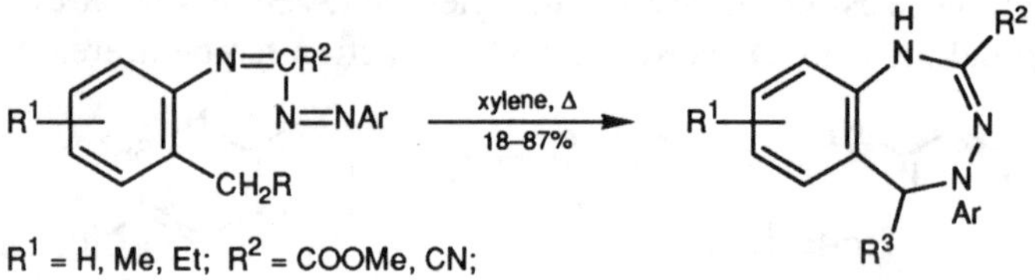

xylene, Δ
18–87%

R^1 = H, Me, Et; R^2 = COOMe, CN;
R^3 = H, Me; Ar = H, 4-$NO_2C_6H_4$,
2,4-$(NO_2)_2C_6H_3$

1,3,4-Benzotriazepine

[3104]

CHAPTER 58

Azide or Azo and Ring-carbon

I. Formation of a Five-membered Ring 1023
1. Pyrrole 1023
2. Imidazole 1026
3. 1,2,3-Triazole 1026
4. Oxazole 1026
5. Thiazole 1026
II. Formation of a Six-membered Ring 1027
1. Pyridine 1027
2. Pyridin-2-one 1028
3. Pyrazine 1028
4. 1,4-Thiazine 1028
5. 1,2-Thiazine 1,1-Dioxide 1028

An example of cyclization of a sulphimide group on to a ring-carbon is included. The chemistry of organic azides has been reviewed [3515].

I. FORMATION OF A FIVE-MEMBERED RING

1. Pyrrole

o-Biphenyl azides decompose in the presence of boron trihalides to give carbazoles, nitrenium ions being probable intermediates [3453]. When the two benzene rings are separated by an azo (N=N) group, this reaction can yield several products according to the relative positions of the azo and azide groups: *o*-azoazides give a triazole ring (see Chapter 55). The point of attack by intermediate nitrene produced thermally is not always predictable; an attempt to synthesize ellipticene produced a mixture consisting mainly of the angular isomer of ellipticene. Migration of a methyl group had occurred [3323]. The synthesis of heterocycles with nitrenes has been reviewed [2598].

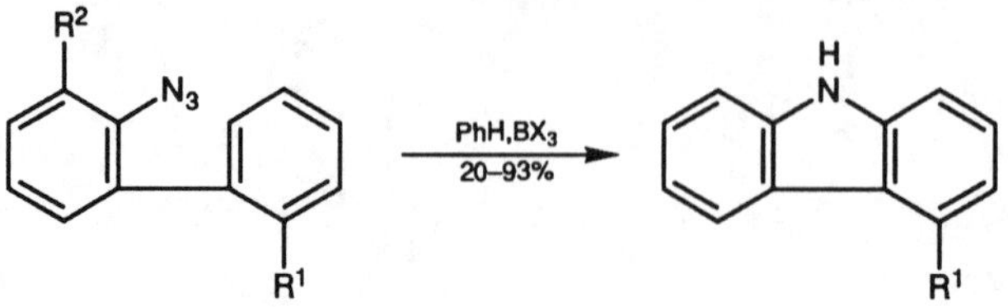

PhH,BX_3 20–93% [3453]

R^1 = H, PhN = N; R^2 = H, NO_2;
X = F, Cl

Carbazole

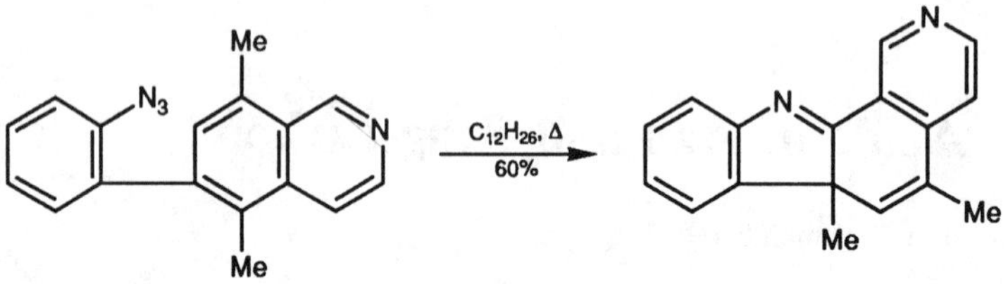

$C_{12}H_{26}$, Δ 60% [3323]

Pyrido[3,4-*a*]carbazole

Another ring system synthesized similarly:

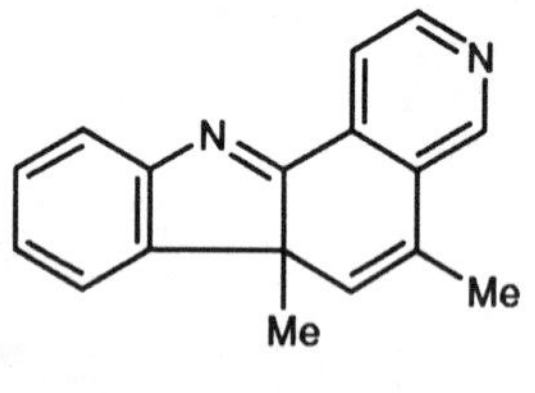

[3323]

Pyrido[4,3-*a*]carbazole

Some azides [such as **(58.1)**] are resistant to thermal cyclization through the nitrene. A strong acid may help the cyclization of such compounds but concentrated sulphuric acid may cause simultaneous sulphonation of a benzene ring.

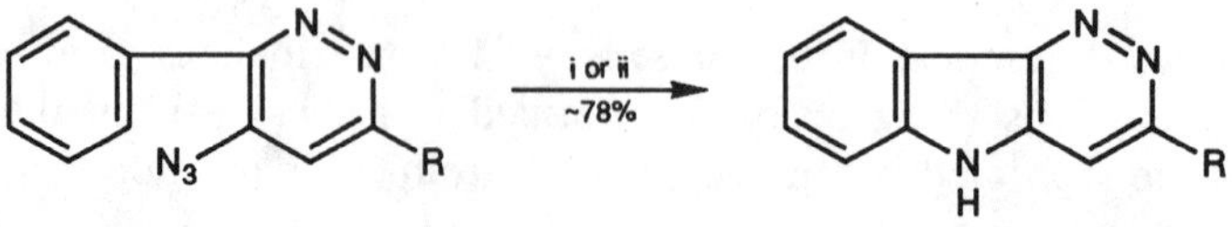

i or ii ~78% [3614]

(58.1)
R = Cl, HO
i, 6NHCl; ii, $MeSO_3H$, Δ

Pyridazino[4,3-*b*]indole

Another ring system synthesized similarly:

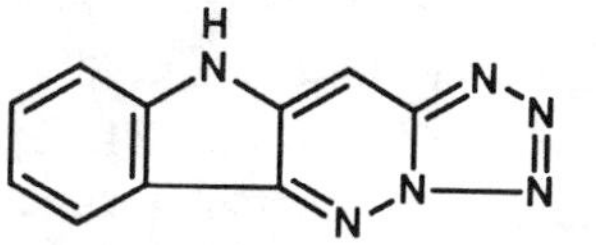

[3614]

Tetrazolo[1',5' : 1,6]pyridazino[4,3-*b*]indole

In the presence of tetracyanoethylene (TCNE), photocyclization of 2-azidobiphenyl gives carbazole as the main product, but an appreciable amount of a spiroazepine is also formed. Reviews of the chemistry [3746] of TCNE and of its use in cycloadditions [3244] are available. The synthesis of an indole by thermolysis of a previously prepared side-chain azide (see p. 301) is a widely used method but such an azide may also be formed *in situ*, as in this conversion of a benzaldehyde to an indole-2-carboxylic ester.

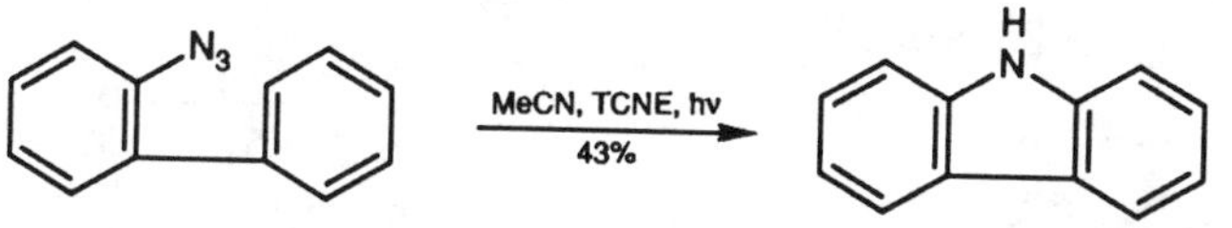

MeCN, TCNE, hν 43% [3064]

Carbazole

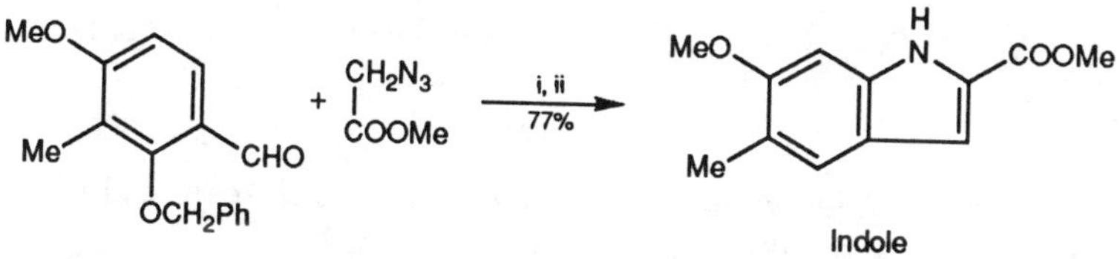

MeO, Me, CHO, OCH_2Ph + CH_2N_3 / COOMe — i, ii 77% → MeO, Me, COOMe [3667]

Indole

i, MeONa, MeOH; ii, xylene, Δ

While ethyl *o*-vinyl-2-azidocinnamate yields isoquinolines on thermolysis (see Chapter 37, Section II.1), its epoxide (**58.2**, R = H or Ph) reacts differently on being heated; the epoxide-containing side-chain is unchanged.

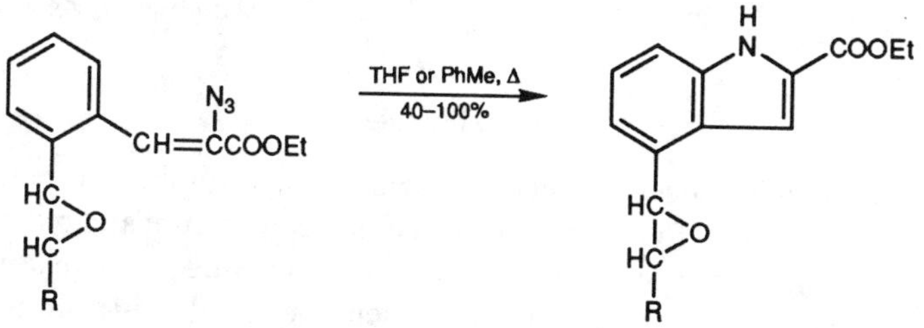

N_3 CH=CCOOEt — THF or PhMe, Δ 40–100% → COOEt [3169]

(58.2)
R = H, Ph

2. Imidazole

Sulphimides, $R^1N{=}SR^2_2$ or $R^1N^+S^-R^2_2$, resemble azides in that there is a possibility that they may form nitrenes by loss of a sulphide, R^2_2S. This occurs when such compounds are photolysed for 3–23 h depending on the substituents; under thermolysis, a mixture of products is formed.

MeCN, hν
38–96%
[2686]

R^1,R^2 = Me, PhCH$_2$, Ph; R^3 = H, Me, Cl

Benzimidazole

3. 1,2,3-Triazole

When a 4-azidocoumarin is warmed with t-butoxide, a fused 1,2,3-triazole ring is formed in low-to-moderate yield by a 1,5-dipolar cyclization. Heating a 3-nitropyridin-2-one with sodium azide–DMF gives a good yield of 1,2,3-triazolo[4,5-*b*]pyridin-5-one [2780].

DMSO, tBuOK, Δ
12–41%
[3328]

R = H, Me, MeO, Cl

[1]Benzopyrano[3-4-*d*]-1,2,3-triazol-4-one

4. Oxazole

Thermolysis of an azide with PPA and an alkanoic acid can lead (through a rearrangement) to a fused oxazole ring being formed.

+ EtCOOH
PPA, Δ
~75%
[2263]

Pyrano[2,3-*g*]benzoxazol-9-one

5. Thiazole

Inter-ring annulation occurs when an azidophenyl sulphide is either thermolysed or photolysed. By varying substituents, temperature and reaction time, it may be possible to improve yields. Photolysis in this study gave relatively poor yields whereas heating in either bromobenzene or 1,2,4-tribromobenzene usually promoted cyclization. A mechanism involving a spirodiene, opening of the sulphur-containing ring and loss of the sulphur atom was proposed.

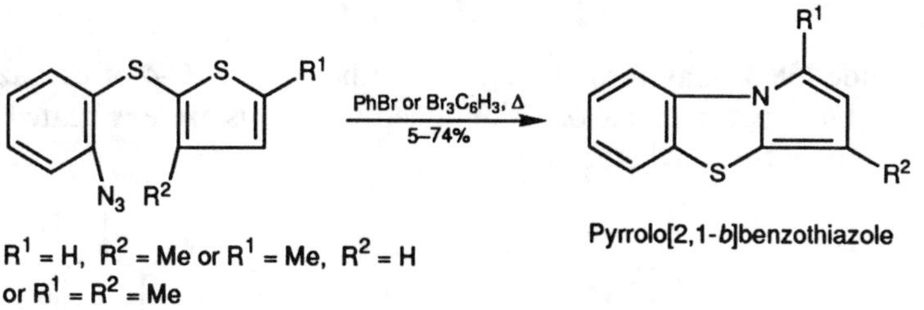

R^1 = H, R^2 = Me or R^1 = Me, R^2 = H
or R^1 = R^2 = Me

PhBr or $Br_3C_6H_3$, Δ
5–74%

Pyrrolo[2,1-*b*]benzothiazole

[2943]

II. FORMATION OF A SIX-MEMBERED RING

1. Pyridine

Thermolysis of *β*-azidovinylarenes to form new pyrrole rings is described in Part 1 (p. 299); a similar reaction on a *γ*-azidoalkadienylcyclohexane gives a mixture of products, a spiropyrrole being a second product which is converted into a cyclohepta[*b*]pyrrole on heating at 200 °C. An unsymmetrical 9-azidovinylacridine may cyclize on to either of the vacant *peri* positions on thermolysis; the 2-methoxy derivative yields a 4:1 mixture in favour of attack by the nitrene on the methoxy-containing ring.

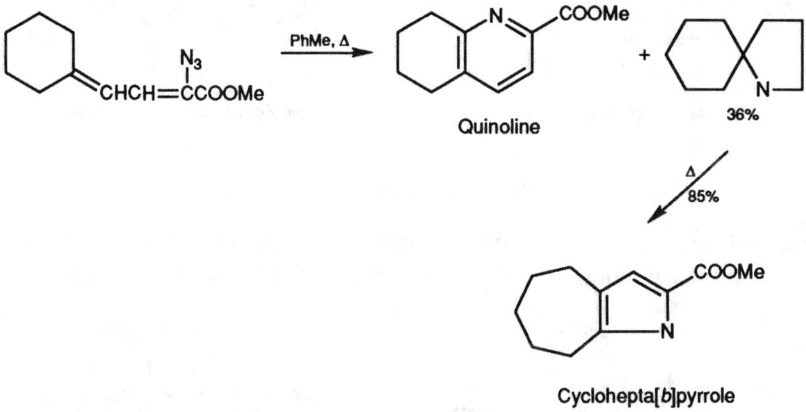

PhMe, Δ

Quinoline

36%

Δ
85%

Cyclohepta[*b*]pyrrole

[3075]

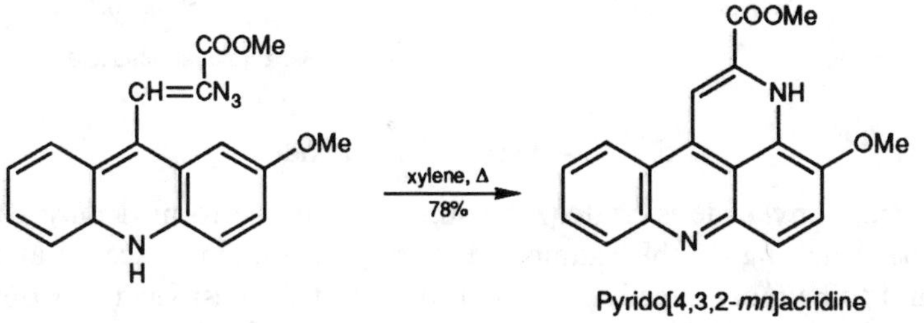

xylene, Δ
78%

Pyrido[4,3,2-*mn*]acridine

[3265]

2. Pyridin-2-one

The acylazide (**58.3**) may form a ring with either C-2 or C-4 of carbazole; in practice, a linear pyridocarbazole is obtained when this azide is heated with a tertiary amine in diphenyl ether for 1 h.

N_3CO CH=CH; Me; 4; 2; Ph_2O, Bu_3N, N_2, Δ; 76%; [3997]

(**58.3**) Pyrido[4,3-*c*]carbazol-1-one

3. Pyrazine

Cleavage of an azo group by hot mineral acid with retention of one nitrogen can be used to form a pyrazine ring in high yield.

AcOH, H_2SO_4, Δ; 83–92%; [2231]

R = H, Br, Cl; Ar = 4-$NO_2C_6H_4$ Pyrimido[4,5-*b*]quinoxalin-4-one

4. 1,4-Thiazine

A rearranged product is obtained when some azido diaryl sulphides are thermolysed in a high-boiling solvent; a spiro intermediate may be involved.

decalin, Δ; 70%; [2074]

Benzo[*a*]phenothiazine

5. 1,2-Thiazine 1,2-Dioxide

When a sulphonylazide is heated, the sulphonylnitrene formed may attack a nearby benzene ring. In this example (R = H), heating in dodecane at 175 °C was found to be more effective that at 150 °C, but the best yield was obtained in cyclohexane in a sealed tube at 120 °C. The presence of a substituent (R = Br) had a strong influence on the reaction, a 5% yield of the thiazine dioxide being obtained in cyclohexane but 24% in chlorobenzene.

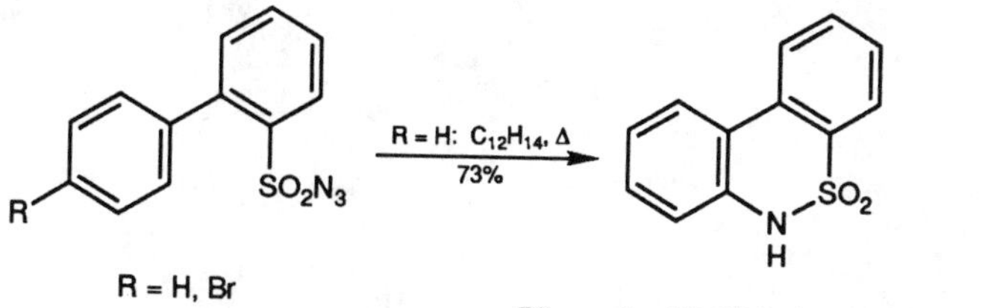

R = H: $C_{12}H_{14}$, Δ; 73%

SO_2N_3 ; R ; SO_2 ; N ; H

R = H, Br

Dibenzo[*c*,*e*][1,2]thiazine 5,5-dioxide

[2907]

CHAPTER 59

Azide and Ring-nitrogen

I.	Formation of a Five-membered Ring	1030
	1. Pyrrole	1030
	2. Pyrazole	1031
	3. Imidazol-4-one	1031
	4. 1,2,4-Triazole	1032
II.	Formation of a Six-membered Ring	1032
	1. Pyrimidin-4-one	1032

This chapter also contains examples of the cyclization of sulphimide and ring-nitrogen and of azide and imine functions. The chemistry of azides has been surveyed [3515].

I. FORMATION OF A FIVE-MEMBERED RING

1. Pyrrole

2-Azido-1-methylbenzimidazole is a relatively stable azide; the azide group survives the addition of DMAD across the 2,3-double bond of the substrate.

MeCN, DMAD, Δ; 87% [2516]

Imidazo[1,2-*a*]benzimidazole

2. Pyrazole

Thermolysis of a 3-azido-imine appears to be a better method of synthesizing the thieno[3,2-*c*]pyrazoles than reduction of the corresponding nitro compound with triethyl phosphite (Chapter 33, Section I.1), but when the isomeric 2-azido-imine was similarly treated, appreciable amounts of the benzothiophene-3-carbonitriles were obtained together with 19–43% yields of the thieno[2,3-*c*]pyrazoles [2493]. A similar conversion is achieved by treatment of the azido-imine with triphenylphosphine at 0–20 °C [3639]. Thermolysis or photolysis of a 2-(azidophenyl)pyridine can give high yields of a fused pyrazole [2654].

N_3 CH=NAr S; $(MeOCH_2CH_2)_2O$, Δ; 66–75%; N NAr S [2493]

[1]Benzothieno[3,2-*c*]pyrazole

Other ring systems synthesized similarly:

S N NAr [2493]

[1]Benzothieno[2,3-*c*]pyrazole

N NN=PPh_3 NHR [3639]

Indazole

N NPh X [2921]

X = O Furo[3,2-*c*]pyrazole
X = S Thieno[3,2-*c*]pyrazole

R N N_3; Δ or *hν*; 43–85%; R N N [2654]

R = NO_2, CF_3, CN

Pyrido[1,2-*b*]indazole

3. Imidazol-4-one

An azidoazole reacts exothermally with diphenyl ketene, and nitrogen is eliminated.

Me N N_3 N + $Ph_2C{=}C{=}O$; 94%; Me N N Ph Ph N O [2516]

Imidazo[1,2-*a*]benzimidazol-3-one

4. 1,2,4-Triazole

Carbamoyl azides (**59.1**) (formed *in situ* from the carbamoyl chloride and azidotrimethylsilane) in boiling toluene cyclize to give good yields of the triazole betaines.

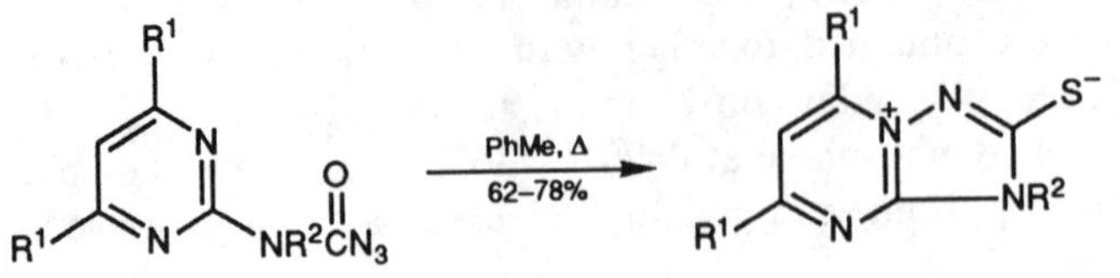

PhMe, Δ 62–78% [3537]

(**59.1**)
R^1 = H, Me; R^2 = alkyl

[1,2,4]Triazolo[1,5-*a*]pyrimidin-8-ium

II. FORMATION OF A SIX-MEMBERED RING

1. Pyrimidin-4-one

Heating picolinoyl azide and diketene yields a pyridopyrimidinedione as the main product, the isocyanate being a likely intermediate.

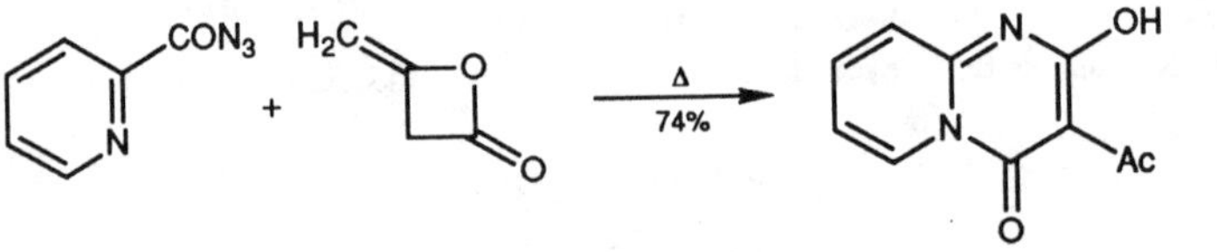

Δ 74% [2666]

Pyrido[1,2-*a*]pyrimidin-4-one

CHAPTER 60

Azo and Carbamate

I. Formation of a Six-membered Ring 1033
1. 1,2,4-Triazin-3-one 1033
2. 1,2,3-Thiadiazin-6-one 1033

Cyclization of an *o*-thiocarboxy-diazonium salt is included in this chapter.

I. FORMATION OF A SIX-MEMBERED RING

1. 1,2,4-Triazin-3-one

Azo compounds containing an *o*-carbamate group may be cyclized by heating either with a mineral acid or in an inert solvent. The azo-carbamate (**60.1**) failed to cyclize under neutral conditions but gave good yields in either an acidic or a weakly basic medium. In contrast, arylazoindol-3-yl compounds readily cyclize on heating to give triazinoindoles [1508].

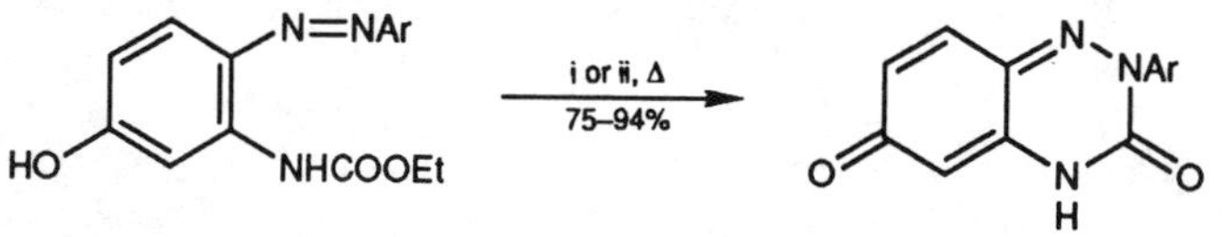

N=NAr, HO, NHCOOEt — i or ii, Δ, 75–94% → N, NAr, O, N, H, O [2865]

(**60.1**)
Ar = Ph, 4-halo-, 4-MeO-, 4-Me-, 4-EtO-C_6H_4
i, aq. Na_2CO_3; ii, aq. HCl

1,2,4-Benzotriazine-3,6-dione

2. 1,2,3-Thiadiazin-6-one

When an attempt is made to diazotize 2-aminothiobenzoic acid with isopentyl nitrite, this ring is formed in high yield; the chemistry of diazine rings containing S and N atoms has been reviewed [3904].

NH_2 / COSH — THF, i (75%) → [2358]

i, Cl_3CCOOH, THF, $Me_2CH(CH_2)_2ONO$

3,1,2-Benzothiadiazin-4-one

CHAPTER 61

Carbamate or Ureide and Ring-carbon or Ring-nitrogen

I.	Formation of a Five-membered Ring	1035
	1. Pyrazole	1035
	2. Imidazol-2-one	1036
	3. 1,2,4-Triazole	1036
	4. Thiazole	1036
	5. 1,2,4-Thiadiazole	1036
II.	Formation of a Six-membered Ring	1037
	1. Pyrimidin-2-one	1037
	2. Pyrimidin-2-one, 4-Thioxo- or Pyrimidin-4-one	1037
	3. 1,2,4-Triazin-5-one	1038
	4. 1,3,5-Triazin-2-one	1038
	5. 1,3,4-Thiadiazine	1039
III.	Formation of a Seven-membered Ring	1039
	1. 1,2,5-Triazepine	1039

I. FORMATION OF A FIVE-MEMBERED RING

1. Pyrazole

1,3-Dipolar addition of a pyridine *N*-ylide to DMAD proceeds at ambient temperature but is susceptible to the nature of the substituent present on the ring; a 4-substituted pyridine usually gives best results and the presence of tetracyanoethylene (TCNE) improves the yield.

Pyridinium N-ylide ($\bar{N}COOEt$) + CCOOMe≡CCOOMe —(MeCN, TCNE; ~44%)→ Pyrazolo[1,5-*a*]pyridine (2,3-di-COOMe) [2294]

2. Imidazol-2-one

Phenylureas undergo oxidative cyclization in high yield when oxidized with LTA at ambient temperature.

R = H, Me, MeO, Cl — $CHCl_3$, LTA, 34–95% — 2-Benzimidazolone [2740]

3. 1,2,4-Triazole

Carbamate and ring-nitrogen form a fused triazole ring when an ethoxycarbonylhydrazinoazine is heated in diphenyl ether.

Ph_2O, Δ, 73% — [1,2,4]Triazolo[3,4-*c*][1,4]benzoxazine-1,4-dione [3070]

4. Thiazole

Ureido side-chains are oxidatively cyclized (p. 311) to a thiazole; heating with hydrochloric acid may also be effective.

dil. HCl, Δ, 58% — Thiazolo[4,5-*b*]pyrazine [3333]

5. 1,2,4-Thiadiazole

A 2-thioureidopyridine is cyclized in a reaction between thioxo and ring-nitrogen functions; no heating is necessary and yields are usually good-to-excellent.

R = HO, EtO, NH_2 — Br_2–AcOH, 68–88% — [1,2,4]Thiadiazolo[2,3-*a*]pyridine [2932]

II. FORMATION OF A SIX-MEMBERED RING

1. Pyrimidin-2-one

Dehydrative cyclization of an acylureidobenzene with PPA gives a 2-quinazolinone, but the yield depends on the temperature, the molar ratio of PPA and the substituent R^2; when the last named is replaced by methyl, a poor yield is obtained. An N^3-phenylureide group attached to a benzene ring reacts with N^4-H of a (reduced) 1,2,4-triazine ring when the compound is heated at about 200 °C; aniline is eliminated. Heating the same compound with PPA produces an isomeric product in which N^2 is attached to the carbonyl.

PPA, Δ 4–80% [2467]

R^1 = alkyl, $PhCH_2$, Ph; R^2 = MeO, AcO; R^3 = H, MeO

2-Quinazolinone

Δ 100% [2652]

[1,2,4]Triazino[4,3-*c*]quinazolin-6-one

PPA, Δ 69%

[1,2,4]Triazino[2,3-*c*]quinazolin-6-one

2. Pyrimidin-2-one, 4-Thioxo- or Pyrimidin-4-one

The thioxopyrimidinone ring may be constructed from a side-chain containing a CH—CS—NH—COOEt; this becomes joined to a ring-NH when the compound is heated in pyridine. Heating the carbamate (**61.1**) in diphenyl ether induces cyclization to a pyrimidin-4-one in good yield.

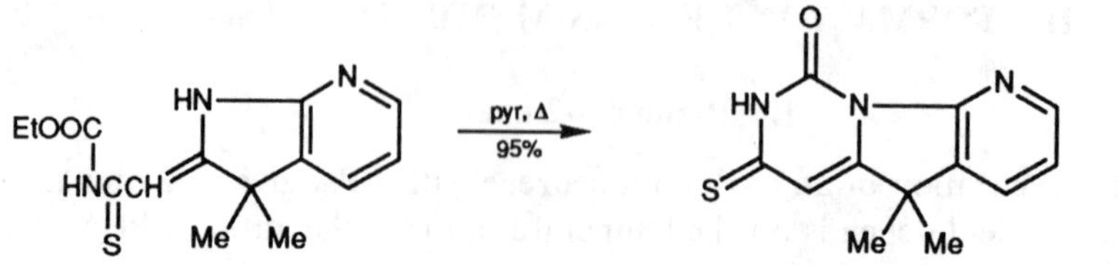

[2578]

Pyrido[3',2' : 4,5]pyrrolo[1,2-*c*]pyrimidin-9-one 7-thioxo-

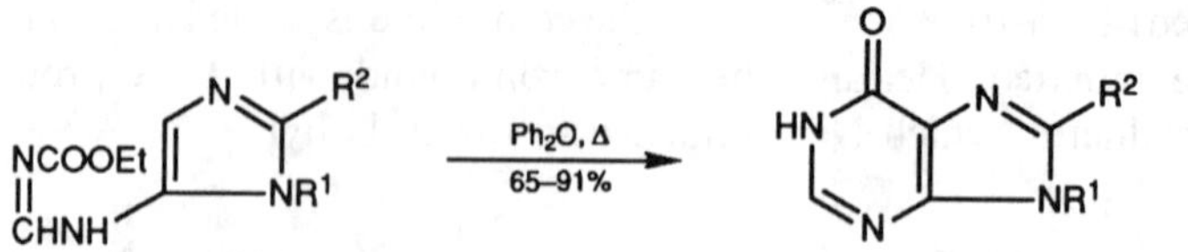

[3668]

(61.1)

R^1 = Me, $HO(CH_2)_2$; R^2 = H, Me, iPr

Imidazo[4,5-*d*]pyrimidin-7-one

3. 1,2,4-Triazin-5-one

A side-chain carbamate group cyclizes on to a neighbouring ring-carbon atom on heating the compound on its own or with phenol under anhydrous conditions.

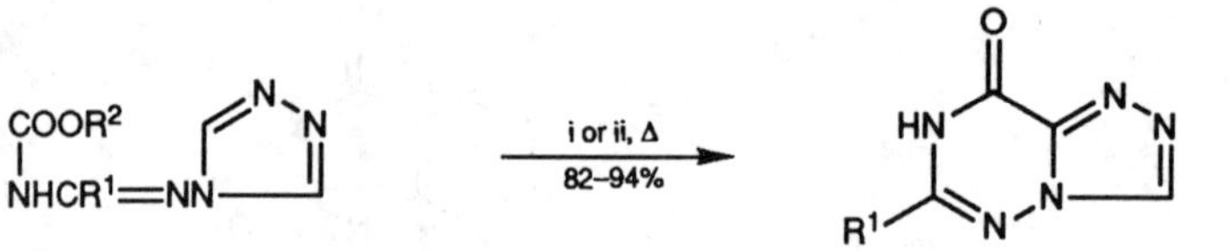

[2124]

R^1 = $PhCH_2$, Cl-, 4-Me-, 4-MeO-C_6H_4, 2,4,6-$Me_3C_6H_2CH_2$; R^2 = $EtOCH_2CH_2$, Bu

i, $(HOCH_2)_2$; ii, PhOH

1,2,4-Triazolo[3,4-*f*][1,2,4]triazin-8-one

4. 1,3,5-Triazin-2-one

Heating an *N*-ethoxycarbonylisothiourea for a short time under reduced pressure is probably the method of choice for the synthesis of this fused triazinone ring [2511], but as an alternative method, prolonged heating in dioxane with an orthoester may be considered [2773]. An acylureide is similarly cyclized by heating in pyridine [2577].

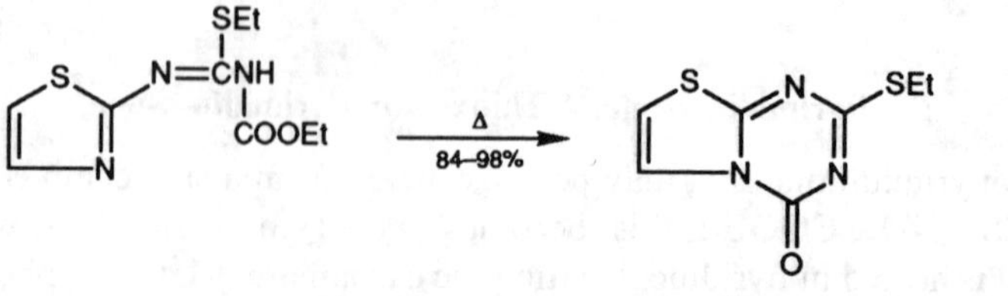

[2511]

Thiazolo[3,2-*a*]-1,3,5-triazin-4-one

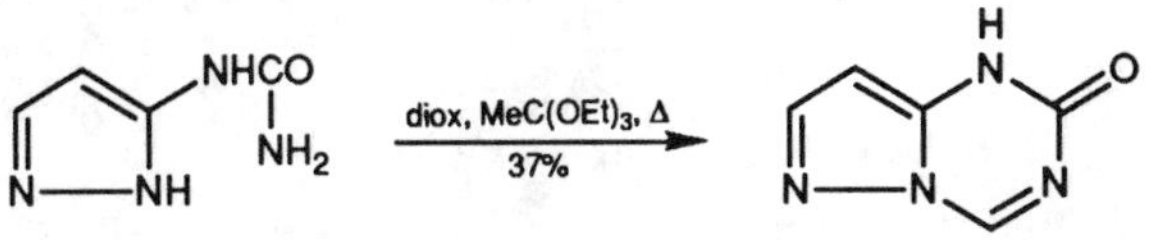

diox, $MeC(OEt)_3$, Δ 37% [2773]

Pyrazolo[1,5-*a*]-1,3,5-triazin-2-one

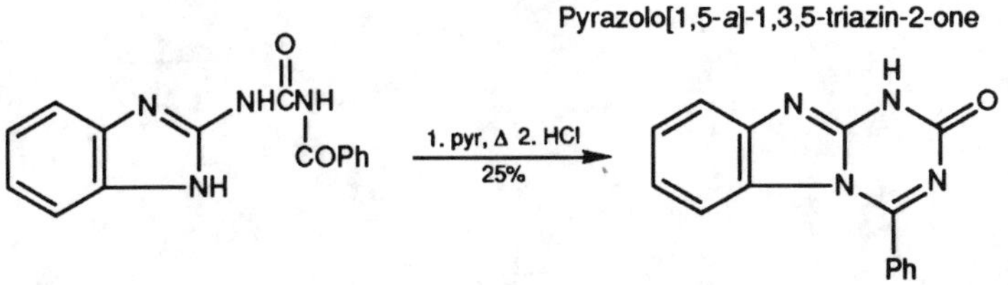

1. pyr, Δ 2. HCl 25% [2577]

1,3,5-Triazino[1,2-*a*]benzimidazol-2-one

5. 1,3,4-Thiadiazine

Heating an N^1-pyridazinylthiosemicarbazide with two molar equivalents of diethyl azodicarboxylate (DEAD) leads to the formation of a new fused thiadiazine ring in good yield.

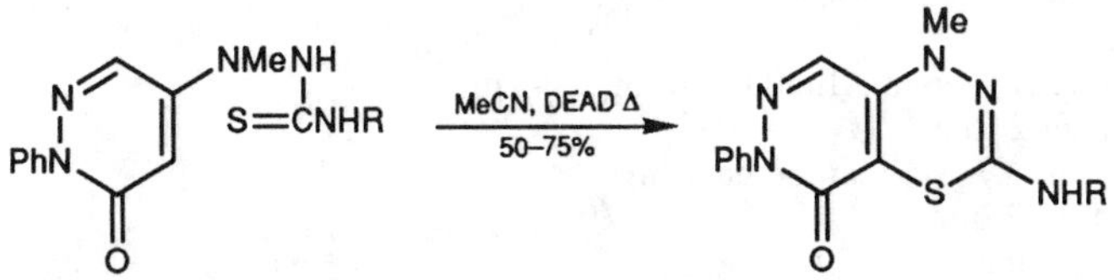

MeCN, DEAD Δ 50–75% [3729]

R = Ph, 4-Br-, 4-Me-C_6H_4, $PhCH_2$ Pyridazino[4,5-*e*][1,3,4]thiadiazin-8-one

III. FORMATION OF A SEVEN-MEMBERED RING

1. 1,2,5-Triazepine

The extended conjugation in compounds which have a *N*-side-chain containing $^+N—\bar{C}H—C{=}N$ enables it to undergo a 1,5-addition cyclization with alkynes.

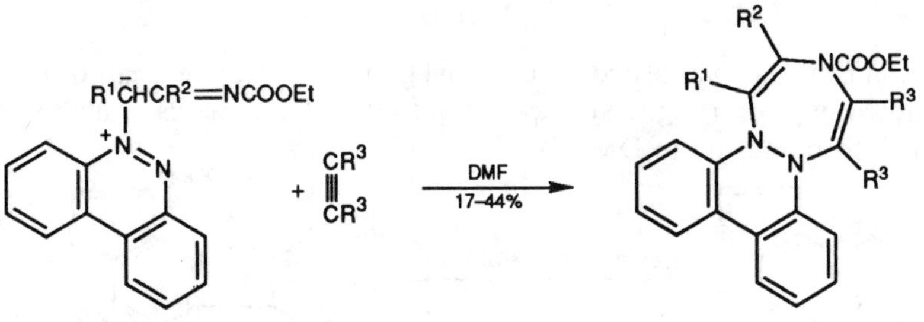

DMF 17–44% [2680]

R^1 = COOMe, COOEt; R^2 = H, COOMe, COOEt; R^3 = COOMe, COOEt

Benzo[*c*][1,2,5]triazepino[1,2-*a*]cinnoline

CHAPTER 62

Carbamate or Carboxamide and another Carboxylic Acid Derivative

I. Formation of a Five-membered Ring 1040
1. Pyrrole 1040
2. Imidazol-2-one or Imidazol-2-one, 4-Thioxo- 1041
3. Oxazol-2-one 1041
II. Formation of a Six-membered Ring 1042
1. Pyridin-2-one 1042
2. Pyrimidine 1042
3. Pyrimidin-4-one, Pyrimidin-4-one, 2-Thioxo or Pyrimidine-2,4-dione 1042
4. 1,3-Oxazin-2-one or 1,3-Oxazine-2,4-dione 1044
III. Formation of a Seven-membered Ring 1044
1. 1,2-Diazepin-3-one 1044
2. 1,4-Diazepine-2,7-dione 1044

I. FORMATION OF A FIVE-MEMBERED RING

1. Pyrrole

Interaction between a nitrile and the carbonyl group of an adjacent carboxamide is promoted by converting the compound into its trifluoromethanesulphonate salt which then adds on to DMAD.

NC, PhCN, O, S — 1. CF_3SO_3H 2. DMAD, DMF, Δ; 17% → COOMe, MeOOC, S, N, Ph [3542]

Thieno[2,3-*g*]indolizine

2. Imidazol-2-one or Imidazol-2-one, 4-thioxo-

The anion of the Reissert compound (**62.1**) reacts at or below room temperature with an aldimine or an isothiocyanate by cycloaddition, and then displacement of the nitrile to form a tricyclic isoquinoline. The chemistry of Reissert compounds has been reviewed [2488, 2959a], as have the synthetic uses of isothiocyanates [3990].

DMF, NaH 48–65% [3699]

(**62.1**) Imidazo[5,1-*a*]isoquinolin-3-one

R^1 = Me, Ph; R^2 = Ph, 4-$MeOC_6H_4$, 3,4,5-$(MeO)_3C_6H_2$;
R^3 = Ph, PhCH=CH, PhC≡C

(**62.1**) DMF, R^4NCS, NaH, N_2 35–58% [3140]

R^4 = Me, Ph, 4-FC_6H_4

Imidazo[5,1-*a*]isoquinolin-3-one, 1-thioxo-

Another ring system synthesized similarly:

[3140]

Imidazo[5,1-*a*]phthalazin-3-one, 1-thioxo-

3. Oxazol-2-one

The ester group of a carbamate is annulated to a neighbouring nitrile (of a Reissert compound) on reaction with an araldehyde under strongly basic conditions (the CN group is eliminated). A better yield of this type of compound was obtained by another method (Chapter 98).

+ ArCHO DMF, NaH 14% [3554]

Ar = 4-$MeOC_6H_4$

Oxazolo[4,3-*a*]phthalazin-3-one

II. FORMATION OF A SIX-MEMBERED RING

1. Pyridin-2-one

Annulation of a cyano-carboxamide is effected in high yield by a strong base at ambient temperature.

EtONa, 74–98% [3111]

Pyrrolo[2,3-*c*]pyridin-7-one
1,7-Naphthyridin-8-one

n = 1,2;
Ar = Ph, 4-Cl-, 3-Me-C_6H_4

2. Pyrimidine

4-Aminoquinazolines are of considerable medicinal interest and may be synthesized by a one-pot reaction from isatoic anhydride (**62.2**) (which, for classification purposes is regarded as a 2-carboxycarbamate) and either ammonia or an amine. The reactions of isatoic anhydride have been reviewed [3008].

i, ii, iii, 44–79% [2820]

(**62.2**) Quinazoline

R^1 = H, Cl; R^2 = H, Me, Ph
i, DMF, $NH_3(g)$; ii, $POCl_3$, ~50 °C; iii, R^2NH_2, Δ

3. Pyrimidin-4-one, Pyrimidin-4-one, 2-Thioxo- or Pyrimidine-2,4-dione

Isatoic anhydride (**62.2**) (review of its reactions [3008]) in a strongly basic environment reacts with cyanamide to give the useful intermediate 2-amino-4-quinazolinone [3281]. Neighbouring thiocarbamate and carboxylic ester groups are converted into one of two pyrimidin-4-one rings on reaction with hydrazine, which behaves as a monofunctional reagent: at room temperature the thioxo group is retained but when heat is applied to the reaction, a 2-hydrazinoquinazolin-4-one is obtained in high yield [3674]. In the presence of a strong base, a carboxamido-carbamate is cyclized in high yields to this dione [2660].

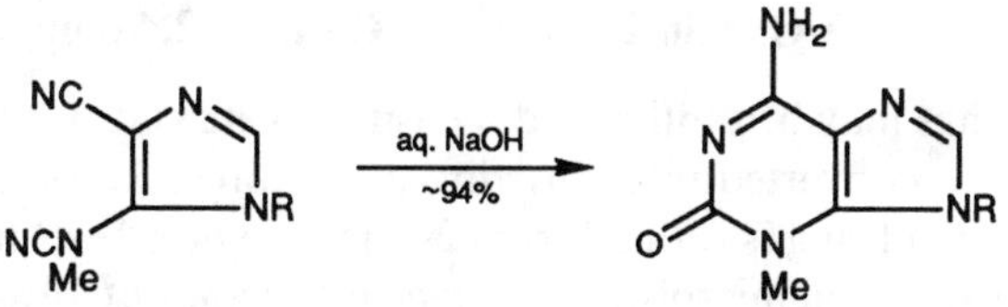

NC, N, NR, NCN, Me

aq. NaOH

~94%

NH_2, N, N, NR, O, N, Me

[3037a]

(62.3)

R = Me, β-D-ribofuranosyl

2-Purinone

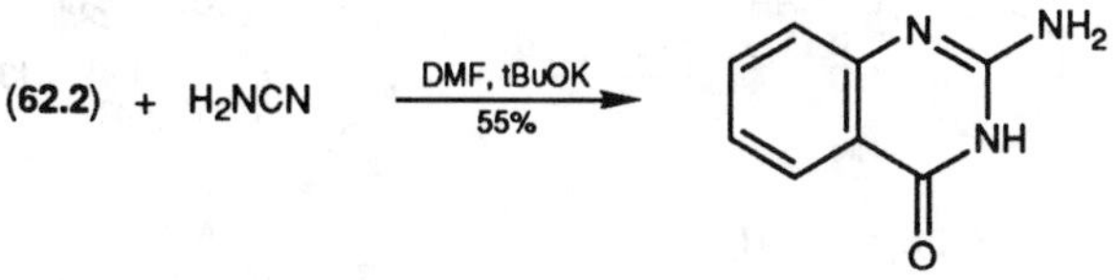

(62.2) + H_2NCN

DMF, tBuOK

55%

N, NH_2, NH, O

[3281]

4-Quinazolinone

S, NHCOEt, N, COOEt

EtOH, N_2H_4, H_2O

20 °C

79%

Δ

93%

H, N, S, NNH_2, N, O

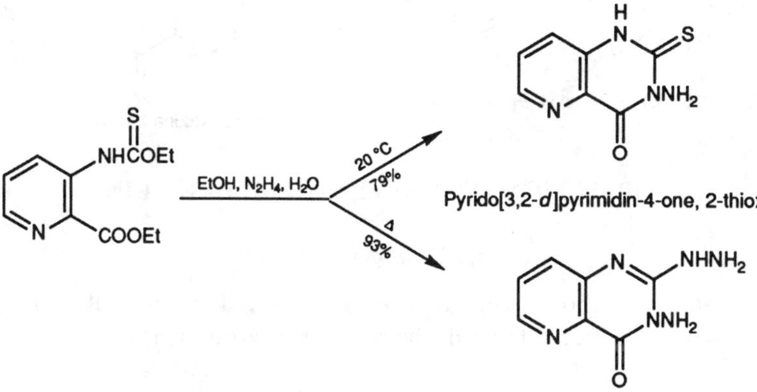

Pyrido[3,2-*d*]pyrimidin-4-one, 2-thioxo- [3674]

N, $NHNH_2$, NNH_2, N, O

Another ring system synthesized similarly:

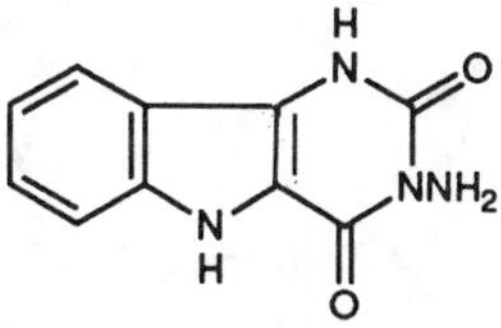

H, N, O, NNH_2, N, H, O

[3206]

Pyrimido[5,4-*b*]indole-2,4-dione

COOEt, Me, NHCOOEt, HN, CONHMe

EtONa, Δ

~100%

H, Me, N, O, HN, NMe, O

[2660]

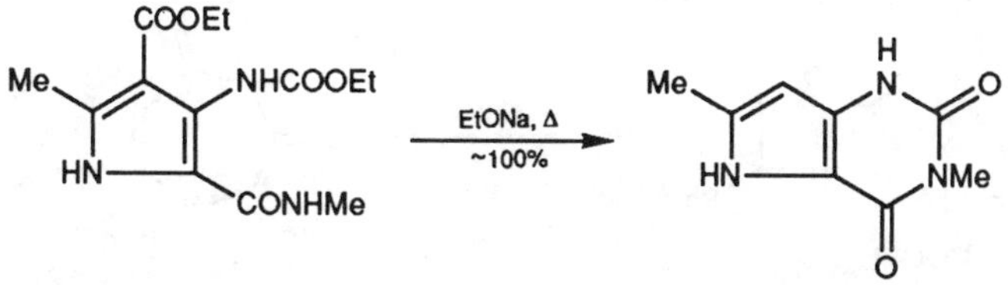

Pyrrolo[3,2-*d*]pyrimidine-2,4-dione

4. 1,3-Oxazin-2-one or 1,3-Oxazine-2,4-dione

Both of these rings may be synthesized by using an enzyme. Carbamoyloxy and nitrile groups are converted into this ring in the presence of baker's yeast or catalase. The second ring is formed from 2-carbamoyloxybenzoic acid esters by oxidation with rat liver microsomal extract in ethanol at pH 7.4 and a temperature of 20–22 °C The process takes about 8 h under aerobic conditions and yields are better than those of some chemical methods.

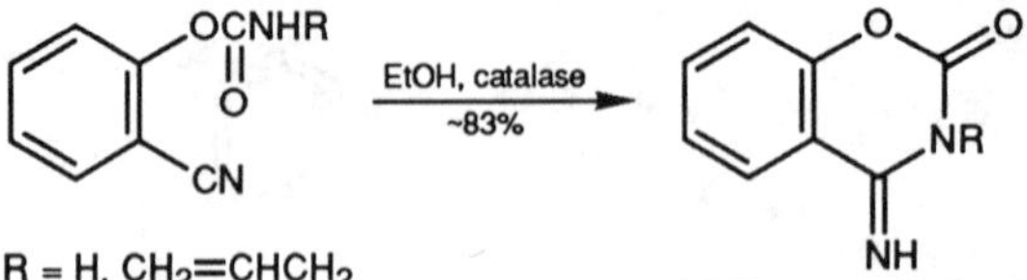

EtOH, catalase, ~83% [3626, 3708, 3791]

R = H, $CH_2{=}CHCH_2$

1,3-Benzoxazin-2-one

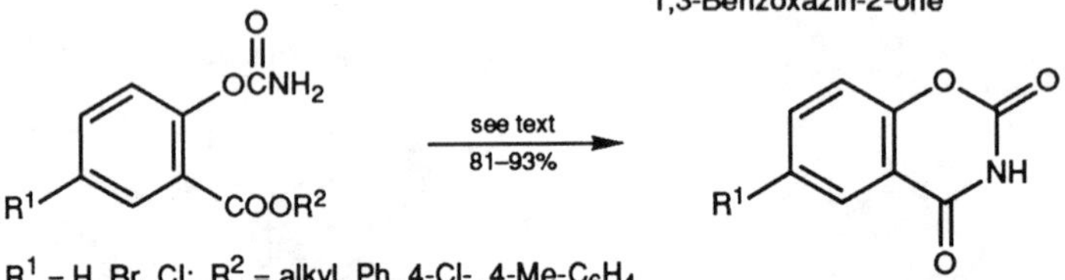

see text, 81–93% [3314]

R^1 = H, Br, Cl; R^2 = alkyl, Ph, 4-Cl-, 4-Me-C_6H_4

1,3-Benzoxazine-2,4-dione

III. FORMATION OF A SEVEN-MEMBERED RING

1. 1,2-Diazepin-3-one

When the t-butoxycarbonylhydrazone (**62.3**) is stirred in ethereal hydrogen chloride, a diazepine ring is formed; the butyl ester group is lost.

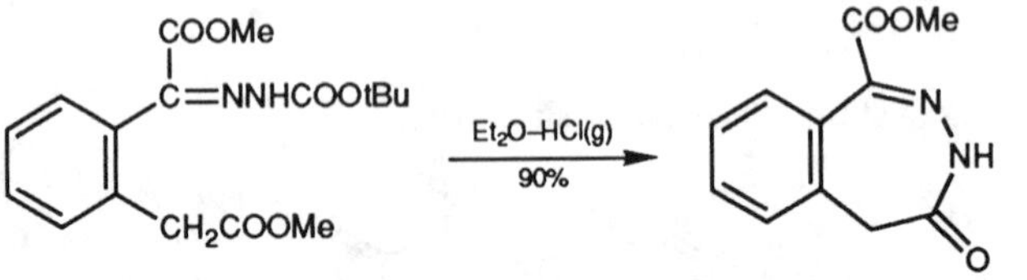

Et_2O–HCl(g), 90% [3010b]

(**62.3**) 2,3-Benzodiazepin-4-one

2. 1,4-Diazepine-2,7-dione

This ring is formed by reaction between a carboxamide and a carboxylic ester under strongly basic conditions.

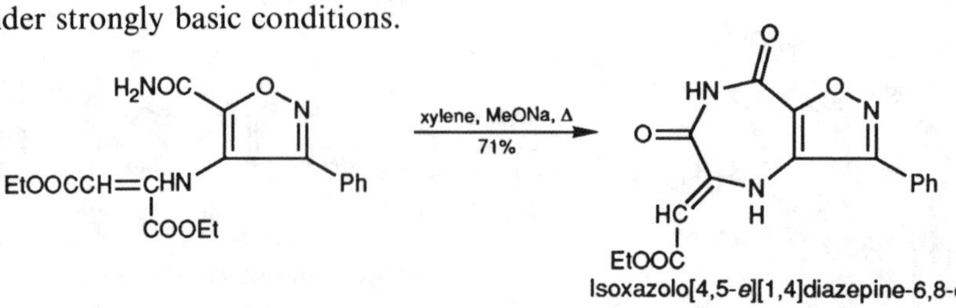

xylene, MeONa, Δ, 71% [3351]

Isoxazolo[4,5-*e*][1,4]diazepine-6,8-dione

CHAPTER 63

Carboxamide or Nitrile and Diazonium Salt

I. Formation of a Six-membered Ring 1045
1. 1,2,3-Triazine 1045
2. 1,2,3-Triazin-4-one 1046
3. 1,2,3-Thiadiazine 1046

The chemistry of heterocyclic diazonium salts has been reviewed [2980].

I. FORMATION OF A SIX-MEMBERED RING

1. 1,2,3-Triazine

Addition of a primary alkylamine to an *o*-cyanodiazonium salt at low temperature leads to an efficient cyclization. A similar ring is formed in high yield when the aminonitrile (**63.1**) is diazotized under standard conditions—a reaction which is not shown by simpler aminonitriles. The product is a useful intermediate.

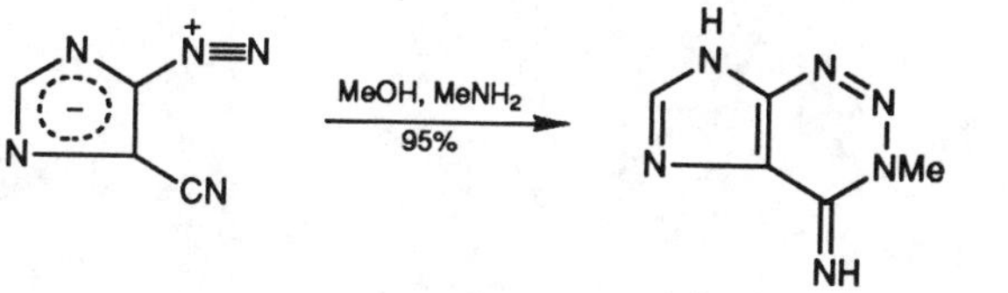

MeOH, $MeNH_2$, 95% [2807]

Imidazo[4,5-*d*]-1,2,3-triazine

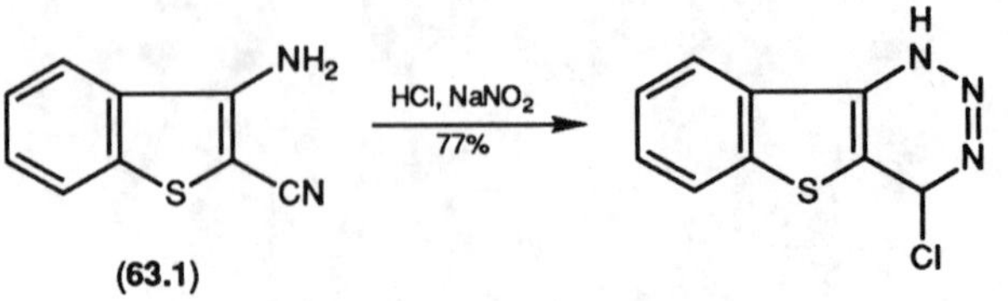

HCl, $NaNO_2$ 77% [2827]

(63.1)

[1]Benzothieno[3,2-*d*]-1,2,3-triazine

2. 1,2,3-Triazin-4-one

When a 2-aminobenzhydrazide in which both the terminal hydrogens have been replaced is diazotized, the hydrazide behaves like a carboxamide and a six-membered ring is formed. The same ring is formed by heating a 2-carboxamidoaryl diazonium salt with PPA (see Chapter 79, section II.4).

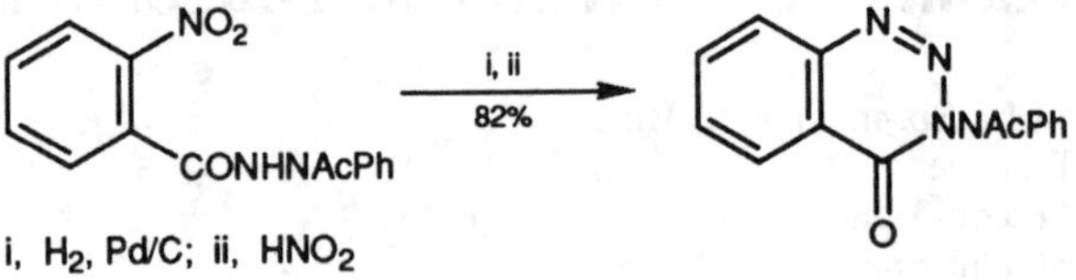

i, ii 82% [2990]

i, H_2, Pd/C; ii, HNO_2

1,2,3-Benzotriazin-4-one

3. 1,2,3-Thiadiazine

Neighbouring thioamide and diazonium groups react at or below room temperature to form a thiadiazine ring.

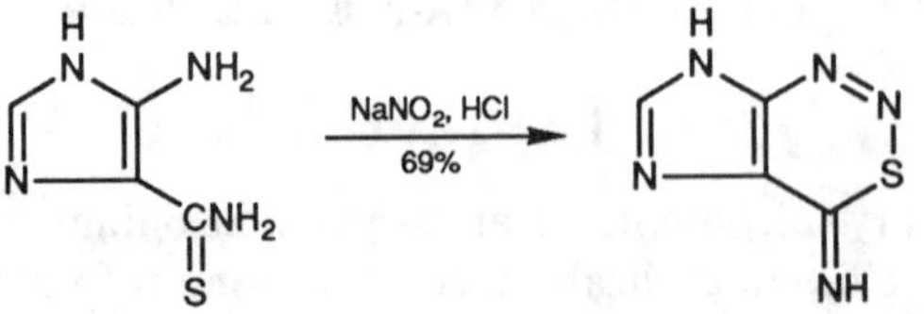

$NaNO_2$, HCl 69% [2719]

Imidazo[4,5-*d*][1,2,3]thiadiazine

CHAPTER 64

Carboxamide, Nitrile or Ureide and Hydroxy or Ether

I. Formation of a Five-membered Ring 1047
1. Pyrazole 1047
2. 1,2,4-Triazole 1048
3. Thiophene 1048
4. Oxazole 1049
5. Isothiazol-3-one 1049
6. 1,2-Oxathiole 2,2-Dioxide 1049
II. Formation of a Six-membered Ring 1049
1. Pyrimidine 1049
2. Pyran-2-one 1050
3. 1,3-Oxazin-4-one 1051
4. 1,2-Oxathiin-4-one 2,2-Dioxide 1052
III. Formation of a Seven-membered Ring 1052
1. 1,4-Diazepine-2,5-dione 1052
2. Oxepine 1053
3. 1,4-Oxazepin-5-one 1053

Examples of cyclizations of a ureido- or thioureido-thiol and of a hydroxy-sulphonamide are given in this chapter.

I. FORMATION OF A FIVE-MEMBERED RING

1. Pyrazole

An *o*-hydroxy-, *o*-methoxy- or *o*-methylthio-nitrile is cyclized in high yield by heating with hydrazine hydrate in an alcohol.

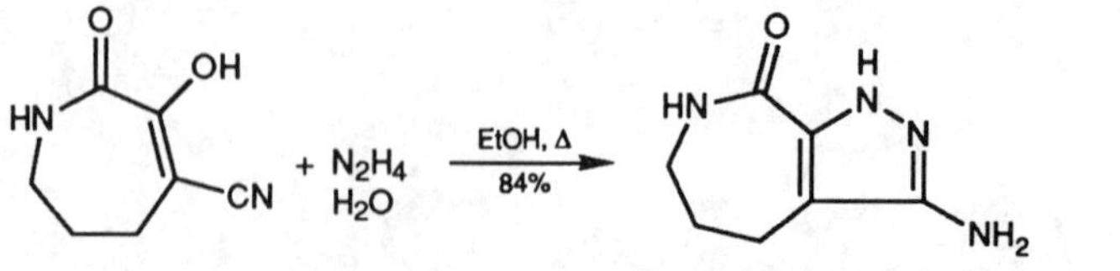

+ N_2H_4 H_2O → EtOH, Δ 84% [2948]

Pyrazolo[3,4-*c*]azepin-8-one

iPrOH, $N_2H_4.H_2O$, Δ 70–90% [2940]

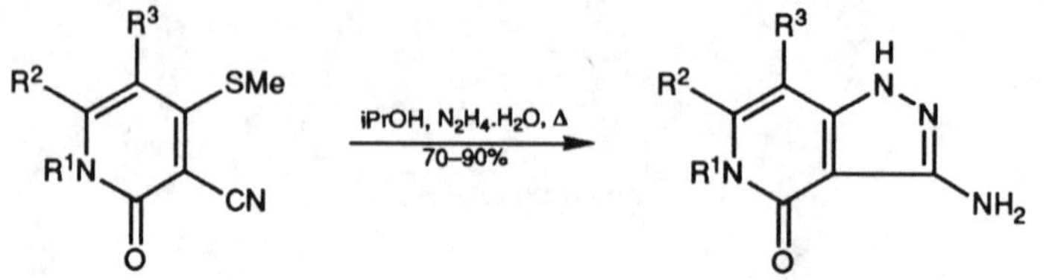

R^1 = H, Me; R^2 = Ph, 4-Me-, 4-Cl-, 4-MeO-, 4-Br-C_6H_4; R^3 = H, Ph

Pyrazolo[4,3-*c*]pyridin-4-one

Another ring system synthesized similarly:

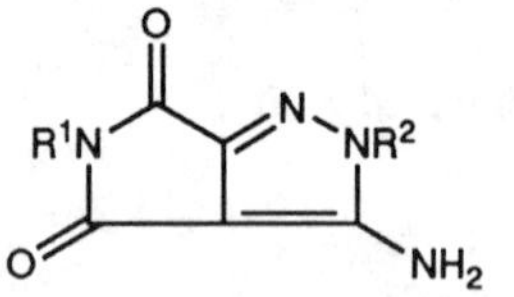

[3378]

Pyrrolo[3,4-*c*]pyrazole-4,6-dione

2. 1,2,4-Triazole

Thioether and ureido groups are joined together to form a triazole ring by heating the compound with iodomethane and then heating the mixture in ethanol.

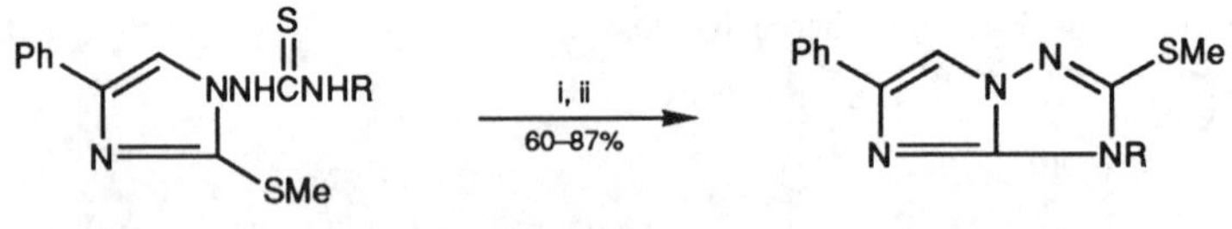

i, ii 60–87% [3616]

R = Me, $PhCH_2$, Ph, 4-Cl-, 4-MeO-, 4-Me-C_6H_4
i, EtOH, MeI, Δ; ii, EtOH, Δ

Imidazo[1,2-*b*][1,2,4]triazole

3. Thiophene

Bromonitromethane is a useful reagent for the cyclization of a mercaptobenzonitrile to a fused 3-aminothiophene.

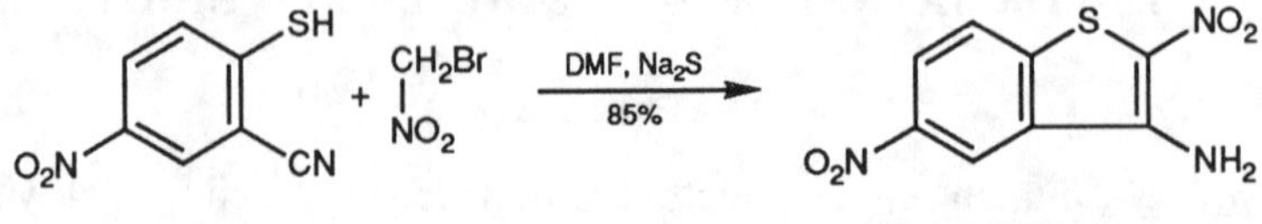

DMF, Na_2S 85% [3170]

Benzo[*b*]thiophene

4. Oxazole

Hydroxy and ester groups (the latter being formed *in situ* from a nitrile) react at ambient temperature with a nucleophile to form a fused oxazole ring.

OH CN NHCHCH=CHAr —MeOH–KOH, 62–85%→ CH$_2$CH$_2$Ar [2446]

Ar = 2-thienyl, Ph, 2-furyl

Benzoxazole

5. Isothiazol-3-one

Successive treatments of 2,2′-dithiobenzamides with bromine and activated basic alumina gives high yields of the isothiazolone.

[S, CNHR, O]$_2$ —1. Br$_2$ 2. Al$_2$O$_3$, 84–97%→ S NR O [3051]

1,2-Benzisothiazol-3-one

R = Et, tBu, cC$_6$H$_{11}$, PhCH$_2$, Ph, 4-Cl-, 4-MeO-, 4-Me-, 4-CN-C$_6$H$_4$

6. 1,2-Oxathiole 2,2-Dioxide

Loss of a t-amine group and cyclization occur in high yield when an *o*-(hydroxymethyl)-sulphonamide is mixed with mineral acid at room temperature.

SO$_2$NMe$_2$ CPh$_2$OH —H$_2$SO$_4$, MeOH, 90%→ O$_2$ S O Ph Ph [2036]

2,1-Benzoxathiole 1,1-dioxide

II. FORMATION OF A SIX-MEMBERED RING

1. Pyrimidine

It is sometimes advantageous to convert an amino group into its ethoxymethylidene derivative (by reaction with an orthoformate); this then cyclizes under mild conditions—in this example, at 0 °C. The ethoxymethylidene derivative does not always need to be isolated, and the two steps may be combined as a 'one-pot' reaction. The chemistry of the medicinally important pteridines has been reviewed [3454, 3594, 3669]. Hydrazinolysis of the ethoxymethylene-nitrile (**64.1**) without external heating is accompanied by cyclization in which only one of the hydrazine nitrogen atoms is part of the ring formed. Analogously, aminolysis gives an *N*-unsubstituted pyrimidine.

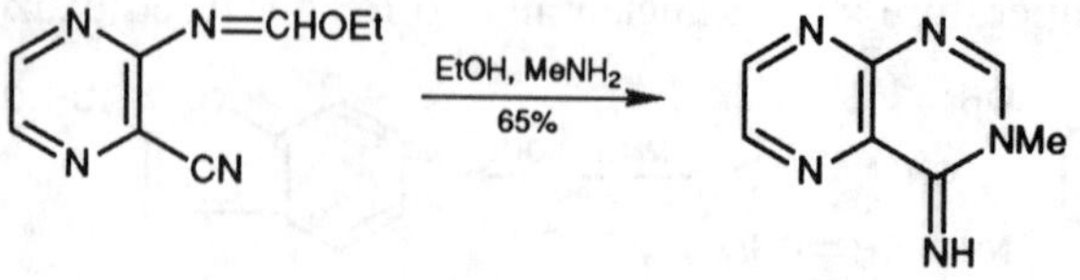

EtOH, $MeNH_2$, 65% [2271]

Pteridine

i, ii, Δ, 48–62% [2323]

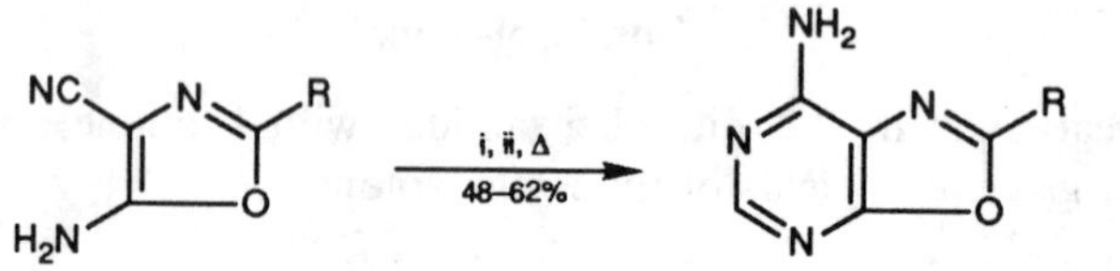

R = Me, Et, Ph
i, $HC(OEt)_3$–Ac_2O; ii, MeOH–NH_3

Oxazolo[5,4-*d*]pyrimidine

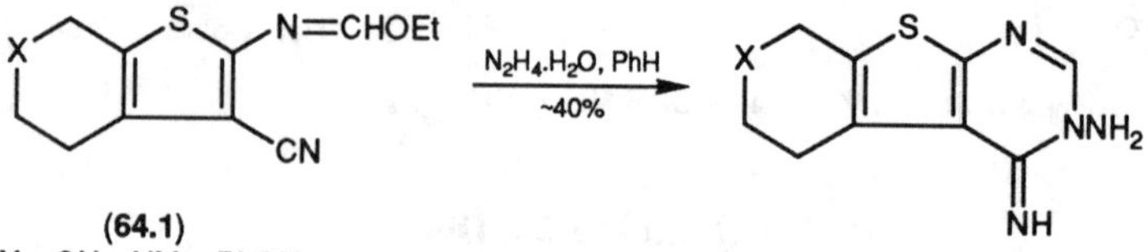

$N_2H_4 \cdot H_2O$, PhH, ~40% [2775]

(64.1)
X = CH_2, NMe, $PhCH_2$

Pyrido[4′,3′ : 4,5]thieno[2,3-*d*]pyrimidine
Benzothieno[2,3-*d*]pyrimidine

Another ring system synthesized similarly:

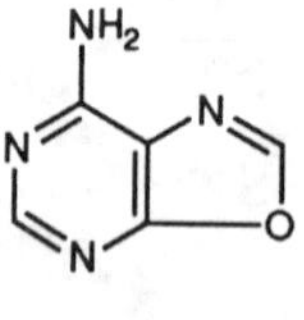

[2230]

Oxazolo[5,4-*d*]pyrimidine

2. Pyran-2-one

One of the byproducts of an attempt to isomerize a *cis*- to *trans*-carboxamide was a cyclized product containing a pyran-2-one ring. In a related cyclization, a phenolic t-carboxamide (the OH being generated *in situ*) in which a C—C double bond was hydrated, cyclized readily by boiling acetic acid in 6 h. The substrate in this reaction was prepared from the *o*-methoxymethoxybenzaldehyde and lithiodimethylacetamide. Several other benzocoumarins were synthesized similarly.

OMe OMe CH=C(Ph)CNMe$_2$ (C=O) —PhNO$_2$, I$_2$, Δ, ~25%→ [2080]

1-Benzopyran-2-one

OCH$_2$OMe, R, CHCH$_2$CNMe$_2$ (OH, O) —AcOH, Δ, 62–91%→ [3411]

R = H, Cl, Me, Ph

Other ring systems synthesized similarly:

[3411] [3411]

Naphtho[2,1-*e*]pyran-2-one Naphtho[1,2-*e*]pyran-3-one

3. 1,3-Oxazin-4-one

o-Hydroxybenzamides are usually cyclized to 4*H*-benzoxazin-4-ones by reaction with aldehydes or ketones in mineral acid, but better yields are often obtained by the use of a milder catalyst such as PPE which functions as catalyst and dehydrating agent [2389]. Another mild method is to use acetic anhydride–perchloric acid followed by TEA at ambient temperature to liberate the free base [2712]. When a cyanophenol is warmed with acetic anhydride and perchloric acid (a potentially dangerous procedure in view of the sensitivity of this acid to heat), the perchlorate salt is obtained in good-to-excellent yield [2808].

OH, CONH$_2$ + R^1CR^2 (=O) —CHCl$_3$, PPE, Δ, 38–92%→ [2389]

(64.2)

R^1 = H, Me; R^2 = alkyl, Ph or R^1R^2 = $(CH_2)_4$ or $(CH_2)_5$

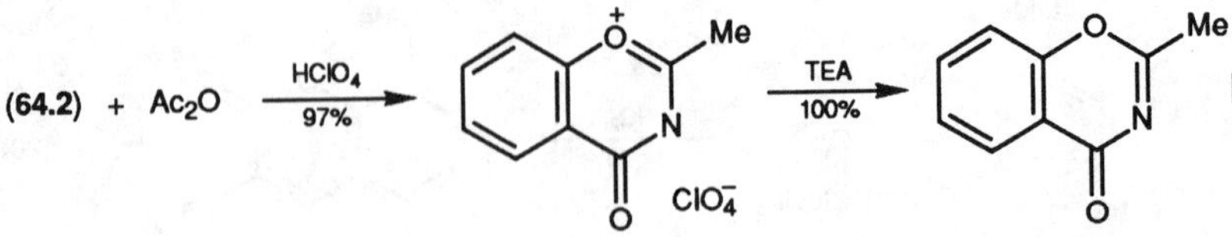

(64.2) + Ac_2O —($HClO_4$, 97%)→ [benzoxazinium ClO_4^-] —(TEA, 100%)→ [2712]

1,3-Benzoxazin-4-one

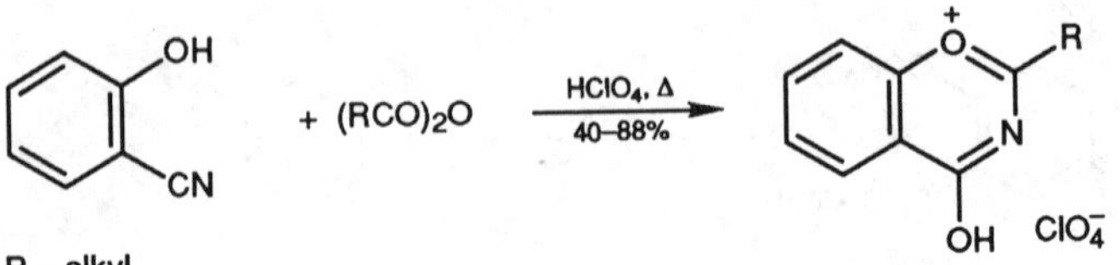

+ $(RCO)_2O$ —($HClO_4$, Δ, 40–88%)→ [2808]

R = alkyl

1,3-Benzoxazin-1-ium

4. 1,2-Oxathiin-4-one 2,2-Dioxide

Reaction between hydroxy and sulphonamide functions (cf. Section I.5) in an acidic medium causes ring closure in high yield without external heating.

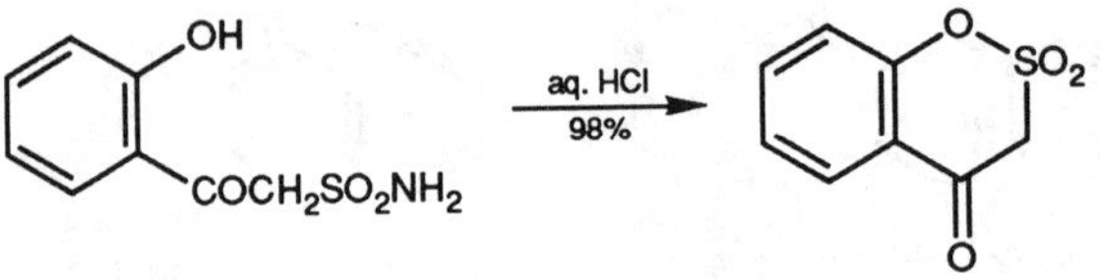

$COCH_2SO_2NH_2$ —(aq. HCl, 98%)→ [241]

1,2-Benzoxathiin-4-one 2,2-dioxide

III. FORMATION OF A SEVEN-MEMBERED RING

1. 1,4-Diazepine-2,5-dione

Base-induced opening of an epoxide ring promotes its reaction with a neighbouring carboxamide function.

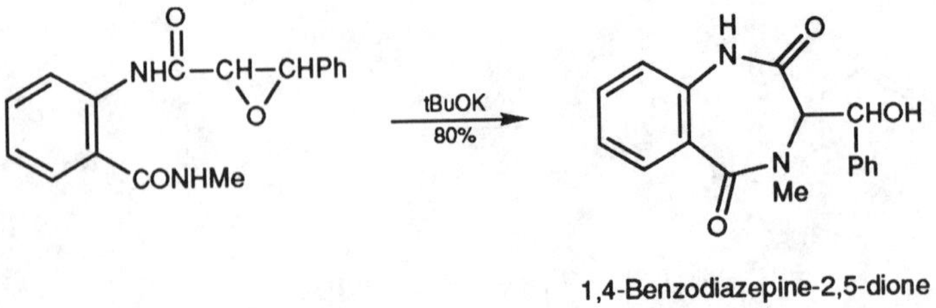

—(tBuOK, 80%)→ [2572]

1,4-Benzodiazepine-2,5-dione

2. Oxepine

This ring is formed by reaction between a side-chain nitrile and a phenolic group; the latter is produced *in situ* by demethylation of an ether with pyridine hydrochloride which also acts as a cyclizing agent and demethylates another methoxy group.

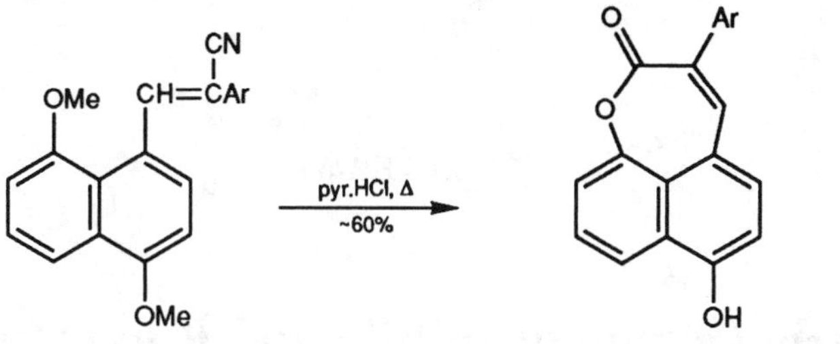

pyr.HCl, Δ
~60%

[2155]

Ar = Ph, 4-ClC_6H_4

Naphth[1,8-*bc*]oxepin-2-one

3. 1,4-Oxazepin-5-one

Salicylamide undergoes *O*-alkylation on heating with bromoacetaldehyde acetal; the product is then cyclized by heating.

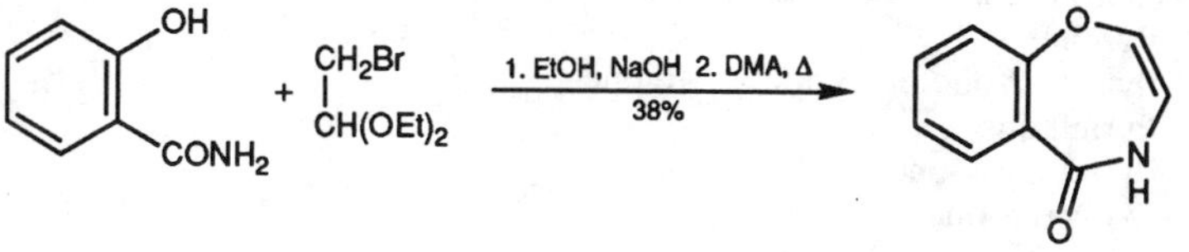

1. EtOH, NaOH 2. DMA, Δ
38%

[3896]

1,4-Benzoxazepin-5-one

CHAPTER 65

Carboxamide or Nitrile and Ring-carbon or Ring-nitrogen

I. Formation of a Five-membered Ring 1054
1. Pyrrole 1054
2. Pyrrol-2- or -3-one 1056
3. Thiophene 1057
4. Isothiazole 1,1-Dioxide 1057
II. Formation of a Six-membered Ring 1057
1. Pyridine 1057
2. Pyridin-2-one or Pyridine-2-thione 1059
3. Pyrimidine 1060
4. Pyrimidin-4-one 1060
5. Pyrazin-4-one 1061
6. Pyran-2-one 1061
7. Thiopyran 1061
8. 1,2-Oxathiin 2,2-Dioxide 1062
III. Formation of an Eight-membered Ring 1062
1. 1,5-Diazocine 1062

Cyclizations of sulphonamide, hydroxamic acid and thiocarboxamide functions with ring-carbon are included in this chapter.

I. FORMATION OF A FIVE-MEMBERED RING

1. Pyrrole

The ylide (**65.1**) is cyclized by reaction with benzyne followed by thermolysis in xylene [2735]. Another malononitrile derivative (**65.2**) cyclizes in a basic medium [2674]. The chemistry of pyrrolizine has been reviewed [3225].

(65.1) + benzyne →(i, ii; 42%) Pyridazino[6,1-*a*]isoindole [2735]

i, BuONO, CH_2Cl_2; ii, xylene, Δ

(65.2) →(EtOH, pip, Δ; 8–90%) Pyrrolizine [2674]

R = MeS, iPrNH, $(CH_2)_5N$, $(CH_2)_4N$, $PhNH_2$, 2-pyrrolyl

The regioselectivity of the addition of a C—N ylide (**65.3**) to an unsymmetrical alkyne (in a sealed system) has been studied. The substituents on both reactants have a strong effect on the course of the reaction and/or on the ratio of regioisomers formed. The equation shows examples of cyclizations which give one isomer [3471]. Reaction of an ylide (from a diazine) with DMAD results in the formation of a new pyrrole ring [2244]. The synthesis of pyrrolopyrimidines has been reviewed [1659, 2593]. DMAD also annulates a nitrile to a ring-carbon in a reaction; the nitrile group is lost in cyclization [3129]. Dibenzoylacetylene forms a ring at the *peri* positions of benzindolizine [3492].

(65.3) + $CR^3{\equiv}CCOOMe$ →(MeCN, Δ; 57–94%) Indolizine [3471]

R^1 = Ac, R^2 = H or R^1R^2 = $(CH{=}CH)_2$; R^3 = H, COOMe

+ $MeOOCC{\equiv}CCOOMe$ →(MeCN; 32–82%) [2244]

R = H, MeO

X = N; Y = Z = CH Pyrrolo[1,2-*a*]pyrimidine
X = Y = CH; Z = N Pyrrolo[1,2-*b*]pyridazine
X = Z = CH; Y = N Pyrrolo[1,2-*a*]pyrazine

+ $R^2COC{\equiv}CCOR^2$ →(PhMe or PhH, Δ; 54–82%) Indolizino[3,4,5-*ab*]isoindole [3129, 3492]

R^1 = H, Me, Ph, PhCO; R^2 = OMe, Ph

When an attempt was made to hydrolyse the nitrile group of **(65.4)**, ring closure on to the pyrazole carbon atom took place in high yield. The well-known reaction of nitriles with gaseous hydrogen chloride (review [2547]) may be used to convert the *N*-cyanoethylpyrrole **(65.5)** into a pyrrolizine.

O_2N, CN; AcOH, H_2SO_4, Δ; 84%; NH; O_2N [2482]

(65.4) Indolo[1,2-*b*]indazole

Br, Ac, NOH, + $AcOCH{=}CH_2$; Li_2PdCl_4; 55%; Br, Ac, N [3311]

Br; Br

Indole

NH_2, COOR

CN, NCHCHCOOR, Ph; Et_2O–HCl; 62–73%; N, Ph [2988, 3225]

Pyrrolizine

(65.5)
R = Me, Et, tBu

2. Pyrrol-2- or -3-one

Nitrenium ions (—N^+—) generated by treatment of *N*-chloro compounds with silver carbonate and trifluoroacetic acid at 0 °C, attack a benzene ring. In this way, good-to-excellent yields of 2-indolones are obtained. When R^2 = NO_2, the yield of indolone is poor, and when R^4 is not a hydrogen, different products are often produced [3550]. A side-chain hydroxamic acid is cyclized on to an adjacent ring-carbon atom on heating the compound with PPA [2239]. A pyrrol-3-one may be built from an *N*-β-cyanoethyl side-chain and zinc chloride [2867].

R^2, R^1, $CR^3CNHOMe$, R^4 O; i, ii; 64–96%; OMe, N, O, R^2, R^1, R^3, R^4 [3550]

2-Indolone

R^1 or R^2 = H, MeO, Cl, Br, AcNH, Me; R^3R^4 = H, Me
i, CH_2Cl_2, tBuOCl; ii, TFA, Ag_2CO_3

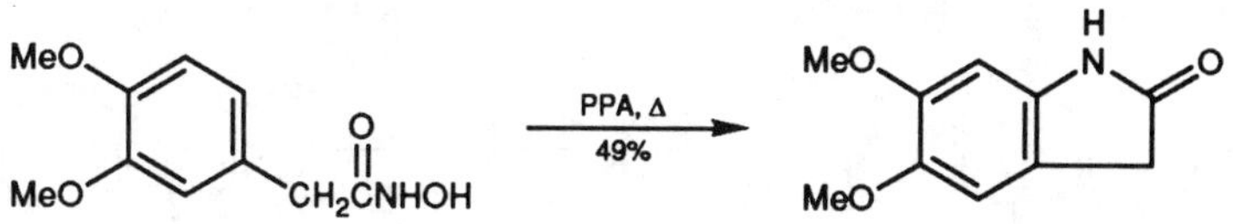

[2239]

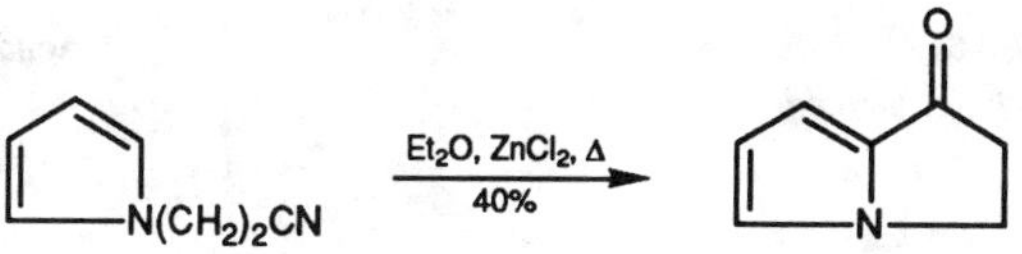

1-Pyrrolizinone

[2867]

3. Thiophene

A malonylidene mononitrile attached to a ring forms a fused thiophene when heated with sulphur and an amine; cyclizations of this type have been reviewed [2977].

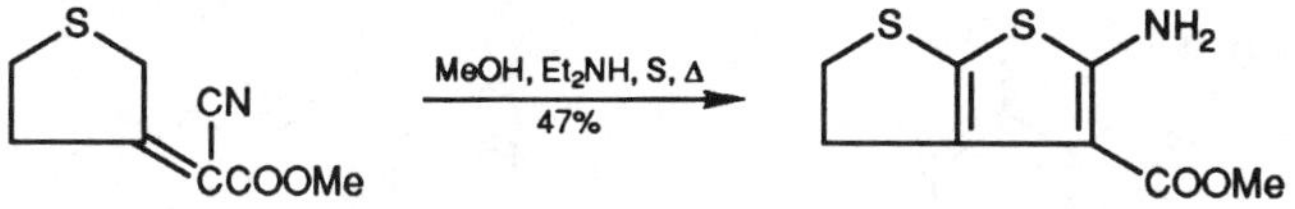

MeOH, Et_2NH, S, Δ 47%

Thieno[2,3-*b*]thiophene

[2555]

4. Isothiazole 1,1-Dioxide

o-Lithiated *N,N*-diphenylbenzenesulphonamides are cyclized by reaction with a nitrile at low temperature with elimination of diphenylamine from an intermediate imine.

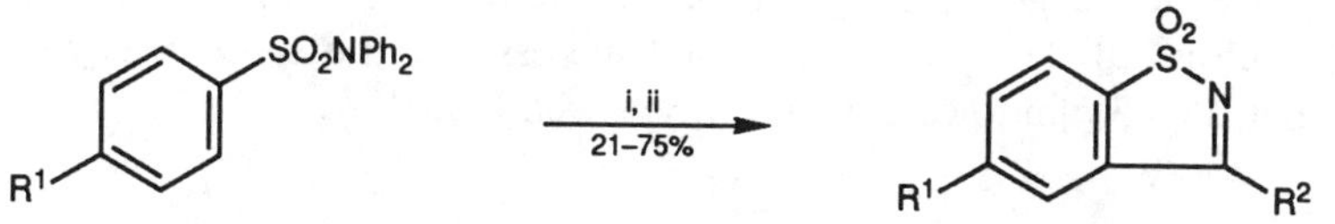

[3604]

R^1 = H, Me, Cl; R^2 = Me_2N, Ph, 4-Me-, 4-Cl-C_6H_4
i, BuLi, THF; ii, R^2CN, THF

1,2-Benzisothiazole 1,1-dioxide

II. FORMATION OF A SIX-MEMBERED RING

1. Pyridine

Interaction between a cyano group and a carbon atom of a cyclic enaminone is facilitated by transition metals. Thus, in the absence of copper(I) chloride, potassium carbonate does not convert the cyano-enamine (**65.6**) into a reduced acridinone. Reduced acridines may also be synthesized by cyclization of *N*-arylbenzamides with phosphorus oxychloride.

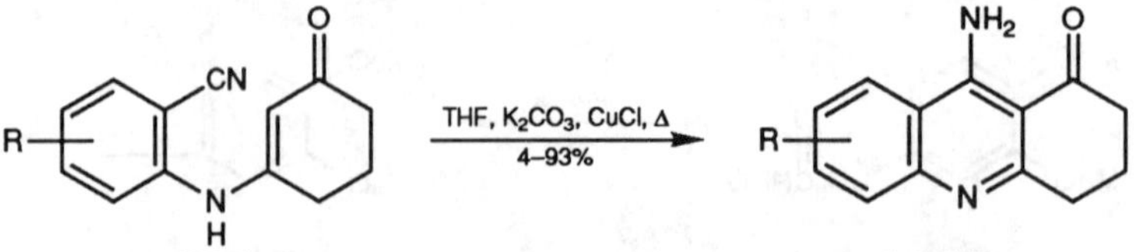

(65.6)
R = H, Cl, F, CF_3, MeO

THF, K_2CO_3, CuCl, Δ
4–93%

1-Acridinone [3524]

$POCl_3$, Δ
47–81%

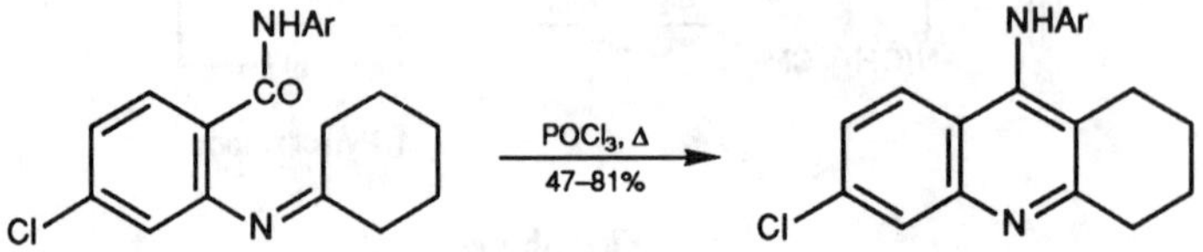

Ar = Ph, Me-, Cl-, Br-, MeO-C_6H_4

Acridine [2537]

Other ring systems synthesized similarly:

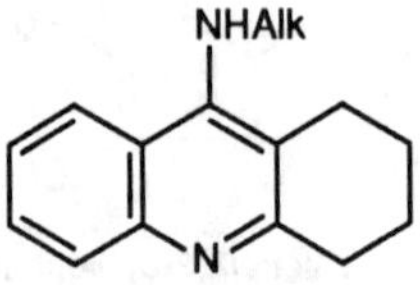

Cyclopenta[*b*]quinoline [2538]

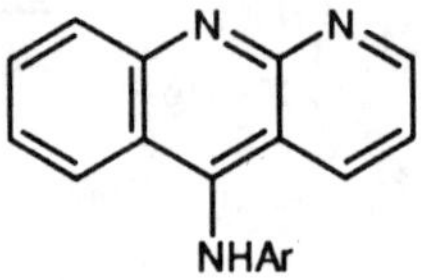

Benzo[*b*][1,8]naphthyridine [2638]

Cyanomethyl groups can be a convenient source of pyridine rings: treatment with acetic anhydride–perchloric acid [3934], acetonitrile–aluminium chloride [2884], Vilsmeier reagent [2727], a 1,3-diketone (or its enol) [2714] and malonaldehyde bisacetal [2714] have been used successfully and sometimes give high yields. Pyrido[4,3-*b*]indoles have been reviewed [2535].

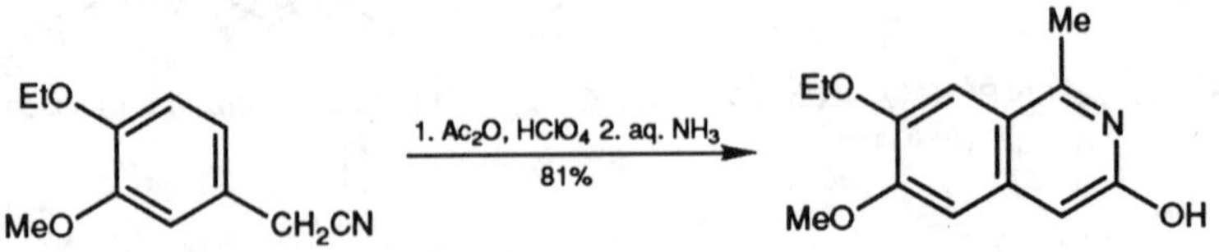

1. Ac_2O, $HClO_4$ 2. aq. NH_3
81%

Isoquinoline [3934]

MeCN, $AlCl_3$, Δ
21%

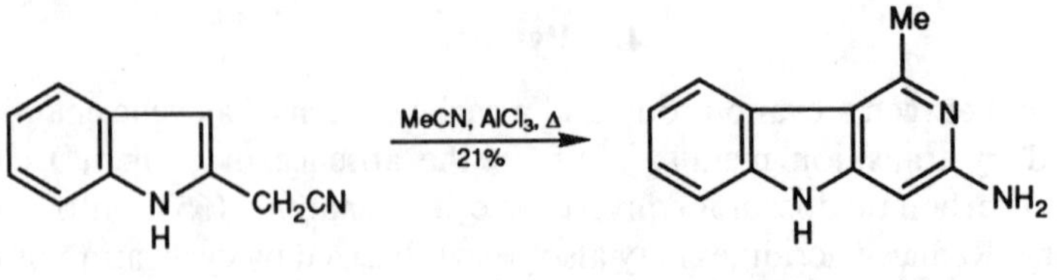

Pyrido[4,3-*b*]indole [2884]

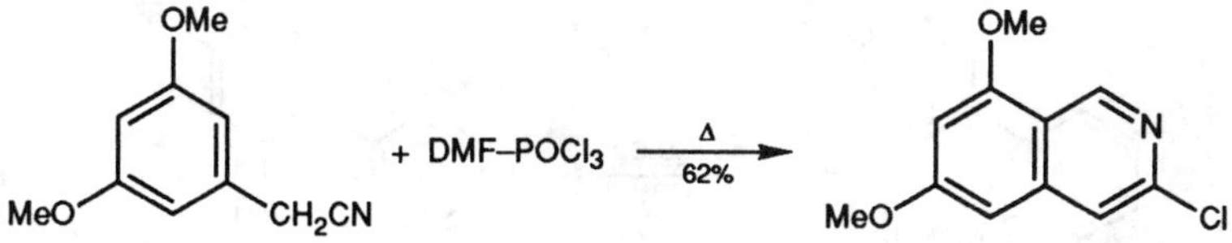

+ DMF–$POCl_3$ $\xrightarrow[62\%]{\Delta}$ [2727]

Isoquinoline

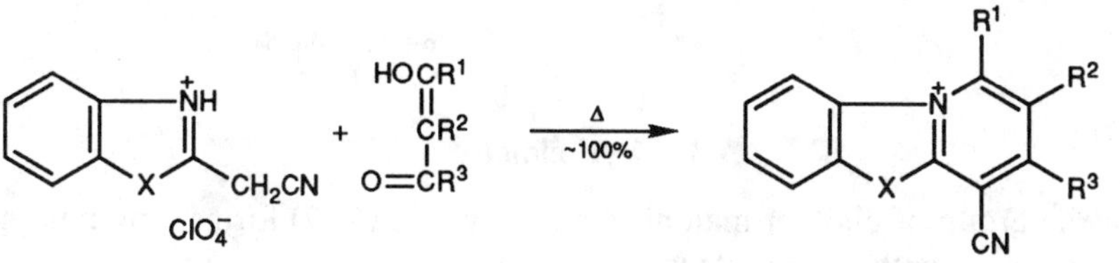

$\xrightarrow[\sim 100\%]{\Delta}$ [2714]

R^1 = H, Me, Ph; R^2 = H, Me, Ph, COOMe;
R^3 = H, Me; X = MeN, O, S

X = NMe Pyrido[1,2-*a*]benzimidazol-10-ium
X = O Pyrido[3,2-*a*]benzoxazol-10-ium
X = S Pyrido[3,2-*a*]benzothiazol-10-ium

Another ring system synthesized similarly

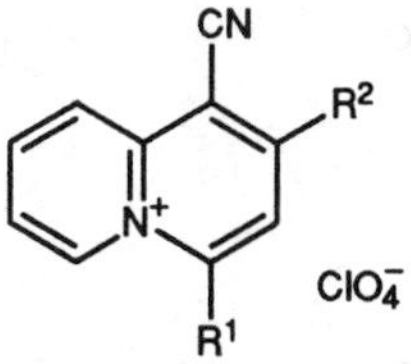

[2714]

Quinolizinium

2. Pyridin-2-one or Pyridine-2-thione

A carboxamide or thiocarboxamide is annulated to the neighbouring ring-carbon atom by heating with a malonic acid derivative and ammonium acetate. Oxidative cyclization of a biphenyl-2-carboxamide with peroxodisulphate (through a radical intermediate) leads to variable yields of a phenanthridinone. The type of substituents on the nitrogen and on the non-carboxamide ring plays an important role in determining the yield, the highest yield being from the *N*-phenylamide (R^2 = H) and the lowest from the substrate where R^1 = H, R^2 = MeO [2412].

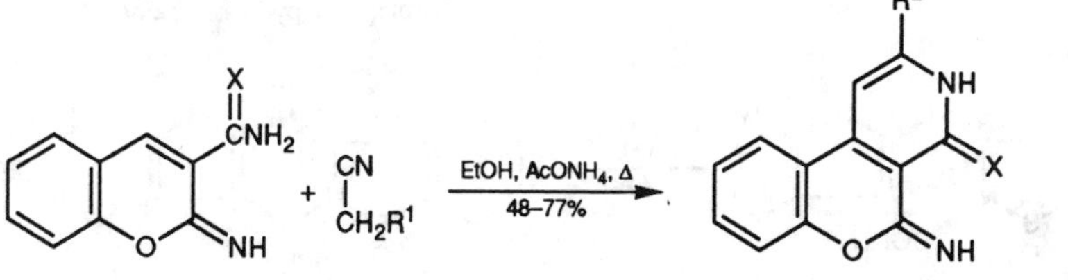

EtOH, $AcONH_4$, Δ; 48–77% [3907]

R^1 = CN, COOEt, PhCO;
R^2 = NH_2, OH, Ph; X = O, S

[1]Benzopyrano[3,4-*c*]pyridin-2-one
[1]Benzopyrano[3,4-*c*]pyridine-2-thione

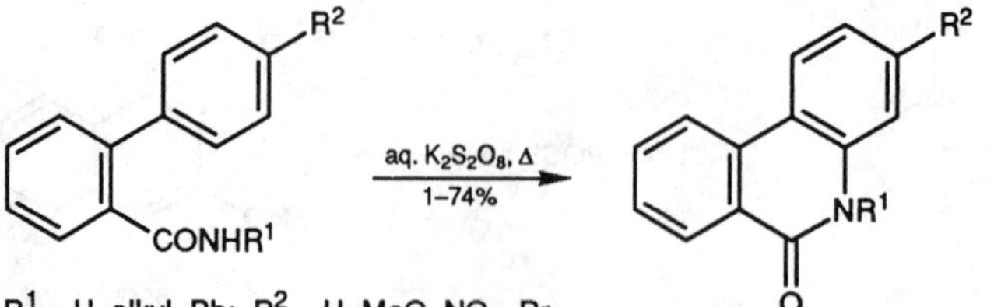

R^1 = H, alkyl, Ph; R^2 = H, MeO, NO_2, Br

aq. $K_2S_2O_8$, Δ; 1–74% [2412]

6-Phenanthridinone

3. Pyrimidine

The nitrile group of another malonic acid derivative (**65.7**) forms a pyrimidine ring by reaction with a ring-nitrogen.

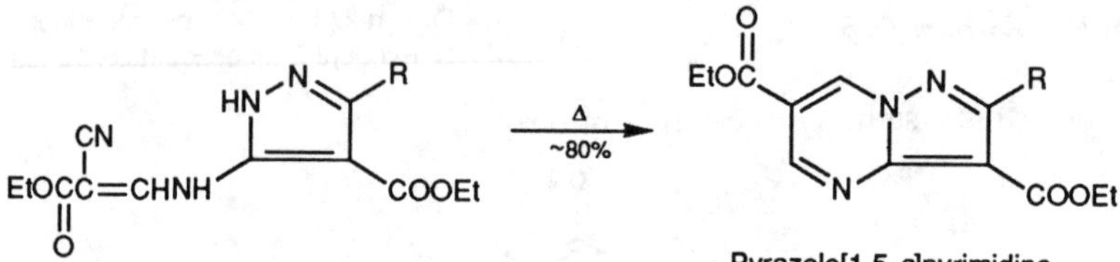

Δ; ~80% [2670]

(**65.7**)
R = H, Me

Pyrazolo[1,5-*a*]pyrimidine

4. Pyrimidin-4-one

The reaction of a side-chain carboxamide with DMFDEA joins the NH_2 to a neighbouring ring-NH and a fused pyrimidinone ring is formed. The same type of reaction may operate in acetic acid when carboxamide and ring-nitrogen are in different rings.

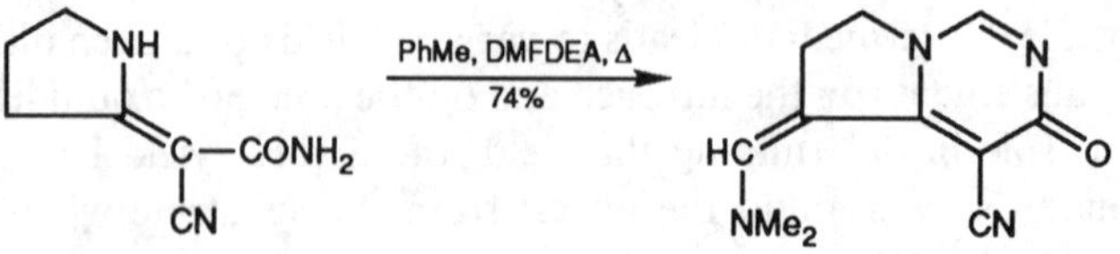

PhMe, DMFDEA, Δ; 74% [2944]

Pyrrolo[1,2-*c*]pyrimidin-3-one

AcOH, Δ; ~40% [2659]

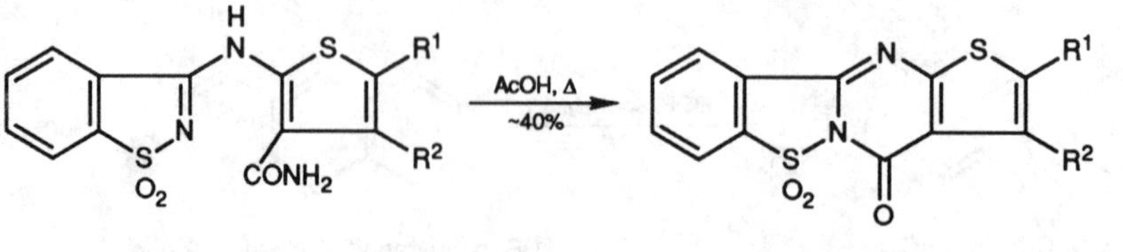

Thieno[2',3' : 4,5]pyrimido[1,2-*b*][1,2]benzisothiazol-4-one 6,6-dioxide

5. Pyrazin-4-one

A carboxamido group may be annulated to a adjoining ring-nitrogen by heating the compound with phenacyl bromide.

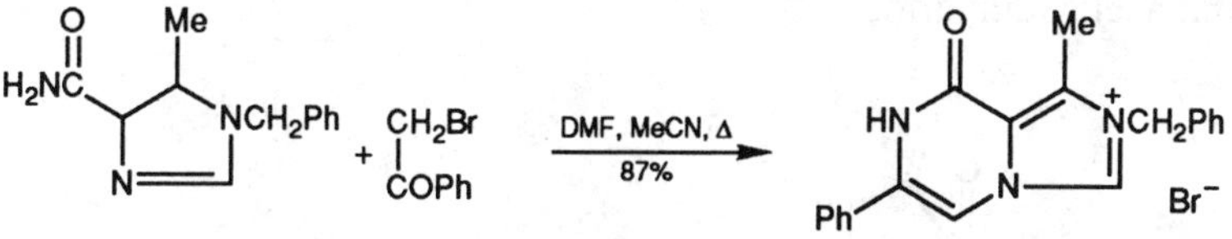

DMF, MeCN, Δ 87% [3288]

Imidazo[1,5-*a*]pyrazin-8-one

Another ring system synthesized similarly:

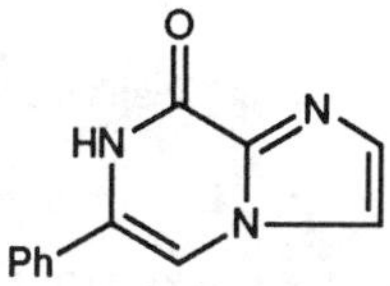

[3288]

Imidazo[1,2-*a*]pyrazin-8-one

6. Pyran-4-one

In the presence of PPA, a cyano group can be partially hydrolysed and a bond is then formed to a nearby ring-carbon atom.

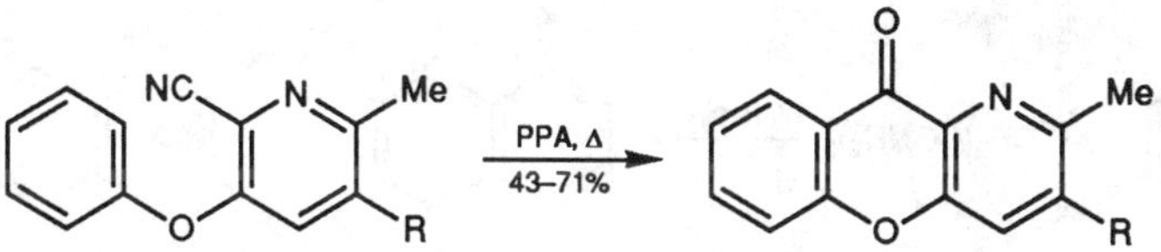

PPA, Δ 43–71% [3032]

R = H, Br, Cl, Me

[1]Benzopyrano[3,2-*b*]pyridin-10-one

7. Thiopyran

A Diels–Alder reaction occurs between the thioxo function of a thiocarboxamide and a neighbouring ring-carbon (as a heterodiene) and the triple bond of DMAD to give a high yield of a fused thiopyran. The usefulness of heterodienes in this type of reaction has been reviewed [2455, 3067a, 3880].

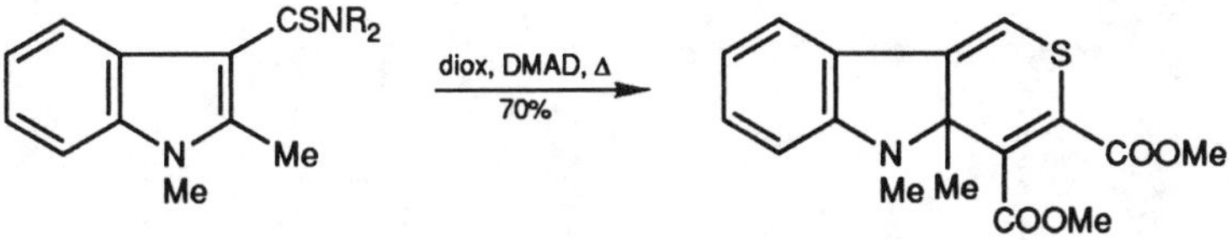

diox, DMAD, Δ 70% [2522]

NR_2 = piperidino, morpholino

Thiopyrano[4,3-*b*]indole

8. 1,2-Oxathiin 2,2-Dioxide

One carbonyl of an unsymmetrical imide is annulated to a ring-carbon atom when the compound is treated at room temperature with a mixture of sulphuric acid and acetic anhydride.

H_2SO_4, Ac_2O 58% [2570]

Naphth[2,1-*c*][1,2]oxathiin 4,4-dioxide

II. FORMATION OF AN EIGHT-MEMBERED RING

1. 1,5-Diazocine

An unexpected product is formed when a chromone-3-nitrile is heated with malonamide and piperidine. 1-Benzopyrans to which a heterocycle is fused at the 2,3-positions have been reviewed [3343, 3346].

+ $CH_2(CONH_2)_2$ EtOH, pip, Δ 80% [3092]

Bis[1]benzopyrano[2,3-*b* : 2',3'-*f*][1,5]diazocin-8,16-dione

CHAPTER 66

Carboxylic Acid or its Derivative and Halogen

I. Formation of a Five-membered Ring 1063
1. Pyrrole 1063
2. Pyrrol-2-one 1064
3. Pyrazole 1065
4. Furan 1065
5. Thiophene 1066
6. Isoxazole or Isoxazol-5-one 1066
7. Thiazole 1067
II. Formation of a Six-membered Ring 1067
1. Pyridine 1067
2. Pyridin-2-one or -4-one 1067
3. Pyridazine 1068
4. Pyrimidine 1068
5. Pyrimidin-4-one 1068
6. Pyrimidine-2,4-dione 1070
7. Pyran-2-one 1070
8. 1,3-Thiazin-4-one or 1,3-Thiazine-2,4-dione 1070
III. Formation of a Seven-membered Ring 1071
1. 1,3-Diazepin-2-one, 3-Thioxo- or 1,3-Thiazepin-4-one 1071
2. 1,4-Diazepine-2,5-dione 1071
IV. Formation of an Eight-membered Ring 1072
1. 1,5-Diazocine-2,8-dione 1072

I. FORMATION OF A FIVE-MEMBERED RING

1. Pyrrole

A primary amine reacts with an *o*-chloro(malonylidenenitrile) function to give a new fused pyrrole ring; when one nitrile group is replaced by ethoxycarbonyl, it is the CN group that reacts with the amine. Suitably positioned bromine and carboxylate ester functions react with a primary amine to form a pyrrol-2-one ring (p. 236), but a malonylidene ester is reported to lead to a pyrrole ring.

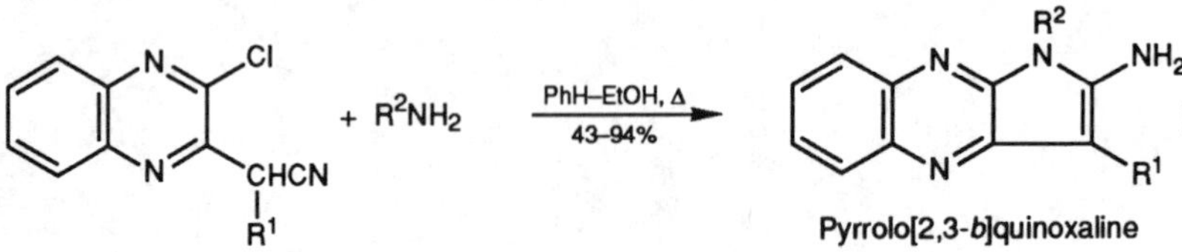

+ R^2NH_2 —PhH–EtOH, Δ; 43–94%→ [2217]

Pyrrolo[2,3-*b*]quinoxaline

R^1 = CN, COOEt; R^2 = Me, $HO(CH_2)_2$, $PhCH_2$, 4-$EtOC_6H_4$

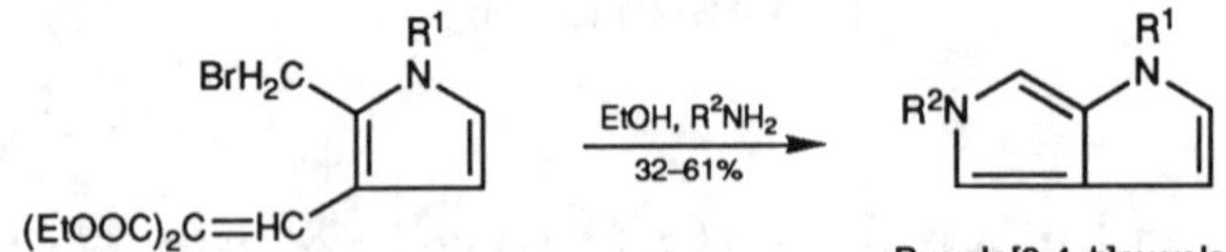

—EtOH, R^2NH_2; 32–61%→ [3792]

Pyrrolo[3,4-*b*]pyrrole

R^1 = 4-$MeOC_6H_4SO_2$; R^2 = H, Me, $PhCH_2$, Ph

Another ring system synthesized similarly:

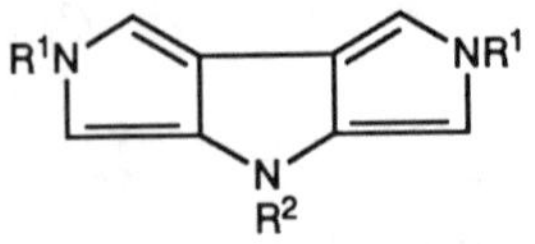

[3792]

Dipyrrolo[3,4-*b* : 3′,4′-*d*]pyrrole

2. Pyrrol-2-one

Thiourea behaves as a monofunctional compound in its reaction with *o*-chloromethylbenzoyl chloride but *N,N′*-thioureas give rise to seven-membered rings (see Section III.1). A fused pyrrol-2-one ring may also be synthesized by treating *o*-*N*-acylbenzamides with sodium hydride and then with butyl-lithium at −78 °C; the lithio-oxy derivative formed is then quenched with hydrochloric acid after standing for 2 h.

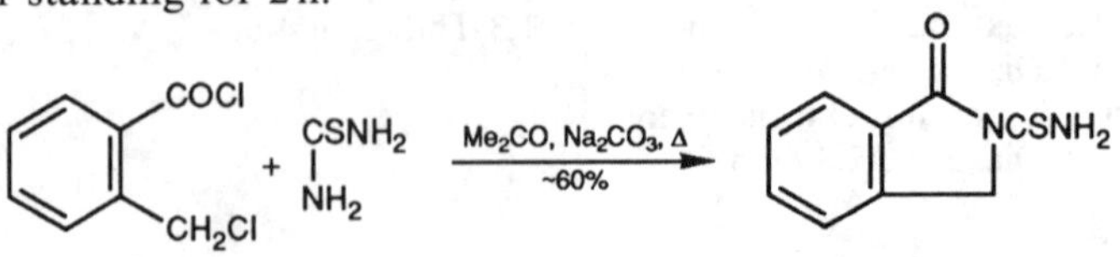

—Me_2CO, Na_2CO_3, Δ; ~60%→ [2731]

1-Isoindolone

—1. NaH, THF, BuLi 2. HCl; 43–93%→ [3623]

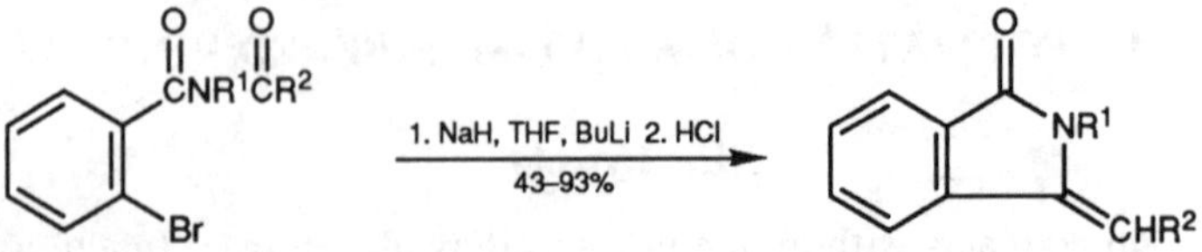

R^1 = H, Me; R^2 = H, Me, Ph

3. Pyrazole

o-Chloro-carbohydrazide (in which the halogen is reactive) is cyclized by heating with copper powder in pyridine [2275]. Chloronitriles are cyclized by heating with hydrazine (p. 338); phenylhydrazine also reacts on heating in dioxane [3569] or on stirring in dichloromethane for 4 h. This latter study showed that it was possible to isolate two isomers from cyclizations with some arylhydrazines whereas others (for example, 4-methyl- or 4-methoxy-phenylhydrazine) produced only the 1-arylpyrazolo-1,3-oxazine. An electron-withdrawing substituent on the phenyl ring may inhibit this cyclization [4003].

pyr, Cu, Δ, 70% [2275]

Ar = 3-$CF_3C_6H_4$

Pyrazolo[3,4-*b*]pyridine

+ $PhNHNH_2$ diox, Δ, ~73% [3569]

R = $PhCH_2NH_2$, PhNH

Pyrazolo[3,4-*d*]pyrimidine

+ $ArNHNH_2$ CH_2Cl_2, 0–52%; CH_2Cl_2, 28–93% [4003]

Ar = Ph, 4-Cl-, 4-MeO-, 4-Me-C_6H_4

(66.1)
Pyrazolo[3,4-*d*][1,3]oxazin-4-one

(66.2)

4. Furan

Neighbouring halogen and ethoxycarbonyl groups are annulated by heating with the sodium salt of ethyl glycolate, but when the latter was replaced by ethyl lactate, prolonged heating was required and decarboxylation also occurred.

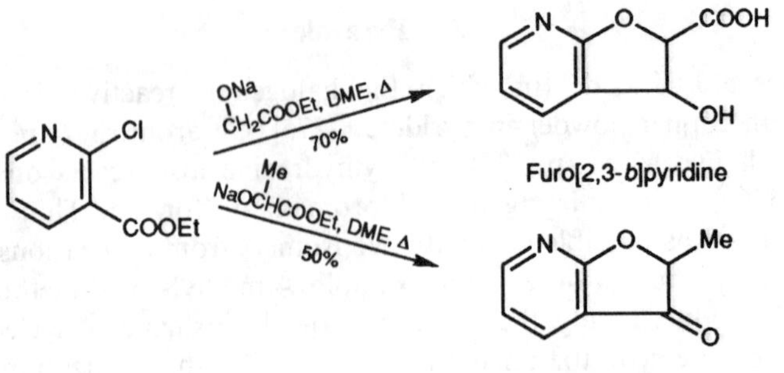

ONa
CH_2COOEt, DME, Δ
70%

Me
NaOCHCOOEt, DME, Δ
50%

Furo[2,3-*b*]pyridine

Furo[2,3-*b*]pyridin-3-one

[3102]

5. Thiophene

A nuclear chlorine and a side-chain nitrile interact when the compound is treated with sodium methoxide and hydrogen sulphide.

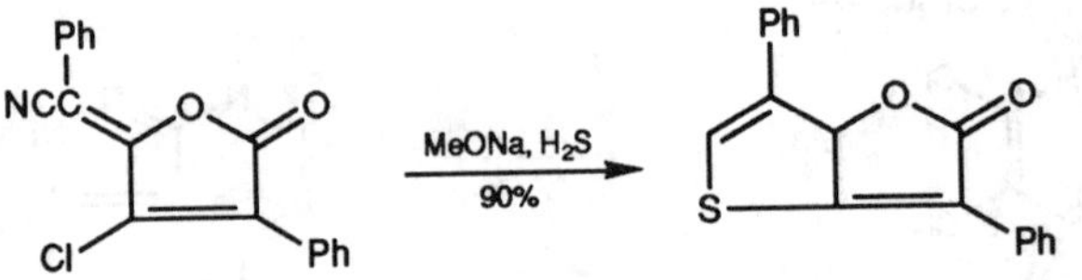

MeONa, H_2S
90%

(66.3)

Thieno[3,2-*b*]furan-2-one

[2619]

6. Isoxazole or Isoxazol-5-one

Advantage is taken of the enhanced reactivity of halogeno and ester substituents on quinones or on a pyridine ring in the synthesis of benzisoxazolediones or isoxazolopyridinones, respectively (see also Chapter 52).

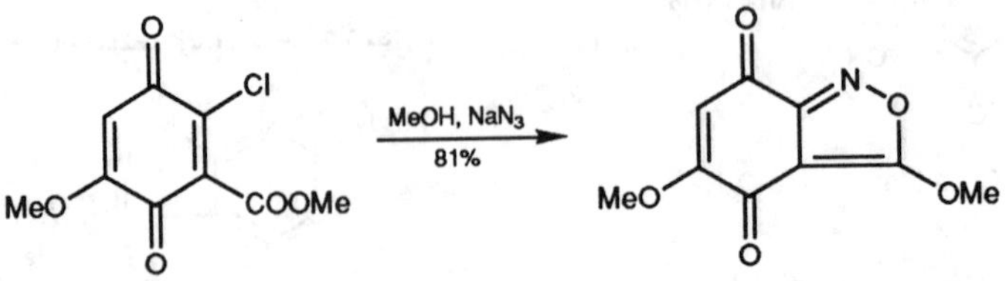

MeOH, NaN_3
81%

2,1-Benzisoxazole-4,7-dione

[2590]

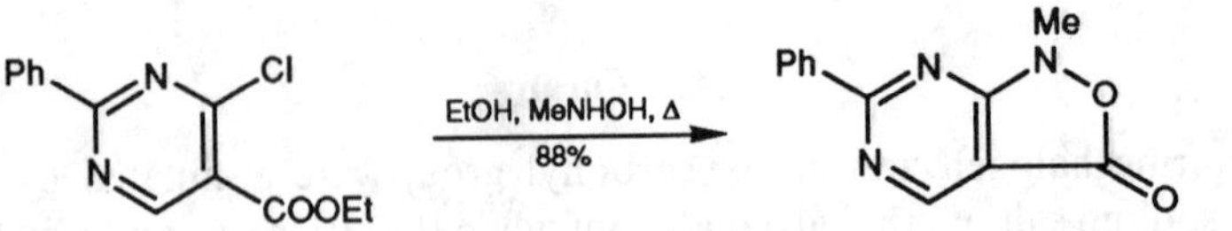

EtOH, MeNHOH, Δ
88%

Isoxazolo[3,4-*d*]pyrimidin-3-one

[2045]

7. Thiazole

Passing a stream of hydrogen sulphide through an isopropanol solution of the 3-chloro-4-cyanoaminopyrroledione (**66.4**) containing pyridine converts it into a pyrrolothiazoledione.

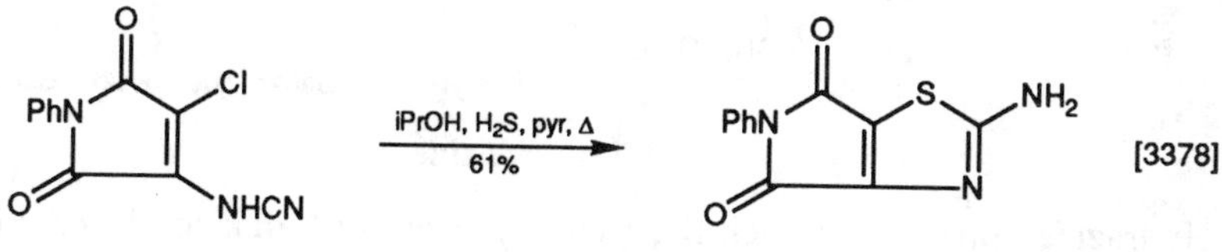

iPrOH, H_2S, pyr, Δ; 61% [3378]

(66.4)

Pyrrolo[3,4-*d*]thiazole-4,6-dione

II. FORMATION OF A SIX-MEMBERED RING

1. Pyridine

3-Cyanovinyl-2-chloroquinolines react with an excess of a cyclic secondary amine to form a 1,8-naphthyridine in good-to-excellent yields.

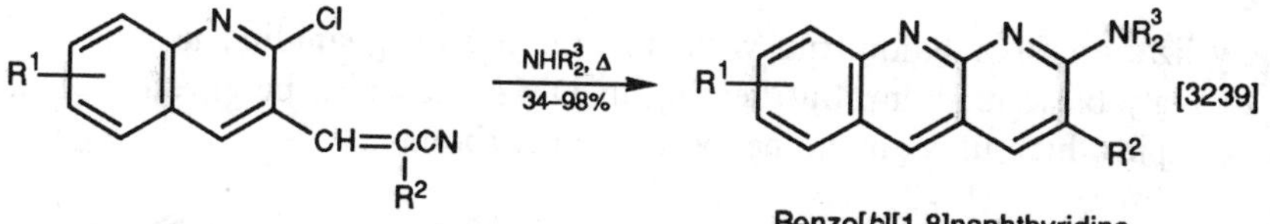

NHR^3_2, Δ; 34–98% [3239]

Benzo[*b*][1,8]naphthyridine

R^1 = H, 6,7-$(MeO)_2$; R^2 = 4-ClC_6H_4, 4-ClC_6H_4S;
NR^3_2 = piperidino, morpholino, 4-Me piperazino

2. Pyridin-2-one or -4-one

2-Bromobenzamides couple with arylboronic acids; thermal deprotection of the amino group is followed by the formation of a fused pyridin-2-one in good yield. The 4-one ring may be constructed in low yield by stirring at a low temperature (−5 °C) a 2-fluorobenzoyl chloride with the lithium derivative of ethyl pyridine-2-acetate.

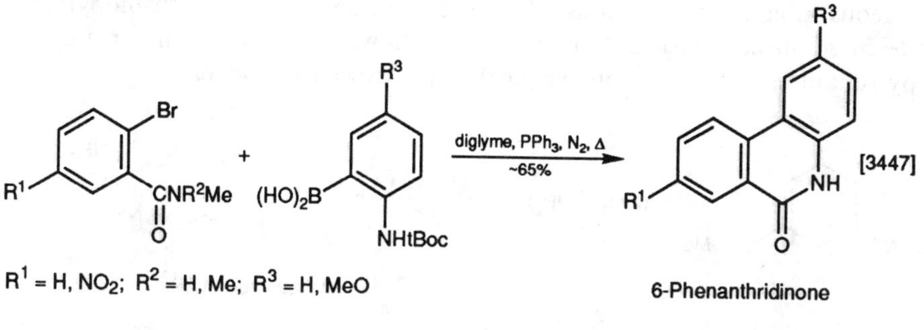

diglyme, PPh_3, N_2, Δ; ~65% [3447]

R^1 = H, NO_2; R^2 = H, Me; R^3 = H, MeO

6-Phenanthridinone

THF, LiN(SiMe$_3$) 25% [3679]

Benzo[*c*]quinolizin-6-one

3. Pyridazine

Hydrazine appears to attack the α-carbon atom of 2-diazocarboxylic esters to give a new fused pyridazine with simultaneous elimination of hydrazoic acid.

EtOH, $N_2H_4.H_2O$ 97% [2605]

Pyrimido[4,5-*c*]pyridazine

4. Pyrimidine

Cyclization of a halonitrile with an amidine or guanidine usually needs a strongly basic medium, but heating in DMA may also be effective as in this example which uses guanidine–potassium carbonate.

DMA, Δ 47–92% [3483]

R = H, Br, Cl, I, MeO, MeS, Me_2N or F_4

Quinazoline

5. Pyrimidin-4-one

A carboxylic acid derivative which also contains an activated neighbouring halogen atom may be cyclized by warming with an amidine (or guanidine) or 2-aminopyridine (a cyclic amidine). In some reactions of this type, it is advantageous to add a base such as TEA. The chlorine atom of the thienopyridine (**66.5**) is unusually resistant to nucleophiles; however, guanidine in *N*-methyl-2-pyrrolidinone gave a reasonable yield of the expected product.

EtOH, AcONa, Δ 86% [2031]

4-Quinazolinone

Other ring systems synthesized similarly:

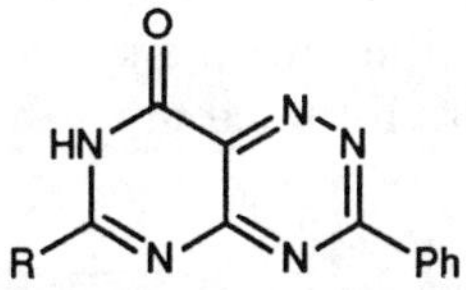

[2438]

Pyrimido[4,5-*e*]-1,2,4-triazin-8-one

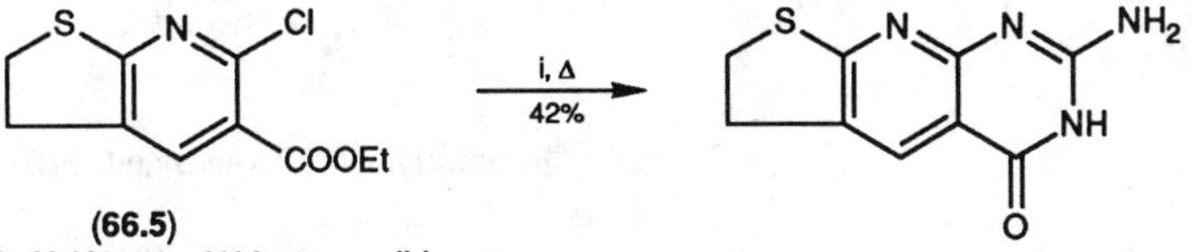

(66.5)

i, $H_2NC(=NH)NH_2$, *N*-Me-2-pyrrolidone

i, Δ 42%

Thieno[3',2' : 5,6]pyrido[2,3-*d*]pyrimidin-4-one

[3490]

Another ring system synthesized similarly:

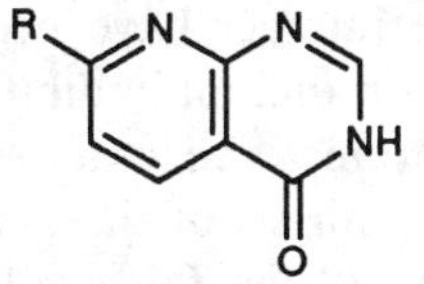

[3551]

Pyrido[2,3-*d*]pyrimidin-4-one

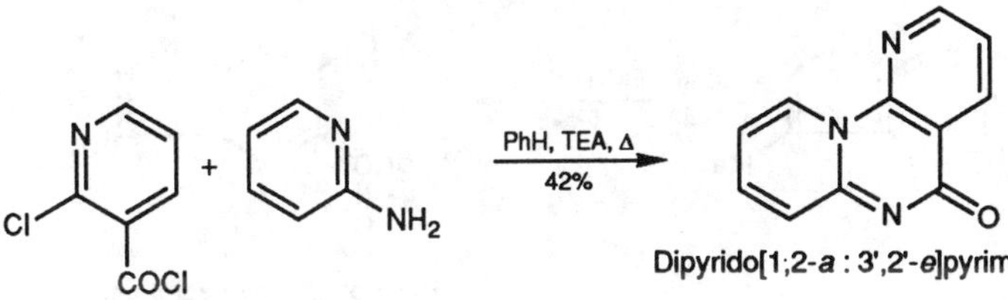

PhH, TEA, Δ 42%

Dipyrido[1;2-*a* : 3',2'-*e*]pyrimidin-5-one

[3130]

Other ring systems synthesized similarly:

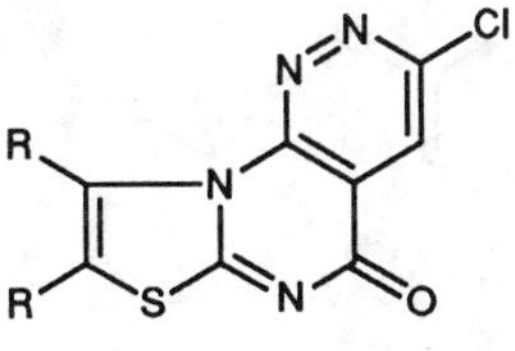

[3390]

R,R = H,H Thiazolo[2',3' : 2,3]pyrimido[4,5-*c*]pyridazin-5-one
R,R = benzo Pyridazino[4',3' : 5,6]pyrimido[2,1-*b*]benzothiazol-5-one

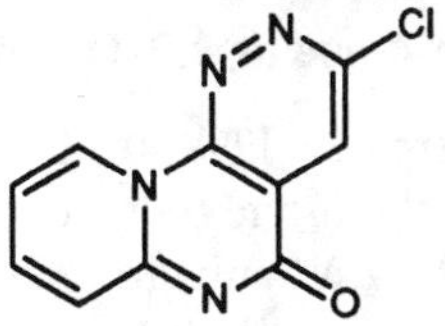

[3390]

Pyrido[2',1' : 2,3]pyrimido[4,5-*c*]pyridazin-5-one

6. Pyrimidine-2,4-dione

A new fused pyrimidinedione ring may be synthesized by heating a reactive *o*-chloro ester with an *N,N'*-disubstituted urea.

+ $(MeNH)_2CO$ Δ 49% [2438]

Pyrimido[4,5-*e*]-1,2,4-triazine-6,8-dione

7. Pyran-2-one

Malonylidene nitrile and a chlorine are transformed into a fused pyran-2-one ring when heated with ethanolic hydrogen chloride [3002]. Reactions in which nitriles and hydrogen chloride interact have been reviewed [2547]. When a 2-bromobenzamide is heated with orcinol in dilute sodium hydroxide containing copper(II) sulphate, the bromo and amino groups are lost; C—C and C—O bonds are formed in a synthesis of the dimethyl ether of alternariol, a compound found in some plants of the *Liliacae* family [2922].

EtOH–HCl, Δ 67% [3002]

Pyrano[2,3-*c*]pyrazol-6-one

NaOH, $CuSO_4$, Δ 25% [2922]

Dibenzo[*b,d*]pyran-6-one

8. 1,3-Thiazin-4-one or 1,3-Thiazine-2,4-dione

Nickel(0)-induced cyclization (review [2998]) of *o*-iodobenzamides gives good yields of 1,3-benzoxazinone or the dione by reaction with a urea or a thiourea, which may be mono-*N*- or di-*N,N'*-substituted. *o*-Iodobenzonitrile behaves similarly. Nickel(II) chloride is complexed with triethylphosphine, and the reaction needs a temperature of about 60 °C.

2-Iodobenzamide ($CONH_2$) + $R^2HNC(=X)NHR^1$ $\xrightarrow[71–85\%]{NiCl_2, PPh_3, \Delta}$ 1,3-Benzothiazin-4-one / 1,3-Benzothiazine [3892]

R^1 = H, Me, Ph; R^2 = H, Me;
X = NH, NMe, NPh, O; Y = O, NH

1,3-Benzothiazin-4-one
1,3-Benzothiazine

III. FORMATION OF A SEVEN-MEMBERED RING

1. 1,3-Diazepin-4-one, 2-Thioxo- or 1,3-Thiazepin-4-one

Either or both of these rings may be formed when a symmetrically substituted thiourea is heated with a 2-halomethylbenzoyl halide; usually the thiazepinone predominates. Under mildly basic conditions, 2-bromomethylbenzoyl bromide was earlier reported to give the 4-one when it reacts with either a thioamide or an *N,N′*-dimethylurea [2571].

2-(CH_2X)benzoyl halide (COX) $\xrightarrow{(RNH)_2CS, Me_2CO, Na_2CO_3, \Delta}$ 2,4-Benzothiazepin-5-one (15–60%) + 2,4-Benzodiazepin-5-one, 3-thioxo- (7–50%) [2731]

R = alkyl, $PhCH_2$, aryl; X = halogen

2,4-Benzothiazepin-5-one

2,4-Benzodiazepin-5-one, 3-thioxo-

2. 1,4-Diazepine-2,5-dione

The benzohydroxamate (**66.6**) cyclizes while it is being stirred in DMF containing TEA.

(**66.6**) $\xrightarrow[\sim 56\%]{DMF, TEA}$ 1,4-Benzodiazepine-2,5-dione [2708]

(**66.6**)
R^1 = H, Cl, NO_2; R^2 = H, Me, $PhCH_2$, 4-ClC_6H_4; X = Cl, Br

1,4-Benzodiazepine-2,5-dione

IV. FORMATION OF AN EIGHT-MEMBERED RING

1. 1,5-Diazocine-2,8-dione

The chlorocarboxamide (**66.7**) reacts with methyl anthranilate under anhydrous conditions to give the diazocinedione in good yield.

NH_2 COOMe + N N Cl Cl $CONH_2$ —THF, AcOH, Δ / 60%→ [3390]

(**66.7**)

Pyridazino[3,4-*b*][1,5]benzodiazocine-5,7-dione

CHAPTER 67

Carboxylic Acid or Ester and Hydrazine or Hydroxylamine Derivative

I. Formation of a Five-membered Ring 1073
1. Pyrazol-3-one 1073
2. Isoxazole 1074
3. Isoxazol-3-one 1074
II. Formation of a Six-membered Ring 1074
1. Pyridazin-3-one 1074
2. Pyrimidine-2,4-dione 1075
III. Formation of a Seven-membered Ring 1075
1. 1,2-Diazepin-3-one 1075
2. 1,4-Diazepin-2-one 1075

I. FORMATION OF A FIVE-MEMBERED RING

1. Pyrazol-3-one

Annulation of hydrazino and carboxylic ester groups may be effected by heating either with acid in DMF or with aqueous alkali.

NC, N, $NHNH_2$, H_2N, COOEt → (HCl, AcOH, DMF, Δ; 65%) → NC, N, H, N, NH, H_2N, O [3336]

Pyrazolo[3,4-*b*]pyridin-3-one

NHNHR COOEt N aq. NaOH, Δ 84% N—NR O N [3518]

R = 2-COOH-5-nPr-thien-3-yl

Pyrazolo[4,3-*c*]quinolin-3-one

2. Isoxazole

A novel synthesis of a fused isoxazole is based on the cyclization of the *O*-(*o*-cyanophenyl)oxime of acetone in an acidic medium and at ambient temperature.

OCH_2OMe $ON{=}CMe_2$ CN HCl, MeOH 69–89% OH O N NH_2 [3586]

1,2-Benzisoxazole

3. Isoxazol-3-one

Base-catalysed cyclization of the *N*-hydroxyureide ester **(67.1)** involves the *N*-hydroxy function rather than the ureide.

CONHMe NOH COOEt PhMe, KOH–EtOH, Δ 79% CONHMe N O O [2078]

(67.1)

2,1-Benzisoxazol-3-one

II. FORMATION OF A SIX-MEMBERED RING

1. Pyridazin-3-one

Interaction of a bishydrazone with a carboxylic ester group in 2*M* sulphuric acid at ambient temperature gives a high yield of pyridazinone; one of the ketones is lost.

O COOEt HN $C{=}NN{=}CCH_2R^1$ R^2 Me Ph THF, H_2SO_4 78–95% O O NH N HN R^2 Me [3243]

R^1 = H, Me; R^2 = Ph, 4-Cl-, 4-Me-C_6H_4

Pyrido[3,4-*d*]pyridazine-1,7-dione

2. Pyrimidine-2,4-dione

An alkyl isocyanate annulates an ester–hydroxamic acid in high yield without the need for external heat.

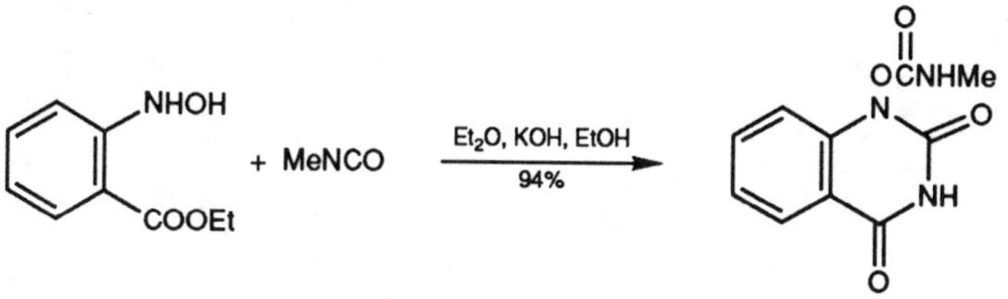

NHOH, COOEt + MeNCO —(Et_2O, KOH, EtOH; 94%)→ OCNHMe (C=O), N, O, NH, O [2078]

2,4-Quinazolinedione

III. FORMATION OF A SEVEN-MEMBERED RING

1. 1,2-Diazepin-3-one

Under mild reaction conditions, this ring is formed from a carboxylic acid and a hydrazone group in the presence of DCC.

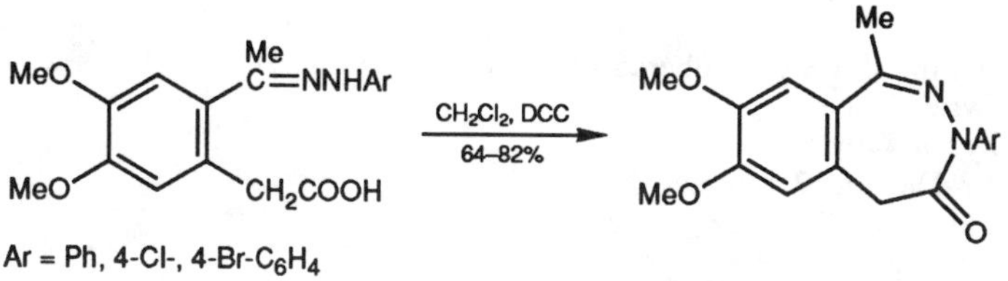

MeO, MeO, Me, C=NNHAr, CH_2COOH —(CH_2Cl_2, DCC; 64–82%)→ MeO, MeO, Me, N, NAr, O [2650]

Ar = Ph, 4-Cl-, 4-Br-C_6H_4

2,3-Benzodiazepin-4-one

2. 1,4-Diazepin-2-one

Reductive cyclization occurs when a suitable oximino ester is treated with zinc and acid.

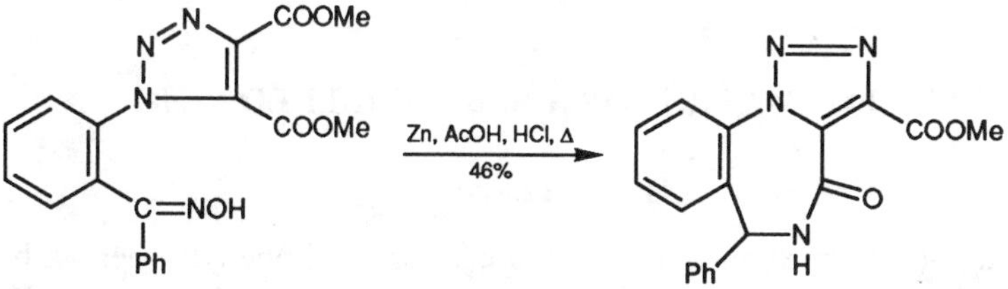

N=N, N, COOMe, COOMe, C=NOH, Ph —(Zn, AcOH, HCl, Δ; 46%)→ N=N, N, COOMe, O, N H, Ph [2742]

[1,2,3]Triazolo[1,5-*a*][1,4]benzodiazepin-4-one

CHAPTER 68

Carboxylic Acid, Acyl Chloride or Ester and Hydroxy or Ether

I. Formation of a Six-membered ring 1076
1. Pyran-2-one 1076
2. Pyran-4-one 1078
3. 1,3-Dioxane-2,4-dione 1079
4. Thiopyran-2-one or -4-one 1079
5. 1,3-Oxazin-4-one 1079
6. 1,3-Thiazin-4-one 1080

An example of the cyclization of an *o*-acyloxy-acyl chloride is included in this chapter.

I. FORMATION OF A SIX-MEMBERED RING

1. Pyran-2-one

Heating a 3-(2-hydroxyphenyl)propionate ester in diphenyl ether with palladium–charcoal induces a combined cyclization and dehydrogenation [3437]. Simultaneous de-*O*-benzylation and cyclization of a 2-(benzyloxy)acrylic acid by heating with acetyl chloride containing phenyltrimethylammonium iodide gives good yields of a fused pyran-2-one [3836]. In contrast, an *o*-methoxybenzylidenemalonic acid (as its potassium salt) cyclizes in the cold when treated with trifluoroacetic acid–trifluoroacetic anhydride [3938].

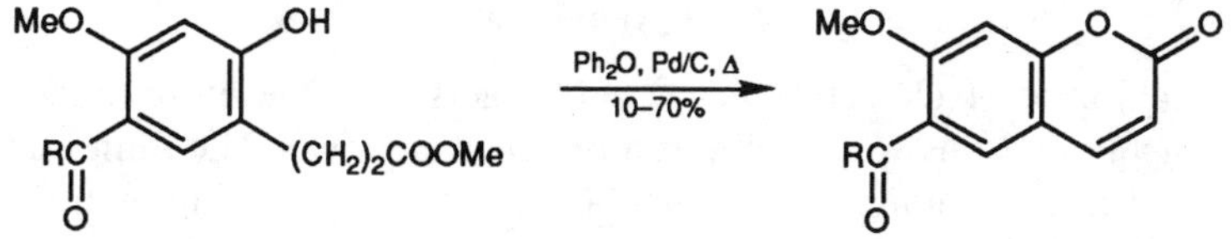

MeO OH RC O (CH$_2$)$_2$COOMe → Ph$_2$O, Pd/C, Δ 10–70% → MeO O O RC O [3437]

R = alkyl, Ph

1-Benzopyran-2-one

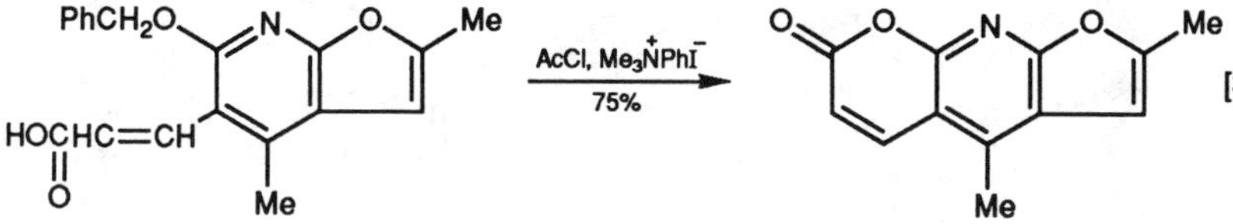

PhCH$_2$O N O Me HOCHC=CH O Me → AcCl, $Me_3\overset{+}{N}Ph\overset{-}{I}$ 75% → O O N O Me Me [3836]

Furo[3,2-*e*]pyrano[2,3-*b*]pyridin-7-one

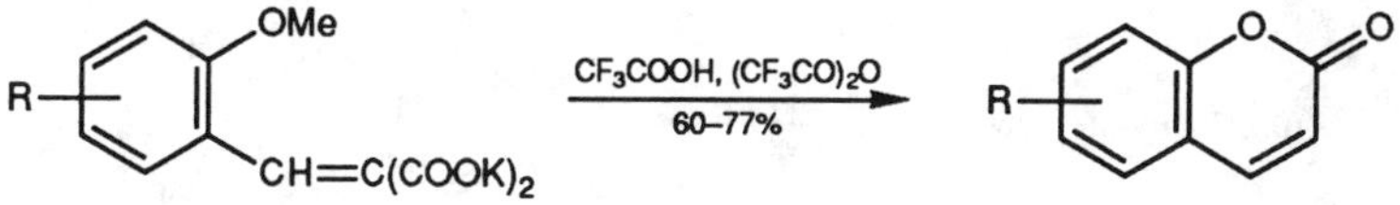

OMe R CH=C(COOK)$_2$ → CF$_3$COOH, (CF$_3$CO)$_2$O 60–77% → O O R [3938]

R = H, MeO, (MeO)$_2$

1-Benzopyran-2-one

Isocoumarin may be synthesized in high yield by heating an *o*-ethoxyvinylbenzoate ester with mineral acid. An *o*-hydroxy (or ethoxy)malonylidene mononitrile gives a high yield of a fused pyran-2-one by heating with hydrochloric acid.

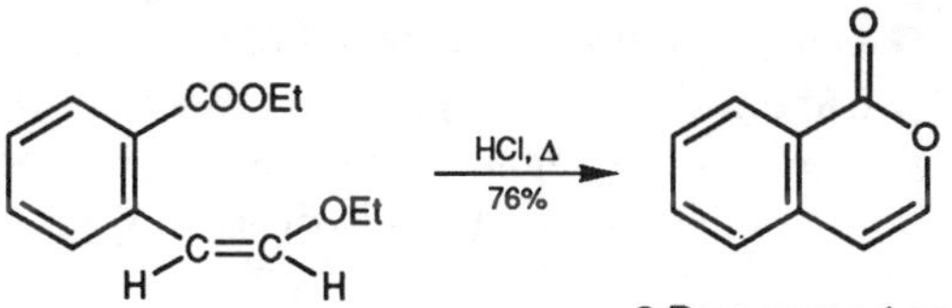

COOEt OEt H C=C H → HCl, Δ 76% → O O [3929]

2-Benzopyran-1-one

Other ring systems synthesized similarly:

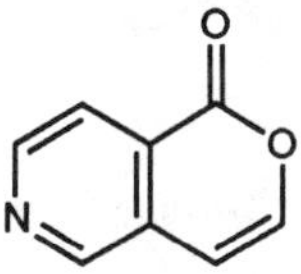

O O N [3929] O O N [3929]

Pyrano[4,3-*c*]pyridin-1-one Pyrano[3,4-*b*]pyridin-8-one

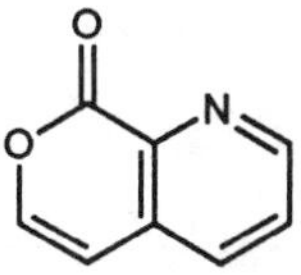

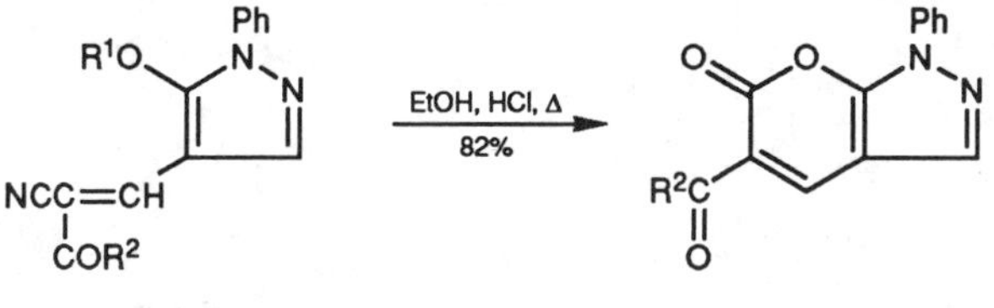

Ph R^{1}O N N NCC=CH COR^2 → EtOH, HCl, Δ 82% → Ph O O N N R^{2}C O [3002]

R^1 = H, Et; R^2 = EtO, NH$_2$

Pyrano[2,3-*c*]pyrazol-6-one

2. Pyran-4-one

The acid chloride of *O*-acetylsalicylic acid reacts with the lithium enolate of acetone or butan-2-one to give a chromone in good yield. The same substrate is converted into a 2-aminochromone-3-carbonitrile by heating with malononitrile (review [2073]) and alkali.

(68.1) $R^1 = H$; $R^2 = H, Me$ + R^2CH_2–MeCOLi —(DME, −70 °C, ~85%)→ 1-Benzopyran-4-one [2939]

(68.1) + $CH_2(CN)_2$ —(aq. NaOH, Δ, 45%)→ [1779]

$R^1 = Br$

Photocyclization (by a radical mechanism) of the mixed anhydride (**68.2**) gives a good yield of the chromanone. An allyloxybenzo radical is probably formed and is cyclized to a benzopyranylmethyl radical. A review of developments in radical, cationic, anionic and metal-promoted cyclizations was recently published [3825]. *O*-Demethylation of a 2-methoxybenzoic acid followed by reaction with 1,3,5-trimethoxybenzene gives a moderate yield of the xanthenone; related xanthenones are similarly synthesized, for example, 2,6-dihydroxybenzoic acid and resorcinol give 1,6-dihydroxy-9-xanthenone [2407].

(68.2) —(PhMe, $AcOCH_2CH{=}CH_2$, *hv*, 70%)→ 1-Benzopyran-4-one [3507]

+ —(i, Δ, ~30%)→ 9-Xanthenone [2272]

i, $AlCl_3$, $ZnCl_2$, $POCl_3$

3. 1,3-Dioxane-2,4-dione

Salicylic acid is cyclized by reaction under anhydrous conditions with phosgene at a low temperature followed by addition of TEA.

OH, COOH + $COCl_2$ —(1. Et_2O 2. TEA; 90%)→ [2408]

1,3-Benzodioxin-2,4-dione

4. Thiopyran-2-one or -4-one

A side-chain acyl chloride cyclizes on to a neighbouring thiol group (formed *in situ* by demethylation of a methylthio ether) under mild conditions. The corresponding thiopyran-4-one is obtained by the reaction of a 2-mercaptobenzoate ester and bromobenzene with lithium *N*-isopropylcyclohexylamide (LICA) at low temperature.

SMe, CH=CHCOCl —(CS_2 or PhH, $AlCl_3$, Δ; 60%)→ [2070]

1-Benzothiopyran-2-one

R, COOMe, SH + Br-Ph —(i, −78 to 20 °C; 22–98%)→ [3659]

R = H, Ac, Me, $CONEt_2$
i, THF, LICA

9-Thioxanthenone

5. 1,3-Oxazin-4-one

Annulation of a hydroxycarboxylic acid is effected at ambient temperature with a carbodi-imide as the reagent and pyridine as solvent.

COOH, OH, N, Ph + RN=C=NR —(pyr; 20–53%)→ [2530]

R = iPr, C_6H_{11}

1,3-Oxazino[6,5-*c*]quinolin-1-one

6. 1,3-Thiazin-4-one

The sulphur analogue of the above oxazin-4-one is prepared from a mercapto-carboxylic acid but with an amino or acyl nitrile in refluxing dioxane or acetic acid.

COOH SH N OH + RCN —(diox or MeOH, Δ; 40–76%)→ [1,3]Thiazino[6,5-*c*]quinolin-1-one [2540]

R = NH_2, $CONH_2$, MeNHCS, PhNHCO, PhNHCS

CHAPTER 69

Carboxylic Acid Derivative and Lactam Carbonyl, Cyanate or Isocyanate

I. Formation of a Five-membered Ring 1081
1. Pyrrol-2-one 1081
2. Thiophene 1082
3. Isothiazol-3-one 1082
II. Formation of a Six-membered Ring 1082
1. Pyridin-2-one 1082
2. Pyrimidin-2-one or Pyrimidine-2-thione 1082
3. Pyrimidin-4-one 1083
4. Pyrimidin-4-one, 2-Thioxo- 1083
5. Pyran-2-one 1084
6. 1,3-Oxazin-4-one or -6-one 1084
7. 1,3-Oxazine-2,4-dione 1085
III. Formation of an Eight-membered Ring 1085
1. Azocin-4-one 1085

I. FORMATION OF A FIVE-MEMBERED RING

1. Pyrrol-2-one

A carboxamide in a side-chain reacts with a neighbouring lactam carbonyl group on heating with a strong base.

$\xrightarrow[62–85\%]{tBuOH–tBuOK, \Delta}$ [2297]

R = HO, NH_2

Pyrrolo[2,3-*b*]pyrrol-2-one

2. Thiophene

Application of the Thorpe cyclization (review [2977]) to mercapto-nitriles yields a fused 3-aminothiophene ring.

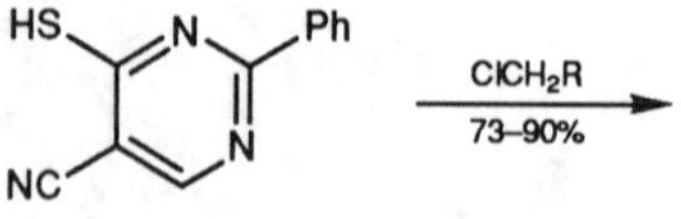

$ClCH_2R$, 73–90% [2977]

R = CN, COOEt, $CONH_2$, Ac, PhCO

Thieno[2,3-*d*]pyrimidine

3. Isothiazol-3-one

Cyano and lactam thiocarbonyl groups are annulated to this fused ring by heating the compound with sulphuric acid.

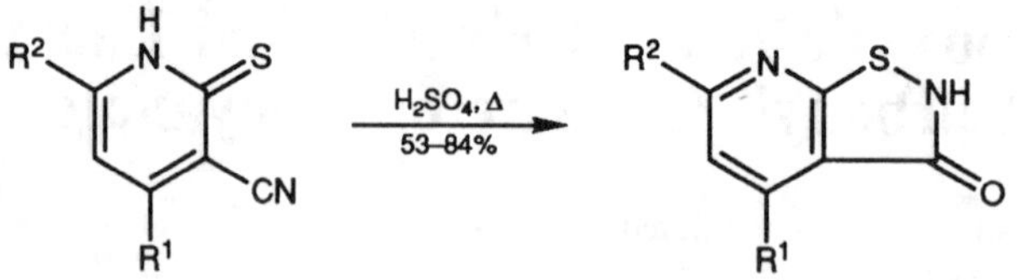

H_2SO_4, Δ, 53–84% [3025]

R^1,R^2 = Me, Ph

Isothiazolo[5,4-*b*]pyridin-3-one

II. FORMATION OF A SIX-MEMBERED RING

1. Pyridin-2-one

Carboxamide and lactam carbonyl functions attached to different rings may be joined together through a pyridinone ring by heating the compound in acetic acid.

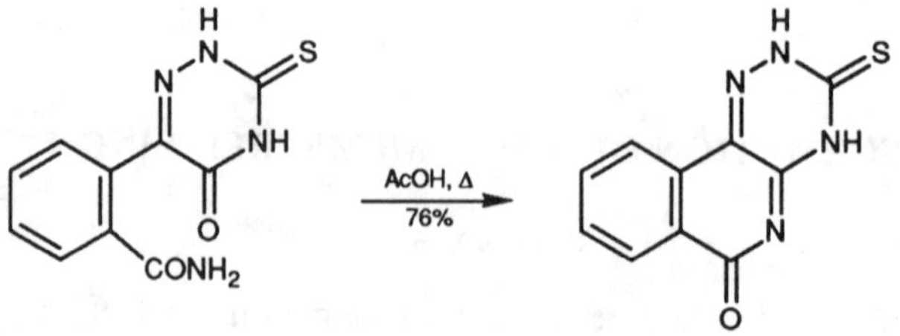

AcOH, Δ, 76% [3061a]

1,2,4-Triazino[5,6-*c*]isoquinolin-6-one, 3-thioxo-

2. Pyrimidin-2-one or Pyrimidine-2-thione

This ring is formed in high yield by heating a 2-isocyanato (or isothiocyanato)-benzonitrile with hydrochloric acid. The use of isothiocyanates in the synthesis of heterocycles has been reviewed [3593, 3990].

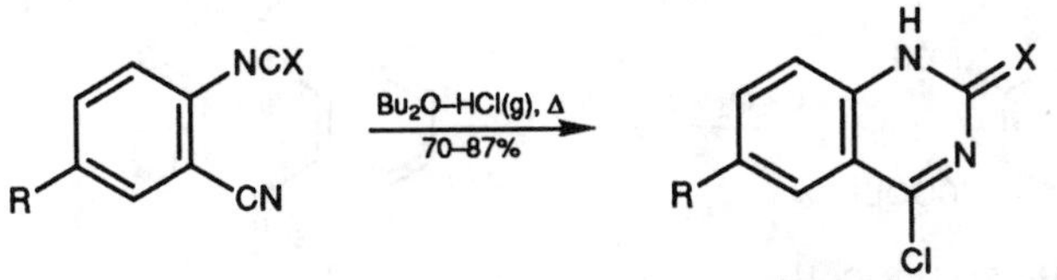

[2209]

R = H, Br; X = O, S

2-Quinazolinone
2-Quinazolinethione

3. Pyrimidin-4-one

Interaction between a carboxamide group attached to one ring and a lactam carbonyl of another ring under dehydrating conditions leads to the formation of a pyrimidin-4-one.

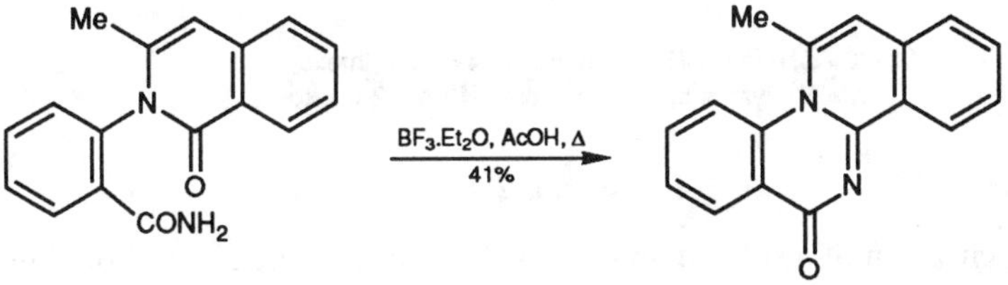

$BF_3.Et_2O$, AcOH, Δ 41% [1113]

Isoquino[2,1-*a*]quinazolin-6-one

4. Pyrimidin-4-one, 2-Thioxo-

Although hydrazines often behave as difunctional reagents, they react with an isothiocyanato-carboxylate ester at room temperature at one nitrogen only. Amines and amino acids yield 3-substituted pyrimidines on being heated with this isothiocyanate (review of isothiocyanates [3990]).

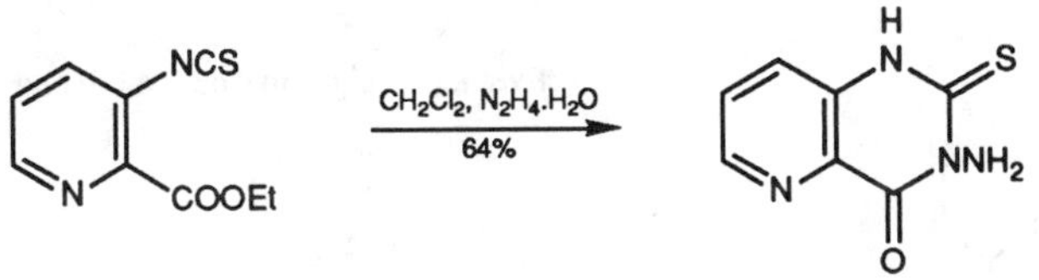

CH_2Cl_2, $N_2H_4.H_2O$ 64% [3674]

Pyrido[3,2-*d*]pyrimidin-4-one, 2-thioxo-

Other ring systems synthesized similarly:

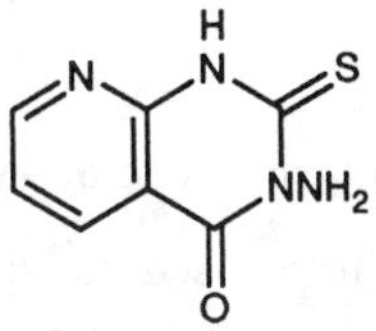

[3681]

Pyrido[2,3-*d*]pyrimidin-4-one, 2-thioxo-

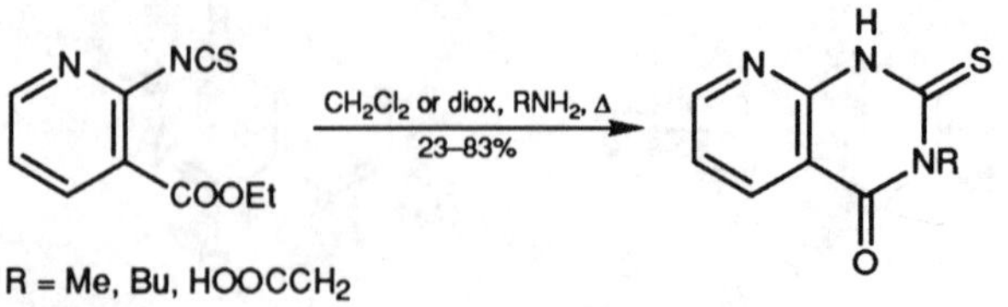

CH_2Cl_2 or diox, RNH_2, Δ
23–83%

[3681]

R = Me, Bu, $HOOCCH_2$

Pyrido[2,3-*d*]pyrimidin-4-one, 2-thioxo-

Other ring systems synthesized similarly:

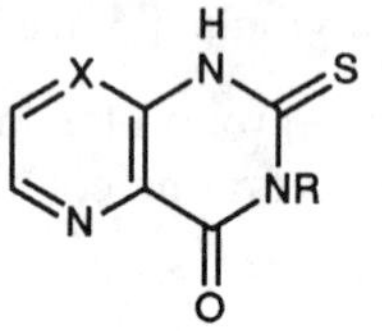

[3675]

X = CH Pyrido[3,2-*d*]pyrimidin-4-one, 2-thioxo-
X = N Pyrazino[2,3-*d*]pyrimidin-4-one, 2-thioxo-

5. Pyran-2-one

Thermal cyclization of a side-chain carboxylic ester to a lactam carbonyl group is an alternative to the photochemical method (p. 359). This cyclization is equally efficient when the two functions are in different rings.

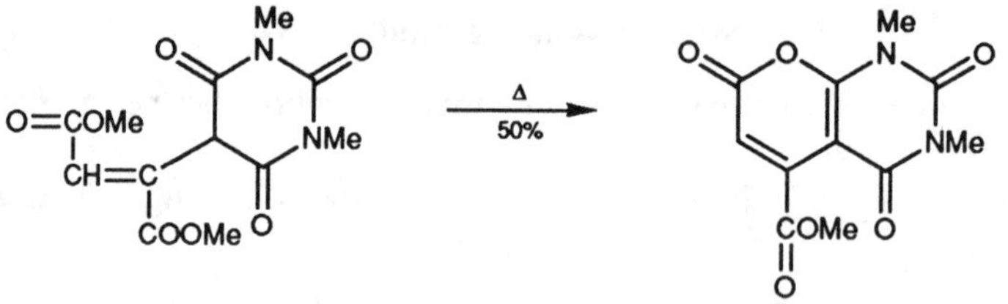

Δ
50%

[2672]

Pyrano[2,3-*d*]pyrimidine-2,4,7-trione

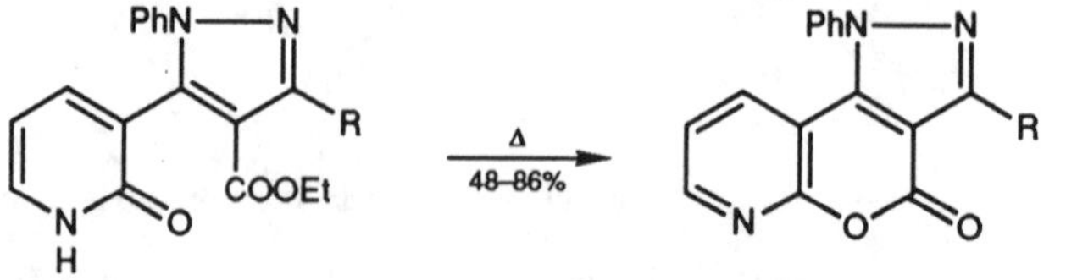

Δ
48–86%

[3486]

R = Me, Ph

Pyrazolo[3',4' : 4,5]pyrano[2,3-*b*]pyridin-4-one

6. 1,3-Oxazin-4-one or -6-one

2-Cyanatobenzoate esters cyclize to a fused oxazin-4-one in the presence of a mild base while the isocyanatobenzoyl chloride is ring-closed by a secondary amine.

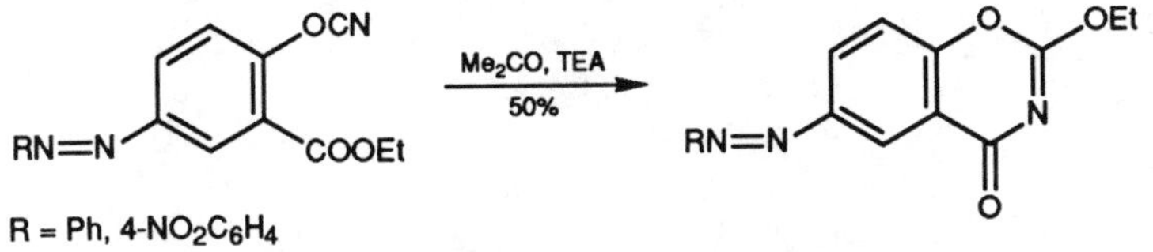

$R = Ph, 4\text{-}NO_2C_6H_4$

Me₂CO, TEA 50% [2047]

1,3-Benzoxazin-4-one

R - alkyl or NR_2 = pyrrolidino, morpholino, piperidino

H_2O 12–81% [3770]

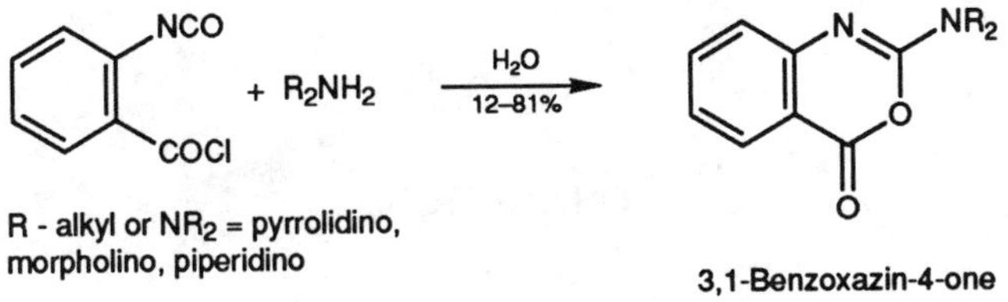

3,1-Benzoxazin-4-one

7. 1,3-Oxazine-2,4-dione

This ring is formed when methyl 2-isocyanatobenzoate is warmed with phosgene.

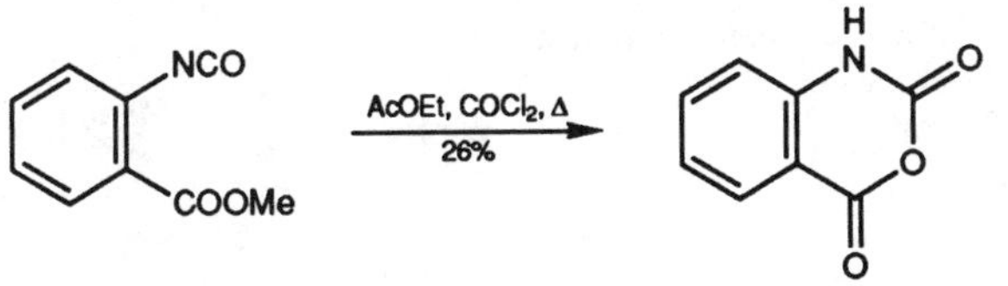

AcOEt, $COCl_2$, Δ 26% [2617]

3,1-Benzoxazine-2,4-dione

III. FORMATION OF AN EIGHT-MEMBERED RING

1. Azocin-4-one

An azocinone ring in the alkaloid magallanesine is formed by reaction of ester and thiocarbonyl groups with DMFDMA.

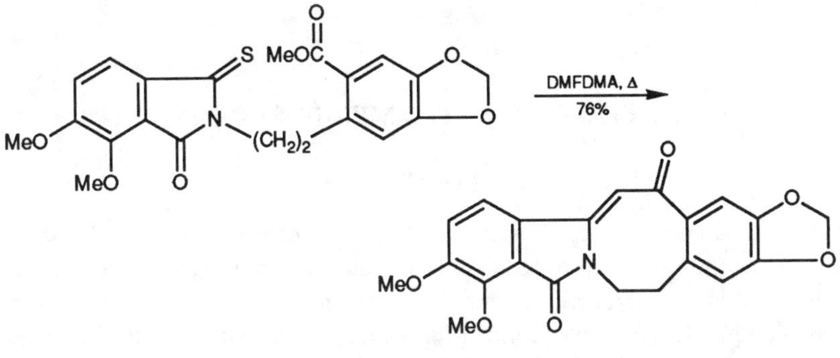

DMFDMA, Δ 76% [3633]

1,3-Dioxolo[4,5-*i*]isoindolo[2,1-*c*][3]benzazocin-8,14-dione

CHAPTER 70

Carboxylic Acid Derivative, Lactam Carbonyl or Isocyanide and Methylene

I. Formation of a Five-membered Ring 1086
1. Pyrrole 1086
2. Pyrrol-3-one 1087
3. Imidazole 1087
4. Furan 1088
5. Furan-3-one 1088
6. Thiophene 1088
II. Formation of a Six-membered Ring 1089
1. Pyridine 1089
2. Pyridin-2-one or -4-one 1090
3. Thiopyran-2-one 1091
4. 1,2-Thiazine 1,1-Dioxide 1091
III. Formation of a Seven-membered Ring 1092
1. Oxepine-3,5-dione 1092
2. 1,4-Oxathiepine or 1,4-Oxathiepin-6-one 1092
IV. Formation of an Eight-membered Ring 1093
1. Azocine or Azocin-2-one 1093

I. FORMATION OF A FIVE-MEMBERED RING

1. Pyrrole

Carboxylic ester and activated *N*-methylene groups react under strongly basic conditions at room temperature to form a pyrrole ring, but attempts to cyclize the *N*-tosyl analogue of (**70.1**) under Dieckmann reaction conditions were unsuccessful; when the *N*-acetyl amine (**70.1**) was treated with sodium hydride, cyclization was achieved in good yield.

THF, tBuOK, 88% — Indole [3520]

(70.1) R = EtO, Et_2N — PhMe, NaH, Δ, 22–63% — Thieno[2,3-*b*]pyrrole [2648]

2. Pyrrol-3-one

The indole-2-carboxylate (**70.2**) undergoes a Dieckmann condensation using sodium in benzene. An excess of lithium bis(trimethylsilyl)amide (LBA) in THF at −78 °C for 5 min cyclizes the thio ester (**70.3**) to the stereoisomer shown as the only product. The same base has been used [3067] to synthesize other biologically promising carbapenems.

(70.2) — PhH, Na, Δ, 58% — Pyrrolo[1,2-*a*]indol-1-one [2237]

(70.3) — THF, LBA, 79% — 1-Azabicyclo[3.2.0]heptan-3,7-dione [3016]

3. Imidazole

Cyanoimino and methylene groups interact in a base-catalysed reaction to form a 2-imidazolamine ring in good-to-excellent yields.

R = CN, Ac, MeOCO, EtOCO, PhCO, Et_2NCO, 4-$NO_2C_6H_4$ — EtONa, EtOH, Δ, 53–93% — Imidazo[2,1-*b*]thiazole [3374]

4. Furan

Base-catalysed cyclization of methylene and nitrile groups gives a fused 3-aminofuran which also has a potentially useful substituent at C-2. This varies according to the base used: potassium carbonate–DMF gives a 2-CN whereas potassium hydroxide produces a 2-$CONH_2$ group [267].

OCH_2CN, CN —EtONa, Δ, ~50%→ OEt, C=NH, O, NH_2 [2556]

Benzofuran

Another ring system synthesized similarly:

O, CN, S, NH_2 [3911]

[1]Benzothieno[3,2-*b*]furan

5. Furan-3-one

An intramolecular Dieckmann reaction may be adapted to produce this ring too. A strong base is needed to convert the *O*-methylene group into its anion, which attacks the 3-methoxycarbonyl function.

R^1, S, OCH_2COOMe, R^2, COOEt —MeONa, or NaH, Δ, 36–87%→ R^1, S, O, COOMe, R^2, O [3178]

R^1 = H, Cl; R^2 = H, Me

Thieno[3,4-*b*]furan-3-one

Another ring system synthesized similarly:

COOEt, O, COOEt, S, O [3178]

Thieno[3,2-*b*]furan-3-one

6. Thiophene

The reactions of various nitriles with sulphur (and sometimes with ammonia) to form a new fused heterocycle have been reviewed [2977]. An *o*-methylcarbonitrile may be cyclized by passing ammonia into a suspension of the substrate and sulphur. t-Butoxide–sodium acetate or TEA are useful promoters of this type of cyclization as is shown in the synthesis of (**70.4**).

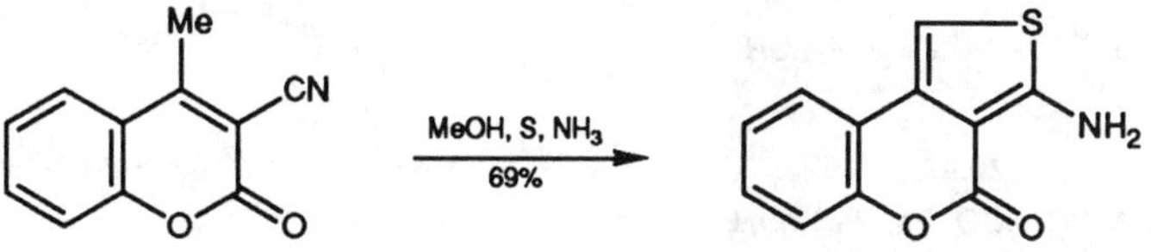

MeOH, S, NH_3 69% [2567]

Thieno[3,4-*c*][1]benzopyran-4-one

Other ring systems synthesized similarly:

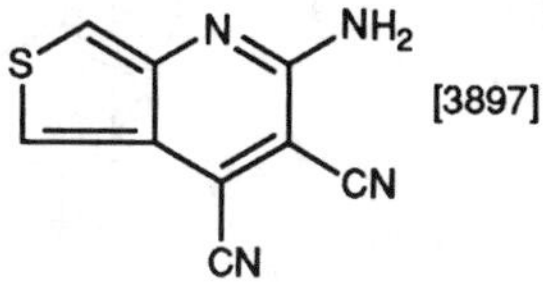

[3897] [3897]

Thieno[3,4-*b*]pyridine

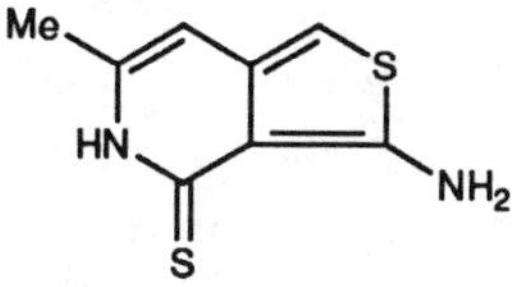

Thieno[3,4-*c*]pyridine-4-thione

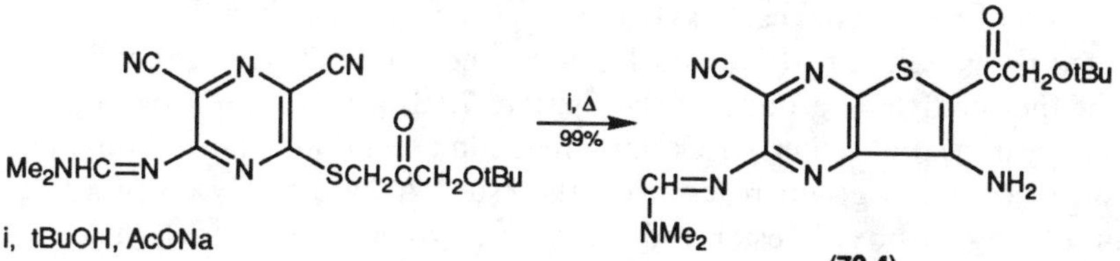

i, Δ 99% [3648]

i, tBuOH, AcONa

(70.4)
Thieno[2,3-*b*]pyrazine

II. FORMATION OF A SIX-MEMBERED RING

1. Pyridine

The anion from (**70.5**) attacks the nitrile group and a 4-aminopyridine ring is formed; in this example, the product is related to the antibacterial agent nalidixic acid. *S*-Phenyl esters are useful in regioselectively directing a Dieckmann condensation of the diesters (**70.6**) to carbacephems. Dieckmann reactions usually require heating but the enaminic ester (**70.7**) undergoes cyclization without external heat.

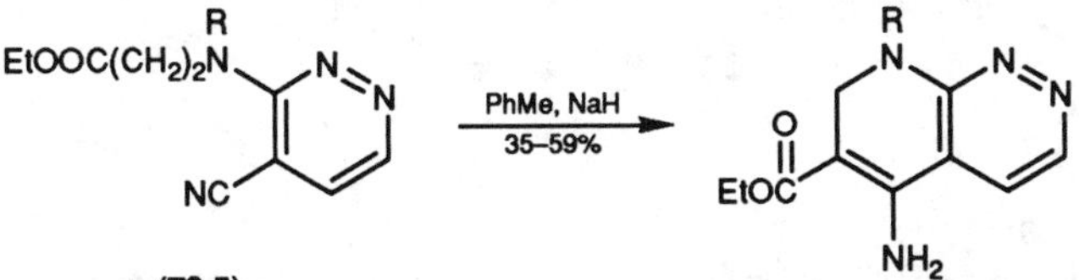

PhMe, NaH 35–59% [3881]

(70.5)
R = alkyl, cyclopropyl

Pyrido[2,3-*c*]pyridazine

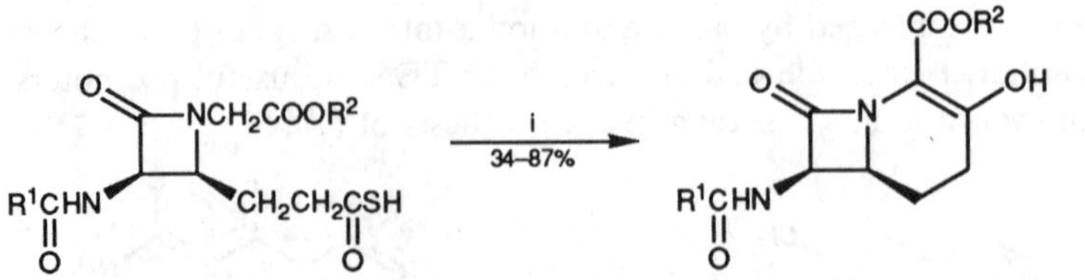

(70.6) → i, 34–87% → 1-Azabicyclo[4.2.0]oct-2-en-8-one [3878]

R^1 = $PhOCH_2CO$, Ph, PhOCO, Boc;
R^2 = 4-$NO_2C_6H_4CH_2$, 4-$MeOC_6H_4CH_2$, Me
i, THF, Li hexamethylsilazide or tBuOK

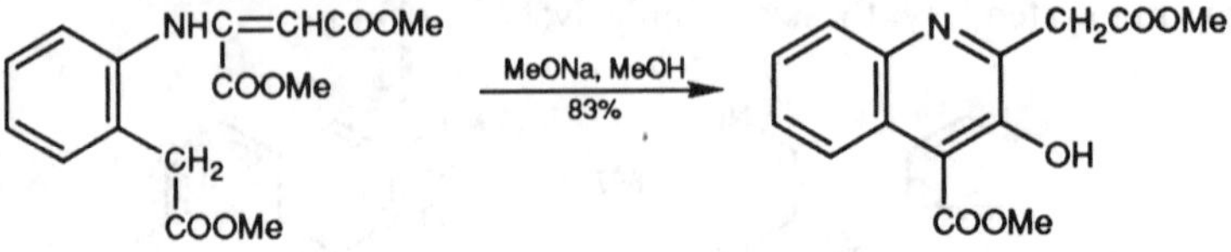

(70.7) → MeONa, MeOH, 83% → [2084]

2. Pyridin-2-one or -4-one

Annulation of methyl and carboxylate ester groups occurs when the compound is treated with sodium hydride and 1,3,5-triazine under anhydrous conditions. Two of the methylene groups of the diester (**70.8**) are activated by adjacent carbonyl groups, but under Dieckmann reaction conditions, the carbonyl next to the carboxamide group reacts with the ester. A base-catalysed procedure for annulating carboxylic ester and methylene groups gives a high yield of a 4-hydroxypyridin-2-one ring (a tautomer of the 2,4-dione). This provides a convenient synthesis of a 1,8-naphthyridin-2-one (or 2,4-dione), which is a leukotriene inhibitor [4002].

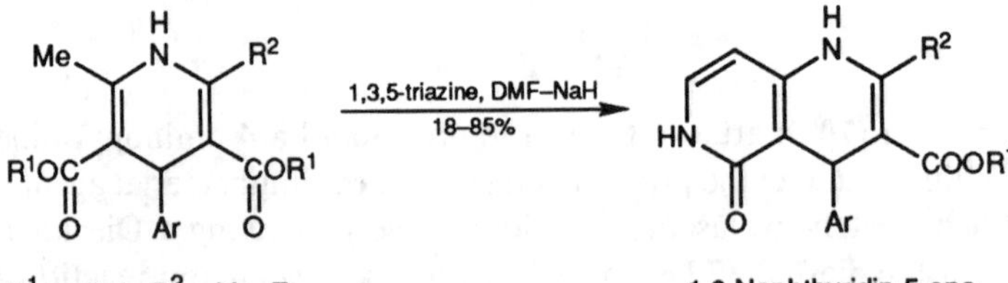

1,3,5-triazine, DMF–NaH, 18–85% → 1,6-Naphthyridin-5-one [3153]

R^1 = alkyl; R^2 = Me, Et

(70.8) → MeONa, MeOH, 30–72% → 2-Quinolinone [2345, 2346]

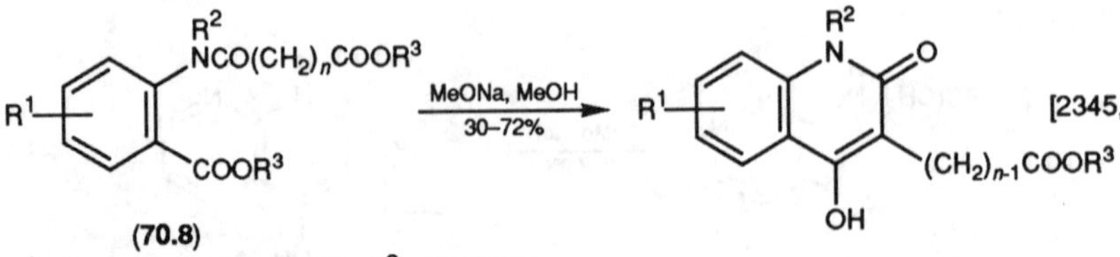

R^1 = H, Cl, NO_2, NH_2, AcNH; R^2 = $EtS(CH_2)_2$, Ph, 4-$MeOC_6H_4$, $PhCH_2$; R^3 = Me, Et; n = 2,3

Ph
NCCH$_2$NR$_2$
O
COOMe
tBuOK, tBuOMe
82%
NR$_2$ = piperidino
Ph
N O
NR$_2$
OH
1,8-Naphthyridin-2-one

[4002]

The microbiologically important 4-oxoquinoline-3-carboxylic acid skeleton is formed by applying Dieckmann conditions to the diester **(70.9)** in which the NH is protected as its acyl derivative. Dehydrogenating agents such as chloranil do not always give good results when applied to the dihydroquinolinones thus produced. Better yields are obtained by electrolysis in acetic acid–t-butanol with a quaternary ammonium salt as electrolyte. With careful choice of conditions, a high yield of the deacylated quinolinone may be obtained.

R^2
R^1 N(CH$_2$)$_2$COOMe
COOMe
(70.9)
R^1 = H, Cl; R^2 = Ac, COOMe
i, NaH, THF; ii, electrolysis, Et$_4$NOTs, tBuOH, AcOH
i, ii
65%
H
R^1 N
COOMe
O
4-Quinolinone

[3927]

3. Thiopyran-2-one

When *N,N*-diethyl-2-methylbenzamide is treated successively with LDA and a thioester (R^{1}COSEt), the sulphur atom is held in the ring of the product.

CONEt$_2$
Me
1. LDA, THF 2. RC(=S)OEt, THF
57–71%
R = Ph, 4-MeO-, 4-Cl-, 4-Me-C$_6$H$_4$
O
S
R
2-Benzothiopyran-1-one

[3731]

4. 1,2-Thiazine 1,1-Dioxide

The use of titanium(IV) chloride as a mediator of organic reactions (review [3045]) has recently been extended to the Dieckmann condensation, which is normally base-catalysed. In this example, an intermediate in the synthesis of piroxicam is synthesized in a reaction which gives the best yield when the substrate : $TiCl_4$: amine ratio is 2.2 : 1 : 3.

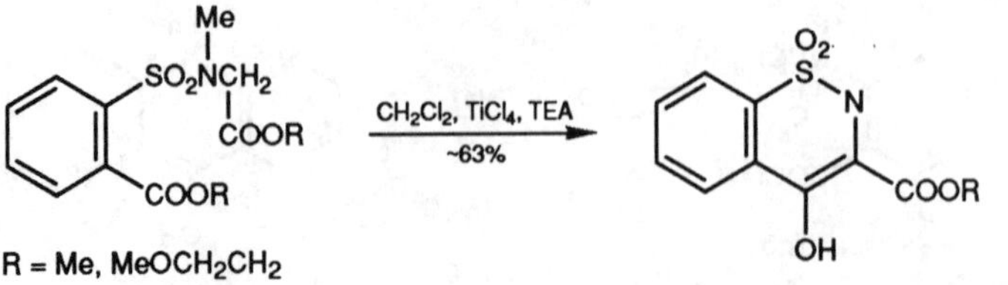

CH_2Cl_2, $TiCl_4$, TEA; ~63% [3782]

R = Me, $MeOCH_2CH_2$

1,2-Benzothiazine 1,1-dioxide

III. FORMATION OF A SEVEN-MEMBERED RING

1. Oxepine-3,5-dione

Attack by an anion derived from an acetyl group on a neighbouring ethoxycarbonylmethoxy of compound **(70.10)** succeeds only when a phenolic group is present adjacent to the acetyl substituent.

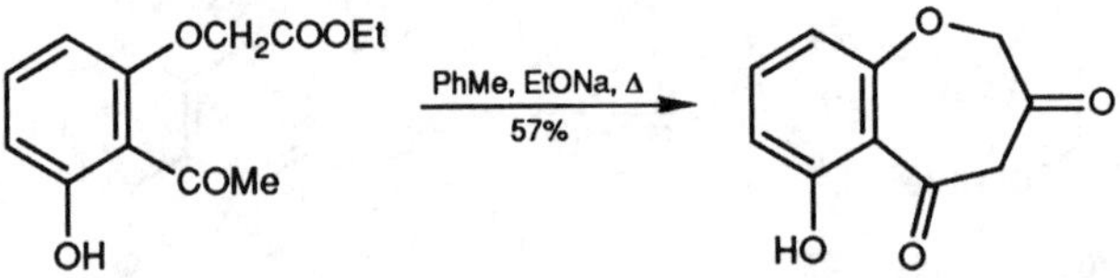

PhMe, EtONa, Δ; 57% [2911]

(70.10) 1-Benzoxepine-3,5-dione

2. 1,4-Oxathiepine or 1,4-Oxathiepin-6-one

The base-catalysed Thorpe–Ziegler reaction (review [1922]) on the dinitrile **(70.11)** can give a good yield of the oxathiepin at room temperature. Replacing one or both nitriles in **(70.11)** by methoxycarbonyl groups enables the oxathiepinone to be prepared under mild conditions.

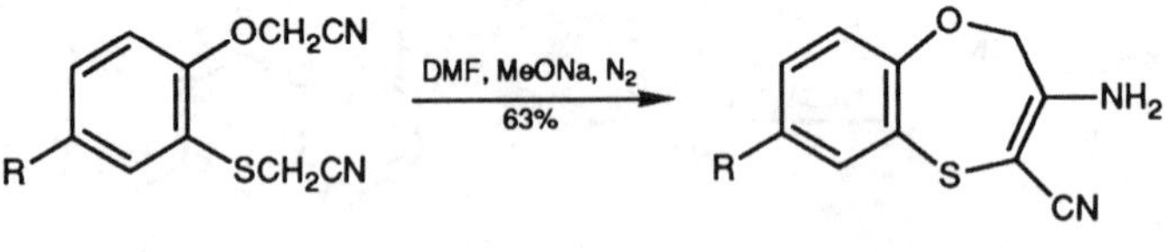

DMF, MeONa, N_2; 63% [3255]

(70.11) 1,5-Benzoxathiepine

MeONa, DMF, N_2; 9–77% [3255]

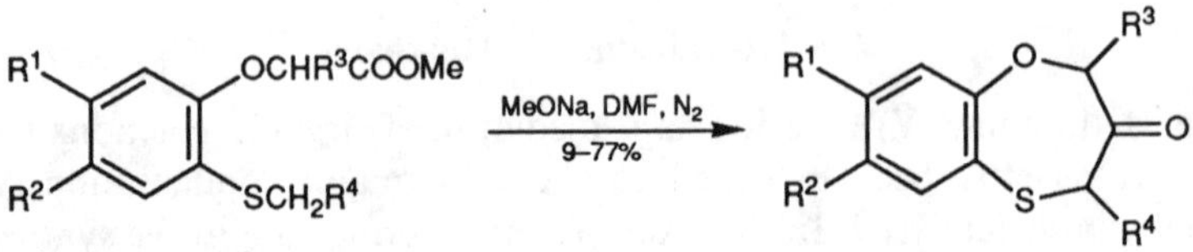

R^1 = H, MeO; R^2 = H, Me, Cl, $PhCH_2O$;
R^3 = H, $MeOOCCH_2CH_2$; R^4 = CN, COOMe

1,5-Benzoxathiepin-3-one

IV. FORMATION OF AN EIGHT-MEMBERED RING

1. Azocine or Azocin-2-one

Bis(tosylmethyl isocyanides) form anions readily; one of these attacks the isocyanide in the neighbouring chain of **(70.12)** to form a fused azocine or azocinone ring depending on the base used to generate the anion.

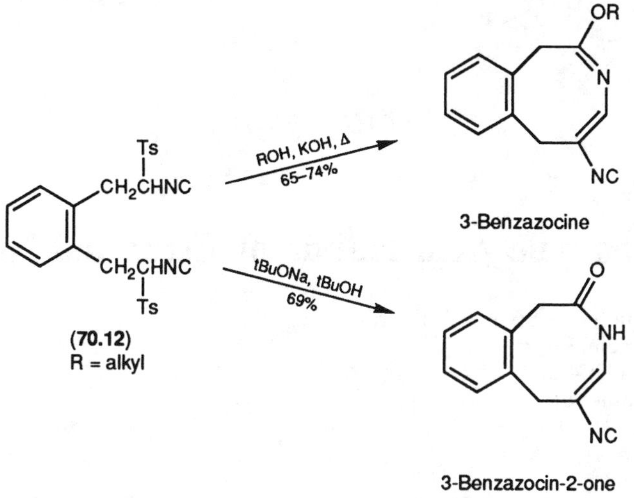

Ts
CH_2CHNC
CH_2CHNC
Ts
(70.12)
R = alkyl

ROH, KOH, Δ
65–74%

OR
N
NC
3-Benzazocine

[3256]

tBuONa, tBuOH
69%

O
NH
NC
3-Benzazocin-2-one

CHAPTER 71

Carboxylic Acid Halide or Ester and Nitrile

I. Formation of a Six-membered ring 1094
1. Pyridin-2-one 1094
2. Pyridin-4-one 1094
3. Pyran-2-one 1095

I. FORMATION OF A SIX-MEMBERED RING

1. Pyridin-2-one

2-Cyanophenylacetic acids are cyclized by first heating with phosphorus pentachloride and then treatment with mineral acid. This appears to give better yields than heating the acid with a mixture of phosphorus pentachloride and oxychloride.

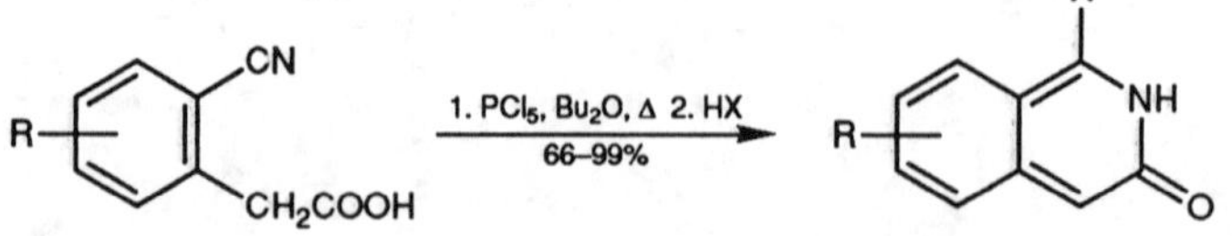

CN, CH_2COOH —— 1. PCl_5, Bu_2O, Δ 2. HX; 66–99% ——> NH, O [2576]

R = H, Cl, Br, MeO, Me; X = Cl, Br, F

3-Isoquinolinone

2. Pyridin-4-one

Interaction between an ester and a nitrile in hot PPA can result in simultaneous cyclization and loss of the nitrile group. An attempt to induce a (base-catalysed) Thorpe–Ziegler cyclization (review [1922], see also Chapter 70, Section IV.1) failed in this instance.

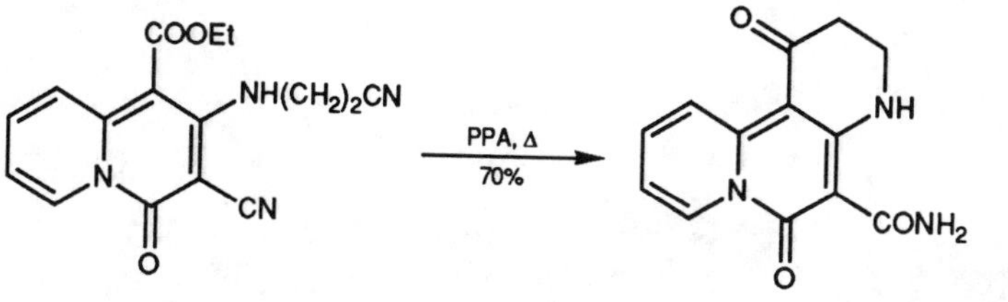

COOEt
NH(CH$_2$)$_2$CN
N
CN
O

PPA, Δ
70%

O
NH
N
CONH$_2$
O

[2199]

Pyrido[3,2-*a*]quinolizine-1,6-dione

3. Pyran-2-one

In a basic medium, *o*-cyanomethylbenzoic acid reacts with acetyl chloride to give isocoumarins instead of the expected mixed anhydrides.

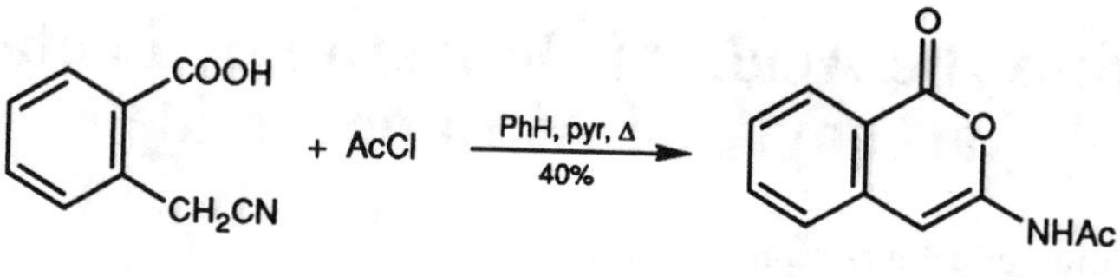

COOH
CH$_2$CN
+ AcCl

PhH, pyr, Δ
40%

O
O
NHAc

[2285]

2-Benzopyran-1-one

CHAPTER 72

Carboxylic Acid, its Derivative or Lactam Carbonyl and Nitro or Ureide

I. Formation of a Five-membered Ring 1096
1. Pyrrol-2-one 1096
2. Imidazol-2-one 1097
3. 1,2,4-Thiadiazole 1097
II. Formation of a Six-membered Ring 1097
1. Pyridine 1097
2. Pyrimidin-2-one 1098
3. Pyrimidine-2,4-dione 1098
4. Pyrimidin-4-one, 2-Thioxo- 1100
5. Pyrazin-2-one or Pyrazine-2,3-dione 1100
6. Pyran-2-one 1101
7. 1,4-Oxazine 1101
8. 1,4-Oxazin-3-one 1102
9. 1,3-Thiazin-6-one 1102
10. 1,4-Thiazine 1103
11. 1,4-Thiazin-3-one 1103
12. 1,2,4-Thiadiazine 1,1-Dioxide 1103
III. Formation of a Seven-membered Ring 1104
1. 1,3-Diazepine-2,4-dione 1104
2. 1,4-Diazepin-5-one 1-Oxide 1104
3. 1,2,4,6-Tetrazepin-5-one or 1,2,4,6-Tetrazepine-5-thione 1104

I. FORMATION OF A FIVE-MEMBERED RING

1. Pyrrol-2-one

The methylene, cyano and nitro groups of compound (**72.2**) are involved in its conversion to an isoindole derivative by heating with a catalytic amount of ethanolic potassium acetate.

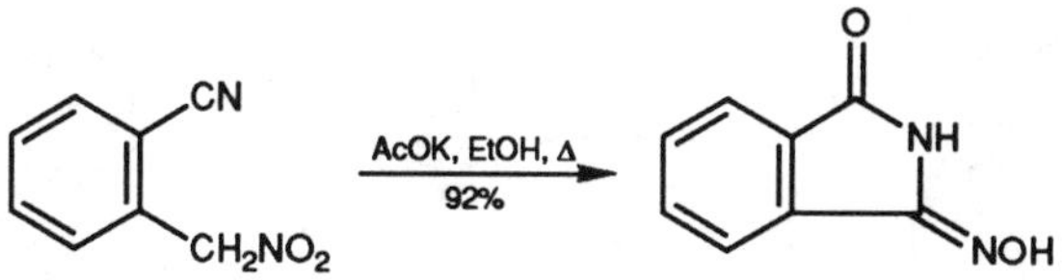

(72.2) → 1-Isoindolone (AcOK, EtOH, Δ; 92%) [2738]

2. Imidazol-2-one

It is sometimes advantageous to avoid alkaline conditions during the synthesis of 6,8-purinediones. This may be achieved by cyclization of 5-ureidopyrimidine-4,6-dione with PPA; hydrochloric acid is usually less efficient. Phosphoric acid (85%) may also be used in the conversion of the ureidopyrimidine (**72.3**) into a 6,8-purinedione.

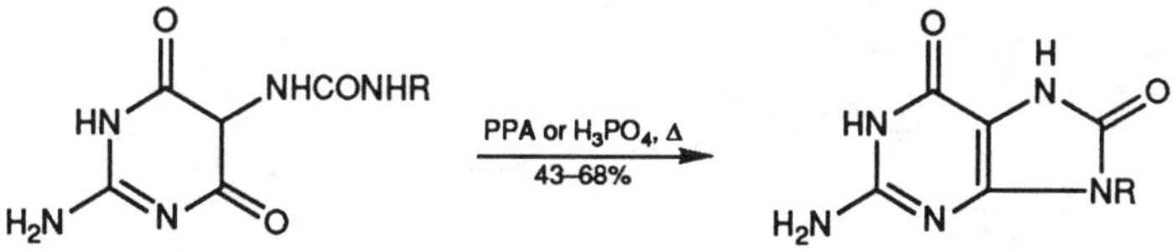

(72.3) R = H, alkyl, Ph → 6,8-Purinedione (PPA or H_3PO_4, Δ; 43–68%) [2173]

3. 1,2,4-Thiadiazole

Oxidative cyclization occurs in high yield when a compound containing lactam thiocarbonyl and thioureido functions is treated with bromine–TEA at ambient temperature.

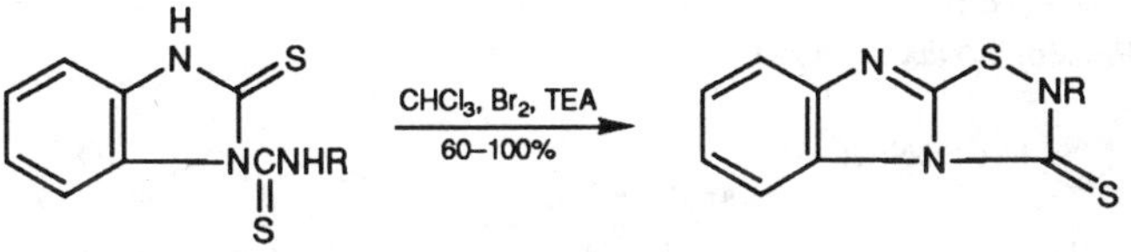

R = Me, Bu, C_6H_{11}, $PhCH_2$ → 1,2,4-Thiadiazolo[4,5-*a*]benzimidazole-3-thione ($CHCl_3$, Br_2, TEA; 60–100%) [3052]

II. FORMATION OF A SIX-MEMBERED RING

1. Pyridine

A nitrene (derived from a nitro group and triethyl phosphite) reacts with an adjacent ethoxycarbonylvinyl group to form a quinoline in moderate yield. Replacing the ester by a keto carbonyl group results in much lower yields.

R^1 = MeO, $PhCH_2O$; R^2 = CN, COOEt

$P(OEt)_3$, Δ ~53% → Quinoline [2238]

2. Pyrimidin-2-one

A naturally occurring purine nucleoside, isoguanosine, is obtained from a 5-benzoylureidoimidazole-4-carbonitrile which is treated with 33% aqueous ammonia for 48 h at room temperature. A ureide is probably an intermediate in the conversion of the dinitrile **(72.4)** into a purinone. The enzyme catalase induces cyclization of *o*-cyanoureides **(72.5)** on incubation at 37 °C for 6 h to give high yields of quinazolinones.

EtOH, aq. NH_3 ~79% → 2-Purinone [3313]

R = H, D-ribofuranosyl

aq. NaOH ~94% [3037a]

(72.4)
R = Me, β-D-ribofuranosyl

EtOH, pH 7.2, catalase 75–95% → 2-Quinazolinone [3626]

(72.5)
R^1 = H, Me, Ph; R^2 = allyl

3. Pyrimidine-2,4-dione

In addition to the methods mentioned on pp. 375–376, 2-ureidobenzoic acids are cyclized by heating with acetic anhydride [2630], alkali [2630] or carbonyldiimidazole [3563]. The carboxylic acid group may be replaced by a carboxamido [3117, 3733] or an *N*-methoxycarboxamido function [2708].

NHCONH$_2$, COOH, R, Me $\xrightarrow{Ac_2O,\ \Delta;\ 40\text{–}70\%}$ 2,4-Quinazolinedione [2630]

R = H, alkyl

2,4-Quinazolinedione

R^3, NHCNHR^2, COOH $\xrightarrow{THF,\ CDI,\ \Delta;\ 20\text{–}83\%}$ Furo[3,4-*d*]pyrimidine-2,4-dione [3563]

R^1 = H, alkyl, acyl, $(CH_2)_3COOEt$;
R^2 = $ArN[(CH_2)_2]_2N(CH_2)_2$; R^3 = H, Me

Furo[3,4-*d*]pyrimidine-2,4-dione

NHCONH$_2$, CONH$_2$ $\xrightarrow{\Delta;\ 98\%}$ Pyrido[3,2-*d*]pyrimidine-2,4-dione [3117, 3734]

Pyrido[3,2-*d*]pyrimidine-2,4-dione

Another ring system synthesized similarly:

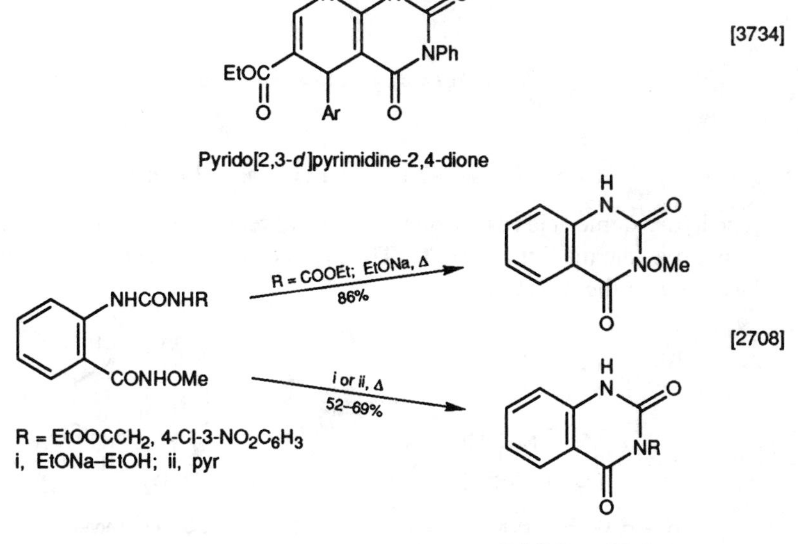

4. Pyrimidin-4-one, 2-Thioxo-

Replacing the ureido function in the reaction described in the previous paragraph [3563] by thioureido provides a route to the title ring. This reaction seems to be particularly efficient when N^3 of the thioureido chain is acylated.

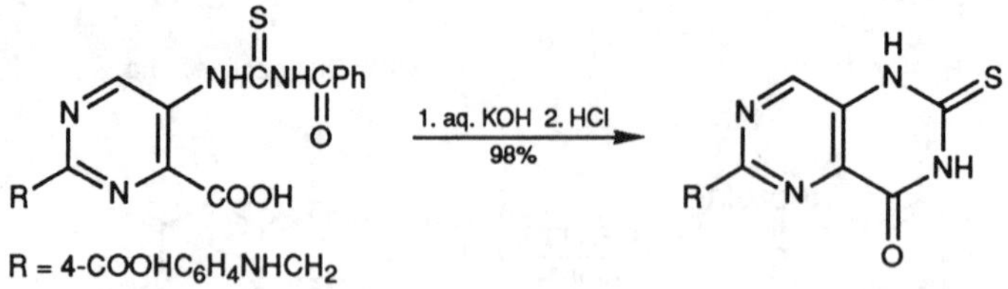

1. aq. KOH 2. HCl, 98% [2172]

R = 4-$COOHC_6H_4NHCH_2$

Pyrimido[5,4-*d*]pyrimidin-4-one, 2-thioxo-

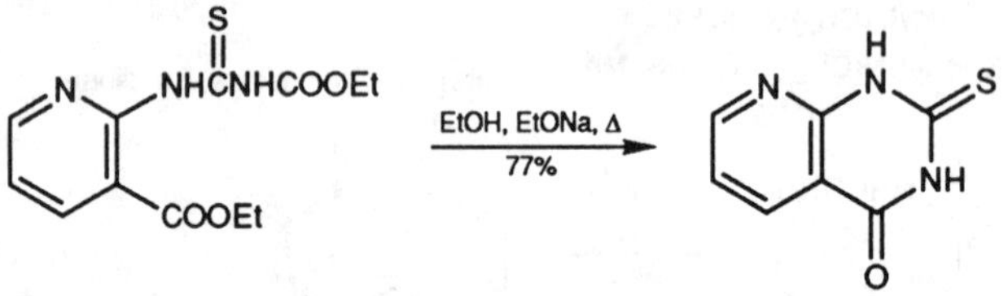

EtOH, EtONa, Δ, 77% [2388]

Pyrido[2,3-*d*]pyrimidin-4-one, 2-thioxo-

Another ring system synthesized similarly:

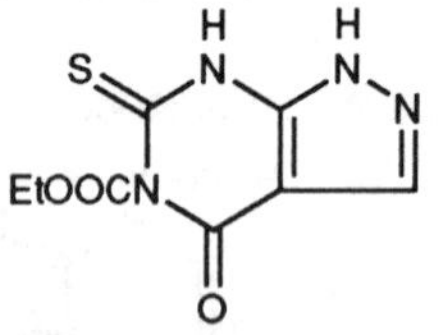

[2339]

Pyrazolo[3,4-*d*]pyrimidin-4-one, 6-thioxo-

5. Pyrazin-2-one or Pyrazine-2,3-dione

Cyclic hydroxamic acids are formed by partial reduction of *N*-(nitrophenyl)glycine in an acidic medium at 60 °C. The glycine ester (**72.6**) may be heated with a base to give the dione.

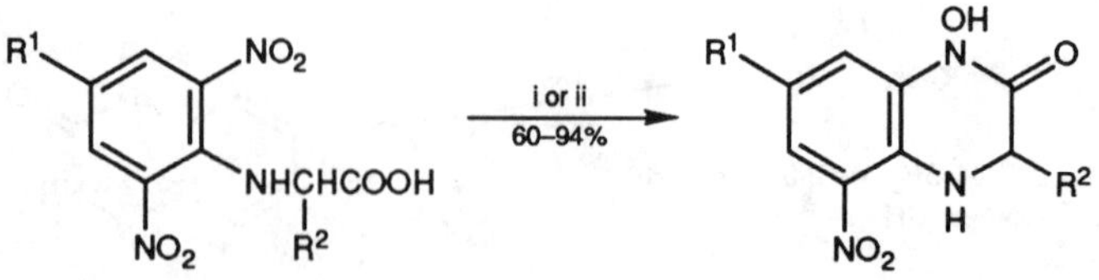

i or ii, 60–94% [2513]

R^1 = H, Cl; R^2 = H, Me

i, MeOH, HCl, $SnCl_2$; ii, Me_2CO, H_2, Pd/C

2-Quinoxalinone

(72.6)

$\xrightarrow[\text{5–44\%}]{\text{EtONa or Na}_2\text{CO}_3,\ \Delta}$ [3320]

X = CH 2,3-Quinoxalinedione
X = N Pyrido[2,3-*b*]pyrazine-2,3-dione

6. Pyran-2-one

2-Nitroethylbenzoic acids, on successive treatment with sodium hydroxide and sulphuric acid–methanol, are converted into 2-benzopyran-1-ones (isocoumarins) in high yield. The nitro acid is prepared from a 3-(1-nitroethyl)-1-isobenzofuranone, and it is possible to obtain a good overall yield of the benzopyranone from this lactone without isolating the nitroethylbenzoic acid. A nitro attached to a benzene ring may be displaced by a nucleophile, especially if the ring is not electron-rich. Thus, a high yield of a coumarin is obtained from the *o*-nitrocinnamic acid (**72.7**); the *trans* isomer cyclizes more readily than the *cis* isomer. When the benzene ring is replaced by thiophene, no cyclization occurs. Nitro group displacement has been reviewed [1398, 1399, 1432].

$\xrightarrow[\text{\textasciitilde 82\%}]{\text{aq. NaOH, H}_2\text{SO}_4\text{–MeOH}}$ [3420]

2-Benzopyran-1-one

R^1 = H, MeO; R^2 = H, Me, Et, 4-$MeOC_6H_4$

$\xrightarrow[\text{78–91\%}]{\text{quinoline, Cu, }\Delta}$ [3254]

1-Benzopyran-2-one

(72.7)
R^1 = H, NO_2; R^2 = Br, Cl; R^3 = COOH, 4-$MeOC_6H_4$; R^4 = H, 4-$MeOC_6H_4$

7. 1,4-Oxazine

A 4-oxacephalosporin is formed in good yield by cyclization of suitably placed carbamate and ester groups using lithium bis(trimethylsilyl)amide at −78 °C under argon.

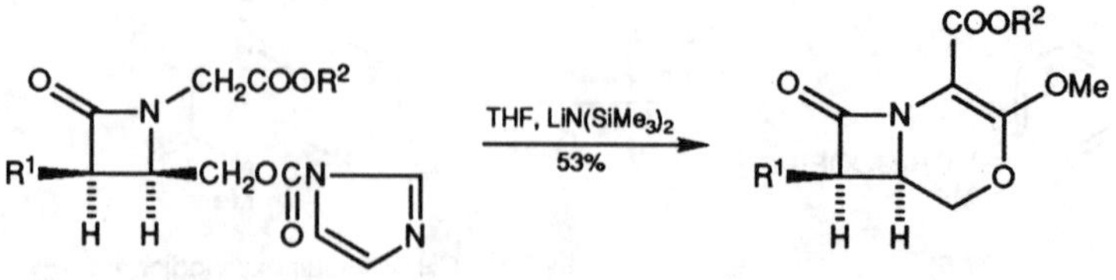
O
N
CH2COOR2
R1
CH2OCN
O
N
H H
THF, LiN(SiMe3)2
53%
COOR2
O
N
OMe
R1
O
H H
R1 = PhCH2CONH; R2 = 4-NO2C6H4CH2

[3738]

4-Oxa-1-azabicyclo[4.2.0]oct-2-en-8-one

8. 1,4-Oxazin-3-one

2-Nitrophenoxyacetates undergo a mild reductive cyclization when treated with zinc-ammonium chloride at 30–40 °C.

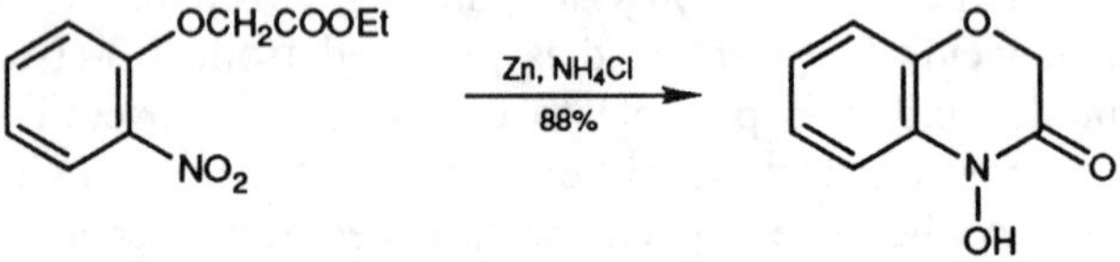
OCH2COOEt
NO2
Zn, NH4Cl
88%
O
N
O
OH

[3599]

1,4-Benzoxazin-3-one

9. 1,3-Thiazin-6-one

This ring may be constructed from neighbouring carboxylic acid or ester and thioureido or acylureido groups by stirring with sulphuric, perchloric or polyphosphoric acid. The last two reagents require heating, and although a high yield is obtainable using perchloric acid–acetic anhydride, the considerable danger of explosions militates against the use of this particular method. The method using sulphuric acid seems the most attractive and efficient [3883]. These 2-aminothiazinones (**72.8**) undergo Dimroth rearrangement under basic conditions even at room temperature.

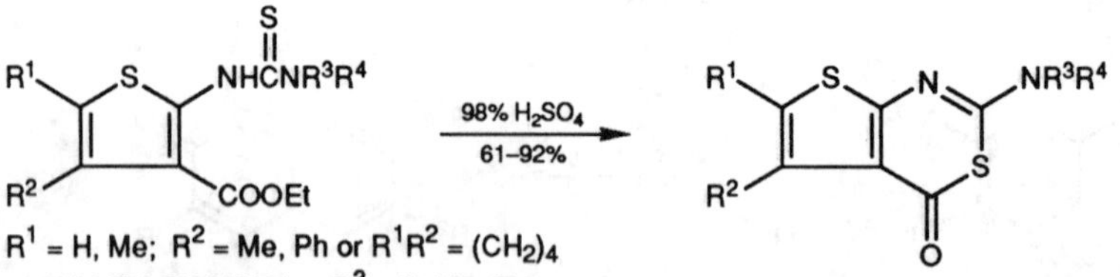
S
R1
S
NHCNR3R4
R2
COOEt
98% H2SO4
61–92%
R1
S
N
NR3R4
S
R2
O

[3237]

R^1 = H, Me; R^2 = Me, Ph or R^1R^2 = $(CH_2)_4$ or $CH_2N(PhCH_2)(CH_2)_2$; R^3 = H, Me, Et; R^4 = H, alkyl, C_6H_{11}, Ph or NR^3R^4 = morphodine

Thieno[2,3-*d*][1,3]thiazin-4-one

Another ring system synthesized similarly:

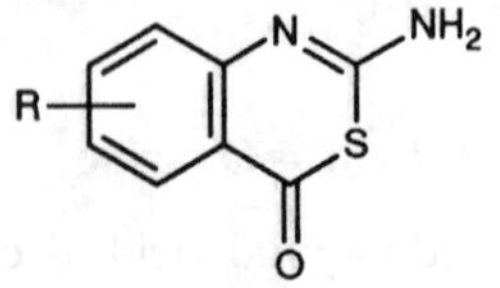
N
NH2
R
S
O

[3883]

(72.8)
3,1-Benzothiazin-4-one

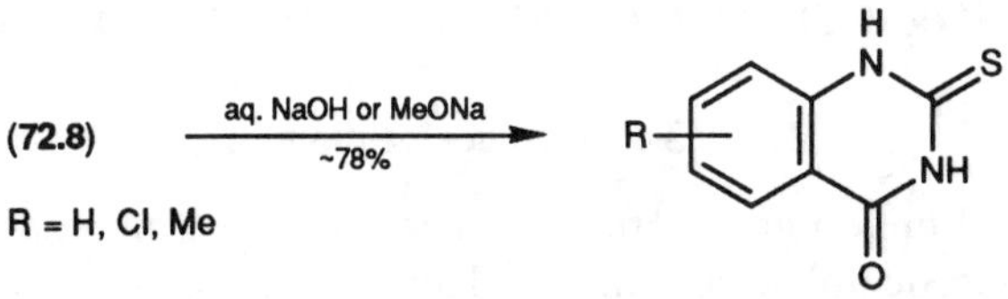

(72.8) —aq. NaOH or MeONa, ~78%→ [3883]

R = H, Cl, Me

H, N, S, NH, O, R

4-Quinazolinone, 2-thioxo-

10. 1,4-Thiazine

Neighbouring ureido and lactam thiocarbonyl groups react in alkali with phenacyl bromides to form a new thiazine ring. The chemistry of α-halo aldehydes and -ketones has been reviewed [B-45].

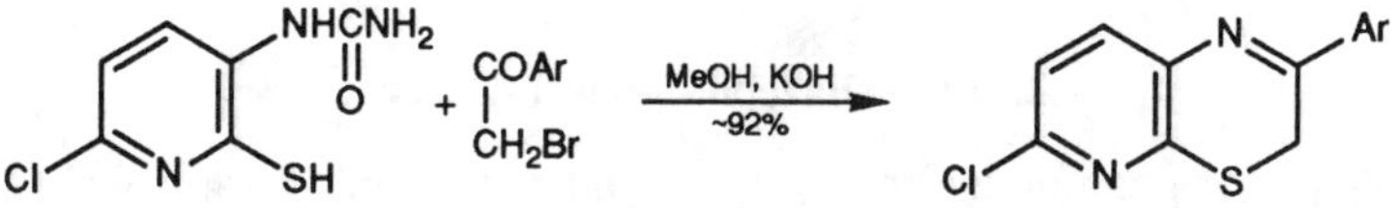

NHCNH$_2$, O, Cl, N, SH + COAr, CH$_2$Br —MeOH, KOH, ~92%→ Cl, N, S, N, Ar [2362]

Ar = Ph, 4-MeOC_6H_4

Pyrido[2,3-*b*][1,4]thiazine

11. 1,4-Thiazin-3-one

Electrolysis of 2-(nitrophenylthio)acetic acid with a mercury cathode in dilute ethanolic sulphuric acid produces a high yield of a thiazin-3-one.

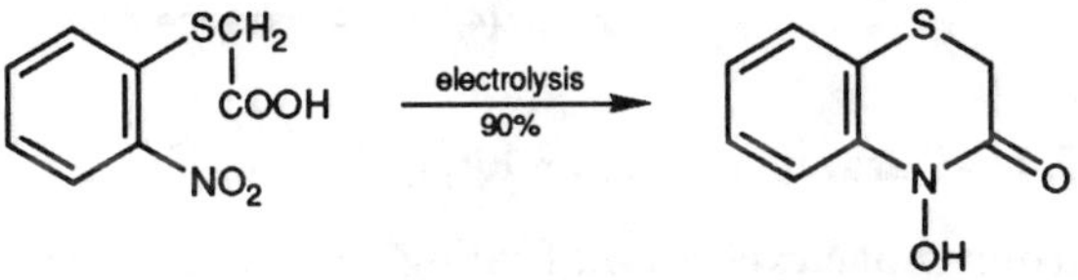

SCH$_2$, COOH, NO$_2$ —electrolysis, 90%→ S, N, O, OH [2374]

1,4-Benzothiazin-3-one

12. 1,2,4-Thiadiazine 1,1-Dioxide

A benzothiadiazine dioxide is formed in high yield when a 2-ureidobenzenesulphonamide is heated for 12–16 h in isopropanol and TEA.

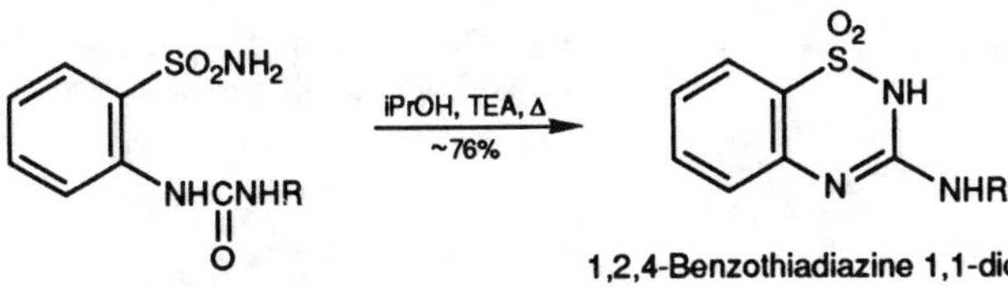

SO$_2$NH$_2$, NHCNHR, O —iPrOH, TEA, Δ, ~76%→ O$_2$, S, NH, N, NHR [3692, 3694]

1,2,4-Benzothiadiazine 1,1-dioxide

R = HO$(CH_2)_n$; n = 2,3

III. FORMATION OF A SEVEN-MEMBERED RING

1. 1,3-Diazepine-2,4-dione

Carboxamide and ureido groups react on heating the compound in a mixture of dioxane and acetic anhydride; the product is a homologous purinedione.

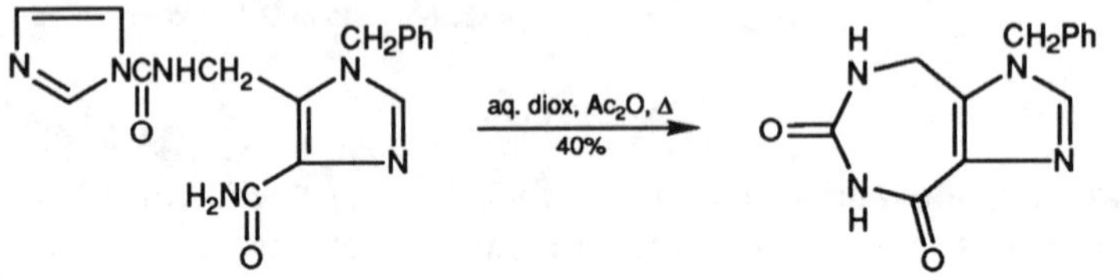

aq. diox, Ac_2O, Δ
40%

[3735]

Imidazo[4,5-*e*][1,3]diazepine-4,6-dione

2. 1,4-Diazepin-5-one 1-Oxide

A homologue of guanine is formed by partial reduction of a nitro-nitrile.

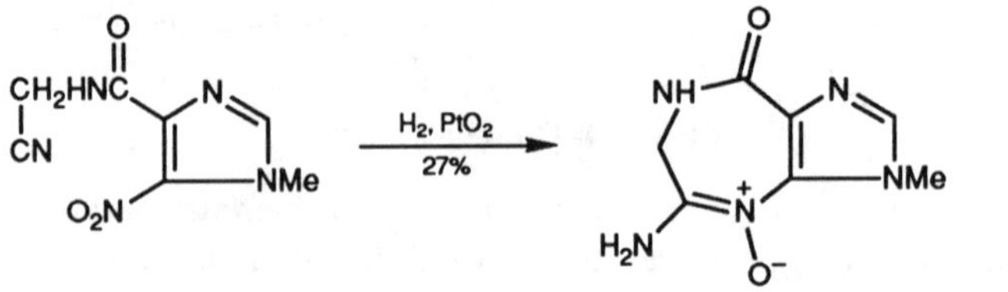

H_2, PtO_2
27%

[3124]

Imidazo[4,5-*e*][1,4]diazepine-8-one 4-oxide

3. 1,2,4,6-Tetrazepin-5-one or 1,2,4,6-Tetrazepine-5-thione

N^3-Ureidoquinazolin-4-one cyclizes on heating with either urea or thiourea to form a fused tetrazepinone ring.

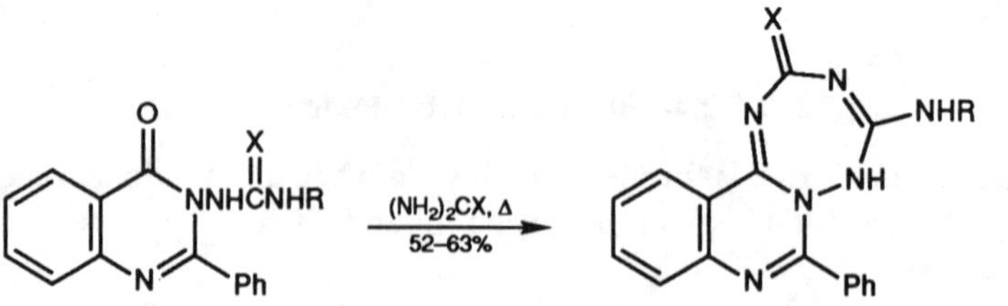

$(NH_2)_2CX$, Δ
52–63%

[3460]

R = H, Ph; X = O, S

X = O 1,2,4,6-Tetrazepino[1,7-*c*]quinazolin-2-one
X = S 1,2,4,6-Tetrazepino[1,7-*c*]quinazoline-2-thione

CHAPTER 73

Carboxylic Acid or Ester and Ring-carbon

I.	Formation of a Five-membered Ring	1105
	1. Pyrrole	1105
	2. Pyrazole	1106
	3. Furan-2-one	1107
II.	Formation of a Six-membered Ring	1107
	1. Pyridine	1107
	2. Pyridine-2-thione	1109
	3. Pyridin-4-one	1109
	4. Pyran-2-one or -4-one	1110
	5. Thiopyran or Thiopyran-4-one	1111

I. FORMATION OF A FIVE-MEMBERED RING

1. Pyrrole

When an *N*-methyl-*N*-phenylamino acid hydrochloride is heated with acetic anhydride and sodium acetate, a high yield of a 3-acetyloxyindole is obtained; this method usually gives good yields, and is not sensitive to oxidizing agents which promote the conversion of 3-indolones (indoxyls) into indigo. The dithiocarboxylic ester ylides (**73.1**) are cyclized at or below room temperature when treated with bromoacetonitrile or a bromoacetic ester and a base, and are then dehydrogenated with chloroanil. The 2-thiol may be obtained by stirring with t-butoxide [3761]. An imidazole *N*-allyl ylide (**73.2**, R^2 = CN, $COOR^4$) is cyclized on heating in xylene to an imidazopyridinone, but when R^2 is a phenylsulphonyl group, a new pyrrole ring is formed [3797].

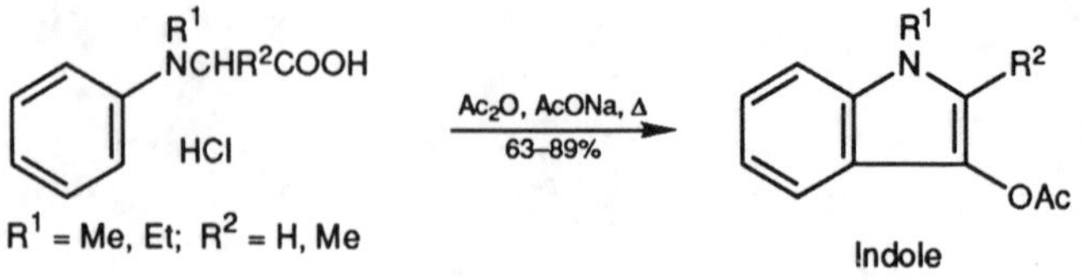

R^1 = Me, Et; R^2 = H, Me

Ac_2O, AcONa, Δ, 63–89%

Indole [3050, 3547]

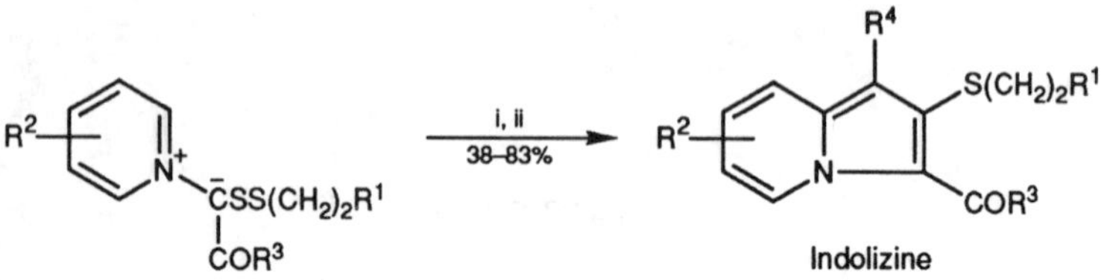

(73.1)

R^1,R^4 = CN, COOEt; R^2 = H, Me; R^3 = EtO, Me

i, $BrCH_2R^1$, $CHCl_3$, DBU; ii, chloranil

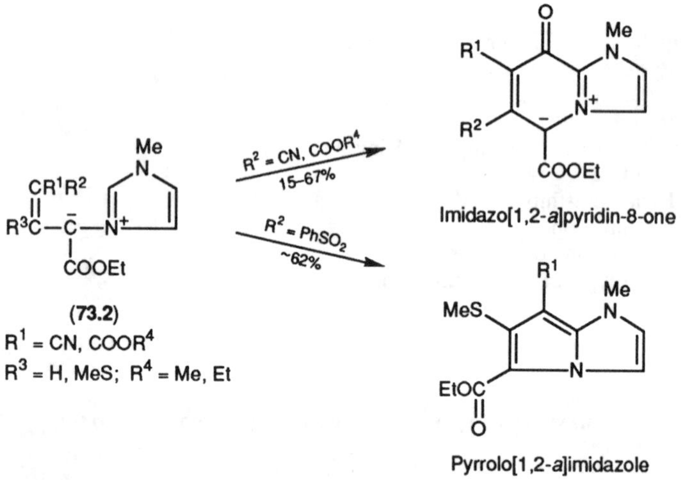

i, ii, 38–83%

Indolizine [3761]

(73.2)

R^1 = CN, $COOR^4$

R^3 = H, MeS; R^4 = Me, Et

R^2 = CN, $COOR^4$, 15–67%

Imidazo[1,2-*a*]pyridin-8-one [3797]

R^2 = $PhSO_2$, ~62%

Pyrrolo[1,2-*a*]imidazole

2. Pyrazole

When a pyridone carboxylic acid is treated with diazomethane, the major product is the pyrazolopyridinone; the carboxylic acid is believed to be esterified and diazomethane may then add across the 3,4-double bond. Replacing diazomethane by diazoethane gave the 1-ethyl-3-methyl analogue (26% yield) of the pyrazolopyridinone shown as the main product of cyclization.

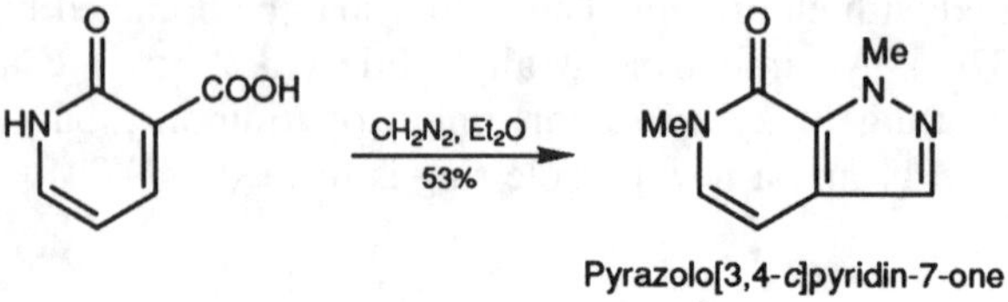

CH_2N_2, Et_2O, 53%

Pyrazolo[3,4-*c*]pyridin-7-one [2792]

3. Furan-2-one

A stereocontrolled ring closure of a cycloalkadienylacetic acid in the presence of a catalytic amount of palladium(II) acetate [B-41] produces either a *cis* or a *trans* product according to the conditions, especially the presence or absence of a small amount of lithium chloride (which blocks the coordination of the carboxylate). Yields vary from 50 to 77%, and the selectivity is greater than 97% under optimum conditions. An attempt to prepare a 3-benzofuranone (see Chapter 88) from the phenoxyarylacetic acid (**73.3**) resulted in a rearrangement and the formation of a 2-furanone.

$(CH_2)_n$ CH_2COOH + RCOOH → (i) / (ii)

AcO $(CH_2)_n$

2-Benzofuranone
Cyclohepta[2,3]furan-2-one [3621]

AcO $(CH_2)_n$

n = 1,2
i, Me_2CO, $Pd(OAc)_2$, benzoquinone, 20 °C;
ii, Me_2CO, $Pd(OAc)_2$, benzoquinone, LiOAc, LiCl, 20 °C

OCHPh COOH —PPA, Δ, 78%→ Ph [3815]

(73.3) 2-Benzofuranone

II. FORMATION OF A SIX-MEMBERED RING

1. Pyridine

When an ester reacts with a ring-carbon atom in the presence of a Friedel–Crafts catalyst, a cyclic ketone (or its tautomeric enol) is usually formed (p. 378), but in boiling phosphorus oxychloride, the oxygen function may be replaced by a chlorine atom [2634]. A carboxylic acid may react with a suitably placed ring-carbon by heating with either PPA or phosphorus oxychloride [2127]; phosphorus pentoxide in xylene is also effective [2347]. When an electron-rich ring such as that of pyrrole is available, heating with acetic anhydride for 1 h suffices [3913].

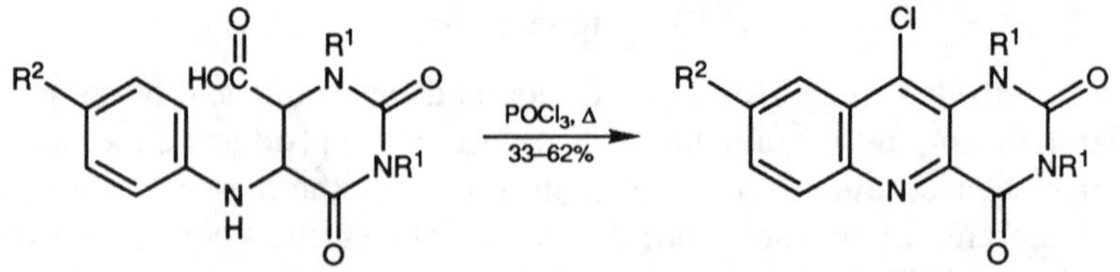

R^1 = H, Me; R^2 = H, Cl, MeO

$POCl_3$, Δ 33–62% [2634]

Pyrimido[5,4-*b*]quinoline-2,4-dione

Another ring system synthesized similarly:

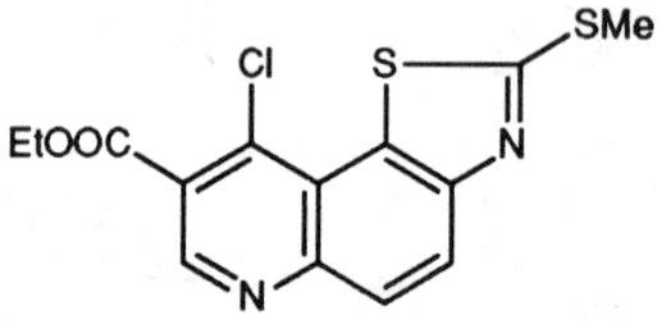

[2783]

Thiazolo[5,4-*f*]quinoline

i or ii, Δ ~90% [2127]

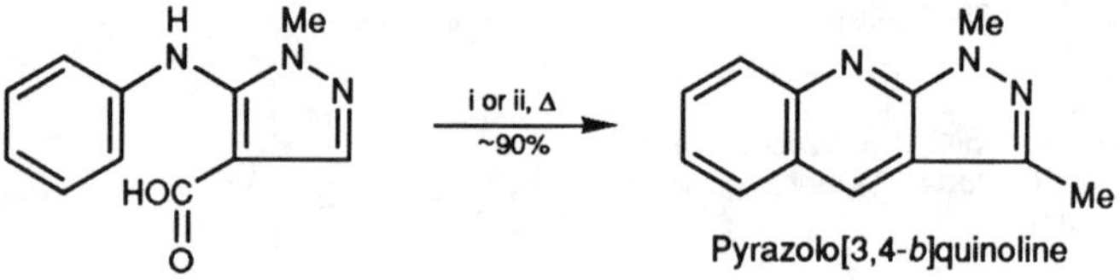

Pyrazolo[3,4-*b*]quinoline

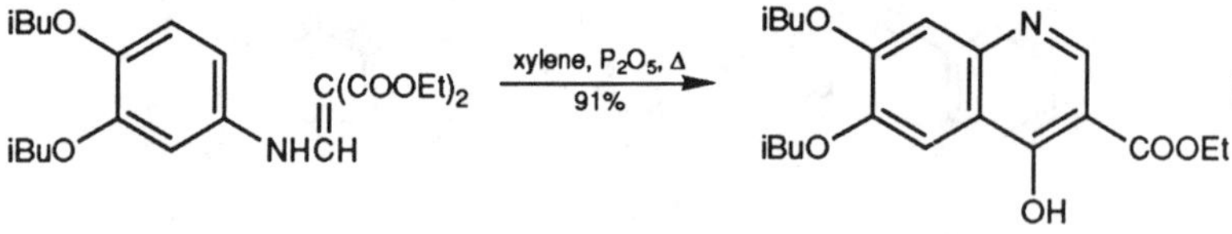

xylene, P_2O_5, Δ 91% [2347]

Quinoline

Ac_2O, Δ 36–58% [3913]

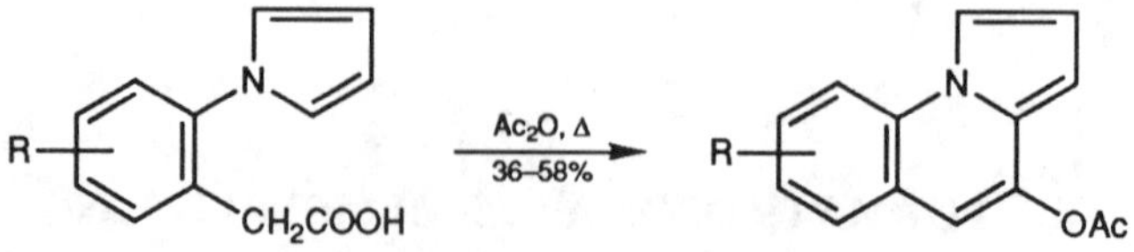

R = H, Br, Cl, Me

Pyrrolo[1,2-*a*]quinoline

Attempts to condense the quinolizine (**73.4**) with methyl propynoate in boiling benzene or toluene containing palladium–charcoal failed, but when nitrobenzene was used, moderate-to-good yields of the pyridoquinolizine were obtained. Formation of a 4-oxo- or 4-chloro-pyridine ring from a carboxylic acid (or ester) or an acyl chloride respectively is well documented (see pp. 378–381, and Section II.3 of this chapter); in this particular example, a fused 4-aminopyridine is formed by successively heating an acyl chloride with an arylamine and phosphorus oxychloride.

CHCOOR^1 + C≡CR^2 / COOR^3 —(PhNO$_2$, Pd/C, Δ; 24–75%)→ Pyrido[2,1,6-*de*]quinolizine [2847]

(73.4)
R^1 = alkyl; R^2,R^3 = H, Ph

—(1. ArNH$_2$, Δ 2. POCl$_3$, Δ; 42–62%)→ Benzo[*b*][1,8]phenanthroline [3807]

Ar = Ph, 4-F-, 4-HO-, 4-Me$_2$N, 4-MeO-, 4-H$_2$NSO$_2$, 4-MeSO$_2$-C$_6$H$_4$

2. Pyridine-2-thione

Thermolysis of the malonic ester (**73.5**) under reduced pressure (~10 mmHg) leads to a high yield of 2-quinolinethione.

NHCCH(COOEt)$_2$ (C=S) —(Δ; 77%)→ 2-Quinolinethione (COOEt, OH) [2134]

(73.4a)

3. Pyridin-4-one

3-(1-Indolyl)propanoic acid (**73.4a**, R = H) forms a third ring by attack of the carboxylic acid group at C-2 of the indole, but a 3-methoxycarbonyl substituent directs the attack to the *peri* (C-7) position [2997, 3015, 3115]. When the chloro ester (**73.6**) is heated with either an amine or a thiophenol, a high yield of the 4-quinolinone is obtained [2441]. Replacement of the chlorine atom by an ethoxycarbonyl group and heating this with phosphorus pentoxide in nitrobenzene is an efficient method of synthesizing a 4-quinolinone-3-carboxylic acid [2380]. An unusual source of carbon atoms in the synthesis of a fused pyridine ring was applied in one stage of the synthesis of a nucleoside; thermolysis of the dilactone (**73.7**) gave a high yield of the pyrazolopyridinone [3757]. Malonylidene derivatives are also cyclized thermally to a fused pyridine-3-carboxylic acid [2240].

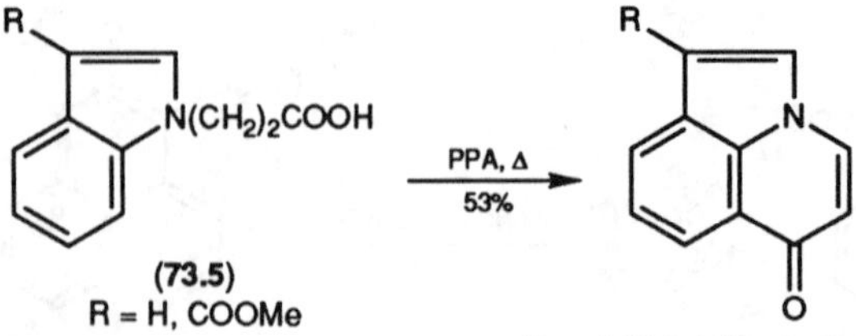

(73.5)
R = H, COOMe

PPA, Δ
53%

Pyrrolo[3,2,1-*ij*]quinolin-6-one [2997, 3015, 3115]

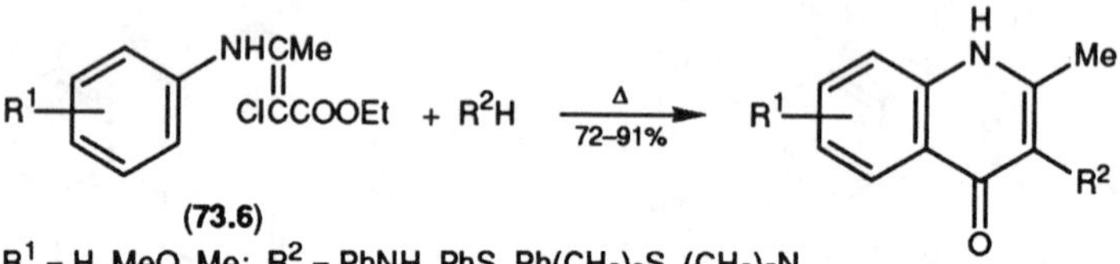

(73.6)
R^1 = H, MeO, Me; R^2 = PhNH, PhS, $Ph(CH_2)_2S$, $(CH_2)_5N$

Δ
72–91%

4-Quinolinone [2441]

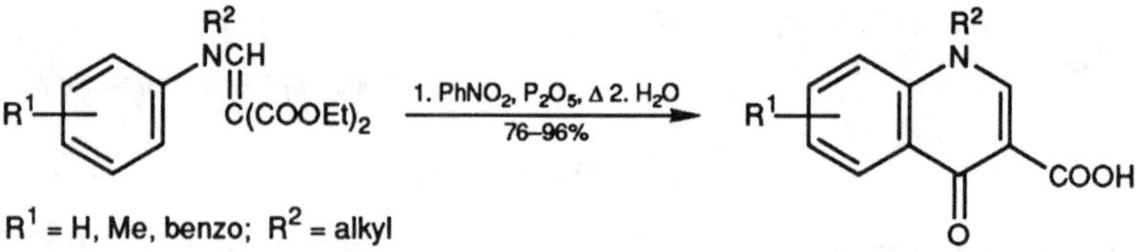

R^1 = H, Me, benzo; R^2 = alkyl

1. $PhNO_2$, P_2O_5, Δ 2. H_2O
76–96%

[2380]

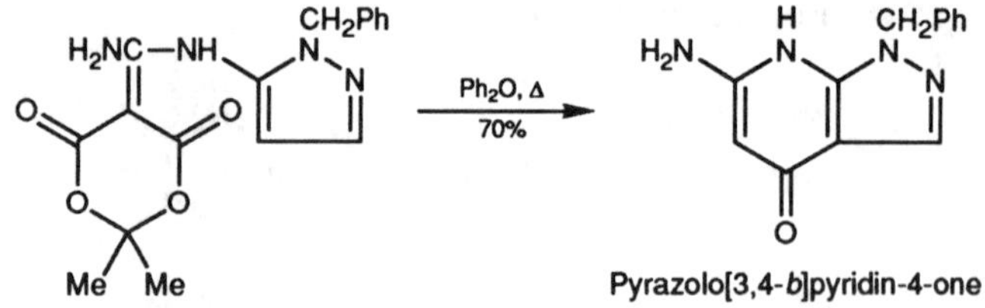

(73.7)

Ph_2O, Δ
70%

Pyrazolo[3,4-*b*]pyridin-4-one [3757]

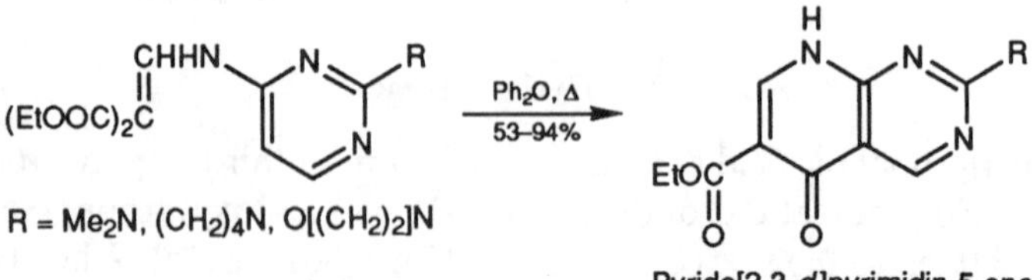

R = Me_2N, $(CH_2)_4N$, $O[(CH_2)_2]N$

Ph_2O, Δ
53–94%

Pyrido[2,3-*d*]pyrimidin-5-one [2240]

4. Pyran-2-one or -4-one

Fairly high yields of coumarins are obtainable by irradiation for 4–6 h of methoxycinnamic acids in acetonitrile–water saturated with oxygen and containing naphthalene-1,4-dinitrile. Further routes to the flavone ring system continue to appear. Flavone-3-carboxylic acids are accessible from β-phenoxybenzylidenemalonic acids and either sulphuric acid or trifluoroacetic acid–trifluoroacetic anhydride. Attempts to cyclize the malonic esters were less successful; of the condensing agents tried, PPA gave the best results. Both PPA and sulphuric acid cyclized a 4-phenoxypyran-2-carboxylic ester [3290]. Cyclization of 2-phenoxybenzoic acids to 9-xanthenones (p. 384) may be efficiently achieved by heating with PPE [3665].

$R = MeO, (MeO)_2, OCH_2O$ — $MeCN–H_2O$, $h\nu$; 60–80% — 1-Benzopyran-2-one [3191]

$R^1 = H, MeO$; $R^2 = H, Et$; Ar = Ph, 2-furyl
i, $R^2 = Et$: PPA, Δ;
ii, $R^2 = H$: $CF_3COOH–(CF_3CO)_2O$ or H_2SO_4
i or ii; 50–65% — 1-Benzopyran-4-one, 2-aryl- [3289]

PPA or H_2SO_4, Δ; 63–72% — Pyrano[4,3-*b*][1]benzopyran-1,10-dione [3290]

5. Thiopyran or Thiopyran-4-one

Both *cis* and *trans*-3-phenoxycinnamic acids cyclize with phosphorus pentoxide to pyran-4-ones but the corresponding phenylthiocinnamic acid isomers behave differently; the *trans* isomer gives a thiopyranone whereas the *cis* compound produces an indenone. Diels–Alder reactions with heterodienes have been reviewed [2455, 3067a, 3880]. Indole-3-dithiocarboxylate esters act as dienes in the Diels–Alder reaction with DMAD (cf. Chapter 65, Section II.7).

$CSSR^3$ + DMAD — diox or DMF, Δ; 65–95% — Thiopyrano[4,3-*b*]indole [2522]

$R^1 = H, Me$; $R^2 = Me, Ph$; $R^3 = Me, CH_2CN$

PhH, P_2O_5, Δ; ~90% — 1-Benzothiopyran-4-one, 2-aryl- [2265]

CHAPTER 74

Carboxylic Acid or Ester and Ring-nitrogen

I. Formation of a Five-membered Ring 1112
1. Pyrrole 1112
2. Pyrrol-2-one 1113
3. Pyrazol-3-one 1113
4. Imidazole-4,5-dione 1113
5. Thiazol-4-one 1114
II. Formation of a Six-membered Ring 1114
1. Pyridin-2-one or Pyridine-2,5-dione 1114
2. Pyrimidin-4-one 1115
3. 1,2,4-Triazin-5-one 1117
4. 1,4-Oxazine or 1,4-Thiazine 1117

I. FORMATION OF A FIVE-MEMBERED RING

1. Pyrrole

The nitrogen of ethyl pyridine-3-acetate becomes a bridgehead atom when the compound is heated with a phenacyl halide. The chemistry of α-halo ketones has been reviewed [B-45].

CH_2COOEt (pyridine) + $BrCH_2C(=O)$ (o-tolyl, Me) $\xrightarrow[60\%]{Me_2CO,\ \Delta}$ Indolizine (CO–OEt, Me) [3817]

2. Pyrrol-2-one

Condensation of Meldrum's acid (review of uses [3949]) with a pyrrole-2-aldehyde yields the pyrrolylidene **(74.1)** which is cyclized efficiently to the pyrrolizinone on flash pyrolysis at 600 °C. No examples of annulation of a carboxylic ester and a ring-nitrogen to a pyrrole or pyrrol-2-one are given in Part 1, but reductive cyclization of **(74.2)** is a step in the synthesis of the alkaloid lennoxamine (see also Chapter 37, Section III.1).

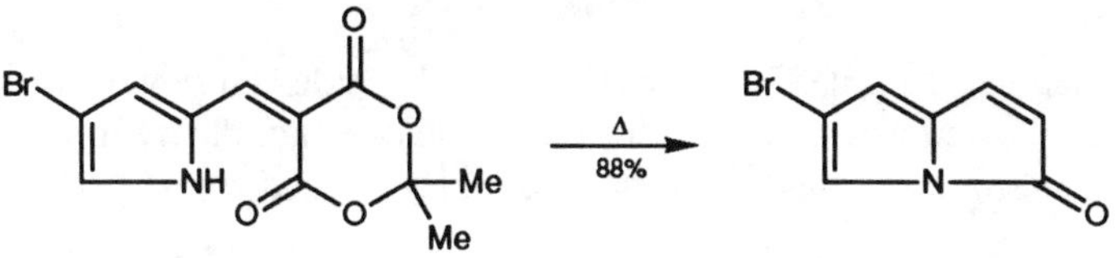

Δ, 88% [3403]

(74.1) 3-Pyrrolizinone

i, 80% [3319]

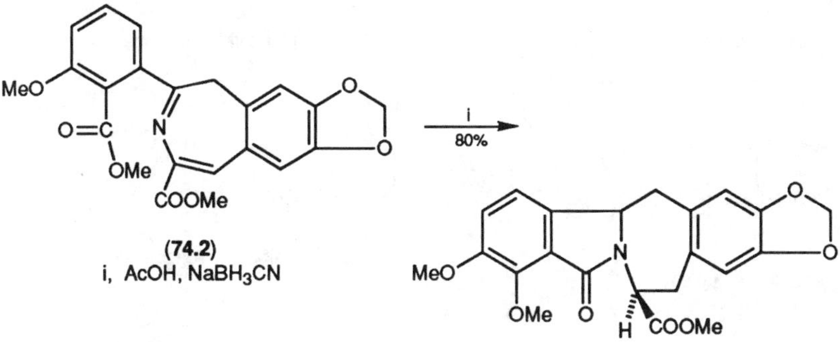

(74.2)
i, AcOH, $NaBH_3CN$

1,3-Dioxolo[4,5-*h*]isoindolo[1,2-*b*][3]benzazepin-8-one

3. Pyrazol-3-one

This ring is formed at room temperature when pyridine-2-acetic acid is stirred with hydroxylamine-*O*-sulphonic acid.

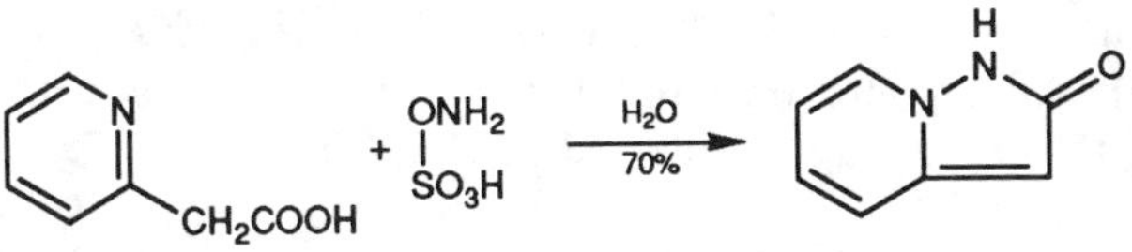

H_2O, 70% [2861]

Pyrazolo[1,5-*a*]pyridin-2-one

4. Imidazole-4,5-dione

2-(Ethoxalylamino)azoles undergo a thermal intramolecular cyclization on to the ring-nitrogen, but the yield may be low.

[2451]

Imidazo[1,2-*b*][1,2,4]triazole-5,6-dione

R = H, NH_2

5. Thiazol-4-one

When azoles carrying a SCH_2COOH or similar side-chain next to the ring-nitrogen are heated with either acetic anhydride or ethanolic hydrogen chloride and 4-toluenesulphonic acid, a thiazolone ring is formed, usually in good yield.

[2464]

R = Ph, $4\text{-}MeOC_6H_4$

Imidazo[2,1-*b*]thiazol-3-one

[3912]

Thiazolo[3,2-*a*]benzimidazol-1-one

Ar = Ph, $4\text{-}MeC_6H_4$

i, EtOH–HCl(g), Δ; ii, EtOH, TsOH, Δ

II. FORMATION OF A SIX-MEMBERED RING

1. Pyridin-2-one or Pyridine-2,5-dione

An acetoacetate side-chain provides the four carbon atoms required to form a pyridine ring by this base-induced cyclization [2034, 2081]; a better yield (90%) was later obtained by heating the acetoacetate **(74.3)** in xylene [2356]. DMSO has also been used as solvent [3252]. A pyridin-2-one ring is formed by heating the ester **(74.4)** with first ammonia and then 4-toluenesulphonic acid. Dehydrogenation with DDQ gives an aromatized ring [3545].

[2034]

(74.3)

R = Me, Et, $PhCH_2$

Pyrido[2,1-*b*]benzothiazol-1-one

DMSO, Δ, 63% [3252]

Pyrido[1,2-*a*]quinoxaline-1,5-dione

(74.4)

i, MeOH, NH_4OH, Δ;
ii, TsOH, diox, PhH; iii, DDQ

i, ii, iii, Δ, 58% [3545]

Indolo[3,2,1-*de*][1,5]naphthyridin-6-one

A high yield of a compound related to ellipticene (which has anticancer properties) and contains a pyridine-2,5-dione ring is obtained by heating the keto acid (**74.5**) with acetic anhydride.

(74.5)
R = H, MeO

Ac_2O, N_2, Δ, 80–96% [3549]

Indolo[1,2-*g*]-1,6-naphthyridine-5,12-dione

2. Pyrimidin-4-one

Cyclization of an ester group on to a ring-nitrogen may be effected by heating the compound under reduced pressure or in diphenyl ether [2211, 2259], but such high temperatures can cause decomposition; for example, the nitroamine (**74.6**) and other compounds give better results when heated at 90 °C with PPA [2924, 3814] or by treatment with t-butoxide–DMF without external heat [3203].

Ph_2O, Δ, 25–99% [2211]

R = Me, Ph, 2-HO-, 4-Me-C_6H_4,
2-Me-, 4-Cl-$C_6H_4OCH_2$

1,3,4-Oxadiazolo[3,2-*a*]pyrimidin-5-one

Another ring system synthesized similarly:

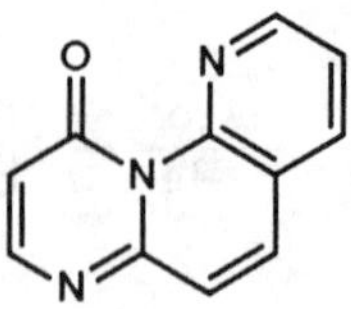

[2053]

Pyrimido[1,2-*a*][1,8]naphthyridin-10-one

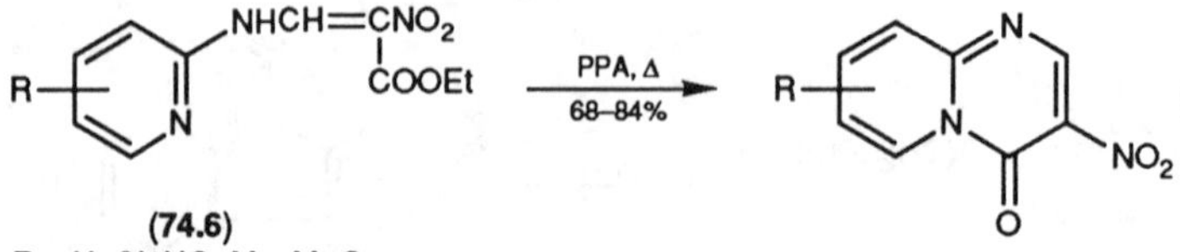

(74.6)
R = H, Cl, HO, Me, MeO

PPA, Δ
68–84%

[2924, 3814]

Pyrido[1,2-*a*]pyrimidin-4-one

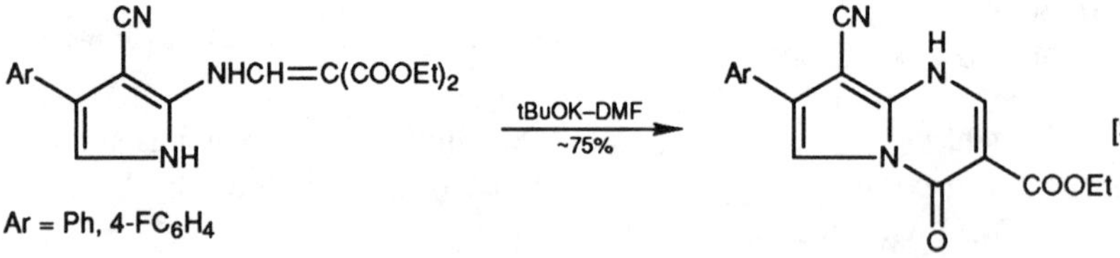

Ar = Ph, 4-FC_6H_4

tBuOK–DMF
~75%

[3203]

Pyrrolo[1,2-*a*]pyrimidin-4-one

Suitably positioned carboxylic acid and ring-nitrogen functions may be annulated by heating the compound with acetic anhydride [3695, 3884].

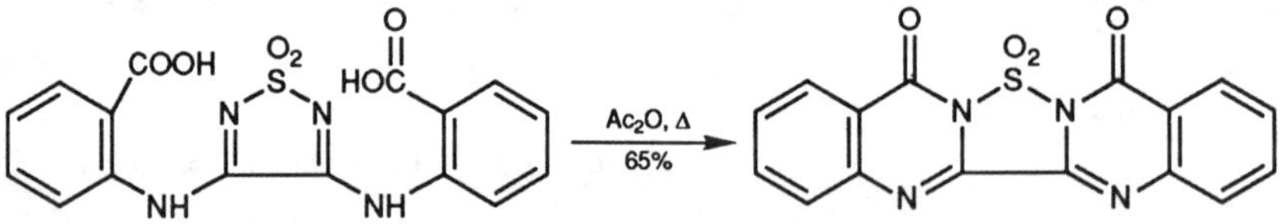

Ac_2O, Δ
65%

[2756]

[1,2,5]Thiadiazolo[3,2-*b* : 4,5-*b'*]diquinazoline-11, 15-dione 13,13-dioxide

Other ring systems synthesized similarly:

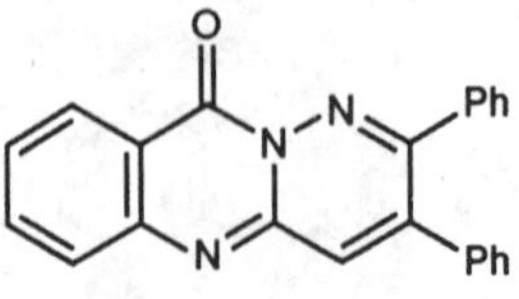

[3884]

Pyridazino[3,2-*b*]quinazolin-10-one

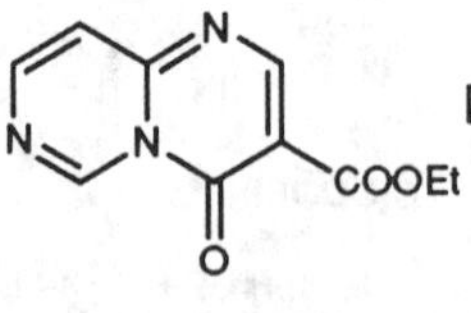

[3695]

Pyrimido[1,6-*a*]pyrimidin-4-one

3. 1,2,4-Triazin-5-one

The carboxylic acid group of the hydrazone (**74.7**) reacts very efficiently with the ring-nitrogen under the influence of hot PPA.

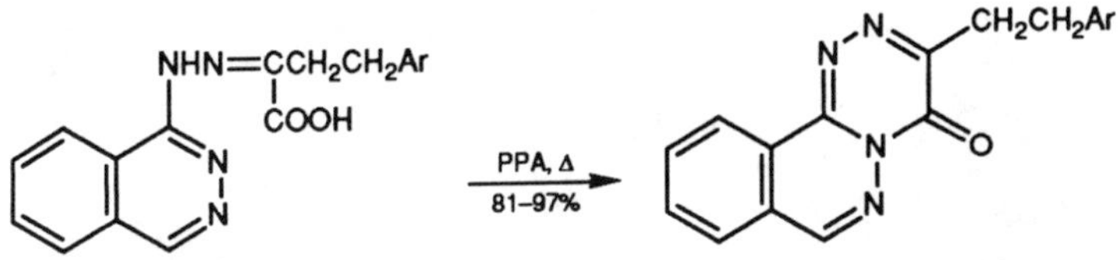

NHHN=C(COOH)CH$_2$CH$_2$Ar

PPA, Δ
81–97%

[3199]

(**74.7**)
Ar = 4-Cl-, 4-Me-C_6H_4

1,2,4-Triazino[3,4-*a*]phthalazin-4-one

4. 1,4-Oxazine or 1,4-Thiazine

The di- or tri-thiocarbonate (**74.8**) is cyclized by first treating it with allyloxalyl chloride to give the *N*-oxalyl derivative which is not isolated in a pure state but is warmed with triethyl phosphite at 40–85 °C to give the ring system of isocephem (or iso-oxacephem) antibiotics which are synthesized by a broadly similar route.

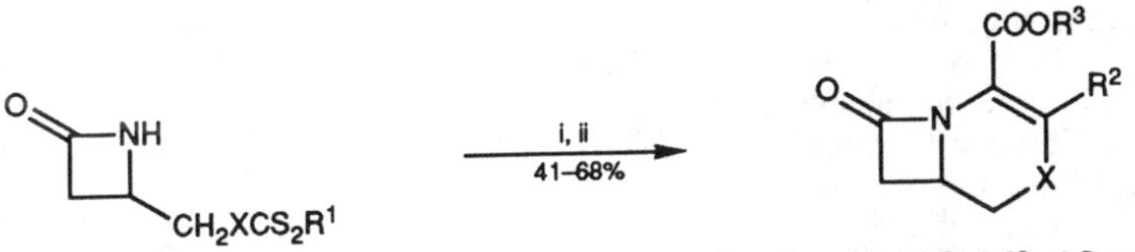

$CH_2XCS_2R^1$

i, ii
41–68%

$COOR^3$

[3184]

(**74.8**)
R^1 = Me, Et; R^2 = SR^1;
R^3 = $CH_2{=}CHCH_2$; X = O, S
i, $ClCOCOOR^3$, iPr_2NEt; ii, CH_2Cl_2, $P(OEt)_3$, Δ

1-Aza-4-oxabicyclo[4.2.0]oct-2-en-8-one
1-Aza-4-thiabicyclo[4.2.0]oct-2-en-8-one

CHAPTER 75

Carboxylic Acid or its Derivative and a Sulphur-containing Group

I. Formation of a Five-membered Ring 1118
1. Isothiazole 1-Oxide 1118
2. Isothiazol-3-one 1119
3. Isothiazol-3-one 1-Oxide or 1,1-Dioxide 1119
II. Formation of a Six-membered Ring 1120
1. 1,2-Thiazine 1,1-Dioxide or 1,2-Thiazin-3-one 1,1-Dioxide 1120
2. 1,2-Thiazin-3-one 1-Oxide 1120
3. 1,3-Thiazin-4-one 1121
4. 1,2,3-Oxathiazine 2,2-Dioxide 1121
5. 1,2,6-Thiadiazine 1,1-Dioxide 1121

I. FORMATION OF A FIVE-MEMBERED RING

1. Isothiazole 1-Oxide

When a 2-sulphinyl-carboxamide is warmed with hydrazoic acid, a 3-iminoisothiazole 1-oxide is formed.

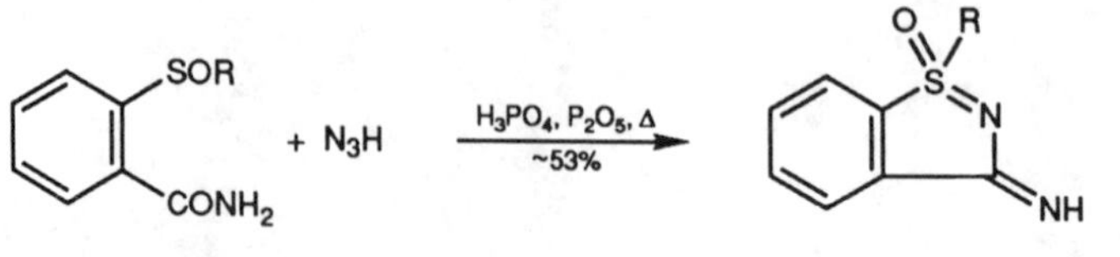

SOR, $CONH_2$ + N_3H —(H_3PO_4, P_2O_5, Δ; ~53%)→ [2705]

R = $HOCH_2CH_2$, Pr, $CH_2{=}CHCH_2$, cC_6H_{11}, $PhCH_2$, Ph, 4-Cl-, 4-MeO-, 2-Me-C_6H_4

1,2-Benzisothiazole 1-oxide

2. Isothiazol-3-one

o-Sulphenamidobenzoates (**75.1**) may be converted into benzisothiazolones in a basic medium and with high yields; the precursors are prepared in two stages from methyl 2-mercaptobenzoates [2749]. A potentially versatile synthesis of the isothiazolone ring is offered by heating *o*-sulphinyl-*N*-unsubstituted benzamides (**75.2**, R = H) with acetic anhydride. However, the scope of this reaction appears to be limited by the fact that a substituent on the carboxamide nitrogen changes the course of the reaction (see Section II.3) [2604]. Isothiazolones may also be synthesized from *o*-sulphinyl esters by successive treatments with sodium hydrosulphide, sodium hydrogencarbonate and hydroxylamine-*O*-sulphonic acid—all at or below room temperature [3689].

SNHR, COOMe → MeOH, MeONa or KOH, 85–93% → S, NR, O [2749]

(75.1)
R = $HOCH_2CH_2$, Pr, allyl, cC_6H_{11}, $PhCH_2$, Ph, 4-MeO-, 2-Me-C_6H_4

1,2-Benzisothiazol-3-one

$SOCH_2Ph$, CONHR → R = H: Ac_2O, Δ, 90% → S, NAc, O

1,2-Benzisothiazol-3-one [2604]

R ≠ H: Ac_2O, Δ, ~77% → S, Ph, NR, O

(75.2)
R = H, Me, Ph

1,3-Benzothiazin-4-one

R^1, X, N, R^2, SOPh, F, COOEt, O → i, ii, iii, ~69% → R^1, X, N, R^2, S, NH, F, O, O [3689]

R^1 = Cl, F; R^2 = cyclopropyl; X = N, CF
i, THF, NaSH; ii, $NaHCO_3$; iii, H_2NOSO_3H

X = N Isothiazolo[5,4-*b*][1,8]naphthyridine-3,4-dione
X = CF Isothiazolo[5,4-*b*]quinoline-3,4-dione

3. Isothiazol-3-one 1-Oxide or 1,1-Dioxide

Neighbouring ester and methylsulphinyl groups may be annulated by heating the compound with sodium azide in the presence of phosphoric acid–phosphoric anhydride. Amination of a 2-chlorosulphonylbenzoate ester [prepared *in situ* by diazotization of the amino ester and then treatment with sulphur dioxide-copper(II) chloride] is a good route to a fused isothiazol-3-one 1,1-dioxide ring.

$R^1 = H, Cl; R^2 = Me, Bu, Ph$ + NaN_3 → (H_3PO_4–P_2O_5, Δ; 14–85%) [2373]

1,2-Benzisothiazol-3-one, 1-dioxide

1. HNO_2, SO_2, $CuCl_2$ 2. NH_3; 50–89% [3098]

1,2-Benzisothiazol-3-one, 1,1-dioxide

II. FORMATION OF A SIX-MEMBERED RING

1. 1,2-Thiazine 1,1-Dioxide or 1,2-Thiazin-3-one 1,1-Dioxide

Cyclization of a sulphonamide and a side-chain carboxylic acid is readily effected by hot PPA; when the carboxyl group is replaced by a nitrile, heating with sulphuric acid has a similar effect, but the carbonyl group is replaced by an imino group.

i or ii; 66–100% [2087]

i, $R^1 = H, Et, Ph; R^2 = COOH$, PPA, Δ; X = O
ii, $R^1 = H; R^2 = CN, H_2SO_4$; X = NH

X = O 1,2-Benzothiazin-3-one 1,1-dioxide
X = NH 1,2-Benzothiazine 1,1-dioxide

2. 1,2-Thiazin-3-one 1-Oxide

o-Sulphinylphenylacetic acid esters give moderately good yields of benzothiazinones when treated with sodium azide and PPA, probably through a nitrone intermediate.

+ P_2O_5 + H_3PO_4 → (NaN_3, $(ClCH_2)_2$, Δ; ~44%) [2450]

R = Me, Et

1,2-Benzothiazin-3-one 1-oxide

3. 1,3-Thiazin-4-one

This ring is formed by acid-catalysed reaction of a 2-thiocyanatobenzoyl chloride with hydrogen chloride, or of the corresponding carboxylic acid with phosphorus pentachloride (review of this kind of reaction [2557]). The chlorine atom in the product is labile.

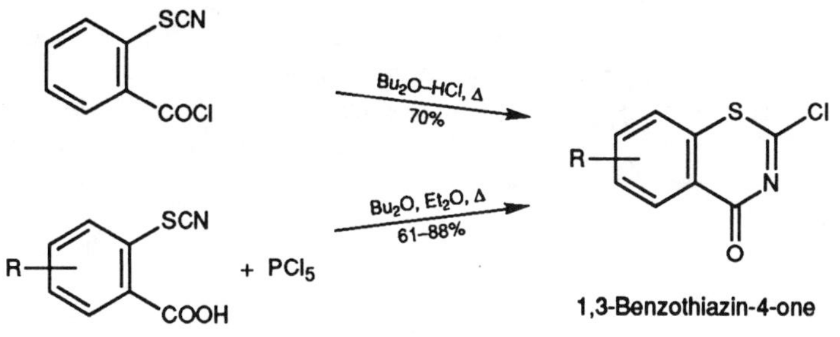
SCN
COCl
Bu2O–HCl, Δ
70%
SCN
R
COOH
+ PCl5
Bu2O, Et2O, Δ
61–88%
R
S
Cl
N
O
[2079]

1,3-Benzothiazin-4-one

R = H, Cl, NO_2

4. 1,2,3-Oxathiazine 2,2-Dioxide

Baker's yeast efficiently cyclizes 2-sulphamoyloxybenzonitrile in a buffer at 37 °C.

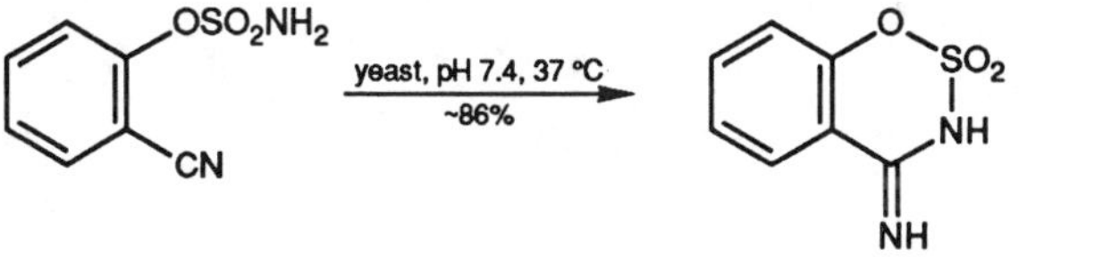
OSO2NH2
CN
yeast, pH 7.4, 37 °C
~86%
O
SO2
NH
NH
[3791]

1,2,3-Benzoxathiazine 2,2-dioxide

5. 1,2,6-Thiadiazine 1,1-Dioxide

Carboxy and sulphamide groups may be annulated by treatment with a carboxylic organic acid.

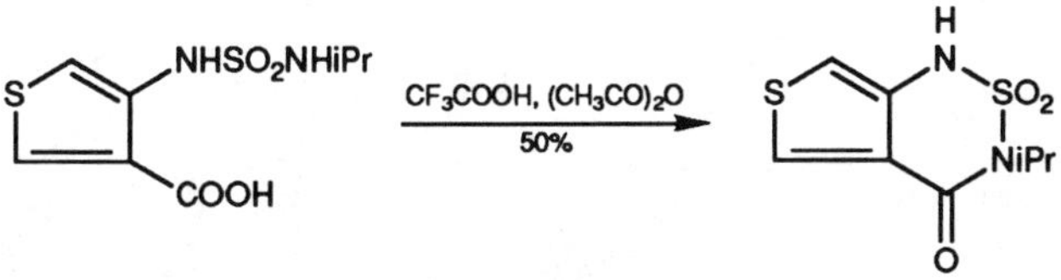
S
NHSO2NHiPr
COOH
CF3COOH, (CH3CO)2O
50%
S
H
N
SO2
NiPr
O
[3601]

Thieno[3,4-*c*][1,2,6]thiadiazin-4-one 2,2-dioxide

Other ring systems synthesized similarly:

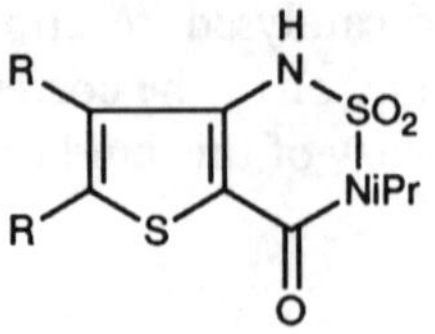

[3601]

R,R = H,H Thieno[3,2-*c*][1,2,6]thiadiazin-4-one 2,2-dioxide

R,R = $(CH{=}CH)_2$ [1]Benzothieno[3,2-*c*][1,2,6]thiadiazin-4-one 2,2-dioxide

CHAPTER 76

1,2-Diamine

I. Formation of a Five-membered Ring 1124
  1. Imidazole 1124
    A. Reaction with a Carboxylic Acid 1124
    B. Reaction with a Carboxylic Acid Derivative 1125
    C. Reaction with an Aldehyde or Ketone 1127
    D. Reaction with Other Reagents 1128
  2. Imidazol-2-one or Imidazole-2-thione 1129
  3. 1,2,3-Triazole 1130
  4. 1,2,4-Triazole 1130
  5. 1,2,5-Thiadiazole 1130
II. Formation of a Six-membered Ring 1131
  1. Pyrazine 1131
  2. Pyrazin-2-one 1134
  3. 1,2,4-Triazine or 1,2,4-Triazin-3-one or -5-one 1136
III. Formation of a Seven-membered Ring 1137
  1. 1,4-Diazepine 1137
  2. 1,4-Diazepin-5-one or 1,4-Diazepine-5-thione 1138
  3. 1,4-Diazepine-5,7-dione 1140
  4. 1,2,4-Triazepin-5-one 1140
  5. 1,2,5-Thiadiazepin-6-one 1-Oxide 1140
  6. 1,3,5-Triazepine 1141
  7. 1,2,4,5-Tetrazepine 1141

The introductory comments on pp. 398–399 also apply to this chapter. The synthesis of heterocycles from 1,2-diamines has been reviewed [3562]. Other cyclizations in which 1,2-diamines take part are mentioned in Chapter 11, Section I.6, Chapter 30, Section II.4 and Chapter 81, Section II.1.

I. FORMATION OF A FIVE-MEMBERED RING

1. Imidazole

A. Reaction with a Carboxylic Acid

The Phillips synthesis of benzimidazoles usually gives lower yields when an aromatic carboxylic acid is used in conjunction with a mineral acid to aid the dehydrating action. A more effective reagent, *N*-diphenylphosphinyl-*N′*-methylpiperazine, is easily prepared (in 95% yield) by adding a solution of *N*-methylpiperazine in dichloromethane to diphenylphosphinic chloride. Cyclization of a diamine in the presence of this reagent and triflic anhydride (review of triflic acid and its derivatives [2920]) proceeds at or below room temperature and yields are usually high [3286]. Most methods for the synthesis of benzimidazoles require acidic conditions or, rarely, a basic environment, but reaction of a diamine with 4-ethoxyphenylacetic acid in the presence of a mildly basic ethyl 2-ethoxyquinoline-4-carboxylate (EEQ) gives a quantitative yield on prolonged heating (about 26 h). Phosphorus pentachloride in chloroform takes only 3 h to produce an 85% yield of the same product [2766]. Photochemical oxidation of *o*-phenylenediamine (**76.7**, R^1 = H) can produce one of several cyclized products depending on the solvent and the presence or absence of air; irradiation in acetic acid with access to oxygen gives 2-methylbenzimidazole in good yield. No cyclization occurred in ethanolic potash [3581].

R^1, NHR^2, NH_2 + $R^3C_6H_4COOH$ —i, 80–95%→ R^1, R^2, N, $C_6H_4R^3$, N [3286]

R^1 = H, COOEt; R^2 = H, Ph; R^3 = H, 2- or 4-Me-

i, CH_2Cl_2, $Ph_2PON(CH_2CH_2)_2NMe$, $(CF_3SO_2)_2O$

Benzimidazole

NHR, O_2N, NH_2 + OEt, CH_2COOH —THF, EEQ, Δ, 100%→ R, N, $CH_2C_6H_4$-4-OEt, N, O_2N [2766]

R = $Et_2NCH_2CH_2$

NH_2, R^1, NH_2 —AcOH, O_2, *hν*, 73%→ H, N, Me, N [3581]

(76.7)
R^1 = H

B. *Reaction with a Carboxylic Acid Derivative*

Conversion of 4,5,6-triaminopyrimidine into purines by heating at 160–240 °C with an anhydride, a carboxamide or a nitrile gives yields which vary from 36 to 96% [2304]. Aroyl halides in DMF give moderate-to-good yields of 2-aryl-benzimidazoles, and some esters cyclize the triamine similarly [2506, 2979].

i or ii, Δ 65–96% → Purine

$(MeCH{=}CHCO)_2O$, Δ 36%

[2304, 2506]

i, $PhCH{=}CHCONH_2$; R^1 = $PhCH{=}CH$
ii, $PhCH_2CN$; R^1 = $PhCH_2$; R^2 = $MeCH{=}CH$

Some difunctional compounds, for example, *N*-(ethoxycarbonyl)thiocarboxamides, may behave as monofunctional reagents in a reaction with a diamine. In this example, an ethoxycarbonyl group is eliminated as ethyl carbamate and the thioamide completes the imidazole ring [2899]. Diamines usually react with 3-oxo esters to give a fused diazepine (p. 423), but when the reactants are heated without a solvent or catalysts at or above 100 °C, a benzimidazole is obtained in high yield [2371].

(76.7) + $R^2C{=}S$ / $NHCOOEt$ —THF or EtOH, Δ, 80–95%→ Benzimidazole [2899]

R^1 = H; R^2 = Et, 4-Me-, 4-MeO-C_6H_4, 2-thienyl

(76.7) + $AcCH_2$ / COOEt —Δ, 79%→ [2371]

R^1 = Me

Dithioate esters react with a diamine at room temperature in methanolic alkali at pH 8 to give high yields of 2-alkyl- or 2-aryl-substituted benzimidazoles [2860]. It is difficult to convert benzimidazole-2-carboxylic acid into its amides by conventional methods as it condenses with itself to form a pentacyclic product, but when *o*-phenylenediamine is heated with an oxamic ester, the

2-carboxamide is formed [3028]. Treatment of nitriles with a strong base converts them into an imidate, which becomes an imidinium hydrochloride on addition of *o*-phenylenediamine dihydrochloride. These two compounds react to form a benzimidazole [3959].

(76.7) + $R^2C(=S)SEt$ —MeOH, aq. NaOH, pH 8; ~88%→ 2-R^2-benzimidazole [2860]

$R^1 = H$; $R^2 = Me, Ph$

(76.7) + $EtOOC{-}CONHR^2$ —DMF, Δ; 20–75%→ 2-$CONHR^2$-benzimidazole [3028]

$R^1 = H$; R^2 = alkyl, allyl, $PhCH_2$, $Ph(CH_2)_2$, Ph, $4\text{-}ClC_6H_4$, $2,4,6\text{-}Me_3C_6H_2$, $2,6\text{-}Cl_2C_6H_3$

Benzimidazole

(76.7) 2HCl + 2-$NCCH_2O$-C_6H_4-MeOOC —MeONa–MeOH; 88%→ 2-(CH_2O-C_6H_4-2-COOMe)-benzimidazole [3959]

$R^1 = H$

Cyanogen bromide in alkali joins together amino and isopropylsulphonylamino groups as a fused 2-aminoimidazole ring without external heating [3887]. The oxazolinone (**76.8**) behaves as a lactone; this is cleaved, and converts the diamine into an imidazole ring; the remainder of the oxazolinone is attached to C-2 of the benzimidazole [3088].

PhC(=O)-C_6H_3($NHSO_2iPr$)(NO_2) —i, ii; 50%→ PhC(=O)-1-(SO_2iPr)-benzimidazole [3887]

i, $Na_2S_2O_4$; ii, THF, BrCN, NaOH

(76.7) + (76.8) [4-(CHR^2=)-2-Ar-oxazol-5(4H)-one] —i, Δ; 66–78%→ 2-[$C(=CHR^2)NHCOAr$]-benzimidazole [3088]

(76.8)

$R^1 = H$; R^2 = Ph, 4-Cl-, 4-MeO-C_6H_4, 2-thienyl;
Ar = Ph, $4\text{-}NO_2C_6H_4$
i, EtOH or AcOH–AcONa

C. *Reaction with an Aldehyde or Ketone*

Several types of products may be obtained from *o*-phenylenediamines and aldehydes (pp. 403–405). An aldehyde converts the triamine (**76.9**) to a fused imidazole or pyrazine according to the temperature, reaction time and ratio of reactants. At room temperature, only the imidazothiadiazine is formed, but a higher temperature, long reaction time and a large excess of aldehyde favour the formation of the pyrazinothiadiazine [3476]. α-Chlorobenzaldoximes react with a diamine (without heating) to give a good yield of 2-arylbenzimidazoles [3468].

(76.9) + RCHO → (DMF or AcOH, 78%) Imidazo[4,5-*d*][1,2,6]thiadiazine 2,2-dioxide

(76.9) + RCHO → (DMF or AcOH, Δ, 51%) Pyrazino[2,3-*c*][1,2,6]thiadiazine 2,2-dioxide [3476]

(76.9)
R = Me, Ph, 4-MeO-, 4-$NO_2C_6H_4$, 2-thienyl-, 2-furyl-, 5-NO_2-2-furyl

(**76.7**) + ArC(Cl)=NOH → (EtOH, 60–90%) 2-arylbenzimidazole [3468]

R^1 = H; Ar = Ph, 4-Cl-, 3-NO_2-C_6H_4

Another carbonyl group derivative which forms an imidazole ring with a diamine is *S,S*-dimethyl-*N*-tosyliminodithiocarbonimidate (**76.10**), which is useful in the synthesis of 8-aminopurines. In an attempt to prepare a 1,5-benzodiazepine, the chloroaldehyde (**76.11**) was stirred or warmed with *o*-phenylenediamine but the main product was the benzimidazole.

Diaminomethylpyrimidine + TsN=C(SMe)$_2$ (**76.10**) → (DMF, Δ, ~60%) Purine (8-NHTs) [2274]

Purine

(76.7) + [2-chloro-3-formyl-1-azaazulene] (76.11), $R^1 = H$ —CH_2Cl_2, 60%→ Benzimidazole [3765]

D. Reaction with Other Reagents

The use of diphenyl cyanocarbonimide [$(PhO)_2C=NCN$] [99] has been extended [3202], and 1,3-bis(methoxycarbonyl)-*S*-methylisothiourea (**76.12**) is a useful reagent for the synthesis of benzimidazole-2-carbamate [3662]. The reagent is easily prepared *in situ* from 2-methyl-2-thiopseudourea sulphate and methyl chloroformate. When a diamine is heated with 2-picoline for several hours, the acidity of the methyl hydrogens allows reaction with even the relatively weakly basic diamines. Addition of sulphur is beneficial in some cyclizations [2877]. A less convenient method of synthesizing a 2-substituted benzimidazole in the average laboratory is to pass a mixture of 2-nitroaniline and alcohol over alumina at 320 °C [2062].

(76.7) + $C(OPh)_2{=}NCN$, R^1 = PrS —MeOH, Δ, 75%→ 5-PrS-2-(NHCN)-benzimidazole [3202]

(76.7) + MeSC(NHCOOMe)=NCOOMe (76.12), R^1 = Me, 2-thienylcarbonyl —diox, HCl, Δ, 51–88%→ R^1-1-Me-2-(NHCOOMe)-benzimidazole [3662]

(76.7) + 2-picoline (Me, N), R^1 = H, Me, Cl, MeO, NO_2, COOEt —S, 170 °C, 10 h, 32–80%→ R^1-2-(2-pyridyl)-benzimidazole [2877]

2-nitroaniline (NH_2, NO_2) + $2R^1OH$ —Al_2O_3, Δ, 30–67%→ 1-R^1-2-R^2-benzimidazole [2062]

R^1 = Me, Et, Pr, Bu; R^2 = H, Me, Et, Pr

Benzimidazole

A benzimidazole carrying a 2-nitromethyl chain may be prepared in good yield and under mild conditions by stirring a diamine at about 10 °C with 1,1-dichloro-2-nitroethene [2723]. β-Nitrostyrene reacts with *o*-phenylenediamine (the only diamine investigated) in boiling ethanol to give a 2-arylbenzimidazole in 30–65% yield [3035]. Heating the diamine with 2-chloro-5-nitropyridine in dilute acid gave a pyridobenzimidazole instead of the expected carboline [2156].

(76.7) + $Cl_2C{=}CHNO_2$ —MeOH, ~75%→ [2723]
R^1 = H, Me

(76.7) + [2-chloro-5-nitropyridine] —2*M* HCl, EtOH, Δ, 58%→ [2156]
R^1 = H

Pyrido[1,2-*a*]benzimidazole

2. Imidazol-2-one or Imidazole-2-thione

Disuccinimido carbonate (DSC) is a useful reagent for the insertion under mild conditions of a carbonyl group in diamines [3019]. A cyclization which leads to imidazo[4,5-*c*]pyridin-2-one is described in Section III. Carbonyldi-imidazole–TEA is another reagent for this type of cyclization [3925]. Potassium ethylxanthate (potassium ethyldithiocarbonate) is an alternative to carbon disulphide in the synthesis of the tautomeric 2-benzimidazolethiones; it was first used many decades ago and usually gives good yields [3259]. The synthesis of an imidazo[4,5-*c*]pyridin-2-one is described in Section III.2 [2327].

—MeCN, DSC, ~84%→ [3019]
X = CH, N

X = CH 2-Benzimidazolone
X = N Imidazo[4,5-*b*]pyridin-2-one

—i, ii, 50%→ [3925]
R = $Me_2NCH_2CH_2$
i, MeOH, H_2, Pd/C; ii, CH_2Cl_2, CDI, TEA, Δ

2-Benzimidazolone

(76.7) —EtOH, EtOCSSK, 83%→ [3259]
R^1 = (1,3-dioxolan-2-yl)CH_2O

2-Benzimidazolethione

Another ring system synthesized similarly:

[3259]

Pyrano[3,4-*e*]benzimidazol-9-one, 2-thioxo-

3. 1,2,3-Triazole

Benzotriazole (review of its properties [3993]) is prepared by diazotization of *o*-phenylenediamine (p. 409); this method may also be used to synthesize 1-substituted benzotriazoles from mono-*N*-substituted diamines.

HNO_2 47% [2105, 2152]

R = Me, N=C(CN)COOEt

Benzotriazole

4. 1,2,4-Triazole

Application of the Vilsmeier reaction (review [1676]) to a *C,N*-diamine gives a good yield of this fused ring. An imino-N^1-amine is cyclized with a dichloromethyleneammonium chloride to form a fused triazole ring in high yield.

DMF–$POCl_3$ 67% [2613]

R = β-D-ribofuranosyl

[1,2,4]Triazolo[2,3-*a*]purin-9-one

+ $Cl_2C{=}\overset{+}{N}Me_2$ Cl^- $CHCl_3$, Δ 95% [2549]

Thiazolo[3,2-*b*][1,2,4]triazole

5. 1,2,5-Thiadiazole

An alternative preparation (cf. p. 412) of a fused thiadiazole consists of heating a diamine and thionyl chloride in toluene rather than in pyridine-dichloromethane. Sulphur monochloride–DMF or sulphur dichloride–DMF also gives good yields.

PhMe, $SOCl_2$, Δ 78% [2315]

R = H, Me

Thieno[3,2-*e*][2,1,3]benzothiadiazole

i or ii 59–75% [2757]

i, SCl, DMF; ii, SCl_2, DMF

[1,2,5]Thiadiazolo[3,4-*c*][1,2,5]thiadiazole

II. FORMATION OF A SIX-MEMBERED RING

1. Pyrazine

Heating a 1,2-dicarbonyl compound with an *o*-phenylenediamine continues to be used and varied for the preparation of quinoxalines. 1,2,3-Propanetrione 1,3-dioxime **(76.13)** and an unsymmetrical diamine give only one of the possible isomers as product [3209, 3946], but a reinvestigation [3557] of the cyclization of a substituted *o*-phenylenediamine with phenylglyoxal has given results which are significantly different from those of an earlier study. The chemistry of pteridines has been reviewed [3454, 3594, 3664]. The 2-oxoacetal **(76.14)** is an oil which exothermally enolizes to a solid on standing at ambient temperature. The solid may be used to cyclize 2,5,6-triaminopyrimidin-4-one to a mixture of 6- and 7-substituted pteridin-4-ones [2987].

+ **(76.13)** i, ii, Δ 82% [3209]

i, EtOH; ii, aq. $NaHCO_3$

2,4-Pteridinedione

+ **(76.13)** dil. HCl 58% [3946]

Pyrazino[2,3-*c*][1,2,6]thiadiazine 2,2-dioxide

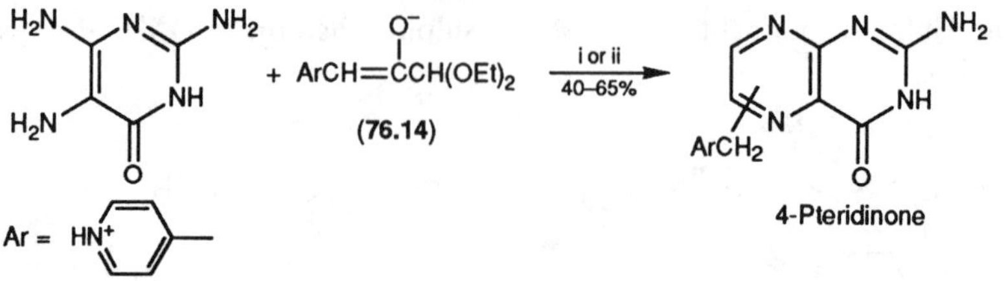

[2987]

Perfluorobutane-2,3-dione (**76.15**), a reactive compound which sometimes shows unexpected properties, behaved predictably with *o*-phenylenediamine to give quinoxalines. The cyclizations proceeded at or below room temperature in good yields [3165]. Transmolecular covalent hydrates were isolated from the reaction of diaminouracils (**76.16**) with the fluorinated diketones [3166]. 3-Phenyl-1,1,1-trifluoropropane-2,3-dione also behaves normally to give 3-(4-tolyl)-2-trifluoromethylquinoxaline in 63% yield [3409].

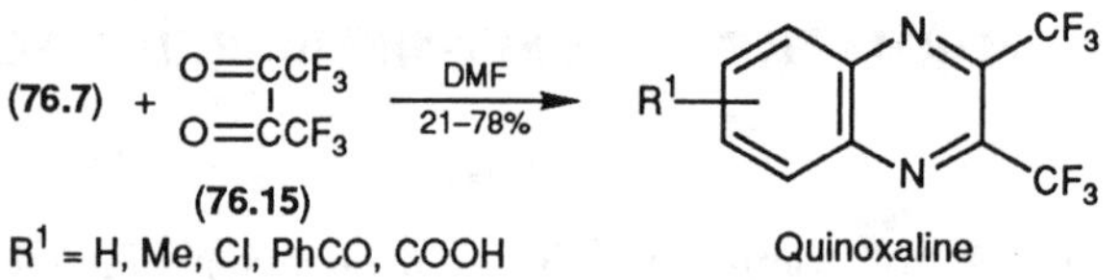

[3165]

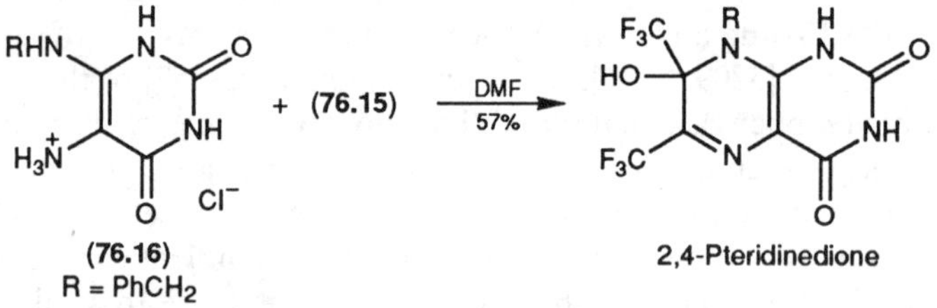

[3166]

α-Oxoaldehydes such as pyruvaldehyde usually react with a diamine on heating, but trifluoropyruvaldehyde in methanol or DMF converts diamines into quinoxalines at ambient temperature. The effect of dissymmetry in the dianion on the ratio of isomers has been studied; the major product is the result of reaction between the aldehyde with the more reactive amino group [3421]. A cyclic diketone readily reacts with *o*-phenylenediamine in acetic acid to give a fused 1,4-dihydropyrazine [2054], and polyglyoxal condenses with a pyrimidine-4,5-diamine (derived *in situ* from the nitrosoamine and hydrosulphite in hydrochloric acid) [2082].

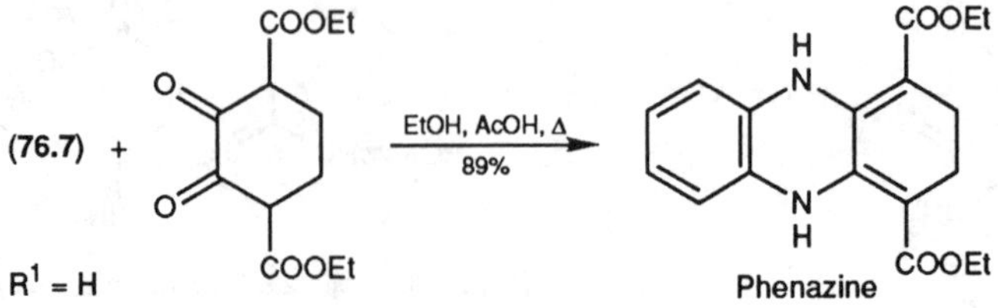

[2054]

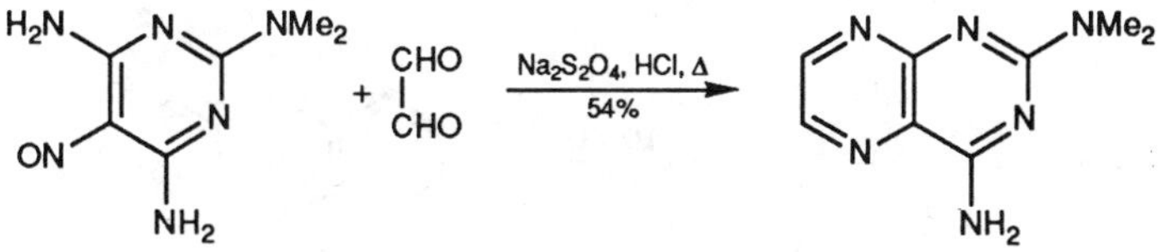

$Na_2S_2O_4$, HCl, Δ; 54% [2082]

Pteridine

A fused pyrazine may be formed from a diamine and a 1,2-diketone [2315] or diiminosuccinonitrile [2694], while a diamine hydrochloride, pentane-2,3,4-trione and sodium acetate yield a 2-methyl-3-acetylquinoxaline [2370].

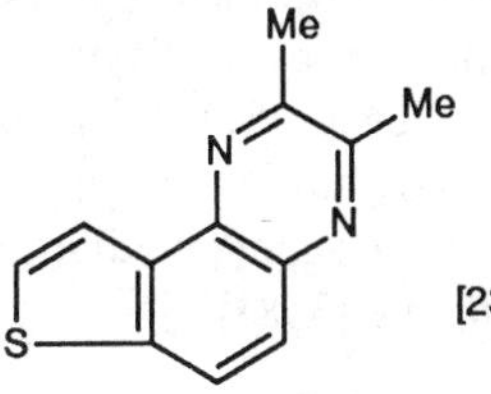

[2315] 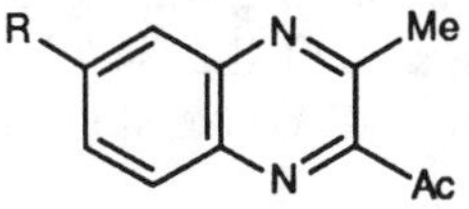 [2370]

Thieno[3,2-*f*]quinoxaline

Quinoxaline

In the presence of a strong base, two methoxy groups of (**76.17**) may be displaced by a diamine [2756]. A keto-(lactam carbonyl) is cyclized with a diamine in the presence of boric acid [2531].

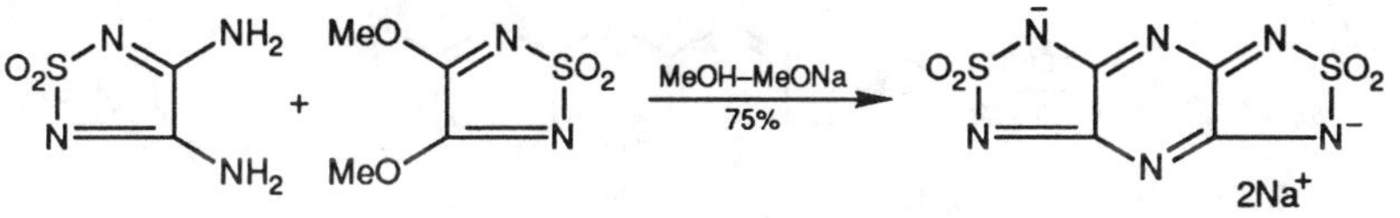

MeOH–MeONa; 75% [2756]

(**76.17**)

Bis[1,2,5]thiadiazolo[3,4-*b* : 3′,4′-*e*]pyrazine 2,2,6,6-tetraoxide

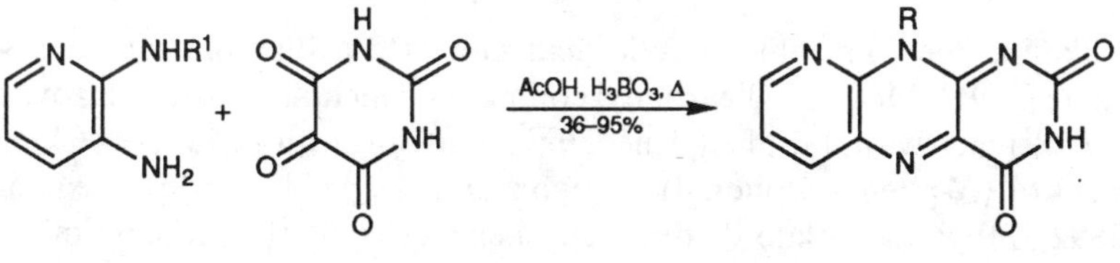

AcOH, H_3BO_3, Δ; 36–95% [2531]

$R^1 = (CH_2)_nXR^2$; R^2 = H, Ac; X = O, NH; n = 2–6

Pyrido[3,2-*g*]pteridine-2,4-dione

Ring opening of epoxy aldehydes or epoxy esters by a diamine at room temperature yields a 2-hydroxymethylquinoxaline. For epoxy esters, heating with lithium hydroxide is necessary and a 3-hydroxymethyl-2-quinoxalinone is formed in good-to-excellent yields [3335]. The chemistry of quinoxalines has been reviewed [3459].

(76.7) + OHC–epoxide–R^2 $\xrightarrow[69–78\%]{MeOH}$ Quinoxaline [3335]

R^1 = H; R^2 = Me, Ph

Quinoxaline

Di(arylazo)quinoxalines, of interest in colour chemistry, may be synthesized from a diamine and 1,2-dichloroglyoxal bisarylhydrazones, which are converted into oxalodinitrile bis(arylimines) on boiling with TEA. The intermediate bis(hydrazone) is converted in high yield to the bisarylazo compound by oxidation at ambient temperature [3612]. Some uses of hypervalent iodine compounds in organic chemistry have been reviewed [3714]. The cyclization of a diamine to a fused pyrazine by an aldehyde under certain conditions is mentioned in Section I.1.C [3476]. Irradiation of *o*-phenylenediamine in ethanol and air promotes its conversion into a phenazine [3581].

ArNHN=C(Cl)–C(Cl)=NNHAr $\xrightarrow{TEA}$ [ArN≡N–C≡C–N≡NAr] $\xrightarrow[60–73\%]{i, ii}$ 2,3-bis(N=NAr)quinoxaline [3612]

Ar = Ph, 2,4-$Cl_2C_6H_3$

i, RC_6H_3-1,2-$(NH_2)_2$, PhH, Δ; ii, CH_2Cl_2, $PhI(OCOCF_3)_2$; R = H or Me

(76.7) $\xrightarrow[60\%]{EtOH, O_2, h\nu}$ 2,3-diaminophenazine [3581]

R^1 = H

Phenazine

2. Pyrazin-2-one

In an alkaline medium, chloroacetic acid converts a diamine into a fused pyrazinone [2191]. Most 1,2-dicarbonyl compounds including ethyl benzoylformate [3776] react with 1,2-phenylenediamines to form quinoxalines. Chloroacetic acid or (*N*-phenylsulphonyl)arylglyoxalamides yield 3-arylquinoxalin-2-ones [2892]; ethyl glyoxalate diethylacetal behaves similarly with pyrimidine-4,5,6-triamine [2083]. Tautomerism in quinoxalinones has been reviewed [3119].

(76.7) + CH_2Cl–COOH $\xrightarrow[50\%]{1. NaOH\ 2. H_2O, Δ}$ 2-Quinoxalinone [2191]

R^1 = H, Me

2-Quinoxalinone

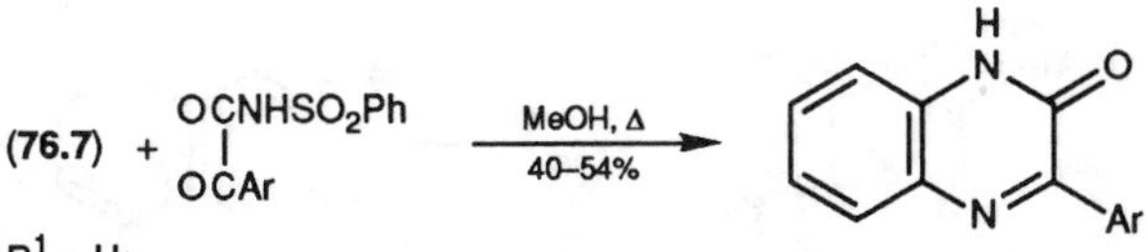

(76.7) + OCNHSO$_2$Ph / OCAr —MeOH, Δ; 40–54%→ [2892]

R^1 = H;
Ar = Ph, 4-halo-, 4-MeO-, 4-NO$_2$-C$_6$H$_4$

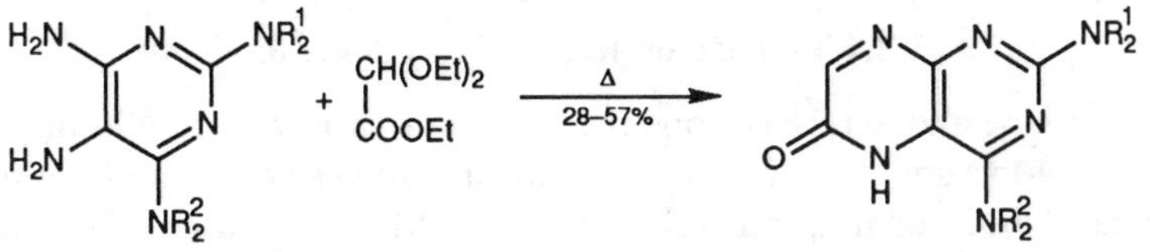

+ CH(OEt)$_2$ / COOEt —Δ; 28–57%→ [2083]

R^1,R^2 = H, Me

6-Pteridinone

Ethyl 7-oxopteridine-6-carboxylate is synthesized from the diamine and diethyl mesoxalate [2299]. A substituted 2-quinoxalinone is obtained from *N*-methylphenylenediamine and DMAD [2678], but under slightly different conditions, two molar amounts of DMAD react with one of diamine [3252]. Stirring a diamine with the oxazolone **(76.18)** in acetic acid under mild conditions leads to the formation of a pyrazinone ring [3250].

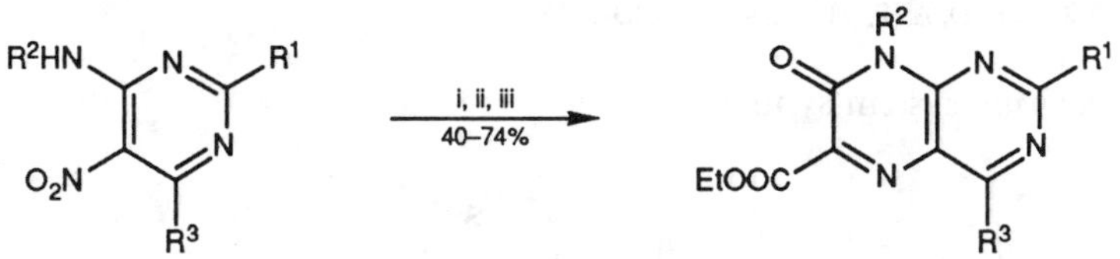

—i, ii, iii; 40–74%→ [2299]

R^1 = Me, Ph, NMe$_2$; R^2 = Me, Et; R^3 = H, Me, Et, Ph
i, POCl$_3$; ii, H$_2$, Pd/C; iii, OC(COOEt)$_2$

7-Pteridinone

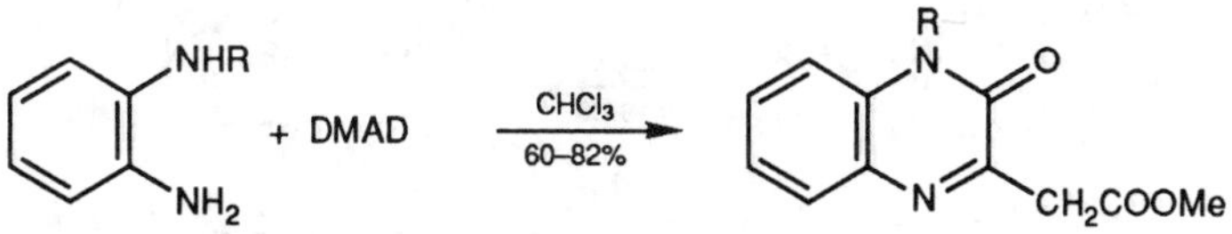

+ DMAD —CHCl$_3$; 60–82%→ [2678]

(76.16)
R = H, Pr, CH$_2$=CHCH$_2$

2-Quinoxalinone

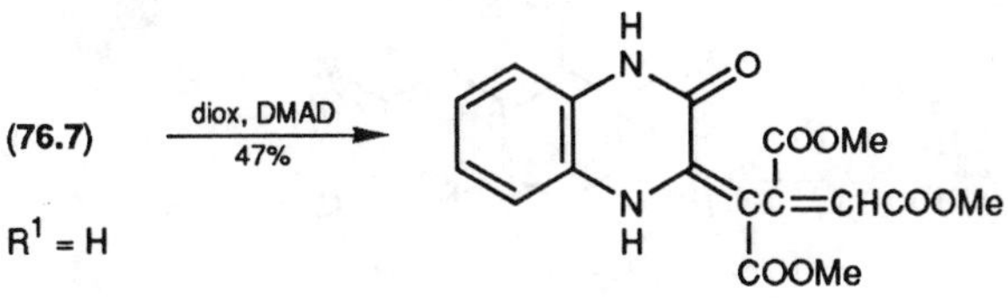

(76.7) —diox, DMAD; 47%→ [3252]

R^1 = H

2-Quinoxalinone

(76.16) + (76.18) $\xrightarrow[\text{30–75\%}]{\text{AcOH}}$ [3250]

R = $PhCH_2$, Ph; Ar = Ph, 4-Me-, 4-MeO-, 4-Cl-C_6H_4

3. 1,2,4-Triazine or 1,2,4-Triazin-3-one or -5-one

A *C,N*-diamine and a 1,2-diketone react under a variety of conditions to form a fused 1,2,4-triazine; this cyclization is undemanding since it can occur in water or ethanol]2612], or in a mineral acid [3423, 3813]. The fused triazin-3-one is formed when the dihydrazine (**76.19**) is treated with ethoxide [2753] while the isomeric 5-one is formed in high yield from the diamine and pyruvic acid [2245].

$\xrightarrow[\text{60\%}]{\text{i or ii}}$ [2612]

R^1 = β-D-ribofuranosyl
i, R^2 = H; H_2O, Δ; ii, R^2 = Me; aq. EtOH

[1,2,4]Triazino[2,3-*a*]purin-10-one

Another ring system synthesized similarly:

[2900]

[1,2,4]Triazino[1,6-*a*]benzimidazole

$\xrightarrow[\text{70–90\%}]{\text{HCl, Δ}}$ [3423]

R^1,R^2 = H, Me

Tetrazolo[1,5-*b*][1,2,4]triazine

Another ring system synthesized similarly:

[3813]

Pyrido[1,2-*b*][1,2,4]triazin-6-one

EtOH–EtONa, 63–81% [2753]

(76.19)
R = H, CHO

Pyrimido[5,4-*e*]-1,2,4-triazine-3,5,7-trione

EtOH, AcCOOH, Δ, ~91% [2245]

R = H, Me

Pyrimido[1,2-*b*][1,2,4]triazine-2,6-dione

III. FORMATION OF A SEVEN-MEMBERED RING

1. 1,4-Diazepine

Nucleophilic addition of diamines to ethynyl ketones is a well used route [2385, 2543, 2782] to 1,5-benzodiazepines (review [2721a]),but when the alkyne is replaced by an equimolar amount of crotonaldehyde (but-2-enal), a complex product is formed [3143].

(76.7) + R^2CO–C≡CR^3 → EtOH, AcOH, Δ, 25–86% [2385, 2543, 2782]

R^1 = H, Me, NO_2; R^2,R^3 = Ph, 4-Me-, 4-Cl-, 4-MeO-C_6H_4, Ph_2C(OMe)

1,5-Benzodiazepine

(76.7) + CHCHO=CHMe → EtOH, Δ, 30% [3143]

R^1 = H

5,13-Ethano[1,5]benzodiazepino[2,3-*b*][1,5]benzodiazepine

Other reagents which react with diamines are as follows: chalcones (1,3-diarylprop-2-en-1-ones) react on heating with a high-boiling tertiary amine [2981], the methylidenepyrrolidinetrione **(76.20)** undergoes a [4 + 3] addition

[3247], the formic acid-catalysed reaction of 1,3-diketone **(76.21)** [2180] and the intriguing cyclization of *o*-phenylenediamine with tetrahydrothiopyran-4-one or its 1,1-dioxide **(76.22)** [2952].

(76.7) + $CH{=}CHAr^1$ / $COAr^2$ —(MeOH, i, Δ; 45–88%)→ 1,5-Benzodiazepine [2981]

R^1 = H; Ar^1 = Ph, 4-Br-, 4-NO_2-, 4-MeO-C_6H_4;
Ar^2 = Ph, 4-Br-, 4-NO_2-, 4-MeO-, 4-Cl-C_6H_4, 4-PhC_6H_4
i, $PhCH_2NMe_2$

(76.7) + **(76.20)** —(EtOH, HCl, Δ; 25–69%)→ Pyrrolo[3,4-*b*][1,5]benzodiazepine-1,3-dione [3247]

R^1 = H; R^2 = Ph, 4-Me-, 3-NO_2-C_6H_4;
R^3 = 4-MeC_6H_4

(76.7) + **(76.21)** —(HCOOH, EtOH, Δ; 35–86%)→ Benz[*b*]indeno[1,2-*c*][1,4]diazepin-12-one [2180]

R^1 = H, alkyl, Ph, 4-ClC_6H_4; R^2 = H, Me, Cl, NO_2

(76.7) + **(76.22)** —(EtOH, BF_3–Et_2O; 46–75%)→ Spiro[thiopyran-4,11'-thiopyrano[4,3-*b*][1,5]benzodiazepine] (2,2-dioxide) [2952]

R^1 = H;
n = 0, 2

2. 1,4-Diazepin-5-one or 1,4-Diazepine-5-thione

More than one 1,4-benzodiazepinone may be formed from an unsymmetrical *o*-phenylenediamine and ethyl acetoacetate. Heating dilute solutions of the reagents in xylene gives a high yield of a 4,7-dimethyl-2-one isomer, but heating the crotonic ester **(76.23)** (formed by mixing the reactants at room temperature) with a strong base yields the 2,7-dimethyl-4-one [2371]. Approximately equal proportions of two regioisomers are also formed when 3,3-dimercapto-1-phenylprop-2-en-1-one reacts with *o*-phenylenediamine [3467]. Replacing ethyl acetoacetate by diketene gives a low yield of a single isomer [2198].

(76.7) + AcCH$_2$ | COOEt

R^1 = Me

xylene, Δ
87%

96%

H N O Me N Me

1,5-Benzodiazepin-2-one [2371]

NHCMe=CH | COOEt
Me NH$_2$

(76.23)

EtOH–Na, Δ
66%

Me N Me N H O

1,5-Benzodiazepin-4-one

C(SH)$_2$ || CHCPh || O

(76.7) + C(SH)$_2$=CHC(O)Ph

xylene, Δ
47%

R^2 R^3 H N S N Ph

[3467]

R^1 = Cl; R^2 or R^3 = Cl

1,5-Benzodiazepine-2-thione

(76.7) + H_2C O O

Δ
20–54%

R H N O N Me

[2198]

R^1 = H, Cl, Me

1,5-Benzodiazepin-2-one

Some pyridodiazepin-4-ones exist as different valence bond tautomers, and may even isomerize to five-membered ring-containing products. Pyridine-3,4-diamine condensed with ethyl acetoacetate in toluene to give a mixture of tautomeric diazepin-4-ones, but in xylene, the product was an imidazopyridine which was also obtained by dry fusion of the diazepin-4-one [2327].

N NH$_2$ NH$_2$ + CH$_2$Ac | COOEt

PhMe, Δ
60%

H N Me N N H O ⇌ N Me N H O

Pyrido[3,4-*b*][1,4]diazepin-4-one

[2327]

xylene, Δ
72%

MeC=CH$_2$ N O N NH

Imidazo[4,5-*c*]pyridin-2-one

3. 1,4-Diazepine-5,7-dione

1,2-Diamines react with either a malonic ester or a malonyl dichloride to form a fused diazepine-5,7-dione.

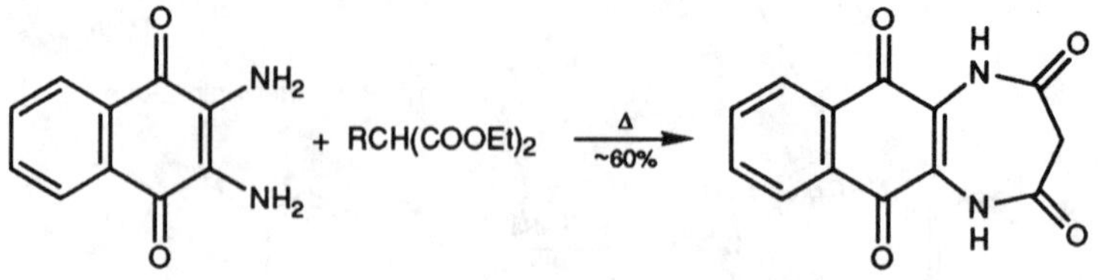

+ $RCH(COOEt)_2$ →(Δ, ~60%) [2215]

R = H, Pr, Bu, $PhCH_2$

Naphtho[2,3-*b*][1,4]diazepine-2,4,6,11-tetrone

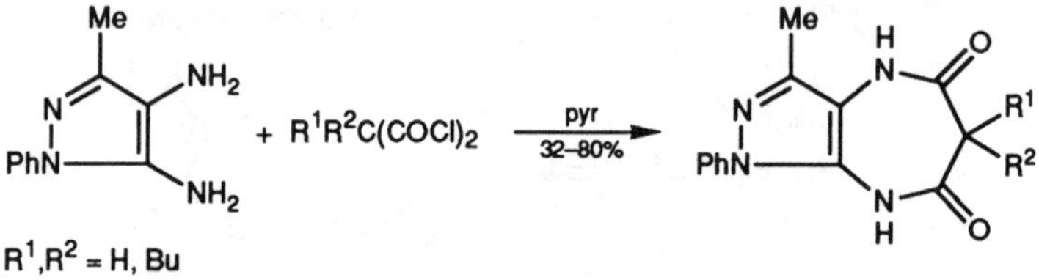

+ $R^1R^2C(COCl)_2$ →(pyr, 32–80%) [3580]

R^1,R^2 = H, Bu

Pyrazolo[3,4-*b*][1,4]diazepine-5,7-dione

4. 1,2,4-Triazepin-5-one

This ring is derived from a *C,N*-diamine and ethyl acetoacetate; although two regioisomers are possible, only the 6-one was obtained in this synthesis.

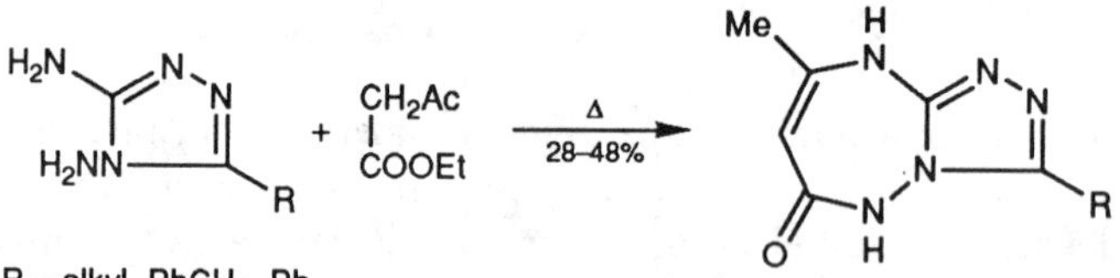

+ CH_2Ac–COOEt →(Δ, 28–48%) [2212]

R = alkyl, $PhCH_2$, Ph

1,2,4-Triazolo[4,3-*b*][1,2,4]triazepin-6-one

5. 1,2,5-Thiadiazepin-6-one 1-Oxide

This interesting ring is obtained from the reaction of ketene and an *o*-diamine in liquid sulphur dioxide at about −70 °C.

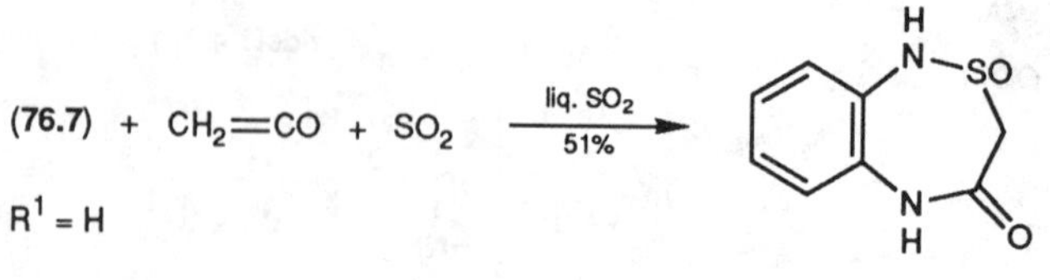

(76.7) + $CH_2{=}CO$ + SO_2 →(liq. SO_2, 51%) [2068]

R^1 = H

2,1,5-Benzothiadiazepin-4-one 2 -oxide

6. 1,3,5-Triazepine

(Dichloromethylene)dimethylammonium salts (review [2549]) are useful cycliz-ing agents for electron-rich bonds or atoms.

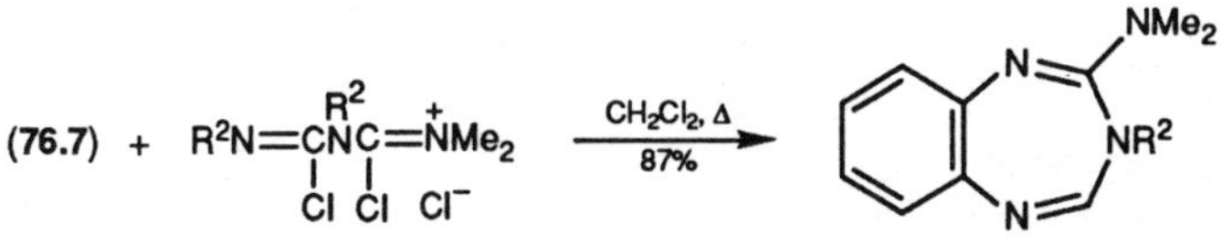

[2918]

R^1 = H; R^2 = cyclohexyl

1,3,5-Benzotriazepine

7. 1,2,4,5-Tetrazepine

When a dihydrazine is heated with orthoacetate, this ring is formed in good yield, but replacing the orthoester with formaldehyde creates a tricyclic compound which contains this ring as well as a 1,3,5-oxadiazine ring.

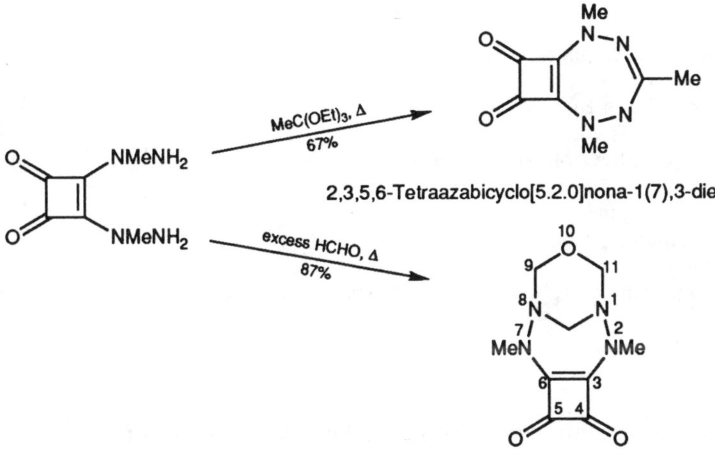

2,3,5,6-Tetraazabicyclo[5.2.0]nona-1(7),3-diene-8,9-dione

[2568]

10-Oxa-1,2,7,8-tetraazatricyclo[6.3.1.0^{3,6}]dodec-3(6)-ene-4,5-dione

CHAPTER 77

1,3-, 1,4- or 1,5-Diamine

I.	Formation of a Six-membered Ring	1142
	1. Pyrimidine	1142
	2. Pyrimidin-2-one or Pyrimidine-2-thione	1143
	3. 1,2,3-Triazine	1144
II.	Formation of a Seven-membered Ring	1144
	1. 1,3-Diazepin-2-one	1144
	2. 1,3,4-Triazepine	1145
	3. 1,4,5-Thiadiazepine 4-Oxide or 1,1-Dioxide	1145
III.	Formation of an Eight-membered Ring	1145
	1. 1,3-Diazocine or 1,3-Diazocine-2-thione	1145

An example of the cyclization of a di-hydroxylamine is included in this chapter.

I. FORMATION OF A SIX-MEMBERED RING

1. Pyrimidine

Orthoesters annulate aminomethyl and amino groups to form a fused pyrimidine ring [2159]. Two amino groups in *peri* positions form a similar ring when treated with benzoic acid and *N,N*-diphenylphosphinyl-*N′*-methylpiperazine [3286] (see also Chapter 76, Section I.1.A). 1,8-Naphthalenediamine is converted into perimidine by heating with 2-acylindanedione and 4-toluenesulphonic acid [2354]. Formic acid similarly cyclizes 4,5-quinolinediamine or the isomeric isoquinolinediamine in about 89% yield [3338].

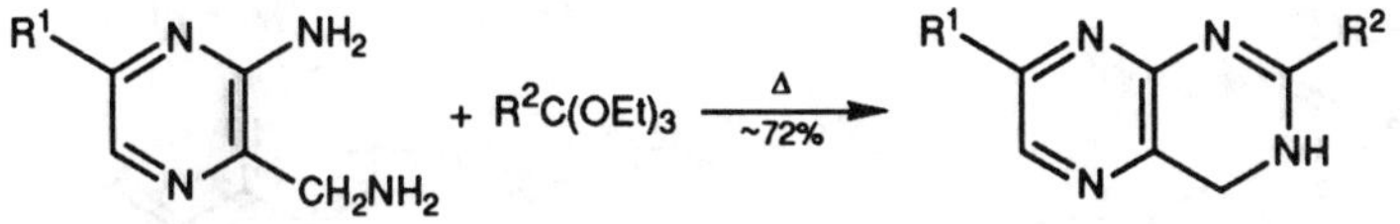

R^1 N NH_2 N CH_2NH_2 + $R^2C(OEt)_3$ →(Δ, ~72%) R^1 N N R^2 NH N [2159]

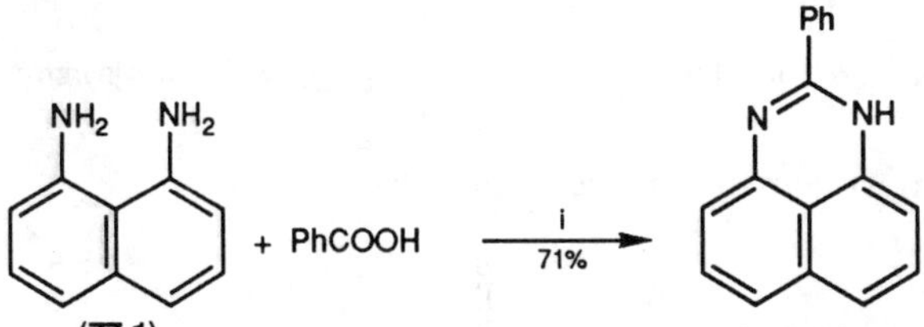

R^1,R^2 = H, Me

Pteridine

NH_2 NH_2 (77.1) + PhCOOH →(i, 71%) Ph N NH [3286]

Perimidine

i, Ph_2PON NMe

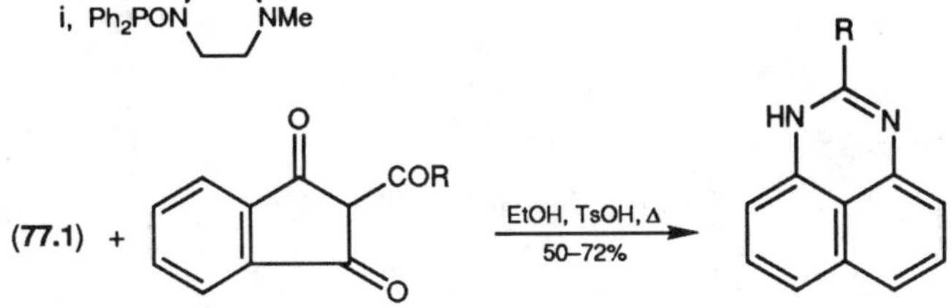

(77.1) + O COR O →(EtOH, TsOH, Δ, 50–72%) R HN N [2354]

R = alkyl, $PhCH_2$, Ph

Other ring systems synthesized similarly:

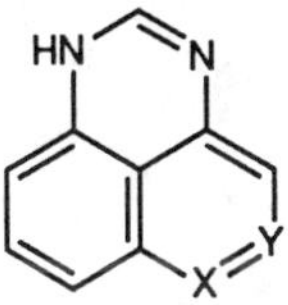

HN N X Y [3338]

X = N; Y = CH Pyrido[4,3,2-*de*]quinazoline
X = CH; Y = N Pyrido[3,4,5-*de*]quinazoline

2. Pyrimidin-2-one or Pyrimidine-2-thione

A fused pyrimidin-2-one may be prepared in good yield by heating a 1,3-diamine with urea. Phosgene reacts at ambient temperature with a 2-imino-amine to form a fused pyrimidin-2-one. Quinoline-1,8-diamine and carbon disulphide react to give a yield of 89% of the pyrimidinethione.

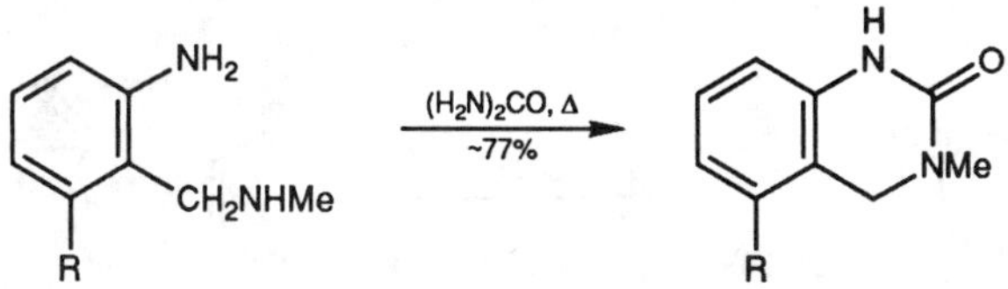

NH_2 CH_2NHMe R →($(H_2N)_2CO$, Δ, ~77%) H N O NMe R [2126]

R = Me, Et

2-Quinazolinone

R^1 = alkyl; R^2 = H, Me, Ph

+ $COCl_2$ → PhH, TEA, 22–68% [2628]

Pyrido[2,3-*d*]pyrimidin-2-one

Another ring system synthesized similarly:

[3338]

Pyrido[3,4,5-*de*]quinazoline-2-thione

3. 1,2,3-Triazine

Diazotization of a 1,4-diamine (**77.2**) can lead to the formation of this ring in high yield.

HCl–$NaNO_2$, 80% [2406]

(**77.2**)

1,2,4-Triazolo[4,3-*c*][1,2,3]benzotriazine

II. FORMATION OF A SEVEN-MEMBERED RING

1. 1,3-Diazepin-2-one

1,4-Diamines are cyclized by heating with phosgene.

PhH, $COCl_2$, Δ, ~33% [2058]

R = Me, Et

1,3-Benzodiazepin-2-one

2. 1,3,4-Triazepine

When the 1,4-diamine (**77.2**) is heated with an orthoester, a new triazepine ring is formed in good yield. Replacing the orthoester by an aldehyde leads to the 6-substituted reduced triazepine.

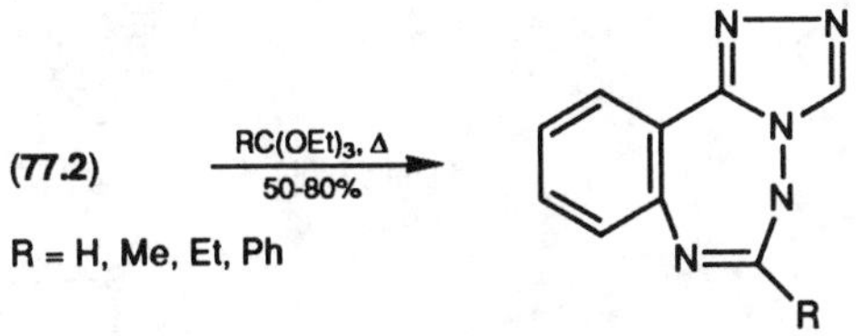

(77.2) —RC(OEt)$_3$, Δ; 50-80%→ [2406]

R = H, Me, Et, Ph

1,2,4-Triazolo[4,3-*d*][1,3,4]benzotriazepine

3. 1,4,5-Thiadazepine 4-Oxide or 1,1-Dioxide

Two suitably positioned hydroxylamine groups undergo oxidative cyclization by low-temperature treatment with peracetic acid to give the 4-oxide, but when the diamino-sulphone was oxidized with iodosobenzene diacetate (review of uses [3714]), an azo group is formed to close the ring.

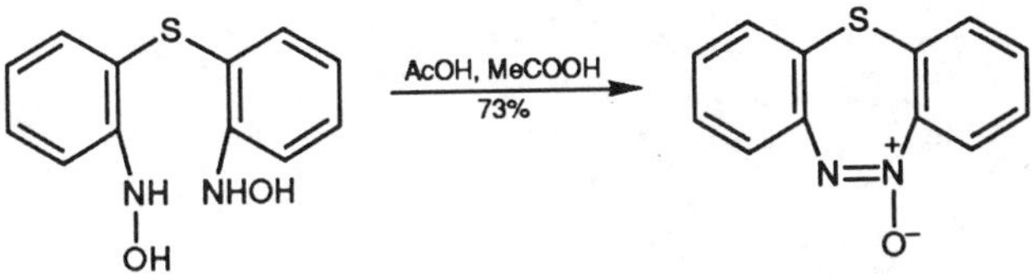

—AcOH, MeCOOH; 73%→ [2290]

Dibenzo[*b,f*][1,4,5]thiadiazepine 5-oxide

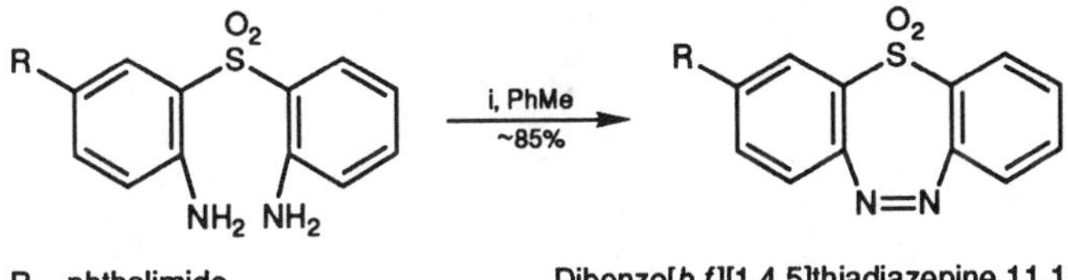

—i, PhMe; ~85%→ [2291]

R = phthalimido

i, PhI(OAc)$_2$

Dibenzo[*b,f*][1,4,5]thiadiazepine 11,11-dioxide

III. FORMATION OF AN EIGHT-MEMBERED RING

1. 1,3-Diazocine or 1,3-Diazocine-2-thione

1,5-Diamines are cyclized by heating with either an orthoester or carbon disulphide (the chemistry of this reagent has been reviewed [B-52].

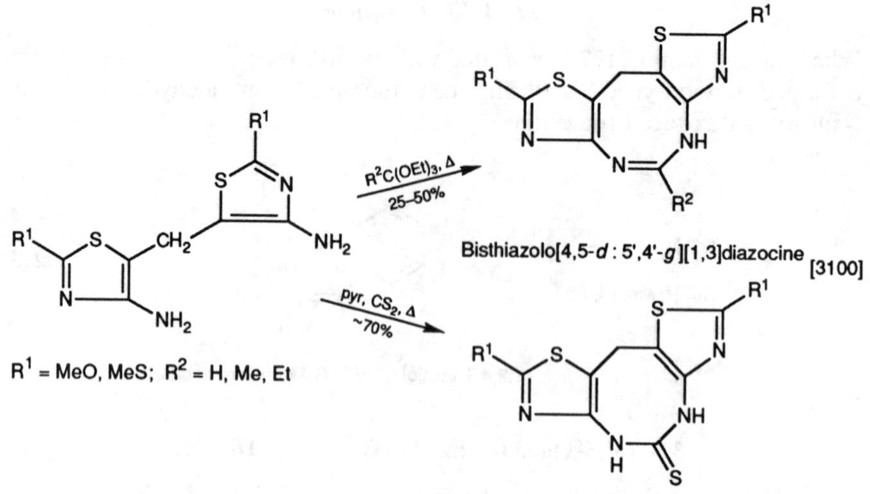

R^1 = MeO, MeS; R^2 = H, Me, Et

$R^2C(OEt)_3$, Δ, 25–50%

Bisthiazolo[4,5-*d* : 5',4'-*g*][1,3]diazocine [3100]

pyr, CS_2, Δ, ~70%

Bisthiazolo[4,5-*d* : 5',4'-*g*][1,3]diazocine-5-thione

CHAPTER 78

Diazo or Diazonium Salt and Halogen or Methylene

I. Formation of a Five-membered Ring 1147
1. 1,2,3-Thiadiazole 1147
II. Formation of a Six-membered Ring 1147
1. Pyridazine or Pyridazin-4-one 1147
III. Formation of a Seven-membered Ring 1148
1. 1,2,4-Triazepin-5-one 1148
2. 1,2,5-Triazepin-4-one 1149

I. FORMATION OF A FIVE-MEMBERED RING

1. 1,2,3-Thiadiazole

Coupling of a diazonium salt with a methylene group of a dithiolium salt yields an S^{IV}-bicyclic compound.

EtOH, $ArN_2^+BF_4^-$, 18–99% [2846]

ClO_4^-, CH_2R^2

R^1 = H, tBu, Ph; R^2 = H, Me, Ph;
Ar = Ph, 4-Br-, 4-MeO-, 4-Ac-, 4-NO_2-, 4-Me-C_6H_4

[1,2]Dithiolo[5,1-*e*][1,2,3]thiadiazole

II. FORMATION OF A SIX-MEMBERED RING

1. Pyridazine or Pyridazin-4-one

In an attempt to displace the chlorine atom of (**78.1**) with hydrazine, the reaction yielded the cyclized pyridazine in high yield provided a 3.6 molar excess of hydrazine hydrate was used. A similar overall reaction occurs when the chloro-diazo ester (**78.1**) is warmed with triphenylphosphine; some of these products are related to the antibacterial nalidixic acid.

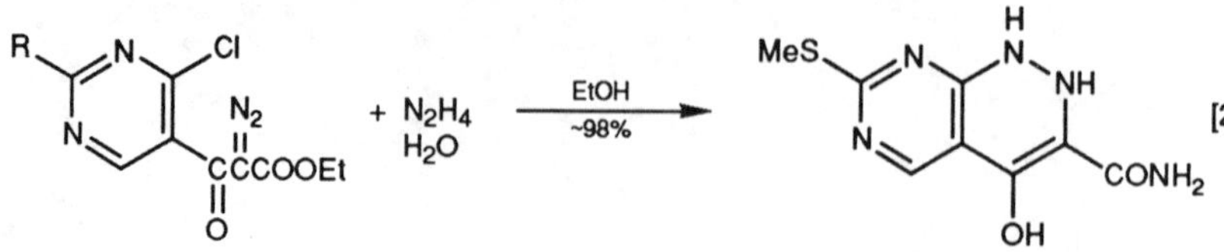

(**78.1**) R = MeS + $N_2H_4 \cdot H_2O$ —EtOH, ~98%→ Pyrimido[4,5-*c*]pyridazine [2795]

(**78.1**) R = Cl, MeS, Ph —iPr$_2$O, Ph$_3$P, Δ, 37–80%→ Pyrimido[4,5-*c*]pyridazin-4-one [2927]

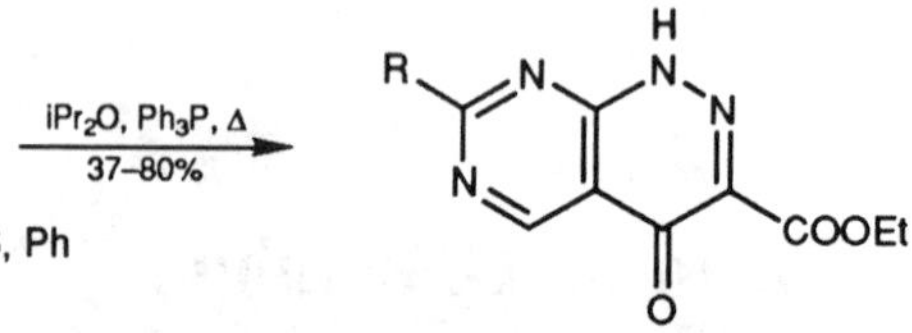

Another ring system synthesized similarly:

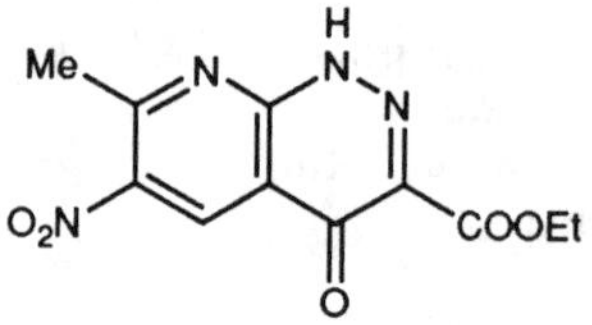

Pyrido[2,3-*c*]pyridazin-4-one [3198]

The diazo group of (**78.2**) reacts with an adjacent acetyl function in preference to the carbon atom of a nearby phenyl ring (cf. Chapter 79, Section II.2). The reactions of diazoazoles have been reviewed [3901].

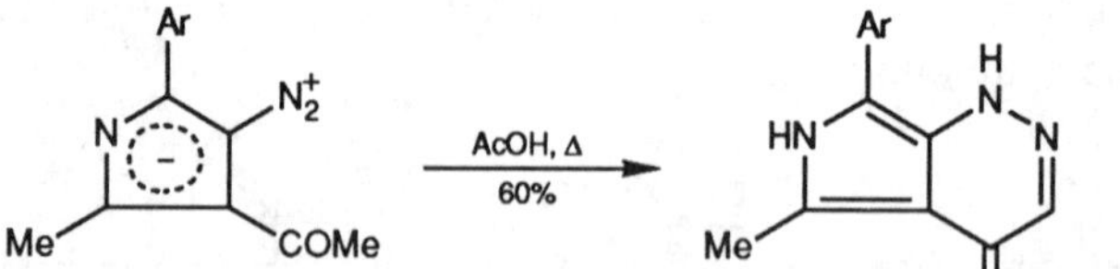

(**78.2**) Ar = 3-MeOC$_6$H$_4$ —AcOH, Δ, 60%→ Pyrrolo[3,4-*c*]pyridazin-4-one [3061]

III. FORMATION OF A SEVEN-MEMBERED RING

1. 1,2,4-Triazepin-5-one

Coupling a diazonium salt with the acidic hydrogen of a side-chain malonate (the Japp–Klingemann reaction) can be an effective way of forming a ring under mildly acidic conditions.

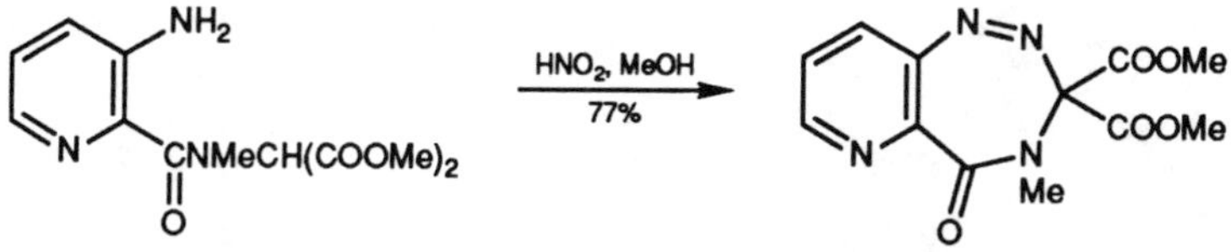

NH$_2$ / CNMeCH(COOMe)$_2$ —HNO$_2$, MeOH, 77%→ [2961]

Pyrido[2,3-*f*]-1,2,4-triazepin-5-one

2. 1,2,5-Triazepin-4-one

An activated side-chain methylene couples with a neighbouring diazonium salt (a Japp–Klingemann reaction) in dilute acid to form this fused ring.

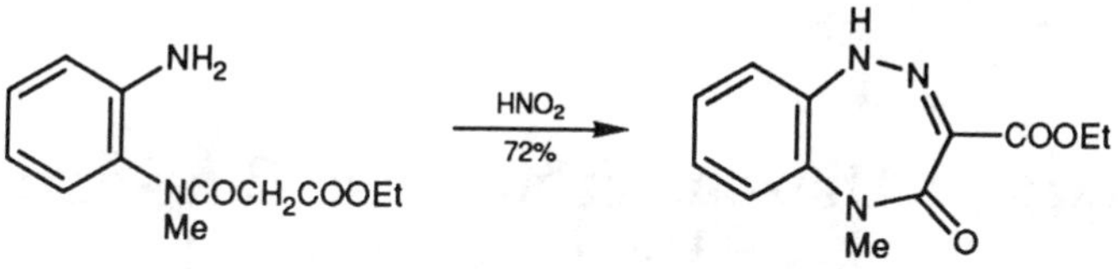

NH$_2$ / NCOCH$_2$COOEt, Me —HNO$_2$, 72%→ COOEt [2029]

1,2,5-Benzotriazepin-4-one

CHAPTER 79

Diazo or Diazonium Salt and Ring-carbon or Ring-nitrogen

I.	Formation of a Five-membered Ring	1150
	1. Pyrrole	1150
	2. Pyrrol-3-one	1151
	3. Furan or Furan-2-one	1151
	4. Thiophene or Thiophene 1-Oxide	1152
II.	Formation of a Six-membered Ring	1152
	1. Pyridine or Pyridin-2-one	1152
	2. Pyridazine	1153
	3. 1,2,3-Triazine	1153
	4. 1,2,3-Triazin-4-one	1154
	5. 1,2,4-Triazine	1154
	6. 1,2,3,5-Tetrazin-4-one	1155
	7. Thiopyran 1,1-Dioxide	1156
	8. 1,3-Oxazine	1156
	9. 1,3,4-Thiadiazine 1,1-Dioxide	1156
III.	Formation of a Seven-membered Ring	1157
	1. 1,2,5-Triazepine	1157
	2. 1,4,5-Oxadiazepine	1157

Examples of the cyclization of a 2-aminomethylaryl diazonium salt and of a 3-triazenopyrazole are included in this chapter

I. FORMATION OF A FIVE-MEMBERED RING

1. Pyrrole

Rhodium-catalysed cyclization of α-diazoacylanilines to 3-acetylindol-2-ols can give high yields. Rhodium octanoate, followed by chlorodiphenyl phosphate and Hünig's base [3060a], catalyses the formation of a pyrrole ring of analogues of the antibacterial asparenomycins. Reactions which are catalysed by rhodium(II) salts have been reviewed [3928].

PhH, $Rh_2(OAc)_4$, Δ ~80% [3837]

R^1 = H, Cl, MeO, Me; R^2 = Me, $PhCH_2$, Ph

Indole

see text 43% [3972]

R^1 = TBS; R^2 = 2-pyridyl

1-Azabicyclo[3.2.0]hept-2-en-7-one

2. Pyrrol-3-one

The diazo compound (**79.1**) is a valuable precursor for the carbene which is formed by heating the compound in benzene with a catalytic amount of rhodium(II) acetate (review [3928]). During this step, ring closure occurs and a high yield of the desired carbapenem is obtained. Photochemical cyclization gives a small amount of the 2-one in addition to the carbapenem. The azetidine nitrogen of (**79.1**) may be protected with a benzyloxy group which is lost during rhodium-catalysed cyclization in about 40% yield [3863].

PhH, $Rh(OAc)_2$, Δ 77–95% [2995, 3010, 3079]

(**79.1**)

R^1 = Et, iPr; R^2 = $4\text{-}NO_2C_6H_4CH_2$

1-Azabicyclo[3.2.0]heptan-3-one

3. Furan or Furan-2-one

A 2-aminodiaryl ether (**79.2**, X = O) is cyclized to the furopyridine (**79.3**, X = O) by using the Pschorr reaction—thermolysis of a mixture of the diazonium salt and copper powder. The use of heterocyclic diazonium salts in organic synthesis has been reviewed [2980]. Rhodium(II) acetate catalyses the room-temperature cyclization of α-diazo-β-oxo esters (**79.4**) by carbenoid insertion into a benzenoid C—H bond.

1. $NaNO_2$, H_2SO_4 2. Cu, Δ; 31% [2461]

(79.2)
X = O, S

(79.3)
X = O Benzofuro[2,3-*b*]pyridine
X = S [1]Benzothieno[2,3-*b*]pyridine

CH_2Cl_2, $Rh_2(OAc)_4$; ~95% [3361]

(79.4)
R = H, Me, Br, OCH_2O

2-Benzofuranone

4. Thiophene or Thiophene 1-Oxide

In the reaction between a diazo group and dichlorosulphine, there is evidence for a 1,3-dipolar addition to give (after loss of nitrogen) an episulphoxide from which an *S*-ring carbon bond is formed. A C—C bond is formed when a diazonium salt of a diaryl thioether is heated with copper powder (a Pschorr reaction); in this example (see above equation), the product is a [1]benzothieno[2,3-*b*]pyridine (**79.3**, X = S, 18% yield).

+ $Cl_2C{=}SO$; C_5H_{12}; 11–55% [2465]

R = H, Me, Cl, MeO; Ar = Ph, Me-, Cl-, MeO-C_6H_4

Benzo[*b*]thiophene 1-oxide

II. FORMATION OF A SIX-MEMBERED RING

1. Pyridine or Pyridin-2-one

As a step in the synthesis of a higher homologue of the antibiotic thienamycin, the closure of the pyridine ring is effected by warming the diazo ester (**79.5**) with rhodium(II) acetate (review of rhodium-catalysed reactions [3928]); the product (**79.6**, R = H) is more easily isolated and purified as its *O*-tosyl derivative (**79.6**, R = Ts). Intramolecular homolytic cyclization (the Pschorr reaction) of the diazonium salt of *N*-methylbenzanilide gives moderate yields which are comparable with those given by the pyrolytic cyclization of the 2-iodo analogue (see Chapter 90, Section II.2 [2303]).

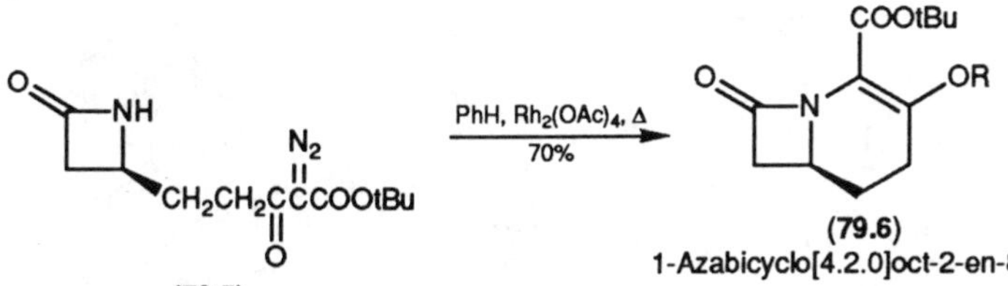

PhH, $Rh_2(OAc)_4$, Δ, 70% [2975]

(79.5)

(79.6)
1-Azabicyclo[4.2.0]oct-2-en-8-one

Me_2CO, Cu, 46% [2303]

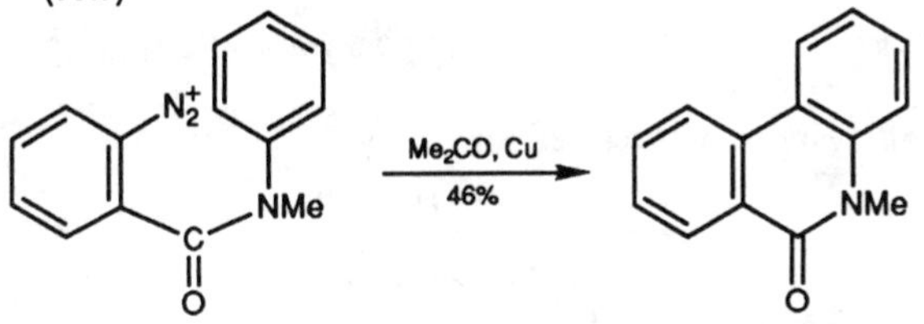

6-Phenanthridinone

Another ring system synthesized similarly:

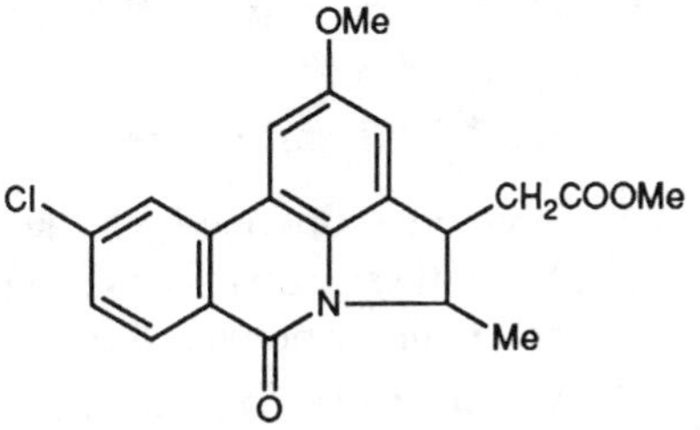

[2627]

Pyrrolo[3,2,1-*de*]phenanthridin-7-one

2. Pyridazine

A diazonium group attacks another benzene ring (especially if the latter is electron-rich) when heated in acetic acid (cf. the reaction in Chapter 78, Section II.1 [3061]).

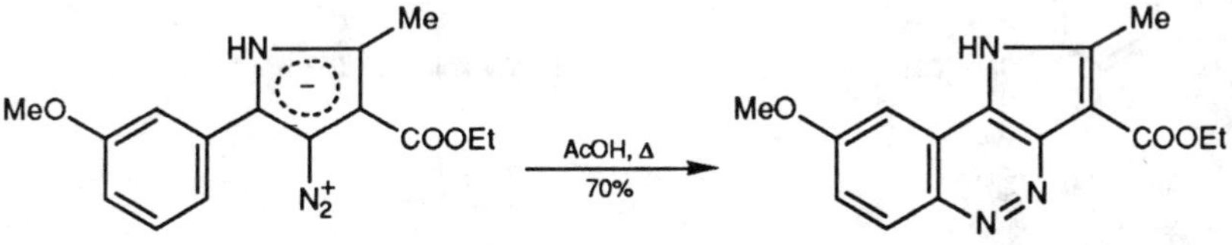

AcOH, Δ, 70% [3061]

Pyrrolo[3,2-*c*]cinnoline

3. 1,2,3-Triazine

An inter-ring cyclization similar to that in the previous section but one in which the diazonium ion attacks a ring-nitrogen atom leads to the formation of a doubly fused triazine ring.

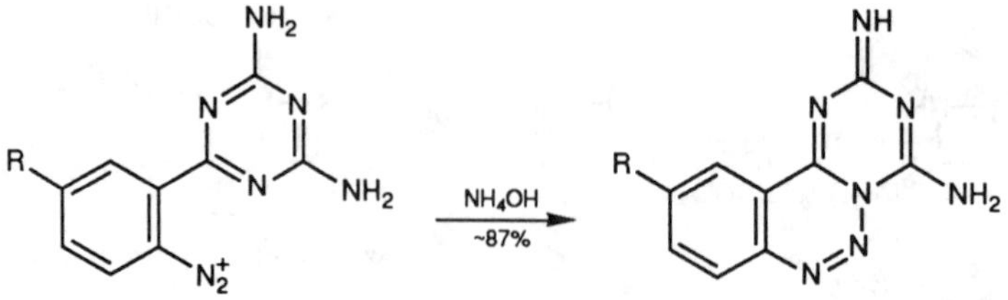

R = H, Me, Br

NH_4OH ~87%

1,3,5-Trazino[1,2-*c*][1,2,3]benzotriazine

[2166]

Another ring system synthesized similarly:

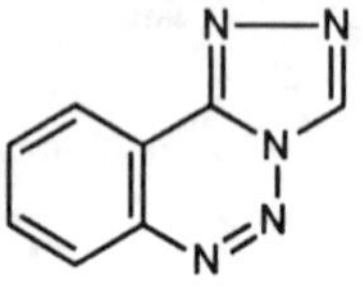

1,2,4-Triazolo[4,3-*c*][1,2,3]benzotriazine

[2406]

4. 1,2,3-Triazin-4-one

Unexpected products are sometimes obtained in the Pschorr reaction; for example, during an attempt to synthesize an alkaloid, thermolysis of the diazonium salt **(79.7)** gave a triazinone, the carbonyl group probably arising from the oxidation of the methylene by nitrous acid.

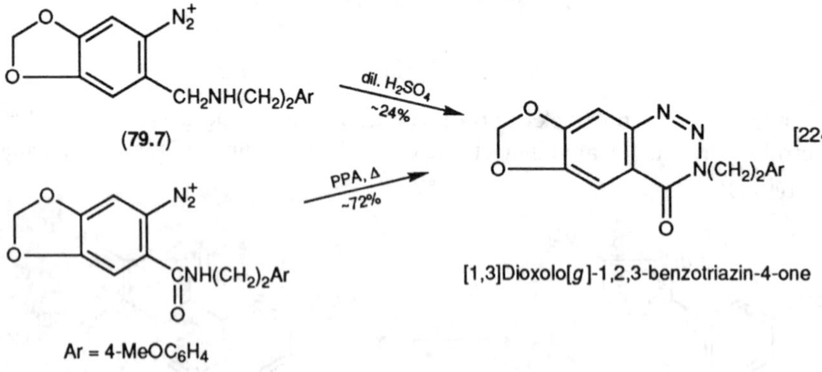

(79.7)

dil. H_2SO_4 ~24%

PPA, Δ ~72%

Ar = 4-$MeOC_6H_4$

[1,3]Dioxolo[*g*]-1,2,3-benzotriazin-4-one

[2242]

5. 1,2,4-Triazine

Diazotization in a nonaqueous medium is useful in synthesis (review [2980]), as is demonstrated by the cyclization of the aminopyrazole **(79.8)** with nitrosyl chloride; amines which lacked the methoxy group failed to cyclize on to the carbon atom [3201]. 2-Diazoazoles [such as **(79.9)**] undergo cycloadditions with 1,2-dimethoxyethene, alkoxyethene, enamines, isocyanates and ketene acetals to form a fused triazine ring at or below ambient temperature. The kinetics and mechanisms of this type of reaction have been studied [3895].

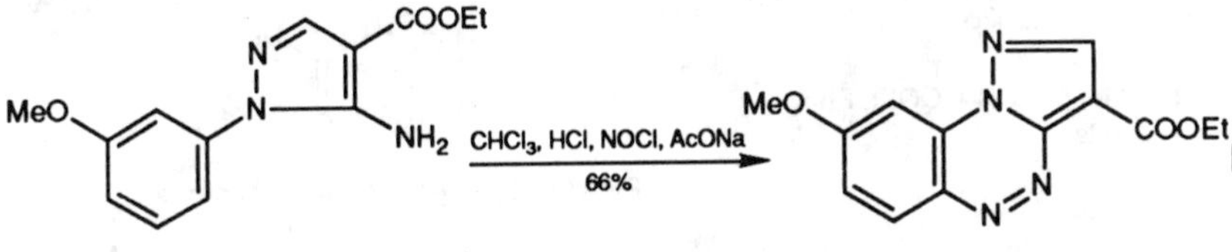

(79.8) — $CHCl_3$, HCl, NOCl, AcONa, 66% → Pyrazolo[5,1-*c*][1,2,4]benzotriazine [3201]

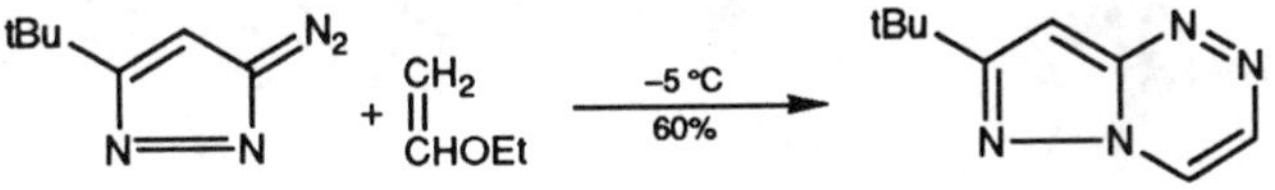

−5 °C, 60% → Pyrazolo[5,1-*c*][1,2,4]triazine [3291]

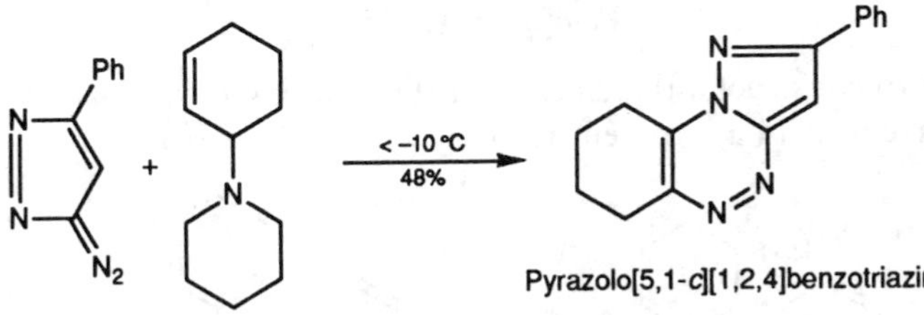

< −10 °C, 48% → Pyrazolo[5,1-*c*][1,2,4]benzotriazine [3291]

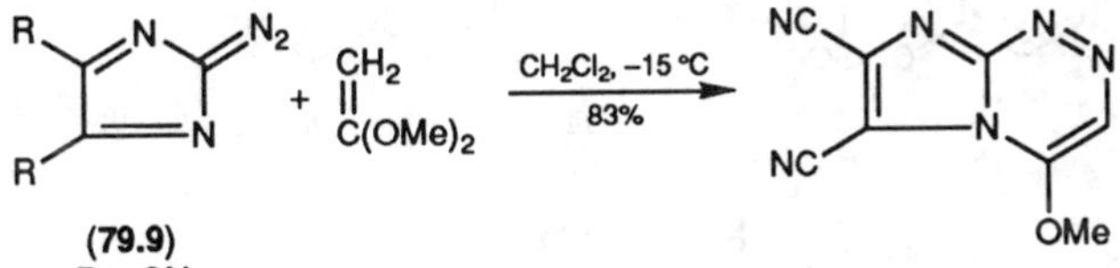

(79.9) R = CN — CH_2Cl_2, −15 °C, 83% → Imidazlo[2,1-*c*][1,2,4]triazine [3291, 3895]

Another ring system synthesized similarly:

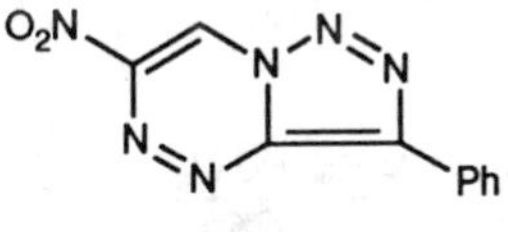

[1,2,3]Triazolo[5,1-*c*][1,2,4]triazine [3291]

6. 1,2,3,5-Tetrazin-4-one

Cyclocarbonylation of an α-triazeno-azole in which the terminal nitrogen becomes attached to the carbonyl group is brought about by phosgene, carbonyldi-imidazole or trichloroethyl chloroformate at ambient or lower temperature. Treatment of 2-diazoimidazole (**79.9**) with isocyanates at −70 °C gives a high yield of the 1,4-cycloaddition product.

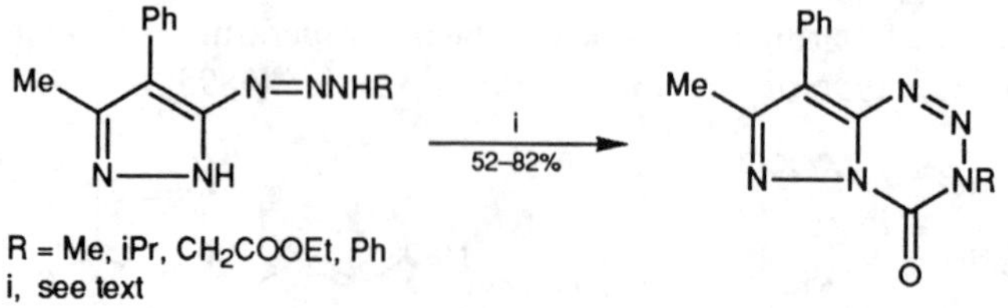

R = Me, iPr, CH_2COOEt, Ph
i, see text

i, 52–82% [3299]

Pyrazolo[5,1-*d*]-1,2,3,5-tetrazin-4-one

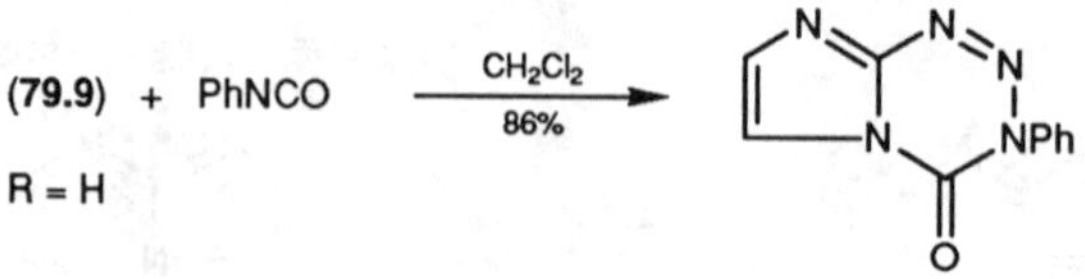

(**79.9**) + PhNCO → (CH_2Cl_2, 86%) [3291]
R = H

Imidazo[2,1-*d*]-1,2,3,5-tetrazin-4-one

7. Thiopyran 1,1-Dioxide

An attempt to construct a thiadiazine 1,1-dioxide ring through diazonium salt coupling gave instead a nitrogen-free ring by a Pschorr reaction.

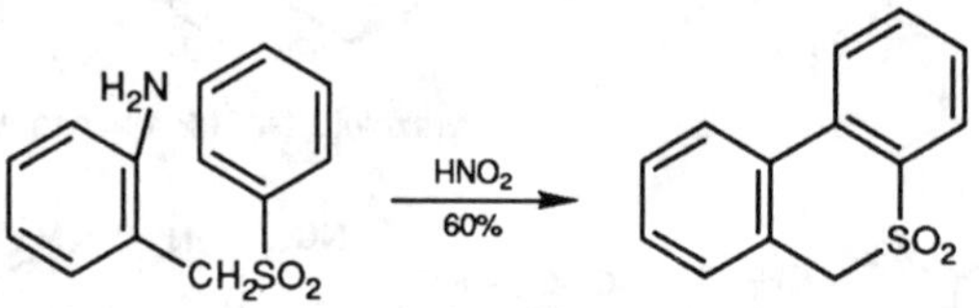

HNO_2, 60% [2938]

Dibenzo[*b,d*]thiopyran 5,5-dioxide

8. 1,3-Oxazine

Annulation of a side-chain diazo group to an azetidinone ring-nitrogen (as described in Section I.2) occurs when the substrate is treated with a catalytic amount of rhodium(II) acetate in warm ethyl acetate.

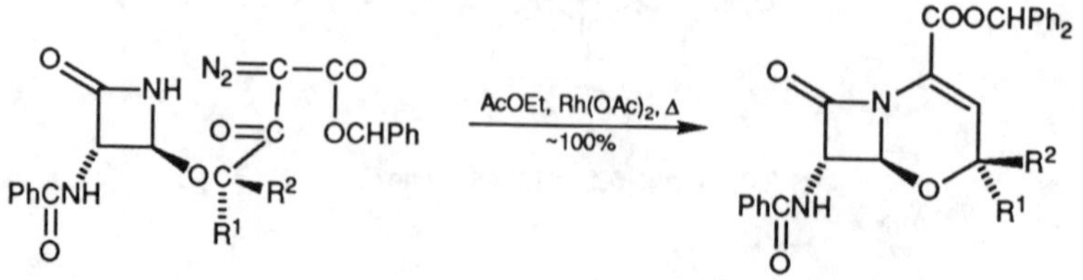

AcOEt, $Rh(OAc)_2$, Δ, ~100% [3794]

R^1,R^2 = H, Me

5-Oxa-1-azabicyclo[4.2.0]oct-2-en-8-one

9. 1,3,4-Thiadiazine 1,1-Dioxide

A diazonium ion may attack a side-chain phenylhydrazono group during diazotization; this fused ring is formed and the phenylazo group is lost.

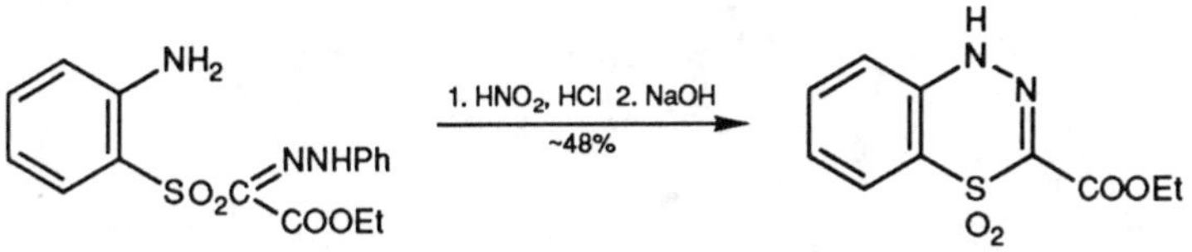

1. HNO_2, HCl 2. NaOH, ~48% [2938]

4,2,1-Benzothiadiazine 4,4-dioxide

III. FORMATION OF A SEVEN-MEMBERED RING

1. 1,2,5-Triazepine

This ring may be formed by a diazonium ion attached to one ring attacking another at one of its carbon atoms; electron-releasing substituents on the latter ring facilitate the cyclization which takes place at room temperature.

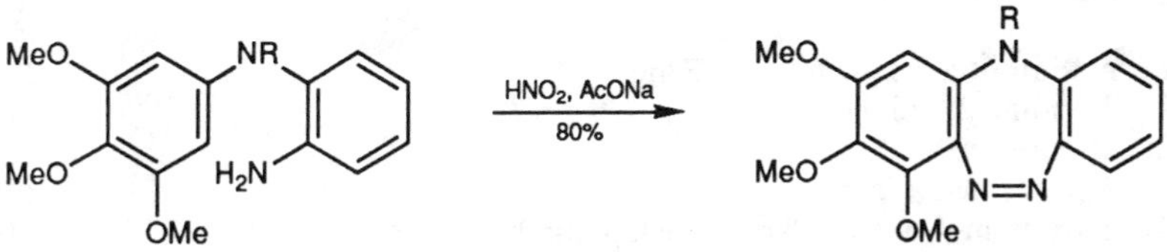

HNO_2, AcONa, 80% [2355]

R = $Me_2N(CH_2)_3$

Dibenzo[*c,f*][1,2,5]triazepine

2. 1,4,5-Oxadiazepine

In a similar reaction, a diazonium ion attached to a diaryl ether attacks a carbon of the other ring to form a doubly fused oxadiazepine. Such reactions have been reviewed [2980].

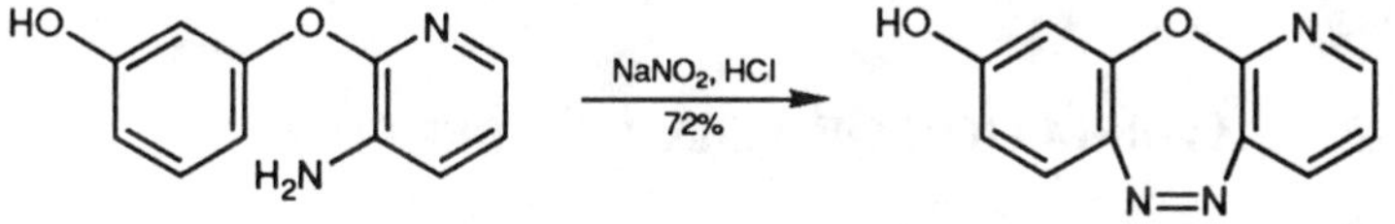

$NaNO_2$, HCl, 72% [2162]

Pyrido[2,3-*b*][1,4,5]benzoxadiazepine

CHAPTER 80

Dicarboxylic Acid or its Derivative or Diazide

I.	Formation of a Six-membered Ring	1158
	1. Pyridine-2,5-dione	1158
	2. Pyridazine	1159
	3. Pyridazine-3,6-dione	1159
	4. Pyrimidin-4-one or Pyrimidine-2,4-dione	1161
	5. Pyran-2-one	1161
II.	Formation of a Seven-membered Ring	1162
	1. 1,3-Diazepine	1162
III.	Formation of an Eight-membered Ring	1162
	1. 1,6,2-Dioxazocine	1162

Cyclization of dinitriles is discussed in Chapter 83.

I. FORMATION OF A SIX-MEMBERED RING

1. Pyridine-2,5-dione

Simultaneous *C*- and *N*-acylation of an imidazole by phthaloyl dichloride occurs in the presence of TEA.

COCl COCl + N R² HN R¹ —MeCN, TEA, Δ / 32–64%→ [2825]

R^1 = H; R^2 = H, Ph or R^1R^2 = (CH=CH)$_2$

Imidazo[1,2-*b*]isoquinoline-5,10-dione
Benzimidazo[1,2-*b*]isoquinoline-6,11-dione

2. Pyridazine

NBS is a better oxidant than LTA for producing the unstable 1,4-phthalazinedione, which then reacts as a dienophile with butadiene. The product can be brominated–dehydrobrominated to give the fully unsaturated product. Irradiation at low temperature of a compound having two suitably placed azide groups creates a new pyridazine ring in high yield.

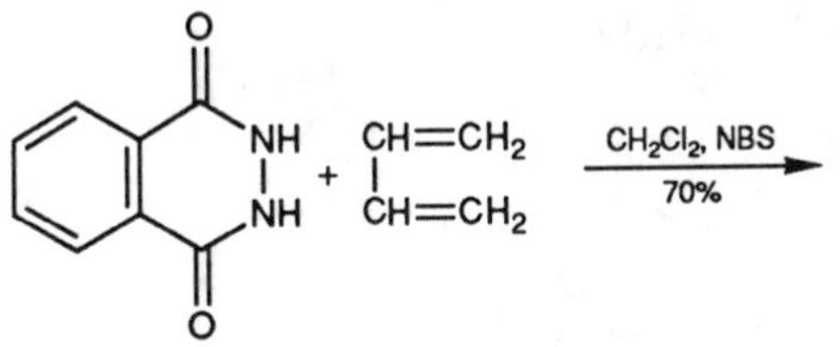

CH_2Cl_2, NBS; 70%

Pyridazino[1,2-*b*]phthalazine-6,11-dione

[3083]

Another ring system synthesized similarly:

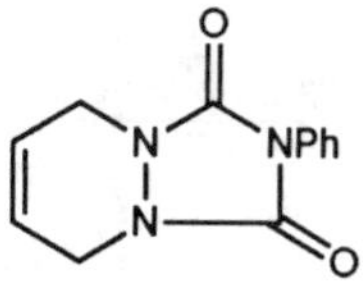

1,2,4-Triazolo[1,2-*a*]pyridazine-1,3-dione

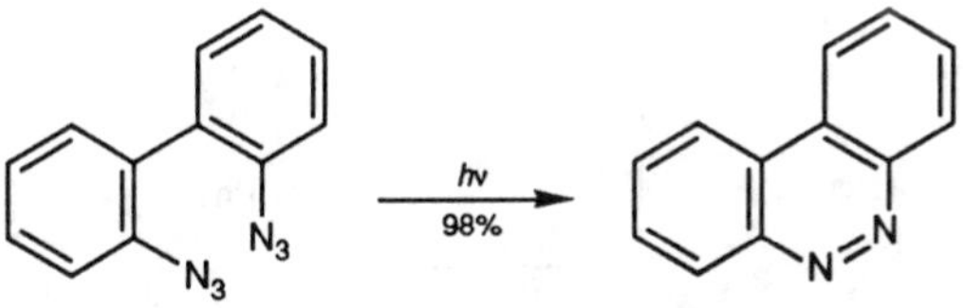

[3083]

hν; 98%

Benzo[*c*]cinnoline

[2763]

3. Pyridazine-3,6-dione

Good yields of a pyridazine (rather than the bishydrazide) are obtained from a dicarboxylic acid by first converting the latter into its anhydride by heating with acetic anhydride, and then adding a hydrazine. A related pyridine diester is cyclized with hydrazine hydrate [2733] (sometimes with hydrochloric acid present [3918]) at ambient temperature to the dihydrazide, which may be converted to the dione by heating with hydrazine [2733], and to the imide on acidification with acetic acid. The pyridazinedione may also be prepared from the imide (**80.3**) by heating with hydrazine hydrate [2797].

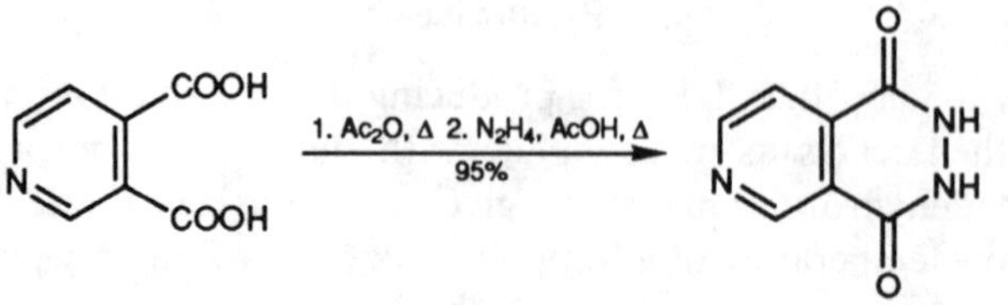

1. Ac_2O, Δ 2. N_2H_4, AcOH, Δ
95%

Pyrido[3,4-*d*]pyridazine-1,4-dione

[2076]

Another ring system synthesized similarly:

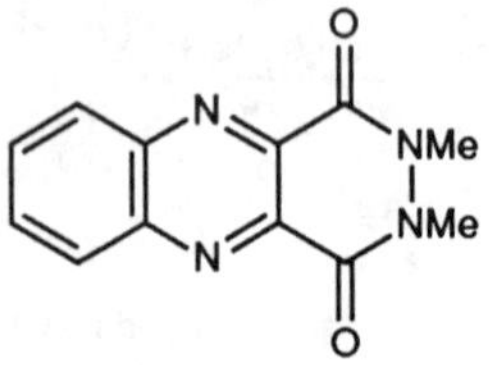

Pyridazino[4,5-*b*]quinoxaline-1,4-dione

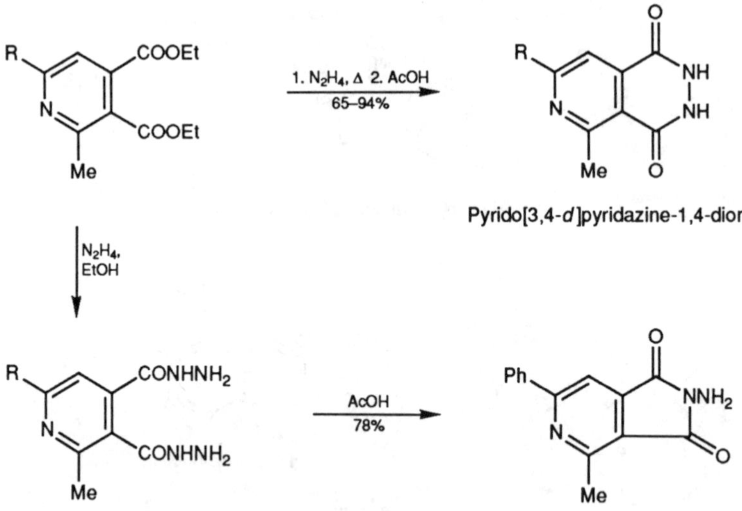

[2539]

1. N_2H_4, Δ 2. AcOH
65–94%

Pyrido[3,4-*d*]pyridazine-1,4-dione

[2733]

N_2H_4, EtOH

AcOH
78%

R = 4-MeO-, 4-Cl-, 3-Cl-, 3-NO_2-C_6H_4, 2,5-$Me_2C_6H_3$

Pyrrolo[3,4-*c*]pyridine-1,3-dione

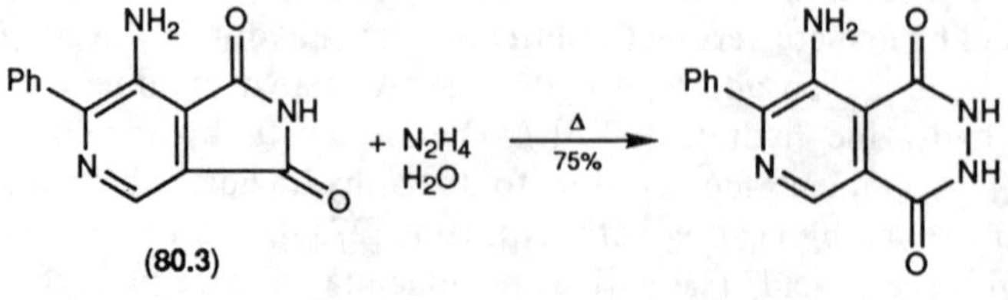

(80.3) + N_2H_4 H_2O → Δ 75%

Pyrido[3,4-*d*]pyridazine-1,4-dione

[2797]

4. Pyrimidin-4-one or Pyrimidine-2,4-dione

The cyclic anhydride (**80.4**) reacts with trimethylsilyl azide under anhydrous conditions to give a mixture of 1,3-oxazine-2,6-dione isomers, which was used without further purification to synthesize several extended purines. Reaction with formamidine acetate in DMF in effect replaced the furandione ring by a pyrimidinone. Cyanamide produced a new fused 2-aminopyrimidin-4-one [in (**80.5**, R = NH_2)], while urea led to a pyrimidinedione. Isatoic anhydride similarly yields 2-aminoquinazolin-4-one on treatment with cyanamide–DMF.

(**80.4**) —i or ii→ (**80.5**) Imidazo[4,5-*g*]quinazolin-8-one [3281]

(**80.4**) —$CO(NH_2)_2$, DMF, Δ, 55%→ Imidazo[4,5-*g*]quinazoline-6,8-dione

i, HC(=NH)NH_2, MeOH, DMF, Δ, R = H (86%);
ii, H_2NCN, tBuOK, DMF, Δ, R = NH_2 (59%)

5. Pyran-2-one

Biphenyl-2,2′-dicarboxylic acids are oxidatively converted into benzocoumarins with loss of carbon dioxide. A more predictable reaction of a dicarboxylic acid occurs on heating with sodium acetate–acetic anhydride; *C*-acetylation first occurs and decarboxylative lactonization follows, but in low yield.

Biphenyl-2,2′-dicarboxylic acid (COOH, COOH) —H_2O_2–H_2SO_4, Δ, 87%→ Dibenzo[*b,d*]pyran-6-one [2204]

OH, COOH, CH_2COOH, Cl —AcONa–Ac_2O, Δ, 10%→ 2-Benzopyran-1-one (OH, O, Me, Cl) [2149]

II. FORMATION OF A SEVEN-MEMBERED RING

1. 1,3-Diazepine

Two iminophosphorane groups separated by four carbon atoms are annulated by stirring the compound with an aroyl chloride and TEA. A review of recent work on seven-membered heterocycles has been published [3987].

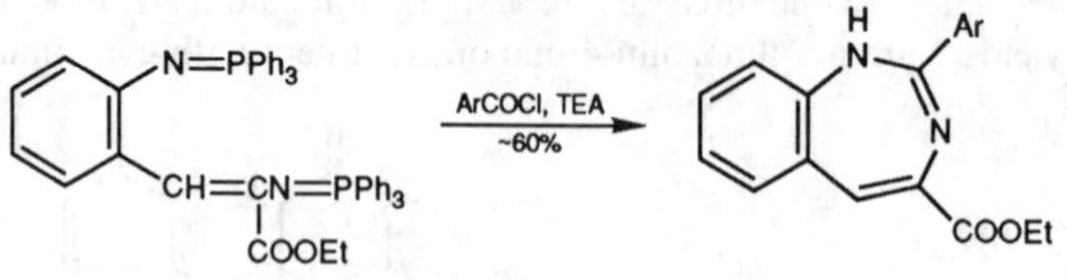

ArCOCl, TEA, ~60% [4005]

Ar = Ph, 4-Cl-, 4-Br-, 4-MeO-, 4-Me-C_6H_4

1,3-Benzodiazepine

III. FORMATION OF AN EIGHT-MEMBERED RING

1. 1,6,2-Dioxazocine

The *N*-substituted phthalimide (**80.6**) undergoes hydrazinolysis over 18 h in boiling ethanol, and the resulting aryloxyethanol cyclizes on heating with hydrochloric acid to this fused eight-membered ring; recent work on such rings has been surveyed [3988].

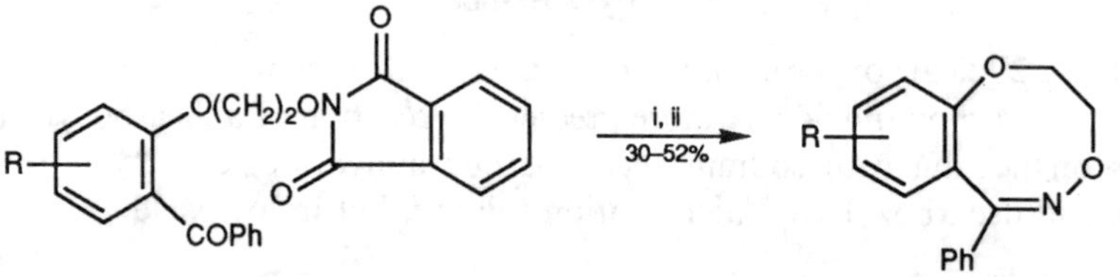

i, ii, 30–52% [3042]

(**80.6**)

R = H, Me, Cl

i, EtOH, $N_2H_4.H_2O$, Δ; ii, HCl, Δ

1,4,5-Benzodioxazocine

CHAPTER 81

Dihalogen

I.	Formation of a Five-membered Ring	1163
	1. Pyrrole	1163
	2. Pyrazole	1164
	3. Imidazole	1165
	4. Thiophene	1165
	5. Thiophene 1,1-Dioxide	1166
	6. Thiazole	1166
II.	Formation of a Six-membered Ring	1167
	1. Pyrazine or 1,4-Thiazine	1167
	2. Pyran-4-one	1167
	3. 1,4-Dioxin or 1,4-Dithiin	1168
	4. 1,3,4-Oxadiazine	1168
	5. 1,3,4-Thiadiazine	1169
III.	Formation of a Seven-membered Ring	1169
	1. 1,3-Thiazepine	1169

I. FORMATION OF A FIVE-MEMBERED RING

1. Pyrrole

Bisbromomethyl derivatives are cyclized by reaction with liquid methylamine at $-80\,°C$ [2061]. Hydrazine and its simple derivatives sometimes behave as monofunctional compounds, for example, in a reaction with the dibromoarylacetic ester (**81.1**) [2658, 2687]. Under these conditions, phenylhydrazine gives a mixture of the isoindole and isoindoline. An unconventional synthesis of a fused pyrrole is that in which a dihalide is heated with ethyl chloroacetate and pyridine in DMF. The pyridine reacts and becomes an integral part of the product [2985].

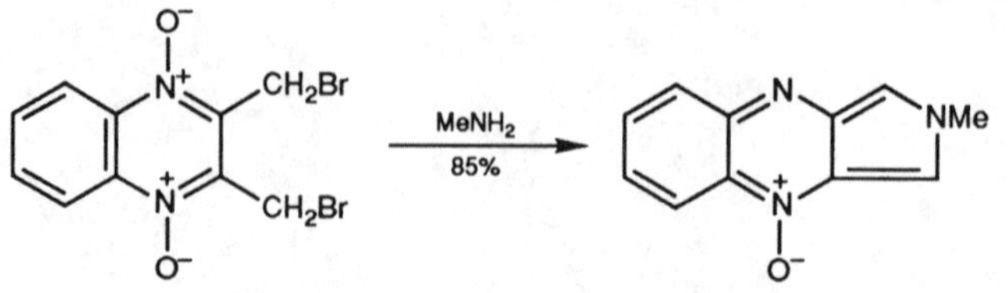

$MeNH_2$
85%

[2061]

Pyrrolo[3,4-*b*]quinoxaline 4-oxide

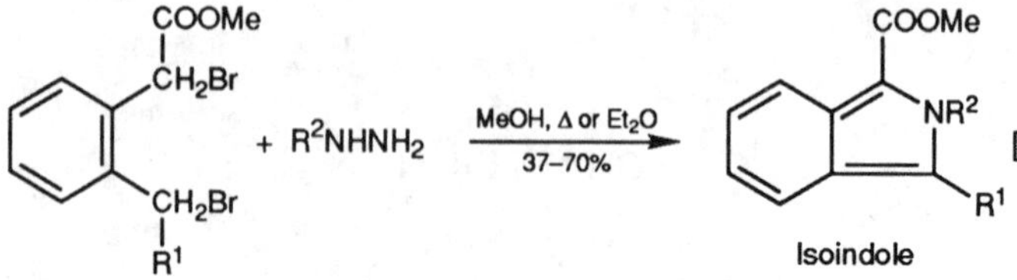

+ R^2NHNH_2 MeOH, Δ or Et_2O 37–70%

[2658, 2687]

Isoindole

(81.1)
R^1 = H, COOMe; R^2 = H, Me

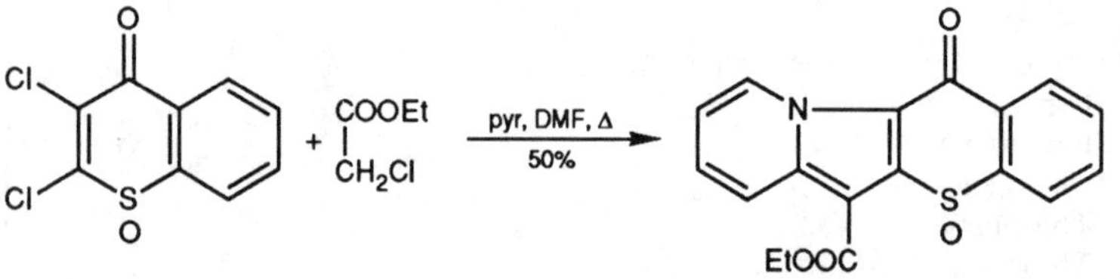

pyr, DMF, Δ
50%

[2985]

[1]Benzothiopyrano[2,3-*b*]indolizin-12-one 5-oxide

2. Pyrazole

Reactive chlorine atoms separated by three carbon atoms are displaced by hydrazine hydrate to give a pyrazole ring in high yield. In an unusual cyclization, hydrazine reacts with two neighbouring functions, a dichloromethyl and a trichloromethyl; the former behaves as a masked formyl group while the latter acts as a leaving group. Pyrazoles fused to other heterocyclic rings have been reviewed [3902].

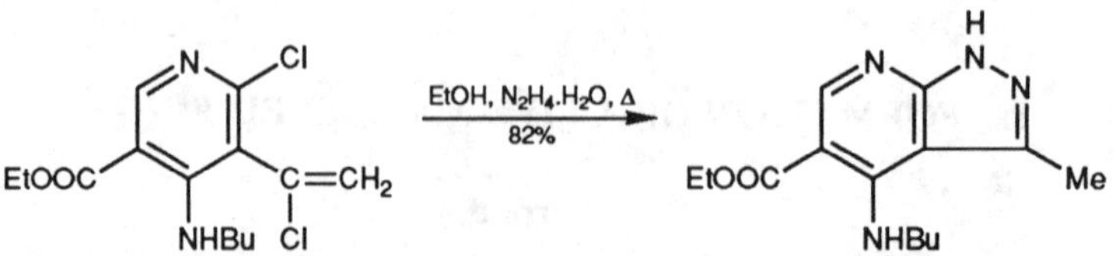

EtOH, $N_2H_4.H_2O$, Δ
82%

[2582]

Pyrazolo[3,4-*b*]pyridine

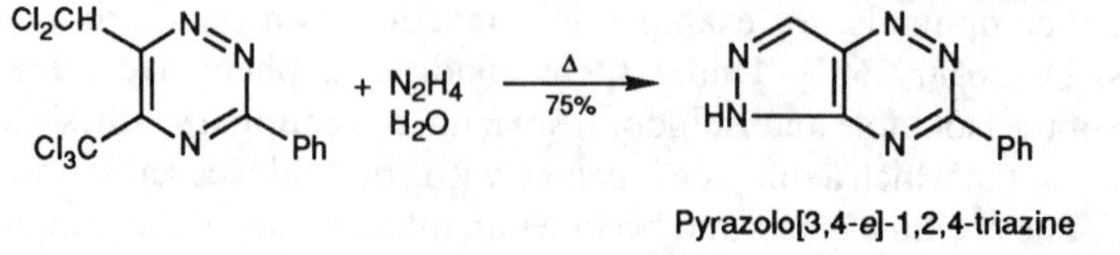

+ N_2H_4 H_2O Δ 75%

[3805]

Pyrazolo[3,4-*e*]-1,2,4-triazine

3. Imidazole

Halogens attached to pyrazine react readily with 2-aminopyridine with the formation of a doubly fused imidazole ring. 2-Aminopyridines carrying alkyl or alkoxy groups react at ambient temperature while two halogen atoms or a nitro group inhibit the cyclization.

diox., 15–91% [3099]

R = alkyl, $PhCH_2O$, Br, Cl

Pyrido[1',2':1,2]imidazo[4,5-*b*]pyrazine

4. Thiophene

Dichloromaleimide is converted at or below ambient temperature to a thienopyrrole by malononitrile–methoxide, and then hydrogen sulphide [3378]. Several dithienothiophenes are synthesized by oxidative cyclization of the lithiated intermediates, but when the latter are unstable, the dibromosulphides may be treated with butyllithium and copper(II) chloride [2287]. The chlorodibromoquinoline (**81.2**) is efficiently converted into the fused thiophene when heated with thiourea in a protic solvent [2873].

i, ii, 29–48% [3378]

R = H, C_6H_{11}, Ph, 2-Me-, 4-Bu-C_6H_4
i, $CH_2(CN)_2$, MeONa, MeOH, 0 °C, 4 h;
ii, AcOH, H_2S, 20–35 °C

Thieno[2,3-*c*]pyrrole-4,6-dione

+ 2BuLi; Et_2O, $CuCl_2$, 19% [2287]

Dithieno[3,2-*b*:3',4'-*d*]thiophene

EtOH, $CS(NH_2)_2$, Δ, 63–97% [2873]

(**81.2**)
R = H, Me, Cl, Br

Thieno[2,3-*b*]quinoline

5. Thiophene 1,1-Dioxide

The conversion of a 1,2-di(bromomethyl) group into the title ring probably proceeds through a dimethide which behaves as a diene towards such dieneophiles as sulphur dioxide to give the Diels–Alder product containing a fused thiophene 1,1-dioxide ring. The product is a stable source of the diene which may be generated by a retro-Diels–Alder reaction.

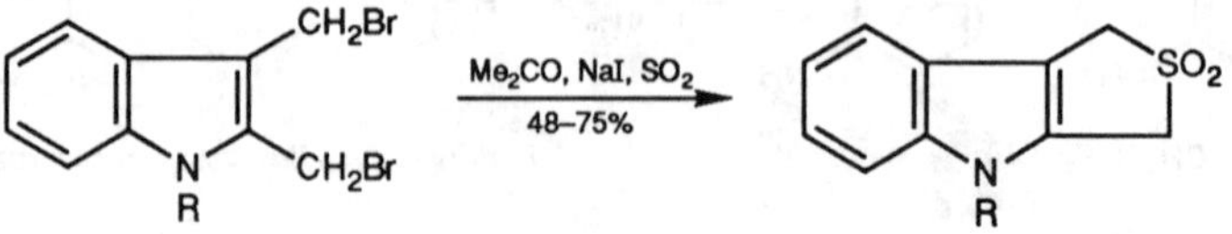

Me_2CO, NaI, SO_2 48–75% [3640]

Thieno[3,4-*b*]indole 2,2-dioxide

6. Thiazole

Dichloroquinoxaline reacts with a mercaptoimidazole in the presence of a mild base or with thiourea in boiling ethanol to form a fused thiazole.

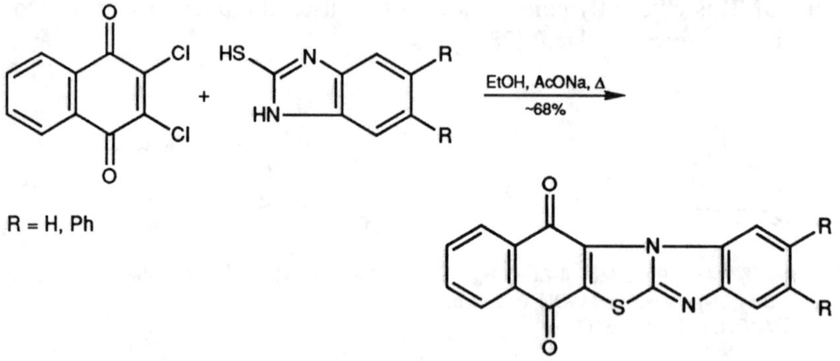

EtOH, AcONa, Δ ~68%

R = H, Ph

[2810]

Naphtho[2',3':4,5]]thiazolo[3,2-*a*]benzimidazole-7,12-dione

Another ring system synthesized similarly:

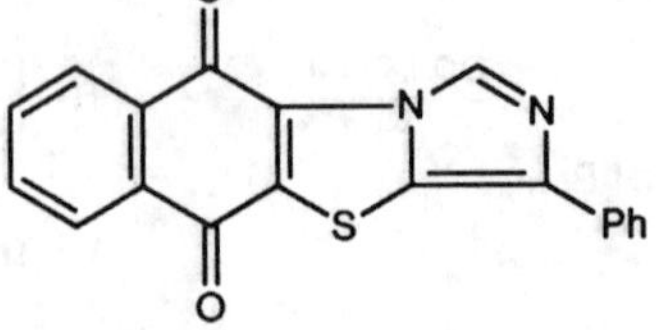

[2810]

Imidazo[5,1-*b*]naphtho[2,3-*d*]thiazole-5,10-dione

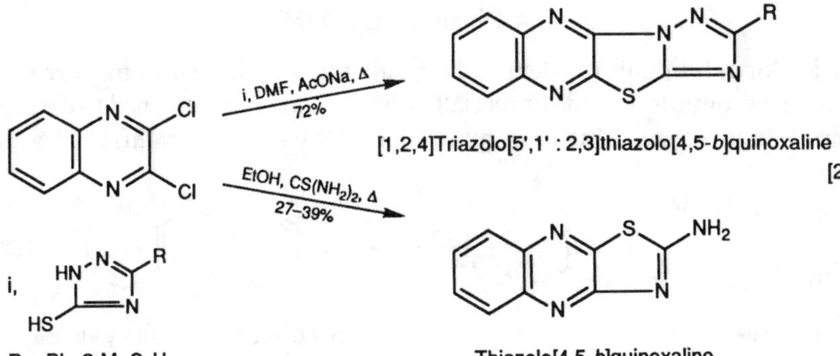

i, DMF, AcONa, Δ 72%

[1,2,4]Triazolo[5',1' : 2,3]thiazolo[4,5-*b*]quinoxaline [2671]

EtOH, $CS(NH_2)_2$, Δ 27–39%

i, R = Ph, 2-MeC_6H_4

Thiazolo[4,5-*b*]quinoxaline

II. FORMATION OF A SIX-MEMBERED RING

1. Pyrazine or 1,4-Thiazine

Dichloro-*N*-phenylmaleimide condenses with *o*-phenylenediamine or *o*-mercaptoaniline to give a high yield of the fused pyrazine or thiazine, respectively. *o*-Aminophenol does not give a cyclized product under similar conditions.

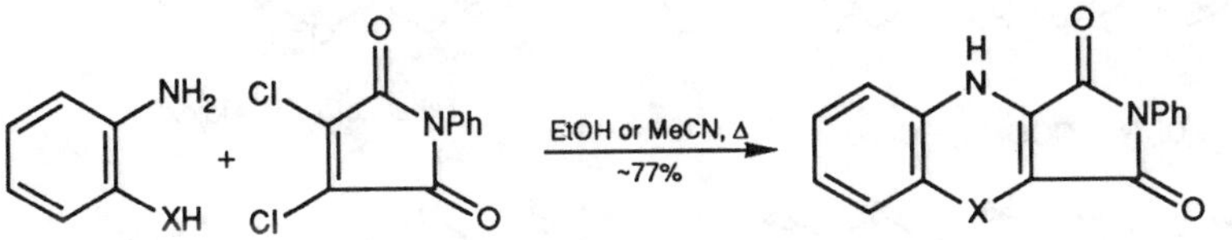

EtOH or MeCN, Δ ~77% [3578]

X = NH, S

X = NH Pyrrolo[3,4-*b*]quinoxaline-1,3-dione
X = S Pyrrolo[3,4-*b*][1,4]benzothiazine-1,3-dione

2. Pyran-4-one

An attempt to replace the bromine atom of the 1,3-diketone (**81.3**) by fluorine led to cyclization with loss of an aryl fluorine and retention of the bromine.

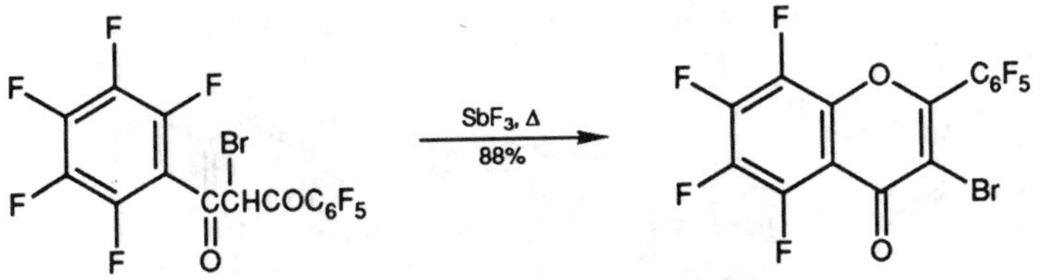

SbF_3, Δ 88% [2077]

(**81.3**)

1-Benzopyran-4-one, 2-aryl-

3. 1,4-Dioxin or 1,4-Dithiin

1,2-Diols or 1,2-dithiols (or their sodium salts) in a basic environment react as nucleophiles towards dihalides, especially when the latter are attached to electron-deficient rings such as pyrazine or pyridazine. HMPT promotes this cyclization.

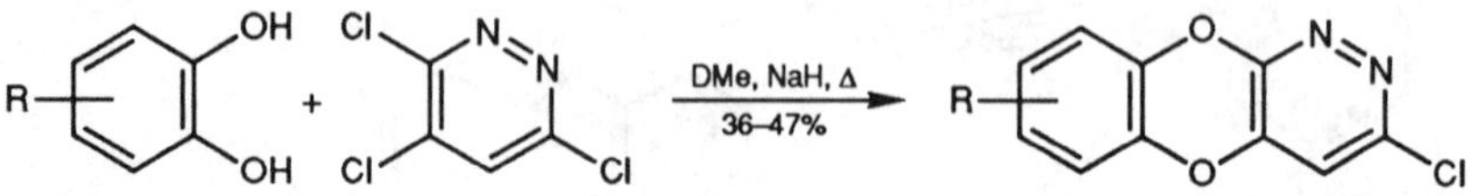

DMe, NaH, Δ
36–47%
[2679]

R = H, NO_2, MeO

[1,4]Benzodioxino[2,3-*c*]pyridazine

Other ring systems synthesized similarly:

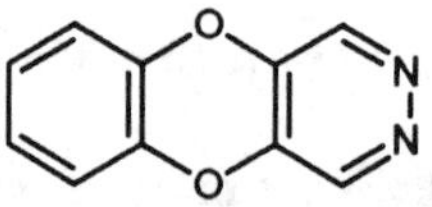

[2679]

[1,4]Benzodioxino[2,3-*d*]pyridazine

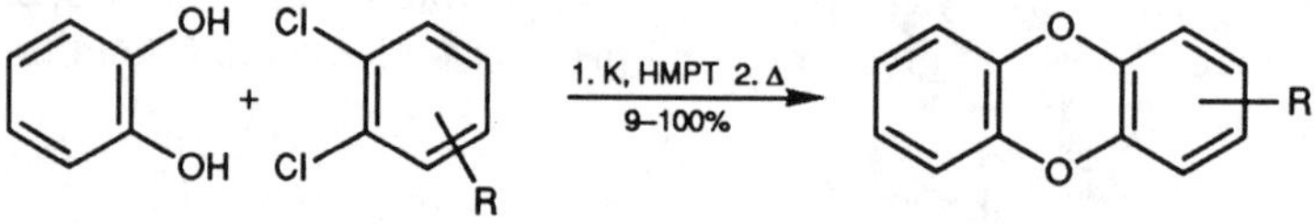

1. K, HMPT 2. Δ
9–100%
[3743]

R = Cl, NO, COOMe, COOiPr

Dibenzo[*b*,*e*][1,4]dioxin

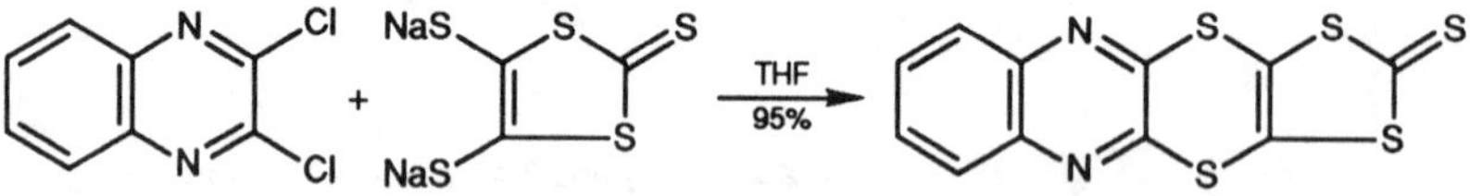

THF
95%
[3752]

1,3-Dithiolo[4′,5′:5,6][1,4]dithiino[2,3-*b*]quinoxaline-2-thione

4. 1,3,4-Oxadiazine

This ring may be formed by heating *o*,α-dibromophenylhydrazones with sodium acetate. The reactions of α-halogenohydrazones have been reviewed [1437, 1753].

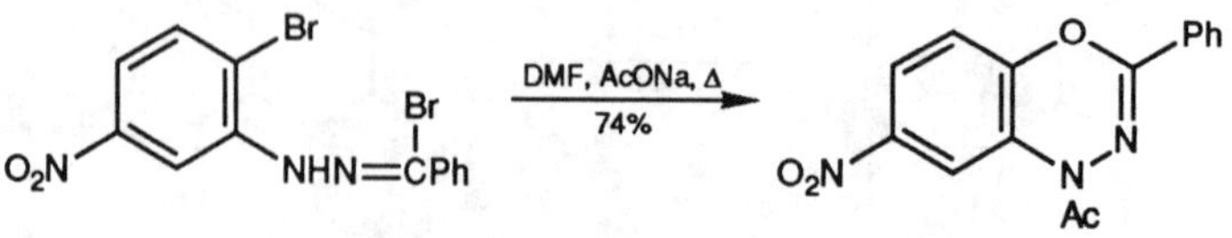

DMF, AcONa, Δ
74%
[2414]

1,3,4-Benzoxadiazine

5. 1,3,4-Thiadiazine

Compounds such as (**81.4**) which have a halogen at the α-position of the hydrazone and another (preferably a fluorine) on the benzene ring are cyclized by heating with potassium thioacetate. Lower yields are obtained when the aryl halide is a bromine or iodine.

(**81.4**) R = halogen; X = F, Br, I —[MeCN, AcSH, Δ; 14–67%]→ 1,3,4-Benzothiadiazine [2164]

III. FORMATION OF A SEVEN-MEMBERED RING

1. 1,3-Thiazepine

Thioamides (review [3458]) cyclize 1,2-di(bromomethyl)benzene to benzothiazepine in good yield.

1,2-di(bromomethyl)benzene + $RC(=S)NH_2$ —[Me_2CO, Na_2CO_3, Δ; 58–77%]→ 2,4-Benzothiazepine [2571]

R = Me, Ph, Me_2N

CHAPTER 82

Dihydroxy

I.	Formation of a Five-membered Ring	1170
	1. Furan	1170
	2. 1,3-Dioxole or 1,3-Dithiole	1171
	3. Thiophene 1,1-Dioxide	1172
II.	Formation of a Six-membered Ring	1173
	1. Pyrazine	1173
	2. Pyran	1173
	3. 1,4-Dioxin	1174
	4. 1,2-Dithiin 1,1-Dioxide	1174
	5. 1,3-Oxazine	1175
	6. 1,4-Oxazine	1175
III.	Formation of a Seven-membered Ring	1176
	1. 1,3,5,2,4-Trithiadiazepine	1176

Cyclizations of a diether, a disulphenyl chloride, a disulphone and of a disulphonyl chloride are mentioned in this chapter.

I. FORMATION OF A FIVE-MEMBERED RING

1. Furan

In contrast to many furan syntheses, this method uses 2-(2-hydroxyalkyl)phenols containing a leaving group at the benzylic carbon atom. Thus, 2-(1-isopropylthio-2-hydroxy)phenol cyclizes on heating with concentrated hydrochloric acid, but when 2-(2-hydroxyethyl)phenol was similarly treated, cyclization did not occur. When R^2 = H, 2-unsubstituted benzofurans are obtained in good yield; minor modifications were made in the synthesis of the 2-phenyl derivatives, and a mechanism for this cyclization involving an episulphonium salt was proposed [3406]. Attempts to hydrolyse the tetrahydropyranyl ether group of (**82.1**) with acid resulted in ring closure to the furan. The product obtained varied with the substituents and the acid used [3322]. When the triether (**82.2**) is stirred with mineral acid and then heated in benzene with 4-toluenesulphonic acid, it cyclizes to benzofuran [3929].

OH
R^1
$CHCHR^2OH$
iPrS
$MeO(CH_2)_2OH$, HCl, Δ
51–67%
R^1 O R^2
Benzofuran
[3406]

R^1 = H, MeO, Cl, Me, C_6H_{11}, NO_2, COOEt; R^2 = H, Me, Ph

MeOOC
R^1
OMe
$CMeCH_2OTHP$
R^2 OH
(**82.1**)
R^1 = $Me_2C{=}CH(CH_2)_2$
R^2 = MeO: TsOH, Me_2CO, Δ
66%
MeOOC
R^1
O
Me
O
4-Benzofuranone [3222]
R^2 = H: THF, H_2SO_4, Δ
55%
MeOOC
R^1
O
Me

OCH_2OMe
OEt
C=C
H H
(**82.2**)
1. EtOH, HCl 2. PhH, TsOH, Δ
73%
O
[3929]

2. 1,3-Dioxole or 1,3-Dithiole

Although several methods of converting a catechol into a 1,3-benzodioxole exist (and a number of them are mentioned [3976]), the use of bromochloromethane in DMF at 110 °C for 2 h in the presence of caesium carbonate provides an improvement on most of them. It does not require harsh conditions and gives high yields in all the examples reported; it gives good results in one instance where other methods had resulted in decomposition of the substrate [3976]. *o*-Diols react with cyanogen bromide at ambient temperature to form a fused 2-iminodioxole ring [2099]. Phenyl isocyanate dichloride annulates a dithiol in the presence of a base to give moderate yields of this ring [2203].

R = Br, CHO + CH_2BrCl → (DMF, Cs_2CO_3, Δ; ~87%) [3976]

1,3-Benzodioxole

Other ring systems synthesized similarly:

[3976]

Naphtho[1,2-*d*]-1,3-dioxol-6-one

1,3-Dioxolo[4,5-*h*][1]benzopyran-8-one [3976]

Naphtho[2,3-*d*]-1,3-dioxole [3976]

+ BrCN → (Me_2CO, TEA; 90%) [2099]

1,3-Benzodioxole

+ PhN=CCl_2 → (DMF, TEA, Δ; ~56%) [2203]

R = Me, $PhCH_2$

1,3-Dithiolo[4,5-*d*]pyridazin-4-one

3. Thiophene 1,1-Dioxide

An intramolecular arylation is believed to occur when the disulphone (**82.3**) is subjected to controlled-potential electrolysis in DMSO. The best yields are recorded in the presence of a two-molar proportion of acetate ion as base.

(82.3) — DMSO, AcO, electrolysis, 95% → Dibenzothiophene 5,5-dioxide [3065]

II. FORMATION OF A SIX-MEMBERED RING

1. Pyrazine

A pyrazine ring is formed rapidly and in high yield when two labile methoxy groups react with *o*-phenylenediamine.

DMF, Δ, 97% → [1,2,5]Thiadiazolo[3,4-*b*]quinoxaline 2,2-dioxide [2756]

2. Pyran

Prolonged heating of the ether-epoxide (**82.4**) in aqueous alkali gave high yields of the pyranoquinoline; several intermediates of this conversion were isolated in separate experiments. Enamines (reviews [3036, 3037]) carrying an electron-withdrawing substituent convert 2-hydroxybenzyl alcohols into benzopyrans in good-to-excellent yields on heating in either acetic anhydride or an acetic anhydride–acetic acid mixture.

(82.4) — DMSO, KOH, Δ, ~76% → Pyrano[3,2-*c*]quinoline [2502]

Ac_2O, Δ, 62–89% → 4*H*-1-Benzopyran [3595]

R^1 = H, MeO; R^2 = H, Me; R^3 = NO_2, CHO, Ac, PhCO, COOEt

3. 1,4-Dioxin

2,3,7,8-Tetrachlorodibenzo-1,4-dioxin (TCDD), a byproduct in the manufacture of the weedkiller 2,4,5-trichlorophenoxyacetic acid, is an exceedingly toxic chemical. It may be synthesized from 4,5-dichlorocatechol dipotassium salt and 1,2,4,5-tetrachlorobenzene [2675]. Other related compounds have been prepared similarly, for example, 2,3,7-trichlorodibenzo-1,4-dioxin. A range of benzonaphthodioxins has been synthesized by heating a catechol with 2,3-dichloro-1,4-naphthoquinones in pyridine, which acts as both solvent and base [3957]. Successful condensation of a catechol with a chloronitrobenzene depends on the presence or absence of substituents and their character, and also on the choice of solvent and base [2831].

DMSO, Δ 19% [2675]

Dibenzo[*b*,*e*][1,4]dioxin

pyr, Δ 25–91% [3957]

R = H, MeO, NO_2, Me, CN, tBu

Benzo[*b*]naphtho[2,3-*e*][1,4]dioxin-6,11-dione

i or ii, Δ 17–44% [2831]

R^1,R^2 = Cl

i, DMSO, K_2CO_3; ii, Me_2CO, K_2CO_3

Dibenzo[*b*,*e*][1,4]dioxin

4. 1,2-Dithiin 1,1-Dioxide

Two suitably placed sulphonyl chloride functions, on mild chemical reduction, form a 1,2-dithiin dioxide ring but heating the substrate (R = H) with hydrogen iodide gives the deoxygenated dibenzodithiin (R = H).

Na_2SO_3, Δ ~42% [2768]

R = H, Cl

Dibenzo[*c*,*e*][1,2]dithiin 5,5-dioxide

5. 1,3-Oxazine

o-Hydroxybenzyl alcohols are cyclized by nitriles in either perchloric acid or sulphuric acid–acetic acid. The usefulness of nitriles in synthesis has been reviewed [3332a].

OH, CR^1_2OH + R^2CN —(i or ii; 32–86%)→ [2182]

R^1 = Me, Et; R^2 = CH_2=CH, $ClCH_2$, $PhCH_2$, $EtOOCCH_2$, $(CH_2)_5NCH_2$

i, $HClO_4$, 0 °C; ii, AcOH, H_2SO_4, 40–50 °C

1,3-Benzoxazine

6. 1,4-Oxazine

Cyclization of the partly protected diol (**82.5**, R^1 = $tBuMe_2Si$) to give the *O*-2-isocephem was achieved by deprotection and treatment with diethyl azodicarboxylate–triphenylphosphine—a useful source of one carbon atom [3350]. Hydrolysis of the t-butyl ester in high yield was difficult, and replacing it with a trimethylsilylethyl group was a worthwhile improvement [3011]. Treatment of the hydroxy ether (**82.6**) with methanesulphonyl chloride and TEA or TEA-hydrogen sulphide results in ring closure in high yield.

(**82.5**) —(i, ii; 57%)→ [3011]

R^1 = $tBuMe_2Si$; R^2 = $PhOCH_2CO$

i, dil. HCl; ii, $(=NCOOEt)_2$. PPh_3

4-Oxa-2-azabicyclo[4.2.0]oct-2-en-8-one

(**82.6**) —(i or ii; ~90%)→ [3820]

R^1 = $PhCH_2$; R^2 = H, $PhCH_2$; X = O, S

i, TEA or $MeSO_2Cl$, CH_2Cl_2; ii, TEA, H_2S

4-Oxa-2-azabicyclo[4.2.0]oct-2-en-8-one
4-Thia-2-azabicyclo[4.2.0]oct-2-en-8-one

III. FORMATION OF A SEVEN-MEMBERED RING

1. 1,3,5,2,4-Trithiadiazepine

Benzene-1,2-disulphenyl dichloride is converted into a trithiadiazepine ring by stirring for several days at room temperature with bis(trimethylsilyl)sulphurdiimide.

$$\text{R-C}_6\text{H}_3(\text{SCl})_2 + (\text{Me}_3\text{SiN}=)_2\text{S} \xrightarrow[\text{50–65\%}]{\text{CH}_2\text{Cl}_2} \text{1,3,5,2,4-Benzotrithiadiazepine} \quad [3260]$$

R = H, Me

1,3,5,2,4-Benzotrithiadiazepine

CHAPTER 83

Dinitrile or Dinitro

I.	Formation of a Five-membered Ring	1177
	1. Pyrrole	1177
II.	Formation of a Six-membered Ring	1178
	1. Pyridine	1178
	2. Pyridazine	1179
	3. Pyridazine 1,2-Dioxide	1179
	4. Pyrimidine	1180

I. FORMATION OF A FIVE-MEMBERED RING

1. Pyrrole

Iron–acetic acid is better as a reducing and cyclizing agent for dinitro compounds (**83.2**) (cf. Chapter 52, Section III.2) than several other reagents, but hydrogenation using Pd–C was a very convenient procedure. An earlier report that a new ring was not formed when phthalonitrile was treated with methoxide was corrected when it was shown that the product was an isoindole.

R, NO_2, $CH{=}CHNO_2$, R → (Fe–AcOH, 64–71%) → R, H, N, R [3812]

(**83.2**) R = H, MeO — Indole

1. H_2O, PhH 2. H_2O ~48% + MeONa → Isoindole [2887]

II. FORMATION OF A SIX-MEMBERED RING

1. Pyridine

o-Cyanomethylbenzonitriles are readily cyclized by heating with ethereal hydrogen bromide or iodide [2129]. This type of reaction has been reviewed [2547]. When a similar reactant derived from pyridine is heated with a strong base (preferably a sodium alkoxide), it behaves differently and gives a methoxy-amine [2762]. A naturally occurring benzo[1,6]naphthyridine called aaptamine is synthesized from the dinitroisoquinoline (**83.3**) by reduction with iron powder and acetic acid (several other methods of reduction were unsuccessful), and heating the product for a short time at 80–90 °C [3760]. A new method [3848] of synthesizing 1,3-dinitriles such as (**83.4**) has enhanced the value of the cyclization of such compounds [48, 1901] (p. 456) to 1,6-naphthyridines and the thieno[3,2-*c*]pyridine shown in the equation [3849].

PhH, Et_2O–HX ~90% [2129]

R^1 = H, MeO, Br, Cl, Me; R^2 = H, Me; X = Br, I

Isoquinoline

RONa, Δ 21–77% [2762]

1,7-Naphthyridine

Another ring system synthesized similarly:

[2762]

2,6-Naphthyridine

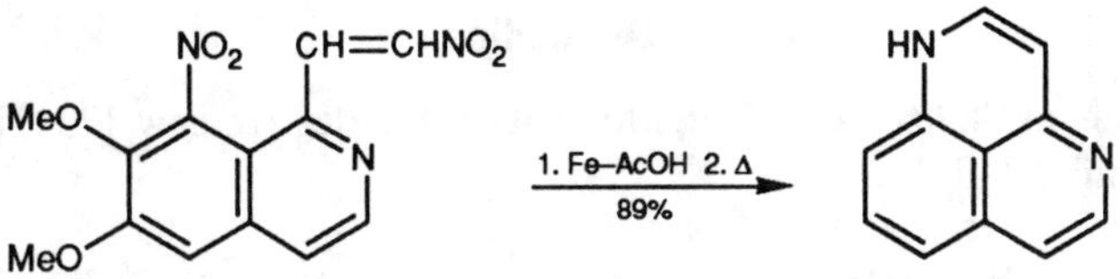

(83.3)

1. Fe–AcOH 2. Δ
89%

Benzo[*de*][1,6]naphthyridine

[3760]

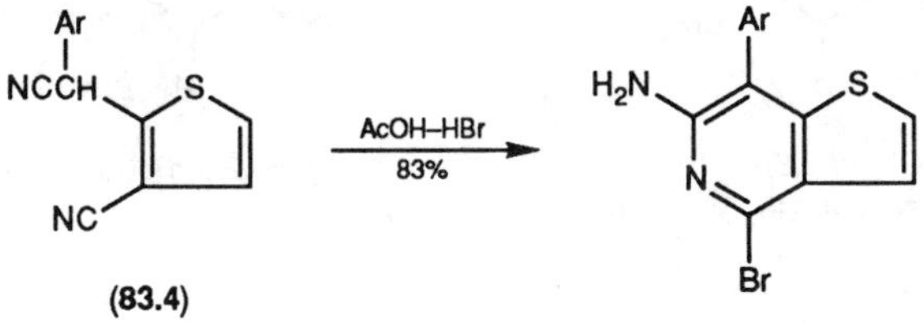

(83.4)

AcOH–HBr
83%

Thieno[3,2-*c*]pyridine

[3849]

2. Pyridazine

Reduction of the dinitro compound (**83.5**) with sodium sulphide gives a high yield of the pyridazine. The [3,4-*c*;4′,3′-*e*]isomer is similarly prepared [2956].

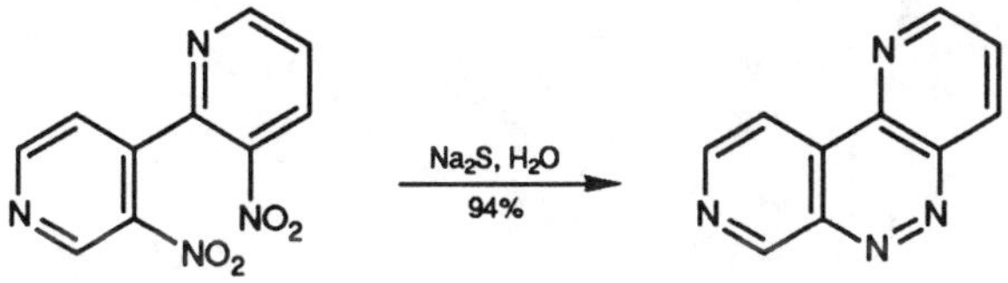

(83.5)

Na_2S, H_2O
94%

Dipyrido[3,4-*c*:2′,3′-*e*]pyridazine

[3062]

3. Pyridazine 1,2-Dioxide

2,2′-Dinitrobiphenyl, on mild reduction in the presence of an aged Raney nickel catalyst, is converted into a benzocinnoline dioxide. The mono-*N*-oxide is formed as a minor product from some dinitrobipyridyls.

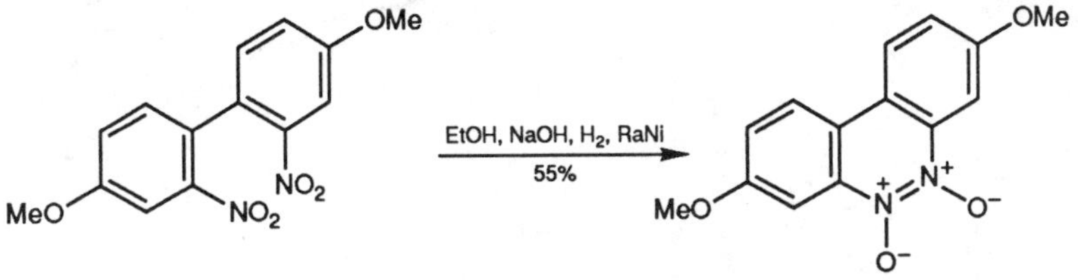

EtOH, NaOH, H_2, RaNi
55%

Benzo[*c*]cinnoline 5,6-dioxide

[2072]

4. Pyrimidine

Dinitriles such as (**83.6**) react with hydrogen halides (review [2547]) to give pyrimidine imino-halides.

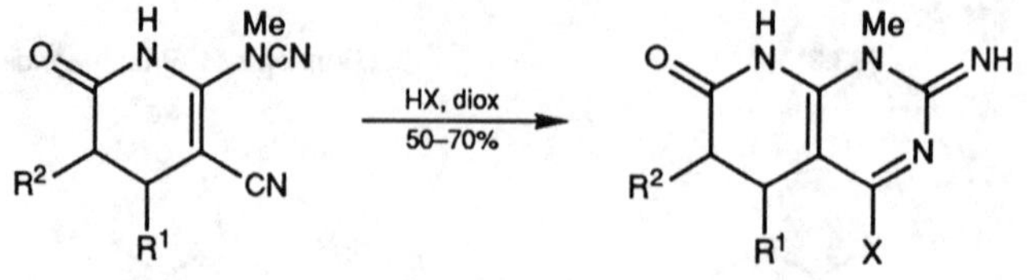

HX, diox
50–70%

[3128]

(83.6)
R^1,R^2 = H, Me; X = Cl, Br, I

Pyrido[2,3-*d*]pyrimidin-7-one

CHAPTER 84

Di-ring-carbon or Di-ring-nitrogen

I. Formation of a Five-membered Ring 1182
1. Pyrrole 1182
2. Pyrrol-2-one 1183
3. Pyrazole 1184
4. Pyrazole-3,5-dione or Pyrazol-1-ium, 3,5-Dioxo- 1186
5. Imidazol-2-one 1186
6. 1,2,3-Triazole 1186
7. 1,2,4-Triazole-3,5-dione 1187
8. Furan 1188
9. Thiophene or Thiophene 1,1-Dioxide 1188
10. Isoxazole 1189
11. 1,2,4-Oxadiazole 1189
II. Formation of a Six-membered Ring 1190
1. Pyridine 1190
2. Pyridin-2-one 1194
3. Pyridazine 1194
4. Pyrazine or Pyrazine 1-Oxide 1195
5. 1,2,4,5-Tetrazine 1195
6. Pyran or Thiopyran 1196
7. Pyran-2-one 1197
8. 1,4-Thiazine 1197
9. 1,3,4-Oxadiazin-6-one 1198
III. Formation of a Seven-membered Ring 1198
1. 1,2,4-Triazepine 1198

Examples of ring formation by joining two methylene groups are included.

I. FORMATION OF A FIVE-MEMBERED RING

1. Pyrrole

A pyrrole ring may be formed from azines and either anhydrous nickel chloride or palladium(II) chloride. Palladium-assisted C—C bond formation connecting one benzene ring to another is also a useful method. A review of transition metal-mediated heterocyclizations was published in 1980 [B-56]. Bis(methylthio)methylenemalonic acid derivatives annulate two methylene groups [2881].

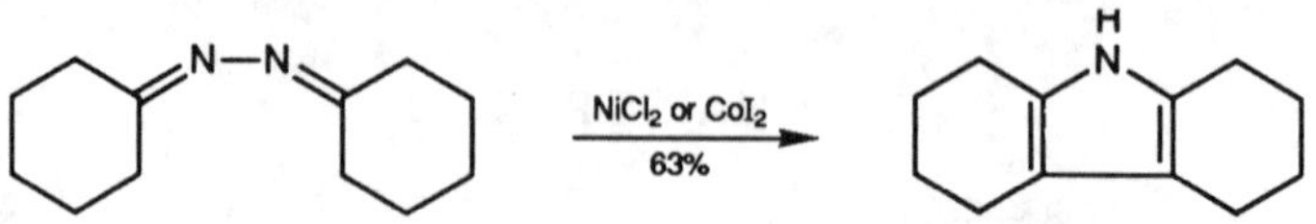
N—N
NiCl2 or CoI2
63%
H
N

[2132]

Carbazole

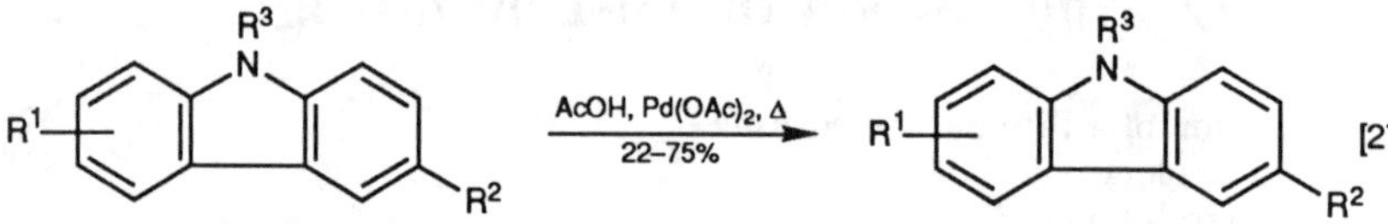
R3
N
R1
R2
AcOH, Pd(OAc)2, Δ
22–75%

[2744a]

R^1 = H, MeO, Br, Cl, Me; R^2 = H, MeO; R^3 = H, Me

Another ring system synthesized similarly:

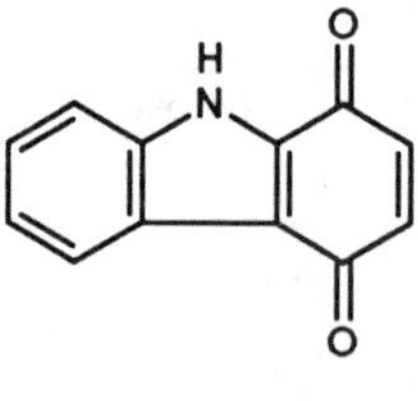
H
N
O
O

[3964]

1,4-Carbazoledione

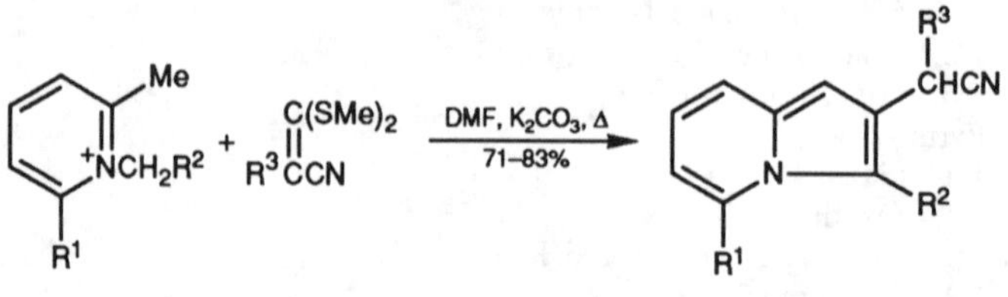
Me
+NCH2R2
R1
C(SMe)2
R3CCN
DMF, K2CO3, Δ
71–83%
R3
CHCN
N
R2

[2881]

R^1 = H, Me; R^2,R^3 = CN, COOEt

Indolizine

The aromatic 10π-electron heterocycle (**84.1**) may be synthesized by reaction of an azaindolizine with DMAD under dehydrogenating conditions [3465]. Another [8 + 2] cycloaddition is that of indolizines with DMAD at the 3,5-positions in the presence of palladium as dehydrogenator [3571]. Very reactive nitriles annulate the *peri* positions of an indene in an [8 + 2] addition [3160].

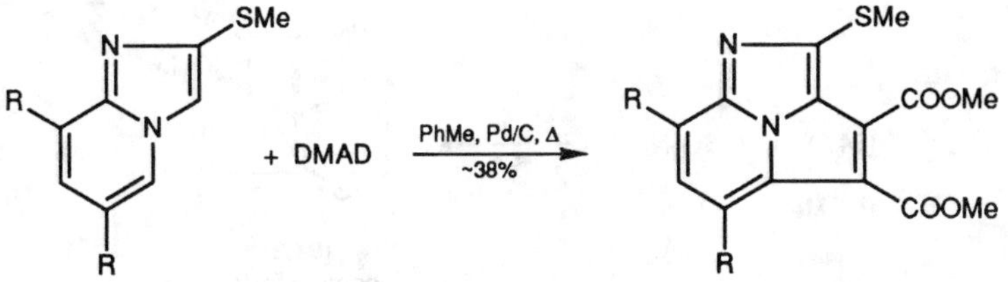

+ DMAD → PhMe, Pd/C, Δ, ~38% [3465]

R = H, Me

(**84.1**)
Imidazo[5,1,2-*cd*]indolizine

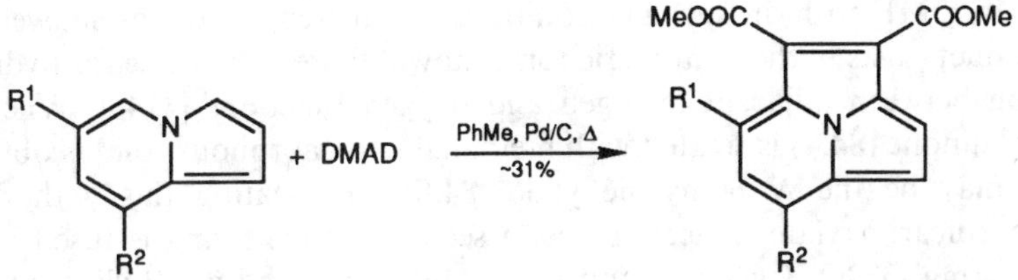

+ DMAD → PhMe, Pd/C, Δ, ~31% [3571]

R^1,R^2 = H, Me

Pyrrolo[2,1,5-*cd*]indolizine

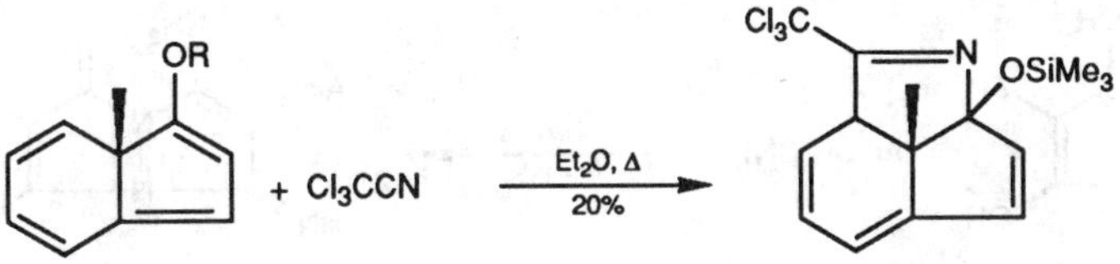

+ Cl_3CCN → Et_2O, Δ, 20% [3160]

(**84.2**)
R = $SiMe_3$

Cyclopent[*cd*]indole

2. Pyrrol-2-one

A C—C bond may be formed between C-2 of indole and that of an *N*-benzoyl group. This dehydrogenating reaction is aided by palladium(II) acetate; one or two byproducts are often obtained. Two *peri* positions of an indene derivative (**84.2**) may be annulated in an [8 + 2] addition by reaction with chlorosulphonyl isocyanate (reviews [3147a, 3340]) which in different experiments gave either pure (**84.3**) or a mixture of this and an isomer.

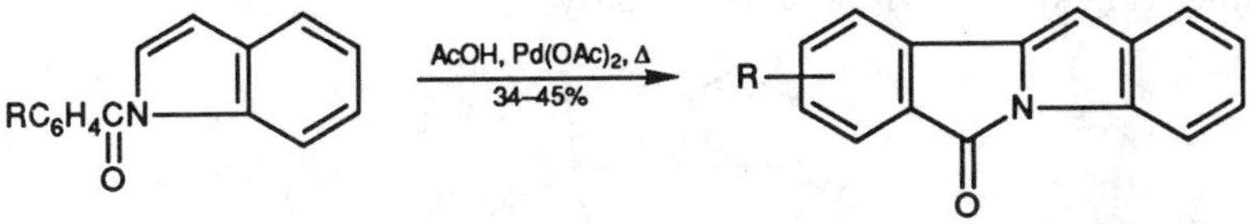

→ AcOH, $Pd(OAc)_2$, Δ, 34–45% [3122]

R = H, Me, benzo

Isoindolo[2,1-*a*]indol-6-one

(84.2) + $ClSO_2NCO$ —45%→ (84.3)

R = Me

(84.3)
Cyclopent[*cd*]indol-2-one

[3160]

3. Pyrazole

The two nitrogen atoms of a pyrazole are annulated by heating the compound briefly in DMF with an arylmalononitrile. Spectroscopic studies showed that the product exists in the tautomeric form shown; there is evidence for hydrogen bonding between adjacent nitrogen and oxygen functions [3145]. When the pyrazolidinone (**84.4**) is treated with methanal, an amorphous solid is obtained which may be the *N*-methylene ylide (**84.5**). On heating this with diallyl acetylenedicarboxylate in acetonitrile, a second pyrazole ring is fused on the original ring [3306]. Cycloaddition of a 1,4-quinone and an *N*-ylide gives an adduct which rapidly aromatizes. When another molecule of ylide is used, it adds on to the first product [3696].

+ $Ar^2CH(CN)_2$ —DMF, Δ, ~35%→

Ar^1 = Ph, 4-Cl-, 4-MeO-C_6H_4; Ar^2 = Ph, 4-$MeOC_6H_4$

Pyrazolo[1,2-*a*]pyrazol-1-one

[3145]

(84.4) + HCHO —MeOH→ (84.5) —i, 49%→

(84.4)
R = CH_2=$CHCH_2$
i, CCOOR ≡ CCOOR

(84.5)

Pyrazolo[1,2-*a*]pyrazol-1-one

[3306]

Another ring system synthesized similarly:

Pyrazolo[1,2-*a*]pyrazol-1-one

[3628]

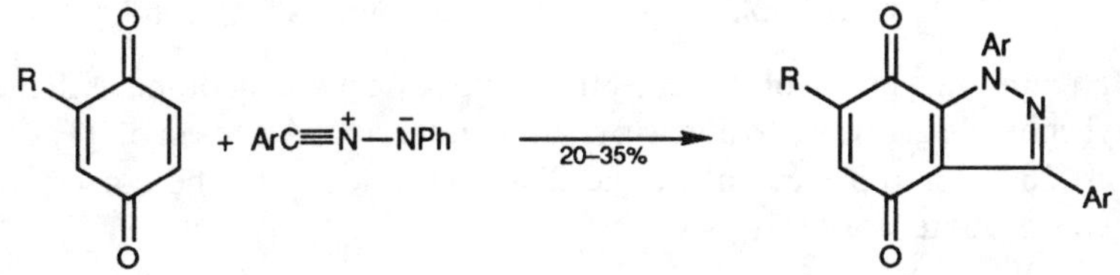

$+ \ ArC{\equiv}\overset{+}{N}{-}\overset{-}{N}Ph \xrightarrow{20\text{–}35\%}$ [3696]

R = Ph, 4-Me-, 4-Me-C_6H_4; Ar = Ph, 4-Cl-, 4-Me-C_6H_4

4,7-Indazoledione

A hydrazidoyl halide (reviews [1437, 1753]) reacts with some C—C double bonds in the presence of TEA or a phase transfer catalyst [3086]. The reactive double bond of thiete sulphone (**84.6**) adds on diazoalkanes under mild conditions to form the fused pyrazole (**84.7**) but diazomethane gives mostly isomer (**84.8**) [2174]. Addition of diazomethanes to the unsubstituted CH=CH group of a 1,4-quinone occurs under mild conditions to give versatile intermediates [2693, 2700]. The chemistry of heterocyclic quinones has been reviewed [2947, 3650].

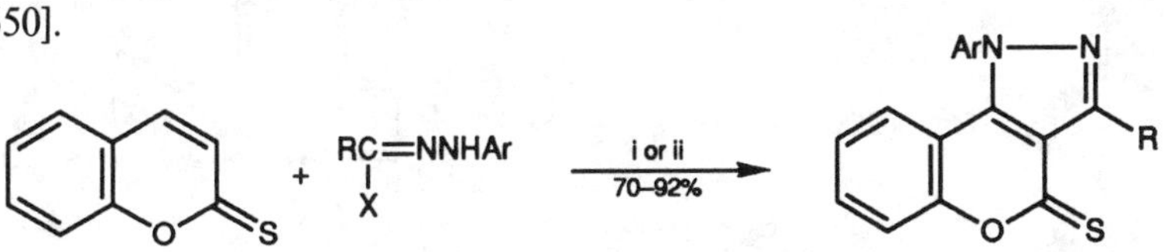

+ RC=NNHAr (X) $\xrightarrow[70\text{–}92\%]{\text{i or ii}}$ [3086]

R = Ac, EtCOO, Ph; Ar = Ph, 4-Br-, 4-Me-C_6H_4;
X = Br, Cl
i, $CHCl_3$, TEA, 0–5 °C; ii, $Bu_4\overset{+}{N}\overset{-}{Br}$, $NaHCO_3$

[1]Benzopyrano[4,3-*c*]pyrazole-4-thione

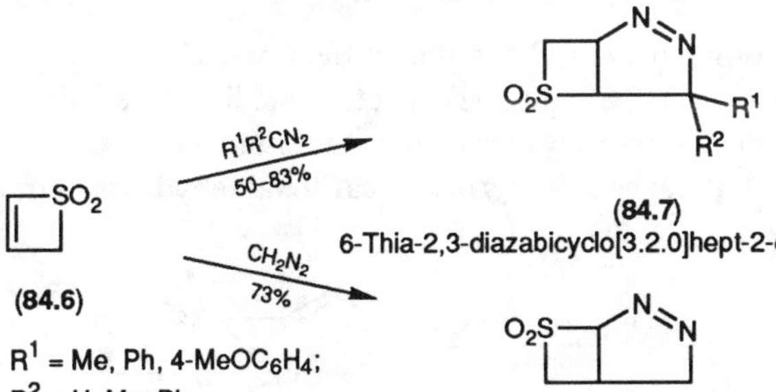

(**84.6**) $\xrightarrow[50\text{–}83\%]{R^1R^2CN_2}$ (**84.7**)

(**84.6**) $\xrightarrow[73\%]{CH_2N_2}$ (**84.8**)

R^1 = Me, Ph, 4-MeOC_6H_4;
R^2 = H, Me, Ph

(**84.7**)
6-Thia-2,3-diazabicyclo[3.2.0]hept-2-ene 6,6-dioxide [2174]

(**84.8**)
7-Thia-2,3-diazabicyclo[3.2.0]hept-2-ene 7,7-dioxide

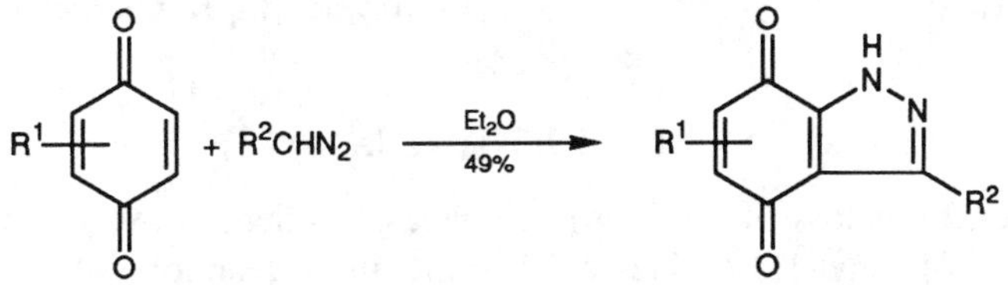

$+ \ R^2CHN_2 \xrightarrow[49\%]{Et_2O}$ [2693, 2700]

R^1 = Cl, NR^3R^4; R^3 = Ph, 4-MeC_6H_4;
R^4 = H, Me; R^2 = H, COOEt

4,7-Indazoledione

4. Pyrazole-3,5-dione or Pyrazol-1-ium, 3,5-Dioxo-

Simultaneous diacylation of the two NH groups of a pyrazoledione with methylmalonyl chloride gives rise to a new pyrazoledione ring. Addition of TEA to this reaction [2697] or replacement of the diacyl chloride with the corresponding ester gives a quaternary salt [3607].

HN, HN, O, O, Me, Me + MeCH(COCl)$_2$ → (THF, Δ, 88%) Me, O, O, N, N, Me, Me, O, O [2141]

Pyrazolo[1,2-*a*]pyrazole-1,3,5,7-tetraone

R^1, N, NH, R^1 + R^{2}CH(COOEt)$_2$ → (Δ, 18–62%) R^1, O, N, N$^+$, R^2, R^1, O [3607]

R^1 = H, Me; R^2 = Me, Ph, COOEt

Pyrazolo[1,2-*a*]pyrazol-4-ium, 1,3-dioxo-

5. Imidazol-2-one

The two rings of the bipyrrole (**84.9**) are joined by a new C—C bond which is formed at about 80 °C in a palladium(II) acetate-mediated reaction (review [B-41]. The salt is supported on polystyrene for maximum yield, but even under optimum conditions, 1,1′-carbonyldipyrrole-2-carboxaldehyde did not cyclize.

NCN, O, R, COOMe → (AcOH, Pd(OAc)$_2$–polymer, Δ, ~93%) N, N, O, R, COOMe [3412]

(**84.9**)
R = H, MeO

Dipyrrolo[1,2-*c*:2′,1′-*e*]imidazol-5-one

6. 1,2,3-Triazole

Vinyl azides (after loss of nitrogen) behave as 1,3-dienes towards the reactive dieneophile 4-phenyl-1,2,4-triazole-3,5-dione in a reaction which proceeds without external heat. Pyrrolo[1,2-*a*]indol-3-one adds on benzyl or phenyl azides.

$$\text{PhN(CO)}_2\text{N=N} + \text{N}_3\text{CR}^1\text{=NCR}^2 \xrightarrow[55\text{–}78\%]{\text{CH}_2\text{Cl}_2} \quad [2829]$$

R^1 = H, Ph, 4-MeO-, 4-NO_2-C_6H_4;
R^2 = H, Me, tBu

[1,2,4]Triazolo[1,2-*a*][1,2,3]triazole-5,7-dione

$$+ \text{RN}_3 \xrightarrow[53\text{–}68\%]{\text{PhH or Me}_2\text{CO}, \Delta} \quad [2273]$$

R = $PhCH_2$, Ph

1,2,3-Triazolo[4',5':3,4]pyrrolo[1,2-*a*]indol-4-one

7. 1,2,4-Triazole-3,5-dione

Adjacent NH functions in a 1,2,4-triazole ring may be annulated by reaction with an isocyanate (two molar proportions) to give a product in which $R^1 = R^2$. When molar proportions of reactants are used, R^1 and R^2 can be different, the second carbonyl group being then supplied by either phosgene or CDI. When the red 4-trimethylsilyl-1,2,4-triazoledione (**84.10**) is left to stand at room temperature for 48 h in moist air (or warmed at 80 °C in a dry box), it is converted into a colourless bicyclic product.

$$\xrightarrow[75\text{–}94\%]{\text{i, ii or iii}} \quad [2923]$$

R^1 = Me, Ph, 4-ClC_6H_4;
R^2 = Me, Ph, 4-ClC_6H_4, C_6H_{11}
i, R^2NCO(2 mol), EtOH, TEA;
ii, R^2NCO(1 mol), PhMe, $COCl_2$;
iii, R^2NCO(1 mol), CDI, Δ

[1,2,4]Triazolo[1,2-*a*][1,2,4]triazole-1,3,5,7-tetraone

$$\text{Me}_3\text{SiN} \ldots \xrightarrow[67\%]{\text{H}_2\text{O}} \quad [3500]$$

(**84.10**)

8. Furan

Diphenyl ether is often used as a high-boiling solvent but it can take part in an intramolecular cyclization under some conditions. The reaction is rapid in the presence of a palladium salt (review [B-41]) and an acid such as methanesulphonic or trifluoroacetic acid or of boron trifluoride. Addition of 4-phenyloxazole to benzyne (generated from 1-aminobenzotriazole and LTA) is more efficient when both reagents are simultaneously introduced through two syringes on opposite sides of the reaction vessel.

CF_3COOH, $Pd(OAc)_2$, Δ; 87% [2744a]

Dibenzofuran

CH_2Cl_2, 0 °C ~100 °C [3424]

1,4-Epoxyisoquinoline

9. Thiophene or Thiophene 1,1-Dioxide

A thiophene ring sandwiched between two other rings may be formed by constructing either a C—C bond or a sulphur bridge. The former method illustrates the use of lithiation followed by oxidation with copper(II) chloride, and the latter is effected with hot fuming sulphuric acid.

Et_2O, BuLi, $CuCl_2$; 52% [2286]

Dithieno[3,2-*b*:2',3'-*d*]thiophene

Other ring systems synthesized similarly:

[2286]

Dithieno[2,3-*b*:3',2'-*d*]thiophene

[2286]

Dithieno[3,4-*b*:3',4'-*d*]thiophene

[2286]

Dithieno[2,3-*b*:3',4'-*d*]thiophene

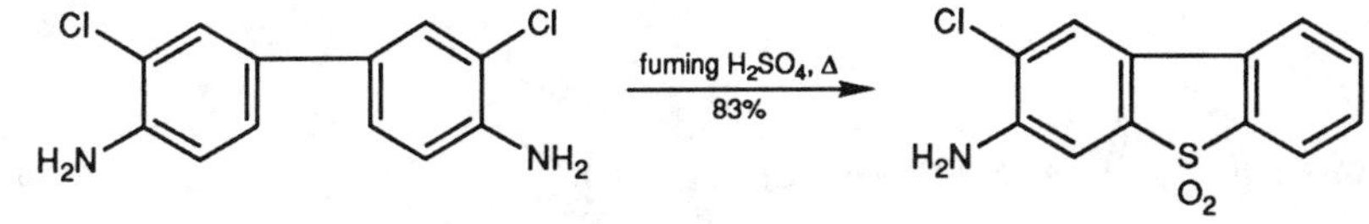

[2689]

Dibenzothiophene 5,5-dioxide

10. Isoxazole

A hydroxamoyl chloride reacts with a C—C double bond of a 1,4-quinone in the presence of silver oxide as base and oxidant. The chemistry of heterocyclic quinones has been reviewed [2947, 3650]. Addition of a nitrile oxide (review of reactions [B-43, 2115]) across the α,β-double bond of a butenolide leads to a new fused isoxazole ring. Cycloheptatriene undergoes a $[6 + 2]\pi$-cycloaddition at the 2,7-positions when it is stirred at room temperature for 4 days with nitrosobenzene [2875].

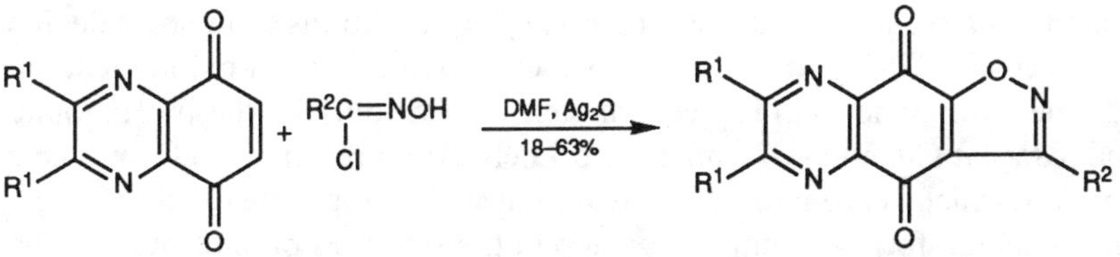

[3356]

R^1 = Cl, 4-MeC_6H_4; R^2 = Ph, 4-Me-, 4-NO_2-C_6H_4 Isoxazolo[4,5-*g*]quinoxaline-4,9-dione

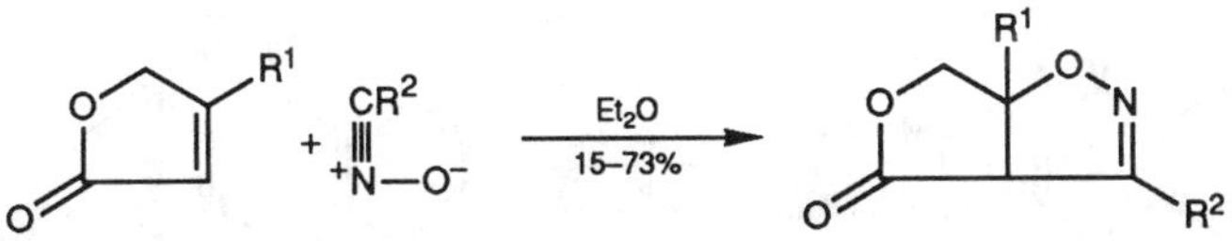

[2116]

R^1 = H, Me, Ph; R^2 = Ph, 2,4,6-$Me_3C_6H_2$ Furo[3,4-*d*]isoxazol-4-one

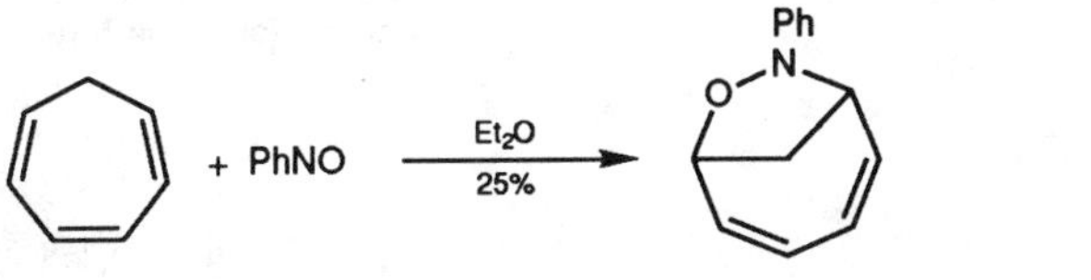

[2875]

7-Oxa-8-azabicyclo[4.2.1]nona-2,4-diene

11. 1,2,4-Oxadiazole

Nitrosobenzene adds to the 2- and 7- positions of ethyl azepine-1-carboxylate and forms an oxadiazole ring under mild conditions (cf. previous section).

COOEt
N
+ PhNO
Et_2O
50%
Ph
O–N
N
COOEt

[2875]

7-Oxa-8,9-diazabicyclo[4.2.1]nona-2,4-diene

II. FORMATION OF A SIX-MEMBERED RING

1. Pyridine

1,4-Quinones undergo a Diels–Alder reaction with azadienes (reviews [B-42, 2455, 3067a, 3880]) such as the *N,N*-dimethylhydrazone of 2-methylacrolein; for example, 5-methoxy-1,4-naphthoquinone reacts with the azadiene under mild conditions to give an easily oxidized intermediate which on warming in ethanol loses hydrogens and the new ring becomes fully aromatized [3279]. Further experiments using related azadienes under varying conditions show some degree of regiospecificity [3866]. In contrast to the behaviour of 5,8-quinolinedione towards the hydrazone (**84.11**) (see Chapter 35, Section I.5), the isoquinoline quinone reacts at C-6 and C-7 (in a hetero-Diels–Alder reaction) to form a new pyridine ring which is aromatized by treatment with manganese dioxide [3973]. A similar reaction forms an important step in the synthesis of analogues of the antibiotic nybomycin [3876].

O
MeO O
+ CMe=CH_2
CH=$NNMe_2$
PhH
62%
O
N
Me
O

[3279]

Benzo[*g*]quinoline-5,10-dione

O
N
O
+ EtOC=CHMe
CH=$NNMe_2$
(**84.11**)
$CHCl_2$, N_2
75%
O
H
N
N
OEt
O Me

[3973]

Pyrido[3,4-*g*]quinoline-5,10-dione

Another ring system synthesized similarly:

[3876]

Pyrido[2,3-*g*]quinoline-2,5,10-trione

N-Acyl-2-cyano-1-azadiene is sufficiently stable for its properties and reactions to be studied. It is a low-melting solid which reacts with the relatively weak dienophile norbornene in a Diels–Alder fashion [B-42] in boiling benzene over 7 h to give one main product and two minor stereoisomers of it [3853]. Similar reactions occur with the *N,N*-dimethylhydrazone of 2-methylacrolein and 1,4-naphthoquinone or its pyridine analogue. In some of these cyclizations, there is good steroselectivity but in others, a mixture of positional isomers is formed [3279]. A Diels–Alder reaction of 2-azabuta-1,3-diene which carries the easily eliminated dimethylamino groups is a convenient route to an azacarboline [3108].

PhH, Δ, 95% [3853]

5,8-Ethano[*g*]quinoline-5,10-dione

1. PhH 2. air, 62% [3279]

R^1 = MeO, then R^3 = MeO;
R^1 = AcO, then R^2 = HO; R^3 = H (52%) + R^2 = H, R^3 = HO (32%);
R^1 = HO, then R^2 = HO (84%)

Benzo[*g*]quinoline-5,10-dione

AcOH, TFA, 41–79% [3108]

R^1 = H, Me; R^2 = H, Me, Cl, NO_2, MeO, $PhCH_2O$;
R^3 = Ph, COOEt

Pyrido[3,4-*b*]indole

Other ring systems synthesized similarly:

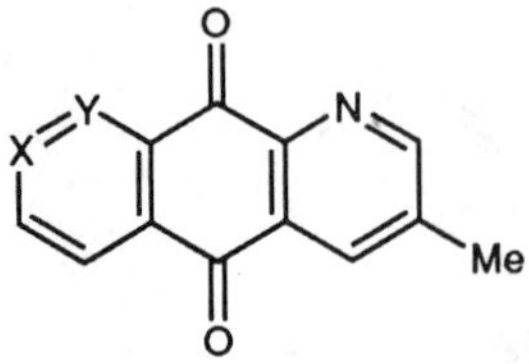

[3279] [3279]

X = CH; Y = N Pyrido[3,2-*g*]quinoline-5,10-dione
X = N; Y = CH Pyrido[4,3-*g*]quinoline-5,10-dione

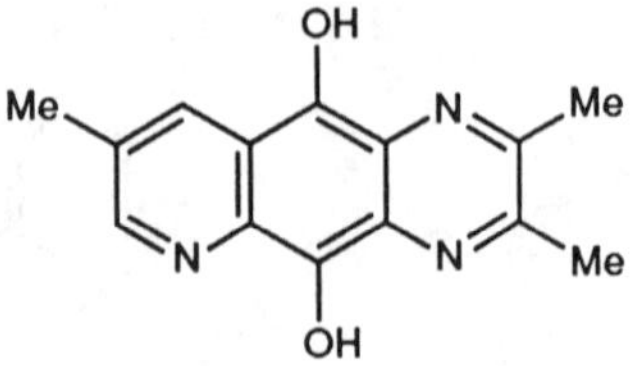

Pyrido[2,3-*g*]quinoxaline

Several methods of forming a bond between two carbon atoms in different aromatic rings are worthy of note. A diphenylamine may be converted into an acridine by heating with benzoic acid in PPA, or using Bernthsen's method, by heating with zinc chloride [2505]. A mixture of valence bond isomers is formed in high yield when the bisindole (**84.12**) is treated at room temperature with trifluoroacetic acid; dehydrogenation of the mixture gives a single product in high yield [3909]. Methyl and methylene groups in a 1,2-dimethylthiazolium salt may be annulated by heating with a 1,2-diketone and TEA; the product is a pyridinium salt [3106]. A C—C bond between different rings is formed when a solution of the *N*-styrylimidazole (**84.13**) containing iodine is irradiated in a Pyrex flask [2845]. In another photocyclization, imines (**84.14**) react in the presence of boron trifluoride etherate to give moderate yields of fused pyridines [3446]. Similarly, the imine (**84.15**) cyclizes on being irradiated in sulphuric acid using a quartz filter to give the marine alkaloid ascididemin [3874].

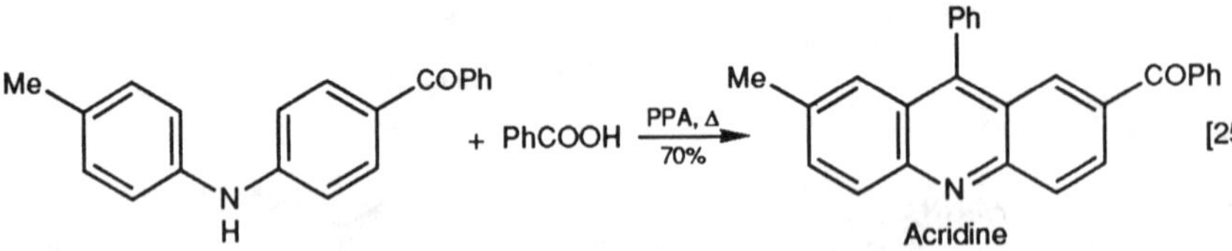

+ PhCOOH PPA, Δ 70% [2505]

Acridine

1. TFA 2. Pd/C, i, Δ 91% [3909]

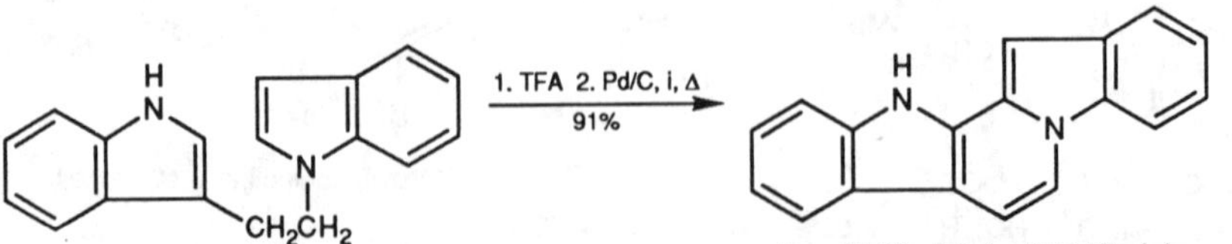

(**84.12**)
i, $(EtOCH_2CH_2)_2O$

Benz[2,3]indolizino[8,7-*b*]indole

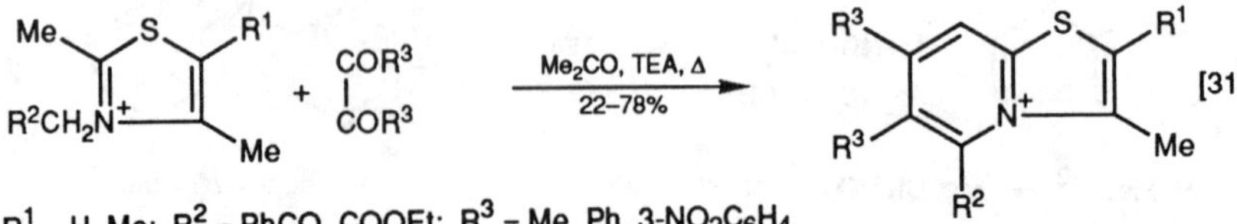

Me₂CO, TEA, Δ 22–78% [3106]

R^1 = H, Me; R^2 = PhCO, COOEt; R^3 = Me, Ph, 3-$NO_2C_6H_4$

Thiazolo[3,2-*a*]pyridinium

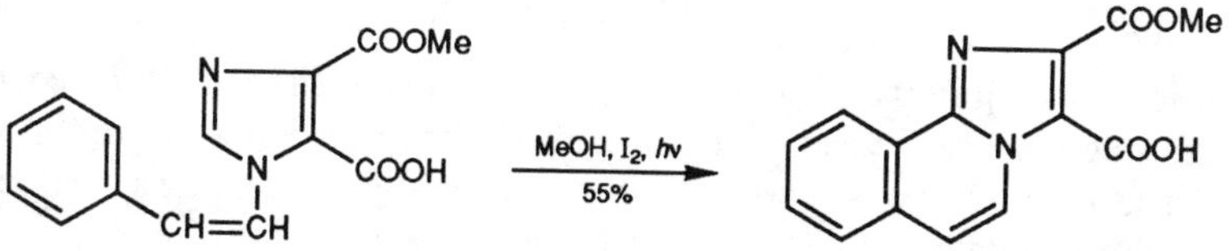

COOMe, N, N, COOH, CH=CH — MeOH, I_2, $h\nu$, 55% → N, COOMe, N, COOH [2845]

(84.13)

Imidazo[2,1-*a*]isoquinoline

Another ring system synthesized similarly:

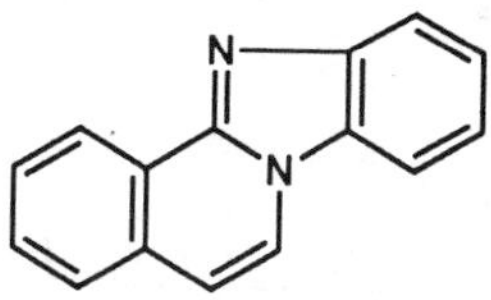

N, N [2845]

Benzimidazo[2,1-*a*]isoquinoline

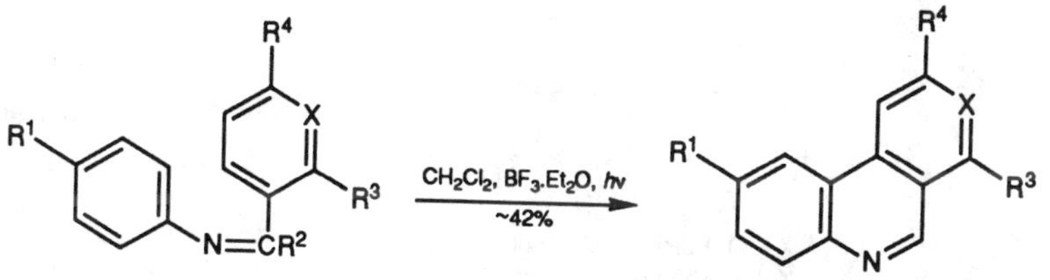

R^4, X, R^1, R^3, $N{=}CR^2$ — CH_2Cl_2, $BF_3.Et_2O$, $h\nu$, ~42% → R^4, X, R^1, R^3, N [3446]

(84.14)
R^1,R^3,R^4 = H, MeO; R^2 = H, Me

X = CH Phenanthridine
X = N Benzo[*c*][2,7]naphthyridine

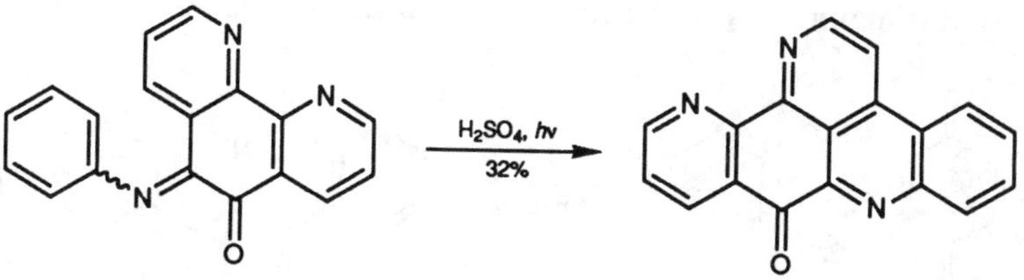

N, N, N, O — H_2SO_4, $h\nu$, 32% → N, N, N, O [3874]

(84.15)

Quino[4,3,2-*de*][1,10]phenanthrolin-9-one

2. Pyridin-2-one

Cycloaddition of a quinone and the protected azadiene (**84.16**) followed by *N*-methylation gives a high yield of the isoquinolinetrione (review of quinonoid heterocycles [3650]). Carbon—carbon bonds between two aromatic rings may be formed photochemically; care should be excercised in choosing the most appropriate solvent [2386].

$R^3OC{=}CH_2$, $N{=}CHOR^3$ (84.16)

1. PhH, Δ 2. HCl ~90%

R^1,R^2 = H, Me, tBu; R^3 = tBuSiMe$_2$

3,5,8-Isoquinolinetrione [3295]

CONHPh

PhH, EtOH, *hν* 77%

Indolo[2,3-*c*]quinolin-6-one [2386]

Other ring systems synthesized similarly:

[2386]

X = O Furo[2,3-*c*]quinolin-4-one
X = S Thieno[2,3-*c*]quinolin-4-one

3. Pyridazine

An activated endocyclic azo group undergoes Diels–Alder reactions with a dienophile on heating the components. Sometimes, the reacting N=N group is formed *in situ* under dehydrogenating conditions. A phenylhydrazone (**84.17**) derived from D-mannose is cyclized by heating with 1,4-benzoquinone for 24 h.

Δ 45–70%

Phthalazino[2,3-*b*]phthalazine-7,12-dione [2600]

HC=CH₂ / HC=CH₂ + HN / HN ... Me → (CH_2Cl_2, LTA, −10 °C; 28%) O, N, N, Me [2318]

Pyrazolo[1,2-*a*]pyridazin-1-one

O, O + ArNHN=CHR (**84.17**) → (PhMe, Δ; ~25%) O, Ar, N, N, O, R [3978]

R = $AcOCH(HCOAc)_2CH_2OAc$;
Ar = Ph, 4-Cl-, 4-MeO-, 4-Br-, 4-Me-C_6H_4

5,8-Cinnolinedione

4. Pyrazine or Pyrazine 1-Oxide

Diethyl azodicarboxylate can join two ring-carbon atoms through a nitrogen atom. An alternative is to treat the compound with nitrous acid so as to connect the two rings through an *N*-oxide function. Both methods can give good yields.

R^1, H, N, R^2, N, O, NR^3, O + NCOOEt=NCOOEt → (Δ; 70–98%) R^1, N, N, R^2, N, O, NR^3, O [2383, 2685]

R^1 = H, Me, Cl, MeO; R^2,R^3 = H, Me

Pyrimido[4,5-*b*]quinoxaline-2,4-dione

R^1, R^2, Me, N, H, N, O, NR^3, O → (AcOH, $NaNO_2$; ~73%) R^1, R^2, Me, N, N, O, NR^3, N^+, O^-, O [2423]

R^1,R^2,R^3 = H, Me

Benzo[*g*]pteridine-2,4-dione 5-oxide

5. 1,2,4,5-Tetrazine

4,4-Disubstituted isopyrazoles and the reactive diene 4-phenyl-1,2,4-triazole-3,5-dione undergo a Diels–Alder reaction at ambient temperature. Reactions involving heterodienes have been reviewed [B-42, 2455, 3067a, 3880].

[2122]

1,4-Methano[1,2,4]triazolo[1,2-*a*][1,2,4,5]tetrazine-6,8-dione

6. Pyran or Thiopyran

A pyran ring is formed from indene and the heterodiene 3-phenylsulphonylbut-3-en-2-one without any external heating [3992]. *N*-Phenylmaleimide reacts under Diels–Alder conditions with diallyl sulphoxide at 110–127 °C over 4–12 h; thioacrolein is released and is trapped by the imide [3877]. Both diaryl and diheteryl ethers or thioethers are ring-closed when stirred with dichloromethyl methyl ether and tin(IV) chloride [2503]. When diaryl ethers are heated with chloroacetyl chloride, a 4-methylidenepyran ring is formed, and if the two aryl rings are substituted differently, the possibility of stereoisomerism arises [2750].

[3992]

Indeno[1,2-*b*]pyran

[3877]

R = H, Me

Thiopyrano[3,4-*e*]pyrrole-5,7-dione

[2503]

X = O, S

Xanthylium
Thioxanthylium

[2750]

$R^1 = R^2 = H$; $R^1 = Br$, $R^2 = F$; $R^1 = 3\text{-}Me$, $R^2 = 6\text{-}Me$;
$R^3, R^4 = H, Cl$

Xanthene

7. Pyran-2-one

A new fused 3-acylaminopyran-2-one ring is generated when 1,3-dimethylbarbituric acid is heated in acetic anhydride with an *N*-acylglycine and a source of one carbon atom such as orthoformic ester, methyl diethoxyacetate or DMFDMA.

NHCOR | CH_2COOH —i, Ac_2O, Δ; 18–65%→ [3826]

R = Ph, pyrazin-2-yl
i, see text

Pyrano[2,3-*d*]pyrimidine-2,4,7-trione

Another ring system synthesized similarly:

[3826]

Indeno[1,2-*b*]pyran-2,5-dione

8. 1,4-Thiazine

Introduction of a sulphur atom as a bridge between two ring-carbon atoms of a diphenylamine may be achieved by heating with sulphur and iodine in 1,2-dichlorobenzene. In a modification of this method, the hydrogen sulphide produced is removed in a current of nitrogen; this gives better results than conducting the reaction under pressure or in an atmosphere of static nitrogen. It is possible for the sulphur bridge to be formed at one of two positions which would lead to either a linear or an angular tetracycle; spectral evidence supports the angular structure.

S, I, N_2, Δ; 70% [3806]

Imidazo[4,5-*c*]phenothiazine

9. 1,3,4-Oxadiazin-6-one

Dimethylketene adds across the N=N bond of cinnolines; the structure of the product depends on whether C-3 has a substituent. In the absence of a bulky substituent, the [4,3-*a*] isomer is formed, but otherwise the [3,4-*a*]cinnolin-3-one is produced.

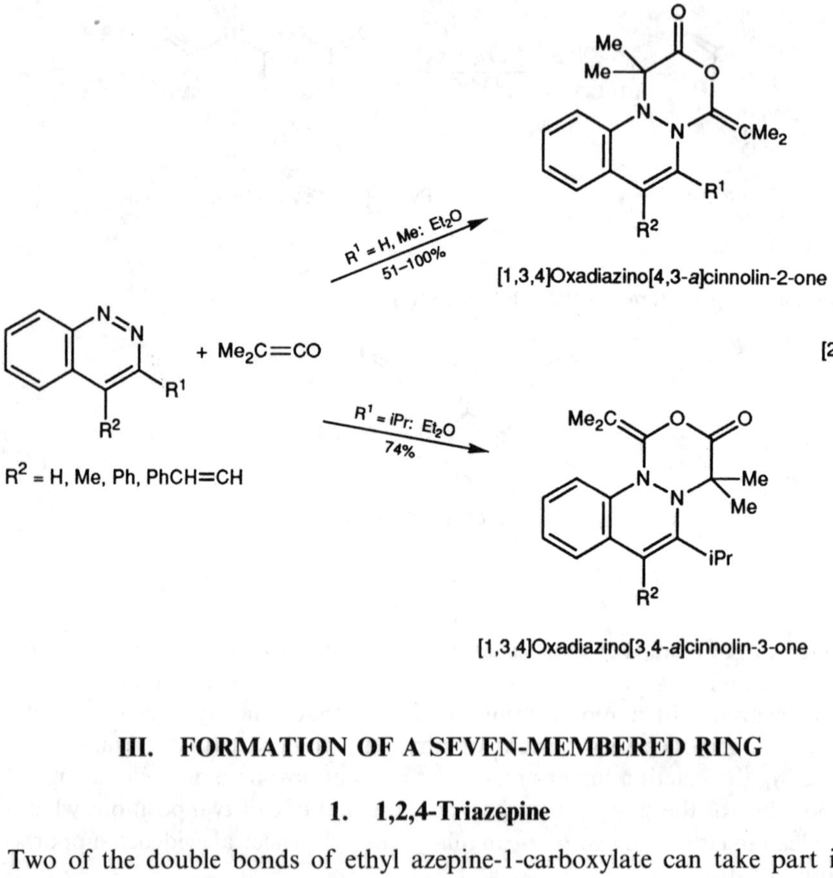

R^1 = H, Me: Et_2O, 51–100%

[1,3,4]Oxadiazino[4,3-*a*]cinnolin-2-one

+ $Me_2C{=}CO$ [2160]

R^2 = H, Me, Ph, PhCH=CH

R^1 = iPr: Et_2O, 74%

[1,3,4]Oxadiazino[3,4-*a*]cinnolin-3-one

III. FORMATION OF A SEVEN-MEMBERED RING

1. 1,2,4-Triazepine

Two of the double bonds of ethyl azepine-1-carboxylate can take part in a Diels–Alder reaction; for example, with heterodienophiles such as diethyl azodicarboxylate or 4-phenyl-1,2,4-triazole-3,5-dione, the addition proceeds at ambient temperature over 12 weeks.

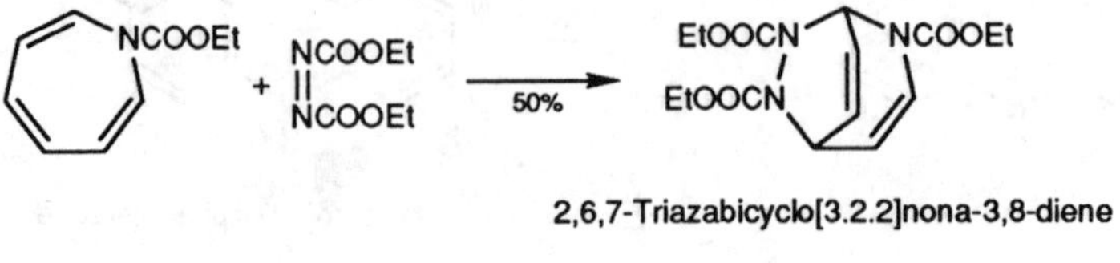

NCOOEt + NCOOEt=NCOOEt → (50%) EtOOCN, EtOOCN, NCOOEt [2871]

2,6,7-Triazabicyclo[3.2.2]nona-3,8-diene

Another ring system synthesized similarly:

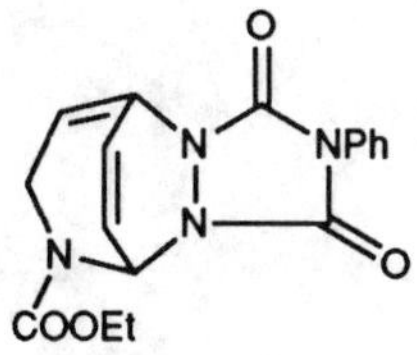

[2871]

5,10-Ethano[1,2,4]triazolo[1,2-*a*][1,2,4]triazepine-1,3-dione

CHAPTER 85

Enamine and Ester or Ketone

I.	Formation of a Five-membered Ring	1200
	1. Pyrrole	1200
II.	Formation of a Six-membered Ring	1201
	1. Pyridine	1201
	2. Pyridine 1-Oxide	1201
	3. Pyridin-2-one	1202
	4. Pyrimidine	1203
	5. Pyran-2-one	1203

Cyclization of enamine and either lactam or lactone carbonyl groups is illustrated.

I. FORMATION OF A FIVE-MEMBERED RING

1. Pyrrole

A 2-enamino-*N*-(2-oxoalkyl)pyridinium salt (review of pyridinium salts [1870]) cyclizes on stirring with TEA to give a high yield of a pyrroloisoquinoline. Heating the oxoenamine (**85.1**) in water induces cyclization to a reduced indolone.

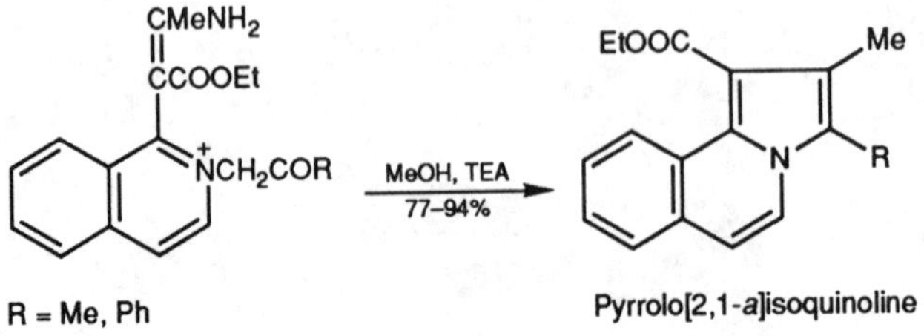

CMeNH$_2$ ‖ CCOOEt; $\overset{+}{N}$CH$_2$COR; MeOH, TEA 77–94% → EtOOC, Me, N, R [3000]

R = Me, Ph

Pyrrolo[2,1-*a*]isoquinoline

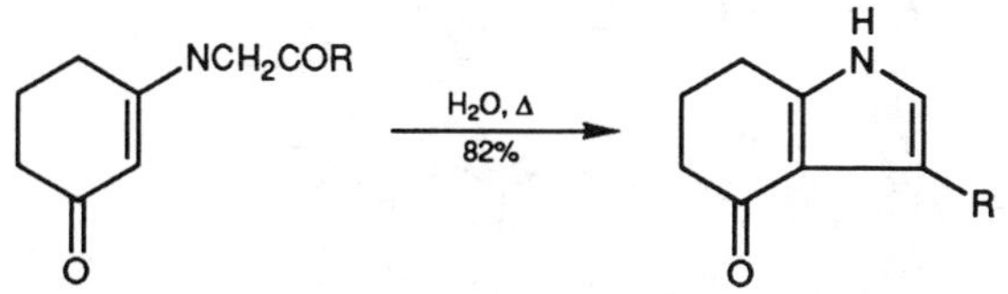

(85.1)
R = $(CHOH)_3CH_2OH$

H_2O, Δ 82%

4-Indolone

[2607]

II. FORMATION OF A SIX-MEMBERED RING

1. Pyridine

Thermally induced annulation of neighbouring acetyl and enamino groups with ammonium acetate in DMF can be a very effective method of building a fused pyridine ring.

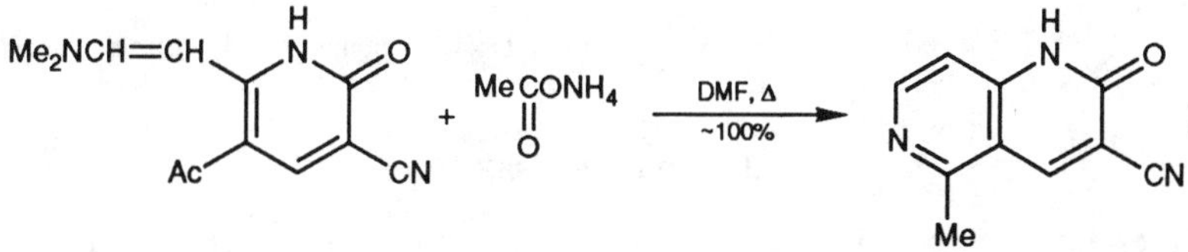

DMF, Δ ~100%

1,6-Naphthyridin-2-one

[3698]

Other ring systems synthesized similarly:

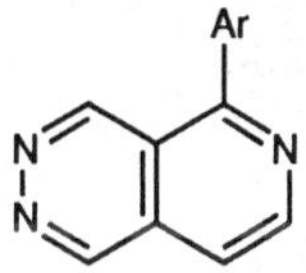

Pyrido[3,4-*d*]pyridazine [3579]

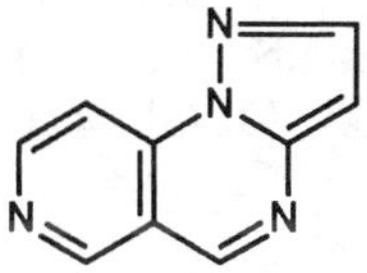

Pyrazolo[1,5-*a*]pyrido[3,4-*e*]pyrimidine [3800]

2. Pyridine 1-Oxide

A side-chain enamine is joined under aqueous alkaline conditions to its neighbouring lactone carbonyl on addition of hydroxylamine. A similar reaction with an acetyl group rather than a lactone carbonyl occurs in acetic acid–sodium acetate to give a high yield of a fused pyridine oxide.

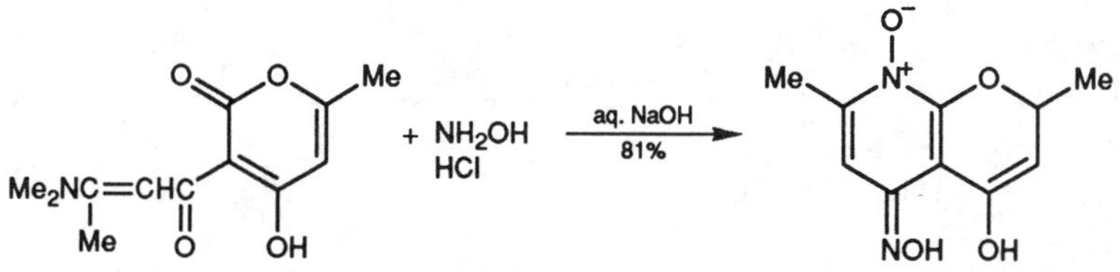

aq. NaOH 81%

Pyrano[2,3-*b*]pyridine 8-oxide

[2951]

Another ring system synthesized similarly:

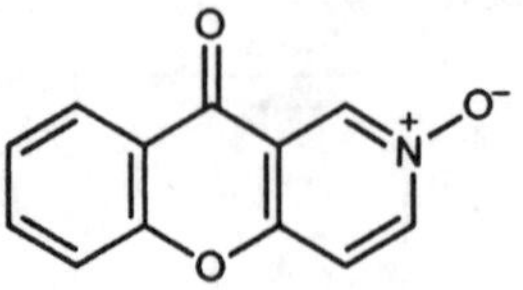

[3400]

[1]Benzopyrano[3,2-*c*]pyridin-10-one 2-oxide

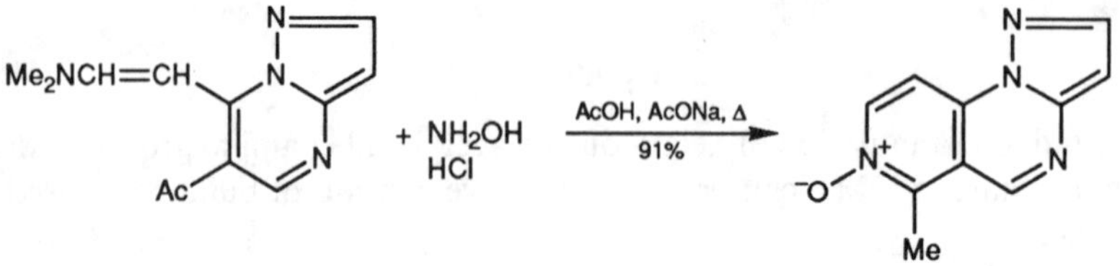

AcOH, AcONa, Δ 91%

[3804]

Pyrazolo[1,5-*a*]pyrido[3,4-*e*]pyrimidine 7-oxide

3. Pyridin-2-one

Enamines may be prepared *in situ* from a ketone and piperidine or similar tertiary amine; in this example, the enamine thus formed reacts with a nearby lactam carbonyl in refluxing acetonitrile. Replacing piperidine by TEA gave a low yield [2662]. A preformed enamine may react with a neighbouring ester group in the presence of a weak base such as benzylamine, sodium acetate or ammonium acetate [3475], or a strong base [3589].

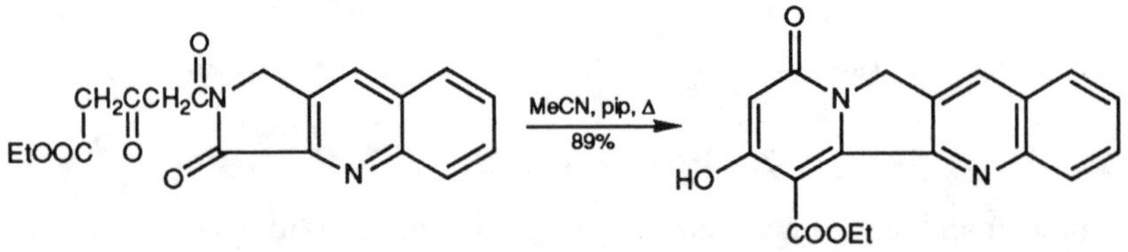

MeCN, pip, Δ 89%

[2662]

Indolizino[1,2-*b*]quinolin-9-one

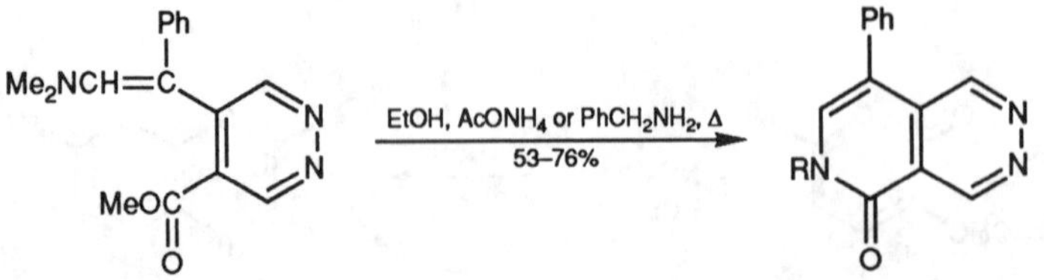

EtOH, $AcONH_4$ or $PhCH_2NH_2$, Δ 53–76%

[3475]

(85.2)
R = H, $PhCH_2NH_2$

Pyrido[3,4-*d*]pyridazin-5-one

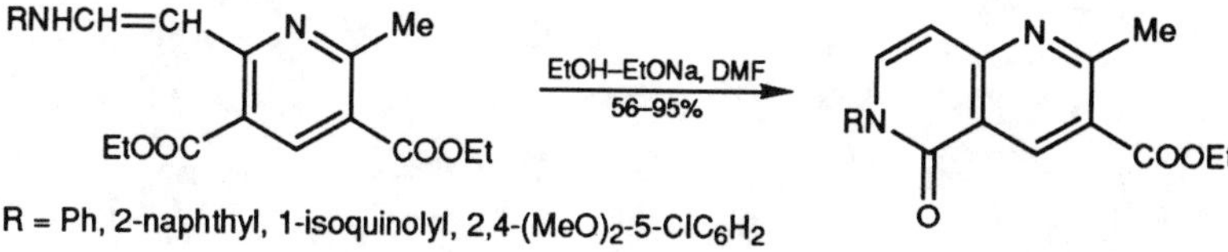

RNHCH=CH ... Me, EtOOC, COOEt → (EtOH–EtONa, DMF; 56–95%) → RN ... Me, COOEt, O [3589]

R = Ph, 2-naphthyl, 1-isoquinolyl, 2,4-$(MeO)_2$-5-ClC_6H_2

1,6-Naphthyridin-5-one

4. Pyrimidine

When an oxo-enamine is heated with an amidine and a strong base, a fused pyrimidine (in this example, a cannabinoid analogue) is formed in high yield.

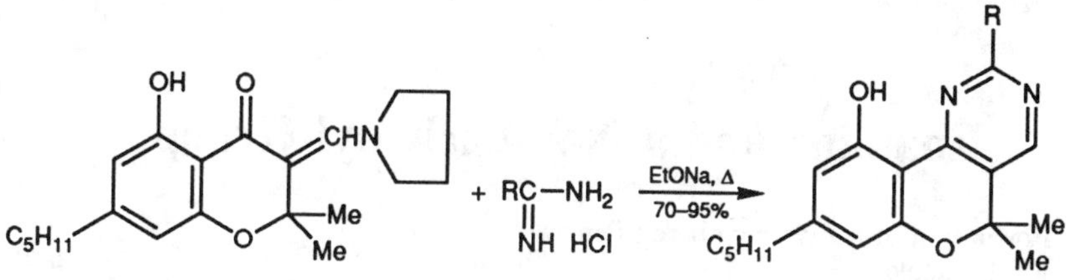

OH, O, CHN, Me, Me, C_5H_{11} + $RC(=NH)NH_2$ HCl → (EtONa, Δ; 70–95%) → R, OH, N, N, Me, Me, C_5H_{11} [3126]

R = H, Me, Ph

[1]Benzopyrano[4,3-*d*]pyrimidine

5. Pyran-2-one

This ring may be formed by stirring the enamino ester (**85.2**) with dilute acetic acid for 20 h or by heating the enaminone (**85.3**) with a malonic ester.

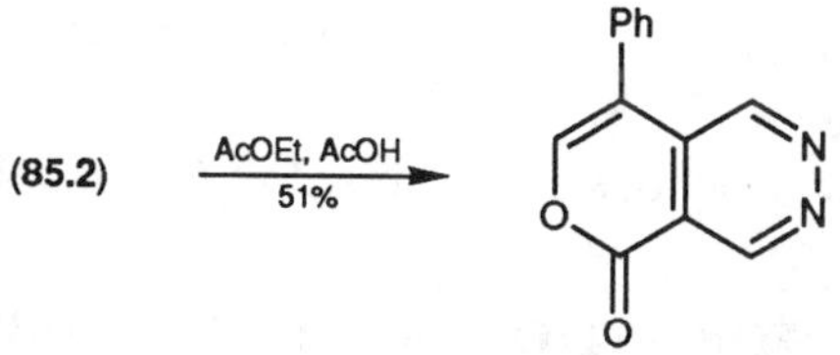

(**85.2**) → (AcOEt, AcOH; 51%) → Ph, O, N, N, O [3475]

Pyrano[3,4-*d*]pyridazin-5-one

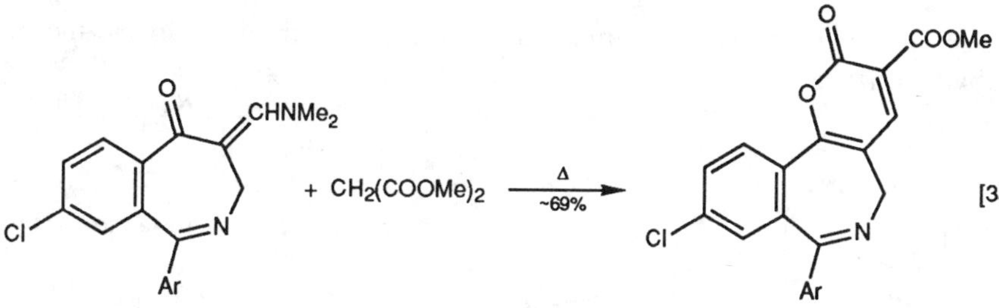

O, $CHNMe_2$, Cl, N, Ar + $CH_2(COOMe)_2$ → (Δ; ~69%) → O, COOMe, O, Cl, N, Ar [3587]

(**85.3**)
Ar = Ph, 2-Cl-, 2-F-C_6H_4

Pyrano[3,2-*d*][2]benzazepin-2-one

CHAPTER 86

Enamine and a Non-carbonyl Group

I. Formation of a Five-membered Ring 1204
 1. Pyrazole 1204
 2. 1,2-Dithiole 1205
 3. Thiazole 1205
II. Formation of a Six-membered Ring 1205
 1. Pyridin-4-one 1205
 2. Pyridazine 1206
 3. Pyran 1206
 4. Pyran-4-one 1206
III. Formation of a Seven-membered ring 1206
 1. Azepine 1206
 2. Azepin-3-one 1207

Amine–enamine reactions are discussed in Chapter 49.

I. FORMATION OF A FIVE-MEMBERED RING

1. Pyrazole

A 2-chloro-enamine may be annulated by warming with either hydrazine or methylhydrazine.

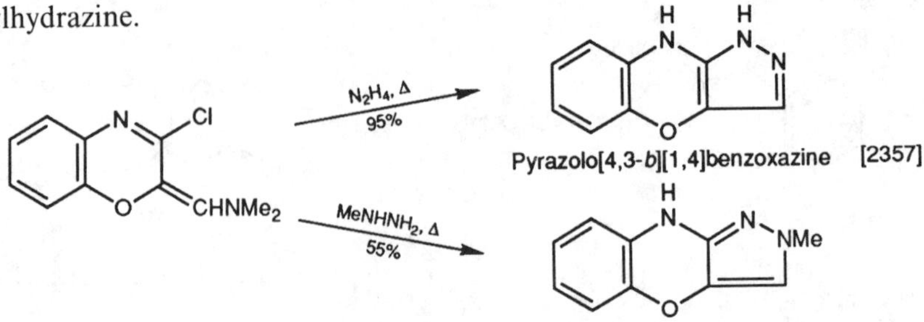

N_2H_4, Δ 95%

Pyrazolo[4,3-*b*][1,4]benzoxazine [2357]

$MeNHNH_2$, Δ 55%

2. 1,2-Dithiole

Treatment of the dithiolium salt (**86.2**) with sodium hydrogensulphide at ambient temperature gives a high yield of the dithiolodithiole.

aq. NaSH, DMF, 77% [2260]

(**86.2**)

[1,2]Dithiolo[1,5-*b*][1,2]dithiole-7-S^{IV}

3. Thiazole

Heating an enamine with sulphur and cyanamide produces a fused 2-aminothiazole.

$(H_2C)_n$ + S + H_2NCN → EtOH, Δ, ~64% [2222]

NR_2 = morpholino
n = 4,5

Benzothiazole
Cyclohepteno[*d*]thiazole

II. FORMATION OF A SIX-MEMBERED RING

1. Pyridin-4-one

When an enaminobenzene solution is saturated with phosgene and then heated, cyclization occurs to a 4-chloropyridinium salt, which is converted into a pyridinone on heating with alkali. A pyridine ring may also be constructed from a ring halogen and an enamine group; yields can be high in this reaction.

i, ii, 51% [2698]

i, $COCl_2$, PhH, Δ; ii, MeOH–NaOH, Δ

4-Quinolinone

i, ii or iii, Δ, 37–91% [3519, 3663]

R^1 = iPr, tBu; R^2 = H, Cl; X = CH, N
i, diox, NaH; ii, diox, MeCN; iii, THF, Bu_4NF

X = CH 4-Quinolinone
X = N 1,8-Naphthyridin-4-one

2. Pyridazine

Cyclization of an enamine function containing the sequence N—N=C—C=C—NMe_2 to a ring-carbon of a 1,2,4-triazole offers an efficient route to a triazolopyridazine.

CHNMe2 / CH / ArC=N—N (triazole, R) → AcOH, Δ, ~91% → [3474]

R = H, Me;
Ar = Ph, 3-MeS-, 4-MeO-,3-NO2-, 3-CF3-C6H4

1,2,4-Triazolo[4,3-*b*]pyridazine

3. Pyran

Intramolecular condensation between enamino and phenolic groups can lead to the formation of a 4*H*-pyran ring. Lewasorb ion-exchange resin gives better yields than Amberlite in this reaction.

OH, CRNMe2, Ph → i, ~100% → [2508]

R = H, Me
i, PhMe, Lewasorb ion exchanger

[1]Benzopyrano[4,3-*b*][1,5]benzodiazepine

4. Pyran-4-one

Heating the phenolic enaminone (**86.3**) with dilute sulphuric acid closes the pyran-4-one ring and gives a good yield of chromone.

OH, CCH=CHNMe2 → dil. H_2SO_4, Δ, 71% → [2334]

(**86.3**) 1-Benzopyran-4-one

III. FORMATION OF A SEVEN-MEMBERED RING

1. Azepine

Flash vacuum pyrolysis of the enamine (**86.4**) gave a low yield of the 4-aza-azulenone. The chemistry of 4-aza-azulenes has been reviewed [3503].

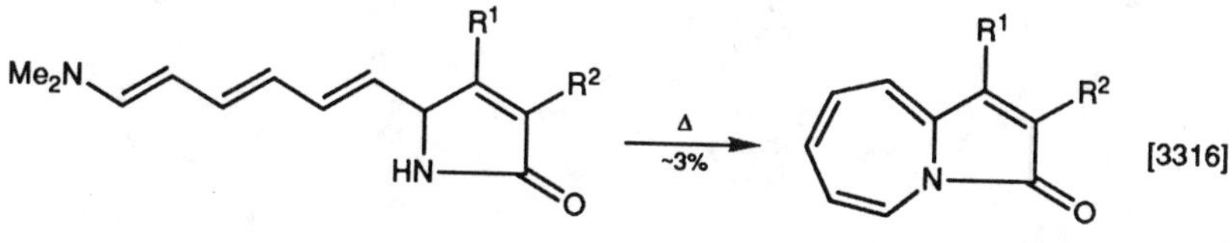

Me_2N R^1 R^2 HN O

(86.4)
$R^1 = R^2$ = H or benzo

$\xrightarrow[\sim 3\%]{\Delta}$

R^1 R^2 N O

Pyrrolo[1,2-*a*]azepin-3-one

[3316]

2. Azepin-3-one

An extended enaminone **(86.5)** ring-closes on to the nitrogen when the compound is heated under reduced pressure; the dimethylamino group is lost.

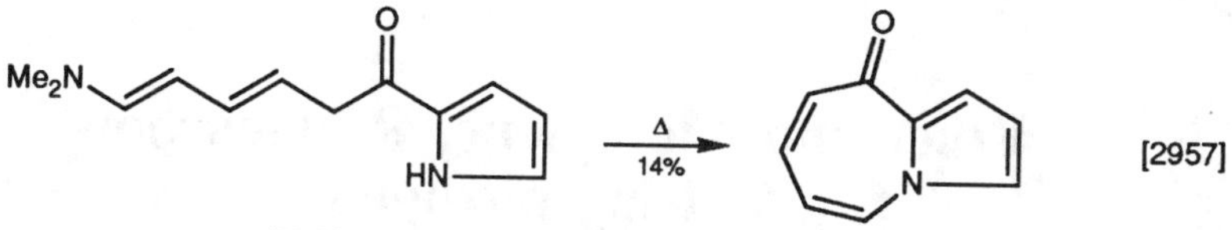

Me_2N O HN

(86.5)

$\xrightarrow[14\%]{\Delta}$

O N

Pyrrolo[1,2-*a*]azepin-9-one

[2957]

CHAPTER 87

Ether and Methylene, Ring-carbon or Ring-nitrogen

I. Formation of a Five-membered Ring 1208
1. Pyrrole 1208
2. Pyrazole 1209
3. Imidazole 1209
4. Furan 1210
5. Thiophene 1210
6. 1,2,4-Oxadiazole 1210
II. Formation of a Six-membered Ring 1211
1. Pyridine 1211
2. Pyridin-2-one or -4-one 1211
3. Pyran-2-one or -4-one 1212

I. FORMATION OF A FIVE-MEMBERED RING

1. Pyrrole

4-Substituted indoles are of biological interest and are not readily synthesized by conventional methods of indole chemistry. Annulation of a nuclear methyl and an α-ethoxyimine (or an imidate) under basic conditions is a promising procedure. The pyridine oxide ester (**87.1**) may be converted in high yields into two kinds of pyrrole carboxylic ester: the potassium salt of the imidate, on heating in DMF, gives the 3-(2-oxocarboxylate) whereas dilute mineral acid leads to the 2-carboxylate ester.

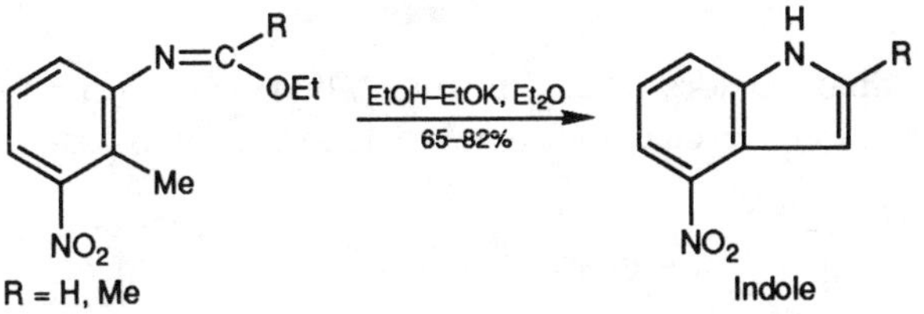

R = H, Me

$EtOH–EtOK, Et_2O$, 65–82%

Indole [3059, 3832]

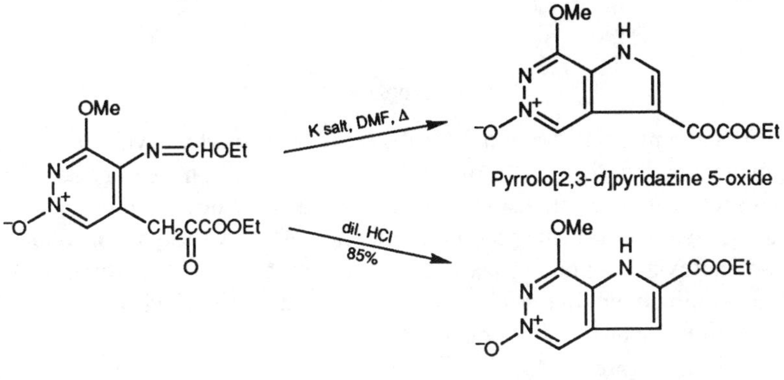

K salt, DMF, Δ

Pyrrolo[2,3-*d*]pyridazine 5-oxide [2563]

dil. HCl, 85%

2. Pyrazole

The *N*-iminopyridinium iodide (**87.2**) reacts with either ethoxymethylenecyanoacetate or 3-(ethoxymethylene)pentane-2,4-dione in chloroform containing potassium carbonate to give high yields of the pyrazolo[1,5-*a*]pyridine.

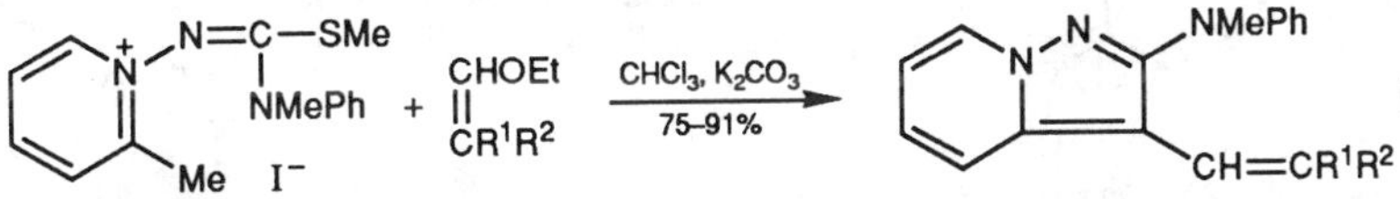

$CHCl_3, K_2CO_3$, 75–91%

(**87.2**)

R^1 = Ac, COOEt; R^2 = Ac, CN

Pyrazolo[1,5-*a*]pyridine [2986]

3. Imidazole

The azetes (**87.3**) undergo addition by nitrile ylides, $Ar\bar{C}H\overset{+}{N}{=}C$ (formed *in situ* from the imidoyl chlorides and TEA) at room temperature.

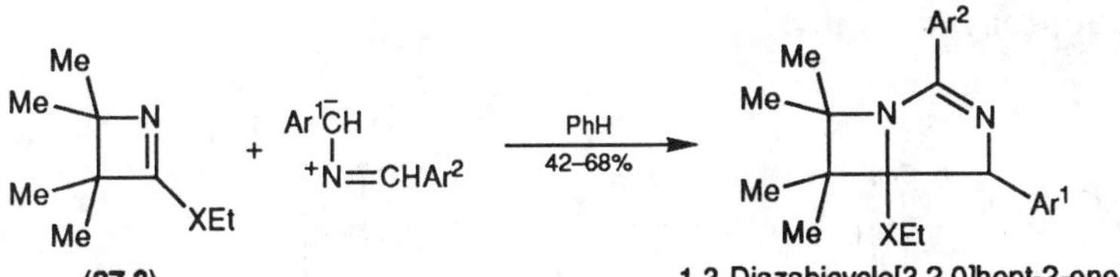

PhH, 42–68%

(**87.3**)

Ar^1 = 4-$NO_2C_6H_4$;

Ar^2 = Ph, 4-NO_2-, 4-Me-C_6H_4; X = O, S

1,3-Diazabicyclo[3.2.0]hept-2-ene [3856]

4. Furan

4-Methoxypyridine 1-oxides react with some alkynes to form a fused furan in good yield; during this cyclization, the *N*-oxide function is lost.

+ PhC≡CR —PhMe, Δ; ~68%→ Furo[3,2-*c*]pyridine [2760]

R = CN, COOMe

5. Thiophene

Lithiation of the methyl of (methylthio)benzene followed by acylation with an acyl chloride and acidification to pH 4–5 yields benzothiophene in good yields, but when aroyl chlorides are used, the mixture has to be heated in benzene. An attempted dehydration of the secondary alcohol (**87.4**) with hydrobromic acid led to *S*-demethylation and cyclization. Similar treatment of the isomeric 2-(2-methylthio-4-nitrophenyl)-1-phenylethanol gave a high yield of 6-nitro-2-phenyl-2,3-dihydrobenzo[*b*]thiophene [2653].

—i, ii, iii; 47–64%→ Benzo[*b*]thiophene [3387]

i, BuLi, Et_2O, TMEDA; ii, RCOCl; iii, HCl
R = alkyl, $PhCH_2$, Ph, 4-MeC_6H_4

—HBr–AcOH, Δ; 48%→ [2653]

(**87.4**)
Ar = 4-$NO_2C_6H_4$

6. 1,2,4-Oxadiazole

Addition of aroylhydroxamoyl chloride (α-chloroaraldehyde oxime) (**87.5**) to the lactim ether (**87.3**) in the presence of a base leads to the formation of the five-membered ring in moderate-to-good yield, the poorest yield being given by the 2-nitrophenyl derivative.

(**87.3**) + ArC(Cl)=NOH (**87.5**) —PhH, TEA; 34–80%→ 4-Oxa-1,3-diazabicyclo[3.2.0]hept-2-ene [3856]

Ar = Ph, 4-MeO, 2-NO_2-C_6H_4;
X = O, S

II. FORMATION OF A SIX-MEMBERED RING

1. Pyridine

The 1,3-oxazinone (**87.6**) is a partly cyclic acetal; the EtO group is displaced by reaction with a conjugated diene in the presence of boron trifluoride; a fused pyridine ring is formed.

O HN EtO O + $R^2C{=}CHR^1$ / $R^3C{=}CHR^4$ → (BF_3–Et_2O, 18–67%) R^1 O R^2 N R^3 O R^4 [2330]

(87.6)
R^1–R^4 = H, Me

Pyrido[2,1-*b*][1,3]benzoxazin-10-one

2. Pyridin-2-one or -4-one

Displacement of an ethoxy attached to a vinylic side-chain occurs in mineral acid at ambient temperature with concurrent formation of a C—C bond [2042]. Ethereal hydrogen chloride annulates nitro and side-chain epoxide groups; intramolecular reduction–oxidation leads to the 4-quinolinone in high yields [2314]. A C—N bond is formed when a methylthiovinylpyrrole is treated with a strong base and a malonic ester. Lactim ethers such as 5-methoxy-2*H*-pyrrole condense with acetonedicarboxylate in a base-induced cyclization at room temperature [3258].

HNCOCH=CHOEt → (H_2SO_4, ~100%) O HN [2042]

Benzo[*h*]quinolin-2-one

Another ring system synthesized similarly:

Me H N O [3664]

2-Quinolinone

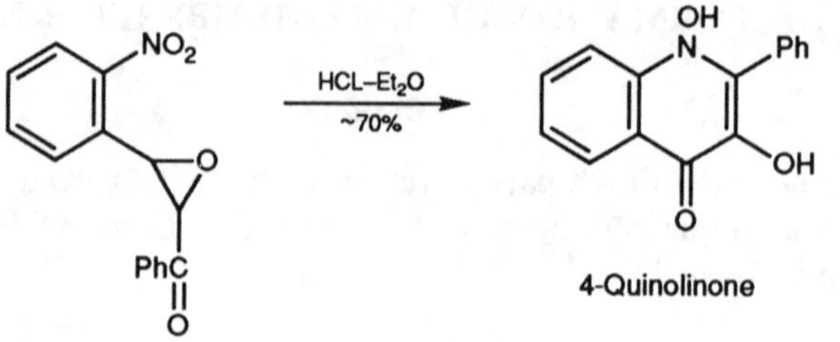

NO_2, PhC, O, O $\xrightarrow[\sim 70\%]{HCL-Et_2O}$ OH, N, Ph, OH, O [2314]

4-Quinolinone

MeC=CH, MeS, N + $CH_2(COOMe)_2$ $\xrightarrow[74\%]{i}$ Me, MeOC, O, N, O [2581]

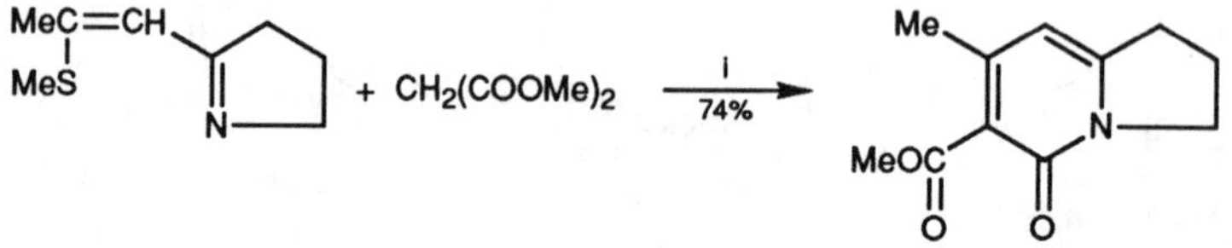

i, tBuOK, tBuOH, Δ

5-Indolizinone

MeO, N + $CO(CH_2COOR)_2$ $\xrightarrow[62-100\%]{TEA}$ COOR, HO, N, O [3258]

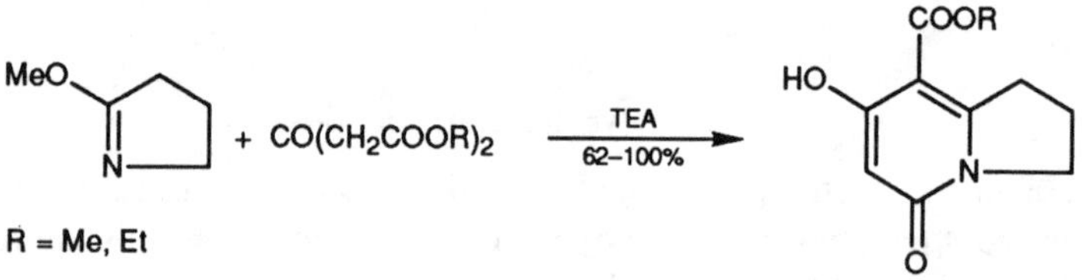

R = Me, Et

3. Pyran-2-one or -4-one

Phenyl 3-ethoxyacrylate is cyclized when treated with sulphuric acid and sulphur trioxide at 0 °C, but electron-withdrawing substituents on the benzene ring inhibit the reaction. Cyclization of a methoxybenzene to a xanthenone is shown in Chapter 68, Section II.2 [2272].

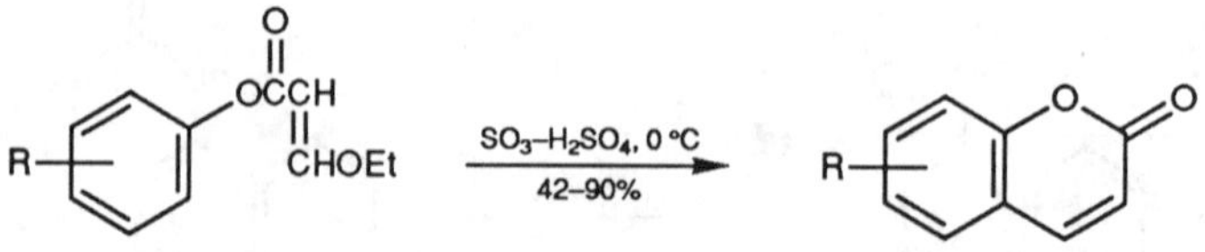

R, O, OCCH, CHOEt $\xrightarrow[42-90\%]{SO_3-H_2SO_4,\ 0\,°C}$ R, O, O [3365]

R = H, MeO, Me

1-Benzopyran-2-one

CHAPTER 88

Halogen and Ether or Hydroxy

I.	Formation of a Five-membered Ring	1213
	1. 1,2,4-Triazole	1213
	2. Furan	1214
	3. Furan-3-one	1215
	4. Thiophene	1215
	5. 1,2,5-Thiadiazole	1216
II.	Formation of a Six-membered Ring	1216
	1. Pyran or Pyran-4-one	1216
	2. 1,4-Dithiin	1217
	3. 1,3-Thiazine	1217
III.	Formation of a Seven-membered Ring	1217
	1. 1,4-Diazepin-2-one	1217
	2. Oxepine	1218
	3. 1,4-Oxazepin-3-one	1218
	4. 1,4-Thiazepine	1218
IV.	Formation of an Eight-membered Ring	1218
	1. 1,7,2,6-Oxathiadiazocine 7,7-Dioxide	1218

I. FORMATION OF A FIVE-MEMBERED RING

1. 1,2,4-Triazole

When the reactive imidoyl chloride in (**88.1**) is treated with an arylamine, it is converted into an amidine which displaces the methylthio group; this reaction proceeds in high yield and under relatively mild conditions.

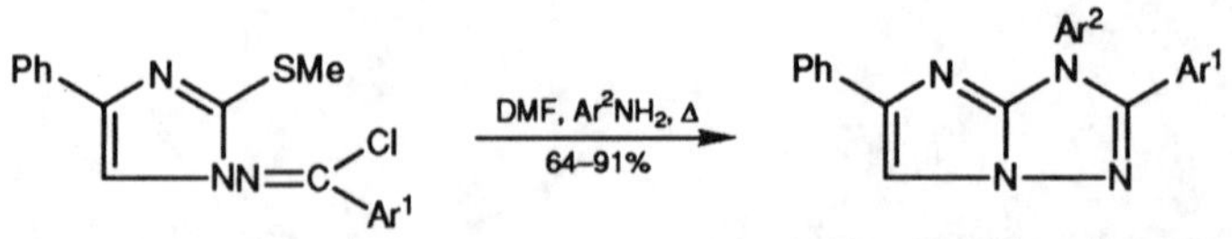

DMF, Ar^2NH_2, Δ, 64–91% [3558]

(88.1)

Ar^1 = 4-Me-, 4-MeO-C_6H_4;
Ar^2 = 4-Me-, 4-MeO-, 4-Cl-C_6H_4

Imidazo[1,2-*b*][1,2,4]triazole

2. Furan

Side-chain halogen and *O*-acylphenol functions interact under basic conditions to give a moderately good yield of benzofuran. An *o*-(2-chloroallyl)phenol is cyclized on warming with mineral acid, but yields are only moderate for 6-chloro- and 6-methoxy-benzofuran. This type of cyclization may give higher yields in a strongly alkaline medium [3943]. This particular synthesis failed in cold sulphuric acid and in hot diethylaniline.

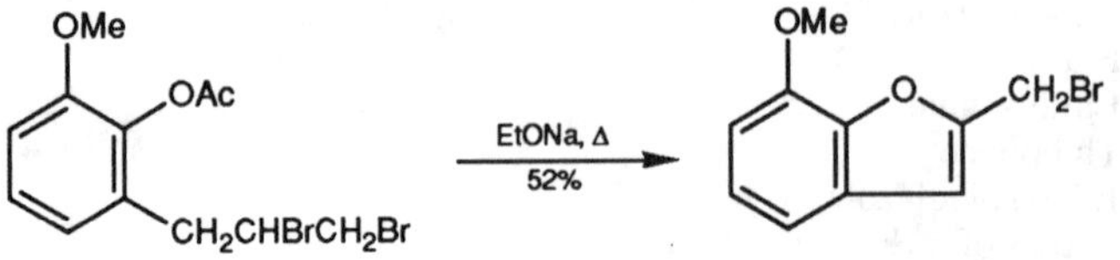

EtONa, Δ, 52% [2796]

Benzofuran

HCl, Δ, 28–77% [2842]

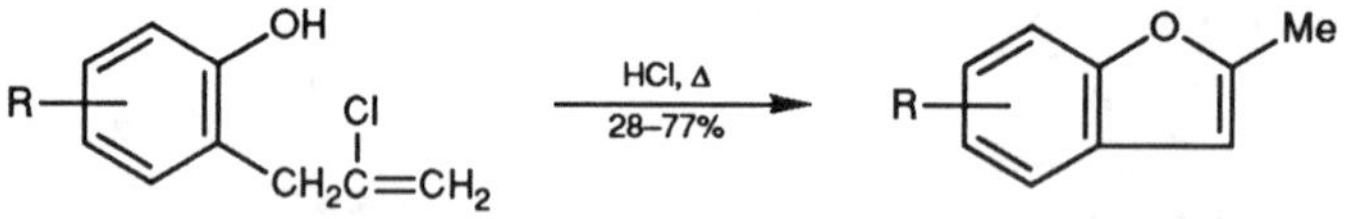

R = H, Cl, MeO, Me

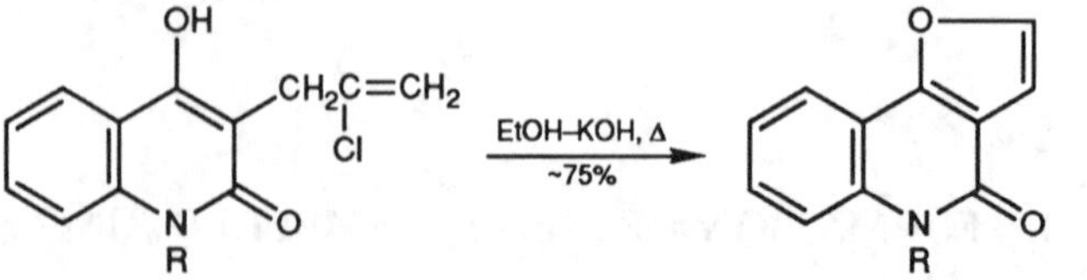

EtOH–KOH, Δ, ~75% [3943]

R = Me, Ph

Furo[3,2-*c*]quinolin-4-one

A palladium-promoted cyclization of 2-halophenols with 1-alkynes gives good yields of benzofurans at ambient temperature. Hydroxy and chloro functions in different rings react to form a furan ring on heating for 18 h with carbonate.

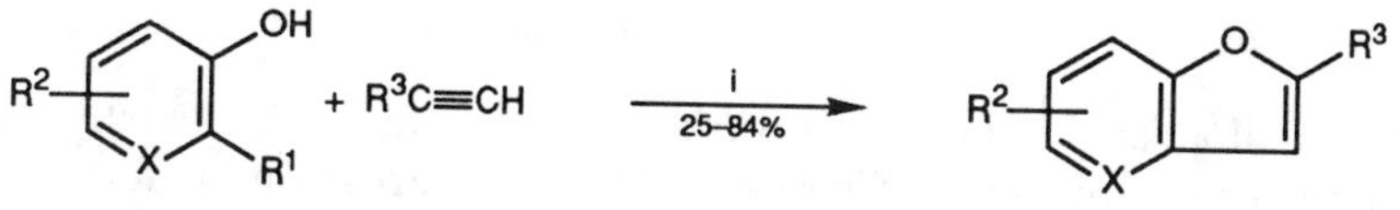

R^1 = Br, I; R^2 = H, MeO, Me, CHO;
R^3 = Bu, Ph, nC_5H_{11}, PhCHOH, CH_2OH, EtCMeOH,
cC_6H_{10}-1-OH, $CH(OEt)_2$, COOEt; X = CH, N
i, $Pd(OAc)_2[PPh_3]_2$, CuI, pip

i, 25–84% [3151]

X = CH Benzofuran
X = N Furo[3,2-*b*]pyridine

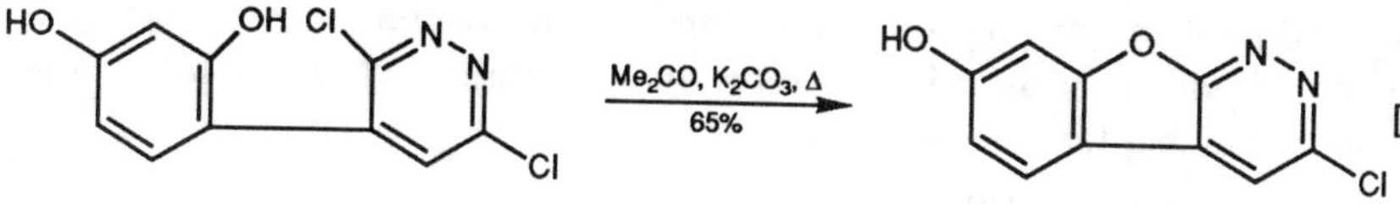

Me_2CO, K_2CO_3, Δ, 65% [3851]

Benzofuro[2,3-*c*]pyridazine

3. Furan-3-one

Cyclization of the α-bromoacetylphenol (**88.2**) (reviews of α-halo ketones [B-45, 2348]) in DMF containing sodium hydrogencarbonate at 40–45 °C gives a low yield of the benzofuran-3-one. The bromo ether (**88.3**) is dealkylated and cyclized with hydrobromic acid in high yield.

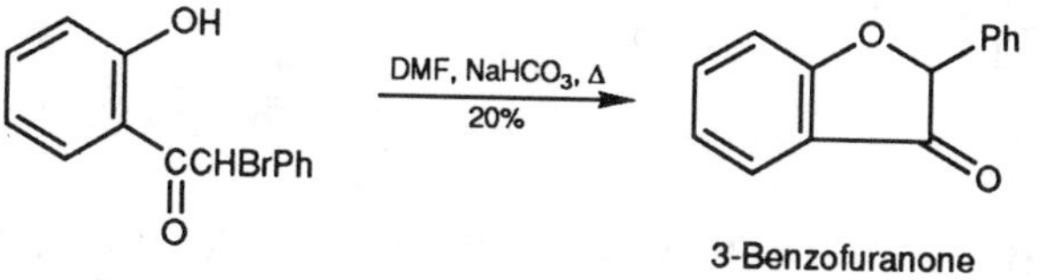

DMF, $NaHCO_3$, Δ, 20% [3815]

(88.2)

3-Benzofuranone

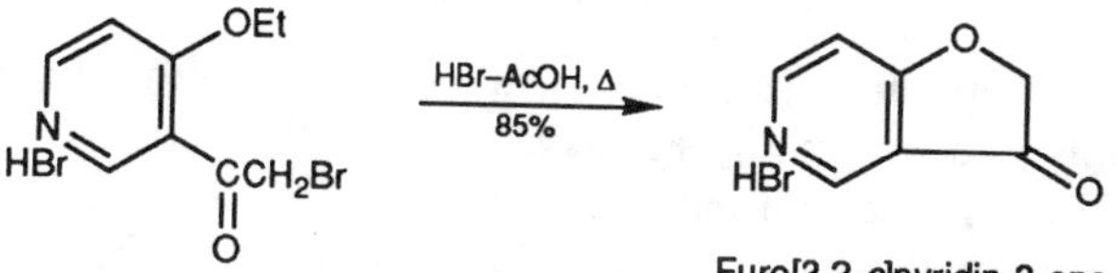

HBr–AcOH, Δ, 85% [2300]

Furo[3,2-*c*]pyridin-3-one

4. Thiophene

Bromine and thiol groups in different rings are annulated by heating in DMF containing copper(I) oxide and potassium hydroxide.

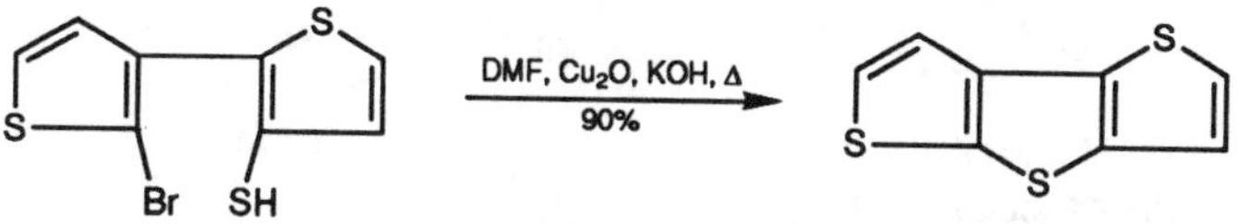

DMF, Cu_2O, KOH, Δ, 90% [2287]

Dithieno[2,3-*b*:2′,3′-*d*]thiophene

5. 1,2,5-Thiadiazole

When a 2-halophenol is heated with (the explosive) tetrasulphur tetranitride, both substituents are usually displaced and a thiadiazole ring is formed, but exceptions to this rule are observed, for example, 4-t-butyl-2-chlorophenol gives 6-t-butyl-4-chloro-2,1,3-benzothiadiazole as the major product (56% yield) and 6-t-butyl-2,1,3-benzothiadiazole as a second product (40%). The ease of displacement of halogen in this reaction decreased in the order I > F > Br > Cl. When this reaction is applied to *o*-halonaphthols, the yields of the thiadiazoles vary considerably according to the positions of the two functions. Sometimes, the main component of the mixture of products contains two fused thiadiazole rings (see also Chapter 99, Section I.7).

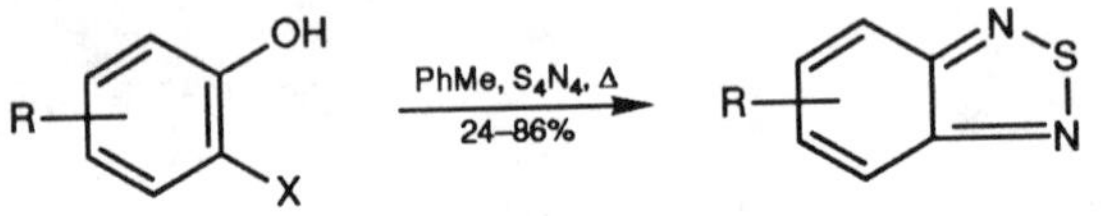
OH
R
X
PhMe, S_4N_4, Δ
24–86%
R
N
S
N

[3077]

R = Me, Br, tBu; X = F, Cl, Br, I   2,1,3-Benzothiadiazole

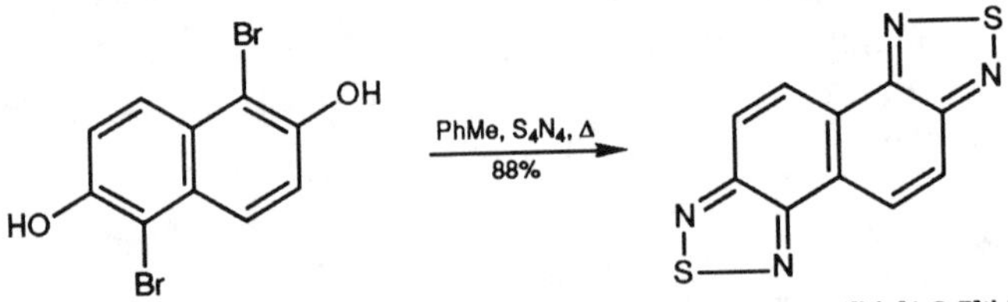
Br
OH
HO
Br
PhMe, S_4N_4, Δ
88%
N
S
N
N
S
N

[3980]

Naphtho[1,2-*c*:5,6-*c*']bis[1,2,5]thiadiazole

II. FORMATION OF A SIX-MEMBERED RING

1. Pyran or Pyran-2-one

2-Methoxy-1-benzopyrans are prepared at or below room temperature from 2-bromophenols and 3-trimethylsilyloxyacroleins. Protection of the phenolic group and lithiation precede treatment with the acrolein; when the protecting group is removed with cold mineral acid, cyclization occurs in good yield. Chalcone dibromides cyclize in a basic medium but yields of pyrans are often better under weakly basic conditions, as is demonstrated in this synthesis of a flavone in dilute alkali at room temperature.

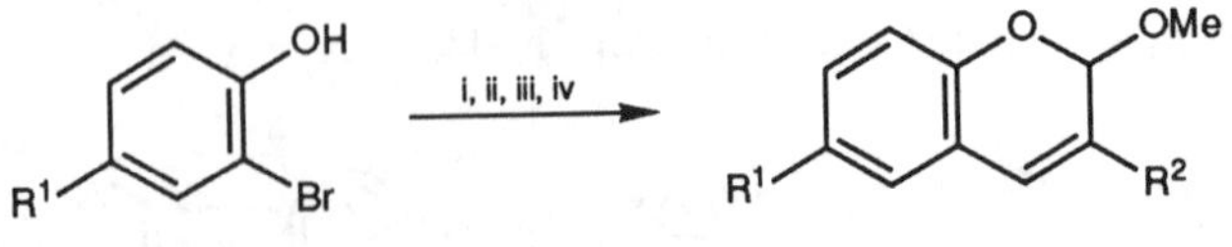
OH
R^1
Br
i, ii, iii, iv
O
OMe
R^1
R^2

[3248]

R^1 = H, MeO, Me; R^2 = alkyl   1-Benzopyran

i, THP, CF_3COOH; ii, BuLi;

iii, $Me_3SiOCH{=}CR^2CHO$; iv, MeOH, HCl

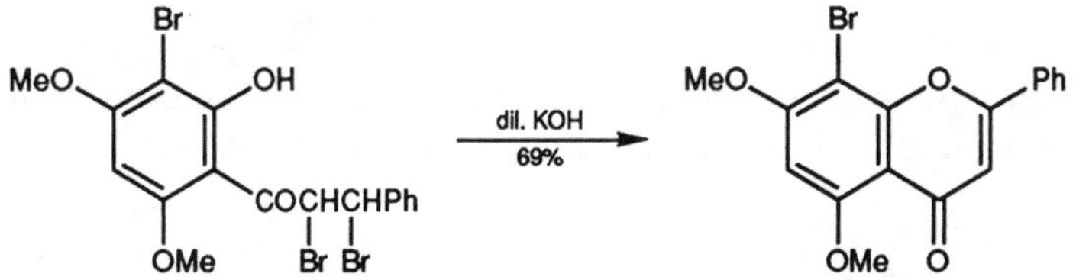

dil. KOH, 69% [2606]

1-Benzopyran-4-one, 2-aryl-

2. 1,4-Dithiin

Self-cyclization of a chlorothiol occurs on prolonged heating in ethanol; varying amounts of the 1,9-dione are also formed.

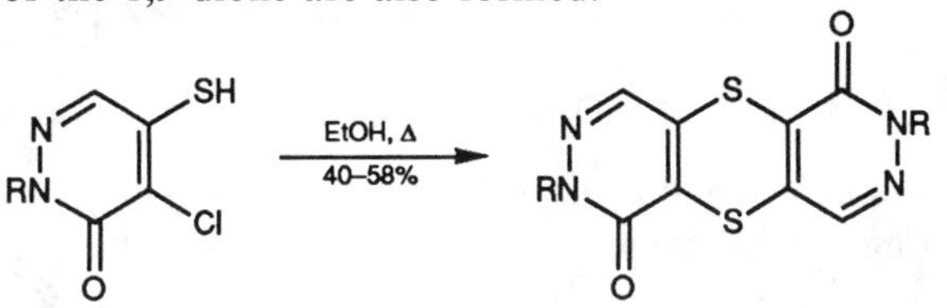

EtOH, Δ, 40–58% [2203]

R = H, Me, $PhCH_2$

[1,4]Dithiino[2,3-*d* : 5,6-*d'*]dipyridazine-1,6-dione

3. 1,3-Thiazine

S-Demethylation of a methylthio ether with aluminium chloride enables cyclization of the resulting chloro-thiol to proceed.

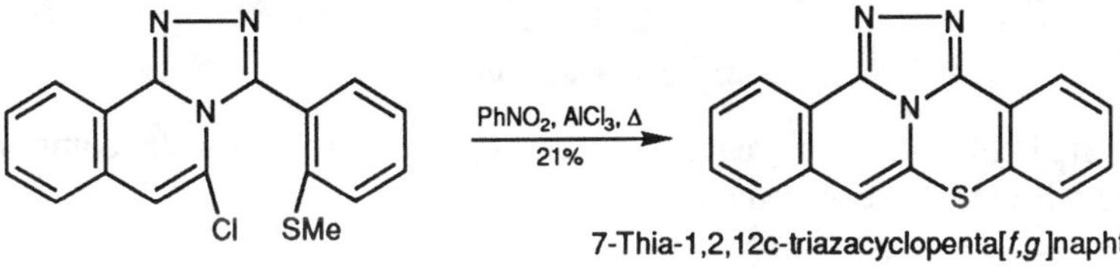

$PhNO_2$, $AlCl_3$, Δ, 21% [2702]

7-Thia-1,2,12c-triazacyclopenta[*f,g*]naphthacene

III. FORMATION OF A SEVEN-MEMBERED RING

1. 1,4-Diazepin-2-one

When an oxime is formed (*in situ*) at the ring carbonyl group of (**88.4**), its anion displaces the chlorine in the basic solution to form a diazepinone ring in good yield.

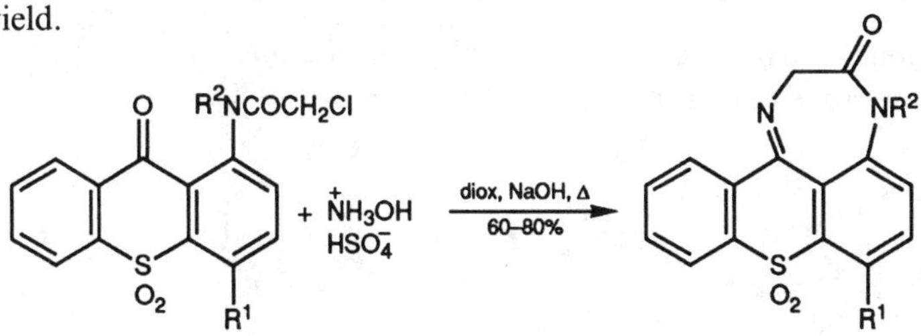

+ $\overset{+}{N}H_3OH$ HSO_4^- — diox, NaOH, Δ, 60–80% [2443]

R^1 = Me, Cl; R^2 = H, alkyl

Benzothiopyrano[4,3,2-*ef*][1,4]benzodiazepin-3-one 8,8-dioxide

2. Oxepine

An Ullman-type reaction between halogen and hydroxy groups attached to different rings can lead to the formation of this fused ring.

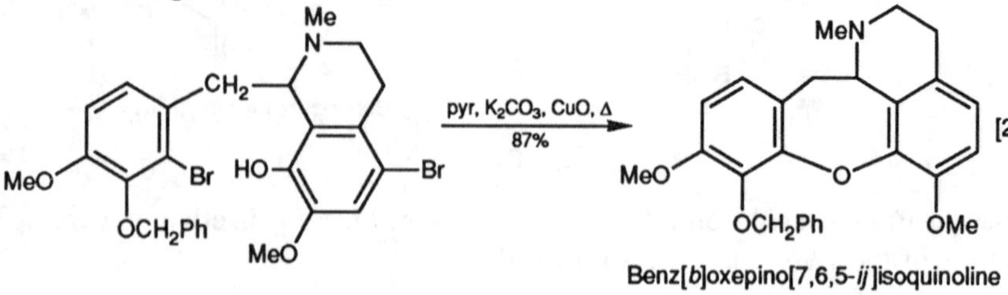

pyr, K_2CO_3, CuO, Δ
87%

Benz[*b*]oxepino[7,6,5-*ij*]isoquinoline

[2421]

3. 1,4-Oxazepin-3-one

The first-formed bromoacetamidobenzyl alcohol from (**88.3**) may be cyclized in high yield by heating wth butoxide.

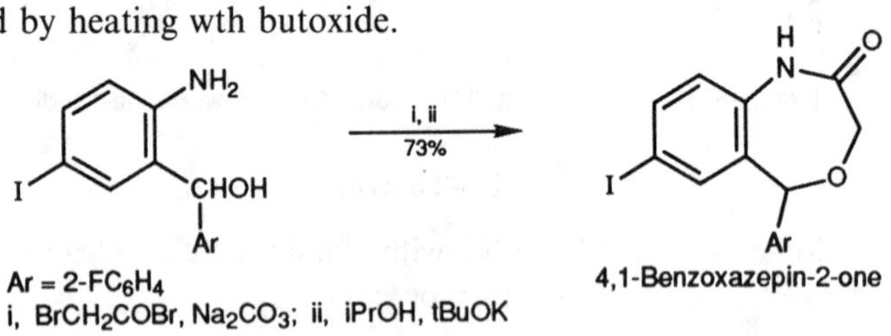

i, ii
73%

Ar = 2-FC_6H_4
i, $BrCH_2COBr$, Na_2CO_3; ii, iPrOH, tBuOK

4,1-Benzoxazepin-2-one

[3925]

4. 1,4-Thiazepine

An activated chlorobenzene reacts with a side-chain thiol when the compound is stirred with ethoxide and then heated in ethanol.

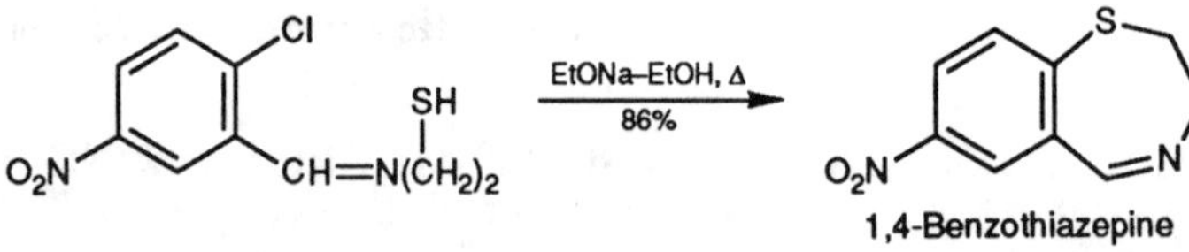

EtONa–EtOH, Δ
86%

1,4-Benzothiazepine

[3367]

IV. FORMATION OF AN EIGHT-MEMBERED RING

1. 1,7,2,6-Oxathiadiazocine 7,7-Dioxide

Suitably positioned oxime (as its sodium salt) and iodomethylsulphonamide functions interact to give this fused ring on heating at 100 °C.

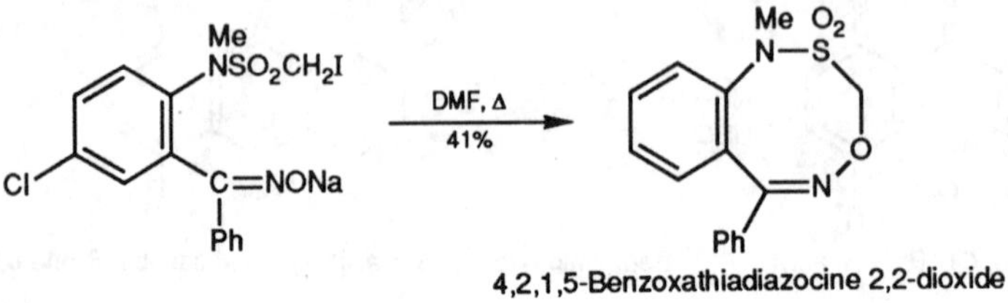

DMF, Δ
41%

4,2,1,5-Benzoxathiadiazocine 2,2-dioxide

[2060]

CHAPTER 89

Halogen and Lactam Carbonyl or a Sulphur-containing Group

I. Formation of a Five-membered Ring 1219
1. Furan 1219
2. Thiophene 1220
3. 1,3-Dithiole 1220
4. Thiazole 1220
II. Formation of a Six-membered Ring 1220
1. 1,3-Thiazine 1220
2. 1,4,2-Dithiazine 1,1-Dioxide 1221
III. Formation of a Seven-membered Ring 1221
1. 1,2,4-Thiadiazepine 1,1-Dioxide 1221

I. FORMATION OF A FIVE-MEMBERED RING

1. Furan

Annulation of 2,2-dichloroethyl and lactam carbonyl functions is an unusual means of forming a fused furan ring. This process occurs when the compound is heated for about 5 h with potassium carbonate in acetone. On the other hand, lactam carbonyl and β-bromovinyl functions require heating with strong base.

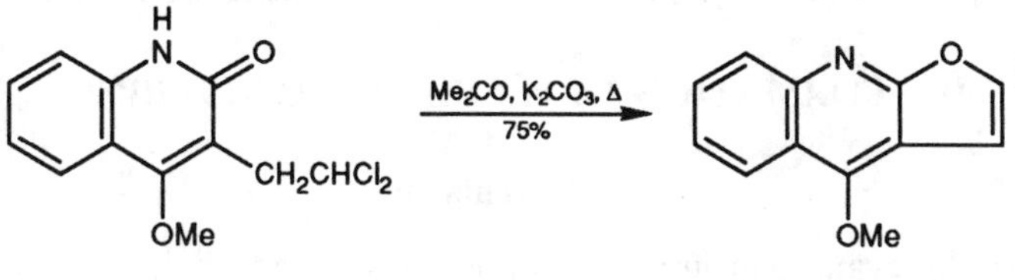

H N O CH_2CHCl_2 OMe $\xrightarrow[75\%]{Me_2CO,\ K_2CO_3,\ \Delta}$ N O OMe [3253]

Furo[2,3-*b*]quinoline

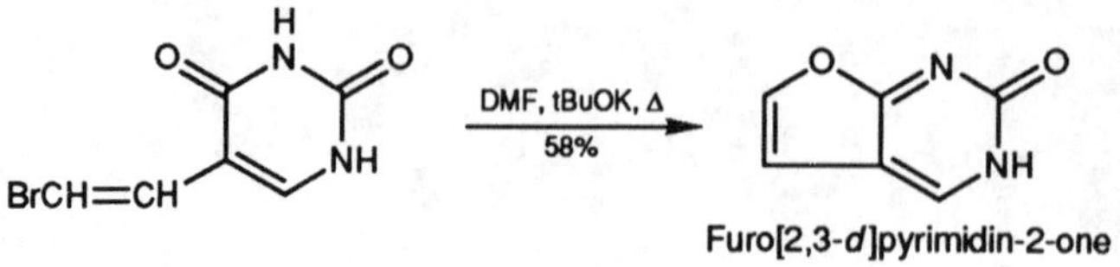
DMF, tBuOK, Δ
58%
BrCH=CH
Furo[2,3-*d*]pyrimidin-2-one

[2859]

2. Thiophene

Nucleophilic displacement of a fluorine atom by a CH_2SCH_2Li side-chain at a low temperature occurs when a pentafluorobenzyl methyl sulphoxide reacts with an excess of butyllithium.

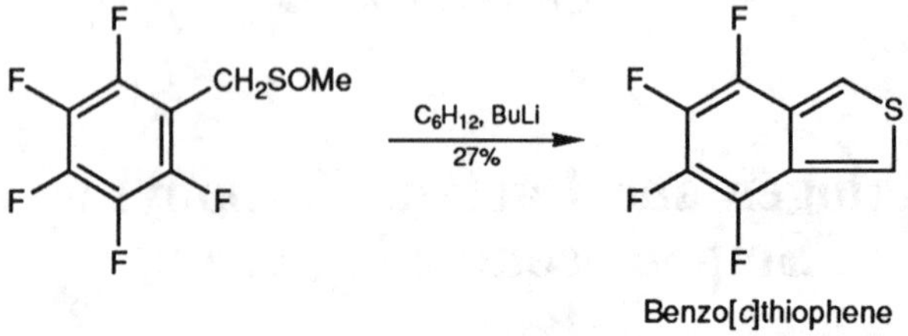
CH_2SOMe
C_6H_{12}, BuLi
27%
Benzo[*c*]thiophene

[3746]

3. 1,3-Dithiole

Neighbouring chloro and thiocyanate groups react to form a dithiole ring on heating with a thiocarboxylic acid.

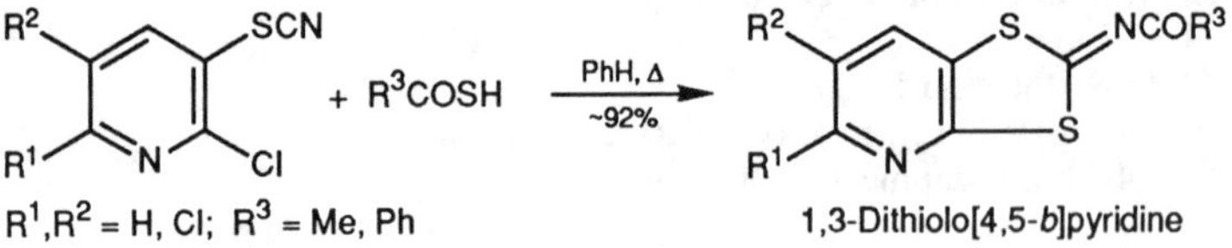
SCN
+ R^3COSH
PhH, Δ
~92%
$NCOR^3$
R^1,R^2 = H, Cl; R^3 = Me, Ph
1,3-Dithiolo[4,5-*b*]pyridine

[3223]

4. Thiazole

An unusual halogen transposition may occur when a halo-isothiocyanate is heated to about 350 °C.

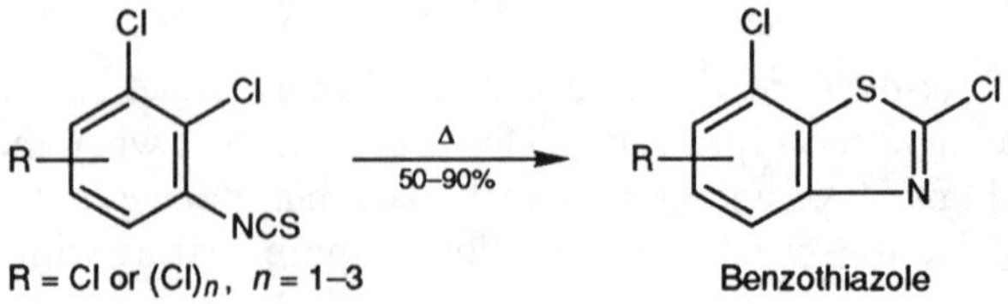
NCS
Δ
50–90%
R = Cl or $(Cl)_n$, n = 1–3
Benzothiazole

[2208]

II. FORMATION OF A SIX-MEMBERED RING

1. 1,3-Thiazine

Iodo and isothiocyanato groups at *peri* positions of naphthalene are reductively annulated by successive treatment at 0 °C with lithium triethylborohydride and copper(II) iodide–DMF. Nucleophilic attack of ammonia or an alkylamine on an *o*-chloromethylene isothiocyanate leads to cyclization at ambient temperature; arylamines have to be heated with the isothiocyanate.

i, $LiBHEt_3$, THF; ii, CuI_2–DMF — i, ii, 86% — Naphtho[1,8-*de*]-1,3-thiazine [3339]

+ R^1NHR^2 — AcOEt, ~55% — 3,1-Benzothiazine [2681]

R^1 = H; R^2 = H, alkyl or R^1R^2N = morpholino, piperidino, pyrrolidino

2. 1,4,2-Dithiazine 1,1-Dioxide

o-Chlorosulphonamides may be converted into the title ring by reaction with alkali in DMF followed by carbon disulphide. The exothermic reaction that follows is allowed to subside before iodomethane is added. The intermediate is believed to contain the group $—SO_2N{=}C(SNa)_2$.

1. NaOH, CS_2 2. MeI — 66% — Pyrazolo[3,4-*e*]-1,4,2-dithiazine 1,1-dioxide [3141]

III. FORMATION OF A SEVEN-MEMBERED RING

1. 1,2,4-Thiadiazepine 1,1-Dioxide

The formation of a nitrilium salt facilitates the formation of this ring from a bromomethyl-sulphonamide and a nitrile.

+ R^2CN — $SnCl_4$, Δ, 10–66% — 1,2,4-Benzothiadiazepine 1,1-dioxide [2434]

R^1 = Me, nC_6H_{13}; R^2 = alkyl, aryl

CHAPTER 90

Halogen and Methylene or Ring-carbon

I. Formation of a Five-membered Ring 1222
 1. Pyrrole 1222
 2. Pyrrol-2-one 1223
 3. Pyrazole 1224
 4. Furan 1224
 5. Thiophene 1225
 6. Oxazol-2-one 1226
 7. 1,2,5-Thiadiazole 1226
II. Formation of a Six-membered Ring 1227
 1. Pyridine 1227
 2. Pyridin-2-one 1228
 3. Pyran 1231
 4. Pyran-4-one 1231
 5. Thiopyran-4-one 1231
 6. 1,3,4-Thiadiazine 1232
III. Formation of a Seven-membered Ring 1232
 1. Azepin-2-one 1232
IV. Formation of an Eight-membered Ring 1232
 1. Azocin-2-one 1232

I. FORMATION OF A FIVE-MEMBERED RING

1. Pyrrole

Indolizines (including 2-hydroxy derivatives) may be synthesized from 2-bromo(or chloro)pyridinium salts (review [1870]) and β-ketoesters in the presence of DBU (review [3364]) [3277]). A fused pyrrole ring is formed in high yield when an *N*-(2-bromoallyl)- or -chloroallyl-arylamine is heated with methanolic boron trifluoride or PPA; isotopic experiments show that the reaction may follow one of two possible courses, but it is not possible to decide between them [2146]. A similar cyclization proceeds by a radical mechanism when a refluxing toluene solution of an *N*-3-haloallylindole containing tributyltin hydride, AIBN and azobis(cyclohexanecarbonitrile) (ACN) is irradiated with a sunlamp for 3 h [3991].

MeCN, DBU, 13–63% [3277]

R^1 = Me, iPr, Ph, EtO;
R^2 = Me, HO, iPr, Ph, 3-$MeOC_6H_4$, 3,4-$Me_2C_6H_3$;
R^3 = Me, iPr, EtO

Indolizine

MeOH–BF_3, Δ, 42–90% [2146]

R^1 = H, Cl, Ph, COOMe; R^2 = H, Me; X = Br, Cl

Indole

Another ring system synthesized similarly:

[2648]

Pyrrolo[3,2-*c*]quinoline

PhMe, *hν*, Bu_3SnH, i, 16–65% [3991]

R^1 = H, COOMe, COOEt; R^2 = H, CH_2COOMe;
R^3 = H, $PhCH_2O$; X = Br, I
i, AIBN, ACN

Pyrrolo[1,2-*a*]indole

2. Pyrrol-2-one

When thallium(I) acetate is added to palladium(0)-catalysed cyclization of the iodobenzoylpyridine (**90.1**), only one of the three isomeric products that are otherwise formed is obtained. Oxindoles may be synthesized from *N*-(α-bromoacyl)-*N*-methylaniline under Friedel–Crafts reaction conditions.

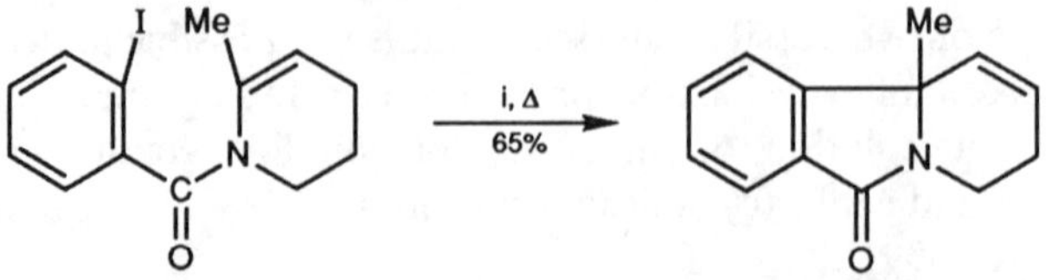

(90.1) → Pyrido[2,1-*a*]isoindol-6-one [3971]

i, MeCN, $Pd(OAc)_2$, PPh_3, TlOAc; i, Δ, 65%

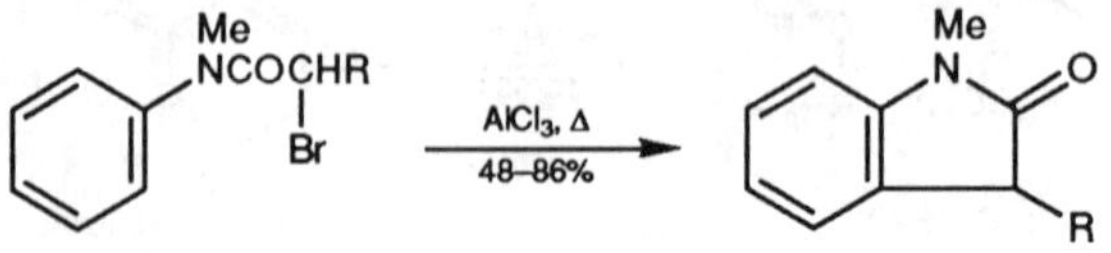

R = alkyl → 2-Indolone ($AlCl_3$, Δ, 48–86%) [2248]

Another ring system synthesized similarly:

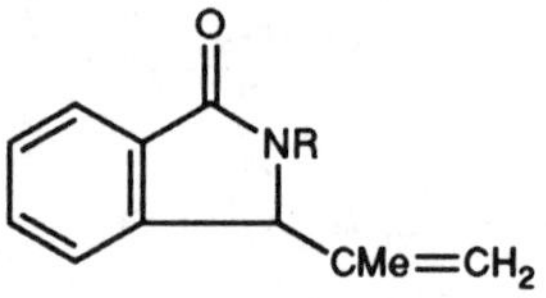

1-Isoindololone [3870]

3. Pyrazole

Phenyldiazomethane adds on to 1,3,5-cyclooctatrien-7-yne (generated *in situ* from the bromide and LDA) at −40 °C.

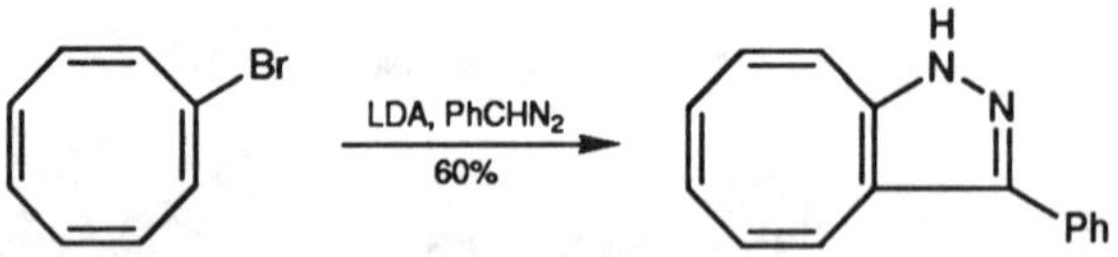

LDA, $PhCHN_2$, 60% → Cycloocta[*c*]pyrazole [3292]

4. Furan

Polyiodinated diphenyl ethers are important thyroid hormones and in order to discover the relationship between activity and planarity of the two benzene rings, iodinated dibenzofurans have been synthesized photochemically. Pyridine *N*-oxides react with alkynes to give furopyridines via a sigmatropic shift; the product varies with the substituents on the alkynes. 3,5-Dihalopyridine oxides give moderate yields of furo[3,2-*b*]pyridines as the major product; smaller amounts of the [3,2-*c*] isomers are also formed. 2,4-Disubstituted pyridine oxides produce the [3,2-*c*] isomers in better yields than for most of the 3,5-isomers.

MeO ... NO_2 —(Me_2CO, $h\nu$; 34%)→ Dibenzofuran [3654]

R^1 ... + $R^2C{\equiv}CR^3$ —(PhMe, Δ; 20–60%)→ Furo[3,2-*b*]pyridine [2760]

R^1 = Br, Cl; R^2 = Ph, CN, COOMe; R^3 = CN, COOMe

Another ring system synthesized similarly:

Furo[3,2-*c*]pyridine [2760]

5. Thiophene

It was not possible to isolate the normal (uncyclized) product of Claisen rearrangement when the allyl thioether (**90.2**) is heated in diethylaniline [2842]. A side-chain bromide may be annulated to a ring-carbon by heating with sulphur (a reaction which the corresponding chloride does not undergo) [2187]. Zirconium complexes of benzyne (generated from bromobenzenes) are promising reagents (review [3502]) as they excercise stereochemical and regiospecific control on some reactions. Substituted benzo[*b*]thiophenes are synthesized in this way from a non-terminal alkyne and a zirconocene [3546].

(**90.2**) —($PhNEt_2$, Δ; 54–80%)→ Benzo[*b*]thiophene [2842]

R = H, 4-Me-, 4-Cl-, 4-MeO-, 2-Cl

PhCHBrCHBrPh —(1,2-$Cl_2C_6H_4$, S, Δ; 40%)→ 2-phenylbenzo[*b*]thiophene [2187]

R^1-benzyne-$ZrCp_2$ + $R^2C{\equiv}CR^3$ —(1. THF 2. SCl_2; 60–92%)→ [3546]

R^1 = H, MeO, Me; R^2,R^3 = Me, Et, Me_3Si; Cp = cyclopentane

6. Oxazol-2-one

Some pyridine *N*-oxides react with isocyanates at the oxide function, but 3-bromopyridine *N*-oxides under the same conditions react differently to give a new oxazolone ring.

+ PhNCO DMF, Δ 11–70% [2588, 2596, 2597]

Oxazolo[4,5-*b*]pyridin-2-one
Oxazolo[4,5-*b*]quinolin-2-one

7. 1,2,5-Thiadiazole

The reaction of bromophenols with tetrasulphur nitride can give one of three types of product: the new ring may result from reaction with a bromine and hydroxy (see Chapter 88, Section I.5), or hydroxy and ring-carbon (Chapter 99, Section I.7) or a mixture of the two products. For example, the table shows the products obtained from variously 4-substituted 1-naphthol.

	Yield (%)	
R	**(90.3a)**	**(90.3b)**
H	9	14
Br	–	27
Cl	1	19

R = H, Cl, Br

PhMe, N_4S_4, Δ

(90.3a)
Naphtho[1,2-*c*][1,2,5]thiadiazole [3980]

(90.3b)
Naphtho[1,2-*c*:3,4-*c*′]bis[1,2,5]thiadiazole

Another ring system synthesized similarly:

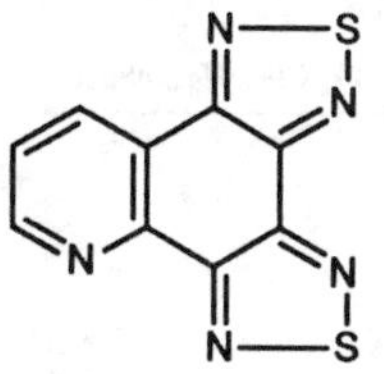

[3980]

Bis[1,2,5]thiadiazolo[3,4-*f*: 3'4'-*h*]quinoline

II. FORMATION OF A SIX-MEMBERED RING

1. Pyridine

The trifluoromethyl-ketimine (**90.4**), on reaction with a strong base, is converted into a 2-phenylquinoline in a novel type of cyclization in which the three fluorine atoms are displaced by nucleophiles. According to the base used, the quinoline may have either an oxygen or a nitrogen substituent at C-4. The t-butyl ether (R = tBu) is converted into the quinolinol (R = H) in high yield [3908]. 5-Deazaflavins are of biological interest and may be synthesized by a simple reaction from a chlorouracil (**90.5**) and *o*-(hydroxymethyl)aniline [3591]. The chlorophenylpyrroles (**90.6**) undergo electrochemical cyclization to the phenanthridine in moderate yield [3758].

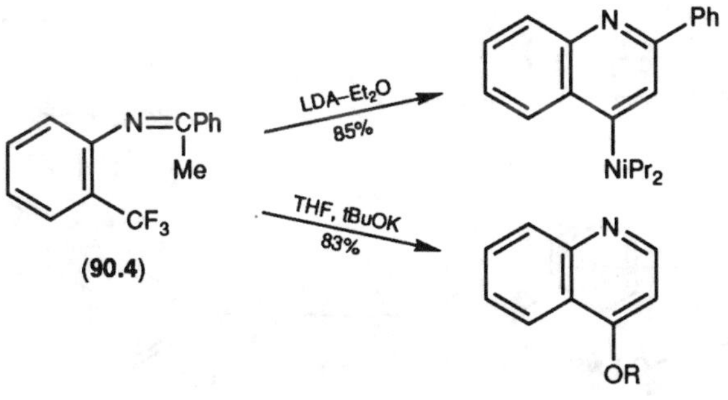

(**90.4**) — LDA–Et_2O, 85% → 2-phenyl-4-($NiPr_2$)quinoline; THF, tBuOK, 83% → 4-OR quinoline [3908]

Quinoline

(**90.5**) + o-(NNR^2)-CH_2OH benzene — DMF, Δ, 16–26% → [3591]

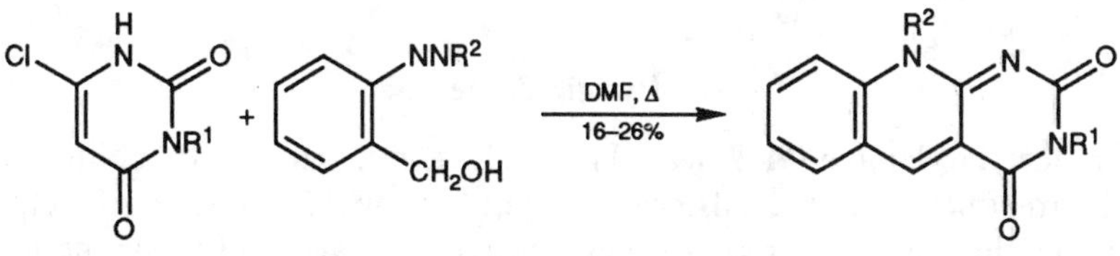

(**90.5**)
R^1 = Et, Ph, 4-$ClC_6H_4CH_2$; R^2 = H, Me

Pyrimido[4,5-*b*]quinoline-2,4-dione

DMF, Pr_4NBF_4, Hg cathode, 30–42% [3758]

(90.6)
R = H, Ph; X = CH, N

X = CH Pyrrolo[1,2-*f*]phenanthridine
X = N Imidazo[1,2-*f*]phenanthridine

Reaction of the trifluoromethyl-imine (**90.7**) with lithium 2-(dimethylamino)-ethylamide or lithium 4-methylpiperazide at −10 °C followed by quenching with either water or dilute mineral acid brings about ring closure [3803]. The C—Br bond of (**90.8**) breaks homolytically when the compound is heated in benzene with tributyltin hydride and AIBN over a prolonged period, and a new C—C bond is formed by attack of the phenyl radical on the arylamine ring. The newly formed ring is dehydrogenated by heating in air. This cyclization is a potentially useful step in the synthesis of lycoricidine [3864].

LiR, Et_2O, 70–88% [3803]

(90.7)
R = $Me_2N(CH_2)_2NH$, $MeN[CH_2CH_2]_2N$

Benz[*c*]acridine

PhH, Bu_3SnH, Δ, ~66% [3864]

(90.8)
R^1–R^3 = H or MeO, or R^1 = H, R^2R^3 = benzo

[1,3]Dioxolo[4,5-*j*]phenanthridine

2. Pyridin-2-one

A chloroacetylarylamine (**90.9**, n = 1) may be converted into quinolin-2-one [2960] by treatment with the Vilsmeier reagent (review [1676]). A novel method of forming this ring was recently described as a stage in the synthesis of pyrrolophenanthridine alkaloids such as ungeremine. A C—C bond is formed by a radical mechanism of debromination by benzyl triethylammonium chloride

—potassium carbonate in hot DMSO. In this example, a 1:1 mixture of two isomers is obtained through radical attacks at two ring-carbon atoms. The alkaloid is later synthesized by chemical modification of one of the isomers (**90.10**) [3967].

R

$NC(CH_2)_nCl$

O

$CHCl_3$, DMF, $POCl_3$

34–49%

R N O Cl

[2960]

(90.9)
R = Me, Ph; n = 1

2-Quinolinone

NO_2 O Br O N O

i

i, DMSO, K_2CO_3, $PhCH_2\overset{+}{N}Et_3Cl^-$

NO_2 O O N C O + NO_2 O O N O

[3967]

(90.10)
[1,3]Dioxolo[4,5-*j*]pyrrolo[3,2,1-*de*]-phenanthridin-7-one

1,3-Dioxolo[4,5-*k*]pyrrolo[3,2,1-*de*]-phenanthridin-6-one

When ultrasonification (US) is applied to the reaction of a bromobenzene with t-butyl isocyanate and butyl lithium in diethyl ether or THF, a considerable improvement in time and yield is obtained; addition of DMF at a low temperature induces cyclization [3184a]. The use of ultrasonification in organic chemistry has been reviewed [B-58, 3615, 3646].

Br Me + tBuNCO

THF, US, BuLi, DMF

40%

O NtBu

[3184a]

1-Isoquinolinone

Photochemical annulation of a chloroarene to a carbon of another ring is mentioned on pp. 490 and 491. The formation of a bond joining thiophene and pyridine rings under different conditions has been developed—irradiation of the substrate in benzene containing TEA through which a stream of air is passed [3217, 3917]. Earlier work had suggested that photolysis of 2-iodobenzanilides in benzene involved the formation of a radical which (among other reactions) forms a bond to the *N*-phenyl ring. The reaction gives results which resemble those for the corresponding 2-diazonium-anilide. *N*-Methyl derivatives give higher yields than the parent compounds [2303].

PhH, TEA, air, *hν*
54%

[3217]

[1]Benzothieno[2,3-*c*][1,8]naphthyridin-6-one

Other ring systems synthesized similarly:

[3217]

X = N; Y = CH [1]Benzothieno[2,3-*c*][1,6]naphthyridin-6-one
X = CH; Y = N [1]Benzothieno[2,3-*c*][1,7]naphthyridin-6-one

[3917]

Benzo[*h*]benzothieno[2,3-*c*][1,6]naphthyridin-6-one

PhH, *hν*
18–36%

[2303]

R = H, Me

6-Phenanthridinone

3. Pyran

3-Bromoprop-2-enyloxybenzenes undergo a combined Claisen rearrangement and cyclodehydrobromination on being heated with *N,N*-diethylaniline in polyethylene glycol (PEG-400) to give 2*H*-chromenes in high yields. The chlorouracil (**90.5**) reacts with 2-hydoxymethylphenol on heating in nitrobenzene at 200 °C for 4 h.

PEG–400, Et_2NPh, Δ; 65–85% [3194]

R^1 = H, Cl, MeO; R^2 = H, Cl, Me, Ph

2*H*-1-Benzopyran

(**90.5**) + 2-$HOCH_2C_6H_4OH$ → $PhNO_2$, Δ; 35–80% [3591]

R^1 = Et, Ph, 4-$ClC_6H_4CH_2$

[1]Benzopyrano[2,3-*d*]pyrimidine-2,4-dione

4. Pyran-4-one

3-Benzoyl-4-fluoropyridine is cyclized by heating with pyridine hydrochloride and then with dilute acid.

1. pyr.HCl, Δ 2. H_2O; 80% [3462]

[1]Benzopyrano[3,2-*c*]pyridin-10-one

5. Thiopyran-4-one

Under strongly basic conditions, *o*-bromoacetophenones and carbon disulphide react to give a high yield of a benzothiopyranone. 2-Bromo-3-pyridyl ketones react with thioesters to produce a fused thiopyranone ring.

+ CS_2 → i, Δ; 80–94% [2963]

R^1 = H, Cl; R^2 = Me, $PhCH_2$

i, NaH, DMSO, R^2Br or R^2I

1-Benzothiopyran-4-one

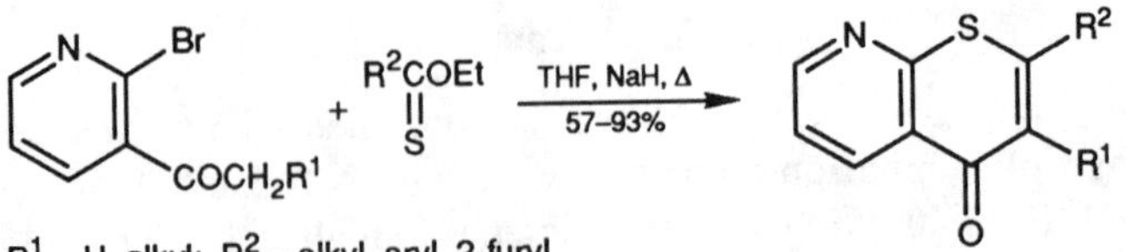

[3605]

R^1 = H, alkyl; R^2 = alkyl, aryl, 2-furyl

THF, NaH, Δ 57–93%

Thiopyrano[2,3-*b*]pyridin-4-one

6. 1,3,4-Thiadiazine

Bromination of the semicarbazide (**90.11**) probably generates an *S*-bromide which attacks the ring and forms a fused thiadiazine.

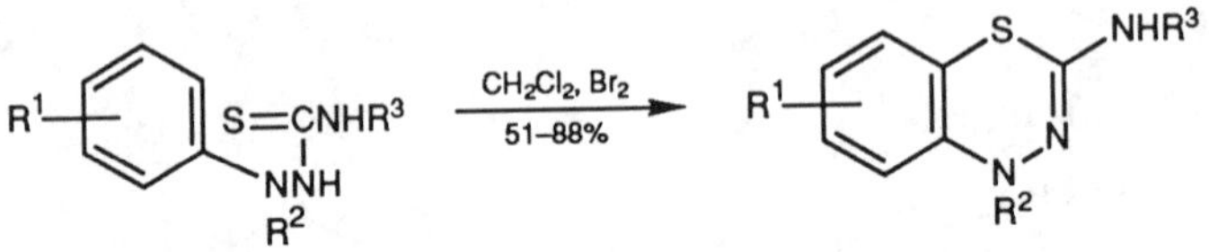

CH_2Cl_2, Br_2 51–88%

[3693]

(**90.11**)

R^1,R^2 = H, Me; R^3 = Me, iPr, tBu, Ph

1,3,4-Benzothiadiazine

III. FORMATION OF A SEVEN-MEMBERED RING

1. Azepin-2-one

Stirring the chlorodiphenylamide (**90.9**, n = 2) with the Vilsmeier reagent (review [1676]) converts it in low yield to a benzazepinone.

(**90.9**) $CHCl_3$, DMF, $POCl_3$ 17–24%

R = Ph; n = 2

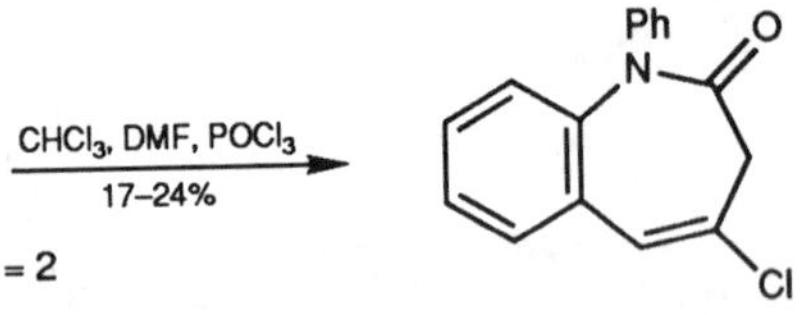

[2960]

1-Benzazepin-2-one

IV. FORMATION OF AN EIGHT-MEMBERED RING

1. Azocin-2-one

The next higher homologue (**90.9**, n = 3) of the chlorodiphenylamide similarly gives the benzazocine.

(**90.9**) $CHCl_3$, DMF, $POCl_3$ 17%

R = Ph; n = 3

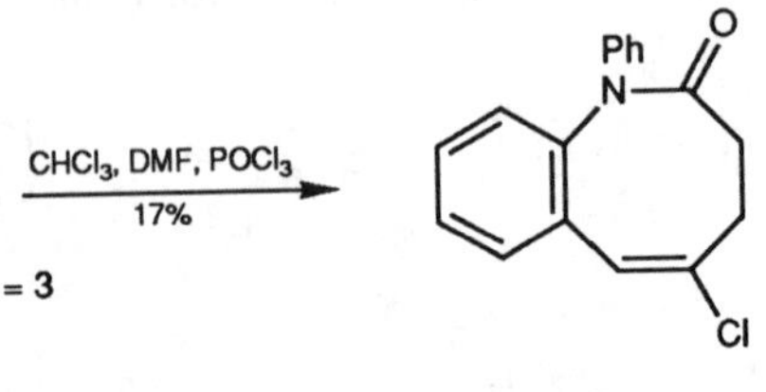

[2960]

1-Benzazocin-2-one

CHAPTER 91

Halogen and Nitro

I.	Formation of a Five-membered Ring	1233
	1. Pyrazole	1233
	2. 1,2,3-Triazole	1234
	3. 1,3-Dithiole	1234
	4. Thiazole	1234
II.	Formation of a Six-membered Ring	1234
	1. 1,4-Dioxin	1234
III.	Formation of a Seven-membered Ring	1235
	1. 1,4-Dithiepine	1235

I. FORMATION OF A FIVE-MEMBERED RING

1. Pyrazole

An exothermic reaction occurs when 2,4-dinitrobenzyl chloride is treated with dimethylamine at 0 °C [2223]. Heating the chloronitropyrimidine (**91.1**) with a hydrazone leads to a good-to-excellent yield of a 3-substituted pyrazole [2351].

2,4-Dinitrobenzyl chloride (O_2N, NO_2, CH_2Cl) + Me_2NH $\xrightarrow[75\%]{Et_2O}$ Indazole (O_2N, NMe) [2223]

(**91.1**) (Me, Cl, MeN, NO_2, O) + $RCH{=}NNH_2$ $\xrightarrow[55\text{–}80\%]{CHCl_3,\ \Delta}$ Pyrazolo[3,4-*d*]pyrimidine-4,6-dione (Me, H, MeN, R, O) [2351]

(**91.1**)
R = Me, Ph

2. 1,2,3-Triazole

Oxidizing conditions (p. 208) in the synthesis of a fused 2-substituted triazole may be avoided by reacting a 2-halonitrobenzene with a hydrazine.

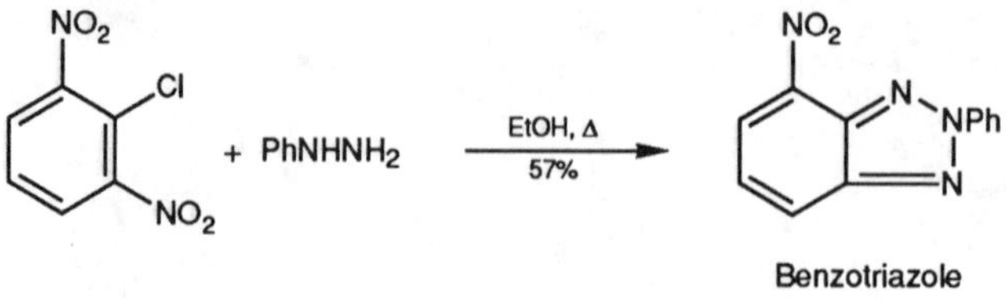

+ PhNHNH$_2$ → (EtOH, Δ; 57%) [2316]

Benzotriazole

3. 1,3-Dithiole

3,3-Dithioacroleins react with 2,4-di- or 2,4,6-tri-nitrobenzenes on warming in ethanol to give high yields of a benzodithiole.

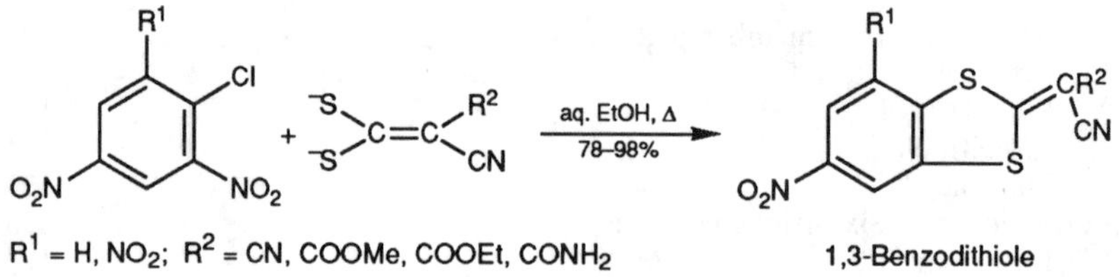

→ (aq. EtOH, Δ; 78–98%) [2171]

R^1 = H, NO$_2$; R^2 = CN, COOMe, COOEt, CONH$_2$

1,3-Benzodithiole

4. Thiazole

A reactive chloronitrobenzene may also be used to synthesize a fused thiazole by heating it in DMF with a compound containing a ring NH and a lactam thiocarbonyl.

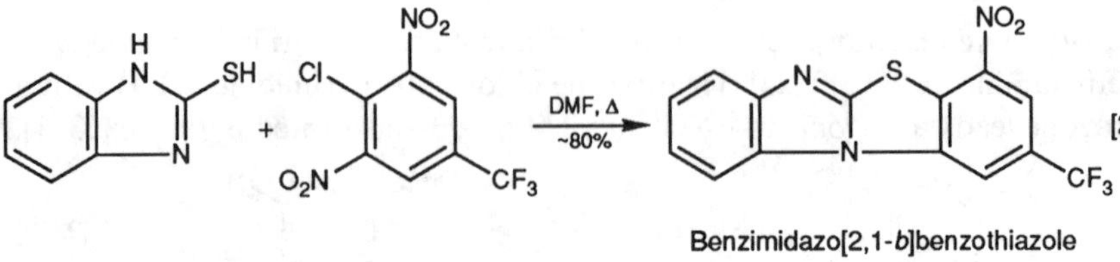

→ (DMF, Δ; ~80%) [2901]

Benzimidazo[2,1-*b*]benzothiazole

II. FORMATION OF A SIX-MEMBERED RING

1. 1,4-Dioxin

Of the three groups in 1,2-dichloro-3-nitrobenzene, a chlorine and a nitro react with the dipotassium salt of catechol and HMPT. This unexpected product is formed in high yield.

Cl OK Cl + OK O_2N 1. HMPT, N_2 2. Δ 96% O Cl O [3743]

Dibenzo[*b*,*e*][1,4]dioxin

III. FORMATION OF A SEVEN-MEMBERED RING

1. 1,4-Dithiepine

Both halo and nitro functions attached to naphthalene are displaced by the anion of benzene-1,2-dithiol in DMF under argon.

S^- + S^- Cl NO_2 DMF, NaH, Ar 45% S S [3572]

Benzo[2,3]naphtho[5,6,7-*ij*][1,4]dithiepine

CHAPTER 92

Halogen and Ring-nitrogen

I. Formation of a Five-membered Ring 1236
1. Pyrrole 1236
2. Imidazole 1237
3. 1,2,4-Triazole 1238
4. 1,2,4-Triazol-3-one 1239
5. Thiazole 1239
6. 1,2,4-Thiadiazole 1239
II. Formation of a Six-membered Ring 1240
1. Pyridine 1240
2. Pyridin-2-one 1240
3. Pyrimidin-4-one 1240
4. Pyrazin-2-one 1242
5. 1,4-Oxazin-2-one 1243
III. Formation of a Seven-membered Ring 1243
1. 1,2,4-Triazepin-7-one 1243

I. FORMATION OF A FIVE-MEMBERED RING

1. Pyrrole

A cheaper method of synthesizing the pyrroloquinolinone (**92.2**) than those previously known (from an indoline) may be the base-catalysed dehalogenation of the dibromide (**92.1**) (prepared by addition of bromine to the 8-allylquinolinone).

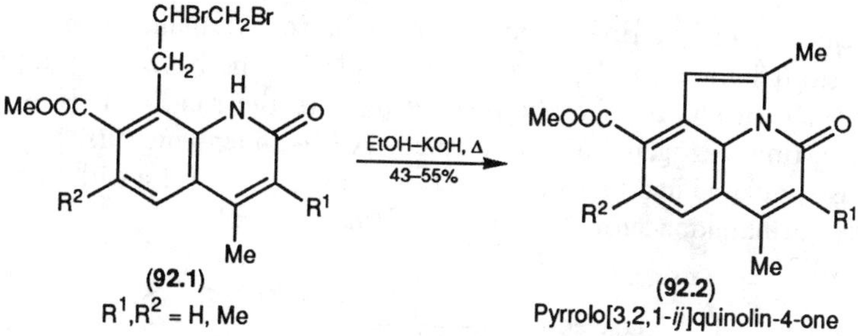

(92.1)
R^1,R^2 = H, Me

EtOH–KOH, Δ
43–55%

(92.2)
Pyrrolo[3,2,1-*ij*]quinolin-4-one

[3955]

2. Imidazole

A C—N bond is formed during nucleophilic attack by a pyridine-type nitrogen on an ω-haloethylamine side-chain [2663]. When an even weaker base is used in this type of cyclization, a partly reduced imidazole ring is formed [2310]. If other easily displaced halogen atoms are present, ring closure with sodium sulphide or an alkylamine may be accompanied by displacement of the second halogen [2310].

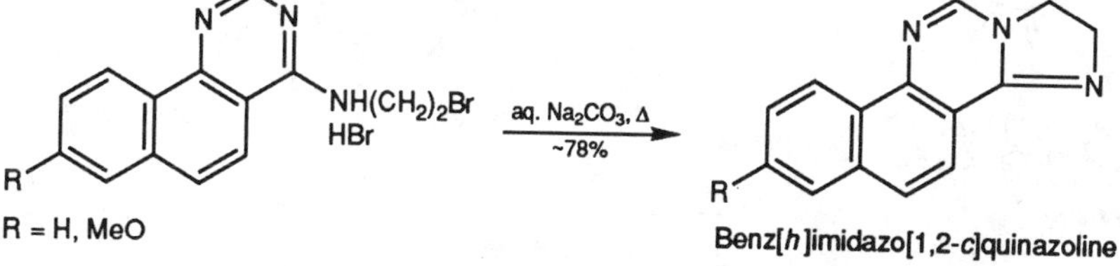

aq. Na_2CO_3, Δ
~78%

R = H, MeO

Benz[*h*]imidazo[1,2-*c*]quinazoline

[2663]

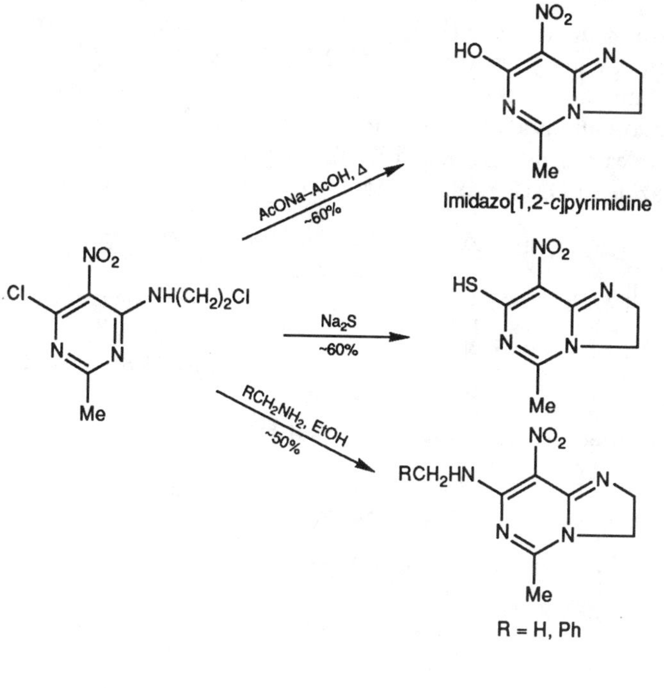

AcONa–AcOH, Δ
~60%

Imidazo[1,2-*c*]pyrimidine

Na_2S
~60%

RCH_2NH_2, EtOH
~50%

R = H, Ph

[2310]

The cation attacks a ring-carbon when 2-(chloroallyl)aniline is heated with a Lewis acid (Chapter 90, Section I.1), but replacing the benzene by a pyridine ring provides a choice of carbon or nitrogen in the ring; with PPA, a new carbon-to-ring nitrogen bond is formed. An α-chloro–ring-nitrogen combination is annulated in a two-stage reaction at low temperature with phosphoryl chloride–formamidoacetonitrile in the presence of TEA.

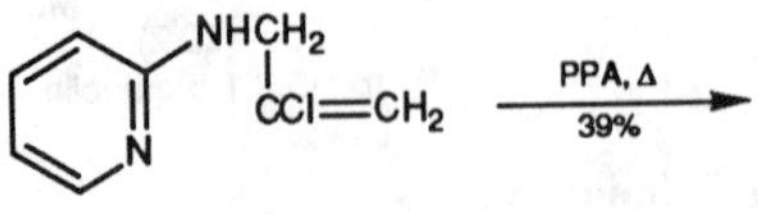

PPA, Δ 39% [2684]

Imidazo[1,2-*a*]pyridine

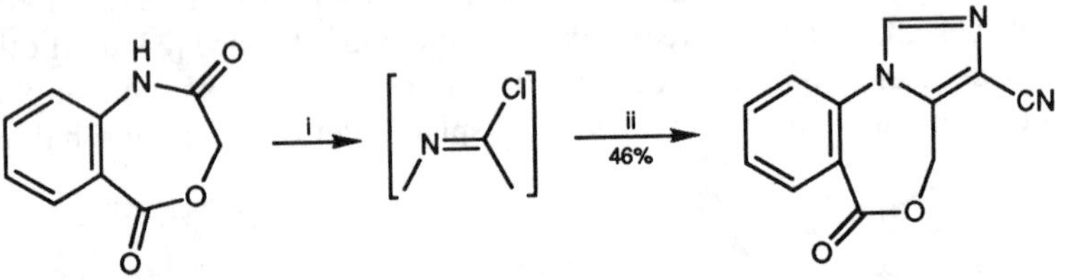

ii 46% [3538]

i, CH_2Cl_2, PhNMe, $POCl_3$;
ii, $OHCNHCH_2CN$ $POCl_3$–TEA

Imidazo[1,5-*a*][4,1]benzoxazepin-6-one

3. 1,2,4-Triazole

α-Halohydrazones (reviews [1437, 1753]) are useful reagents for several types of cyclizations. When this function (—NHN=CX—) is adjacent to a ring-nitrogen, annulation occurs on treatment with a base or even under mild solvolytic conditions at ambient temperature [2410]. The synthesis of condensed 1,2,4-triazoles has been reviewed [2889]. Reaction of a 2-chloroazine with an acylhydrazine in DME produces good yields of triazoloazines [3070].

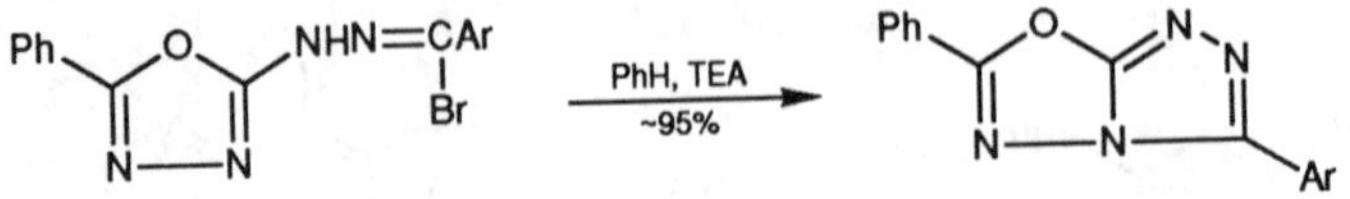

PhH, TEA ~95% [2403, 2410]

Ar = Ph, 4-Me-, 4-Br-, 4-Cl-, 3-NO_2-C_6H_4

1,2,4-Triazolo[3,4-*b*][1,3,4]oxadiazole

Another ring system synthesized similarly:

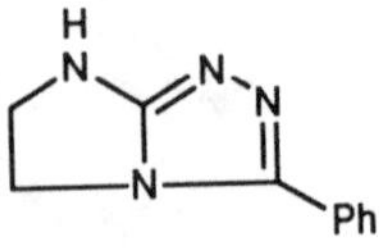

[2410]

Imidazo[2,1-*c*]-1,2,4-triazole

R^1, N, Cl, X, O + NHNH$_2$, R^2C=O → (DME, Δ; 31–73%) → R^2, N, N, N, R^1, X, O [3070]

R^1 = H, Cl, Me; R^2 = H, Me; X = O, S

X = O [1,2,4]Triazolo[3,4-*c*][1,4]benzoxazin-4-one
X = S [1,2,4]Triazino[3,4-*c*][1,4]benzothiazin-4-one

4. 1,2,4-Triazol-3-one

Semicarbazide hydrochloride or ethyl carbazate cyclize 2-chloroazines or 2-chloroazepines on heating the compounds in ethanol to form this fused ring in moderate yield.

Cl, N, N, Cl + NHNH$_2$, CONH$_2$, HCl → (EtOH, HCl, Δ; 48%) → N, NH, N, N, Cl, O [2322]

1,2,4-Triazolo[4,3-*b*]pyridazin-3-one

Cl, N, N, OMe + COOEt, NHNH$_2$ → (EtOH, Δ; 36%) → N—NH, N, N, O, OMe [3139]

Pyrido[2,3-*c*]-1,2,4-triazolo[4,3-*a*]azepin-3-one

5. Thiazole

Ammonium thiocyanate annulates chloromethyl and ring-nitrogen functions to a thiazole ring on warming in methanol. Under mild conditions, the intermediate thiocyanatomethyl derivative may be isolated.

R^1, CHCl, N, NH, R^2 → (MeOH, NH_4SCN, Δ; 19–59%) → R^1, N, S, N, NH, R^2 [2350, 2614]

R^1 = H, Me; R^2 = H, Cl, Me

Thiazolo[3,4-*a*]benzimidazole

6. 1,2,4-Thiadiazole

When a 2-trichloromethanesulphenamidothiazole is stirred with a primary arylamine and TEA, a thiadiazole ring is formed and the chlorine atoms are lost.

CHCl$_3$, RNH$_2$, TEA
34–68%
[3585]

R = 4-Cl-, 3- or 4-NO$_2$-C$_6$H$_4$, Cl$_2$C$_6$H$_3$, 2-pyridyl, 5-iodo-2-pyridyl

[1,2,4]Thiadiazolo[3,4-*b*]benzothiazole

II. FORMATION OF A SIX-MEMBERED RING

1. Pyridine

When 4-(bromobutenyl)azetidin-2-one is heated in DMF with activated copper powder, cyclization to the carbacephem occurs. The yield depends on the proportion of *Z*-isomer (which decreases during cyclization) in the enantiomeric mixture of reactant.

DMF, Δ
20–33%
[3624]

R = H, COOEt

1-Azabicyclo[4.2.0]oct-2-en-8-one

2. Pyridin-2-one

Cross-coupling of 7-haloindoles with an *o*-formylboronic acid (**92.3**) using modified Suzuki reaction conditions [3609] gives compounds which are of interest in the synthesis of the alkaloids ungerimine and hippadine.

i
~45%
[3862]

(**92.3**)

R = H, MsO; X = Br, I
i, DME, PPh$_3$, Na$_2$CO$_3$, Δ

[1,3]Dioxolo[4,5-*j*]pyrrolo[3,2,1-*de*]phenanthridin-7-one

3. Pyrimidin-4-one

Two bromine atoms are lost during the conversion of the 2-(2,3-dibromopropanoylamino)pyrimidine into the bicyclic product by heating in DMF. Thermolysis of the 3-chloro analogue in DMSO gives an even higher yield of the same compound, but heating the 3-chloropropenoylaminopyridazine

(**92.4**), which already contains the required C—C double bond, proceeds in lower yield.

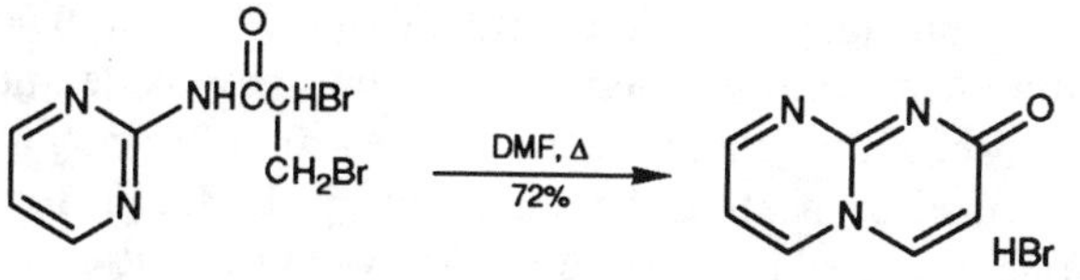

DMF, Δ 72% HBr [2279]

Pyrimido[1,2-*a*]pyrimidin-2-one

DMSO, Δ ~95% HCl [2296]

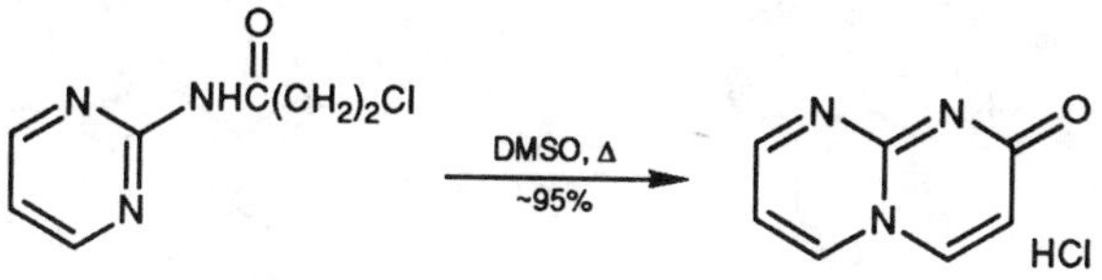

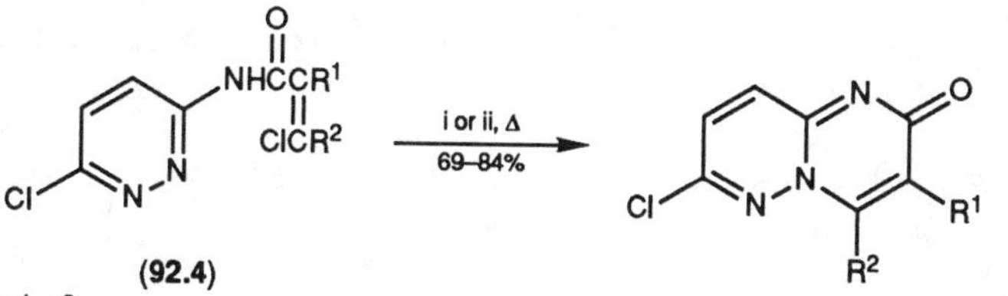

(**92.4**)
R^1,R^2 = H, Cl
i, 1,2,4-$Cl_3C_6H_3$; ii, xylene

i or ii, Δ 69–84% [2296]

Pyrimido[1,2-*b*]pyridazin-2-one

A useful selectivity was observed when the trichloroacryloylaminopyridazine (**92.5**) was thermally cyclized at 140 °C and at 210 °C. No such distinction was apparent with the dichloroacryloylaminothiadiazole (**92.6**).

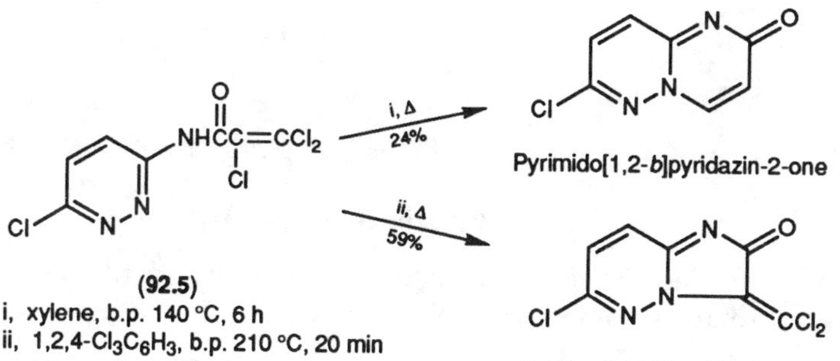

(**92.5**)
i, xylene, b.p. 140 °C, 6 h
ii, 1,2,4-$Cl_3C_6H_3$, b.p. 210 °C, 20 min

i, Δ 24% Pyrimido[1,2-*b*]pyridazin-2-one [2296]

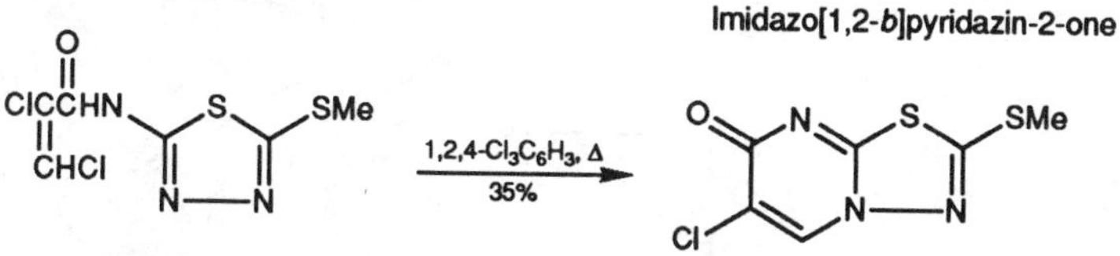

ii, Δ 59% Imidazo[1,2-*b*]pyridazin-2-one

(**92.6**)

1,2,4-$Cl_3C_6H_3$, Δ 35% [2296]

1,3,4-Thiadiazolo[3,2-*a*]pyrimidin-7-one

When a 3-chloropyridazine is heated with anthranilic acid, a high yield of a quinazolinone is obtained [2219]. A halogen attached to one ring is displaced by reaction with a ring-nitrogen in another and a carbonyl group placed between them in a palladium-catalysed synthesis with carbon monoxide under pressure [3280]. A bond between nitrogen in one ring and carbon (bearing a chlorine) in another ring is formed by heating the compound in DMA [3130]; pyrido[1,2-*a*]quinazolin-6-one has been synthesized in 83% yield by heating 2-chloro-*N*-2-pyridylbenzamide in tetralin and then basifying the mixture [3697].

EtOH, Δ 82% [2219]

Pyrido[2′,3′:4,5]pyridazino[6,1-*b*]quinazolin-8-one

i 35–65% [3280]

Pyrido[2,1-*b*]quinazolin-11-one

R^1 = HO, $(CH_2)_5R^2$; R^2 = imidazol-1-yl, pyridin-3-yl

i, tBuOH, PPh_3, $(Ph_3P)_2PdCl_2$, Bu_3N, CO

DMA, Δ ~63% [3130]

R = H, Cl

Pyrido[3′,2′:5,6]pyrimido[1,2-*a*]quinazolin-5-one

4. Pyrazin-2-one

A bromoacetylamino group is annulated to a *peri*-placed ring-nitrogen by heating with t-butoxide in isopropanol.

iPrOH, tBuOK 66% [3925]

Pyrrolo[1,2,3-*de*]quinoxalin-5-one

5. 1,4-Oxazin-2-one

An unexpected product was isolated when 2-trichloroacetylpyrrole was stirred at room temperature with chloroacetone.

$COCCl_3$ NH + $COCH_2Cl$ Me $\xrightarrow[58\%]{Me_2CO}$ O O N Me [3418]

Pyrrolo[2,1-*c*][1,4]oxazin-1-one

III. FORMATION OF A SEVEN-MEMBERED RING

1. 1,2,4-Triazepin-7-one

The ω-chlorohydrazide (**92.7**) cyclizes on to the ring-nitrogen when the compound is stirred with pyridine.

S $NHNHC(CH_2)_2Cl$ O N Ar $\xrightarrow[62\text{–}75\%]{pyr}$ S N–NH =O N Ar [2863]

(92.7)
Ar = Ph, 4-Br-, 4-Cl-, 4-Et-, 4-MeO-, 4-Me-C_6H_4

Thiazolo[2,3-*c*][1,2,4]triazepin-3-one

CHAPTER 93

Hydrazide, Hydrazine or Hydrazone and Nitro

I. Formation of a Five-membered Ring 1244
1. Pyrazole 1244
2. 1,2,3-Triazole 1-Oxide 1244
II. Formation of a Six-membered Ring 1245
1. 1,2,3-Triazin-4-one 1-Oxide 1245
2. 1,2,4-Triazin-3-one 1245
3. 1,2,4-Triazin-6-one 1246
4. 1,3,4-Thiadiazine 1246

I. FORMATION OF A FIVE-MEMBERED RING

1. Pyrazole

Annulation of a *peri*-positioned nitro and hydrazine functions of the quinoline (**93.1**) occurs when it is refluxed in acetic acid, the 5-nitro group being displaced (cf. Section I.2). *Peri*-annulated compounds have been reviewed [3905].

NO_2 NHNHPh MeO NO_2 N —AcOH, Δ 50%→ PhN—NH MeO NO_2 N [3004]

(**93.1**) Pyrazolo[3,4,5-*de*]quinoline

2. 1,2,3-Triazole 1-Oxide

o-Nitrohydrazines such as (**93.1**) are cyclized by heating with either aqueous alkali or sulphuric acid, apparently in better yield under acidic conditions, but as no variants of these substrates were studied, no firm conclusion can be drawn (but see previous paragraph).

(93.1) aq. NaOH, Δ 57% [3004]

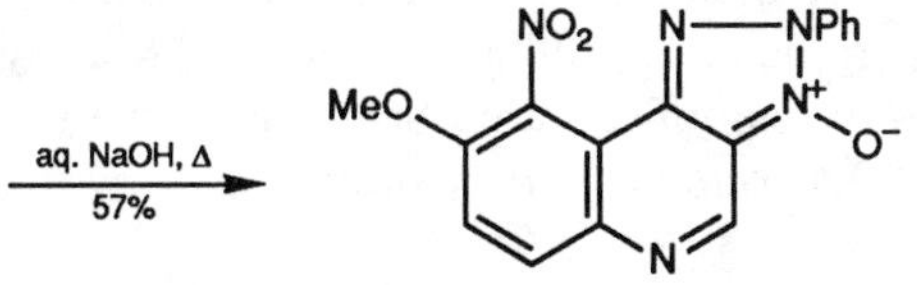

1,2,3-Triazolo[4,5-*c*]quinoline 3-oxide

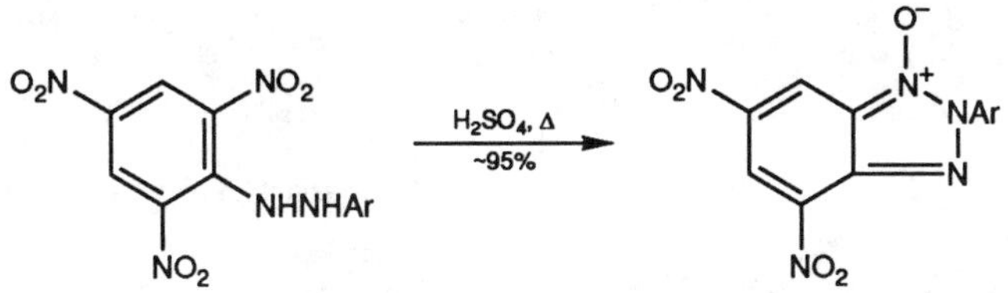

[3041]

$Ar = 2,4,6\text{-}(NO_2)_3C_6H_2$ H_2SO_4, Δ ~95%

Benzotriazole 1-oxide

II. FORMATION OF A SIX-MEMBERED RING

1. 1,2,3-Triazin-4-one 1-Oxide

2-Nitrobenzaldehyde methylhydrazone undergoes oxidative cyclization to a betaine *N*-oxide in high yield. Although the yield in this particular conversion is very high, it is worth considering the use of the phenyliodo diesters (review of the use of these and similar reagents [3714]). The reactions of hydrazones have been reviewed [2384].

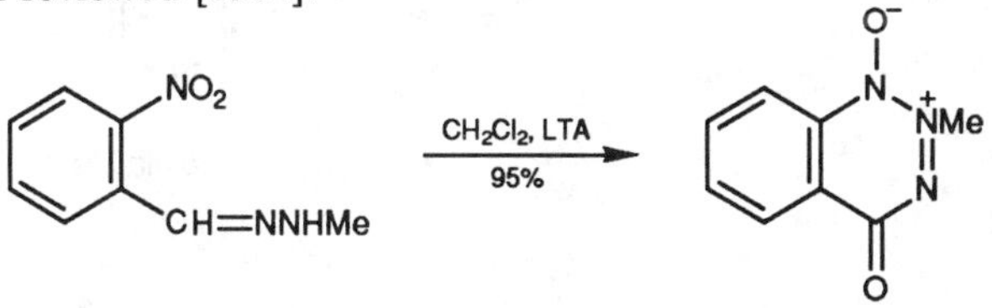

CH_2Cl_2, LTA 95% [2621]

1,2,3-Benzotriazin-4-one 1-oxide

2. 1,2,4-Triazin-3-one

A hydrazono-carbamate may be cyclized efficiently by heating it in decalin.

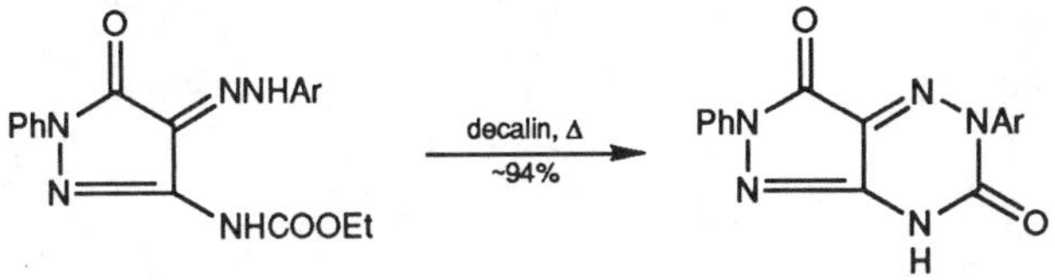

decalin, Δ ~94% [2440]

$Ar = Ph, 4\text{-halo-}C_6H_4$

Pyrazolo[3,4-*e*]-1,2,4-triazine-3,7-dione

3. 1,2,4-Triazin-6-one

A hydrazide may be linked to a ring-nitrogen atom by heating the compound with an orthoester; the same product is formed by heating the ω-acetylhydrazide with phosphorus oxychloride, but in such reactions, it is possible that cyclization to an isomeric 1,3,4-oxadiazole may occur.

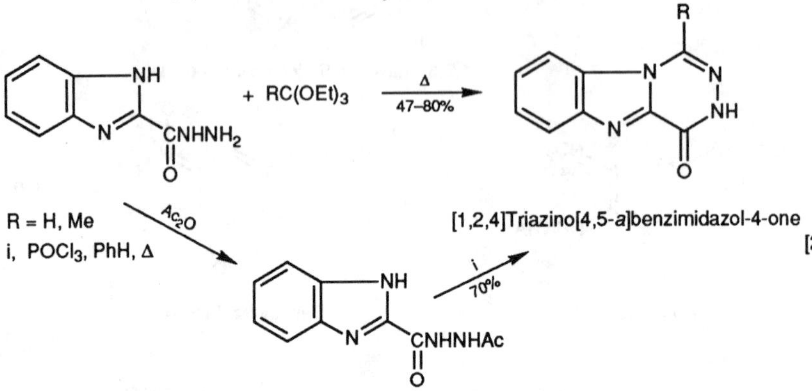

+ $RC(OEt)_3$ —Δ, 47–80%→

$R = H, Me$

i, $POCl_3$, PhH, Δ

Ac_2O

i, 70%

[1,2,4]Triazino[4,5-*a*]benzimidazol-4-one

[2185]

4. 1,3,4-Thiadiazine

A nitro group attached to a benzene ring is readily displaced by a neighbouring hydrazono group (joined to the ring in this example through a sulphur) on heating the compound with a weak base.

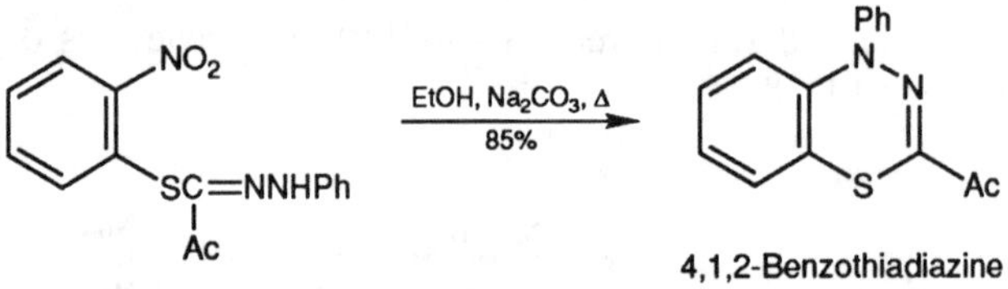

EtOH, Na_2CO_3, Δ, 85%

4,1,2-Benzothiadiazine

[2938]

CHAPTER 94

Hydrazine and Ring-carbon or Ring-nitrogen

I.	Formation of a Five-membered Ring	1247
	1. Pyrrole	1247
	2. Pyrazole	1253
	3. 1,2,4-Triazole	1253
	4. 1,2,4-Triazol-3-one or 1,2,4-Triazole-3-thione	1256
	5. Furan	1256
II.	Formation of a Six-membered Ring	1257
	1. Pyridazine	1257
	2. 1,2,4-Triazine	1257
	3. 1,2,4-Triazin-5-one or 1,2,4-Triazine-5,6-dione	1257
	4. Pyran-2-one	1258
III.	Formation of a Seven-membered Ring	1259
	1. 1,2-Diazepine	1259

As in Part 1, Fischer indolization of arylhydrazines or of hydrazones is discussed in this chapter. Similar cyclizations in which the arylhydrazine is replaced by either an *O*-arylhydroxylamine or a 2-oximinoazine are also included.

I. FORMATION OF A FIVE-MEMBERED RING

1. Pyrrole

Fischer indolization is usually a Lewis acid-catalysed reaction but thermal cyclization of 4-pyridylhydrazone *N*-oxides gave moderate-to-good yields of indoles. The 3-pyridylhydrazones yielded lower yields of pyrrolo[3,2-*b*]pyridines together with small amounts of the [2,3-*c*] isomers. A comparison of yields obtained using various catalysts on 2-pyridylhydrazones of cyclohexanone showed wide variations (from 21 to 98%); in this reaction, 4-toluenesulphonic acid gave the best yield. The effect of varying the substituent on the pyridine ring was also investigated [2150]. Heating a hydrazine with the carbonyl compound in either tetralin or ethylene glycol gave moderate-to-high yields of pyrrolo[2,3-*d*]pyrimidines [2664]. 4-Quinolinylhydrazones are difficult to cyclize under the usual conditions but thermolysis gives good results [2851].

NHNH$_2$ + $R^1C{=}O$ / CH_2R^2 —DEG, Δ, 10–77%→ [2150]

R^1 = alkyl, Ph; R^2 = H, Me, Ph or R^1R^2 = $(CH_2)_3$

Pyrrolo[3,2-*c*]pyridine

Other ring systems synthesized similarly:

Pyrrolo[3,2-*b*]pyridine [2150]

Pyrido[2,3-*b*]indole [2184]

Pyrrolo[2,3-*d*]pyrimidine-2,4-dione [2664]

Early attempts to prepare indole itself by Fischer's method failed, possibly because the expected product may have been unstable or reactive under the experimental conditions employed. By removing the product from the reaction environment (zinc chloride-coated beads at 290–300 °C) with a stream of nitrogen, a moderate yield of indole may be obtained. This method gives high yields of 2- and 3-methylindoles from the relevant hydrazones [2828]. The mechanism of a variant of the Fischer reaction has been demonstrated by the use of ^{15}N-labelled phenylhydrazine; 2-methyltryptamine was the product of this indolization [2188].

NHN=CR^1 / CH_2R^2 —$ZnCl_2$, N_2, Δ, 36–79%→ [2828]

R^1,R^2 = H, Me

Indole

NHNH$_2$ + $(CH_2)_3Cl$ / COMe —Δ, 81%→ (Me, $(CH_2)_2NH_2$) [2188]

Using α-acetyl-γ-butyrolactone as the carbonyl component in the Fischer synthesis, decarboxylation occurs and it is thus possible to prepare a 1,2-disubstituted indol-3-ylethanol [2637]. In a related reaction, a γ-haloaldehyde or a γ-haloketone reacts with an arylhydrazine to give a tryptamine in which the side-chain nitrogen is derived from the terminal nitrogen of the reacting phenylhydrazine; this and related reactions were reviewed in 1974 and 1988 [2641, 3461]. The course of Fischer indolization of the preformed monophenylhydrazone of cyclohexane-1,3-dione changes when the reaction is allowed to proceed under conditions which encourage ketal formation [2565].

iPrOH, HCl, Δ; 41–58% [2637]

R = Me, iPr, $PhCH_2$, Ph

$(CH_2OH)_2$, TsOH, PhMe, Δ; 34–54% [2565]

R^1 = H, Me; $R^2 = R^1$; R^3 = H; X = O; Y = H_2 or R^2 = H; $R^3 = R^2$; X = H_2; Y = O

X = O; Y = H_2 1-Carbazolone
X = H_2; Y = O 2-Carbazolone

The different electronic effect of the sulphur function in tetrahydrothiopyran-3-one and its 1,1-dioxide is reflected in the different products obtained [2193, 2194]. The readily prepared dihydrochalcones (1,3-diphenylpropan-2-ones) give moderate yields of 3-benzyl-2-arylindoles in the Fischer reaction using ethanolic hydrogen chloride as catalyst [2721]. The use of phosphorus trichloride as catalyst, mentioned in a preliminary publication (p. 508), has recently been described in full and illustrated with further examples [3396], and the usefulness of boron trifluoride–acetic acid (not mentioned in Part 1) has been demonstrated [2524].

n = 0: EtOH, Δ; 10–65% → Thiopyrano[3,2-*b*]indole

n = 2: EtOH–HCl, Δ; ~85% → Thiopyrano[3,4-*b*]indole

R = H, Me [2193, 2194]

R^1 or R^2 = Me + PhNHNH$_2$ $\xrightarrow[72\%]{PhMe, (CH_2OH)_2, \Delta}$ [3222]

Cyclopent[*b*]indole

When the Fischer synthesis is applied to an unsymmetrical ketone, either one of two isomers or a mixture of them may be produced. (+)-3-Methylcyclopentanone gives a mixture of 1- and 2-methylcyclopent[*b*]indoles, and the relative amounts of these formed under various conditions are analysed [3222]. Further work has recently been published on the Fischer indolization of *N*-methoxyphenyl-*N*-phenylhydrazones of an unsymmetrical ketone (ethyl pyruvate). Cyclization in acid occurs mostly on to the more electron-rich benzene ring whereas under nonacidic (for example, thermal) conditions there is less regioselectivity [3539]. 2-Methoxyphenylhydrazine sometimes behaves anomalously and does not yield the expected 7-methoxyindole, but when *o*-4-toluenesulphonyl- or *o*-4-trifluoromethylsulphonyl-phenylhydrazine is used to prepare the hydrazone, the main product is the 7-sulphonyloxyindole which may be hydrolysed to the 7-hydroxyindole with alkali [3629].

$\xrightarrow[36–52\%]{PPA, \Delta}$ [3629]

R^1 = 4-MeC$_6$H$_4$SO$_2$, CF$_3$SO$_2$; R^2 = H, Br

Indole

Replacement of methoxy groups by chlorine atom(s) in the phenylhydrazine ring also leads to a mixture of products, some of which are anomalous. The equation shows the main product; when 2,6-dichlorophenylhydrazine reacts with ethyl pyruvate, no indole is formed in the presence of ethanolic hydrogen chloride, but zinc chloride–acetic acid gives a low yield (4.9%) of ethyl 5,7-dichloroindole-2-carboxylate [3764]. A study of the indolization of α-phenyl-α-4-substituted phenylhydrazine shows that isomeric indoles are formed in ratios which depend on the 4-substituent. The study supports the premise that a C—C bond is formed in the second step of this reaction [B-29, 2818]. *N*-Benzyl-*N*-phenylhydrazine reacts with 2-methyl-4-oxohexanoic acid to give an appreciable amount of *N*-unsubstituted indoles [3841]. Fischer cyclizations using cyclic phenylhydrazines (such as *N*-phenylpyrazolidine) and cycloalkanones were not expected to lose one nitrogen atom as ammonia. An indole was obtained in this reaction, the second nitrogen being in a side-chain [2478].

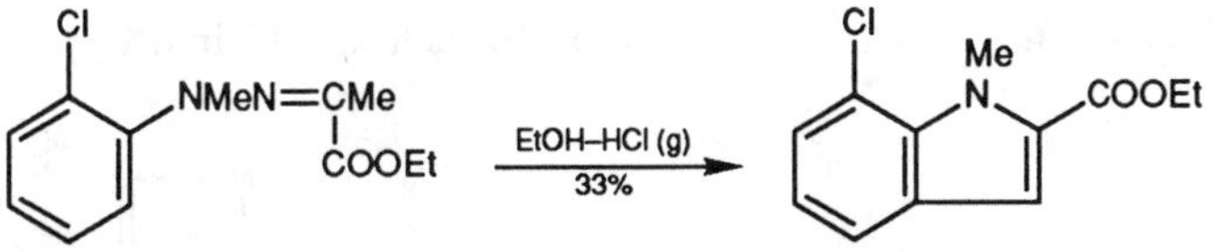

[3764]

When a dihydrazinobenzene is converted into a ketone dihydrazone (such as (**94.4**), two pyrrole rings may be formed [2893]. Complementary to this kind of reaction is the double indolization of bis(phenylhydrazones) from diketones to form indolo[2,3-*a*]pyrrolo[3,4-*c*]carbazoles in good yield [3541]. An indoloindole is formed when an *N*-aryl-*N*-alkylhydrazone of *N*-acetylindoxyl (**94.5**) is subjected briefly to Fischer reaction conditions. Similarly, the simple hydrazone of *N*-acetylindoxyl reacts with a $COCH_2$-containing ketone to form a dihydrazone, and the latter is cyclized to form another pyrrole ring (the Piloty–Robinson reaction) on stirring at 20 °C in acetic acid [3021, 3022]. The resulting pyrrolo[3,2-*b*]indole is a component of several alkaloids.

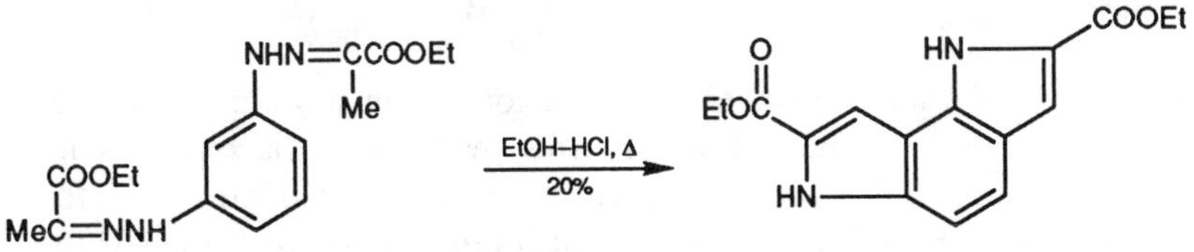

(**94.4**) → Pyrrolo[2,3-*e*]indole [2893]

Another ring system synthesized similarly:

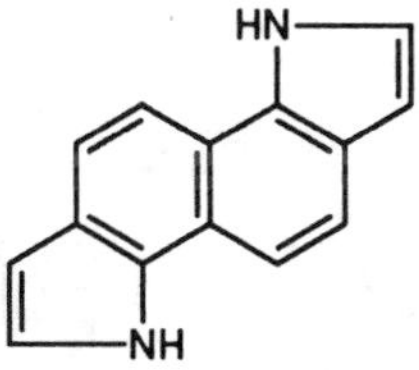

Indolo[7,6-*g*]indole [2896]

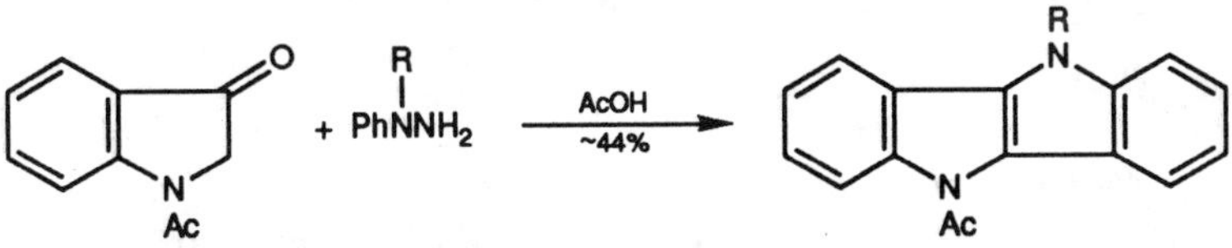

(**94.5**) R = Me, $PhCH_2$ → Indolo[3,2-*b*]indole [3021]

A combined Fischer indolization, debenzylation and aromatization occurs when ketones of type (**94.6**) are heated for several hours with an arylhydrazine. Examples of reactions in which hydrazines exhibit dehydrogenative properties are given in this paper [3287]. 2,3-Dihydrofuran behaves as would be expected of 4-hydroxybutanal in a Fischer reaction to yield a 3-(2-hydroxyethyl)indole [3700].

(94.6)
R = H, Cl

+ $NHNH_2$–C_6H_4-4-R, Δ, ~69%

Pyrido[3,2-*b*:5,4-*b*']diindole [3287]

i, Δ, 48%

i, THF, $(HOCH_2)_2$, $ZnCl_2$

Indole [3700]

Application of the Japp–Klingemann reaction to an arylamine can be a convenient synthesis of some indoles. In this reaction, a diazonium salt is formed *in situ* and is coupled with the active methylene group of ethyl 2-methyl-3-oxobutanoate under weakly basic or neutral conditions. Heating the resulting hydrazone gives an indole [3932]. When a phenylhydrazone is treated with iodomethane for about 15 h (with or without a weak base), the overall effect is similar to that of the Fischer and Piloty cyclizations [2622].

HNO_2, HCl, Δ, 95% [3932]

MeI, 28–100%

R = H, tBu, cC_6H_{11}

Carbazole [2622]

PrOH, Δ, 87% [3834]

R^1 = H, Me; R^2 = Ph, 2-furyl, 2-thienyl;
R^3 = CN, COOEt, $CONH_2$

Pyrazolo[3,4-*d*]pyrimidine-4,6-dione

2. Pyrazole

In the cyclization of a hydrazino group on to a neighbouring ring-carbon to form a new fused pyrazole ring, the required carbon atom may be supplied by the arylidene part of a malonic acid derivative. Dichloromethylene ammonium salts (review [2549]) are readily attacked by nucleophiles such as phenylhydrazines which displace the chlorine atoms and form a fused pyrazole in high yield.

Ph NNH$_2$ + $Cl_2C{=}\overset{+}{N}Me_2$ Cl^- —($CHCl_3$, Δ; 83%)→ Indazole (Ph, NMe_2) [2548]

3. 1,2,4-Triazole

A hydrazinopyrrole (**94.7**) may be cyclized to a pyrrolotriazole by heating with cyanogen bromide [3078]. Numerous examples of similar cyclizations using a carboxylic acid are known (see p. 513); aliphatic acids react at about 100 °C but reactions of their aromatic and heteroaromatic analogues give better yields when heated briefly at 180–200 °C [2113]; addition of DCC followed by heating in ethylene glycol can promote cyclization [3521]. When oxalic acid is used, decarboxylation also occurs; the acid may be replaced by diphenylcarbodi-imide [2114]. The hydrazinocarboxamide (**94.8**) could cyclize in more than one way when treated with formic acid or ethyl orthoformate, but the product is the fused triazole [2420].

(94.7) (COOEt, NHNH$_2$) —(aq. MeOH, BrCN; 83%)→ Cyclohepta[4,5]pyrrolo[2,1-*c*]-1,2,4-triazole (COOEt, NH_2) [3078]

Me, NHNH$_2$, Cl + COOH/COOH —(Δ; 70%)→ Furo[3,2-*d*]-1,2,4-triazolo[4,3-*b*]pyridazine (Me, Cl) [2113]

NH_2, Cl, NHNH$_2$ + RCOOH —(i, ii; 66%)→ 1,2,4-Triazolo[4,3-*b*]pyridazine (NH_2, Cl, R) [3521]

R = tribenzoyl-D-ribofuranosyl

i, CH_2Cl_2, DCC; ii, $(CH_2OH)_2$, Δ

NHNH$_2$ + RN=C=NR → (xylene, Δ; 47–95%) [2114]

R = alkyl, Ph

1,2,4-Triazolo[3,4-*a*]isoquinoline

CONH$_2$ NHNH$_2$ Cl → (HCOOH or HC(OEt)$_3$, Δ; 74%) [2420]

(**94.8**)

1,2,4-Triazolo[4,3-*b*]pyridazine

In the cyclization with an orthoester (see p. 515), addition of silica gel can improve the yield [3925]. It is possible for the product of such a cyclization to have undergone a Dimroth rearrangement; under milder reaction conditions, the primary product can often be isolated [3573, 3787]. To the list of numerous reagents which annulate a 2-hydrazinoazine (p. 513) may be added ethyl acetamidothioate which reacts without the need for external heating [3382]. *N,N*-Dimethylphosgenimmonium chloride converts a hydrazinothiazole into a thiazolotriazole in high yield [2945]. The synthesis of fused triazoles has been reviewed [2889, 3903].

NHNH$_2$ I Ar → (PhMe, RC(OEt)$_3$, SiO$_2$, Δ; 68–97%) [3925]

Ar = 2-F- or 2-Cl-C_6H_4; R = H, Me, Et

[1,2,4]Triazolo[4,3-*a*][1,4]benzodiazepine

R^1 R^2 NHNH$_2$ → (HC(OEt)$_3$, Δ; 30–80%) [ ] → [3573, 3787]

R^1 = H, MeO; R^2 = MeO, NO$_2$

[1,2,4]Triazolo[1,5-*c*]pyrimidine

S HN NHNH$_2$ Ph + MeC(=NH_2^+)SEt → (DMF; 46%) S HN Me [3382]

1,2,4-Triazolo[4,3-*d*][1,2,4]triazine-6-thione

PhCl, Δ, 82–100% [2945]

R = H, EtO

1,2,4-Triazolo[3,4-*b*]benzothiazole

A useful distinction between two possible modes of cyclization is illustrated by the behaviour of 3-amino-2-hydrazinopyrimidin-4-one (**94.9**). In boiling butanol or briefly in cold acetic acid, it reacts with an orthoester to give a triazolopyrimidinone, but in acetic acid at room temperature for 1 h, the same reagent converts it into the pyrimidotetrazine [3278]. Substitution on either N-1 or N-3 of the 4-quinazolinone (**94.10**) or (**94.11**) with a methyl or amino controls the direction of cyclization [3732].

+ $R^2C(OEt)_3$ — BuOH, Δ, 29–67%; AcOH, 23–51% [3278]

(**94.9**)
R^1 = Me; R^2 = H, Me, Ph

1,2,4-Triazolo[4,3-*a*]pyrimidin-7-one

Pyrimido[1,2-*b*]-1,2,4,5-tetrazin-6-one

CS_2 or HCOOH, ~80% [3732]

(**94.10**)
R^1 = Me, NH_2;
R^2 = HS(from CS_2) or H(from HCOOH)

[1,2,4]Triazolo[4,3-*a*]quinazolin-5-one

DMF–CS_2 or HCOOH, Δ, ~76% [3732]

(**94.11**)
R = HS(from CS_2) or H(from HCOOH)

1,2,4-Triazolo[4,3-*b*]quinazolin-5-one

4. 1,2,4-Triazol-3-one or 1,2,4-Triazole-3-thione

Additional methods of cyclizing 2-hydrazino-azoles (**94.7**) or -azines (pp. 516, 517) include the use of carbonyl di-imidazole or carbon disulphide and a base (reviews [B-52, 3582]).

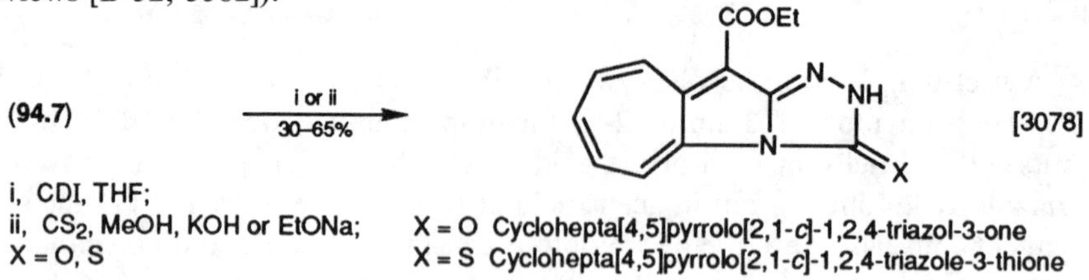

(94.7) → (i or ii, 30–65%) [3078]

i, CDI, THF;
ii, CS_2, MeOH, KOH or EtONa;
X = O, S

COOEt

X = O Cyclohepta[4,5]pyrrolo[2,1-*c*]-1,2,4-triazol-3-one
X = S Cyclohepta[4,5]pyrrolo[2,1-*c*]-1,2,4-triazole-3-thione

Another ring system synthesized similarly:

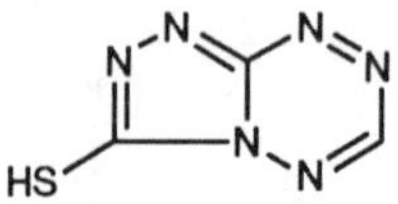

[2682]

[1,2,4]Triazolo[4,3-*b*]-1,2,4,5-tetrazine

5. Furan

Replacing phenylhydrazine by an *O*-arylhydroxylamine derivative (p. 511) leads to benzofurans; phenylhydroxylamine hydrochloride and a ketone may be similarly cyclized in acetic acid–sulphuric acid or in 98% sulphuric acid.

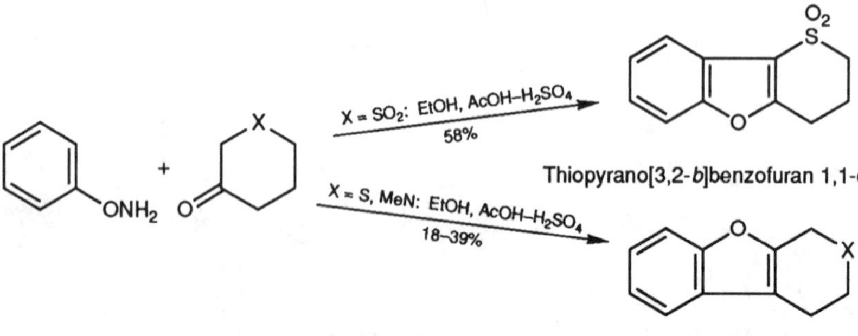

X = SO_2: EtOH, AcOH–H_2SO_4, 58%

Thiopyrano[3,2-*b*]benzofuran 1,1-dioxide

X = S, MeN: EtOH, AcOH–H_2SO_4, 18–39%

[2536]

X = MeN Benzofuro[2,3-*c*]pyridine
X = S Thiopyrano[3,4-*b*]benzofuran

Other ring systems synthesized similarly:

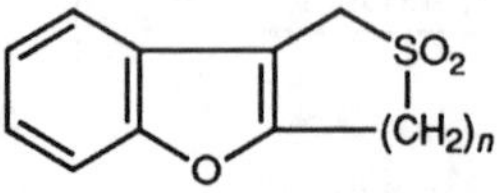

[3230]

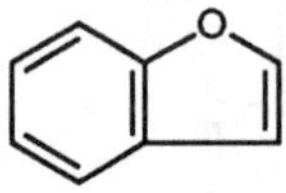

n = 2 Thiopyrano[4,3-*b*]benzofuran 2,2-dioxide
n = 3 Thiepino[4,3-*b*]benzofuran 2,2-dioxide

[3280]

Benzofuran

II. FORMATION OF A SIX-MEMBERED RING

1. Pyridazine

An unexpected product was obtained in high yield when the hydrazine (**94.12**) was stirred with tetrafluorobenzoquinone in an acidic medium.

(**94.12**) + tetrafluorobenzoquinone → EtOH, HCl, 88% → Pyrimido[4,5-*c*]cinnoline-1,3-dione [2632]

2. 1,2,4-Triazine

When conventional methods of cyclization discussed above are applied to the hydrazine (**94.13**), a triazine ring is formed.

(**94.13**) R = H, Cl + $HC(OEt)_3$ → TsOH, Δ, 9–96% → 2a,4,5-Triazabenz[*cd*]azulene [3334]

3. 1,2,4-Triazin-5-one or 1,2,4-Triazine-5,6-dione

Reaction of 1-hydrazinophthalazine (**94.14**) (hydralazine, an antihypertensive compound) with a 2,4-dioxoalkanoic ester did not yield the expected triazinophthalazine but either a 1,2,4-triazinophthalazinone or a 1,2,4-triazolophthalazine, depending on the conditions and/or the substituents in the dioxo ester. Under nonacidic conditions, the triazinophthalazinone is obtained but hydralazine hydrochloride gives the triazole except when a nitrophenyl group is present in the dioxo ester; a mechanism is suggested [3344]. Hydralazine and oxalic acid react in DMF at 30 °C for 24 h to form the dione in high yield whilst mesoxalate (diethyl oxomalonate) in refluxing ethanol converts the hydrazine efficiently into the triazinone.

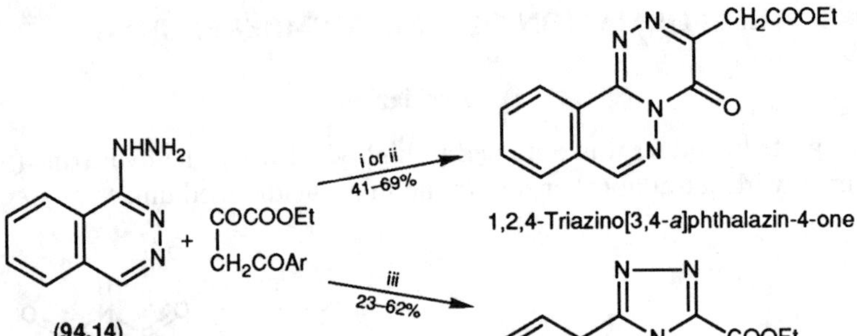

NHNH2 + COCOOEt / CH2COAr

(94.14)

i or ii, 41–69% → 1,2,4-Triazino[3,4-*a*]phthalazin-4-one [3344]

iii, 23–62% → 1,2,4-Triazino[3,4-*a*]phthalazine

i, Δ; Ar = MeO-, Me-, NO_2-C_6H_4;
ii, EtOH, HCl, Δ; Ar = 4-$NO_2C_6H_4$;
iii, EtOH, HCl, Δ; Ar = 3- or 4-MeO-C_6H_4

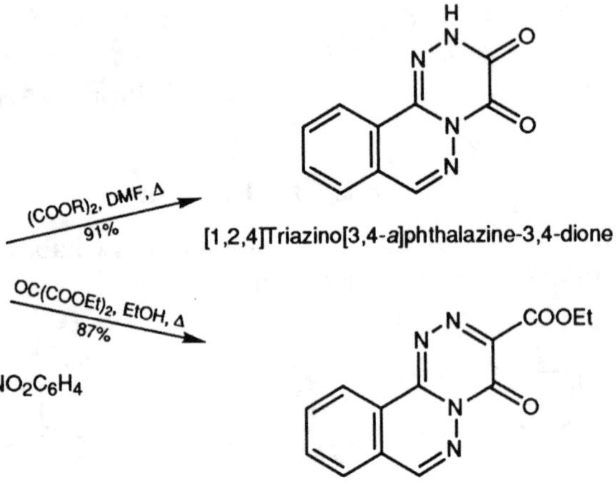

(94.14)

$(COOR)_2$, DMF, Δ, 91% → [1,2,4]Triazino[3,4-*a*]phthalazine-3,4-dione [2758]

$OC(COOEt)_2$, EtOH, Δ, 87% → [1,2,4]Triazino[3,4-*a*]phthalazin-4-one

R = H, Et, 4-$NO_2C_6H_4$

4. Pyran-2-one

O-Phenylhydroxylamines (see also Section I.5) react at room temperature with DMAD to give an enamine which undergoes a Cope rearrangement; this cleaves the N—O bond and a pyranone is obtained in high yield. When R = NO_2, additional heating of an intermediate with ethanolic hydrochloric acid is necessary.

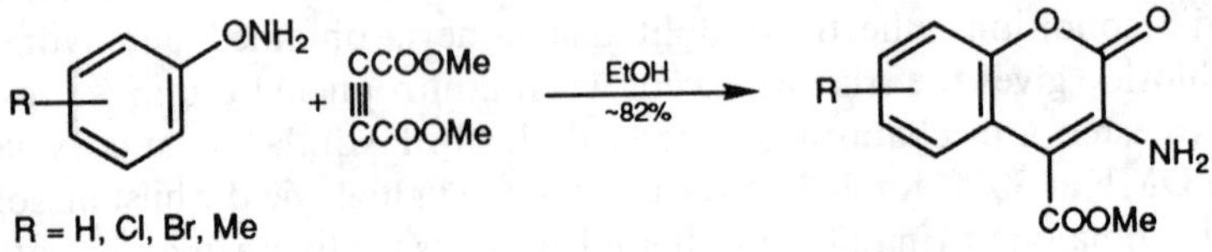

R–C_6H_4–ONH2 + CCOOMe ≡ CCOOMe → EtOH, ~82% → [2872]

R = H, Cl, Br, Me

1-Benzopyran-2-one

III. FORMATION OF A SEVEN-MEMBERED RING

1. 1,2-Diazepine

3-Alkoxyacroleins (3-alkoxyprop-2-enals) are useful cyclizing agents (review [3224], for example, in the formation of a 1,2-diazepine ring. The chemistry of hydrazinopyrimidines has been reviewed [3026].

$NMeNH_2$ + CHO—CR=CHOEt —(aq. AcOH, Δ; 35–62%)→ [2934]

R = alkyl

Pyrimido[4,5-*c*]-1,2-diazepine-6,8-dione

CHAPTER 95

Hydrazone or Oxime and Ring-carbon or Ring-nitrogen

I. Formation of a Five-membered Ring 1260
   1. Pyrrol-2-one or Pyrrole-2,3-dione 1260
   2. Pyrazole 1261
   3. 1,2,3-Triazole 1262
   4. 1,2,4-Triazole 1262
   5. 1,2,4-Triazole 1-Oxide 1263
   6. 1,2,4-Oxadiazol-5-one 1264
   7. 1,2,3,5-Oxathiadiazole 2-Oxide 1264
II. Formation of a Six-membered Ring 1264
   1. Pyridine 1264
   2. 1,2,4-Triazine or 1,2,4-Triazine 4-Oxide 1265
   3. 1,2,4-Oxadiazine 1266
III. Formation of a Seven-membered ring 1266
   1. 1,2-Diazepine 1266

Fischer indole-type cyclization is discussed in Chapter 94.

I. FORMATION OF A FIVE-MEMBERED RING

1. Pyrrol-2-one or Pyrrole-2,3-dione

N-Acylhydrazones undergo cyclization to a neighbouring carbon on heating in pentanol containing hydrogen chloride; both nitrogen atoms are retained but the arylidene group migrates from nitrogen to carbon. Isatin is formed in good yield by annulation of an aldoxime to a neighbouring ring-carbon (a variation of Sandmeyer's isatin synthesis).

Ar = Ph, 4-Cl-, 4-MeO-, 4-Me-C_6H_4 — $C_5H_{11}OH$–HCl, Δ; 60–85% — Pyrrolo[3,4-*b*]indol-3-one [2452]

R = $Pr_2N(CH_2)_2$ — 1. H_2SO_4, Δ 2. $NaHCO_3$; 58% — 2,3-Indoledione [3358]

2. Pyrazole

Thermolysis of the sodium derivative of the tosylhydrazone (**95.1**) at 280–340 °C gives the fused pyrazole; it is possible that a diazo derivative is an intermediate [3292]. The monohydrazone of cyclohexane-1,3-dione is cyclized by heating with ethyl orthoacetate and an acidic catalyst [2565]. Annulation of a hydrazone [such as (**95.2**)] to a ring-carbon of the original hydrazine may be achieved by heating the compound in DMF with an aryl aldehyde [2912].

(**95.1**) R = H, Me, Ph — NaH, Δ; 76–99% — Cycloocta[*c*]pyrazole [3292]

+ $MeC(OEt)_3$ — TsOH, PhMe, Δ; 29% — 4-Indazolone [2565]

(**95.2**) — DMF, R^3CHO, Δ; 32–82% — Pyrazolo[3,4-*d*]pyrimidine-4,6-dione [2912]

R^1 = Ph, 4-Cl-, 4-MeO-, 4-Me_2N-C_6H_4, MeCH=CH, PhCH=CH; R^2 = H;
R^3 = Ph, 4-Cl-, 4-MeO-, 4-Me_2N-, 4-Cl-C_6H_4

3. 1,2,3-Triazole

Nitrosation of the uracil (**95.2**) with a mixture of *N*-nitrosodimethylamine and phosphorus oxychloride in benzene leads to the formation of a fused 1,2,3-triazole ring, but the reaction takes a different course when R^2 = H (see Section II.2) [2793]. Oxidation of a hydrazone of a 2-acylpyridine at ambient temperature with activated manganese dioxide gives a fused triazole [2107].

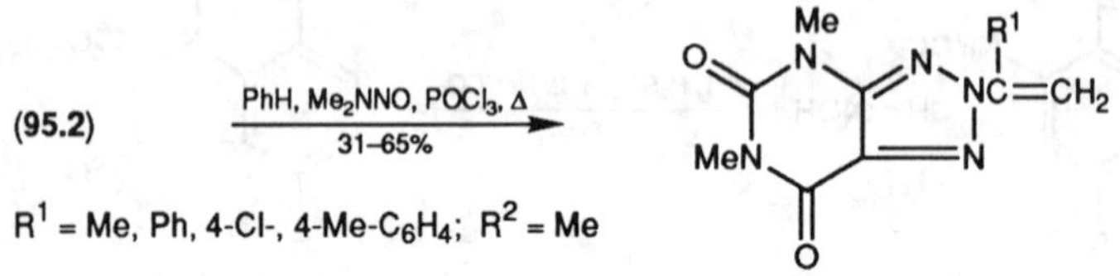

(**95.2**) —PhH, Me_2NNO, $POCl_3$, Δ; 31–65%→ [2793]

R^1 = Me, Ph, 4-Cl-, 4-Me-C_6H_4; R^2 = Me

1,2,3-Triazolo[4,5-*d*]pyrimidine-4,6-dione

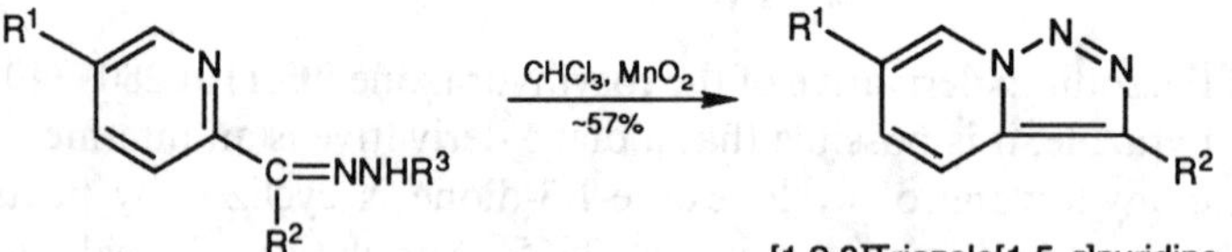

—$CHCl_3$, MnO_2; ~57%→ [2107]

R^1 = H, Me; R^2 = PhCO; R^3 = H, Ts

[1,2,3]Triazolo[1,5-*a*]pyridine

Another ring system synthesized similarly:

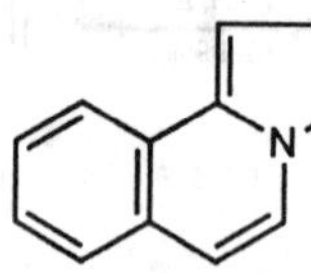

[2703]

[1,2,3]Triazolo[5,1-*a*]isoquinoline

4. 1,2,4-Triazole

2-Benzylidenehydrazino-1,3,5-triazines are cyclized by LTA to either of the adjacent nitrogen atoms. An unsymmetrically substituted triazine then gives two different products, often in unequal amounts. However, 1-methyl-1,3,5-triazin-2-ones yield one product [2145]. Hydrazones of azines may also be cyclized by heating with potassium nitrate in acetic acid—conditions which encourage retention of the nitro group in the substrate [2732]. The synthesis of fused 1,2,4-triazoles has been reviewed [2889]. Orthoesters annulate a hydrazono group of (**95.3**) to a neighbouring ring-nitrogen, but when the *N*-trifluoroacetyl hydrazone (**95.3**, R^1 = CF_3CO) is heated in DMSO, a Dimroth rearrangement occurs to produce an isomer of the first product [3673].

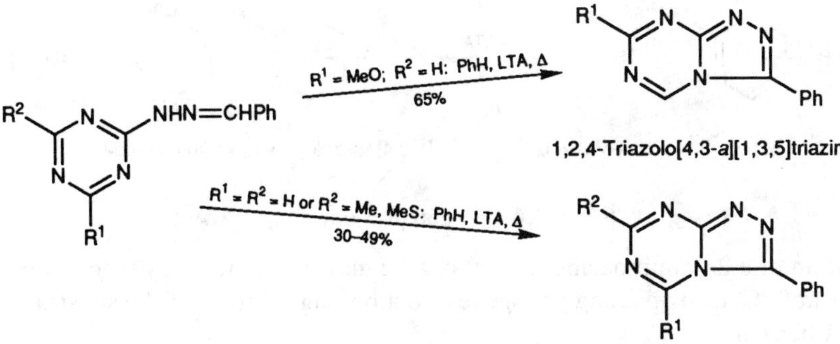

R¹ = MeO; R² = H: PhH, LTA, Δ 65%

1,2,4-Triazolo[4,3-*a*][1,3,5]triazine

[2145]

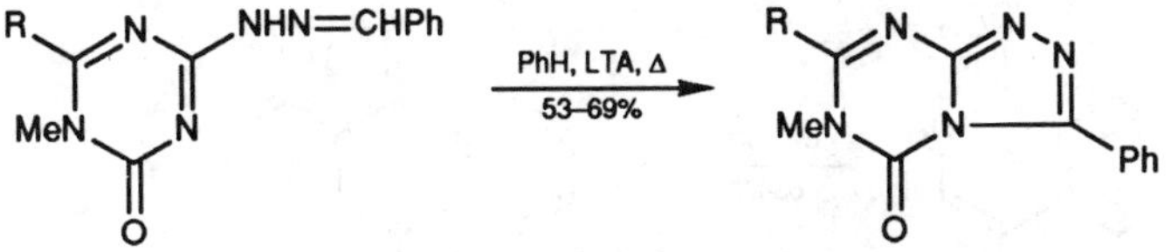

R¹ = R² = H or R² = Me, MeS: PhH, LTA, Δ 30–49%

PhH, LTA, Δ 53–69%

[2145]

R = H, Me

1,2,4-Triazolo[4,3-*a*][1,3,5]triazin-5-one

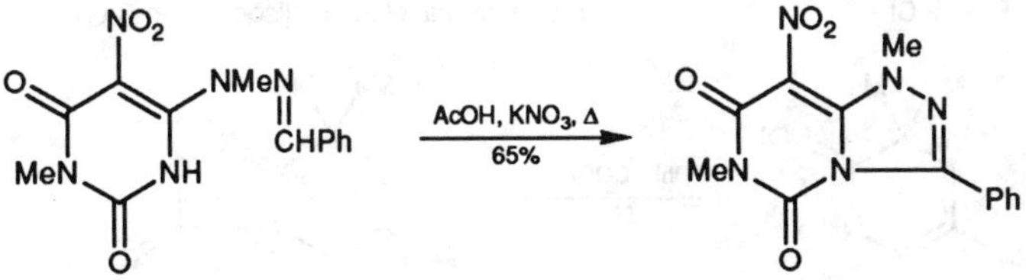

AcOH, KNO_3, Δ 65%

[2732]

1,2,4-Triazolo[4,3-*c*]pyrimidine-5,7-dione

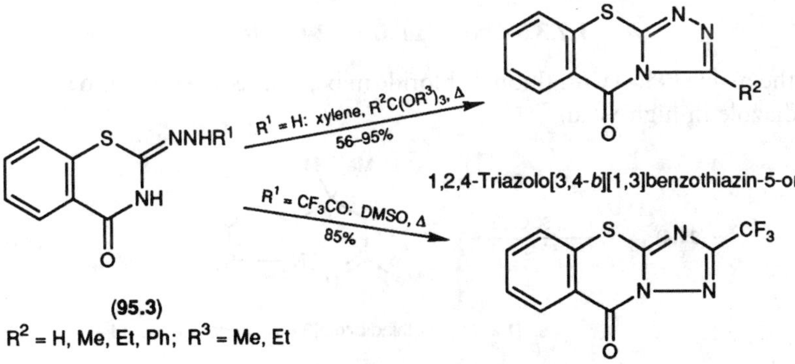

R¹ = H: xylene, $R^2C(OR^3)_3$, Δ 56–95%

1,2,4-Triazolo[3,4-*b*][1,3]benzothiazin-5-one

[3673]

R¹ = CF_3CO: DMSO, Δ 85%

(95.3)

R^2 = H, Me, Et, Ph; R^3 = Me, Et

1,2,4-Triazolo[5,1-*b*][1,3]benzothiazin-5-one

5. 1,2,4-Triazole 1-Oxide

An amidoxime forms a ring with an adjacent ring-nitrogen when the compound is oxidized with LTA at or below room temperature.

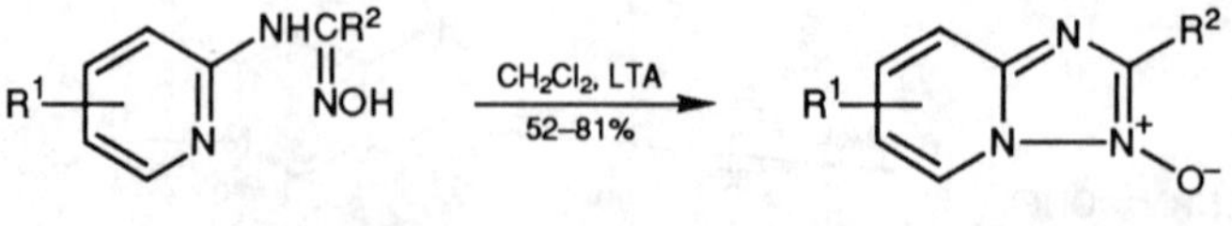

CH_2Cl_2, LTA 52–81% [2853]

R^1 = H, Me; R^2 = Ph, 4-MeC_6H_4

[1,2,4]Triazolo[1,5-*a*]pyridine 3-oxide

6. 1,2,4-Oxadiazol-5-one

Cyclization of a 2-oximinoazine with phosgene may occur either by reaction in pyridine at 0 °C, or by passing phosgene into a boiling solution of the substrate (**95.4**) in benzene.

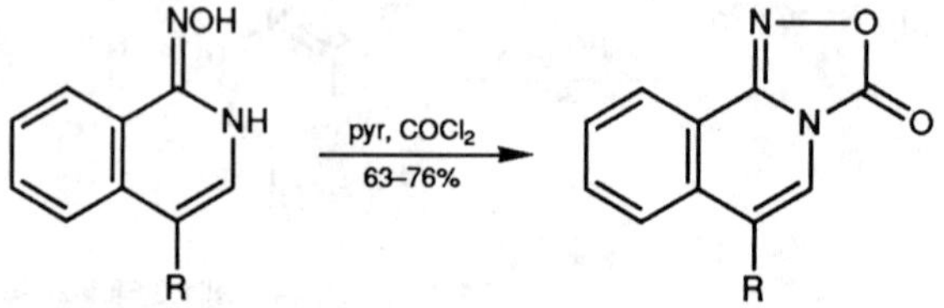

pyr, $COCl_2$ 63–76% [2704]

R = H, Cl

[1,2,4]Oxadiazolo[3,4-*a*]isoquinolin-3-one

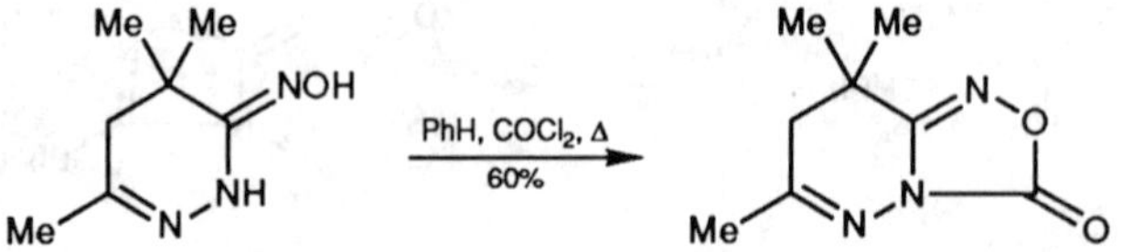

PhH, $COCl_2$, Δ 60% [2841]

(**95.4**)

[1,2,4]Oxadiazolo[4,3-*b*]pyridazin-3-one

7. 1,2,3,5-Oxathiadiazole 2-Oxide

Boiling the oxime (**95.4**) with thionyl chloride in benzene converts it into a fused oxathiadiazole in high yield.

(**95.4**) PhH, $SOCl_2$, Δ 83% [2841]

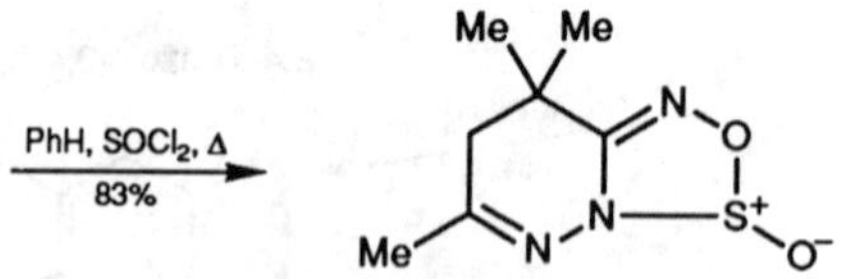

[1,2,3,5]Oxathiadiazolo[3,4-*b*]pyridazine 3-oxide

II. FORMATION OF A SIX-MEMBERED RING

1. Pyridine

The hydrazine part of a phenylhydrazone is sometimes displaced during cyclization, for example, with cinnamaldehyde or arylidenemalononitriles.

$Ar^1NHN{=}C(COOEt)$–(1,3,4-oxa/thiadiazole, 5-Ph) + $CHAr^2{=}CR^1R^2$ $\xrightarrow{\text{pyr, }\Delta,\ \sim58\%}$ [3818]

$R^1 = H$; $R^2 = CHO$ then $R^3 = R^4 = H$ or
$R^1 = R^2 = CN$ then $R^3 = CN$; $R^4 = NH_2$;
$Ar^1, Ar^2 = Ph, 4\text{-}ClC_6H_4$; $X = O, S$

[1,3,4]Oxadiazolo[4,5-*a*]pyridine
[1,3,4]Thiadiazolo[4,5-*a*]pyridine

2. 1,2,4-Triazine or 1,2,4-Triazine 4-Oxide

The product of the cyclization of hydrazones under nitrosating conditions (as described in Section I.3) is a triazine (rather than a triazole) when the hydrazone is derived from a benzaldehyde instead of an acetophenone [2793]. A mixture of nitrous and acetic acids also gives a triazine [2247, 2646]. Further variations in the product are possible by varying the nitrosating reagent; when diethyl azodicarboxylate–acetic acid–nitrous acid is used, triazine-4-oxides are formed [2247, 2849]. Recent work on polyazines has been reviewed [3985].

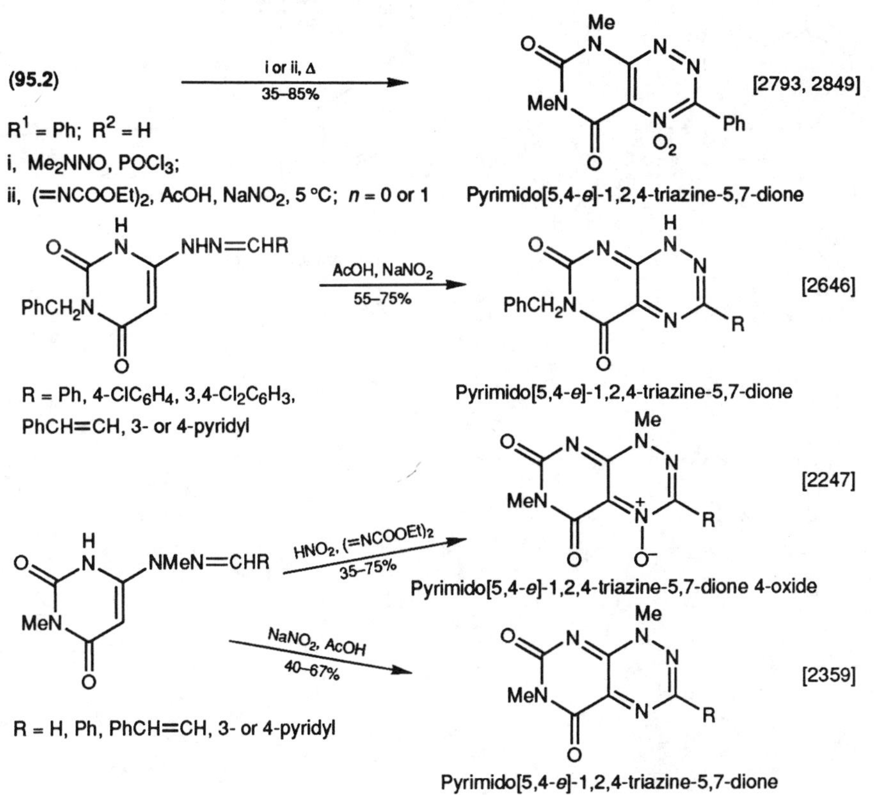

3. 1,2,4-Oxadiazine

Oxidative cyclization of an amidoxime at room temperature with LTA gives a 3-substituted oxadiazine in good-to-high yield.

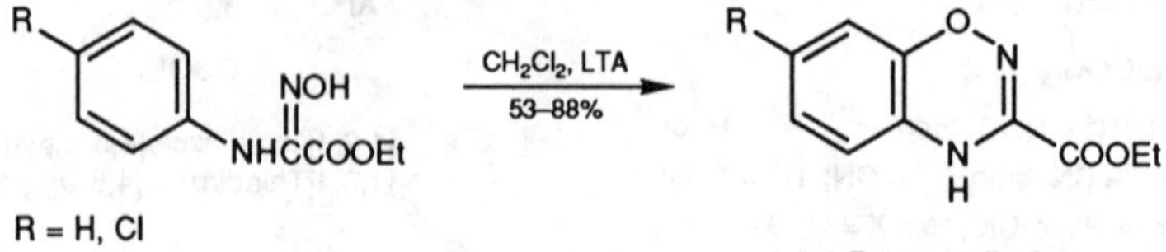

R = H, Cl

CH_2Cl_2, LTA
53–88%

1,2,4-Benzoxadiazine [2852]

III. FORMATION OF A SEVEN-MEMBERED RING

1. 1,2-Diazepine

1,3-Cycloaddition of a 2-chloroacrylonitrile to the hydrazone (**95.5**) gives a high yield of a diazepinoquinoline.

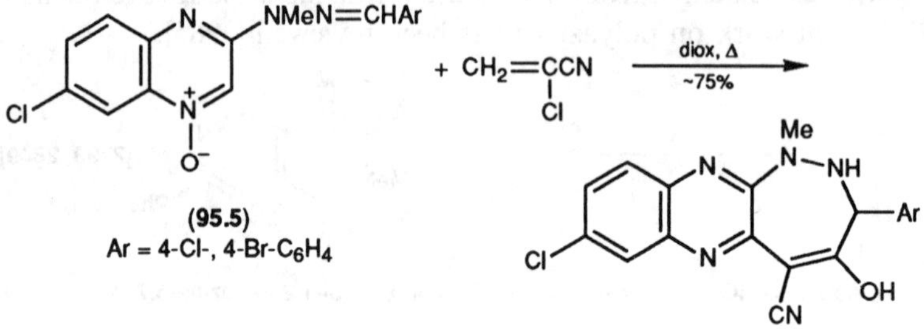

(**95.5**)
Ar = 4-Cl-, 4-Br-C_6H_4

+ CH_2=C(Cl)CN

diox, Δ
~75%

1,2-Diazepino[3,4-*b*]quinoxaline [3683, 3701]

CHAPTER 96

Hydroxy or Ether and Hydrazone or Oxime

I.	Formation of a Five-membered Ring	1267
	1. Pyrazole	1267
	2. Isoxazole	1268
	3. Isoxazole 2-Oxide	1269
	4. Oxazole	1269
	5. Thiazole	1270
II.	Formation of a Six-membered Ring	1270
	1. Pyridazine or 1,2-Oxazine	1270
	2. 1,3-Oxazin-4-one	1271

Cyclizations of ureidophenol and of a hydroxy-hydroxamic acid are described in this chapter.

I. FORMATION OF A FIVE-MEMBERED RING

1. Pyrazole

2-Hydrazonophenols react to form this fused ring on heating the compound in xylene containing a catalytic amount of 4-toluenesulphonic acid, but a positional isomer of this product is obtained when the hydrazone is heated with the same arylhydrazine that was used to prepare the substrate. In an attempt to prepare an acridine derivative, a tetrahydroxybenzophenone was heated with a hydrazine (either under reflux or in a sealed tube), but an indazole was isolated in good yield [3469].

xylene, TsOH, Δ 28–58%

$ArNHNH_2$, HCl, AcOH, Δ 30–65%

R = Br, NO_2;
Ar = Ph, 3-Br-, 3-Me-C_6H_4, 3,5-$Me_2C_6H_3$

[1]Benzopyrano[4,3-*c*]pyrazol-4-one [3950]

Δ 52–75%

Ar = 2,4-$(HO)_2C_6H_3$; R = H, Me, Ph

Indazole [3469]

Another ring system synthesized similarly:

[3211, 3793]

(96.2)
X = NR Cyclohepta[*c*]pyrazol-8-one
X = O Cyclohept[*d*]isoxazol-8-one

2. Isoxazole

Hydroxymethylene and keto groups are annulated by heating the compound with hydroxylamine (without the necessity of isolating the intermediate oxime) [2883]. Another fused isoxazole (**96.3**) is synthesized by heating a 2-hydroxyketoxime with thionyl chloride, or in even better yield, with trichloroacetyl isocyanate. However, chlorosulphonyl isocyanate gave poor yields [3355]. An isoxazole ring is formed by heating a 2-hydroxy ketone with hydroxylamine in pyridine [2055], and, in contrast, another fused isoxazole (**96.2**, X = O) is synthesized by heating a hydroxy ketone with hydroxylamine in methanolic hydrochloric acid [3211]. 3-Hydroxypyridine-2-carboxaldehyde oxime gives unstable isoxazoles but the parent hydroxy ketone reacts with hydroxylamine-*O*-sulphonic acid to give approximately equal proportions (about 45% of each) of the isoxazolopyridine (**96.3**) and the isomeric oxazolo[4,5-*b*]pyridine; the latter is probably formed as a result of a competing Beckmann rearrangement [3454a].

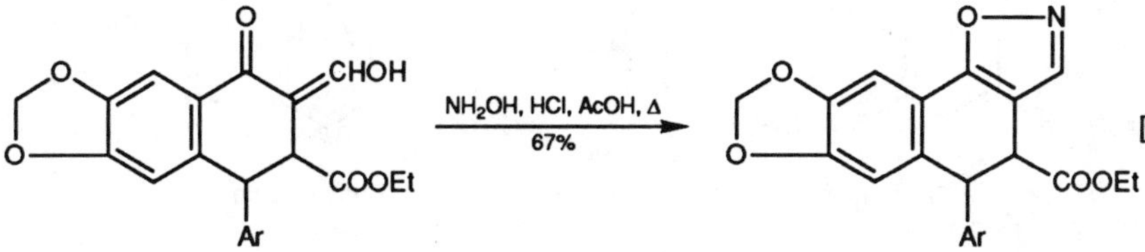

NH_2OH, HCl, AcOH, Δ
67%

[2883]

Ar = 3,4-methylenedioxyphenyl

[1,3]Dioxolo[6,7]naphth[2,1-*d*]isoxazole

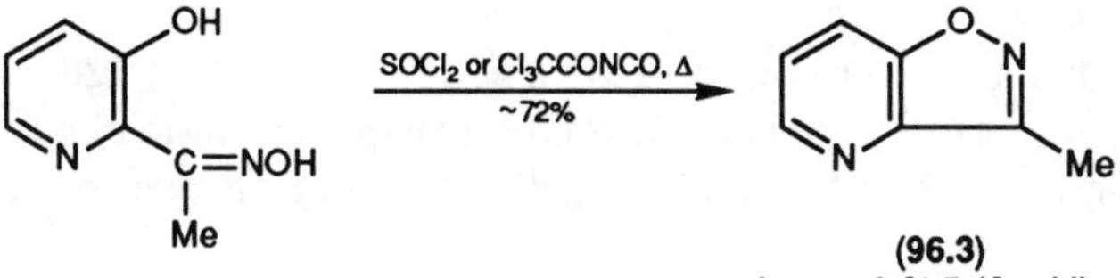

$SOCl_2$ or $Cl_3CCONCO$, Δ
~72%

[3355]

(96.3)
Isoxazolo[4,5-*b*]pyridine

+ NH_2OH

EtOH–pyr, Δ
33%

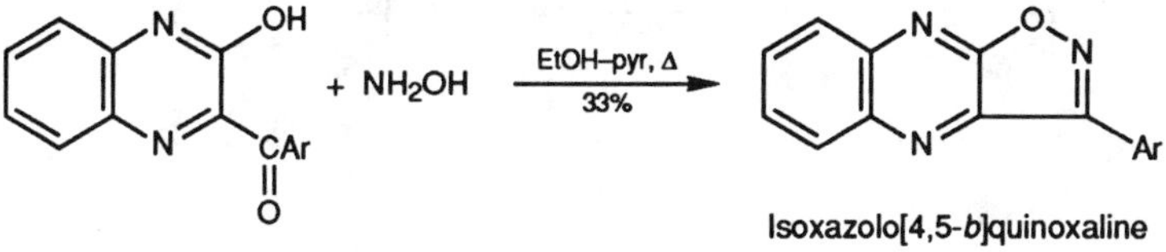

[2055]

Isoxazolo[4,5-*b*]quinoxaline

3. Isoxazole 2-Oxide

This ring is produced by subjecting a 2-hydroxy-oxime to either sodium hypochlorite or LTA [3355].

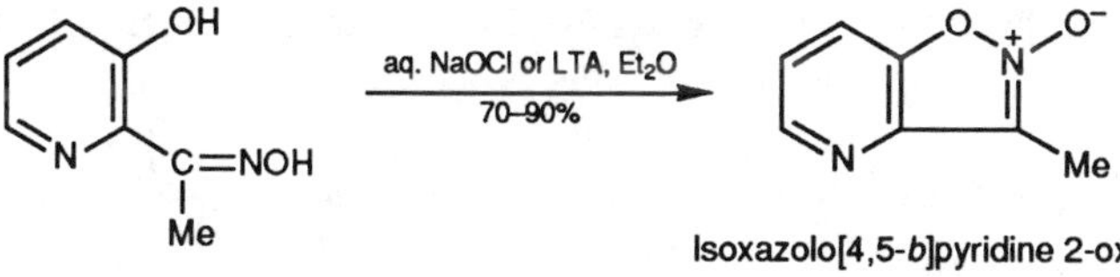

aq. NaOCl or LTA, Et_2O
70–90%

[3355]

Isoxazolo[4,5-*b*]pyridine 2-oxide

Other ring systems synthesized similarly:

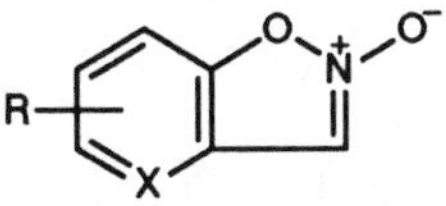

[3174, 3355]

X = CH 1,2-Benzisoxazole 2-oxide
X = $\overset{+}{N}$–O^- Isoxazolo[4,5-*b*]pyridine 2,4-dioxide

4. Oxazole

A high yield of benzoxazole is usually obtained when a 2-thioureidophenol is warmed with ammoniacal silver nitrate.

$$\text{5-}H_2NO_2S\text{-2-}(RNHC(=S)NH)C_6H_3OH \xrightarrow[75\text{–}91\%]{NH_3\text{–}AgNO_3} \text{5-}H_2NO_2S\text{-2-}(NHR)\text{benzoxazole}$$ [2804]

Benzoxazole

R = alkyl, $PhCH_2$, Ph

5. Thiazole

Heating the oxime of 2-mercaptoacetophenones with PPA at 120–130 °C gives mainly a benzothiazole, but as a Beckmann rearrangement may occur to some extent during the reaction, some benzisothiazole may also be formed.

$$R\text{-}C_6H_3(SH)(C(Me){=}NOH) \xrightarrow[30\text{–}70\%]{PPA,\ \Delta} R\text{-2-methylbenzothiazole}$$ [2497]

Benzothiazole

R = Cl, NO_2

II. FORMATION OF A SIX-MEMBERED RING

1. Pyridazine or 1,2-Oxazine

Peri-positioned hydrazono (formed *in situ*) and hydroxy groups on naphthalene react on heating the compound in ethane-1,2-diol; similarly, 8-acyl-1-naphthol oxime yields a 1,2-oxazine. Both reactions proceed in high yield.

$$\text{8-COR-1-naphthol} + N_2H_4 \cdot H_2O \xrightarrow[\sim 89\%]{(CH_2OH)_2,\ \Delta} \text{3-R-1H-benzo[de]cinnoline}$$ [2312]

R = Me, Ph

Benzo[*de*]cinnoline

Another ring system synthesized similarly:

[2312]

Naphth[1,8-*de*]-1,2-oxazine

2. 1,3-Oxazin-4-one

Acid-catalysed cyclization of a salicylaldehyde oxime with acetic anhydride gives high yields of an oxazinone as its salt.

OH
CH=NOH
+ Ac_2O $\xrightarrow[95\%]{HClO_4}$
O Me
N
O
[2611]

1,3-Benzoxazin-4-one

CHAPTER 97

Hydroxy and Methylene

I. Formation of a Five-membered Ring 1272
1. Furan or Furan-3-one 1272
2. Thiophene 1273
3. Thiazole-5-thione 1273
II. Formation of a Six-membered Ring 1273
1. Pyran 1273
2. Pyran-2-one or Pyran-2-thione 1274
3. Pyran-4-one 1274

I. FORMATION OF A FIVE-MEMBERED RING

1. Furan or Furan-3-one

An activated methylene group interacts with an adjacent acetyloxy function when the compound is heated with acetic acid–acetic anhydride. The trimethylsilyl enol ethers (**97.2**) of 2-hydroxyacetophenones are useful precursors for the synthesis of heterocycles. They react with nitrosobenzene without external heat under oxidative conditions to form benzofuran-3-ones in one step without isolating intermediates, some of which are shown to be present by spectroscopic evidence.

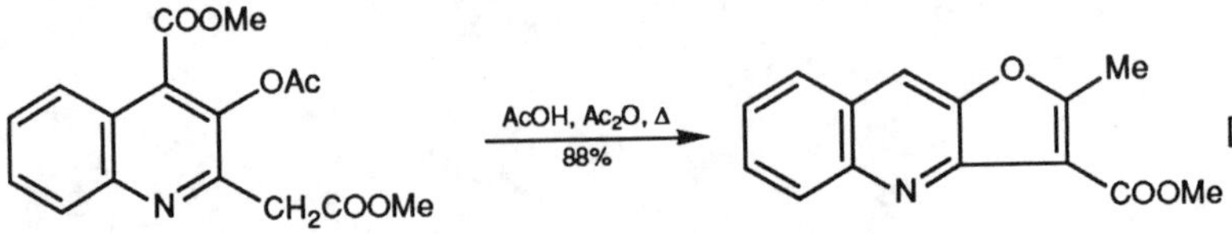

COOMe, OAc, N, CH_2COOMe —[AcOH, Ac_2O, Δ; 88%]→ O, Me, N, COOMe [2335]

Furo[3,2-*b*]quinoline

(97.2)
R = H, Cl, Me, 1-adamantyl
i, $CHCl_3$; ii, HCl, THF; iii, O_2, TEA

+ PhNO → (i–iii, ~35%) 3-Benzofuranone [3723]

2. Thiophene

Catalytic dehydrogenation of a (2-ethyl)thiophenol at about 450 °C in the presence of a copper–chromium catalyst gives the benzo[*b*]thiophene in good yields.

R = H, NH_2 → (Cu–Cr–C, Δ, ~60%) Benzo[*b*]thiophene [3049]

3. Thiazole-5-thione

The penem-3-thione (**97.3**) may be synthesized from the silver thiolate by the use of thiocarbonyldi-imidazole (TDI).

R = $CH_2{=}CHCH_2$, $Me_3Si(CH_2)_2$ → (CH_2Cl_2, TDI, ~80%) (97.3) [3058]

1-Aza-4-thiabicyclo[3.2.0]heptan-7-one, 3-thioxo-

II. FORMATION OF A SIX-MEMBERED RING

1. Pyran

A substituted *o*-hydroxyacetophenone (**97.4**) may be converted in high yield into a 2*H*-chromene by reaction at −78 °C with phenyl vinyl sulphoxide. Recent developments in the chemistry of fused pyrans (and thiopyrans) have been reviewed [3986].

(97.4) + $PhS(O)CH{=}CH_2$ → (i, ii, ~90%) 2*H*-1-Benzopyran [4001]

R = 5-Me, 5-Et
i, THF, NaH; ii, EtOH–EtONa, Δ

2. Pyran-2-one or Pyran-2-thione

Annulation of methylene and phenolic groups usually needs strongly basic conditions; sodium dissolved in 2-ethylbutan-2-ol was chosen for a low-temperature synthesis of a chromene-2-thione but sodium hydride has also been successful in a similar synthesis of the chromenone.

(97.4) + CS_2 —(Et_2CMeOH, Na; 56%)→ [2306]

R = 5-Me

1-Benzopyran-2-thione

HCOOEt, NaH; 71% [2661, 2894]

1-Benzopyran-2-one

3. Pyran-4-one

The Kostanecki reaction is a long-established method of synthesizing chromones [B-12, p. 515], but one of its disadvantages is that occasionally several other benzopyran-4-ones and/or benzopyran-2-ones may be formed as minor or major byproducts [B-12, p. 516]. A recent report [3660] describes the formation of 3-benzoyl-7-hydroxyflavone as the main product of reacting resacetophenone, sodium benzoate and benzoic anhydride. A better synthesis of the 7-hydroxyflavone is described in Chapter 29, Section II.3 [3660]. Cyclization of resacetophenone (**97.4**, R = OH) by reaction with a benzaldehyde, ethyl orthoformate and perchloric acid gives high yields of 4-ethoxybenzopyrylium salts which are easily hydrolysed to the benzopyranones with water. When R = Cl, good yields are still obtained provided electron-releasing substituent(s) are attached to the Ar ring.

(97.4) + ArCHO + $HC(OEt)_3$ —(1. $HClO_4$ 2. H_2O, Δ; 12–98%)→ [2890]

R = HO, Cl;
Ar = Ph, HO-, MeO-C_6H_4, HO-Me-C_6H_3, $OCH_2OC_6H_3$

1-Benzopyran-4-one, 2-aryl-

An *o*-hydroxyphenylpropan-1-one reacts with ethyl acetate in the presence of a strong anhydrous base to give a 1,3-diketone which cyclizes in hot mineral acid [B-12, p. 496] (see also p. 534 of Part 1). Sodium alkoxide is frequently used but sodium may give better yields with some less reactive esters [3436]. One of the simplest methods of synthesizing chromones is treatment of 2-hydroxyacetophenones with ethyl orthoformate and perchloric acid. The main limitation imposed by this method is that the reacting hydroxy ketone should not have an electron-withdrawing substituent [2457]. Heating an activated methylene-containing phenol with acetic anhydride–TEA, acetic anhydride–orthoformate or acetic formic anhydride–sodium formate also seems to give good results in the absence of a deactivating substituent on the benzene ring [2460, 2661, 2894]. The chemistry of acetic formic anhydride has been reviewed [3823]. A method which proceeds in the absence of bases uses boron trifluoride etherate and methanesulphonyl chloride [2779].

1. Na, MeCOOEt 2. HCl, Δ; 75% [3436]

1-Benzopyran-4-one

(**97.4**) + $HC(OEt)_3$ — 1. $HClO_4$ 2. H_2O, Δ; 30–85% [2457]

R = H, MeO, Me

R^1 = 2-furyl: Ac_2O, TEA, Δ; ~96% [2460]

R^1 = COOEt; R^2 = H: AcOCHO, HCOONa; 76% [2661]

R^1 = 2-furyl; R^2 = H, OH; R^3 = H, AcO

+ DMF — $BF_3.Et_2O$, $MeSO_2Cl$, Δ; 82% [2779]

1-Benzopyran-4-one, 3-aryl-

CHAPTER 98

Hydroxy and Nitro or Ring-nitrogen

I. Formation of a Five-membered Ring 1276
 1. Pyrrole 1276
 2. Imidazole 1277
 3. Furan 1277
 4. Isoxazole 1277
 5. Oxazole or Oxazol-2-one 1277
 6. 1,3,4-Dithiazole 1278
II. Formation of a Six-membered Ring 1278
 1. 1,2,4-Triazine 1278
 2. 1,4-Oxazin-2-one 1279
III. Formation of a Seven-membered Ring 1279
 1. 1,4-Thiazepin-3-one 1279

Examples of the conversion of hydroxy and nitroso groups into a heterocyclic ring are given in this chapter.

I. FORMATION OF A FIVE-MEMBERED RING

1. Pyrrole

Acid-catalysed elimination of water from a side-chain hydroxy and ring-NH functions leads to the formation of a C—N bond of a fused pyrrole.

N NH CHOH Ph —xylene, TsOH, Δ / 54%→ N N Ph Imidazo[2,1-*a*]isoindole [2200]

2. Imidazole

Cyclodehydration of a β-hydroxyethylamine group to a neighbouring ring-nitrogen is induced by PPA (p. 537), but phosphorus oxychloride may be a more efficient reagent and was recently used in the synthesis of the first member of a tetracyclic ring system.

$POCl_3$, Δ, ~89% [3966]

R^1,R^2 = H, Me

[1]Benzoxepino[4,5-*e*]imidazo[1,2-*c*]pyrimidine

3. Furan

Displacement of a nitro group by a hydroxy (as its anion) is a well known reaction (reviews [1398, 1399, 1432]. In this example, a 2-hydroxy-2-nitrodiphenyl is efficently cyclized under strongly basic conditions at room temperature.

HMPT, NaH, 82–98% [3040]

R = H, NO_2

Dibenzofuran

4. Isoxazole

Treatment of a 3-benzoylpyridin-4-ol with hydroxylamine leads directly to the isoxazole in good yields.

pyr, NH_2OH, Δ, 44–87% [3472]

Ar = Cl-, F-, MeO-, Me-C_6H_4

Isoxazolo[5,4-*d*]pyrazolo[3,4-*b*]pyridine

5. Oxazole or Oxazol-2-one

A very efficient method of converting a 1-nitroso-2-naphthol (**98.2**) into a naphthoxazole is to stir the precursor with an *N*-phenacylpyridinium salt at

$-10\,°C$. A bromine atom in the phenyl ring reduces the yield considerably. Phosgene in the presence of TEA annulates a hydroxymethyl group to a ring-nitrogen at low temperature.

NO, OH, (98.2) + pyridinium N^+–CH_2COPh —EtOH, NaOH, ~100%→ Naphth[1,2-*d*]oxazole (COPh) [2136]

CHArOH phthalazine —Et_2O, $COCl_2$, TEA, ~43%→ Oxazolo[4,3-*a*]phthalazin-3-one [3554]

Ar = Ph, 4-$MeOC_6H_4$

6. 1,3,4-Dithiazole

A dithiazole ring fused to an azetidine is formed by treating the thioacyl derivative with *N*-chlorosuccinimide and a tertiary amine. The *cis* : *trans* isomer ratio varies with temperature.

R^1OHCMe, NH, $SC(S)CH_2COOtBu$ —i, ~90%→ R^1OHCMe, COOtBu [4004]

R^1 = blocking group

2,4-Dithia-1-azabicyclo[3.2.0]heptan-7-one

II. FORMATION OF A SIX-MEMBERED RING

1. 1,2,4-Triazine

Acid-catalysed annulation of a phenolic hydroxy to a nitrogen which is a member of another ring can give high yields of a triazine when the two ring systems are joined by an azo group.

X—Y, HN—Z, N=N, OH —H^+, Δ, 62–95%→ [2655]

Y = Z = CH, X = N Imidazo[2,1-*c*]naphtho[2,1-*e*][1,2,4]triazine
X = Y = CH, Z = N Naphtho[2,1-*e*]pyrazolo[5,1-*c*][1,2,4]triazine
X = Y = Z = N Naphtho[2,1-*e*]tetrazolo[5,1-*c*][1,2,4]triazine

Another ring system synthesized similarly:

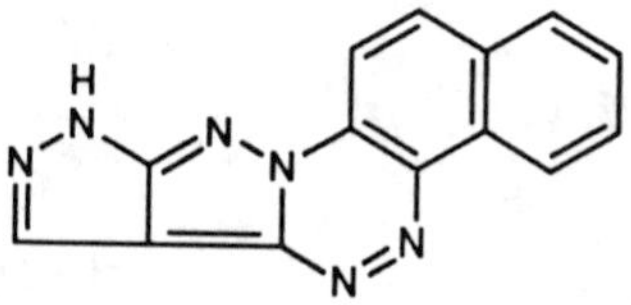

[3780]

Naphtho[2,1-*e*]pyrazolo[3',4':3,4]pyrazolo[5,1-*c*][1,2,4]triazine

2. 1,4-Oxazin-2-one

The two carbon atoms required to form this ring from an *o*-nitrosophenol may be supplied by a phosphorane (cf. Chapter 35, Section I.1).

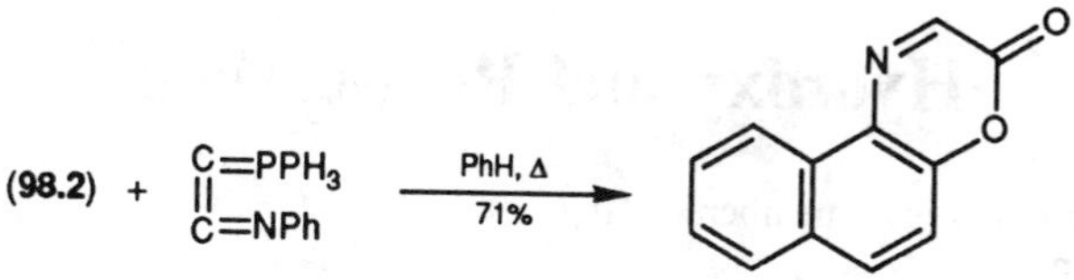

(98.2) + C=PPH$_3$ / C=NPh —PhH, Δ, 71%→ [2777]

Naphth[2,1-*b*][1,4]oxazin-3-one

III. FORMATION OF A SEVEN-MEMBERED RING

1. 1,4-Thiazepin-3-one

Thiol and ring-nitrogen functions in different rings react with methyl 2-chloroalkanoate in a basic environment to form a doubly fused thiazepinone ring.

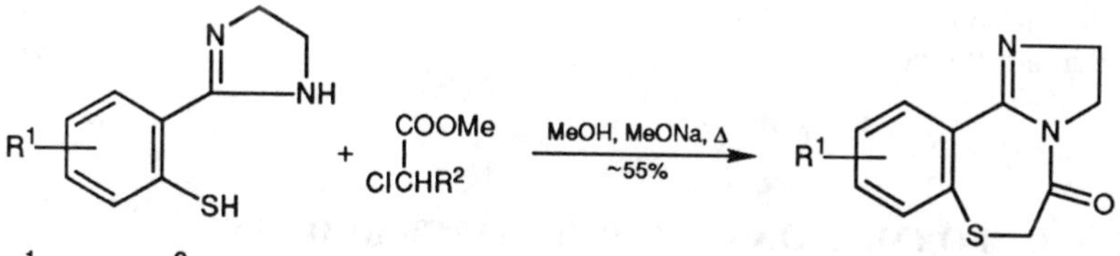

+ COOMe / ClCHR^2 —MeOH, MeONa, Δ, ~55%→ [2709]

R^1 = H, Cl; R^2 = H, Me

Imidazo[1,2-*d*][1,4]benzothiazepin-5-one

CHAPTER 99

Hydroxy and Ring-carbon

I. Formation of a Five-membered Ring 1280
  1. Furan 1280
  2. Furan-2-one 1281
  3. Thiophene 1282
  4. Thiophene-2,3-dione 1282
  5. 1,3-Dithiole 1283
  6. Oxazole 1283
  7. 1,2,5-Thiadiazole 1283
II. Formation of a Six-membered Ring 1284
  1. Pyridine 1284
  2. Pyridin-4-one 1285
  3. Pyran 1286
  4. Pyran-2-one 1287
  5. Pyran-4-one 1290
  6. Pyran-4-thione 1292

I. FORMATION OF A FIVE-MEMBERED RING

1. Furan

α-Haloalkynes cyclize hydroxyquinolin-2-ones in a weakly basic medium to give variously 2,3-disubstituted furoquinolinones [3943]. Cyclodehydrogenation of a suitably structured hydroxy compound is accomplished by heating it in a high-boiling solvent [2390, 2669]. An α-hydroxy ketone such as benzoin annulates a hydroxy group and its neighbouring ring-carbon in a reaction which is acid-catalysed [3091].

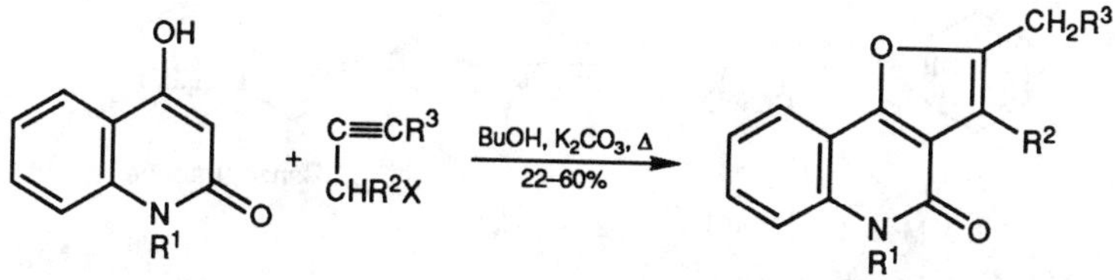

OH

$C{\equiv}CR^3$ | CHR^2X ; BuOH, K_2CO_3, Δ ; 22–60% ; [3943]

R^1 = Me, Ph; R^2 = H, Me;
R^3 = $PhOCH_2$, 4-Br-, 4-Cl-, 2-Cl-C_6H_4; X = Br, Cl

Furo[3,2-*c*]quinolin-4-one

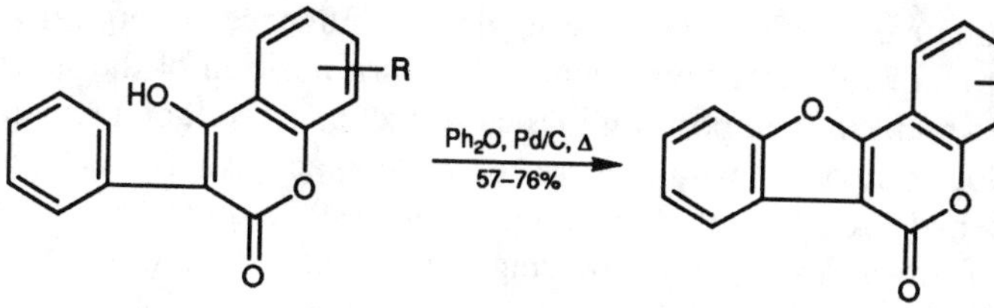

Ph_2O, Pd/C, Δ ; 57–76% ; [2390, 2669]

R = H, Me, Me

Benzofuro[3,2-*c*][1]benzopyran-6-one

Another ring system synthesized similarly:

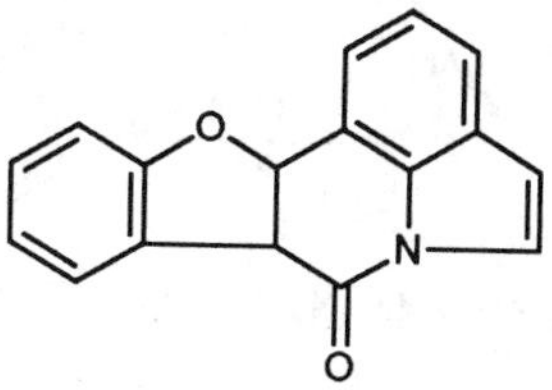

[2390]

Benzofuro[3,2-*c*]pyrrolo[3,2,1-*ij*]quinolin-7-one

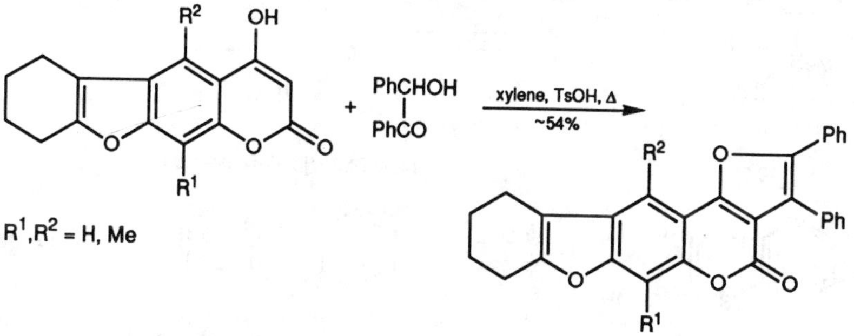

PhCHOH | PhCO ; xylene, TsOH, Δ ; ~54% ; [3091]

R^1,R^2 = H, Me

Benzofuro[3,2-*g*]furo[3,2-*c*][1]benzopyran-4-one

2. Furan-2-one

Although this ring may be formed from *o*-hydroxyphenylalkanoic acids, the latter are usually less readily available than the corresponding phenol (**99.1**), which cyclizes on treatment with ethyl pyruvate in the presence of a catalytic amount of titanium tetrachloride and a reducing agent.

(99.1) + COMe / COOEt —i, 50–80%→ 2-Benzofuranone [3076]

R^1 = H, Me, MeO, MeS, $PhCH_2$, Ph, 2,4-F_2,C_6H_3
i, CH_2Cl_2, $TiCl_4$, AcOH, Zn

3. Thiophene

The arylsulphonate of an alcohol cyclizes under the influence of boron trifluoride at ambient temperature; a substituent in the *para* position of the substrate may appear in an unexpected position in the product as a result of a rearrangement which does not occur with a *meta*- or *ortho*-placed substituent [2170]. Chloroacetaldehyde reacts with the uracil-6-thiol (**99.2**) under mildly basic conditions to form a fused thiophene ring (review of recent work on such compounds [3981]) unsubstituted at the 2 and 3 positions [3682]. A fused 2-chlorothiophene ring is formed when the t-alcohol (**99.3**) is heated with thionyl chloride [3914].

—CH_2Cl_2, BF_3, ~90%→ Benzo[*b*]thiophene [2170]

R = Ph, 4-MeC_6H_4; Ar = 2,4,6-$(NO_2)_3C_6H_2$

(99.2) + CH_2Cl / CHO —AcONa, 66%→ Thieno[2,3-*d*]pyrimidine-4,6-dione [3682]

(99.3) —$SOCl_2$, Δ, 35–53%→ Thieno[3,2-*b*][1]benzothiophene [3914]

R = iPr, tBu

4. Thiophene-2,3-dione

Heating a thiophenol with oxalyl dichloride in diethyl ether gives a good yield of a benzo[*b*]thiophene-2,3-dione; its chemistry has been reviewed [3514].

(COCl)$_2$, Et$_2$O, Δ, 68% → Benzo[*b*]thiophene-2,3-dione [3017a]

5. 1,3-Dithiole

1,3-Benzodithioles, which have interesting electronic and antimicrobial properties, can be synthesized by stirring a thioacyl derivative of a thiophenol with sulphuric acid at 0–20 °C. When sulphuric acid is replaced by either hydrochloric acid or PPA, no dithiole was isolated.

S—C(=S)CH(COOEt)$_2$ → H_2SO_4, 36–82% → C(COOEt)$_2$ [3865]

R = H, 4-Me, Br, Cl, 2-6-Cl$_2$, 3-MeO

1,3-Benzodithiole

6. Oxazole

Some hydroxyindoles are susceptible to oxidation, and in the presence of a primary amine and silver(II) oxide, lead(IV) oxide or manganese dioxide, a 6-hydroxyindole is converted into a fused benzoxazole.

+ RCH$_2$NH$_2$ → DME, Ar, MnO$_2$, 25–61% [3422]

R = Bu, CH$_2$=CH, MeSCH$_2$CH$_2$, MeO$_2$C, nC$_7$H$_{15}$, HOCH$_2$, HO(CH$_2$)$_2$, Ph

Pyrrolo[2,3-*e*]benzoxazole

7. 1,2,5-Thiadiazole

Phenols which are unsubstituted at C-2 and/or C-6 are cyclized by heating with the explosive tetrasulphur tetranitride. 2-Naphthol also gives a high yield of the 1,2-fused naphthol but 1-naphthol tends to produce a mixture of 1,2-mono- and 1,2,3,4-di-fused naphthalenes (see also Chapter 88, Section I.5).

(99.1) + N_4S_4 → PhMe, Δ, 12–84% [3077]

R^1 = H, alkyl

2,1,3-Benzothiadiazole

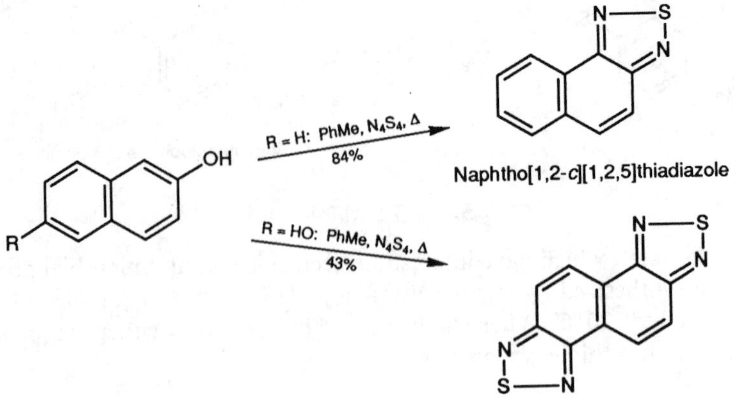

R = H: PhMe, N_4S_4, Δ 84%

Naphtho[1,2-*c*][1,2,5]thiadiazole [3980]

R = HO: PhMe, N_4S_4, Δ 43%

Naphtho[1,2-*c*:5,6-*c*']bis[1,2,5]thiadiazole

Other ring systems synthesized similarly:

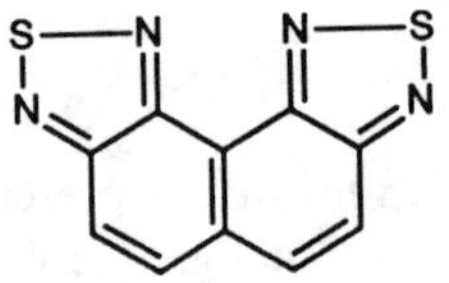

[3980]

Naphtho[1,2-*c*:7,8-*c*']di[1,2,5]thiadiazole

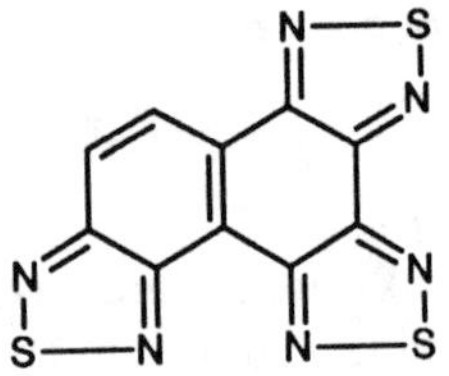

[3980]

Naphtho[1,2-*c*:3,4-*c*':5,6-*c*"]tris[1,2,5]thiadiazole

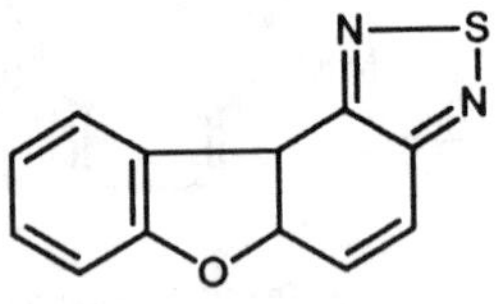

[3980]

Benzofuro[3,2-*e*]-2,1,3-benzothiadiazole

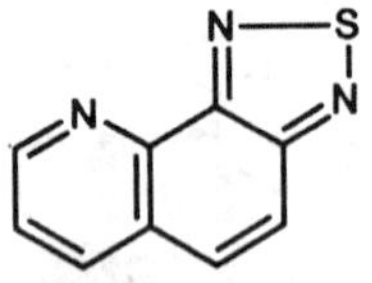

[3980]

[1,2,5]Thiadiazolo[3,4-*h*]quinoline

II. FORMATION OF A SIX-MEMBERED RING

1. Pyridine

An attempt to prepare reduced 5-quinolinones from cyclohexane-1,3-diones (or their tautomers) and 3-amino-2-methylacrolein gave 1,8-acridinediones with the loss of one carbon atom (a mechanism for this reaction has been suggested). The acridinediones were also synthesized (in 40–79% yield) from the cyclohexanediones, propanal and ammonium acetate. Further investigation of the former synthesis showed that quinolinones are also produced in the reaction but as they are oils, they had to be isolated as their salts (see also Chapter 53).

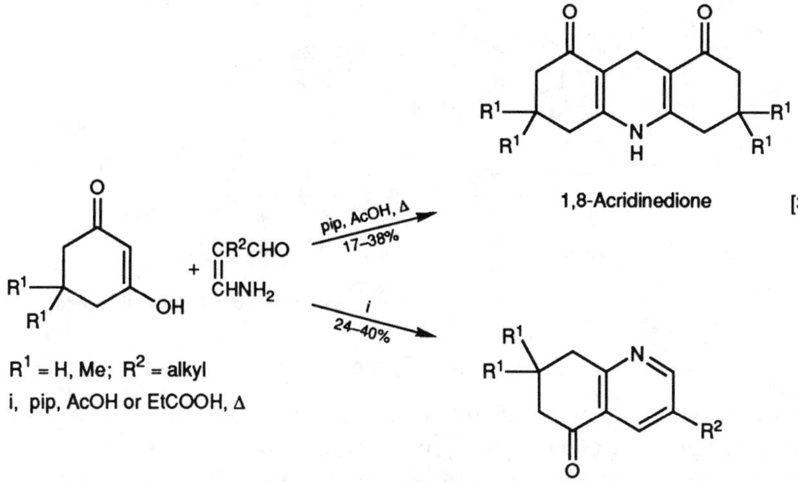

pip, AcOH, Δ 17–38%

i 24–40%

R^1 = H, Me; R^2 = alkyl

i, pip, AcOH or EtCOOH, Δ

1,8-Acridinedione [3012]

5-Quinolinone

AcOH, H_2SO_4, Δ 60–84%

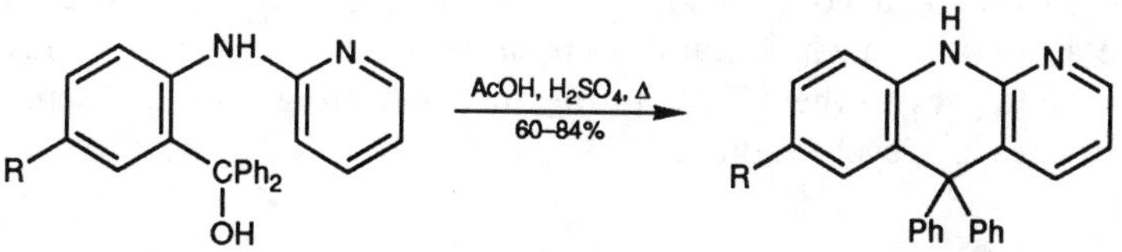

(99.4)
R = H, Me, MeO, Br, Cl

Benzo[*g*]-1,8-naphthyridine [2633]

A side-chain t-alcohol group as in **(99.4)** undergoes cyclodehydration when heated with sulphuric acid. Annulation of a phenylethanol with a nitrile and sulphuric acid can give fairly high yields of an isoquinoline; the usefulness of nitriles in the synthesis of heterocycles has been reviewed [3332a].

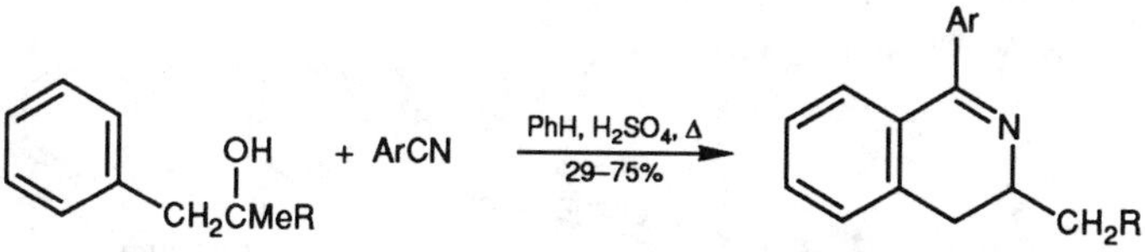

PhH, H_2SO_4, Δ 29–75%

R = alkyl, $PhCH_2$; Ar = Ph, pyridyl

Isoquinoline [2086]

2. Pyridin-4-one

Acylation of a hydroxy group followed by displacement of the acyloxy by an amino group occurs when anthranilic acid is heated under strictly anhydrous conditions with a resorcinol derivative.

BuOH, $ZnCl_2$, Δ
30%

[2480]

1-Benzopyrano[7,6-*b*]quinolin-6-one

3. Pyran

A 2-hydroxyquinone is converted into a benzopyran by heating with 3-methylbut-2-enal in the presence of TEA [3435]. Heterocyclic quinones have been reviewed [3650]. A one-step synthesis of xanthene is based on the reaction between a phenoxymagnesium bromide and triethyl orthoformate [2592]; when the latter is replaced by cinnamaldehyde and the mixture is heated for 15–24 h, it is possible to obtain a good yield of a flavene [2254]. Shorter reaction times produce a mixture which contains a larger proportion of the 2*H*-benzopyran, but this isomerizes to the 4*H*-pyran on further heating. In the absence of the 3-substituent (R^2), cyclization fails.

PhH, TEA, Δ
30%

[3435]

R^1 or R^2 = HO

Naphtho[2,3-*b*]pyran-5,10-dione

R^2 = H: PhH, $HC(OEt)_3$, Δ
15–86%

[2592]

Xanthene

R^2 ≠ H: CHPh=CHCHO, PhH, Δ
25–75%

[2254]

R^1 = H, alkyl; R^2 = Me, iPr

4*H*-1-Benzopyran, 2-aryl-

A hydroxy group may be joined to a carbon atom of an adjacent ring by heating with an anhydride [2573]. On the other hand, a side-chain enol acetate is cyclized by heating with PPA followed by perchloric acid to precipitate the product as its salt [2803]. A photochemical cyclization of a 3-arylallyl alcohol in the presence of 1,4-naphthalenedinitrile (DCN) gives a good yield of a 2*H*-benzopyran but oxygen must be excluded during the reaction [3413].

$(RCO)_2O$ (RCOCMe), Δ; 23–45% [2573]

R = H, Me; X = NH, NAc, NCOOMe, S

[1]Benzopyrano[4,3-*b*][1,5]benzodiazepine
[1]Benzopyrano[3,4-*c*][1,5]benzothiazepine

Other ring systems synthesized similarly:

[2573]

X = S [1]Benzopyrano[3,4-*b*][1,4]benzothiazine
X = NH [1]Benzopyrano[3,4-*b*]quinoxaline

i, ii; 56–75% [2803]

Ar = 3-MeO-, 3-EtO-C_6H_4, 3,4-$(MeO)_2C_6H_3$
i, PPA, Δ; ii, $HClO_4$

2-Benzopyrylium

MeCN–H_2O, DCN, *hν*; 50–65% [3413]

2*H*-1-Benzopyran

R^1 = MeO or 3,4-OCH_2O; R^2,R^3 = H, Me

4. Pyran-2-one

Examples of the conversion of phenols to coumarins by the Pechmann reaction demonstrate the effectiveness of trichloroacetic acid, which gives results comparable with those with sulphuric acid [2103, 2181, 2908]. Hydrogen chloride has also been used but whether the resulting product is a linear or an angular thienobenzopyranone may depend on the presence or absence of substituents, as is demonstrated by the behaviour of the 5-hydroxybenzothiophene [2038]. Phenols with electron-rich rings condense with propynoic esters to form coumarins; some of these reactions do not need external heating [2313, 2858]. Pechmann synthesis of coumarins from 2-acylphloroglucinol and a 3-oxocarboxylic ester using 5% sulphuric acid in acetic acid at ambient temperature gives a separable mixture of 6- and 8-acylcoumarins in the ratio of 3:2. Yields in the benzoylacetate reaction deteriorated as the scale of the reaction increased, especially above 20 mmol. Variations in the condensing agent and in the ratio of acetic and sulphuric acids did not improve yields. Adding the benzoylacetate in three portions over 20 days gave the best results [3261].

CCl_3COOH or H_2SO_4, Δ; 68–84% [2103, 2181, 2908]

R^1 = H, HO, Ac; R^2 = Me, HO; R^3,R^4 = H, HO

1-Benzopyran-2-one

R = H: HCl–MeOH, 100%: Thieno[3,2-*f*][1]benzopyran-7-one [2038]

R = Me: HCl–MeOH, 80%: Thieno[2,3-*g*][1]benzopyran-6-one

$ZnCl_2$, Δ; 90% [2313, 2858]

1-Benzopyran-2-one

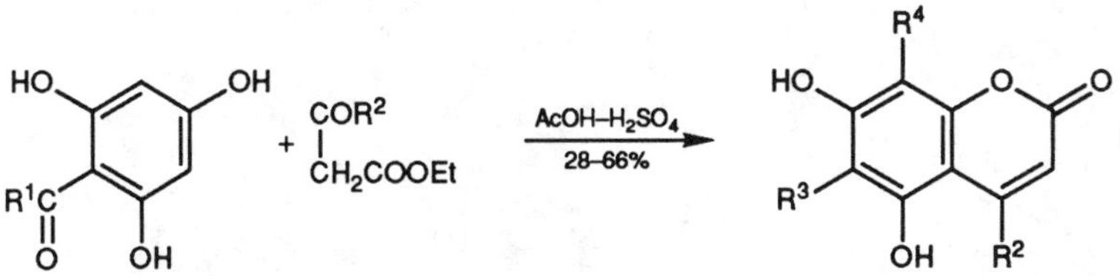

[3261]

R^1 = alkyl; R^2 = alkyl; R^3 = H; R^4 = R^1CO
or R^3 = R^1CO; R^4 = H

1-Benzopyran-2-one

The product of the Pechmann reaction between resorcinol and 2-allylaceto-acetate varies with reaction time but the yields of both products are low [2183], as is that from ethyl 2-acetamidoacetoacetate and zinc chloride [3749]. A mixture of citric acid and sulphuric acid when heated with a phenol gives a benzopyran-2-one such as (**99.5**) or (**99.6**) in moderate yield [2262].

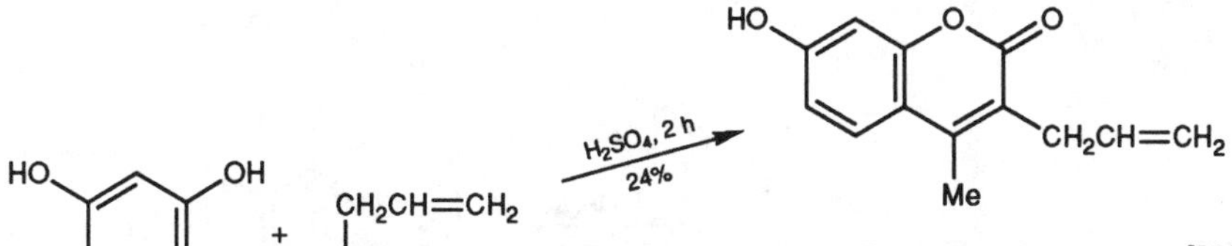

1-Benzopyran-2-one [2183]

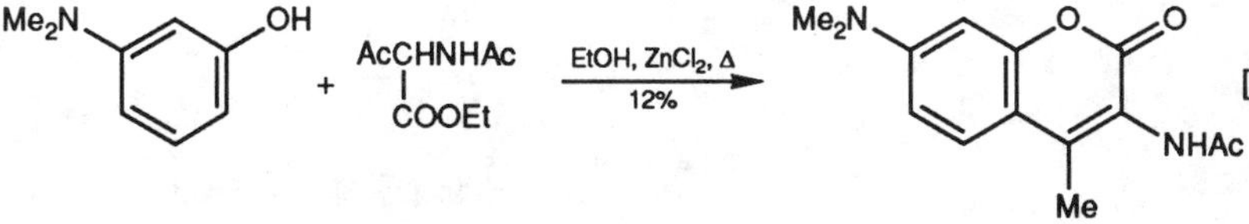

[3749]

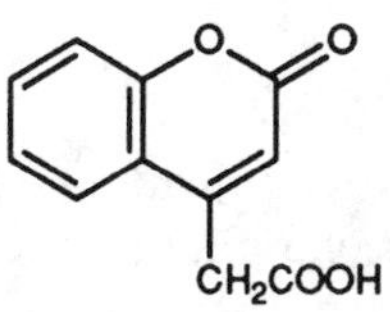

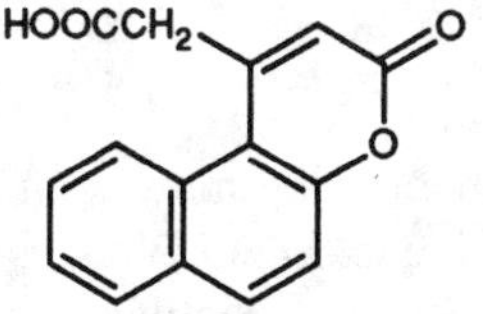

[2262]

(**99.5**)
1-Benzopyran-2-one

[2262]

(**99.6**)
Naphtho[2,1-*b*]pyran-3-one

Derivatives of malonic acid convert phenols into coumarins; malonyl chloride at ambient temperature and with titanium(IV) chloride leads to a 4-hydroxycoumarin [2283] whereas pyridinols condense with monosubstituted malonic esters and with diketene or 3-aminocrotonic ester to give a similar product [2437, 2439].

MeO OH + $CH_2(CCl)_2$ (O) → CH_2Cl_2, $TiCl_4$, 60% → MeO, O, O, OH [2283]

OH, N + COOAr, RCHCOOAr → Δ, 50–60% → O, O, N, R, OH [2439]

R = alkyl, $PhCH_2$, Ph; Ar = 2,4,6-$Cl_3C_6H_2$

Pyrano[3,2-*b*]pyridin-2-one

OH, N, O → O, CH_2, O, AcOH, Δ, 60% → Me, O, O, N, O [2437]

Pyrano[3,2-*a*]quinolizine-3,6-dione

→ COOEt, CH=CMeNH$_2$, 65% → O, O, N, O, Me [2437]

Pyrano[2,3-*b*]quinolizine-2,5-dione

5. Pyran-4-one

An activated phenol ring may be annulated by reaction with an alkenoic acid and a Lewis acid, e.g., PPA [2302], zinc chloride or phosphorus oxychloride–zinc chloride [2479, 3930]. A similar substrate undergoes thermal cyclization to a flavone by reaction with ethyl benzoylacetate [2499]. Another variation of the acylation method of synthesizing benzopyran-4-ones is that which uses an electron-rich phenol and a propynoic acid in the presence of phosphorus pentoxide–methanesulphonic acid (under argon). Varying amounts of the coumarins are often produced alongside the chromone, depending to a large extent on the number of methoxy groups present [3845].

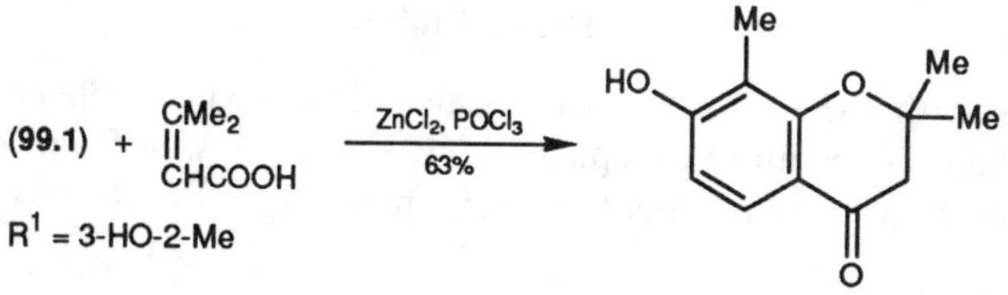

(99.1) + $CMe_2{=}CHCOOH$ —($ZnCl_2$, $POCl_3$; 63%)→ [2479, 3930]

R^1 = 3-HO-2-Me

1-Benzopyran-4-one

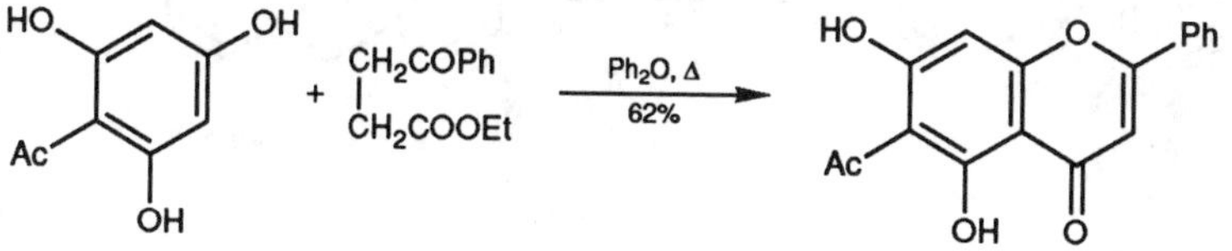

+ CH_2COPh / CH_2COOEt —(Ph_2O, Δ; 62%)→ [2499]

1-Benzopyran-4-one, 2-aryl-

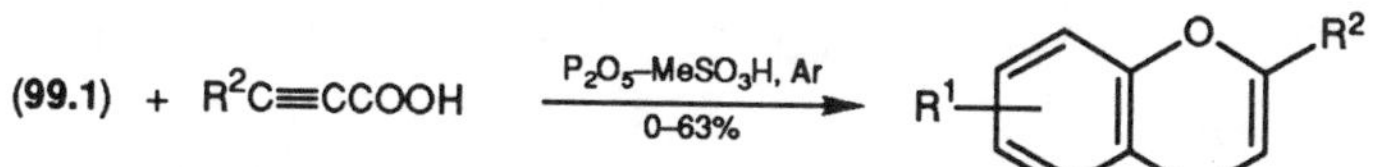

(99.1) + $R^2C{\equiv}CCOOH$ —(P_2O_5–$MeSO_3H$, Ar; 0–63%)→ [3845]

R^1 = HO, $(MeO)_n$; n = 1–3;
R^2 = Me, Ph, 4-$MeOC_6H_4$, 3,4-$(MeO)_2C_6H_3$

Suitably hydroxylated benzophenones are oxidatively coupled to give xanthones by manganese dioxide, DDQ or potassium ferricyanide [2151, 2941]. 2′-Halo-2-hydroxybenzophenones, which are capable of forming benzyne intermediates, are converted into xanthenones by potassamide. Some 3′-halo-2-hydroxybenzophenones also yield xanthenones as one of several products [2676].

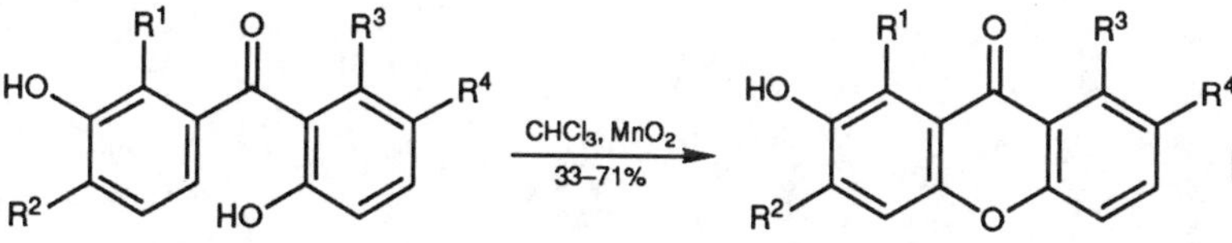

—($CHCl_3$, MnO_2; 33–71%)→ [2151, 2941]

R^1 = H; R^2 = HO or R^1 = HO; R^2 = H;
R^3 = H, HO; R^4 = H, COOH

9-Xanthenone

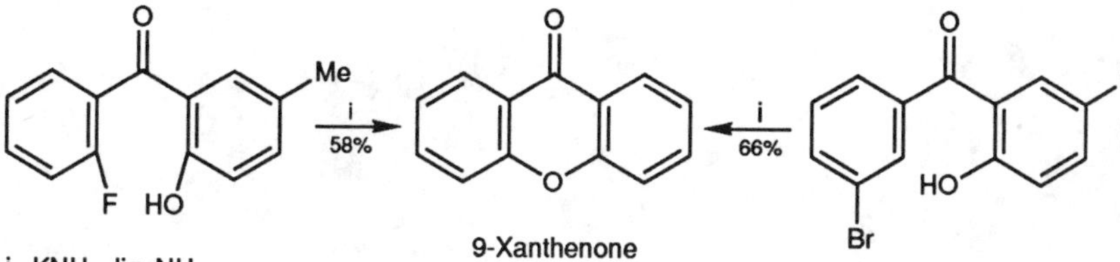

—(i; 58%)→ ←(i; 66%)— [2676]

9-Xanthenone

i, KNH_2–liq. NH_3

6. Pyran-4-thione

By analogy with the cyclization of 2′-halo-2-hydroxybenzophenone (above) [2676], 2′-halo-2-mercaptobenzophenone (formed *in situ* from the methyl thioether) reacts with potassamide–liquid ammonia to give 9-thioxanthenone.

KNH_2, liq. NH_3, 17% [2676]

9-Thioxanthenone

CHAPTER 100

Hydroxy, Ring-carbon or Ring-nitrogen and Imine or Iminophosphorane

I. Formation of a Five-membered Ring 1293
1. Pyrrole 1293
2. Imidazole 1294
3. Oxazole or Thiazole 1294
II. Formation of a Six-membered Ring 1295
1. Pyridine 1295
2. Pyridin-2-one or Pyridine-2-thione 1298
3. Pyridin-3-one 1298
4. Pyrimidine 1298
5. Pyrimidin-2-one or Pyrimidine-2-thione 1299
6. Pyrimidin-4-one 1299
7. 1,2,4-Triazine 1301
8. 1,3-Oxazine, 1,3-Oxazin-2-one or 1,3-Oxazine-2-thione 1301
9. 1,3,4-Thiadiazine 1301
10. 1,3,2,4-Dithiadiazine 1302
III. Formation of a Seven-membered Ring 1302
1. Azepine 1302
2. 1,3-Oxazepine 1302

An example of ring formation from a sulphimide and a ring-carbon atom is included in this chapter.

I. FORMATION OF A FIVE-MEMBERED RING

1. Pyrrole

The iminophosphorane (**100.3**) is a versatile intermediate. It reacts with α-bromoketones (reviews [B-45]) to give moderate yields of reduced 4-indolones (see Section II.1).

$N{=}PPh_3$ + $RC{=}O$ / CH_2Br → (diox, K_2CO_3, Δ; 28–56%) [3762]

(100.3)
R = Me, Ph, 4-Cl-, 4-Br-, 4-Me-C_6H_4

4-Indolone

2. Imidazole

Some *o*-alkylideneaminopyridines are cyclized at ambient temperature by reaction with trimethylsilyl cyanide, which supplies a ring-carbon double bonded to an exocyclic nitrogen.

$N{=}C(CF_3)_2$ → (Me_3SiCN; 91%) → CF_3, CF_3, $NSiMe_3$ [3370]

Imidazo[1,2-*a*]pyridine

3. Oxazole or Thiazole

An efficient conversion of a hydroxy-imine into a fused oxazole ring under relatively mild conditions is effected by either silver oxide or thionyl chloride. Barium manganate is claimed to be more efficient than manganese dioxide in this cyclization.

XH, R^1, $N{=}CHR^2$ → (CH_2Cl_2, Ag_2O; 69–92%) → O, R^3, N, R^1 [2930]

(100.4)
R^1 = H, Cl, Me;
R^2 = Ph, 4-MeO-, 4-NO_2-C_6H_4, PhCH=CH; X = O

Benzoxazole

Me, O, N, OH, MeN, $N{=}CHC_6H_4R$, O → ($SOCl_2$, Δ; 69–78%) → Me, O, N, O, C_6H_4R, MeN, N, O [2929]

R = H, halo, Me, NO_2, MeO

Oxazolo[5,4-*d*]pyrimidine-4,6-dione

(100.4) $\xrightarrow[60-80\%]{PhH, BaMnO_4}$ [3457]

R^1 = H; X = O, S;
R^2 = Ph, 4-Me-, 4-MeO-, 4-Me_2N-, 3- or 4-NO_2-C_6H_4

X = O Benzoxazole
X = S Benzothiazole

II. FORMATION OF A SIX-MEMBERED RING

1. Pyridine

A ring is formed between an iminophosphorane group and a ring-carbon by reaction with an isocyanate, an isothiocyanate or an α,β-unsaturated ketone. The iminophosphorane may be either preformed or generated *in situ* from an azide and triphenylphosphine [3227, 3401, 3762, 3854]. The product in some instances needs to be treated with a dehydrogenating agent in order to form the fully aromatized ring [3739, 3941]. There is evidence that a carbodi-imide is an intermediate in this cyclization; for some iminophosphoranes, it was possible to isolate oily carbodi-imides when the normal time for cyclization (52 h) to the isoquinoline was reduced to 24 h. However, a differently substituted isoquinoline was obtained on heating the carbodi-imides at 230 °C under pressure [3745].

EtOOCC=CH, Ph_3P=N $\xrightarrow[62-98\%]{PhMe, ArNCO, \Delta}$ [3227]

Ar = Ph, 4-Cl-, 4-MeO-, 4-Me-C_6H_4; X = O, S

Furo[3,2-*c*]pyridine
Thieno[3,2-*c*]pyridine

Other ring systems synthesized similarly:

Thieno[2,3-*c*]pyridine [3227]

Isoquinoline [3745]

N=PPh_3, CH=CCOOEt $\xrightarrow[70-84\%]{PhMe, ArNCS, \Delta}$ [3401]

Ar = Ph, 4-Cl-, 4-MeO-, 4-Me-C_6H_4

Pyrido[3,4-*b*]indole

NHAr, N, COOEt, N Me [3401]

Pyrido[4,3-*b*]indole

NHPh, MeO, N, COOEt, OCH_2Ph [3854]

Isoquinoline

(100.3) + $CR^2{=}CHR^3$ / COR^1 (100.5) —PhMe or diox, Δ; 18–60%→ [3762]

R^1 = Me, Ph; R^2 = H, Me; R^3 = H, Me, Ph

5-Quinolinone

$N{=}PBu_3$ (100.6) + (100.5) —PhH or PhMe, Pd/C, Δ; 21–89%→ [3739]

R^1 = Me, Ph; R^2 = H, Me; R^3 = H, Me, Ph

Cyclohepta[*b*]pyridine

N_3 —i, ii; 49–85%→ [3941]

R^1 = Bu, Ph; R^2,R^3 = H, Me, Ph

i, PhH or PhMe, PR^1_3; ii, R^3 = $R^3C(=O)CH{=}CHR^2$, Pd/C, N_2

Indeno[1,2-*b*]pyridine

Arylideneanilines are converted into fused pyridines by heating with an alkyne and an oxidizing agent such as di-isopropyl peroxydicarbonate (DP); when the aniline is unsymmetrically substituted, a mixture of isomers may be formed [3173]. α-Halo ketones (and, less efficiently, simple aliphatic ketones [2639]) react with imines to give fused pyridines in variable yields [2715]. An imine formed by reaction of a 6-aminopyrimidinedione with DMFDMA undergoes cycloaddition with an electron-deficient alkene; the first product is dehydrogenated in hot nitrobenzene [3620].

$N{=}CHPh$, MeO (100.7) + PhC≡CH —PhH, DP, Δ; ~40%→ [3173]

Quinoline

$N{=}CHC_6H_4R$ + CH_2X / COMe → diox, $BF_3.Et_2O$, 35–80% → Me, XCH_2, C_6H_4R, N [2715]

$R = H, Cl, MeO, NO_2$; $X = Cl, Br$

Benzo[*f*]quinoline

$Me_2NCH{=}N$, Me, N, O, NMe, O + CHR^1 / CHR^2 → 1. PhMe, Δ 2. $PhNO_2$, Δ, 35–60% → Me, N, N, O, NMe, R^2, R^1, O [3620]

$R^1 = H, COOMe, COOEt$;
$R^2 = CN, Ac, COOMe, COOEt$

Pyrido[2,3-*d*]pyrimidine-2,4-dione

The 1-azadiene (**100.8**) is cyclized in a hetero-Diels–Alder reaction (reviews [B-42, 2455, 3067a, 3880]) with either ethoxyethene or methoxyallene [3649]. The former reagent gives better results in its reaction with an imine in the presence of a tetracarbonylcobaltate [2133]. Prolonged heating of the iminoethenylaniline (**100.9**) in dioxane–aluminium chloride results in the formation of a quinoline and loss of terminal imino nitrogen [3228]. A survey of recent work on quinolines and isoquinolines has been published [3984].

$CMe{=}NSO_2Ph$

(**100.8**)

$CH_2{=}CHOEt$, PhMe, Δ, 79% → Me, NSO_2Ph, OEt, H

Isoquinoline [3649]

$CH_2{=}C{=}CHOMe$, mesitylene, Δ, 28% → Me, NSO_2Ph, OMe, H, CH_2

(**100.7**) + $EtOCH{=}CH_2$ → $HCo(CO)_4$, MeOH, 86% → N, Ph, MeO [2133]

Quinoline

diox, $AlCl_3$, Δ; 58–90% [3228]

(100.9)

R^1 = H, Me; R^2 = Me, $PhCH_2$; R^3 = Ph, 4-MeC_6H_4, C_6H_{11}

2. Pyridin-2-one or Pyridine-2-thione

When an iminophosphorane is heated with either carbon dioxide or carbon disulphide in benzene, annulation to a convenient ring-carbon occurs in high yield.

CO_2 or CS_2, PhH, Δ; 83–94% [3635]

X = O, S

Indolo[3,2-*c*]quinolin-5-one
Indolo[3,2-*c*]quinoline-5-thione

3. Pyridin-3-one

A side-chain containing a C=C—N=C sequence attached to a benzene ring cyclizes on to the *o*-position when the compound is irradiated together with equimolar amounts of perchloric acid and sodium carbonate.

$HClO_4$, Na_2CO_3, *hν*; 23–65% [3536]

4-Isoquinolinone

R^1 = Me, Ph; R^2 = Ph, 4-MeO-, 4-CN-, 4-Me-C_6H_4;
R^3 = H, Me, MeO, CN

4. Pyrimidine

The triphenylphosphoamidine (**100.10**) reacts with an aldehyde in refluxing xylene to give either the quinazoline (**100.11**) or the dihydroquinazoline (**100.12**). The proportion of these two compounds is dependent on the nature of R: (**100.11**) is the only product (71% yield) when R = Ph, but the dihydro derivative is formed in 46–70% yield when R = 2- or 4-nitrophenyl; non-aryl aldehydes give a mixture. An iminophosphorane group and a ring-nitrogen in compound (**100.13**) react with an aryl isocyanate at room temperature to form a pyrimidine ring.

(100.10) + RCHO —xylene, Δ (see text)→ (100.11) + (100.12) [3858]

R = C_6H_{13}, $PhCH_2CH_2$, Ph, 2- or 4-$NO_2C_6H_4$

Quinazoline

(100.13) —CH_2Cl_2, ArNCO, 74–86%→ [3441]

Ar = Ph, 4-MeO-, 3-Me-C_6H_4

Benzimidazo[1,2-*c*]quinazoline

5. Pyrimidin-2-one or Pyrimidine-2-thione

The benzimidazole-phosphorane (**100.13**) is readily cyclized by reaction with either carbon dioxide or carbon disulphide.

(100.13) —CH_2Cl_2, CX_2, ~73%→ [3441, 3499]

X = O, S

Benzimidazo[1,2-*c*]quinazolin-6-one
Benzimidazo[1,2-*c*]quinazoline-6-thione

6. Pyrimidin-4-one

A fused 5-chloropyrimidin-4-one ring is formed from an imine and ring-nitrogen by heating with either trichloroacetic anhydride in boiling toluene

[2882], or dichloroacetic acid and phosphorus oxychloride in DMF [3464]. Tautomeric 2-aryliminoimidazolines react with an enamine-acid chloride (**100.14**) to produce either the 5- or 7-one according to the solvent used [2534]. An iminophosphorane (formed *in situ* from an azide) reacts at room temperature with a lactam carbon atom to form a doubly fused pyrimidinone in excellent yield [3956].

N=CHAr; $(CCl_3CO)_2O$, PhMe, Δ; 23–37%; Ar; Cl; O [2882]

Ar = Ph, 4-Cl-, 4-MeO-C_6H_4

Pyrido[1,2-*a*]pyrimidin-4-one

Me; SMe; ArCH=N; + $CHCl_2$ COOH; DMF, $POCl_3$, Δ; 96%; O; Cl; Me; SMe; Ar [3464]

Ar = 3,4,5-$(MeO)_3C_6H_2$

[1,2,4]Triazolo[1,5-*a*]pyrimidin-7-one

Another ring system synthesized similarly:

R^1; Ar; Cl; O; R^2 [3464]

1,2,4-Triazolo[4,3-*a*]pyrimidin-5-one

R^1N; H; HN; + R^2C=C(Cl)NMe$_2$; COCl; (**100.14**); MeOH, TEA; 15–61%; PhH, TEA; 34–69%; Me_2N; R^1; O; NMe_2 [2534]

R^1 = aralkyl, aryloxyalkyl, aryl; R^2 = alkyl

Imidazo[1,2-*a*]pyrimidin-5-one

Imidazo[1,2-*a*]pyrimidin-7-one

N_3 O C N $(CH_2)_n$ C O

$n = 5,6$

PhH, PBu_3, N_2
84–96%

N N $(CH_2)_n$ O [3956]

$n = 5$ Azepino[2,1-*b*]quinazolin-7-one
$n = 6$ Azocino[2,1-*b*]quinazolin-8-one

7. 1,2,4-Triazine

An isocyanate reacts readily at ambient temperature with an iminophosphorane derivative of an amidine to form a triazine ring. Recent research on polyazines has been reviewed [3985].

SMe N $NN{=}CR^1$ $N{=}PPh_3$ Ph

R^1 = 4-MeO-, 4-Me-C_6H_4;
R^2 = Me-, Cl-, MeO-C_6H_4

CH_2Cl_2, R^2NCO
66–81%

SMe N N N R^1 N Ph NHR^2 [3616]

Imidazo[5,1-*f*][1,2,4]triazine

8. 1,3-Oxazine, 1,3-Oxazin-2-one or 1,3-Oxazine-2-thione

Cyclization of iminophosphorane and hydroxy groups with an isocyanate leads to the formation of benzoxazines; replacing the isocyanate by carbon dioxide or carbon disulphide gives a good yield of the corresponding oxazinone or oxazinethione.

$N{=}NPPh_3$ CH_2OH

RNCX, PhMe, Δ
40–65%

N NHR O

3,1-Benzoxazine [3999]

CX_2, PhMe, Δ
53–74%

H N X O

R = $PhCH_2$, 4-MeO-, 4-halo-, 4-Me-C_6H_4;
X = O, S

3,1-Benzoxazin-2-one
3,1-Benzoxazine-2-thione

9. 1,3,4-Thiadiazine

An aza-Wittig reaction may be adapted to the synthesis of a fused thiadiazine ring under mild conditions with a two-molar proportion of an isothiocyanate. The reactions of isothiocyanates have been reviewed [3990]. In this way, an iminophosphorane is annulated to a neighbouring sp^3 ring-carbon atom.

PhN NN=PPh$_3$ —MeOH, RNCS, 40–54%→ [3829]

R = Me, PhCH$_2$, Ph, 4-Me-, 4-MeO-, 4-F-C$_6$H$_4$

Imidazo[1,5-*d*][1,3,4]thiadiazin-5-one, 7-thioxo-

10. 1,3,2,4-Dithiadiazine

A novel and electron-rich ring is formed when the trimethylsilylsulphur di-imide (**100.15**) is treated with sulphur dichloride without external heating.

—CH$_2$Cl$_2$, SCl$_2$, 12%→ [3053]

(**100.15**) 1,3,2,4-Benzodithiadiazine

III. FORMATION OF A SEVEN-MEMBERED RING

1. Azepine

The nitrogen ylide (**100.16**), produced *in situ* from an imidoyl chloride and a strong base, cyclizes at room temperature or lower to the benzazepine. This and similar methods of synthesizing azepines have been reviewed [3962].

—THF, tBuOK, 73%→ [–CH=N$^+$=C$^-$Ph] → [3312]

(**100.16**) 2-Benzazepine

2. 1,3-Oxazepine

Side-chain iminophosphorane and acyloxy groups react in refluxing benzene with the formation of a C—N bond of an oxazepine ring.

—PhH, P(OEt)$_3$, Δ, 30–74%→ [3715]

R = H, Me, cC$_3$H$_5$, Ph 1,3-Benzoxazepine

CHAPTER 101

Isocyanate and Methylene, Ring-carbon or Ring-nitrogen

I.	Formation of a Five-membered Ring	1303
	1. Thiazole	1303
II.	Formation of a Six-membered Ring	1304
	1. Pyridine	1304
	2. Pyridin-2-one	1304
	3. Pyridine-2-thione	1306
	4. Pyrimidin-4-one, 2-thioxo-	1306

An example of the cyclization of a thiocyanate and a ring-nitrogen is included in this chapter.

I. FORMATION OF A FIVE-MEMBERED RING

1. Thiazole

Warming a 2-thiocyanatomethylbenzimidazole in methanol converts it into the thiazolo derivative in high yield. The reactions of isothiocyanates have been reviewed [3990].

R^2 CHSCN, N, NH, R^1 —— MeOH, Δ / ~80% ——> R^1, N, S, N, NH, R^2 [2350]

R^1 = H, Cl, Me; R^2 = H, Me

Thiazolo[3,4-*a*]benzimidazole

II. FORMATION OF A SIX-MEMBERED RING

1. Pyridine

Reaction of an ynamine with an isocyanate (prepared *in situ* by rearrangement of an azide) leads to the formation of a pyridine ring [2333]. Intramolecular cyclization of an (isothiocyanatoethyl)benzene is effected by heating with triethyloxonium tetrafluoroborate or with PPA. With the former, the salt may be converted into its free base with aqueous carbonate [2843]. Photolysis of a vinyl isocyanate (formed *in situ*) gives a good yield of a 1-isoquinolinone. The corresponding isothiocyanate ion produces a mixture of products of which the 1-isoquinolinethione is usually a minor component [3839].

NCO; + $CNEt_2$ ≡ CMe; PhH, 54–68%; OH, Me, NEt_2, Ar [2333]

R = H, MeO, Br; Ar = Ph, 4-Cl, 4-Br-C_6H_4

Benzo[*h*]-1,6-naphthyridine

NCS, $CR^2R^3CHR^4$; i, ii, 45–80%; SEt, R^4, R^2, R^3 [2843]

R^1 = H, Me, Br, MeO; R^2,R^3,R^4 = H, Me, Ph

i, CH_2Cl_2, $Et_3O^+.BF_4^-$, Δ; ii, aq. Na_2CO_3

Isoquinoline

C_6H_4–4–R^2, C=CBr, C_6H_4–4–R^1 + KCNX; i, 15–93%; X, NH, C_6H_4–4–R^2, C_6H_4–4–R^1 [3839]

R^1 = H, MeO; R^2 = H, MeO, Me

i, CH_2Cl_2, $Bu_4N^+Cl^-$, *hν*

X = O 1-Isoquinolinone

X = S 1-Isoquinolinethione

2. Pyridin-2-one

Vinylic isocyanates react with enamines such as 1-pyrrolidino-1-cyclohexene under mild conditions to form a new pyridinone ring and simultaneous loss of pyrrolidine [3540]. Reaction of an ynaminone with phenyl isocyanate in refluxing THF gives good yields of 3-acyl-4-dimethylamino-2-quinolinone [2056]. A perchloroynamine behaves similarly [2838].

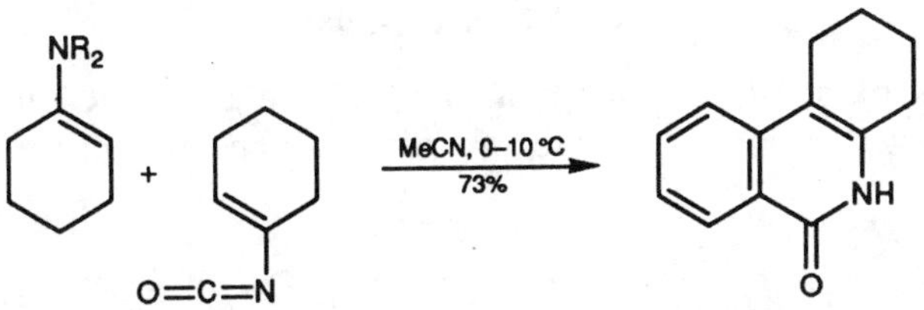

NR2 = pyrrolidino

MeCN, 0–10 °C, 73%

6-Phenanthridinone [3540]

Other ring systems synthesized similarly:

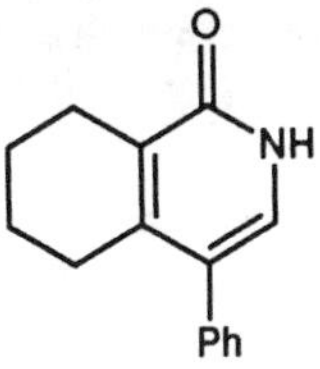

1-Isoquinolinone [3540]

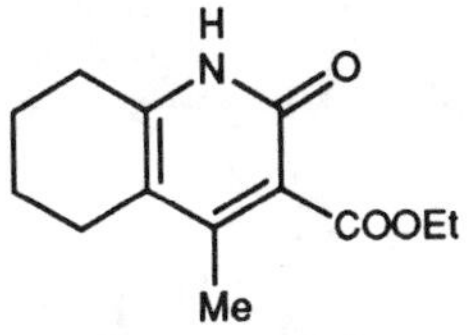

2-Quinolinone [3540]

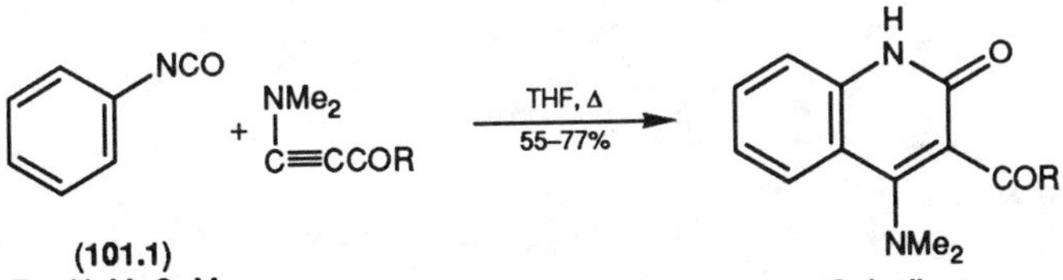

(101.1) R = H, MeO, Me

THF, Δ, 55–77%

2-Quinolinone [2056]

(101.1) + $C{\equiv}CC{=}CCl_2$ (Cl, NR_2)

THF, Δ, ~30%

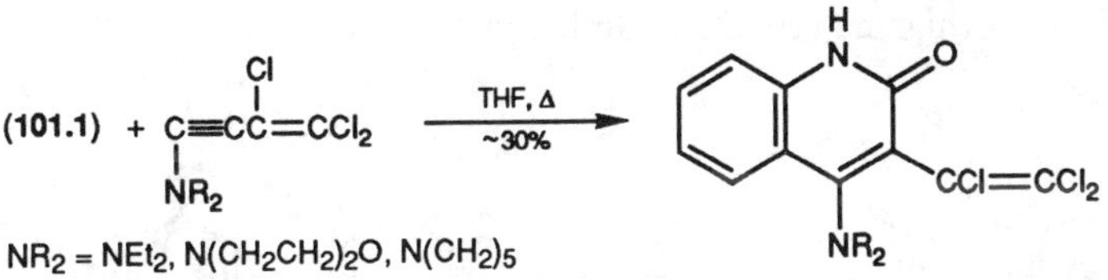

$NR_2 = NEt_2, N(CH_2CH_2)_2O, N(CH_2)_5$ [2838]

When (+)-1-methyl-2-phenylethyl isocyanate is heated with triethyloxonium tetrafluoroborate or boron trifluoride diethyl etherate, a high yield of a dihydroisoquinolin-1-one is obtained, but heating with methyl fluorosulphonate leads to 3,4-dihydro-1-methoxy-3-methylisoquinoline [2935].

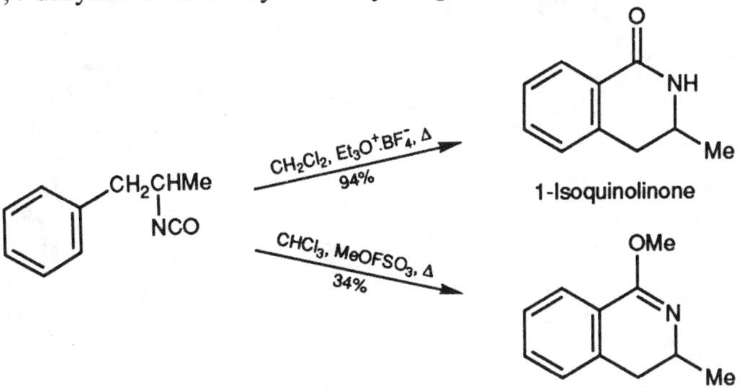

$CH_2Cl_2, Et_3O^+BF_4^-$, Δ, 94%

1-Isoquinolinone

$CHCl_3, MeOFSO_3$, Δ, 34%

[2935]

3. Pyridine-2-thione

Thermolysis of a 3-(β-isothiocyanatovinyl)indole at 170 °C under reduced pressure converts it into the 2-thioxopyrido analogue in high yield.

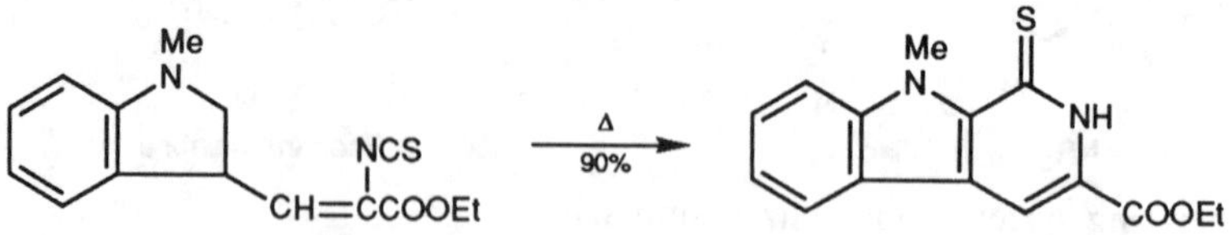

Δ, 90% [3401]

Pyrido[3,4-*b*]indole-1-thione

Another ring system synthesized similarly:

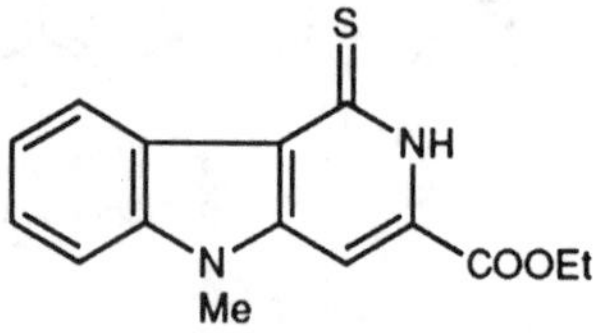

[3401]

Pyrido[4,3-*b*]indole-1-thione

4. Pyrimidin-4-one, 2-thioxo-

Another high-yielding thermolysis is the *peri*-cyclization of (isothiocyanatocarbonyl)phenothiazine (**101.2**) in diphenyl ether.

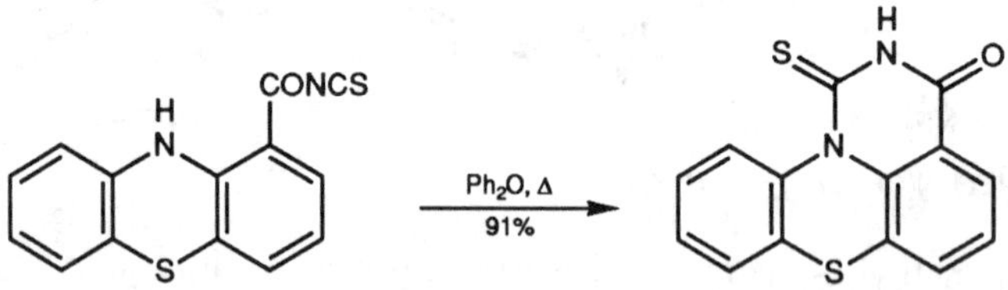

Ph_2O, Δ, 91% [2748]

(**101.2**) Pyrimido[5,6,1-*kl*]phenothiazin-3-one, 1-thioxo-

CHAPTER 102

Ketone and Lactam Carbonyl

I. Formation of a Five-membered Ring 1307
1. Pyrrole or Pyrrole 1-Oxide 1307
2. Pyrazole 1308
3. Furan 1308
4. Thiophene 1309
5. Thiazole 1309
II. Formation of a Six-membered Ring 1309
1. Pyridine 1309
2. Pyrimidin-4-one 1310
3. Pyrazine 1310
4. 1,2,4-Triazine 1310
5. Pyran or Thiopyran 1310
III. Formation of a Seven-membered Ring 1311
1. Azepine 1311

The use of lactams (and lactones) in heterocyclic synthesis has been reviewed [2387]. Examples of the cyclization of aldehyde, aldimine or thiosemicarbazide and lactam carbonyl (or thiocarbonyl) functions are included in this chapter.

I. FORMATION OF A FIVE-MEMBERED RING

1. Pyrrole or Pyrrole 1-Oxide

A side-chain oxime becomes joined to a lactam carbonyl when the compound is treated at room temperature or lower with sodium borohydride. The product varies even within this limited temperature range, and is partially dependent on the substituents on the reactant. For example, at $-10\,°C$ and when $R^1 = H$,

R^2 = Me, the *N*-hydroxy derivative (**102.1**) is the sole product isolated in 55% yield; at ambient temperature, the *N*-oxide (**102.2**) is obtained in 95% yield.

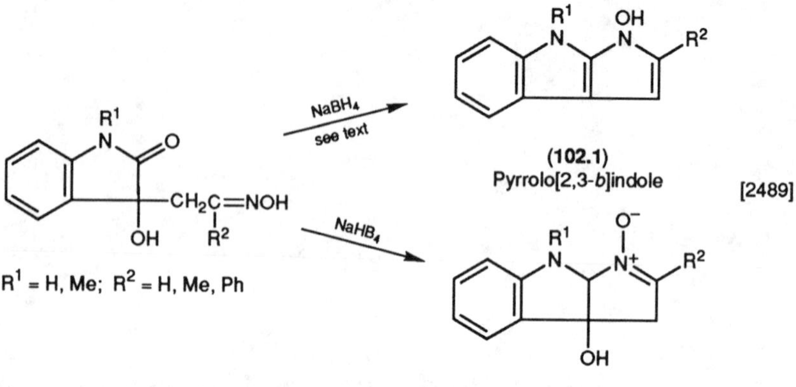

R^1 = H, Me; R^2 = H, Me, Ph

$NaBH_4$, see text

$NaHB_4$

(**102.1**)
Pyrrolo[2,3-*b*]indole

[2489]

(**102.2**)
Pyrrolo[2,3-*b*]indole 1-oxide

2. Pyrazole

An acetyl group and an adjacent lactam thiocarbonyl react with hydrazine hydrate to form a fused pyrazole.

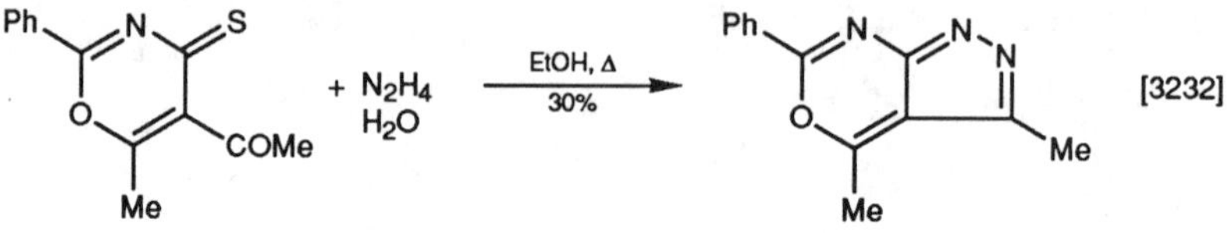

+ $N_2H_4 \cdot H_2O$ — EtOH, Δ, 30%

Pyrazolo[3,4-*d*][1,3]oxazine

[3232]

3. Furan

Cyclization of a compound containing a side-chain ketone and a lactam carbonyl occurs at room temperature in mineral acid [2050] or by heating it with PPA [2491]. When the lactam nitrogen is substituted, the reaction takes a different course although a furan ring is still formed (see Chapter 29, Section I.2).

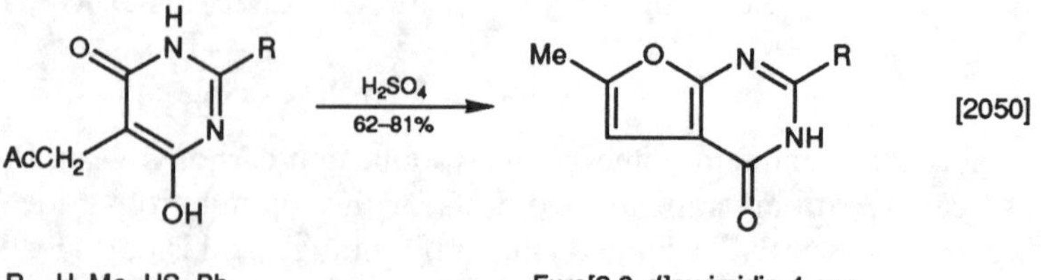

H_2SO_4, 62–81%

R = H, Me, HS, Ph

Furo[2,3-*d*]pyrimidin-4-one

[2050]

4. Thiophene

S-Alkylation of a lactam thiocarbonyl by a phenacyl halide is followed by cyclization with the neighbouring aldehyde group. The intermediate quaternary salt is converted into the free base by heating it above its melting point and at reduced pressure.

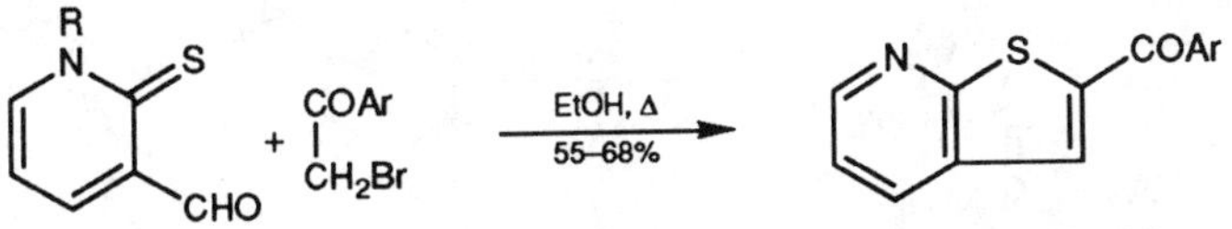

EtOH, Δ; 55–68% [3147]

R = iPr, Ph; Ar = Ph, 4-Br-, 4-NO_2-, 4-Ph-C_6H_4

Thieno[2,3-*b*]pyridine

5. Thiazole

Cyclodehydration of a side-chain ketone and lactam thiocarbonyl requires heating with phosphorus oxychloride, PPA or sulphuric acid.

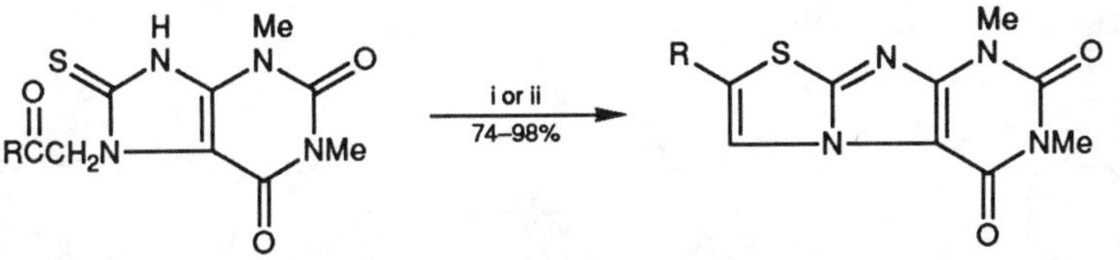

i or ii; 74–98% [2718]

R = Me, tBu, Ph, 4-Br-, 4-MeO-, 4-Me-C_6H_4
i, $POCl_3$, Δ; ii, H_2SO_4

Thiazolo[2,3-*f*]purine-2,4-dione

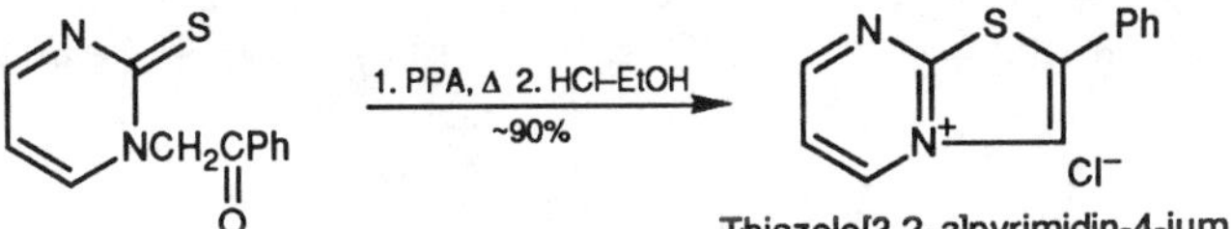

1. PPA, Δ 2. HCl–EtOH; ~90% [3383]

Thiazolo[3,2-*a*]pyrimidin-4-ium

II. FORMATION OF A SIX-MEMBERED RING

1. Pyridine

A 3-oxopropyl side-chain reacts with a lactam carbonyl [as in (**102.3**] and a source of ammonia such as ammonium acetate to form a fused dihydropyridine ring.

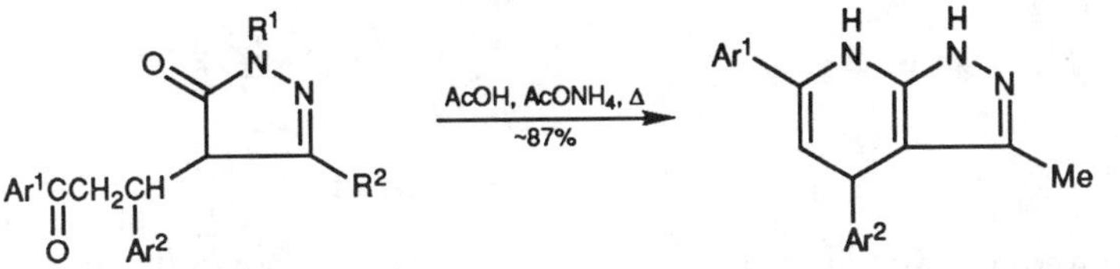

AcOH, $AcONH_4$, Δ; ~87% [3113]

(**102.3**)
Ar^1 = 4-MeC_6H_4, 1- or 2-naphthyl;
Ar^2 = 4-ClC_6H_4; R^1 = H; R^2 = Me

Pyrazolo[3,4-*b*]pyridine

2. Pyrimidin-4-one

In studies related to the synthesis of camptothecin, lactam carbonyl and side-chain keto groups reacted on heating the compound with ammonium acetate under dehydrating conditions.

PhMe, AcOH, AcONH$_4$, Δ; 87% [2662]

Pyrimido[1',2':1,2]pyrrolo[3,4-*b*]quinolin-4-one

3. Pyrazine

A ketone and its neighbouring lactam carbonyl form a pyrazine ring when the compound is heated with an *o*-phenylenediamine.

i, ii; 55% [3341]

i, Fe, PhH; ii, dil. HCl

Pyrimido[4,5-*b*]quinoxaline-2,4-dione

4. 1,2,4-Triazine

A thiosemicarbazone and its neighbouring lactam carbonyl react to form a triazine when the compound is heated with carbonate.

Na_2CO_3, Δ; 47–96% [2154]

R = Me, Et

Pyrazolo[3,4-*e*]-1,2,4-triazine

5. Pyran or Thiopyran

The compound (**102.3**), used in the synthesis of a fused pyridine in Section II.1, may also be cyclized to pyrano- or thiopyrano-pyrazole by heating with phosphorus pentoxide–PPA or phosphorus pentasulphide. The presence of ethanol or acetic acid retards the reaction.

(102.3) $\xrightarrow[54-94\%]{P_2O_5-PPA \text{ or } P_2S_5\text{-xylene}, \Delta}$ [3023, 3113]

Ar^1 = 4-MeC_6H_4, Ph; Ar^2 = 4-ClC_6H_4;
R^1,R^2 = Me, Ph; X = O, S

Pyrano[2,3-*c*]pyrazole
Thiopyrano[2,3-*c*]pyrazole

III. FORMATION OF A SEVEN-MEMBERED RING

1. Azepine

Rhodium acetate-assisted cyclization (review [3928]) of an aldehyde (protected as an aminoaziridinehydrazone) which also has a lactam in another ring gives a high yield of the azepine.

$\xrightarrow[76\%]{PhMe, Rh(OAc)_2, \Delta}$ [3634]

R = (2,3-diphenylaziridin-1-yl)

1,3-Dioxolo[4,5-*h*]isoindolo[1,2-*b*]benzazepin-8-one

CHAPTER 103

Methylene and Nitro or Nitroso

I. Formation of a Five-membered Ring 1312
  1. Pyrrole 1312
  2. Pyrrol-2-one 1312
  3. Imidazole 1313
  4. Imidazole 3-Oxide 1314
II. Formation of a Six-membered Ring 1315
  1. 1,2,5-Thiadiazine 1,1-Dioxide 1315

I. FORMATION OF A FIVE-MEMBERED RING

1. Pyrrole

Mild hydrogenation of the nitro ketone (**103.2**) followed by heating the resulting cyclized solid in toluene gives the 1-hydroxyindole.

NO_2; $CH_2C(=O)CH_2SO_2Ph$ (**103.2**) —1. AcOEt, H_2, PtO_2 2. PhMe, Δ; 77%→ OH, N, CH_2SO_2Ph Indole [3273]

2. Pyrrol-2-one

Irradiation of *o*-t-butylnitrobenzene (**103.3**) in alkali leads to a good yield of the indolone; the percentage conversion in the reaction is low.

MeOH, aq. NaOH, *hν* 66%

(103.3) → 2-Indolone [2336]

3. Imidazole

A side-chain methylenamino group reacts with a nitroso group (introduced *in situ*) when the compound is heated with mineral acid [2466]. An unusual reaction occurred when the nitro-t-amine (**103.4**) was refluxed with acetic anhydride and zinc chloride; the course of this reaction may involve an *N*-oxide [2039]. When an *o*-nitroalkylamine is heated with ethoxide, the two functions are converted into a 1-hydroxyimidazole ring [3004]. Nitrosation of the pyrimidinone (**103.5**) involves several steps including a Fischer–Hepp rearrangement of an *N*-nitroso to a ring-C-nitroso; the final product is a 6-(4′-nitrophenylamino)purin-2-one [2667]. In a weakly basic medium, the nitroamine (**103.6**) is cyclized to either the benzimidazole (when R = H), or the benzimidazolone (when R ≠ H).

1. nPrONO, HCl 2. Δ 64%

(103.3) → 2,6-Purinedione [2466]

$ZnCl_2$, Δ 15–87%

(103.4) + Ac_2O → [2039]

X = bond, CH_2, $(CH_2)_2$, O

Pyrrolo[1,2-*a*]benzimidazole
Pyrido[1,2-*a*]benzimidazole
Azepino[1,2-*a*]benzimidazole
[1,4]Oxazino[1,2-*a*]benzimidazole

EtONa, EtOH, Δ 54–67%

R = HO, $HO(CH_2)_2$, Ph → Imidazo[4,5-*c*]quinoline [3004]

NHAr / MeN / O / N / Me / NCH_2Ph — AcOH, $NaNO_2$, Δ, 53% → NHAr / MeN / N / Ph / N / O / N / Me [2667]

(103.5)
$Ar = 4\text{-}NO_2C_6H_4$

2-Purinone

R / NCH_2CN / NO_2 — R = H: EtOH, Na_2CO_3, Δ, 77% → OH / N / CN / N — Benzimidazole [2550]

R ≠ H: EtOH, Na_2CO_3, Δ, 60–75% → R / N / O / NOH — 2-Benzimidazolone

(103.6)
R = H, Me, $PhCH_2$, Ph

4. Imidazole 3-Oxide

A recent improvement in the synthesis of benzimidazole *N*-oxides depends on the presence of a 2-carboxylic ester group at the time the *N*-oxide ring is formed; the carboxylate function may be easily removed if desired. Stirring or gentle warming of a nitrophenylglycine ester or nitrile with sodium ethoxide produces a good yield of the *N*-oxide [3395, 3506]. A method of preparing an analogous benzimidazole but under less basic conditions is to heat the nitrophenylamino-acetonitrile (**103.7**) with carbonate in ethanol [3402]. A purine 7-oxide (guanine 7-oxide) which has antitumour action may be synthesized by stirring the nitro-phenacylamine (**103.8**) with dilute sodium hydroxide; the benzoyl group is lost as benzoate [3310].

R^1 / $NHCH_2R^2$ / NO_2 — EtONa–EtOH, 31–71% → H / N / R^2 / R^1 / N^+ / O^- [3395, 3506]

R^1 = H, MeO, Me, AcNH; R^2 = COOEt, CN

Benzimidazole 3-oxide

NHAc / NO_2 / $NHCH_2CN$ — EtOH, K_2CO_3, Δ, 78% → NHAc / O^- / N^+ / CN / NH [3402]

(103.7)

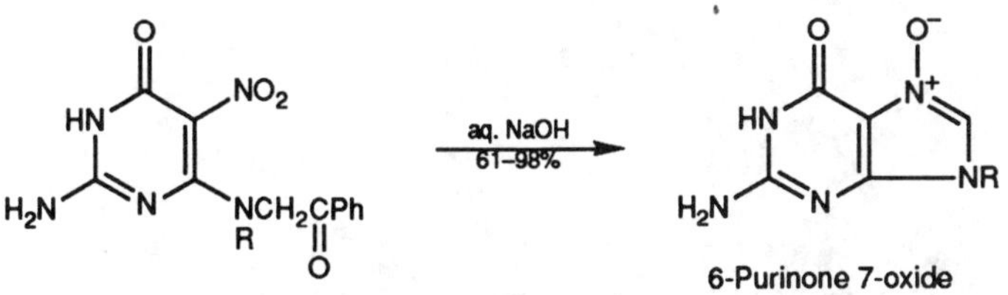

aq. NaOH
61–98%

[3310]

6-Purinone 7-oxide

(103.8)
R = Me, Pr, allyl, $PhCH_2$, $4\text{-}MeOC_6H_4CH_2$, cC_6H_{11}

I. FORMATION OF A SIX-MEMBERED RING

1. 1,2,5-Thiadiazine 1,1-Dioxide

A reactive side-chain methylene condenses with a nitroso group in a basic medium; in this example, the methylene is joined to the ring through a sulphonamide group. The chemistry of thiadiazines containing adjacent S and N atoms has been reviewed [3904].

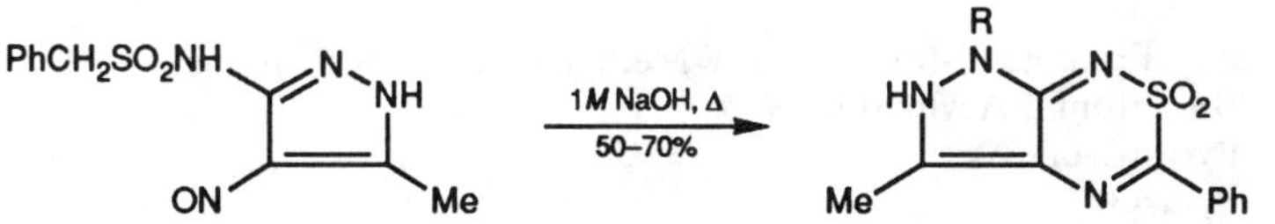

1 *M* NaOH, Δ
50–70%

[3575]

R = Me, Ph, $3\text{-}ClC_6H_4$

Pyrazolo[3,4-*c*][1,2,5]thiadiazine 2,2-dioxide

CHAPTER 104

Methylene and Ring-carbon or Ring-nitrogen

I. Formation of a Five-membered Ring 1316
   1. Pyrrole 1316
      A. From an *N*-Methyl- or *N*-Methylene-substituted Azine 1317
      B. From 2-Acylpyridines 1320
   2. Pyrrol-2-one 1320
   3. Pyrazole 1320
   4. Imidazole 1321
   5. Thiophene 1321
   6. 1,2-Dithiol-3-one 1322
   7. Thiazole 1322
II. Formation of a Six-membered Ring 1323
   1. Pyridine 1323
   2. Pyridin-2-one 1324
   3. Pyrimidine-2,4-dione or Pyrimidin-2-one, 4-Thioxo- 1326
   4. 1,2,3-Triazine 1-Oxide 1326
   5. Pyran 1326
   6. 1,4-Thiazin-3-one 1327
III. Formation of a Seven-membered Ring 1327
   1. Azepine 1327

An example of the formation of a pyran ring from a side-chain phosphorane is included in this chapter.

I. FORMATION OF A FIVE-MEMBERED RING

1. Pyrrole

Many references to the synthesis of a fused pyrrole are mentioned in this section; for convenience, these are divided into two types.

A. From an N-Methyl- or N-Methylene-substituted Azine

Quaternized pyridines which have an R^1R^2CH group on the nitrogen are useful compounds because of the ease with which they react with bases to give reactive ylides; the latter add on to some alkynes such as DMAD under mild conditions. The quaternized pyridines form ylides more readily when R^1 is an electron-attracting group such as acyl, 4-nitrophenyl, tosyl, nitrile or carboxylic ester. When both R^1 and R^2 are electrophilic, as in $(NC)_2CH$, ylide formation is very facile [2280, 2284, 2419, 2422, 2970]. The chemistry of *N*-substituted pyridines [1870] and the use of azomethine ylides [2509] have been reviewed.

R = H, MeO, EtO

DMAD, MeCN, Δ 50–70%

HC≡CCN, MeCN, Δ 30–80%

Pyrrolo[1,2-*b*]pyridazine [2280]

+ DMAD MeCN 27%

Indolizine [2419, 2422]

PhMe, Δ 32–73% [2970]

R = H, Me, $PhCH_2$, Ph, CN

The reactions shown in this section illustrate the versatility of *N*-substituted pyridines in the synthesis of condensed heterocycles containing pyrrole. Some of the cyclizations proceed without the need for externally applied heat [2419, 2422, 2830, 2980a], and for most of the cyclizations, the conditions are relatively mild. The preformed ylide is used in some of the reactions while in others, it is generated *in situ*. It is even possible to react a 2-methylpyridine with an *N*-alkylating agent, such as bromoacetone, and then cyclize the intermediate *N*-alkyl derivative by heating with sodium acetate-acetic anhydride or carbonate without isolating either the intermediate *N*-alkylpyridine or the ylide [2157, 2295, 2393, 3512].

COOMe
COOMe
R— CF$_3$SO$_3^-$ $^+$NCH$_2$SO$_2$Ts + DMAD —CHCl$_3$, TEA, 64–88%→ R— N [2830]

R = H, CN, tBu, Me, Cl, PhCO

R^2
$^+$NCH$_2$COAr + R^{2}CH=CH$_2$ —MeCN or DMF–TEA, 5–29%→ N COAr [2980a]
R^1 R^1

R^1 = H, Br, Me, COOEt; R^2 = 2- or 4-pyridyl; Ar = Ph, 4-NO$_2$C$_6$H$_4$

COOR^1
CH$_2$COOR^1 COOEt OH
N + BrCHR^2 —N$_2$, Δ, ~50%→ N R^2 [2157]

R^1 = Me, Et; R^2 = Me, Ph

Indolizine

Amongst the most commonly used reagents are DMAD [2280, 2419, 2422, 2523, 2830], bis(methylthio)methylenemalonic acid derivatives [3491, 3571], propiolic acid nitrile [2280], 3-methoxyacrylic ester or nitrile [3704], 2- or 4-vinylpyridine [2980a] and an anhydride [2458, 2876]. This kind of cyclization is not confined to pyridinium salts; pyridazine ylides react similarly and in good yields [2295, 2523].

R^1
R^{1}CH$_2$N$^+$–N CR^2 R^2 N–N
X$^-$ + ≡CCOOMe —i, ii, 28–84%→ MeOOC [2523]

R^1 = H, Ac, Ph, 4-NO$_2$C$_6$H$_4$; R^2 = H, COOMe; X = Br, Cl
i, PhH or DMF, TEA; ii, chloranil, Δ

Pyrrolo[1,2-*b*]pyridazine

Another ring system synthesized similarly:

N COOMe [2728]

Pyrrolo[2,1-*a*]isoquinoline

OMe
CO
O H
MeOC N S Me
N [3542]
Me
O

Pyrrolo[1,2-*a*]thieno[2,3-*d*]pyrimidin-4-one

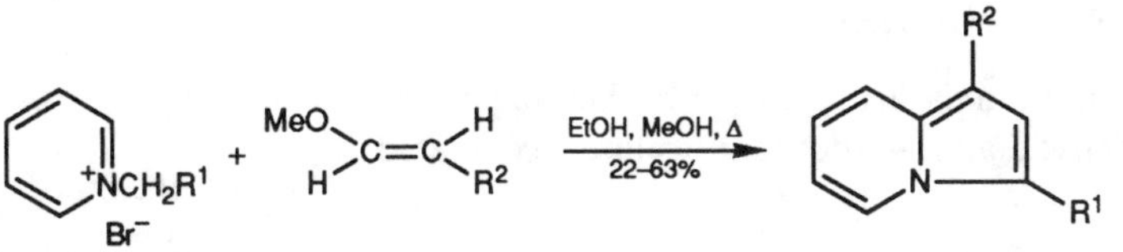

$$\text{Pyridinium}^{+}NCH_2R^1\ Br^- + MeO(H)C{=}C(H)R^2 \xrightarrow[22\text{–}63\%]{EtOH,\ MeOH,\ \Delta} \text{indolizine } (R^1, R^2)$$ [3704]

R^1 = COOEt, CN; R^2 = Ac, PhCO, 4-BrC_6H_4CO, 4-MeOC_6H_4CO, CN, COOMe

Other ring systems synthesized similarly:

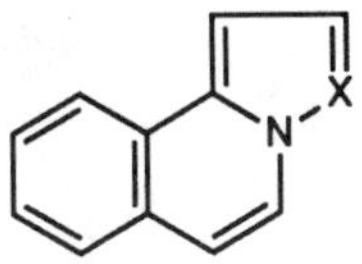

X = CH Pyrrolo[2,1-*a*]isoquinoline
X = N Pyrazolo[5,1-*a*]isoquinoline [3704]

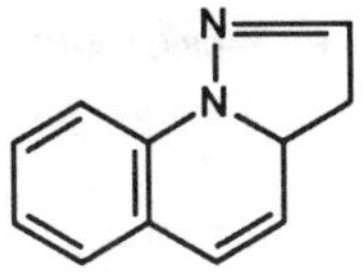

Pyrazolo[1,5-*a*]quinoline [3704]

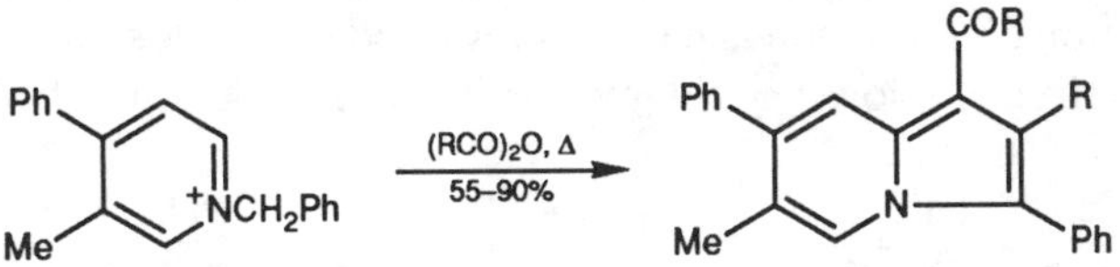

$$\xrightarrow[55\text{–}90\%]{(RCO)_2O,\ \Delta}$$ [2458]

R = Me, Et, Ph

Indolizine

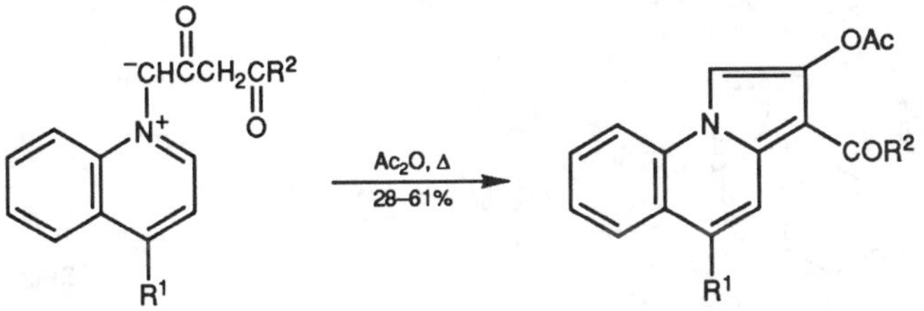

$$\xrightarrow[28\text{–}61\%]{Ac_2O,\ \Delta}$$ [2876]

R^1 = H, Me; R^2 = PhNH, EtO

Pyrrolo[1,2-*a*]quinoline

Another ring system synthesized similarly:

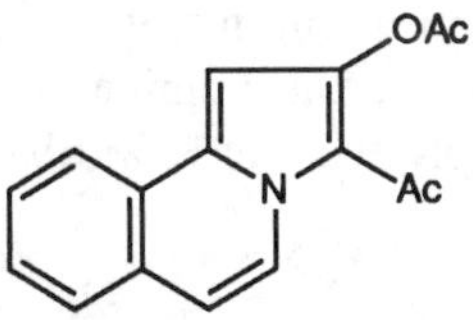

Pyrrolo[2,1-*a*]isoquinoline [2876]

$$+\ COR^2CH_2Br \xrightarrow[40\text{–}60\%]{1.\ Me_2CO,\ \Delta\ \ 2.\ NaHCO_3\text{ or }Na_2CO_3}$$ [2295]

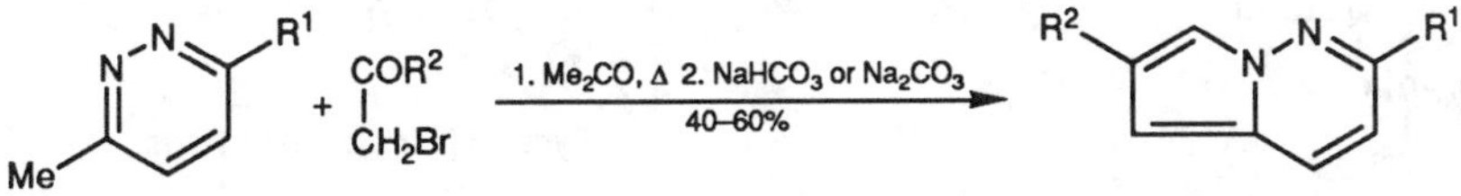

R^1 = H, Me; R^2 = H, Me, Ph

Pyrrolo[1,2-*b*]pyridazine

B. From 2-Acylpyridines

These compounds are cyclized by heating with two molar proportions of a benzaldehyde and ammonium acetate–acetic acid.

2-Pyridyl-$COCH_2R$ + ArCHO + NH_4OAc —(AcOH, Δ; 37–58%)→ Indolizine (N=CHAr, R, Ar substituted) [2337, 2369]

R = H, Me, Ph; Ar = Ph, 4-MeO-, 4-Cl-C_6H_4

Indolizine

2. Pyrrol-2-one

A pyrrolone ring is formed from a methylene and a ring-carbon in a low-temperature cyclo-oxidation of the amide (**104.2**) which is promoted by [bis(trifluoroacetyloxy)iodo]benzene. The application of hypervalent iodine compounds to the formation of C—C bonds has been reviewed [3714]. A similar compound (**104.3**) is cyclized in lower yield by stirring it at 0 °C with trifluoroacetic anhydride.

PhN(C=O)CH_2SMe (**104.2**) —($(CH_2Cl)_2$, $PhI(OCOCF_3)_2$; 63%)→ 2-Indolone (Ph, SMe) [3198]

(**104.2**) 2-Indolone

Cyclohexenyl-$N(Me)C(O)CH_2SOMe$ (**104.3**) —(CH_2Cl_2, $(CF_3CO)_2O$; 22%)→ (Me, SMe) [3037b]

(**104.3**)

3. Pyrazole

A side-chain phosphonate and a ring-nitrogen react with allyloxalyl chloride and a twofold proportion of a tertiary amine at a low temperature to form a fused pyrazole; some of the products have antibacterial activity.

Pyrazolidinone (NHBoc; N-$CH_2CH(CN)PO_3Et$) + COCl–COOR —(CH_2Cl_2, iPr_2NEt; 25%)→ Pyrazolo[1,2-*a*]pyrazol-1-one (ROC(O), NC, NHBoc) [3868]

Pyrazolo[1,2-*a*]pyrazol-1-one

R = CH_2=$CHCH_2$

4. Imidazole

When the dipyridylimines (**104.4**) are treated first with LDA at −78 °C and then with benzophenone, a new imidazole ring is formed in good yield [3190]. An isomeric imidazopyridine is produced by reacting a perfluoroalkanoic nitrile with a pyridinium salt (**104.5**) at low temperature in the presence of a strong base [3175].

Ar C=NR N → THF, LDA, Ph_2CO 66–88% → Ar N N R [3190]

(**104.4**) R = Et, iPr; Ar = 2-pyridyl — Imidazo[1,5-*a*]pyridine

$^{+}NCH_2COOR$ ClO_4^- + $CF_3(CF_2)_nCN$ → MeCN, NaH 19–25% → N $(CF_2)_nCF_3$ N COOR [3175]

(**104.5**) R = Et, tBu; *n* = 0,3 — Imidazo[1,2-*a*]pyridine

Annulation of a methylene and a carbon atom in a different ring with butyl-lithium, DMF and tetramethylethylenediamine (TMEDA) gives a low yield of product [2743].

Me N + BuLi + DMF → THF, TMEDA, N_2 14% → N N OHC [2743]

Imidazo[2,1,5-*cd*]indolizine

5. Thiophene

Heating 3,3-diphenylpropanoic acid with thionyl chloride is a reaction which does not produce the expected product; this gives a nearly theoretical yield of a benzo[*b*]thiophene-2-carboxylic acid. When Vilsmeier reaction conditions are applied to the pyrimidylthioacetic acid (**104.6**), a fused thiophene ring is formed.

COOH CHCH₂ Ph → $SOCl_2$, Δ ~99% → S COOH Ph [2473]

Benzo[*b*]thiophene

(104.6) → DMF–$POCl_3$, 38% → Thieno[2,3-*d*]pyrimidine-4,6-dione [3682]

6. 1,2-Dithiol-3-one

When thionyl chloride is heated in an atmosphere of argon with 4-methylquinoline for 20 h, the methyl group and the pyridine ring are chlorinated. The mechanism appears to be complex but the product has a fused dithiolone ring. However, treatment of 2-chloro-4-methylquinoline with thionyl chloride does not yield a dithioloquinolinone.

$SOCl_2$, Δ, 1–10% [3504]

R = H, Me → [1,2]Dithiolo[3,4-*c*]quinolin-1-one

7. Thiazole

The sulphur atom of thiazole is derived from thionyl chloride in this unusually mild reaction in which a reactive methylene is joined to a ring-carbon through sulphur. An *S*-acylmethyl side-chain may be annulated to a ring-nitrogen by heating with a mixture of a sodium alkanoate and its anhydride.

$SOCl_2$, pyr, 90% [2065]

Thiazolo[4,5-*d*]pyrimidine-5,7-dione

+ R^4COONa + $(R^4CO)_2O$ → Δ, 52–98% [2363]

R^1 = Me, Ph, 4-Br-, 4-NO_2-C_6H_4; R^2 = H, Ph;
R^3 = Ph, 4-$NO_2C_6H_4$; R^4 = Me, Et

Imidazo[2,1-*b*]thiazole

II. FORMATION OF A SIX-MEMBERED RING

1. Pyridine

Vilsmeier reagent supplies the required carbon to form a new fused pyridine by joining a methylene to a carbon of a pyrrole ring [3220]. Heating a similar compound with DMFDEA is an alternative method [2586]. Annulation of a reactive methyl attached to one ring to a ring-nitrogen in another may be achieved by heating the compound with iodine and cyclohexene oxide [3116].

PhN—N; CH_2CN; $POCl_3$–DMF, 39%; PhN—N; CN [3220]

Pyrazolo[4,3-*e*]pyrrolo[1,2-*a*]pyridine

Me; NO_2; DMFDEA, DMF, Δ, 60%; NO_2 [2586]

Pyrrolo[1,2-*a*]quinoline

H N; COMe; i, Δ, 72%; H N; I^-; OH [3116]

i, I_2, cyclohexene oxide, CCl_4

Indolo[2,3-*a*]quinolizine

A variation on the cyclization of α-cyanomethylpyridines with a malonic ester (p. 566) is the ring closure of the acetate (**104.7**) with ethoxymethylenemalononitrile and of the quinolinone (**104.8**) with ethyl (ethoxymethylene)cyanoacetate. A methylene group in the succinic acid (**104.9**) readily forms an anion which attacks the electron-rich pyrrole ring on heating the compound in piperidine.

O; CH_2COOEt; NH; N; OAr + $EtOCH{=}C(CN)_2$ Δ, 73% → COOEt; O; N; CN; N; NH_2; OAr [3284]

(**104.7**) Ar = 4-MeC_6H_4

Benzo[*f*]pyrido[1,2-*a*][1,8]naphthyridin-12-one

(104.8) + $NCC(=CHOEt)COOEt$ →(Δ, 90%) Benzo[c]quinolizin-6-one [3284]

(104.9) →(py, pip, Δ, 51%) Azepino[2,1,7-cd]indolizine [2701]

2. Pyridin-2-one

A reactive methylene may be joined to a ring-nitrogen by heating with one of several malonic acid derivatives: methyl 2-cyano-3,3-di(methylthio)acrylate [2071], a 2,4,6-trichlorophenyl malonate (**104.10**) [2382, 3258] or diethyl ethoxymethylenemalonate [3566]. The last named reagent was successful where several others failed.

R = Me, Et; + $(MeS)_2C{=}C(CN)COOMe$ →(Δ, ~49%) 4-Quinolizinone [2071]

(104.10) $R = PhCH_2, Ph$; $Ar = 2,4,6\text{-}Cl_3C_6H_2$; X = O, S →(Δ, 50–100%) Pyrido[2,1-b]benzoxazol-1-one; Pyrido[2,1-b]benzothiazol-1-one [2382]

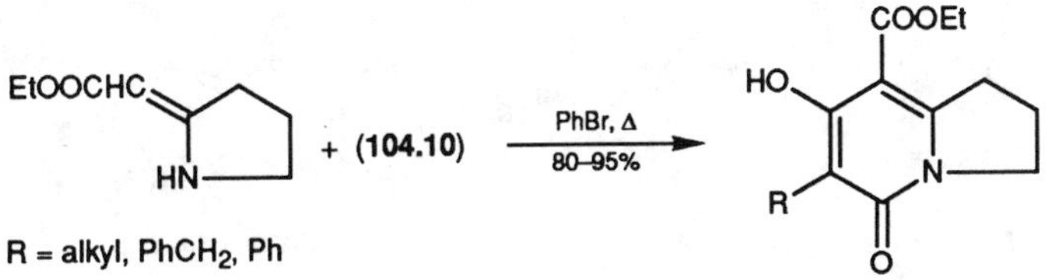

EtOOCHC= (pyrrolidin-2-ylidene, HN) + (104.10) —PhBr, Δ / 80–95%→ [3258]

R = alkyl, $PhCH_2$, Ph

5-Indolizinone

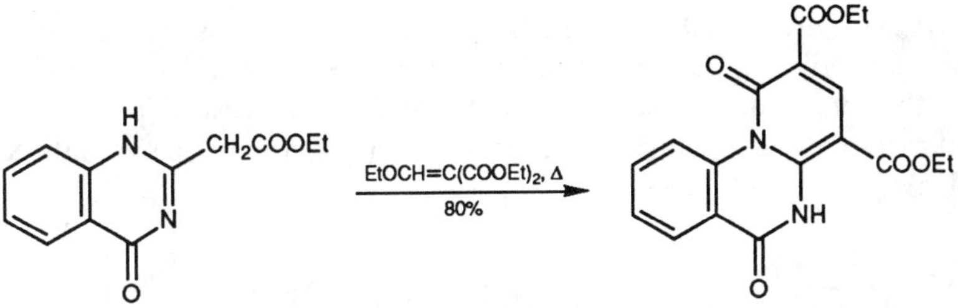

2-(CH_2COOEt)quinazolin-4-one —EtOCH=C(COOEt)$_2$, Δ / 80%→ [3566]

Pyrido[1,2-*a*]quinoxaline-1,6-dione

Dimethyl oxalate [2955], diketene [2789], an acetoacetate [3489], 3-aminocrotonic ester [3489] and the Vilsmeier reagent [2994] have also been demonstrated to be useful synthons.

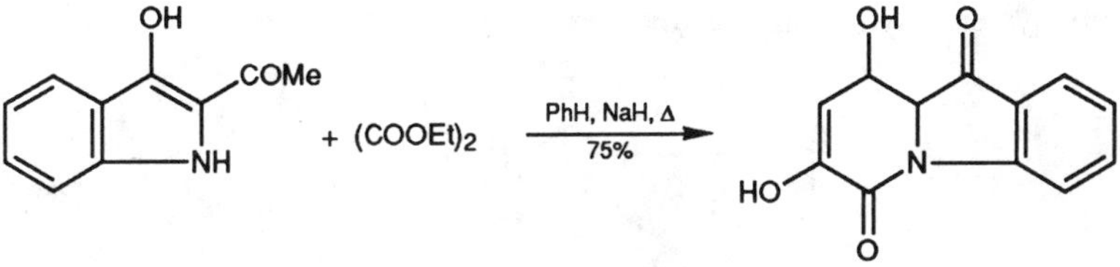

3-hydroxy-2-acetylindole + $(COOEt)_2$ —PhH, NaH, Δ / 75%→ [2955]

Pyrido[1,2-*a*]indole-6,10-dione

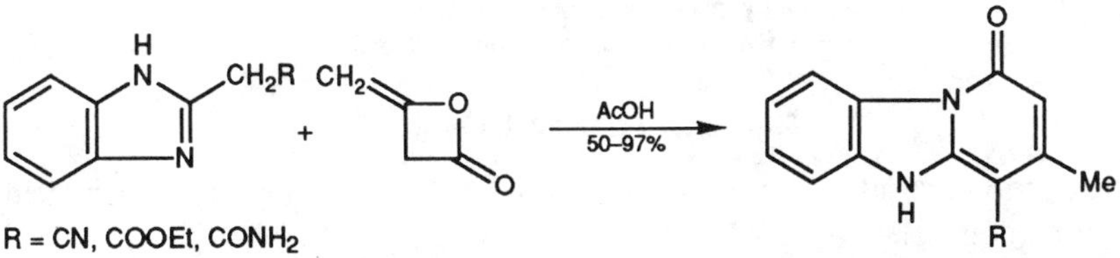

2-(CH_2R)benzimidazole + diketene —AcOH / 50–97%→ [2789]

R = CN, COOEt, $CONH_2$

Pyrido[1,2-*a*]benzimidazol-1-one

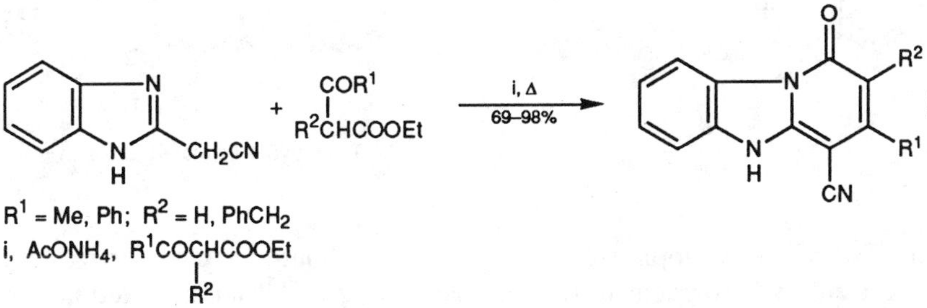

2-(CH_2CN)benzimidazole + $R^2CH(COR^1)COOEt$ —i, Δ / 69–98%→ [3489]

R^1 = Me, Ph; R^2 = H, $PhCH_2$

i, $AcONH_4$, $R^1COCH(R^2)COOEt$

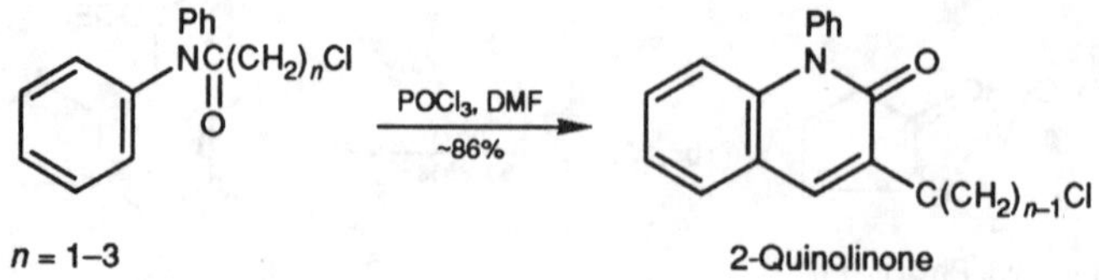
Ph
NC(CH2)nCl
O
POCl3, DMF
~86%
Ph
N
O
C(CH2)n-1Cl
n = 1–3
2-Quinolinone

[2994]

3. Pyrimidine-2,4-dione or Pyrimidin-2-one, 4-Thioxo-

The synthetically useful ethoxycarbonyl isocyanate reacts with 2-benzimidazolylacetonitrile or its corresponding acetate ester. Alkyl isocyanates also cyclize the benzimidazol-2-acetate ester [3570, 3588]. Acyliso(thio)cyanates and a mild base convert 2-methylindolenines (3*H*-indoles) to a fused pyrimidinedione ring [2695].

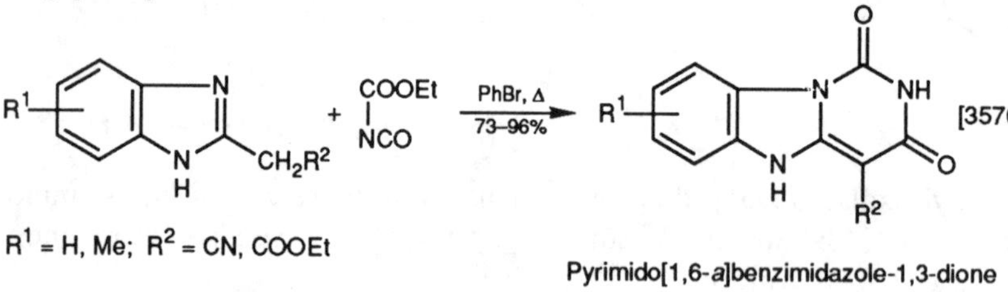
R1
N
N
H
CH2R2
+
COOEt
NCO
PhBr, Δ
73–96%
O
N
NH
O
R2
R1 = H, Me; R2 = CN, COOEt
Pyrimido[1,6-a]benzimidazole-1,3-dione

[3570, 3588]

Other ring systems synthesized similarly:

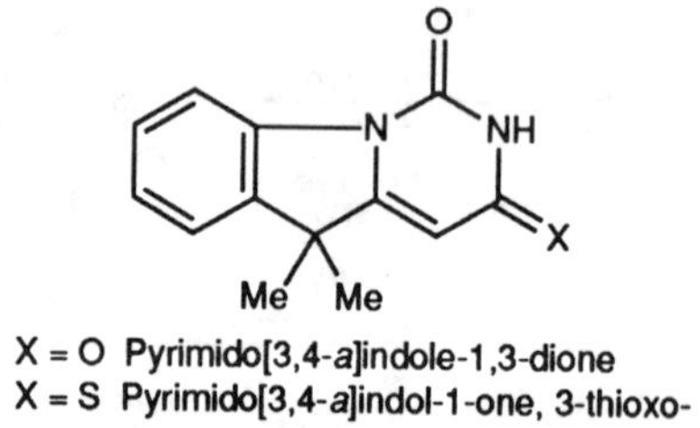
O
N
NH
X
Me
Me
X = O Pyrimido[3,4-a]indole-1,3-dione
X = S Pyrimido[3,4-a]indol-1-one, 3-thioxo-

[2695]

4. 1,2,3-Triazine 1-Oxide

An attempt to nitrosate an indole-2-acetic ester at C-3 gave instead a high yield of the triazinoindole.

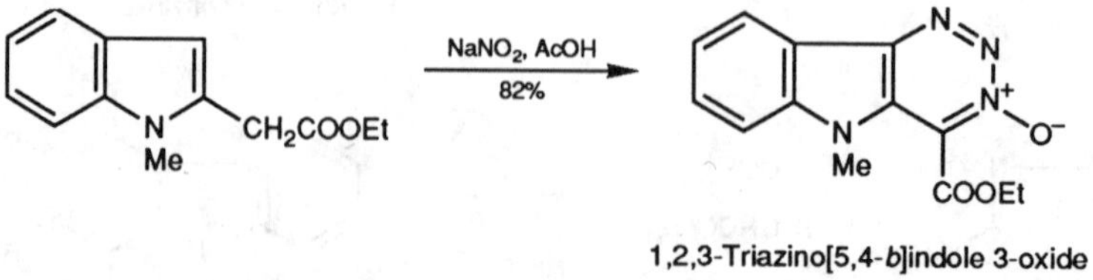
N
Me
CH2COOEt
NaNO2, AcOH
82%
N
N
N+
O−
N
Me
COOEt
1,2,3-Triazino[5,4-b]indole 3-oxide

[3430]

5. Pyran

Thermolysis of the cyanophosphorane (**104.11**) at about 265 °C under reduced pressure resulted in tandem intramolecular Wittig and Claisen reactions; in some reactions, a mixture of a benzopyran and a benzofuran is obtained. This may be difficult to separate.

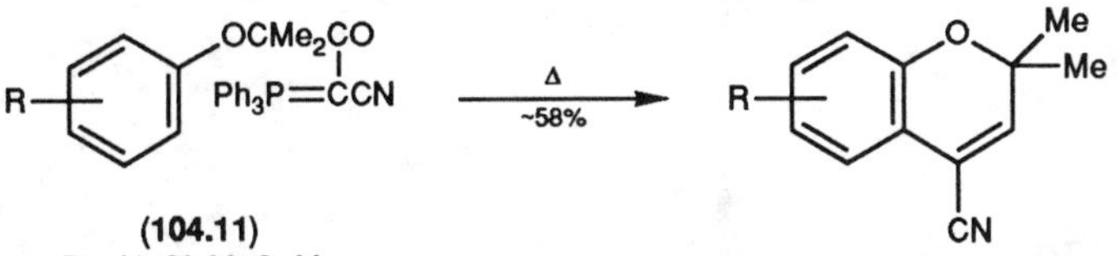

(104.11)
R = H, Cl, MeO, Me

Δ, ~58%

2*H*-1-Benzopyran [3293]

6. 1,4-Thiazin-3-one

Warming an *N*,*N*-diarylacetamide with thionyl chloride at about 55 °C converts it into a 2,2-dichloro-4-arylthiazinone.

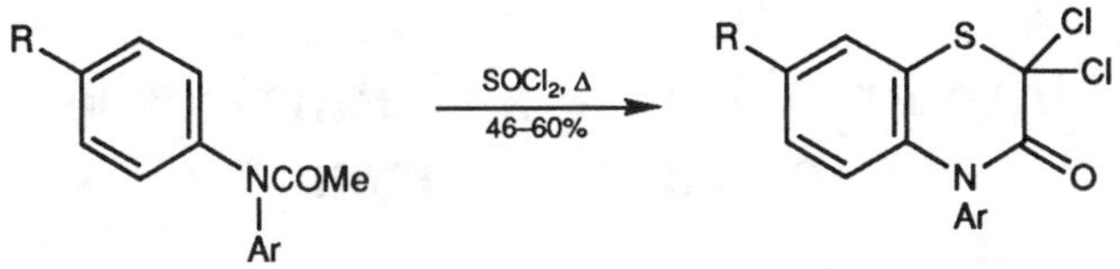

$SOCl_2$, Δ, 46–60%

R = H, Me, MeO; Ar = 4-RC_6H_4

1,4-Benzothiazin-3-one [3112]

III. FORMATION OF A SEVEN-MEMBERED RING

1. Azepine

Addition of a 2-methylquinazolin-4-one to DMAD over a long time at the boiling point of acetonitrile leads to the formation of a new fused azepine ring. A two- or three-fold excess of methyl propiolate reacts with 2-methylbenzothiazole in a similar fashion.

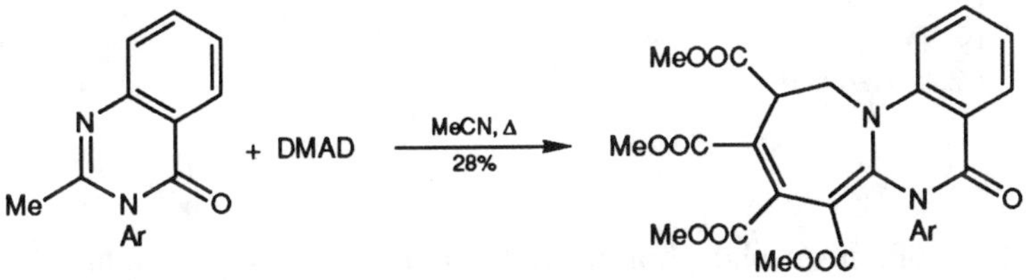

+ DMAD; MeCN, Δ, 28%

Ar = 2-MeC_6H_4

Azepino[1,2-*a*]quinazolin-5-one [2401]

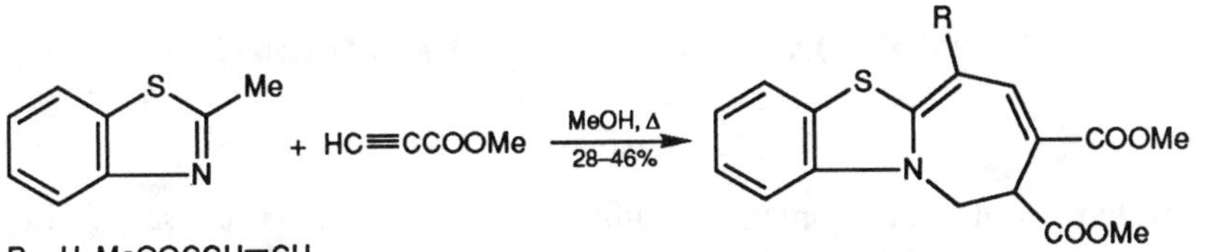

+ HC≡CCOOMe; MeOH, Δ, 28–46%

R = H, MeOOCCH=CH

Azepino[2,1-*b*]benzothiazole [3555]

CHAPTER 105

Nitro or *N*-Oxide and Ring-carbon or Ring-nitrogen

I. Formation of a Five-membered Ring 1328
1. Pyrrole 1328
2. Pyrrol-2-one or Pyrrole-2,3-dione 1330
3. Pyrazole 1330
4. Imidazole 1331
5. Imidazole 3-Oxide 1331
6. 1,2,3-Triazole 1331
7. Furan 1332
8. Isoxazole 1332
9. 1,2,4-Oxadiazol-5-one 1333
II. Formation of a Six-membered Ring 1333
1. Pyridazine 1-Oxide 1333
2. 1,2-Oxazine 1333
3. 1,4-Thiazine 1334
4. 1,2,4-Oxadiazine 1334

Cyclizations of a nitro-phosphorane and of a nitroso-ring-carbon are shown in this chapter.

I. FORMATION OF A FIVE-MEMBERED RING

1. Pyrrole

7-Substituted indoles are sometimes difficult to synthesize because of the tendency of the Fischer (and some other) reactions to produce the 5-substituted isomers. A method which avoids this problem is the treatment of *o*-substituted nitrobenzenes with a vinylmagnesium bromide at −40 °C; 3- and 4-halonitrobenzenes, on the other hand, give poor yields of 4-, 5- and 6-haloindoles by this method. 7-Methoxy- (and 7-hydroxy-)indoles are biologically significant compounds [3630] but there is little information as to how *o*-methoxynitrobenzene would behave in this reaction. Deoxygenation of the *β*-nitro-*α*-phenylstyrenes with triethyl phosphite at 150 °C usually produces high yields of 3-phenylindoles [3951]. Flash vacuum pyrolysis (fvp) (at 800 °C and 1–6 × 10^{-4} Torr) of 2- and 4-benzylpyrimidine 1-oxide (review of pyrimidine oxides [3795]) gives pyrimidoindoles [3059a]. 4-Nitrophenylthiazole undergoes a cyclodeoxygenation when treated with triethyl phosphite, the nitrene intermediate attacking C-4 of the thiazole [2823].

R, NO_2 + $3CH_2{=}CHMgBr$ —THF, 41–67%→ R, H, N [3630]

R = Me, Cl, Br, Me_3SiO Indole

Other ring systems synthesized similarly:

HN, N [3630] H, N [3630]

Pyrrolo[2,3-*f*]quinoline Indeno[1,7-*fg*]indole

NO_2, C=CR, Ph —$(EtO)_3P$, Δ, 32–99%→ H, N, R, Ph [3951]

R = tBuS, PhS, NO_2

$PhCH_2$, N, N^+, ^-O —fvp, Δ, 48%→ N, N [3059a]

Pyrimido[1,2-*a*]indole

Another ring system synthesized similarly:

N, N [3059a]

Pyrimido[1,6-*a*]indole

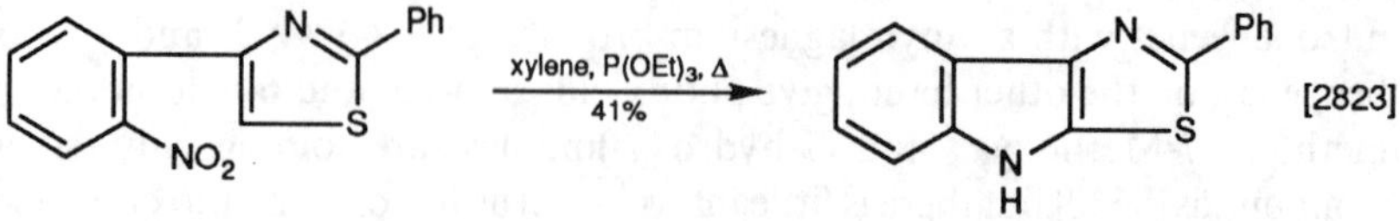

xylene, $P(OEt)_3$, Δ 41% [2823]

Thiazolo[5,4-*b*]indole

2. Pyrrol-2-one or Pyrrole-2,3-dione

The lability of the nitro group of 4-nitroquinoline 1-oxide is utilized in the reaction with an enamine at ambient temperature to form a new pyrrolone ring. When the reaction time is extended, the product lacks the *N*-hydroxy group. The lability of nitrobenzenes has been reviewed [1398, 1432]. A *C,N*-diphenylnitrone [*Chemical Abstracts* name: *N*-(phenylmethylene)benzamine *N*-oxide] reacts with two moles of dichloroketene (review [2348]) (prepared *in situ* from chloroacetyl chloride and TEA at about 5 °C) to give a high yield of the 3,3-dichloroindol-2-one, which is easily converted into the biologically and analytically interesting isatin.

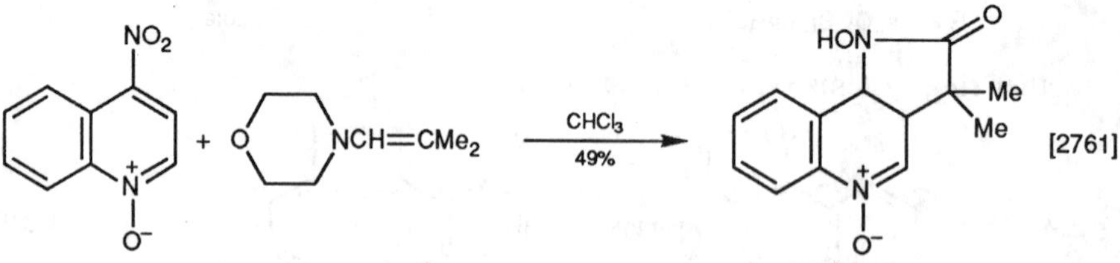

$CHCl_3$ 49% [2761]

Pyrrolo[3,2-*c*]quinolin-2-one 5-oxide

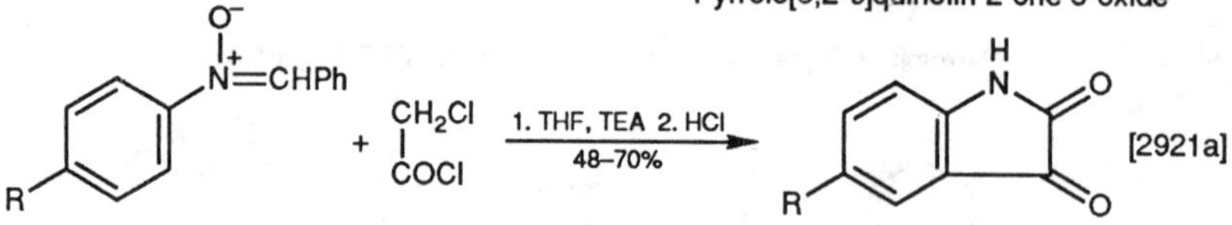

1. THF, TEA 2. HCl 48–70% [2921a]

R = H, Cl, Br, Me

2,3-Indoledione

3. Pyrazole

When the isoxazole (**105.1**) is treated with an excess of diazomethane at laboratory temperature, a [1,5]sigmatropic shift occurs and the benzoyl group migrates from the bridgehead position to an adjacent carbon (that derived from diazomethane)—a move which resembles the Van Alphen–Häuttel rearrangement [B-4].

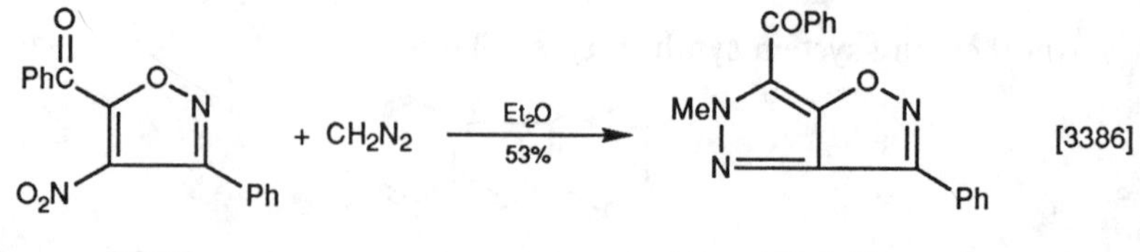

+ CH_2N_2 Et_2O 53% [3386]

(**105.1**)

Pyrazolo[3,4-*d*]isoxazole

4. Imidazole

Hot sulphuric acid converts a 2-nitroenamino-oxazine into an imidazoazine with loss of the nitro group. A different kind of annulation occurs when the nitro is attached to a phenyl ring, and the compound is heated in a high boiling solvent to form an intramolecular C—N bond with loss of the nitro group.

MeOH, H_2SO_4, Δ; 70–90% [3110]

X = CH, N

Imidazo[1,2-*a*]pyridine
Imidazo[1,2-*a*]pyrazine

xylene, Δ; 78% [3097]

Pyrido[1,2-*a*]benzimidazole

5. Imidazole 3-Oxide

N,α-Diphenyl-α-nitrosonitrone, an explosive compound, spontaneously cyclizes at room temperature to a benzimidazole 3-oxide.

CH_2Cl_2; 86% [3428]

Benzimidazole 3-oxide

6. 1,2,3-Triazole

A nitro group attached to a pyrimidinone or pyridinone ring may be displaced by a nucleophile such as an azide ion. On heating the azide in DMF, a fused triazole ring is formed.

DMF, NaN_3, Δ; 50–87% [2176]

R^1 = H, Me; R^2 = NH_2, Me_2N, EtO;
X = CH, N

1,2,3-Triazolo[4,5-*b*]pyridin-5-one
1,2,3-Triazolo[4,5-*d*]pyrimidin-5-one

Other ring systems synthesized similarly:

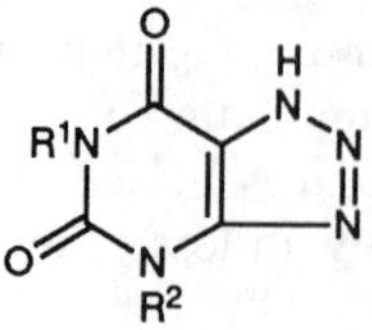

[2176] [2176]

1,2,3-Triazolo[4,5-*d*]pyrimidine-5,7-dione

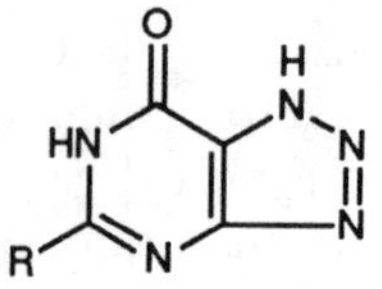

1,2,3-Triazolo[4,5-*d*]pyrimidin-7-one

7. Furan

4-Nitropyridine 1-oxide reacts with 1-cyano-2-phenylethyne in refluxing toluene to form a 3-cyano-2-phenylfuran ring fused to the pyridine.

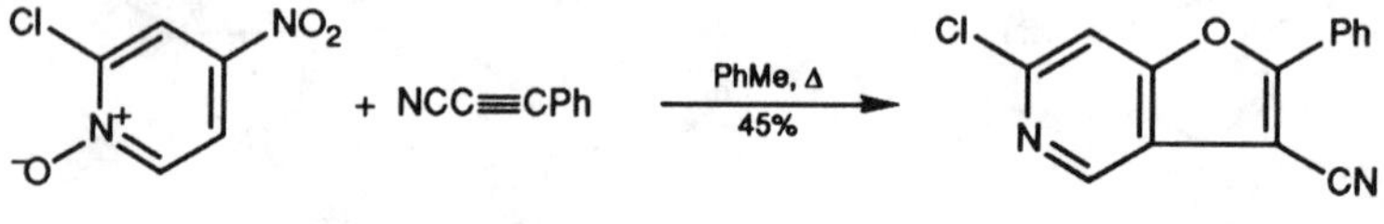

+ NCC≡CPh —PhMe, Δ, 45%→ [2760]

Furo[3,2-*c*]pyridine

8. Isoxazole

The regioselectivity of 1,3-dipolar cyclization is affected by steric and electronic factors. Addition of cyclic *N*-oxides to propynoic esters shows regioselectivity as only the 1-carboxylic ester is obtained when the temperature is kept at 60 °C. When the phosphorane (**105.2**) is heated in methanol, the isoxazole is formed instead of the expected diarylmethane.

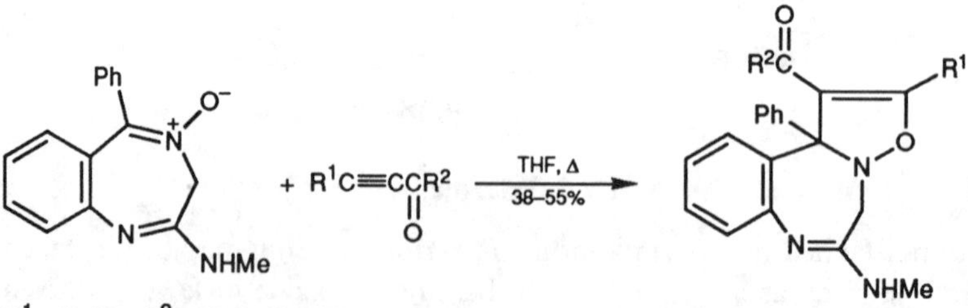

+ $R^1C{\equiv}CC(O)R^2$ —THF, Δ, 38–55%→ [3073, 3131]

R^1 = H, Ph; R^2 = MeO, EtO, Me

Isoxazolo[2,3-*d*][1,4]benzodiazepine

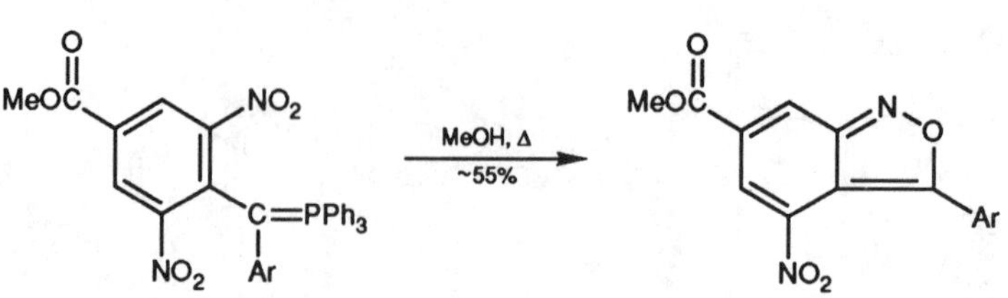

—MeOH, Δ, ~55%→ [2526]

(**105.2**)
Ar = 4-$MeOC_6H_4$

2,1-Benzisoxazole

9. 1,2,4-Oxadiazol-5-one

Annulation of an electron-enriched pyridine *N*-oxide with phenyl isocyanate at 110 °C or lower gives an oxadiazolone, but 3-bromopyridine *N*-oxide behaves differently (see Chapter 90); an unsymmetrically substituted *N*-oxide usually leads to a mixture of isomers.

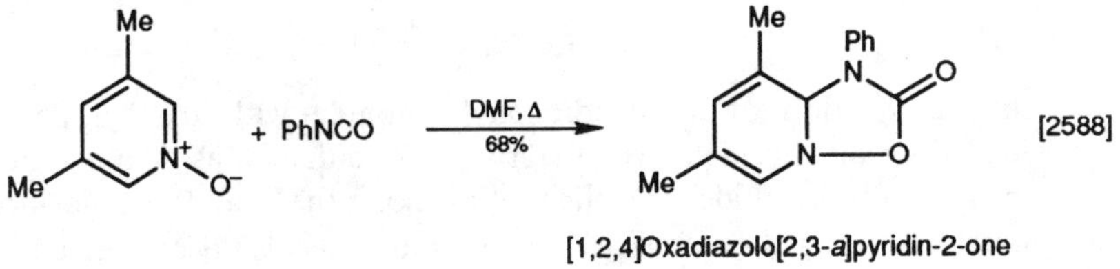

+ PhNCO —DMF, Δ, 68%→ [2588]

[1,2,4]Oxadiazolo[2,3-*a*]pyridin-2-one

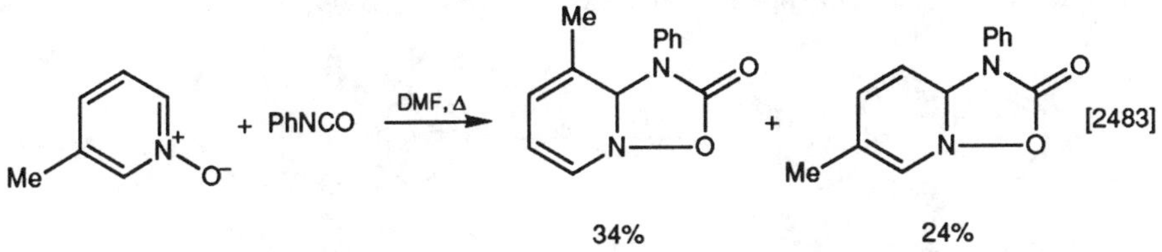

+ PhNCO —DMF, Δ→ 34% + 24% [2483]

II. FORMATION OF A SIX-MEMBERED RING

1. Pyridazine 1-Oxide

Nitro and ring-nitrogen in different rings may be joined together by heating with iodine–sodium acetate in ethanol. The *N*-oxide was formed instead of the expected dehydrogenated pyridine.

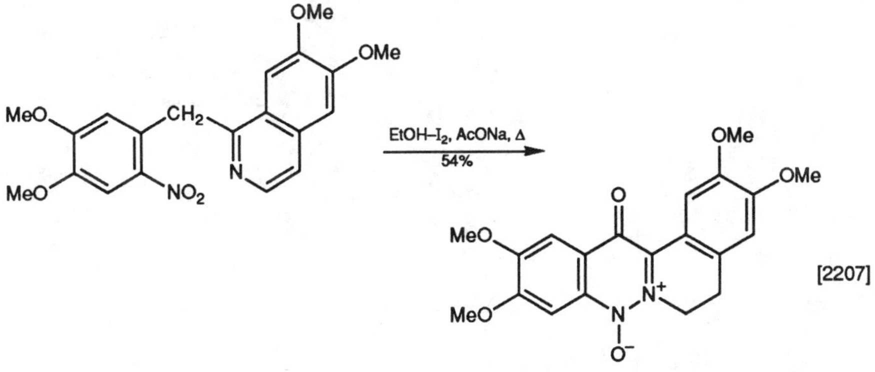

—EtOH–I_2, AcONa, Δ, 54%→ [2207]

Isoquino[2,1-*b*]cinnolin-13-one 8-oxide

2. 1,2-Oxazine

Irradiating a *β*-nitrovinylfuran in acetone results in the formation of a fused 1,2-oxazine.

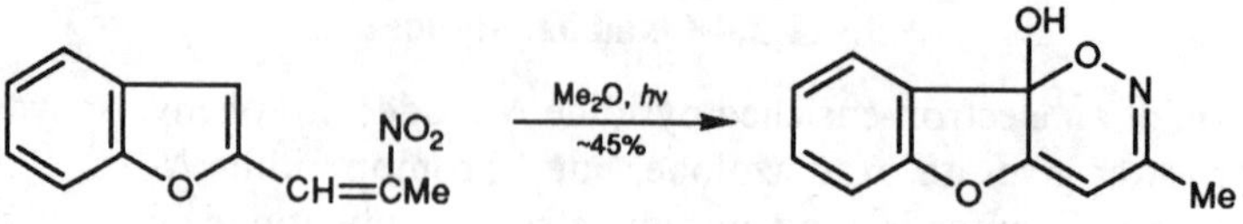

NO_2 CH=CMe → (Me_2O, $h\nu$; ~45%) → OH, O, N, Me

Benzofuro[2,3-*e*]-1,2-oxazine

[2448]

3. 1,4-Thiazine

Reductive cyclization of a nitrodiphenyl sulphide with triethyl phosphite proceeds by a rearrangement (via a spirocyclohexadiene), as demonstrated by 4-monosubstituted sulphide. The effect of blocking the 2- and/or 6-positions of the unnitrated ring with methyl or chloro groups has also been studied [2256]. The chemistry of phenothiazines has been reviewed [B-44, 3332].

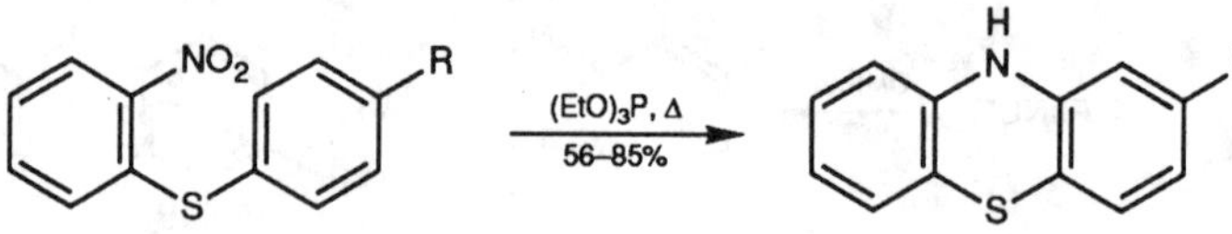

NO_2, S, R → ($(EtO)_3P$, Δ; 56–85%) → H, N, S, R

R = Me, tBu, Cl

Phenothiazine

[2168]

4. 1,2,4-Oxadiazine

Nitrosobenzene and ethyl cyanoformate *N*-oxide react at or below room temperature to give a benzoxadiazine.

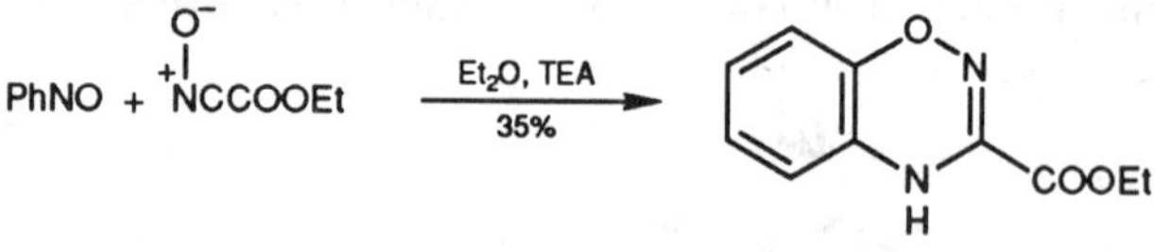

PhNO + $^{+}$NCCOOEt (O^-) → (Et_2O, TEA; 35%) → O, N, N, H, COOEt

1,2,4-Benzoxadiazine

[3428]

CHAPTER 106

Ring-carbon or Ring-sulphur and Ring-nitrogen

I. Formation of a Five-membered Ring 1335
1. Pyrrole 1335
2. Pyrrol-2-one or Pyrrole-2,3-dione 1336
3. Pyrazole 1337
4. Imidazole 1338
5. 1,2,4-Triazole 1338
6. 1,2,4-Oxadiazole 1338
7. 1,2,4-Thiadiazol-3-one 1339
II. Formation of a Six-membered Ring 1339
1. Pyridine 1339
2. Pyridin-2-one 1340
3. Pyrimidin-4-one 1340
4. 1,3-Oxazin-6-one 1341
III. Formation of a Seven-membered Ring 1341
1. Azepine 1341

I. FORMATION OF A FIVE-MEMBERED RING

1. Pyrrole

Addition of a 1,2,4-triazine to a DMAD (two molar proportions) results in the formation of a new fully substituted pyrrole ring [2915a]. DMAD also reacts with *N*-substituted pyridazines under basic conditions to give a new fused pyrrole ring [3304]. Reaction of the cyclopropenone (**106.1**) with 4-picoline and then with pivoloyl chloride gives a high yield of an indolizine [3553].

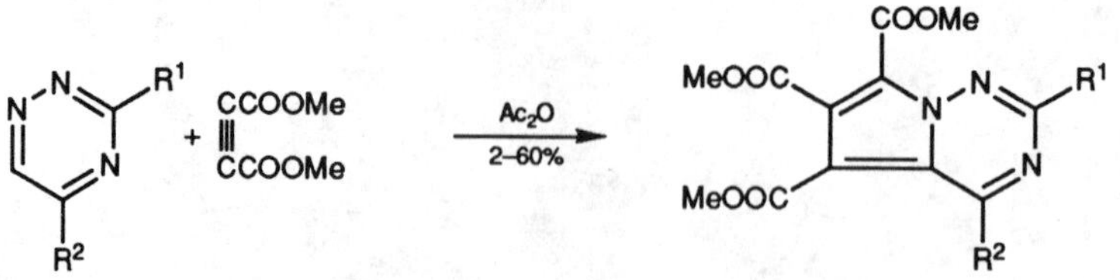

Ac_2O, 2–60%

[2915a]

R^1 = MeO, Ph, 4-MeC_6H_4; R^2 = H, MeO, Ph

Pyrrolo[2,1-*f*][1,2,4]triazine

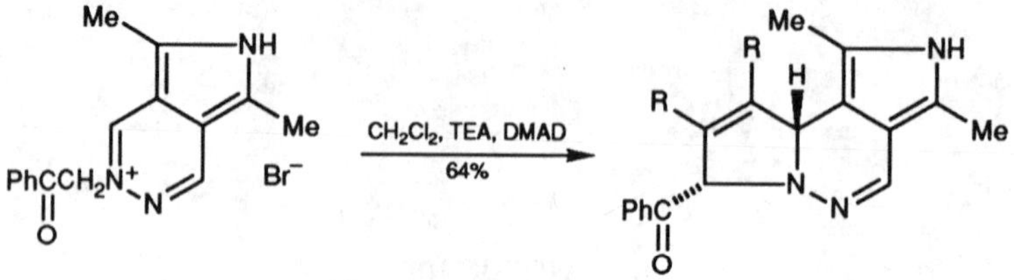

CH_2Cl_2, TEA, DMAD, 64%

[3304]

R = COOMe

Dipyrrolo[1,2-*b*:3′,4′-*d*]pyridazine

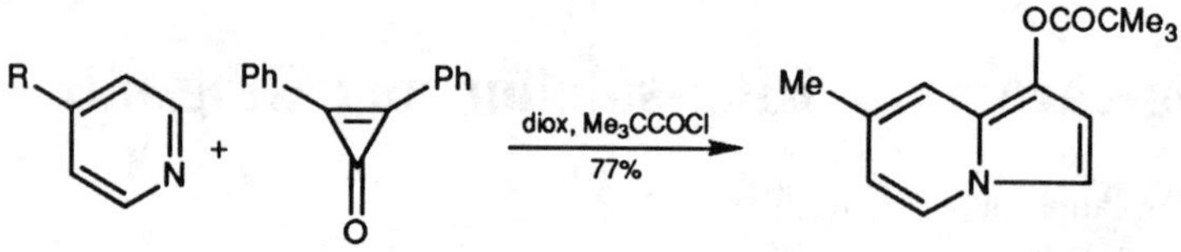

diox, Me_3CCOCl, 77%

[3553]

(106.1) (106.2) Indolizine

R = Me

2. Pyrrol-2-one or Pyrrole-2,3-dione

1,2-Diarylcyclopropenone (**106.1**) reacts with argon-purged pyridine to give labile intermediates (3-indolizinols) which are simultaneously oxidized and dimerized. When similar reactants were heated in dioxane or when 4,4′-bipyridyl replaced pyridine, bisindolizin-1-ones were obtained.

(106.1) + (106.2) — pyr, $Cu(OAc)_2$, Δ, ~50%

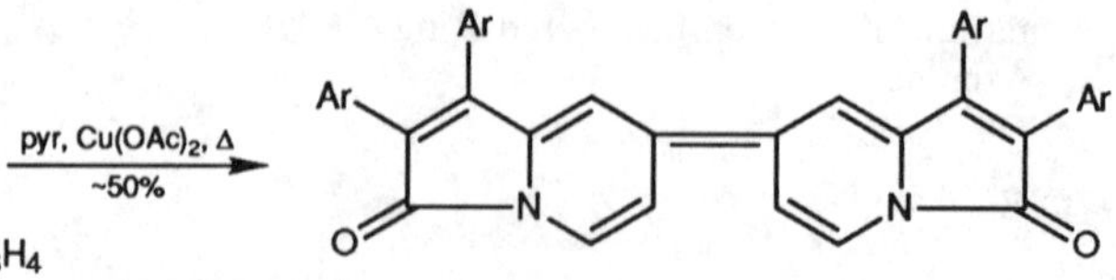

R = H; Ar = Ph, 4-$tBuC_6H_4$

[3552]

3-Indolizinone

Diethyl methyloxalacetate (**106.2**, R^1 = Me) adds to tetrahydropyridine and forms a new pyrroledione ring, but diethyl oxalacetate gives a product which is shown to be the enol form of the corresponding dione [2744]. The 1,2-double bond of 3*H*-indoles undergoes a cycloaddition on treatment at ambient temperature with either diethyl oxalacetate or oxalopropionate. When R^1 = H, the product may exist as its 2-hydroxy tautomer [3394]. Successive *N*- and

C-acylation of a 9,10-dihydroacridine with oxalyl chloride can lead to the formation of a *peri*-positioned pyrrole-2,3-dione ring being formed in good-to-excellent yield [2321].

3. Pyrazole

Peri-annulation of the benzotriazole ylide (**106.4**) occurs when it is heated in benzene with an alkyne. Arylethynes produce a mixture of positional isomers but the propynoate ester gives a preponderance of (**106.5**, R^2 = $COOR^4$).

4. Imidazole

Doubly fused imidazoles may be synthesized by trapping the benzyne formed by lithiation and loss of lithium fluoride from the amidine (**106.6**). By limiting the amount of t-butyllithium used to about two equivalents, monosubstitution of the intermediate with an electrophile (such as water or iodomethane) is possible.

1. tBuLi, THF 2. H_2O or MeI; 30–74% [3948]

(106.6)
n = 0–2; R = H, Me

Pyrrolo[1,2-*a*]benzimidazole
Pyrido[1,2-*a*]benzimidazole
Azepino[1,2-*a*]benzimidazole

Other ring systems synthesized similarly:

[3948]

X = O [1,3]Oxazino[3,2-*a*]benzimidazole
X = S [1,3]Thiazino[3,2-*a*]benzimidazole

5. 1,2,4-Triazole

The dipolarophilic nature of the C=N bond of the 3*H*-indole (**106.7**) is demonstrated by its reaction (without external heating) with a nitrile imine.

PhH, TEA; ~89% [3989]

(106.7)
Ar = Ph, 4-Me-, 4-Cl-C_6H_4

[1,2,4]Triazolo[4,3-*a*]indole

6. 1,2,4-Oxadiazole

Benzohydroxamic acid chloride (the oxime of benzoyl chloride) adds on to the benzodiazepin-2-one (**106.8**) at ambient temperature with loss of chlorine.

(106.8)
R = Me, Ph

PhCCl(=NOH); CH_2Cl_2, TEA; 45–87%

[1,2,4]Oxadiazolo[4,5-*a*][1,5]benzodiazepin-5-one [3672]

7. 1,2,4-Thiadiazol-3-one

A ring S(II) is converted into a bridgehead S(IV) during the cyclization of the thiadiazole-3-thione (**106.9**) by heating it with an isocyanate in chloroform for between 6 and 24 h.

(106.9) + R^2NCO; $CHCl_3$, Δ; 24–98%

2aλ⁴-Thia-2,3,4a,7a-tetraazacyclopent[*cd*]indene-1,4-dione [3656]

R^1 = Me, Ph;
R^2 = Me, $Cl(CH_2)_2$, Ph, $CH_2=CMeCOOCH_2CH_2$

II. FORMATION OF A SIX-MEMBERED RING

1. Pyridine

The NH and neighbouring 2-CH of imidazole are annulated when the compound is heated with cyclopropyl aryl ketone and PPA in decalin or heated without a solvent. When the 2-position is blocked, cyclization at C-5 of the imidazole occurs [3276]. 3*H*-Indole adds on to DMAD at room temperature to form a pyridoindole in moderate-to-good yield [3394]. 2-Styrylbenzimidazole is photocyclized in good yield in methanol containing iodine [2845].

R = H: decalin, PPA, Δ; 17–79% → Imidazo[1,2-*a*]pyridine

R = Me: Δ; 36% → Imidazo[1,5-*a*]pyridine [3276]

R = H, Me;
Ar = Ph, MeO-, Cl-, F-, MeO-C_6H_4

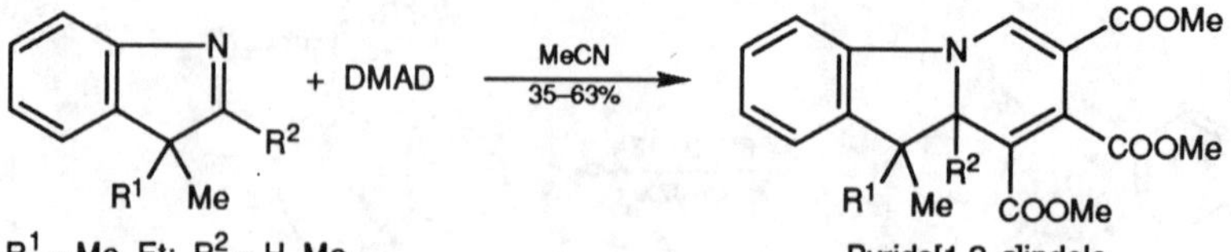

+ DMAD → MeCN, 35–63% [3394]

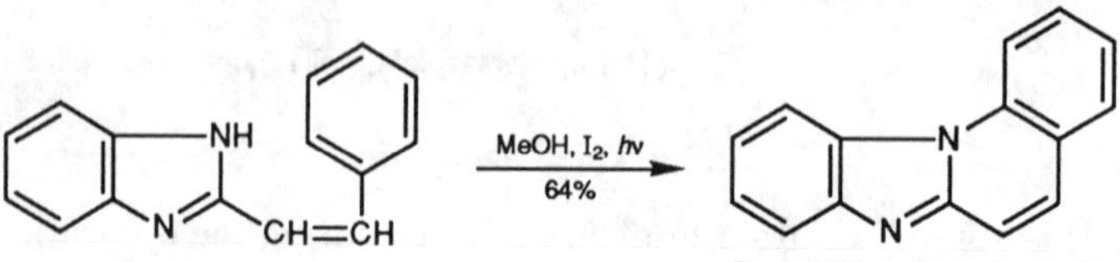

R^1 = Me, Et; R^2 = H, Me

Pyrido[1,2-*a*]indole

MeOH, I_2, *hν*, 64% [2845]

Benzimidazo[1,2-*a*]quinoline

2. Pyridin-2-one

During an attempt to formylate the carbazole (**106.10**) by heating it with *N*-methylformamide and phosphorus oxychloride, ring closure gave the pentacyclic compound instead.

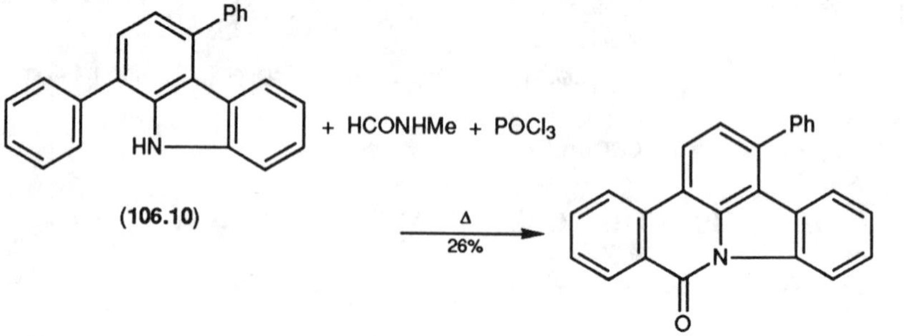

(**106.10**) + HCONHMe + $POCl_3$ → Δ, 26% [2214]

Indolo[3,2,1-*de*]phenanthridin-9-one

3. Pyrimidin-4-one

A new pyrimidinone ring is formed when the 6-benzoyladenosine (**106.11**) is dissolved in acetonitrile and irradiated under argon and in the presence of iodine, tetracyanoethene or 1,4-dinitrobenzene. This may be the first report of an oxidative photocyclization of a 1,3-diazahexatriene system, N=C—N=C—C=C.

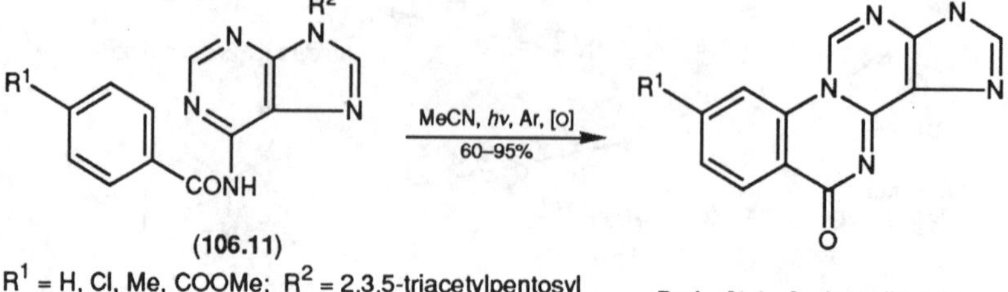

(**106.11**) → MeCN, *hν*, Ar, [O], 60–95% [3305]

R^1 = H, Cl, Me, COOMe; R^2 = 2,3,5-triacetylpentosyl

Purino[1,6-*a*]quinazolin-5-one

4. 1,3-Oxazin-6-one

Phthalazines and dimethylketene (in a molar ratio of 1:2) give a high yield of an adduct without external heating. Quinazolines need four molar equivalents of diketene and form an adduct which contains two oxazinone rings [2161].

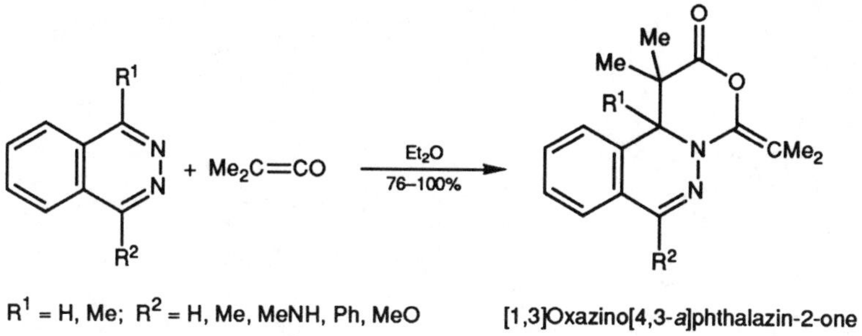

R^1 = H, Me; R^2 = H, Me, MeNH, Ph, MeO + $Me_2C{=}CO$ → (Et_2O, 76–100%) [2161]

[1,3]Oxazino[4,3-*a*]phthalazin-2-one

III. FORMATION OF A SEVEN-MEMBERED RING

1. Azepine

Cycloaddition of cyclohepta[*b*]pyrrole to DMAD in boiling benzene gives a product which contains two molecules of DMAD and a minor product which contains one DMAD molecule. Cyclohept[*d*]imidazole gives lower yields.

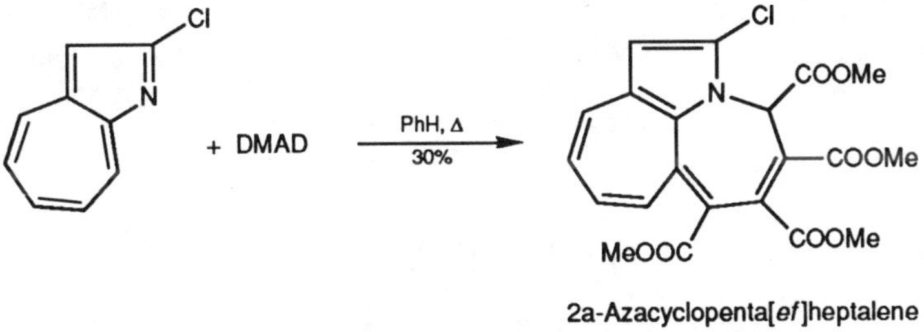

+ DMAD → (PhH, Δ, 30%) [2936]

2a-Azacyclopenta[*ef*]heptalene

Another ring system synthesized similarly:

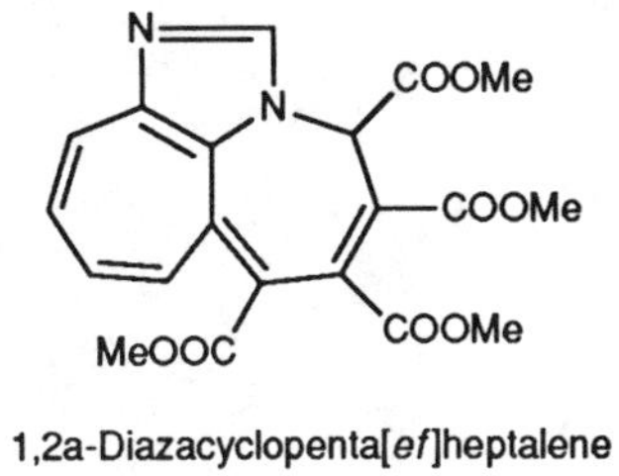

[2936]

1,2a-Diazacyclopenta[*ef*]heptalene

List of Books and Monographs

This list is a continuation of that on pages 581 and 582 of Part 1 (1987). In addition to additional volumes in this Series on isoquinolines [B-17] and quinolines [B-13], the following are some recently published books that are of interest in synthetic heterocyclic chemistry.

B-40. Barton, D. and Ollis, W.D. (eds), *Comprehensive Organic Chemistry*, Pergamon Press, Oxford, 1979.

B-41. Heck, R.F., *Palladium Reagents in Organic Synthesis*, Academic Press, London, 1985.

B-42. Boger, D.L. and Weinreb, S.M., *Hetero Diels–Alder Methodology in Organic Synthesis*, Academic Press, San Diego, 1987.

B-43. Torsell, K.B.G., *Nitrile Oxides, Nitrones and Nitronates in Organic Synthesis*, Academic Press, San Diego, 1987.

B-44. Gupta, R.R. (ed.),*Bioactive Molecules, Vol. 4: Phenothiazines and Benzothiazines*, Elsevier, Amsterdam, 1988.

B-45. De Kimpe, R. and Verhé, R., *Chemistry of α-Haloketones, α-Haloaldehydes and α-Haloimines*, Wiley, Chichester, 1988.

B-46. Patai, S. (ed.), *Chemistry of Sulphones and Sulphoxides*, Wiley, Chichester, 1988.

B-47. Wakefield, B.J., *Organolithium Methods*, Academic Press, London, 1988.

B-48. Pelter, A., Smith, K. and Brown, H.C., *Borane Reagents*, Academic Press, London, 1988.

B-49. Patai, S. and Rappoport, Z. (eds), *Chemistry of Enones*, Wiley, Chichester, 1989.B-50 Ninomiya, I. and Naito, T. Photochemical Synthesis, Academic Press, London, 1989.

B-50. Ninomiya, I. and Naito, T., *Photochemical Synthesis*, Academic Press, London, 1989.

B-51. Breuer, E., Aurich, H.G. and Nielsen, A.T., *Nitrones, Nitronates and Nitroxides*, Wiley, Chichester, 1989.

B-52. Dunn, A.D. and Rudorf, W.-D., *Carbon Disulphide in Organic Chemistry*, Ellis Horwood, Chichester, 1989.

B-53. Davies, H.G., Green, R.H., Kelly, D.R. and Roberts, S.M., *Biotransformations in Preparative Organic Chemistry*, Academic Press, London, 1989.

B-54. Patai, S. and Rappoport, Z. (eds), *Chemistry of Sulphinic Acids, Esters and Derivatives*, Wiley, Chichester, 1990.

B-55. Patai, S. (ed.), *Chemistry of Sulphenic Acids and their Derivatives*, Wiley, Chichester, 1990.

B-56. Harrington, P.J., *Transition Metals in Total Synthesis*, Wiley, New York, 1990.
B-57. Fringuelli, F. and Taticchi, A., *Dienes in the Diels-Alder Reaction*, Wiley, Chichester, 1990.
B-58. Mason, T.J. (ed.), *Sonochemistry*, Royal Society of Chemistry, Cambridge, 1990.
B-59. Patai, S: and Rappoport, Z. (eds), *Chemistry of Sulphonic Acids, Esters and their Derivatives*, Wiley, Chichester, 1991.
B-60. Hesse, M., *Ring Enlargement in Organic Chemistry*, VCH, Weinheim, 1991.

References

The first 31 references are from Part 1 and are cited in this volume.

48. Koitz, G., Thierrichter, B. and Junek, H., *Heterocycles*, 1983, **20**, 2405.
100. Kappe, T. and Ziegler, E., *Angew. Chem., Int. Ed. Engl.*, 1974, **13**, 491.
116. Kato, T., *Acc. Chem. Res.*, 1974. **7**, 265
182. Nakanishi, S. and Massett, S.S., *Org. Prep. Proced. Int.*, 1980, **12**, 219.
241. Uno, H. and Kurokawa, M., *Chem. Pharm. Bull.*, 1982, **30**, 333.
267. Beck, J.R. and Suhr, R.G., J. *Heterocycl. Chem.*, 1974, **11**, 227.
324. Szabo, W.A., *Aldrichim. Acta*, 1977, **10**, 23.
345. Shridhar, D.R., Sarma, C.R., Krishna, R.R., Prasad, R.S. and Sachdeva, *Org. Prep. Proced. Int.*, 1978, **10**, 163.
348. Acheson, R.M., Prince, R.J. and Proctor, G., *J. Chem. Soc., Perkin Trans. 1*, 1979, 595.
526. Markovac, A., Wu, G.S., LaMontagne, M.P., Blumbergs, P. and Ao, M.S., *J. Heterocycl. Chem.*, 1982, **19**, 829.
551. Van der Plas, H.C., *Acc. Chem. Res.*, 1978, **11**, 462.
562. Daisley, R.W. and Hanbali, J.R., *J. Heterocycl. Chem.*, 1983, **20**, 999.
1064. MacKensie, N.E., Thomson, R.H. and Greenhalgh, C.W., *J.Chem. Soc., Perkin Trans. 1*, 1980, 2923.
1096. Elnagdi, M.H., Elfahham, H.A.E. and Elgemeie, G.E.H., *Heterocycles*, 1983, **20**, 519.
1113. Ozaki, K.Y., Yamada, Y. and Oine, T., *Chem. Pharm. Bull.*, 1984, **32**, 2160.
1150. Hermecz, I., Meszaros, Z., Simon, K., Saszlo, L. and Pal, Z., *J. Chem. Soc., Perkin Trans. 1*, 1984, 1795.
1221. McCord, T.J., DuBose, C.E., Shafer, P.L. and Davis, A.L., *J. Heterocycl. Chem.*, 1984, **21**, 643.
1398. Migachev, G.I. and Danilenko, V.A., *Chem. Heterocycl. Compds.*, 1982, **18**, 649.
1399. Beck, J.R., *Tetrahedron*, 1978, **34**, 2057.
1432. Preston, P.N. and Tennant, G., *Chem. Rev.*, 1972, **72**, 627.
1437. Hafez, E.A.A., Abed, N.M., Elmoghayar, M.R.H. and El-Agamey, A.G.A., *Heterocycles*, 1984, **22**, 1821.
1491. Perera, R.C., Smalley, R.K. and Rogerson, L.G., *J. Chem. Soc. (C)*, 1971, 1348.
1508. Slouka, J., Bekárek, V. and Sternberk, V., *Collect. Czech. Chem. Commun.*, 1978, **43**, 960.
1510. Sakamoto, T., Kondo, Y. and Yamanaka, H., *Heterocycles*, 1984, **22**, 1347.
1618. Meth-Cohn, O. and Suschitzky, H., *Adv. Heterocycl. Chem.*, 1972, **14**, 211.
1753. Shawali, A.S. and Parkanyi, C., *J. Heterocycl. Chem.*, 1980, **17**, 833.
1779. Bevan, P.S., Ellis, G.P., Hudson, H.V., Romney-Alexander, T.M. and Williams, J.M., *J. Chem. Soc., Perkin Trans. 1*, 1986, 1643.

1870. Sliwa, W., *Heterocycles*, 1986, **24**, 181.
1922. Schaefer, J.P. and Bloomfield, J.J., *Org. React.*, 1967, **15**, 1.
2014. McNaught, A.D., *Adv. Heterocycl. Chem.*, 1976, **20**, 175.
2018. Clemens, R.J., *Chem. Rev.*, 1986, **86**, 241.
2028. Takamizawa, A., Hamashima, Y., Hayashi, S. and Sakai, S., *Chem. Pharm. Bull.*, 1968, **16**, 2195.X
2029. Rossi, S., Pirola, O. and Selva, F., *Tetrahedron*, 1968, **24**, 6395.
2030. Kiffer, D. and Levy, R., *C.R. Hebd. Seances Acad. Sci.*, 1968, **267C**, 1730.
2031. Gupta, C.M., Bhaduri, A.P. and Khanna, N.M., *Indian J. Chem.*, 1968, **6**, 758.
2032. Schmitt, J., Suquet, M., Comoy, P., Clim, T. and Callet, G., *Bull. Soc. Chim. Fr.*, 1968, 4575.
2033. Grunsbergs, F., Arens, A. and Vanags, G., *Latv. PSR Zinat. Akad. Vestis, Kim. Ser.*, 1968, 324.
2034. Hawlitzky, N., Haller, R. and Merz, K.W., *Arch. Pharm. (Weinheim)*, 1968, **301**, 17.
2035. Tuchtenhagen, G. and Rühlmann, K., *Liebigs Ann. Chem.*, 1968, **711**, 174.
2036. Watanabe, H., Schwarz, R.A. and Hauser, C.R., *J. Chem. Soc., Chem. Commun.*, 1969, 287.
2037. Barrett, G.C., Khokhar, A.R. and Chapman, J.R., *J. Chem. Soc., Chem. Commun.*, 1969, 818.
2038. Chakrabarti, P.M., Chapman, N.B. and Clarke, K., *J. Chem. Soc.(C)*, 1969, 1.
2039. Grantham, R.K. and Meth-Cohn, O., *J. Chem. Soc.(C)*, 1969, 70.
2040. Jones, D.W., *J. Chem. Soc.(C)*, 1969, 1729.
2041. Stockelmann, G., Specker, H. and Riepe, W., *Chem. Ber.*, 1969, **102**, 455.
2042. Effenberger, F. and Hartmann, W., *Chem. Ber.*, 1969, **102**, 3260.
2043. Regitz, M. and Schwall, H., *Liebigs Ann. Chem.*, 1969, **728**, 99.
2044. Schefczik, E., *Liebigs Ann. Chem.*, 1969, **729**, 97.
2045. Kim, D.H. and Santilli, A.A., *Chem. Ind.(London)*, 1969, 458.
2046. Ishiwata, S. and Shiokawa, Y., *Chem. Pharm. Bull.*, 1969, **17**, 1153.
2047. Hedayatullah, M., Nunes, A., Binick, A. and Denivelle, L., *Bull. Soc. Chim. Fr.*, 1969, 2729.
2048. Royer, R., Colin, G., Demerseman, P., Combrisson, S. and Cheutin, A., *Bull. Soc. Chim. Fr.*, 1969, 2785.
2049. Bisagni, E., Marquet, J.-P. and Andre-Louisfert, J., *Bull. Soc. Chim. Fr.*, 1969, 4338.
2050. Marquet, J.-P., Andre-Louisfert, J. and Bisagni, E., *Bull. Soc. Chim. Fr.*, 1969, 4344.
2051. Undheim, K. and Borka, L., *Acta Chem. Scand.*, 1969, **23**, 1715.
2052. Guarneri, M. and Giori, P., *Gazz. Chim. Ital.*, 1969, **99**, 463.
2053. Carboni, S., da Settimo, A., Bertini, D. and Biagi, G., *Gazz. Chim. Ital.*, 1969, **99**, 677.
2054. Schweizer, H.R., *Helv. Chim. Acta*, 1969, **52**, 322.
2055. Dahn, H. and Nussbaum, J., *Helv. Chim. Acta*, 1969, **52**, 1661.
2056. Gais, H.-J., Hafner, K. and Neuenschwander, M., *Helv. Chim. Acta*, 1969, **52**, 2641.
2057. Weber, K.H., *Arch. Pharm. (Weinheim)*, 1969, **302**, 584.
2058. Hromatka, O., Knollmüller, M. and Desehler, H., *Monatsh. Chem.*, 1969, **100**, 469.
2059. Hromatka, O., Knollmüller, M. and Krenmüller, F., *Monatsh. Chem.*, 1969, **100**, 941.
2060. Hromatka, O., Knollmüller, M. and Binder, D., *Monatsh. Chem.*, 1969, **100**, 1434.
2061. Anderson, R.C. and Fleming, R.H., *Tetrahedron Lett.*, 1969, 1581.
2062. Kozlov, N.S. and Stepanova, M.N., *Dokl. Akad. Nauk BSSR*, 1969, **13**, 541.
2063. Shvedov, V.I., Altukhova, L.B., Chernyshkova, L.A. and Grinev, A.N., *Khim.-Farm. Zh.*, 1969, **3**, 15.
2064. Davis, M. and White, A.W., *J. Org. Chem.*, 1969, **34**, 2985.
2065. Goldman, I.M., *J. Org. Chem.*, 1969, **34**, 3285.
2066. Freeman, F. and Ito, T.I., *J. Org. Chem.*, 1969, **34**, 3670.

2067. Venturella, V.S. and Watson, E.J., *J. Heterocycl. Chem.*, 1969, **6**, 671.
2068. Gomes, A., de S, and Jullie, M.M., *J. Heterocycl. Chem.*, 1969, **6**, 729.
2069. Paolini, J.P. and Lendvay, L.J., *J. Med. Chem.*, 1969, **12**, 1031.
2070. Ruwet, A. and Renson, M., *Bull. Soc. Chim. Belg.*, 1969, **78**, 449.
2071. Kobayashi, G., Furukawa, S., Matsuda, Y. and Matsunaga, S., *Yakugaku Zasshi*, 1969, **89**, 203.
2072. Kempter, F.E. and Castle, R.N., *J. Heterocycl. Chem.*, 1969, **6**, 523.
2073. Kim, D.H. and Santilli, D.H., *Chem. Ind. (London)*, 1969, 458.
2074. Messer, M. and Farge, D., *Bull. Soc. Chim. Fr.*, 1969, 4395.
2075. Petersen, J.B. and Lakowitz, K.H., *Acta Chem. Scand.*, 1969, **23**, 971.
2076. Matsuura, I. and Okui, K., *Chem. Pharm. Bull.*, 1969, **17**, 2206.
2077. Osadchii, S.A. and Barkhash, V.A., *J. Org. Chem. USSR*, 1970, **6**, 1639.
2078. Capuano, L., Ebner, W. and Schrepfer, J., *Chem. Ber.*, 1970, **103**, 82.
2079. Simchen, G. and Wenzelburger, J., *Chem. Ber.*, 1970, **103**, 413.
2080. Fuks, R. and Viehe, H.G., *Chem. Ber.*, 1970, **103**, 564.
2081. Achenbach, H., Haller, R. and Hawlitzky, N., *Chem. Ber.*, 1970, **103**, 677.
2082. Konrad, G. and Pfleiderer, W., *Chem. Ber.*, 1970, **103**, 722.
2083. Konrad, G. and Pfleiderer, W., *Chem. Ber.*, 1970, **103**, 735.
2084. Winterfeldt, E. and Nelke, J.M., *Chem. Ber.*, 1970, **103**, 1183.
2085. Schulte, K.E., von Weissenborn, V. and Tittel, G.L., *Chem. Ber.*, 1970, **103**, 1250.
2086. Seeger, E., Engel, W., Teufel, H. and Machleidt, H., *Chem. Ber.*, 1970, **103**, 1674.
2087. Sianesi, E., Reduelli, R., Bertani, M. and Da Re, P., *Chem. Ber.*, 1970, **103**, 1992.
2088. Ruangsiyanand, C., Rimek, H.-J. and Zymalkowski, F., *Chem. Ber.*, 1970, **103**, 2403.
2089. Alles, H., Bormann, D. and Musso, H., *Chem. Ber.*, 1970, **103**, 2526.
2090. Schmidt, R.R., Schwille, D. and Wolf, H., *Chem. Ber.*, 1970, **103**, 2760.
2091. Capuano, L. and Ebner, W., *Chem. Ber.*, 1970, **103**, 3104.
2092. Reimlinger, R., Peiren, M.A. and Merényi, R., *Chem. Ber.*, 1970, **103**, 3252.
2093. Reimlinger, H. and Peiren, M.A., *Chem. Ber.*, 1970, **103**, 3266.
2094. Goerdeler, J. and Lüdke, H., *Chem. Ber.*, 1970, **103**, 3393.
2095. Capuano, L. and Ebner, W., *Chem. Ber.*, 1970, **103**, 3459.
2096. Richter, R. and Ulrich, H., *Chem. Ber.*, 1970, **103**, 3525.
2097. Hetzheim, H., Pusch, H. and Beyer, H., *Chem. Ber.*, 1970, **103**, 3533.
2098. Reimlinger, H., Lingier, W.R.F. and Merényi, R., *Chem. Ber.*, 1970, **103**, 3817.
2099. Hedayatullah, M., Binick, A. and Denivelle, L., *C.R. Hebd. Seances Acad. Sci.*, 1970, **271C**, 1599.
2100. Petracek, F.J., Sugisaka, N., Klohs, M.W., Parker, R.G., Borner, J. and Roberts, J.D., *Tetrahedron Lett.*, 1970, 707.
2101. Junek, H. and Wrtilek, I., *Monatsh. Chem.*, 1970, **101**, 1130.
2102. Holland, J.W. and Jones, D.W., *J. Chem. Soc.(C)*, 1970, 530.
2103. Reichel, L. and Hampel, W., *Liebigs Ann. Chem.*, 1970, **733**, 1.
2104. Dickoré, K., Sasse, K. and Bode, K.-D., *Liebigs Ann. Chem.*, 1970, **733**, 70.
2105. Kamel, M., Ali, M.I. and Kamel, M.M., *Liebigs Ann. Chem.*, 1970, **733**, 115.
2106. Ried, W., Piechaczek, D. and Vollberg, E., *Liebigs Ann. Chem.*, 1970, **734**, 13.
2107. Eistert, B. and Endres, E., *Liebigs Ann. Chem.*, 1970, **734**, 56.
2108. Flitsch, W. and Krämer, U., *Liebigs Ann. Chem.*, 1970, **735**, 35.
2109. Hünig, S. and Fleckenstein, E., *Liebigs Ann. Chem.*, 1970, **738**, 192.
2110. Mustafa, A., Fleifel, A.M., Ali, M.I. and Hassam, N.M., *Liebigs Ann. Chem.*, 1970, **739**, 75.
2111. Kanaoka, Y., Hamada, T. and Yonemitsu, O., *Chem. Pharm. Bull.*, 1970, **18**, 587.
2112. Kametani, T., Yagi, H., Kawamura, K. and Kohno, T., *Chem. Pharm. Bull.*, 1970, **18**, 645.
2113. Yoshina, S. and Maeba, I., *Chem. Pharm. Bull.*, 1970, **18**, 842.
2114. Reimlinger, H., Billiau, F. and Lingier, W.R.F., *Synthesis*, 1970, 260.

2115. Grundmann, C., *Synthesis*, 1970, 344.
2116. Metelli, R. and Bettinetti, G.F., *Synthesis*, 1970, 365.
2117. Eiden, F. and Dobinsky, H., *Synthesis*, 1970, 365.
2118. Singh, S., Singh, H., Singh, M. and Narang, K.S., *Indian J. Chem.*, 1970, **8**, 230.
2119. Modi, S.K., Gakhar, H.K. and Narang, K.S., *Indian J. Chem.*, 1970, **8**, 389.
2120. Nair, M.D., *Indian J. Chem.*, 1970, **8**, 949, 950.
2121. Chadha, V.K. and Pujari, H.K., *Indian J. Chem.*, 1970, **8**, 1039.
2122. Evnin, A.B., Arnold, D.R., Karnischky, L.A. and Strom, E., *J. Am. Chem. Soc.*, 1970, **92**, 6218.
2123. Mühlstädt, M., Krausmann, H. and Fischer, G., *J. Prakt. Chem.*, 1970, **312**, 254.
2124. Becker, H.G.O., Bayer, D., Israel, G., Müller, R., Riediger, W. and Timpe, H.-J., *J. Prakt. Chem.*, 1970, **312**, 669.
2125. van der Burg, W.J., Bonta, I.L., Delobelle, J., Ramon, C. and Vargraftig, B., *J. Med. Chem.*, 1970, **13**, 35.
2126. Boots, M.R., Boots, S.G. and Moreland, D.E., *J. Med. Chem.*, 1970, **13**, 144.
2127. Stein, R.G., Biel, J.H. and Singh, T., *J. Med. Chem.*, 1970, **13**, 153.
2128. Snyder, H.R. and Benjamin, L.E., *J. Med. Chem.*, 1970, **13**, 164.
2129. Neumeyer, J.L. and Weinhardt, K.K., *J. Med. Chem.*, 1970, **13**, 613.
2130. Bell, M.R., Zalay, A.W., Oesterlin, R., Schane, P. and Potts, G.O., *J. Med. Chem.*, 1970, **13**, 665.
2131. Arnall, P. and Dace, G.R., *Manuf. Chem.*, 1970, **41**, 21.
2132. Stapfer, C.H. and D'Andrea, R.W., *J. Heterocycl. Chem.*, 1970, **7**, 651.
2133. Joh, T. and Hagihara, N., *Nippon Kagaku Zasshi*, 1970, **91**, 383.
2134. Junek, H., Metallides, A. and Ziegler, E., *Org. Prep. Proced. Int.*, 1970, **2**, 161.
2135. Claret, P.A. and Osborne, A.G., *Org. Prep. Proced. Int.*, 1970, **2**, 305.
2136. Lown, J.W. and Moser, J.P., *Can. J. Chem.*, 1970, **48**, 2227.
2137. Ishikawa, T., Yonemoto, M., Isagawa, K. and Fushizaki, Y., *Bull. Chem. Soc. Jpn.*, 1970, **43**, 1839.
2138. Sliwa, H., *Bull. Soc. Chim. Fr.*, 1970, 631.
2139. Cadogan, J.I.G. and Kulik, S., *J. Chem. Soc., Chem. Commun.*, 1970, 233.
2140. Dorofeenko, G.N. and Korobkova, V.G., *J. Gen. Chem. USSR*, 1970, **40**, 230.
2141. Zinner, G. and Böse, D., *Pharmazie*, 1970, **25**, 309.
2142. Bourdais, J., *Tetrahedron Lett.*, 1970, 2895.
2143. Desimoni, G. and Minoli, G., *Tetrahedron*, 1970, **26**, 1393.
2144. Bisagni, E., Bourzat, J.-D. and Andre-Louisfert, J., *Tetrahedron*, 1970, **26**, 2087.
2145. Kobe, J., Stanovnik, B. and Tisler, M., *Tetrahedron*, 1970, **26**, 3357.
2146. George, C., Gill, E.W. and Hudson, J.A., *J. Chem. Soc.(C)*, 1970, 74.
2147. Ollis, W.D., Ormand, K.L., Redman, B.T., Roberts, R.J. and Sutherland, I.O., *J. Chem. Soc.(C)*, 1970, 125.
2148. Albert, A., *J. Chem. Soc.(C)*, 1970, 230.
2149. Roberts, J.C. and Woollven, P., *J. Chem. Soc.(C)*, 1970, 278.
2150. Kelly, A.H. and Parrick, J., *J. Chem. Soc.(C)*, 1970, 303.
2151. Locksley, H.D. and Murray, I.G., *J. Chem. Soc.(C)*, 1970, 392.
2152. Rees, C.W. and West, D.E., *J. Chem. Soc.(C)*, 1970, 583.
2153. Clarke, K., Hughes, C.G., Humphries, A.J. and Scrowston, R.M., *J. Chem. Soc.(C)*, 1970, 1013.
2154. Lister, J.H., Manners, D.S. and Timmis, G.M., *J. Chem. Soc.(C)*, 1970, 1313.
2155. Buu-Hoï, N.P., Saint-Ruf, G. and Perche, J.-C., *J. Chem. Soc.(C)*, 1970, 1327.
2156. Stephenson, L. and Warburton, W.K., *J. Chem. Soc.(C)*, 1970, 1355.
2157. Bowers, R.J. and Brown, A.G., *J. Chem. Soc.(C)*, 1970, 1434.
2158. Adembri, G., De Sio, F., Nesi, R. and Scotton, M., *J. Chem. Soc.(C)*, 1970, 1536.
2159. Albert, A. and Ohta, K., *J. Chem. Soc.(C)*, 1970, 1540.

2160. Shah, M.A. and Taylor, G.A., *J. Chem. Soc.(C)*, 1970, 1642.
2161. Shah, M.A. and Taylor, G.A., *J. Chem. Soc.(C)*, 1970, 1651.
2162. Eatough, L.L., Fuller, L.S., Good, R.H. and Smalley, R.K., *J. Chem. Soc.(C)*, 1970, 1874.
2163. Sword, I.P., *J. Chem. Soc.(C)*, 1970, 1916.
2164. Callaghan, P.D. and Gibson, M.S., *J. Chem. Soc.(C)*, 1970, 2106.
2165. Kirby, P., Soloway, S.B., Davies, J.H. and Webb, S.B., *J. Chem. Soc.(C)*, 1970, 2250.
2166. Mackenzie, S.M. and Stevens, M.F.G., *J. Chem. Soc.(C)*, 1970, 2298.
2167. Chapman, N.B., Hughes, C.G. and Scrowston, R.M., *J. Chem. Soc.(C)*, 1970, 2431.
2168. Cadogan, J.I.G., Kulik, S., Thomson, C. and Todd, M.J., *J. Chem. Soc.(C)*, 1970, 2437.
2169. Baker, J.A. and Chatfield, P.V., *J. Chem. Soc.(C)*, 1970, 2478.
2170. Capozzi, G., Mellon, G. and Modena, G., *J. Chem. Soc.(C)*, 1970, 2621.
2171. Yokoyama, M., *J. Org. Chem.*, 1970, **35**, 283.
2172. Kim, D.H. and McKee, R.L., *J. Org. Chem.*, 1970, **35**, 455.
2173. Perini, F. and Tieckelmann, H., *J. Org. Chem.*, 1970, **35**, 812.
2174. Dittmer, D.C. and Glassman, R., *J. Org. Chem.*, 1970, **35**, 999.
2175. Mosher, W.A. and Piesch, S., *J. Org. Chem.*, 1970, **35**, 1026.
2176. Blank, H.U., Wempen, I. and Fox, J.J., *J. Org. Chem.*, 1970, **35**, 1131.
2177. Pilgram, K., *J. Org. Chem.*, 1970, **35**, 1165.
2178. Brossi, A. and Teitel, S., *J. Org. Chem.*, 1970, **35**, 1684.
2179. Potts, K.T. and Armbruster, R., *J. Org. Chem.*, 1970, **35**, 1965.
2180. Mosher, W.A. and Piesch, S., *J. Org. Chem.*, 1970, **35**, 2109.
2181. Still, W.C. and Goldsmith, D.J., *J. Org. Chem.*, 1970, **35**, 2282.
2182. Lopatina, K.I., Klyuev, S.M. and Zagorevskii, V.A., *Chem. Heterocycl. Compds. (Engl. Transl.)*, 1970, **6**, 39.
2183. Zaitsev, I.A., Shestaeva, M.M. and Zagorevskii, V.A., *Chem. Heterocycl. Compds. (Engl. Transl.)*, 1970, **6**, 136.
2184. Yakhontov, L.N., Pronina, E.V. and Rubstov, M.V., *Chem. Heterocycl. Compds. (Engl. Transl.)*, 1970, **6**, 170.
2185. Pankina, Z.A. and Shchukina, M.N., *Chem. Heterocycl. Compds. (Engl. Transl.)*, 1970, **6**, 228.
2186. Sycheva, T.P., Pankina, Z.A. and Shchukina, M.N., *Chem. Heterocycl. Compds. (Engl. Transl.)*, 1970, **6**, 406.
2187. Voronkov, M.G. and Udre, V.É., *Chem. Heterocycl. Compds. (Engl. Transl.)*, 1970, **6**, 421.
2188. Grandberg, I.I., Przheval'skii, N.M., Vysotskii, V.I. and Khmel'nitskii, R.A., *Chem. Heterocycl. Compds. (Engl. Transl.)*, 1970, **6**, 441.
2189. Aryuzina, V.M. and Shchukina, M.N., *Chem. Heterocycl. Compds. (Engl. Transl.)*, 1970, **6**, 486.
2190. Bokaldere, R.P. and Grinshtein, V.Y., *Chem. Heterocycl. Compds. (Engl. Transl.)*, 1970, **6**, 522.
2191. Koshel, N.G. and Postovskii, I.Y., *Chem. Heterocycl. Compds. (Engl. Transl.)*, 1970, **6**, 633.
2192. Simonov, A.M., Anisimova, V.A. and Grushina, L.E., *Chem. Heterocycl. Compds. (Engl. Transl.)*, 1970, **6**, 778.
2193. Aksanova, L.A., Sharkova, L.M. and Zagorevskii, V.A., *Chem. Heterocycl. Compds. (Engl. Transl.)*, 1970, **6**, 864.
2194. Borisova, L.N., Kucherova, N.F. and Zagorevskii, V.A., *Chem. Heterocycl. Compds. (Engl. Transl.)*, 1970, **6**, 868.
2195. Safonova, T.S. and Levkovskaya, L.G., *Chem. Heterocycl. Compds. (Engl. Transl.)*, 1970, **6**, 1023.

2196. Abramenko, P.I. and Zhiryakov, V.G., *Chem. Heterocycl. Compds. (Engl. Transl.)*, 1970, **6**, 1515.
2197. Belyaev, E.Y., Kumarev, V.P., Kondrat'eva, L.E. and Shakhova, E.S., *Chem. Heterocycl. Compds. (Engl. Transl.)*, 1970, **6**, 1576.
2198. Motoki, S., Urakawa, C., Kano, A., Fushimi, Y., Hirano, T. and Murata, K., *Bull. Chem. Soc. Jpn.*, 1970, **43**, 809.
2199. Kobayashi, G., Furukawa, S., Matsuda, Y., Natsuki, R. and Matsunaga, S., *Chem. Pharm. Bull.*, 1970, **18**, 124.
2200. Chaykovsky, M., Benjamin, L., Fryer, R.I. and Metlesics, N., *J. Org. Chem.*, 1970, **35**, 1178.
2201. Keane, D.D., Marathe, K.G., O'Sullivan, W.I., Philbin, E.M., Simons, R.M. and Teague, P.C., *J. Org. Chem.*, 1970, **35**, 2286.
2202. McOmie, J.F.W. and West, D.E., *J. Chem.Soc.(C)*, 1970, 1084.
2203. Kaji, K., Kuzuya, M. and Castle, R.N., *Chem. Pharm. Bull.*, 1970, **18**, 147.
2204. Ota, E. and Okazaki, M., *Yuki Gosei Kagaku Kyokai Shi*, 1970, **28**, 341.
2205. Cadogan, J.I.G., Marshall, R., Smith, D.M. and Todd, M.J., *J. Chem.Soc.(C)*, 1970, 2441.
2206. Challand, S.R., Herbert, R.B. and Holliman, F.G., *J. Chem. Soc., Chem. Commun.*, 1970, 1423.
2207. Cava, M.P., Mitchell, M.J. and Hill, D.T., *J. Chem. Soc., Chem. Commun.*, 1970, 1601.
2208. Degener, E., Beck, G. and Holtschmidt, H., *Angew. Chem., Int. Ed. Engl.*, 1970, **9**, 65.
2209. Simchen, G., Entenmann, G. and Zondler, R., *Angew. Chem., Int. Ed. Engl.*, 1970, **9**, 523.
2209a. Künzle, F. and Schmutz, J., *Helv. Chim. Acta*, 1970, **53**, 798.
2210. Nakagawa, S., Okumura, J., Sakai, F., Hoshi, H. and Naito, T., *Tetrahedron Lett.*, 1970, 3719.
2211. Gehlen, H. and Simon, B., *Arch. Pharm. (Weinheim)*, 1970, **303**, 501.
2212. Gehlen, H. and Drohla, R., *Arch. Pharm. (Weinheim)*, 1970, **303**, 709.
2213. Dell, H.-D. and Kamp, R., *Arch. Pharm. (Weinheim)*, 1970, **303**, 785.
2214. Teitei, T., *Aust. J. Chem.*, 1970, **23**, 185.
2215. Winterfeldt, E. and Wildersohn, M., *Pharm. Acta Helv.*, 1970, **45**, 323.
2216. Pratt, R.E., Welstead, W.J. and Lutz, R.E., *J. Heterocycl. Chem.*, 1970, **7**, 1051.
2217. Otomasu, H., Ohmiya, S., Sekiguchi, T. and Takahashi, H., *Chem. Pharm. Bull.*, 1970, **18**, 2065.
2218. Bagli, J.F. and Immer, H., *J. Org. Chem.*, 1970, **35**, 3499.
2219. Völcker, C.E. and Bress, H.J., *Z. Chem.*, 1970, **10**, 399.
2220. Petrov, V.P. and Barkhash, V.A., *Chem. Heterocycl. Compds. (Engl. Transl.)*, 1970, **6**, 357.
2221. Petrov, V.P. and Barkhash, V.A., *Chem. Heterocycl. Compds. (Engl. Transl.)*, 1970, **6**, 573.
2222. Gewald, K., Spies, H. and Mayer, R., *J. Prakt. Chem.*, 1970, **312**, 776.
2223. Patey, A.L. and Waldron, N.M., *Tetrahedron Lett.*, 1970, 1442.
2224. Fielden, R., Meth-Cohn, O. and Suschitzky, H., *Tetrahedron Lett.*, 1970, 1229.
2225. Yoneda, F., Ogiwara, K., Kanahori, M. and Nishigaki, S., *J. Chem. Soc., Chem. Commun.*, 1970, 1068.
2226. Julia, M., Hurion, N. and Tam, H.D., *Chim. Ther.*, 1970, **5**, 343.
2227. Lau, P.T.S. and Gompf, T.E., *J. Org. Chem.*, 1970, **35**, 4103.
2228. Yale, H.L. and Pluscec, J., *J. Org. Chem.*, 1970, **35**, 4254.
2229. Abushanab, E., *J. Org. Chem.*, 1970, **35**, 4279.
2230. Ohtsuka, Y., *Bull. Chem. Soc. Jpn.*, 1970, **43**, 3909.

2231. Yoneda, F., Ichiba, M., Ogiwara, K. and Nishigaki, S., *J. Chem. Soc., Chem. Commun.*, 1971, 23.
2232. Tamura, Y., Tsujimoto, N. and Uchimura, M., *Chem. Pharm. Bull.*, 1971, **19**, 143.
2233. Nishigaki, S., Fukazawa, S., Ogiwara, K. and Yoneda, F., *Chem. Pharm. Bull.*, 1971, **19**, 206.
2234. Kugita, H., Inoue, H., Ikezaki, M., Kondon, M. and Takeo, S., *Chem. Pharm. Bull.*, 1971, **19**, 595.
2235. Kato, T., Niitsuma, T. and Maeda, K., *Chem. Pharm. Bull.*, 1971, **19**, 832.
2236. Shirai, H., Hayazaki, T. and Aoyama, T., *Chem. Pharm. Bull.*, 1971, **19**, 892.
2237. Takada, T., Kunugi, S. and Ohki, S., *Chem. Pharm. Bull.*, 1971, **19**, 982.
2238. Kametani, T., Nyu, K. and Yamanaka, T., *Chem. Pharm. Bull.*, 1871, **19**, 1321.
2239. Kametani, T. and Nemoto, N., *Chem. Pharm. Bull.*, 1971, **19**, 1325.
2240. Minami, S., Shono, T. and Matsumoto, J., *Chem. Pharm. Bull.*, 1971, **19**, 1426.
2241. Hamada, Y. and Takeuchi, I., *Chem. Pharm. Bull.*, 1971, **19**, 1857.
2242. Kametani, T., Seino, C. and Nakano, T., *Chem. Pharm. Bull.*, 1971, **19**, 1959.
2243. Sugasawa, T., Toyota, T., Sasakura, K. and Hidaka, T., *Chem. Pharm. Bull.*, 1971, **19**, 1971.
2244. Kobayashi, Y., Kutsuma, T. and Morinaga, K., *Chem. Pharm. Bull.*, 1971, **19**, 2106.
2245. Tsuji, T. and Ueda, T., *Chem. Pharm. Bull.*, 1971, **19**, 2530.
2246. Hisano, T. and Ichikawa, M., *Chem. Pharm. Bull.*, 1971, **19**, 2625.
2247. Yoneda, F., Nishigaki, S. and Shinomura, K., *Chem. Pharm. Bull.*, 1971, **19**, 2647.
2248. Daisley, R.W. and Walker, J., *J. Chem. Soc.(C)*, 1971, 1375.
2249. Sutherland, D.R. and Tennant, G., *J. Chem. Soc.(C)*, 1971, 2156.
2250. Bowden, K. and Brown, T.H., *J. Chem. Soc.(C)*, 1971, 2163.
2251. Mair, A.C. and Stevens, M.F.G., *J. Chem. Soc.(C)*, 1971, 2317.
2252. Rogers, G.T. and Ulbricht, T.L.V., *J. Chem. Soc.(C)*, 1971, 2364.
2253. Davis, M., Knowles, P., Sharp, B.W., Walsh, R.J.A. and Wooldridge, K.R.H., *J. Chem. Soc.(C)*, 1971, 2449.
2254. Casiraghi, G., Casnati, G. and Salerno, G., *J. Chem. Soc.(C)*, 1971, 2546.
2255. Brown, D.J. and Sugimoto, T., *J. Chem. Soc.(C)*, 1971, 2616.
2256. Cadogan, J.I.G. and Kulik, S., *J. Chem. Soc.(C)*, 1971, 2621.
2257. Birnie, J.M. and Campbell, N., *J. Chem. Soc.(C)*, 1971, 2634.
2258. Cheeseman, G.W.H. and Rafiq, M., *J. Chem. Soc.(C)*, 1971, 2732.
2259. Landquist, J.K., *J. Chem. Soc.(C)*, 1971, 2735.
2260. Duguay, G., Reid, D.H., Wade, K.O. and Webster, R.G., *J. Chem. Soc.(C)*, 1971, 2829.
2261. Cullen, W.P., Donnelly, D.M.X., Keenan, A.K., Levin, T.P., Melody, D.P. and Philbin, E.M., *J. Chem. Soc.(C)*, 1971, 2848.
2262. Bradley, E., Cotterile, W.D., Livingstone, R. and Walshaw, M., *J. Chem. Soc.(C)*, 1971, 3028.
2263. Chohan, M.I., Fitton, A.O., Hatton, B.T. and Suschitzky, H., *J. Chem. Soc.(C)*, 1971, 3079.
2264. Wermuth, C.G. and Flammang, M., *Tetrahedron Lett.*, 1971, 4293.
2265. Buggle, K., Delahunty, J.J., Philbin, E.M. and Ryan, N.D., *J. Chem. Soc.(C)*, 1971, 3168.
2266. Marshall, R. and Smith, D.M., *J. Chem. Soc.(C)*, 1971, 3510.
2267. Zalkow, L.A., Nabors, J.B., French, K. and Bisarya, S.C., *J. Chem. Soc.(C)*, 1971, 3551.
2268. Ahmad, S., Whalley, W.B. and Jones, D.F., *J. Chem. Soc.(C)*, 1971, 3590.
2269. Brooke, G.M., Musgrave, W.K.R. and Thomas, T.R., *J. Chem. Soc.(C)*, 1971, 3596.
2270. Blackshire, R.B. and Sharpe, C.J., *J. Chem. Soc.(C)*, 1971, 3602.
2271. Albert, A. and Ohta, K., *J. Chem. Soc.(C)*, 1971, 3727.

2272. Locksley, H.D., Quillinan, A.J. and Scheinmann, F., *J. Chem. Soc.(C)*, 1971, 3804.
2273. Franck, R.W. and Auerbach, J., *J. Org. Chem.*, 1971, **36**, 31.
2274. Rodricks, J.V. and Rapoport, H., *J. Org. Chem.*, 1971, **36**, 46.
2275. Bellani, P., Lauria, E. and Zoni, G., *Farmaco, Ed. Sci.*, 1971, **26**, 872.
2276. René, L. and Royer, R., *Bull. Soc. Chim. Fr.*, 1971, 4329.
2277. Dutta, N.L., Wadia, M.S. and Bindra, A.P., *J. Indian Chem. Soc.*, 1971, **48**, 873.
2278. Frydman, B., Los, M. and Rapoport, H., *J. Org. Chem.*, 1971, **36**, 450.
2279. Greco, C.V. and Warchol, J.F., *J. Org. Chem.*, 1971, **36**, 604.
2279a. Derieg, M.E., Fryer, R.I., Hillery, S.S., Metlesics, W. and Silverman, G., *J. Org. Chem.*, 1971, **36**, 782.
2280. Sasaki, T., Kanematsu, K., Yukimoto, Y. and Ochiai, S., *J. Org. Chem.*, 1971, **36**, 813.
2281. White, J.D., Mann, M.E., Kiershenbaum, H.D. and Mitra, A., *J. Org. Chem.*, 1971, **36**, 1048.
2282. Sheradsky, T. and Saleminck, G., *J. Org. Chem.*, 1971, **36**, 1061.
2283. Büchi, G., Klambert, D.H., Shank, R.C., Weinreb, S.M. and Wogan, G.N., *J. Org. Chem.*, 1971, **36**, 1143.
2284. Douglass, J.E. and Wesolosky, J.M., *J. Org. Chem.*, 1971, **36**, 1165.
2285. Wells, J.N., Wheeler, W.J. and Davisson, L.M., *J. Org. Chem.*, 1971, **36**, 1503.
2286. de Jong, F. and Janssen, M.J., *J. Org. Chem.*, 1971, **36**, 1645.
2287. de Jong, F. and Janssen, M.J., *J. Org. Chem.*, 1971, **36**, 1998.
2288. Potts, K.T. and Sorm, M., *J. Org. Chem.*, 1971, **36**, 8.
2289. Pollak, A., Stanovnik, B. and Tisler, M., *J. Org. Chem.*, 1971, **36**, 2457.
2290. Szmant, H.H. and Chow, Y.L., *J. Org. Chem.*, 1971, **36**, 2887.
2291. Szmant, H.H. and Chow, Y.L., *J. Org. Chem.*, 1971, **36**, 2889.
2292. Ahmad, Y. and Smith, P.A.S., *J. Org. Chem.*, 1971, **36**, 2972.
2293. Temple, C., Kussner, C.L. and Montgomery, J.A., *J. Org. Chem.*, 1971, **36**, 2974.
2294. Sasaki, T., Kanematsu, K. and Kakehi, A., *J. Org. Chem.*, 1971, **36**, 2978.
2295. Fraser, M., *J. Org. Chem.*, 1971, **36**, 3087.
2296. Kuderna, J.G., Skiles, R.D. and Pilgram, K., *J. Org. Chem.*, 1971, **36**, 3506.
2297. Blanton, C.D., Whidby, J.F. and Briggs, F.H., *J. Org. Chem.*, 1971, **36**, 3929.
2298. Raines, S., Chai, S.Y. and Palopoli, F.P., *J. Org. Chem.*, 1971, **36**, 3992.
2299. Taylor, E.C., Thomson, M.J., Perlman, K., Mengel, R. and Pfleiderer, W., *J. Org. Chem.*, 1971, **36**, 4012.
2300. Llommet, G., Sliwa, H. and Maitté, P., *C.R. Hebd. Seances Acad. Sci.*, 1971, **272C**, 2197.
2301. Kawase, Y., Takata, S. and Hitikishima, E., *Bull. Chem. Soc. Jpn.*, 1971, **44**, 749.
2302. Munro, H.D., Musgrave, O.C. and Templeton, R., *J. Chem. Soc.(C)*, 1971, 95.
2303. Hey, D.H., Jones, G.H. and Perkins, M.J., *J. Chem. Soc.(C)*, 1971, 116.
2304. Giner-Sorolla, A. and Brown, D.M., *J. Chem. Soc.(C)*, 1971, 126.
2305. Horspool, W.M., Smith, P.I. and Tedder, J.M., *J. Chem. Soc.(C)*, 1971, 138.
2306. Dean, F.M., Frankham, D.B., Hatam, N. and Hill, A.W., *J. Chem. Soc.(C)*, 1971, 218.
2307. Kessar, S.V., Jit, P., Mundra, K.P. and Lumb, A.K., *J. Chem. Soc.(C)*, 1971, 266.
2308. Cattenach, C.J. and Rees, R.G., *J. Chem. Soc.(C)*, 1971, 53.
2309. Birchall, J.M., Haszeldine, R.N., Nikokavouras, J. and Wilks, E.S., *J. Chem. Soc.(C)*, 1971, 562.
2310. Clark, J. and Ramsden, T., *J. Chem. Soc.(C)*, 1971, 679.
2311. Sutherland, D.R. and Tennant, G., *J. Chem. Soc.(C)*, 1971, 706.
2312. Lacy, P.H. and Smith, D.C.C., *J. Chem. Soc.(C)*, 1971, 747.
2313. Crombie, L. and Ponsford, R., *J. Chem. Soc.(C)*, 1971, 788.
2314. Sword, I.P., *J. Chem. Soc.(C)*, 1971, 820.
2315. Chapman, N.B., Clarke, K. and Sharma, K.S., *J. Chem. Soc.(C)*, 1971, 919.

2316. Boulton, A.J., Fletcher, I.J. and Katritzky, A.R., *J. Chem. Soc.(C)*, 1971, 1193.
2317. Zupan, M., Stanovnik, B. and Tisler, M., *J. Heterocycl. Chem.*, 1971, **8**, 1.
2318. Farnum, D.G., Au, A.T. and Rasheed, K., *J. Heterocycl. Chem.*, 1971, **8**, 25.
2319. Lewis, A. and Shepherd, R.G., *J. Heterocycl. Chem.*, 1971, **8**, 41.
2320. Lewis, A. and Shepherd, R.G., *J. Heterocycl. Chem.*, 1971, **8**, 47.
2321. Hess, B.A. and Corbino, S., *J. Heterocycl. Chem.*, 1971, **8**, 161.
2322. Francavilla, P. and Lauria, F., *J. Heterocycl. Chem.*, 1971, **8**, 415.
2323. Patil, V.D. and Townsend, L.B., *J. Heterocycl. Chem.*, 1971, **8**, 503.
2324. Carr, J.B., *J. Heterocycl. Chem.*, 1971, **8**, 511.
2325. Lhommet, G., Sliwa, H. and Maitté, P., *J. Heterocycl. Chem.*, 1971, **8**, 517.
2326. Wright, W.B. and Brabander, H.J., *J. Heterocycl. Chem.*, 1971, **8**, 711.
2327. Israel, M. and Jones, L.C., *J. Heterocycl. Chem.*, 1971, **8**, 797.
2328. Blazevic, N. and Kajfez, F., *J. Heterocycl. Chem.*, 1971, **8**, 845.
2329. Mosher, W.A. and Soeder, R.W., *J. Heterocycl. Chem.*, 1971, **8**, 855.
2330. Ben-Ishay, D. and Warshawsky, A., *J. Heterocycl. Chem.*, 1971, **8**, 865.
2331. Wu, M.T. and Lyle, R.E., *J. Heterocycl. Chem.*, 1971, **8**, 989.
2332. Mosher, W.A. and Banks, T.E., *J. Heterocycl. Chem.*, 1871, **8**, 1005.
2333. Heindel, N.D., Hsia, R. and Holcombe, F.O., *J. Heterocycl. Chem.*, 1971, **8**, 1047.
2334. Föhlisch, B., *Chem. Ber.*, 1971, **104**, 348.
2335. Winterfeldt, E., *Chem. Ber.*, 1971, **104**, 677.
2336. Döpp, D., *Chem. Ber.*, 1971, **104**, 1035.
2337. Kröck, F.W. and Kröhnke, F., *Chem. Ber.*, 1971, **104**, 1629.
2338. Reimlinger, H., *Chem. Ber.*, 1971, **104**, 2801.
2339. Capuano, L. and Schrepfer, H.J., *Chem. Ber.*, 1971, **104**, 3039.
2340. Wagner, H., Maurer, G., Hörhammer, L. and Farkas, L., *Chem. Ber.*, 1971, **104**, 3357.
2341. Reimlinger, H., Lingier, W.R.F. and Mernyi, R., *Chem. Ber.*, 1971, **104**, 3965.
2342. Ried, W. and Wiedemann, P., *Chem. Ber.*, 1971, **104**, 3329.
2343. Eiden, F. and Kuckländer, U., *Arch. Pharm. (Weinheim)*, 1971, **304**, 57.
2344. Roth, H.J. and Hagen, H.-E., *Arch. Pharm. (Weinheim)*, 1971, **304**, 73.
2345. Hörlein, U., *Arch. Pharm. (Weinheim)*, 1971, **304**, 81.
2346. Hörlein, U. and Geiger, W., *Arch. Pharm. (Weinheim)*, 1971, **304**, 130.
2347. Harris, N.D., *Synthesis*, 1971, 256.
2348. Brady, W.T., *Synthesis*, 1971, 415.
2349. Bugge, A., *Acta Chem. Scand.*, 1971, **25**, 27.
2350. Hauzwitz, R.D., Maurer, B.V. and Narayan, V.L., *J. Chem. Soc., Chem. Commun.*, 1971, 1100.
2351. Maki, Y., Izuta, K. and Suzuki, M., *J. Chem. Soc., Chem. Commun.*, 1971, 1442.
2352. Leyshon, L.J. and Saunders, D.G., *J. Chem. Soc., Chem. Commun.*, 1971, 1608.
2353. Wick, A.E., Bartlett, P.A. and Dolphin, D., *Helv. Chim. Acta*, 1971, **54**, 513.
2354. Mosher, W.A. and Banks, T.E., *J. Org. Chem.*, 1971, **36**, 1477.
2355. Kirchner, E. and Bretschneider, H., *Monatsh. Chem.*, 1971, **102**, 162.
2356. Hawlitzky, N. and Haller, R., *Monatsh. Chem.*, 1971, **102**, 718.
2357. Mazharuddin, M. and Thyagarajan, G., *Tetrahedron Lett.*, 1971, 307.
2358. Fanning, A.T. and Roberts, T.D., *Tetrahedron Lett.*, 1971, 805.
2359. Yoneda, F., Shinomura, K. and Nishigaki, S., *Tetrahedron Lett.*, 1971, 851.
2360. Yamamoto, I., Tabo, Y., Gotoh, H., Minami, T., Ohshivo, Y. and Agawa, T., *Tetrahedron Lett.*, 1971, 2295.
2361. Sagitullin, R.S., Borisov, N.N., Kost, K.N. and Simonova, N.A., *Chem. Heterocycl. Compds. (Engl. Transl.)*, 1971, **7**, 58.
2362. Safonova, T.S. and Leykovskaya, L.G., *Chem. Heterocycl. Compds. (Engl. Transl.)*, 1971, **7**, 65.

2363. Mazur, I.A., Kochergin, P.M. and Tromsa, V.G., *Chem. Heterocycl. Compds. (Engl. Transl.)*, 1971, **7**, 359.
2364. Kost, A.N., Stankevichus, A.P., Zhukauskaite, L.N. and Shulyakene, I.I., *Chem. Heterocycl. Compds. (Engl. Transl.)*, 1971, **7**, 474.
2365. Simonov, A.M. and Uryukina, I.G., *Chem. Heterocycl. Compds. (Engl. Transl.)*, 1971, **7**, 536.
2366. Lozinskii, M.O., Shivanyuk, A.F. and Pel'kis, P.S., *Chem. Heterocycl. Compds. (Engl. Transl.)*, 1971, **7**, 869.
2367. Knysh, E.G., Krasovshii, A.N. and Kochergin, P.M., *Chem. Heterocycl. Compds. (Engl. Transl.)*, 1971, **7**, 1059.
2368. Priimenko, B.A. and Kochergin, P.M., *Chem. Heterocycl. Compds. (Engl. Transl.)*, 1971, **7**, 1168.
2369. Prostakov, N.S. and Baktibaev, O.B., *Chem. Heterocycl. Compds. (Engl. Transl.)*, 1971, **7**, 1302.
2370. Titov, V.V. and Kozhokiina, L.F., *Chem. Heterocycl. Compds. (Engl. Transl.)*, 1971, **7**, 1328.
2371. Kost, A.N., Solomky, Z.F., Prikhod'ko, N.M. and Teteryuk, S.S., *Chem. Heterocycl. Compds. (Engl. Transl.)*, 1971, **7**, 1447.
2372. Gruzdev, V.G., *J. Org. Chem. USSR*, 1971, **7**, 1113.
2373. Stoss, P. and Satzinger, G., *Angew. Chem., Int. Ed. Engl.*, 1971, **10**, 76.
2374. Tallec, A., Mennereau, G. and Robic, G., *C.R. Hebd. Seances Acad. Sci.*, 1971, **273C**, 1378.
2375. McCorkindale, N.J. and McCulloch, A., *Tetrahedron*, 1971, **27**, 4653.
2376. Hlubucek, J., Ritchie, E. and Taylor, W.C., *Aust. J. Chem.*, 1971, **24**, 2347.
2377. Mazharuddin, M. and Thyagarajan, G., *Chem. Ind. (London)*, 1971, 178.
2378. Bevis, M.J., Forbes, E.J., Naik, N.N. and Uff, B.C., *Tetrahedron*, 1971, **27**, 1253.
2379. Bradbury, S., Keating, M., Rees, C.W. and Storr, R.C., *J. Chem. Soc., Chem. Commun.*, 1971, 827.
2380. Markees, D.G. and Schwab, L.S., *Helv. Chim. Acta*, 1972, **55**, 1319.
2381. Bertelson, R.C. and Glanz, K.D., *J. Org. Chem.*, 1972, **37**, 2207.
2382. Kappe, T., Chirazi, M.A.A. and Ziegler, E., *Monatsh. Chem.*, 1972, **103**, 234.
2383. Yoneda, F. and Fukazawa, S., *J. Chem. Soc., Chem. Commun.*, 1972, 503.
2384. Kitaev, Y.P. and Buzykin, B.I., *Russ. Chem. Rev. (Engl. Transl.)*, 1972, **41**, 495.
2385. Ried, W. and König, E., *Liebigs Ann. Chem.*, 1972, **755**, 24.
2386. Kanaoka, Y. and Itoh, K., *Synthesis*, 1972, 36.
2387. Wamhoff, H. and Korte, F., *Synthesis*, 1972, 151.
2388. Stanovnik, B. and Tisler, M., *Synthesis*, 1972, 308.
2389. Irvine, J.L. and Piantadosi, C., *Synthesis*, 1972, 568.
2390. Kappe, T. and Schmidt, H., *Org. Prep. Proced. Int.*, 1972, **4**, 233.
2391. Potts, K.T. and Elliott, A.J., *Org. Prep. Proced. Int.*, 1972, **4**, 269.
2392. Clark-Lewis, J.W., Dainis, I., McGarry, E.J. and Baig, M.I., *Aust. J. Chem.*, 1972, **25**, 857.
2393. Dainis, I., *Aust. J. Chem.*, 1972, **25**, 1003.
2394. Dainis, I., *Aust. J. Chem.*, 1972, **25**, 1025.
2395. Demerae, S., Dalton, L.K. and Elmes, B.I., *Aust. J. Chem.*, 1972, **25**, 2651.
2396. Hirata, T., Twanmoh, L.M., Wood, H.B., Goldin, A. and Driscoll, J.S., *J. Heterocycl. Chem.*, 1972, **9**, 99.
2397. Israel, M., Jones, L.C. and Modest, E.J., *J. Heterocycl. Chem.*, 1972, **9**, 255.
2398. Lambardino, J.G., *J. Heterocycl. Chem.*, 1972, **9**, 315.
2399. Schapira, C.B. and Lamdan, S., *J. Heterocycl. Chem.*, 1972, **7**, 569.
2400. Montgomery, J.A. and Laseter, A.G., *J. Heterocycl. Chem.*, 1972, **9**, 1077.
2401. Taylor, J.B., Harrison, D.R. and Fried, F., *J. Heterocycl. Chem.*, 1972, **9**, 1227.
2402. Kwasnik, H.R., Oliver, J.E. and Brown, R.T., *J. Heterocycl. Chem.*, 1972, **9**, 1429.

2403. Butler, R.N., Lambe, T.M. and Scott, F.L., *J. Chem. Soc., Perkin Trans. 1*, 1972, 269.
2404. Albert, A. and Taguchi, H., *J. Chem. Soc., Perkin Trans. 1*, 1972, 449.
2405. Albert, A., *J. Chem. Soc., Perkin Trans. 1*, 1972, 461.
2406. Bowie, R.A. and Thomason, D.A., *J. Chem. Soc., Perkin Trans. 1*, 1972, 1842.
2407. Finnegan, R.A. and Patel, J.K., *J. Chem. Soc., Perkin Trans. 1*, 1972, 1896.
2408. Dean, F.M., Hindley, K.B. and Small, S., *J. Chem. Soc., Perkin Trans. 1*, 1972, 2007.
2409. Hastings, J.S. and Heller, H.G., *J. Chem. Soc., Perkin Trans. 1*, 1972, 2128.
2410. Scott, F.L., O'Halloran, J.K., O'Driscoll, J. and Hegarty, A.F., *J. Chem. Soc., Perkin Trans. 1*, 1972, 2224.
2411. Earl, R.A., Panzica, R.P. and Townsend, L.B., *J. Chem. Soc., Perkin. Trans. 1*, 1972, 2672.
2412. Forrester, A.R., Ingram, A.S. and Thomson, R.H., *J. Chem. Soc., Perkin Trans. 1*, 1972, 2847.
2413. Heaney, H., Jablonski, J.M. and McCarty, C.T., *J. Chem. Soc., Perkin Trans. 1*, 1972, 2903.
2414. Elliott, A.J. and Gibson, M.S., *J. Chem. Soc., Perkin Trans. 1*, 1972, 2915.
2415. Fujihara, M. and Kawazu, M., *Chem. Pharm. Bull.*, 1972, **20**, 88.
2416. Ogura, H., Sakaguchi, M. and Takeda, K., *Chem. Pharm. Bull.*, 1972, **20**, 404.
2417. Maki, Y. and Taylor, E.C., *Chem. Pharm. Bull.*, 1972, **20**, 605.
2418. Aki, O. and Nakagawa, Y., *Chem. Pharm. Bull.*, 1972, **20**, 1325.
2419. Kutsuma, T., Fujiyama, K., Sekine, Y. and Kobayashi, Y., *Chem. Pharm. Bull.*, 1972, **20**, 1558.
2420. Yanai, M., Kinoshita, T., Takeda, S., Nishimura, M. and Kuraishi, T., *Chem. Pharm. Bull.*, 1972, **20**, 1617.
2421. Kametani, T., Fukumoto, K. and Fujihara, M., *Chem. Pharm. Bull.*, 1972, **20**, 1800.
2422. Kutsuma, T., Fujiyama, K. and Kobayashi, Y., *Chem. Pharm. Bull.*, 1972, **20**, 1809.
2423. Yoneda, F., Sakuma, Y., Ichiba, M. and Shinomura, K., *Chem. Pharm. Bull.*, 1972, **20**, 1833.
2424. Uchimaru, F., Okada, S., Kosasayama, A. and Konno, T., *Chem. Pharm. Bull.*, 1972, **20**, 1834.
2425. Higashino, T., Amano, T., Tamura, Y., Katsumata, N., Washizu, Y., Ono, T. and Hayashi, E., *Chem. Pharm. Bull.*, 1972, **20**, 1874.
2426. Watanabe, T., Hamaguchi, F. and Ohki, S., *Chem. Pharm. Bull.*, 1972, **20**, 2123.
2427. Hisano, T., Ichikawa, M., Kito, G. and Nishi, T., *Chem. Pharm. Bull.*, 1972, **20**, 2575.
2428. Born, J.L., *J. Org. Chem.*, 1972, **37**, 3952.
2429. Swett, L.R., Stein, R.G. and Kimura, E.T., *J. Med. Chem.*, 1972, **15**, 42.
2430. Heindel, N.D., Ko, C.C.H., Birrer, R.C. and Markel, J.R., *J. Med. Chem.*, 1972, **15**, 118.
2431. Montgomery, J.A. and Thomas, H.J., *J. Med. Chem.*, 1972, **15**, 182.
2432. Ghosh, P.B., Ternai, B. and Whitehouse, M.W., *J. Med. Chem.*, 1972, **15**, 255.
2433. Kochkar, M.M. and Williams, B.B., *J. Med. Chem.*, 1972, **15**, 332.
2434. Fernandez-Tomé, M.P., Madronero, R., del Rio, J. and Vega, S., *J. Med. Chem.*, 1972, **15**, 887.
2435. Ogura, H., Takayanagi, H., Yamazahi, Y., Yonezawa, S., Takagi, H., Kobayashi, S., Kamioka, T. and Kamoshita, K., *J. Med. Chem.*, 1972, **15**, 923.
2436. Fisher, M.H., Schwartzkopf, G. and Hoff, D.R., *J. Med. Chem.*, 1972, **15**, 1168.
2437. Kappe, T. and Linnau, Y., *Liebigs Ann. Chem.*, 1972, **761**, 25.
2438. Brugger, M., Wamhoff, H. and Korte, F., *Liebigs Ann. Chem.*, 1972, **758**, 173.
2439. Kappe, T., Chirazi, M.A.A. and Ziegler, E., *Monatsh. Chem.*, 1972, **103**, 426.
2440. Slouka, J. and Pec, P., *Monatsh. Chem.*, 1972, **103**, 1444.
2441. Böhme, H. and Braun, R., *Arch. Pharm. (Weinheim)*, 1972, **305**, 93.

2442. Kreutzberger, A. and Uzbeck, M.U., *Arch. Pharm. (Weinheim)*, 1972, **305**, 171.
2443. Eiden, F. and Dusemund, J., *Arch. Pharm. (Weinheim)*, 1972, **305**, 324.
2444. Manecke, G. and Zerpner, D., *Chem. Ber.*, 1972, **105**, 1943.
2445. Hug, R., Hansen, H.-J. and Schmid, H., *Helv. Chim. Acta*, 1972, **55**, 1675.
2446. Walia, J.S. and Walia, P.S., *J. Chem. Soc., Chem. Commun.*, 1972, 108.
2447. Yoneda, F., Matsumura, T. and Senga, H., *J. Chem. Soc., Chem. Commun.*, 1972, 606.
2448. Hunt, R., Reid, S.T. and Taylor, K.T., *Tetrahedron Lett.*, 1972, 2861.
2449. Kreutzberger, A. and Meyer, B., *Chem. Ber.*, 1972, **105**, 1810.
2450. Stoss, P. and Satzinger, G., *Chem. Ber.*, 1972, **105**, 2575.
2451. Kreutzberger, A. and Meyer, B., *Chem. Ber.*, 1972, **105**, 3974.
2452. Kogan, N.A. and Vlassova, M.I., *Chem. Heterocycl. Compds. (Engl. Transl.)*, 1972, **8**, 252.
2453. Shvedov, V.I., Altukhova, L.B. and Grinev, A.N., *Chem. Heterocycl. Compds. (Engl. Transl.)*, 1972, **8**, 310.
2454. Knysh, E.G., Krasovskii, A.N., Kochergin, P.M. and Shabel'nik, P.M., *Chem. Heterocycl. Compds. (Engl. Transl.)*, 1972, **8**, 364.
2455. Zelenin, K.N., Bezhan, I.P. and Matveeva, Z.M., *Chem. Heterocycl. Compds. (Engl. Transl.)*, 1972, **8**, 525.
2456. Safonova, T.S., Nemeryuk, M.P., Myshkina, L.A. and Traven, N.I., *Chem. Heterocycl. Compds. (Engl. Transl.)*, 1972, **8**, 859.
2457. Dorofeenko, G.N. and Tkachenko, V.V., *Chem. Heterocycl. Compds. (Engl. Transl.)*, 1972, **8**, 935.
2458. Prostakov, N.S. and Baktibaev, O.B., *Chem. Heterocycl. Compds. (Engl. Transl.)*, 1972, **8**, 1102.
2459. Sazonov, N.V. and Safonova, T.S., *Chem. Heterocycl. Compds. (Engl. Transl.)*, 1972, **8**, 1163.
2460. Khilya, V.P., Grishko, L.G. and Szabo, V., *Chem. Heterocycl. Compds. (Engl. Transl.)*, 1972, **8**, 1189.
2461. Abramenko, P.I. and Zhiryakov, V.G., *Chem. Heterocycl. Compds. (Engl. Transl.)*, 1972, **8**, 1392.
2462. Bogat-skii, A.V., Rudenko, O.P., Andron, S.A. and Chumachenko, T.K., *Chem. Heterocycl. Compds. (Engl. Transl.)*, 1972, **8**, 1547.
2463. Sindelár, K., Svatek, E. and Protiva, M., *Collect. Czech. Chem. Commun.*, 1972, **37**, 3660.
2464. Mustafa, A., Ali, M.O., Abou-State, M.A. and Hamman, A.E.G., *J. Prakt. Chem.*, 1972, **314**, 785.
2465. Thijs, L., Strating, J. and Zwenenberg, B., *Recl. Trav. Chim. Pays-Bas*, 1972, **91**, 1345.
2466. Rybar, A. and Stibranyi, L., *Collect. Czech. Chem. Commun.*, 1972, **37**, 2630.
2467. Budesinsky, Z. and Lederer, P., *Collect. Czech. Chem. Commun.*, 1972, **37**, 2779.
2468. Spagnolo, P., Tiecco, M., Tundo, A. and Martelli, G., *J. Chem. Soc., Perkin Trans. 1*, 1972, 256.
2469. Tamura, Y., Tsujimoto, N., Sumida, Y. and Ikeda, M., *Tetrahedron*, 1972, **28**, 21.
2470. Dore, G., Bonhomme, M. and Robba, M., *Tetrahedron*, 1972, **28**, 2553.
2471. Tamura, Y., Kanematsu, K., Kakehi, A. and Ito, G., *Tetrahedron*, 1972, **28**, 4947.
2472. Sheradsky, T. and Lewinter, S., *Tetrahedron Lett.*, 1972, 3941.
2473. Krubsack, A.J. and Higa, T., *Tetrahedron Lett.*, 1972, 4823.
2473a. Omar, A.M.M.E., *Pharmazie*, 1972, **27**, 798.
2474. Yoneda, F., Higuchi, M., Matsumura, T. and Senda, K., *Bull. Chem. Soc. Jpn.*, 1973, **46**, 1863.
2475. Dhaka, K.S., Chadha, V.K. and Pujari, H.K., *Indian J. Chem.*, 1973, **11**, 554.
2476. Kamel, M., Sherif, S., Issa, R.M. and Abd-el-Hay, F.I., *Tetrahedron*, 1973, **29**, 221.

2477. Fuks, R., *Tetrahedron*, 1973, **29**, 2153.
2478. Eberle, M.K., Kahle, G.G. and Talati, S.M., *Tetrahedron*, 1973, **29**, 4045.
2479. Lahey, F.N. and Stick, R.V., *Aust. J. Chem.*, 1973, **26**, 2307.
2480. Lahey, F.N. and Stick, R.V., *Aust. J. Chem.*, 1973, **26**, 2311.
2481. Mock, J., Ritchie, E. and Taylor, W.C., *Aust. J. Chem.*, 1973, **26**, 2315.
2482. Gale, D.J. and Wilshire, J.F.K., *Aust. J. Chem.*, 1973, **26**, 2683.
2483. Hisano, T., Yoshikawa, S. and Muraoka, K., *Org. Prep. Proced. Int.*, 1973, **5**, 95.
2484. Hisano, T., *Org. Prep. Proced. Int.*, 1973, **5**, 145.
2485. Davis, M., Homfeld, E. and Paproth, T., *Org. Prep. Proced. Int.*, 1973, **5**, 197.
2486. Ninomiya, I., Takasugi, H. and Naito, T., *Heterocycles*, 1973, **1**, 17.
2487. Kametani, T. and Fukumoto, K., *Heterocycles*, 1973, **1**, 129.
2488. Popp, F.D., *Heterocycles*, 1973, **1**, 165.
2489. Shoji, N., Kondo, Y. and Takemoto, T., *Heterocycles*, 1973, **1**, 251.
2490. Denzel, T. and Höhn, H., *Arch. Pharm. (Weinheim)*, 1973, **306**, 746.
2491. Collins, J.F., Gray, G.A., Grundon, M.F., Harrison, D.M. and Spyropoulos, C.G., *J. Chem. Soc., Perkin Trans. 1*, 1973, 94.
2492. Chippendale, K.E., Iddon, B. and Suschitzky, H., *J. Chem. Soc., Perkin Trans. 1*, 1973, 125.
2493. Chippendale, K.E., Iddon, B. and Suschitzky, H., *J. Chem. Soc., Perkin Trans. 1*, 1973, 129.
2494. Perche, J.-C., Saint-Ruf, G. and Buu-Hoï, N.P., *J. Chem. Soc., Perkin Trans. 1*, 1973, 260.
2495. Foster, H.E. and Hurst, J., *J. Chem. Soc., Perkin Trans. 1*, 1973, 319.
2496. Ackrell, J. and Boulton, A.J., *J. Chem. Soc., Perkin Trans. 1*, 1973, 351.
2497. Clarke, K., Hughes, C.G. and Scrowston, R.M., *J. Chem. Soc., Perkin Trans. 1*, 1973, 356.
2498. Partridge, M.W. and Smith, A., *J. Chem. Soc., Perkin Trans. 1*, 1973, 453.
2499. Agasimundin, Y.S. and Siddappa, S., *J. Chem. Soc., Perkin Trans. 1*, 1973, 503.
2500. Glover, E.E., Rowbottom, K.T. and Bishop, D.C., *J. Chem. Soc., Perkin Trans. 1*, 1973, 842.
2501. Sutherland, D.R., Tennant, G. and Vevers, R.J.S., *J. Chem. Soc., Perkin Trans. 1*, 1973, 943.
2502. Bowman, R.M., Grundon, M.F. and James, K.J., *J. Chem. Soc., Perkin Trans. 1*, 1973, 1055.
2503. Ashby, J., Ayed, M. and Meth-Cohn, O., *J. Chem. Soc., Perkin Trans. 1*, 1973, 1104.
2504. Hirata, T., Wood, H.B. and Driscoll, J.S., *J. Chem. Soc., Perkin Trans. 1*, 1973, 1209.
2505. Birchall, J.M. and Clark, M.T., *J. Chem. Soc., Perkin Trans. 1*, 1973, 1259.
2506. Smith, C.V.Z., Robins, R.K. and Tolman, R.L., *J. Chem. Soc., Perkin Trans. 1*, 1973, 1855.
2507. Hecht, S.M. and Werner, D., *J. Chem. Soc., Perkin Trans. 1*, 1973, 1903.
2508. Eiden, F. and Héja, G., *Synthesis*, 1973, 148.
2509. Stuckwisch, C.G., *Synthesis*, 1973, 469.
2510. Uno, H., Irie, A. and Hino, K., *Chem. Pharm. Bull.*, 1973, **21**, 34.
2511. Nagano, M., Matsui, T., Tobitsuka, J. and Oyamada, K., *Chem. Pharm. Bull.*, 1973, **21**, 74.
2512. Uno, H., Irie, A. and Hino, K., *Chem. Pharm. Bull.*, 1973, **21**, 256.
2513. Otomatsu, H., Ohmiya, S., Takahashi, H., Yoshida, K. and Sato, S., *Chem. Pharm. Bull.*, 1973, **21**, 353.
2514. Otomatsu, H., Takahashi, H. and Yoshida, K., *Chem. Pharm. Bull.*, 1973, **21**, 492.
2515. Kobayashi, G., Matsuda, Y. and Natsuki, R., *Chem. Pharm. Bull.*, 1973, **21**, 921.
2516. Shiokawa, Y. and Ohki, S., *Chem. Pharm. Bull.*, 1973, **21**, 98.
2517. Yanagisawa, H., Nakas, H. and Ando, A., *Chem. Pharm. Bull.*, 1973, **21**, 1080.

2518. Tamura, Y., Sumida, Y., Tamada, S. and Ikeda, M., *Chem. Pharm. Bull.*, 1973, **21**, 1139.
2519. Suzue, S., Hirobe, M. and Okamoto, T., *Chem. Pharm. Bull.*, 1973, **21**, 2146.
2520. Murata, T., Sugawara, T. and Ukawa, K., *Chem. Pharm. Bull.*, 1973, **21**, 2571.
2521. Matsuura, S., Kunii, T. and Matsuura, A., *Chem. Pharm. Bull.*, 1973, **21**, 2757.
2522. Tominaga, Y., Natsuki, R., Matsuda, Y. and Kobayashi, G., *Chem. Pharm. Bull.*, 1973, **21**, 2770.
2523. Masaki, Y., Otsuka, H., Nakayama, Y. and Hioki, M., *Chem. Pharm. Bull.*, 1973, **21**, 2780.
2524. Hino, T., Suzuki, T. and Nakagawa, M., *Chem. Pharm. Bull.*, 1973, **21**, 2786.
2525. Butler, D.E., Myer, R.F., Alexander, S.M., Bass, P. and Kennedy, J.A., *J. Med. Chem.*, 1973, **16**, 49.
2526. Tripp, S.L., Block, F.B. and Barile, G., *J. Med. Chem.*, 1973, **16**, 60.
2527. Makriyannis, A., Frazee, J.S. and Wilson, J.W., *J. Med. Chem.*, 1973, **16**, 118.
2528. Feit, P.W., Nielsen, O.B.T. and Rastrup-Andersen, N., *J. Med. Chem.*, 1973, **16**, 127.
2529. Nakanishi, M., Tahara, T., Araki, K., Shiroki, M., Tsumagari, T. and Takigawa, Y., *J. Med. Chem.*, 1973, **16**, 214.
2530. March, L.C., Romanchick, W.A., Bajiva, G.S. and Joullie, M.M., *J. Med. Chem.*, 1973, **16**, 337.
2531. Israel, M. and Tirosh, N., *J. Med. Chem.*, 1973, **16**, 520.
2532. Jones, J.H., Holtz, J. and Cragoe, E.J., *J. Med. Chem.*, 1973, **16**, 537.
2533. Willenbrock, H.J., Wamhoff, H. and Korte, F., *Liebigs Ann. Chem.*, 1973, 103.
2534. Stähle, H. and Köppe, H., *Liebigs Ann. Chem.*, 1973, 1275.
2535. Kost, A.N., *Chem. Heterocycl. Compds. (Engl. Transl.)*, 1973, **9**, 267.
2536. Aksanova, L.A., Sharkova, L.M., Kucherova, N.F. and Zagorevskii, V.A., *Chem. Heterocycl. Compds. (Engl. Transl.)*, 1973, **9**, 289.
2537. Konshin, M.E., *Chem. Heterocycl. Compds. (Engl. Transl.)*, 1973, **9**, 488.
2538. Konshin, M.E. and Uvarov, D.I., *Chem. Heterocycl. Compds. (Engl. Transl.)*, 1973, **9**, 490.
2539. Koksharova, T.G., Lipatova, L.F., Konyukhov, V.N. and Smotrina, G.N.P.Z.V., *Chem. Heterocycl. Compds. (Engl. Transl.)*, 1973, **9**, 513.
2540. Kretov, A.G., Momsenko, A.P. and Levin, Y.A., *Chem. Heterocycl. Compds. (Engl. Transl.)*, 1973, **9**, 595.
2541. Dorofeenko, G.N. and Luk'yanov, S.M., *Chem. Heterocycl. Compds. (Engl. Transl.)*, 1973, **9**, 812.
2542. Kucherova, N.F., Aksanova, L.A., Sharkova, L.M. and Zagorevskii, V.A., *Chem. Heterocycl. Compds. (Engl. Transl.)*, 1973, **9**, 835.
2543. Korshunov, S.P., Kazantseva, V.M., Vopilina, L.A., Pisareva, V.S. and Utckhina, N.V., *Chem. Heterocycl. Compds. (Engl. Transl.)*, 1973, **9**, 1287.
2544. Dorofeenko, G.N., Krivan, S.V. and Korobkova, V.G., *Chem. Heterocycl. Compds. (Engl. Transl.)*, 1973, **9**, 1318.
2545. Shvedov, V.I., Vasil'eva, V.K., Trofimkin, Y.I. and Grinev, A.N., *Chem. Heterocycl. Compds. (Engl. Transl.)*, 1973, **9**, 1473.
2546. Sazonov, N.V. and Safonova, T.S., *Chem. Heterocycl. Compds. (Engl. Transl.)*, 1973, **9**, 1532.
2547. Simchen, G. and Entenmann, G., *Angew. Chem., Int. Ed. Engl.*, 1973, **12**, 119.
2548. Hervens, F. and Viehe, H.G., *Angew. Chem., Int. Ed. Engl.*, 1973, **12**, 405.
2549. Viehe, H.G. and Janousek, Z., *Angew. Chem., Int. Ed. Engl.*, 1973, **12**, 806.
2550. Livingstone, D.B. and Tennant, G., *J. Chem. Soc., Chem. Commun.*, 1973, 96.
2551. Shanmugan, P., Lakshminarayana, P. and Palaniappar, R., *Monatsh. Chem.*, 1973, **104**, 633.
2552. Hromatka, O. and Binder, D., *Monatsh. Chem.*, 1973, **107**, 704.

2553. Elliott, A.J., Gibson, M.S., Kayser, M.M. and Pawelchak, G.A., *Can. J. Chem.*, 1973, **51**, 4115.
2554. Petersen, H.J., *Acta Chem. Scand.*, 1973, **27**, 2655.
2555. Gewald, K. and Schael, J., *J. Prakt. Chem.*, 1973, **315**, 39.
2556. Gewald, K. and Jnsch, H.J., *J. Prakt. Chem.*, 1973, **315**, 779.
2557. Yanagida, S. and Komori, S., *Synthesis*, 1973, 189.
2558. Garcia, E.E., *Synth. Commun.*, 1973, **3**, 397.
2559. Omar, A.M.M.E., El-Dine, S.A.S. and Hazzaa, A.A.B., *Pharmazie*, 1973, **28**, 682.
2560. Jen, T., Bender, P., VanHoeven, H., Dienel, B. and Loev, B., *J. Med. Chem.*, 1973, **16**, 407.
2561. Nielsen, O.B.T., Nielsen, C.K. and Feit, P.W., *J. Med. Chem.*, 1973, **16**, 1170.
2562. Sonveaux, E. and Ghosez, L., *J. Am. Chem. Soc.*, 1973, **95**, 5417.
2563. Cook, P.D. and Castle, R.N., *J. Heterocycl. Chem.*, 1973, **10**, 551.
2564. Beck, J.R. and Yahner, J.A., *J. Org. Chem.*, 1973, **38**, 2450.
2565. Borch, R.F. and Newell, R.G., *J. Org. Chem.*, 1973, **38**, 2729.
2566. Reid, W. and Merkel, W., *Liebigs Ann. Chem.*, 1973, 122.
2567. Reid, W. and Nyiondi-Bonguan, E., *Liebigs Ann. Chem.*, 1973, 134.
2568. Seitz, G. and Morck, H., *Synthesis*, 1973, 355.
2569. Häfel, J. and Breitmaier, E., *Angew. Chem., Int. Ed. Engl.*, 1973, **12**, 922.
2570. Paull, K.D. and Cheng, C.C., *J. Heterocycl. Chem.*, 1973, **10**, 137.
2571. Reinhoudt, D.N., *Recl. Trav. Chim. Pays-Bas.*, 1973, **92**, 20.
2572. Richter, H., Winter, K., Elkousy, S. and Luckner, M., *Pharmazie*, 1974, **29**, 506.
2573. Curtze, J. and Thomas, K., *Liebigs Ann. Chem.*, 1974, 328.
2574. Röchling, H. and Hörlein, G., *Liebigs Ann. Chem.*, 1974, 504.
2575. Reid, W. and Ochs, W., *Liebigs Ann. Chem.*, 1974, 1248.
2576. Simchen, G. and Häfner, M., *Liebigs Ann. Chem.*, 1974, 1802.
2577. Capuano, L., Schrepfer, H.J., Jaeschke, M.E. and Porsche, H., *Chem. Ber.*, 1974, **107**, 62.
2578. Capuano, L., Schrepfer, H.J., Müller, K. and Roos, H., *Chem. Ber.*, 1974, **107**, 929.
2579. Körösi, J. and Lang, T., *Chem. Ber.*, 1974, **107**, 3883.
2580. Yoneda, F., Higuchi, M. and Nagamatsu, T., *J. Am. Chem. Soc.*, 1974, **96**, 5607.
2581. Fuhrer, W., Hobi, R., Pfaltz, A. and Schneider, P., *Helv. Chim. Acta*, 1974, **57**, 1498.
2582. Denzel, T., *Arch. Pharm. (Weinheim)*, 1974, **307**, 177.
2583. Otto, H.H., *Arch. Pharm. (Weinheim)*, 1974, **307**, 444.
2584. Eiden, F. and Dusemund, J., *Arch. Pharm. (Weinheim)*, 1974, **307**, 701.
2585. Lebenstedt, E. and Schunack, W., *Arch. Pharm. (Weinheim)*, 1974, **307**, 894.
2586. Garcia, E.E., *Org. Prep. Proced. Int.*, 1974, **6**, 11.
2587. Claret, P.A. and Osborne, A.G., *Org. Prep. Proced. Int.*, 1974, **6**, 149.
2588. Hisano, T., Matsuoka, T. and Ichikawa, M., *Org. Prep. Proced. Int.*, 1974, **6**, 243.
2589. Bennett, O.F., Johnson, J. and Tramondozzi, J., *Org. Prep. Proced. Int.*, 1974, **6**, 287.
2590. Schäfer, W., Moore, H.W. and Aguado, A., *Synthesis*, 1974, 30.
2591. Omar, A.M.M.E., *Synthesis*, 1974, 41.
2592. Casiraghi, G., Casnati, G., Castellani, M. and Corina, M., *Synthesis*, 1974, 564.
2593. Amaranth, Y. and Madhav, R., *Synthesis*, 1974, 837.
2594. Tsuge, O. and Samura, H., *Heterocycles*, 1974, **2**, 27.
2595. Ninomiya, I., *Heterocycles*, 1974, **2**, 105.
2596. Hisano, T., Noda, H. and Aoyama, M., *Heterocycles*, 1974, **2**, 163.
2597. Hamana, M., Noda, H. and Aoyama, M., *Heterocycles*, 1974, **2**, 167.
2598. Kametani, T., Ebetino, F.F., Yamanaka, T. and Nyu, K., *Heterocycles*, 1974, **2**, 209.
2599. Meguro, K. and Kuwala, Y., *Heterocycles*, 1974, **2**, 335.
2600. Lopez, B., Lora-Tamayo, M., Novarro, P. and Soto, J.L., *Heterocycles*, 1974, **2**, 649.

2601. Bakke, J., Heikman, H. and Hellgren, E.B., *Acta Chem. Scand.*, 1974, **28B**, 393.
2602. Pedersen, E.B. and Lawesson, S.D., *Acta Chem. Scand.*, 1974, **28B**, 1045.
2603. Augustine, M. and Kuppe, K.R., *Tetrahedron*, 1974, **30**, 3533.
2604. Oae, S. and Numata, T., *Tetrahedron*, 1974, **30**, 2641.
2605. Minami, S., Kimura, Y., Miyamoto, T. and Matsumoto, J., *Tetrahedron Lett.*, 1974, 3893.
2606. Donnelly, J.A. and Doran, H.J., *Tetrahedron Lett.*, 1974, 4083.
2607. Sanchez, A.G., Toledano, E. and Guillén, M.G., *J. Chem. Soc., Perkin Trans. 1*, 1974, 1237.
2608. Peek, M.E., Rees, C.W. and Storr, R.C., *J. Chem. Soc., Perkin Trans. 1*, 1974, 1260.
2609. Birch, A.J., Jackson, A.H. and Shannon, P.V.R., *J. Chem. Soc., Perkin Trans. 1*, 1974, 2185.
2610. Chimenti, F., Vomero, S., Nacci, V., Sealzo, M., Giuliano, R. and Artico, M., *Farmaco, Ed. Sci.*, 1974, **29**, 589.
2611. Dorofeenko, G.N., Ryabukhin, Y.I. and Mezheritskii, W., *J.Org.Chem. USSR*, 1974, **10**, 2251.
2612. Coffen, D.L., DeNoble, J.P., Evans, E.L., Field, G.F., Fryer, R.I., Katonak, D.A., Mandel, B.J., Sternbach, L.H. and Zally, W.J., *J. Org. Chem.*, 1974, **39**, 167.
2613. Anderson, G.L., Rizkalla, B.H. and Brown, A.D., *J. Org. Chem.*, 1974, **39**, 937.
2614. Haugwitz, R.D., Maurer, B.V. and Narayanam, V.L., *J. Org. Chem.*, 1974, **39**, 1359.
2615. Kakehi, A. and Ito, S., *J. Org. Chem.*, 1974, **39**, 1542.
2616. Fortuna, D., Stanovnik, B. and Tisler, M., *J. Org. Chem.*, 1974, **39**, 1833.
2617. Peet, N.P. and Sunder, S., *J. Org. Chem.*, 1974, **39**, 1931.
2618. DeBoer, C.D., *J. Org. Chem.*, 1974, **39**, 2426.
2619. Weinstock, J., Bank, J.E. and Sutton, B.M., *J. Org. Chem.*, 1974, **39**, 2454.
2620. Papadopoulos, E.P., *J. Org. Chem.*, 1974, **39**, 2540.
2621. McKillop, A. and Kobylecki, R.J., *J. Org. Chem.*, 1974, **39**, 2710.
2622. Posvic, H., Dombro, R., Ito, H. and Tetinski, T., *J. Org. Chem.*, 1974, **39**, 2575.
2623. Pilgram, K. and Skiles, R.D., *J. Org. Chem.*, 1974, **39**, 3277.
2624. Danishefsky, S. and Etheredge, S.J., *J. Org. Chem.*, 1974, **39**, 3430.
2625. Maguire, J.H. and McKee, R.L., *J. Org. Chem.*, 1974, **39**, 3434.
2626. Wikel, J.H. and Paget, C.J., *J. Org. Chem.*, 1974, **39**, 3506.
2627. Olson, D.R., Wheeler, W.J. and Wells, J.N., *J. Med. Chem.*, 1974, **17**, 167.
2628. Hardtmann, G.E., Huegi, B., Koletar, G., Kroin, S., Ott, H., Perrine, J.W., and Takesue, E.I., *J. Med. Chem.*, 1974, **17**, 636.
2629. Zweig, J.S. and Castagnoli, N., *J. Med. Chem.*, 1974, **17**, 747.
2630. DeGraw, J.I., Brown, V.H., Colwell, W.T. and Morrison, N.E., *J. Med. Chem.*, 1974, **17**, 762.
2631. Buchman, R., Heinstein, P.F. and Wells, J.N., *J. Med. Chem.*, 1974, **17**, 1168.
2632. Wright, G.E. and Brown, N.C., *J. Med. Chem.*, 1974, **17**, 1277.
2633. Konshin, M.E. and Chesnokov, V.P., *Chem. Heterocycl. Compds. (Engl. Transl.)*, 1974, **10**, 105.
2634. Britikova, N.E., Belova, L.A., Mogidson, O.Yu. and Elina, A.S., *Chem. Heterocycl. Compds. (Engl. Transl.)*, 1974, **10**, 117.
2635. Sumlivenko, N.V. and Dvorko, G.F., *Chem. Heterocycl. Compds. (Engl. Transl.)*, 1974, **10**, 121.
2636. Sharanina, L.G. and Baranov, S.N., *Chem. Heterocycl. Compds. (Engl. Transl.)*, 1974, **10**, 171.
2637. Grandberg, I.I., *Chem. Heterocycl. Compds. (Engl. Transl.)*, 1974, **10**, 179.
2638. Chesnokov, V.P. and Konshin, M.E., *Chem. Heterocycl. Compds. (Engl. Transl.)*, 1974, **10**, 217.
2639. Kozlov, N.S., Zhikhareva, O.D. and Stremok, I.P., *Chem. Heterocycl. Compds. (Engl. Transl.)*, 1974, **10**, 219.

2640. Dubenko, R.G. and Gorbenko, E.F., *Chem. Heterocycl. Compds. (Engl. Transl.)*, 1974, **10**, 434.
2641. Grandberg, I.I., *Chem. Heterocycl. Compds. (Engl. Transl.)*, 1974, **10**, 501.
2642. Petyunin, P.A., Bulgakov, V.A. and Petyunin, G.P., *Chem. Heterocycl. Compds. (Engl. Transl.)*, 1974, **10**, 527.
2643. Takahashi, M., Onizawa, S. and Satoh, T., *Bull. Chem. Soc. Jpn.*, 1974, **47**, 2724.
2644. Saint-Ruf, G., *J. Heterocycl. Chem.*, 1974, **11**, 13.
2645. Coppola, G.M., Hardtmann, G.E. and Huegi, B.S., *J. Heterocycl. Chem.*, 1974, **11**, 51.
2646. Yoneda, F., Nagamatsu, T. and Ichiba, M., *J. Heterocycl. Chem.*, 1974, **11**, 83.
2647. Ning, R.Y., Madan, P.B. and Sternbach, L.H., *J. Heterocycl. Chem.*, 1974, **11**, 107.
2648. Crochet, R.A., Boatright, J.T., Blanton, C.D., Wier, C.T. and Hochholzer, W.E., *J. Heterocycl. Chem.*, 1974, **11**, 143.
2649. Chesney, D.R. and Castle, R.N., *J. Heterocycl. Chem.*, 1974, **11**, 167.
2650. Sotiriadis, A., Catsoulacos, P. and Theodoropoulos, D., *J. Heterocycl. Chem.*, 1974, **11**, 401.
2651. Alcalde, E., de Mendoza, J., Garcia-Marquina, J., Almera, C. and Elguero, J., *J. Heterocycl. Chem.*, 1974, **11**, 423.
2652. Trepanier, D.L., Sunder, S. and Braun, W.H., *J. Heterocycl. Chem.*, 1974, **11**, 747.X
2653. Fletcher, T.L., Pan, H.L., Cole, C.A. and Namkung, M.J., *J. Heterocycl. Chem.*, 1974, **11**, 815.
2654. Abramovitch, R.A. and Kalinowski, J., *J. Heterocycl. Chem.*, 1974, **11**, 857.
2655. Vilarrasa, J. and Granados, R., *J. Heterocycl. Chem.*, 1974, **11**, 867.
2656. Grol, C.J., *J. Heterocycl. Chem.*, 1974, **11**, 953.
2657. Paull, K.D. and Cheng, C.C., *J. Heterocycl. Chem.*, 1974, **11**, 1027.
2658. Cignarella, G., Savelli, F., Cerri, R. and Sanna, P., *J. Heterocycl. Chem.*, 1974, **11**, 1049.
2659. Sauter, F. and Deinhammer, W., *Monatsh. Chem.*, 1974, **105**, 1249.
2660. Murata, T. and Ukawa, K., *Chem. Pharm. Bull.*, 1974, **22**, 240.
2661. Okumura, K., Kondo, K., Oine, F. and Inoue, I., *Chem. Pharm. Bull.*, 1974, **22**, 331.
2662. Sugasawa, T., Toyoda, T. and Sasakura, K., *Chem. Pharm. Bull.*, 1974, **22**, 771.
2663. Koyama, T., Hirota, T., Yoshida, T., Hara, H. and Ohmori, S., *Chem. Pharm. Bull.*, 1974, **22**, 1451.
2664. Senda, S. and Hirota, K., *Chem. Pharm. Bull.*, 1974, **22**, 1459.
2665. Hamana, M. and Kumadaki, S., *Chem. Pharm. Bull.*, 1974, **22**, 1506.
2666. Kato, T. and Masuda, S., *Chem. Pharm. Bull.*, 1974, **22**, 1542.
2667. Yoneda, F. and Matsumoto, S., *Chem. Pharm. Bull.*, 1974, **22**, 1652.
2668. Senda, S. and Hirota, K., *Chem. Pharm. Bull.*, 1974, **22**, 2921.
2669. Kappe, T. and Branduer, A., *Z. Naturforsch.*, 1974, **29B**, 292.
2670. Saito, K., Hori, I., Igarashi, M. and Midorikawa, H., *Bull. Chem. Soc. Jpn.*, 1974, **47**, 476.
2671. Dhaka, K.S., Mohau, J., Chadha, V.K. and Pyari, H.K., *Indian J. Chem.*, 1974, **12**, 966.
2672. Rao, A.S. and Mitra, R.M., *Indian J. Chem.*, 1974, **12**, 1028.
2673. Zbiral, E., *Synthesis*, 1974, 775.
2674. Hartke, K. and Radau, S., *Liebigs Ann. Chem.*, 1974, 2110.
2675. Kende, A.S., Wade, J.J., Ridge, D. and Poland, A., *J. Org. Chem.*, 1974, **39**, 931.
2676. Gibson, M.S., Vines, S.M. and Walthew, J.M., *J. Chem. Soc., Perkin Trans. 1*, 1975, 155.
2677. Baldwin, D. and van den Broek, P., *J. Chem. Soc., Perkin Trans. 1*, 1975, 375.
2678. Suschitzky, H., Wakefield, B.J. and Whittaker, R.A., *J. Chem. Soc., Perkin Trans. 1*, 1975, 401.
2679. Ames, D.E. and Ward, R.J., *J. Chem. Soc., Perkin Trans. 1*, 1975, 534.

2680. Gait, S.F. and Rance, M.J., Rees, C.W., Stephenson, R.W. and Storr, R.C., *J. Chem. Soc., Perkin Trans. 1*, 1975, 556.
2681. Hull, R., van den Broek, P.J. and Swain, M.L., *J. Chem. Soc., Perkin Trans. 1*, 1975, 922.
2682. Dickinson, R.G. and Jacobson, N.W., *J. Chem. Soc., Perkin Trans. 1*, 1975, 975.
2683. Callaghan, P.D., Gibson, M.S. and Elliott, A.J., *J. Chem. Soc., Perkin Trans. 1*, 1975, 1386.
2684. McDonald, B.G. and Proctor, G.R., *J. Chem. Soc., Perkin Trans. 1*, 1975, 1446.
2685. Yoneda, F., Matsumoto, S., Sakuma, Y. and Fukazawa, S., *J. Chem. Soc., Perkin Trans. 1*, 1975, 1907.
2686. Gilchrist, T.L., Moody, C.J. and Rees, C.W., *J. Chem. Soc., Perkin Trans. 1*, 1975, 1964.
2687. Cignarella, G., Savelli, F. and Sanna, P., *Synthesis*, 1975, 252.
2688. Esmail, R. and Kurzer, F., *Synthesis*, 1975, 301.
2689. Arnold, F.E., *Org. Prep. Proced. Int.*, 1975, **7**, 123.
2690. Hisano, T., Shoji, K. and Ichikawa, M., *Org. Prep. Proced. Int.*, 1975, **7**, 271.
2691. Friedrichsen, W. and Oeser, H.G., *Chem. Ber.*, 1975, **108**, 31.
2692. Hartke, K., Krampetz, D. and Uhde, W., *Chem. Ber.*, 1975, **108**, 128.
2693. Eistert, B., Pfleger, K., Arackal, T.J. and Holzer, G., *Chem. Ber.*, 1975, **108**, 693.
2694. Rothkopf, H.W., Wöhrle, D., Müller, R. and Kossmehl, G., *Chem. Ber.*, 1975, **108**, 875.
2695. Capuano, L. and Müller, K., *Chem. Ber.*, 1975, **108**, 1541.
2696. Seebach, D., Ehrig, V., Leitz, H.F. and Henning, R., *Chem. Ber.*, 1975, **108**, 1946.
2697. Dubau, F.-P. and Zinner, G., *Chem. Ber.*, 1975, **108**, 2189.
2698. Ahlbrecht, H. and Vonderheid, C., *Chem. Ber.*, 1975, **108**, 2300.
2699. Aurich, H.G. and Stork, K., *Chem. Ber.*, 1975, **108**, 2764.
2700. Eistert, B., Golbran, L.S.B., Vamvakaris, C. and Arackal, T.J., *Chem. Ber.*, 1975, **108**, 2941.
2701. Flitsch, W., Gurke, A. and Müter, R., *Chem. Ber.*, 1975, **108**, 2969.
2702. Reimlinger, H. and Lingier, W.R.F., *Chem. Ber.*, 1975, **108**, 3787.
2703. Reimlinger, H., Lingier, W.R.F. and Merényi, R., *Chem. Ber.*, 1975, **108**, 3794.
2704. Reimlinger, H. and Billiau, F., Lingier, W.R.F. and Peiren, M.A., *Chem. Ber.*, 1975, **108**, 3799.
2705. Stoss, P. and Satzinger, G., *Chem. Ber.*, 1975, **108**, 3855.
2706. Reid, W., Merkel, W. and Park, S.W., *Liebigs Ann. Chem.*, 1975, 79.
2707. Reichardt, C. and Halbritter, K., *Liebigs Ann. Chem.*, 1975, 470.
2708. Wolf, E. and Kohl, H., *Liebigs Ann. Chem.*, 1975, 1245.
2709. Hagen, H. and Fleig, H., *Liebigs Ann. Chem.*, 1975, 1994.
2710. Rogal'chenko, G.K., Mazur, I.A. and Kochergin, P.M., *Chem. Heterocycl. Compds. (Engl. Transl.)*, 1975, **11**, 81.
2711. Kovtunenko, V.A., Bubnovskaya, Y., and Babichev, F.S., *Chem. Heterocycl. Compds. (Engl. Transl.)*, 1975, **11**, 120.
2712. Ryabukhin, Y.I., Dorofeenko, G.N. and Mezheritskii, V.V., *Chem. Heterocycl. Compds. (Engl. Transl.)*, 1975, **11**, 243.
2713. Popov, I.I., Tkachenko, P.V. and Simonov, A.M., *Chem. Heterocycl. Compds. (Engl. Transl.)*, 1975, **11**, 347.
2714. Chuiguk, V.A. and Vohrenko, Y.M., *Chem. Heterocycl. Compds. (Engl. Transl.)*, 1975, **11**, 467.
2715. Kozlov, N.S. and Vorob'eva, G.V., *Chem. Heterocycl. Compds. (Engl. Transl.)*, 1975, **11**, 708.
2716. Povstyanoi, M.V., Klybor, M.A., Gorban, N.M. and Kochergin, P.M., *Chem. Heterocycl. Compds. (Engl. Transl.)*, 1975, **11**, 751.

2717. Grineva, V.S., Buevich, V.A. and Rudchenko, V.V., *Chem. Heterocycl. Compds. (Engl. Transl.)*, 1975, **11**, 752.
2718. Krasovskii, A.N., Yurchenko, M.I., Kochergin, P.M. and Soroka, I.I., *Chem. Heterocycl. Compds. (Engl. Transl.)*, 1975, **11**, 979.
2719. Pushkareva, Z.V., Ofitserov, V.I., Mokrushin, V.S. and Aglitskaya, K.V., *Chem. Heterocycl. Compds. (Engl. Transl.)*, 1975, **11**, 995.
2720. Shabarov, Y.S., Mochalov, S.S., Fedotov, A.N. and Kalashnikov, V.V., *Chem. Heterocycl. Compds. (Engl. Transl.)*, 1975, **11**, 1038.
2721. Yakovenko, V.I., Mikhlina, E.E., Oganesyan, E.T. and Yakhontov, L.N., *Chem. Heterocycl. Compds. (Engl. Transl.)*, 1975, **11**, 1104.
2721a. Solomko, Z.F. and Kost, A.N., *Chem. Heterocycl. Compds. (Engl. Transl.)*, 1975, **11**, 1231.
2722. Bachkovskii, I.P. and Chuiguk, V.A., *Chem. Heterocycl. Compds. (Engl. Transl.)*, 1975, **11**, 1272.
2723. Rudchenko, V.V., Buevich, V.A., Grineva, V.S. and Perekalin, V.V., *Chem. Heterocycl. Compds. (Engl. Transl.)*, 1975, **11**, 1341.
2724. Astrakhantseva, N.I., Zhiryakov, V.G. and Abramenko, P.I., *Chem. Heterocycl. Compds. (Engl. Transl.)*, 1975, **11**, 1364.
2725. Uno, H., Irie, A. and Hino, K., *Chem. Pharm. Bull.*, 1975, **23**, 450.
2726. Kato, T. and Masuda, S., *Chem. Pharm. Bull.*, 1975, **23**, 452.
2727. Koyama, T., Hirota, T., Shinohara, Y., Yamato, M. and Ohmori, S., *Chem. Pharm. Bull.*, 1975, **23**, 497.
2728. Kobayashi, Y., Kumadaki, I., Sekine, Y., Naito, Y. and Kutsuma, T., *Chem. Pharm. Bull.*, 1975, **23**, 566.
2729. Kurihara, T., Sano, H. and Hirano, H., *Chem. Pharm. Bull.*, 1975, **23**, 1153.
2730. Okamoto, Y. and Ueda, T., *Chem. Pharm. Bull.*, 1975, **23**, 1391.
2731. Fugita, H. and Sato, Y., *Chem. Pharm. Bull.*, 1975, **23**, 1764.
2732. Yoneda, F. and Nagamatsu, T., *Chem. Pharm. Bull.*, 1975, **23**, 1885.
2733. Oka, Y., Omura, K., Miyake, A., Itoh, K., Tominoto, M., Tada, N. and Yurugi, S., *Chem. Pharm. Bull.*, 1975, **23**, 2239.
2734. Kobayashi, G., Matsuda, Y., Tominaga, Y., Maseda, C., Awaya, H. and Kurata, K., *Chem. Pharm. Bull.*, 1975, **23**, 2759.
2735. Igeta, H., Arai, H., Hasegawa, H., and Tschuchiya, T., *Chem. Pharm. Bull.*, 1975, **23**, 2791.
2736. Schweizer, E.E. and DeVoe, S.V., *J. Org. Chem.*, 1975, **40**, 144.
2737. Abushanab, E. and Alteri, N.D., *J. Org. Chem.*, 1975, **40**, 157.
2738. Feuer, H. and Friedman, H., *J. Org. Chem.*, 1975, **40**, 187.
2739. Macco, A.A., Godefroi, E.F. and Drouen, J.J.M., *J. Org. Chem.*, 1975, **40**, 252.
2740. Cooley, J.H. and Jacobs, P.T., *J. Org. Chem.*, 1975, **40**, 552.
2741. Hoogenboom, B.E., Fink, S.C., Ihrig, P.J., Langsjoen, A.N., Linn, C.J. and Machling, K.L., *J. Org. Chem.*, 1975, **40**, 880.
2742. Coffen, D.L., Fryer, R.I., Katonak, D.A. and Wong, F., *J. Org. Chem.*, 1975, **40**, 894.
2743. Fuentes, O. and Paudler, W.W., *J. Org. Chem.*, 1975, **40**, 1210.
2744. Bender, D.R., Bjeldanes, L.F., Knapp, D.R. and Rapoport, R.,*J. Org. Chem.*, 1975, **40**, 1264.
2744a. Akermark, B., Eberson, L., Jonsson, E. and Petterson, E., *J. Org. Chem.*, 1975, **40**, 1365.
2745. Nair, M.G., Campbell, P.T. and Baugh, C.M., *J. Org. Chem.*, 1975, **40**, 1745.
2746. Earl, R.A., Pugmire, R.J., Revankar, G.R. and Townsend, L.B., *J. Org. Chem.*, 1975, **40**, 1822.

2747. Danishefsky, S., Bryson, T.A. and Puthenpurayil, J., *J. Org. Chem.*, 1975, **40**, 1846.
2748. Weinstock, J., Gaitanopoulos, D.E. and Sutton, B.M., *J. Org. Chem.*, 1975, **40**, 1914.
2749. Grivas, J.C., *J. Org. Chem.*, 1975, **40**, 2029.
2750. Granoth, I. and Pownall, H.J., *J. Org. Chem.*, 1975, **40**, 2088.
2751. Temple, C., Kussner, C.L. and Montgomery, J.A., *J. Org. Chem.*, 1975, **40**, 2205.
2752. Bousquet, E.W., Moran, M.D., Harmon, J., Johnson, A.L. and Summers, J.C., *J. Org. Chem.*, 1975, **40**, 2208.
2753. Taylor, E.C. and Sowinski, F., *J. Org. Chem.*, 1975, **40**, 2321.
2754. Taylor, E.C. and Sowinski, F., *J. Org. Chem.*, 1975, **40**, 2329.
2755. Potts, K.T. and Kane, J., *J. Org. Chem.*, 1975, **40**, 2600.
2756. Wen, R.Y., Komin, A.P., Street, R.W. and Carmack, M., *J. Org. Chem.*, 1975, **40**, 2743.
2757. Komin, A.P., Street, R.W. and Carmack, M., *J. Org. Chem.*, 1975, **40**, 2749.
2758. Zimmer, H., Kokosa, J.M. and Shah, K.J., *J. Org. Chem.*, 1975, **40**, 2901.
2759. Higa, T. and Krubsack, A.J., *J. Org. Chem.*, 1975, **40**, 3037.
2760. Abramovitch, R.A. and Shinkai, I., *J. Am. Chem. Soc.*, 1975, **97**, 3227.
2761. Yamamori, T., Noda, H. and Hamana, M., *Tetrahedron*, 1975, **31**, 945.
2762. Alhaique, F., Riccieri, F.M. and Santucci, E., *Tetrahedron Lett.*, 1975, 173.
2763. Yabe, A. and Honda, K., *Tetrahedron Lett.*, 1975, 1079.
2763a. de Groot, A. and Jansen, B.J.M., *Tetrahedron Lett.*, 1975, 3407.
2764. Nielsen, O.B.T., Bruun, H., Bretting, C. and Feit, P.W., *J. Med. Chem.*, 1975, **18**, 41.
2765. Kavai, I., Mead, L.H., Drobniak, J. and Zakrzewski, S.F., *J. Med. Chem.*, 1975, **18**, 272.
2766. Carroll, F.I. and Coleman, M.C., *J. Med. Chem.*, 1975, **18**, 318.
2767. Buckle, D.R., Cantello, B.C.C., Smith, H. and Spicer, B.A., *J. Med. Chem.*, 1975, **18**, 726.
2768. Srivastava, P.R., Field, L. and Grenan, M.M., *J. Med. Chem.*, 1975, **18**, 798.
2769. Grol, C.J. and Rollema, H., *J. Med. Chem.*, 1975, **18**, 857.
2770. Dunwell, D.W., Evans, D. and Hicks, T.A., *J. Med. Chem.*, 1975, **18**, 1158.
2771. Reichardt, C. and Halbritter, K., *Angew. Chem., Int. Ed. Engl.*, 1975, **14**, 86.
2772. Metallidis, A., Sotiriadis, A. and Theodoropoulos, D., *J. Heterocycl. Chem.*, 1975, **12**, 359.
2773. Vogel, A. and Troxler, F., *Helv. Chim. Acta*, 1975, **58**, 761.
2774. Edenhofer, A., *Helv. Chim. Acta*, 1975, **58**, 2192.
2775. Sauter, F. and Stanetty, P., *Monatsh. Chem.*, 1975, **106**, 1111.
2776. Mitscher, L.A., Shiphchandler, M. and Showalter, H.D.H., *Heterocycles*, 1975, **3**, 7.
2777. Bestmann, H.J., Schmid, G. and Sandmeier, D., *Angew. Chem., Int. Ed. Engl.*, 1976, **15**, 115.
2778. Hollitzer, O., Seewald, A. and Steglish, W., *Angew. Chem., Int. Ed. Engl.*, 1976, **15**, 444.
2779. Bass, R.J., *J. Chem. Soc., Chem. Commun.*, 1976, 78.
2780. Lynch, B.M. and Sharma, S.C., *Can. J. Chem.*, 1976, **54**, 1029.
2781. Farnier, M., Soth, S. and Fournari, P., *Can. J. Chem.*, 1976, **54**, 1066.
2782. Amey, R.L. and Heindel, N.D., *Org. Prep. Proced. Int.*, 1976, **8**, 306.
2783. Dohmori, R., Kadoya, S., Takamura, I. and Suzuki, N., *Chem. Pharm. Bull.*, 1976, **24**, 130.
2784. Makisumi, Y. and Takada, S., *Chem. Pharm. Bull.*, 1976, **24**, 770.
2785. Furukawa, Y., Miyashita, O. and Shima, S., *Chem. Pharm. Bull.*, 1976, **24**, 970.
2786. Ito, I., Oda, N. and Kato, T., *Chem. Pharm. Bull.*, 1976, **24**, 1189.
2787. Itoh, T., Honma, M. and Ogura, H., *Chem. Pharm. Bull.*, 1976, **24**, 1390.
2788. Kato, T., Tabei, K. and Kawashima, E., *Chem. Pharm. Bull.*, 1976, **24**, 1544.

2789. Kato, T. and Daneshtalab, M., *Chem. Pharm. Bull.*, 1976, **24**, 1640.
2790. Akiba, M., Kosugi, Y. and Takada, T., *Chem. Pharm. Bull.*, 1976, **24**, 1731.
2791. Takeuchi, I. and Hamada, Y., *Chem. Pharm. Bull.*, 1976, **24**, 1813.
2792. Kametani, T., Kigawa, Y., Takahashi, T., Nemoto, H. and Fukumoto, K., *Chem. Pharm. Bull.*, 1976, **24**, 1870.
2793. Senga, K., Kanamori, Y., Nishigaki, S. and Yoneda, F., *Chem. Pharm. Bull.*, 1976, **24**, 1917.
2794. Takada, K., Haginiwa, J. and Murakoshi, I., *Chem. Pharm. Bull.*, 1976, **24**, 2265.
2795. Kimura, Y., Miyamoto, T., Matsumoto, J. and Minami, S., *Chem. Pharm. Bull.*, 1976, **24**, 2637.
2796. Hirose, N., Kuriyama, S. and Toyoshima, S., *Chem. Pharm. Bull.*, 1976, **24**, 2661.
2797. Omura, K., Tada, N., Tomimoto, M., Usni, Y., Oka, Y. and Yurugi, F., *Chem. Pharm. Bull.*, 1976, **24**, 2699.
2798. Higashino, T., Iwai, Y. and Hayashi, E., *Chem. Pharm. Bull.*, 1976, **24**, 3120.
2799. Petersen, U. and Heitzer, H., *Liebigs Ann. Chem.*, 1976, 1659.
2800. Schäfer, W. and Falkner, C., *Liebigs Ann. Chem.*, 1976, 1809.
2801. Draber, W., Dickoré, K., Donner, W. and Timmler, H., *Liebigs Ann. Chem.*, 1976, 2206.
2802. Heesing, A., Kappler, U. and Rauh, W., *Liebigs Ann. Chem.*, 1976, 2222.
2803. Kuznetsov, E.V., Shcherbakova, I.V. and Dorofeenko, G.N., *Chem. Heterocycl. Compds. (Engl. Transl.)*, 1976, **12**, 47.
2804. Simov, D. and Davidkov, K., *Chem. Heterocycl. Compds. (Engl. Transl.)*, 1976, **12**, 151.
2805. Chernykh, V.J., Goya, P. and Ochoa, C., *Chem. Heterocycl. Compds. (Engl. Transl.)*, 1976, **12**, 81.
2806. Grishchenko, A.S. and Kost, A.N., *Chem. Heterocycl. Compds. (Engl. Transl.)*, 1976, **12**, 411.
2807. Mokrushin, V.S., Ofitserov, V.I., Rapakova, T.V., Tsaur, A.G. and Pushkareva, Z.Y., *Chem. Heterocycl. Compds. (Engl. Transl.)*, 1976, **12**, 465.
2808. Dorofeenko, G.N., Ryabukhin, Y.I. and Mezheritskii, V.V., *Chem. Heterocycl. Compds. (Engl. Transl.)*, 1976, **12**, 621.
2809. Sych, E.D., Mikitenko, E.K. and Kornilov, M.Y., *Chem. Heterocycl. Compds. (Engl. Transl.)*, 1976, **12**, 650.
2810. Simomov, A.M. and Komissarov, V.N., *Chem. Heterocycl. Compds. (Engl. Transl.)*, 1976, **12**, 654.
2811. Skortsova, G.G., Kim, D.G. and Sigalov, M.V., *Chem. Heterocycl. Compds. (Engl. Transl.)*, 1976, **12**, 717.
2812. Grinev, A.N., Lyubchanskaya, V.M., Uretskaya, G.Ya. and Vlasova, T.F., *Chem. Heterocycl. Compds. (Engl. Transl.)*, 1976, **12**, 733.
2813. Mochalov, S.M., Surikova, T.P. and Shabarov, Y.S., *Chem. Heterocycl. Compds. (Engl. Transl.)*, 1976, **12**, 738.
2814. Kost, A.N., Sagitullin, R.S. and Gromov, S.P., *Chem. Heterocycl. Compds. (Engl. Transl.)*, 1976, **12**, 766.
2815. Tkachenko, P.V., Popov, I.I., Simonov, A.M. and Medvedev, Yu. V., *Chem. Heterocycl. Compds. (Engl. Transl.)*, 1976, **12**, 805.
2816. Acheson, R.M., *Chem. Heterocycl. Compds. (Engl. Transl.)*, 1976, **12**, 837.
2817. Abramenko, P.I. and Zhiryaker, V.G., *Chem. Heterocycl. Compds. (Engl. Transl.)*, 1976, **12**, 860.
2818. Przheval'skii, N.M., Grandberg, I.I. and Klyuer, N.A., *Chem. Heterocycl. Compds. (Engl. Transl.)*, 1976, **12**, 880.
2819. Grinev, A.N., Druzhinina, A.A. and Sorokina, I.K., *Chem. Heterocycl. Compds. (Engl. Transl.)*, 1976, **12**, 1048.
2820. Foster, C.H. and Elam, E.U., *J. Org. Chem.*, 1976, **41**, 2646.

2821. Ferris, J.P. and Trimmer, R.W., *J. Org. Chem.*, 1976, **41**, 13.
2822. Ferris, J.P. and Trimmer, R.W., *J. Org. Chem.*, 1976, **41**, 19.
2823. Potts, K.T. and Marshall, J.L., *J. Org. Chem.*, 1976, **41**, 129.
2824. Moschel, R.C. and Leonard, N.J., *J. Org. Chem.*, 1976, **41**, 294.
2825. Johnson, A.L., *J. Org. Chem.*, 1976, **41**, 836.
2826. Suzuki, M. and Yoneda, N., *J. Org. Chem.*, 1976, **41**, 1482.
2827. Beck, J.R. and Yahner, J.A., *J. Org. Chem.*, 1976, **41**, 1733.
2828. Nakazaki, M. and Yamamoto, K., *J. Org. Chem.*, 1976, **41**, 1877.
2829. Hassner, A., Tang, D. and Keogh, J., *J. Org. Chem.*, 1976, **41**, 2102.
2830. Abramovitch, R.A. and Alexanian, V., *J. Org. Chem.*, 1976, **41**, 2144.
2831. Gray, A.P., Cepa, S.P., Salomon, I.J. and Aniline, O., *J. Org. Chem.*, 1976, **41**, 2435.
2832. Benati, L. and Montevechi, P.C., *J. Org. Chem.*, 1976, **41**, 2639.
2833. Ning, R.Y., Fryer, R.I., Madan, P.B. and Sluboski, B.C., *J. Org. Chem.*, 1976, **41**, 2724.
2834. Sunder, S., Peet, N.P. and Trepanier, D.L., *J. Org. Chem.*, 1976, **41**, 2732.
2835. Reimlinger, H., Billiau, F. and Lingier, W.R.F., *Chem. Ber.*, 1976, **109**, 118.
2836. Ried, W. and Schweitzer, R., *Chem. Ber.*, 1976, **109**, 1643.
2837. Stoss, P. and Satzinger, G., *Chem. Ber.*, 1976, **109**, 2097.
2838. Roedig, A. and Fouré, M., *Chem. Ber.*, 1976, **109**, 2159.
2839. Böhme, H. and Weisel, K.H., *Chem. Ber.*, 1976, **109**, 2908.
2840. Capuano, L., Tammer, T. and Zander, R., *Chem. Ber.*, 1976, **109**, 3497.
2841. Mack, W., *Chem. Ber.*, 1976, **109**, 3564.
2842. Anderson, W.K., LaVoie, E.J. and Bottaro, J.C., *J. Chem. Soc., Perkin Trans. 1*, 1976, 1.
2843. Gittos, M.W., Robinson, M.R., Verge, J.P., Davies, R.N., Iddon, B. and Suschitzky, H., *J. Chem. Soc., Perkin Trans. 1*, 1976, 33.
2844. Barton, D.H.R., Ducker, J.W., Lord, W.A. and Magnus, P.D., *J. Chem. Soc., Perkin Trans. 1*, 1976, 38.
2845. Cooper, G. and Irwin, W.J., *J. Chem. Soc., Perkin Trans. 1*, 1976, 75.
2846. Christie, R.M. and Reid, D.H., *J. Chem. Soc., Perkin Trans. 1*, 1976, 228.
2847. Farquhar, D., Gough, T.T. and Leaver, D., *J. Chem. Soc., Perkin Trans. 1*, 1976, 341.
2848. Johnston, D. and Smith, D.M., *J. Chem. Soc., Perkin Trans. 1*, 1976, 399.
2849. Yoneda, F., Nagamatsu, T. and Shinomura, K., *J. Chem. Soc., Perkin Trans. 1*, 1976, 713.
2850. Adams, J.H., Gupta, P., Khan, M.S., Lewis, J.R. and Watt, R.A., *J. Chem. Soc., Perkin Trans. 1*, 1976, 2089.
2851. Parrick, J. and Wilcox, R., *J. Chem. Soc., Perkin Trans. 1*, 1976, 2121.
2852. Gilchrist, T.L., Harris, C.J., King, F.D., Peek, M.E. and Rees, C.W., *J. Chem. Soc., Perkin Trans. 1*, 1976, 2161.
2853. Gilchrist, T.L., Harris, C.J., Hawkins, D.G., Moody, C.J. and Rees, C., *J. Chem. Soc., Perkin Trans. 1*, 1976, 2166.
2854. Eiden, F. and Wendt, R., *Arch. Pharm. (Weinheim)*, 1976, **309**, 70.
2855. Röder, E. and Franke, U., *Arch. Pharm. (Weinheim)*, 1976, **309**, 131.
2856. Franke, U. and Röder, E., *Arch. Pharm. (Weinheim)*, 1976, **309**, 185.
2857. Böhme, H. and Weisel, K.-H., *Arch. Pharm. (Weinheim)*, 1976, **309**, 959.
2858. Chatterjee, A., Ganguly, D. and Sen, R., *Tetrahedron*, 1976, **32**, 2407.
2859. Bleackley, R.C., Jones, A.S. and Walker, R.T., *Angew. Chem., Int. Ed. Engl.*, 1976, **32**, 2795.
2860. Suzuki, N., Yamabayashi, T. and Izawa, Y., *Bull. Chem. Soc. Jpn.*, 1976, **49**, 353.
2861. Ochi, H., Miyasaka, T., Kanada, K. and Arakawa, K., *Bull. Chem. Soc. Jpn.*, 1976, **49**, 1980.

2862. Mahajan, M.P., Sondhi, S.M. and Rahlan, N.K., *Bull. Chem. Soc. Jpn.*, 1976, **49**, 2609.
2863. Mahajan, M.P., Sondhi, S.M. and Ralhan, N.K., *Bull. Chem. Soc. Jpn.*, 1976, **49**, 2651.
2864. Hosokawa, T., Yamashita, S., Murahashi, S.I. and Sonoda, A., *Bull. Chem. Soc. Jpn.*, 1976, **49**, 3662.
2865. Slouka, J., Srámkova, M. and Bekárek, V., *Collect. Czech. Chem. Commun.*, 1977, **42**, 3449.
2866. Böhme, H. and Weisel, K.H., *Arch. Pharm. (Weinheim)*, 1977, **310**, 26.
2867. Schnekenburger, J. and Breit, E., *Arch. Pharm. (Weinheim)*, 1977, **310**, 152.
2868. Dusemund, J., *Arch. Pharm. (Weinheim)*, 1977, **310**, 417.
2869. Liu, K.C., Tuan, J.Y., Shih, B.-J. and Lee, L.-C., *Arch. Pharm. (Weinheim)*, 1977, **310**, 522.
2870. Jackson, A.H., Jenkins, P.R. and Shannon, P.V.R., *J. Chem. Soc., Perkin Trans. 1*, 1977, 1698.
2871. Murphy, W.S. and Raman, K.P., *J. Chem. Soc., Perkin Trans. 1*, 1977, 1824.
2872. Sheradsky, T., Nov, E., Segal, S. and Frank, A., *J. Chem. Soc., Perkin Trans. 1*, 1977, 1827.
2873. Shanmugam, P., Soundarajan, N. and Kanakarajan, K., *J. Chem. Soc., Perkin Trans. 1*, 1977, 2024.
2874. Kametani, T., Luc, C.V., Higa, T., Ihara, M. and Fukumoto, K., *J. Chem. Soc., Perkin Trans. 1*, 1977, 2347.
2875. Murphy, W.S., Raman, K.P. and Hathaway, B.J., *J. Chem. Soc., Perkin Trans. 1*, 1977, 2521.
2876. Kato, T., Chiba, T. and Kimura, H., *Chem. Pharm. Bull.*, 1977, **25**, 203.
2877. Ichikawa, M. and Hisano, T., *Chem. Pharm. Bull.*, 1977, **25**, 358.
2878. Senga, K., Shimiza, K. and Nishigaki, S., *Chem. Pharm. Bull.*, 1977, **25**, 495.
2879. Takada, T., Kosugi, Y. and Akiba, M., *Chem. Pharm. Bull.*, 1977, **25**, 543.
2880. Tominaga, Y., Fujito, H., Mizuyama, K., Matsuda, Y. and Kobayashi, G., *Chem. Pharm. Bull.*, 1977, **25**, 1519.
2881. Tominaga, Y., Miyake, Y., Fujito, H., Kurata, K., Awaya, H., Matsudo, Y. and Kobayashi, G., *Chem. Pharm. Bull.*, 1977, **25**, 1528.
2882. Morimoto, T. and Sekiya, M., *Chem. Pharm. Bull.*, 1977, **25**, 1607.
2883. Horii, Z., Tsujiuchi, M., Kanai, K. and Momose, T., *Chem. Pharm. Bull.*, 1977, **25**, 1803.
2884. Takeda, K., Ohta, T., Shudo, K., Okamoto, T., Tsugi, K. and Kosage, T., *Chem. Pharm. Bull.*, 1977, **25**, 2145.
2885. Hara, T., Fujimori, H., Kayama, Y., Mori, T., Itoh, K. and Hashimoto, Y., *Chem. Pharm. Bull.*, 1977, **25**, 2584.
2886. Sakamoto, M., Miyazawa, K. and Tomimatsu, Y., *Chem. Pharm. Bull.*, 1977, **25**, 3360.
2887. Pankratova, E.V, Rodinova, G.N., Zaitsev, B.E. and Titkov, V.A., *Chem. Heterocycl. Compds. (Engl. Transl.)*, 1977, **13**, 54.
2888. Boyarintseva, O.N., Kurilo, G.N., Anisimova, O.S. and Grinev, A.N., *Chem. Heterocycl. Compds. (Engl. Transl.)*, 1977, **13**, 70.
2889. Babichev, F.S. and Kovtunenko, V.A., *Chem. Heterocycl. Compds. (Engl. Transl.)*, 1977, **13**, 117.
2890. Dorofeenko, G.N., Tkachenko, V.V., Yakovenko, V.I., Mezheritskii, V.V. and Oganeoyan, E.T., *Chem. Heterocycl. Compds. (Engl. Transl.)*, 1977, **13**, 149.
2891. Royer, R., *Chem. Heterocycl. Compds. (Engl. Transl.)*, 1977, **13**, 463.
2892. Rovinskii, M.S., Dolyuk, V.G., Zhukhovitskii, V.B., Shtamburg, V.S. and Kremlev, M.M., *Chem. Heterocycl. Compds. (Engl. Transl.)*, 1977, **13**, 667.
2893. Samsoniya, S.A., Targamadze, N.L., Tret'yakova, L.G., Efimova, T.K., Turchin,

K.F., Gverdtsileli, I.M. and Suvorov, N.N., *Chem. Heterocycl. Compds. (Engl. Transl.)*, 1977, **13**, 757.
2894. Khilya, V.P., Kupchevskaya, I.P., Salikhova, A.J., Grishko, L.G., Babichev, F.S. and Kirillova, L.G., *Chem. Heterocycl. Compds. (Engl. Transl.)*, 1977, **13**, 948.
2895. Bednyagina, N.P., Karavaeva, E.S., Lipunova, G.N., Medvedeva, L.I. and Buzykin, B.I., *Chem. Heterocycl. Compds. (Engl. Transl.)*, 1977, **13**, 1024.
2896. Samsoniya, S.A., Trapaidze, M.V., Gverdtsileli, I.M. and Suvorov, N.N., *Chem. Heterocycl. Compds. (Engl. Transl.)*, 1977, **13**, 1035.
2897. Britikova, N.E. and Novitskii, K.Y., *Chem. Heterocycl. Compds. (Engl. Transl.)*, 1977, **13**, 1338.
2898. Kohn, H. and Davis, R.E., *J. Org. Chem.*, 1977, **42**, 72.
2899. George, B. and Papadopoulos, E.P., *J. Org. Chem.*, 1977, **42**, 441.
2900. Zeiger, A.V. and Joullie, M.M., *J. Org. Chem.*, 1977, **42**, 542.
2901. D'Amico, J.J., Tung, C.C. and Dahl, W.E., *J. Org. Chem.*, 1977, **42**, 600.
2902. Tamura, Y., Tsunekawa, M., Miyamoto, T. and Ikeda, M., *J. Org. Chem.*, 1977, **42**, 602.
2903. Pizzorno, M.T. and Albonico, S.M., *J. Org. Chem.*, 1977, **42**, 909.
2904. Leir, C.M., *J. Org. Chem.*, 1977, **42**, 911.
2905. Kakehi, A., Ito, S, Manabe, T., Maeda, T. and Imai, K., *J. Org. Chem.*, 1977, **42**, 2514.
2906. Thummel, R.P. and Kohli, D.K., *J. Org. Chem.*, 1977, **42**, 2742.
2907. Abramovitch, R.A., Chellathurai, T., McMaster, I.T., Takaya, T., Azogu, C.I. and Vanderpool, D.P. *J. Org. Chem.*, 1977, **42**, 2914.
2908. Padwa, A. and Owens, W., *J. Org. Chem.*, 1977, **42**, 3076.
2909. Gassman, P.G. and Schenk, W.N., *J. Org. Chem.*, 1977, **42**, 3240.
2910. Rhee, R.P. and White, J.D., *J. Org. Chem.*, 1977, **42**, 3650.
2911. Wasson, B.K., Hamel, P. and Rooney, C.S., *J. Org. Chem.*, 1977, **42**, 4265.
2912. Yoneda, F., Nagamatsu, T., Nagamura, T. and Senga, K., *J. Chem. Soc., Perkin Trans. 1*, 1977, 765.
2913. Keir, W.F., MacLennan, A.H. and Wood, H.C.S., *J. Chem. Soc., Perkin Trans. 1*, 1977, 1321.
2914. Yoneda, F., Higuchi, M. and Matsumoto, S., *J. Chem. Soc., Perkin Trans. 1*, 1977, 1754.
2915. Park, S.W., Reid, W. and Schuckman, W., *Liebigs Ann. Chem.*, 1977, 106.
2915a. Ewald, H., Lehman, B. and Neunhoeffer, H., *Liebigs Ann. Chem.*, 1977, 1718.
2916. Linke, S., *Liebigs Ann. Chem.*, 1977, 1787.
2917. Grohe, K. and Heitzer, H., *Liebigs Ann. Chem.*, 1977, 1947.
2918. Elgavi, A. and Viehe, H.G., *Angew. Chem., Int. Ed. Engl.*, 1977, **16**, 181.
2919. Hagemann, H., *Angew. Chem., Int. Ed. Engl.*, 1977, **16**, 743.
2920. Howells, R.D. and McCown, J.D., *Chem. Rev.*, 1977, **77**, 69.
2921. Gronowitz, S., Westerlund, C. and Hörnfeldt, A.B., *Chem. Scr.*, 1977, **12**, 1.
2921a. Baker, A.D., Wong D., Lo, S., Bloch, M., Horozoglu, G., Goodman, N.L., Engel, R. and Liotta, D.C., Tetrahedron Lett., 1978, 415.
2922. Soti, F., Incze, M., Kajtar-Peredy, M., Baitz-Gacs, E., Imre, L. and Farkas, L., *Chem. Ber.*, 1977, **110**, 979.
2923. Capuano, L. and Miller, L., *Chem. Ber.*, 1977, **110**, 1691.
2924. Wolfbeis, O.S., *Chem. Ber.*, 1977, **110**, 2480.
2925. Braun, M., Hanel, G. and Heinisch, G., *Monatsh. Chem.*, 1978, **109**, 63.
2926. Brunskill, J.S.A., De, A., Elagbar, Z., Jeffrey, H. and Ewing, D.F., *Synth. Commun.*, 1978, **8**, 533.
2927. Miyamoto, T., Kimura, Y., Matsumoto, J. and Minami, S., *Chem. Pharm. Bull.*, 1978, **26**, 14.

2928. Yoneda, F., Nagamatsu, T., Ogiwara, K., Kanahori, M., Nishigaki, S. and Taylor, E.C., *Chem. Pharm. Bull.*, 1978, **26**, 367.
2928a. Tanabe, S. and Sakaguchi, T., *Chem. Pharm. Bull.*, 1978, **26**, 423.
2929. Senga, K., Sato, J. and Nishigaki, S., *Chem. Pharm. Bull.*, 1978, **26**, 765.
2930. Yoshifuji, M., Nagase, R., Kawashima, T. and Inamoto, N., *Heterocycles*, 1978, **10**, 57.
2931. Havakawa, I., Yamazaki, K., Dohmori, R. and Koga, N., *Heterocycles*, 1978, **10**, 241.
2932. Vercek, B., Stanovnik, B. and Tisler, M., *Heterocycles*, 1978, **11**, 313.
2933. Shafiee, A., *J. Heterocycl. Chem.*, 1978, **15**, 473.
2934. Waid, K. and Breitmaier, E., *Synthesis*, 1978, 748.
2935. Davies, R.V., Iddon, B., Suschitzky, H. and Gittos, M.W., *J. Chem. Soc., Perkin Trans. 1*, 1978, 180.
2936. Abe, N., Tanaka, Y. and Nishiwaki, T., *J. Chem. Soc., Perkin Trans. 1*, 1978, 429.
2937. Garcia-Lopez, M.T., de las Heras, F.G. and Stud, M., *J. Chem. Soc., Perkin Trans. 1*, 1978, 483.
2938. Ames, D.E., Chandraskhar, S. and Hansen, K.J., *J. Chem. Soc., Perkin Trans. 1*, 1978, 539.
2939. Watanabe, T., Katayama, S., Nakashito, Y. and Yamauchi, M., *J. Chem. Soc., Perkin Trans. 1*, 1978, 726.
2940. Kumar, A., Ila, H. and Junjappa, H., *J. Chem. Soc., Perkin Trans. 1*, 1978, 857.
2941. Graham, R. and Lewis, J.R., *J. Chem. Soc., Perkin Trans. 1*, 1978, 876.
2942. Chakrabarti, J.K., Hicks, T.A., Hotten, T.M. and Tupper, D.E., *J. Chem. Soc., Perkin Trans. 1*, 1978, 937.
2943. Lindley, J.M., Meth-Cohn, O. and Suschitzky, H., *J. Chem. Soc., Perkin Trans. 1*, 1978, 1198.
2944. Azimov, V.A., Granik, V.G., Glushkov, R.G. and Yakhontov, L.N., *Chem. Heterocycl. Compds. (Engl. Transl.)*, 1978, **14**, 289.
2945. Gorelik, M.V., Kuleshova, N.D. and Arinich, L.V., *Chem. Heterocycl. Compds. (Engl. Transl.)*, 1978, **14**, 332.
2946. Loseva, T.S., Goizman, M.S., Alekseeva, L.M., Shvarts, O.R., Mikhlina, E.E. and Yakhontov, L.N.,*Chem. Heterocycl. Compds. (Engl. Transl.)*, 1978, **14**, 657.
2947. Tsizin, Y.S., *Chem. Heterocycl. Compds. (Engl. Transl.)*, 1978, **14**, 925.
2948. Glushkov, R.G. and Stezhko, T.V., *Chem. Heterocycl. Compds. (Engl. Transl.)*, 1978, **14**, 1013.
2949. Granik, V.G., Marchenko, N.B. and Glushkov, R.G., *Chem. Heterocycl. Compds. (Engl. Transl.)*, 1978, **14**, 1261.
2949a. Montgomery, J.A. and Thomas, H.J., *J. Labelled Compd. Radiopharm.*, 1978, **15(S)**, 727.
2950. Fenner, H. and Grauert, R.W., *Arch. Pharm. (Weinheim)*, 1978, **311**, 303.
2951. Löwe, W., *Arch. Pharm. (Weinheim)*, 1978, **311**, 414.
2952. Eiden, F. and Schmiz, E., *Arch. Pharm. (Weinheim)*, 1978, **311**, 799.
2953. Pigulla, J. and Röder, E., *Arch. Pharm. (Weinheim)*, 1978, **311**, 822.
2954. Eiden, F. and Schmiz, E., *Arch. Pharm. (Weinheim)*, 1978, **311**, 867.
2955. Görlitzer, K. and Engler, E., *Arch. Pharm. (Weinheim)*, 1978 **311**, 960.
2956. Kanoktanaporn, S. and MacBride, J.A.H., *J. Chem. Soc., Perkin Trans. 1*, 1978, 1126.
2957. Flitsch, F., Kappenberg, F. and Schmitt, H., *Chem. Ber.*, 1978, **111**, 2407.
2958. Lauer, R.F. and Zenchoff, G., *J. Heterocycl. Chem.*, 1979, **16**, 339.
2959. Mori, M. and Ban, Y., *Tetrahedron Lett.*, 1979, 1133.
2959a. Popp, F.D., *Adv. Heterocycl. Chem.*, 1979, **24**, 187.
2960. Schulte, K.E. and Bergenthal, D., *Arch. Pharm. (Weinheim)*, 1979, **312**, 265.

2961. Dannhardt, G. and Obergrusberger, R., *Arch. Pharm. (Weinheim)*, 1979, **312**, 896.
2962. Hengartner, U., Batcho, A.D., Blount, J.F., Leimgruber, W., Larshied, M.E. and Scott, J.W.,*J. Org. Chem.*, 1979, **44**, 3748.
2963. Rudorf, W.D., Schierhorn, A. and Augustine, M., *Tetrahedron*, 1979, **35**, 551.
2964. Tashiro, M. and Yamato, T., *Synthesis*, 1979, 49.
2965. Mignot, A., Moskowitz, H. and Micoque, M., *Synthesis*, 1979, 52.
2966. Tamura, Y., Kawasaki, T., Tanio, M. and Kita, Y.,*Synthesis*, 1979, 120.
2966a. Blazevic, N., Kolbah, D., Belin, B., Sunjic, V. and Kajfig, F., *Synthesis*, 1979, 161.
2967. Skötsch, C. and Breitmaier, E., *Synthesis*, 1979, 370.
2968. Skötsch, C., Kohlmeyer, I. and Breitmaier, E., *Synthesis*, 1979, 449.
2969. Narasimhan, N.S. and Chandrachod, P.S., *Synthesis*, 1979, 589.
2970. Matsumoto, K., Uchida, T. and Paquette, L.A., *Synthesis*, 1979, 746.
2971. Tashiro, M., *Synthesis*, 1979, 921.
2972. Hercouet, A. and Le Corre, M., *Tetrahedron Lett.*, 1979, 2145.
2973. Begasse, B., Hercouet, A. and Le Corre, M., *Tetrahedron Lett.*, 1979, 2149.
2974. Horino, H. and Inoue, N., *Tetrahedron Lett.*, 1979, 2403.
2975. Salzmann, T.N., Ratcliffe, R.W. and Christensen, B.G., *Tetrahedron Lett.*, 1980, **21**, 1193.
2976. Khamal, A., Rao, K.R. and Sattur, P.B., *Synth. Commun.*, 1980, **10**, 799.
2977. Gewald, K., *Chimia*, 1980, **34**, 101.
2978. Grasheva, I.N. and Tochilkin, A.T., *Chem. Heterocycl. Compds. (Engl. Transl.)*, 1980, **16**, 275.
2979. Afanasiadi, L.S., Tur, I.N. and Bolotin, B.M., *Chem. Heterocycl. Compds. (Engl. Transl.)*, 1980, **16**, 298.
2980. Tisler, M. and Stanovnik, B., *Chem. Heterocycl. Compds. (Engl. Transl.)*, 1980, **16**, 443.
2980a. Tereat'ev, P.B., Vinogradova, S.M. and Kost, A.N., *Chem. Heterocycl. Compds. (Engl. Transl.)*, 1980, **16**, 506.
2981. Orlov, V.D., Kolos, N.N., Yaremenko, F.G. and Lavrushin, V.F.,*Chem. Heterocycl. Compds. (Engl. Transl.)*, 1980, **16**, 547.
2982. Kuz'menko, V.V., Komissarov, V.N. and Simonov, A.M., *Chem. Heterocycl. Compds. (Engl. Transl.)*, 1980, **16**, 634.
2983. Glushkov, R.G. and Stezhko, T.V., *Chem. Heterocycl. Compds. (Engl. Transl.)*, 1980, **16**, 853.
2984. Kost, A.A., *Chem. Heterocycl. Compds. (Engl. Transl.)*, 1980, **16**, 903.
2985. Christie, R.M., Shand, C.A, Thomson, R.H. and Greenhalgh, C.W., *J. Chem. Res.(S)*, 1980, 8.
2986. Kakehi, A., Ito, S., Watanabe, K., Ono, T. and Miyazima, T., *J. Chem. Res.(S)*, 1980, 18.
2987. Walsh, R.J.A. and Wooldridge, K.R.H., *J. Chem. Res.(S)*, 1980, 38.
2988. Neidlein, R. and Jeromin, G., *J. Chem. Res.(S)*, 1980, 232.
2989. Neidlein, R. and Jeromin, G., *J. Chem. Res.(S)*, 1980, 233.
2990. Paterson, T. McC. and Smalley, R.K., *J. Chem. Res.(S)*, 1980, 246.
2991. Houghton, P.G. and Rees, C.W., *J. Chem. Res.(S)*, 1980, 303.
2992. Clarke, G.M., Lee, J.B., Swinbourne, F.J. and Williamson, B., *J. Chem. Res.(S)*, 1980, 398.
2993. Clarke, G.M., Lee, J.B., Swinbourne, F.J. and Williamson, B., *J. Chem. Res.(S)*, 1980, 400.
2994. Hayes, R., Meth-Cohn, O. and Tarnowski, B., *J. Chem. Res.(S)*, 1980, 414.
2995. Ratcliffe, R.W., Salzman, T.N. and Christensen, B.G., *Tetrahedron Lett.*, 1980, **21**, 31.
2996. Cuny, E., Lichtenthaler, F.W. and Moser, A., *Tetrahedron Lett.*, 1980, **21**, 3029.
2997. Nakatsuka, S., Miyazaki, H. and Goto, T., *Tetrahedron Lett.*, 1980, **21**, 2817.

2998. Tolstikov, G.A. and Dzhemiler, U.M., *Chem. Heterocycl. Compds. (Engl. Transl.)*, 1980, **16**, 19.
2999. Grigg, R. and Gunaratne, H.Q.N., *J. Chem. Soc., Chem. Commun.*, 1980, 661.
3000. Iwao, M. and Kuraishi, T., *Bull. Chem. Soc. Jpn.*, 1980, **53**, 297.
3001. Kawase, Y., Yamaguchi, S., Morita, M. and Uesugi, T., *Bull. Chem. Soc. Jpn.*, 1980, **53**, 1057.
3002. Otto, H.H. and Schmalz, H., *Monatsh. Chem.*, 1980, **111**, 53.
3003. Wolfeis, O.S., Ziegler, E., Knierzinger, H., Wipfler, H. and Trummer, I., *Monatsh. Chem.*, 1980, **111**, 93.
3004. Abbasi, M.M., Nasr, M. and Zoorob, H.H., *Monatsh. Chem.*, 1980, **111**, 963.
3005. Smith, R.G., Lucas, R.A. and Wasley, J.W.F., *J. Med. Chem.*, 1980, **23**, 952.
3006. Garanti, L. and Zecchi, G., *J. Org. Chem.*, 1980, **45**, 4767.
3006a. Zelinsky, W., *Synthesis*, 1980, 70.
3007. Senga, K., Furukawa, K. and Nishigaki, S., *Synthesis*, 1980, 479.
3008. Coppola, G.M., *Synthesis*, 1980, 505.
3009. Walsh, A.D., *Synthesis*, 1980, 677.
3009a. Garanti, L. and Zecchi, G., *J. Chem. Soc., Perkin Trans. 1*, 1980, 116.
3010. Kametani, T., Honda, T., Nakayama, A., Sasakai, Y., Mochizuki, T. and Fukumoto, K., *J. Chem. Soc., Perkin Trans. 1*, 1981, 2228.
3010a. El'tsov, A.V., Kukushin, V.Y. and Bykova, L.M., *J. Gen. Chem. USSR*, 1981, **51**, 1822.
3010b. Sanna, P., Savelli, F. and Cignarella, G., *J. Heterocycl. Chem.*, 1981, **18**, 475.
3011. Mitsunobu, O., *Synthesis*, 1981, 1.
3012. Greenhill, J.V., Moteu, M.A., Hanke, R. and Breitmaier, E., *J. Chem. Res.(S)*, 1981, 66.
3013. Abdou, S.E., Fahmy, S.M., Sadek, K.U. and Elnagdi, M.H., *Heterocycles*, 1981, **16**, 2177.
3014. Corey, E.J. and Tramontano, A., *J. Am. Chem. Soc.*, 1981, **103**, 5599.
3015. Nakatsuka, S., Miyazaki, H. and Goto, T., *Chem. Lett.*, 1981, 407.
3016. Hatanaka, M., Nitta, H. and Ishimaru, T., *Tetrahedron Lett.*, 1981, **22**, 3886.
3017. Cheng, C.C. and Yang, S.-J., *Org. React.*, 1982, **28**, 37.
3017a. Hannoun, M., Blazevic, N., Kolbah, M., Mihalic, M. and Kajfez, F., *J. Heterocycl. Chem.*, 1982, **19**, 1131.
3018. Kamal, A. and Sattur, P.B., *Synth. Commun.*, 1982, **12**, 157.
3019. Takeda, K. and Ogawa, H., *Synth. Commun.*, 1982, **12**, 213.
3020. Lupo, B. and Tarrago, G., *Synth. Commun.*, 1982, **12**, 381.
3020a. Gelin, S., Chantegrel, B. and Chabannet, M., *Synth. Commun.*, 1982, **12**, 431.
3021. Grinev, A.N. and Ryabova, S.Y., *Chem. Heterocycl. Compds. (Engl. Transl.)*, 1982, **18**, 153.
3022. Grinev, A.N. and Ryabova, S.Y., *Chem. Heterocycl. Compds. (Engl. Transl.)*, 1982, **18**, 155.
3023. Mityurina, K.V., Kharchenko, V.G. and Charkesova, L.V., *Chem. Heterocycl. Compds. (Engl. Transl.)*, 1982, **18**, 237.
3024. Mikitenko, E.K. and Romanov, N.N., *Chem. Heterocycl. Compds. (Engl. Transl.)*, 1982, **18**, 248.
3025. Krauze, A.A., Bomika, Z.A., Pelcher, Yu.E., Mazheika, I.B. and Dubur, G.Ya., *Chem. Heterocycl. Compds. (Engl. Transl.)*, 1982, **18**, 385.
3026. Ivashchencko, A.V. and Garicheva, O.N., *Chem. Heterocycl. Compds. (Engl. Transl.)*, 1982, **18**, 429.
3027. Petrichenko, V.M. and Konshin, M.E., *Chem. Heterocycl. Compds. (Engl. Transl.)*, 1982, **18**, 459.
3028. Petyunin, P.A. and Choudry, A.M., *Chem. Heterocycl. Compds. (Engl. Transl.)*, 1982, **18**, 519.

3029. Petrichenko, V.M. and Konshin, M.E., *Chem. Heterocycl. Compds. (Engl. Transl.)*, 1982, **18**, 605.
3030. Abramova, N.D., Trzhtsinskaya, B.V., Skortsov, Yu.M., Mal'kov, A.G., Albanov, A.I. and Skortsova, G.G., *Chem. Heterocycl. Compds. (Engl. Transl.)*, 1982, **18**, 800.
3031. Kofman, T.P., Kirpenko, Z.V. and Pevzner, M.S., *Chem. Heterocycl. Compds. (Engl. Transl.)*, 1982, **18**, 854.
3032. Petrichenko, V.M. and Konshin, M.E., *Chem. Heterocycl. Compds. (Engl. Transl.)*, 1982, **18**, 946.
3033. Pakhurova, T.F., Paulin'sh, Y.Y., Gudrinietse, E.Yu., Dashkevich, S.V. and Ablovatskaya, M.V., *Chem. Heterocycl. Compds. (Engl. Transl.)*, 1982, **18**, 1217.
3034. Melik-Ogandzhanyan, R.G., Gapoyan, A.S., Khathatryan, V. and Mirzajan, V.S., *Chem. Heterocycl. Compds. (Engl. Transl.)*, 1982, **18**, 1305.
3035. Latif, N., Mishriky, N., Assad, F.M. and Meguid, S.A., *Indian J. Chem.*, 1982, **21B**, 872.
3036. Hickmott, P.W., *Tetrahedron*, 1982, **38**, 1975.
3037. Hickmott, P.W., *Tetrahedron*, 1982, **38**, 3363.
3037a. Itaya, T. and Harada, T., *Tetrahedron Lett.*, 1982, **23**, 2203.
3037b. Tamura, Y., Maeda, H., Akai, S. and Ishibashi, H., *Tetrahedron Lett.*, 1982, **23**, 2209.
3038. Barlin, G.B., *Aust. J. Chem.*, 1983, **36**, 983.
3039. Barlin, G.B., Brown, D.J., Kadunc, Z., Petric, A., Stanovnik, B. and Tisler, M., *Aust. J. Chem.*, 1983, **36**, 1215.
3040. Sierakowski, A.F., *Aust. J. Chem.*, 1983, **36**, 1281.
3041. Renfrow, R.A., Strauss, M.J., Cohen, S. and Buncel, E., *Aust. J. Chem.*, 1983, **36**, 1843.
3042. Browne, E.J., Engelhardt, L.M. and White, A.H., *Aust. J. Chem.*, 1983, **36**, 2555.
3043. Cambie, R.C., Hume, B.A., Rutledge, P.S. and Woodgate, P.D., *Aust. J. Chem.*, 1983, **36**, 2569.
3044. Adams, D.R., Dominguez, J.N. and Perez, J.A., *Tetrahedron Lett.*, 1983, **24**, 517.
3045. Weidmann, B. and Seebach, D., *Angew. Chem., Int. Ed. Engl.*, 1983, **22**, 31.
3046. Neunhoeffer, H. and Schaberger, F.-.D., *Liebigs Ann. Chem.*, 1983, 1845.
3047. Girgis, N.S., Elgemeie, G.E.H., Nawar, G.A.M. and Elnagdi, M.H., *Liebigs Ann. Chem.*, 1983, 1469.
3048. Risitano, F., Grossi, G., Foti, F. and Carnso, F., *J. Chem. Res.(S)*, 1983, 52.
3049. Sauter, F., Jordis, U. and Cai, G., *J. Chem. Res.(S)*, 1983, 276.
3050. Daisley, R.W. and Elagbar, Z.A., *Org. Prep. Proced. Int.*, 1983, **15**, 278.
3051. Kamigata, N., Hashimoto, S. and Kobayashi, M., *Org. Prep. Proced. Int.*, 1983, **15**, 315.
3052. Martin, D. and Tittelbach, F., *Tetrahedron*, 1983, **39**, 2311.
3053. Koenig, H. and Oakley, R.T., *J. Chem. Soc., Chem. Commun.*, 1983, 73.
3054. Wallace, T.W., *J. Chem. Soc., Chem. Commun.*, 1983, 535.
3055. Dean, F.M., Al-Sattar, M. and Smith, D.A., *J. Chem. Soc., Chem. Commun.*, 1983, 535.
3056. Zanirato, P., *J. Chem. Soc., Chem. Commun.*, 1983, 1065.
3057. Rajappa, S., Sreenivasan, R. and Rane, A.V., *Tetrahedron Lett.*, 1983, **24**, 3155.
3058. Girijavallabhan, V.M., Ganguly, A.K., Pinto, P. and Versace, R., *Tetrahedron Lett.*, 1983, **24**, 3179.
3059. Bergman, J., Sand, P. and Tilstam, U., *Tetrahedron Lett.*, 1983, **24**, 3665.
3059a. Ohsawa, A., Kawaguchi, T. and Igeta, H., *Synthesis*, 1983, 1037.
3060. Marchand, G., Decroix, B. and Morel, J., *J. Heterocycl. Chem.*, 1984, **21**, 877.
3060a. Shih, D.H., Baker, F., Cama, L. and Christensen, B.G., *Heterocycles*, 1984, **21**, 29.
3061. Dattolo, G., Cirricione, G., Almerico, A.M., Presti, G. and Aiello, E., *Heterocycles*, 1984, **22**, 2269.

3061a. Hejsek, M. and Slouka, J., *Pharmazie*, 1984, **39**, 186.
3062. Hull, R., MacBride, J.A.H. and Wright, P.M., *J. Chem. Res.(S)*, 1984, 328.
3063. Euerby, M.R. and Waigh, R.D., *J. Chem. Soc., Chem. Commun.*, 1984, 127.
3064. Murata, S., Sugawara, T. and Iwamera, A., *J. Chem. Soc., Chem. Commun.*, 1984, 1198.
3065. Novi, M., Garbarino, G., Dell'Erba, C. and Petrillo, G., *J. Chem. Soc., Chem. Commun.*, 1984, 1205.
3066. Nair, V. and Turner, G.A., *Tetrahedron Lett.*, 1984, **25**, 247.
3067. Hatanaka, M., Nitta, H. and Ishimaru, T., *Tetrahedron Lett.*, 1984, **25**, 2387.
3067a. Yoshida, A., Tajima, Y., Takeda, N. and Oida, S., *Tetrahedron Lett.*, 1984, **25**, 2793.
3067b. Ciganek, E., *Org. React.*, 1984, **32**, 1.
3068. Watanabe, M., Kurosaki, A and Furukawa, S., *Chem. Pharm. Bull.*, 1984, **32**, 1264.
3068a. Harris, R.L.N. and McFadden, H.G., *Aust. J. Chem.*, 1984, **37**, 1473.
3069. Hegedus, L.S., *Tetrahedron*, 1984, **40**, 2415.
3070. Musser, J.H., Georgiev, V.S., Mack, R., Loev, B., Brown, R.E. and Huang, F., *Heterocycles*, 1985, **23**, 871.
3070a. Bitter, I., *Heterocycles*, 1985, **23**, 1167.
3071. Carotti, A., Casini, G., Gavuzzo, E. and Mazza, F., *Heterocycles*, 1985, **23**, 1659.
3072. Burke, J.M. and Stevenson, R., *J. Chem. Res.(S)*, 1985, 34.
3073. Capozzi, G., Liguori, A., Ottana, R., Romeo, G., Russo, N. and Uccella, N., *J. Chem. Res.(S)*, 1985, 96.
3074. Ames, D.E., Mitchell, J.C. and Takundwa, C.C., *J. Chem. Res.(S)*, 1985, 144.
3075. Moody, C.J., Rees, C.W., Rodrigues, J.A.R. and Tsoi, S.C., *J. Chem. Res.(S)*, 1985, 238.
3076. Piccolo, O., Filippini, L., Tinucci, L., Valoti, E. and Citterio, A., *J. Chem. Res.(S)*, 1985, 258.
3077. Mataka, S., Takahashi, K., Shiwaku, S. and Tashiro, M., *J. Chem. Res.(S)*, 1985, 346.
3078. Nishiwaki, T. and Kunishige, N., *J. Chem. Res.(S)*, 1985, 363.
3079. Fetter, J., Lempert, K., Kajtár-Peredy, M., Simig, G., Hornyák, G. and Horváth, Z., *J. Chem. Res.(S)*, 1985, 368.
3080. Anderson, R.M., *J. Chem. Res.(S)*, 1985, 376.
3081. Haider, N., Heinisch, G., Kurzmann-Rauscher, I. and Wolf, M., *Liebigs Ann. Chem.*, 1985, 167.
3082. Neunhoeffer, H., Hausa, J. and Hammann, H., *Liebigs Ann. Chem.*, 1985, 640.
3083. Sheradsky, T. and Moshenberg, R., *J. Org. Chem.*, 1985, **50**, 5604.
3084. Prudhomme, M., Dauphine, G. and Jeminet, G., *J. Antibiot.*, 1986, **39**, 922.
3084a. Sha, C.-K. and Tsou, C.-P., *J. Chem. Soc., Chem. Commun.*, 1986, 310.
3085. Haider, N. and Heinisch, G., *Arch. Pharm. (Weinheim)*, 1986, **319**, 850.
3086. Baruah, A.K., Prajapati, D. and Sandhu, J.S., *Heterocycles*, 1986, **24**, 1527.
3087. Varma, R.S., Kadkhodayan, M. and Kabalka, G.W., *Heterocycles*, 1986, **24**, 1647.
3088. Harb, A.F.A., Zayed, S.E., El-Maghraby, A.M. and Metwally, S.A., *Heterocycles*, 1986, **24**, 1873.
3089. Munavalli, S., Hsu, F.L. and Poziomek, E.J., *Heterocycles*, 1986, **24**, 1893.
3090. Urleb, U., Stanovnik, B., Stibilj, V. and Tisler, M., *Heterocycles*, 1986, **24**, 1899.
3091. Ahluwalia, V.K., Adhikari, R. and Singh, R.P., *Heterocycles*, 1986, **24**, 1919.
3092. Ukawa, K., Ishiguro, T., Wada, Y. and Nohara, A., *Heterocycles*, 1986, **24**, 1931.
3093. Lal, B., Krishnan, L. and de Souza, N.J., *Heterocycles*, 1986, **24**, 1977.
3094. Harris, R.L.N. and McFadden, H.G., *Aust. J. Chem.*, 1986, **39**, 887.
3095. Heinisch, G., Holzer, W. and Nawwar, G.A.M., *J. Heterocycl. Chem.*, 1986, **23**, 93.
3096. D'Amico, J.J., Fuhrhop, R.W., Bollinger, F.G. and Dahl, W.E., *J. Heterocycl. Chem.*, 1986, **23**, 641.
3097. Anderson, D.J. and Taylor, A.J., *J. Heterocycl. Chem.*, 1986, **23**, 1091.

3098. Saari, W.S. and Schwering, J.E., *J. Heterocycl. Chem.*, 1986, **23**, 1253.
3099. Suzuki, T., Nagae, Y. and Mitsuhashi, K., *J. Heterocycl. Chem.*, 1986, **23**, 1419.
3100. Rahman, L.K.A., *J. Heterocycl. Chem.*, 1986, **23**, 1435.
3101. Takagi, K., Aotsuka, T., Morita, H. and Okamoto, Y., *J. Heterocycl. Chem.*, 1986, **23**, 1443.
3102. Morita, H. and Shiotani, S., *J. Heterocycl. Chem.*, 1986, **23**, 1465.
3103. Hénichart, J.P., Houssin, R. and Bernier, J.-L., *J. Heterocycl. Chem.*, 1986, **23**, 1531.
3103a. Mitsuhashi, K., Nagae, Y. and Suzuki, T., *J. Heterocycl. Chem.*, 1986, **23**, 1741.
3104. Fusco, R., Marchesimi, A. and Sannicolo, F., *J. Heterocycl. Chem.*, 1986, **23**, 1795.
3105. Press, J.B., Bandurco, V.T., Wong, E.M., Hajos, Z.G., Kanojia, R.M., Mallory, R.A., Deegan, G., McNally, J.J., Roberts, J.R., Cotter, M.L., Gradon, D.W. and Lloyd, J.R., *J. Heterocycl. Chem.*, 1986, **23**, 1821.
3106. Galera, C., Vaquero, J.J., Navio, J.L.G. and Alvarez-Buillon, J., *J. Heterocycl. Chem.*, 1986, **23**, 1889.
3107. Andersen, K.E., Hammad, M. and Pedersen, E.B., *Liebigs Ann. Chem.*, 1986, 1255.
3108. Biere, H., Russe, R. and Seelen, W., *Liebigs Ann. Chem.*, 1986, 1749.
3109. Hanold, N., Kalbitz, H., Pieper, M., Zimmer, O. and Meier, H., *Liebigs Ann. Chem.*, 1986, 1344.
3110. Krowczynski, A. and Kozerski, L., *Heterocycles*, 1986, **24**, 1209.
3111. Takahata, H., Suzuki, T. and Yamazaki, T., *Heterocycles*, 1986, **24**, 1247.
3112. Konishi, H., Takishita, M., Koketsu, H., Okawa, T. and Kiji, J., *Heterocycles*, 1986, **24**, 1557.
3113. Metwally, S.W., Younes, M.I. and Nour, A.M., *Heterocycles*, 1986, **24**, 1631.
3114. Abdel-Galil, F.M., Sherif, S.M. and Elnagdi, M.H., *Heterocycles*, 1986, **24**, 2023.
3115. Nakatsuka, S., Asano, H. and Goto, T., *Heterocycles*, 1986, **24**, 2109.
3116. Johnson, D.A. and Gribble, G.W., *Heterocycles*, 1986, **24**, 2127.
3117. Fahmy, A.F., Youssef, M.S.K., Halim, M.S.A., Hassan, M.A. and Sauer, J., *Heterocycles*, 1986, **24**, 2201.
3118. Sakamoto, T., Kondo, Y., Miura, N., Hayashi, K. and Yamanaka, H., *Heterocycles*, 1986, **24**, 2311.
3119. Kurasawa, Y., Wakefield, B.J. and Whittaker, R.A., *Heterocycles*, 1986, **24**, 2321.
3120. Garcia, H., Iborra, S., Miranda, M.A. and Primo, J., *Heterocycles*, 1986, **24**, 2511.
3121. Metz, W.A. and Wepplo, P.J., *Heterocycles*, 1986, **24**, 2353.
3122. Itahara, T., *Heterocycles*, 1986, **24**, 2557.
3123. Varma, R.S. and Kabalka, G.W., *Heterocycles*, 1986, **24**, 2645.
3124. Hosmane, R.S., Bhan, A. and Rauser, M.E., *Heterocycles*, 1986, **24**, 2743.
3125. Kawahara, N., Shimamori, T., Itoh, T. and Ogura, H., *Heterocycles*, 1986, **24**, 2803.
3126. Bellassoued-Fargeau, M.C., Graffe, B., Sacquet, M.C. and Maitté, *Heterocycles*, 1986, **24**, 2831.
3127. Takahashi, M., Orihara, T., Sasaki, T., Yamayera, T., Yamazaki, K. and Yoshida, A., *Heterocycles*, 1986, **24**, 2857.
3128. Victory, P. and Garriga, M., *Heterocycles*, 1986, **24**, 3053.
3129. Tominaga, Y., Shiroshita, Y., Gotou, H. and Matsuda, Y., *Heterocycles*, 1986, **24**, 3071.
3130. Kamal, A., Rao, M.V. and Sattur, P.B., *Heterocycles*, 1986, **24**, 3075.
3131. Capozzi, G., Ottana, R., Romeo, G. and Uccella, N., *Heterocycles*, 1986, **24**, 3087.
3132. Dal Piaz, V., Ciciani, G. and Chimichi, S., *Heterocycles*, 1986, **24**, 3143.
3133. Peet, N.P. and Sunder, S., *Heterocycles*, 1986, **24**, 3213.
3134. Abdel-Galil, F.M., Sallam, M.M., Sherif, S.M. and Elnagdi, M.H., *Heterocycles*, 1986, **24**, 3341.
3135. Mohiuddin, G., Reddy, P.S., Ahmed, K. and Raman, C.V., *Heterocycles*, 1986, **24**, 3489.
3136. Molina, P., Alajarin, M. and Ferao, A., *Heterocycles*, 1986, **24**, 3633.

3137. Bartsch, H., Schwarz, O. and Neubauer, G., *Heterocycles*, 1986, **24**, 3483.
3138. Capozzi, G., Ottana, R., Romeo, G. and Matsuda, Y., *J. Chem. Res.(S)*, 1986, 200.
3139. Patel, D.I., Smalley, R.K. and Scopes, D.I.C., *J. Chem. Res.(S)*, 1986, 204.
3140. Uff, B.C., Budhram, R.S., Ghaem-Maghami, G., Mallard, A.S., Harutunian, V., Calinghen, S. and Chaudhury, N., *J. Chem. Res.(S)*, 1986, 206.
3141. Deane, C.C., *J. Chem. Res.(S)*, 1986, 388.
3142. Doad, G.J.S., Jordis, U., Rudolf, M., Sauter, F. and Scheinmann, F., *J. Chem. Res.(S)*, 1986, 10.
3143. Aversa, M.C., Gianetto, P. and Krohnke, F.H., *J. Chem. Res.(S)*, 1986, 430.
3144. Mizutani, M., Sanemitsu, Y., Tamaru, Y. and Yoshida, Z., *Tetrahedron*, 1986, **42**, 305.
3145. Zvilichovsky, G. and David, M., *Synthesis*, 1986, 239.
3146. Fatiadi, A.J., *Synthesis*, 1986, 249.
3147. Molina, P., Alajarin, M. and de Vega, M.J.P., *Synthesis*, 1986, 342.
3147a. Dhar, D.N. and Murthy, K.S.K., *Synthesis*, 1986, 437.
3148. Molina, P., Tarraga, A. and Lidon, M.J., *Synthesis*, 1986, 635.
3149. Wojciechowski, K. and Makosza, M., *Synthesis*, 1986, 651.
3150. Aotsuka, T., Morita, H., Takagi, J. and Okamoto, Y., *Synthesis*, 1986, 668.
3151. Arcadi, A., Marinelli, F. and Cacchi, S., *Synthesis*, 1986, 749.
3152. Bansal, R.K. and Jain, J.K., *Synthesis*, 1986, 840.
3153. Kleenschroth, J., Mannhardt, K., Hartenstein, J. and Satzinger, G., *Synthesis*, 1986, 859.
3154. Phadke, R.C. and Rangnekar, D.W., *Synthesis*, 1986, 860.
3155. Haider, N. and Heinisch, G., *Synthesis*, 1986, 862.
3156. Lewis, C.N., Spargo, P.L. and Staunton, J., *Synthesis*, 1986, 944.
3157. Yoshida, Y., Nagai, S., Oda, N. and Sakakibara, J., *Synthesis*, 1986, 1026.
3158. Potts, K.T. and Kane, J.M., *Synthesis*, 1986, 1027.
3159. Awasthi, A.K. and Tewari, R.S., *Synthesis*, 1986, 1061.
3160. Gibbard, A.C., Moody, C.J. and Rees, C.W., *J. Chem. Soc., Perkin Trans. 1*, 1986, 145.
3161. Gairns, R.S., Grant, R.D, Moody, C.J., Rees, C.W. and Tsoi, S.C., *J. Chem. Soc., Perkin Trans. 1*, 1986, 483.
3162. Gairns, R.S., Grant, R.D., Moody, C.J., Rees, C.W. and Tsoi, S.C.,*J. Chem. Soc., Perkin Trans. 1*, 1986, 491.
3163. Gairns, R.S., Moody, C.J. and Rees, C.W., *J. Chem. Soc., Perkin Trans. 1*, 1986, 501.
3164. Saenghantara, S.T. and Wallace, T.W., *J. Chem. Soc., Perkin Trans. 1*, 1986, 789.
3165. Cushman, M., Wong, W.C. and Bacher, A., *J. Chem. Soc., Perkin Trans. 1*, 1986, 1043.
3166. Cushman, M., Wong, W.C. and Bacher, A., *J. Chem. Soc., Perkin Trans. 1*, 1986, 1051.
3167. Branch, C.L. and Pearson, M.J., *J. Chem. Soc., Perkin Trans. 1*, 1986, 1077.
3168. Ardakani, M.A., Alkhader, M.A., Lippiatt, J.H., Patel, D.D., Smalley, R.K. and Higson, S., *J. Chem. Soc., Perkin Trans. 1*, 1986, 1107.
3169. Hickey, D.M.B., Moody, C.J. and Rees, C.W., *J. Chem. Soc., Perkin Trans. 1*, 1986, 1113.
3170. Fishwick, B.R., Rowles, D.K. and Stirling, C.J.M., *J. Chem. Soc., Perkin Trans. 1*, 1986, 1171.
3171. Upton, C., *J. Chem. Soc., Perkin Trans. 1*, 1986, 1225.
3172. Brindley, J.C., Gillon, D.G. and Meakins, G.D., *J. Chem. Soc., Perkin Trans. 1*, 1986, 1255.
3173. Leardini, R., Nanni, D., Pedulli, G.F., Tundo, A. and Zanardi, G., *J. Chem. Soc., Perkin Trans. 1*, 1986, 1591.

3174. Boulton, A.J., Tsoungas, P.G. and Tsiamis, C., *J. Chem. Soc., Perkin Trans. 1*, 1986, 1665.
3175. Banks, R.E., Pritchard, R.G. and Thomson, J., *J. Chem. Soc., Perkin Trans. 1*, 1986, 1769.
3176. Molina, P., Alajarin, M., Saez, J.R., Foces-Foces, M. de la C., Cana, F.H., Claramunt, R.M. and Elguero, J., *J. Chem. Soc., Perkin Trans. 1*, 1986, 2037.
3177. Giannola, L.I., Palazzo, S. , Lamartina, L., di Sanseverino, L.R. and Sabatino, P., *J. Chem. Soc., Perkin Trans. 1*, 1986, 2095.
3178. Banks, M.R., Barker, H.M. and Huddleston, P.R., *J. Chem. Soc., Perkin Trans. 1*, 1986, 2223.
3179. Mackenzie, N.E., Surendrakumar, S., Thomson, R.H., Cowe, H.J. and Cox, P.J., *J. Chem. Soc., Perkin Trans. 1*, 1986, 2233.
3180. Taylor, E.C. and Davies, H.M.L., *J. Org. Chem.*, 1986, **51**, 1537.
3181. Taylor, E.C., Katz, A.H. and Alvarado, S.I., *J. Org. Chem.*, 1986, **51**, 1607.
3182. Kumar, C.V., Gopidas, K.R., Bhattacharyya, K., Das, P.K. and George, M.V., *J. Org. Chem.*, 1986, **51**, 1967.
3183. Wamhoff, H. and Schupp, W., *J. Org. Chem.*, 1986, **51**, 2787.
3184. McCombie, S.W., Metz, W.A. and Afonso, A., *Tetrahedron Lett.*, 1986, **27**, 305.
3184a. Einhorn, J. and Luche, J.L., *Tetrahedron Lett.*, 1986, **27**, 501.
3185. Baudin, J.-B. and Julia, S.A., *Tetrahedron Lett.*, 1986, **27**, 837.
3186. Sharp, J.T. and Skinner, C.E.D., *Tetrahedron Lett.*, 1986, **27**, 869.
3187. Shibuya, M., Sakurai, H., Maeda, T., Nishiwaki, E. and Saito, M., *Tetrahedron Lett.*, 1986, **27**, 1351.
3188. Alcaide, B., Perez-Ossorio, R., Plumet, J. and Sicora, M.A., *Tetrahedron Lett.*, 1986, **27**, 1627.
3189. Tischler, A.N. and Lanza, T.J., *Tetrahedron Lett.*, 1986, **27**, 1653.
3190. Krapcho, A.P. and Powell, J.R., *Tetrahedron Lett.*, 1986, **27**, 3713.
3190a. Dieter, R.K. and Fishpaugh, J.R., *Tetrahedron Lett.*, 1986, **27**, 3823.
3191. Pandey, G., Krishna, A. and Rao, J.M., *Tetrahedron Lett.*, 1986, **27**, 4075.
3192. Endo, Y., Namikawa, K. and Shudo, K., *Tetrahedron Lett.*, 1986, **27**, 4209.
3193. Chong, R.J., Siddiqui, M.A. and Snieckus, V., *Tetrahedron Lett.*, 1986, **27**, 5323.
3193a. LeFloc'h, Y. and Lefeuvre, M., *Tetrahedron Lett.*, 1986, **27**, 5503.
3194. Gopal, D. and Rajagopalan, K., *Tetrahedron Lett.*, 1986, **27**, 5883.
3195. Gronowitz, S., Hörnfeldt, A.-B. and Yang, Y.-H., *Chem. Scr.*, 1986, **26**, 311.
3196. Gronowitz, S., Hörnfeldt, A.-B. and Yang, Y.-H., *Chem. Scr.*, 1986, **26**, 383.
3197. Gronowitz, S., Hörnfeldt, A.-B. and Yang, Y., *Croat. Chem. Acta*, 1986, **59**, 313.
3198. Tamara, Y., Yakura, T., Shirouchi, Y. and Haruta, J., *Chem. Pharm. Bull.*, 1986, **35**, 1061.
3199. El-Ashry, E.H., Amer, A., Labib, G.H., Ralman, M.M.A. and El-Massry, A.M., *J. Heterocycl. Chem.*, 1987, **24**, 63.
3200. Egawa, H., Kataoka, M., Shibamori, K., Miyamoto, T., Nakano, J. and Matsumoto, T., *J. Heterocycl. Chem.*, 1987, **24**, 181.
3201. Beck, J.R., Gajewski, R.P., Lynch, M.P. and Wright, F.L., *J. Heterocycl. Chem.*, 1987, **24**, 267.
3202. Webb, R.L., Eggleton, D.S., Labaw, C.S., Lewis, J.J. and Wert, K., *J. Heterocycl. Chem.*, 1987, **24**, 275.
3203. Abdalla, G.M. and Sowell, J.W., *J. Heterocycl. Chem.*, 1987, **24**, 297.
3204. Tita, T.T. and Kornet, M.J., *J. Heterocycl. Chem.*, 1987, **24**, 409.
3205. Eger, K., Pfahl, J.G., Folkers, G. and Roth, H.J., *J. Heterocycl. Chem.*, 1987, **24**, 425.
3206. Monge, A., Palop, J.A., Parrado, P., Perizo-Ilarbe, C. and Fernandez-Alvarez, E., *J. Heterocycl. Chem.*, 1987, **24**, 437.
3207. Chu, D.T.W. and Maleczka, R.E., *J. Heterocycl. Chem.*, 1987, **24**, 453.

3208. El-Kerdawy, M.M., Bayomi, S.M., Shebata, I.A. and Glennor, R.A., *J. Heterocycl. Chem.*, 1987, **24**, 501.
3209. Kang, Y., Soyka, R. and Pfleiderer, W., *J. Heterocycl. Chem.*, 1987, **24**, 597.
3210. Sanchez, J.P., Trehan, A.K. and Nichols, J.B., *J. Heterocycl. Chem.*, 1987, **24**, 55.
3211. Li, Z.H., Jin, Z.T. and Yin, B.Z., *J. Heterocycl. Chem.*, 1987, **24**, 779.
3212. Basha, F.Z. and DeBernardis, J.F., *J. Heterocycl. Chem.*, 1987, **24**, 789.
3213. Tseng, S.S., Epstein, J.W., Brabender, H.J. and Francisco, G., *J. Heterocycl. Chem.*, 1987, **24**, 837.
3214. Epling, G.A. and Lin, K.Y., *J. Heterocycl. Chem.*, 1987, **24**, 853.
3215. McKenzie, T.C. and Rolfes, S.M., *J. Heterocycl. Chem.*, 1987, **24**, 859.
3216. Horaguchi, T., Matsuda, S., Tanemura, K. and Suzuki, T., *J. Heterocycl. Chem.*, 1987, **24**, 965.
3217. Kudo, H., Takahashi, K., Castle, R.N. and Lee, M.L., *J. Heterocycl. Chem.*,]1987, **24**, 1009.
3218. Cheeseman, G.W.H. and Varvounis, G., *J. Heterocycl. Chem.*, 1987, **24**, 1157.
3219. Eweiss, N.F. and Bahajaj, A.A., *J. Heterocycl. Chem.*, 1987, **24**, 1173.
3220. Stefancich, G., Corelli, F., Massa, S., Silvestri, R. and di Santo, R., *J. Heterocycl. Chem.*, 1987, **24**, 1199.
3221. Rodios, N.A., *J. Heterocycl. Chem.*, 1987, **24**, 1275.
3222. Robinson, B., *J. Heterocycl. Chem.*, 1987, **24**, 1321.
3223. Chen, T.K. and Flowers, W.T., *J. Heterocycl. Chem.*, 1987, **24**, 1569.
3224. Breitmaier, E., Ullrich, F.W., Potthoff, B., Böhm, R. and Bastien, H., *Synthesis*, 1987, 1.
3225. Hall, G., Sugden, J.K. and Waghela, M.B., *Synthesis*, 1987, 10.
3226. Vaquero, J.J., Fuentes, L., Del Castillo, J.C., Perez, M.I., Garcia, J.L. and Soto, J.L., *Synthesis*, 1987, 33.
3227. Molina, P., Fresneda, P.M. and Hurtado, F., *Synthesis*, 1987, 45.
3228. Barluenga, J., Cuervo, H., Fustero, S. and Gotor, V., *Synthesis*, 1987, 82.
3229. Molina, P., Tirraga, A., Espinosa, A. and Lidon, M.J., *Synthesis*, 1987, 128.
3230. Alemagna, A., Baldoli, C., Del Buttero, P., Licandro, E. and Maiorana, S., *Synthesis*, 1987, 192.
3231. Imafuku, K., Honda, M. and McOmie, J.F.W., *Synthesis*, 1987, 199.
3232. Mohareb, R.M., Habashi, A., Ibrahim, N.S. and Sherif, S.M., *Synthesis*, 1987, 228.
3233. Baldoli, C., Licandro, E., Maiorana, S., Menta, E. and Papagni, A., *Synthesis*, 1987, 288.
3234. Asaad, F.M., Becher, J., Moller, S. and Varma, K.S., *Synthesis*, 1987, 301.
3235. Van den Goorbergh, J.A.M., van der Steeg, M. and van der Gen, A., *Synthesis*, 1987, 314.
3236. Brady, W.T., Marchand, A.P., Giang, Y.F. and Wu, A.-H., *Synthesis*, 1987, 395.
3237. Leistner, S., Gütschow, M. and Wagner, G., *Synthesis*, 1987, 466.
3238. Phadke, R.C. and Rangnekar, D.W., *Synthesis*, 1987, 484.
3239. Srivastava, R.P., Neelima, and Laduri, A.P., *Synthesis*, 1987, 512.
3240. L'abbé, G., *Synthesis*, 1987, 525.
3241. Ayyangar, N.R., Kumar, S.M. and Srinivasan, K.V., *Synthesis*, 1987, 616.
3242. Madroñero, R. and Vega, S., *Synthesis*, 1987, 628.
3243. Barluenga, J., Iglesias, M.J. and Gotor, V., *Synthesis*, 1987, 662.
3244. Fatiadi, A.J., *Synthesis*, 1987, 749.
3245. Reinaud, O., Capdevielle, P. and Maumy, M., *Synthesis*, 1987, 790.
3246. Yokoyama, M., Watanabe, S. and Hatanaka, H., *Synthesis*, 1987, 846.
3247. Augustin, M. and Jeschke, P., *Synthesis*, 1987, 937.
3248. Ulrich, F.-W. and Breitmaier, E., *Synthesis*, 1987, 951.
3249. Benoit, R., Dupas, G., Bourguignon, J. and Quśgniner, G., *Synthesis*, 1987, 1124.
3250. Subhashini, N.J.P. and Hanumenthu, P., *Indian J. Chem.*, 1987, **26B**, 32.

3251. Kodato, S., Wada, H., Saito, S., Takeda, M., Nishibata, Y., Aoe, K., Date, T., Omada, Y. and Tamaki, H., *Chem. Pharm. Bull.*, 1987, **35**, 80.
3252. Kawahara, N., Shimamori, T., Itoh, T., Takayanagi, H. and Ogura, H., *Chem. Pharm. Bull.*, 1987, **35**, 457.
3253. Saito, M., Kawakami, K. and Kaneko, C., *Chem. Pharm. Bull.*, 1987, **35**, 1319.
3254. Oda, N., Yoshida, Y., Nagai, S., Ueda, T. and Sakakibara, J., *Chem. Pharm. Bull.*, 1987, **35**, 1796.
3255. Sugihara, H., Mabuchi, H. and Kawamatsu, Y., *Chem. Pharm. Bull.*, 1987, **35**, 1919.
3256. Sasaki, H. and Kitagawa, T., *Chem. Pharm. Bull.*, 1987, **35**, 3475.
3257. N'Diaye, I., Mayrargue, J., Farnoux, C., Miocque, M. and Gayral, P., *Eur. J. Med. Chem,.*, 1987, **22**, 403.
3258. Dannhardt, G., Meindl, W., Gussmann, S., Ajili, S. and Kappe, T., *Eur. J. Med. Chem.*, 1987, **22**, 505.
3259. Cereda, E., Turconi, M., Ezhaya, A., Bellora, E., Brambilla, A., Pagani, F. and Donetti, A., *Eur. J. Med. Chem.*, 1987, **22**, 527.
3260. Morris, J.L. and Rees, C.W., *J. Chem. Soc., Perkin Trans. 1*, 1987, 211.
3260a. Mitchell, G. and Rees, C.W., *J. Chem. Soc., Perkin Trans. 1*, 1987, 403.
3261. Crombie, L., Jones, R.C.F. and Palmer, C.J., *J. Chem. Soc., Perkin Trans. 1*, 1987, 317.
3262. Johnstone, G., Smith, D.M., Shepherd, T. and Thompson, D., *J. Chem. Soc., Perkin Trans. 1*, 1987, 495.
3263. Huber, I., Fülöp, G., Dombi, G., Bernáth, G., Hermecz, I. and Mészáros, Z., *J. Chem. Soc., Perkin Trans. 1*, 1987, 909.
3264. Hickey, D.M.B., Mankenzie, A.R., Moody, C.J. and Rees, C.W., *J. Chem. Soc., Perkin Trans. 1*, 1987, 921.
3265. Labarca, C.V., Mackenzie, A.R., Moody, C.J., Rees, C.W. and Vaquero, J.J., *J. Chem. Soc., Perkin Trans. 1*, 1987, 927.
3266. Rees, C.W. and Smith, D.I., *J. Chem. Soc., Perkin Trans. 1*, 1987, 1159.
3267. Iddon, B., Khan, N. and Lim, B.L., *J. Chem. Soc., Perkin Trans. 1*, 1987, 1457.
3268. Takahata, H., Anazawa, A., Moriyama, K. and Yamagaki, T., *J. Chem. Soc., Perkin Trans. 1*, 1987, 1501.
3269. Barton, J.W. and Pearson, N.D., *J. Chem. Soc., Perkin Trans. 1*, 1987, 1541.
3270. Molina, P., Alajarin, A., de Vega, M.J.P., Foces-Foces, M.de la C., Cano, F.H., Claramunt, R.M. and Elguero, J., *J. Chem. Soc., Perkin Trans. 1*, 1987, 1853.
3271. Molina, P., Alajarin, M., Ferao, A. and de Vega, M.J.P., *J. Chem. Soc., Perkin Trans. 1*, 1987, 2667.
3272. Langley, D.R. and Thurston, D.E., *J. Org. Chem.*, 1987, **52**, 91.
3273. Parpani, P. and Zecchi, G., *J. Org. Chem.*, 1987, **52**, 1417.
3274. Lau, C.K., Bélanger, P.C., Dufresne, C. and Scheigetz, J., *J. Org. Chem.*, 1987, **52**, 1670.
3275. Tsuji, Y., Huh, K.-T. and Watanabe, Y., *J. Org. Chem.*, 1987, **52**, 1673.
3276. Davey, D.D., *J. Org. Chem.*, 1987, **52**, 1863.
3277. Nugent, R.A. and Murphy, M., *J. Org. Chem.*, 1987, **52**, 2206.
3278. Bitha, P., Hlavka, J.J. and Lin, Y., *J. Org. Chem.*, 1987, **52**, 2220.
3279. Potts, K.T., Walsh, E.B. and Bhattacharjee, D., *J. Org. Chem.*, 1987, **52**, 2285.
3280. Tilley, J.W., Coffen, D.L., Schaer, B.H. and Lind, J., *J. Org. Chem.*, 1987, **52**, 2469.
3281. Leonard, N.J., Kazamierczak, F. and Rykowski, A., *J. Org. Chem.*, 1987, **52**, 2933.
3282. Bock, M.G., DiPardo, R.M., Evans, B.E., Rittle, K.E., Veber, D.F., Fredinger, R.M., Hirshfield, J. and Springer, J.P., *J. Org. Chem.*, 1987, **52**, 3232.
3283. Brady, W.T., Giang, Y.F., Marchand, A.P. and Wu, A., *J. Org. Chem.*, 1987, **52**, 3457.
3284. Deady, L.W. and Werden, D.M., *J. Org. Chem.*, 1987, **52**, 3930.

3285. Sanicanin, Z., Juric, A., Tabalovic, I. and Trinjstic, N., *J. Org. Chem.*, 1987, **52**, 4053.
3286. Hendrickson, J.B. and Hussoin, M.S., *J. Org. Chem.*, 1987, **52**, 4137.
3287. Trudell, M.L., Fukade, N. and Cook, J.M., *J. Org. Chem.*, 1987, **52**, 4293.
3288. Davey, D.D., *J. Org. Chem.*, 1987, **52**, 4379.
3289. Hormi, O.E.O., Moisio, M.R. and Sund, B.C., *J. Org. Chem.*, 1987, **52**, 5272.
3290. Hormi, O.E.O. and Paakkanen, A.M., *J. Org. Chem.*, 1987, **52**, 5275.
3291. Magee, W.L., Rao, C.B., Glinka, J., Hui, H., Amick, T.J., Fiscus, D., Kakodkar, S., Nair, M. and Shechter, H., *J. Org. Chem.*, 1987, **52**, 5538.
3292. Sanders, D.C., Marczak, A., Melendez, J.L. and Shechter, H., *J. Org. Chem.*, 1987, **52**, 5622.
3293. Rehman, H. and Rao, J.M., *Tetrahedron*, 1987, **43**, 5335.
3294. Kendall, H.E. and Luscombe, D.K., *Prog. Med. Chem.*, 1987, **24**, 249.
3295. McKillop, A. and Brown, S.P., *Synth. Commun.*, 1987, **17**, 657.
3296. Imafuku, K., Kikuchi, Y. and Yin, B.-Z., *Bull. Chem. Soc. Jpn.*, 1987, **60**, 185.
3297. Gonda, J., Kristian, P. and Imrich, J., *Collect. Czech. Chem. Commun.*, 1987, **52**, 2508.
3298. Graf, H. and Klebe, G., *Chem. Ber.*, 1987, **120**, 965.
3299. Ege, G., Gilbert, K. and Maurer, K., *Chem. Ber.*, 1987, **120**, 1375.
3300. Ming, Y., Horlemann, N. and Wamhoff, H., *Chem. Ber.*, 1987, **120**, 1427.
3301. Fetter, J., Lempert, K., Kajtar-Peredy, M., Bujtás, G. and Simig, G., *J. Chem. Res.(S)*, 1987, 28.
3302. Euerby, M.R. and Waigh, R.D., *J. Chem. Res.(S)*, 1987, 36.
3303. Euerby, M.R. and Waigh, R.D., *J. Chem. Res.(S)*, 1987, 38.
3304. Fuentes-Rodriguez, F., Sepulveda-Arques, J., Jones, R.A., Bates, P.A. and Hursthouse, M.B., *J. Chem. Res.(S)*, 1987, 354.
3305. Shimada, K., Sako, M., Hirota, K. and Maki, Y., *Tetrahedron Lett.*, 1987, **28**, 207.
3306. Jungheim, L.N., Sigmund, S.K., Jones, N.D. and Swartzendruber, J.K., *Tetrahedron Lett.*, 1987, **28**, 289.
3307. Atanes, N., Guitian, E., Saa, C., Castelo, L. and Saá, J.M., *Tetrahedron Lett.*, 1987, **28**, 817.
3308. Bock, M.G., DiPardo, R.M., Evans, B.E., Rittle, K.E., Veber, D.F. and Freidinger, R.M., *Tetrahedron Lett.*, 1987, **28**, 939.
3309. Lash, T.D., Bladel, K.A. and Johnson, M.C., *Tetrahedron Lett.*, 1987, **28**, 1135.
3310. Nohara, F., Nishii, M., Ogawa, H., Isano, K., Ubukata, M., Fujii, T., Itaya, T. and Saito, T., *Tetrahedron Lett.*, 1987, **28**, 1287.
3311. Martin, P., *Tetrahedron Lett.*, 1987, **28**, 1645.
3312. Groundwater, P.W. and Sharp, J.T., *Tetrahedron Lett.*, 1987, **28**, 2069.
3313. Chern, J.-W., Lee, H.Y., Hung, M. and Shish, F.-J., *Tetrahedron Lett.*, 1987, **28**, 2151.
3314. Kamal, A., Rao, A.B. and Sattur, P.B., *Tetrahedron Lett.*, 1987, **28**, 2425.
3315. Lim, M.I. and Patil, D.G., *Tetrahedron Lett.*, 1987, **28**, 3775.
3316. Flitsch, W., Jones, R.A. and Hohenhorst, M., *Tetrahedron Lett.*, 1987, **28**, 4397.
3317. Molina, P., Arques, A., Vinader, M.V., Becher, J. and Broudum, K., *Tetrahedron Lett.*, 1987, **28**, 4451.
3318. Randles, K.R. and Storr, R.C., *Tetrahedron Lett.*, 1987, **28**, 5555.
3319. Moody, C.J. and Warrellow, G.R., *Tetrahedron Lett.*, 1987, **28**, 6089.
3320. McFarlane, M.D. and Smith, D.M., *Tetrahedron Lett.*, 1987, **28**, 6363.
3321. Kawamoto, I., Endo, R., Suzaki, K. and Hata, T., *Heterocycles*, 1987, **25**, 123.
3322. Castanedo, R., Zetina, C.B. and Maldonado, L.A., *Heterocycles*, 1987, **25**, 175.
3323. Miller, R.B., Dugar, S. and Epperson, J.R., *Heterocycles*, 1987, **25**, 217.
3324. Shah, N.V. and Cama, L.D., *Heterocycles*, 1987, **25**, 221.

3325. Golic, L., Stropnik, C., Stanovnik, B. and Tisler, M., *Heterocycles*, 1987, **25**, 347.
3326. Bobbitt, J.M. and Bourque, A.J., *Heterocycles*, 1987, **25**, 601.
3327. Nagahara, T. and Kametani, T., *Heterocycles*, 1987, **25**, 729.
3328. Ito, K. and Hariya, J., *Heterocycles*, 1987, **26**, 35.
3329. Capozzi, G., Ottana, R. and Romeo, G., *Heterocycles*, 1987, **26**, 39.
3330. Miki, Y., Kinoshita, H., Yoshimara, T., Takemura, S. and Ikeda, M., *Heterocycles*, 1987, **26**, 199.
3331. Riad, B.Y., Negm, A.M., Abdou, S.E. and Dabun, H.A., *Heterocycles*, 1987, **26**, 205.
3332. Saraf, S., Al Omran, F. and Al Saleh, B., *Heterocycles*, 1987, **26**, 239.
3332a. Elnagdi, M.H., Sherif, S.M. and Mohareb, R.M., *Heterocycles*, 1987, **26**, 497.
3333. Koren, B., Stanovnik, B. and Tisler, M., *Heterocycles*, 1987, **26**, 689.
3334. Abe, N. and Gomi, A., *Heterocycles*, 1987, **26**, 767.
3335. Murata, S., Sugimoto, T. and Matsuura, S., *Heterocycles*, 1987, **26**, 883.
3336. Mohamed, M.H., Ibrahim, N.S. and Elnagdi, M.H., *Heterocycles*, 1987, **26**, 899.
3337. Hafez, E.A.A., Elnagdi, M.H., Elagamay, A.G.A and El-Tawesl, F.M.A.A., *Heterocycles*, 1987, **26**, 903.
3338. Woodgate, P.D., Herbert, J.M. and Denny, W.A., *Heterocycles*, 1987, **26**, 1029.
3339. Herbert, J.M., Woodgate, P.D. and Denny, W.A., *Heterocycles*, 1987, **26**, 1037.
3340. Kamal, A. and Sattur, P.B., *Heterocycles*, 1987, **26**, 1051.
3341. Radzicjewski, C., Ghosh, S. and Kaiser, E.T., *Heterocycles*, 1987, **26**, 1227.
3342. Garin, J., Loscertales, M.P., Melandez, E., Merchán, F.L., Rodriguez, R. and Tejero, T., *Heterocycles*, 1987, **26**, 1303.
3343. Ghosh, C.K., Bandyopadhyay, C. and Haiti, J., *Heterocycles*, 1987 **26**, 1623.
3344. Amer, A., Weisz, K. and Zinner, H., *Heterocycles*, 1987, **26**, 1851.
3345. Hibino, S. and Sugino, E., *Heterocycles*, 1987, **26**, 1883.
3346. Romney-Alexander, T.M., *Heterocycles*, 1987, **26**, 1899.
3347. Molina, P., Tarraga, A., Lorenzo-Pena, M. and Espinosa, A.,*Heterocycles*, 1987, **26**, 2183.
3348. Alonso, R. and Brossi, A., *Heterocycles*, 1987, **26**, 2209.
3349. Garin, J., Guillen, C., Melendez, E., Merchán, F.L., Ordune, I., and Tejero, T., *Heterocycles*, 1987, **26**, 2371.
3350. Hrytsak, M. and Durst, T., *Heterocycles*, 1987, **26**, 2393.
3351. Nesi, R., Giomi, D., Quartara, L., Papaleo, S. and Tedeschi, T., *Heterocycles*, 1987, **26**, 2419.
3352. Zecchini, G., Torrini, I. and Paradisi, M.P., *Heterocycles*, 1987, **26**, 2443.
3353. Martin, N., Pascual, C., Seoane, C. and Soto, J.L., *Heterocycles*, 1987, **26**, 2811.
3354. Bashir, M. and Kingston, D.G.I., *Heterocycles*, 1987, **26**, 2877.
3355. Tagawa, Y. and Goto, Y., *Heterocycles*, 1987, **26**, 2921.
3356. Venugopalan, B., Iyer, S.S., Karnik, P.J. and de Souza, N.J., *Heterocycles*, 1987, **26**, 3173.
3357. Ming, Y. and Boykin, D.W., *Heterocycles*, 1987, **26**, 3229.
3358. Weinstock, J., Gaitanopoulos, D.E., Stringer, O.D., Frane, R.G., Hiebler, J.P., Kinter, L.B., Mann, W.A., Flaim, K.E and Glessner, G., *J. Med. Chem.*, 1987, **30**, 1166.
3359. Evans, B.E., Rittle, K.F., Bock, M.G., DiPardo, R.M., Friedinger, R.M., Whitter, W.L., Gould, N.P., Lundell, G.F., Hommick, C.F., Veher, D.F., Anderson, P.S.,Chang, R.S.L., Lotti, V.J., Cerini, D.J., Chen, T.B., King, P.J., Kunkel, K.A., Springer, J.P. and Hirschfield, J., *J. Med. Chem.*, 1987, **30**, 1229.
3360. Elnagdi, M.H., Elgemeie, G.E.H. and Elmoghayer, M.R.H., *Adv. Heterocycl. Chem.*, 1987, **41**, 319.
3361. Hrytsak, M. and Durst, T., *J. Chem. Soc., Chem. Commun.*, 1987, 1150.
3362. Molina, P., Vilaplana, M.J., Anfreu, P.L. and Moller, J., *J. Heterocycl. Chem.*, 1987, **24**, 1281.

3363. Hiremath, S.P., Hiremath, D.M. and Purohit, M.G., *Indian J. Chem.*, 1987, **26B**, 1042.
3364. Hermesz, I., *Adv. Heterocycl. Chem.*, 1987, **42**, 83.
3365. Ziegler, T., Möhler, T., Bosehi, G., Reny-Palasse, V. and Rips, R., *Chem. Ber.*, 1987, **120**, 373.
3366. Grasso, S., Zappala, M. and Chimirri, A., *Heterocycles*, 1987, **26**, 2477.
3367. Besanty, D., Mayrarque, J., Moussa, G.E., Shaaban, M.E., Gayral P. and Miocque, M., *Eur. J. Med. Chem.*, 1988, **23**, 403.
3368. Reddy, A.V.N., Kamal, A. and Sattur, P.B., *Synth. Commun.*, 1988, **18**, 525.
3369. René, L., El Cherif, S., Boschi, G., Reny-Palasse, V. and Rips, R., *Eur. J. Med. Chem.*, 1988, **23**, 592.
3370. Burger, K., Huber, E., Kahl, T., Partscht, H. and Ganzer, M., *Synthesis*, 1988, 44.
3371. Arnoldi, A. and Carughi, M., *Synthesis*, 1988, 155.
3372. Hosmane, R.S. and Lim, B.B., *Synthesis*, 1988, 242.
3373. Latif, N., Asaad, F.A. and Grant, N., *Synthesis*, 1988, 246.
3374. Iwata, C., Watanabe, M., Okamoto, S., Fujimoto, M., Sakae, M., Katsurada, M. and Imanishi, T., *Synthesis*, 1988, 261.
3375. Gupta, A.K., Ila, H. and Junjappa, H., *Synthesis*, 1988, 284.
3376. Sastry, C.V.R., Rao, K.S., Krishnan, V.S.H., Rastogi, K. and Jain, M.L., *Synthesis*, 1988, 336.
3377. Booth, B.L., Coster, R.D., Fernanda, M. and Proenca, J.R.P., *Synthesis*, 1988, 389.
3378. Etzbach, K.H. and Eilingsfeld, H., *Synthesis*, 1988, 449.
3379. Tanabe, Y. and Sanemitsu, Y., *Synthesis*, 1988, 482.
3380. Sadek, K.U. and Elnagdi, M.H., *Synthesis*, 1988, 483.
3381. Mitschker, A. and Wedermeyer, K., *Synthesis*, 1988, 517.
3382. Neunhoeffer, H., Dostert, M. and Hammann, H., *Synthesis*, 1988, 778.
3383. Liebscher, J. and Hassoun, A., *Synthesis*, 1988, 816.
3384. Chakrasali, R.R., Ila, H. and Junjappa, H., *Synthesis*, 1988, 851.
3385. Neunhoeffer, H. and Reichel, D., *Synthesis*, 1988, 877.
3386. Nesi, R., Giomi, D., Papaleo, S., Bracci, S. and Dapporto, P., *Synthesis*, 1988, 884.
3387. Cabiddu, S., Cancellu, D., Florio, C., Gelli, G. and Melis, S., *Synthesis*, 1988, 888.
3388. Wamhoff, H. and Muhr, J., *Synthesis*, 1988, 919.
3389. Gewald, K., Hain, U. and Gruner, M., *Chem. Ber.*, 1988, **121**, 573.
3390. Ried, W. and Eichhorn, T.A., *Chem. Ber.*, 1988, **121**, 2049.
3391. Hofmann, H. and Fischer, H., *Chem. Ber.*, 1988, **121**, 2147.
3392. Maini, P.N., Sammes, M.P. and Katritzky, A.R., *J. Chem. Soc., Perkin Trans. 1*, 1988, 161.
3393. Haider, N. and Heinisch, G., *J. Chem. Soc., Perkin Trans. 1*, 1988, 401.
3394. Le Count, D.J. and Marson, A.P., *J. Chem. Soc., Perkin Trans. 1*, 1988, 451.
3395. Harvey, I.W., McFarlane, M.D., Moody, D.J. and Smith, D.M., *J. Chem. Soc., Perkin Trans. 1*, 1988, 681.
3396. Baccolini, G. and Dalpozzo, R., *J. Chem. Soc., Perkin Trans. 1*, 1988, 971.
3397. Landor, S.R., Landor, P.D., Johnson, A., Fomum, Z.T., Mbafor, J.T. and Nkengfack, A.E., *J. Chem. Soc., Perkin Trans. 1*, 1988, 975.
3398. Girling, I.R. and Widdowson, A., *J. Chem. Soc., Perkin Trans. 1*, 1988, 1317.
3399. Moody, C.J. and Shah, P., *J. Chem. Soc., Perkin Trans. 1*, 1988, 1407.
3400. Ghosh, C.K., Pal, C., Maiti, J., and Sarkar, M., *J. Chem. Soc., Perkin Trans. 1*, 1988, 1489.
3401. Molina, P. and Fresneda, P.M., *J. Chem. Soc., Perkin Trans. 1*, 1988, 1819.
3402. Harvey, I.W., McFarlane, M.D., Moody, D.J. and Smith D.M., *J. Chem. Soc., Perkin Trans. 1*, 1099, 1939.
3403. Blake, A.J., McNab, H. and Morrison, R., *J. Chem. Soc., Perkin Trans. 1*, 1988, 2145.

3404. Hasiguchi, S., Natsugari, H. and Ochiai, M., *J. Chem. Soc., Perkin Trans. 1*, 1988, 2345.
3405. Doad, G.J.S., Okor, D.I., Scheinmann, F., Bates, P.A. and Hursthouse, M.B., *J. Chem. Soc., Perkin Trans. 1*, 1988, 2993.
3406. Ota, T., Hasegawa, S., Inoue, S. and Sato, K., *J. Chem. Soc., Perkin Trans. 1*, 1988, 3029.
3407. Ananda, G.D.S. and Stoodley, R.J., *J. Chem. Soc., Perkin Trans. 1*, 1988, 3359.
3408. King, F.D., *J. Chem. Soc., Perkin Trans. 1*, 1988, 3381.
3409. Kamitori, Y., Hojo, M., Masuda, R., Fujitani, T., Ohara, S. and Yokoyama, T., *J. Org. Chem.*, 1988, **53**, 129.
3410. Krolski, M.E., Renaldo, A.F., Rudisill, D.E. and Stille, J.K., *J. Org. Chem.*, 1988, **53**, 1170.
3411. Harvey, R.G., Cortez, C., Ananthanaryan, T.P. and Schmolka, S., *J. Org. Chem.*, 1988, **53**, 3936.
3412. Boger, D.L. and Patel, M., *J. Org. Chem.*, 1988, **53**, 1405.
3413. Pandey, G. and Krishna, A., *J. Org. Chem.*, 1988, **53**, 2364.
3414. Estel, L., Marsais, F. and Queguiner, G., *J. Org. Chem.*, 1988, **53**, 2740.
3415. Kumar, S. and Leonard, N.J., *J. Org. Chem.*, 1988, **53**, 3959.
3416. Reed, M.W. and Moore, H.W., *J. Org. Chem.*, 1988, **53**, 4166.
3417. Qiang, L.Q. and Baine, N.H., *J. Org. Chem.*, 1988, **53**, 4218.
3418. Dumas, D.J., *J. Org. Chem.*, 1988, **53**, 4650.
3419. Molina, P., Arques, A., Vinader, M.V., Becher, J. and Brondum, K., *J. Org. Chem.*, 1988, **53**, 4654.
3420. Hauser, F.M. and Baghdanov, V.M., *J. Org. Chem.*, 1988, **53**, 4676.
3421. Cushman, M., Patel, H. and McKenzie, A., *J. Org. Chem.*, 1988, **53**, 5088.
3422. Boger, D.L., Carbone, L.R. and Yohannes, D., *J. Org. Chem.*, 1988, **53**, 5163.
3423. Willer, R.L. and Henry, R.A., *J. Org. Chem.*, 1988, **53**, 5371.
3424. Whitney, S.E. and Rickborn, B., *J. Org. Chem.*, 1988, **53**, 5596.
3425. Al Nakib, T., Tyndall, D.V. and Meegan, M.J., *J. Chem. Res.(S)*, 1988, 10.
3426. Golec, J.M.C. and Scrowston, R.M., *J. Chem. Res.(S)*, 1988, 46.
3427. McCarthy, E.T., Tyndall, D.V. and Meegan, M.J., *J. Chem. Res.(S)*, 1988, 145.
3428. Gilchrist, T.L., Gordon, P.F. and Rees, C.W., *J. Chem. Res.(S)*, 1988, 148.
3429. Korbonits, D. and Kolonits, P., *J. Chem. Res.(S)*, 1988, 209.
3430. Monge, A., Palop, J.A., Recalde, J.I. and Fernandez-Alvarez, E., *J. Chem. Res.(S)*, 1988, 344.
3431. Fitton, A.O., Muzanila, C.N., Odusanya, O.M., Oppong-Boachic, F.K., Duckworth, S.J. and Hadi, A.H.B.A., *J. Chem. Res.(S)*, 1988, 352.
3432. Yang, Y., Hörnfeldt, A.-B. and Gronowitz, S., *Chem. Scr.*, 1988, **28**, 275.
3433. Gronowitz, S., Hörnfeldt, A.-B. and Yang, Y., *Chem. Scr.*, 1988, **28**, 281.
3434. Dittami, J.P. and Ramanathan, H., *Tetrahedron Lett.*, 1988, **29**, 45.
3435. de Oliveira, A.B., Ferreira, D.T. and Poslan, D.S., *Tetrahedron Lett.*, 1988, **29**, 155.
3436. Alonso, R. and Brossi, A., *Tetrahedron Lett.*, 1988, **29**, 735.
3437. Cairns, N., Harwood, L.M. and Astles, D.P., *Tetrahedron Lett.*, 1988, **29**, 1311.
3438. Iritani, K., Matsubara, S. and Utimoto, K., *Tetrahedron Lett.*, 1988, **29**, 1799.
3439. Ariamala, G. and Balasubramanian, K.K., *Tetrahedron Lett.*, 1988, **29**, 3487.
3440. Qiang, L.G. and Baine, N.H., *Tetrahedron Lett.*, 1988, **29**, 3517.
3441. Molina, P., Alajarin, M. and Vidal, A., *Tetrahedron Lett.*, 1988, **29**, 3849.
3442. Burns, B., Grigg, R., Sridharan, V. and Worakun, T., *Tetrahedron Lett.*, 1988, **29**, 4325.
3443. Larock, R.C. and Stinn, D.E., *Tetrahedron Lett.*, 1988, **29**, 4687.
3444. Srikrishna, A. and Krishna, K., *Tetrahedron Lett.*, 1988, **29**, 4995.
3445. Jones, R.C.F. and Smallridge, M.J., *Tetrahedron Lett.*, 1988, **29**, 5005.

3446. Thompson, C.M. and Docter, S., *Tetrahedron Lett.*, 1988, **29**, 5213.
3447. Siddiqui, M.A. and Snieckus, V., *Tetrahedron Lett.*, 1988, **29**, 5463.
3448. Burns, B., Grigg, R., Ratananukul, P., Sridharan, V., Stevens, P., Sukirthalingam, S. and Worakun, T., *Tetrahedron Lett.*, 1988, **29**, 5565.
3449. Vicentini, C.B., Veronese, A.C., Giori, P. and Guarneri, M., *Tetrahedron Lett.*, 1988, **29**, 6171.
3450. Dieter, R.K. and Fishpaugh, J.R., *Tetrahedron Lett.*, 1988, **29**, 6633.
3451. Subramanian, R.S. and Balasubramanian, K.K., *Tetrahedron Lett.*, 1988, **29**, 6797.
3452. Bergman, J., Tilstam, U. and Törnoos, K.-W., *J. Chem. Soc., Perkin Trans. 1*, 1988, 519.
3453. Spagnolo, P. and Zanirato, P., *J. Chem. Soc., Perkin Trans. 1*, 1988, 2615.
3454. Palmer, D.C., Skotnicki, J.S. and Taylor, E.C., *Prog. Med. Chem.*, 1988, **25**, 85.
3454a. Gawley, R.E., *Org. React.*, 1988, **35**, 1.
3455. Armstrong, V., Soto, O., Valderama, J.A. and Tapia, R., *Synth. Commun.*, 1988, **18**, 717.
3456. Taylor, E.C. and Yoon, C.-M., *Synth. Commun.*, 1988, **18**, 1187.
3457. Srivastava, R.G. and Venkatarmani, P.S., *Synth. Commun.*, 1988, **18**, 1537.
3458. Takahata, H. and Yamazaki, T., *Heterocycles*, 1988, **27**, 1953.
3459. Sakata, G.. Makino, K. and Kurasawa, Y., *Heterocycles*, 1988, **27**, 2481.
3460. Abdel-Megeed, M.F. and Teniou, A., *Collect. Czech. Chem. Commun.*, 1988, **53**, 329.
3461. Pindur, U. and Adam, R., *J. Heterocycl. Chem.*, 1988, **25**, 1.
3462. Marsais, F., Trecort, F., Bréant, P. and Quéguiner, G., *J. Heterocycl. Chem.*, 1988, **25**, 81.
3463. Haider, N., Heinisch, G. and Lassnigg, D., *J. Heterocycl. Chem.*, 1988, **25**, 119.
3464. Reiter, J., Pongó, L., Solár., P. and Dvortsák, P., *J. Heterocycl. Chem.*, 1988, **25**, 173.
3465. Tominaga, Y., Shiroshita, Y., Kurokawam T., Matsuda, Y. and Hosomi, A., *J. Heterocycl. Chem.*, 1988, **25**, 185.
3466. Noguchi, M., Sakamoto, K., Nagata, S. and Kajigaeshi, S., *J. Heterocycl. Chem.*, 1988, **25**, 205.
3467. Pennini, R., Cerri, A., Tajana, A. and Nardi, D., *J. Heterocycl. Chem.*, 1988, **25**, 305.
3468. Abdelhamid, A.O., Parkányi, C., Rashid, S.M.K. and Lloyd, N.D., *J. Heterocycl. Chem.*, 1988, **25**, 403.
3469. Krishnan, R. and Long, S.A., *J. Heterocycl. Chem.*, 1988, **25**, 447.
3470. Reisch, J. and Scheer, M., *J. Heterocycl. Chem.*, 1988, **25**, 677.
3471. Matsumoto, K., Okemi, Y., Konishi, H., Shi, X., Uchida, T. and Aoyama, K., *J. Heterocycl. Chem.*, 1988, **25**, 689.
3472. Shutske, G.M. and Huger, F.P., *J. Heterocycl. Chem.*, 1988, **25**, 703.
3473. Bitha, J., Hlavka, J.J. and Lin, Y., *J. Heterocycl. Chem.*, 1988, **25**, 739.
3474. Lieberman, D.F. and Albright, J.D., *J. Heterocycl. Chem.*, 1988, **25**, 827.
3475. Boamah, P.Y., Haider, N., Heinisch, G. and Moshuber, J., *J. Heterocycl. Chem.*, 1988, **25**, 879.
3476. Herrero, A. and Ochoa, C., *J. Heterocycl. Chem.*, 1988, **25**, 891.
3477. Katritzky, A.R. and Fan, W.-Q., *J. Heterocycl. Chem.*, 1988, **25**, 901.
3478. Hamer, R.L., Sekarek, D., Effland, R.C. and Klein, J.T., *J. Heterocycl. Chem.*, 1988, **25**, 991.
3479. Benincori, T. and Sannicolo, F., *J. Heterocycl. Chem.*, 1988, **25**, 1029.
3480. Al Kathlan, H. and Zinner, H., *J. Heterocycl. Chem.*, 1988, **25**, 1047.
3481. Sutherland, R.G., Abd-Et-Aziz, A.S. Piórko, A., Gill, U.S. and Lee, C.C., *J. Heterocycl. Chem.*, 1988, **25**, 1107.
3482. Eid, M.M., Kadry, A.M. and Hassan, R.A., *J. Heterocycl. Chem.*, 1988, **25**, 1117.

3483. Hynes, J.B., Pathak, A., Panos, C.H. and Okeke, C.C., *J. Heterocycl. Chem.*, 1988, **25**, 1173.
3484. Kurasawa, Y., Okiyama, M., Kamigaki, Y., Kano, L.M., Takada, A. and Okamoto, Y., *J. Heterocycl. Chem.*, 1988, **25**, 1259.
3485. Plé, P.A. and Marnett, L.J., *J. Heterocycl. Chem.*, 1988, **25**, 1271.
3486. Melani, F., Cecchi, L., Colotta, V., Palazzino, G. and Filacchioni, G., *J. Heterocycl. Chem.*, 1988, **25**, 1367.
3487. Nan'ya, S., Katsuranaya, K.,Okamoto, Y., Maekawa, E. and Ueno, Y., *J. Heterocycl. Chem.*, 1988, **25**, 1373.
3488. Press, J.B. and McNally, J.J., *J. Heterocycl. Chem.*, 1988, **25**, 1571.
3489. Rida, S.M., Soliman, F.S.G., Badawey, E.H.M. and Kappe, T., *J. Heterocycl. Chem.*, 1988, **25**, 1725.
3490. Taylor, E.C., Pont, J.L. and Warner, J.C., *J. Heterocycl. Chem.*, 1988, **25**, 1733.
3491. Tominaga, Y., Shiroshita, Y. and Hosmane, A., *J. Heterocycl. Chem.*, 1988, **25**, 1745.
3492. Matsumoto, K., Uchida, T., Aoyama, K., Nishikawa, M., Kuroda, T. and Okamoto, T., *J. Heterocycl. Chem.*, 1988, **25**, 1793.
3493. Bartoli, G., Bosco, M., Dalpozzo, R. and Todesco, P.E., *J. Chem. Soc., Chem. Commun.*, 1988, 807.
3494. Sha, C.-K., Tsou, C.-P., Li, Y.-C., Lee, R.S., Tsai, F.-Y. and Yeh, R.-H., *J. Chem Soc., Chem. Commun.*, 1988, 1081.
3495. Haider, N. and Heinisch, G., *Arch. Pharm. (Weinheim)*, 1988, **321**, 309.
3496. Sakamoto, T., Annaka, M., Kondo, Y., Araki, T. and Yamanaka, H., *Chem. Pharm Bull.*, 1988, **36**, 1890.
3497. L'abbé, G., Vanderdreisshe, H. and Weyns, N., *Bull.Soc. Chim. Belg.*, 1988, **97**, 85
3498. Kelly, T.R., Field, J.A. and Li, Q., *Tetrahedron Lett.*, 1988, **29**, 3545.
3499. Sha, C.-K., Tsou, C.-P. and Wang, S.L., *J. Chem. Soc., Chem. Commun.*, 1988, 320
3500. Nachbaur, E., Faleschimi, G., Belay, F. and Janoschek, R., *Angew. Chem., Int. Ed Engl.*, 1988, **27**, 701.
3501. Hegedus, L.S., *Angew. Chem., Int. Ed. Engl.*, 1988, **27**, 1113.
3502. Buchwald, S.L. and Nielsen, R.B., *Chem. Rev.*, 1988, **88**, 1047.
3503. Flitsch, W., *Adv. Heterocycl. Chem.*, 1988, **43**, 35.
3504. Al-Shaar, A.H.M., Lythgoe, D.J., McClenaghan, I.M. and Ramsden, C.A., *J. Chem Soc., Perkin Trans. 1*, 1988, 3025.
3505. Sakamoto, T., Kondo, Y. and Yamanaka, H., *Heterocycles*, 1988, **27**, 2225.
3506. McFarlane, M.D., Moody, D.J. and Smith, D.M., *J. Chem. Soc., Perkin Trans. 1*, 1988, 681.
3507. Pelduc, P., Tailhan, C. and Zard, S.Z., *J. Chem. Soc., Chem. Commun.*, 1988, 308.
3508. Rodrigo, R., *Tetrahedron*, 1988, **44**, 2093.
3509. Beecher, J. and Stidsen, C.E., *Sulfur Rep.*, 1988, **8**, 105.
3510. Aran, V.J., Goya, P. and Ochoa, C., *Adv. Heterocycl. Chem.*, 1988, **44**, 81.
3511. Heitmann, W., Liepmann, H., Mätzel, U., Zeugner, H., Fuchs, A.M., Krähling, H., Ruhlan,, M., Mol, F. and Tulp, M.T.M, *Eur. J. Med. Chem.*, 1988, **23**, 249.
3512. Cingolani, G.M., Claudi, F. and Venturi, F., *Eur. J. Med. Chem.*, 1988, **23**, 291.
3513. Lamas, C., Castelo, L. and Dominguez, D., *Tetrahedron Lett.*, 1988, **29**, 3865.
3514. Rajopadhye, M. and Popp, F.D., *Heterocycles*, 1988, **27**, 1489.
3515. Scriven, E.F.V. and Turnbull, K., *Chem. Rev.*, 1988, **88**, 297.
3516. Galli, C., *Chem. Rev.*, 1988, **88**, 765.
3517. Kagichika, H., Kawachi, E., Hashimoto, Y. and Shudo, K., *J. Med. Chem.*, 1989, **32**, 834.
3518. Shindo, H., Takada, S., Murata, S., Eigyo, M. and Matsushita, A., *J. Med. Chem.*, 1989, **32**, 1213.
3519. Moran, D.B., Ziegler, C.B., Dunne, T.S., Kuck, N.K. and Lin, Y., *J. Med. Chem.*, 1989, **32**, 1313.

3520. Unangst, P.C., Connors, D.T., Stabler, S.R., Weckert, R.J., Carethers, M.E., Kennedy, J.A., Thomson, D.O., Chestnut, J.C., Adolphson, R.L. and Conroy, M.C., *J. Med. Chem.*, 1989, **32**, 1360.
3521. Kang, Y., Larson, S.B., Robins, R.K. and Ravenkar, G.R., *J. Med. Chem.*, 1989, **32**, 1547.
3522. Girard, G.R., Bondmill, W.E., Hillegass, L.M., Holben, K.G., Pendleton, R.G. and Uzinskes, I., *J. Med. Chem.*, 1989, **32**, 1566.
3523. Engel, W.W., Eberlin, W.G., Mihn, G., Hammer, R. and Trummlitz, G., *J. Med. Chem.*, 1989, **32**, 1718.
3524. Shutske, G.M., Pierrat, F.A., Kapples, K.J., Corndeldt, M.L., Szewczak, M.R., Huger, F.P., Bores, G.M., Harontanioan, V. and Davis, K.L., *J. Med. Chem.*, 1989, **32**, 1805.
3525. Chakrabarti, J.K., Hotten, T.M., Pullar, I.A., and Steggles, R.T., *J. Med. Chem.*, 1989, **32**, 2375.
3526. Reich, M.F., Fabio, F.F., Lee, V.J., Kuck, N.A. and Testa, R.T., *J. Med. Chem.*, 1989, **32**, 2474.
3527. Bare, T.M., McLaren, C.D., Campbell, J.B., Firor, J.W., Resch, J.F., Walters, C.P., Salama, A.I., Meiners, B.A. and Patel, J.B., *J. Med. Chem.*, 1989, **32**, 2561.
3528. Nitta, M., Iino, Y., Hara, E. and Kobayashi, T., *J. Chem. Soc., Perkin Trans. 1*, 1989, 51.
3529. Molina, P., Alajarin, M. and Vidal, A., *J. Chem. Soc., Perkin Trans. 1*, 1989, 247.
3530. Landor, S.R., Fomum, Z.T., Asobo, P.F., Landor, P.D. and Johnson, A., *J. Chem. Soc., Perkin Trans. 1*, 1989, 251.
3531. Pant, C.M., Stoodley, R.J., Witing, A. and Williams, D.J., *J. Chem. Soc., Perkin Trans. 1*, 1989, 297.
3532. Boulton, A.J., Deri, P., Henderson, N., Jarrar, A.A. and Kiss, M., *J. Chem. Soc., Perkin Trans. 1*, 1989, 543.
3533. Meakins, G.D., Musk, S.R.R., Robertson, C.A. and Woodhouse, L.S., *J. Chem. Soc., Perkin Trans. 1*, 1989, 643.
3534. Altamura, M., Cesti, P., Francalanci, P., Marchi. M. and Cambiaghi, S., *J. Chem. Soc., Perkin Trans. 1*, 1989, 1225.
3535. Majumdar, K.C., Chattopadhyay, S.K. and Khan, A.T., *J. Chem. Soc., Perkin Trans. 1*, 1989, 1285.
3536. Armesto, D., Gallego, M.G., Ortiz, M.J., Romano, S. and Horspool, W.M., *J. Chem. Soc., Perkin Trans. 1*, 1989, 1343.
3537. Marley, H., Wright, S.H.B. and Preston, P.N., *J. Chem. Soc., Perkin Trans. 1*, 1989, 1727.
3538. Ashwood, M.S., Gourlay, J.A., Houghton, P.C., Nagel, M.A. and Wright, S.H.B., *J. Chem. Soc., Perkin Trans. 1*, 1989, 1889.
3539. Ishii, H., Takeda, H., Hagiwara, T., Sakamoto, M., Kogusuri, K. and Murakami, Y., *J. Chem. Soc., Perkin Trans. 1*, 1989, 2407.
3540. Rigby, J.H. and Balasubramanian, N., *J. Org. Chem.*, 1989, **54**, 224.
3541. Bergman, J. and Peleman, B., *J. Org. Chem.*, 1989, **54**, 824.
3542. New, J.S., Christopher, W.L. and Jass, P.A., *J. Org. Chem.*, 1989, **54**, 990.
3543. Eisch, J.J. and Dluzniewski, T., *J. Org. Chem.*, 1989, **54**, 1269.
3544. Taylor, E.C., Pont, J.L. and Warner, J.C., *J. Org. Chem.*, 1989, **54**, 1456.
3545. Hagen, T.J., Nagayanan, K., Names, J. and Cook, J.M., *J. Org. Chem.*, 1989, **54**, 2170.
3546. Buchwald, S.L. and Fang, Q., *J. Org. Chem.*, 1989, **54**, 2793.
3547. Brady, W.T. and Gu, Y.Q., *J. Org. Chem.*, 1989, **54**, 2834.
3548. Padwa, A., Bullock, W.H., Kline, D.N. and Perumattam, J., *J. Org. Chem.*, 1989, **54**, 2862.
3549. Gribble, G.W., Fletcher, G.L., Ketcha, D.M. and Rajadhye, M., *J. Org. Chem.*, 1989, **54**, 3264.

3550. Kawase, M., Kitamura, T. and Kikugawa, Y., *J. Org. Chem.*, 1989, **54**, 3394.
3551. Taylor, E.C. and Wong, G.S.K., *J. Org. Chem.*, 1989, **54**, 3618.
3552. Wadsworth, D.H., Weidner, C.H., Bender, S.L., Nuttall, R.H. and Lewis, H.R., *J. Org. Chem.*, 1989, **54**, 3652.
3553. Weidner, C.H., Wadsworth, D.H., Bender, S.L. and Bettman, D.J,, *J. Org. Chem.*, 1989, **54**, 3660.
3554. Uff, B.C., Ho, Y.P., Hussain, F. and Haji, M.S. *J. Chem. Res.(S)*, 1989, 24.
3555. Letcher, R.M., Chung, K.K. and Sin, D.W.M., *J. Chem. Res.(S)*, 1989, 115.
3556. Barker, J.M., Huddleston, P.R., Heath, D.J., Jackson, E. and Holmes, D., *J. Chem. Res.(S)*, 1989, 196.
3557. Loriga, M., Nuvole, A. and Paglietti, G., *J. Chem. Res.(S)*, 1989, 202.
3558. Molina, P., Lorenzo, M.A. and Aller, E., *J. Chem. Res.(S)*, 1989, 262.
3559. Gorvin, J.H., *J. Chem. Res.(S)*, 1989, 294.
3560. Gabriel, G., Pickles, R. and Tyman, J.H.P., *J. Chem. Res.(S)*, 1989, 348.
3561. Aubert, T., Tabyaoui, B., Farnier, M. and Guilard, R., *J. Chem. Soc., Perkin Trans. 1*, 1989, 1369.
3562. Kempter, G. and Spindler, J., *Z. Chem.*, 1989, **29**, 276.
3563. Press, J.B., McNally, J.J., Keiser, J.A., Offord, S.J., Katz, L.B., Giardino, E., Falotico, R. and Tobia, A.J., *Eur. J. Med. Chem.*, 1989, **24**, 627.
3564. Tsukayama, M., Kawamura, Y., Tamaki, H., Kabo, T. and Horie, T., *Bull. Chem. Soc. Jpn.*, 1989, **62**, 826.
3565. Babin, P. and Devaux, G., *J. Chem. Educ.*, 1989, **66**, 522.
3566. Deady, L.W., Mackay, M.F. and Werden, D.M., *J. Heterocycl. Chem.*, 1989, **26**, 161.
3567. Bartsch, H., Erker, T. and Neubeuer, G., *J. Heterocycl. Chem.*, 1989, **26**, 205.
3568. Yamaguchi, S., Tsuzuki, K., Kinoshita, M., Oh-hira, Y. and Kawase, Y., *J. Heterocycl. Chem.*, 1989, **26**, 281.
3569. El Reddy, A.M., Ali, A.S. and Ayad, A.O., *J. Heterocycl. Chem.*, 1989, **26**, 313.
3570. Badawey, E.A.M., Rida, S.M., Soliman, F.S.G. and Kappe, T., *J. Heterocycl. Chem.*, 1989, **26**, 405.
3571. Tominaga, Y., Kurokawa, T., Gotou, H., Matsuda, Y. and Hosomi, A., *J. Heterocycl. Chem.*, 1989, **26**, 477.
3572. Musmar, M.J., Khan, S.R., Zaktzer, A.S., Martin, G.E., Lynch, V.M., Simmonsen, S.H. and Smith, K., *J. Heterocycl. Chem.*, 1989, **26**, 667.
3573. Sliwa, H. and Blondeau, D. and Rousseau, O., *J. Heterocycl. Chem.*, 1989, **26**, 687.
3574. Robey, R.L., Copley-Merriman, C.R. and Phelps, M.A., *J. Heterocycl. Chem.*, 1989, **26**, 779.
3575. Vincentini, C.B., Veronese, A.C., Poli, T., Guarneri, M., Giori, P. and Ferretti, V., *J. Heterocycl. Chem.*, 1989, **26**, 797.
3576. Kurasawa, Y., Kamigaki, Y., Kim, H.S., Watanabe, C., Kanoh, M., Okiyama, M., Takada, A. and Okamoto, Y., *J. Heterocycl. Chem.*, 1989, **26**, 861.
3577. Yang, Y. and Hörnfeldt, A.-B. and Gronowitz, S., *J. Heterocycl. Chem.*, 1989, **26**, 865.
3578. Katritzky, A.R., Fan, W.-Q., Li, Q.-L. and Bayyuk, S., *J. Heterocycl. Chem.*, 1989, **26**, 885.
3579. Boamah, P.Y., Haider, N. and Heinisch, G., *J. Heterocycl. Chem.*, 1989, **26**, 933.
3580. Colombo, A., Frigola, J., Parés, J. and Andaluz, B., *J. Heterocycl. Chem.*, 1989, **26**, 949.
3581. Crank, G. and Makin, M.I.H., *J. Heterocycl. Chem.*, 1989, **26**, 1163.
3582. Tominaga, Y., *J. Heterocycl. Chem.*, 1989, **26**, 1167.
3583. Hirayama, R., Kawase, M., Kimachi, T., Tanaka, K. and Yoneda, F., *J. Heterocycl. Chem.*, 1989, **26**, 1255.
3584. Stanovnik, B., Svete, J. and Tisler, M., *J. Heterocycl. Chem.*, 1989, **26**, 1273.
3585. Potts, K.T. and Kane, J.M., *J. Heterocycl. Chem.*, 1989, **26**, 1289.

3586. Shutske, G.M. and Kapples, K.J., *J. Heterocycl. Chem.*, 1989, **26**, 1293.
3587. Walser, A. and Todaro, L., *J. Heterocycl. Chem.*, 1989, **26**, 1299.
3588. Badawey, E.A.M., Rida, S.M., Soliman, F.S.G. and Kappe, T., *J. Heterocycl. Chem.*, 1989, **26**, 1401.
3589. Balogh, M., Hermecz, I., Simon, K. and Pusztay, L., *J. Heterocycl. Chem.*, 1989, **26**, 1755.
3590. L'abbé, G. and Vanderstede, E., *J. Heterocycl. Chem.*, 1989, **26**, 1811.
3591. Chen, X., Nagata, M., Tanaka, K. and Yoneda, F., *J. Chem. Soc., Chem. Commun.*, 1989, 44.
3592. Reiter, J. and Pongo, L., *Org. Prep. Proced. Int.*, 1989, **21**, 163.
3593. Sharma, S., *Sulfur Rep.*, 1989, **8**, 327.
3594. Rosowsky, A., *Prog. Med. Chem.*, 1989, **26**, 1.
3595. René, L., *Synthesis*, 1989, 69.
3596. Yang, Y., Hörnbeldt, A.-B. and Gronowitz, S., *Synthesis*, 1989, 130.
3597. Rao, V.S. and Darbarwar, M., *Synthesis*, 1989, 139.
3598. Gupta, A.K., Chakrasali, R.T., Ila, H. and Junjappa, H., *Synthesis*, 1989, 141.
3599. Sicker, D., Prätorius, B., Mann, G. and Meyer, L., *Synthesis*, 1989, 211.
3600. Rossi, E. and Stradi, R., *Synthesis*, 1989, 214.
3601. Goya, P., Lissavetzky, J. and Rozas, I., *Synthesis*, 1989, 280.
3602. Black, D. St. C., Ivory, A.J., Keller, P.A. and Kumar, N., *Synthesis*, 1989, 322.
3603. Eitel, M. and Pindur, U., *Synthesis*, 1989, 364.
3604. Hellwinkel, D. and Karle, R., *Synthesis*, 1989, 394.
3605. Couture, A., Grandclaudon, P. and Huguerre, E., *Synthesis*, 1989, 456.
3606. Maruyama, K., Osuka, A., Nakagawa, K., Nabeshima, T. and Tabuchi, K., *Synthesis*, 1989, 628.
3607. Kappe, T. and Kos, C., *Synthesis*, 1989, 629.
3608. Lee, K.-J., Kim, S., Um, H. and Park, H., *Synthesis*, 1989, 638.
3609. Sato, M., Miyaura, N. and Suzuki, A., *Bull. Chem. Soc, Jpn.*, 1989, 1405.
3610. Hesek, D., Tegza, M., Rybar, A. and Povazance, F., *Synthesis*, 1989, 681.
3611. Ried, W. and Laoutidis, J., *Synthesis*, 1989, 739.
3612. Gallos, J.K. and Corobili, E.E., *Synthesis*, 1989, 751.
3613. Subramanian, M., Prasad, K.J.R. and Shanmugam, P., *Synthesis*, 1989, 777.
3614. Stadlbauer, W., Pfaffenschlager, A. and Kappe, T., *Synthesis*, 1989, 781.
3615. Einhorn, C., Einhorn, J. and Luche, J.L., *Synthesis*, 1989, 787.
3616. Molina, P., Lorenzo, A. and Aller, E., *Synthesis*, 1989, 843.
3617. Sicker, D., *Synthesis*, 1989, 875.
3618. Molina, P. and Fresneda, P.M., *Synthesis*, 1989, 878.
3619. Molina, P., Arques, A., Fresneda, P.M., Vinader, M.V. and Foces-Foces, M. de la C., *Chem. Ber.*, 1989, **122**, 307.
3620. Walsh, E.B. and Wamhoff, H., *Chem. Ber.*, 1989, **122**, 1673.
3621. Bäckvall, J.E., Andersson, P.C. and Vägberg, J.O., *Tetrahedron Lett.*, 1989, **30**, 137.
3622. Motoki, S., Watanabe, T. and Saito, T., *Tetrahedron Lett.*, 1989, **30**, 187.
3623. Hendi, M.S., Natalie, K.S., Hendi, S.B., Campbell, J.H., Greenwood, T.D. and Wolfe, J.F., *Tetrahedron Lett.*, 1989, **30**, 275.
3624. Joyeau, R., Kobaiter, R., Sadet, J. and Wakselman, M., *Tetrahedron Lett.*, 1989, **30**, 337.
3625. Gourcy, J.-G., Prudhomme, M., Dauphin, G. and Jeminet, G., *Tetrahedron Lett.*, 1989, **30**, 351.
3626. Kamal, A. and Sattur, P.B., *Tetrahedron Lett.*, 1989, **30**, 1133.
3627. Doad, G.J.S., Barltrop, J.A., Petty, C.M. and Owen, T.C., *Tetrahedron Lett.*, 1989, **30**, 1597.
3628. Jungheim, L.N., *Tetrahedron Lett.*, 1989, **30**, 1889.

3629. Murakami, Y., Takahashi, H., Nakazawa, Y., Koshimizu, M., Watanabe, T. and Yoloyama, Y., *Tetrahedron Lett.*, 1989, **30**, 2099.
3630. Bartoli, G., Palmieri, G., Bosco, M. and Dalpozzo, R., *Tetrahedron Lett.*, 1989, **30**, 2129.
3631. Arcadi, A., Cacchi, S. and Marinelli, F., *Tetrahedron Lett.*, 1989, **30**, 2581.
3632. Jones, K. and McCarthy, C., *Tetrahedron Lett.*, 1989, **30**, 2657.
3633. Fang, F.G., Feigelson, G.B. and Danishefsky, S.J., *Tetrahedron Lett.*, 1989, **30**, 2743.
3634. Fang, F.G. and Danishefsky, S.J., *Tetrahedron Lett.*, 1989, **30**, 2747.
3635. Molina, P., Alazarin, M. and Vidal, A., *Tetrahedron Lett.*, 1989, **30**, 2847.
3636. Takeuchi, H. and Eguchi, S., *Tetrahedron Lett.*, 1989, **30**, 3313.
3637. Chen, H.G., Hoechstetter, C. and Knochel, P., *Tetrahedron Lett.*, 1989, **30**, 4795.
3638. Desai, M.C. and Thadeio, P.F., *Tetrahedron Lett.*, 1989, **30**, 5223.
3639. Molina, P., Arques, A. and Vinader, M.V., *Tetrahedron Lett.*, 1989, **30**, 6237.
3640. Vice, S.F., de Carvalho, H.N., Taylor, N.G. and Dmitrienko, G.J., *Tetrahedron Lett.*, 1989, **30**, 7289.
3641. Bartsch, H. and Erker, T., *Liebigs Ann. Chem.*, 1989, 177.
3642. Grabenwöger, M., Haider, N. and Heinisch, G., *Liebigs Ann. Chem.*, 1989, 481.
3643. Sadek, K.U., Abouhadid, K. and Elghandour, A.H.H., *Liebigs Ann. Chem.*, 1989, 501.
3644. Kalcheva, V., Stoyanova, D. and Simova, S., *Liebigs Ann. Chem.*, 1989, 1251.
3645. Haider, N., Heinisch, G. and Volf, I., *Heterocycles*, 1989, **29**, 481.
3646. Gol'dberg, Y., Sturckovich, R. and Lukevics, E., *Heterocycles*, 1989, **29**, 597.
3647. Haider, N., Heinisch, G. and Volf, I., *Heterocycles*, 1989, **29**, 1309.
3648. Taylor, E.C. and Rieter, L.H., *J. Am. Chem. Soc.*, 1989, **111**, 285.
3649. Boger, D.L. and Kaspar, A.M., *J. Am. Chem. Soc.*, 1989, **111**, 1517.
3650. Tisler, M., *Adv. Heterocycl. Chem.*, 1989, **45**, 37.
3651. Yang, Y., *Synth. Commun.*, 1989, **19**, 1001.
3652. Mylari, B.L., Scott, P.J. and Zombrocoski, W.J., *Synth. Commun.*, 1989, **19**, 2921.
3653. Deady, L.W., Ganakas, A.M. and Ong, B.H., *Aust. J. Chem.*, 1989, **42**, 1029.
3654. Chandler, C.J., Craik, D.J. and Waterman, K.J., *Aust. J. Chem.*, 1989, **42**, 1407.
3655. Perlmutter, H.D., *Adv. Heterocycl. Chem.*, 1989, **46**, 1.
3656. Matsumura, N., Mort, O., Tomura, M. and Yoneda, S., *Chem. Lett.*, 1989, 39.
3657. Himbert, G., Diehl, K. and Schlindwein, H.-J., *Chem. Ber.*, 1989, **122**, 1691.
3658. Huang, Z.-T. and Zhang, P.-C., *Chem. Ber.*, 1989, **122**, 2011.
3659. Watanabe, M., Date, M., Tsakazaki, M. and Furukawa, S., *Chem. Pharm. Bull.*, 1989, **37**, 36.
3660. Wu, E.S.C., Cole, T.E., Davidson, T.A., Darley, M.A., Doering, K.G., Fedorchuk, U., Loh, J.T., Thomas, T.L., Blosser, J.C., Borelli, A.R., Kinsolving, C.R., Parker, R.B., Strand, J.C. and Watkins, B.E ., *J. Med. Chem.*, 1989, **32**, 183.
3661. Kelley, J.L., Linn, J.A. and Selway, J.W.T., *J. Med. Chem.*, 1989, **32**, 218.
3662. Kruse, L.J., Ladd, D.L., Harrsch, P.B., McCabe, F.L., Mong, S.-M., Faucette, L. and Johnson, R., *J. Med. Chem.*, 1989, **32**, 409.
3663. Bonzard, D., Di Cesare, P., Essiz, M., Jacquet, J.P., Remuzon, P., Weber, A., Oki, T. and Masuyoshi, M., *J. Med. Chem.*, 1989, **32**, 537.
3664. Alabaster, C.T., Bell, A.S., Campbell, S.F., Ellis, P., Henderson, C.G., Morris, D.S., Roberts, D.A., Ruddock, K.S., Samuels, G.M.R. and Stefaniak, M.H., *J. Med. Chem.*, 1989, **32**, 575.
3665. Rewcastle, G.W., Atwell, G.J., Baguley, B.C., Calveley, S.B. and Denny, W.A., *J. Med. Chem.*, 1989, **32**, 793.
3666. Padwa, A., Dean, D.C. and Zhi, L., *J. Am. Chem. Soc.*, 1989, **111**, 6451.
3667. Jones, G.B. and Moody, C.J., *J. Chem. Soc., Chem. Commun.*, 1989, 186.

3668. Al-Shaar, A.H., Gilmour, D.W., Lythgoe, D.J., McClenaghan, I. and Ramsden, C.A., *J. Chem. Soc., Chem. Commun.*, 1989, 851.
3669. Taylor, E.C., *J. Heterocycl. Chem.*, 1990, **27**, 1.
3670. Shishoo, C.J., Devani, M.B., Bhatti, V.S., Jain, K.S. and Ananthan, S., *J. Heterocycl. Chem.*, 1990, **27**, 119.
3671. Freedman, J. and Huber, E.W., *J. Heterocycl. Chem.*, 1990, **27**, 343.
3672. Chimirri, A., Grasso, S., Ottana, K., Romeo, G. and Zappala, M., *J. Heterocycl. Chem.*, 1990, **27**, 371.
3673. Liu, K.-C., Shih, B.J. and Chern, J.-W., *J. Heterocycl. Chem.*, 1990, **27**, 391.
3674. Urleb, U., Stanovnik, B. and Tisler, M., *J. Heterocycl. Chem.*, 1990, **27**, 407.
3675. Urleb, U., Neidlein, R. and Kramer, W., *J. Heterocycl. Chem.*, 1990, **27**, 433.
3676. Schober, B.D., Megyeri, G. and Kappe, T., *J. Heterocycl. Chem.*, 1990, **27**, 471.
3677. Katner, A.S. and Brown, R.F., *J. Heterocycl. Chem.*, 1990, **27**, 563.
3678. Jin, R.-H., Yin, B.-Z. and Jin, Z.T., *J. Heterocycl. Chem.*, 1990, **27**, 583.
3679. Ziegler, C.B., Moran, D.B., Fenton, T.J. and Lin, Y., *J. Heterocycl. Chem.*, 1990, **27**, 587.
3680. Bock, M.G., DiPardo, R.M., Carson, K.G. and Freidinger, R.M., *J. Heterocycl. Chem.*, 1990, **27**, 631.
3681. Urleb, U., Stanovnik, B. and Tisler, M., *J. Heterocycl. Chem.*, 1990, **27**, 643.
3682. Hirota, K., Shirahashi, M., Senda, S. and Yogo, M., *J. Heterocycl. Chem.*, 1990, **27**, 717.
3683. Kim, H.S., Kurasawa, Y., Yoshii,C., Masuyama, M., Takada, A. and Okamoto, Y., *J. Heterocycl. Chem.*, 1990, **27**, 819.
3684. Tsuji, T. and Takenaka, K., *J. Heterocycl. Chem.*, 1990, **27**, 851.
3685. Quijano, M.L., Nogueras, M., Sanchez, A., de Cienfueges, G.A. and Melgarejo, M., *J. Heterocycl. Chem.*, 1990, **27**, 1079.
3686. Licandro, E., Maioran, S., Papagni, A., Tavallo, D., Slawin, A.M.Z. and Williams, D.J., *J. Heterocycl. Chem.*, 1990, **27**, 1103.
3687. Gronowitz, S. and Timari, G., *J. Heterocycl. Chem.*, 1990, **27**, 1127.
3688. Gronowitz, S. and Timari, G., *J. Heterocycl. Chem.*, 1990, **27**, 1159.
3689. Chu, D.T.W. and Claiborne, A.K., *J. Heterocycl. Chem.*, 1990, **27**, 1191.
3690. Tominaga, Y., Michioka, T., Moriyama, K. and Hosomi, A., *J. Heterocycl. Chem.*, 1990, **27**, 1217.
3691. Izumi, T., Nishimoto, Y., Kohei, K. and Kasahara, H., *J. Heterocycl. Chem.*, 1990, **27**, 1419.
3692. Chern, J.-W., Wu, Y.-H. and Liu, K.-C., *J. Heterocycl. Chem.*, 1990, **27**, 1485.
3693. Sanemitsu, Y., Manabe, A., Kawamura, S., Satoh, J. and Tanabe, Y., *J. Heterocycl. Chem.*, 1990, **27**, 1517.
3694. Chern, J.W., Ho, C.P., Wu, Y.H., Rong, J.G., Liu, K.C., Cheng, M.C. and Wang, Y., *J. Heterocycl. Chem.*, 1990, **27**, 1909.
3695. Dennin, F., Blondeau, D. and Sliwa, H., *J. Heterocycl. Chem.*, 1990, **27**, 1963.
3696. Argyropoulos, N.G., Mentzafos, D. and Terzis, A., *J. Heterocycl. Chem.*, 1990, **27**, 1983.
3697. Timari, G., Hajos, G. and Messmer, A., *J. Heterocycl. Chem.*, 1990, **27**, 2005.
3698. Singh, B. and Lesker, G.Y., *J. Heterocycl. Chem.*, 1990, **27**, 2085.
3699. Kant, J., *J. Heterocycl. Chem.*, 1990, **27**, 2129.
3700. McKittrick, B., Failli, A., Steffan, R.J., Soll, R.M., Hughes, P., Schmid, J., Asselin, A.A., Shaw, C.C., Noureldin, R. and Gavin, G., *J. Heterocycl. Chem.*, 1990, **27**, 2151.
3701. Kim, S.C., Kurasawa, Y., Yoshii, C., Masuyama, M. and Takada, A., *J. Heterocycl. Chem.*, 1990, **27**, 2197.
3702. Buisson, J.P. and Demersemann, P., *J. Heterocycl. Chem.*, 1990, **27**, 2213.

3703. Matyus, P., Malek, N., Tegeles, A., Kosary, J., Kasztreiner, E., Podany, B., Rabloczky, G. and Kürthy, M., *J. Heterocycl. Chem.*, 1990, **27**, 151.
3704. Tominaga, Y., Ichihara, Y., Mori, T., Kanio, C. and Hosomi, A., *J. Heterocycl. Chem.*, 1990, **27**, 263.
3705. Huang, J. and Haigh, D.S., *J. Heterocycl. Chem.*, 1990, **27**, 331.
3706. Vega, S., Alonso, J., Diaz, J.H. and Junquera, F., *J. Heterocycl. Chem.*, 1990, **27**, 269.
3707. Goldstein, S.W. and Dambek, P.J., *J. Heterocycl. Chem.*, 1990, **27**, 335.
3708. Servi, S., *Synthesis*, 1990, 1.
3709. Flouzat, C. and Guillaumet, G., *Synthesis*, 1990, 64.
3710. Thurston, D.E., Murty, V.S., Langley, D.R. and Jones, G.B., *Synthesis*, 1990, 81.
3711. White, J. and Baker, D.C., *Synthesis*, 1990, 151.
3712. Kolb, S., *Synthesis*, 1990, 171.
3713. Sambaiah, T. and Reddy, K.K., *Synthesis*, 1990, 422.
3714. Moriarty, R.M. and Vaid, R.K., *Synthesis*, 1990, 431.
3715. Lee, K.-J., Choi, D.O., Kim, J., Jeong, J.U. and Park, H., *Synthesis*, 1990, 455.
3716. Molina, P., Arques, A. and Vinader, M.V., *Synthesis*, 1990, 469.
3717. Molina, P. and Vilaplana, M.J., *Synthesis*, 1990, 474.
3718. Hojo, M., Masuda, R. and Okada, E., *Synthesis*, 1990, 481.
3719. Alcaide, B., Dominguez, G., Plumet, J., Sierra, M.A., Monge, A. and Perez-Garcia, V., *Synthesis*, 1990, 485.
3720. Patonay-Péli, E., Litkei, G. and Patonay, T., *Synthesis*, 1990, 511.
3721. Smalley, R.K. and Teguiche, M., *Synthesis*, 1990, 654.
3722. Volz, W. and Voss, J., *Synthesis*, 1990, 670.
3723. Ohno, M., Ido, M., Shimizu, S. and Eguchi, S., *Synthesis*, 1990, 703.
3724. Younes, M.I., Metwally, S.A.M. and Atta, A.H., *Synthesis*, 1990, 704.
3725. Laus, G. and Klötzer, W., *Synthesis*, 1990, 707.
3726. Russ, T., Bats, J.W. and Reid, W., *Synthesis*, 1990, 721.
3727. Jacobsen, N. and Kolind-Andersen, H., *Synthesis*, 1990, 911.
3728. Jakobsen, M.H., Buchardt, O., Holm, A. and Meldal, M., *Synthesis*, 1990, 1008.
3729. Kihara, Y., Kabashima, S., Uno, K., Okawara, T., Yamasaki, T. and Furukawa, M., *Synthesis*, 1990, 1020.
3730. Hosmane, R.S., Bhadti, V.S. and Lim, B.B., *Synthesis*, 1990, 1095.
3731. Couture, A., Cornet, H. and Grandclaudon, P., *Synthesis*, 1990, 1133.
3732. Badawy, M.A., Abdel-Hady, S.A., Mahmoud, A.H. and Ibrahim, Y.A., *Liebigs Ann. Chem.*, 1990, 815.
3733. Wamhoff, H., Wintersohl, H., Stölben, S., Pasach, J., Naijue, Z. and Fang, G., *Liebigs Ann. Chem.*, 1990, 901.
3734. Wamhoff, H. and Paasch, J., *Liebigs Ann. Chem.*, 1990, 995.
3735. Bridson, P.K. and Lambert, S.J., *J. Chem. Soc., Perkin Trans. 1*, 1990, 173.
3736. Jones, R.C.F., Smallbridge, M.J. and Chapleo, C.B., *J. Chem. Soc., Perkin Trans. 1*, 1990, 385.
3737. Jong, T.-T. and Len, S.J., *J. Chem. Soc., Perkin Trans. 1*, 1990, 423.
3738. Nitta, H., Hatanaka, M. and Ueda, I., *J. Chem. Soc., Perkin Trans. 1*, 1990, 432.
3739. Nitta, M. and Iino, Y., *J. Chem. Soc., Perkin Trans. 1*, 1990, 435.
3740. Kang, W.-B., Sekiya, T., Toru, T. and Ueno, Y., *J. Chem. Soc., Perkin Trans. 1*, 1990, 441.
3741. Moody, C.J. and Rahimtoola, K.F., *J. Chem. Soc., Perkin Trans. 1*, 1990, 673.
3742. Johnson, P.M. and Moody, C.J., *J. Chem. Soc., Perkin Trans. 1*, 1990, 681.
3743. Lee, H.H. and, and Denny, W.A., *J. Chem. Soc., Perkin Trans. 1*, 1990, 1071.
3744. Alves, M.J., Booth, B.L. and Proenc, M.F.J.R.P., *J. Chem. Soc., Perkin Trans. 1*, 1990, 1705.
3745. Molina, P., Tarraga, A. and Lidon, M.J., *J. Chem. Soc., Perkin Trans. 1*, 1990, 1727.

3746. Brooke, G.M. and Mawson, S.D., *J. Chem. Soc., Perkin Trans. 1*, 1990, 1919.
3747. Kubo, Y., Kuwana, M., Okamoto, K. and Yoshida, K., *J. Chem. Soc., Perkin Trans. 1*, 1990, 1975.
3748. Boyle, P.H., Hughes, E.M., Khattab, H.A. and Lockhart, R.J., *J. Chem. Soc., Perkin Trans. 1*, 1990, 2071.
3749. Corrie, J.E.T., *J. Chem. Soc., Perkin Trans. 1*, 1990, 2151.
3750. Aldersley, M.F., Chishti, S.H., Dean, F.M., Douglas, M.E. and Ennis, D.S., *J. Chem. Soc., Perkin Trans. 1*, 1990, 2160.
3751. Trécourt, F., Marsais, F., Güngör, T. and Quéguiner, G., *J. Chem. Soc., Perkin Trans. 1*, 1990, 2409.
3752. Varma, K.S., Edge, S., Underhill, A.E., Becher, J. and Bojesen, G., *J. Chem. Soc., Perkin Trans. 1*, 1990, 2563.
3753. Bode, M.L. and Kaye, P.T., *J. Chem. Soc., Perkin Trans. 1*, 1990, 2612.
3754. Kamal, A., Rao, M.V. and Rao, A.B., *J. Chem. Soc., Perkin Trans. 1*, 1990, 2755.
3755. Jackson, P.M., Moody, C.J. and Shah, P., *J. Chem. Soc., Perkin Trans. 1*, 1990, 2909.
3756. Moody, C.J. and Warrellow, G.J., *J. Chem. Soc., Perkin Trans. 1*, 1990, 2929.
3757. Sanghvi, Y.S., Larson, S.B., Robins, R.K. and Ravenkar, G.R., *J. Chem. Soc., Perkin Trans. 1*, 1990, 2943.
3758. Grimshaw, J. and Newitt, S.A., *J. Chem. Soc., Perkin Trans. 1*, 1990, 2995.
3759. Conti, C., Desideri, N., Orsi, N., Sestili, I. and Stein, M.L., *Eur. J. Med. Chem.*, 1990, **25**, 725.
3760. Balczewski, P., Makkon, M.K.J., Street, J.D. and Joule, J.A., *J. Chem. Soc., Perkin Trans. 1*, 1990, 3193.
3761. Kakehi, A., Ito, S., Yamada, N. and Yamaguchi, K., *Bull. Chem. Soc. Jpn.*, 1990, **63**, 829.
3762. Nitta, M., Soeda, H. and Iino, Y., *Bull. Chem. Soc. Jpn.*, 1990, **63**, 932.
3763. Sato, R., Endoh, H., Abe, A., Yamouchi, S., Goto, T. and Saito, M., *Bull. Chem. Soc. Jpn.*, 1990, **63**, 1160.
3764. Ishii, H., Murakami, Y. and Ishikawa, T., *Chem. Pharm. Bull.*, 1990, **38**, 597.
3765. Abe, N. and Emoto, Y., *Bull. Chem. Soc. Jpn.*, 1990, **63**, 1543.
3766. Abe, N., Ishikawa, N., Hayashi, T. and Miyra, Y., *Bull. Chem. Soc. Jpn.*, 1990, **63**, 1617.
3767. Elnagdi, M.H. and Erian, A.W., *Bull. Chem. Soc. Jpn.*, 1990, **63**, 1854.
3768. Abe, N. and Ueno, T., *Bull. Chem. Soc. Jpn.*, 1990, **63**, 2121.
3769. Krantz, A., Spencer, R.A., Tam, T.F., Liak, T.J., Copp, L.J., Thomas, E.M. and Rafferty, S.P., *J. Med. Chem.*, 1990, **33**, 464.
3770. Reiffen, M., Eberlin, W., Müller, P., Psiorz, M., Noll, K., Heider, J., Lilliec, C., Kobinger, W. and Luger, P., *J. Med. Chem.*, 1990, **33**, 1496.
3771. Jiang, J.B., Hesson, D.P., Dusak, B.A., Dexter, D.L., Kang, G.J. and Hamel, E., *J. Med. Chem.*, 1990, **33**, 1721.
3772. McNamara, D.J., Berma, E.M., Fry, D.W. and Werbel, L.M., *J. Med. Chem.*, 1990, **33**, 2045.
3773. Turconi, M., Nicola, M., Quintero, M.G., Maiocchi, L., Micheletti, R., Giraldo, E. and Donetti, A., *J. Med. Chem.*, 1990, **33**, 2101.
3774. Roberts, D.A., Bradbury, R.H., Brown, D., Faull, A., Griffiths, D., Major, J.S., Oldham, A.A., Pearce, R.J., Radcliffe, A.H., Revill, J. and Waterson, D., *J. Med. Chem.*, 1990, **33**, 2326.
3775. Ting, P.C., Kaminski, J.J., Sherlock, M.H., Tan, W.C., Loc, J.F., Bryant, R.W., Watnick, A.S. and McPhail, A.T., *J. Med. Chem.*, 1990, **33**, 2697.
3776. Temple, C. and Rener, G.A., *J. Med. Chem.*, 1990, **33**, 3044.
3777. Hughes, L.R., Jackman, A.L., Oldfield, J., Smit, R.C., Barrows, K.D., Marsham, P.R., Bishop, J.A.M., Jones, T.R., O'Connor, B.M. and Calvert, A.H., *J. Med. Chem.*, 1990, **33**, 3060.

3778. Peet, N.P., Lentz, N.L., Meng, E.C., Dudley, M.W., Ogden, A.M.L., Demeter, D.A., Weintraub, H.J.R. and Bey, P., *J. Med. Chem.*, 1990, **33**, 3127.
3779. Krutosikova, A., *Collect. Czech. Chem. Commun.*, 1990, **55**, 597.
3780. Deeb, A., El Mobayed, M., Essawy, A.N., El-Hamid, A.H. and Hrubantova, M., *Collect. Czech. Chem. Commun.*, 1990, **55**, 728.
3781. Gonda, J. and Barnikol, M., *Collect. Czech. Chem. Commun.*, 1990, **55**, 753.
3782. Patek, M., *Collect. Czech. Chem. Commun.*, 1990, **55**, 1223.
3783. Sindelár, K., Holubek, J., Koruna, I. and Hrubantova, M., *Collect. Czech. Chem. Commun.*, 1990, **55**, 1586.
3784. Tominaga, Y., Norisue, H., Kamio, C., Masunari, T., Miyashiro, Y. and Hosomi, A., *Heterocycles*, 1990, **31**, 1.
3785. Campagna, F., Carotti, A., Casini, G. and Macripo, M., *Heterocycles*, 1990, **31**, 97.
3786. Romano, C., de la Cuesta, E. and Avondano, C., *Heterocycles*, 1990, **31**, 267.
3787. Rousseau, O., Blondeau, D. and Sliwa, H., *Heterocycles*, 1990, **31**, 277.
3788. Batori, S., Sandor, P. and Messmer, A., *Heterocycles*, 1990, **31**, 289.
3789. Hilmy, K.M.H., Mogensen, J., Jorgenson, A. and Pedersen, E.B., *Heterocycles*, 1990, **31**, 367.
3790. Friary, R., Schwerdt, J.A., Seidl, V. and Topliss, J.G., *Heterocycles*, 1990, **31**, 419.
3791. Kamal, A., Rao, M.V. and Rao, A.B., *Heterocycles*, 1990, **31**, 577.
3792. Sha, C.-K., Liu, J.-M., Chiang, R.-K. and Wang, S.-L., *Heterocycles*, 1990, **31**, 603.
3793. Kishi, I., Imafuku, K., Ogawa, K. and Matsushita, Y., *Heterocycles*, 1990, **31**, 677.
3794. Okonogi, T., Shibahara, S., Murai, Y., Inouge, S., Kondo, S. and Christensen, B.G., *Heterocycles*, 1990, **31**, 791.
3795. Yamanaka, H., Sakamoto, T. and Niitsuma, S., *Heterocycles*, 1990, **31**, 923.
3796. Kamal, A. and Hashim, R., *Heterocycles*, 1990, **31**, 969.
3797. Matsuda, Y., Goto, H., Katou, K., Matsumoto, H., Yamashita, M., Takahashi, K. and Ide, S., *Heterocycles*, 1990, **31**, 977.
3798. Aso, M., Sakamoto, M., Urakawa, N. and Kanematsu, K., *Heterocycles*, 1990, **31**, 1103.
3799. Mayor, J.P., Cassady, J.M. and Nichols, D.E., *Heterocycles*, 1990, **31**, 1035.
3800. Bruni, F., Chimichi, S., Cosimelli, B., Costanzo, A., Guerrini, G. and Selleri, S., *Heterocycles*, 1990, **31**, 1141.
3801. Nagaoka, H., Hara, H. and Mase, T., *Heterocycles*, 1990, **31**, 1241.
3802. Kamal, A., *Heterocycles*, 1990, **31**, 1377.
3803. Strekowski, L., Wydra, R.L., Harden, D.B. and Honkau, V.A., *Heterocycles*, 1990, **31**, 1565.
3804. Bruni, F., Chimichi, S., Cosimelli, B., Costanzo, A., Guerrini, G. and Selleri, S., *Heterocycles*, 1990, **31**, 1635.
3805. Konno, S., Sagi, M., Yokoyama, M. and Yamanaka, H., *Heterocycles*, 1990, **31**, 1933.
3806. Lopez-Alvarado, P., Avendaño, C., Grande, M.T. and Menéndez, J.C., *Heterocycles*, 1990, **31**, 1983.
3807. Sanchez, E., del Campo, C., Avendaño, C., and Llama, E., *Heterocycles*, 1990, **31**, 2003.
3808. Muzaffar, A., Hamel, E. and Brossi, A., *Heterocycles*, 1990, **31**, 2037.
3809. Gogoi, P.C. and Kataky, J.C.S., *Heterocycles*, 1990, **31**, 2147.
3810. Kasztreiner, E., Rabloczky, G., Makk, N., Mátyus, P., Diesler, E., Tegeles, A., Kosáry, J., Czakó, K., Gyürki, S., Cseh, G., Kürthy, M. and Jaszlitz, L., *Eur. J. Med. Chem.*, 1990, **25**, 333.
3811. Ram, V.J., Singha, U.K. and Guru, P.Y., *Eur. J. Med. Chem.*, 1990, **25**, 533.
3812. Ijaz, A.S., Parrick, J. and Yahya, A., *J. Chem. Res.(S)*, 1990, 116.
3813. Atanasov, K.D., Lafuente, P., Quintero, M., Seoane, C. and Soto, J.L., *J. Chem. Res.(S)*, 1990, 186.

3814. Parrick, J. and Rami, H.K., *J. Chem. Res.(S)*, 1990, 308.
3815. Hanaya, K., Muramatsu, T. and Hasgawa, E., *Chem. Ind. (London)*, 1990, 802.
3816. Reddy, N.A.V., Maiti, S.N. and Micetich, R.G., *J. Chem. Res.(S)*, 1990, 32.
3817. Parrick, J. and Rami, H.K., *J. Chem. Res.(S)*, 1990, 64.
3818. Elnagdi, M.H., Erian, A.W., Sadek, K.U. and Selim, M.A., *J. Chem. Res.(S)*, 1990, 148.
3819. Martin-Leon, N., Quinteiro, M., Seoane, C. and Soto, J.L., *J. Chem. Res.(S)*, 1990, 156.
3820. Greff, Z., Horvath, Z., Nyitrai, J., Kajtar-Peredy, M. and Brlik, J., *J. Chem. Res.(S)*, 1990, 170.
3821. Ariamala, G. and Balasubramanian, K.K., *Tetrahedron*, 1990, **46**, 309.
3822. Thomas, A., Chakraborty, M., Ila, H. and Junjappa, H., *Tetrahedron*, 1990, **46**, 577.
3823. Strazzolini, P., Giumanini, A.G. and Cauci, S., *Tetrahedron*, 1990, **46**, 1081.
3824. Bartoli, G., Palmieri, G., Petrini, M., Bosco, M. and Dalpozzo, R., *Tetrahedron*, 1990, **46**, 1379.
3825. Thebtaranonth, C. and Thebtaranonth, Y., *Tetrahedron*, 1990, **46**, 1385.
3826. Kepe, V., Kocevar, M., Polanc, S., Vercek, B. and Tisler, M., *Tetrahedron*, 1990, **46**, 2081.
3827. Verbrüggen, H., Bohn, B.D. and Kroliewicz, K., *Tetrahedron*, 1990, **46**, 3489.
3828. Rossi, E., Stradi, R. and Visentin, P., *Tetrahedron*, 1990, **46**, 3581.
3829. Molina, P., Arques, A., Alies, M.A., Llamas-Saiz, A. and Foces-Foces, C., *Tetrahedron*, 1990, **46**, 4353.
3830. Junjappa, H., Ida, H. and Asokan, C.V., *Tetrahedron*, 1990, **46**, 5423.
3831. Vicentini, C.B., Veronese, A.C., Giori, P., Lumachi, B. and Guarneri, M., *Tetrahedron*, 1990, **46**, 5777.
3832. Bergman, J. and Sand, P., *Tetrahedron*, 1990, **46**, 6085.
3833. Molina, P., Vilaplana, M.J. and Perez, J., *Tetrahedron*, 1990, **46**, 7855.
3834. Bhuyan, P., Boruah, R.C. and Sandhu, J.S., *J. Org. Chem.*, 1990, **55**, 568.
3835. Tsuji, Y., Kotachi, S., Huh, K.T. and Watanabe, Y., *J. Org. Chem.*, 1990, **55**, 580.
3836. VanSickle, A.P. and Rapoport, H., *J. Org. Chem.*, 1990, **55**, 895.
3837. Etkin, N., Babu, S.D., Fooks, C.J. and Durst, T., *J. Org. Chem.*, 1990, **55**, 1093.
3838. Donnelly, J.A. and Farrell, D.F., *J. Org. Chem.*, 1990, **55**, 1757.
3839. Kitamura, T., Kobayashi, S. and Taniguchi, H., *J. Org. Chem.*, 1990, **55**, 1801.
3840. Sha, C.-K. and Tsou, C.-P., *J. Org. Chem.*, 1990, **55**, 2446.
3841. Conn, R.S.E., Douglas, A.W., Karady, S., Corby, E.G., Lovell, A.V. and Shankai, I., *J. Org. Chem.*, 1990, **55**, 2908.
3842. Taylor, E.C. and Harrington, P.M., *J. Org. Chem.*, 1990, **55**, 3222.
3843. Echavarren, A.M., *J. Org. Chem.*, 1990, **55**, 4255.
3844. Couture, A., Huguerre, E. and Grandclaudon, P., *J. Org. Chem.*, 1990, **55**, 4337.
3845. McGarry, L.W. and Detty, M.R., *J. Org. Chem.*, 1990, **55**, 4349.
3846. Molina, P., Arques, A. and Vinader, M.V., *J. Org. Chem.*, 1990, **55**, 4724.
3847. Turner, J.A., *J. Org. Chem.*, 1990, **55**, 4744.
3848. Sommer, M.B., Begtrup, M. and Bogeso, K.P., *J. Org. Chem.*, 1990, **55**, 4817.
3849. Sommer, M.B., Begtrup, M. and Bogeso, K.P., *J. Org. Chem.*, 1990, **55**, 4822.
3850. Barrett, A.G.M. and Sakadarat, S., *J. Org. Chem.*, 1990, **55**, 5110.
3851. Coates, W.J. and McKillop, A., *J. Org. Chem.*, 1990, **55**, 5418.
3852. Floyd, D.M., Moquin, R.V., Atwal, K.S., Ahmed, S.Z., Spergel, S.H., Gongontas, J.Z. and Malley, M.F., *J. Org. Chem.*, 1990, **55**, 5572.
3853. Teng, M. and Fowler, F.W., *J. Org. Chem.*, 1990 **55**, 5646.
3854. Molina, P., Alajarin, M. and Vidal, A., *J. Org. Chem.*, 1990, **55**, 6140.
3855. Harvey, R.G., Haha, J.-T., Bukowska, M. and Jackson, H., *J. Org. Chem.*, 1990, **55**, 6161.

3856. Luheshi, A.B.N., Smalley, R.K., Kennewell, P.D. and Westwood, R., *Tetrahedron Lett.*, 1990, **31**, 123.
3857. Jain, S., Jain, R., Singh, J. and Anand, N., *Tetrahedron Lett.*, 1990, **31**, 131.
3858. Rossi, E., Celentano, G., Stradi, R. and Strada, A., *Tetrahedron Lett.*, 1990, **31**, 903.
3859. Reid, J.G. and Runge, J.M.R., *Tetrahedron Lett.*, 1990, **31**, 1093.
3860. Doise, M., Dennin, F., Blondeau, D. and Sliwa, H., *Tetrahedron Lett.*, 1990, **31**, 1155.
3861. Majundar, K.C., De, R.N. and Saha, S., *Tetrahedron Lett.*, 1990, **31**, 1207.
3862. Siddiqui, M.A. and Snieckus, V., *Tetrahedron Lett.*, 1990, **31**, 1523.
3863. Williams, M.A. and Miller, M.J., *Tetrahedron Lett.*, 1990, **31**, 1807.
3864. Rosa, A.M., Prabhakar, S. and Lobo, A.M., *Tetrahedron Lett.*, 1990, **31**, 1881.
3865. Yeung, C.M. and Klein, L.L., *Tetrahedron Lett.*, 1990, **31**, 2121.
3866. Nebois, P., Barret, R. and Fillion, H., *Tetrahedron Lett.*, 1990, **31**, 2569.
3867. Linderman, R.J. and Kirollos, K.S., *Tetrahedron Lett.*, 1990, **31**, 2689.
3868. Ternansky, R.J. and Draheim, S.E., *Tetrahedron Lett.*, 1990, **31**, 2805.
3869. Khanapure, S.P., Bhawal, B.M. and Biehl, E.R., *Tetrahedron Lett.*, 1990, **31**, 2869.
3870. Grigg, R., Dorrity, M.J.R., Malone, J.F., Mongkolaussavaratana, T., Norbert, W.D.J.A. and Sridharan, V., *Tetrahedron Lett.*, 1990, **31**, 3075.
3871. Veronese, A.C., Callegari, R. and Salah, S.A.A., *Tetrahedron Lett.*, 1990, **31**, 3485.
3872. Machiguchi, T. and Yamabe, S., *Tetrahedron Lett.*, 1990, **31**, 4169.
3873. Barker, A.J., Campbell, M.M. and Jenkins, M.J., *Tetrahedron Lett.*, 1990, **31**, 4359.
3874. Moody, C.J., Rees, C.W. and Thomas, R., *Tetrahedron Lett.*, 1990, **31**, 4375.
3875. Liebskind, L.S., Johnson, S.A. and Melallum, J.S., *Tetrahedron Lett.*, 1990, **31**, 4397.
3876. Lee, H. and Anderson, W.K., *Tetrahedron Lett.*, 1990, **31**, 4405.
3877. Block, E. and Zhao, S.H., *Tetrahedron Lett.*, 1990, **31**, 5003.
3878. Jackson, B.G., Gardner, J.P. and Heath, P.C., *Tetrahedron Lett.*, 1990, **31**, 6317.
3879. Grigg, R., Loganathan, V., Sukirthalingan, S. and Sridharan, V., *Tetrahedron Lett.*, 1990, **31**, 6573.
3880. Barluenga, J., Aznar, F., Fustero, S. and Tomas, M., *Pure Appl. Chem.*, 1990, **62**, 1957.
3881. Boamah, P.Y., Haider, N. and Heinisch, G., *Arch. Pharm. (Weinheim)*, 1990, **323**, 207.
3882. Görlitzer, K. and Vogt, R., *Arch. Pharm. (Weinheim)*, 1990, **323**, 837.
3883. Leistner, S., Gütschow, M. and Stach, J., *Arch. Pharm. (Weinheim)*, 1990, **323**, 857.
3884. Lalifa, F.A.K., *Arch. Pharm. (Weinheim)*, 1990, **323**, 883.
3885. Verboom, W. and Reinhoudt, D.N., *Recl. Trav. Chim. Pays-Bas*, 1990, **109**, 311.
3886. Verboom, W., Verboom, C., Eissiak, I.M., Lammerink, B.H.M and Reinhoudt, D.W., *Recl. Trav. Chim. Pays-Bas*, 1990, **109**, 481.
3887. Dominianni, S.J., *Org. Prep. Proced. Int.*, 1990, **22**, 106.
3888. Korodi, F. and Cziaky, Z., *Org. Prep. Proced. Int.*, 1990, **22**, 579.
3889. Kunz, K.R., Taylor, E.W., Hutton, H.M. and Blackburn, B.J., *Org. Prep. Proced. Int.*, 1990, **22**, 613.
3890. Fischer, G., *Z. Chem.*, 1990, **30**, 305.
3891. Brehme, R., *Chem. Ber.*, 1990, **123**, 2039.
3892. Takagi, K., *Chem. Lett.*, 1990, 2205.
3893. Sundberg, R.J., *Prog. Heterocycl. Chem.*, 1990, **2**, 70.
3894. Bird, C.W., *Prog. Heterocycl. Chem.*, 1990, **2**, 87.
3895. Cabre, M., Farrás, Sanz, J.F. and Vilarrasa, J., *J. Chem. Soc., Perkin Trans. 2*, 1990, 1943.
3896. Hofmann, H. and Fischer, H., *Liebigs Ann. Chem.*, 1990, 917.
3897. Elnagdi, M.H. and Erian, A.W., *Liebigs Ann. Chem.*, 1990, 1215.
3898. Khalil, Z.H., Abdel-Hafez, A.A., Geics, A.A. and El-Dean, A.M.K., *Bull. Chem. Soc. Jpn.*, 1990, **63**, 668.

3899. Yamaguchi, S., Oh-hira, Y., Yamada, M., Michitani, H. and Kawase, Y., *Bull. Chem. Soc. Jpn.*, 1990, **63**, 952.
3900. Sato, R., Endoh, H., Abe, A., Yamaichi, S., Goto, T. and Saito, M., *Bull. Chem. Soc. Jpn.*, 1990, **63**, 1160.
3901. Cirricione, G., Almerico, A.M., Aiello, E. and Dattolo, G., *Adv. Heterocycl. Chem.*, 1990, **48**, 65.
3902. Elnagdi, M.H., Elmoghayer, M.R.H. and Sadek, K.U., *Adv. Heterocycl. Chem.*, 1990, **48**, 223.
3903. Shaban, M.A.E. and Nazr, A.Z., *Adv. Heterocycl. Chem.*, 1990, **49**, 277.
3904. Busby, R.E., *Adv. Heterocycl. Chem.*, 1990, **50**, 255.
3905. Mezheritskii, V.V. and Tkachenko, V.V., *Adv. Heterocycl. Chem.*, 1990, **51**, 1.
3906. Acheson, R.M., *Adv. Heterocycl. Chem.*, 1990, **51**, 105.
3907. Elgemeie, G.E.H. and Elgdandour, A.H.H., *Bull. Chem. Soc. Jpn.*, 1990, **63**, 1230.
3908. Strekowski, L., Wydra, R.L., Cegla, M.T., Czarny, A., Harden, D.B., Patterson, S.E., Battiste, M.A. and Coxon, J.M., *J. Org. Chem.*, 1990, **55**, 4777.
3909. Pelcman, B. and Gribble, G.W., *Tetrahedron Lett.*, 1990, **31**, 2381.
3910. Younes, M.I., Atta, A.H. and Metwally, S.A.M., *Gazz. Chim. Ital.*, 1991, **121**, 185.
3911. Sasaki, K., Tashima, Y., Nakayama, T. and Hirota T., J. Heterocycl. Chem., 1991, 28, 269.
3912. Abdelhamid, A.O. and Attaby, F.A., *J. Heterocycl. Chem.*, 1991, **28**, 41.
3913. Bouyazza, L., Lancelot, J.-C., Raoult, S. and Robba, M., *J. Heterocycl. Chem.*, 1991, **28**, 77.
3914. Jackson, J.L., *J. Heterocycl. Chem.*, 1991, **28**, 109.
3915. Yamaguchi, S., Saitoh, M. and Kawase, Y., *J. Heterocycl. Chem.*, 1991, **28**, 125.
3916. Date, M., Kawanishi, K., Akiyoshim, R. and Furukawa, S., *J. Heterocycl. Chem.*, 1991, **28**, 173.
3917. Musmar, M.J. and Castle, R.N., *J. Heterocycl. Chem.*, 1991, **28**, 203.
3918. Luo, J.K. and Castle, R.N., *J. Heterocycl. Chem.*, 1991, **28**, 205.
3919. Hirota, T., Sasaki, K., Tashima, Y. and Nakayama, T., *J. Heterocycl. Chem.*, 1991, **28**, 263.
3920. Giammona, G., Neri, M., Caarlisi, B., Palazzo, A. and LaRosa, C., *J. Heterocycl. Chem.*, 1991, **28**, 325.
3921. Tanemura, K., Suzuki, T., Horagualin, T. and Sudo, M., *J. Heterocycl. Chem.*, 1991, **28**, 305.
3922. Varella, E.A. and Nicolaides, D.N., *J. Heterocycl. Chem.*, 1991, **28**, 311.
3923. Venugopalan, B., Bagat, C.P., de Souza, E.P. and de Souza, N.J., *J. Heterocycl. Chem.*, 1991, **28**, 337.
3924. Bailey, S., Harnden, M.R., Jarvest, R.L., Parkin, A. and Boyd, M.R., *J. Med. Chem.*, 1991, **34**, 57.
3925. Walser, A., Flynn, T., Mason, C., Crowley, H., Maresca, C., Yarenko, B. and O'Donnell, M., *J. Med. Chem.*, 1991, **34**, 1209.
3926. Giannetti, D., Viti, G., Sbracci, P., Pestellini, V., Volterra, G., Borsini, F., Lecc, A., Meli, A., Dapporto, P. and Paoli, P., *J. Med. Chem.*, 1991, **34**, 1356.
3927. Torii, S., Okumoto, H. and Xu, L.H., *Tetrahedron*, 1991, **47**, 665.
3928. Adams, J. and Spero, D.M., *Tetrahedron*, 1991, **47**, 1765.
3929. Sakamoto, T., Kondo, K., Yasahara, A. and Yamanaka, A., *Tetrahedron*, 1991, **47**, 1877.
3930. Dike, S.Y., Merchant, J.R. and Sapre, N.Y., *Tetrahedron*, 1991, **47**, 4775.
3931. Rival, Y., Grassy, G., Taudou, A. and Escalle, R., *Eur. J. Med. Chem.*, 1991, **26**, 13.
3932. Cruces, M.A., Elorriega, C. and Fernandez-Alvarez, E., *Eur. J. Med. Chem.*, 1991, **26**, 33.
3933. Jaguelin, S., Robert, A. and Gayral, P., *Eur. J. Med. Chem.*, 1991, **26**, 51.

3934. Kanojia, R.M., Press, J.B., Lever Jr, O.W., Williams, L., Werblood, H.M., Falotico, R., Moore, J.M. and Tobia, A.J., *Eur. J. Med. Chem.*, 1991, **26**, 137.
3935. Nakib, T.A., Bezjak, V., Raschid, S., Fullam, B. and Meegan, M.J., *Eur. J. Med. Chem.*, 1991, **26**, 221.
3936. Bruché, L., Garanti, L. and Zecchi, G., *J. Chem. Res.(S)*, 1991, 2.
3937. Bradshaw, D.P., Jones, D.W. and Tideswell, J., *J. Chem. Soc., Perkin Trans. 1*, 1991, 169.
3938. Hormi, O.E.O., Peltonen, C. and Bergström, R., *J. Chem. Soc., Perkin Trans. 1*, 1991, 219.
3939. O'Daly, M.A., Hopkinson, C.P., Meakins, G.D. and Raybould, A.J., *J. Chem. Soc., Perkin Trans. 1*, 1991, 855.
3940. Yamasaki, T., E., Kawaminami, E., Yamada, T., Okawa, T. and Furukawa, M., *J. Chem. Soc., Perkin Trans. 1*, 1991, 991.
3941. Nitta, M., Ohnuma, M. and Iino, Y., *J. Chem. Soc., Perkin Trans. 1*, 1991, 1115.
3942. Nitta, M., Mori, S. and Iino, Y., *Heterocycles*, 1991, **32**, 23.
3943. Majumdar, K.C. and Choudhury, P.K., *Heterocycles*, 1991, **32**, 73.
3944. Gogoi, P.C. and Kataky, J.C.S., *Heterocycles*, 1991, **32**, 231.
3945. Gogoi, P.C. and Kataky, J.C.S., *Heterocycles*, 1991, **32**, 237.
3946. Alkorta, I., Goya, P. and Páez, J.A., *Heterocycles*, 1991, **32**, 279.
3947. Bradbury, R.H., *Heterocycles*, 1991, **32**, 449.
3948. Caroon, J.M. and Fisher, L.E., *Heterocycles*, 1991, **32**, 459.
3949. Chen, B.-C., *Heterocycles*, 1991, **32**, 529.
3950. Coletta, V., Cecchi, L., Melani, F., Filacchioni, G., Martini, C., Gelli, S. and Lucacchini, A., *J. Pharm. Sci.*, 1991, **80**, 276.
3951. Russell, G.A, Yao, C.F., Tashtoush, H., Russell, J.E. and Dedolph, D.F., *J. Org. Chem.*, 1991, **56**, 663.
3952. Shaw, K.J. and Vartanian, M., *J. Org. Chem.*, 1991, **56**, 858.
3953. Boger, D.L. and Nakahara, S., *J. Org. Chem.*, 1991, **56**, 880.
3954. Laufer, D.A. and Al-Farhan, E., *J. Org. Chem.*, 1991, **56**, 891.
3955. Chilin, A., Rodighiero, P., Pastorini, G. and Guiotto, A., *J. Org. Chem.*, 1991, **56**, 908.
3956. Takeuchi, H., Matsushita, Y. and Eguchi, S., *J. Org. Chem.*, 1991, **56**, 1535.
3957. Czekanski, T., Hanack, M., Becker, J.Y., Bernstein, J., Bittner, S., Kaufman-Orenstein, L. and Peleg, D., *J. Org. Chem.*, 1991, **56**, 1569.
3958. Atkinson, J., Morand, P., Arnason, J.T., Niemeyer, H.N. and Brown, H.R., *J. Org. Chem.*, 1991, **56**, 1788.
3959. Nabulsi, N.A.R. and Gandour, R.D., *J. Org. Chem.*, 1991, **56**, 2260.
3960. Robl, J.A., *Synthesis*, 1991, 56.
3961. Birkett, P.R., Chapleo, C.B. and Mackenzie, G., *Synthesis*, 1991, 157.
3962. Zecchi, G., *Synthesis*, 1991, 181.
3963. Wermann, K. and Hartmann, M., *Synthesis*, 1991, 189.
3964. Bittner, S., Krief, P. and Massil, T., *Synthesis*, 1991, 215.
3965. Villemin, D. and Alloum, A.B., *Synthesis*, 1991, 301.
3966. Nagamatsu, T., Tsurubayashi, S., Sasaki, K. amd Hirota, T., *Synthesis*, 1991, 303.
3967. Lauk, U., Dürst, D. and Fischer, W., *Tetrahedron Lett.*, 1991, **32**, 65.
3968. Toril, S., Okumoto, H. and Xu, L.H., *Tetrahedron Lett.*, 1991, **32**, 237.
3969. Larock, R.C. and Kuo, M.-Y., *Tetrahedron Lett.*, 1991, **32**, 359.
3970. Billeret, D., Blondeau, D. and Sliwa, H., *Tetrahedron Lett.*, 1991, **32**, 627.
3971. Grigg, R., Loganathan, V., Santhakumar, V. Sridharan, V. and Teasdale, A., *Tetrahedron Lett.*, 1991, **32**, 687.
3972. Bouthillier, G., Mastalerz, H. and Menard, M., *Tetrahedron Lett.*, 1991, **32**, 1023.
3973. Nebois, P. and Fillion, H., *Tetrahedron Lett.*, 1991, **32**, 1307.
3974. Inanaga, J., Ujikawa, O. and Yamaguchi, M., *Tetrahedron Lett.*, 1991, **32**, 1737.

3975. Yang, Z., Hon, P.M., Chui, K.Y., Xu, Z.L., Chang, H.M., Lee, C.M., Chi, Y.X., Wong, H.N.C., Poon, C.D. and Fang, B.M., *Tetrahedron Lett.*, 1991, **32**, 2061.
3976. Zelle, R.E. and McClellan, W.J., *Tetrahedron Lett.*, 1991, **32**, 2461.
3977. Anastasiou, D., Chaouk, H. and Jackson, W.R., *Tetrahedron Lett.*, 1991, **32**, 2499.
3978. Avalos, M., Babiano, R., Cintas, P., Jiménez, J.L., Molina, M.M., Palacios, J.C. and Sanchez, J.B., *Tetrahedron Lett.*, 1991, **32**, 2513.
3979. Molina, P., Arques, A., Cartagena, I. and Obón, R., *Tetrahedron Lett.*, 1991, **32**, 2521.
3980. Mataka, S., Takahashi, K., Ikezaki, Y., Hatta, T., Tori-i, A. and Tashiro, M., *Bull. Chem. Soc. Jpn.*, 1991, **64**, 68.
3981. Press, J. and Russell, R.K., *Prog. Heterocycl. Chem.*, 1991, **3**, 70.
3982. Sundberg, R.J., *Prog. Heterocycl. Chem.*, 1991, **3**, 90.
3983. Bird, C.W., *Prog. Heterocycl. Chem.*, 1991, **3**, 109.
3984. Keay, J.G. and Sherman, A.R., *Prog. Heterocycl. Chem.*, 1991, **3**, 187.
3985. Hurst, D.T., *Prog. Heterocycl. Chem.*, 1991, **3**, 224.
3986. Hepworth, J.D., *Prog. Heterocycl. Chem.*, 1991, **3**, 253.
3987. Kane, J.M. and Peet, N., *Prog. Heterocycl. Chem.*, 1991, **3**, 277.
3988. Newkome, G.R., *Prog. Heterocycl. Chem.*, 1991, **3**, 317.
3989. Coutouli-Argyropoulou, E., and Malamidou-Xenikaki, E., *J. Chem. Res.(S)*, 1991, 106.
3990. Mukerjee, A.K. and Ashare, R., *Chem. Rev.*, 1991, **91**, 1.
3991. Ziegler, F.E. and Jeroncic, L.O., *J. Org. Chem.*, 1991, **56**, 3479.
3992. Weichert, A. and Hoffmann, H.M.R., *J. Org. Chem.*, 1991, **56**, 4098.
3993. Katritzky, A.R., Rachwal, S. and Hitchings, G.J., *Tetrahedron*, 1991, **47**, 2683.
3994. Francis, J.E., Cash, W.D., Barbaz, B.S., Bernard, P.S., Lovel, R.A., Mazzenga, G.C., Friedmann, R.C., Hyun, J.L., Braunwalder, A.F., Loo, P.S. and Bennett, D.A., *J. Med. Chem.*, 1991, **34**, 281.
3995. Saito, T., Ayukawa, H., Sumizawa, N., Shizuta, T., Utoloki, S. and Kobayashi, K., *J. Chem. Soc., Perkin Trans. 1*, 1991, 1405.
3996. Howson, A.T., Hughes, K., Richardson, S.K., Sharpe, D.A. and Wadsworth, A.H., *J. Chem. Soc., Perkin Trans. 1*, 1991, 1565.
3997. Karmaker, A.C., Kar, G.K. and Ray, J.K., *J. Chem. Soc., Perkin Trans. 1*, 1991, 1997.
3998. Larock, R.C. and Yum, E.K., *J. Am. Chem. Soc.*, 1991, **113**, 6689.
3999. Molina, P., Arques, A. and Molina, A., *Synthesis*, 1991, 21.
4000. Ye, F.-C., Chen, B.-C. and Huang, X., *Synthesis*, 1991, 317.
4001. Solladie, G. and Girardin, A., *Synthesis*, 1991, 569.
4002. Nyce, P.L., Gala, D. and Steinman, M., *Synthesis*, 1991, 571.
4003. Neidlein, R. and Sui, Z., *Synthesis*, 1991, 658.
4004. Jasys, V.J., Kellogg, M.S. and Volkmann, R.A., *Tetrahedron Lett.*, 1991, **32**, 3771.
4005. Molina, P., Arques, A., Alies, A. and Vinader, M.V., *Tetrahedron Lett.*, 1991, **32**, 4404.
4006. Meyer, M., Deschamps, J.C. and Molho, D., *Bull. Soc. Chim. Fr.*, 1991, 91.

General Index

A topic may be mentioned on several pages following those which are printed in bold. (R) means that a reference to a review is included on one or more of the pages cited.

Aaptamine, 1178
Abbreviations, 669
Algar–Flynn–Oyamada reaction, 902
Alternariol, 1070
t-Amino effect (R), 916
2-Aminobenzophenones (R), **973**
Aminoethylthiolation, 873
Ascididemin, 1192
Asparenomycins, 1151
Auromomycin, 826
4-Azaazulenes (R), 1206

Baker–Venkataraman reaction, 824
Batcho–Leimgruber reaction, 957
Beckmann rearrangement (R), 807, 1268, 1270
Benzazepines (R), 1302
1,4-Benzodiazepines, tricyclic (R), 710
1,5-Benzodiazepines (R), 1137
Benzodiazocines (R), 816
Benzofurans (R), 676, 703, 800, 818
1-Benzopyrans, 2,3-fused (R), 1062
Benzopyrroles (R), 702, 969
  1-hydroxy- (R), 885
Benzo[*b*]thiophene-2,3-dione (R), 1282
Bernthsen's synthesis of acridines, 1192
Beyer–Combes reaction, 989
Biguanides, 830
Bischler–Napieralski isoquinoline synthesis, 770
t-Butyl groups as blockers (R), 982
Butotenin, 697

Carbolines—*see* Pyridoindoles
Camptothecin, 1310
Chalcone dibromide, 1216
Colchineinamide, 972
Combes quinoline synthesis, 866, 990
Cope rearrangement, 1258
Curtius reaction, 933
Cyclization, cobalt-induced, 1297
  copper-induced, 711, 737, 864, 1057, 1065, 1070, 1240,
  copper–chromium-induced, 1273
  homolytic (R), 894, 1078, 1078, 1152, 1228, 1229
  iron complex-mediated, 813
  mercury(II)-assisted, 778, 885
  metal-mediated (R), 885, 992, 1057
  nickel-induced, 1070, 1182
  palladium-induced (R), 702, **706**, 735, 885, **893**, 900, 915, 992, 1076, **1106**, 1214, 1223, 1242
  photochemical (R), 737, 740, 884, 900, 992, 1016, **1025**, 1031, 1032, 1078, 1110, 1124, 1134, 1151, 1159, **1182**, **1223**, 1230, 1287, 1298, 1304, 1313, 1333, **1339**
  rhodium-mediated (R), 768, 890, **1150**, 1311,
  ruthenium-mediated, 713, 981, 992
  sonochemically promoted (R), 886, 1229
  thallium derivatives in, 769, 818, 886, 897, 902, 1223
  transition metal-mediated (R), 1070, 1182
  zirconium-mediated (R), 1225
Cyclization of functional groups, acetal and aldehyde, 831
  acetal or thioacetal and amine, **671**
  acetal and carboxamide, 674
  acetal or thioacetal and ring-C, **675**
  acetal and ring-N, **677**
  acyl halide and acyloxy, **1078**
  acyl halide and halogen, 1068
  acyl halide and isocyanate, 1084
  acyl halide and nitrile, **1094**
  acyl halide and acyloxy, 1078

GENERAL INDEX

Cyclization of functional groups (*cont.*)
acyl halide and ring-C or ring-N, **783**,
acyl halide and thiocyanate, 1121
acyl halide and thioether, 1079
O-acylamidroxamine and amine, **691**
acylamine and alkene, 733
acylamine and alkoxyamine (or hydroxyamine), 684
acylamine and amidroxamine, 691
acylamine and amine or imine, 683
acylamine and carboxamide, **691**
acylamine and carboxylic acid, **693**
acylamine and carboxylic ester, 694
acylamine and ether or thioether, 697
acylamine and halogen, **701**
acylamine and hydrazine, 687
acylamine and hydroxy or thiol, **712**
acylamine and imine, 688
acylamine and ketone, **680**
acylamine and lactam carbonyl or thiocarbonyl, **726**
acylamine and methylene, **733**
acylamine and phosphorane, **760**
acylamine and nitrile, 692
acylamine and nitroso, 756
acylamine and oximine, 691
acylamine or acylimine and ring-C, **767**
acylamine and ring-N, **775**
acylamine and thiocyanate, **781**
acylamine and thioureide, **778**
acylamine and trimethylsilylmethyl, 734
acylhydrazine or thioacylhydrazine and amine, 686
acylhydrazine and carboxylic acid, 694
acylhydrazine and lactam carbonyl, 731
acylhydrazine and phosphorane, 764
acylhydrazine and ring-C, **767**
acylhydrazine and ring-N, 775
acyloxy and azide, 1019
acyloxy and methylene, 1272
acyloxy or acylthio and phosphorane, **760**
acylhydroxyamine and ring-C, 1056
aldehyde and amine, **670**
aldehyde and azide, **790**
aldehyde and carboxylic acid, **800**
aldehyde and carboxylic ester, **802**
aldehyde and halogen, **809**
aldehyde and hydroxy, **817**
aldehyde and ketone, **827**
aldehyde or ketone and lactam carbonyl or thiocarbonyl, **1308**
aldehyde and methylene, 839
aldehyde and nitrile, **845**
aldehyde and nitro, **848**
aldehyde or ketone and phosphorane, **852**
aldehyde and ring-C, **859**
aldehyde and ring-N, **872**
aldehyde and triazene, 790
aldoxime and alkyne, 787
aldoxime and amine, 673
aldoxime and lactam carbonyl, 1308
alkene and amine or imine, **884**
alkene and azide, **884**
alkene and carboxamide, 891
alkene and carboxylic acid, 891
alkene and diazonium, 889
alkene and halogen, 893
alkene and hydroxy or thiol, **898**
alkene and iminophosphorane, **888**
alkene and ketone or thioketone, **786**
alkene and lactam carbonyl, **904**
alkene and lactone carbonyl, 907
alkene and methylene, 920
alkene and nitro, 889
alkene and ring-C, 908
alkene and ring-N, **910**
alkene and thioether, 898
alkyne and amine, 884
alkyne and azide, **884**
alkyne and carbamate, 884
alkyne and ether, 899
alkyne and halogen, 893
alkyne and hydroxy, 899
alkyne and methylene, 911
alkyne and nitrile, 891
alkyne and oxime, 787
alkyne and ring-C, **910**
alkyne and ring-N, 913
amidine and amine, **921**
amidine and carboxylic ester, 924
amidine and lactam carbonyl, 727
amidine and methylene, 922
amidine and methylthio, 922
amidine and nitrile, 923
amidine and nitro, 754
amidine and ring-C or ring-N, **927**
amidroxamine and amine, 690
amine and alkene, **733**
amine and *N*-acylsulphonamide, 688
amine or imine and azide, 933
amine and azo, 932, 934
amine and carbamate, **685**
amine and carboxamide, 935
amine or imine and carboxylic acid, **940**

Cyclization of functional groups (*cont.*)
amine and carboxylic ester, **946**
amine and cyanosulphonamide, 749
amine and disulphide, 718
amine and dithiocarbamate, 718
amine and enamine, **957**
amine and epoxide, 697
amine and ether or thioether, **697**
amine and halogen, **701**
amine and hydrazide or hydrazine, **959**
amine and hydrazone, **964**
amine and hydroxy or thio, **712**
amine and *N*-hydroxysulphonamide, 779
amine and imine, **965**
amine and iminoether, 740, 748
amine or imine and ketone, **969**
amine or imine and lactam carbonyl or thiocarbonyl, **726**
amine or imine and nitrile, **739**
amine or imine and nitro, **751**
amine and nitroso or *N*-oxide, **756**
amine and oxime, 958
amine and oximinoacylamine, 673, 934
amine or imine and phosphorane, 761, 763
amine and ring-C, **980**
amine and ring-N, **997**
amine or imine and sulphonamide, **778**
amine and sulphoximide, 780
amine and thiocarbamate, 699
amine and thiocyanate, **781**
amine and thioureide, 778
aminoxy and ring-C, **984**
arylidenimino and lactam carbonyl, 731
azide and azo, 1015
azide and carboxamide, 1019
azide and carboxylic ester, 1019
azide and halogen, 1018
azide and imide, 936
azide and imine, 1031
azide and ketone, 791
azide and methyl or methylene, **1020**
azide and nitrile, 1019
azide and nitro, 1016
azide and ring-C, **1023**
azide and ring-N, **1030**
azide and sulphimide, 1026
azide and thioether, 1021
azo and carbamate, 1033
azo and ring-C, 1028
carbamate and carboxamide, 1042
carbamate and carboxylic acid, 1042
carbamate and carboxylic ester, **1042**
carbamate and dithiocarbamate, 718
carbamate and halogen, **708**
carbamate and nitrile, **1041**
carbamate and ring-C, 1035
carbamate and ring-N, 1036
carbamate and trimethylsilylmethylamine, 782
carbamoyl azide and ring-N, 1032
carbodiimide and nitro, 1016
carboxamide and carboxylic ester, 1044
carboxamide and disulphide, 1049
carboxamide and diazonium salt, 1046
carboxamide and epoxide, 1052
carboxamide and ether or thioether, 1051
carboxamide and halogen, **1064**
carboxamide and hydroxy, **1051**
carboxamide and iminophosphorane, 936
carboxamide and ketone, **794**
carboxamide and lactam carbonyl, **1081**
carboxamide and methylene, 1091
carboxamide and nitrile, 1042
carboxamide and phosphonium salt, 935
carboxamide and ring-C, **1056**
carboxamide and ring-N, **1060**
carboxamide and sulphenamide, 1118
carboxamide and sulphinyl, 1119
carboxamide and ureide, **1098**
carboxylic acid chloride—*see* acyl chloride
carboxylic acid and ether, **1076**
carboxylic acid and hydrazone, 1075
carboxylic acid and hydroxy or thiol, **1076**
carboxylic acid and ketone, **799**
carboxylic acid and nitrile, 1095
carboxylic acid and nitro, **1096**
carboxylic acid and ring-C, **1105**
carboxylic acid and ring-N, **1112**
carboxylic acid and sulphamide, 1121
carboxylic acid and sulphonamide, 1120
carboxylic acid and thiocyanate, 1121
carboxylic acid and thioureide or ureide, **1098**
carboxylic acid anhydride and ether, 1078
carboxylic ester and cyanate, 1085
carboxylic ester and ether or thioether, **1076**
carboxylic ester and halogen, **1063**
carboxylic ester and hydrazine, **1073**
carboxylic ester and hydrazone, **1073**
carboxylic ester and hydroxy, 1077

Cyclization of functional groups (*cont.*)
carboxylic ester and hydroxylamine, **1073**
carboxylic ester and iminophosphorane, **853**
carboxylic ester and isothiocyanate, 1083
carboxylic ester and ketone, **799**
carboxylic ester and lactam carbonyl, 1084
carboxylic ester and methylene, **1086**
carboxylic ester and nitrile, **1094**
carboxylic ester and nitro, **1097**
carboxylic ester and hydroxyureide, 1074
carboxylic ester and phosphorane, **853**
carboxylic ester and ring-C, **1105**
carboxylic ester and ring-N, **1112**
carboxylic ester and sulphenamide, 1119
carboxylic ester and sulphinyl, 1120
carboxylic ester and sulphonylamine, 696
carboxylic ester and thiocarbamate, 1043
carboxylic ester and thioureide or ureide, 1102
chlorodithioformate and ring-C, 785
diacyl chloride, 1158
dialdehyde, **827**
1,2-diamine, **1123**
1,3-, 1,4- or 1,5-diamine, **1142**
diazide, 1159
diazo and halogen, 1148
diazo and ring-C or ring-N, 1150
diazonium and hydrazide, 1046
diazonium and hydrazone, 1156
diazonium and methylene, 1148, 1149
diazonium and nitrile, **1045**
diazonium and ring-C, **1150**
diazonium and ring-N, 1154
diazonium and thioamide, 1046
diazonium and thiocarboxylic acid, 1034
dicarbohydrazide, 1160
dicarboxamide (cyclic), 1159
dicarboxylic acid, **1158**
dicarboxylic acid anhydride, 1161
dicarboxylic acid imide, **1159**
dicarboxylic ester, **1160**
diether, **1171**
dihalogen, **1163**
dihydroxy, **1170**
di(hydroxyamino), 1145
di(iminophosphorane), 1162
diketone, **827**
di(methylene), 1194
dinitrile, **1177**
dinitro, **1177**
diol—*see* dihydroxy
dioxime, 832, 833
di(ring-C), **1181**
di(ring-N), **1181**
di(sulphenyl chloride), 1174
di(sulphone), 1173
di(sulphonyl chloride), 1174
dithioacetal and sulphonylamine, 818
dithioacetal and ketone, 834, 837
dithiocarboxylic acid and ring-C, **1105**
dithiol, 1172
di(trimethylsilylimine), 833
enamine and carbonyl, **1200**
enamine and non-carbonyl, **1204**
ether or thioether and ketone, **806**
ether or thioether and ketoxime, 808
ether or thioether and methylene, 1208
ether or thioether and nitrile, **1047**
ether or thioether and ring-C, **1210**
ether or thioether and ring-N, 1209
formazan and ring-N, 929
guanidine and nitro, 752, 755
halogen and acyloxy or hydroxy, **1213**
halogen and ether or thioether, **1213**
halogen and hydrazide, 1065
halogen and hydrazine, **707**
halogen and hydrazone, 704
halogen and hydroxylamine diacetate, 703
halogen and isocyanate or isothiocyanate, 704, 1220
halogen and ketone, **809**
halogen and lactam carbonyl, 1219
halogen and methylene, **1222**
halogen and nitrile, **1063**
halogen and nitro, **1233**
halogen and oxime, 1218
halogen and ring-C, **1222**
halogen and ring-N, **1236**
halogen and semicarbazide, 708
halogen and sulphinyl, 1220
halogen and sulphonamide, 1221
halogen and thiocyanate, 1220
halogen and thiol or thioether, **1214**
halogen and ureide, **704**
hydrazide and nitro, **1244**
hydrazine and nitro, **1244**
hydrazine and ring-C, **1247**
hydrazine and ring-N, **1247**
hydrazone and ketone, 797
hydrazone and nitro, **1244**

Cyclization of functional groups (*cont.*)
hydrazone and phosphorane, 761
hydrazone and ring-C, **1260**
hydrazone and ring-N, **1262**
hydroxamic acid and hydroxy, 1269
hydroxamic acid and ring-C, 1057
hydroxy and hydrazone, 1267
hydroxy or thiol and imine, 1294
hydroxy or thiol and ketone, **817**
hydroxy and lactam carbonyl or thiocarbonyl, **817**
hydroxy or thiol and methylene, 1272
hydroxy or thiol and nitrile, **1047**
hydroxy and nitro, **1276**
hydroxy and nitroso, 1277
hydroxy or thiol and oxime, **1269**
hydroxy or thiol and phosphonium, 762
hydroxy or thiol and phosphorane, **762**, 1301
hydroxy and ring-C, **1280**
hydroxy or thiol and ring-N, **1276**
hydroxy and sulphonamide, **1049**
hydroxy and thioureide, 1270
hydroxyamino and ring-C, 991
imidate and methylene, 733
imine and methylene, **733**
imine and ring-C or ring-N, **1294**
isocyanate and nitrile, **1082**
isocyanate or isothiocyanate and ketone, **792**
isocyanate or isothiocyanate and ring-C or ring-N, **1303**
isocyanide and methylene, 1093
ketone and ketoxime, 831
ketone and lactam carbonyl or thiocarbonyl, **1307**
ketone and methylene, 839
ketone and nitrile, **845**
ketone and nitro, **848**
ketone and nitroso, **848**
ketone and phosphorane, 855
ketone and phosphorodiamidothioate, 854
ketone and ring-C, **859**
ketone and ring-N, **872**
ketone and sulphonamide, 795
ketone and sulphoximine, 808
ketone and thiocarbamate, 797
ketone and thioether, 807
ketone and thioketal, 834
ketone and thioureide or ureide, 792
ketone and tosylhydrazone, 681
ketone and trichloromethyl, 814
ketoximine and lactam carbonyl, 1308
lactam carbonyl or thiocarbonyl and methylene, **904**
lactam carbonyl or thiocarbonyl and nitrile, 1082
lactam carbonyl or thiocarbonyl and ring-C, **906**
lactam carbonyl or thiocarbonyl and ring-N, **872**
lactam carbonyl and thiosemicarbazide, 1310
lactam carbonyl or thiocarbonyl and thioureide or ureide, **1096**
lactone and ketone, 793
methylene and nitrile, **1086**
methylene and nitro, **1312**
methylene and nitroso, 1315
methylene and ring-C, **1316**
methylene and ring-N, **1318**
methylene and ring-S, 1147
nitrile and lactam thiocarbonyl, 1082
nitrile and nitro, **1104**
nitrile and oxime, 1074
nitrile and ring-C or ring-N, **1055**
nitrile and sulphamoyloxy, 1121
nitrile and sulphonamide, 1120
nitrile and thioether, 1048
nitrile and thiol, 1082
nitrile and ureide, 1098
nitro and phosphorane, 1332
nitro or nitroso and ring-C, **1329**
nitro and ring-N, 1331
N-oxide and ring-C, 1329
oxime and ring-N, 1261
phosphorane and ring-C or ring-N, 1293
phosphorane and thiocarboxylic ester, or thiocyanate, **853**
phosphorane and trithiocarbonate, 854
ring-C and ring-N, **1335**
ring-C and sulphimide, 1026
ring-C and sulphinamide, 912
ring-C and sulphonamide, 1057
ring-C and sulphonylimino, **768**
ring-C and sulphonylazide, 1029
ring-C and thioacylamine, 767
ring-C and thioacylhydrazine, 772
ring-C and thiocarboxamide, **1059**
ring-C and thioketone, 865
ring-C and thioureide or ureide, **1036**
ring-C and trimethylsilyl, 1302
ring-C and *N*-ylide, 768

Cyclization of functional groups (*cont.*)
ring-N and di- or tri-thiocarbonate, 1117
ring-N and ring-S, 1339
ring-N and sulphenamide, 776
ring-N and sulphimide, 929
ring-N and thiocyanate, 782
ring-N and thioureide or ureide, 1036
sulphonamide and ureide, 1103
thiocarboxylic acid and methylene, 1090
thioether and ureide, 1048
Cyclizing reagents, acetals, 862, 881, 912, 1003, 1058, 1134
acethydrazide, 875
acetic anhydride–perchloric acid, 874
acetic formic anhydride (R), 718, 923, 924, 1275
radiolabelled, 923
acetonitrile–aluminium chloride, 1058
acrylonitrile, 743
(acylamino)amidroxamine, 691
3-alkoxyacroleins, 1259
alkynes (R), 707, 736, 877, 882, 941, 962, 982, 998, 1009, 1012, 1039, 1055, 1108, 1147, 1210, 1214, 1224, 1230, **1288**, 1296, 1318, 1332
allenes, 863, 882, 917, 949, 991, 1004, 1154, 1297
amidines, **807, 919**, 974, 1068, 1203
amidrazones, 925
aminoacetals (R), 674
3-aminoacroleins, 990, 1284
3-aminocrotonic ester or nitrile, 1003, 1290, 1325
2-aminonitrile, 723
N-aminopyridinium salts (R), 747, 768
5-aminotetrazole, 835
ammonium acetate, 801, 816, 827, 835, 842, 881, 973, 1059, 1201, 1284, **1309**, 1320
ammonium carbonate, 801, 816
ammonium thiocyanate, 1239
aryl chloroformate, 688
aroylhydroxamoyl chloride, 878, 1210
arylazochloroacetyl chloride, 878
arylidenemalononitrile, 742, 878, 908, 1055, 1264
azadienes—*see* Heterodienes
azides (R), 941, 1015, 1018, 1186
baker's yeast, 1121—*see also* catalase
barium manganate as dehydrogenator, 1294
benzohydroxamic acid chloride, 1338
benzonitrile oxide (R), 983
benzoyl isothiocyanate, 930
benzoylcyanamide, 951
biguanides, 829
bis(acetonitrile)dichloropalladium, 884
1,3-bis(methoxycarbonyl)-*S*-methylisothiourea, 937, 1128
bis(methylthio)methylenemalonic acid derivatives, 1318
bis(trichloromethyl) carbonate ('triphosgene'), 715
bis(trimethylsilyl)sulphurdiimide, 1176
boronic acids, 705, 812, 1067
bromoacetyl bromide, 978, 1218
bromoacetaldehyde acetal, 1053
bromoacetonitrile, 1105
2-bromoacrylic acid or its nitrile, 877
bromochloromethane, 1171
bromonitromethane, 1048
(+)-10-camphorsulphonic acid, 727
carbenes, 862, 872, 1151
carbodiimides, 698, 1079
carbon disulphide (R), 763, 1001, **1143**, 1221, 1231, 1256, 1274, **1298**
carbon suboxide, (R), 1011
carbonyl sulphide, 715, 936
carbonyldiimidazole, 820, 939, 1098, 1129, 1155, 1187, 1256
catalase (R), 736, 1044, 1098
chalcones, 918, 1137
chloral hydrate–hydroxylamine, 713
chloroacetyl chloride, 1000, 1196
2-chloroacrylonitrile, 1266
chlorobenzaldoximes, 1127
chlorocarbonyl isocyanate (R), 1014
chlorocarbonylsulphenyl chloride, 985
3-chlorocrotonic acid, 882
chlorodiphenyl phosphate, 1150
chloroformamidine hydrochloride, 951
ω-chlorohydrazide, 1243
chloroiminium chloride, 681
2-chloronitrobenzene, 1174
N-chlorosuccinimide, 905, 1278
chlorosulphonyl isocyanate (R), 746, 825, 974, 1183, 1268
chlorothioformic acid chloride, 1001
phenyl ester, 837
2-chloro-1,1,1-triethoxyethane, 714
crotonaldehyde, 1137
crotonic anhydride, 738
cyanamide, 1042, 1161
1-cyano-2-phenylethyne, 1332

Cyclizing reagents (*cont.*)
cyanoacetamide (R), 804, 822, 823, 834
cyanogen, 942
cyanogen bromide, 729, 757, 780, 943, 1126, 1171, 1253
cyanohydrins, 714
cyanothioacetamide (R), 671, 821, 822, 905
DBN, 672
DBU (R), 803, 824, 852, 855, 880, 916
DEAD, 941, 90, 1039, 1211
(diacetoxyiodo)benzene (R)—*see* Iodo compounds, hypervalent
diallyl acetylenedicarboxylate, 1184
diallyl sulphoxide, 1196
1,3-diazahexatriene, 1340
diazoalkanes, 1106, 1185
dibenzoylacetylene, 1012, 1055
o,α-dibromophenylhydrazone, 1168
dibromotriphenylphosphorane, 952
1,2-dicarbonyl compounds, 1132, 1133, 1136, 1282
1,3-dicarbonyl compounds, 722, 908, **953**, 993, 994, 1005, 1008, 1058, **1138**, 1143, 1209
dichloroacetic acid, 1300
dichloroketene (R), 1330
dichloromethyl methyl ether, 1196
N-(dichloromethylene)carboxamides, 1013
(dichloromethylene)dimethylammonium salts (R), 718, 1130, 1141
1,1-dichloro-2-nitroethene, 714, 1129
dichlorosulphine, 1152
diethoxymethyl acetate, 684, 744
diethyl azodicarboxylate, 998, 1175, 1195, 1265
diethyl chlorophosphate, 874
diethyl mesoxalate, 1135, 1257
diethyl methyloxalacetate, 1336
diethyl oxalacetate, 1336
diethyl or dimethyl oxalate, 723, 921, 925, 939, 1012, 1325
diethyl oxalopropionate, 1336
di-isopropyl peroxydicarbonate, 1296
di-isopropylethylamine, 977
diketene, 918, 930, 983, 984, 1032, 1138, 1290, 1325, 1341
dimedone, 877
dimethyl allenedicarboxylate, 949
dimethylacetamide diethyl (or dimethyl)acetal, 973
dimethylketene, 1198, 1341
dimorpholinophosphoric chloride, 874
O-(2,4-dinitrophenyl)hydroxylamine, 741
diphenyl cyanocarbonimide, 1128
diphenyl ether (as reactant), 1188
diphenyl sulphoacetate, 823
diphenyl thiomalonate, 724
diphenylketene, 865, 1031
N-diphenylphosphinyl-*N*′-methylpiperazine, 1124, 1142
diphenylphosphine chloride, 1124
diphenylphosphorylation, 872
diphosphorane, 1162
disuccinimido carbonate, 716, 719, 1129
3,3-dithioacroleins, 1234
dithioate esters, 713, 1125
dithiocarbamate, 718, 951, 1127
ditosyldiazomethane, 864
DMAD, 707, 721, 779, 877-879, 882, 917, 918, 949, 962, 995, 1012, **1030**, 1035, 1055, 1111, 1135, **1182**, **1258**, **1317**, 1327, **1335**
DMFDEA, 924, 925, 937, 949, 1060, 1323
DMFDMA, 966, 1085, 1197, 1296
enamines, 821, 908, 929, 957, 985, 990, 995, 1004, 1154, 1173, 1258, 1300, 1304, 1330
epoxides, 878, 885, 998
epoxyaldehydes, 1133
3-ethoxyacrolein diethylacetal, 993
3-ethoxyacrolein, 990
ethoxycarbonyl isothiocyanate (R), 951, 1326
N-ethoxycarbonylthioamide, 717
N-(ethoxycarbonyl)thiocarboxamides, 1125
ethoxyethene, 1297
ethoxymethylenecyanoacetate, 1209, 1323
ethoxymethylenemalonate, 909, 965, 1009, 1324
ethoxymethylenemalononitrile, 1003, 1323
ethyl (ethoxymethylene)cyanoacetate, 1323
ethyl acetimidate, 998
ethyl benzoylacetate, 1290
ethyl 2-benzoylglycinate, 713
ethyl carbamate, 974
ethyl carbazate, 875, 1239
ethyl 2-chloroacetoacetate, 721, 863, 1008

Cyclizing reagents (*cont.*)
ethyl or phenyl chloroformate, 831, 938, 943, 963, 837, 993, 1128
ethyl cyanoacetate, 673, 967
ethyl cyanoformate *N*-oxide, 1334
ethyl glycolate, 1065
ethyl glyoxalate diethylacetal, 1134
ethyl hippurate, 713
ethyl isocyanoacetate, 874
ethyl lactate, 1065
ethyl pyruvate, 1250, 1281
N-ethylmorpholine, 783
formaldehyde, 939, 943, 1141
formamide, 744, 847, 974
N-methyl-, 745
formamidine, 1161
formamidoacetonitrile, 1238
formic acid, 718, 854, 1143
glycidylaldehyde, 998
glyoxals, 1000, 1132, 1134
guanidine, 750, 807, 1068
α-haloacetals, 881, 912
α-haloaldehydes and -ketones (R), 723, 731, 839, 877, 878, 981, 997, 998, 1103, 1215, 1243, **1294**, 1309, 1317
α-halo-acids or -esters, 721, 821, 878, 1105, 1112, 1134, **1163**, 1279
α-halocarboxamides, 818
α-halohydrazones, **1168,** 1238
halomalondialdehydes (R), 1003
heterodienes (R), 869, 1061, 1111, 1191
hexamethylenetetramine (hexamine) (R), 815, 816
hydrazide, 759
hydrazidoyl halide (R), 1185
hydrazine, ring enlargement of thiophenes, 947
hydrazines, 728, 740, 745, 800, 804, 810, 814, 829, 834, 846, 947, 969, 1019, 1042, **1049**, 1065, 1068, 1147, 1159, **1162**, 1204, 1234, **1248**, 1308
hydrazoic acid, 1118
hydrazones (R), 756, 836, 1194, 1233, 1245, **1247**, **1261**
hydroxamoyl chlorides, 717, 1189, 1210
hydroxylamines, 768, 795, 799, 801, 831, 971, 972, 983, 991, 1145, 1201, 1256, 1268, 1277
hydroxylamine-*O*-sulphonic acid, 1113, 1268
imidoyl chlorides, 728, 1214
iminium halides, 681, 718, 1253
imino-ester or -ether, 942, 951
iodo compounds, hypervalent (R), 753, 1134, 1145, 1320
iron (cyclopentadienyl)hexafluoro-phosphate complexes, 813
iron and acid, 958, 977, 1177
iron(II) sulphate and ammonia, 850
isatoic anhydride (R), 907, 925, 1042, 1161
isocyanates (R), 748, **761,** 828, 856, 861, 881, 887, 925, 931, **933**, 970, 1014, 1075, 1082, 1154, 1155, 1183, 1187, 1226, 1229, **1295**, 1304, 1326, 1333, 1339
isothiocyanates (R), 792, 856, 877, 881, 930, 951, 995, 1041, 1082, **1295**, 1303
ketene acetals, 913, 1154
ketene dithioacetals (R), 675, 834, 910, 1004
ketenes, 725, 800, 836, 865, 1140
lactams (R), 724, 727, 1307
lactim ethers, 1211
lactones (R), 742, 1307
Lawesson's reagent, 727
Lewasorb ion-exchange resin, 1206
lithium bis(trimethylsilyl)amide (LBA), 1087
lithium 2-(dimethylamino)ethylamide, 1228
lithium 4-methylpiperazide, 1228
lithium *N*-isopropylcyclohexylamide, 950, 1079
LTA, 927, 928, 933, 961, 1017, 1036, 1159, 1188, 1262, 1263, 1266, 1269
maleic anhydride, 913
malondiamide, 804, 1062
malonic acid derivatives, 743, 818, 823, 879, 880, 948, 973, 990, 992, 1003, 1059, 1076, 1109, 1110, 1140, 1203, 1211, 1253, 1289, **1323**
malononitrile (R), 672, 730, 787, 788, 906, 908, 967, 1070, 1165, 1184
malonyl dichloride, 1140, 1186
mercury(II) chloride, 674, 778
mercury(II) oxide (R), 743, 778
3-methoxyacrylic ester or nitrile, 1318
methyl acrylate, 768
methyl diethoxyacetate, 1197
methyl fluorosulphonate, 1305
methyl propiolate, 1327
2-methylacrolein, *NN*-dimethylhydrazone, 1190

Cyclizing reagents (*cont.*)
N-methylformamide, 745, 1340
methylmalonyl chloride, 1186
NBS as antioxidant, 1159
NBS-sulphuric acid, 905
nitrenes (R), **1020**, 1024, 1028
nitrile-HCl (R), 743, 942, 1056, 1070, 1178
nitrile imine, 1338
nitrile oxides (R), 995, 1000, 1189
nitriles (R), 713, 730, 1003, 1175, 1285
unsaturated (R), 742, 1004
with sulphur (R), 1088
nitrilium salt, 772
nitroalkenes (R), 821
nitro group displacement (R), 753, 1101, 1277
4-nitrophenyl chloroformate, 688
nitrosation, 1265
nitrosobenzene, 1189, 1272
N-nitrosodimethylamine, 1262
β-nitrostyrene, 1129
nitrosyl chloride, **1154**
orthoesters, 714, 720, 735, 737, 747, 779, 875, 924, 960, 961, 967, 973, 1038, 1049, 1141, 1142, 1145, 1197, 1246, **1253, 1261, 1274**, 1286
oxalyl dichloride, 707, 945, 967, 1282, 1320, 1337
oxazolinone as lactone, 716, 1126, 1135
perchloroynamine, 1304
perfluorodiketones, 1132
phase transfer catalysis, 1016
phenoxycarbonyl isocyanate, 1014
phenyl chloroformate, 837
phenyl isocyanate dichloride, 925, 1171
phenyl vinyl sulphoxide, 1273
1,2-phenylenediamine, 805, 815, 835, 1167, 1173, 1310
phenylglyoxal, 1131
O-phenylhydroxylamine, 984, 1258
N-phenylmaleimide, 1196
3-phenylsulphonylbut-3-en-2-one,
4-phenyl-1,2,4-triazole-3,5-dione, 1273
phosgene or thiophosgene, 719, 722, 771, 801, 815, 819, 837, 887, 931, 938, 972, 1007, 1079, 1085, 1143, 1144, 1155, 1187, 1205, 1264
phosphonium salts, 762, 820, 935
phosphoranes (R), 718, **760**, 809, 819, 837, **852**, 873, **885**, 899, 936, 953
phosphorus pentasulphide, 728, 729, 815, 830, 1310
polyphosphoric acid, ethyl ester—*see* PPE
trimethylsilyl ester, 704
potassium cyanide, **751**
potassium ethylxanthate, 719, 1129
potassium thiocyanate, 930
PPE, 713, 874, 1000, 1051, 1110
prop-2-ynyltriphenylphosphonium bromide, 716
propiolic acid nitrile, 1318
pyridine (as reactant), 703, 840, 1163, 1231
pyridine hydrochloride as demethylating agent (R), 819, 899, 1053
pyridinium salts, *N*-substituted (R), 839, **910**, 946, 1200, 1277
pyruvaldehyde, 1132
pyruvic acid, 1136
samarium di-iodide, 894
semicarbazide, 834, 1239
silver(I) perchlorate, 704
sodium borohydride–nickel boride–hydrazine hydrate, 957
sodium imidazolide, 820
sodium methylsulphinylmethanide, 954
sodium hydrogen sulphide, 1205
sulphamide, 837
sulphenyl halides, 1001
sulphonium salts, 738, 863
sulphonylnitrene, 1028
sulphur, replacement in ring by hydrazine, 947
sulphur dichloride, 833, 1131, 1225, 1301
sulphur monochloride (disulphur dichloride), 914, 986, 1131
tetracyanoethylene (TCNE) (R), 882, 1003, 1025, 1035
tetrahydrothiopyran-4-one or its 1,1-dioxide, 1138
tetrasulphur tetranitride, 1216, 1226, 1283
thioacetic acid, 944
thioamides (R), 717, 812, 1071, 1169
thiocarbamoyl isothiocyanate, 995
thiocarbonyldiimidazole, 1273
thiosemicarbazide hydrochloride, 750, 926
thiourea, 812, 865, 937, 807, 1104, 1165, 1166
titanium(IV) chloride (R), 1091
tosyl azide, 913
1,3,5-triazine, 942

Cyclizing reagents (*cont.*)
tributyltin derivatives, 894, 895
trichloroacetyl isocyanate, 1268
trichloroethyl or trichloromethyl chloroformate, 685, 1007, 1155
trichloromethanesulphenyl chloride, 986
triethyloxonium tetrafluoroborate, 912, **1304**,
triflic acid or anhydride (R), 1124
trifluoroacetic anhydride, 718, 1076, 1110, 1320, 1299
1,1,1-trifluorobut-3-yn-2-one, 988
triformamidomethane, 937
trimethylaluminium, 956
trimethylsilyl cyanide, 1294
trimethylsilyl derivatives, 781, 943, 1161, 1272
trimethylsilyl polyphosphoric ester, 704
3-trimethylsilyloxyacroleins, 1216
triphenylphosphine thiocyanate, 975
triphenylmethyl perchlorates, 869
urea, 746, 788, 807, 1000, 1070, 1071, 1143, 1161
δ-valerolactone, 742
Vilsmeier reagent (R), 703, 730, 735, 745, 756, 771, 916, 1058, 1130, 1228, 1232, **1321**
vinyl azides, 1186
N-ylide, 747, 768, 807, 911, 946, 1035, **1054**, 1105, 1184, 1209, 1302, **1317**, 1337
ynamine or ynamine, **1304**
zirconocenes, 1225

Demethylation with pyridine HCl (R)—*see* Cyclizing reagents, pyridine hydrochloride
Diazaquinomycin A, 867
Diazotization of heterocyclic primary amines (R), **1148**, **1150**
Dieckmann condensation, **1086, 1151**
Diels-Alder reaction (R)—*see also* Hetero–Diels–Alder reaction, 769, 828, 882, 1111, 1166
Diltiazem (R), 955
Dimroth rearrangement, 745, 946, 1102, 1254, 1262
C,*N*-Diphenylnitrone, 982
Doebner–Miller synthesis, **987**

Electrochemical cyclization, 1103, 1172, 1227
Ellipticene, 678, 1023, 1115
Enamides, photolytic cyclization (R), 900
Enzymes as chemical catalysts, 982, 1044
Evodone, 896

Fischer indolization (R), 980, 984, **1247**, 1328
Fischer–Hepp rearrangement, 1313
Flash vacuum photolysis, 921
Flash vacuum pyrolysis, 1113, 1329
Formulae, alignment and numbering, 666
Friedel–Crafts cyclization, 772, 773, 784, 785, 1107, 1223
Friedländer reaction, 671
Furans, condensed (R), 819

Gattermann aldehyde synthesis, 866

Herz reaction, 986
Heterocycles, quinonoid (R), 717, 788, 803, 806, 811, 964, 970, 1185, 1189, 1194, 1286
Hetero–Diels–Alder reaction (R), 869, 907, 1190, 1191, **1194**, 1297
Hippadine, 1240
Hydralazine, 1257
Hydrazones (R), 1245

Indoles (R), 675—*see also* Benzopyrroles
1-hydroxy- (R), 885
Irradiation—*see* Cyclization, photochemical
Isatin, 803
Isobenzofurans (R), 830
Isocyanates, 6-membered heterocyclic (R), 1082
Isophellopterin, 803
Isoquinoline alkaloids (R), 900
IUPAC nomenclature rules, 666

Japp–Klingemann reaction, 1148, 1149, 1252

Knoevenagel reaction, 804
Knorr reaction, 867
Knorr synthesis of pyrroles, 860
Knorr synthesis of quinolines, 973
Kostanecki reaction, 1274

Lennoxamine, 1113
Lycoricidine, 1228

Madelung synthesis of indoles, 733
Magallanesine, 1085

Mannich reaction, 861
Meldrum's acid and its derivatives (R), 824, 1011, 1112

Nalidixic acid analogues, 1089, 1148
Nenitzescu synthesis, 859
Newman rearrangement, 699
Newman–Kwart rearrangement, 719
Niementowski synthesis of quinazolines (R), 942
Nitroxides (R), 848, 1329
Nomenclature of heterocycles, IUPAC, 666
Nybomycin, 1190

Pechman reaction, **1287**
Peri-annulation, 702, 821, 831, 841, 868, 879, 911, 974, 1000, 1027, 1055, 1109, 1113, 1142, 1182, 1220, 1242, 1244, 1270, 1306, 1337
Phase transfer catalysis, 903, 1185
Phenothiazines (R), 709, 722, 1306, 1334
Phillips synthesis of benzimidazoles, 1124
Photocyclization—*see* Cyclization, photochemical
Pictet–Gams synthesis, 770
Piloty–Robinson reaction, 1251
Pomeranz–Fritsch reaction, 675, 677, 678
Pschorr reaction, **1151**
Pteridines (R), 758, 923, 990, 1131
Pyrazoles, fused (R), 703, 741, 1164
Pyrazolopyrimidines (R), 763, 958
Pyrido[4,3-*b*]indoles (γ-carbolines) (R), 1058
Pyrimidine *N*-oxides (R), 1329
Pyrimidines (R), hydrazino-, 1259
Pyrrolizines (R), 873, 1054
Pyrrolo[1,2-*a*]azepines (4-azaazulenes) (R), 1206
Pyrrolopyrimidines (R), 1055

Quinones, heterocyclic—*see* Heterocycles, quinonoid
Quinoxalines (R), 1133
Quinoxalinediones, tautomerism (R), 1134

Reissert compounds (R), 768, 900, 1041
Ring system names and their numbering, 666

Schmidt reaction, 820, 866
Sincitine, 900
Skraup synthesis of quinolines, 989
Sonification (R), 886, 1229
Stereochemistry of vinyl azides in cyclization, 888
Stereoselectivity during cyclization, 707, 893, 965, 1087, 1106, 1196, 1225
Sulphur, extrusion in cyclization, 912
Suzuki reaction, 1240

TCDD, 1174
1,2,*x*-Thiadiazines, fused (R), 779, 1034, 1315
Thienamycin, 873, 1152
Thioridazine, 709
Thorpe cyclization (R), 1082
Thorpe–Ziegler reaction (R), 1092, 1094
Timmis reaction, 758
1,2,4-Triazoles, fused (R), 776, 1238, 1254, 1262
Trimethylsilyl group, utility of, 781, 1175
Tschitschibabin synthesis, 998
Tyrosinase, 982

Ullman-type reaction, 1218
Ultrasonification (R), 886, 1229
Ungeremine, 1228, 1240

Van Alphen–Hüttel rearrangement, 1330
Vilsmeier reaction (R)—*see* under Cyclizing reagents

Willgerodt reaction, 942
Wittig reaction, 820, **853**, 1326—*see* Cyclizing reagents, phosphoranes,
Wolff–Kinshner reduction, 829

N-Ylides, 747, 768, 807, 911, 1035, **1054**, 1105, 1184, 1209, 1302, **1317**, 1337

Zirconium complexes (R), 1225

GENERAL INDEX

Index of Ring Systems

The following abbreviations are used in this index (see p. 663): azep, azepine; im, imidazole; Nox, N-oxide; O_2, dioxide; ox, oxazine; oxdazol, oxadiazole; oxep, oxepine; p, pyridine; paz, pyrazine; pdaz, pyridazine; pm, pyrimidine; prr, pyrrole; pzol, pyrazole; tetrazep, tetrazepine; thzn, thiazine; thzol, thiazole; triazep, triazepine; trzn, triazine.

Page numbers in bold indicate that the ring system is mentioned on later pages of the chapter.

Acenaphtho[1,2-*c*][1,2,5]thiadazole, 833
Acridine, 973, 1058, 1192
1,8-Acridinedione, 1285
1-Acridinone, 1058
9-Acridinone, 699, 950
1-Azabicyclo[3.2.0]heptan-3,7-dione (prr3one), 1087
1-Azabicyclo[3.2.0]heptan-3-one (prr3one), 1151
1-Azabicyclo[3.2.0]hept-2-en-7-one (prr), 853, 1151
1-Azabicyclo[4.2.0]oct-2-en-8-one (p), 1090, 1153, 1240
2a-Azacyclopenta[*ef*]heptalene (azep), 1341
1-Aza-4-oxabicyclo[4.2.0]oct-2-en-8-one (1,4ox), 1117
1-Aza-4-thiabicyclo[3.2.0]heptan-7-one, 3-thioxo (thzol), 1273
1-Aza-4-thiabicyclo[4.2.0]oct-2-en-8-one (1,4thzn), 1117
Azepino[1,2-*a*]benz**imidazole**, 1313, 1338
Azepino[2,1-*b*]benzothiazole, 1327
Azepino[3,4-*b*]indole-1,3-**dione**, 804
Azepino[2,1,7-*cd*]**indolizine** (p), 1324
Azepino[1,2-*a*]quinazolin-5-one, 1327
Azepino[2,1-*b*]**quinazolin**-7-**one**, 1301
Azocino[2,1-*b*]**quinazolin**-8-**one**, 1301

Benz[*a*]acridine, 672
Benz[*c*]acridine, 1228
2-Benzazepine, 682, 1302
3-Benzazepine, 679, **888**
1-Benzazepin-2-one, 844, 1232
1-Benzazepin-4-one, 982
3-Benzazepin-2-one, 679
3-Benzazocine, 1093
1-Benzazocin-2-one, 1232
3-Benzazocin-2-one, 1093
Benzimidazo[1,2,3-*ij*]benzo[*c*][1,8]-**naphthyridine**-3,10-di**one** (p2one), 879
Benzimidazo[2,1-*b*]benzo**thiazole**, 1234
Benz[*g*]**imidazo**[2,1-*a*]isoindole, 727
Benzimidazo[2,1-*a*]**isoquinoline**, 1193
Benzimidazo[1,2-*b*]**isoquinoline**-6,11-**dione**, 1158
Benzimidazole, 684, 685, 690, 737, 740, 752, 778, 922, 928, 933, 959, 1026, **1124**, 1314
Benzimidazole 3-oxide, 753, 984, 1314, 1331
2-Benzimidazolethione, 779, 1129
2-Benzimidazolone, 685, 1036, 1129, 1314
Benz[*a*]imidazo[4,5-*c*]**phenazin**-2-one, 835
Benz[*h*]**imidazo**[1,2-*c*]quinazoline, 1237
Benzimidazo[1,2-*c*]**quinazoline**,1299
Benzimidazo[1,2-*c*]**quinazoline**-6-**thione**, 1299
Benzimidazo[1,2-*c*]**quinazolin-6-one, 1299**
Benz[*g*]indazole, 970
Benz[*f*]indazol-3-one, 1019
Benz[*b*]indeno[1,2-*c*][1,4]diazepin-12-one, 1138
Benz[*b*]indeno[1,2-*e*][1,4]thiazin-11-one, 723
Benz[*e*]indole, 981
Benz[*f*]indole-4,9-dione, 806

INDEX OF RING SYSTEMS

Benz[*f*]indolizine (prr), 694
Benz[2,3]**indolizino**[8,7-*b*]indole (p), 1192
Benz[*f*]isoquinoline, 770
2,1-Benzisothiazole, 737
1,2-Benzisothiazole 1,1-dioxide, 795, 972, 1057
1λ^4,2-Benzisothiazole 1-oxide, 1118
1,2-Benzisothiazol-3-one, 1049, 1119
1,2-Benzisothiazol-3-one 1,1-dioxide, 1118, 1120
1,2-Benzisothiazol-3-one 1-oxide, 1120
1,2-Benzisoxazole, 831, 1074
2,1-Benzisoxazole, 849, 1332
1,2-Benzisoxazole 2-oxide, 1269
2,1-Benzisoxazole-4,7-dione, 972, 1066
2,1-Benzisoxazol-3-one, 1074
Benzo[*a*]-1,3-benzodioxolo[4,5-*g*]-**quinolizin**-14-**one** (p2one), 900
Benzo[*h*]benzothieno[2,3-*c*][**1,6**]-**naphthyridin**-6-**one**, 1230
Benzo[*c*]cinnoline, 1159
Benzo[*de*]cinnoline, 1270
Benzo[*c*]cinnoline 5,6-dioxide, 1179
1,2-Benzodiazepine, 897
1,3-Benzodiazepine, 1162
1,4-Benzodiazepine, 774, 815
1,5-Benzodiazepine, 750, **1137**
2,3-Benzodiazepine, 797
1,4-Benzodiazepine-2,5-dione, 955, 1052, 1071
1,5-Benzodiazepine-2-thione, 1139
1,3-Benzodiazepin-2-one, 1144
1,3-Benzodiazepin-4-one, 939
1,4-Benzodiazepin-2-one, 791, 857, **977**
1,4-Benzodiazepin-5-one, 976
1,5-Benzodiazepin-2-one, 1139
2,3-Benzodiazepin-4-one, 804, 1044, 1075
2,4-Benzodiazepin-1-one, 976
2,4-Benzodiazepin-5-one, 3-thioxo-, 1071
1,5-Benzodiazocine, 774
1,6-Benzodiazocine, 890
1,5-Benzodiazocin-2-one, 816
1,4,5-Benzodioxazocine, 1162
1,3-Benzodioxin-2,4-dione, 1079
[**1,4**]Benzo**dioxino**[2,3-*c*]pyridazine, 1168
[**1,4**]Benzo**dioxino**[2,3-*d*]pyridazine, 1168
1,3-Benzodioxole, 1172
Benzo[1,2-*b*:5,4-*b*′]**dipyran-4**,6-**dione**, 902
1,3,2,4-Benzodithiadiazine,1302
1,2-Benzodithiol-4-one, 3-thioxo-, 985
1,3-Benzodithiole, 1234, 1283
Benzofuran, 676, 762, 800, 812, 818, 819, 841, 863, 895, 899, 985, 1088, 1171, **1214**, 1256
2-Benzofuranone, 1107, 1152, 1282
3-Benzofuranone, 784, 900, 902, 1215, 1273
4-Benzofuranone, 863, 895, 1171
4-Benzofurazanone, 832
Benzo**furo**[3,2-*c*][1]benzopyran-6-one, 1281
Benzofuro[3′,2′:6,7][1]benzopyrano[4,3-*b*]-**pyrrol**-4-one, 981
Benzofuro[3,2-*e*]-2,1,3-benzo**thiadiazole**, 1284
Benzofuro[3,2-*g*]**furo**[3,2-*c*][1]benzopyran-4-one, 1281
Benzofuro[2,3-*e*]-**1,2-oxazine**,1334
Benzo**furo**[2,3-*c*]pyridazine, 1215
Benzo**furo**[2,3-*b*]pyridine, 1152
Benzo**furo**[2,3-*c*]pyridine, 1256
Benzo**furo**[3,2-*c*]pyridine, 984
Benzofuro[3,2-*c*]**pyridine**, 770
Benzofuro[3,2-*d*]**pyrimidine**, 745
Benzofuro[2,3-*c*]**pyrrole**, 894
Benzo**furo**[3,2-*c*]pyrrolo[3,2,1-*ij*]quinolin-7-one, 1281
Benzo**furo**[2,3-*b*]quinoline, 819
Benzo**furo**[2,3-*b*]quinolin-11-one, 819
Benzofuro[2,3-*c*]**quinolin-6-one**, 743
Benzo[*b*]naphtho[2,3-*e*][1,4]dioxin-6,11-dione, 1174
Benzo[2,3]naphtho[5,6,7-*ij*][1,4]**dithiepine**, 1235
Benzo[*b*][1,8]naphthyridine, 1058, 1067
Benzo[*c*][1,8]naphthyridine, 705
Benzo[*c*][2,7]naphthyridine, 1193
Benzo[*de*][1,6]naphthyridine, 1179
Benzo[*g*]-1,8-naphthyridine, 1285
Benzo[*h*]-1,6-naphthyridine,1304
Benzo[*b*][1,8]phenanthroline, 1109
Benzo[*a*]phenazine, 772
Benzo[*a*]phenothiazine, 1028
Benzo[*g*]pteridine (paz), 759
Benzo[*g*]pteridine-2,4-dione 5-oxide, (pazNox) 1195
1*H*-2-Benzopyran, 765
2*H*-1-Benzopyran, 821, 901, 920, 1216, 1231, 1273, 1287, 1327
4*H*-Benzopyran, 1173
2*H*-1-Benzopyran, 2-aryl-, 821
2*H*-1-Benzopyran, 3-aryl-, 857
4*H*-1-Benzopyran, 2-aryl-, 1286
2*H*-1-Benzopyran-5,8-dione, 788

[1]Benzopyrano[2,3-*d*]**azepin**-11-one, 844
[1]Benzo**pyrano**[3,4-*b*][1,5]benzodiazepine, 1206
[1]Benzo**pyrano**[4,3-*c*][1,5]benzodiazepine, 1287
[1]Benzopyrano[6,5,4-*def*][1]benzopyran, 821
[1]Benzo**pyrano**[3,4-*b*][1,5]-benzothiazepine, 1287
[1]Benzo**pyrano**[3,4-*b*][1,4]benzothiazine, 1287
[1]Benzopyrano[3,4-*b*][**1,4**]benzo**thiazin**-6-one, 722
[1]Benzopyrano[3,2-*e*]-1,4-**diazepine-2,3,5,**11-**tetraone**, 939
[1]Benzo**pyrano**[2,3-*b*]indol-11-**one**, 825
1-Benzopyran-2-one, 804, 815, **822**, 909, 1051, 1077, 1101, 1111, 1212, 1258, 1274, 1288
1-Benzopyran-4-one, 765, 766, 824, 1078, 1206, 1275, 1291
2-Benzopyran-1-one, 802, 1077, 1095, 1095, 1101, 1161
2-Benzopyran-3-one, 803, 837
2*H*-1-Benzopyran-5-one, 870
1-Benzopyran-4-one, 3-aroyl-, 1274
1-Benzopyran-4-one, 2-aryl-, 825, **901**, 1111, 1167, 1217, 1274, 1291
1-Benzopyran-4-one, 3-aryl-, 1275
[1]Benzopyrano[4,3-*c*]**pyrazole**-4-thione, 1185
[1]Benzopyrano[4,3-*c*]**pyrazol**-4-one, 1268
[1]Benzopyrano[2,3-*d*]**pyridazin**-10-one, 815
[1]Benzo**pyrano**[2,3-*b*]pyridine, 870
[1]Benzopyrano[3,4-*c*]**pyridine-2-thione**, 1059
[1]Benzopyrano[2,3-*b*]**pyridin**-5-one, 672, 823
[1]Benzo**pyrano**[3,2-*b*]pyridin-10-**one**, 1061
[1]Benzopyrano[3,2-*c*]**pyridin**-10-one, 842, 1202, 1231
[1]Benzopyrano[3,4-*c*]**pyridin-2-one**, 1059
[1]Benzopyrano[3,4-*c*]**pyridin**-5-one, 769
[1]Benzopyrano[3,2-*c*]**pyridin**-10-one 2-**oxide**, 1202
[1]Benzopyrano[4,3-*d*]**pyrimidine**, 1203
[1]Benzo**pyrano**[2,3-*d*]pyrimidine-2,4-dione, 1231
[1]Benzopyrano[3,2-*d*]**pyrimidine-4,10-dione**, 939
1-Benzopyrano[7,6-*b*]**quinolin**-6-**one**, 1286
[1]Benzo**pyrano**[3,4-*b*]quinoxaline, 1287
[1]Benzopyrano[4,3-*c*][**1,2,6**]**thiadiazine** 2,2-**dioxide**, 837
[1]Benzopyrano[3,4-*d*]-**1,2,3**-**triazol**-4-one, 1026
1-Benzopyran-2-thione, 822, 1274
Benzo[*f*]**pyrido**[1,2-*a*][1,8]naphthyridin-12-one, 1323
2-Benzopyrylium, 869, 1287
Benzo[*h*]quinazoline, 847
Benzo[*f*]quinoline, 1297
Benzo[*g*]quinoline-5,10-dione, **1190**
Benzo[*h*]quinolin-2-one, 796, 1211
Benzo[*b*]quinolizin-6-one (p2one), 868
Benzo[*c*]quinolizin-6-one (p4one), 1068
Benzo[*c*]quinolizin-6-one (p), 1324
1,2,4-Benzothiadiazepine 1,1-dioxide, 1221
1,2,4-Benzothiadiazepin-3-one 1-oxide, 780
2,1,5-Benzothiadiazepin-4-one 2-oxide, 1140
1,3,4-Benzothiadiazine, 1169, 1232
4,1,2-Benzothiadiazine, 710, 772, 1246
1,2,4-Benzothiadiazine 1,1-dioxide, 688, 749, **779**, 925, 1103
4,1,2-Benzothiadiazine 4,4-dioxide, 1157
1,2,4-Benzothiadiazine 1-oxide, 780
3,1,2-Benzothiadiazin-4-one, 1034
1,2,3-Benzothiadiazole, 720, 986
2,1,3-Benzothiadiazole, 833, 1216, 1283
2,1,3-Benzothiadiazole 1-oxide, 833
1,4-Benzothiazepine, 1218
1,5-Benzothiazepine, 750
2,4-Benzothiazepine, 1169
1,4-Benzothiazepin-5-one, 674
1,5-Benzothiazepin-4-one, 955
2,4-Benzothiazepin-5-one, 1071
1,2-Benzothiazine, 1021
1,3-Benzothiazine, 1071
3,1-Benzothiazine, 1221
1,2-Benzothiazine 1,1-dioxide, 1092, 1120
1,3-Benzothiazine-4-thione, 995
1,3-Benzothiazin-4-one, 1071, **1119**
1,4-Benzothiazin-3-one, 723, 1103, 1327
3,1-Benzothiazin-4-one, 944, 1102
1,2-Benzothiazin-3-one 1,1-dioxide, 1120
2,1-Benzothiazin-4-one 2,2-dioxide, 696
1,2-Benzothiazin-3-one 1-oxide, 1120
2,1-Benzothiazin-5-one 2-oxide, 808
Benzothiazole, 699, **713**, 769, 865, 1205, 1220, 1270, 1295
2-Benzothiazolone, 699, 716, 719, 986

Benzo**thiazolo**[2,3-*b*]quinazolin-12-one, 905
[1]Benzothieno[3,2-*b*][**1,5**]benzo**diazepine** 5,5-dioxide, 976
[1]Benzothieno[3,2-*b*]**furan**, 1088
[1]Benzothieno[3,2-*e*]**indolizine** (p), 916
[1]Benzothieno[2,3-*c*][1,6]**naphthyridin**-6-**one**, 1230
[1]Benzothieno[2,3-*c*][1,7]**naphthyridin-6-one**, 1230
[1]Benzothieno[2,3-*c*][1,8]**naphthyridin**-6-**one**, 1230
[1]Benzothieno[2,3-*c*]**pyran-3-one**, 803
[1]Benzothieno[2,3-*c*]**pyrazole**, 1031
[1]Benzothieno[3,2-*c*]**pyrazole**, 849, 1031
[1]Benzothieno[2,3-*d*]**pyridazine**, 835
[1]Benzothieno[2,3-*b*]**pyridine**, 1003, 1152
[1]Benzothieno[3,2-*c*]**pyridine**, 770
[1]Benzothieno[2,3-*b*]**pyridin-4**-one, 950
[1]Benzothieno[2,3-*d*]**pyrimidine**, 1050
[1]Benzothieno[3,2-*d*]**pyrimidine-2,4-dione**, 938
[1]Benzothieno[2,3-*d*]**pyrimidine-2**-thione, 793
[1]Benzothieno[2,3-*c*]**quinolizine** (p), 916
[1]Benzothieno[3,2-*c*][**1,2,6**]**thiadiazin-4-one 2,2-dioxide**, 1122
[1]Benzothieno[2,3-*d*]**thiazole**, 719
[1]Benzothieno[3,2-*d*]-**1,2,3-triazine**, 1046
Benzo[*b*]thiophene, 676, 762, 784, 797, 854, 914, 1048, 1210, 1225, 1273, 1282, 1321
Benzo[*c*]thiophene, 831, 865, 1220
Benzo[*b*]thiophene 1-oxide, 915, 1152
2,3-Benzo[*b*]thiophenedione, 1283
2*H*-1-Benzothiopyran, 765, 920
[1]Benzothiopyrano[4,3-*b*][**1,4**]-benzo**diazepin-7-one**, 805
[1]Benzothiopyrano[4,3,2-*ef*][**1,4**]-benzo**diazepin-3-one** 8,8-dioxide, 1217
[1]Benzothiopyrano[2,3-*b*]**indolizin**-12-one 5-oxide (prr), 1164
1-Benzothiopyran-2-one, 1079
1-Benzothiopyran-4-one, 797, 892, 1231
2-Benzothiopyran-1-one, 1091
1-Benzothiopyran-4-one, 2-aryl-, 1111
[1]Benzothiopyrano[4,3-*b*]**pyran**, 788
[2]Benzothiopyrano[3,4-*e*]pyrazolo[1,5-*a*]-**pyrimidine**, 1005
[1]Benzo**thiopyrano**[2,3-*d*]pyridazin-10-**one**, 815
[1]Benzothiopyrano[4,3-*b*]**pyridine**, 787
[1]Benzothiopyrano[4,3-*b*]**pyridin-2-one**, 787
[1]Benzothiopyrano[4,3,2,-*de*]**quinazoline**, 974
[1]Benzothiopyrano[4,3,2-*de*]**quinazoline 1**,7,7-tri**oxide**, 691
1,2,4-Benzotriazepine, 858
1,3,4-Benzotriazepine, 1022
1,3,5-Benzotriazepine, 1141
1,3,4-Benzotriazepine-2,5-dione, 963
Benzo[*c*][**1,2,5**]**triazepino**[1,2-*a*]cinnoline, 1039
1,2,5-Benzotriazepin-4-one, 1149
1,3,4-Benzotriazepin-5-one, 926, 962
1,2,3-Benzotriazine, 1017
1,2,4-Benzotriazine, 686
1,2,4-Benzotriazine-3,6-dione, 1033
1,2,3-Benzotriazin-4-one, 943, 1046
1,2,3-Benzotriazin-4-one 1-oxide, 1231
1,2,3-Benzotriazin-4-one 2-oxide, 747
Benzotriazole, 934, 941, 1015, 1130, 1234
Benzotriazole 1-oxide, 1245
4-Benzotriazolone, 984
1,3,5,2,4-Benzotrithiadiazepine, 1176
1,2,4-Benzoxadiazine, 890, 1266, 1334
1,3,4-Benzoxadiazine, 710, 1168
2,1,3-Benzoxadiazol-4-one 1-oxide, 832, 833
2,1,3-Benzoxadiazole 1-oxide, 1016
4,2,1,5-Benzoxathiadiazocine 2,2-dioxide, 1218
1,2,5-Benzoxathiazepin-4-one 2-oxide, 725
1,2,3-Benzoxathiazine 2,2-dioxide, 1121
1,5-Benzoxathiepine, 1092
1,5-Benzoxathiepin-3-one, 1092
1,2-Benzoxathiin 2,2-dioxide, 824
1,2-Benzoxathiin-4-one 2,2-dioxide, 1052
1,5-Benzoxathiocin, 789
2,1-Benzoxathiole 1,1-dioxide, 1049
1,3-Benzoxazepine, 738, 1302
1,5-Benzoxazepine-2,4-dione, 724
1,4-Benzoxazepin-5-one, 798, 1053
4,1-Benzoxazepin-2-one, 1218
1,3-Benzoxazine, 720, 1175
1,4-Benzoxazine, 749
3,1-Benzoxazine, 709, 738, 890, 1301
1,3-Benzoxazine-2,4-dione, 1044
3,1-Benzoxazine-2,4-dione, 1085
1,3-Benzoxazine-2-thione, 903
1,3-Benzoxazin-1-ium, 1052
1,3-Benzoxazin-2-one, 826, 903, 1044
1,3-Benzoxazin-4-one, 1052, 1085, 1271

1,4-Benzoxazin-2-one, 721
1,4-Benzoxazin-3-one, 826, 953,1102
3,1-Benzoxazin-2-one,1301
3,1-Benzoxazin-4-one, 695, 943, 1085
3,1-Benzoxazin-4-one, 2-thioxo-, 943
Benzoxazole, 704, **713**, 762, 769, 808, 1019, 1049, 1270, **1294**
2-Benzoxazolone, 716
1-Benzoxepine-3,5-dione, 1092
[1]Benzoxepino[4,5-*e*]**imidazo**[1,2-*c*]-pyrimidine, 1277
Benz[*b*]**oxepino**[7,6,5-*ij*]isoquinoline, 1218
[1]Benzoxepino[5,4-*d*]**pyrimidine**, 847
Biphenyleno[2,3-*c*][1,2,5]oxadiazole, 832
Biphenyleno[2,3-*c*][1,2,5]oxadiazole 3-oxide, 832
Bisbenzopyrano[2,3-*b*:2′,3′-*f*][**1,5**]-**diazocine**-8,16-dione, 1062
Bis[1,2,5]thiadiazolo[3,4-*b*:3′,4′-*e*]**pyrazine** 2,2,6,6-tetraoxide, 1133
Bis[**1,2,5**]**thiadiazolo**[3,4-*f*:3′,4′-*h*]-quinoline, 1227
Bisthiazolo[4,5-*d*:5′,4′-*g*][**1,3**]**diazocine**, 1146

Carbazole, 886, **1024**, 1182
1,4-Carbazoledione, 1182
3-Carbazolone, 1249
Cinnoline, 889
Cinnoline 1-oxide, 754
5,8-Cinnolinedione, 1195
4-Cinnolinone, 763
Cyclohepta[*b*]furan-5-one, 1107
Cyclohepta[*b*]furan-8-one, 899
Cyclohepta[*c*]pyrazole, 807
Cyclohepta[*c*]pyrazol-8-one, 1268
Cyclohepta[*e*]pyrazolo[1,5-*a*]**pyrimidine**, 1005
Cyclohepta[*b*]pyridine, 1296
Cyclohepta[*b*]pyrrole, 810, 861, 1027
Cyclohepta[*d*]pyrrolo[1,2-*a*]benz**imidazole**, 999
Cyclohepta[*d*]pyrrolo[2,3-*b*][1,5]-benzo**diazepine**, 815
Cyclohepta[*d*]pyrrolo[2,3-*b*][1,5]-benzo**diazepin-**12-**one**, 954, 999
Cyclohepta[4,5]pyrrolo[2,1-*c*]-**1,2,4-triazole-3-thione**, 1256
Cyclohepta[4,5]pyrrolo[2,1-*c*]-**1,2,4-triazol-3-one**, 1256
2-Cyclohepta[*b*]thiophenone, 865
Cyclohepta[*e*]-1,2,4-triazine, 837
Cyclohepteno[*d*]thiazole, 1205
Cyclohept[*d*]imidazole, 741, 807
Cyclohept[1,2,3-*cd*]isoindol-2-one, 862
Cyclohept[*d*]isoxazol-8-one, 1268
Cyclohept[*d*]oxazole-2-thione, 972
Cyclohept[*d*]oxazol-2-one, 972
Cyclohept[*d*]oxazol-8-one, 820
Cycloocta[*c*]pyrazole, 1224, 1261
Cyclopenta[*b*][1,4]benzothiazine, 722
Cyclopenta[*d*]pyran-3-one, 837
Cyclopenta[*e*]pyrazolo[1,5-*a*]**pyrimidine**, 1005
Cyclopenta[*b*]pyridine, 855
Cyclopenta[*b*]pyridin-5-one, 987
Cyclopenta[*d*]pyrrolo[2,3-*b*][1,5]-benzo**diazepine**, 815
Cyclopenta[4,5]pyrrolo[2,1-*c*]-1,2,4-**triazole**, 1253
Cyclopenta[4,5]pyrrolo[2,1-*c*]-**1,2,4-triazole -3-thione, 1256**
Cyclopenta[*b*]quinoline, 1058
Cyclopent[*d*]imidazole, 741
Cyclopent[*b*]indole, 1250
Cyclopent[*cd*]indol-2-one, **1183**
Cyclopent[*cd*]isoindole, 1183

2a,5-Diazabenz[*cd*]azulene (paz), 1012
1,3-Diazabicyclo[3.2.0]hept-2-ene (im), 1209
1,3-Diazabicyclo[3.2.0]hept-3-en-7-one (im), 886
1,3-Diazabicyclo[4.2.0]oct-3-en-8-one (pm), 890
1,6-Diazabicyclo[4.2.0]oct-3-en-8-one (pdaz), 856
1,2a-Diazacyclohepta[*ef*]heptalene (azep), 1341
4a,7b-Diazacyclopent[*cd*]indene (pzol), 913
1,2-Diazepino[3,4-*b*]quinoxaline, 1266
Dibenzo[*b,e*][1,4]thiazepine, 996
Dibenzo[*b,f*][1,4]thiazepin-11-one, 945
Dibenzo[*b,e*][1,4]dioxin, 1168, 1174, 1235
Dibenzo[*c,e*][1,2]dithiin 5,5-dioxide, 1174
Dibenzofuran, 864, 1188, 1225, 1277
Dibenzo[*b,d*]pyran-6-one, 1070, 1161
Dibenzo[*c,f*]**pyrazino**[1,2-*a*]azepine-3,4-**dione**, 1013
Dibenzo[*b,f*][1,4,5]thiadiazepine 11,11-dioxide, 1145
Dibenzo[*c,f*][1,2,5]thiadiazepine **5,5-**dioxide, 711
Dibenzo[*b,f*][1,4,5]thiadiazepine 5-oxide, 1145

Dibenzo[*b,e*][1,4]diazepine, 996, 1022
Dibenzo[*b,f*][1,5]diazocine-6,12-dione, 945
Dibenzo[*c,e*][1,2]thiazine 5,5-dioxide, 1029
Dibenzothiophene 5,5-dioxide, 1173, 1189
Dibenzo[*b,d*]thiopyran 5,5-dioxide, 1156
Dibenzo[*b,d*]thiopyran-6-thione, 785
Dibenzo[*c,f*][1,2,5]triazepine, 1157
Dibenz[*b,e*][1,4]oxazepine, 996
Dicyclopenta[*b,e*]pyridine, 973
Diimidazo[1,5-*a*:1′,5′-*d*]pyrazine-5,10-dione, 785
1,3-Dioxolo[4,5-*h*][1]benzopyran-8-one, 1172
1,3-Dioxolo[4,5-*h*]isoindolo[1,2-*b*][3]-benz**azepin**-8-one, 1311
1,3-Dioxolo[4,5-*h*]**isoindolo**[1,2-*b*][3]-benzazepin-2-one, 1113
1,3-Dioxolo[4,5-*i*]isoindolo[2,1-*c*][3]-benz**azocine**-8,14-dione, 1085
[1,3]Dioxolo[6,7]naphth[2,1-*d*]**isoxazole**, 1269
[1,3]Dioxolo[4,5-*j*]**phenanthridine**, 813, 1228
[1,3]Dioxolo[4,5-*j*]pyrrolo[3,2,1-*de*]-**phenanthridin**-7-**one**, 1229, 1240
1,3-Dioxolo[4,5-*k*]pyrrolo[3,2,1-*de*]-**phenanthridin-6-one**, 1229
Di**pyrido**[1,2-*b*:3′,2′-*d*]pyrazole, 741
Di**pyrido**[1,2-*b*:3′,2′-*d*]pyrazol-2-one, 949
Dipyrido[3,4-*c*:2′,3′-*e*]**pyridazine**, 1179
Dipyrido[3,4-*c*:4′,3′-*e*]**pyridazine**, 1179
Dipyrido[1,2-*a*:4′,3-′-*d*]**pyrimidin**-11-**ium**, 1009
Dipyrido[1,2-*a*:3′,2′-*e*]**pyrimidin-5-one**, 1069
Dipyrido[1,2-*a*:2′,1′-*d*][**1,3,5**]**triazin**-5-**ium**, 1013
Dipyrimido[1,6-*a*:6′,1′-*d*][1,3,5]**triazin**-10-ium, 1013
Dipyrrolo[1,2-*c*:2′,1′-*e*]**imidazol**-5-one, 1186
Di**pyrrolo**[1,2-*b*:3′,4′-*d*],pyridazine, 1336
Dipyrrolo[2,3-*b*:2′,3′-*d*]**pyridine**, 770
Dipyrrolo[3,4-*b*:3′,4′-*d*]pyrrole, 1064
2,4-Dithia-1-azabicyclo[3.2.0]heptan-7-one, (1,3,4-dithiazole) 1278
Dithieno[3,2-*b*:2′,3′-*f*][**1,4**]**oxazepine**, 851
Dithieno[2,3-*b*:2′,3′-*d*]**pyridine**, 705
Dithieno[3,2-*b*:2′,3′-*d*]**pyridine**, 705
Dithieno[3,2-*b*:3′,2′-*d*]**pyridine**, 705
Dithieno[3,4-*b*:2′,3′-*d*]**pyridine**, 705
Dithieno[3,4-*b*:3′,2′-*d*]**pyridine**, 705
Dithieno[3,4-*b*:3′,4′-*d*]**pyridine**, 705
Dithieno[3,2-*b*:3′,2′-*d*]**pyridine-4-oxide**, 850
Dithieno[3,2-*b*:3′,4′-*d*]**pyridine 4-oxide**, 850
Dithieno[3,4-*b*:3′,2′-*d*]**pyridine-4-oxide**, 850
Dithieno[2,3-*b*:2′,3′-*e*][**1,4**]**thiazine**, 710
Dithieno[2,3-*b*:2′,3′-*d*]thiophene, 1215
Dithieno[2,3-*b*:3′,2′-*d*]thiophene, 1188
Dithieno[3,2-*b*:2′,3′-*d*]thiophene, 1188
Dithieno[3,2-*b*:3′,4′-*d*]thiophene, 1165
Dithieno[3,4-*b*:3′,4′-*d*]thiophene, 1188
1,3-Dithiolo[4′,5′:5,6][**1,4**]**dithiino**[2,3-*b*]-quinoxaline-2-thione, 1168
[1,2]Dithiolo[1,5-*b*]dithiole-7-S^{iv}, 1205
1,3-**Dithiolo**[4,5-*d*]pyridazin-4-one, 1172
1,3-**Dithiolo**[4,5-*b*]pyridine, 1220
[1,2]**Dithiolo**[3,4-*c*]quinolin-1-**one**,1322
[1,2]Dithiolo[5,1-*e*][**1,2,3**]**thiadiazole**, 1147

1,4-Epoxyisoquinoline (furan), 1188
5,13-Ethano[1,5]benzo**diazepino**[2,3-*b*]-[1,5]benzodiazepine, 1137
5,8-Ethanoquinoline, 1191
5,10-Ethano[1,2,4]triazolo[1,2-*a*][**1,2,4**]-**triazepine**-1,3-dione, 1199

Fluoreno[1,9,9*a*-*cd*]**pyridin-3-one**, 802
Furo[3,4-*d*][**1,3**]benzo**diazepin-3-**one, 688
Furo[4,3,2-*de*][1]benzopyran, 841
Furo[2,3-*h*][1]benzo**pyran-2**-one, 803
Furo[3,2-*c*][1]benzopyran-4-one, 863
Furo[3,2-*g*][1]benzopyran-5-one, 841
Furo[3,4-*d*]**isoxazol**-4-one, 1189
Furo[2,3-*d*][**1,3**]**oxazin-4-one**, 953
Furo[3,2-*e*]**pyrano**[2,3-*b*]pyridin-7-**one**, 1077
Furo[3,2-*c*]**pyrazole**, 1031
Furo[3,4-*c*]**pyrazol-4-**one, 829
Furo[3,4-*d*]**pyridazine**, 834
Furo[3,4-*d*]pyridazin-5-**one**, 801
Furo[2,3-*b*]pyridine, 899, 1066
Furo[3,2-*b*]pyridine, 1215, **1224**
Furo[3,2-*c*]pyridine, 1210, **1224**, 1295, 1332
Furo[2,3-*b*]pyridin-**3**-one, 1066
Furo[3,2-*c*]pyridin-3-**one**, 1215
Furo[2,3-*d*]**pyrimidine**, 744
Furo[3,4-*d*]**pyrimidine**-2,4-**dione**, 1099
Furo[2,3-*d*]pyrimidin-2-one, 1220
Furo[2,3-*d*]pyrimidin-4-one, 1308
Furo[2,3-*b*]quinoline, 1219
Furo[3,2-*b*]quinoline, 1272

Furo[5,6-*f*]quinoline, 864
Furo[2,3-*c*]**quinoline 5-oxide**, 850
Furo[3,2-*c*]**quinoline** 5-**oxide**, 850
Furo[2,3-*c*]**quinolin-4-one**, 1194
Furo[2,3-*h*]**quinolin-2-one**, 992
Furo[2,3-*h*]quinolin-2-one, 992
Furo[3,2-*c*]quinolin-4-one, 818, 914, 1214, 1281
2λ^4-Furo[3,2-*c*][1,2]**thiazine**, 1021
Furo[3,2-*d*]-**1,2,4-triazolo**[4,3-*b*]pyridazine, 1253

Imidazo[1,2-*a*]benzimidazole, 736, 971
Imidazo[5,1-*b*]benzimidazole, 775
Imidazo[1,2-*a*]benzimidazole-3-**one**, 1030
Imidazo[1,2-*b*]benzimidazol-2-**one**, 1000
Imidazo[2,1-*b*][1,3]benzo**diazepine**, 700
Imidazo[4,5-*b*][1,5]benzo**diazepine, 749**
Imidazo[1,2-*a*][1,5]benzodiazepin-5-one, 998
Imidazo[1,2-*d*][**1,4**]benzo**thiazepine**, 883
Imidazo[1,2-*d*][**1,4**]benzo**thiazepin-5-one**, 1279
Imidazo[2,1-*b*]benzothiazol-2-**one**, 1000
Imidazo[4,5-*f*]2,1,3-benzoxadiazole 1-**oxide**, 1016
Imidazo[1,5-*a*][4,1]benzoxazepin-6-one, 1238
Imidazo[5,1-*c*][1,4]benzoxazine, **740**, **875**
Imidazo[1,2-*c*][1,3]benzoxazine-5-thione, 886
Imidazo[1,2-*c*][1,3]benzoxazin-5-one, 886
Imidazo[5,1-*b*]benzoxazole, 775
Imidazo[2,1-*b*]benzoxazol-2-one, 1000
Imidazo[4,5-*e*][1,3]**diazepine**-4,6-**dione**, 1104
Imidazo[4,5-*e*][1,4]**diazepin**-8-**one**, 674
Imidazo[4,5-*e*][1,4]**diazepin-8-one** 4-**oxide**, 1104
Imidazo[1,2-*a*]imidazole, 811
Imidazo[4,5-*d*]imidazole, 727
Imidazo[1,2-*a*]imidazol-2-**one**, 948
Imidazo[5,1,2-*cd*]**indolizine** (prr), 1183
Imidazo[2,1,5-*cd*]indolizine, 1321
Imidazo[4,5-*b*]indol-**2**-**one**, 933
Imidazo[2,1-*a*]**isoindole**, 873, 1276
Imidazo[2,1-*a*]**isoquinoline**, 1193
Imidazo[1,2-*b*]**isoquinoline**-5,10-**dione**, 1158
Imidazo[5,1-*a*]isoquinolin-3-**one**, 1041
Imidazo[5,1-*a*]isoquinolin-3-**one**, 1-**thioxo**-, 1041
Imidazo[5,1-*b*]naphtho[2,3-*d*]**thiazole**-5,10-dione, 1166
Imidazo[2,1-*c*]naphtho[2,1-*e*][**1,2,4**]-**triazine**, 1278
Imidazo[1,2-*c*][**1,3**]**oxazine**, 882
Imidazo[1,2-*f*]**phenanthridine**, 1228
Imidazo[4,5-*c*]pheno**thiazine**, 1197
Imidazo[5,1-*a*]phthalazin-3-**one**, 1-**thioxo-**, **1041**
Imidazo[1,2-*a*]purin-9-one, 999
Imidazo[1,2-*a*]pyrazine, 1331
Imidazo[1,5-*a*]pyrazine-1-**thione**, 1001
Imidazo[1,2-*a*]**pyrazin-8-one**, 1061
Imidazo[1,2-*a*]pyrazin-3-**one**, 1000
Imidazo[1,5-*a*]**pyrazin-8-one**, 1061
Imidazo[4,5-*b*]pyrazin-**2**-**one**, 960
Imidazo[4,5-*c*]pyrazole, 757
Imidazo[1,2-*b*]pyridazin-2-**one**, 1241
Imidazo[1,2-*b*]pyridazin-3-**one**, 1000
Imidazo[1,2-*a*]pyridine, 874, 913, 1238, 1294, 1321, 1331
Imidazo[1,2-*a*]**pyridine**, 879, 1339
Imidazo[1,5-*a*]pyridine, 874, 1339
Imidazo[1,5-*a*]**pyridine**, 1321
Imidazo[4,5-*b*]**pyridine**, 741
Imidazo[1,2-*a*]pyridin-3-**one**, 1000
Imidazo[1,2-*a*]**pyridin-5-one**, 917
Imidazo[1,2-*a*]**pyridin-7-one**, 918
Imidazo[4,5-*b*]pyridin-**2**-**one**, 1129
Imidazo[4,5-b]**pyridin-2-one**, 948
Imidazo[4,5-*c*]pyridin-**2**-**one**, 1139
Imidazo[1,2-a]pyrimidine, 874
Imidazo[1,2-*c*]pyrimidine, 998, 1237
Imidazo[1,2-*a*]**pyrimidin-4**-one, 1009
Imidazo[1,2-*a*]**pyrimidin-5-one**, 1300
Imidazo[1,2-*a*]**pyrimidin-7-one**, 1300
Imidazo[1,2-*a*]pyridin-7-one, 1106
Imidazo[1,2-*c*]quinazolin-5-one, 886
Imidazo[4,5-*f*]quinazolin-7-one, 684
Imidazo[4,5-*g*]**quinazolin-8-one**, 1161
Imidazo[1,5-*a*]quinoline, 998
Imidazo[4,5-*c*]quinoline, 1313
Imidazo[4,5-*f*]quinoline, 981, 989
Imidazo[4,5-*f*]quinolin-7-one, 684
Imidazo[1,2-*a*]**quinoxaline**, 975
Imidazo[1,5,4-*de*]quinoxaline, 776
Imidazo[1,5,4-*de*]quinoxaline-**2**,5-di**one**, 1000
Imidazo[1,5,4-*de*]quinoxalin-5-one, 1000
Imidazo[2,1-*d*]-**1,2,3,5-tetrazin-4-one**, **1156**
Imidazo[2,1-*b*][**1,3,4**]**thiadiazine**, 731
Imidazo[4,5-*d*][**1,2,3**]**thiadiazine, 1046**

INDEX OF RING SYSTEMS

Imidazo[4,5-*d*][1,2,6]thiadiazine 2,2-dioxide, 1127
Imidazo[1,5-*d*][**1,3,4**]**thiadiazin**-5-one 7-thioxo-, 1302
Imidazo[2,1-*b*][**1,3**]**thiazine**-2,5-dione, 883
Imidazo[2,1-*b*]thiazole, 941, 1087
Imidazo[2,1-*b*]**thiazole**, 1322
Imidazo[2,1-*b*]**thiazol-4-ium**, 876
Imidazo[2,1-*b*]**thiazol-3-one**, 878, 1114
Imidazo[4,5-*e*][**1,2,4**]**triazepine**-5,8-**dione**, 689
Imidazo[1,2-*b*][**1,2,4**]**triazine**, 686
Imidazo[2,1-*c*]-**1,2,4-triazine**, 1155
Imidazo[4,5-*d*]-**1,2,3-triazine**, **1045**
Imidazo[5,1-*c*][**1,2,4**]**triazine**, 814
Imidazo[5,1-*f*][**1,2,4**]**triazine**, 1301
Imidazo[5,1-*f*][1,2,4]triazine-**2**,4,7-**trione**, 948
Imidazo[1,2-*d*][1,2,4]triazin-4-ium, 8-oxo-, 874
Imidazo[4,5-*d*]**1,2,3-triazin-4-one 2-oxide,** 747
Imidazo[1,2-*b*][1,2,4]triazole, 998, 1048
Imidazo[1,2-*b*]-[**1,2,4**]**triazole**, 685, 1214
Imidazo[1,5-*b*][1,2,4]triazole, 874
Imidazo[2,1-*c*]-**1,2,4-triazole**, 1238
Imidazo[1,2-*b*][1,2,4]triazole-5,6-**dione**, 1114
Imidazo[1,2-*c*]1,2,3-triazolo[4,5-*e*]-**pyrimidine**, 1006
Indazole, 752, 761, 790, 1031, 1233, 1353, 1268
Indazole 1-oxide, 752
4,7-Indazoledione, 965, 1185
3-Indazolone, 694
4-Indazolone, 1261
Indeno[2,1-*c*][2]benzopyran-5,11-dione, 804
Indeno[1,2-*e*]-1,4-diazepin-6-one, 838, 976
Indeno[2,1-*d*]-1,3-dioxin-**2**-one, 837
Indeno[2,1-*d*]-1,3-dioxin-**2**-thione, 837
Indeno[1,2-*c*]-1,2-dithiole-3-thione, 985
Indeno[1,2-*d*][1,2]oxathiin 3,3-dioxide, 920
Indeno[2,1-*g*]pteridin-6-one (paz), 758
Indeno[1,2-*b*]pyran, 1196
Indeno[1,2-*b*]pyran-2,5-dione, 1197
Indeno[1,2-*c*]pyrazole, 829
Indeno[1,2-*c*]pyrazol-4-one, 829, 970
Indeno[1,2-*b*]pyridine, 1296
Indeno[1,2-*b*]pyrido[3,2-*e*]**pyrazin-6-**one, 758
Indeno[1,2-*d*]pyrimidine-2,5-dione, 788
Indeno[1,2-*d*]pyrimidin-5-one, 808
Indeno[2′,1′:4,5]**pyrrolo**[2,1-*a*]isoquinolin-12-one, 840
Indeno[2′,1′:4,5]**pyrrolo**[2,1-*a*]phthalazin-12-one, 840
Indeno[1,2-*d*]tetrazolo[1,5-*a*]**pyrimidin**-9-one, 835
Indole, 676, 698, 702, 713, 734, 768, 781, 818, 828, 848, 854, 860, 861, 885, 886, 893, 894, 912, 922, 957, 958, 1018, 1020, 1025, 1056, 1087, 1106, 1151, 1177, 1209, 1223, **1248**, 1312, 1329
2,3-Indoledione, 1261, 1330
Indolizine (prr), 675, 694, 839, 840, **911**, 1055, 1106, 1112, 1182, 1223, **1317,** 1336
2,3-Indolizinedione (prr2,3dione), 1337
Indolizino[3,4,5-*ab*]isoindole (prr), 1055
3-Indolizinone (prr2one), 1336
5-Indolizinone (p2one), 879, 917, 1212, 1325
5-Indolizinone (prr), 929
Indolizino[1,2-*b*]quinolin-9-**one** (p2one), 1202
Indolo[2,1-*a*]isoquinoline, 786
Indolo[1,2-*g*]-**1,6-naphthyridine**-5,12-**dione** (p2,5dione), 1115
Indolo[3,2,1-*de*][1,5]**naphthyridin-6-one**, 880, 1115
Indolo[2,1-*b*]**quinazolin**-12-**one**, 907
Indolo[2,3-*c*]**quinolin-6-one**, 1194
Indolo[1,2-*b*]indazole, 1056
Indolo[3,2-*b*]indole, 1251
Indolo[7,6-*g*]indole, 1251
2-Indolone, 782, 784, 895, 947, 1056, 1224, 1313, 1320
3-Indolone, 1021
3*H*-4-Indolone, 1294
4-Indolone, 983, 1201, 1207
Indolo[3,2,1-*de*]**phenanthridin-9-one**, 1340
Indolo[2,3-*a*]**pyrrolo**[3,4-*c*]carbazole, 1251
Indolo[3,2-*c*]**quinoline**-5-**thione**, 1298
Indolo[3,2-*c*]**quinolin-5-one**, 1298
Indolo[2,3-*a*]**quinolizine**, 1323
Isobenzofuran, 704, 830
Isoindole, 828, 1164, 1178
Isoindolo[2,1-*a*]indol-6-one, 1183
1-Isoindolone, 795, 828, 862, 947, 1064, 1097, 1224
Isoquino[2,1-*b*]**cinnolin**-13-**one** 8-**oxide**, 1333

Isoquinoline, 677, 678, 743, **770**, 855, 866, 888, **1058**, 1178, 1285, **1295**, 1304
Isoquinoline 2-oxide, 673, 788
1-Isoquinolinethione, 1304
3,5,8-Isoquinolinetrione, 1194
1-Isoquinolinone, 802, 868, 891, 895, 1229, **1384**
3-Isoquinolinone, 802, 1094
4-Isoquinolinone, 1298
Isoquino[2,1-*a*]**quinazolin-6-one, 1083**
Isothiazolo[5,4-*b*][1,8]naphthyridine-**3**,4-di**one**, 1119
Isothiazolo[5,4-*b*]**pyridine**, 987
Isothiazolo[5,4-*b*]pyridin-**3-one**, 1082
Isothiazolo[3,4-*d*]pyrimidine-4,6-dione, 985
Isothiazolo[5,4-*b*]quinoline-**3**,4-di**one,** 1119
Isoxazolo[3,4,5-*kl*]acridine, 831
Isoxazolo[2,3-*d*][1,4]benzodiazepine, 1332
Isoxazolo[4,5-*b*][**1,4**]**diazepine**, 976
Isoxazolo[4,5-*e*][**1,4**]**diazepine**-6,8-**dione**, 1044
Isoxazolo[5,4-*d*][**1,3**]**oxazin-4-one**, 953
Isoxazolo[5,4-*d*]pyrazolo[3,4-*b*]pyridine, 1277
Isoxazolo[4,5-*b*]pyridine, 1269
Isoxazolo[5,4-*b*]**pyridine**, 987
Isoxazolo[4,5-*b*]pyridine 2,4-dioxide, 1269
Isoxazolo[4,5-*b*]pyridine **2-oxide**, 1269
Isoxazolo[3,4-*d*]pyrimidin-3-one, 1066
Isoxazolo[4,5-*b*]quinoxaline, 795, 801, 845, 1269
Isoxazolo[4,5-*g*]quinoxaline-4,9-dione, 1189
Isoxazolo[3,2-*c*][**1,2,4**]**thiadiazole**, 776

1,4-Methano[1,2,4]triazolo[1,2-*a*][**1,2,4,5**]-**tetrazine**-6,8-dione, 1196

Naphth[1,2-*d*]imidazol-2-one, 971
Naphth[1,2-*d*]imidazol-5-one, 928
Naphth[2′,1′:4,5]imidazo[2,1-*b*]**thiazole**, 677
Naphth[1,2-*d*]imidazo[2,1-*b*]**thiazol-8-one**, 878
Naphth[1,2-*c*]isoxazole, 972
Naphth[2,1-*d*]isoxazole, 972
Naphtho[1,2-*c*:3,4-*c*′]bis[1,2,5]thiadiazole, 1226
Naphtho[1,2-*c*:5,6-*c*′]bis[1,2,5]thiadiazole, 1216, 1284
Naphtho[1,2-*c*:7,8-*c*′]bis[1,2,5]thiadiazole, 1284
Naphtho[2,3-*b*][1,4]diazepine-2,4,6,11-tetraone, 1140
Naphtho[1,2-*d*]-1,3-dioxole, 1172
Naphtho[2,3-*d*]-1,3-dioxole, 1172
Naphtho[1′,2′:4,5]**furo**[2,3-*e*][1,2,4]triazine, 819
Naphtho[2,3-*b*]pyran-5,10-dione, 1286
Naphtho[2,3-*c*]pyran-5,10-dione, 843
Naphtho[1,2-*b*]pyran-4-one, 902
Naphtho[1,2-*e*]pyran-2-one, 1051
Naphtho[2,1-*b*]pyran-3-one, 822, 1289
Naphtho[2,1-*e*]pyran-2-one, 1051
Naphtho[2,3-*b*]pyran-4-one, 919
Naphtho[2,1-*e*]pyrazolo[3′,4′:3,4]-pyrazolo[5,1-*c*][**1,2,4**]**triazine**, 1279
Naphtho[2,1-*e*]pyrazolo[5,1-*c*][**1,2,4**]-**triazine**, 1278
Naphtho[2,1-*e*]tetrazolo[5,1-*c*][**1,2,4**]-**triazine**, 1278
Naphtho[1,2-*c*][1,2,5]thiadiazole, 1226, 1284
Naphtho[1,8-*de*]-1,3-thiazine, 1221
Naphtho[2,3-*d*]thiazole-4,9-dione, 717
Naphtho[2′,3′:4,5]**thiazolo**[3,2-*a*]-benzimidazole-7,12-dione, 1166
Naphtho[1,2-*c*:3,4-*c*′:7,8-*c*″]tris[1,2,5]-thiadiazole, 1284
Naphth[2,1-*c*][1,2]oxathiin 4,4-dioxide, 1062
Naphth[1,8-*de*]-1,2-oxazine 1270
Naphth[2,1-*b*][1,4]oxazin-3-one, 1279
Naphth[1,2-*d*]oxazole, 1278
Naphth[1,8-*bc*]oxepin-2-one, 1053
1,5-Naphthyridine, 989
1,6-Naphthyridine, 989
1,7-Naphthyridine, 1178
1,8-Naphthyridine, 989
2,6-Naphthyridine, 1178
1,5-Naphthyridine 1-oxide (p), 989, 990
1,5-Naphthyridin-2-one (p2one), 949
1,6-Naphthyridin-2-one (p), 1201
1,6-Naphthyridin-2-one (p2one), 694
1,6-Naphthyridin-5-one (p2one), 1090, 1203
1,6-Naphthyridin-7-one (p2one), 846
1,7-Naphthyridin-2-one (p2one), 694, 949
1,7-Naphthyridin-8-one (p2one), 1042
1,8-Naphthyridin-2-one (p2one), 694, 1091
1,8-Naphthyridin-4-one (p4one), 737, 1205
2,7-Naphthyridin-1-one (p2one), 846

7-Oxa-8-azabicyclo[4.2.1]nona-2,4-diene (isoxazole), 1189
4-Oxa-1-azabicyclo[4.2.0]oct-2-en-8-one (1,4ox), 1102
4-Oxa-2-azabicyclo[4.2.0.]oct-2-en-8-one (1,4ox), 1175
5-Oxa-1-azabicyclo[4.2.0]oct-2-en-8-one (1,3ox), 1156
4-Oxa-1,3-diazabicyclo[3.2.0]hept-2-ene (1,2,4oxdazol), 1210
7-Oxa-8,9-diazabicyclo[4.2.1]nona-2,4-diene (1,2,4oxdazol), 1190
[1,3,4]Oxadiazino[3,4-*a*]cinnolin-3-**one**, 1198
[1,3,4]**Oxadiazino**[4,3-*a*]cinnolin-2-**one**, 1198
[1,2,4]Oxadiazolo[4,5-*a*][1,5]-benzodiazepin-5-one, 1339
[1,2,4]Oxadiazolo[3,4-*c*][1,4]benzoxazin-1-one, 832
[1,2,4]Oxadiazolo[3,4-*a*]isoquinolin-3-**one**, 1264
[1,2,4]Oxadiazolo[4,3-*b*]pyridazin-3-**one**, 1264
[1,3,4]Oxadiazolo[4,5-*a*]pyridine, 1265
[1,2,5]**Oxadiazolo**[3,4-*d*]pyridine 1-oxide, 1016
[1,2,5]Oxadiazolo[4,5-*b*]pyridine **3-oxide**, 753
1,2,4-Oxadiazolo[2,3-*a*]pyridin-2-**one**, 1333
[1,3,4]Oxadiazolo[3,2-*a*]**pyrimidin-4-ium**, 1005
1,3,4-Oxadiazolo[3,2-*a*]**pyrimidin-5-one**, 930, 1115
1,2,4-Oxadiazolo[2,3-*a*]quinoline, 757
[1,2,5]Oxadiazolo[3,4-*b*]quinoxaline, 832
10-Oxa-1,2,7,8-tetraazatricyclo[6.3.1.0^{3,6}]-dodec-3(6)-ene-4,5-dione (1,2,5,6tetrazep), 1141
[1,2,3,5]Oxathiadiazolo[3,4-*b*]pyridazine 3-**oxide**, 1264
1,3-Oxathiolo[4,5-*b*]pyridin-**2-one**, 820
[1,3]Oxazino[3,2-*a*]benz**imidazole**, 1338
[1,4]Oxazino[1,2-*a*]benz**imidazole**, 1313
[1,3]Oxazino[4,5-*c*]isoquinolin-1-one, 952
[1,4]Oxazino[2,3,4-*kl*]phenothiazine, 882
[1,3]**Oxazino**[4,3-*a*]phthalazin-2-**one**, 1341
1,3-Oxazino[6,5-*c*]quinolin-1-one, 1079
Oxazolo[5,4-*d*]**[1,3]oxazin-4-one**, 953
Oxazolo[4,3-*a*]phthalazin-3-**one**, 1041, 1278
Oxazolo[4,5-*b*]pyridine, 715
Oxazolo[5,4-*b*]**pyridine**, 704, 906
Oxazolo[3,2-*a*]**pyridin-5-one,** 918
Oxazolo[4,5-*b*]pyridin-**2-one**, 716, 1226
Oxazolo[4,5-*d*]pyrimidine, 715
Oxazolo[5,4-*d*]**pyrimidine**, 1050
Oxazolo[5,4-*d*]pyrimidine-4,6-dione, 1294
Oxazolo[5,4-*d*]pyrimidine-**2**,5,7-tri**one**, 704
Oxazolo[3,2-*a*]pyrimidin-4-ium, 807
Oxazolo[5,4-*d*]pyrimidin-7-one, 714
Oxazolo[4,5-*b*]quinolin-2-**one**, 1226

Perimidine, 1143
Phenanthridine, 705, 812, 1193
Phenanthridine 5-oxide, 850
6-Phenanthridinone, 1060, 1067, 1153, 1230, 1305
Phenanthro[9,10-*d*]imidazole, 830
Phenanthro[9,10-*c*][1,2,5]thiadiazole 2-oxide, 833
Phenazine, 836, 994, 1132, 1134
Phenothiazine, 709, 1334
1-Phenothiazinone, 722
Phenoxazine, 721
3-Phenoxazinone, 721
Phthalazine, 681
1-Phthalazinone, 707, 796, 846
Phthalazino[2,3-*b*]phthalazine-7,12-dione, 1194
Pteridine (paz), 758, 1133
Pteridine (pm), 923, 938, 1050, 1143
2,4-Pteridinedione (paz), **1131**
2,4-Pteridinedione (pm2,4dione), 952
2-Pteridinone (pm2one), 685
4-Pteridinone (paz), 1132
6-Pteridinone (pazone), 1135
7-Pteridinone (paz), 1135
Purine (im), 684, 756, **1125**
Purine (pm), 746, 923
2,6-Purinedione (im), 757, 932, 999, 1313
6,8-Purinedione (im2one), 1097
2-Purinone (im), 1314
2-Purinone (pm2one), 1043, 1098
6-Purinone (im), 684, 757, 933
6-Purinone (pm6one), 692, 1038
6-Purinone 7-oxide (imNox), 1315
Pyrano[3,2-*d*][2]benzazepin-**2-one**, 1203
Pyrano[3,4-*e*]benz**imidazol**-9-one, **2-thioxo-**, 1130
Pyrano[4,3-*b*][1]benzo**pyran**-1,10-di**one**, 1111
Pyrano[3,2-*c*][1]benzopyran-**2**,5-di**one**, 793
Pyrano[4,3-*b*][1]benzopyran-10-one, 869

Pyrano[2,3-*g*]benz**oxazol**-9-one, 1026
Pyrano[2,3-*b*:6,5-*b'*]dipyridin-5-one, 823
Pyrano[2,3-*b*]indol-**4-one**, 808
Pyrano[3,4-*b*]indol-3-**one**, 802
Pyrano[4,3-*b*]indol-3-**one**, 803
Pyrano[3',2':5,6]pyrano[4,3-*b*]benzopyran-7-one, 869
Pyrano[2,3-*c*]pyrazole, 908, 1311
Pyrano[2,3-*c*]pyrazol-6-**one**, 720, 909, 1070, 1077
Pyrano[2,3-*d*]pyridazine, 870
Pyrano[3,4-*d*]pyridazin-5-**one**, 1203
Pyrano[2,3-*b*]**pyridine**, 742
Pyrano[2,3-*c*]pyridine, 821
Pyrano[3,2-*b*]pyridine, 821
Pyrano[2,3-*b*]**pyridine** 8-**oxide**, 1201
Pyrano[2,3-*b*]pyridin-**2**-one, **823**
Pyrano[2,3-*b*]pyridin-**4-one**, 825
Pyrano[3,2-*b*]pyridin-**2-one**, 1201
Pyrano[3,2-*c*]pyridin-**4-one**, 825
Pyrano[3,4-*b*]pyridin-8-**one**, 1077
Pyrano[4,3-*b*]pyridin-5-**one**, 891
Pyrano[4,3-*c*]pyridin-1-**one, 1077**
Pyrano[2,3-*d*]pyrimidine-2,4-dione, 908
Pyrano[2,3-*d*]pyrimidine-**2**,4,7-tri**one**, 908, 1084, 1197
Pyrano[2,3-*d*]**pyrimidin-4-one**, 857
Pyrano[2,3-*b*]**quinoline**, 742
Pyrano[3,2-*c*]quinoline, 995, 1173
Pyrano[3,2-*c*]quinoline-**2**,5-di**one**, 823
Pyrano[2,3-*b*]quinolizine-**2**,5-di**one**, 1290
Pyrano[3,2-*a*]quinolizine-3,6-di**one**, 1290
Pyrazino[1,2-*a*]benzimidazole, 836, 881
Pyrazino[1,2-*a*][**1,4**]benzo**diazepine**, 773
Pyrazino[3,2,1-*kl*]phenothiazine-1,2-**dione**, 1012
Pyrazino[2,3-*d*]**pyrimidin-4-one**, **2-thioxo-**, 1084
Pyrazino[2,3-*c*][1,2,6]thiadiazine 2,2-dioxide, 1127, 1131
Pyrazino[2,3-*b*]-**1,4-thiazine**, 749
Pyrazolo[3,4-*c*]azepin-8-one, 1048
Pyrazolo[3,4-*b*][1]benzazepine, 811
Pyrazolo[1,5-*a*]benzimidazole, 736
Pyrazolo[4,3-*b*][1,4]benzothiazine, 811
Pyrazolo[5,1-*c*][**1,2,4**]benzo**triazine**, 1155
Pyrazolo[5,1-*c*][1,2,4]**benzotriazine** 5-**oxide**, 755
Pyrazolo[3,4-*b*][**1,4**]benz**oxazepin-5-one**, 955
Pyrazolo[4,3-*b*][1,4]benzoxazine, 703, 1204
Pyrazolo[3',4,:3,4]cyclopenta[1,2-*b*]-pyridin-4-one, 970
Pyrazolo[3',4':3,4]cyclopenta[1,2-*c*]-pyridin-4-one, 970
Pyrazolo[3,4-*b*][**1,4**]**diazepine-5,7-dione**, 1140
Pyrazolo[1,5-*a*:3,4-*d'*]di**pyrimidine**, 1006
Pyrazolo[3,4-*e*]-**1,4,2-dithiazine 1,1-dioxide**, 1221
Pyrazolo[3,4-*e*]**indolizine** (p), 855
Pyrazolo[5,1-*a*]isoindol-8-one, 800
Pyrazolo[5,1-*a*]isoquinoline, **1319**
Pyrazolo[3,4-*d*]isoxazole, 1330
Pyrazolo[3,4-*d*][1,3]oxazine, **1065**, **1308**
Pyrazolo[5,1-*c*][**1,4**]**oxazine**, 903
Pyrazolo[3,4-*d*][1,3]oxazin-4-one, 1065
Pyrazolo[3,4-*a*]phen**oxazin-4-**one, 871
Pyrazolo[5,1-*f*]**purin-2-one** (im2one), 960
Pyrazolo[3',4':4,5]**pyrano**[2,3-*b*]pyridin-4-**one**, 1084
Pyrazolo[4'',3'':5',6']pyrano[2',3':4,5]-pyrimido[1,6-*b*][1,2,4]**triazine**, 967
Pyrazolo[4'',3'':5',6']pyrano[2',3':4,5]-pyrimido[1,6-*b*][**1,2,4**]**triazine**-2,3-**dione**, 965
Pyrazolo[4',3':5,6]pyrano[3,2-*e*][**1,2,4**]-**triazolo**[1,5-*c*]pyrimidine, 967
Pyrazolo[3,4-*b*]**pyrazine**, 759
Pyrazolo[3,4-*c*]pyrazole, 740
Pyrazolo[1,2-*a*]pyrazole-1,**3**,**5**,7-tetra**one**, 1186
Pyrazolo[1,2-*a*]**pyrazol-4-**ium, 1,3-**dioxo-**, 1186
Pyrazolo[1,2-*a*]pyrazol-1-one, 1184, 1320
Pyrazolo[1,2-*a*]**pyridazin**-1-one, 1195
Pyrazolo[3,4-*c*]pyridazin-**3-one**, 947
Pyrazolo[1,5-*a*]pyridine, 727, 736, 983, 1035, 1209
Pyrazolo[3,4-*b*]pyridine, 1065, 1164
Pyrazolo[3,4-*b*]**pyridine**, 671, 741, 887, 1003, 1309
Pyrazolo[4,3-*c*]pyridine, 970
Pyrazolo[1,5-*a*]pyridin-2-**one**, 1113
Pyrazolo[3,4-*b*]**pyridin-3-one**, 1002, 1073
Pyrazolo[3,4-*b*]**pyridin-4-one**, 1110
Pyrazolo[3,4-*c*]pyridin-7-one, 1106
Pyrazolo[4,3-*b*]pyridin-**3-one**, 752
Pyrazolo[4,3-*c*]pyridin-4-one, 1048
Pyrazolo[1,5-*a*]**pyrido**[3,4-*e*]pyrimidine, 1201

Pyrazolo[1,5-*a*]**pyrido**[3,4-*e*]pyrimidine 7-**oxide**, 1202
Pyrazolo[1,5-*a*]**pyrimidine**, 744, 919, **1003**
Pyrazolo[3,4-*d*]**pyrimidine**, 745, 746, 763, 764, 772
Pyrazolo[3,4-*d*]pyrimidine, 1065
Pyrazolo[3,4-*d*]**pyrimidine** 5-**oxide**, 924, 958
Pyrazolo[3,4-*d*]pyrimidine-4,6-dione, 1233, 1252, 1261
Pyrazolo[3,4-*d*]**pyrimidine-6-thione**, 764
Pyrazolo[1,5-*a*]**pyrimidin-5-one**, 1010
Pyrazolo[1,5-*a*]**pyrimidin-7-one**, 1010
Pyrazolo[3,4-*d*]**pyrimidin-4-one**, 938
Pyrazolo[3,4-*d*]**pyrimidin-6-one**, 764
Pyrazolo[3,4-*d*]**pyrimidin-4-one** 6-**thioxo**-, 952, 1100
Pyrazolo[3′,4′:4,5]pyrimido[2,1-*b*]-benzo**thiazol**-4-one, 906
Pyrazolo[1′,5′:1,6]**pyrimido**[4,5-*d*]-pyridazin-4-one, 1006
Pyrazolo[4,3-*e*]pyrrolo[1,2-*a*]**pyridine**, 1323
Pyrazolo[1,5-*a*]**quinazoline**, 1005
Pyrazolo[4,3-*g*]**quinazolin-5-one**, 942
Pyrazolo[1,5-*a*]quinoline, 898, 1319
Pyrazolo[3,4,5-*de*]quinoline, 1244
Pyrazolo[3,4-*b*]**quinoline**, 1108
Pyrazolo[4,3-*c*]quinolin-3-**one**, 1074
Pyrazolo[3,4-*b*]quinoxaline, 800, 810, 862
Pyrazolo[5,1-*d*]-**1,2,3,5-tetrazin-4**-one, 1156
Pyrazolo[3,4-*c*][**1,2,5**]**thiadiazine**, 1315
Pyrazolo[3,2-*c*][1,4]thiazine, 811
Pyrazolo[1,5-*a*]-**1,3,5-triazine**, 925
Pyrazolo[3,4-*e*]-**1,2,4-triazine**, 1310
Pyrazolo[3,4-*e*]-1,2,4-triazine, 1164
Pyrazolo[5,1-*c*][**1,2,4**]**triazine**, 1155
Pyrazolo[3,4-*e*]-**1,2,4-triazine**-3,7-**dione**, 1245
Pyrazolo[5′,1′:3,4][1,2,4]triazino[5,6-*b*]-[1,5]benzoxazepine, 724
Pyrazolo[5′,1′:3,4][1,2,4]triazino[5,6-*b*]-[1,5]benzo**xazepin**-6-one, 724
Pyrazolo[1,5-*a*]-**1,3,5-triazin-2-one**, 1039
Pyrazolo[3,2-*c*][1,2,4]triazin-4-one, 983
Pyrazolo[3,4-*d*]-**1,2,3-triazine-4-one** 2-oxide, 747
Pyrazolo[5′,1′:3,4][1,2,4]triazino[6,5-*f*]-[**1,3,4**]**thiadiazepine**, 750, 926
Pyrazolo[2′,3′:2,3][**1,2,4**]**triazolo**[1,5-*a*]-pyridine, 1001
Pyridazino[3,4-*b*][**1,5**]benzo**diazocine**-5,7-**dione**, 1072
Pyridazino[1,6-*b*]**indazole**, 768
Pyridazino[4,3-*b*]**indole**, 1024
Pyridazino[6,1-*a*]**isoindole**, 1055
Pyridazino[1,2-*b*]phthalazine-6,11-dione, 1159
Pyridazino[4,5-*d*]pyridazine, 834
Pyridazino[4,5-*c*]pyridazin-8-one, 707
Pyridazino[4′,3′:5,6]**pyrimido**[2,1-*b*]-benzothiazol-5-one, 1069
Pyridazino[3,2-*b*]**quinazolin**-10-**one**, 1116
Pyridazino[4,5-*b*]**quinolin**-10-**one**, 706, 813
Pyridazino[4,5-*b*]quinoxaline-1,4-**dione**, 1160
Pyridazino[4,5-*e*][**1,3,4**]**thiadiazin**-8-one, 1039
Pyrido[4,3,2-*mn*]acridine, 1027
Pyrido[1,2-*a*]benz**imidazole**, 703, 999, 1129, 1313, 1331, 1338
Pyrido[1,2-*a*]benzimidazol-10-**ium**, 1059
Pyrido[1,2-*a*]benzimidazol-1-**one**, 1325
Pyrido[1,2-*b*][**2,4**]benzo**diazepin-6-one**, 953
Pyrido[2,3-*b*][**1,4**]benzo**diazepin-6-one**, 711
Pyrido[3,2-*b*][1,4]benzothiazine, 907
Pyrido[3,2-*a*]benzothiazol-10-**ium**, 1059
Pyrido[2,1-*b*]benzothiazol-1-**one**, 1114, 1324
Pyrido[2,3-*b*][**1,4,5**]benz**oxadiazepine**, 1157
Pyrido[3,2-*b*][**4,1**]benz**oxazepine**, 711
Pyrido[2,1-*b*][1,3]benzoxazin-10-one, 1211
Pyrido[3,2-*a*]benzoxazol-10-ium, 1059
Pyrido[2,1-*b*]benzoxazol-1-**one**, 1324
Pyrido[2,3-*c*]carbazole, 989
Pyrido[3,2-*b*]carbazole, 989
Pyrido[3,4-*a*]**carbazole**, 1024
Pyrido[4,3-*a*]**carbazole**, 1024
Pyrido[4,3-*b*]carbazole, 678
Pyrido[4,3-*c*]carbazol-1-**one**, 1028
Pyrido[3,4-*b*][**1,4**]**diazepin**-2-**one**, 954
Pyrido[3,4-*b*][1,4]**diazepin-4-one**, 1139
Pyrido[3,2-*b*:5,4-*b*′]di**indole**, 1252
Pyrido[1′,2′:1,2]imidazo[4,5-*f*]-**2,1,3**-benz**oxadiazole** 1-**oxide**, 1016
Pyrido[1′,2′:1,2]**imidazo**[4,5-*b*]pyrazine, 703, 984, 1165
Pyrido[1,2-*b*]**indazole**, 1031

Pyrido[1,2-*a*]indole, 1340
Pyrido[2,3-*b*]indole, 887, 1248
Pyrido[3,4-*b*]indole, 1191, 1295
Pyrido[4,3-*b*]indole, 1058, 1296
Pyrido[1,2-*a*]indole-6,10-di**one**, 1325
Pyrido[3,2-*a*]indolizin-**2-one**, 949
Pyrido[3,2-*b*]indol-**4-one**, 941
Pyrido[2,1-*a*]**isoindol**-6-**one**, 1224
Pyrido[2,3-*h*]-1,6-naphthyridine, 770
Pyrido[3,2-*g*]**pteridine**-2,4-dione (paz), 1133
Pyrido[2,3-*b*]**pyrazine**, 1133
Pyrido[2,3-*b*]**pyrazine-2,3-dione**, 1101
Pyrido[2,3-*b*]**pyrazin**-3-**one**, 961
Pyrido[2′,3′:3,4]pyrazolo[1,5-*a*]quinolin-9-**one**, 949
Pyrido[2,3-*c*]pyridazine, 1089
Pyrido[2,3-*d*]**pyridazine**, 846
Pyrido[3,4-*d*]pyridazine, 1201
Pyrido[2,3-*d*]pyridazine-**2**,5-di**one**, 918
Pyrido[3,4-*d*]**pyridazine**-1,4-**dione**, 1160
Pyrido[3,4-*d*]**pyridazine**-1,7-di**one**, 1074
Pyrido[2,3-*b*]**pyridazin-3-one**, 961
Pyrido[2,3-*c*]**pyridazin-4**-one, 1148
Pyrido[2,3-*d*]pyridazin-2-**one**, 681, **973**
Pyrido[3,4-*d*]pyridazin-5-**one**, 1202
Pyrido[2′,3′:4,5]pyridazino[6,1-*b*]-**quinazolin-8-one**, 1242
Pyrido[2,3-*d*]**pyrimidin-4-one**, **2-thioxo**-, 1100
Pyrido[1,2-*a*]**pyrimidine**, 1005
Pyrido[2,3-*d*]pyrimidine, 705
Pyrido[2,3-*d*]pyrimidine-2,4-dione, 887, 1003, 1297
Pyrido[3,2-*d*]pyrimidine-2,4-dione, 966
Pyrido[3,2-*d*]**pyrimidine-2,4-dione**, 1099
Pyrido[3,4-*d*]pyrimidine-2,4-dione, 842
Pyrido[1,2-*a*]**pyrimidin-2-one**, 1010, 1011
Pyrido[1,2-*a*]**pyrimidin-4-one**, **930, 1007**, 1032, 1116
Pyrido[2,3-*d*]**pyrimidin-2-one**, 975, 1144
Pyrido[2,3-*d*]**pyrimidin-4-one**, 1069
Pyrido[2,3-*d*]pyrimidin-4-one, 857, 991
Pyrido[2,3-*d*]pyrimidin-5-**one**, 950, 1110
Pyrido[2,3-*d*]**pyrimidin**-7-one, 1180
Pyrido[3,4-*d*]**pyrimidin-4-one**, 951
Pyrido[2,3-*d*]**pyrimidin-4-one**, 2-**thioxo**-, 695, 1100
Pyrido[3,2-*d*]**pyrimidin-4-one**, 2-**thioxo**-, 952, 1043, **1083**
Pyrido[2′,1′:2,3]**pyrimido**[4,5-*c*]pyridazin-5-**one**, 1069
Pyrido[3′,2′:5,6]**pyrimido**[1,2-*a*]-quinazolin-5-**one**, 1242
Pyrido[2,3-*e*]pyrrolo[1,2-*a*]**pyrazin**-6-**one**, 994
Pyrido[3′,2′:4,5]pyrrolo[1,2-*c*]**pyrimidin**-9-**one**, 7-**thioxo-**, 1038
Pyrido[3,4,5-*de*]**quinazoline**, 1141
Pyrido[4,3,2-*de*]**quinazoline**, 1141
Pyrido[3,4,5-*de*]**quinazoline**-2-**thione**, 1144
Pyrido[1,2-*a*]**quinazolin**-6-**one**, 1242
Pyrido[2,1-*b*]**quinazolin**-11-**one**, 1242
Pyrido[2,1-*b*]**quinazolin**-5-**one**, 1007
Pyrido[3,2-*g*]quinoline-5,10-dione, 1192
Pyrido[3,4-*g*]quinoline-5,10-dione, 1190
Pyrido[4,3-g]quinoline-5,10-dione, 1192
Pyrido[3,2-*g*]quinoline-**2**,5,8,10-tetra**one**, 867
Pyrido[2,3-*g*]quinoline-2,5,10-trione, 1191
Pyrido[2,3-*b*]**quinolin**-5-**one**, 706
Pyrido[2,1,6-*de*]quinolizine, 1109
Pyrido[3,2-*a*]quinolizine-1,6-di**one**, 1095
Pyrido[2,3-*g*]quinoxaline, 1192
Pyrido[1,2-*a*]quinoxaline-1,5-dione, 1115
Pyrido[1,2-*a*]quinoxaline-1,6-di**one**, 1325
Pyrido[2,3-*b*][**1,4**]**thiazepin**-2-**one**, 711
Pyrido[2,3-*b*][**1,4**]**thiazine**, 731, 749, 1103
Pyrido[2,3-*d*]**thiazolo[3,2**-*a*]pyrimidin-5-one, 876
Pyrido[3,2-*b*]thieno[2,3-*f*][**1,4**]**oxazepine**, 851
Pyrido[4′,3′:4,5]thieno[2,3-*d*]**pyrimidine**, 1050
Pyrido[2,3-*f*]-**1,2,4-triazepin-5-one**, 1149
Pyrido[1,2-*b*][**1,2,4**]**triazine**, 994
Pyrido[3,4-*e*]-**1,2,4-triazine**, 686, 961
Pyrido[4,3-*e*]-**1,2,4-triazine** **1-oxide**, 785
Pyrido[2,1-*f*][**1,2,4**]**triazin**-9-**ium**, 748
Pyrido[2,1-*f*][**1,2,4**]**triazin**-9-**ium**, 4-oxo-, 748
Pyrido[1,2-*a*]-**1,3,5-triazin-4-one**, 1013
Pyrido[1,2-*b*][**1,2,4**]**triazin**-6-one, 1136
Pyrido[1,2-*a*]-**1,3,5-triazin-4-one, 2-thioxo**-, 687
Pyrido[2,3-*c*]-**1,2,4-triazolo**[4,3-*a*]azepin-**3-one**, 1239
Pyrido[2,3-*e*][**1,2,4**]**triazolo**[4,3-*a*]-pyrimidin-5-one, 960
Pyrido[3,2-*d*][**1,2,4**]**triazolo**[1,5-a]-pyrimidin-5-one, 960
Pyrimido[4,5-*b*]azepin-**4-one**, 857
Pyrimido[1,2-*a*]benzimidazole, 1005

Pyrimido[1,6-*a*]benzimidazole-1,3-**dione**, 1326
Pyrimido[1,2-*a*]benzimidazol-2-**one**, 1010
Pyrimido[1,2-*a*]benzimidazol-**4-one**, 930, **1008**
Pyrimido[4,5-*b*][**1,5**]benzo**diazepin-2-**one, 749
Pyrimido[5,4-*b*][**1,4**]benzo**thiazine**-2,4-dione, 731
Pyrimido[1,2-*a*]benzothiazol-2-**one**, 1010
Pyrimido[2,1-*b*]benzothiazol-**4-one**, 1008
Pyrimido[2,1-*b*]benzoxazol-**4-one**, 930
Pyrimido[4,5-*c*]**cinnoline**-1,3-dione, 1257
Pyrimido[4,5-*c*]-**1,2-diazepine**-6,8-dione, 1259
Pyrimido[4,5-*b*][**1,4**]**diazepin-4-one**, 977
Pyrimido[1,2-*b*]indazole, 1004
Pyrimido[1,2-*a*]**indole**, 1329
Pyrimido[1,6-*a*]**indole**, 1329
Pyrimido[3,4-*a*]indole-1,3-**dione**, 1326
Pyrimido[5,4-*b*]indole-**2,4-dione**, 1043
Pyrimido[4,5-*b*]indol-**2-one**, 856
Pyrimido[3,4-*a*]indol-1-**one**, 3-**thioxo**-, 1326
Pyrimido[4,5-*c*]isoquinolin-1-**one**, 924
Pyrimido[1,2-*a*][1,8]naphthyridin-10-**one**, 1116
Pyrimido[5,4-*b*][**1,4**]**oxazine**, 721
Pyrimido[5,4-*b*][**1,4**]**oxazine**-4,7-di**one**, 944
Pyrimido[5,4-*b*][**1,4**]**oxazin**-7-**one**, 721, 722
Pyrimido[5,6,1-*kl*]phenothiazin-3-**one**, 1-**thioxo**-, 1306
Pyrimido[1,2-*e*]purine-2,4-dione, 889
Pyrimido[2,1-*f*]purine-2,**4**,8-tri**one**, 1011
Pyrimido[1,2-*a*]purin-10-one, 1004
Pyrimido[4′,5′:6,5]pyrano[3,2-*h*]quinoline, 745
Pyrimido[4′,5′:6,5]pyrano[3,2-*h*]quinolin-8-**one**, 747
Pyrimido[4,5-*c*]**pyridazine**, 1068, 1148
Pyrimido[1,2-*b*]pyridazin-2-**one**, 1007, 1068, 1241
Pyrimido[1,2-*b*]pyridazin-**4-one**, 1009
Pyrimido[4,5-*c*]**pyridazin-4**-one, 1148
Pyrimido[4,5-*d*]pyridazin-**2-one**, 793
Pyrimido[4,5-*d*]pyrimidine, 764
Pyrimido[4,5-*d*]pyrimidine-2,4-dione, 764
Pyrimido[4,5-*d*]pyrimidine-2,**4**,5-tri**one**, 856
Pyrimido[1,2-*a*]**pyrimidin-2-one**, 920, 1011, 1241
Pyrimido[1,2-*a*]**pyrimidin-4-one**, 1009
Pyrimido[1,6-*a*]**pyrimidin-4-one**, 1116
Pyrimido[4,5-*d*]**pyrimidin-4-one**, 937
Pyrimido[5,4-*d*]**pyrimidin-4-one**, **2-thioxo**-, 1100
Pyrimido[4′,5′:4,5]pyrrolo[2,3-*c*]azepine-**4**,6-di**one**, 924
Pyrimido[1′,2′:1,2]pyrrolo[3,4-*b*]quinolin-**4-one**, 1310
Pyrimido[4,5-*b*]**quinoline**-2,4-dione, 813, 1227
Pyrimido[5,4-*b*]**quinoline**-2,4-dione, 1108
Pyrimido[2,1,6-*de*]quinolizine, 738
Pyrimido[4,5-*b*]**quinoxaline**-2,4-dione, 1195, 1310
Pyrimido[4,5-*b*]**quinoxaline**-2,4-dione **5-oxide**, 754
Pyrimido[4,5-*b*]**quinoxalin**-4-one, 1028
Pyrimido[1,2-*b*]-**1,2,4,5-tetrazin**-6-one, 1255
Pyrimido[4,5-*b*][**1,4**]**thiazine**, 748
Pyrimido[1,6-*b*][**1,2,4**]**triazine**, 967
Pyrimido[5,4-*e*]-**1,2,4-triazine**, 961
Pyrimido[1,2-*b*][**1,2,4**]**triazine**-2,6-di**one**, 1137
Pyrimido[1,2-*b*][**1,2,4**]**triazine**-2,6-di**one**, 962
Pyrimido[4,5-*e*]-1,2,4-triazine-6,8-**dione**, 1070
Pyrimido[4,5-*e*]-**1,2,4-triazine**-6,8-dione, 759
Pyrimido[5,4-*e*]-**1,2,4-triazine**-5,7-dione, 1265
Pyrimido[5,4-*e*]-**1,2,4-triazine**-5,7-dione **4-oxide**, 1265
Pyrimido[1,6-*a*]-**1,3,5-triazine-2,4**,6-tri**one**, 1014
Pyrimido[4,5-*e*]-**1,2,4-triazine-3**,6,8-tri**one**, 962
Pyrimido[5,4-*e*]-**1,2,4-triazine-3**,5,7-tri**one**, 1137
Pyrimido[5,4-*e*]-**1,2,4-triazine-3**,5,7-tri**one**, 687, 709
Pyrimido[4,5-*e*]-1,2,4-triazin-8-**one**, 1069
Pyrrolizine, 873, **1055**
1-Pyrrolizinone (prr3one), 1057
3-Pyrrolizinone (prr2one), 783, 873, 1113
Pyrrolizino[2,1-*b*]quinolin-5-**one**, 860
Pyrrolo[3,2,1 -*de*]acridine-1,2-**dione**, 1337
Pyrrolo[1,2-*a*]**azepin**-3-one, 1207
Pyrrolo[1,2-*a*]**azepin-9-one**, 1207
Pyrrolo[1,2-*a*]benz**imidazole**, 1313, 1338

Pyrrolo[1,2-*a*]benzimidazole, 890, 911, 1030
Pyrrolo[3,4-*b*][**1,5**]benzo**diazepine**-1,3-dione, 1138
Pyrrolo[2,1-*c*][**1,4**]benzo**diazepin-5-one**, 674
Pyrrolo[1,2-*b*][**1,2,5**]benzo**thiadiazepine** 5,5-**dioxide**, 773
Pyrrolo[1,2-*a*][**3,1,6**]benzo**thiadiazocin**-6-**one**, 956
Pyrrolo[3,4-*c*][**1,5**]benzo**thiazepine**-1,3-dione, 724
Pyrrolo[3,4-*b*][**1,4**]benzo**thiazine**-1,3-dione, 1167
Pyrrolo[2,1-*b*]benzo**thiazole**, 1027
Pyrrolo[2,3-*e*]benz**oxazole**, 1283
Pyrrolo[3,2-*c*]**cinnoline**, 1153
Pyrrolo[1,2-*b*]**cinnolin**-10-**one**, 708
Pyrrolo[1,2-*a*]imidazole, 1106
Pyrrolo[1,2-*a*]**imidazole**, 862
Pyrrolo[1,2-*c*]imidazole-3,5-di**one**, 873
Pyrrolo[1,2-a]imidazol-5-**one**, 913
Pyrrolo[1,2-*a*]**indole**, 861
Pyrrolo[1,2-*a*]indole, 1223
Pyrrolo[2,3-*b*]indole, 1308
Pyrrolo[2,3-*e*]indole, 1251
Pyrrolo[1,2-*a*]indole-2,3-**dione**, 1337
Pyrrolo[2,1,5-*cd*]indolizine, 1183
Pyrrolo[2,1,5-*cd*]indolizine, 905
Pyrrolo[1,2-*a*]indol-1-**one**, 861, 1087
Pyrrolo[3,4-*b*]indol-3-**one**, 1261
Pyrrolo[2,1-*a*]isoquinoline, 1200, **1318**
Pyrrolo[1,2-*b*]**isoquinolin**-5-**one**, 868
Pyrrolo[3,2-*c*][**1,8**]**naphthyridine**, 770
Pyrrolo[2,1-*c*][**1,4**]**oxazin**-1-**one** 1243
Pyrrolo[2,3-*d*][**1,3**]**oxazin**-4-**one**, 952, 953
Pyrrolo[3,2,1-*de*]**phenanthridin**-7-**one**, 1153
Pyrrolo[1,2-*f*]**phenathridine**, 1228
Pyrrolo[1,2-*a*]pyrazine, 1055
Pyrrolo[3,4 -*b*]pyrazine, 894
Pyrrolo[1,2-*a*]**pyrazin**-1-**one**, 814
Pyrrolo[3,4-*c*]**pyrazole**-4,6-dione, 1048
Pyrrolo[3,4-*c*]**pyrazol**-4-one, 829
Pyrrolo[1,2-*b*]pyridazine, 1055, **1317**
Pyrrolo[1,2-*b*]**pyridazine**, 993
Pyrrolo[2,3-*d*]pyridazine 5-oxide, 1209
Pyrrolo[1,2-*b*]**pyridazin**-2-**one**, 868
Pyrrolo[3,4-*c*]**pyridazin-4-one**, 1148
Pyrrolo[3,4-*d*]pyridazin-5-**one**, 799
Pyrrolo[2,3-*b*]pyridine, 681
Pyrrolo[2,3-*b*]**pyridine**, 1003
Pyrrolo[3,2-*b*]pyridine, 734, 1248
Pyrrolo]3,2-*c*]pyridine, 681, 1248
Pyrrolo[3,4-*c*]pyridine, 894
Pyrrolo[3,4-*c*]pyridine-1,3-**dione**, 1160
Pyrrolo[3,2-*c*]pyridine 5-oxide, 885
Pyrrolo[2,3-*b*]pyridin-**2**-**one**, 736
Pyrrolo[2,3-*c*]**pyridin-**7-**one**, 679, 1042
Pyrrolo[3,2-*c*]**pyridin**-6-**one**, 929
Pyrrolo[1,2-*a*]pyrimidine, 1055
Pyrrolo[1,2-*c*]**pyrimidine**, 880
Pyrrolo[2,3-*d*]**pyrimidine**, 744
Pyrrolo[2,3-*d*]pyrimidine, 810
Pyrrolo[3,2-*d*]pyrimidine, 670
Pyrrolo[2,3-*d*]**pyrimidine-2,4-dione**, 1019
Pyrrolo[2,3-*d*]pyrimidine-2,4-dione, 860, 1248
Pyrrolo[3,2-*d*]**pyrimidine**-**2,4**-**dione**, 1043
Pyrrolo[2,3-*d*]pyrimidine-**2**,4,6-tri**one**, 947
Pyrrolo[1,2-*a*]**pyrimidin-4-one**, 1116
Pyrrolo[1,2-*c*]**pyrimidin**-3-**one**, 1060
Pyrrolo[2,3-*d*]pyrimidin-4-one, 810, 981
Pyrrolo[2,3-*d*]**pyrimidin**-**4**-**one**, 746, 857
Pyrrolo[3,2-*d*]pyrimidin-4-one, 735
Pyrrolo[1,2-*a*]pyrrole, 768
Pyrrolo[3,4-*b*]pyrrole, 853, 1064
Pyrrolo[2,3-*b*]**pyrrol-2**-one, 1080
Pyrrolo[3,2-*b*]pyrrol-5-one, **2-thioxo**-, 905
Pyrrolo[2,1-*b*]**quinazolin**-9-**one**, 936
Pyrrolo[1,2-*a*]**quinoline**, 1108, 1323
Pyrrolo[1,2-*a*]quinoline, 1319
Pyrrolo[2,3-*f*]**quinoline**, 988
Pyrrolo[2,3-*f*]quinoline, 1329
Pyrrolo[3,2,1-*ij*]quinoline, 912
Pyrrolo[3,2-*b*]**quinoline**, 672
Pyrrolo[3,2-*b*]quinoline, 735
Pyrrolo[3,2-*c*]quinoline, 1223
Pyrrolo[3,4-*b*]**quinoline**, 672
Pyrrolo[3,2,1-*ij*]quinolin-**2-one**, 702
Pyrrolo[3,2,1-*ij*]**quinolin**-**4**-**one**, 880
Pyrrolo[3,2,1-*ij*]quinolin-4-one, 1237
Pyrrolo[3,2,1-*ij*]**quinolin**-6-**one**, 1110
Pyrrolo[3,4-*b*]quinolin-3-**one**, 768
Pyrrolo[3,2-*c*]quinolin-**2**-**one** 5-oxide, 1330
Pyrrolo[2,1,5-*de*]**quinolizin**-3-**one** (p3one), 843
Pyrrolo[2,3-*b*]quinoxaline, 1064
Pyrrolo[3,4-*b*]**quinoxaline**-1,3-dione, 1167
Pyrrolo[3,4-*b*]quinoxaline 4-oxide, 1164
Pyrrolo[1,2,3-*de*]**quinoxalin**-5-**one**, 1242
Pyrrolo[2,1-*b*]thiazole, 840, 912
Pyrrolo[3,4-*d*]**thiazole**-4,6-dione, 1067
Pyrrolo[2,1-*b*]thiazol-5-**one**, 913

Pyrrolo[1,2-*c*]thieno[3,2-*e*]pyrimidine, 907
Pyrrolo[1,2 -*a*]thieno[2,3-*d*]pyrimidin-4-one, 1318
Pyrrolo[2,1-*f*][1,2,4]triazine, 1336

Quinazoline, 682, 744, 746, 919, 929, 1042, 1068, 1299
Quinazoline 3-oxide, 691, 792, 958
2,4-Quinazolinedione, 938, 1075, 1099
5,8-Quinazolinedione, 974c
2-Quinazolinethione, 975, 1083
4-Quinazolinethione, 930
Quinazolino[2,3-*c*][1,4]benzothiazin-12-**one**, 880
2-Quinazolinone, 692, 746, 918, 975, 1037, 1083, 1098, 1143
4-Quinazolinone, 692, 695, 857, 924, 930, 935, 937, 942, 951, 1043, 1068
4-Quinazolinone, 2-thioxo-, 936, 1103
Quino[2,3-*b*][**1,5**]benz**oxazepine**, 816
Quinoline, 671, 681, 706, 737, 742, 770, 834, 866, 867, 887, 916, 917, 1027, 1090, 1098, 1108, 1227, **1296**
Quinoline 1-oxide, 849, 889
2-Quinolinethione, 1109
2-Quinolinone, 673, 694, 743, 771, 842, 867, 941, 1090, 1211, 1229, 1305, 1326
4-Quinolinone, 706, 707, 737, 753, 935, 950, 1091, 1110, 1205, 1212
5-Quinolinone, 988, 1285, 1296
Quinolizinium, 1059
4-Quinolizinone (p4one), 679, 1324
Quino[4,3,2-*de*][1,10]phenanthrolin-9-one, 1193
Quinoxaline, 708, **1132**
2,3-Quinoxalinedione, 1101
2-Quinoxalinone, 836, 1100, **1134**
2-Quinoxalinone 4-oxide, 836

Spiro[thiopyran-4,11′-thiopyrano[4,3-*b*]-[**1,5**]benzo**diazepine**], 1138
Spiro[thiopyran-4,11′-thiopyrano[4,3-*b*]-[**1,5**]benzo**diazepine**] 3,3-dioxide, 1138

2,3,5,6-Tetraazabicyclo[5.2.0]nona-1(7),3-diene-8,9-dione, (1,2,4,5tetrazep), 1141
1,2,4,6-Tetrazepino[1,7-*c*]quinazoline-2-**thione**, 1104
1,2,4,6-Tetrazepino[1,7-*c*]quinazolin-2-**one**, 1104
Tetrazolo[1,5-*b*]pyridazine, 1019
Tetrazolo[1′,5′:1,6]pyridazin[4,3-*b*]**indole**, 1025
Tetrazolo[1,5-*b*][**1,2,4**]**triazine**, 1136
4-Thia-1-azabicyclo[3.2.0]heptan-7-one (thzol), 916
4-Thia-1-azabicyclo[3.2.0]hept-2-en-7-one (thzol), 719, 854
4-Thia-2-azabicyclo[4.2.0]oct-2-en-8-one (1,4thzn), 1175
5-Thia-1-azabicyclo[4.2.0]oct-2-en-8-one 5,5-dioxide (1,3thznO_2), 843
6-Thia-2,3-diazabicyclo[3.2.0]hept-2-ene 6,6-dioxide (pzol), 1185
7-Thia-2,3-diazabicyclo[3.2.0]hept-2-ene 7,7-dioxide (pzol), 1185
1,2,4-Thiadiazolo[4,5-*a*]benzimidazole-**3-thione**, 1097
[**1,2,4**]**Thiadiazolo**[3,4-*b*]benzothiazole, 1002, 1240
[1,2,5]Thiadiazolo[3,2-*b*:4,5-*b*′]-di**quinazoline**-11,15-di**one** 13,13-dioxide, 1116
[**1,2,3**]**Thiadiazolo**[5,4-*h*]isoquinoline, 720
[**1,2,4**]**Thiadiazolo**[3,4-*b*][1,3,4]oxadiazole, 776
[**1,2,4**]**Thiadiazolo**[4,3-*a*]pyrazine, 1002
[**1,2,4**]**Thiadiazolo**[2,3-*a*]pyridine, 1036
[**1,2,4**]**Thiadiazolo**[4,3-*a*]pyridine, 986, 1002
[1,3,4]Thiadiazolo[4,5-*a*]**pyridine**, 1265
[**1,2,4**]**Thiadiazolo**[4,3-*c*]pyrimidine, **1002**
1,3,4-Thiadiazolo[3,2-*a*]pyrimidin-**4**-ium, 730
[**1,2,4**]**Thiadiazolo**[4,3-*a*]pyrimidin-**3-one**, 1002
[1,2,4]**Thiadiazolo**[4,3-*c*]pyrimidin-**3-one**, 1002
1,3,4-Thiadiazolo[3,2-*a*]pyrimidin-5-one, 729
1,3,4-Thiadiazolo[3,2-*a*]**pyrimidin**-5-**one**, 930, 1241
1,3,4-Thiadiazolo[3,2-*a*]**pyrimidin**-7-**one**, 1011
[**1,3,4**]**Thiadiazolo**[2,3-*b*]quinazolin-9-one, 729
[**1,2,5**]**Thiadiazolo**[3,4-*h*]quinoline, (3980.) 1284
[1,2,5]Thiadiazolo[3,4-*b*]**quinoxaline** 2,2-dioxide, 1173
[1,2,5]Thiadiazolo[3,4-*c*][1,2,5]thiadiazole, 1131

[1,3,4]Thiadiazolo[2,3-*c*][1,2,4]thiadiazole, 776, 1002
[1,3,4]Thiadiazolo[3,2-*a*][**1,3,5**]**triazine**-5,7-**dione**, 1014
1,2,3-Thiadiazolo[5,4-*c*][1,2,4]triazolo[1,2-*a*]**pyridazine**-7,9-dione, 907
2aλ^4-Thia-2,3,4a,7a-tetraazacyclopent[*cd*]-indene-1,4-dione (1,2,4-thiadiazole), 1339
7-Thia-1,2,12c-triazacyclopenta[*fg*]-naphthacene (1,3thzn), 1217
1,2,3,5-Thiatriazolo[4,5-*c*][1,4]-benzoxazine 1-**oxide**, 777
[1,3]Thiazino[3,2-*a*]benz**imidazole**, 1338
[**1,3**]**Thiazino**[3,2-*a*]benzimidazol-**4-one**, 883
[**1,3**]**Thiazino**[6,5-*c*]quinolin-1-**one**, 1080, 1075
Thiazolo[3,2-*a*]benzimidazole, 876, 877, 879
Thiazolo[3,4-*a*]benzimidazole, 1239, 1303
Thiazolo[3,2-*a*]benzimidazol-3-**one**, 878, 1114
Thiazolo[3,2-*a*][**1,3,5**]benzo**triazepine**, 968
Thiazolo[3,2-*a*][1,3]diazepine, 877
Thiazolo[5,4-*b*]**indole**, 1330
Thiazolo[5,4-*b*]indole, 728
Thiazolo[5,4-*h*]isoquinolin-2-**one**, 720
Thiazolo[3,2-*a*]perimidine, 876
Thiazolo[3,2-*e*]purine, 878
Thiazolo[2,3-*f*]purine-2,4-dione, 1309
Thiazolo[2,3-*f*]purin-4-one, 878
Thiazolo[4,5-*b*]pyrazine, 1036
Thiazolo[5,4-*c*]pyridine, 718
Thiazolo[3,2-*a*]pyridin-4-ium, 1309
Thiazolo[3,2-*a*]pyridinium, 877, 916
Thiazolo[3,2-*a*]**pyridinium**, 1192
Thiazolo[2,1-*f*]pyrido[2,3-*c*][1,2,4]triazin-10-ium, 5-oxo-, 876
Thiazolo[3,2-*a*]pyrimidine, 877
Thiazolo[4,5-*d*]pyrimidine, 782
Thiazolo[5,4-*d*]**pyrimidine**, 691
Thiazolo[4,5-*d*]pyrimidine-5,7-dione, 1322
Thiazolo[4,5-*d*]py**rimidine**-5-**thione**, 764
Thiazolo[3,2-*a*]**pyrimidin**-5-**one**, 1008, 1011, 1012
Thiazolo[3,2-*a*]**pyrimidin-7-one**, 1011
Thiazolo[4,5-*d*]**pyrimidin**-5-**one**, 856
Thiazolo[2′,3′:2,3]**pyrimido**[4,5-*c*]-pyridazin-5-**one**, 1069
Thiazolo[2,3-*b*]quinazolin-5-one, 877
Thiazolo[5,4-*f*]**quinoline**, 1108
Thiazolo[4,5-*b*]quinoxaline, 1167
Thiazolo[2,3-*c*][**1,2,4**]**thiadiazole**, 776, 1002
Thiazolo[4,3-*b*]-**1,3,4-thiadiazole**-5-thione, 729
Thiazolo[2,3-*c*][**1,2,4**]**thiadiazol**-3-**one**, 1002
Thiazolo[2,3-*c*][**1,2,4**]**triazepin**-3-**one**, 1243
Thiazolo[3,2-*a*]-**1,3,5-triazine-2,4-dione**, 931
Thiazolo[3,2-*a*]-**1,3,5-triazin-4-one**, 1038
Thiazolo[2,3-*c*]-1,2,4-triazole, 915
Thiazolo[3,2-*b*][**1,2,4**]**triazole**, 1130
Thiazolo[3,2-*b*][1,2,4]triazole, 878, 915
Thiazolo[3,2-*b*][**1,2,4**]**triazole**, 741
Thiazolo[3,2-*c*]-1,2,4-triazole, 877
Thiazolo[3,2-*c*][**1,2,3**]**triazole**, 913
Thiazolo[3,4-*b*][**1,2,4**]**triazole**-5-thione, 728
Thiazolo[3,2-*b*][1,2,4]triazol-4-ium, 876
Thieno[2,3-*b*][**1,5**]benzo**diazepin**-4-**one**, 944, 954
Thieno[2,3-*g*][1]benzo**pyran**-6-**one**, 1288
Thieno[3,2-*f*][1]benzo**pyran**-7-**one**, 1288
Thieno[3,4-*c*][1]benzopyran-4-one, 1089
Thieno[3,2-*e*][**2,1,3**]benzo**thiadiazole**, 1131
Thieno[2,3-*b*][**1,4**]benzo**thiazine**, 710
Thieno[3,2-*b*][**1,4**]benzo**thiazine**, 710
Thieno[3,4-*b*][**1,4**]benzo**thiazine**, 710
Thieno[2,3-*b*][1]benzothiophene, 676, 865
Thieno[3,2-*b*][1]benzothiophene, 676, 1282
Thieno[3,2-*b*][**1,4**]benz**oxazepine**, 851
Thieno[2,3-*e*]-**1,4-diazepin-2-one**, 978
Thieno[3,4-*e*]-**1,4-diazepin-2-one**, 978
Thieno[3,4-*d*][**1,3**]**dioxol-2-one** 5,5-dioxide, 820
Thieno[3,2-*b*]furan-2-one, 1066
Thieno[3,2-*b*]**furan-3-one**, 1088
Thieno[3,4-*b*]**furan-3-one**, 1088
Thieno[3,4-*b*]indole **2,2-dioxide**, 1166
Thieno[2,3-*g*]**indolizine** (prr), 905, 1040
Thieno[3,2-*e*]**indolizine** (p), 917
Thieno[3,2-*c*][**1,8**]**naphthyridine**, 705
Thieno[2,3-*c*]**pyran**-5-**one**, 803
Thieno[3,2-*c*]**pyran-6-one**, 803
Thieno[2,3-*b*]pyrazine, 1089
Thieno[3,2-*c*]**pyrazole**, 1031
Thieno[2,3-*b*]pyridine, 1309
Thieno[2,3-*b*]**pyridine**, 742, 1003
Thieno[3,2-*b*]**pyridine**, 1003
Thieno[3,2-*c*]**pyridine**, 866, 1179, 1295
Thieno[3,4-*b*]**pyridine**, 950
Thieno[3,4-*b*]pyridine, 1089

Thieno[3,4-*c*]pyridine-4-thione, 1089
Thieno[2,3-*b*]pyridin-7-ium, 787
Thieno[2,3-*b*]**pyridin**-4-**one**, 950
Thieno[2,3-*c*]**pyridin**-7-**one**, 791
Thieno[3,2-*c*]**pyridin**-4-**one**, 791
Thieno[2′,3′:3,4]**pyrido**[1,2-*c*]pyrimidin-7-one, 906
Thieno[3′,2′:3,4]**pyrido**[1,2-*c*]pyrimidin-4-one, 906
Thieno[3′,2′:5,6]pyrido[2,3-*d*]**pyrimidin-4-one**, 1069
Thieno[2,3-*d*]pyrimidine, 812, 1082
Thieno[2,3-*d*]**pyrimidine**, 744
Thieno[2,3-*d*]pyrimidine-2,4-dione, 865
Thieno[2,3-*d*]pyrimidine-4,6-dione, 1282, 1322
Thieno[2′,3′:4,5]**pyrimido**[1,2-*b*][1,2]-benzisothiazol-**4-one** 6,6-dioxide, 1060
Thieno[2,3-*b*]**pyrrole**, 740, **1087**
Thieno[2,3-*b*]pyrrole, 863
Thieno[2,3-*c*]**pyrrole**, 885, 894
Thieno[3,2-*b*]pyrrole, 863
Thieno[3,2-*b*]**pyrrole**, 885
Thieno[3,4-*c*]**pyrrole**, 885
Thieno[2,3-*c*]**quinoline**, 705
Thieno[3,4-*c*]**quinoline**, 705
Thieno[2,3-*c*]**quinoline** 5-**oxide**, 850
Thieno[3,2-*c*]**quinoline** 5-**oxide**, 850
Thieno[3,4-*c*]**quinoline** 5-oxide, 850
Thieno[2,3-*b*]quinoline-4,6-dione, 1165
Thieno[3,2-*c*]**quinolin**-4-one, 1194
Thieno[2,3-*c*]**quinolizine**, (917
Thieno[2,3-*b*]quinoxaline, 897
Thieno[3,2-*f*]**quinoxaline**, 1133
Thieno[3,2-*c*][**1,2,6**]**thiadiazin-4-one 2,2-dioxide**, 1122
Thieno[3,4-*c*][**1,2,6**]**thiadiazin-4-one 2,2-dioxide**, 1121
Thieno[3,4-*d*]-**1,2,3**-thiadiazole, 812
Thieno[2,3-*e*][**1,4**]**thiazepin-2**-one, 945
2λ^4-Thieno[2,3-*e*][**1,2**]**thiazine**, 1021
2λ^4-Thieno[3,2-*c*][**1,2**]**thiazine**, **1021**
Thieno[2,3-*d*][**1,3**]**thiazin**-4-**one**, 1102
Thieno[2,3-*d*]**thiazole**, 717
Thieno[2′,3′:4,5]thieno[2,3-*d*]**thiazole**, 717
Thieno[2,3-*b*]thiophene, 842, 1057
Thieno[2,3-*e*][**1,2,4**]**triazolo[3,4-*c*][1,4]**-thiazepine, 875
Thiepino[4,3-*b*]benzo**furan** 2,2-dioxide, 1256
Thiopyrano[3,2-*b*]benzo**furan**, 1256
Thiopyrano[3,2-*b*]benzo**furan** 1,1-dioxide, 1256
Thiopyrano[4,3-*b*]benzo**furan** 2,2-dioxide, 1256
Thiopyrano[3,2-*c*][1]benzopyran-4,5-**dione**, 903
Thiopyrano[3,2-*b*]benzothiophen-4-**one**, 870
Thiopyrano[3,2-*b*]**indole**, 1249
Thiopyrano[3,4-*b*]**indole**, 1249
Thiopyrano[4,3-*b*]indole, 1061, 1111
Thiopyrano[2,3-*b*]indol-1-ium, 909
Thiopyrano[3,2-*b*]indol-4-**one**, 870
Thiopyrano[2,3-*c*]pyrazole, 1311
Thiopyrano[2,3-*b*]pyridin-**4**-one, 1232
Thiopyrano[2,3-*d*]pyrimidine-2,**4**,5-tri**one**, 909
Thiopyrano[3,4-*e*]pyrrole-5,7-dione, 1196
9-Thioxanthenone, 1079, 1292
Thioxanthylium, 1196
2a,4,5-Triazabenz[*cd*]azulene (1,2,4trzn), 1257
2a,6b,10c-Triazabenzo[2,3]pentalen[1,6-*ab*]indene (pzol), 1337
2,6,7-Triazabicyclo[3.2.2]nona-3,8-diene (1,2,4triazep), 1198
1,3,4-Triazabicyclo[4.2.0]oct-2-en-8-one (1,2,4trzn), 708
1,3,7b-Triazacyclopent[*c,d*]inden-4-one (im4one), 948
2,3a,6a-Triazaphenalen-6-one, 3-thioxo-(pm2thione), 782
[**1,2,4**]**Triazino**[1,6-*a*]benzimidazole, 1136
[**1,2,4**]**Triazino**[4,3-*a*]benzimidazole, 814
1,3,5-Triazino[1,2-*a*]benzimidazole, 748, 931
[**1,2,4**]**Triazino**[4,5-*a*]benzimidazol-4-one, 1246
1,3,5-Triazino[5,6-*b*]benzimidazol-2-**one**, 748, 1039
1,2,4-Triazino[5,6-*b*][**1,4**]benzo**diazepin-10-one**, 732
1,3,5-Triazino[1,2-*b*]benzothiazole-**2,4-dione**, 931
1,3,5-Triazino[1,2-*c*][**1,2,3**]benzo**triazine**, 1154
1,2,4-Triazino[5,6-*b*][**1,4**]**diazepin-8-one**, 732
1,2,4-Triazino[4,3-*b*]indazole, 881
1,2,3-Triazino[5,4-*b*]indole 3-**oxide**, 1326
1,3,5-Triazino[2,1-*a*]isoquinolin-4-**one**, 931, **1013**
1,2,4-Triazino[5,6-*c*]**isoquinolin-6-one**, 3-thioxo-, 1082

[1,2,4]Triazino[5,6-*e*][**1,3,4**]**oxadiazinium**, 731
[**1,2,4**]**Triazino**[**3,4-***a*]phthalazine-3,4-**dione**, 1258
[**1,2,4**]**Triazino**[**3,4-***a*]phthalazin-4-**one**, **1117**, 1258
[**1,2,4**]**Triazino**[**2,3-***a*]purin-10-one, 1136
[1,2,4]Triazino[2,3-*c*]**quinazolin-6-one**, 1037
[1,2,4]Triazino[4,3-*c*]**quinazolin-6-one**, 1007, 1037
1,3,5-Triazino[1,2-*a*]quinolin-1-one, 1014
[1,2,4]Triazino[4,3-*b*]-**1,2,4,5-tetrazin**-6-one, 764
[1,2,4]Triazino[5,6-*e*][**1,3,4**]**thiadiazine**, 688
[1,2,4]Triazino[6,5-*e*]-1,2,4-triazine, 967
[**1,3,5**]-**Triazino**[1′,2′:1,5][1,2,4]triazolo[3,4-*a*]isoquinolin-11-**one**, 1014
1,2,3-Triazolo[4,5-*b*][**1,5**]benzo**diazepine**, 750
[1,2,3]Triazolo[1,5-*a*][**1,4**]benzo**diazepin**-4-**one**, 1075
[1,2,3]Triazolo[1,5-*a*][**1,5**]benzo**diazepin**-5-**one**, 954
[**1,2,4**]**Triazolo**[3,4-*c*][1,4]benzothiazin-4-one, 1239
1,2,4-Triazolo[3,4-*b*][**1,3**]benzothiazin-5-one, 1263
1,2,4-Triazolo[5,1-*b*][**1,3**]benzothiazin-5-one, 1263
1,2,4-Triazolo[3,4-*b*]benzothiazole, 1255
[1,2,4]Triazolo[5,1-*b*]benzo**thiazol**-5-one, 877
1,2,4-Triazolo[4,3-*d*][**1,2,4**]-benzo**triazepine**, 1145
1,2,4-Triazolo[4,3-*d*][**1,3,4**]benzo**triazine**, 1144, 1154
[**1,2,4**]**Triazolo**[3,4-*c*][1,4]benzoxazine-1,4-**dione**, 1036
[**1,2,4**]**Triazolo**[3,4-*c*][1,4]benzoxazin-4-one, 1239
[**1,2,4**]**Triazolo**[5,1-*b*]benzoxazole, 928
[**1,2,4**]**Triazolo**[1,5-*b*]indazole, 761
[**1,2,4**]**Triazolo**[4,3-*a*]indole, 1338
[**1,2,3**]**Triazolo**[5,1-*a*]isoquinoline, 1262
[**1,2,4**]**Triazolo**[5,1-*a*]isoquinoline, 928
1,2,4-Triazolo[3,4-*a*]isoquinoline, 1254
[1,2,3]Triazolo[1,5-*d*][**1,3,4**]**oxadiazin**-4-**one**, 696
1,2,4-Triazolo[3,4-*b*][1,3,4]oxadiazole, 728
[1,2,4]Triazolo[3,4-*b*]**purin**-7-**one** (im2one), 933
1,2,4-Triazolo[3,4-*b*][1,3,4]oxadiazole, 1238
1,2,4-Triazolo[3,4-*a*]phthalazine, 1258
[**1,2,4**]**Triazolo**[2,3-*a*]purin-9-one, 1130
1,2,4-Triazolo[4,3-*a*]pyrazine, 776
[**1,2,4**]**Triazolo**[1,5-*a*]pyrazine **3-oxide**, 1001
[1,2,4]Triazolo[1,5-*a*]**pyrazin**-1-ium, 8-**oxo**-, 881
1,2,4-Triazolo[4,3-*a*]**pyrazin**-8-**one**, 3-thioxo-, 673
1,2,4-Triazolo[4,3-*b*]**pyridazine**, 1206
1,2,4-Triazolo[4,3-*b*]pyridazine, 1206, 1239, **1253**
1,2,4-Triazolo[1,2-*a*]**pyridazine**-1,3-dione, 1159
1,2,4-Triazolo[4,3-*b*]pyridazin-**3-one**, **1239**
[**1,2,3**]**Triazolo**[1,5-*a*]pyridine, 1262
[**1,2,4**]**Triazolo**[1,5-*a*]pyridine, 728, 769, 929
1,2,3-Triazolo[4,5-*b*]**pyridine**, 672
[**1,2,4**]**Triazolo**[1,5-*a*]pyridine **3-oxide**, 1001, 1264
1,2,3-Triazolo[4,5-*b*]pyridin-5-one, 1026, 1331
1,2,3-Triazolo[4,5-*b*]**pyridin**-**6**-**one**, 673
[1,2,3]Triazolo[1,5-*a*]**pyrimidine**, 1006
[1,2,4]Triazolo[1,5-*a*]**pyrimidine**, 1006
[**1,2,4**]**Triazolo**[1,5-*c*]pyrimidine, 1254
1,2,3-Triazolo[4,5-*d*]**pyrimidine**, 764, 923, 934
1,2,3-Triazolo[4,5-*d*]**pyrimidine**-4,6-**dione**, 1262
1,2,3-Triazolo[4,5-*d*]pyrimidine-5,7-dione, 934, 1332
1,2,4-Triazolo[4,3-*c*]pyrimidine-5,7-dione, 1263
[**1,2,4**]**Triazolo**[1,5-*a*]pyrimidine **1-oxide**, 1001
[**1,2,4**]**Triazolo**[1,5-*a*]pyrimidin-8-ium, 1032
1,2,3-Triazolo[4,5-*d*]**pyrimidin-4-one**, 692
1,2,3-Triazolo[4,5-*d*]pyrimidin-5-one, 1331
1,2,3-Triazolo[4,5-*d*]pyrimidin-7-one, 1332
1,2,3-Triazolo[4,5-*d*]**pyrimidin**-7-**one**, 938
1,2,4-Triazolo[4,3-*a*]**pyrimidin**-5-**one**, 1300
1,2,4-Triazolo[4,3-*a*]pyrimidin-5-one, 929
[1,2,3]Triazolo[3,4-*a*]**pyrimidin**-7-**one**, 1008

[1,2,4]Triazolo[1,5-*a*]**pyrimidin**-7-**one**, 1300
[1,2,4]**Triazolo**[1,5-*c*]pyrimidin-5-one, 730
1,2,4-Triazolo[4,3-*a*]pyrimidin-7-one, 1255
1,2,4-Triazolo[4,3-*a*]**pyrimidin**-7-**one**, 1010
1,2,3-Triazolo[4′,5′:3,4]pyrrolo[1,2-*a*]-indol-4-one, 1187
[1,2,4]Triazolo[5,1-*b*]**quinazolin**-9-one, 1009
[1,2,4]Triazolo[1,5-*a*]**quinazolin**-5-**one**, 1009
[**1,2,4**]**Triazolo**[4,3-*a*]quinazolin-5-one, 1255
1,2,4-Triazolo[4,3-*b*]quinazolin-5-one, 1255
1,2,4-Triazolo[4,3-*c*]**quinazolin**-5-**one**, 1007
1,2,3-Triazolo[4,5-*f*]quinoline, 934
1,2,3-Triazolo[4,5-*c*]quinoline **3-oxide**, 1245
[**1,2,4**]**Triazolo**[4,3-*b*]-1,2,4,5-tetrazine, 1256
[1,2,3]Triazolo[5,1-*b*][**1,3,4**]**thiadiazine**, 723
1,2,4-Triazolo[3,4-*b*][**1,3,4**]**thiadiazine**, 732
[1,2,4]Triazolo[5,1-*b*][**1,3,4**]**thiadiazin-8-ium**, 732
1,2,4-Triazolo[3,4-*b*][**1,3,4**]**thiadiazin**-1-ium, 731
1,2,4-Triazolo[3′,4′:2,3][**1,3,4**]-**thiadiazino**[5,6-*b*]quinoxaline, 732
1,2,4-Triazolo[3,4-*b*][**1,3,4**]**thiadiazole**, 729
[1,2,4]Triazolo[5,1-*b*][**1,3**]**thiazine**, 882
[1,2,4]Triazolo[5,1-*b*][**1,3**]**thiazin**-7-**one**, 883
1,2,4-Triazolo[3,4-*b*][**1,3**]**thiazin**-5-**one**, 883
[1,2,4]Triazolo[5′,1′:2,3]**thiazolo**[4,5-*b*]-quinoxaline, 1167
1,2,4-Triazolo[4,3-*b*][**1,2,4**]**triazepin**-6-**one**, 1140
[1,2,3]Triazolo[5,1-*c*][**1,2,4**]**triazine**, 1155
[1,2,4]Triazolo[1,5-*a*][**1,3,5**]**triazine**, 925
[1,2,4]Triazolo[5,1-*c*][**1,2,4**]**triazine**, 814
1,2,4-Triazolo[4,3-*a*][1,3,5]triazine, 1263
[1,2,4]Triazolo[1,5-*a*][**1,3,5**]**triazine**-5,7-**dione**, 687
1,2,4-Triazolo[4,3-*d*][1,2,4]triazine-3,8-di**one**, 875
1,2,4-Triazolo[4,3-*d*][1,2,4]triazine-6-thione, 1254
[**1,2,4**]**Triazolo**[5,1-*c*][1,2,4]triazin-4-one, 698, 761
1,2,4-Triazolo[3,4-*f*][**1,2,4**]**triazin**-8-**one**, 1038
1,2,4-Triazolo[4,3-*a*][1,3,5]triazin-5-one, 1263
[1,2,4]Triazolo[1,5-*a*][**1,3,5**]**triazin**-5-**one**, 7-**thioxo-**, 687
[1,2,4]Triazolo[3,4-*c*]-1,2,4-triazole, 965
[1,2,4]Triazolo[1,2-*a*][**1,2,3**]**triazole**-5,7-dione, 1187
[1,2,4]Triazolo[1,2-*a*][**1,2,4**]**triazole-1**, **3**,5,7-**tetraone**, 1187
1,2,4-Triazolo[4,3-*b*][1,2,4]triazole-3-thione, 922
1,2,4-Triazolo[4,3-*b*][1,2,4]triazol-3-one, 922

Xanthene, 1196, 1286
9-Xanthenone, 1078